# THE PHYSICS OF ELECTRONIC AND ATOMIC COLLISIONS

XIX International Conference

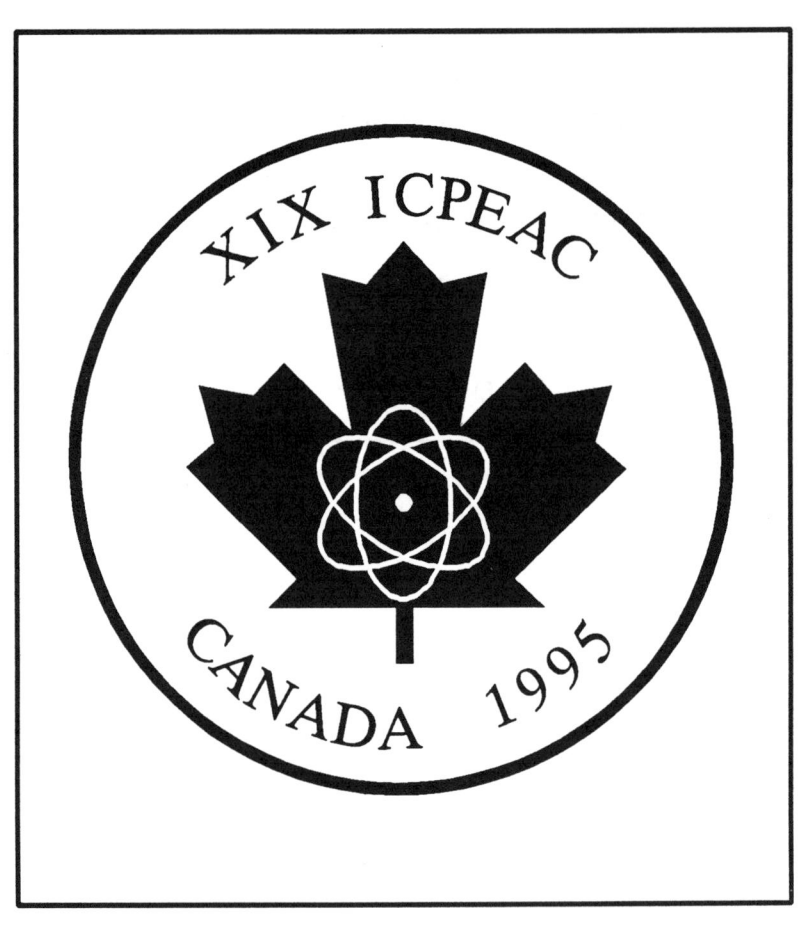

# THE PHYSICS OF ELECTRONIC AND ATOMIC COLLISIONS

XIX International Conference

Whistler, Canada  July–August 1995

EDITORS
Louis J. Dubé
*Université Laval, Québec*

J. Brian A. Mitchell
*University of Western Ontario, Ontario*

J. William McConkey
*University of Windsor, Ontario*

Chris E. Brion
*University of British Columbia,
British Columbia*

AIP CONFERENCE
PROCEEDINGS 360

**American Institute of Physics**   **Woodbury, New York**

Authorization to photocopy items for internal or personal use, beyond the free copying permitted under the 1978 U.S. Copyright Law (see statement below), is granted by the American Institute of Physics for users registered with the Copyright Clearance Center (CCC) Transactional Reporting Service, provided that the base fee of $6.00 per copy is paid directly to CCC, 222 Rosewood Drive, Danvers, MA 01923. For those organizations that have been granted a photocopy license by CCC, a separate system of payment has been arranged. The fee code for users of the Transactional Reporting Service is: 1-56396-440-6/ 95 /$6.00.

© 1995 American Institute of Physics

Individual readers of this volume and nonprofit libraries, acting for them, are permitted to make fair use of the material in it, such as copying an article for use in teaching or research. Permission is granted to quote from this volume in scientific work with the customary acknowledgment of the source. To reprint a figure, table, or other excerpt requires the consent of one of the original authors and notification to AIP. Republication or systematic or multiple reproduction of any material in this volume is permitted only under license from AIP. Address inquiries to Office of Rights and Permissions, 500 Sunnyside Boulevard, Woodbury, NY 11797-2999; phone 516-576-2268; fax: 516-576-2499; e-mail: rights@aip.org.

L.C. Catalog Card No. 95–83671
ISBN 1-56396-440-6
DOE CONF- 950706

Printed in the United States of America

## Contents

Preface .................................................................... xiii
Sponsors and Exhibitors .................................................. xiv
Executive Committee ICPEAC 1993-95 ....................................... xv
Local Committee XIX ICPEAC ............................................... xv
General Committee ICPEAC 1993-95 ......................................... xvi

### PLENARY

Mesoscopic Systems and Quantum Corrals ..................................... 3
    E.J. HELLER, M.F. Crommie, C.P. Lutz, and D.M. Eigler
Ion-Atom Collision Measurements
Relevant to Fusion Plasmas ................................................ 19
    H.B. GILBODY
Synchrotron Radiation Sources:
Present Status, Anticipated Developments
and Applications to Photon Collision Physics .............................. 39
    B. SONNTAG

### PHOTON COLLISIONS

Dissociation Dynamics of Superexcited Molecules ........................... 67
    Y. HATANO
Inner Shell Excitation, Relaxation
and Fragmentation of Cluster Beams and Molecules .......................... 89
    A.P. HITCHCOCK and E. Rühl
Anisotropy and Kinetic Energy Release
in Ionic Fragmentation of Molecules Photoexcited
to Inner-Shell Hole States ............................................... 105
    I.H. SUZUKI and N. Saito

### PHOTOIONIZATION

Double Photoionization of He ............................................. 117
    N. BERRAH
Signatures of Strong Correlations in Photoionization ..................... 127
    C. H. GREENE
Energy and Angular Resolved Studies
of Double Photoionization of Helium and Rare Gases ....................... 139
    A. HUETZ, L. Andric, A. Jean,
    P. Lablanquie, P. Selles, and J. Mazeau

## ELECTRON-ATOM COLLISIONS

Electron Collisions with Atoms in Excited States .................. 155
    C.C. LIN

Recent Progress in Polarized-Electron Scattering
from Atoms and Molecules ......................................... 163
    S. MAYER

The Role of Lasers in Electron-Atom Collision Physics ............ 173
    M. STANDAGE

Laser-Assisted Electron-Atom Collisions .......................... 189
    B. WALLBANK

## ELECTRON-MOLECULE COLLISIONS: THEORY

Electron Collisions with Oriented Molecules ...................... 201
    C.M. FULLERTON

Low-Energy Electron Scattering from Polyatomic Molecules:
Recent Theoretical Results ....................................... 211
    F.A. GIANTURCO

Electronic Excitation in Electron-Molecule Scattering
using the R-Matrix Method ........................................ 233
    J. TENNYSON

## ELECTRON-MOLECULE COLLISIONS: EXPERIMENT

High Resolution Studies of the Dissociative Attachment
of Low Energy Electrons to State-Selected Sodium Dimers ....... 247
    K. BERGMANN, M. Keil, M. Külz,
    A. Kortyna, D. Weyh, and W. Meyer

Electron Scattering from Vibrationally Excited Molecules ........ 257
    P.D. BURROW

Studies of Low Energy Electron Collisions at
Sub-meV Resolution ............................................... 267
    H. HOTOP, D. Klar, J. Kreil,
    M.-W. Ruf, A. Schramm, and J.M. Weber

Absolute Cross Section Measurements for Electron Collisions
with Polyatomic Molecules of Plasma Chemistry ................... 279
    H. TANAKA and L. Boesten

## ELECTRON-ION COLLISIONS

Experimental Studies of Electron Impact Dissociation
of Molecular Ions ................................................ 297
    N. DJURIĆ, Y.S. Chung, B. Wallbank, and G.H. Dunn

New Mechanisms for Dissociative Recombination .................. 307
    S.L. GUBERMAN
Electron-Ion Collisions in Storage Rings .......................... 317
    A. MÜLLER
Dissociative Recombination of Molecular Ions
in a Cooler Ring ................................................ 329
    T. TANABE, I. Katayama, H. Kamegaya,
    K. Chida, T. Watanabe, Y. Arakaki, M. Yoshizawa,
    M. Saito, Y. Haruyama, K. Hosono, K. Hatanaka,
    T. Honma, K. Noda, S. Ohtani, and H. Takagi

## ELECTRON COLLISIONS: IONIZATION

Ionization of Atoms as a Three-Body Process .................... 341
    S. JONES and D.H. Madison
Basis Spline Method for $e^- + H$ Collisions ......................... 347
    S.V. PASSOVETS, J.H. Macek, and S.Yu Ovchinnikov
(e, 2e) Momentum Spectroscopy of Thin Films .................... 357
    P. STORER, Y.Q. Cai, S.A. Canney, R. Caprari,
    S.A.C. Clark, A.S. Kheifets, I.E. McCarthy,
    S. Utteridge, M. Vos, and E. Weigold

## POSITRON AND MUON COLLISIONS

Why Positron Physics is Fun ..................................... 369
    R.J. DRACHMAN
Collisions of Positrons and Positronium
with Atoms and Molecules ....................................... 385
    G. LARICCHIA
Overview of the Present Theoretical Status
of Positron-Atom Collisions ...................................... 397
    H.R.J. WALTERS, A.A. Kernoghan, and M.T. McAlinden
Spin Relaxation Phenomena in Muon Collision
Processes in Gases .............................................. 413
    D.G. FLEMING, M. Senba, J.J. Pan, and D.J. Arseneau

## ION-ATOM COLLISIONS: THEORY

Time Dependent Description of Inner Shell Excitation
and Transfer in Ion-Atom Collisions ............................. 435
    B. FRICKE, P. Kürpick, and W.-D. Sepp
Molecular Treatment of Ion-Atom Collisions
at Intermediate Energies ........................................ 445
    C. HAREL, H. Jouin, B. Pons,
    L.F. Errea, L. Mendez, and A. Riera

Time-Dependent, Lattice Approach to Atomic Collisions ......... 455
    D. R. SCHULTZ
Theory of Low Energy Ion-Atom Collisions ....................... 471
    E.A. SOLOV'EV

## ION-ATOM COLLISIONS: EXPERIMENT

Non-Statistically Populated Autoionizing Levels
of Li-like Carbon: Hidden Crossings ................................ 485
    E.F. DEVENEY, H.F. Krause, N.L. Jones,
    J.M. Sanders, C.R. Vane, W. Wu, S.Datz,
    M. Breinig, D.Desai, S.Yu Ovchinnikov,
    Q.C. Kessel, and S.M. Shafroth
Cold Target Recoil Ion Momentum Spectroscopy .................. 495
    R. DÖRNER, V. Mergel, L. Spielberger,
    O. Jagutzki, M. Unverzagt, W. Schmitt,
    J. Ullrich, R. Moshammer, H. Khemliche,
    M. Prior, R.E. Olson, L. Zhaoyuan, W. Wu,
    C.L. Cocke, and H. Schmidt-Böcking
Electron-Electron and Electron-Nuclear Interactions
in Dressed Ion-Atom Collisions ..................................... 505
    R.D. DUBOIS
Projectile Electron Excitation and Loss in Ion-Atom Collisions .. 515
    E.C. MONTENEGRO, W.E. Meyerhof,
    J.H. McGuire, and C.L. Cocke
Radiative and Resonant Electron Capture
Studies for High-Z Projectiles ....................................... 525
    Th. STÖHLKER

## ION-ATOM COLLISIONS: CHARGE TRANSFER

Electron Spectroscopy of Rydberg States
Produced in Capture Processes ...................................... 537
    A. BORDENAVE-MONTESQUIEU and P. Moretto-Capelle
Gathering and Evaluation of State Selective Capture
Cross Sections for Plasma Diagnostics ............................. 547
    R. HOEKSTRA
Double Electron Capture: Complex Amplitudes
from Auger Anisotropy Measurements ............................. 557
    M.H. PRIOR and H. Khemliche

## ION-MOLECULE COLLISIONS

Coupled Wavepackets Study of Ion-Molecule Collisions .......... 569
    F. AGUILLON

Direct Ionization in Diatomic and Triatomic
Quasimolecules ..................................................... 579
      Y.S. GORDEEV and G.N. Ogurtsov
Wavefunction Overlap Effects in
Collisional Excitation of Molecules ................................. 587
      D. MATHUR and M. Krishnamurthy

## RYDBERG COLLISIONS

Charge Transfer between Rydberg Atoms and
Polar Molecules or Clusters ........................................ 599
      C. DESFRANÇOIS,
      H. Abdoul-Carime, and J.P. Schermann
Application of Coherent Rydberg States
for Collision Studies ............................................... 609
      E. HORSDAL-PEDERSEN, J.C. Day,
      B. DePaola, T. Ehrenreich, S.B. Hansen, Y. Leontiev,
      K.B. MacAdam, and K.S. Mogensen
Atomic Scattering from Oriented Rydberg Atoms ................. 619
      J. WANG, J.H. McGuire, and R.E. Olson

## COLLISIONS WITH SURFACES AND CLUSTERS

Interaction of Highly Charged Ions
with Metal and Insulator Surfaces ................................. 631
      F. AUMAYR
Inelastic Ion Surface Collisions .................................... 647
      V.A. ESAULOV, L. Guillemot, S. Lacombe,
      and V. Ngoc Tuan
Molecular Collisions on Large van der Waals Clusters ............. 657
      J.P. VISTICOT, J. Berlande, X. Biquard,
      M.A. Gaveau, A. Lallement, J.M. Mestdagh,
      and O. Sublemontier

## COLD ATOMS AND ION-ION COLLISIONS

Photoassociation in Ultracold Collisions: High Resolution
Spectroscopy from the Collision Continuum ....................... 667
      P.D. LETT
Differential Scattering in Ion-Ion Collisions of He ................ 677
      S. KRÜDENER
Collision Processes Involving Negative Hydrogen Ions ............ 687
      D.B. USKOV

## NOVEL TECHNIQUES

Recent Results from the Super EBIT .............................. 705
    R.E. MARRS

Attaining Low Electron Temperatures in Electron Coolers ........ 721
    H. DANARED

Production of Bright, Cold Beams
for Atomic Collision Experiments .................................... 731
    K.A.H. VAN LEEUWEN, E.J.D. Vredenbregt, P.G.M. Sebel,
    J.P.J. Driessen, M.D. Hoogerland, and H.C.W. Beijerinck

"Complete" Measurement of
Molecular Coulomb-Explosions ....................................... 741
    U. Werner and H.O. LUTZ

## SELECTED TOPICS: PHOTONS

Photoionization of $Sr^+$ Ions in the 3d Ionization Region .......... 755
    Y. ITOH, T. Koizumi, Y. Awaya, S.D. Kravis,
    M. Oura, M. Sano, T. Sekioka, and F. Koike

Angular Distributions and Retardation
in Photoionization of Two Electrons in Helium ..................... 763
    M.A. KORNBERG and J.E. Miraglia

Experimental Separation of Photoabsorption
and Compton Scattering Contributions
to He Single and Double Ionization ................................. 773
    L. SPIELBERGER, O. Jagutzki, R. Dörner,
    J. Ullrich, U. Meyer, V. Mergel, M. Unverzagt,
    M. Damrau, T. Vogt, I. Ali, Kh. Khayyat, D. Bahr,
    H.G. Schmidt, R. Frahm, and H. Schmidt-Böcking

## SELECTED TOPICS: ELECTRONS

Theoretical and Experimental Investigation of
Electron-Helium Scattering .......................................... 785
    I. BRAY, D.V. Fursa, D.T. McLaughlin,
    B.P. Donnelly, and A. Crowe

Spin Effects in (e,2e) Collisions ..................................... 795
    X. Guo, J. Hurn, J. LOWER, S. Mazevet,
    Y. Shen, I.E. McCarthy, and E. Weigold

Measurement of Exchange and Spin-Orbit Effects
and their Interference in Elastic e-Cs Scattering ................... 805
    M. TONDERA, G. Baum, P. Baum, L. Grau,
    B. Leuer, R. Niemeyer, and W. Raith

Studies of Electron-Molecule Scattering
at Microelectronvolt Energies
Using Very-High-$n$ Rydberg Atoms .................................. 815
    M.T. FREY, S.B. Hill, K.A. Smith,
    F.B. Dunning, and I.I. Fabrikant

On the Ionisation Mechanism of Reflection (e,2e) Events ......... 825
    S. IACOBUCCI, P. Luches, L. Marassi, R. Camilloni,
    B. Marzilli, S. Nannarone, and G. Stefani

Recombination of $H_3^+$ and $D_3^+$ Ions with Electrons ................. 835
    R. JOHNSEN, T. Gougousi, and M.F. Golde

## SELECTED TOPICS: ATOMS

Generalizations of Distorted-Wave Capture Theories
to Relativistic Atomic Collisions ..................................... 847
    J. EICHLER and N. Toshima

X-Ray Emission in Relativistic Ion-Atom Collisions ............... 857
    J.F. McCANN, J.T. Glass, and D.S.F. Crothers

Measurements of Charge Transfer Cross Sections
for $Ar^{q+}$ (q=6,7,8,9 and 11) Collisions
with He and $H_2$ Targets at Low Energies .......................... 867
    K. OKUNO, H. Saitoh, K. Soejima,
    S. Kravis, and N. Kobayashi

Projectile Charge Dependence of Helium Single Excitation
at Medium and High Impact Energies ............................. 877
    F. Martín and A. SALIN

Hyperspherical Elliptic Coordinates:
New Approximate Symmetry of
Three-Body Coulomb Problem ....................................... 887
    O.I. TOLSTIKHIN, S. Watanabe, and M. Matsuzawa

Author Index ........................................................... 897

# PREFACE

The Nineteenth International Conference on the Physics of Electronic and Atomic Collisions was held at the Conference Center in Whistler, British Columbia, Canada, from July 26 to August 1, 1995.

The conference was attended by 609 delegates from 35 countries. Of these, 135 were students who had not yet completed their PhD thesis. In addition, 138 accompanying persons participated in the social programme. There were 880 contributed papers submitted to the conference. The 868 abstracts received before June 1, 1995 were printed in the Book of Abstracts.

This book contains the written versions of the invited talks presented at the XIX ICPEAC. Of the 79 invited speakers, six could not submit a manuscript. The present contributions include 3 plenary lectures, 9 review talks, 47 progress reports, and 14 selected topics chosen among the submitted abstracts by the International ICPEAC committee.

We are grateful for financial support from the sponsors listed on the following page which made it possible to support the participation of 92 delegates. Special emphasis was put on supporting young scientists. We appreciate the contributions of the exhibitors and the competent assistance of "International Conference Services Inc." .

Louis J. Dubé
J. Brian A. Mitchell
J. William McConkey
Chris E. Brion

SPONSORS

The University of Western Ontario

The University of Windsor

The International Union of Pure and Applied Physics (IUPAP)

Lawrence Livermore National Laboratory

Air Canada

Avis Rent-A-Car

EXHIBITORS

The Institute of Physics (IOP)

Springer Verlag

International Conference on the Physics
of Electronic and Atomic Collisions

- Executive Committee ICPEAC 1993-95

    Chairman        FRANÇOIS J. WUILLEUMIER, France

    Vice-Chairman   GORDON H. DUNN, USA

    Secretary       J. NORMAN BARDSLEY, USA

    Treasurer       REINHARD MORGENSTERN, The Netherlands

    Members
                    T. ANDERSEN, Denmark
                    K. BARTSCHAT, USA
                    C.E. BRION, Canada
                    S. DATZ, USA
                    B. FASTRUP, Denmark
                    Y. ITIKAWA, Japan
                    A. LAHMAM-BENNANI, France
                    M. MATSUZAWA, Japan
                    J.W. MCCONKEY, Canada
                    J.B.A. MITCHELL, Canada
                    E. SALZBORN, Germany
                    A.S. SCHLACHTER, USA
                    H.-P. WINTER, Austria
                    L. ROOS, adm. secr., The Netherlands

- Local Committee XIX ICPEAC, Canada

    Chairmen  J. BRIAN A. MITCHELL, University of Western Ontario
              J. WILLIAM MCCONKEY, University of Windsor
              CHRIS E. BRION, University of British-Columbia

    Secretary  MARITA C. CHIDICHIMO, University of Waterloo

    Treasurer  WILLIAM N. LENNARD, University of Western Ontario

# General Committee ICPEAC 1993-95

**Argentina**
R.D. Rivarola

**Australia**
W.R. MacGillivray

**Belgium**
P. DeFrance

**Brazil**
M.A.P. Lima

**Canada**
C.E. Brion
J.W. McConkey
J.B.A. Mitchell

**Denmark**
T. Andersen
B. Fastrup

**Finland**
T. Åberg

**France**
D. Dowek
R. Gayet
A. Lahmam-Bennani
J. Pascale
F.J. Wuilleumier

**Germany**
H.J. Korsch
K.H. Meiwes-Broer
H. Loesch
E. Salzborn
R. Schinke

**Hungary**
J. Palinkas

**India**
D.P. Dewangan
S. Sathyamurty

**Israel**
M. Shapiro

**Italy**
M. Zarcone

**Japan**
Y. Awaya
Y. Itikawa
M. Matsuzawa
S. Ohtani

**Latvia**
E.G. Karule

**Malaysia**
W.C. Fon

**Mexico**
I. Alvarez

**The Netherlands**
R. Morgenstern
W.J. van der Zande

**Russia**
S. Ovchinnikov
B.M. Smirnov

**Spain**
G. Delgado-Bario

**Sweden**
H. Cederquist

**Ukraine**
O.B. Shpenik

**United Kingdom**
A. Crowe
J. Geddes
J. Humberston

**USA**
J.N. Bardsley
K. Bartschat
K. Becker
J.E. Burgdörfer
C.L. Cocke
S. Datz
G.H. Dunn
D.H. Jaecks
K. Kirby
S.T. Manson
A.S. Schlachter

# PLENARY

Mesoscopic Systems and Quantum Corrals ........................  3
    E.J. HELLER, M.F. Crommie, C.P. Lutz, and D.M. Eigler
Ion-Atom Collision Measurements
Relevant to Fusion Plasmas ............................................ 19
    H.B. GILBODY
Synchrotron Radiation Sources:
Present Status, Anticipated Developments
and Applications to Photon Collision Physics ....................... 39
    B. SONNTAG

# Mesoscopic Systems and Quantum Corrals

E. J. Heller*, M. F. Crommie[+], C. P. Lutz[†], and D. M. Eigler[†]

*Department of Physics, Harvard University and
Harvard-Smithsonian Center for Astrophysics
Cambridge, Massachusetts 02138
[†]IBM Research Division, Almaden Research Center
650 Harry Road, San Jose, California 95120
[+]Department of Physics, Boston University,
Boston Massachusetts 02215

> Mesoscopic systems are under intense theoretical and experimental development. New theoretical concepts are required which patch the statistical assumptions of random and incoherent systems to coherent quantum mechanics. The field is made even more compelling because of the potential for devices which are so small that quantum coherences and ballistic transport are important. We focus here on the quantum corral experiments involving Fe adatoms placed one at a time on the surface of Cu(111).

## INTRODUCTION TO MESOSCOPIC SYSTEMS

We define "mesoscopic" systems as exhibiting clear evidence of coherent quantum mechanics (or coherent wave equations) while at the same time being amenable to statistical approximation. They need not be small! Examples include nuclear scattering (the birthplace of Random Matrix Theory, an important tool in mesoscopic physics), sound scattering in a concert hall with obstacles and absorbers, certain microwave cavities with chaotic classical motion or random scatterers, some small and carefully constructed semiconductor microstructures, Rydberg atoms in external fields, excited many electron atoms, collisions involving complex partners, polyatomic molecules (e.g. their spectra), and quantum corrals. In this paper we shall give a few examples of the systems and progress in mesoscopics. There is no hope that even a lengthy article be comprehensive in this burgeoning field, so we shall finally focus on a particular example of a beautiful system which admits of a completely quantum mechanical treatment: the STM measurements of Crommie, Lutz, and Eigler [1,2] involving Fe adatoms placed individually on a Cu(111) surface. Complexity enters through multiple scattering which is handled in a conventional way. We will examine the problem of STM images of surface electron waves in the presence of the Fe adatom "quantum corrals", which were originally constructed to produce quantum chaos and which turned out to deliver a quite different lesson. We will see that the quantum corrals are ideal quasiatom-quasiparticle scattering systems in two dimensions, with some unique properties.

One of the themes of this work is the importance of coherent atomic physics and scattering theory in mesoscopic systems: In two dimensions, instead of three! Certain experiments that we can only dream of in three dimensions become feasible in two, including scattering experiments where the "targets" can be arranged like iron ducks in a shooting gallery.

## The Mesoscopic Interface

The road from the microscopic quantum world to macroscopic properties is well traveled. Quantum statistical mechanics of bulk properties of materials is the prime example. However this is not mesoscopic physics. On the other hand the properties of a metal cluster of 10,000 atoms may very well be mesoscopic: the mean free path for quasiparticle collisions may be on the order of the cluster size, there may be only a few impurities or defects in the cluster, or the cluster boundaries may be irregular. Perhaps decreasing temperature causes a transition from bulk to mesoscopic behavior.

A fully quantum treatment would solve all the problems of the cluster. However there are several problems with this seemingly ultimate goal. First, a full *ab initio* is not feasible, even in the foreseeable future, due to computational complexity. Second, the experiment producing the cluster is is likely churning out a *distribution* of clusters of varying numbers of particles and varying shapes. Third, we actually do not want the precise many body eigenfunctions or even the eigenvalues of one or all the clusters; rather, we want certain measurable properties such as magnetic susceptibilities, conductance fluctuations, and scattering cross sections. We need a theory which can handle some explicit coherence as well as take advantage of statistical ideas.

Two of the primary tools in the mesoscopic theory toolchest are random matrix theory and semiclassical methods. Connecting these two tools is the theory of quantum chaos, which has built many links between underlying classical chaos and random matrix ideas, as developed by Wigner, Dyson, Mehta and others. Bohigas,Giannoni and Schmit made a conjecture of the connection between random matrix theories and spectral fluctuations [3] and classically chaotic ststems. The discovery of eigenstate scarring by the least unstable periodic orbits was explained in terms of stability of periodic orbits [4]. One of the most significant discoveries was a modification to random matrix energy level correlations due to periodic orbits [5]. However there is no complete semiclassical theory eigenvalues and eigenfunctions. The semiclassical methods are especially impotent when it comes to individual eigenfunctions and nearest neighbor level spacings, and especially powerful when considering certain averaged properties, or spectral properties involving long range energy correlations.

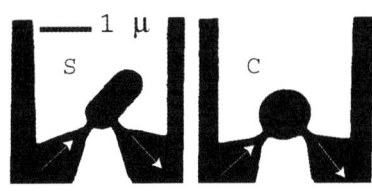

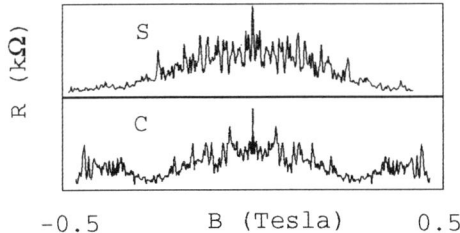

**FIG. 1.** Semiconductor microstructures constructed by Marcus and Westervelt to measure conductance fluctuations in integrable (circle, C) and chaotic (stadium, S) shaped billiard cavities

## Conductance Fluctuations

*Landauer formula.* One of the first mesoscopic formulas is due to Landauer [7], who connected the seemingly macroscopic idea of conductance with the microscopic quantum mechanics of scattering theory. He expressed the conductance in terms of the complex T-matrix amplitudes $t_{n,m}$ connecting incoming and outgoing electrical leads (channels), considering the "device" to be merely a source of conventional scattering:

$$G = \frac{2e^2}{h} \sum_{n,m} |t_{n,m}|^2 \qquad (1)$$

Later Jalabert, Baranger and Stone [8] further approximated the T-matrix elements semiclassically, following Blümel and Smilansky [9], who in turn had used Miller's classical S-matrix theory [10]. Jalabert, Baranger and Stone considered specifically the experiments of Marcus and Westervelt [6], which measured the conductance fluctuations of micron-sized semiconductor heterostructures, as a function of applied magnetic field (Fig. (1)).

Using gate voltages, the structures were shaped into paradigm billiard systems: a circle, which is and example of a completely integrable classical dynamics and simple eigenstates, and the stadium billiard, which is probably the most popular billiard paradigm for global classical chaos. The microstructures were pure enough that the mean free path and coherence lengths are larger that the size of the billiard enclosure, and ballistic transport is possible. The number of open transverse channels in the leads can be varied from one to several. Complexity comes from the extremely chaotic motion of the stadium billiard.

Using the semiclassical scattering theory modified to include the additional semiclassical phase due to the magnetic flux $B$ penetrating the billiard, Jalabert *et. al* were able to derive an expression for the conductance autocorrelation,

$$\langle \delta G(B+\Delta B)\delta G(B)\rangle \propto \left| \int_{-\infty}^{\infty} dA e^{i2\pi \Delta B A/\phi_0} P(A) \right|^2 \qquad (2)$$

where $A$ is the areas enclosed by classical trajectories contributing to the semiclassical scattering, $\phi_0$ is the fundamental flux quantum, and $P(A)$ is the distribution of areas $A$ enclosed by the trajectories. Some averaging over the semiclassical scattering expressions was required. Numerical simulations suggested that these area could be exponentially distributed,

$$P(A) \propto e^{-\alpha A} \qquad (3)$$

so that

$$\langle \delta G(B+\Delta B)\delta G(B)\rangle \propto 1/[1+(2\pi\Delta B/\alpha\phi_0)^2]^2 \qquad (4)$$

This Lorenzian behavior has indeed been observed by Marcus and Westervelt [6].

This example is a paradigm for mesoscopic studies: A macroscopic quantity (conductance) is related to a microscopic one (scattering theory T- matrix); then semiclassical and statistical ideas are applied to obtain a formula which is successfully related to experiment and shown to contain new physical insight (areas of trajectories).

The earlier work of Blümel and Smilanski [9] addressed the question of Ericson fluctuations in scattering cross sections, which occur when resonances widths exceed resonance spacings, and multiple overlap results. Making certain statistical assumptions, Ericson found Lorenzian correlations in S-matrix autocorrelation functions. Using semiclassical arguments, Blümel and Smilanski also obtained a Lorenzian correlation,

$$\langle S^*_{nn'}(E) S_{nn'}(E+\epsilon)\rangle_E \propto \frac{1}{1-i\epsilon/(\gamma\hbar)} \qquad (5)$$

thus linking the semiclassical formulae (based on chaotic scattering, which is characterized typically by self-similar fractal "strange repellors") and statistical, random matrix ideas.

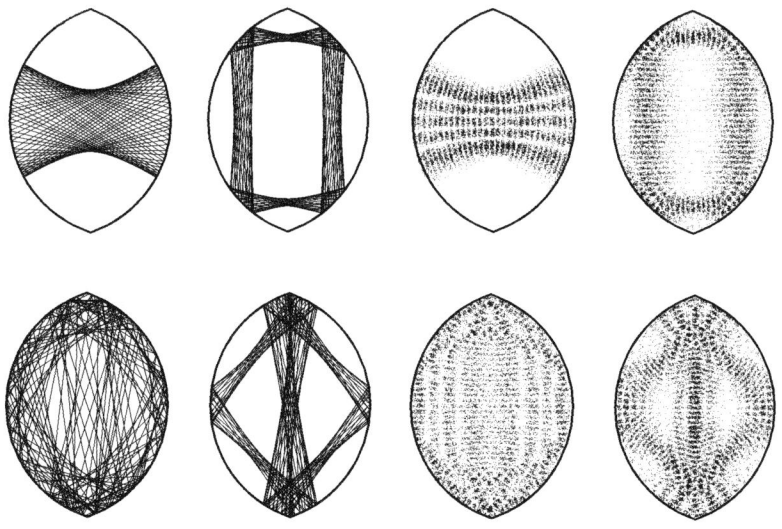

**FIG. 2.** Classical trajectories and quantum eigenstates of the lemon billiard.

## BILLIARDS

The semiconductor heterostructure ballistic billiard systems just considered are one of perhaps a dozen or more distinct experimental recent realizations of wave motion in billiard enclosures. Examples include thin microwave cavities (both superconducting and room temperature) in various shapes, water waves, acoustic waves in air confined by walls and in blocks of metal, tunnel junctions, and quantum corrals. Billiards have played a prominent role for several reasons. First, theory is cleaner for certain billiard systems (e.g. the classical motion can be proved to be completely chaotic for some shapes. Billiards are scaling systems, in which only the speed of the motion, but not the shape of the trajectories changes as energy is changed); this has attracted experimental realizations. Second, in microstructures, waveguides and acoustical experiments it is easier to construct cavities than smooth potentials.

Billiards of different shapes generate integrable (e.g. a rectangle or a circle) motion, mixed motion (e.g. "lemon" billiard; see below), and completely chaotic motion (the stadium shape). By far the most common type of motion is mixed, where region of chaotic and quasiperiodic motion co-exist at the same energy (Fig.2).

## WHAT'S AT STAKE?

There are many issues under consideration in connection with the field of mesoscopic physics. The question of practical devices with new properties or smaller scale looms very large, but we do not consider it further here.

Many of the issues fall broadly under the heading of quantum correspondence to classical chaos. They include 1) as we have seen, fluctuation and susceptibilities obtained or understood semiclassically; 2) eigenvalue distributions and the relation to random matrix theories, 3) eigenfunction distributions and scarring by periodic orbits; 4) wavefunction propagation in the time domain; and 5) the role of tunneling and diffraction in the presence of chaos. In some cases quantum and semiclassical calculations as well as the experiments can all be done and compared; in others, such as conductance fluctuations a good understanding of the experiments is obtained semiclassically or by random matrix assumptions without resorting (or without even the possibility) of quantum calculations. With the exception of the nearest neighbor level spacing distributions, semiclassical methods have proven to be astonishing in their insight and utility for ballistic systems. This situation might be termed "self generated" chaos, as opposed to say a set of random impurity scatterers which generate explicitly random paths. The latter case is the traditional one in condensed matter physics; we do not address it here.

## QUANTUM CORRALS

It is against this backdrop that Crommie, Lutz, and Eigler [1] constructed circular and stadium shaped "quantum corrals" made out of Fe adatoms individually placed on the surface of Cu(111). The placement is done by sliding the Fe atoms, one at a time, using the tip of the STM. To initiate this "sliding" process, the tip is centered over an Fe atom and lowered until the tunnel junction resistance is approximately $100k\Omega$ ($V \approx 0.01$ volt, $I \approx 100nA$). This creates a "bond" between the tip and surface atom. To slide the atom, the tip is moved laterally to the desired final resting spot, and then retracted. The tip-adatom bond is thus broken, and the atom remains at the final spot. Such a capability has already been demonstrated for a variety of adsorbate/surface combinations [2].

A schematic of the tip/adatom/surface geometry is shown in Fig. (3).

A strong scattering occurs between the surface state electrons known to exist on Cu(111) and the Fe adatoms. The surface state electrons obey almost perfect free-electron like quadratic $E$ vs. $k$, with an effective mass of close to $0.4m_e$. Because the Fe atoms scatter the surface state electrons, it is possible to disturb the 2D electron gas at the atomic scale by rearranging the Fe atoms to fixed sites, something that is impossible in the gas phase.

The STM experiments are performed at 4K once the Fe adatoms have been arranged. The measured quantity is dI/dV, or the rate of change of tunnel current with tip bias voltage. For fixed bias voltage, the tunnel current is due to all occupied bands which can tunnel to unoccupied bands across the vacuum

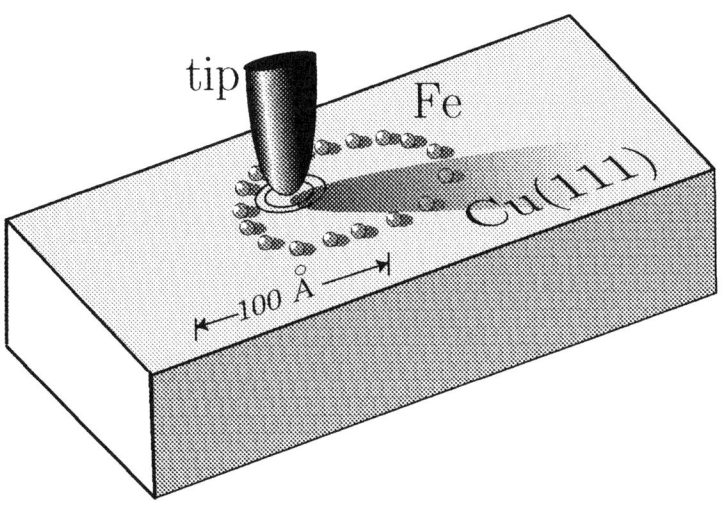

**FIG. 3.** An STM tip scanning a previously constructed corral.

gap. Therefore the *change* in tunnel current with bias voltage measures the current due to electrons or holes tunneling at the new voltage.

The Fe adatoms act as corral posts; the corrals act to confine the electrons, although this confinement is partial as we shall see. Seeing that some sort of confinement was possible, Crommie *et.al* set out to investigate the possibility of seeing eigenfunction scars and other manifestations of quantum mechanics of the classically chaotic stadium billiard. It was initially clear that a simple particle in the box model explained some qualitative features of a circular enclosure, but the model lacked many features, including the ability to describe "open" structures which were not really corrals. The pattern seen inside a stadium shaped enclosure also needed explanation.

By placing the tip at the center of a 60 atom circle one gets the $dI/dV$ curve shown in Fig.(4). If the circular corral were a very good box with a high Q, the effect of reflection from the walls would lead to no net enhancement or de-enhancement of the current, except at the eigenvalues of the box. The fact that instead the resonances are moderately broad indicates that the box instead is rather leaky; i.e. it has a low Q.

It is evident that a better, microscopic picture is needed to understand the patterns seen in the $dI/dV$ curve [11]. The need is even more compelling when we try to understand the pattern of intensity across a corral at fixed bias voltage, or indeed the changes in the pattern with bias voltage. Such a theory is provided by two dimensional electron-atom scattering! (Really quasiparticle-quasiatom scattering; by quasiatom we mean the Fe atom as

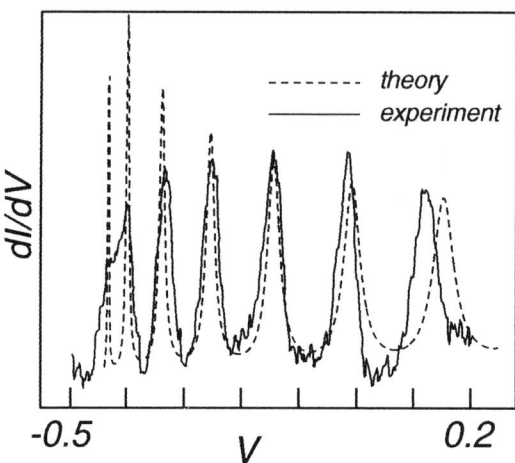

**FIG. 4.** 60 atom circular corral dI/dV for tip in the center of the circle. Solid:experiment; dashed:theory.

modified by the contact with the Cu(111) surface.)

The problem is perhaps best seen within a scattering theory approach [11]. A multiple scattering formalism is ideal because the Fe adatoms are presumably outside each others interaction radius for the quasiparticles. It is convenient and suggestive to use a time dependent context initially. Consider Fig.(5). It shows electron amplitude leaving the tip and returning after colliding with something on the surface. Since the electron flux (current) is proportional to something like $\psi^*\nabla\psi$, and $\psi$ is the sum of direct and scattered parts, it is clear that the current will fluctuate with the strength of the scattered returning amplitude and its phase. In the time domain the tip acts like a Green function source of quasiparticle amplitude: $\langle \vec{r}|e^{-iHt\hbar}|\vec{r}\rangle \equiv \langle \vec{r}|\vec{r}(t)\rangle$ ). The $dI/dV$ signal is proportional to the Fourier transform (i.e., the local density of states [1])

$$dI/dV = \int_{-\infty}^{\infty} e^{i\epsilon t/\hbar} \langle \vec{r}|\vec{r}(t)\rangle \, dt \qquad (6)$$

The position $\vec{r}$ is the tip location. The state $|r(t)\rangle$ can be thought of as electron amplitude initially localized beneath the tip, propagated in time. The Fourier transform energy is $\epsilon = E_F + V - E_0$, where $V$ is the bias voltage and $E_0$ is the surface state band-edge. If electron amplitude expands into a perfect, clean Cu(111) surface then it simply leaves the region, never to return. The

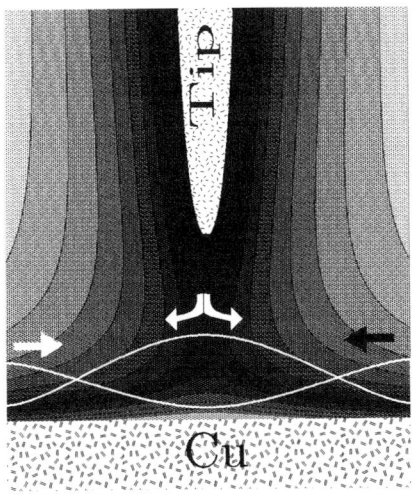

**FIG. 5.** Schematic of the operation of the STM instrument in the mode where it is sensitive to reflection and interference of surface waves

function $\langle \vec{r}|e^{-iHt\hbar}|\vec{r}\rangle$ is then diagonal element of the two dimensional *free* particle Green function. Moving $\vec{r}$ around on such a perfect surface would bring about no change in $dI/dV$.

If there are obstacles on the surface, e.g., Fe adatoms, the electron amplitude coming from $\vec{r}$ will scatter off them, either elastically or inelastically. Inelastic scattering includes any process that decreases the amount of reflected electron amplitude. Amplitude which returns to the tip at $\vec{r}$ (from single or multiple encounters with Fe adatoms) contributes to position and time dependent recurrences in the correlation $\langle \vec{r}|e^{-iHt\hbar}|\vec{r}\rangle$; tunnel current will thus vary with position and with voltage. The amplitude arriving from the tip to another position is given by the Fourier transform of the free part of the amplitude

$$\psi_0 \equiv \int_{-\infty}^{\infty} e^{i\epsilon t/\hbar} \langle \vec{r}'|\vec{r}(t)\rangle_0 \, dt$$
$$= \text{Re}\left[ J_0(k|\vec{r}-\vec{r}'|) + iY_0(k|\vec{r}-\vec{r}'|) \right] \tag{7}$$

where $J_0$ and $Y_0$ are the usual Bessel functions of the first and second kind, respectively and $k = \sqrt{2m\epsilon/\hbar^2}$. The asymptotic form of the tip wave is given in terms of the asymptotic forms of $J_0$ and $Y_0$ and is

$$\psi_0 \to \text{Re}\left[\sqrt{\frac{2}{\pi k r}} e^{ikr - i\pi/4}\right] \tag{8}$$

This wave impinges on the Fe adatoms. For an incident plane wave along the x-axis, the asymptotic form of the scattering from a target in two dimensions is $\psi(x,y) = e^{ikx} + f(\theta)e^{ikr}/\sqrt{r}$.

Since the wavelength of the surface state electrons is large near the Fermi energy in Cu(111) (about 50 atomic units) it is reasonable to assume only s-wave scattering. In this case the asymptotic form becomes

$$\psi(x,y) = e^{ikx} + \sqrt{\frac{2}{\pi k}} e^{i\eta_0 + i\pi/4} \sin\eta_0 \frac{e^{ikr}}{\sqrt{r}} \tag{9}$$

where $\eta_0$ is the phase shift of the scattered s-wave component of $\psi$. If $\eta_0$ is real, the cross section (really a length) is $\sigma = \frac{4}{k}\sin^2\eta_0$.

To account for the presence of an arbitrary number of atoms we let $a_T(r_j)$ be the amplitude arriving from the tip to the $j^{th}$ atom a distance $r_j$ away from the tip; from Eq. (8) this is $a_T(r_j) = \sqrt{\frac{2}{\pi k r_j}} e^{ikr_j - i\pi/4}$ The scattered wave from the $j^{th}$ atom is $a_T(r_j)\,a(r)$ where $r_j$ is the distance to the $j^{th}$ atom and

$$a(r) = \sqrt{\frac{2}{\pi k}} e^{i\pi/4} \frac{(\alpha_0 e^{2i\delta_0} - 1)}{2i} \frac{e^{ikr}}{\sqrt{r}} \tag{10}$$

The amplitude arriving at atom $j'$ having come from the tip and scattered off atom $j$ is then $a_T(r_j)\,a(r_{jj'})$, where $r_{jj'}$ is the distance between atoms $j$ and $j'$. The total amplitude for scattering once and returning to the tip at $\vec{r}$ is $g_1(\vec{r}) = \sum_j a_T(r_j)\,a(r_j)$ The amplitude having scattered once from atom $j$ and thence from $j'$ before returning to the tip is $g_2(\vec{r}) = \sum_j \sum_{j'} a_T(r_j)\,a(r_{jj'})\,a(r'_j)$.

The complete multiple scattering for all orders yields

$$\frac{dI((\vec{r},\epsilon)}{dV} \propto \text{Re}\left[\sum_{n=0}^{\infty} g_n(\vec{r})\right]$$
$$= \text{Re}\left[1 + \mathbf{a_T}\left(1 + \mathbf{A} + \mathbf{A}^2 + \cdots\right)\mathbf{a}\right]$$
$$= \text{Re}\left[(1 + \mathbf{a_T}[1 - \mathbf{A}]^{-1}\mathbf{a})\right] \tag{11}$$

where $\mathbf{a_T}$ and $\mathbf{a}$ are vectors whose length is N, the number of Fe adatoms, and $\mathbf{A}$ is an N × N matrix. At each new position the distances from tip to the adatoms changes, so the vectors $\mathbf{a_T}$ and $\mathbf{a}$ change, but the matrix of atom-to-atom amplitudes $\mathbf{A}$ is fixed for constant bias voltage.

Expression (11) was used in all the comparisons with experiment which follow. At fixed energy there are just two parameters to specify: the s-wave scattering phase shift $\delta_0$, and the inelastic attenuation $\alpha_0$. The data we examined are (1) STM topographs of single Fe atoms on Cu(111), (2) the bias

**FIG. 6.** dI/dV near a single Fe atom. Solid:experiment; dashed:theory, with a phase shift of $\pi/2$.

voltage dependence of the differential conductivity (dI/dV) of the STM tunnel junction at the center of a 60 atom radius = 88.7Å circle of Fe atoms (Fig. 4), (3) STM topographs of a 60 atom 88.7Å circle of Fe atoms for several voltages near the Fermi energy, and (4) the STM topograph of a stadium shaped "corral" consisting of 76 Fe atoms.

By adjusting the phase shift, and assuming no attenuation, it is possible to get a fairly good fit to the $dI/dV$ curve for the center of the 60 atom circle. However, the phase shift one must use to do this is incompatible with the single atom data ((Fig.(6), which requires a phase shift of near $\pm\pi/2$. Since the atoms are almost certainly nearly independent of one another, spaced as they are by a minimum of approximately 10 Angstroms, the single atom phase shift must be used in the many atom algorithm. If one uses a phase shift of $\pm\pi/2$ in the multiatom corrals, however the results are very poor. Essentially, the scattering is much too strong; the resonance structure in dI/dV is too narrow, giving too high a Q compared to experiment.

The dilemma can be removed by realizing that the scattering need not be completely coherent. If the iron atoms couple the surface with the (degenerate) bulk states, then scattering into the bulk Bloch states will occur, causing attenuation on the scattered flux in the surface states. If such inelastic processes remove electrons from the surface states, the scattering is modified by an inelastic factor $\alpha_0$, so that $\exp[2i\eta_0] \equiv \alpha_0 \exp[2i\delta_0]$ and $\delta_0$ is real. The single atom cross section is then

$$\sigma = \frac{|\alpha_0 e^{2i\delta_0} - 1|^2}{k} + \frac{(1-\alpha_0^2)}{k} \equiv \sigma_{\text{elastic}} + \sigma_{\text{inelastic}} \qquad (12)$$

Suppose we make the very simple assumption that the Fe atom completely absorbs the incoming s-wave, but leaves the p-wave and all higher partial waves alone. The scattering wavefunction would then be

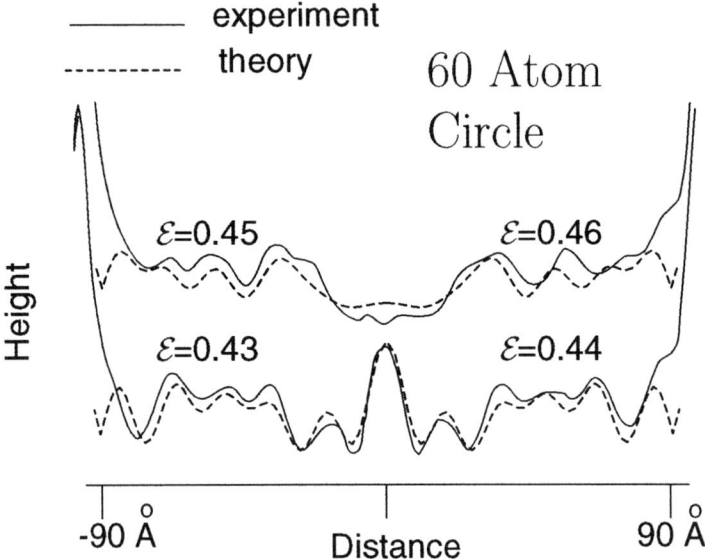

**FIG. 7.** dI/dV slices across the center of the 60 atom circle at two different bias voltages.

$$\psi(\vec{r}) = (\text{plane wave}) - (\text{outgoing s-wave}) =$$
$$= e^{ikx} - \frac{1}{2}H_0(kr)$$
$$\rightarrow e^{ikx} - \frac{1}{2}\sqrt{\frac{2}{\pi k}}\frac{e^{-i\pi/4+ikr}}{\sqrt{r}}$$
$$= e^{ikx} + \sqrt{\frac{2}{\pi k}}\frac{e^{i\pi/4}(e^{2i\delta_0}-1)}{2i}\frac{e^{ikr}}{\sqrt{r}}$$
$$\Rightarrow \delta_0 = a + i\infty \tag{13}$$

giving an effective phase shift of $\pi/2$. Thus we say that the Fe adatoms absorb all the incoming s-wave amplitude and act as "black dots".

Using this black dot assumption, we have very little else to adjust. The phase shift is locked at $\pi/2$, and the other parameters ($\alpha_0 = 0$, energy fixed by bias voltage, effective mass, position of the atoms) are determined by other experiments or are very well known. We do find it best to adjust the energy of the surface state band by 0.01 V; this however is within the experimental error (earlier photoemission experiments) and probably represents a better determination of the surface state band energy. This adjustment is indicated in Fig.(7).

Finally we show the theory-experiment comparison for the 76 atom stadium with no adjustment of parameters.

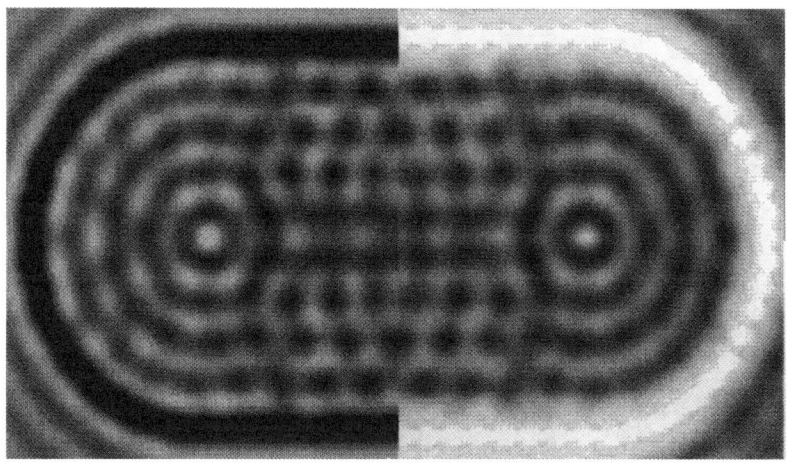

**FIG. 8.** Direct comparison of theory and experiment for the 76 atom stadium corral. (Right side shows experimental results, left side shows the results of a multiple scattering calculation in the black dot limit)

## CONCLUSION

It is remarkable how many assumptions and approximations have gone into the theory of the STM quantum corral images. We list some of them now.

1. independent quasiparticle model for the electrons

2. purely quadratic $E$ vs. $k$.

3. s-wave scattering only

4. multiple scattering theory used

5. black dot model for Fe-quasiparticle interaction

6. no quasiparticle-phonon scattering

Quasiparticle-phonon scattering would have caused additional damping for which there is no evidence. The same holds true for quasiparticle-quasiparticle scattering.

The theory presented for the quantum corrals is completely quantum mechanical. In this sense the "mesoscopic" potential of the corrals has not yet been realized. This will become possible as new kinds of structures are built.

The hope that the quantum corrals would shed some light on the question of chaos and quantum mechanics has been somewhat dimmed by the fact of the low Q's for the corrals. There is no time for chaos to play a role. Various possibilities suggest themselves for raising the Q values, including trying different adatoms and using thin films (removing the bulk states altogether).

One of the most glaring difficulties with the theory is seen in Fig.(4). Above a voltage of 0, (at the Fermi energy), there is an obvious peak shift. In the positive voltage range, electrons are tunneling off the tip onto the surface into empty surface Bloch modes. These are not exactly equivalent to holes below the Fermi level, due to electron correlation effects. This is but one of very many questions and opportunities for further study that the STM quantum corral experiments raise.

## ACKNOWLEDGMENTS

This research was supported by the National Science Foundation under grant number CHE-9014555. EJH acknowledges helpful conversations with S. Chan.

## REFERENCES

1. M.F. Crommie, C.P. Lutz, and D. M. Eigler, *Nature* **363**, 524 (1993); *Phys. Rev.* **B 48**, 2851 (1993); *Science* **262**, 218 (1993).
2. D. M. Eigler and E. K. Schweitzer, *Nature* **344**, 524 (1990); J.A. Stroscio and D. M. Eigler, *Science* **254**, 1319 (1991).

3. O. Bohigas, M.-J. Giannoni and C. Schmit, J. Physique. Lett. **45**, (1984) L-1015.
4. E. J. Heller, Phys. Rev. Lett. **53**, 1515 (1984).
5. M. V. Berry, Proc. R. Soc. Lond. **400** 229 (1985).
6. C.M. Marcus, A.J. Rimberg, R.M. Westervelt, R.F. Hopkins, and A.C.Gossard, *Phys. Rev. Lett.* bf 69 506 (1992).
7. R. Landauer, *Phil. Mag.* **21** 863-7 (1970).
8. R.A. Jalabert, H.U. Baranger, and A. D. Stone, *Phys. Rev. Lett.* **65**, 2442 (1990).
9. R. Blümel and U. Smilansky,*Phys. Rev. Lett.* **60**, 477 (1988).
10. W. H. Miller *J. Chem. Phys.* **53** 5(1970).
11. E. J. Heller, M. F. Crommie, C. P. Lutz and D. M. Eigler, *Nature* **369**, 464 (1994).

# Ion-atom collision measurements relevant to fusion plasmas

## H B Gilbody

*Department of Pure and Applied Physics,*
*The Queen's University of Belfast, Belfast, United Kingdom.*

**Abstract.** Some of the ion-atom processes relevant to plasma heating by fast neutral beams, modelling of edge plasmas and plasma diagnostics in current and next-step Tokamak devices are considered. Results of measurements on a number of selected processes involving charge transfer, ionization and excitation are discussed in relation to theoretical predictions.

## INTRODUCTION

Controlled thermonuclear fusion of the isotopes of hydrogen in a high temperature plasma offers the prospect of a major source of energy with low environmental impact and virtually inexhaustible supplies of fuel. Over the past few decades we have seen steady progress towards practical realisation based on schemes in which the plasma is heated during magnetic confinement. Attention has also been directed towards heating and inertial confinement by energetic laser or particle beams. While both these general approaches involve many atomic physics problems, in this short review the scope will be limited to aspects of magnetic confinement fusion.

The preferred nuclear reaction is

$$D + T \rightarrow (^4He + 3.52 \text{ MeV}) + (n + 14.06 \text{ MeV})$$

Plasma ion temperatures $T_i > 10$ keV are required with Lawson number $Nt > 10^{14}$ s cm$^{-3}$ where N is the plasma density and t is the confinement time. In the schemes envisaged, the fast neutrons are then trapped in a lithium wall or blanket where kinetic energy is converted to recoverable heat and tritium breeding occurs through the reaction

$$^6Li + n \rightarrow {}^4He + T + 4.80 \text{ MeV}$$

The fast alpha particles give up their energy with the plasma constituents and sustain plasma heating in the 'ignition' phase.

Of the many experimental magnetic confinement schemes, measurements utilising toroidal plasmas in Tokamak devices have been generally the most successful. In recent years, most attention has been focused on experiments in three large Tokamaks; the Joint European Torus (JET) in the UK; the Tokamak Fusion Test Reactor (TFTR) in the USA and JT60-U in Japan. Most of the measurements to date have been carried out with deuterium plasmas but preliminary measurements with DT plasmas commenced in 1991. However the establishment of reliable methods of impurity control is a major outstanding problem relevant to the design of next-step devices such as the proposed International Thermonuclear Experimental Reactor (ITER). This device, the design of which is to be finalised in 1998, is intended to demonstrate feasibility by generating at least 1000 MW of power under ignition conditions. This would provide the fusion technology for the subsequent construction of a demonstration reactor (DEMO) to be followed by a prototype reactor probably by the middle of the 21st century.

Among a number of general reviews, a NATO ASI Series publication[1] provides a general discussion of the role of atomic and molecular processes in Tokamak devices. Current and future cross section needs are considered in a summary report by Janev[2] and in a recent review by Summers[3].

The range of relevant collision processes is very wide. Apart from the partially or fully ionized primary (hydrogen and helium) constituents in the different regions of the plasma, a detailed understanding of the role of plasma impurities, continues to be of major importance. These impurities, which originate mainly from plasma facing materials, can include low Z species such as Be, B, C and O ranging from neutral to fully ionized and high Z species such as $Fe^{q+}$, $Mo^{q+}$ and $W^{q+}$ in respective charge states q up to about 23, 32 and 34 in current devices. Reliable data are required on collisions between heavy particles including charge exchange, ionization and excitation; electron collisions relevant to step-wise excitation, de-excitation, electron-ion recombination processes and bremsstrahlung production; the interaction of plasma constituents with facing materials including problems of erosion, heat dissipation and plasma contamination.

The scope of this short review will be limited to a discussion of some of the results of measurements of heavy particle collision processes important to current and future Tokamak devices. Some selected processes involving charge transfer, ionization and excitation will be considered which are relevant to

(a) plasma heating by fast neutral beams of the central region of the plasma where fusion is intended to occur,
(b) modelling of the comparatively cool edge of the plasma where effective control and removal of impurities is of crucial importance, and
(c) diagnostics of plasma parameters by neutral beam probes.

Accurate experimental measurements over an energy range ~ 1 eV to 1 MeV, particularly on processes involving H atoms, provide a substantial challenge.

Theoretical descriptions are also difficult but reliable measurements can provide important benchmarks for tests of the approximations used. In some cases, the measurements can also provide a basis for general semi-empirical scaling relations to predict approximate cross sections for a wide range of different species.

Although the D and T isotopes of hydrogen are of greater relevance than H, most experimental measurements have been made with ions or atoms of H. For this reason and, since isotope effects do not arise for most of the processes considered, reference will generally be made to H in this discussion.

# RELEVANT ION-ATOM COLLISION PROCESSES

## Plasma heating by fast neutral beams

In a Tokamak device, direct ohmic heating of ions via the induced toroidal current becomes ineffective at temperatures above a few keV. Supplementary heating by injection of fast neutral beam or by rf heating is essential. In the present large Tokamaks, heating by injection of 40-80 keV amu$^{-1}$ hydrogen isotope beams at the tens of MW power level has been extremely effective. The heating beams are produced by charge transfer neutralization of fast positive hydrogen ions (comprising $H^+$, $H_2^+$ and $H_3^+$) in hydrogen thereby providing full, half and one third energy neutral components. The spacial and energy profiles of these components can be assessed[4], through a knowledge of the appropriate cross sections, by observing the Doppler shifted Balmer alpha emissions from the fast excited neutrals.

In ITER, where higher energy beams are required, injectors being developed based on neutralization of fast $H^-$ ions are expected to provide higher neutralization efficiencies. Fast helium beams, which have also been used for heating in present devices may also be more effective for energy deposition in the much larger ITER torus. However, a significant metastable content in the fast helium beams prepared by charge transfer neutralization of $He^+$ ( see [5] ) can modify the energy deposition profile.

For heating by hydrogen neutral beams, the fast hydrogen atoms pass through the magnetic confining fields and undergo electron removal in collisions with the plasma constituents. Charge transfer

$$H^+ + H \rightarrow H + H^+ \qquad (1)$$

and ionization $\qquad H^+ + H \rightarrow H^+ + H^+ + e \qquad (2)$

in collisions with plasma protons are important in this context. The resulting fast protons are then trapped in the confining field and give up their energy in further collisions with the plasma constituents.

The effectiveness of neutral beam heating can be strongly influenced by collisions with multiply charged impurity ions. High cross sections for both charge transfer

$$X^{q+} + H \rightarrow X^{(q-1)+} + H^+ \qquad (3)$$

and ionization $\qquad X^{q+} + H \rightarrow X^{q+} + H^+ + e \qquad (4)$

can seriously modify the energy deposition profiles. In addition, electron capture into high $n$ states which then decay radiatively can lead to substantial cooling of the plasma.

Estimates of beam penetration and energy deposition with H beams have normally assumed that the beam particles are all in the ground state. However for beam energies above 200 keV amu$^{-1}$, at high plasma densities, the radiative and collision times of excited beam particles become comparable. It is then necessary to consider processes involving excited beam atoms which will lead to the enhancement of beam attenuation.

Similar considerations apply to heating by fast helium beams but, in this case, processes involving both one and two electron removal must be considered. Transfer ionization processes of the type

$$X^{q+} + He \rightarrow X^{(q-1)+} + He^{2+} + e \qquad (5)$$

may include contributions from two-electron capture into states which subsequently undergo autoionization. Account must also be taken of processes relevant to an initial admixture of metastable atoms in fast helium beams.

## Plasma edge processes

A detailed understanding of collision processes in the comparatively cool plasma edge regions is essential for the efficient design of divertors for the control and removal of impurities including the helium 'ash'[2]. Charge transfer processes which are near-resonant or moderately exothermic are of particular concern since they may have large cross sections in the ~ 1 eV - 500 eV energy range of interest. In present devices, additional magnetic fields allow the edge plasma to flow into the divertor region where it undergoes neutralization in collisions with the divertor plates. The resulting neutrals are either pumped away or undergo further collisions. Charge transfer collisions are important in determining the degree of recycling of hydrogen and other species together with the associated energy losses.

While hydrogen atoms and molecules and their ions together with helium atoms and ions are primary constituents of the edge plasma, impurities such as Be, B, C and O in all stages of ionization are of interest. Some of these impurities arise from the use of low Z plasma facing materials in the divertor region. However, in ITER, erosion and power dissipation considerations may dictate the use of high Z metals such as W, Mo, Nb or V in spite of the fact that the permitted influx into the plasma of such species must be kept very small to avoid unacceptable cooling of the plasma core [6]. Molecular impurities such as $H_2O$ and CO are also relevant. The possible future use of carbon fibre composites as plasma facing materials may also give to a range of $C_nH_m$ hydrocarbon impurities.

Of particular importance in modelling edge plasmas are symmetrical resonant charge transfer processes such as $H^+$-H collisions. Moderately exothermic charge transfer processes involving partially ionized impurity ions may also take place very effectively through curve crossings leading selective population of a limited number of excited states. Excited product states decay radiatively thereby cooling the plasma. The introduction of cold hydrogen or impurities such as Ne, Ar or Kr by 'gas puffing' may provide enhanced cooling in the divertor region and also diagnostic information. This may be important in ITER where there is difficulty in reducing the anticipated power dissipation at the divertor plates to practicable levels.

## Neutral beam diagnostics

### *Charge exchange recombination spectroscopy*

In this technique, which has been used extensively with both H heating beams and dedicated H probes[7], for the diagnostics of fully ionized species (eg $He^{2+}$, $C^{6+}$, $O^{8+}$), spectroscopic observations are made of the decay of the H-like excited products formed in state selective electron capture

$$X^{z+} + H \rightarrow X^{(z-1)+}(n,l) + H^+ \qquad (6)$$

Local ion densities, ion temperatures and plasma rotational velocities can be deduced from studies of the Doppler-broadened line profiles. The main excited product channels generally radiate in the VUV. However, observations of the non-dominant high n capture channels which radiate in the visible region, while much weaker, are preferable because of the convenience of fibre optics coupling to the spectrometer and ease of calibration of the latter. Account has to be taken of the effect of Zeeman and motional Stark effects and $l$ mixing collisions on the observed line intensities and profiles.

Similar measurements have also been carried out with fast He beams[3] in which observations of $He^+$ ($n = 4 \rightarrow n = 3$) decay at 468.6 nm following one-electron capture have been utilised for diagnostics of $He^{2+}$ populations in different regions of the plasma. Corrections arising from electron capture by even small fractions of fast metastable helium atoms may be substantial.

Lithium beam charge exchange spectroscopy has proved effective in probing edge plasmas. For example Schorn et al[8] have used a low energy Li beam probe to measure temperatures of $C^{6+}$ impurity ions through observations of the Doppler broadened $C^{5+}$ emissions.

## *Beam emission spectroscopy*

This technique is based on spectroscopic analysis of the emissions from fast excited beam atoms formed in collisions with the plasma ions and electrons. Accurate cross sections for direct excitation of H or He by $H^+$, $He^{2+}$ or $X^{q+}$ impurity ions are important in this context. In one example of the use of this approach, Boileau et al[9] have shown that observations of Balmer alpha radiation emitted from fast hydrogen heating beams can provide local proton densities and measurement of toroidal and poloidal magnetic fields via Zeeman and motional Stark effects.

## *Fast alpha particle diagnostics*

Techniques appropriate to measurements of the density and slowing-down velocity distributions of fast alpha particles in an ignited plasma are important. Some years ago Post (see[1]) proposed the use of a 6 MeV Li beam 'merged' with the fast alpha particles to collide at low cm energies and then record the escaping He atom products of two-electron capture. The use of two-electron capture with fast He beams now seems likely together with observations of the emissions from an appropriate excited state of He.

# EXPERIMENTAL METHODS FOR STUDIES OF COLLISIONS INVOLVING H ATOMS

A brief summary of the main experimental techniques used in collision studies involving H atoms is appropriate since, unlike stable gases such as $H_2$ or He, special well characterised targets of atomic hydrogen have had to be developed to facilitate accurate cross section measurements. Few of the measurements are absolute and most rely on normalisation of relative cross sections to other data. The main techniques are:-
(a) *Crossed beam methods employing highly dissociated thermal energy hydrogen beams.*
The modulated crossed beam technique pioneered by Fite et al[10] provided much of the early data on charge transfer and ionization but the accuracy is severely limited by poor signal to background ratios. The crossed beam coincidence counting technique incorporating TOF spectroscopy developed

by Shah and Gilbody[11] provides high sensitivity and accuracy. Studies of electron capture into excited states in $X^{q+}$-H collisions have been carried out using photon emission spectroscopy (PES) first used by Ciric et al[12] using intense ion beams and hydrogen beams of high density from a Wood's tube discharge source. While of low sensitivity, the method provides results of moderate to high accuracy.

(b) *Furnace target methods.*

In this approach, first used by Lockwood and Everhart[13], the primary ion beam is passed through highly dissociated hydrogen contained within a tungsten tube furnace. Many measurements of total electron capture cross sections have since been based on this approach. Energy loss studies by Park et al[14] have provided data on direct excitation and ionization of H while translational energy spectroscopy (TES) (involving measurements of either energy gain or loss), first used by McCullough et al[15], has provided data on state-selective electron capture in $X^{q+}$- H collisions.

(c) *Fast intersecting beam methods.*

In this approach, a target beam of fast H atoms usually produced by electron capture neutralisation of an $H^+$ beam, is arranged to collide with the primary ion beam in an ultra-high vacuum region. While the use of keV energy beams facilitates collision product detection through particle counting, low signal levels and high signal to background ratios make such measurements difficult. Account must also be taken of the excited state population of the H beams. Kim and Meyer[16] arranged the beams to intersect at right angles and used electric field ionization to define and control the H($n$) state populations. Several experiments using a merged beam configuration have provided data for collisions at very low c.m. energies. These include measurements by Newman et al[17] of $H^+$- H($1s$) charge transfer down to c.m. energies of 0.1eV and studies by Koch and Bayfield[18] of electron removal in $H^+$- H($n$) collisions for $n$ = 44 to 50 in the c.m. energy range 0.4 - 61eV.

# SELECTED EXPERIMENTAL DATA

## Charge transfer and ionization in $H^+$-H and $He^{2+}$-H collisions

Processes (1) and (2) have been extensively studied since the late 1950's and data are available for H($1s$) atoms over wide energy range. The most reliable of the experimental data within the range 2 -1500 keV amu$^{-1}$ are shown in Fig 1 together with a few theoretical predictions. The absolute total charge transfer cross sections measured by McClure[19] using a furnace target method are of particular importance since these have been used to normalise (either directly or indirectly) the results of many other measurements on H atoms. The ionization cross sections shown

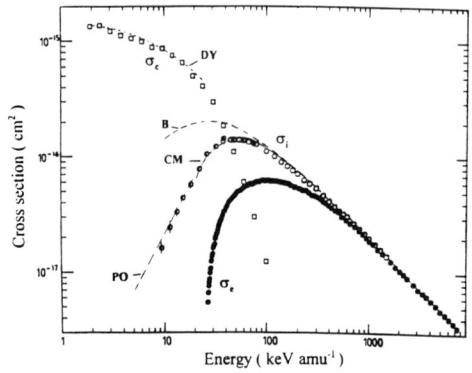

**Fig 1.** Cross sections $\sigma_c$ for charge transfer and $\sigma_i$ for ionization in $H^+$ - H($1s$) collisions together with equivelocity cross sections $\sigma_e$ for e-H($1s$) ionization.
$\sigma_c$ : □, McClure[19]; DY, theory-PSS method[25]
$\sigma_i$ : O, ●, Shah & Gilbody[11,20]; B, theory-first Born[22]; CM, theory-CDW-EIS method[23]; PO, theory, - non-adiabatic treatment[24].
$\sigma_e$ : ●, Shah et al[22]

for $H^+$ impact and for electron impact[21] are based on variants of a crossed beam technique with coincidence counting of the collision products. These $H^+$-H ionization cross sections[11,20] were normalised to the first Born approximation prediction[22] at 1500 keV. However in the subsequent measurements[21] of ionization by $H^+$ and equivelocity electrons, the low energy cross sections were also effectively normalised to the absolute charge transfer cross sections of McClure[19].

For ionization, the first Born values can be seen to diverge greatly from the lower energy experimental values particularly near the cross section peak. Improved agreement down to about 20 keV is provided by CDW-EIS calculations[23] in which distortion is accounted for in the exit channel by the continuum distorted wave approximation and in the entrance channel by the eikonal approximation. Below 20 keV, the calculations of Pieksma and Ovchinnikov[24] within the framework of non-adiabatic transitions in slow collisions are in reasonable accord with experiment. In the case of charge transfer, the perturbed stationary state calculations of Dalgarno and Yadav[25] are in remarkably good agreement with experiment.

While the high energy data in Fig 1 are important to modelling of fast H atom beam penetration, the large low energy charge transfer cross sections, which are dominated by resonant electron capture, are important in descriptions of the plasma edge. Lower energy measurements include absolute $H^+$- H($1s$) charge transfer cross sections[17] in the range 0.2 - 300 eV based on a merged beam method.

Charge transfer in $He^{2+}$ - H collisions has been the subject of many experimental and theoretical studies (see reviews [26], [27]). Total electron capture cross sections $\sigma_{21}$ are dominated by the accidentally resonant product channels $He^+$ (*2s* or *2p*) + $H^+$ and (due to the Coulomb repulsion between the charged collision products) peak at about 10 keV amu$^{-1}$ with a large value of about $2 \times 10^{-15}$ cm$^2$.

Cross sections for $He^+(2p)$ and $He^+(2s)$ formation which have been based on observations of 30.4 nm emission following either spontaneous or electric-field induced decay of fast $He^+(2p)$ or $He^+(2s)$ atoms are described reasonably well by calculations based on both AO and MO approaches. While electron capture into $He^+(n = 2)$ states accounts for more than 90% of $\sigma_{21}$ at energies below about 50 keV amu$^{-1}$, the need to carry out plasma diagnostics using charge exchange recombination spectroscopy in the visible region has led to recent studies of $He^+$ ($n = 4 \rightarrow n = 3$) 468.4 nm emission by Hoekstra et al[28] and by Frieling et al[24] in the energy range 2 - 132 keV amu$^{-1}$. These measurements (Fig 2) are based on a crossed beam technique in which the separate $He^+(4l)$ contributions are

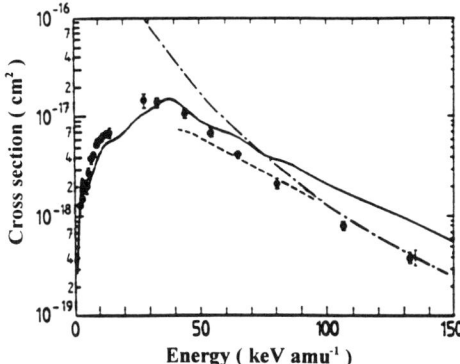

Fig 2. Cross sections for $He^+$ ($n = 4 \rightarrow n = 3$) emission following electron capture in $He^{2+}$ - H collisions. ● : Experiment [28], [29]. —— : Theory, AO method [30]. – – – : Theory, CTMC method [31]. —·— : Theory, CDWA method [32]. ( adapted from [28] )

determined by deconvoluting the spacial distribution of intensities along the beam path arising from the different lifetimes of the $4l$ states. In Fig 2 the measured cross sections can be seen to be in good general accord with the close coupling A.O. calculations of Fritsch[30] below 50 keV amu$^{-1}$ and, at higher energies, approach values based on the distorted wave approximation[32]; at intermediate energies, classical-trajectory-Monte-Carlo (CTMC) calculations[31] are in good agreement with experiment.

## Electron capture in collisions of $He^{2+}$ ions with He and $H_2$

Cross sections $\sigma_{21}$ and $\sigma_{20}$ for one and two electron capture by $He^{2+}$ ions in the He and $H_2$ have been the subject of many experimental and theoretical studies over a very wide energy range (see summary in [33]). In He, values of $\sigma_{20}$ decrease with increasing energy and, at energies above about 10 keV amu$^{-1}$, are exceeded by $\sigma_{21}$. In $H_2$, at energies above about 3 keV amu$^{-1}$, where $\sigma_{21}$ is more than an order of magnitude greater than $\sigma_{20}$, both cross sections rise to peak values at about 20 keV amu$^{-1}$ and thereafter decrease. Okuno et al[33]., using a beam attenuation technique, which employs an octupole ion beam guide, have extended available data in both He and $H_2$ downwards in energy from 667 to 0.3 eV amu$^{-1}$. In He, values of $\sigma_{20}$ derived from these measurements exhibit a steady increase with decreasing energy characteristic of resonant two-electron capture while values of $\sigma_{21}$ show a corresponding rapid decrease; at 0.3 eV amu$^{-1}$ $\sigma_{20}$ exceeds $10^{-15}$ cm$^2$. In the case of H$_2$, values their measured values of $\sigma_{20}$ also exhibit an increase with decreasing energy (Fig 3) becoming larger than corresponding values in He. This result was, at first, surprising since the He + H$^+$ + H$^+$ product channel would be about 28 eV exothermic for a He($1s^2$) ground state product assuming Franck-Condon transitions.

Shimakura et al[34] have considered both one and two-electron capture in He$^{2+}$ - H$_2$ collisions in terms of the adiabatic potential energy curves of the (HeH$_2$)$^{2+}$ system. A series of avoided crossings between about 3.5 and 8 au is shown to determine both the magnitude and energy dependence of $\sigma_{21}$ and $\sigma_{20}$. At the lower impact energies, they show that the adiabacity of these crossings increases leading to a large probability for formation of He($1s3l$) excited products through two-electron capture. At higher impact energies, the curves become more diabatic resulting in the dominance of one-electron capture. Their calculated cross sections (Fig 3) for both $\sigma_{20}$ and $\sigma_{21}$ can be seen to be in reasonable accord with experiment. More recently, Sato et al[35] have confirmed the importance of the accidentally resonant He($1s3l$) + H$^+$ + H$^+$ two-electron capture product channel by puffing cold H$_2$ into a low temperature helium plasma and observing that very strong He ($n = 3 \rightarrow n = 2$) emissions then arise.

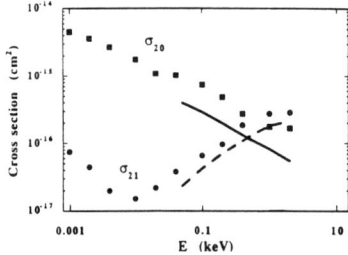

Fig 3. Cross sections for $\sigma_{21}$ and $\sigma_{20}$ for one and two-electron capture in He$^{2+}$-H$_2$ collisions.

■, ● : $\sigma_{20}$, $\sigma_{21}$, experiment [33].

—— , – – : $\sigma_{20}$, $\sigma_{21}$, theory [34].

( adapted from [34] )

In the case of one-electron capture in He$^{2+}$ - H$_2$ collisions, recent TES studies by Hodgkinson et al[36] in the range 0.5 - 2 keV amu$^{-1}$ have confirmed indications in PES studies by Hoekstra et al[37] that the He$^+$($1s$) + H$^+$ + H($n$ = 2) product channel is dominant at low energies. Non-dissociative electron capture (mainly into the He$^+$ ($n$ = 2) state) accounts for only 1% of $\sigma_{21}$ at 0.5 keV amu$^{-1}$ rising to about 25% at 2 keV amu$^{-1}$.

## Charge transfer and ionization in collisions involving multiply charged ions of velocity v > 1 au

The late 1970's and early 1980's saw a major upsurge of interest in measurements of charge transfer and ionization in collisions between multiply charged ions and H, $H_2$ and He. This interest was stimulated by the CTMC calculations of Olson and Salop[38] for collisions of bare nuclei with $H(1s)$ which predicted large cross sections for both (3) and (4) which increased as q increased. This general theoretical approach has since proved very effective in describing experimental data at intermediate velocities. Most of the early experimental studies of (3) and (4) for $q \geq 3$ and the corresponding processes in $H_2$ and He were carried out mainly at velocities $v > 1$ au (correspond to 25 keV amu$^{-1}$) because of the limited availability of suitable sources of low energy multiply charged ions. Measurements of charge transfer cross sections $\sigma_{q,q-1}$ for (3) have been based almost entirely on the furnace target technique.

Fig 4 shows a typical set of experimental data for (3) and (4) for collisions of $Ar^{(2-9)+}$ ions with $H(1s)$ atoms which indicate the relative importance of charge transfer and ionization over a wide energy range. In this case, the low energy measurements[39] of $\sigma_{q,q-1}$ using the furnace target technique can be seen to be in good accord with the higher energy measurements[40] based on the crossed beam coincidence technique. For $q \geq 3$, values of $\sigma_{q,q-1}$ for a given velocity can be seen to increase with q. At velocities beyond the cross section peak, (which occurs at higher velocities as q increases) ionization cross sections at a given velocity also increase with q.

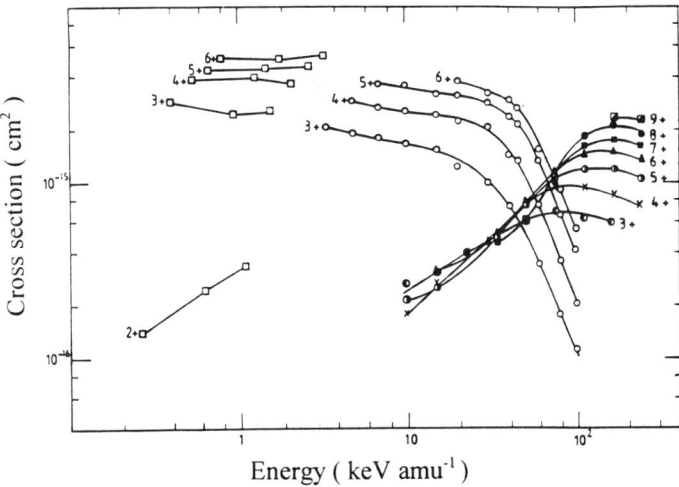

**Fig 4.** Cross sections for charge transfer and ionization in $Ar^{q+}$ - $H(1s)$ collisions.
□ : charge transfer - furnace target method [38].
○ : charge transfer - crossed beam coincidence method [40].
◢, ●, ■, ▲, ◐, ×, ◑ : ionization - crossed beam coincidence method [40].

The results in Fig 4 indicate that for $q \geq 3$, the total cross sections for one-electron removal from H atoms decrease quite slowly with increasing energy over a wide energy range. This is typical of data for many other collision systems (see review [41]). In addition, at velocities beyond the peak values, cross sections for either charge transfer or ionization (for a given q) for q > 3 are found to scale according to simple relations of the form

$$\sigma = \sigma_0 q^n \tag{7}$$

where the scaling parameters $\sigma_0$ and n depend on velocity. Values of n for $\sigma_{q,q-1}$ vary from about 1.5 to 3 over the velocity range $3 - 6.5 \times 10^8$ cm s$^{-1}$. The more limited data for ionization of H atoms are consistent with n ≅ 1.5. At these high velocities, cross sections for both charge transfer and ionization are only weakly dependent on species. Measured cross sections $\sigma_{q,q-1}$ for the fully stripped ions He$^{2+}$, Li$^{3+}$, B$^{5+}$ and C$^{6+}$ in H at high velocities[42] appear to approach a common curve in close agreement with the relation $\sigma_{q,q-1} = 25.14$ q$^3$ a$_0^2$v$^{-7}$ au predicted by Crothers and Todd[43] on the basis of the Bohr-Lindhard model of electron capture. The limited data available for ionization of H by fast bare nuclei[44] are consistent with the Z$^2$ scaling predicted by the first Born approximation but at velocities which are observed to increase with Z.

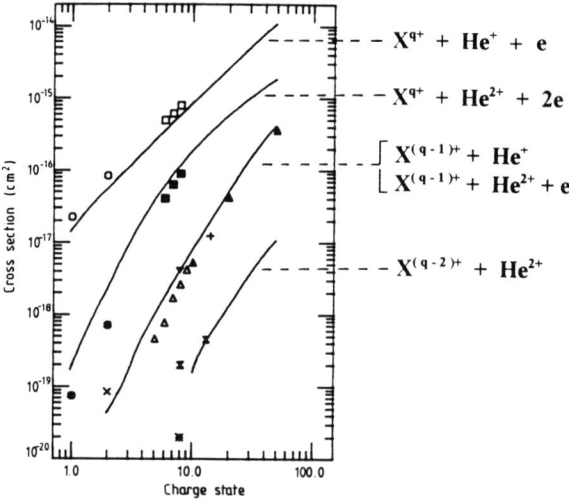

**Fig 5.** Cross section for electron capture and ionization in X$^{q+}$ - He collisions calculated by Olson et al [48] (solid curves) compared with experimental data: O, ●, [49]; □, ■, [50]; ×, [51]; Δ, [52];▼, [53]; ▲, [54]; +, [55]; ✕, [56]; ✗, [57]; ✻, [58]. ( from [48] )

Measured total one-electron capture cross sections and ionization cross sections for multiply charged ions in H$_2$ and He exhibit the same general characteristics as the corresponding results for H. On the basis of these data, there have been a number of attempts to develop general scaling relations (see for example [45], [46], [47]) to allow approximate cross sections to be predicted for any high q primary ion in H, H$_2$ or He.

For He, a useful insight into the relative importance of single and double electron capture, single and double ionization and transfer ionization (process 5) at high velocities is provided by the calculations of Olson et al[48] who have used a semi-classical description of the atom which allows the use of the CTMC method. In Fig 5, their theoretical predictions for the different electron capture and ionization channels are shown for 1 MeV amu$^{-1}$ X$^{q+}$ ions where q = 1 to 50. The calculations can be seen to be in reasonable general accord with the limited experimental data available. Single ionization is dominant with cross sections scaling roughly as q$^2$. For q < 10, double ionization cross sections scale as q$^3$ while, for q > 10 an approximately q$^2$ scaling is predicted. In the case of both one and two-electron capture, a q$^3$ scaling relation is indicated.

## Charge transfer in collisions involving multiply charged ions at velocities v < 1 a.u.

In recent years, low energy measurements of total electron capture cross sections for (3) in H and the corresponding processes in H$_2$ and He have been facilitated by the increasing availability of ECR and similar intense sources of multiply charged ions. However data for many partially ionized fusion impurity species is still very limited. At velocities V < 1 a.u. for q ≥ 3, exothermic one-electron capture may take place very effectively through avoided crossings of the adiabatic potential energy curves describing the initial and final molecular systems. In many reactions, electron capture is dominated by a limited number of such crossings. These occur at internuclear separations R$_c$ ≅ (q-1)/ΔE where ΔE is the energy defect for a collision channel leading to the formation of product ions in specified excited states ( $n,l$ ). The low energy values of σ$_{q, q-1}$ evident in Fig 4 for Ar$^{q+}$ - H electron capture for q ≥ 3 are indicative of a number of curve crossings which result in large total cross sections weakly dependent on energy.

Simple q scaling relations which can provide an approximate description of charge transfer at higher velocities do not, in general, apply. Accurate theoretical descriptions are difficult (see review [59]) and generally involve solution of the coupled channel equations with a molecular basis expansion of the total electron wave function. Approximate descriptions such as those based on the multi-channel Landau-Zener (MCLZ) model are sometimes useful in predicting the dominant product channels. However, in order to obtain a better understanding of state-selective electron capture and the range of validity of the theoretical models, experimental measurements based on the complementary techniques of TES and PES are very important. Both techniques, in principle, allow identification of the main excited product channels and determination of the corresponding cross sections. While the PES approach provides much greater energy resolution than TES, optical calibration, particularly in the VUV, is difficult. In TES, calibration with respect to known total electron capture cross sections is much simpler and the method (unlike PES) also provides an unambiguous indication of product channels associated with any possible admixture of metastable ions in the primary beam.

In collisions between slow fully stripped ions and H, H$_2$ and He, one-electron capture is highly selective and, most cases, dominated by a single avoided crossing. For example both PES and TES measurements [60], [61] show that, for $C^{6+}$-H, $N^{7+}$-H and $O^{8+}$-H collisions, the dominant products are $C^{5+}(n=4)$, $N^{6+}(n=5)$ and $O^{7+}(n=5)$ respectively in accord with theoretical predictions. For partially ionized species, where a number of avoided crossings may be important, the main product channels are less easy to predict.

The $C^{3+}$-H($ls$) process has been studied by both the TES and PES methods [62], [63] in the overlapping energy ranges 1.5-1.8keV and 9-60keV respectively. In the TES energy change spectrum shown in Fig. 6, the three main peaks

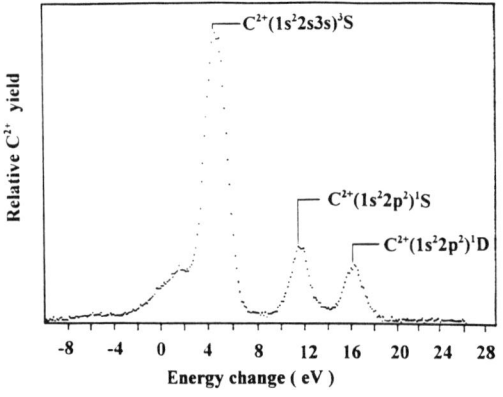

Fig 6. Energy change spectrum for one-electron capture in $C^{3+}$-H($ls$) collisions at 15 keV [62]

can be identified with the $C^{2+}(ls^22s3s)^3S$, $C^{2+}(ls^22p^2)^1S$ and $C^{2+}(ls^22p^2)^1D$ product channels arising from curve crossings at internuclear separations $R_c \cong 11.5$ a.u., 4.7 a.u. and 16.2 a.u. respectively. In Fig 7, cross sections for these three products channels obtained from TES studies are compared with values derived from PES measurements. The most recent theoretical predictions by Errea et al [64] which are also shown, are based on an expansion with 22 molecular states modified with common translation factors. The limited degree of agreement between the TES and PES measurements and with the theoretical predictions indicates the difficulty of carrying out both accurate measurements and calculations.

Rough calculations based on the MCLZ approach have been shown to provide a useful indication of the main product channels in many different processes but the quantitative agreement with experiment is usually poor. For example, in a recent TES study of one-electron capture in $N^{4+}$- He collisions[65], MCLZ calculations satisfactorily predict the main $N^{3+}(ls^2 2p^2)^1S$ product channel for $R_c \cong$ 3.4 a.u. However, the calculations fail to account for three other channels in the range $R_c = 2.8 - 1.8$ a.u. which together account for 37% of the total cross section at 16keV.

Many of the TES measurements reveal the presence of collision channels associated with metastable as well as ground state primary ions. However a detailed analysis of the energy change spectra is precluded unless the metastable fractions are known.

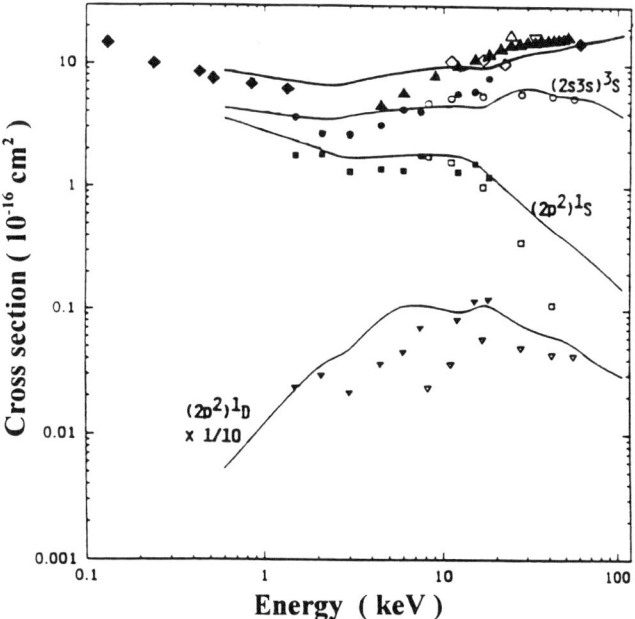

**Fig 7.** Cross sections for $C^{2+}(n,l)$ formation in $C^{3+}$-$H(ls)$ one-electron capture collisions: ●, ■, ▼, TES measurements [62]; ○, □, ▽, PES measurements [63]; ◆, ▽, ▲, ◊ total electron capture cross sections (see [62]); —, theory [64]

For example, TES studies [66] of one-electron capture in $C^{2+}$- $H(ls)$ collisions show that, at 8 keV, about 65% of the $C^+$ products are associated with product channels involving $C^{2+}(2s2p)^3P^0$ metastable primary ions. Some of the partially ionized metallic species of relevance to fusion have metastable states of high statistical weight indicating their likely influence on measured cross sections. This is illustrated by the TES energy change spectrum shown in Fig 8 for $Fe^{3+}$-He collisions at 12 keV.

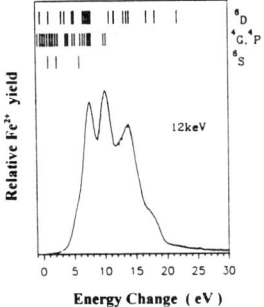

**Fig 8.** Energy change spectrum [67] for one-electron capture in $Fe^{3+}$ - He collisions at 12 keV. Energy defects corresponding to possible collision channels involving $^6S$ ground state and $^4G$, $^4P$ and $^6D$ metastable primary ions are indicated.

In this case, the product channels correlate with $^6$D, $^4$G and $^4$P metastable rather than the $^6$S ground state of the primary ions. Measurements of this type indicate that many published cross sections based on measurements with primary ion beams of unspecified metastable content should be interpreted with care. In future work the development of techniques to allow measurements with state selected primary ions is essential for the resolution of the present ambiguities. Measurements based on the use of double translational energy spectroscopy [68] or ion traps[69] would seem to have considerable potential in this regard.

For helium targets, two-electron capture is significant. For example, PES studies [70] of the $O^{8+}$ - He system show that the cross sections for one and two-electron capture are of comparable magnitude. In this case, double capture leads to the formation of doubly excited helium-like $O^{6+}(n=3,n'=4)$ and $O^{6+}(n=3,n'=3)$ ions. For $C^{4+}$ - He collisions, measured cross sections for two-electron capture exceed those for one-electron capture at energies below about 2.5keV amu$^{-1}$ ( see [71] ). In this case, curve crossing considerations result in the dominance of ground state $C^{2+}(1s^22s^2)^1S$ products. Electron spectroscopy can provide detailed information about the decay of autoionizing states formed in two-electron capture[72]. In the case of slow $N^{7+}$ - He and $O^{7+}$ - He collisions, TES studies [73] have shown that two-electron capture into autoionizing states of $N^{5+}$ and $O^{5+}$ ( the decay of which provide contributions to (5) ) are about one order of magnitude smaller than total cross sections for one- electron capture.

## Excitation of H and He atoms

Experimental studies of direct excitation of the type

$$X^{q+} + H(1s) \rightarrow X^{q+} + H(n,l) \tag{8}$$

have not been extensive. For H$^+$ impact these include the 2.3-26 keV amu$^{-1}$ crossed beam studies of H(2p) and H(2s) excitation by Morgan et al[74], energy loss measurements by Park et al[75] and 40-800 keV amu$^{-1}$ crossed beam measurements for H(np) excitation by Detleffsen et al[76]. Theoretical descriptions of $n = 2$ excitation, which have been based mainly on M.O. and A.O. expansions in a number of different variants, have been discussed by Fritsch and Lin[27].

Cross sections $\sigma(H_\alpha)$ for Balmer alpha emission as a result of the direct excitation process (8), which are of direct relevance to fusion diagnostics, have been measured by Donnelly et al[77]. In terms of the separate cross sections $\sigma(n,l)$ for excitation of the separate fine structure states, $\sigma(H_\alpha) = \sigma(3s) + 0.12\ \sigma(3p) + \sigma(3d)$ where the factor 0.12 is the branching ratio for 3p decay via Balmer alpha radiation. These measured cross sections (Fig 9) can be seen to be in surprisingly poor accord with predictions based on close-coupling calculations [78], [79], [80]. In addition, rough estimates of $\sigma(H_\alpha) = \sigma(n=3) - 0.88\sigma(3p)$ by Detleffsen et al[76] using their values of σ(*3p*) together with renormalised values of σ(*n=3*)due to Park et al[75] can be seen to be in better accord with theory.

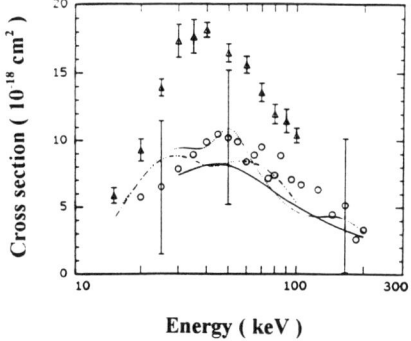

**Fig 9.** Cross sections σ(H$_\alpha$) for Balmer alpha emission in H$^+$-H collisions.
▲, Donnelly et al[77]
o, Detleffsen et al[76]
——— , theory [79]
········ , theory [78]
——-- , theory [80]
( from [76] )

Further measurements seem desirable. However, it is interesting to note that the corresponding cross sections σ(H$_\alpha$) measured by Donnelly et al[77] for excitation in 17-67 keV amu$^{-1}$ He$^{2+}$-H(*1s*) collisions are in good agreement with the close coupling calculations of Fritsch et al[81]. At present there are no measurements for excitation of H by heavier bare nuclei.

The measurements of H(*np*) excitation by Detleffsen et al[76] include data for the multiply charged primary ions He$^{2+}$, Si$^{(2-9)+}$ and Cu$^{(3-11)+}$ which have been considered in terms of a scaling relation derived by Janev and Presnyakov[82]

$$\sigma / q = f(v^2 / q) \qquad (9)$$

for the excitation cross section as a function of the scaled velocity v. A universal curve based on this relation is found to hold for the 2p state for q ≥ 3 and for higher n values for q ≥ 2 in the scaled specific energy E/mq range 15 to 220 keV amu$^{-1}$.

While there have been many studies of excitation of He($1^1S$) by H$^+$ ions over a wide energy range (see review [83]) measurements with multiply charged ions are sparse. Anton et al[84] have shown that measured cross sections for excitation of $n^1P$, $n^1S$ and $n^1D$ states of He for a number of ions which include He$^{2+}$, Si$^{(3-10)+}$ and Cu$^{(6-13)+}$ can be described by the scaling relation (9) to within about 30% for q between 2 and 12 in the E/mq range between 20 and 150 keV amu$^{-1}$.

## CONCLUSIONS

Over the past twenty five years the amount of data from experimental studies on fusion-relevant ion-atom collision processes, particularly on those involving multiply charged ions, has been impressive. Cross sections for many of the basic processes relating to the penetration of fast H and He beams are now well established although data are still needed for excited species. At high velocities, the measurements on charge transfer and ionization in collisions involving highly charged species have provided the basis for the development of general scaling relations to allow the prediction of approximate cross sections. At low energies, where many charge transfer processes have been shown to be important for the accurate modelling of edge plasmas, there is a need for measurements in which both the initial and final excited states are well defined. For neutral beam diagnostics, there is also a need for more accurate measurements on processes involving the excitation of non-dominant levels.

# REFERENCES

1. Joachain, C.J. and Post, D.E. (editors), *Atomic and Molecular Physics of Controlled Thermonuclear Fusion*, New York, Plenum, 1982.
2. Janev, R.K., Summary report of IAEA Technical Committee Meeting on Atomic and Molecular Data for Fusion Reactor Technology, Vienna, *IAEA report* INCD(NDS)-277, 1993.
3. Summers, H.P., *Advances in Atomic, Molecular and Optical Physics*, Vol. 33 (editor,M.Inokuti), New York, Academic Press, 1994, pp. 275-319.
4. Bonnal, J.F., Bracco, G., Breton, C., de Michelis, C.,Bruaux, J., Mattioli, M., Oberson, R. and Ramette, J., *J. Phys. D: Appl. Phys.* **15**, 805, (1982).
5. Gilbody, H.B., *Inst. Phys. Conf.* Ser. No. **38**, 156, (1978).
6. Meade, D.M. *Nucl. Fusion* **14**, 289 (1974).
7. Fonck, R.J., Goldston, R.J., Kaita, R. and Post, D.E., *Appl. Phys. Lett.* **42**, 239, (1983).
8. Schorn, R.P., Wolfrum, E., Aumayr, F., Hintz, E., Rusbüldt, H., and Winter, H., *Nuclear Fusion* **32**, 35, (1992).
9. Boileau, A., von Hellermann, M., Mandel, W., Summers, H.P., Weisen, H., and Zinoviev, A., *J. Phys. B: At. Mol. Opt. Phys.* **22**, L145, (1989).
10. Fite, W.L., Brackmann, T.R. and Snow, W.R., *Phys. Rev.* **112**, 1161 (1958).
11. Shah, M.B. and Gilbody, H.B., *J. Phys. B: At. Mol. Phys.* **14**, 2361, (1981).
12. Ciric, D., Brazuk, A., Dijkkamp, D., de Heer, F.J., and Winter, H., *J. Phys. B* **18**, 3629 (1985).
13. Lockwood, G.J., and Everhart, E., *Phys. Rev.* **125**, 567, (1962).
14. Park, J.T., Aldag, J.E. George, J.M., *Phys. Rev. Lett.* **34**, 1253, (1975).
15. McCullough R.W., Wilkie, F.G. and Gilbody, H.B. *J. Phys. B: At. Mol. Phys.* **16**, 1573, (1984).
16. Kim, H.J. and Meyer, F.W., *Phys. Rev. A* **26**, 1310, (1982).
17. Newman, J.H., Cogan, J.D., Zeigler, D.L., Nitz, D.E., Rundel, R.D., Smith, K.A. and Stebbings, R.F., *Phys. Rev. A* **25**, 2976, (1982).
18. Koch, P.M. and Bayfield, J.E., *Phys. Rev. Lett.* **34**, 448, (1975).
19. McClure, G.W., *Phys. Rev A* **148**, 47, (1966).
20. Shah, M.B., Elliott, D.S. and Gilbody, H.B., *J. Phys. B: At. Mol. Phys.* **20**, 2481 (1987).
22. Bates, D.R. and Griffing, G., *Proc. Phys. Soc. A* **66**, 961, (1953).
23. Crothers, D.S.F. and McCann, J.F., *J. Phys. B: At. Mol. Phys.* **16**, 3229, (1983).
24. Pieksma, M. and Ovchinnikov, S.Y., *J. Phys. B: At. Mol. Opt. Phys.* **24**, 2699, (1991).
25. Dalgarno, A. and Yadav, H.N., *Proc. Roy. Soc. A* **66**, 173, (1953).
26. Gilbody, H.B., in *Advances in Atomic, Molecular and Optical Physics*, Vol. 33 (eds. B. Bederson and H. Walther), Academic, 1994, pp. 149-180.
27. Fritsch, W. and Lin, C.D., *Phys Reports* **202**, 2, (1991).

28. Frieling, G.J., Hoekstra, R., Smulders, E., Dickson, W.J., Zinoviev, A.N., Kuppens, S.J. and de Heer, F.J., *J. Phys. B: At. Mol. Opt. Phys.* **25**, 1245 (1992).
29. Hoekstra, R., de Heer, F.J. and Morgenstern, R., *J. Phys. B: At. Mol. Opt. Phys.* **24**, 4025, (1991).
30. Fritsch, W., *J. Physique. Coll.* **50**, 87, (1989).
31. Schultz, D.R., Meng, L., Reinhold, C.O. and Olson, R.E., *Phys. Scr. T* **37**, 89, (1991).
32. Belkic, D., Gayet, R., and Salin, A. cited in Ref. 1.
33. Okuno, K., Soejima, K. and Kaneko, Y., *J. Phys. B* **25**, L105, (1992).
34. Shimakura. N., Kimura, M. and Lane, N.F., *Phys. Rev. A* **47**, 709, (1993).
35. Sato. K., Takiyama, K., Oda, T., Furukane, U., Akiyama, R., Mimura, M., Otsuka, M. and Tawara, H. *J. Phys. B: At. Mol. Opt. Phys.* **27**, L651, (1994).
36. Hodgkinson, J.M., McLaughlin, T.K., McCullough, R.W., Geddes, J. and Gilbody, H.B. *J. Phys. B: At. Mol. Opt. Phys.* (1995) - in course of publication.
37. Hoekstra, R., Folkerts, H.O., Beijers, J.P.M., Morgenstern, R. and de Heer, F.J., *J. Phys. B: At. Mol. Opt. Phys.* **27**, 2021, (1994).
38. Olson, R.E. and Salop, A., *Phys. Rev. A* **16**, 531 (1977).
39. Crandall, D.H., Phaneuf, R.A. and Meyer, F.W., *Phys. Rev. A* **22**, 379 (1980).
40. Shah, M.B. and Gilbody, H.B., *J. Phys. B: At. Mol. Phys.* **16**, 4395, (1983).
41. Gilbody, H.B., in *Advances in Atomic and Molecular Physics*, Volume 22 (Editors, D.R. Bates and B. Bederson), New York, Academic Press, 1986, pp. 143-192.
42. Goffe, T.V., Shah, M.B. and Gilbody, H.B., *J. Phys. B: At. Mol. Phys.* **12**, 3763, (1979).
43. Crothers, D.S.F. and Todd, N.R., *J. Phys. B: At. Mol. Phys.* **13**, 2277, (1980).
44. Shah, M.B. and Gilbody, H.B., *J. Phys. B: At. Mol. Phys.* **16**, L449, (1983).
45. Janev, R.K. and Hvelplund, P., *Comments At. Mol. Phys.* **11**, 75, (1981).
46. Gillespie, G.H., *J. Phys. B: At. Mol. Phys.* **15**, L729, (1982).
47. Gillespie, G.H., *Phys. Lett.* **93A**, 327, (1983).
48. Olson, R.E., Wetmore, A.E. and McKenzie, M.L., *J. Phys. B: At. Mol. Phys.* **19**, L629 (1986).
49. Shah, M.B. and Gilbody, H.B., *J. Phys. B: At. Mol. Phys.* **18,** 899, (1985).
50. Hvelplund, P., Haugen, H.K. and Knudsen, H., *Phys. Rev. A* **22**, 1930, (1980).
51. Hvelplund, P., Heinemeier, J., Horsdal-Pedersen, E. and Simpson, F.R., *J. Phys. B: At. Mol. Phys.* **9**, 491 (1976).
52. Guffy, J.A., Ellsworth, L.D. and MacDonald, J.R., *Phys. Rev. A* **15**, 1863, (1977).
53. Knudsen, H., Haugen, H.K., and Hvelplund, P., *Phys. Rev. A* **23**, 597, (1981).
54. Schlachter, A.S., Stearns, J.W., Graham, W.G., Berkner, K.H., Pyle, R.V. and Tanis, J.A., *Phys Rev A* **27**, 3372 (1983).
55. Schiebel, U., Doyle, B.L., MacDonald, J.R. and Ellsworth, L.D., *Phys Rev A* **16**, 1089, (1977).
56. MacDonald, J.R. and Martin, F.W., *Phys. Rev. A* **16**, 41, (1977).
57. Tanis, J.A., et al. *Bull. Am. Phys. Soc.* **31**, 955 (1986).
58. Hippler, R., Datz, S., Miller, P.D. and Pepmiller, P.L., *Abstracts of Proc 14th ICPEAC* (Palo Alto), p. 505, (1985).
59. Janev, R.K. and Winter, H., *Phys. Reports* **117**, 266, (1985).
60. Dijkkamp, D., Ciric, D. and de Heer F.J., *Phys. Rev. Lett.* **54**, 1004, (1985).
61. Kimura, M., Kobayashi, N., Ohtani, S. and Tawara, H., *J. Phys. B: At. Mol. Phys.* **20**, 3873 (1987).
62. Wilkie, F.G., McCullough, R.W. and Gilbody, H.B., *J. Phys. B: At. Mol. Phys.* **19**, 239, (1986).
63. Ciric, D., Brazuk, A., Dijkkamp, D., de Heer, F.J. and Winter, H., *J. Phys. B: At. Mol. Phys.* **18**, 3639, (1985).
64. Errea, L.F., Herrero, B., Mendez, L. and Riera, A., *J. Phys. B: At. Mol. Opt. Phys.* **24,** 4061, (1991).

65. McLaughlin, T.K., Tanuma, H., Hodgkinson, J., McCullough, R.W. and Gilbody, H.B., *J. Phys. B: At. Mol. Opt. Phys.* **26**, 3871, 1993.
66. McCullough, R.W., Wilkie, F.G. and Gilbody, H.B., *J. Phys. B: At. Mol. Phys.* **17**, 1373, (1984).
67. McLaughlin, T.K., Hodgkinson, J.M., Tawara, H., McCullough, R.W. and Gilbody H.B., *J. Phys. B: At. Mol. Opt. Phys.* **26**, 3587 (1993).
68. Huber, B.A., Kahlert, H.J. and Wiesemann, K., *J. Phys. B: At. Mol. Phys.* **17**, 2883, (1984).
69. Church, D.A. and Holzscheiter, H.M., *Phys. Rev. A* **40**, 54, (1989).
70. Bliman, S., Hitz, D., Jacquot, B., Hard, C. and Salin, A., *J. Phys. B: At. Mol. Phys. B* **16**, 2349 (1983).
71. Kimura, M. and Olson, R.E., *J. Phys. B: At. Mol. Phys.* **17**, L713 (1984).
72. Bordenave-Montesquieu, A.,Benoit-Cattin, P., Boudjema, M., and Gleizes, A. in *Invited papers of 15th ICPEAC*, Brighton; Amsterdam, North Holland, 1987, pp. 643-653.
73. Tsurubachi, S., Iwai, T., Kaneko, Y., Kimura, M., Kobayashi, N., Matsumoto, A., Ohtani, S., Okuno, K., Takagi, S. and Tawara, H., *J. Phys. B: At. Mol. Phys.* **15,** L733 (1982).
74. Morgan, T.J., Geddes, J. and Gilbody, H.B., *J. Phys. B: At. Mol. Phys.* **6**, 2118, (1973).
75. Park, J.T., Aldag, J.E., George, M. and Peacher, J.L., *Phys Rev A* **14**, 608, (1976).
76. Detleffsen, D., Anton, A., Werner, A. and Schartner, K-H., *J Phys B: At. Mol. Opt. Phys.* **27**, 4195 (1994).
77. Donnelly, A., Geddes, J. and Gilbody, H.B., *J. Phys. B: At. Mol. Opt. Phys.* **24**, 165 (1991).
78. Shakeshaft, R., *Phys. Rev. A* **18**, 1930 (1978).
79. Ermolaev, A.M., *J. Phys. B: At. Mol. Opt. Phys.* **24**, L49, (1991).
80. Slim, H.A., *J. Phys. B: At. Mol. Opt. Phys.* **26**, L743, (1993).
81. Fritsch, W., Shingal, R. and Lin, C.D., *Phys. Rev. A* **44**, 5686 (1991).
82. Janev, R.K. and Presnyakov, L.P., *J. Phys. B: At. Mol. Phys.* **13**, 4233 (1980).
83. de Heer, F.J., Hoekstra, R. and Summers. H.P., *Atomic and Plasma-Material Interaction Data for Fusion*, (Supplement to Nuclear Fusion), **3**, 47, (1992).
84. Anton, M., Detleffsen, D. and Schartner, K-H., *Atomic and Plasma-Material Interaction Data for Fusion* (Supplement to Nuclear Fusion), **3**, 51, (1992).

# Synchrotron Radiation Sources: Present Status, Anticipated Developments and Applications to Photon Collision Physics

B. Sonntag

*II. Institut für Experimentalphysik, Universität Hamburg,*
*Luruper Chaussee 149, 22761 Hamburg, Germany*

**Abstract.** The advent of third generation synchrotron radiation sources has opened up exciting possibilities for research. The characteristic features of these sources will be presented and the road towards diffraction limited storage ring sources and Free Electron Lasers for the soft- and hard X-ray regime will be outlined. The large impact of the new sources on atomic physics will be demonstrated by the discussion of most recent experiments.

## INTRODUCTION

Over the past 30 years synchrotron radiation has turned into a most powerful research tool applied in many different fields of science: in physics, chemistry, and biology, in material sciences, geophysics and medicine. Synchrotron radiation is the electromagnetic radiation emitted by relativistic electrons or positrons travelling in curved paths through the bending magnets, wigglers and undulators of modern storage rings. Synchrotron radiation spans the electromagnetic spectrum from the infrared to the hard X-rays. The intense radiation is well collimated and polarized. The light generated in an ultrahigh-vacuum environment is emitted in short pulses with a precisely known repetition rate. The unique features of synchrotron radiation have lead to an explosive growth of synchrotron radiation research and facilities over the last decades. At present more than thirty storage rings are in operation around the world as synchrotron radiation sources for basic and applied research. The number will definitely grow in the near future because there are several rings in the construction or proposal stage. In a joint effort driven by the facilities and their user communities the sources have undergone dramatic improvements. At the same time new beamline concepts, optical components and monochromators have been developed together with sophisticated instrumentation. With the new 3rd generation synchrotron radiation facilities in operation, ESRF in Grenoble, ELETTRA in Trieste, SRRC in Hsinchu, PLS in Pohang, ALS in Berkeley and Max II in Lund, as well as facilities under construction, APS in Argonne, SPring 8 in Himeji and BESSY II in Berlin, new highlights are in store. Recent advances in the development of linear accelerators for large linear colliders considered as the next generation of machines for particle physics, new breakthroughs in the performance of low emittance electron guns and the successful operation of very precise undulators open the exciting possibility of building Free Electron Lasers for the vacuum ultraviolet range for soft and even hard X-rays [1-9 and references therein].

# THIRD GENERATION STORAGE RING SOURCES

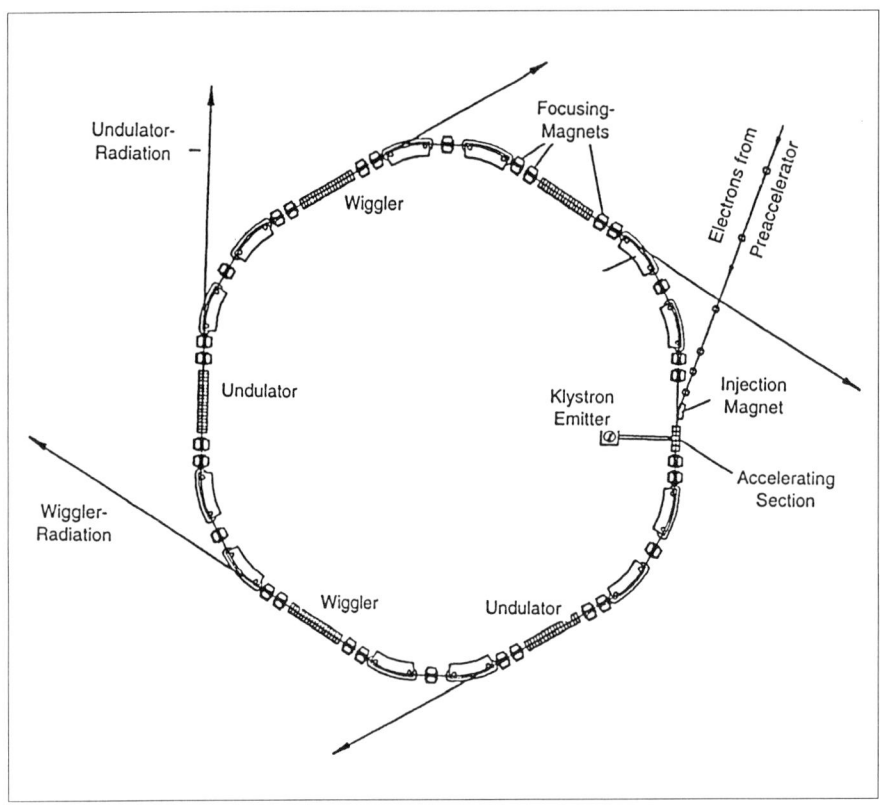

**FIGURE 1.** Scheme of a third generation storage ring radiation source.

In Figure 1, the general layout of a third generation storage ring light source with long straight sections for wigglers and undulators is depicted. In Table 1, the salient parameters for a representative low-energy and a representative high-energy third generation storage ring are summarized.

The radiation emitted by storage ring sources is quantitatively characterized by brilliance (photons per second per mm$^2$ source area per mrad$^2$ source divergence per 0.1 % spectral bandwidth), by brightness (photons per second per mrad$^2$ source divergence and per 0.1 % spectral bandwidth) and by the spectral flux (photons per second per 0.1% spectral bandwidth). The key quantity is brilliance from which brightness is obtained by integrating over the source area and flux by furthermore integrating over all angles. Depending on the source parameters all three quantities display a characteristic dependence on the photon energy. (Note that the European nomenclature used here differs from that used in the US). In order to increase the brilliance of the photon beam great efforts have been made to decrease the emittance of the electron beam; the ultimate limit is set by diffraction effects.

**TABLE 1.** Salient parameters of 3rd generation electron storage ring light sources (ALS: Advanced Light Source Berkeley; ESRF: European Snychrotron Radiation Facility Grenoble).

|  | ALS | ESRF |
|---|---|---|
| **Energy** | 1.5 GeV | 6 GeV |
| **Circumference** | 197 m | 844 m |
| **Emittance** | > 10 nmrad | > 4 nmrad |
| **Critical Photon Energy** | 1.56 keV | 19.2 keV |
| **Beam Current** | 400 mA | 200 mA |
| **Beam Lifetime** | > 10 h | > 10 h |
| **Straight Sections** | 10 | 32 |
| **Undulator Length** | 4.5 m | 5 m |
| **Bunch Length** | 30-50 ps | 50 ps |

The diffraction-limited emittance, given by $\lambda/4\pi$ where $\lambda$ is the radiation wavelength, has still not been reached in the soft and hard X-ray regime. Low emittance imposes stringent requirements on stability and reproducibility of the stored beam, since the source of the synchrotron radiation beam must be kept stable to a fraction of the beam diameter (at present > 10 μ). With the 3rd generation storage ring sources the interest is focused on the radiation generated in wigglers and undulators installed in the straight sections of the rings. But for a considerable number of applications the radiation emitted from the bending magnets provides an attractive alternative. The characteristic emission patterns of these sources are displayed in Figure 2. In all three cases the radiation is emitted in a narrow cone the angular aperture of which is given by $\gamma = E/m_0 c_2$ where E is the electron energy.

The spectrum of electrons passing through bending magnets is continuous with a high energy cut off above the critical photon energy

$$\varepsilon_c \ [keV] = 0.665 \cdot B \ [T] \cdot E^2 \ [GeV] \qquad (1)$$

where B is the magnetic field. The total power radiated is:

$$P[kW] = 0.2654 \cdot B \ [T] \cdot E^3 \ [GeV] \cdot I \ [A] \qquad (2)$$

where I is the current circulating in the ring. In wigglers or undulators the electrons travel through a periodic magnetic structure which produces a sinusoidally varying field. The magnetic structure and the sinusoidal electron trajectory is given in the schematic diagram in Figure 3.

Insertion devices are characterized by the parameter

$$K = 0.934 \ B_0 \ [T] \cdot \lambda_0 \ [cm] \qquad (3)$$

where $B_0$ is the peak magnetic field and $\lambda_0$ the period length. K can also be expressed in terms of the maximum deflection angle of the electrons.

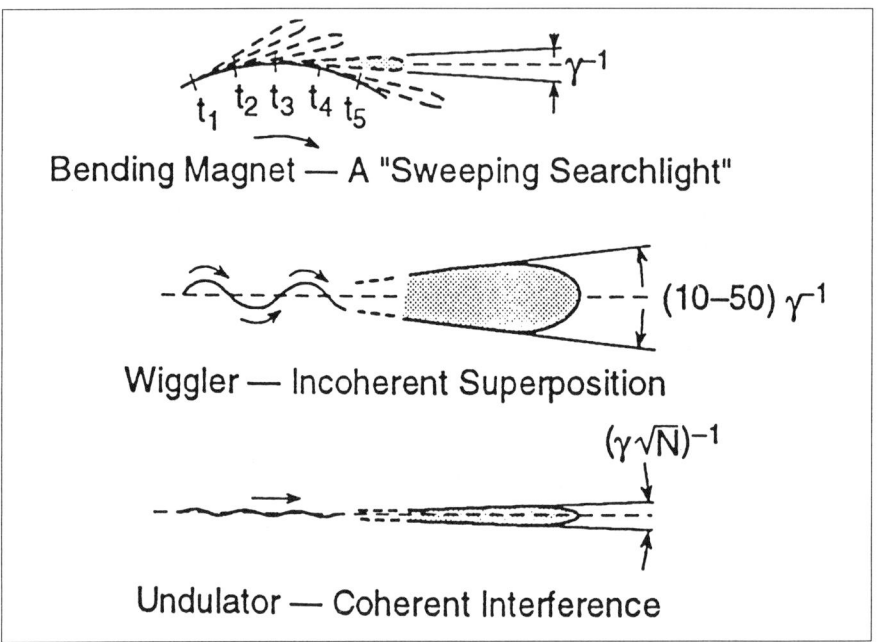

**FIGURE 2.** Emission pattern of radiation from bending magnets, wiggler magnets and undulator magnets.

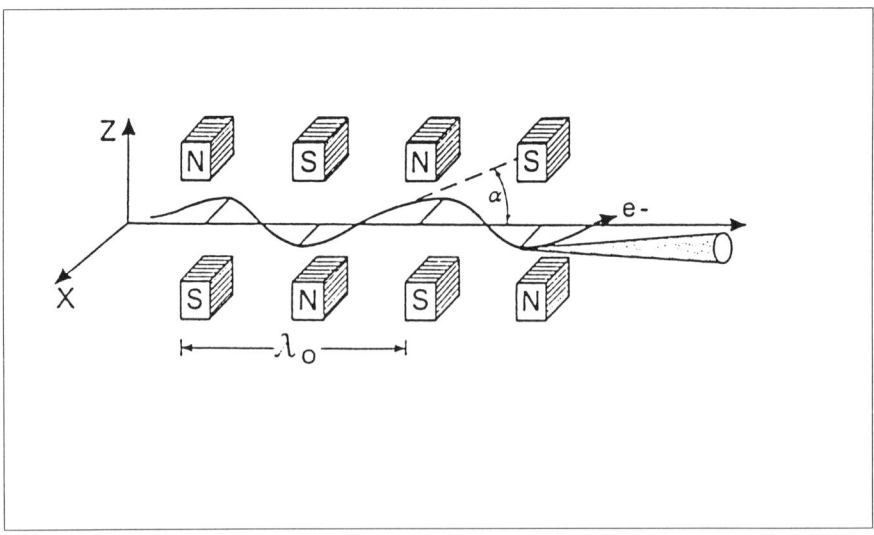

**FIGURE 3.** Schematic view of an insertion device magnetic structure. Also shown are the electron trajectory, the maximum angle of deviation $\alpha$ and one radiation cone.

$$K = \alpha \cdot \gamma \qquad (4)$$

## Wigglers

For wigglers ($K \gg 1$) the radiation emitted at different parts of the electron trajectory adds up incoherently. The spectrum resembles that of a bending magnet, the total flux being nearly 2 N times higher where N is the number of periods.

## Undulators

Undulators produce angular deviations of the electron beam which are small compared to $\gamma^{-1}$. Therefore the emission cones of successive bends of the trajectory overlap. Interference effects concentrate the radiation in a series of lines with $\Delta\lambda/\lambda \sim 1/[N \cdot n]$. The wavelengths of the fundamental and the $n$th harmonics are given by

$$\lambda_n = \frac{\lambda_0}{2n\gamma^2}(1 + \frac{1}{2}K^2 + \gamma^2\theta^2) \qquad (5)$$

where $\theta$ is the angle between the undulator axis and the direction of observation.

The odd harmonics reach their maximum intensity on axis, whereas the intensity of the even harmonics is zero on axis. The wavelengths of the undulator lines can be shifted by changing the magnetic field for example by varying the magnetic gap.

Figure 4 shows the average brilliance of several second and third generation synchrotron radiation sources. For comparison the expected average brilliance of two projected Free Electron Lasers is included. The typical total photon flux achieved with undulators mounted in third generation storage rings lies between $10^{15}$ to $10^{16}$ photons per sec per 0.1% bandwidth.

The well defined degree of polarization is one of the big advantages of synchrotron radiation widely exploited in a great variety of experiments. Bending magnet radiation is linearly polarized in the plane of the electron orbit but has a circularly polarized component above and below the orbit plane. This component has often been used in dichroism experiments, though the intensity drops considerably for off plane observation. The periodic magnet structures of standard wigglers and undulators give rise to linearly polarized radiation. Intense elliptically or circularly polarized radiation can be produced by special insertion devices: asymmetric devices, which have positive and negative magnetic poles of different magnitude; helical devices (bifilar solenoids, planar helical magnet structures); interference devices, in which the radiation from several devices is mixed coherently. For many systems the polarization of the radiation can be switched from linearly polarized to left or right circularly polarized by changing the magnetic field.

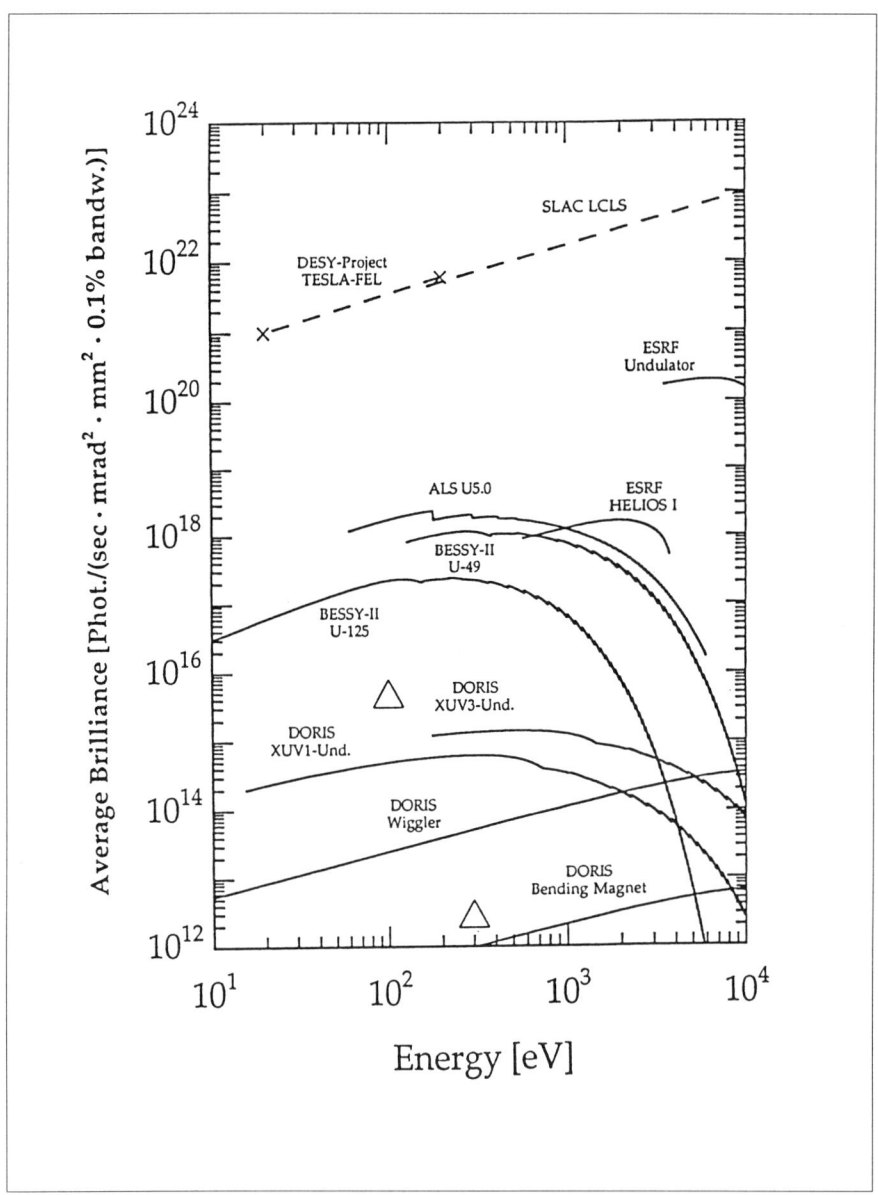

**FIGURE 4.** Average spectral brilliance of several second and third generation synchrotron radiation sources and of Free Electron Lasers proposed at DESY (TESLA-FEL, Hamburg) and SLAC (SLAC LCLS, Stanford). The open triangles represent values achieved with plasma lasers under the optimistic assumption that they can fire once a second.

# PHOTON-ATOM COLLISION PHYSICS

Synchrotron radiation has made a great impact on atomic physics. For an overview of the state of the field the reader is referred to recent reviews [2,10,11]. However, in the following examples are presented which demonstrate the considerable advances achieved within the last years.

## ULTRA-HIGH RESOLUTION SPECTROSCOPY OF HELIUM

Helium is the archetypical system in which the effects of electron-electron correlation can be studied. Such effects are most readily observed in multiply excited atoms and hence the study of doubly excited states in He provides important information on the correlated motion of a pair of electrons in the field of the doubly charged nucleus. Since the pioneering photoabsorption experiment by Madden and Codling [12] the double excitation Rydberg series of He have been the focus of many experimental and theoretical investigations [see 13-15 and references therein]. Recent interest has been stimulated by the substantially improved experimental data obtained due to advances in spectral resolution and photon flux. The best data reported so far have been obtained at the Advanced Light Source in Berkeley [16,17]. Figure 5 shows the optical layout of the spherical grating monochromator beamline.

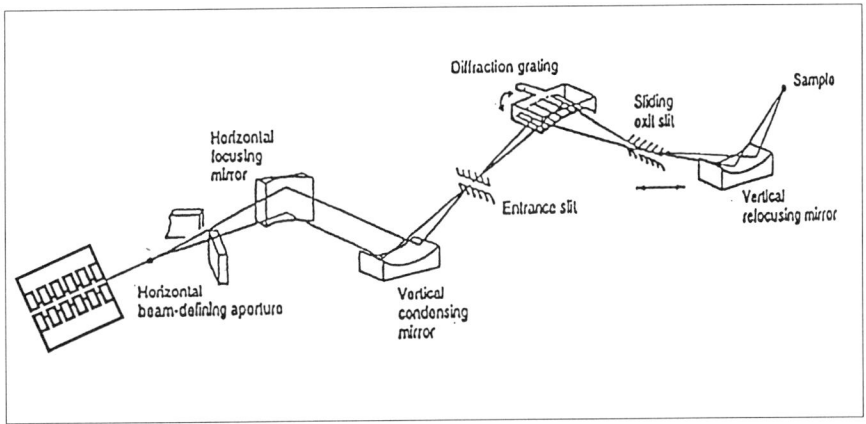

**FIGURE 5**. Optical layout of the spherical grating monochromator beamline at the Advanced Light source.

In the range of the He double excitations the beamline provides approximately $10^{13}$ photons per sec per 0.01% bandwidth. The total photoionization yield was determined by using a gas cell with two parallel charge collecting plates. In Figure 6 the higher members of the (sp, 2n$^\pm$) Rydberg series are shown The principal (sp, n+) series could be resolved up to the n = 26 member. Two members of the "hidden" (2p, nd) series are barely visible.

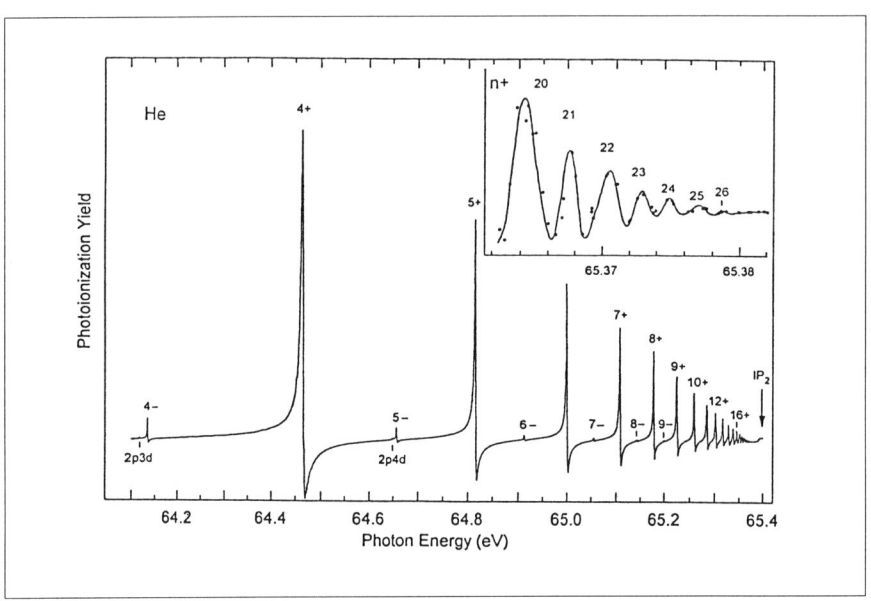

**FIGURE 6.** The three optical allowed Rydberg series in the double-excitation region of He below the N = 2 ionization threshold of He, IP2. The inset shows the "+" Rydberg series from n = 20 up to n = 26.

The (2p, 3d) and the (sp, 4-) lines are given on an enlarged scale in Figure 7. This spectrum clearly demonstrates the extremely high resolving power achieved. Note that the intrinisic width of the (2p, 3d) line is expected to be less than 3 µeV [14].

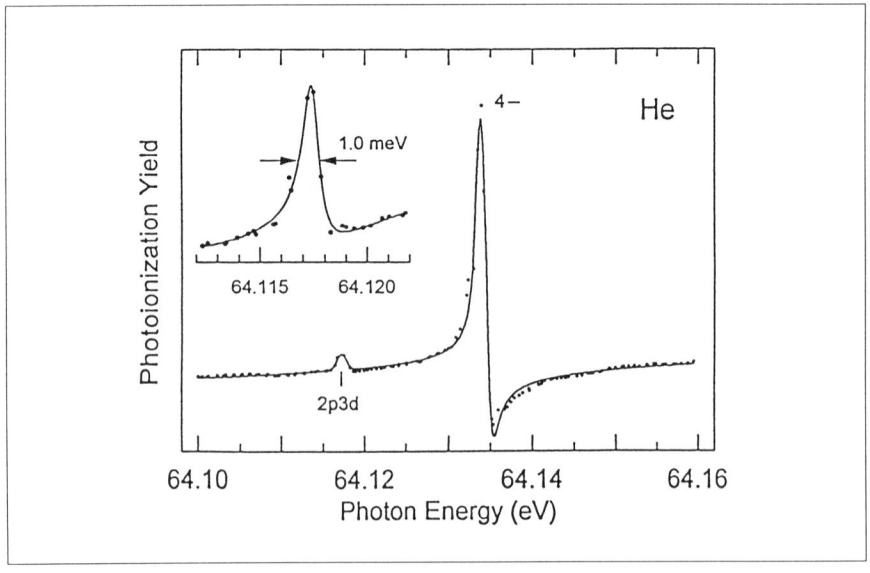

**FIGURE 7.** (sp, 4n-) and (2p, 3d) double excitation line of He.

# DOUBLE PHOTOIONIZATION OF HELIUM

The double photoionization of He has attracted considerable interest because of its fundamental nature and because of the challenge it poses to experiment and theory [see 18-27 and references therein]. For high photon energies photoabsorption and Compton scattering contribute to the double ionization. For the ratio of double to single ionization

$$\mathbf{R}_{Ph} = \frac{\sigma^{2+}}{\sigma^{+}} \qquad (6)$$

experimental and theoretical results converged towards a value of 1.67% for the high photon energy limit. For Compton scattering the value of the high energy asymptotic ratio

$$\mathbf{R}_{c} = \frac{\sigma^{2+}}{\sigma^{+}} \qquad (7)$$

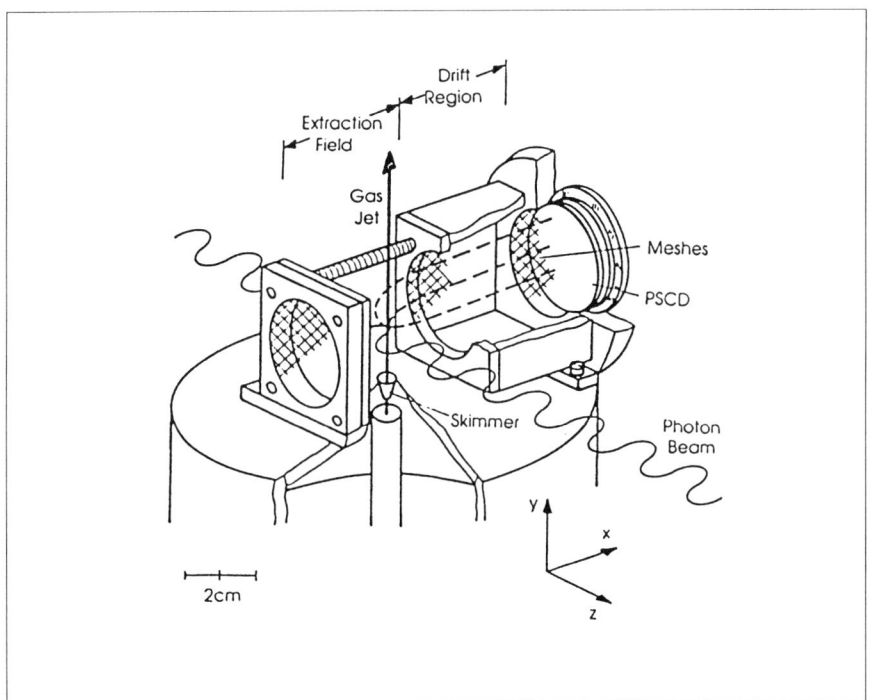

**FIGURE 8.** Recoil-ion momentum spectrometer with supersonic He gas jet. The recoil ions have been measured in coincidence with the beam pulse, the channel-plate detector is two-dimensional position sensitive. The electric field vector of the linear polarized light is parallel to the extraction field.

still controversial. The situation is aggravated by the fact that most experiments are unable to discriminate between photoabsorption and Compton scattering. Spielberger et al. [27] recently succeeded to separate the two contributions for high energy photon impact by measuring the full momentum vector of the He$^+$ and He$^{++}$ recoil icons. In the photoabsorption process the ions acquire large momenta whereas in Compton scattering most of the momentum is transferred to the scattered electron and therefore the ions have small momenta. The recoil-ion momentum spectrometer used for the determination of the charge state and the momentum of the outgoing ions is shown in Figure 8.

The data, obtained at the BW1 undulator line of HASYLAB at DESY in Hamburg using an approximately 3 keV broad photon band centred at the third harmonic of the undulator, are given in Figure 9. For photoabsorption $R_{Ph}$ was found to be 1.72 ± 0.12 % at ~7.0 keV, in good agreement with earlier experimental and theoretical values. For Compton scattering the value obtained for $R_c$ of 1.22 ± 0.06 % corroborates the prediction of Andersson and Burgdörfer [24].

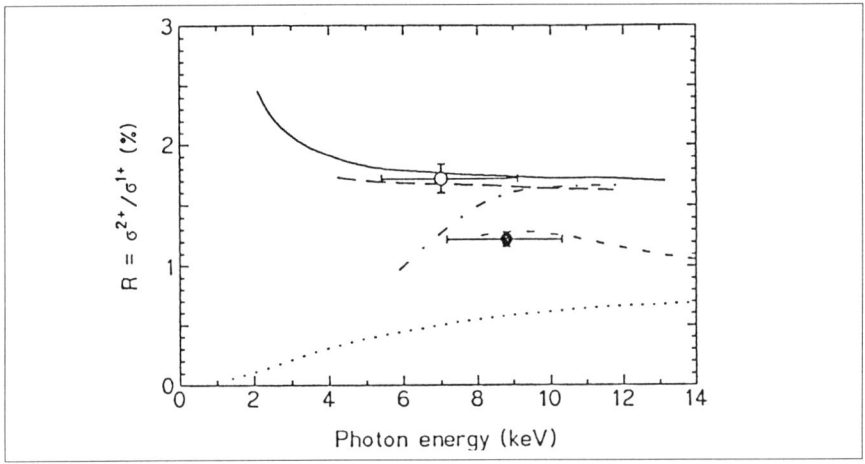

**FIGURE 9.** Ratios of double to single ionization. Full circle: experimental ratio for Compton scattering ($R_c$); open circle: experimental ratio for photoabsorption ($R_{Ph}$). The horizontal error bars indicate the energy region from which 67% of all He$^{1+}$ ions result. Full line: $R_{Ph}$ from Andersson and Burgdürfer [20]; broken line: $R_{Ph}$ from Hino et al.[25]; dashed line: $R_c$ from Andersson and Burgdörfer [24]; dotted line: $R_c$ from Suric et al. [26]; dash-dotted line: $R_c$ from Hino et al. [25].

# ELECTRON ANGULAR CORRELATION IN DOUBLE PHOTOIONIZATION

The angular correlation of the two outgoing electrons in dependence on the excess energy and the energy sharing between the two electrons provides the most stringent tests for the theoretical description of the double photoionization. The angular correlation is contained in the triple differential cross section (TDCS), a quantity which can be determined if, after energy and angle selection, both electrons are detected in coincidence. Theoretical calculations of the TDCS have been performed on the basis of extended Wannier models [28-30 and references therein] and by using correlated wavefunctions in the initial and final state treating all three two-particle Coulomb interactions equally [31 and references therein]. Because of the extreme weakness of the coincidence signal experiments on He have become feasible only within the last few years [28,30,32,33 and references therein]. For linearly polarization radiation the TCDS can be written as

$$\mathrm{TDCS} = C(E_1, E_2, \theta_{12}) |\cos\theta_1 + z(E_1, E_2, \theta_{12}) \cdot \cos\theta_2|^2 \tag{8}$$

where $\theta_1$ and $\theta_2$ are the polar angles of the ejected electrons with respect to the electric field vector, and $\theta_{12}$ is the relative angle between both electrons. $E_1$ and $E_2$ are the energies of the two electrons. The three particle Coulomb interaction is included in the $\theta_{12}$ dependence of the parameters $C$ and $z$. For equal energy sharing ($E_1 = E_2$) the parameter $z$ assumes the constant value $z = 1$. In accordance with equation 8 the TDCS is zero at $\theta_{12} = 180°$ for $E_1 = E_2$. For unequal energy sharing the TDCS still displays a marked minimum at $\theta_{12} = 180°$ for low excess energies. This minimum disappears for high excess energies and significantly different electron energies. These theoretical predictions are corroborated by the results obtained by Schwarzkopf et al. [33,34] and given in Figure 10. The upper TDCS has been obtained for 99 eV photon energy and equal energy sharing whereas the lower curve represents the TDCS taken at 131.9 eV photon energy and unequal energy sharing $E_2 = 5$ eV, $E_2 = 47.9$ eV.

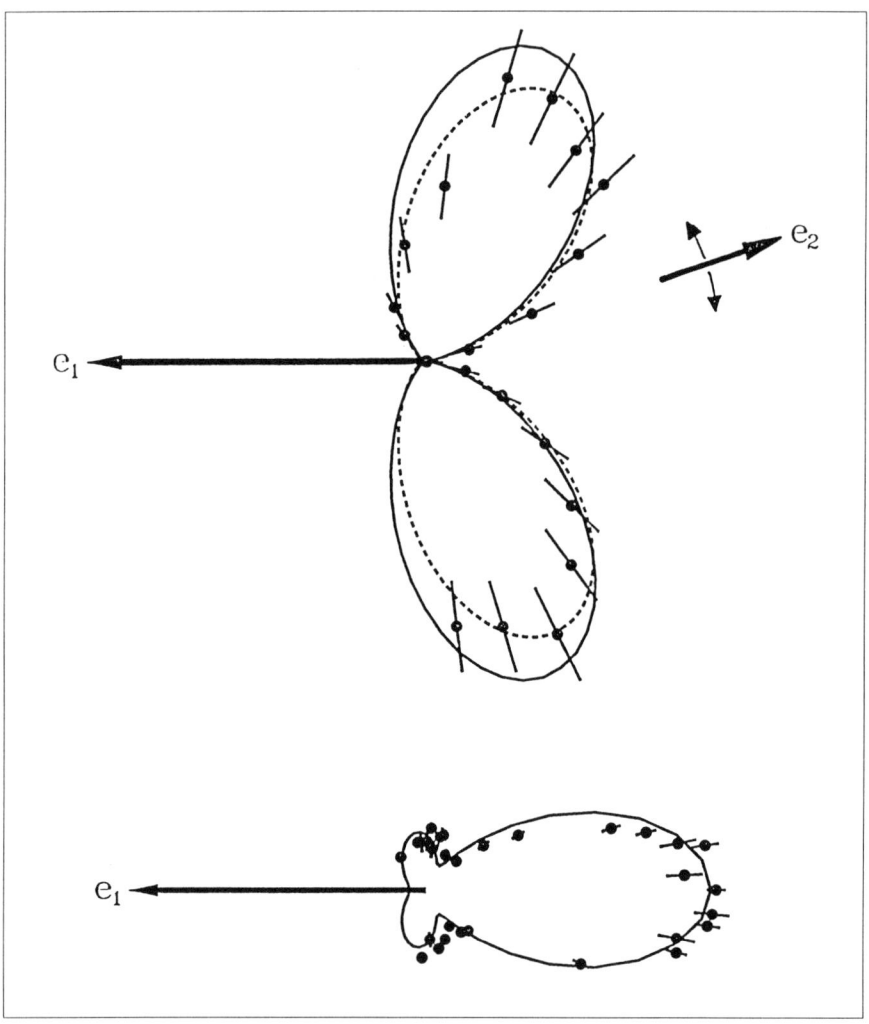

**FIGURE 10.** Experimental and theoretical (solid and dashed lines) TDCS for double ionization of He of equal (upper curve) and unequal energy sharing, measured in a plane perpendicular to the photon beam. The direction of electron $e_1$ is parallel to the electric field vector while the direction of $e_2$ is varied with respect to $e_1$.

## TRIPLY EXCITED STATES OF LITHIUM

Lithium is the simplest open-shell many-electron atom. Lithium, a four-body Coulombic system, provides an opportunity for investigating highly correlated multiply excited states where all three electrons reside outside the K-shell. In addition, whereas Helium only exhibits direct double photoionization, Lithium offers the potential for the study of both direct and resonant double photoionization. Until recently triply excited states of Li were only observed in collisions of Li$^+$ ions with solids, gases and electrons [35-38 and references therein]. In addition to these measurements a number of calculations for double K-shell resonance positions and widths have been performed [39-43 and references therein].

Using a dual laser plasma technique Kiernan et al. [44] measured for the first time the photoabsorption spectrum of atomic Li at energies corresponding to the simultaneous excitation of both 1s electrons. Shortly afterwards high resolution photoion spectra taken at undulator beamlines at HASYLAB in Hamburg [45] and at the Photon Factory in Tsukuba [46] were reported. In both experiments a beam of Li atoms was crossed with monochromatized undulator radiation. The resulting singly or double charged photoions were measured as a function of photon energy with a time-of-flight (TOF) analyzer. Figure 11 (upper part) shows the Li$^+$ photoion yield spectra in the 148 eV - 155 eV region where many $1s^22s \rightarrow 2l2l'nl''$ transitions are expected to lie. The lowest triply excited transitions $1s^22s(^2S) \rightarrow 2p^3(^2P)$ appears as a weak feature at 148.77 eV while the spectrum is dominated by the asymmetric resonance $1s^22s(^2S) \rightarrow 2p^2(^3P)3p(^2P)$ at 152.46 eV. In the lower part of Figure 11 the Li$^+$ photoion yield spectrum in the 160-163 eV region, where $1s^22s \rightarrow 2l3l'nl''$ transitions are expected to contribute, is presented. The asymmetric resonance at 161.66 eV has been tentatively assigned to overlapping $1s^22s(^2S) \rightarrow 2s3s(^3S, {}^1S)3p(^2P)$ transitions [45]. The dominating resonance $1s^22s$ $^2S \rightarrow 2s^22p$ $^2P$ has been studied in great detail in Li$^+$, Li$^{++}$ ion yield spectroscopy [45,46,47] photoelectron spectroscopy [48].

In the Li ion spectrum this resonance can be well described by an asymmetric Fano profile for a single discrete state interacting with many continua. The parameters reported by Kiernan et al. [45], resonance energy $E_0 = 142.32 \pm 0.02$ eV, width $\Gamma = 0.14 \pm 0.02$ eV and asymmetry parameter $q = -2.0 \pm 0.3$ are in respect to energy in best agreement with the theoretical value of 142.26 eV given by Chung [41] and with respect to width with the value of 0.13 eV calculated by Simons et al. [40]. Due to lower spectral resolution the width $\Gamma$ was overestimated in the dual plasma experiment [44] and to a lesser extent by Azuma et al. [46]. The coupling of the $1s^22s \rightarrow 2s^22p$ $^2P$ resonance to the $1s2s$ $^{1,3}S\varepsilon l$ and $1s2p$ $^{1,3}P\varepsilon l$ continua differs. This is borne out by the different Fano profiles in the constant ionic state (CIS) photoelectron spectra determined at SUPER ACO in Paris [48]. The Li$^+$1s2s $^1S$ partial cross section displays an asymmetric Fano profile whereas the Li1s2p $^3P$ partial cross section shows a symmetric Lorentzian profile. Both spectra are given together with the results of a R-matrix calculation in length (L) and velocity (V) form [48-50] in Figure 12.

It is interesting to note that the resonance profiles also differ in the Li ion spectra, being asymmetric Fano type for Li$^+$ while Lorentzian for Li$^{++}$ [46,47].

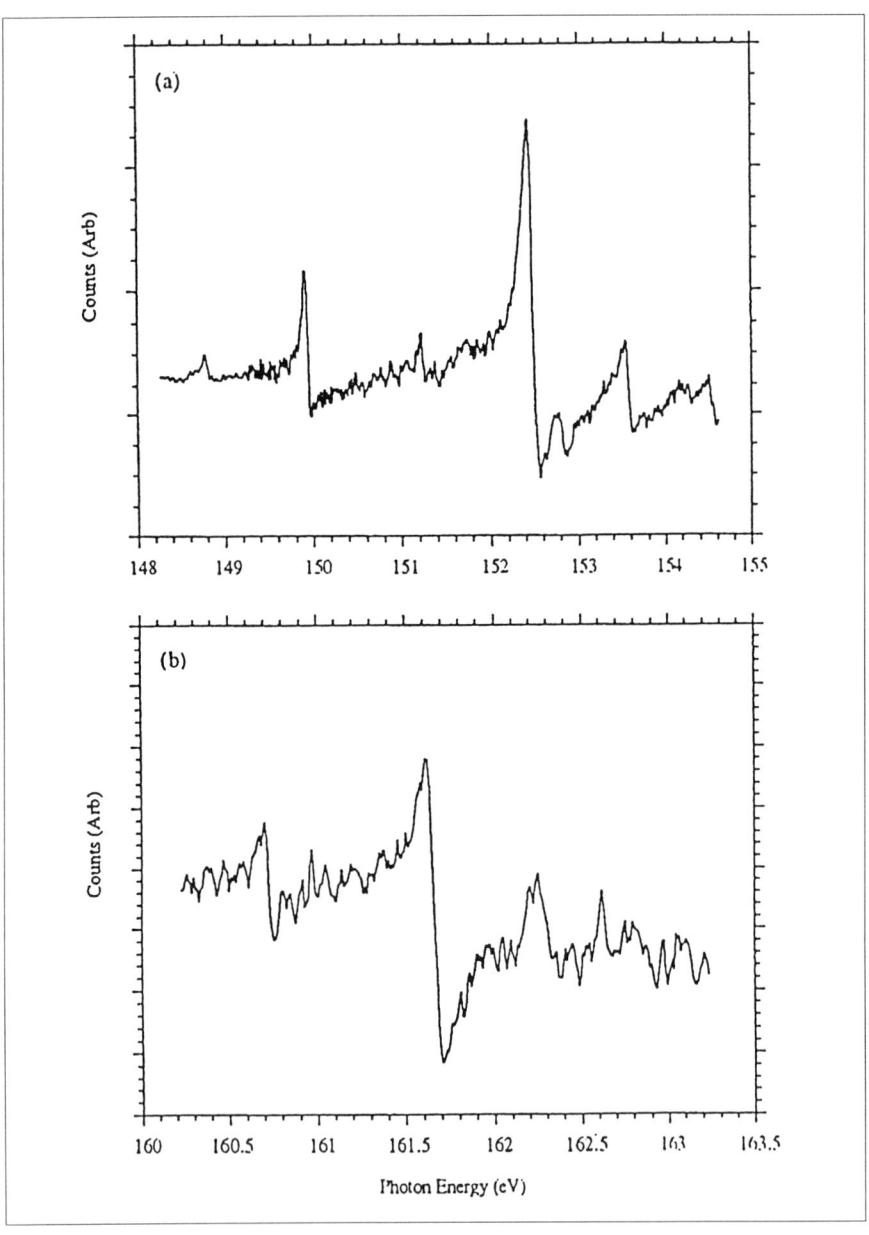

**FIGURE 11.** Li$^+$ ion yield spectra in the range of the $1s^22s$ - $2l2l'nl''$ (a) and the $1s^22s$ -> $2l3l'nl''$ (b) transitions.

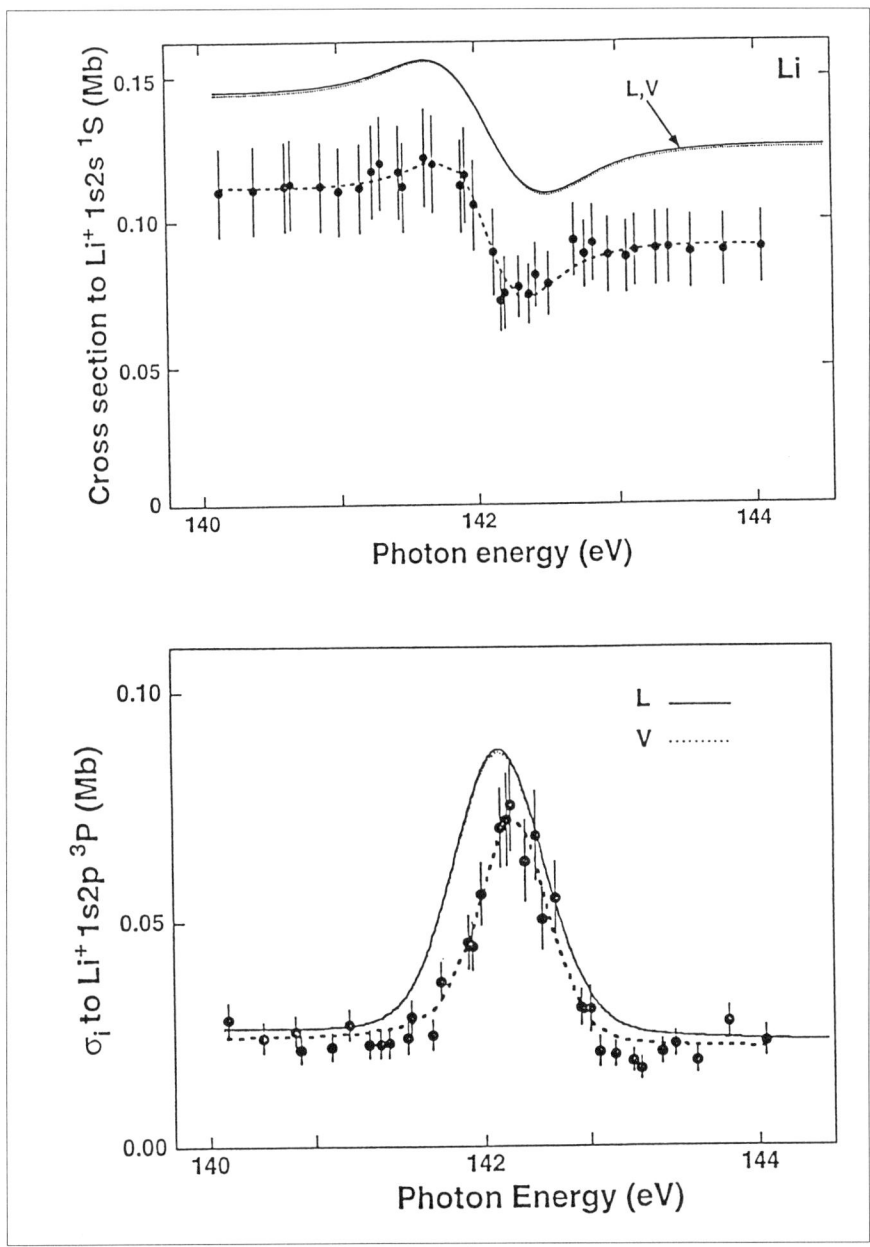

**FIGURE 12.** Partial photoionization cross sections for the $Li^+ 1s2s\ ^1S$ (upper part) and for the $Li^+\ 1s2p\ ^3P$ (lower part) final ionic states in the range of the $Li\ 2s^22p\ ^2P$ resonance.

## RESONANT RAMAN AUGER

Close to threshold the excitation of atomic inner-shell hole states and the radiative or radiationless decay cannot be treated as distinct processes. Near the threshold, photoionization and radiative or radiationless de-excitation occur in a single quantum process, the resonant Raman scattering. This results in a linear dispersion and a resonance narrowing of spectator Auger-electron lines [51,52 and reference therein]. These effects were first observed in the X-ra range [53,54] and only very recently in the vacuum ultraviolet range (VUV) [55,56]. In order to exploit the resonant Raman Auger effect for atomic physics in the VUV one needs a very narrow bandwidth photon beam and a very high resolution electron spectrometer. Both requirements are met by the Finish beamline in MAX-lab, which is equipped with a modified SX-700 plane grating monochromator and a hemispherical electron energy analyzer [57]. For Xe and Kr resonant Auger lines Kivimäki et al. [55] could nicely demonstrate that the width of the resonant Auger lines was determined by the bandwidth of the photons and did not depend on the lifetime width of the core hole. The linear shift of the lines with photon energy was also clearly to be seen. Apart from the interesting physics of the resonant Raman effect itself the

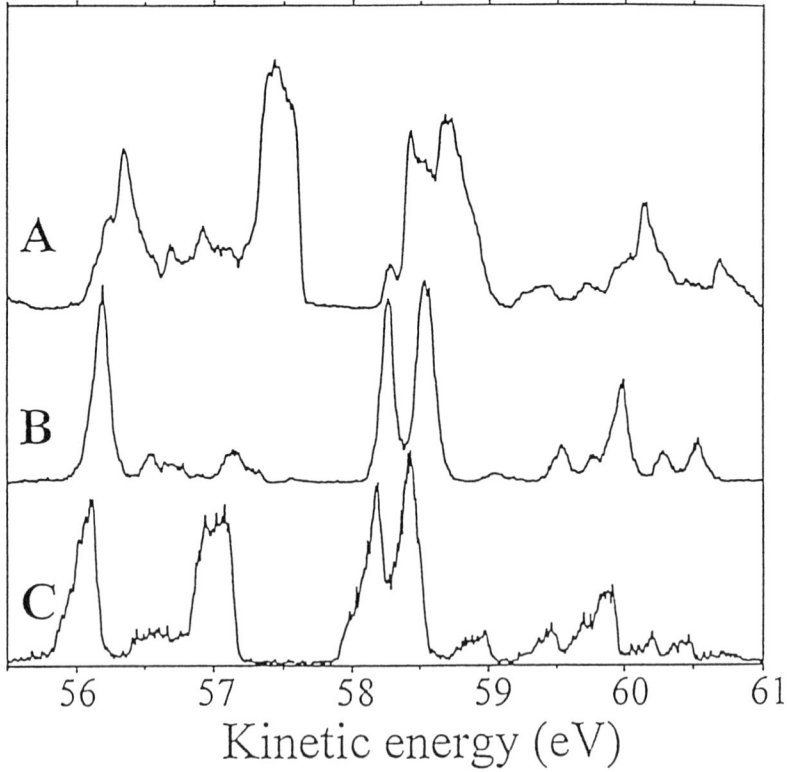

**FIGURE 13**. Kr $3d^{-1}_{5/2}$ 5p -> $4p^{-2}$5p resonance Auger spectra, excited at 91.45 eV (A), 91.20 eV (B) and 90.91 eV (C) mean photon energy.

resonant Raman effect can be used to enhance the experimental resolution of the resonant Auger spectra and thus disentangle much finer details. But there is a caveat one has to take into account before entering upon a very detailed analysis of the resonant Auger spectra and the shapes of the lines. Aksela et al. [58] have given an impressive example for the crucial importance of the exciting photon energy distribution for the line shape of resonance Auger lines. The Kr $3d^{-1}_{5/2}$ 5p -> $4p^{-2}$5p resonance Auger spectrum was selected for the demonstration. Figure 13 shows the spectra taken with a photon band width of about 280 meV.

The spectra were taken with the average photon energy tuned exactly to the resonance energy (B), and when it was detuned by 0.25 eV to higher (A) or lower (C) energies. The shape of the double peak structure between 58 and 59 eV, which is ascribed to the $3d^{-1}_{5/2}$5p -> $4p^{-2}(^1D)$5p transitions, is dramatically changed and the peak maxima are shifted. The distortion of the resonance Auger-spectra by the photon energy distribution is obvious from the small extra peak at the low energy side of the doublet in spectrum A. Aksela et al. [58] were able to prove that this was due to a weak tail on the low energy side of the photon band.

## Linear Dichroism in the Photoelectron Spectra of Laser-aligned Chromium Atoms

Transition metal atoms embedded in solids are of great basic and practical importance. This rests with the partially unoccupied outer d or f-shells which strongly influence the optical, electrical and magnetical properties of the solids. Due to the local character of these orbitals the adequate treatment of the intra- atomic interactions in many cases is essential for description of transition metals / rare earths and transition metal/rare earths compounds [59]. Recently inner shell angular resolved photoelectron spectra of magnetically oriented solids, multilayer systems and surface layers excited with linearly or circularly polarized VUV radiation (linear/circular magnetic dichroism in the angular distribution of photoelectrons LMDAD/CMDAD) were proved to be very sensitive probes for the electronic structure [60-63 and references therein]. Atomic models have been successfully invoked for explaining the origin of the dichroism [64-66]. Experiments on the corresponding free atoms can substantially help to disentangle the intra-atomic and the inter-atomic effects. By laser pumping the atoms can be prepared in a well defined aligned or oriented state. Determining the energy and angular distribution of the electrons excited by linearly polarized VUV radiation provides very detailed information on the states and the many electron dynamics of the open shell transition metal atoms. Test experiments on alkali atoms have successfully been performed [11, 67, 68] by combining laser and VUV undulator radiation. In a first experiment on transition metal atoms the resonant 4s- and 3d- photoionization of laser aligned Cr-atoms was investigated in the range of the 3p - 4s, 3d inner-shell excitations [69, 70]. The experimental arrangement is shown schematically in Figure 14. Cr atoms emanating from a resistively heated furnace are prepared in an aligned ground state by pumping the $4s^7S_3$ -> $4p^7P_2$ transition with a cw-dye laser. The 3p-core resonances are excited by monochromatized and linearly polarized undulator radiation which propagates antiparallel to the laser beam. The 4s and 3d electrons emitted upon the $3p^5 3d^5 4s^2 \rightarrow 3p^6 3d^5 4s\, \varepsilon l$, $3p^5 3d^5 nd\, 4s \rightarrow 3p^6 3d^4 4s\, \varepsilon l$ autoionization of the core resonances are detected by a cylindrical mirror analyser (CMA). Only electrons emitted with polar angles with respect to the electric vector of the undulator radiation which are close to the magic angle of 54,7 ° are registered. The azimuthal angle extends over $\pi$.

The asymmetric 3p - 4s (lines 1-3), 3p - 3d (lines 12 - 17) and 3p - nd (lines above 44 eV) lines are clearly to be seen in Figure 15. The asymmetry of the lines is caused by the interference between the core resonances and the direct photoionization channels. Most features of the spectra are well described by model calculations performed by Dolmatov within the Spin Polarized Random Phase Approximation with Exchange (SPRPAE) [70, 71].
In order to achieve agreement with the experimental results it was essential to take the relaxation of the nd, especially 3d (anti-collapse), orbitals upon creation of the 3p - hole into account. In the following we will concentrate on the dependence of the lines $3p^6 3d^5 4s^7\, S_3 \rightarrow 3p^5 3d^5 4s^2\ ^7P_4(1), ^7P_3(2), ^7P_2(3)$ on the relative orientation of the electric vector of the laser radiation with respect to the electric

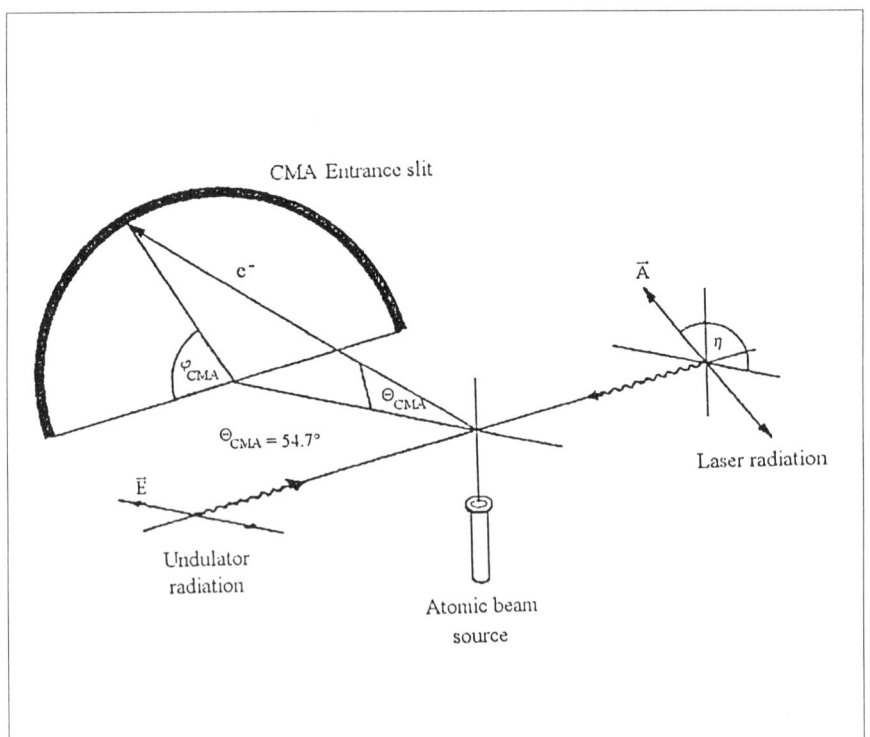

vector of the undulator radiation given by the angle $\eta$ (Figure 14).

For the geometry shown in Figure 14, neglecting higher atomic state multipoles, the Linear Dichroism in the Angular Distribution of the electrons (LDAD) can be described by [69, 72 -74] :

$$A_{20}\beta_{LASER} = 2\left[\frac{I(\eta = 0°) - I(\eta = 90°)}{I(\eta = 0°) + 2I(\eta = 90°)}\right] \quad (9)$$

where $\beta_{laser}$ is given by :

$$\beta_{LASER} = \frac{3\hat{J}_0}{\sqrt{5}} \frac{\sum_J (-1)^{1+J_0+J} \begin{Bmatrix} J_0 & 1 & J \\ 1 & J_0 & 2 \end{Bmatrix} \sigma_J}{\sum_J \sigma_J} \quad (10)$$

For the transitions Cr $3p^6\,3d^5\,4s\ ^7S_3$ ($J_0 = 3$) to $3p^5\,3d^5\,4s^2\ ^7P_{2,3,4}$ (J = 2, 3, 4) this reduces to

$$\beta_{Laser} = \sqrt{3}\,\frac{-\frac{2}{5}\sigma_2 + \frac{1}{2}\sigma_3 - \frac{1}{6}\sigma_4}{\sigma_2 + \sigma_3 + \sigma_4} \quad (11)$$

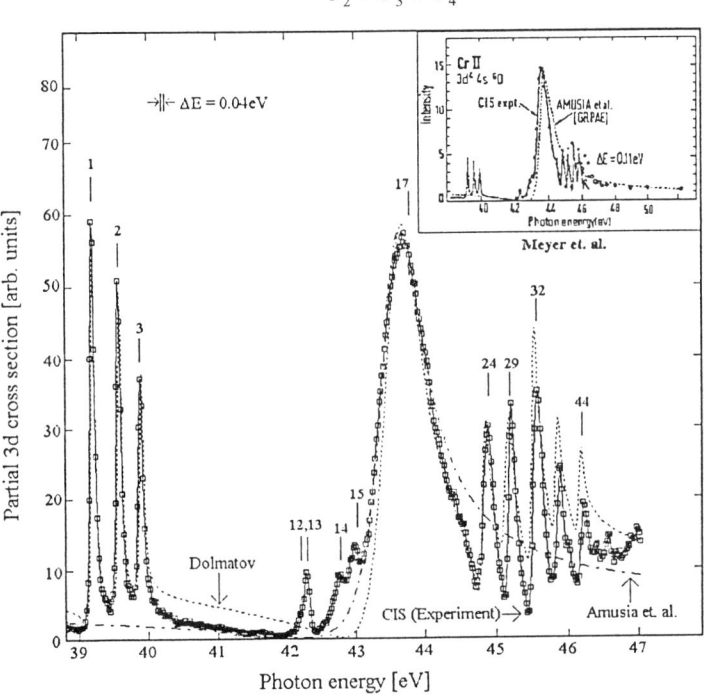

**Figure 15.** Experimental and theoretical partial 3d iron reaction of atomic Cr in the region of the 3p -> 4s, 3p-nd resonances.

The partial cross sections for the individual channels characterized by the total angular momentum J of the final state are given by $\sigma_J$. The intensities obtained with the electric vectors of both radiation fields parallel I ($\eta = 0°$) and perpendicular to each other I ($\eta = 90°$) are presented in the upper part of Figure 16. There is a marked dependence on $\eta$. The experimental LDAD signal $A_{20}$ $\beta_{laser}$ calculated according to formula 9 is given in the outer part of Figure 16. Inserting the theoretical partial iron sections given in the lower part of Figure 16 [70] into equation 11 a value for $\beta_{laser}$ can be obtained. The solid line in the center part of Figure 16 which is in fair agreement with the experimental results is based on $A_{20} = 0,45$. This value for $A_{20}$ is consistent with the value obtained in rate equation calculations for the pumping process. The experimental LDAD corroborates the theoretical predictions for the 3 p -> 4 s resonances. This is not the case for the 3 p - nd resonances which display only very weak LDAD. This indicates that overlapping transitions to final states with different angular momenta I contribute to these resonances.

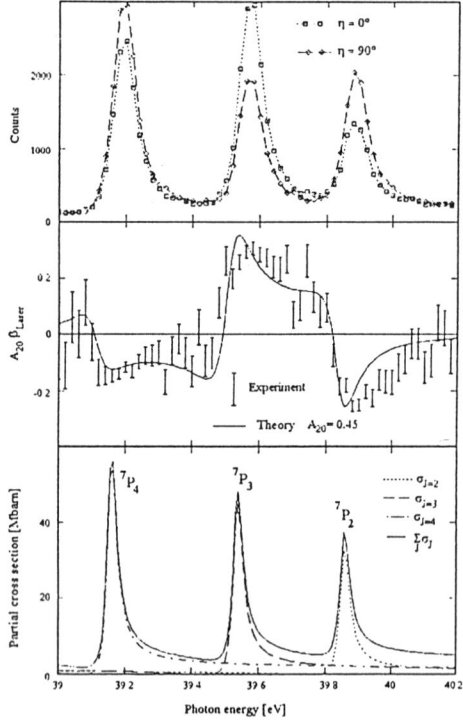

**Figure 16.** Experimental strength of the Cr $3p^6$ $3d^5$ $4s^7$ $S_3$ -> $3p^5$ $3d^6$ $4s^2$ $^7P_{2, 3, 4}$ resonances in the 3d photoionization channel for both radiation parallel ($\eta = 0°$) and perpendicular ($\eta = 90°$) to each other (upper part). Linear Dichroism in the Angular Distribution (LDAD) (center part) and theoretical partial cross sections (lower part).

## Perspectives

Progress in photon collision physics is intimately related to the quality of the light sources that are available. Key parameters are the brilliance, the flux, the polarization, the time structure and the coherence of the photon beam. Lasers providing outstanding values for all these parameters have revolutionized photon physics in the infrared and visible part of the electromagnetic spectrum. Free Electron Lasers (FEL) are very promising candidates for bringing this revolution to the vacuum ultraviolet (VUV) and even the X-ray part of the spectrum [1, 4-8]. As an example the average photon flux expected to be emitted by the TESLA-TEST-FACILITY FEL (TTF-FEL) [5] is presented in Figure 17. The average brilliance has already been shown in Figure 4. The characteristic parameters of the photon beam are summarized in Table 2.

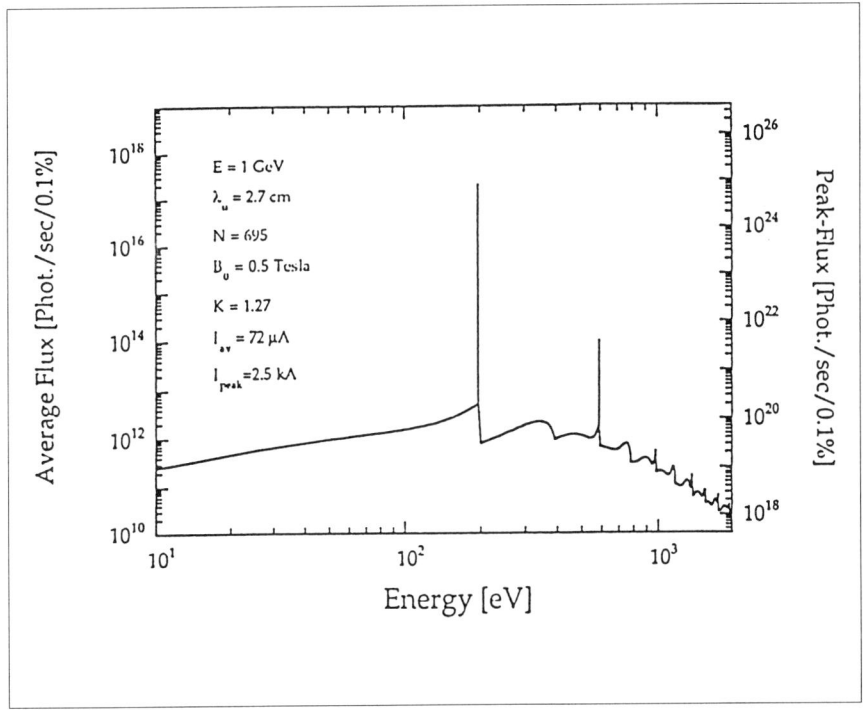

**Figure 17.** Angle integrated photon flux of the TTF FEL calculated for its design parameters. The ordinate on the left shows the time averaged flux, the ordinate on the right the peak flux. Note the small spectral width of the delta function like laser radiation compared with the ordinary, spontaneous emission of the undulator.

**TABLE 2.** Typical parameters expected for the photon beam of the TTF-FEL.

| | |
|---|---|
| Energy | 30 eV - 200 eV first harmonic |
| Divergence (1σ) | 28 μrad |
| Spectral bandwidth at saturation (FWHM) | 1 % |
| Pulse length | 350 fms |
| Photons per bunch | $4 \times 10^{13}$ |
| Average flux | $3 \times 10^{18}$ photons/sec |
| Peak flux | $1 \times 10^{26}$ photons/sec |
| Average brilliance | $6 \times 10^{21}$ photons/sec/mm$^2$/mrad$^2$/0.1% |
| Peak brilliance | $2 \times 10^{29}$ photons/sec/mm$^2$/mrad$^2$/0.1% |

The TTF-FEL will be a unique light source which emits radiation up to photon energies of about 200 eV in the first harmonic and up to 600 eV in the 3rd harmonic. The average values of brilliance and photon flux are three orders of magnitude higher than those of third generation synchrotron sources. The realization of such a FEL is a very difficult and challenging task but the exciting perspectives for basic and applied research opened up by the outstanding properties of the radiation justify the effort.

## Acknowledgements

The author is indebted to many colleagues for stimulating discussions, for providing figures, and making results available prior to publication.

# REFERENCES

1. Winnick, H. ed., *Synchrotron Radiation Sources a Primer*, Singapore, New Jersey, London, Honkong, World Scientifice, 1994.
2. Schlachter ,A.S., and Wuilleumier, F.J., eds. *New Directions in Research with Third-Generation Soft X-ray Synchrotron Radiation Sources,* Dordrecht, Boston, London, Kluwer Academic Publishers, 1994.
3. Ebashi , S., Koch, M., and Rubenstein, E., *Handbook of Synchrotron Radiation,*Vol 4, Amsterdam, Oxford, New York, Tokyo, North-Holland, 1991.
4. Arthur, J., Materlik, G., and Winnick, H., eds., *Workshop on Scientific Applications of Coherent X-Rays,* SLAC Report 437, 1994.
5. *A VUV Free Electron Laser at the TESLA Test Facility at DESY,* Conceptual Design Report, Hamburg, DESY Print, TESLA-FEL 95-03, 1995.
6. Report of the Committee of the National Academy of Sciences on*, Free Electron Laser and Other Advanced Sources of Light,* Washington DC, National Academy Press, 1994.
7. Spicer, W., Arthur, J., and Winnick, H., eds., *Workshop on Scientific Applications of Short Wavelength Coherent Light Sources,* Stanford, SLAC Report 414, 1992.
8. Ben-Zvi, I., and Winnick, H., eds., *Report on the Workshop Towards Short Wavelength Free Electron Lasers,* Brookhaven National Laboratory, BNL Report 49651, 1993.
9. Elleaume, P. J., Synchrotron Rad.**1**, 19 (1994).
10. .Schmidt, V., Rep. Progr. Phys. **55**, 1483 (1992).
11. Sonntag, B., and Zimmermann, P., Rep. Progr. Phys. **55**, 911 (1992).
12. Madden, R.P., and Codling, K., Phys. Rev. Lett. **10**, 516 (1963 ).
13. Tang, J.Z., Watanabe, S., Matsuzawa, M., and Lin, C.D., Phys. Rev. Lett. **69**, 1633 (1992).
14. Domke, M., Remmers, G., and Kaindl, G., Phys. Rev. Lett. **69**, 1171 (1992).
15. Domke, M., Schulz, K., Remmers, G., Gutlierrez, A., Kaindl, G., and Wintgen, D., Phys. Rev. **A51**, R4309 (1995).
16. Kaindl G., Schulz, K., Heimann., P.A., Bozek, J.D., and Schlachter, A.S., Technical Report Advanced Light Source, Berkeley, 1995.
17. Kaindl, G., Schulz, K., Bozek, J.D., Schlachter, A.S., and Heimann, P.A.,XIX, ICPEAC , Scientific Programm and Abstracts of Contributed Papers,p.325 (1995).
18. Dalgarno, A., and Sadeghpour. H.R., Comments At. Mol. Phys. **30**, 143 (1994).

19. Samson, J.A.R., He, Z., Bartlett, R., and Sagurton, M, Phys. Rev. Lett. **72**, 3329 (1994).
20. Andersson, L., and Burgdörfer, J., Phys. Rev. Lett. **71**, 50 (1993).
21. Levin, J.C., Sellin, I.A., Johnson, B.M., Lindle, D.W., Miller, R.D., Berrah, N., Azuma, Y., Berry, H.G., and Lee, H.D., Phys. Rev. **A47**, R16 (1993).
22. McGuire, J.H., Berrah, N., Bartlett, R.J., Samson, J.A.R., Tanis, J.A., Cocke, C.L., and Schlachter, A.S., J. Phys. **B28**, 913 (1995).
23. Bergström, P., Hino, K., and Macek, J., Phys. Rev. **A51**, 3044 (1995).
24. Andersson, L., and Burgdörfer, J., Phys. Rev. **A50**, R2810 (1994).
25. Hino, K., Bergström, P., and Macek, J., Phys. Rev. **72**, 1620 (1994).
26. Suric', T., Pisk, K., Logan, B.A., and Pratt, R., Phys. Rev. Lett. **73**, 790 (1994).
27. Spielberger, L., Jagutzki, O., Dörner, R., Ullrich, J., Meyer, U., Mergel, V., Unverzagt, M., Damrau, M., Vogt, T., Ali, I., Khayyat, Kh., Bahr, D., Schmidt, H.G., Frahm, R., and Schmidt-Böcking, H., Phys. Rev. Lett. **74**, 4615 (1995).
28. Lablanquie, P., Mazeau, J., Andric, L., Selles, P., and Huetz, A., Phys. Rev. Lett. **74**, 2192 (1995).
29. Kazansky, A.K., and Ostrowsky, V.N., J. Phys. **B28**, 1453 (1995).
30. Dawber, G., Avaldi, L., McConkey, A.G., Rojast, H., MacDonald, M.A., and King, G.C., J. Phys. **B28**, L271 (1995).
31. Maulbetsch, F., and Briggs J.S., J. Phys. **B27**, 4095 (1994).
32. Schwarzkopf, O., Krässig, B., Elmiger, J., and Schmidt, V., Phys. Rev. Lett. **70**, 3008 (1995).
33. Schwarzkopf, O., Krässig, B., Schmidt, V., Maulbetsch, F., and Briggs, J.S., J. Phys.**B27**, L347 (1994).
34. Schwarzkopf, O., and Schmidt, V., J. Phys. in press.
35. Bruch, R., Paul, G., Andrä, J., and Lipsky, L., Phys. Rev. **A12**, 1808 (1975).
36. Rodbrö, M., Bruch, R., and Bisgaard, J. Phys. **B12**, 2413 (1979).
37. Agentoft, M., Andersen, T., and Chung, K.T., J. Phys. **B17**, L433 (1984).
38. Müller, A., Hofmann, G., Weissbecker, B., Stenke, M., Tinschert, K., Wagner, M., and Salzborn, E., Phys Rev. Lett. **63**, 758 (1989).
39. Safranova, U.I., and Senashenko, V.S., J. Phys. **B11**, 2623 (1978).
40. Simons, R.L., Kelly, H.P., and Bruch, R., Phys. Rev. **A19**, 682 (1979).
41 Chung, K.T., Phys. Rev. **A25**, 1596 (1982).
42. Nicolaides, C.A., J. Phys. **B26**, L291 (1993)

43. Piangos, N.A., and Nicolaides, C.A., Phys Rev. **A48**, 4142 (1993).
44. Kiernan, L.M., Kennedy, E.T., Mosnier, J.P., Costello, J.T., and Sonntag, B.F., Phys. Rev. Lett. **72**, 2359 (1994).
45. Kiernan, L.M., Lee, K.M., Sonntag, B.F., Sladeczek, P., Zimmermann, P., Kennedy, E.T., Mosnier, J.P., and Costello, J.T., J. Phys. **B28**, L161 (1995).
46. Azuma, Y., Hasegawa, S., Koike, F., Kutluk, G., Nagata, T., Shigemasa, E., Yagishita, A., and Sellin, I.A., Phys. Rev. Lett. **74**, 3768 (1995).
47. Kiernan, L.M., Lee, K.M., Sonntag, B.F., Zimmermann, P., Kennedy, E.T., Mosnier, J.P., and Costello, J.T., to be published.
48. Journel M.L., These de Doctorat de l' Universite Paris XI, 1995.
49. Voky, L., private communication.
50. Journel, L., Cubaynes, D., Bizau, J.M., AlMoussalami, S., Rouvellou, B., Wuillemier, F.J., Voky, L., Faucher, P.,and Hibbert, A., to be published.
51. Aberg, T., Phys. Scr. **T41**, 71 (1992).
52. Aberg, T., and Crasemann, B., *Radiative and radiationless resonant Raman scattering,* in *Resonant Anomalous X-Ray Scattering,* Materlik, G., Sparks, C., and Fischer, K., eds., Amsterdam, Elsevier 1994, page 430.
53. Brown, G.S., Chen, M.H., Crasemann, B., and Ice, G.E., Phys. Rev. Lett.**45**, 1937 (1980).
54. Armen, G.B., Aberg,T., Levin,J.C., Crasemann, B., Chen, M.H., Ice, G.E., and Brown, G.S., Phys. Rev. Lett. **54**, 1142 (1985).
55. Kivimäki, A., Naves de Brito, A., Aksela, S., Aksela, H.,Sairanen,O.P., Ausmees,A., Osborne, S.J., Dantas, L.B., and Svensson, S., Phys. Rev. Lett. **71**, 4307 (1993).
56. Liu,Z.F., Bancroft, G.M., Tan, K.H., and Schachter, M., Phys. Rev. Lett. **72**, 621 (1994).
57. Aksela, S., Kivimäki, A., Naves de Brito, A., Sairanen, O.P., Svensson , S., and Väyrynen, J., Rev. Sci. Instrum. **65**, 831 (1994).
58. Aksela, S., Kukk, E., Aksela, H., and Svensson, S., Phys. Rev. Lett. **74**, 2917 (1995).
59. Davis, L.C., J. Appl. Phys. **59**, 25 (1986).
60. Baumgarten, L., Schneider, C.M., Petersen, H., Schäfers, F., and Kirschner, J., Phys. Rev. Lett. **65**, 492 (1990).
61. Roth, Ch., Hillebrecht, F.U., Rose, H.B., and Kisker, E., Phys. Rev. Lett. **70**, 3479 (1993).
62. Sirotti, F., and Rossi, G., Phys. Rev. **B49**, 15682 (1994).
63. Hillebrecht F.U., Roth, Ch., Jungblut, R., Kisker, E., and Bringer, A., Europhys. Lett. **19**, 711 (1992).

64. Thole, B.T., and van der Laan, G., Phys. Rev. **B44**, 12424 (1991).
65. Cherepkov, N.A., Phys. Rev. **B50**, 13831 (1994).
66. Thole, B.T., Dürr, H.A., and van der Laan, G., Phys Rev. Lett. **74**, 2371 (1995).
67. Pahler, M., Lorenz, C., v. Raven, E., Rüder, J., Sonntag B., Baier, S., Müller, B., Schulze, M., Staiger, H., Zimmermann, P., and Kabachnik, N.M., Phys. Rev. Lett. **68**, 2285 (1992).
68. Baier, S., Schulze, M., Staiger, H., Zimmermann, P., Lorenz, C., Pahler, M., Rüder, J., Sonntag, B., Costello, J.T., and Kiernan, L., J. Phys. **B27**, 1341 (1994).
69. Dohrmann, T., Thesis, Universität Hamburg, 1995.
70. Dohrmann, T., v.d. Borne, A., Sonntag, B., Wedowski, M., Weisbarth, F., Zimmermann, P., and Dolmatov, V.K., to be published.
71. Dolmatov, V.K., J. Phys. **B26**, L585 (1993).
72. Cherepkov, N.A., Kuznetsov, V.V., Verbitskii, V.A., J. Phys. **B28**, 1221 (1995).
73. Baier, S., Grum-Grzhimailo, A.N., and Kabachnik, N.M., J. Phys. **B27**, 3363 (1994).
74. Grum-Grizhimailo, A.N., private communication.

# PHOTON COLLISIONS

Dissociation Dynamics of Superexcited Molecules .................. 67
    Y. HATANO
Inner Shell Excitation, Relaxation
and Fragmentation of Cluster Beams and Molecules ............... 89
    A.P. HITCHCOCK and E. Rühl
Anisotropy and Kinetic Energy Release
in Ionic Fragmentation of Molecules Photoexcited
to Inner-Shell Hole States ............................................ 105
    I.H. SUZUKI and N. Saito

# Dissociation Dynamics of Superexcited Molecules

## Yoshihiko Hatano

*Department of Chemistry, Tokyo Institute of Technology,*
*Meguro-ku, Tokyo 152, Japan*

**Abstract.** A survey is given of the spectroscopy and dynamics studies of superexcited molecules by means of electron- and photon-impact methods with particular emphasis on their dissociation dynamics. Topics chosen from recent progress in these studies are 1) comparison of electron- and photon-impact studies with each other, 2) dissociation dynamics of $H_2$ in doubly excited states, 3) dissociation dynamics of inner core excited states of molecules, e.g., $C_2H_2$ and $O_2$, and 4) absolute photoabsorption, photoionization, and photodissociation cross sections, and photoionization quantum yields of molecules.

### INTRODUCTION

Platzman [1, 2] considered theoretically the interactions of ionizing radiation with atoms and molecules, and pointed out that atoms and molecules receive a large fraction of energy from ionizing radiation with a spectrum determined by their optical oscillator strength distributions according to the optical approximation. He also pointed out the following important features, although very few experimental data were available at that time.

a) The value of an oscillator strength distribution shows generally its maximum at the energy of 10-30 eV, which is larger than the first ionization potential (Ip).
b) Ionization efficiency ($\eta$) values of molecules are much smaller than unity in the energy range just above Ip.
c) There exists a hydrogen isotope effect in $\eta$ values.

By combining all available information, (a)-(c), Platzman indicated an important role of highly excited electronic states in the primary action of ionizing radiation as follows:

$$
\begin{aligned}
AB &\rightarrow AB^+ + e^- \quad \text{direct Ionization} &(1)\\
&\rightarrow AB' \quad \text{superexcitation} &(2)\\
&\rightarrow AB^{**} \quad \text{excitation} &(3)\\
\\
AB' &\rightarrow AB^+ + e^- \quad \text{autoionization} &(4)\\
&\rightarrow A + B \quad \text{dissociation} &(5)\\
&\rightarrow AB'' \quad \text{processes other than (4) and (5).} &(6)
\end{aligned}
$$

When a molecule AB receives energy which is larger than its Ip, AB may be directly ionized (1) and may be excited (2) to form AB' which was named by Platzman a "superexcited" molecule. The superexcited AB' can ionize (4) or dissociate into neutral fragments (5). Since the ionization efficiency, $\eta$, is defined as the ratio between the cross section for the total ionization ($\sigma i$) and the cross section for the total energy absorption ($\sigma t$), then the value of $1-\eta$ shows the importance of the dissociation process in the total decay channels of AB' because the process (6) seems not to be important in the decay of such extremely highly excited states. As one of the process (6), ion-pair formation, AB'$\rightarrow$A$^+$+B$^-$, has been extensively studied to clarify the dynamics of superexcited molecules although its cross section is much smaller than those of the processes (4) and (5) [3-7]. The value of $\eta$ may therefore be smaller than unity in the energy range even above Ip, which means that the molecules excited into this energy range are not always ionized but use their energy for processes other than ionization.

The Platzman's concept of the superexcited state has made a profound influence on the science of excited states and motivated researchers in a wide field to find new objectives of research. In order to study experimentally the superexcited states or to substantiate Platzman's ideas, scientists had to find the way to produce such superexcited states. Electron beams were mainly used in 1960's and 70's for this purpose, whereas after 80's laser multiphotons and synchrotron radiation (SR) have also been used. Details of the electronic states of superexcited molecules and the mechanism of autoionization and dissociation have been quickly clarified. The theory of superexcited states has also been advanced greatly [8-11]. This paper gives a survey of recent experimental studies of dynamics of superexcited molecules as a summary of the first comprehensive survey by the present author [12] together with further new information on the subject, and places particular emphasis on the recent progress in understanding their dissociation dynamics and corresponding superexcited electronic states.

# ELECTRON IMPACT STUDIES
# OF SUPEREXCITED MOLECULES

## Production of Excited Fragments by Electron Impact

There have been three major experimental studies of superexcited molecules by means of an electron beam as an excitation source to form superexcited states. In the following a brief survey is given of these studies with particular emphasis on a comparison with photon impact studies.

Ionization and excitation of molecules in collisions with electrons in the energy range higher than about $10^2$ eV are well elucidated by the Born-Bethe theory [13-15] and the cross section $Q_s(T)$ at the electron energy $T$ to form the state $s$ at least for optically-allowed transitions is given by

$$Q_s(T) = \frac{4\pi a_o^2 R}{T} M_s^2 \ln \frac{4C_s T}{R} \quad (7)$$

where $a_o$, $R$, and $C_s$ are the Bohr radius, the Rydberg energy, and a constant, respectively; $M_s^2$ is the dipole matrix element squared for the state $s$ formation. In general, it is difficult to measure $M_s^2$ experimentally for every excited state $s$ of interest. However, when a superexcited molecule produces an electronically excited, in some cases optically emissive, dissociation fragments, $M_s^2$ may be obtained experimentally from the measurement of optical emission cross section as a function of $T$. de Heer and his coworkers [16-20] have extensively investigated the dissociative excitation of molecules by electron impact and obtained the values of $M_s^2$ from the slope of a plot of $Q_s T/4\pi a_o^2$ vs $\ln T$ (the Fano plot).

## Absolute cross sections for molecular photoabsorption and photoionization processes by fast electron impact

In electron energy loss experiments of molecules extrapolation of energy loss intensities for incident electron energies much higher than $10^2$ eV to zero momentum transfer, i.e., to the intensity at the forward scattering, should give, according to the Bethe theory, a corresponding optical oscillator strength [13]. Brion and his coworkers [21-23] have extensively investigated such an experimental approach using fast electrons as the virtual photon source, which M.J. van der Wiel has named "the poor man's synchrotron". He has pointed out some expected characteristics of SR in comparison with virtual photons in understanding ionization and excitation of molecules and made clear some necessary assumptions to virtual photons instead of real photons. It should be noted here, as described in detail later in this section, that these two methods, i.e., real- and virtual photon experiments, have complementary roles with each other to substantiate the dynamics of superexcited molecules.

Brion's approach to real-photon experiments using the virtual photon source, which is called the electron impact dipole-simulation method, is summarized in Table 1 and compared with real-photon experiments. These simulation techniques have provided a

TABLE 1. Photon and electron-impact experiments [23]

| Photon experiment | Equivalent electron-impact experiment |
|---|---|
| total photoabsorption | electron-energy-loss spectroscopy, dipole (e,e) |
| total photoionization | dipole (e,2e) or (e,e+ion) (from sums of partial cross sections) |
| photoelectron spectroscopy | electron energy loss-ejected electron coincidence, dipole (e,2e) |
| photoionization mass spectrometry | electron-ion coincidence, dipole (e,e+ion) |

large body of data for comparison with photon experiments of absolute cross sections [23]. The photon experiments using SR, which are greatly in progress, can be compared with the simulation measurements.

Examples are chosen in the following for this comparison. The absolute $\sigma t$ values for $H_2S$ were measured [24] using low resolution dipole (e,e) spectroscopy in the equivalent photon energy range up to 90 eV and compared with that obtained from SR experiments [25]. Good agreement was obtained in the $\sigma t$ values between the two different experiments, providing confirmation that the necessary assumptions, such as the sum-rule normalization used in the simulation experiments, are reasonable at least for the present molecule. However, a large difference was observed in the wavelength or energy resolution between the two experiments. This is not an essential comparison between the simulation and the real photon experiments to understand superexcited states. In an SR experiment in the vacuum ultraviolet (VUV) region, a high energy-resolution of about 10 meV is easily obtained, while in most of the simulation experiments the energy resolution was about 1 eV. However, Brion's group has developed, as was expected by the present author [12], a dipole (e,e) simulation experiment with a higher energy resolution of about 50 meV [26-31].

Another example is shown in Fig.1 of a comparison between a dipole (e,e) simulation experiment and a real-photon experiment. There existed a marked discrepancy between the two in the absolute $\sigma t$ values [12, 32-35]. In a recent high resolution dipole (e,e) experiment of $SiH_4$ [31], however, as shown in Fig.1, good agreement has been obtained with a real-photon SR experiment

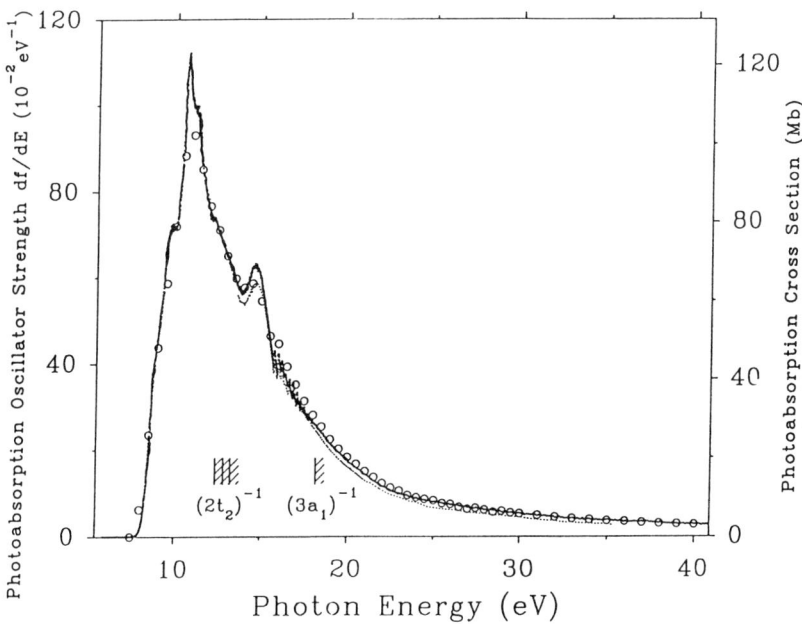

FIGURE 1. Comparison of the photoabsorption cross sections ($\sigma t$) of $SiH_4$. ○ Low resolution dipole (e,e) experiment [31], − high resolution dipole (e,e) experiment [31], and ......SR experiment [33].

[33]. The former discrepancy has been ascribed to the interaction of a target gas with the oxide cathode of the electron gun in the simulation experiment [31]. Although this simulation experiment has still somewhat lower energy resolution in the $\sigma t$ curve, it has a distinctive advantage in that it shows the gross features of photo-absorption, -ionization, and -ionic fragmentation of a molecule over a wide energy range, and provides information about superexcited states complementary to that obtained from SR experiments.

## Translational spectroscopy of dissociation fragments produced by electron-impact excitation of simple molecules

For deeper understanding of the electronic structure of super-excited molecules and their dissociation processes, it is indispensable to measure the kinetic energy of dissociation fragments and their angular distribution as well as to measure the threshold energy of dissociation [36]. The former measurement, which is called translational spectroscopy of molecular dissociation, is described in this section, while the latter one in a preceding section typically represented by the Fano plot. The former one has been made not only for charged particles, for long lived metastables or high Rydberg atoms by means of time-of-flight spectroscopy, but also for emissive excited atoms with short lifetimes by means of Doppler-profile spectroscopy [36].

The Doppler experiments on $H_2$, for example, determined the kinetic energy of $H^*(n)$, where $n$ is the principal quantum number of a hydrogen atom formed from doubly excited states $(2p\sigma_u)(nl\lambda)$ as well as from vibrationally excited states $(1s\sigma_g)(nl\lambda)$ [36-40]. This is typical evidence for Platzman's prediction of the dissociation of superexcited states of molecules. For molecular hydrogen excess energies beyond Ip result from vibrational or double excitation. For other molecules one can easily estimate another type of excess energy resulting from inner valence excitation. The same experimental technique has been applied to $H_2O$, $D_2O$, $CH_4$, $NH_3$, HF and other more complex molecules, giving direct evidence for the dissociation potential of inner valence excited states as well as doubly excited states [36, 41-44]. In a series of these investigations the internal energies of these highly excited states have been found to be in good agreement with the energies of corresponding molecular ions obtained from photoelectron or ESCA spectra and (e, 2e) experiments [36, 43]. The agreement shows that superexcited states are molecular high-

Rydberg states converging to corresponding ionic states, and supports the core-ion model proposed for high-Rydberg atomic dissociation fragments.

# PHOTON IMPACT STUDIES OF SUPEREXCITED MOLECULES

### General

Excitation photon sources which have been used for photon impact studies of superexcited molecules are classified into discharge lamps, lasers, and SR sources. Photochemistry of gas-phase molecules [45-47] has been investigated by means mainly of discharge lamps up to about 1980 and of lasers since then, and has been restricted almost to consequences of photoabsorption below Ip. In research fields other than photochemistry, however, there are several useful sources of information on photon impact studies of superexcited molecules.

Berkowitz [48] compiled the experimental data on photoabsorption, photoionization and photoelectron spectroscopy in the photon energy range much wider than that in photochemistry, i.e., from absorption thresholds to the VUV-SX region in some cases even to hard X-ray region, summarized theoretical backgrounds, and briefly introduced a general scheme of the dynamics of superexcited states. The most important point in this book from the viewpoint of superexcited states is a comprehensive survey of the data on $\eta$ values up to late 1970's.

Recently published books [49-51] included only fragmentary information on studies of superexcited states. Most of the papers therein were not concerned with the dynamics of superexcited molecules, but with either the spectroscopy of formed ions and electrons or the dissociation of excited molecules below Ip. In summaries of photophysics and photochemistry in the VUV range an important role of SR was pointed out as a promising excitation source in the experiments of molecular photoionization and dissociative ionization, giving typical examples chosen mainly from the experimental results obtained at Orsay [52, 53].

The first comprehensive survey of the photon impact studies of the dynamics of superexcited molecules has been given by the present author [12]. In this section, therefore, a summary of this survey article is presented together with further new information on this subject.

## Laser-multiphoton studies of superexcited molecules

Since NO is a simple diatomic molecule with a relatively low ionization potential, this molecule has been most intensively studied in the entitled experiments. The autoionization dynamics of NO has been studied in detail using photoelectron spectroscopy combined with laser multiphoton ionization, providing conclusive evidence for the electronic autoionization due to continuum-continuum interaction via dissociative valence-excited states [54-56].

Neutral fragments, $N(^2D)$ and $N(^4S)$, formed from a superexcited NO molecule in Rydberg states have been also detected recently with a resonance-enhanced multiphoton ionization (REMPI) technique [57-60]. The use of lasers for detecting products as well as for producing superexcited states has provided precise information on the spectroscopic features of these states, which is certainly helpful also to understand their dynamic features.

## Synchrotron radiation studies of superexcited molecules

### General

Spectroscopy and dynamics studies of superexcited molecules have greatly progressed by applying SR as a new excitation source to form superexcited states.

As pointed out in preceding sections, there have been relatively few studies on superexcited molecules using SR in comparison with those using electron impact or discharge-lamp photon impact even in the case of $H_2$, and much less in the case of chemically important complex molecules. Recently, however, such studies have been quickly motivated and accelerated by the development of new dedicated SR facilities [12].

As selected topics in this section, we will discuss, first of all, recent progress in SR research on the dissociation dynamics of superexcited molecular hydrogen, particularly doubly excited states [12, 36, 61-63], and secondly in SR research on superexcited states of chemically important complex molecules. An approach to complex molecules is the choice of molecules to be studied, which are in a stereo-isomer series [64-68], e.g., cyclopropane and propylene for $C_3H_6$. To clarify the neutral dissociation of superexcited molecules, we need to compare in detail an absolute $\eta$ value with $\sigma t$, $\sigma i$, and cross sections for optical emission from neutral dissociation fragments as a function of photon energy. An example in such comparison is presented for $C_2H_2$ [69] as chosen from our recent research on $N_2$, $O_2$, CO, $CO_2$, $N_2O$, simple hydrocarbons, ethers, and Si-containing molecules [70-77]. Finally, a newly developed experimental method for the entitled subject, that is, the two dimensional observation of the neutral dissociation of superexcited $O_2$ is briefly described [78].

*Dissociation of superexcited states of $H_2$*

Dissociation of superexcited states of $H_2$ has been extensively studied both experimentally and theoretically [17, 36, 37-40, 53, 61-63, 79, 80]. The observation of fluorescence from excited fragments is highly informative as a method for studying these states and their dissociation processes. It has been known, as summarized in a preceding section, from the translational spectroscopy [36] of H* formed by electron impact on $H_2$ that the slow and fast H* atoms are produced from vibrationally excited Rydberg states $(1s\sigma_g)(nl\lambda)$ and doubly excited states $(2p\sigma_u)(nl\lambda)$, respectively. The former states have been extensively investigated also in photon-impact experiments [81-86], while the latter states have not been investigated in detail.

The first observation of the neutral dissociation of doubly excited molecular hydrogen has been successful by the measurement of Lyman-$\alpha$ emission in the photodissociation of $H_2$ using SR [12, 61-63] and compared with the theoretically predicted potential curves [12, 87-91]. The Lyman-$\alpha$ excitation spectrum has been observed in the energy region from the threshold to about 35 eV. The spectrum observed in the lower energy region is attributed to slow H(2$l$) atoms produced from vibrationally excited Rydberg states, while the spectrum observed in the energy region higher than 26.4 eV is attributed to fast H(2$l$) atoms produced from the neutral fragmentation of optically formed doubly excited states. The cross section of the Lyman-$\alpha$ fluorescence produced from doubly excited states at 30 eV is very small, about $10^{-20}$cm$^2$, as compared with that, about $10^{-17}$cm$^2$ at the main peaks in the excitation spectrum, from vibrationally excited Rydberg states (single-electron excited states). It is concluded, therefore, that the cross section for the photon-impact neutral fragmentation of doubly excited states of $H_2$ is much smaller than that of the single-electron excited states. Since, in the case of electron-impact excitation at low energies, however, both cross sections do not differ so much from each other [17], the doubly excited states, which are optically forbidden from the ground state, seem to have a significant role in the neutral fragmentation of $H_2$.

In the Lyman-$\alpha$ excitation spectrum [62] in the higher energy region, there have been observed three thresholds which discriminate some theoretical results from the others.

To clarify further the doubly excited states of $H_2$, Lyman-$\alpha$ Lyman-$\alpha$ coincidence has been measured in the dissociation process of

$$H_2 + h\nu \rightarrow H_2^{**} \rightarrow H(2p) + H(2p) \qquad (8)$$

where $H_2^{**}$ is doubly excited molecular hydrogen formed by optically allowed transition [61]. The excitation spectrum of this process obtained from coincidence signals by changing the photon energy has the threshold at the energy just below 29 eV. The $Q_2{}^1\Pi_u(1)$ state is a precursor of this process because it is the lowest $Q_2$ state and is expected to dissociate into the lowest limit of the $Q_2$ state, i.e., H(2p)+H(2p) at 24.9 eV.

The time-dependent intensity of Balmer-$\alpha$ and Lyman-$\alpha$ emissions produced in the photodissociation of $H_2$ has been measured using pulsed SR with the single bunch operation of a storage ring to obtain the angular momentum population of $H^*(3l)$ and $H^*(2l)$ as dissociation fragments, respectively [92, 93]. The population has shown a strong dependence on the incident photon wavelength, from which detailed information has been obtained on the dissociation dynamics of superexcited molecular hydrogen.

*Photoabsorption, Photoionization, and Photodissociation Cross Sections, and Photoionization Quantum Yields of Molecules*

It should be noted first of all that a major part of both $\sigma t$ and $\sigma i$ curves for almost all molecules should excist in the VUV region [12, 94, 95]. The data of these cross section values obtained from real-photon experiments have not been sufficient in general because suitable photon sources and window materials have been very few in this wavelength region. Recently, however, the situation has been quickly improved by the development of new dedicated SR facilities. This section summarizes the present status of the measurements of the entitled cross sections particularly in an absolute scale for chemically important molecules.

The values of $\sigma t$ or the values of oscillator strength $f$ or $df/dE$ have already been measured for various molecules in the wavelength region at least longer than the near UV region, whereas the values in the wavelength region shorter than the LiF cutoff at 105 nm in particular are very few. When the Thomas-Kuhn-Reiche (TKR) sum rule is compared with the sum of the obtained $f$ values for a molecule in the wavelength region longer than 105 nm which corresponds, roughly speaking, to about the Ip of most molecules, it is found that this sum of the obtained $f$ values corresponds to less than a few percent of $Z$, the number of electrons in the molecule. It is, therefore, concluded that the interaction of a photon with a molecule in the VUV region is predominant over all the other wavelength regions.

With this idea, therefore, $\sigma t$ and $\sigma i$ values have been systematically measured for molecules in several stereo-isomer series e.g., $C_3H_6$ (cyclopropane and propylene) and $C_2H_6O$ (ethyl alcohol and dimethyl ether) and compared with each other [64-68] to see how $df/dE$ changes with changing molecular structure, giving the common new features of $\sigma t$ or $df/dE$ values summarized as follows [12, 94, 95]:

(a) The $\sigma t$ values show a maximum at about 70-80 nm (about 16-18 eV) for each molecule.
(b) In the wavelength region shorter than that at the maximum the $\sigma t$ values are almost the same among the isomer molecules and equal to the sum of the cross sections for the constituent atoms.
(c) In the low energy side, the $\sigma t$ values have different peaks and shoulders depending on the isomer, i.e., on its molecular structure. The sum of the $\sigma t$ values in this energy region is, however, almost equal among the isomer molecules.

The photoionization quantum yield $\eta$ is a key quantity of considerable importance, as pointed out in the beginning of this article, to serve as an index for the degree of competition between direct ionization and excitation to the superexcited states opening to autoionization and dissociation. Several experimental efforts have been devoted to the measurement of $\eta$. Even for simple molecules, however, serious conflicts exist not only between real-photon and virtual-photon experiments but also between real-photon experiments themselves. The problems have originated mainly from lack of an intense light source and a suitable window material in the VUV region particularly in the wavelength region shorter than the LiF cutoff at 105 nm.

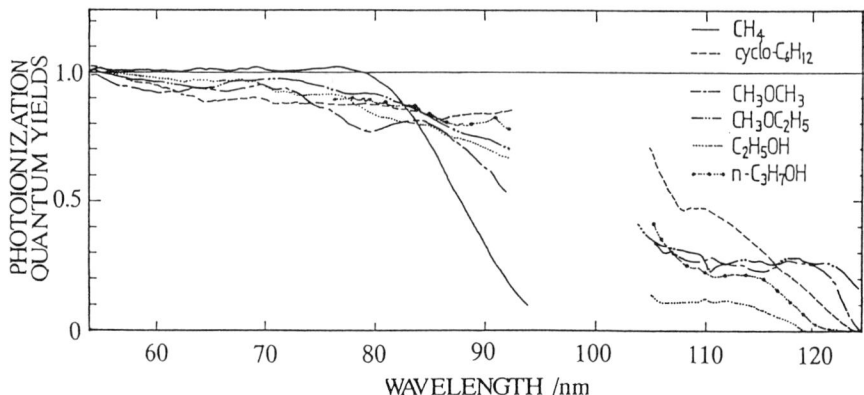

FIGURE 2. Photoionization quantum yields of several molecules in the wavelength regions shorter and longer than the LiF cutoff at 105 nm [96].

Experiments with differential pumping and without a window for the entrance of a photon beam into an ionization chamber also present several significant problems, such as an effusion of sample gases into the beam transportation region or a contribution from a diffracted photon beam in the higher order, making it difficult to determine correct, absolute, and comprehensive values of $\eta$. In most cases, the $\eta$-values have been assumed to be unity in the energy region far above the first ionization potential.

Recently systematic measurements have been reported also of the absolute $\eta$ values for molecules in several stereo-isomer series using a multiple-staged photoionization chamber with metal thin foil windows and an SR light source in the wavelength region from their respective ionization potentials at about 120-140 nm (9-10 eV) to the wavelength region of 54-92 nm (13-23 eV) [64-77, 96]. The obtained results, some of which are shown in Fig.2, have common features summarized as follows.

(a) $\eta$-values in the region close to the first ionization potential are much less than unity, which means that the molecules studied here are not easily ionized even when they have received enough energy to ionize.

(b) $\eta$-values do not reach unity even in the energy range more than about 10 eV above Ip.
(c) $\eta$-values increase with increasing photon energy and reach unity at around 23 eV (or 54 nm).
(d) $\eta$-curves show considerable structures.

It is concluded, therefore, that non-ionizing processes, such as the neutral fragmentation of superexcited molecules, have an important role in their decay channels and that molecules can not be easily ionized even when they absorb, energies larger than Ip.

*Comparative studies of ionization quantum yields with excitation spectra of optical emission from dissociation fragments: Acetylene*

As concluded experimentally in the last section, the magnitude of the internal energy of superexcited states of a molecule is much larger than its bond energy. Neutral fragments formed from these states should have excess large energies in their electronically, vibrationally, rotationally, and/or translationally excited states. It is, therefore, of great interest to observe optical emissions from excited fragments as a function of wavelength and to compare with the structures in $\eta$-curves. In this section, an example of such a comparison is presented for acetylene [69] to clarify further the dynamics of superexcited molecules.

In the $\eta$ curve of $C_2H_2$, as shown in Fig.3, obvious deviations from unity with interesting structures of at least three minima indicate the neutral dissociation of a superexcited acetylene molecule strongly competing with autoionization. Combining the $\eta$ values with the $\sigma t$ values in Fig.3, the cross sections for non-ionizing decay processes, the $\sigma d$ values, are obtained. The excitation spectra of the dispersed fluorescence from excited fragments produced in the dissociation of a superexcited acetylene molecule are further observed and compared with the calculated thresholds of related dissociation processes. The neutral dissociation of a superexcited acetylene molecule as a function of wavelength is, therefore, summarized as follows.

$$C_2H_2 + h\nu \rightarrow C_2(d^3\Pi_g) + 2H \qquad (9)$$
$$C_2(C^1\Pi_g) + 2H \qquad (10)$$
$$CH(A^2\Delta) + CH \text{ (or } CH(A^2\Delta)) \qquad (11)$$
$$H(2p) + C_2H \text{ (or } C_2 + H) \qquad (12)$$

The three minima in the $\eta$ curve correlate well with the structures observed in the excitation spectra. By further comparison with the $\sigma t$ and $\sigma i$ curves, the precursor superexcited states for the dissociation processes (9)-(12) have been assigned [69] in detail to high Rydberg states and/or innervalence excited states which have been investigated theoretically. The dissociation process (11) is of particular interest from chemical viewpoints because of a predominant breaking of the triple bond of acetylene in a specific energy range. This is explained, by observing in detail a hydrogen isotope effect of the dissociation processes, in terms of the structure-deformation of an acetylene molecule in superexcited states to loosen the triple bond [97].

## CONCLUDING REMARKS AND COMMENTS ON FUTURE PROBLEMS

A survey has been given of recent progress in the spectroscopy and dynamics studies of superexcited molecules with particular emphasis on the observation of the neutral dissociation of superexcited molecules and their electronic states. The molecules studied are simple molecules ranging from $H_2$ ($D_2$), $N_2$, $O_2$, and CO to $CO_2$ and $N_2O$ and complex molecules as hydrocarbons, alcohols, ethers, and some Si-containing molecules. It has been emphasized experimentally [12], as well as theoretically [98], to measure *absolute* photoabsorption, photoionization, and photodissociation cross sections, and photoionization quantum yields of these molecules.

It is concluded, therefore, that molecular superexcited states are, in most cases, vibrationally (or/and rotationally) excited, inner-core excited, or doubly excited Rydberg states converging to corresponding ionic states. In some cases, they are also valence-excited non-Rydberg states. It is also concluded in particular from measured *absolute* photoionization quantum yields that the cross section for the neutral dissociation in competing with autoionization is considerably large especially in the energy range near the ionization threshold. In this energy range photoabsorption cross sections show their maximum values. This conclusion means that molecules are not easily ionized even when they absorb the energy being larger than their ionization thresholds.

Some future problems needing more work are not comprehensively summarized here, but are chosen in the following, for example, from our research programs in progress.

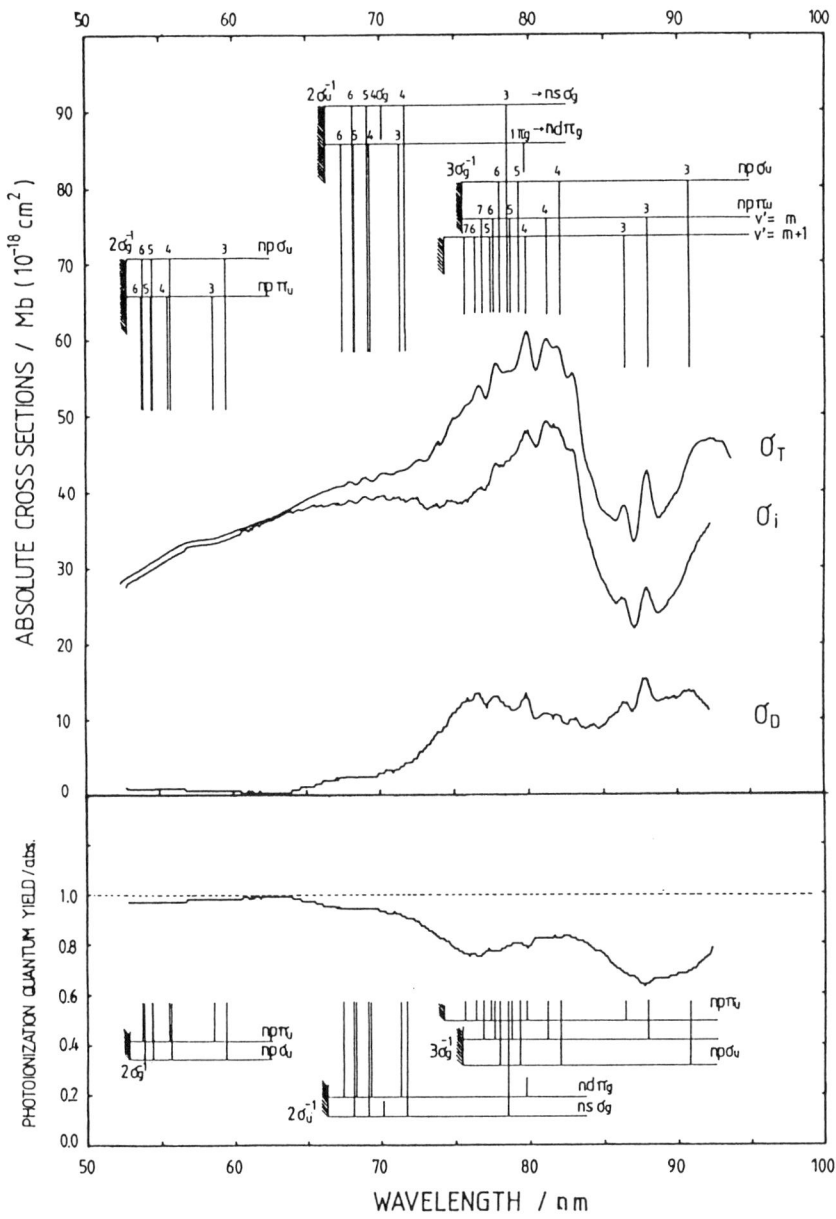

**FIGURE 3.** Photoabsorption ($\sigma t$), photoionization ($\sigma i$), photodissociation cross sections ($\sigma d$) (upper), and photoionization quantum yield (lower) of $C_2H_2$ [69].

It is informative, as described in the last section, for a clear understanding of the entitled subject to measure the excitation spectrum of fluorescence from excited dissociation fragments as compared with a photoionization quantum yield curve. Recently, a new experimental method has been developed in this measurement by the two-dimentional observation of the neutral dissociation of superexcited molecules, e.g., $O_2$ as shown in Fig. 4, as a function of both excitation photon wavelength and fluorescence wavelength [78].

A new experimental method has been developed also for the observation of ionic satellite states, to which superexcited states converge, by the coincidence detection of, e.g., $N^+$ and the optical emission from $N^*$ which are produced from $N_2^{+*}$ [99].

It is greatly needed to obtain the detailed information of superexcited states produced also by the optically forbidden transition from the ground state of a molecule. Another new experimental method has been further developed to obtain such information by means of the coincidence detection of fluorescence from a dissociation fragment, e.g., a Lyman-$\alpha$ photon, and an electron energy loss spectrum of $H_2$ with a relatively low incident electron energy [100]. The coincidence yield has been measured as a function of the energy loss, from which an additional peak is clearly observed due to an optically forbidden doubly excited molecular hydrogen as compared with optically allowed ones.

As a final comment obtained not from our research programs but from recent publications by other authors, one should notice a number of recent papers on experiments and theories of the ionization of large molecules such as tryptophan excited with laser photons into the energy range above their ionization thresholds [101-104]. The delayed ionization has been observed in some molecules, from which it has been emphasized that large molecules are not easily ionized and that a new theory explaining this result is developed. We are interested at least in the experimental result itself, but, as far as we understand, we should not yet fully rely on its theory [105]. It may be concluded from their feature article itself [101] and the references cited in their papers [101-104] that the authors have almost no information on the dynamics of superexcited molecules obtained in both experimental and theoretical investigations.

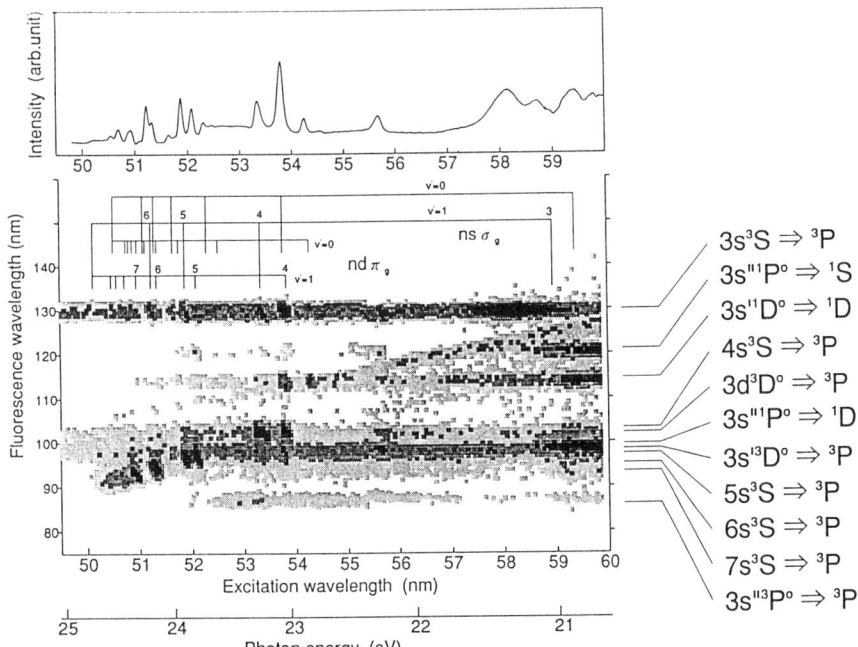

**FIGURE 4.** Two-dimensional yield spectrum of the fluorescence radiation emitted from excited oxygen atoms produced in the neutral dissociation of superexcited molecular oxygen as a function of both the excitation wavelength and the fluorescence wavelength (lower panel). The yields presented by the grey rectangle plots increase from white to black linearly with 8 steps of the color. Yield spectrum of nondispersed fluorescence for 105<$\lambda_f$<180 nm (upper panel) is shown together [78].

## ACKNOWLEDGMENTS

The author wishes to thank Drs. M. Inokuti, F.J. de Heer, and C.E. Brion for helpful discussion and comments. He is indebted to Drs. N. Kouchi, M. Ukai, K. Kameta, H. Koizumi, S. Arai, A. Ehresmann, and graduate students, M. Kitajima, S. Machida, and T. Odagiri, in his group for their excellent collaboration. The author's synchrotron radiation research described herein at the photon factory has been supported scientifically by Drs. K. Tanaka, K. Ito, and T. Hayaishi, and financially by the Ministry of Education, Science and Culture.

# REFERENCES

1. R.L. Platzman, *Vortex*, 23, 372 (1962).
2. R.L. Platzman, *Radiat.Res.*, 17, 419 (1962).
3. H. Oertel, H. Schenk, and H. Baumgartel, *Chem.Phys.Lett.*, 46, 251 (1980).
4. A. Dadouch, G. Dujardin, L. Hellner, M.J. Besnard-Ramage, and B.J. Olsson, *Phys.Rev.A* 43, 6057 (1991).
5. K. Mitsuke, S. Suzuki, T. Imamura, and I. Koyano, *J.Chem. Phys.*, 92, 6556(1990); 93, 1710 (1990); 93, 8717 (1990); 94, 6003 (1991).
6. K. Mitsuke, H. Yoshida, and H. Hattori, *Z.Phys.D* 27, 267 (1993).
7. H. Yoshida and K. Mitsuke, *J.Chem.Phys.*, 100, 8817 (1994).
8. H. Nakamura, *Int.Rev.Phys.Chem.*, 10, 123 (1991).
9. H. Nakamura, *J.Phys.Chem.*, 88, 4812 (1984).
10. S. Lee, M. Iwai, and H. Nakamura, *Molecules in Laser Fields*, ed. A.D. Bandrauk, Marcel Decker, New York, 1994, p.217.
11. H. Nakamura, *J. Chinese Chem.Soc.*, 42, 359 (1995).
12. Y. Hatano, *Dynamics of Excited Molecules*, ed. K. Kuchitsu Elsevier, Amsterdam, 1994, chapter 6.
13. M. Inokuti, *Rev.Mod.Phys.*, 43, 297 (1971).
14. M. Inokuti, *Applied Atomic Collision Physics*, Vol.4, ed. S. Datz, Academic Press, New York, 1983, p.179.
15. F.J. de Heer and M. Inokuti, *Electron Impact Ionization*, eds. T.D. Mark and G.H. Dunn, Springer-Verlag, Vienna, 1985, p.232.
16. F.J. de Heer, *Int.J.Radiat.Phys.Chem.*, 7, 137 (1975).
17. F.J. de Heer, H.A. van Sprang and F.J. Mohlmann, *J.Chim. Phys.*, 77, 773 (1980).
18. D.A. Vroom and F.J. de Heer, *J.Chem.Phys.*, 50, 573; 580; 1883 (1969).
19. F.J. de Heer, H.R.M. Moussa and M. Inokuti, *Chem.Phys.Lett.*, 1, 484 (1967).
20. C.I.M. Beenaker and F.J. de Heer, *Chem.Phys.*, 7, 130 (1975).
21. C.E. Brion and A. Hamnett, *Adv.Chem.Phys.*, 45, 1 (1981).
22. C.E. Brion, *Comm.At.Mol.Phys.*, 16, 249 (1985).
23. J.W. Gallagher, C.E. Brion, J.A.R. Samson and P.W. Langhoff, *J.Phys.Chem.Ref. Data*, 17, 9 (1988).
24. C.E. Brion, Y. Iida and J.P. Thomson, *Chem.Phys.*, 101, 449 (1986).
25. T. Ibuki, H. Koizumi, T. Yoshimi, M. Morita, S. Arai, K. Hironaka, K. Shinsaka, Y. Hatano, Y. Yagishita and K.Ito, *Chem.Phys.Lett.*, 119, 327 (1985).

26. W.F. Chan, G. Cooper, and C.E. Brion, *Phys.Rev.*, **A44**, 186 (1991).
27. W.F. Chan, G. Cooper, X. Guo, G.R. Burton, and C.E. Brion, *Phys.Rev.*, **A45**, 1420 (1992).
28. W.F. Chan, G. Cooper, and C.E. Brion, *Chem.Phys.*, **168**, 375 (1992).
29. G. Cooper, G.R. Burton, and C.E. Brion, *J. Electron Spectr. Related Phenomena*, **73**, 139 (1995).
30. G. Cooper, T.N. Olney, and C.E. Brion, *Chem.Phys.*, **194**, 175 (1995).
31. G. Cooper, G.R. Burton, W.F. Chan, and C.E. Brion, *Chem. Phys.*, **196**, 293 (1995).
32. G. Cooper, T. Ibuki and C.E. Brion, *Chem.Phys.*, **140**, 133 (1990).
33. K. Kameta, M. Ukai, R. Chiba, K. Nagano, N. Kouchi, Y. Hatano and K. Tanaka, *J.Chem.Phys.*, **95**, 1456 (1991).
34. M. Suto and L.C. Lee, *J.Chem.Phys.*, **84**, 1160 (1986).
35. U. Itoh, Y. Toyoshima, H. Onuki, N. Washida and T. Ibuki, *J. Chem.Phys.*, **85**, 4867 (1986).
36. Y. Hatano, *Comm.At.Mol.Phys.*, **13**, 259 (1983).
37. R.S. Freund, J.A. Schiavone and D.F. Brader, *J.Chem.Phys.*, **64**, 1122 (1976).
38. K. Ito, N. Oda, Y. Hatano and T. Tsuboi, *Chem.Phys.*, **17**, 35 (1976); **21**, 203 (1977).
39. M. Higo, S. Kamata, and T. Ogawa, *Chem.Phys.*, **66**, 243 (1982).
40. T. Ogawa, M. Taniguchi, K. Nakashima, and H. Kawazumi, *Chem. Phys.*, **137**, 323 (1989).
41. N. Kouchi, K. Ito, Y. Hatano, N. Oda and T. Tsuboi, *Chem. Phys.*, **36**, 239 (1979).
42. M. Ohno, N. Kouchi, K. Ito, N. Oda and Y. Hatano, *Chem. Phys.*, **58**, 45 (1981).
43. N. Kouchi, M.Ohno, K. Ito, N. Oda and Y. Hatano, *Chem.Phys.*, **67**, 287 (1982).
44. N. Kouchi, K. Ito, Y. Hatano and N. Oda, *Chem.Phys.*, **70**, 105 (1982).
45. H. Okabe, *Photochemistry of Small Molecules*, Wiley-Interscience Publ., New York, 1978.
46. K.P. Lawley (ed.), *Photodissociation and Photoionization*, Adv.Chem.Phys. Vol.LX, Wiley-Interscience, New York, 1985.
47. M.N.R. Ashfold and J.E. Baggott (eds.), *Molecular Photodissociation Dynamics*, Adv. Gas Phase Photochemistry and Kinetics, Royal Soc.Chem. 1987, and references cited therein.
48. J. Berkowitz, *Photoabsorption, Photoionization, and Photoe-*

*lectron Spectroscopy*, Academic Press, New York, 1979.
49. S.P. McGlynn, G.L. Findley and R.H. Huebner (eds.), *Photophysics and Photochemistry in the Vacuum Ultraviolet*, D. Reidel Publ., Dordrecht, 1985.
50. F. Lahmani (ed.), *Photophysics and Photochemistry above 6 eV*, Elsevier, Amsterdam, 1985.
51. C.Y. Ng (ed.), *Vacuum Ultraviolet Photoionization and Photodissociation of Molecules and Clusters*, World Scientific, Singapore, 1991.
52. P.M. Guyon and I. Nenner, *Appl.Opt.*, **19**, 4068 (1980).
53. I. Nenner and J.A. Beswick, *Handbook on Synchrotron Radiation*, **2**, 355 (1987).
54. Y. Achiba and K. Kimura, *Chem.Phys.*, **129**, 11 (1989).
55. K. Nakashima, H. Nakamura, Y. Achiba and K. Kimura, *J.Chem.Phys.*, **91**, 1603 (1989).
56. K. Kimura, *Int.Rev.Phys.Chem.*, **6**, 195 (1987).
57. G.E. Gadd, L.E. Jusinski, and T.G. Slanger, *J.Chem.Phys.*, **91**, 3378 (1989).
58. A. Fujii and N. Morita, *Chem.Phys.Lett.*, **182**, 304 (1991).
59. A. Fujii and N. Morita, *SPIE*, **1858**, 184 (1993).
60. A. Fujii and N. Morita, *J.Chem.Phys.*, **97**, 327 (1992).
61. S. Arai, T. Kamosaki, M. Ukai, K. Shinsaka, Y. Hatano, Y. Ito, H. Koizumi, A. Yagishita, K. Ito and K. Tanaka, *J.Chem.Phys.*, **88**, 3016 (1988).
62. S. Arai, T. Yoshimi, M. Morita, K. Hironaka, T. Yoshida, H. Koizumi, K. Shinsaka, Y. Hatano, A. Yagishita and K. Ito, *Z.Phys.*, **D4**, 65 (1986), and references cited therein.
63. M. Glass-Maujean, *J.Chem.Phys.*, **85**, 4830 (1986); **89**, 2839 (1988).
64. H. Koizumi, T. Yoshimi, K. Shinsaka, M. Ukai, M. Morita, Y. Hatano, A. Yagishita and K. Ito, *J.Chem.Phys.*, **82**, 4856 (1985).
65. H. Koizumi, K. Hironaka, K. Shinsaka, S. Arai, H. Nakazawa, A. Kimura, Y. Hatano, Y. Ito, Y.W. Zhang, A. Yagishita, K. Ito and K. Tanaka, *J.Chem.Phys.*, **85**, 4276 (1986).
66. H. Koizumi, K. Shinsaka, T. Yoshimi, K. Hironaka, S. Arai, M. Ukai, M. Morita, H. Nakazawa, A. Kimura, Y. Hatano, Y. Ito, Y.W. Zhang, A. Yagishita, K. Ito and K. Tanaka, *Radiat.Phys.Chem.*, **32**, 111 (1988).
67. H. Koizumi, K. Shinsaka and Y. Hatano, *Radiat.Phys.Chem.*, **34**, 87 (1989).
68. H. Koizumi, *J.Chem.Phys.*, **95**, 5846 (1991).
69. M. Ukai, K. Kameta, R. Chiba, K. Nagano, N. Kouchi, K. Shinsaka, Y. Hatano, H. Umemoto, Y. Ito and K. Tanaka, *J.*

Chem.Phys., **95**, 4142 (1991).
70. M. Ukai, K. Kameta, K. Shinsaka, Y. Hatano, T. Hirayama, S. Nagaoka, and K. Kimura, Chem.Phys.Lett., **167**, 334 (1990).
71. K. Kameta, M. Ukai, N. Terazawa, K. Nagano, Y. Chikahiro, N. Kouchi, Y. Hatano, and K. Tanaka, J.Chem.Phys., **95**, 6188 (1991).
72. K. Kameta, M. Ukai, T. Kamosaki, K. Shinsaka, N. Kouchi, Y. Hatano, and K. Tanaka, J.Chem.Phys., **96**, 4911 (1992).
73. M. Ukai, N. Kouchi, K. Kameta, N. Terazawa, Y. Chikahiro, Y. Hatano, and K. Tanaka, Chem.Phys.Lett., **195**, 298 (1992).
74. M. Ukai, K. Kameta, N. Kouchi, K. Nagano, Y. Hatano, and K. Tanaka, J.Chem.Phys., **97**, 2835 (1992).
75. M. Ukai, K. Kameta, N. Kouchi, Y. Hatano, and K. Tanaka, Phys.Rev., **A46**, 7019 (1992).
76. K. Kameta, M. Ukai, T. Numazawa, N. Terazawa, Y. Chikahiro, N. Kouchi, Y. Hatano, and K. Tanaka, J.Chem.Phys., **99**, 2487 (1993).
77. M. Ukai, K. Kameta, S. Machida, N. Kouchi, Y. Hatano, and K. Tanaka, J.Chem.Phys., **101**, 5473 (1994).
78. M. Ukai, S. Machida, K. Kameta, M. Kitajima, N. Kouchi, Y. Hatano, and K. Ito, Phys.Rev.Lett., **74**, 239 (1995).
79. J.W.J. Verschuur and H.B. van Linden van den Heuvell, Chem. Phys., **129**, 1 (1989).
80. C.Y.R. Wu, T.S. Chien and D.L. Judge, J.Chem.Phys., **92**, 1713 (1990).
81. J.E. Mentall and E.P. Gentieu, J.Chem.Phys., **52**, 5641 (1970).
82. P. Borrell, P.M. Guyon, M. Glass-Maujean, J.Chem.Phys., **66**, 818 (1977).
83. J.E. Mentall and P.M. Guyon, J.Chem.Phys., **67**, 3845 (1977).
84. M. Glass-Maujean, J. Breton and P.M. Guyon, Phys.Rev.Lett., **40**, 181 (1978).
85. P.M. Guyon, J. Breton and M. Glass-Maujean, Chem.Phys.Lett., **68**, 314 (1979).
86. J. Breton, P.M. Guyon and M. Glass-Maujean, Phys.Rev., **A21**, 1909 (1980).
87. C. Bottcher, J.Phys., **B7**, L352 (1974).
88. A.U. Hazi, J.Phys., **B8**, L262 (1975).
89. S.L. Guberman, J.Chem.Phys., **78**, 1404 (1983).
90. H. Takagi and H. Nakamura, Phys.Rev., **A27**, 691 (1973).
91. J. Tennyson, C.J. Noble and S. Salvini, J.Phys., **B17**, 905 (1984).
92. N. Kouchi, N. Terazawa, Y. Chikahiro, M. Ukai, K. Kameta, Y. Hatano, and K. Tanaka, Chem.Phys.Lett., **190**, 319 (1992).

93. N. Terazawa, N. Kouchi, M. Ukai, K. Kameta, Y. Hatano, and K. Ito, *J.Chem.Phys.*, **100**, 7036 (1994).
94. Y. Hatano, *Radiation Research*, eds. E.M. Fielden, J.F. Fowler, J.H. Hendry and D. Scott, Taylor & Francis, London, 1987, p.35.
95. Y. Hatano and M. Inokuti, *Atomic and Molecular Data for Radiotherapy and Radiation Research*, ed. M. Inokuti, IAEA, 1995, chapter 5.
96. M. Ukai, K. Kameta, N. Kouchi, and Y. Hatano, *Comm.At.Mol.Phys.*, to be published.
97. T. Ibuki, Y. Horie, A. Kamiuchi, Y. Morimoto, M.C.K. Tinone, K. Tanaka, and K. Honma, *J.Chem.Phys.*, **102**, 5301 (1995).
98. H. Lefevre-Brion and A. Suzor-Weiner, *Comm.At.Mol.Phys.*, **29**, 305 (1994).
99. M. Kitajima, M. Ukai, S. Machida, K. Kameta, N. Kouchi, Y. Hatano, T. Hayaishi, and K. Ito, *J.Phys.B: At.Mol.Opt.Phys.*, **28**, L185 (1995).
100. T. Odagiri, N. Uemura, K. Koyama, M. Ukai, N. Kouchi, and Y. Hatano, *J.Phys.B: At.Mol.Opt.Phys.*, **28**, L465 (1995); ibid., to be published.
101. E.W. Schlag and R.D. Levine, *J.Phys.Chem.*, **96**, 10608 (1992) (Feature Article).
102. W.G. Scherzer, H.L. Selzle, E.W. Schlag, and R.D. Levine, *Phys.Rev.Lett.*, **72**, 1435 (1994).
103. R. Weinkauf, P. Aicher, G. Wesley, J. Grotemeyer, and E.W. Schlag, *J.Phys.Chem.*, **98**, 8381 (1994).
104. E. Rabani, R.D. Levine, and U. Even, *J.Phys.Chem.*, **98**, 8834 (1994).
105. M. Inokuti and Y. Hatano , Personal communication (1994).

# Inner-shell Excitation, Relaxation and Fragmentation of Cluster Beams and Molecules

## Adam P. Hitchcock[*] and Eckart Rühl[+]

[*]*Dept. of Chemistry, McMaster University, Hamilton, Ont. L8S 4M1 Canada.*
[+]*Institut für Physikalische Chemie, Freie Universität Berlin, Berlin D-14195 Germany*

**Abstract.** Recent synchrotron radiation studies of inner-shell excitation and associated ionic decay spectroscopies of molecular beams of clusters of rare gas atoms and molecules are reviewed. The dependence of Ar(2p) spectra on the size of argon clusters provides information on cluster structure and the length scale over which the cluster properties change from atom-like to solid-like. Ion-ion and photoelectron-photoion-photoion coincidence studies (charge separation mass spectrometry) probe charge separation mechanisms in rare gas and molecular clusters. The inner-shell spectroscopy and ionic fragmentation of three isomeric *closo*-carboranes species are reported.

## INTRODUCTION

Inner-shell excitation and associated decay spectroscopies are useful site specific probes of electronic and geometrical structure and photoionisation dynamics. In recent years synchrotron radiation, X-ray optics and cluster beam technologies have advanced to the point where it is practical to carry out detailed studies of homogeneous and heterogeneous neutral clusters of atoms and molecules. [1-7]

Clusters are a novel state of matter, intermediate between the isolated units (atoms or molecules) and the condensed phase. Studies as a function of (average) cluster size help to understand the relationship of properties of the isolated species to those of the condensed phase. Free clusters, in the form of stable molecular clusters or supersonic beams of condensed atoms and molecules, avoid the complications of cluster-substrate or cluster-matrix interactions which may occur in supported clusters. Vacuum ultraviolet and X-ray absorption spectroscopies can be used to investigate the structural and electronic properties of atomic and molecular cluster beams. Associated spectroscopies based on the optical, electronic and/or ionic decay products provide information on the relaxation of the initially produced inner-shell state as well as the fragmentation dynamics of neutral and singly or multiply charged clusters.

Investigations of homogeneous Van der Waals clusters of argon, krypton and neon give excellent insight into the utility of inner-shell spectroscopy, motivating future studies of more exotic metallic and semiconductor clusters. The cluster size dependence of the near edge and extended fine structure illustrates the structural sensitivity, the spatial range of which depends on what aspect of the spectrum is being probed. Comparisons among various yield techniques (partial ion yield, Auger electron yield, threshold electron yield and total ion or electron yield) can be very useful to probe subsets of the cluster size distribution and different regions (surface versus bulk) of individual clusters. The electronic relaxation of the core excited/ionized clusters may involve a number of intermediate steps which can be detected through electron [3,6] and optical (luminescence [8] and fluorescence) spectroscopic studies.

The Auger decay of inner-shell excited and ionised states generated by synchrotron radiation photoionization is an efficient source of doubly charged ions. The charge separation and fragmentation of these species can be studied by photoelectron-photoion-photoion coincidence (PEPIPICO) (also called charge separation mass spectrometry - CSMS) and related techniques. Such studies provide detailed information on the fragmentation dynamics which in turn reveals insights into the bonding and electronic structure. In molecules, the dependence of the fragmentation process on the X-ray energy can reveal cases of site and/or state selective fragmentation [9-11].

Heterogeneous clusters composed of mixtures of different rare gases, mixtures of rare gases and molecules [5], or mixtures of molecules are being studied. The ability to examine different sites through use of their unique inner-shell edge is a particularly powerful aspect of this methodology. Thus the ion-ion charge separation signals for $Ar:N_2$ mixtures differ significantly for Ar 2p and N 1s ionization, revealing information about the interplay between bond strengths and ionization potentials of the cluster components [5]. Inner-shell excitation, relaxation and fragmentation studies of stable cluster molecules such as multi-metal organometallic complexes [12] or icosahedral carboranes [13,14] are also underway. The excitation spectra of these species are dominated by valence rather than Rydberg transitions. Extensive fragmentation is observed following inner-shell excitation and ionization, notwithstanding the much stronger bonding in these species as compared to the weak Van der Waals bonding in the rare gas clusters.

## EXPERIMENTAL

The compact molecular beam apparatus used for these studies has been described earlier in detail [1,3,15]. A supersonic jet expansion of homogeneous or heterogeneous clusters is prepared by expanding appropriate source gases through a 50 μm conical nozzle. Variations in stagnation pressures ($p_0$ up to 5 bar) and nozzle temperatures ($T_0$ of 150-300 K) allow systematic variation of the average size of the cluster distribution. For rare gas expansions, average cluster sizes are estimated on the basis of the reduced scaling parameter, $\Gamma^*$ of the expansion [16-18].

We have carried out cluster experiments using this compact souce at the BESSY storage ring, primarily at the HE-TGM-2 beam line [19], super-ACO at LURE and at HASYLAB. Experiments on molecular gases were carried out at BESSY, as well as at the Canadian grasshopper beamline at the Synchrotron Radiation Centre (U. Wisconsin-Madison) and beamlines 9.0.1 and 6.3.2 of the Advanced Light Source (LBNL, Berkeley). Total electron yield spectra are recorded using a biased channeltron electron detector placed close to the intersection of the X-ray beam with the cluster beam or gas jet. Time-of-flight mass spectra are recorded using either pulse extraction, which provides more quantitative yields, or photoelectron coincidence, which is best suited for using electron-ion-ion correlation techniques to investigate charge separation mechanisms. Because of relatively limited extraction fields there is significant kinetic energy discrimination in the ion and electron detectors such that lower energy species are emphasized. The time correlations among cations produced in a single core excitation event are detected using multistop time-to-digital converters [20].

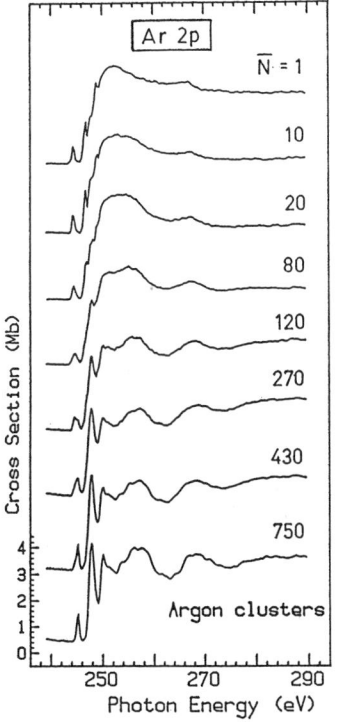

**FIGURE 1** Ar(2p) total electron yield spectra of argon clusters. The average cluster sizes ($\bar{N}$) are deduced from a reduced scaling parameter derived from the expansion conditions [16-18].

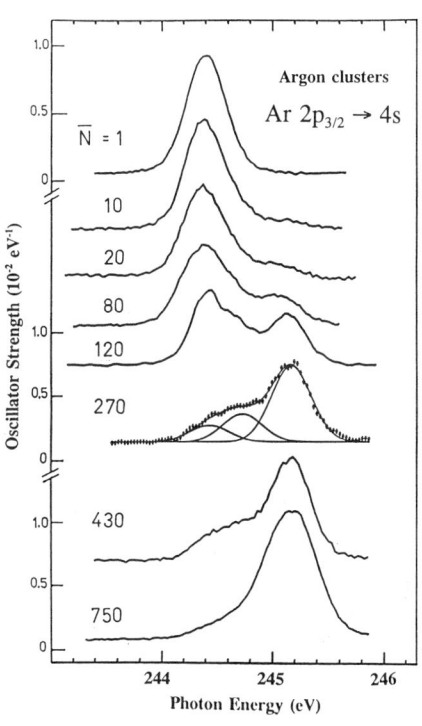

**FIGURE 2** Expansion of the Ar(2p)→4s excitation. A typical curve fit is shown. The scales of these spectra have been calibrated at the energy of the atomic component.

## RESULTS AND DISCUSSION

### Near edge spectroscopy

Fig. 1 plots the total electron yield spectra in the region of Ar(2p) excitation of argon cluster beams as a function of average cluster size. The spectral features of the Ar(2p) TEY spectrum evolve from those of the atom to those of the solid [21] with increasing cluster size. Broad oscillations develop in the Ar(2p) continua for clusters of an average size of 20 or more. These oscillations are the first-shell extended X-ray absorption fine structure (EXAFS) signal. Analysis using multiple scattering phase shifts [1] indicates that, independent of cluster size, the first shell distance in Ar clusters is indistinguishable from that of bulk solid argon, to within the precision of the analysis (±0.05 Å). A similar result was obtained at the Ar 1s edge [2], where a longer data range provided improved analysis. In addition to the first shell backscattering, a higher frequency X-ray absorption near edge structure (XANES) develops in the 255-275 eV range for clusters of average size greater than 300. In the largest clusters this XANES signal becomes very similar to that in solid argon [21]. It arises from longer path length multiple scattering. In principle a combined analysis of the XANES and EXAFS measured on a narrow size distribution of clusters would provide quantitative structural information on clusters. At present the wide distribution of cluster sizes greatly complicate this analysis.

Fig. 2 is an expanded presentation of the Ar(2p) → 4s transition as a function of cluster size. The energy scales of these spectra are set by aligning the fitted atomic component to 244.39 eV. This differs from the original report [1], since we have subsequently determined that there were instrumental energy shifts among the as-recorded data. As the cluster size increases, the Ar(2p)→4s Rydberg states of the atom shift up in energy to those of the exciton states of solid Ar [21]. Similar shifts are observed for the Ar(2p)→3d [5] and the Ar(1s)→4p Rydberg transitions [2]. In general the Ar(2p)→ 4s line shape consists of three distinct components (see sample curve fit in Fig. 2). The lowest energy peak is associated with free atoms (244.390 eV) [22]; the middle signal, which is strongest in intermediate values of $\overline{N}$ and thus small clusters, is associated with surface Ar atoms; and the highest energy signal is associated with fully coordinated Ar atoms in the inner-regions which dominate large clusters. There are systematic shifts in the energy of the Ar(2p)→4s transition of the surface Ar atoms. The energy of the bulk cluster signal (245.14(3) eV) provides an accurate evaluation of the 4s exciton energy of solid argon. Our value is in good agreement with the best recent measurements of the solid [23] as well as recent higher resolution measurements of argon clusters [6].

Recently Björneholm et al [6] have used a high flux, high resolution undulator beam line at HASYLAB to carry out more detailed studies of the size dependence of the Ar(2p) spectrum of argon clusters. Measuring the yield of $Ar_2^+$ excluded signal from residual atomic argon in the beam. This, along with higher flux and energy resolution, provided more precise values for the cluster size dependence of the surface and bulk Ar(2p) → 4s transition energies. The energy of the bulk 4s component, which begins to appear for $\overline{N}$ of 5 does not shift significantly whereas the energy of the surface component increases by 0.12 eV. Björneholm et al [6] associate the shift in the energy of the surface Ar(2p) → 4s transition with a systematic increase in first shell co-ordination number of the surface atoms, while the absence of any shift in the energy of the bulk component is related to a constant first shell co-ordination environment for those atoms.

In contrast to the *upward* shift in the energy of the 4s transition in Ar surface atoms, threshold photoelectron spectroscopy [3] and later, X-ray photoelectron spectroscopy (XPS) [6], measurements have shown that there is a *downward* shift in the Ar $2p_{3/2}$ ionization potential (IP) with increasing cluster size. In addition to providing reliable values for the Ar 2p IP of solid argon, the IP shifts provide information about cluster structure. Björneholm et al. [6] assumed that the energy shift of the surface IP is linearly related to surface co-ordination number. By scaling this shift against a value of 12 for the atom-solid energy shift, Björneholm et al estimated the effective surface co-ordination number as a function of average cluster size. These values were then compared to predictions for closed shell structures of either multiicosahedra [24] or cuboctahedra (fcc bulk termination). The trend in the experimental values is in somewhat better agreement with that predicted for closed-shell cuboctahedral structures rather than that for completed shells of the multishell icosahedral structure. However, this is not considered a conclusive structural result, because of limitations of the model relating energy shifts to surface co-ordination number, and also because many of the argon clusters will have an incomplete shell structure. A proper consideration of the reduced surface co-ordination of open shell icosahedral clusters would bring the predicted trend in the average co-ordination number into better agreement with experiment. This impressive study is the first XPS measurement of a free cluster beam. While it did not give a definitive structural answer, it does indicate the potential of inner-shell spectroscopies of clusters to provide quantitative structural information. Photo-yield spectroscopy as a function of average cluster size will provide substantial ability to distinguish among structural models in systems where the electronic and geometric structure changes appreciably with size, such as metal and semiconductor clusters. Studies in this area are in progress.

In contrast to rare gas clusters, where relatively large spectral changes occur with increasing cluster size, the inner-shell excitation spectra of clusters of covalently bound molecules [5,25,26] tend to differ relatively little from those of the isolated molecule. This is not surprising since the spectra of condensed molecular solids are dominated by core → valence transitions which are highly spatially localised and thus not much affected by alterations of the molecular surroundings. An exception to this is the core → Rydberg transitions, where the large spatial extent of the Rydberg orbital makes these excitations sensitive to cluster size. Fig. 3 presents the time-of-flight mass spectrum of a beam of ethylene clusters (stagnation conditions, $P_o$=5bar, $T_o$=230K) recorded at 285 eV. The H$^+$,

$CH_x^+$ and $C_2H_x^+$, $x \leq 3$ signals arise mostly from fragmentation of residual molecules in the beam, but the $C_2H_4^+$ parent ion and larger ions are associated with clusters. Fragmentations which leave intact ethylene units dominate, although there are small yields of the $C_3H_x^+$, $C_5H_x^+$ and $C_7H_x^+$ odd-carbon species which may be associated with ion-molecule reactions in the cluster.

Comparison of the total yield C 1s spectrum of this ethylene cluster beam with partial yield spectra of various cluster ions provides a means to qualitatively investigate the size dependence of the C 1s spectrum since the yield of a given cluster ion can only arise from clusters of larger size. Fig. 4 plots the C 1s spectrum of molecular ethylene [27] with that for the total ion yield, the $(C_2H_4)_2^+$ dimer yield, the $(C_2H_4)_4^+$ tetramer yield, and that of solid ethylene (multilayer condensed on a noble metal) [28]. While the total ion yield spectrum is very similar to that of the free molecule, suggesting there is appreciable unclustered ethylene under these expansion conditions, the dimer and tetramer spectra differ in the 287 to 290 eV region. In particular the spectral trend is consistent with a progressive filling in of the dip at 289 eV as the cluster size increases.

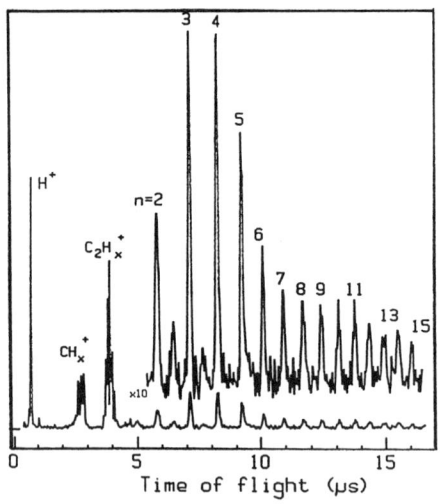

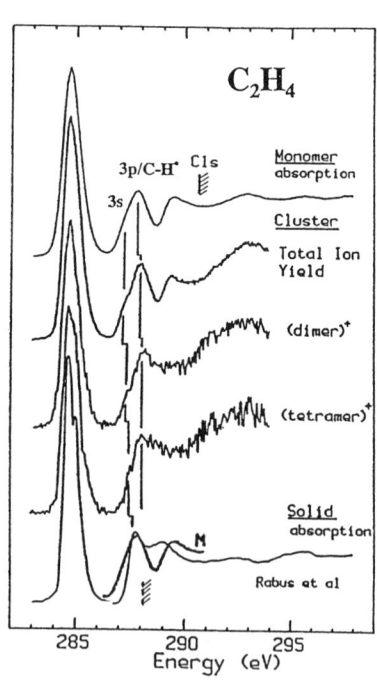

**FIGURE 3** Time-of-flight mass spectrum of a cluster beam of ethylene (generated with $p_o$= 5 bar, $T_o$ = 230 K) recorded with 285 eV photons.

**FIGURE 4** C(1s) spectra of the ethylene cluster beam recorded using total ion yield and dimer and tetramer cation yields. The C(1s) spectra of the free molecule [27], and a condensed multilayer [28] are also plotted.

This is consistent with a shift of the ionization threshold to lower energy in large clusters and the condensed state which is expected to occur because there is better polarisation shielding of the core hole when there are surrounding molecules. The same effect is operative in rare gas clusters and solids - the Ar $2p_{3/2}$ IP decreases by 1 eV from 248.63 eV in the atom to 247.7(1) eV in the solid, as determined from threshold [3] and high-energy [6] photoelectron spectra of Ar clusters. In addition to spectral changes consistent with a reduced IP, the C(1s)→3s and C(1s)→3p Rydberg transitions shift to lower energy, converging towards the first peak at 287.8 eV in solid ethylene. The size dependence of inner-shell spectra of beams of clustered $N_2$ [5, 25], CO [25], $N_2O$ [25], benzene and methanol [26] molecules has also been recorded, with analogous results.

## Fragmentation of Rare Gas Clusters

Non-radiative Auger, which dominates core hole decay in the soft X-ray region, produces doubly charged or more highly charged species. Since it is difficult to stabilize two positive charges in a small spatial region, there is a large dissociative double ionization yield in core ionization, as well as an appreciable one for core excitation. Photoion-photoion coincidence (PIPICO) and particularly PEPIPICO (CSMS) are useful tools to study the fragmentation of the initially formed multiply charged species. A detailed lineshape analysis provides much information about the photoionization fragmentation dynamics [29,30]. Auger decay of the Ar(2p) core hole results mainly in $(Ar)_n^{2+}$ doubly charged cluster ions. A cluster of more than 90 argon atoms is required for this species to be stable on the experimental time scale (µs). Since it is energetically more favourable to break the relatively weak Van der Waals bonds than to localise two charges on a single Ar ion, decay of a doubly ionized argon cluster leads almost exclusively to the eventual production of several smaller, singly charged cluster ions (or $Ar^+$), possibly along with a number of neutral argon species. This reaction could occur either in a concerted or a multi-step process.

Information about the competition between various charge separation mechanisms as well as the role of pre- or post-separation neutral evaporation can be deduced from a consideration of intensities and shapes of PEPIPICO signals. Fig. 5 presents a PEPIPICO spectrum recorded at 260 eV from an argon cluster beam ($p_o$ = 5 bar, $T_o$ = 193 K, $\overline{N}$~200). The pairs of cations giving rise to these signals are readily identified by their measured flight time relative to the photoelectron start signal. In addition to $Ar_n^+/Ar_m^+$, m>n pairs arising from asymmetric charge separation, there are also strong signals near the diagonal of this plot associated with symmetric charge separation processes producing $Ar_n^+/Ar_n^+$ pairs. The ratio of asymmetric to symmetric charge separation increases as the average cluster size increases [30], probably because the relative number of asymmetric channels gets larger in larger clusters. In general, large singly charged fragments are not formed. Instead pairs of small cations are produced, along with one or more neutral products. The PEPIPICO results suggest the neutrals are generated in steps before or after charge separation rather than in a concerted process. The energetics of neutral evaporation are discussed in detail in [30].

Mechanisms leading to particular cation pairs can be inferred from the detailed PEPIPICO peak shapes. Fig. 6 plots the shape of $Ar^+/Ar_2^+$ signals at three different expansion conditions corresponding to $\overline{N}$ values of 4, 10 and 60. If the dominant process producing an ion pair is a simple two-body dissociation the PEPIPICO signal should have a slope of -1 in the $t_1/t_2$ presentation [31] because the momenta of the two ions must be equal and opposite to conserve linear momentum. The width of the signal is related to the kinetic energy release (KER). On the other hand, if there is loss of neutrals after the charge separation, the slope (m) will differ from -1, with $0 > m > -1$ if the neutral is lost by the heavier of the initially produced ions and $m < -1$ if the neutral is lost by the lighter ion. There is a distinct evolution in peak shapes with cluster size, as illustrated in Fig. 6. For $\overline{N} = 4$ the contours have a slope of -0.66(4). This can be rationalised by a decay mechanism in which $Ar_4^{2+}$ carries out a two-step decay. Note that $Ar_4^{2+}$ may be the direct product of Auger decay of Ar(2p) ionized $Ar_4$, or it may arise from larger clusters by pre-evaporation of neutrals.

$$Ar_4^{2+} \rightarrow Ar^+ + Ar_3^+ \rightarrow Ar^+ + Ar_2^+ + Ar \qquad (1)$$

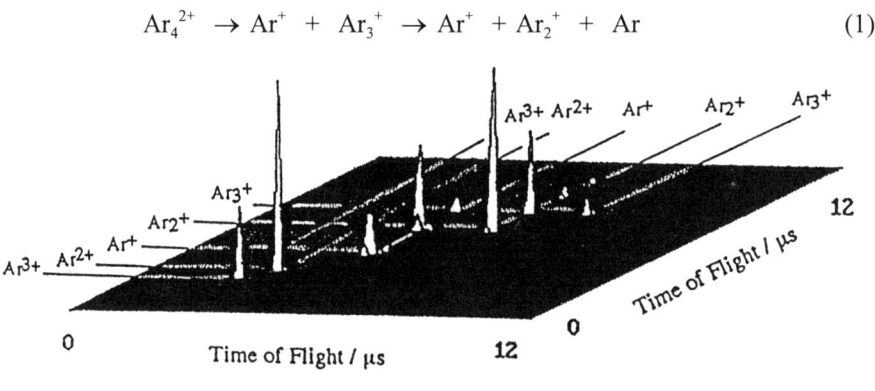

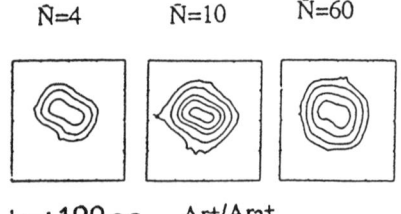

$\overline{N}=4$ $\quad$ $\overline{N}=10$ $\quad$ $\overline{N}=60$

**FIGURE 5** PEPIPICO spectrum of argon cluster beam ($p_o = 5$ bar, $T_o = 190$ K) [30]. The signals from multiply charge atomic ions arise from false coincidences associated with atoms in the beam [1].

⊢⊣ 100 ns $\quad$ Ar+/Ar2+

**FIGURE 6** Peak shapes of the $Ar^+/Ar_2^+$ PEPIPICO signal at different expansion conditions corresponding to estimated average cluster sizes of 4, 10 and 60 [30].

Consideration of momentum conservation shows that the slope expected for this process is -0.67, the ratio of the mass of the final dimer cation to the trimer cation intermediate. As the average cluster size increases the base of the PEPIPICO signal retains a slope of -0.70(5) but the peak of the signal 'twists' to a slope of -0.40(3). Twisted PEPIPICO signals have been observed in molecules [32] and interpreted either as the result of two (or more) competing processes, or from a semi-concerted process in which there is secondary decay in the Coulomb field of the separating ions. In the case of cluster beams, where the sample is a mixture rather than a unique species, it seems most reasonable to assume different cluster sizes give different PEPIPICO signal slopes. The base, which corresponds to high kinetic energy release events and thus fragmentation of smaller doubly charged cluster ions, is likely associated with the same process which we have attributed to sequential $Ar_4^{2+}$ decay in the $\overline{N} = 4$ cluster signal. The shallower slope at the base of the $Ar^+/Ar_2^+$ signal is consistent with a mechanism analogous to (1) for an $Ar_6^{2+}$ species:

$$Ar_6^{2+} \rightarrow Ar^+ + Ar_5^+ \rightarrow Ar^+ + Ar_2^+ + 3\ Ar \qquad (2)$$

since a slope of -0.4 eV is expected ($m(Ar_2^+)/m(Ar_5^+)$). At even larger cluster sizes (($\overline{N} = 60$) the PEPIPICO signal becomes a 'roundish' square. This can be interpreted in terms of prior and post evaporation of neutrals occurring along with the charge separation in heavier larger argon clusters. This results in the monocations separating at a wide range of angles, depending on the size and mass ratio of the intermediate and final cation (cf. mechanisms 1 and 2 above). The limiting result (for $\overline{N}>50$ in the case of argon) is a square with similar signals in horizontal and vertical directions of the $t_1/t_2$ PEPIPICO plot from which little information can be deduced.

Examples of PEPIPICO studies of charge separation in homogeneous clusters of molecular $N_2$ and heterogeneous mixed $N_2$/Ar clusters are described in ref. [5]. Differences are observed between the PEPIPICO signals for the same ion pair produced with N(1s) ionization relative to those for Ar(2p) ionization. These differences are interpretable in terms of preferential directions of charge transfer within the heterocluster which can be rationalised in terms of the relative ionization potentials of $N_2$ and Ar.

## Excitation and Fragmentation of *closo*-Carborane Clusters

It is interesting to compare the spectroscopy and fragmentation of rare gas clusters described above with that of stable, covalently bound cluster compounds such as organometallic complexes [10,12,33] and borane and carborane cluster compounds [13,14]. Because of the much stronger bonding in the stable cluster species the ionic fragmentation might be expected to lead to characteristic patterns of heavier charged ions. In addition there is the possibility of detecting examples of state and/or site selective fragmentation processes [9], which are of interest with regard to potential synthetic applications of X-ray photochemistry.

*closo*-1,2-orthocarborane is being used as a source compound for chemical vapour deposition (CVD) of boron carbide thin film semiconductors [34-36]. Understanding the electronic excitations and the preferred fragmentation pathways of this species and its meta and para isomers should help in understanding electron- and photon-induced decomposition at the molecular level. The B 1s and C 1s spectra of ortho-, meta-, and paracarborane are compared in Fig. 7. These results, obtained with total ion yield detection at the ALS undulator beamline 9.0.1, are much better resolved but otherwise similar to dipole-regime electron impact spectra of the same species [14].

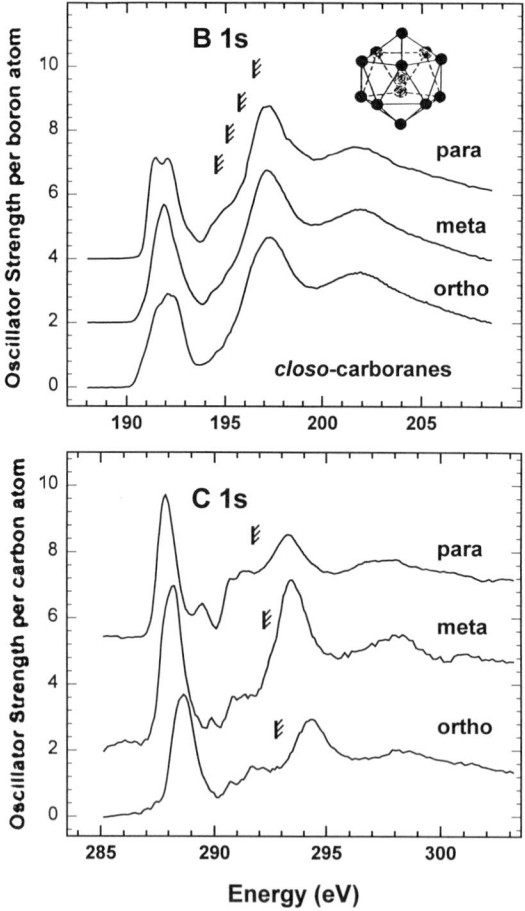

**FIGURE 7** Total ion yield spectra of ortho- meta- and para-closo carborane ($B_{10}C_2H_{12}$). The underlying pre-edge signals have been subtracted and the intensities converted to an approximate absolute oscillator strength scale by normalisation to atomic intensity in the high energy limit.

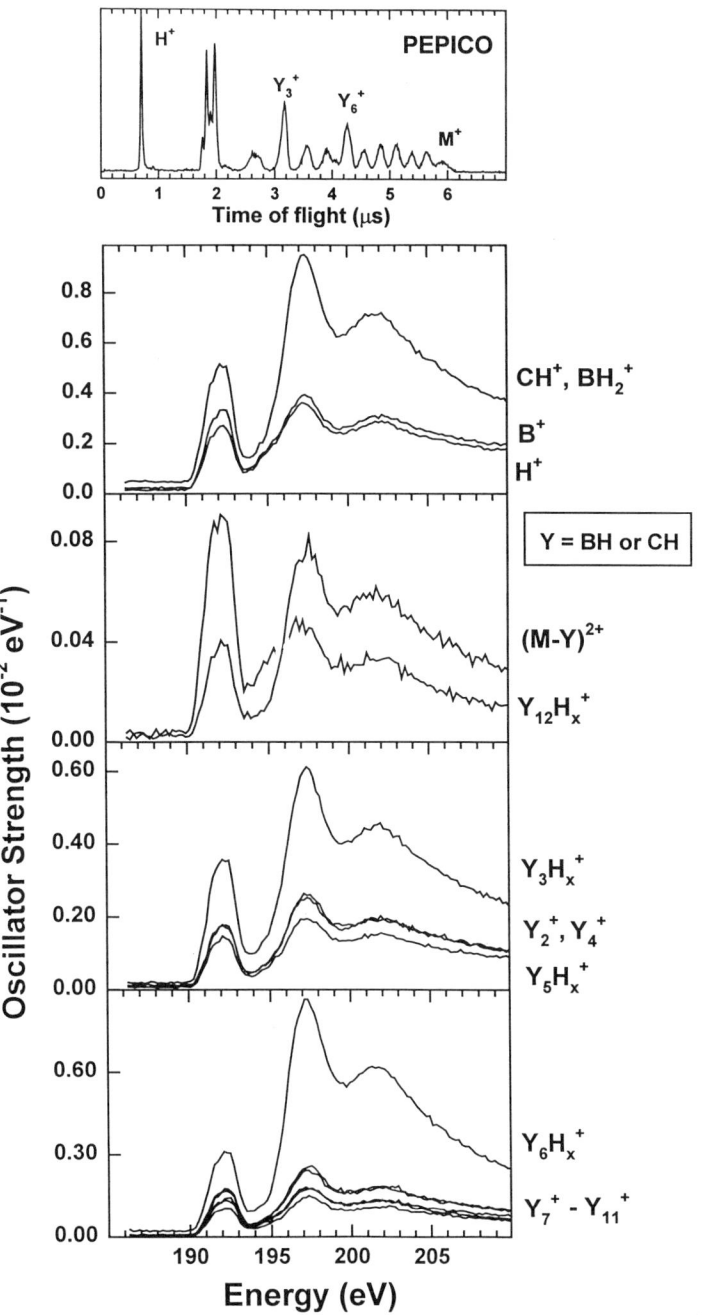

**FIGURE 8** (top) PEPICO time-of-flight mass spectrum (250 V/cm extraction field) of orthocarborane recorded at 192 eV. (bottom) Partial ion yields for photoionization of orthocarborane in the B 1s region. $Y_n$ symbolises a $B_nH_n$, $B_{n-1}CH_n$ or $B_{n-2}C_2H_n$ unit.

The B 1s spectra are quite similar, with only minor variations in the fine structure of the lowest energy discrete excitation, which is believed to arise from a complex overlap of many transitions [13]. There is a significant isomeric dependence of the C 1s spectra, consistent with a larger change in the carbon than the boron environment. The main effects are a systematic shift to higher energy in the series p→m→o, as monitored by the position of the sharp low energy feature, and an enhanced intensity of the 294 eV peak in the meta isomer. The energy shift is primarily associated with variations in the C 1s ionization potential [13] since the overall shift from para to ortho in excitation (0.7 eV) is similar to that in the C 1s IP (1.0 eV estimated; the meta IP is 0.6 eV above para [14]). While it had been postulated that the sharp feature in orthocarborane was a type of local $\pi^*$ orbital [13], such an explanation is not applicable to the meta or para species where the two carbon atoms are not adjacent. Rather MNDO calculations [14] suggest that the low-lying C 1s feature arises from C 1s excitation to an extra-atomic linear combination of atomic B 2s and B 2p orbitals which make up the $7b_1$ (ortho, para) or $7b_2$ (meta) lowest unoccupied molecular orbital (LUMO). This extra-atomic delocalised excitation is symmetry allowed in all three species.

Partial ion yields and ion pair yield spectra in the B 1s and C 1s regions have been measured for all three species [14]. A sample PEPICO mass spectrum of metacarborane is given in Fig. 8 along with the ion yields of orthocarborane in the B 1s region. The relatively strong signals of $Y_3^+$ and $Y_6^+$ (where Y indicates a cluster vertex, either a CH or a BH species) in the PEPICO mass spectrum can be rationalised by a greater stability of planar trigonal and octahedral structures which arise by rearrangement of the ions initially formed. With the PEPICO method used for these measurements the yield of high mass fragments is very low, with at most only a few percent yield of the parent ion. This is in sharp contrast to the conventional electron impact mass spectrum of these species, which is dominated by parent ion production [37]. This is consistent with core hole decay leading to extensive ionic fragmentation, including large amounts of multiple ionisation, almost all of which ends up as ion pairs. The only doubly charged ion detected is a weak $B_9C_2H_x^{2+}$ signal at 4.1 μs flight time. To illustrate the extensive dissociative double ionization processes Fig. 9 presents a comparison of the PIPICO and PEPIPICO spectra of metacarborane, recorded at 192 eV, the discrete B 1s excitation feature. The PIPICO spectrum is very complex because of superposition of many ion pair signals. The PEPIPICO $t_1/t_2$ plot is very effective in separating these overlapping signals, with each peak in this plot readily identified by the flight times of each ion involved.

Overall there is little difference among the partial yields of the isomeric species and relatively little change in individual ion or ion pair yields aside from a major step up or step down in specific channels at the onsets of B 1s core excitation and ionization. Thus any selectivity among these three species which may exist with regard to properties of boron carbide films prepared by X-ray assisted CVD is more likely to be associated with specificity of the chemistry of fragments rather than selectivity in the initial excitation.

Morin et al [29] have noted that while state selective ionic fragmentation can be demonstrated in smaller molecules, larger molecules (with some exceptions [38]) rarely exhibit any strong dependence of their final fragmentation products on the initial inner-shell state that is created. In large molecules, this can be rationalised by a very fast and extensive internal energy redistribution following Auger decay which puts energy into many different vibrational and rotational modes. Stepwise decay processes generally dominate and a variety of different pathways leading to common final products results in little state, bond or site selectivity [10]. In contrast inner-shell induced ionic fragmentation of small molecules can be much influenced by the specific inner-shell excited or ionized state which is produced, particularly in cases where bond-breaking occurs on a time scale comparable to the core hole decay [39]. Recently the impulsive model [29], which uses a classical treatment of the forces on individual atoms associated with geometry changes between the ground and core excited states, has successfully explained the core state dependence of shapes of PEPIPICO peaks corresponding to atomization events in $N_2O$ and $CO_2$. In these cases the strong coupling between the properties of the inner-shell excitation and the final ion products results in ion angular distributions which are characteristic of the symmetry of the resonances as excited by linearly polarised light [40].

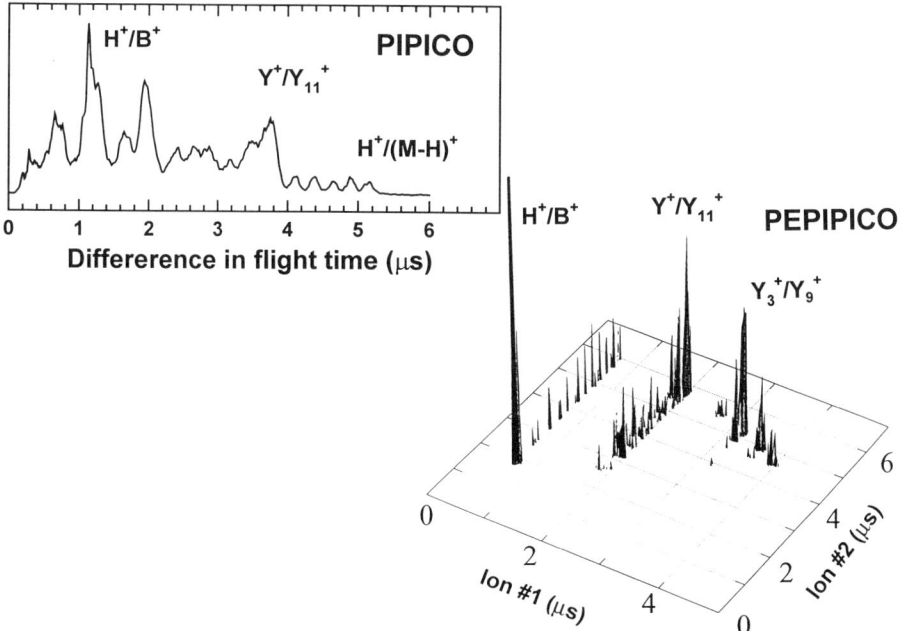

**FIGURE 9** Comparison of PIPICO (ion auto-correlation) and PEPIPICO signals (multi-stop PEPICO) of B 1s ionised metacarborane. The peak of the $H^+/B^+$ signal is off-scale by 30%.

## SUMMARY

Considerable progress has been made over the past 4-5 years in inner-shell excitation and ionic fragmentation spectroscopic studies not only of molecules but also of supersonic beams of clusters of rare gases and molecules. Several groups are presently attempting to apply these techniques to sources of metal or semiconductor clusters. This will produce interesting insights into the links between the geometric and electronic structure of atoms, small molecules and the solid state.

## ACKNOWLEDGEMENTS

Adam Hitchcock thanks the Advanced Light Source, Lawrence-Berkeley National Laboratory, for their hospitality during 1994-95. The expert assistance of Dr. J.D. Bozek with the experiments at ALS is gratefully acknowledged. Special thanks are given to A.L.D. Kilcoyne and T. Tylisczak who developed the time-of-flight instrument and acquisition, and to P.D Dowben who motivated the carborane project and provided samples. The research is funded by NSERC (Canada), BMBF (Germany) grant 05-5KEFXB5-TP3, and NATO. It is conducted at SRC (U. Wisconsin-Madison, supported by NSF), ALS (LBL, supported by DOE), BESSY (Berlin) and HASYLAB (Hamburg).

## REFERENCES

1. Rühl, E., Heinzel, C., Hitchcock A.P., and Baumgärtel, H., *J. Chem. Phys.* **98**, 2653 (1993).
2. Rühl, E., Heinzel, C., Hitchcock, A.P., Schmeltz, H., Reynaud, C., Baumgärtel, H., Drube W., and Frahm, R., *J. Chem. Phys.* **98**, 6820 (1993).
3. Knop, A., Jochims, H.W., Kilcoyne, A.L.D., Hitchcock, A.P., and Rühl, E., *Chem. Phys. Lett.* **223**, 553 (1994).
4. Federmann, F., Björneholm, O., Beutler, A. and Möller, T., *Phys. Rev. Lett.* **73**, 1549 (1994).
5. Rühl, E., Hitchcock, A.P., Morin, P. and Lavollée, M., *J. Chim. Phys.* **92**, 521 (1995).
6. Björneholm, O., Federmann, F., Fossing, F. and Möller, T., *Phys. Rev. Lett.* **74**, 3017 (1995).
7. Rühl, E., Knop, A., Hitchcock, A.P., Dowben, P.A. and McIlroy, D.N., *Surf. Rev. Lett.* (1995) in press.
8. Rühl, E, Heinzel C., and Jochims, H.W., *Chem. Phys. Lett.* **211**, 403 (1993); Meyer, M., Lacoursière, J., Simon, M., Morin P. and Larzillière, M., *Chem. Phys.* **187**, 143 (1994)
9. Hanson, D.M. *Adv. Chem. Phys.* **77**, 1 (1990).
10. Simon, M., Thèse d'Université, Orsay, France (1992)
11. Simon, M., LeBrun, T., Martins, R., de Souza, G.G.B., Nenner, I., Lavolée, M. and Morin, P., *J. Phys. Chem.* **97**, 5228 (1993).
12. Hitchcock, A.P., McGlinchey, M.J., Johnson, A.L., Walter, W.K., Perez-Jigato, M., King, D.A., Norman, D.L., Rühl, E., Heinzel, C., and Baumgärtel, H., *Trans. Faraday Soc.* **89**, 3331 (1993).
13. Hitchcock, A.P., Wen, A.T., Lee, S., Glass, J.A., Spencer, J.T. Jr. and Dowben, P.A., *J. Phys. Chem.* **97**, 8171 (1993).
14. Hitchcock, A.P. et al. (manuscript in preparation)
15. Rühl, E., Schmale, C., Jochims, H.W., Biller, E., Simon M., and Baumgärtel, H., *J. Chem. Phys.* **95**, 6544 (1991).

16. Hagena, O.F., *Z. Phys. D* **4**, 291 (1987).
17. Farges, J., deFeraudy, M.F., Raoult B., and Torchet, G., *J. Chem. Phys.* **84**, 3491 (1986).
18. Wörmer, J., Ph.D. thesis, University of Hamburg, (1990)
19. Bernstorff, S., Braun, W., Mast, M., Peatman W., and Schroeter, T., *Rev. Sci. Inst.* **60**, 2097 (1989).
20. In the BESSY epxeriments a LeCroy 4208 and a MIPSYS FLY-TDC are used for coincidence detection. In the ALS experiments a Tolmar TDC is used.
21. Haensel R., Keitel, G., Kosuch, N., Nielsen, U., and Schreiber, P. *J. Phys. (Paris)* C **4**, 236 (1971); Haensel R., Kosuch, N., Nielsen, U., Rössler, U., and Sonntag, B. *Phys. Rev. B* **7**, 1577 (1973). Niemann W., Malzfeldt, W., Rabe, P., Haensel, R. and Lübcke, M. *Phys. Rev. B* **35**, 1099 (1987).
22. King, G.C., Tronc, M., Read F.H., and Bradford, R.C., *J. Phys. B* **10**, 2479 (1977).
23. Scheuerer, R., Feulner, P., Rocker, G., Lin Z., and Menzel, D., *DIET IV*, Betz, G. and Varga, P. (eds). 235 (1990).
24. Farges J., *Surf. Sci.* **106**, 95 (1981).
25. Rühl, E., *Ber. Bunsenges Phys. Chem.* **96** 1172 (1992); Rühl, E., Heinzel, C., Jochims, EH.W., Biller, E., Locht, R., Hitchcock, A.P. and Baumgärtel H., *Synchrotron Radiation and Dynamic Phenomena*, Grenoble, Beswick, A., ed. (AIP, NY 1992) 230.
26. Geiger, J., Rabe, S., Heinzel, C., Baumgärtel H., and Rühl, E., *NATO Conf. Proc.* Patras, L.G. Christopherou, ed. (1994) in press.
27. McLaren, R., Clark, S.A.C., Ishii I., and Hitchcock, A.P., *Phys. Rev. A* **36**, 1683 (1987).
28. Rabus, H., Arvanitis, D., Domke M., and Babershke, K., *J. Chem. Phys.* **96**, 1560 (1992).
29. Morin, P., Lavollée, M., Meyer M., and Simon, M., *AIP Conf. Proc.* **295**, 139 (1993).
30. Rühl, E., Heinzel, C., Baumgärtel, H., Lavollée M., and Morin, P., *Z. Phys. D* **31**, 245 (1994).
31. Eland, J.H.D., *Mol. Phys.* **61**, 725 (1987); ibid, *Acc. Chem. Res.* **22**, 381 (1989).
32. Eland J.H.D., and Tewes-Bowen, B.J., *Synchrotron Radiation and Dynamic Phenomena*, Grenoble, Beswick, A. ed. (1992) 100.
33. Rühl, E., Heinzel, C., Baumgärtel, H., and Hitchcock, A.P., *Chem. Phys.* **169**, 243 (1993).
34. S. Lee, J. Mazurowski, G. Ramseyer, and P.A. Dowben, *J. Appl. Phys.* **72**, 4925 (1992).
35. Kim, Y.-G, Dowben, P.A., Spencer, J.T., Ramseyer, G.O., *J. Vac. Sci. Tech. A* **7**, 2796 (1989).
36. Mazurowski, J., Baral-Tosh, S., Ramseyer, G., Spencer, J.T., Kim, Y-.G., Dowben, P.A., *Mat. Res. Soc. Symp. Proc.*, **190**, 101 (1991); Mazurowski, J., Lee, S., Ramseyer, G., Dowben, P.A., *Mat. Res. Soc. Symp. Proc.*, **242**, 637 (1992).
37. N.I. Vasyukova, Y.S. Nekrasov, Y.N. Sukharev, V.A. Mazunov and Y.L. Sergeev, *Izv. Akad. Nauk SSSR, Ser Khim.* **6**, 1337 (1985). (transl. p. 1223).
38. Schmelz, H.C., Reynaud, C. Simon, M. and Nenner, I, *J. Chem. Phys.* **101**, 3742 (1994).
39. Morin P., and Nenner, I. *Phys. Rev. Lett.* **56**, 1913 (1986); Aksela, H., Aksela, S., Hotokka, M., Yagashita, A. and Shigemasa, E. *J. Phys. B* **25**, 3357 (1992).
40. Bozek, J.D., Saito N., and Suzuki, I.H., *Rev. Phys. A,* **51**, 4563 (1995).

# Anisotropy and Kinetic Energy Release in Ionic Fragmentation of Molecules Photoexcited to Inner-Shell Hole States

Isao H. Suzuki and Norio Saito

*Electrotechnical Laboratory, Umezono, Tsukuba-shi, Ibaraki 305, Japan*

**Abstract.** Angular distributions of fragment ions produced from linear molecules photoexcited to inner-shell excited states reflect symmetry of these excited states because the production of the ions is very fast compared to the rotational motion of the molecules. Some examples have been shown for the π* state, Rydberg orbital excited states and the shape resonance of diatomic and linear triatomic molecules using a rotatable time-of-flight mass spectrometer and monochromatized synchrotron radiation. Kinetic energy distributions of fragment ions from inner-shell hole states have been determined at several photon energies using time-of-flight spectra and a simulation fitting calculation. These distributions give a considerable difference between neutral excited and ionized states, indicating that the dissociation pathway from the π* excited state is different from that at the ionized state.

## INTRODUCTION

Inner-shell excitation and decay processes of small molecules attract much attention in these years because intense and highly monochromatic soft X-rays become available in many laboratories of synchrotron radiation. Vibrationally resolved photoabsorption spectra on inner-shell excitation have been observed for some molecules, giving information on geometrical structure of inner-shell hole states.[1-4] Molecules photoexcited using soft X-rays usually eject electrons through Auger type transitions, turning into molecular ions with valence holes,[5-8] although Cl $K$-shell excited molecules emit fluorescent X-rays considerably.[9] Electron spectra from inner-shell excited neutral states have been measured, and these spectra were found to mainly consist of two types of Auger electrons, spectator Auger and participator Auger. The strength of the two components largely depends on character of the inner-shell excited states, and thus close examination of these spectra gives deep understanding on overlap and interaction of related electron orbitals.

An electron energy loss technique has given much information on characteristics of inner-shell hole states of many molecules, together with photoabsorption spectroscopy.[10-12] Those studies reported that inner-shell spectra of unsaturated linear molecules are characterized by two kinds of resonances ($\pi^*$ and $\sigma^*$) and by transitions to Rydberg orbitals. After an Auger transition from the inner-shell hole state, a highly valence-excited molecular ion decomposes into fragment ions. The Auger transition and subsequent ionic fragmentation usually occur much faster than the rotational motion of the molecule because the highly valence-excited states have repulsive potential curves in many cases. Therefore the angular distribution of the fragment ion can be a powerful tool for direct determination of the transition symmetry.[13-21]

Since inner-shell excitation of a molecule confines excess energy to a particular site in the molecule, reactions with site specific character are expected to take place around the atom initially excited. In order to examine the relation between the excited site and the final products, several measurements were performed, showing a site-selective feature in the yield spectrum of a few fragment ions.[22,23] In this series of studies, it is very important to understand the dissociation pathways and intermediate states in ionic decomposition. Coincidence studies of Auger electrons with fragment ions were carried out for a few molecules, which clearly demonstrated intermediate states of molecular ions.[24-26] Since kinetic energy of product ions gives an energy difference between intermediate states of molecular ions and states of final products, determination of this energy should serve to clarify the dissociation mechanism in detail.[27-34] Further, the kinetic energy of the fragment ion is a critical factor in the angular distribution measurement described above, because the dissociation is required to proceed faster than the rotational motion of the molecule.

In the present study, angular distributions and kinetic energy distributions (KED) of fragment ions from small molecules have been determined using a rotatable time-of-flight (TOF) mass spectrometer and monochromatic soft X-rays. Asymmetry parameters ($\beta$) of inner-shell excitation and dissociation pathways of excited molecules are discussed in diatomic molecules and $CO_2$.

## EXPERIMENTAL

Synchrotron radiation from the TERAS electron storage ring (750 MeV) at the Electrotechnical Laboratory was monochromatized with a Grasshopper monochromator.[13,29] In order to improve spectral purity of the photon, the electron energy in the ring was occasionally lowered down to 400 MeV and several thin foils for filtration were inserted into the photon beam. A rotatable TOF mass spectrometer consists of a mutually orthogonal arrangement of the soft X-ray beam, a nozzle for the introduction of an effusive jet of the sample gas, and a microchannel plate (MCP) detector and a flight tube with an MCP detector mounted opposite each other to detect the electrons and ions (see Fig.1). Charged particles generated in the interaction region of the sample gas with the photon beam were accelerated to the MCP detectors using a series of grids biased to the appropriate potentials. The TOF spectra were obtained using signals from no energy-analyzed electrons as a start for a time-to-amplitude converter, and photoion-photoion coincidence (PIPICO) spectra were measured using signals from two fragment ions. The spectra were observed at angles of $0°$, $55°$, and $90°$ with respect to the direction of polarization of the synchrotron radiation.

Peaks in the TOF and PIPICO spectra have been fitted to a model, in order to obtain $\beta$ parameters and KED's for individual ions from the experimental data.[13-18] The shape of these peaks in the spectra is determined by the KED and angular distribution of the ions, the characteristics and orientation of the TOF spectrometer and the applied electric fields. Trajectories of ions in the spectrometer are first calculated by assuming that (1) the electrical properties of the spectrometer meet the design specifications, (2) the space focusing of the spectrometer is perfect, and (3) the density of the sample gas in the interaction region is uniform. These trajectory calculations provide the collection efficiencies of ions with different kinetic energies in the spectrometer at the experimental conditions and thus the spectral profiles for ions of different kinetic energies. Next, assuming that the incident photon beam is 90% linearly polarized, KED's of the ions are determined by fitting the TOF spectra measured at $55°$ with the profiles calculated in the trajectory calculations. Angular distribution effects on the observed TOF spectra are supposed to be zero at $55°$. Angular distributions of the ions are then determined by fitting the KED for the isotropic distribution of ions to the experimental spectra measured at $0°$ and $90°$ allowing the peak shape to change by varying only the value of $\beta$.

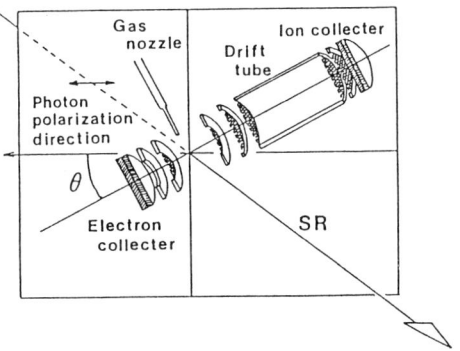

**FIGURE 1**. A rotatable time-of-flight mass spectrometer.

# RESULTS AND DISCUSSION

## Anisotropy in Ionic Fragmentation

### Diatomic molecules

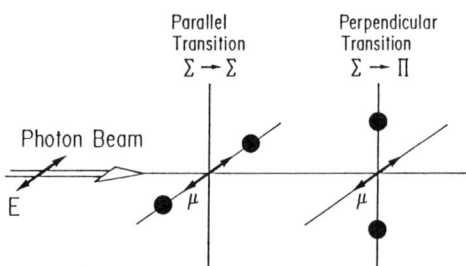

In linear molecules having $\Sigma$ symmetry in the ground state, electronic transitions are allowed to states having $\Sigma$ and $\Pi$ symmetries. The direction of the transition moment is parallel or perpendicular to the molecular axis (see Fig.2). In the instance of photoabsorption using linearly polarized light, the molecule oriented parallel to the photon polarization can induce a parallel type transition ($\Sigma$ type) and the molecule perpendicular can do a perpendicular type transition ($\Pi$ type). When the angle between the photon polarization and the molecular axis is expressed with $\theta$,[13-21] the transition intensity of the molecule oriented to a direction of $\theta$ is given by

**FIGURE 2.** Schematic illustration of electronic transitions in diatomic molecules. E denotes the photon polarization direction, and $\mu$ is the transition moment.

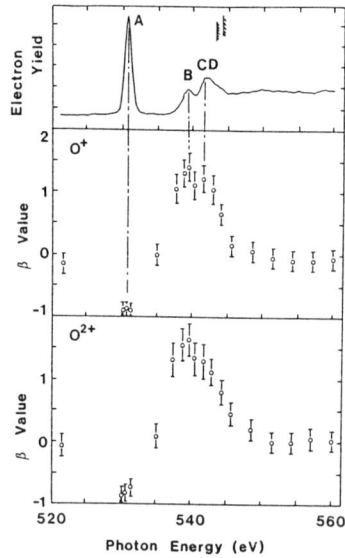

$$f(\theta) = \frac{1}{4\pi}\left[1+\frac{\beta}{2}\left(3\cos^2\theta-1\right)\right]. \quad (1)$$

In the $\Sigma$ type transition, $\beta$ is equal to 2, making the angular distribution be $\cos^2\theta$. On the other hand, the angular distribution becomes $\sin^2\theta$ in the $\Pi$ type transition, $\beta = -1$. Dissociation following inner-shell excitation usually happens very fast, and then the angular distribution of fragment ions is essentially the same as the molecular orientation at the instant of photoabsorption.

**FIGURE 3.** Asymmetry parameters of $O^+$ and $O^{2+}$ from $O_2$ as a function of photon energy in the O$K$-edge region.

Figure 3 shows asymmetry parameters ($\beta$) for the $O^+$ and $O^{2+}$ ions from $O_2$ as a function of photon energy around the O$K$-edge.[14] The electron yield spectrum shown at the top of Fig.3 exhibits the same feature as found in the electron energy

**TABLE 1.** Asymmetry parameters for transitions to discrete orbitals below the K-edge and the shape resonance in $N_2$, $O_2$, CO, and NO.

| Orbital | $N_2$ | $O_2$ | CO(CK) | NO(NK) |
|---|---|---|---|---|
| $\pi^*$ | -0.85 | -0.89 | -0.83 | -0.83 |
| 3s | 1.4 | ------ | 1.2 | 1.0 |
| 3p | -0.65 | ------ | -0.7 | -0.6 |
| $\sigma^*$ | 0.45 | 1.4 | 0.6 | 0.5 |

Uncertainties in the listed values are approximately ±0.15.

loss spectra.[10] The β parameters obtained have been corrected for the contribution from ionization of valence electrons. The β values for $O^{2+}$ show the same feature as those for $O^+$ within the experimental uncertainty. The β parameter at 531 eV is -0.89, which is in good agreement with the expected value, -1, for the 1s→π* transition. At 539 eV (1s→3sσ), β is 1.41, being close to the expected value of 2.

At 541 eV (1s→3p), β is 1.23. If the transition to the 3pπ orbital dominates that to the 3pσ for $O_2$ as well as $N_2$,[35] the β parameter should show a negative value at this energy. The broad structure around this energy in the photoabsorption spectrum had been interpreted to result from the 1s→σ* shape resonance transition.[10,12] This type of transition is expected to give a positive value for β parameter. The present positive value of β indicates that the σ* resonance transition takes place at this energy, which directly supports the previous assignment.

The β parameters for inner-shell transitions of diatomic molecules are listed in Table 1.[13-16] At the transition of 1s electron to the π* orbital, the β values for all molecules are close to -1 within experimental uncertainties. The β parameters at the 1s→3sσ transition range from 1.0 to 1.4, showing Σ transition. As mentioned above, $O_2$ shows the 3sσ transition overlapping with the σ* shape resonance and then the oscillator strength for the latter dominates the former transition. The transition to the 3p orbital is mainly contributed from the 1s→3pπ transition, i.e. Π transition. The parameter values are about -0.65, being in good agreement with the expectation. The broad band structures above the ionization thresholds observed previously in the inner-shell excitation spectra provided positive β values in the present study. The feature in the inner-shell spectra is very close to that in the β parameter curves except for $O_2$. The previous interpretation that this feature originates from the σ* shape resonance has been directly confirmed by the feature of the β parameters obtained here. The energies for the maximum β values are about 10 eV above the ionization threshold for $N_2$, 10 eV for CO and 4 eV for NO.

## *Linear triatomic molecules*

Correspondence between the angular distribution of fragment ions and the molecular orientation at the instant of photoabsorption seems to be possibly obscured because triatomic molecules are able to be bent.[17,18,36] The ion angular distribution largely depends on the bending motion as well as the symmetry of the inner-shell hole states. In triatomic molecules, the excited state including $\pi^*$ orbital becomes stable at the bent structure, Renner-Teller effect. Then the $\beta$ parameter at the $1s \to \pi^*$ excitation can deviate from the value of -1.0 expected for the $\Pi$ transition. On the other hand, the bending vibration holds a symmetry of $\pi$-character in these molecules, and hence the $\beta$ parameter has a possibility to show a negative value at the energy where the $\Sigma$ transition takes place in the frame of only electronic states concerned.

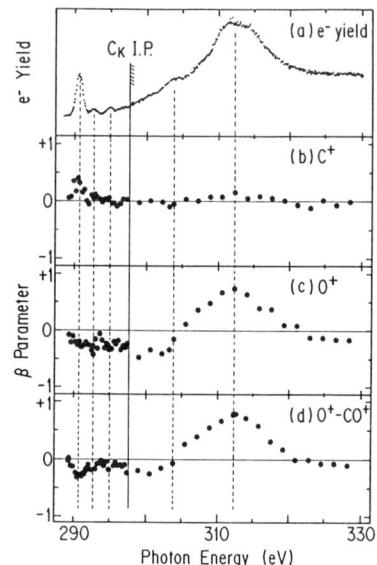

FIGURE 4. Asymmetry parameters of $C^+$, $O^+$, and the ion pair of $O^+$-$CO^+$ from $CO_2$ as a function of photon energy in the $CK$-edge region.

Figure 4 shows asymmetry parameters for $C^+$, $O^+$, and PIPICO signals ($O^+$-$CO^+$) from $CO_2$ in the region of the $CK$-edge.[18] Since the PIPICO signal can specify a bond to be broken and remove a dissociation of two steps type, the PIPICO angular distribution reflects the direction of the specified bond in the molecule. The electron yield spectrum at the top approximates a photoabsorption spectrum measured at a moderate resolution,[1,11] which exhibits some peaks of discrete transitions to $\pi^*$, $3s$, and $3p$ orbitals and broad structures owing to shake-up states and the $\sigma^*$ shape resonance. The $\beta$ parameter at 290.7 eV ($1s \to \pi^*$) is +0.4 for $C^+$, -0.3 for $O^+$, and -0.3 for $O^+$-$CO^+$. The photoexcitation at this energy should be $\Pi$ type, but the $\beta$ values differ from -1 for the three species. This finding, however, is consistent with the bent structure in the $\pi^*$ excited state, as mentioned above. The present value for the $\beta$ parameter gives a bond angle of about 130°. The finding of the positive $\beta$ value for $C^+$ means that the $C^+$ is ejected to the direction of the photon polarization, indicating that the bent structure of the excited state is responsible for the formation of $C^+$.

At 292.7 eV ($C1s^{-1}3s$), the parameters for $O^+$ and $O^+$-$CO^+$ are negative although that for $C^+$ is near to zero. Since the electronic transition from $2\sigma_g$ ($C1s$) to $3s\sigma_g$ is forbidden, the coupling with $\pi$ type vibrational motion enables the molecule to be excited at this energy. This vibronic state possibly interferes with another electronic state having the same symmetry and then shares transition strength. When the bending motion is excited as well as borrowing transition strength, it is reasonable that the fragment ions are ejected to the perpendicular direction. At 295.0 eV (C1s$^{-1}$3p), the parameters are negative for O$^+$ and O$^+$-CO$^+$. Around 312 eV, broad positive structures are seen for the β spectra of O$^+$ and O$^+$-CO$^+$. These structures exhibit that the σ$_u$ shape resonance takes place in this energy region. Although another shape resonance, $1s \rightarrow \sigma_g^*(5\sigma_g)$, is forbidden in the C1s region, this resonance is allowed in the O1s region, i.e. $O1s(1\sigma_u) \rightarrow 5\sigma_g$. This transition has been confirmed to happen near the O1s ionization threshold. The β parameters showed a broad positive structure for O$^+$ and O$^+$-CO$^+$.

## Kinetic Energy Distribution

### *Nitrogen molecule*

Inner-shell hole states of a molecule usually decay to valence hole states of the molecular ion accompanied with Auger electron emission and the molecular ion turns into fragment ions, although some exceptions were reported so far, in which molecular dissociation happens faster than the Auger decay.[8,37] The Auger electron spectrum is supposed to reflect states of the molecular ion just before dissociation. When the energy levels of product species are given, kinetic energies of the fragment ions are a critical factor for clarification of dissociation pathways.[27-32]

The KED's of N$^+$ produced from N$_2$ are shown at several photon energies in Fig.5.[27] The N$^+$ ions produced at 414 eV above the K-edge (409.9 eV) have a broad maximum around 4 eV kinetic energy, which differs from those at other photon energies. The intensity of the ions with kinetic energies below 2 eV is considerably smaller at 414 eV than that at 401 eV. The peak of 0.5 eV at the N1s$^{-1}\pi^*$ state seems to shift slightly to higher energy at the excitation to the Rydberg orbitals. The broad peak of 5 eV at the π* state becomes wider at the Rydberg states. These results indicate that the molecular ion states just before dissociation are different among the π* state, the Rydberg states, and the K-shell ionized state. The KED's of doubly charged ions (N$_2^{2+}$) show a broad maximum around 8 eV in all photon energies, which moves slightly to higher energy with increasing photon energy.

A coincidence technique between the fragment ion and the Auger electron was

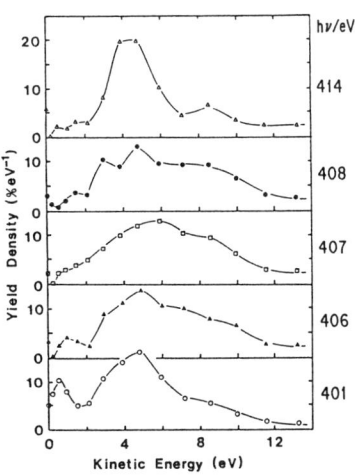

**FIGURE 5.** Kinetic energy distributions of N$^+$ (and N$_2^{2+}$) at several photon energies indicated in the right end.

utilized for dissociation from the K-shell ionized state, which clarified the relation between two valence hole states of the molecular ion and fragment charge states.[24] The present authors have compared the obtained KED of $N^+$ with the Auger electron spectrum on the assumption that the products exist in their ground states.[29] This comparison provided that most $N^+$ ions are produced in the ground state ($^3P$) or low excited states. The total KED of $N^++N^+$ obtained from PIPICO spectrum was also compared with the Auger electron spectrum, which supported the present postulation.[38] A comparison was made between the KED of $N^+$ at 401 eV($\pi^*$ state) and a resonant Auger spectrum.[29] The component of low kinetic energies was well reproduced with the binding energies from this Auger spectrum, but the structure around 5 eV was not obtained. This finding suggests that many $N^+$ ions are produced through double Auger processes at this photon energy. This fact is consistent with the experimentally obtained branching ratio in the fragment ion formation.

## Oxygen molecule

Similar examination has been made for dissociation pathways of $O_2$ in the K-shell hole states. The KED's of $O^+$ from $O_2$ depend considerably on the photon energy, indicating that main dissociation pathways at the 1s ionized state differ from those at inner-shell excited neutral states.[28,30] The lower panel of Fig.6 shows comparison of the total KED of $O^++O^+$ obtained here with a binding energy spectrum derived from the Auger electron spectrum.[39,40] If all states of $O_2^{2+}$ dissociates to the ground state of two atomic ions, $O^+(^4S)$, the two curves should agree each other within the experimental uncertainty. The lower panel in Fig.6 exhibits a considerable difference between the solid and broken curves, although maximum positions are close each other. This finding indicates that many $O^+$ ions are not formed in the ground state. In consideration of the energy levels of $O_2^{2+}$ and $O^+$, dissociation pathways are proposed here as shown in Table 2. The expected KED is shown with the broken curve at the upper panel in Fig.6. The essential feature of the expected curve is close to that of the measured KED, which postulates that the present assignment on the dissociation pathways is appropriate.

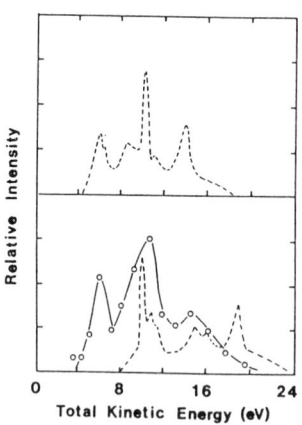

FIGURE 6. Total kinetic energy distributions of $O^++O^+$ from $O_2$ at a photon energy of 560 eV. The solid curve indicates the experimental result, and the broken curves are total kinetic energy distributions calculated from an Auger electron spectrum with different assumptions.

**TABLE 2.** Proposed dissociation pathways from $O_2^{2+}$ states into $O^+ + O^+$.

| $O_2^{2+}$ state[#] | Energy (eV) | $O^+ + O^+$ state (Energy / eV) | Released energy (eV) |
|---|---|---|---|
| w $^3\Delta_u$(A $^3\Sigma_u^+$) | 42.6 | $^4S + {}^4S$ (32.35) | 10.2 |
| b $^3\Pi_g$ | 43.4 | $^4S + {}^2P$ (37.35) | 6.0 |
| b' $^3\Sigma_u^-$ | 43.8 | $^4S + {}^2P$ (37.35) | 6.4 |
| $^3\Delta_g$ | 47.3 | $^2D + {}^2D$ (39.03) | 8.3 |
| c $^3\Pi_u$ | 48.5 | $^4S + {}^2P$ (37.35) | 11.1 |
| $^3\Pi_g$(c $^3\Pi_u$) | 51.2 | $^4S + {}^2P$ (37.35) | 13.8 |

[#]Assignment by Ref.39. Parentheses indicate Ref.40.

## SUMMARY

Asymmetry parameters for photoexcitation of *K*-shell electrons in linear molecules have been determined using monochromatic synchrotron radiation and a rotatable TOF spectrometer. The obtained parameters of diatomic molecules directly confirmed symmetry of inner-shell hole states and shape resonances. The β parameters for triatomic molecules depended often on the detected species, indicating importance of structures in excited states and of bending motion. Kinetic energy distributions of fragment ions from molecules in inner-shell hole states have been obtained at several photon energies around the K-edges. The KED's obtained at the ionized states were compared with the energy states of doubly charged molecules on the assumption that the final product ions exist in their ground states. Ion-ion coincidence spectra were demonstrated to be useful for clarifying correlation between the molecular ion states after Auger transitions and the states of the final products. For deeper understanding of the dissociation pathways, it becomes very important to carry out precise coincidence measurements of multi-dimensional type among an energy-selected Auger electron and fragment ions.

## ACKNOWLEDGMENT

The authors are grateful to the staff of the accelerator group at ETL for their continued operation of the TERAS electron storage ring. We are indebted to Prof. Becker, Prof. Samson, and Prof. Hitchcock for fruitful comments on the present study, and to Dr.Bozek for cooperation in performing a part of this work.

## REFERENCES

1. Ma Y., Chen C.T., Meigs G., Randall K., and Sette F., Phys. Rev. A **44**, 1848 (1991).
2. Domke M., Remmers G., and Kaindl G., Nucl. Instrum. Meth. B **87**, 173 (1994).

3. Kosugi N., Shigemasa E., and Yagishita A., Chem. Phys. Lett. **190**, 481 (1992).
4. Saito N., Heiser F., Hemmer O., Hempelmann A., Wieliczek K., Vielhaus J., and Becker U., Phys. Rev. A **51** R4313 (1995).
5. Eberhardt W., Rubensson J-E., Randall K.J., Feldhaus J., Kilcoyne A.L.D., Bradshaw A.M., Xu Z., Johnson P.D., and Ma Y., Physica Scripta, T**41**, 143 (1992).
6. Hemmer O., Heiser F., Eiben J., Wehlitz R., and Becker U., Phys. Rev. Lett. **71**, 987 (1993).
7. Schmidbauer M., Kilcoyne A.L.D., Koppe H.M., Feldhaus J., and Bradshaw A.M., Chem. Phys. Lett., **199**, 119 (1992).
8. Caldwell C.D., Schaphorst S.J., Krause M.O., and Jimenez-Mier J., J. Electron Spectrosc. **67**, 243 (1994).
9. Southworth S.H., Lindle D.W., Mayer R., and Cowan P.L., Phys. Rev. Lett. **67**, 1098 (1991).
10. Hitchcock A.P. and Brion C.E., J. Electron Spectrosc. **18**, 1 (1980).
11. Wight G.P. and Brion C.E., J. Electron Spectrosc. **3**, 191 (1974).
12. Sette F., Stohr J., and Hitchcock A.P., J. Chem.Phys. **81**, 4906 (1984).
13. Saito N. and Suzuki I.H., Phys.Rev.Lett. 61, 2740 (1988).
14. Saito N. and Suzuki I.H, J. Phys. B**22**, L517 (1989).
15. Saito N. and Suzuki I.H, Phys. Rev. A**43**, 3662 (1991).
16. Bozek J.D., Saito N., and Suzuki I.H., J. Chem. Phys. **100**, 393 (1994).
17. Bozek J.D., Saito N., and Suzuki I.H., J. Chem. Phys. **98**, 4652 (1993).
18. Bozek J.D., Saito N., and Suzuki I.H., Phys. Rev. A**51**, 4563 (1995).
19. Lee K., Kim D.Y., Ma C.I., and Hanson D.M., J. Chem. Phys. **100**, 8550 (1994).
20. Kim D.Y., Lee K., Ma C.I., Mahalingam M., Hanson D.M., and Hulbert S.L., J. Chem. Phys. **97**, 5915 (1992).
21. Shigemasa E., Hayaishi T., Sasaki T. and Yagishita A., Phys. Rev. A**47**, 1824 (1993).
22. Eberhardt W., Sham T.K., Carr R., Krummacher S., Strongin M., Weng S.L., and Wesner D., Phys. Rev. Lett. **50**, 1038 (1983).
23. Suzuki I.H., Saito N., and Bozek J.D., Bull. Chem. Soc. Jpn. **68**, 1119 (1995).
24. Eberhardt W., Plummer E.W., Lyo I.W., Murphy R., Carr R., and Ford W.K., J. de Physique, C**9**, 679 (1987).
25. Ueda K., Chiba H., Sato Y., Hayaishi T., Shigemasa E.,and Yagishita A., Phys.Rev. A**46**, R5 (1992).
26. Lindle D.W., Manner W.L., Steinbeck L., Villalobos E., Levin J.C., and Sellin I.A., J. Electron Spectrosc. **67**, 373 (1994).
27. Suzuki I.H. and Saito N., J. Chem. Phys. **91**, 5324 (1989).
28. Saito N. and Suzuki I.H., J. Chem. Phys. **91**, 5329 (1989).
29. Saito N. and Suzuki I.H., Int. J. Mass Spectrom. Ion Proces. **82**, 61 (1988).
30. Saito N. and Suzuki I.H., J. Chem. Phys. **93**, 4073 (1990).
31. Suzuki I.H. and Saito N., Laser Chem. **16**, 5 (1995).
32. Saito N., Bozek J.D., and Suzuki I.H., J. Phys. B (in press).
33. Hanson D.M., Lapiano-Smith D.A., Lee K., Ma C.I., and Kim D.Y., Chem. Phys. **162**, 439 (1992).
34. LeBrun T., Lavollee M., Simon M,. and Morin P., J. Chem. Phys. **98**, 2534 (1993).
35. Dehmer J.L. and Dill D., J. Chem. Phys. **65**, 5327 (1976).
36. Adachi J., Kosugi N., Shigemasa E., and Yagishita A., J. Chem. Phys. (in press).
37. Morin P. and Nenner I., Phys. Rev. Lett. **56**, 1913 (1986).
38. Saito N. and Suzuki I.H., J. Phys. B**20**, L785 (1987).
39. Sambe H. and Ramaker D.E., Chem. Phys. **104**, 331 (1986).
40. Moddeman W.E., Carlson T.A., Krause M.O., Pullen B.P., Bull W.E., and Schweitzer G.K,. J. Chem. Phys. **55**, 2317 (1971).

# PHOTOIONIZATION

Double Photoionization of He .................................. 117
    N. BERRAH
Signatures of Strong Correlations in Photoionization .......... 127
    C. H. GREENE
Energy and Angular Resolved Studies
of Double Photoionization of Helium and Rare Gases ......... 139
    A. HUETZ, L. Andric, A. Jean,
    P. Lablanquie, P. Selles, and J. Mazeau

# Double Photoionization of He

## Nora Berrah

Physics Department, Western Michigan University,
Kalamazoo, MI 49008.

Abstract. Various theoretical studies have investigated correlations between the two continuum electrons in photo-double-ionization of He. Near-threshold, intermediate and high energy studies have been performed using different experimental methods to test the corresponding predictions. Recent developments in studies of double photoionization of He are presented from below the double-ionization threshold to 12 keV.

### INTRODUCTION

Photo-double-ionization is one of the fundamental processes of physics because it requires a solution of the three-body Coulomb problem where the boundary conditions for the two continuum electrons must be included. There has been much interest in the study of photo-double-ionization of He because it is a system that is dominated by electron-electron correlations. Because the independent electron model failed to provide adequate agreement with measurements, new theoretical approaches have had to be developed. In this progress report, we present recent developments in measurements and calculations in some aspects of the field of photo-double-ionization.

### PHOTOIONIZATION NEAR THE HE DOUBLE-IONIZATION THRESHOLD

Since the classic work of Wannier (1), numerous theoretical studies have been made on near-threshold ionization (2-4). The various theories yield predictions for three different observable situations: (a) the energy dependence of the cross section (b) the energy sharing of the two outgoing electrons and (c) the angular correlation of these electrons. Wannier theory predicts, in the energy range just above threshold, that $\sigma^{++} = \sigma_0 E^{\alpha}_{exc}$. Kossmann et al. (5) made an extensive study of the threshold law for the cross section of double ionization in helium. Ion analysis was carried out with a pulsed-field time-of-flight (TOF) e/m analyzer. Their results provide quantitative information about the Wannier exponent, $\alpha=1.05(2)$ which agrees with the theoretical prediction of 1.056; a threshold value $\sigma^0=1.02(4) \times 10^{-21}$ cm$^2$ and $E_{th}=79.013(10)$ eV. Furthermore, their experimental results find the range of validity of the cross-section threshold law to be approximately 2-eV excess energy above threshold. Lablanquie et al. (6) used coincidence measurements between low energy electrons and doubly charged ions to study the dynamics of double photoionization and confirmed the range of validity of the Wannier theory. They found that the energy distribution of the two outgoing electrons is flat, within 20%, in agreement with the theoretical prediction, but in a 15 eV energy range above threshold. Photoionization phenomena near the double-ionization threshold has also been extensively studied by Hall et al (7). Using a photoelectron/photoion coincidence technique they find the value of the exponent $\alpha$ to be consistent with the Wannier prediction. They also investigate the behavior of the asymmetry parameter, ß, near threshold and obtain a nearly constant value close to -0.4. Their result is in disagreement with the prediction of the Wannier theory which appears to underestimate the angular correlation between the two electrons (8). Dawber et al. (9) have exploited the photoelelectron-photoelectron coincidence technique to measure the triple differential cross section (TDCS) at very low excess energies E (0.6 eV <E< 2 eV) for both equal and unequal energy sharing between the two outgoing electrons. The measured data are compared with the Wannier(1) predictions and also with recent ab initio calculations (10-12) that are not based on a Wannier-like treatment. Their measurements suggest a departure from the predictions of the Wannier model at the largest excess energy studied, E=2 eV. Lablanquie et al. (13) have also very recently studied the effect of electron energy sharing near the double photoionization threshold. In their energy and angle resolved measurements, they observed that although the angular distributions do not depend much on the energy sharing of the two electrons at 4 eV above threshold, a strong effect is measured at E=18.6 eV. We have used a zero-volt spectrometer (14) to study with higher resolution than 50 meV (15) photoionization phenomena near the double-ionization threshold. Fig. 1 shows a preliminary spectrum taken with photons from an undulator beamline coupled with a spherical-grating monochromator of the Advanced Light Source at Lawrence Berkeley

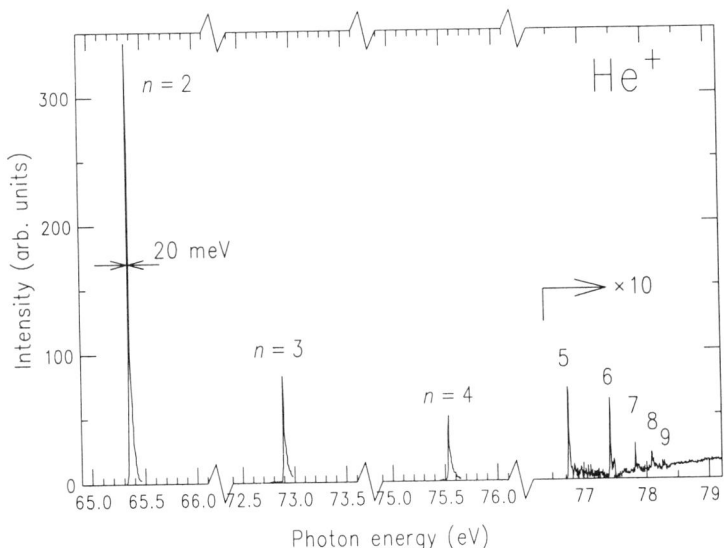

FIG. 1. Threshold ($E_e$=0 eV) photoelectron spectrum obtained with a monochromator bandpass of 6 meV and and photon energy increment of 5 meV.

Laboratory. The 1.5 GeV storage ring was filled to 40 mA in the two bunch mode at injection. The monochromator bandpass was 6 meV near 79 eV with 31 µm entrance and 27 µm exit slit widths. The scan in Fig.1 shows eight satellite lines which are the result of electron correlations. These satellite lines originate from an ionization process with additional excitation leaving the ion in a $He^+$ nl (n>1) state. The linewidths are about 20 meV, an improvment by a factor of 2.5 over previous measurements (15).

## ANGULAR DISTRIBUTIONS OF HELIUM SATELLITES $HE^+$ Nl (N=2-7)

The angular distributions of the satellite n=2 have been extensively studied in the low energy range (16-19) and especially in the region of autoionization (19,20). Calculations have concentrated on the satellite n=2 (21-24) but do not exist for the angular distributions over a wider energy range for the higher members. The angular distributions for n>2 were calculated only at threshold (25). This study (26) aimed to determine the angular distribution parameters, $\beta_n$, for the satellites n=2-7 as a function of photon energy, in the high energy region, to determine whether the ß-values converge to ß=2 and at which photon energy this limit is reached. Furthermore, this study makes it possible to compare the energy dependence of the angular distribution, for the satellites n=3-7, with respect to the principal quantum number n.

Helium photoelectron satellites He⁺ nl (n=2-7) were measured and the angular distributions were determined in the photon energy range from 80 eV to 603 eV (26). The experiment was performed at HASYLAB and some of the measurements were carried out on the BW3 undulator beamline equipped with an SX-700 monochromator. Two angle-resolved TOF spectrometers were used to analyze the kinetic energy and direction of the photoelectrons. Both analyzers were mounted perpendicular to the photon beam on a rotatable chamber in order to determine the angular distribution. Fig. 2 shows the angular

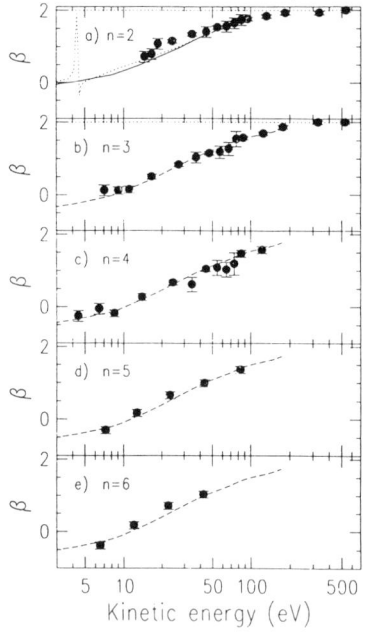

FIG. 2. Angular distribution parameters ß for the satellites n=2-6 (filled circles) on a logarithmic scale. The dotted line in a) was calculated by Salomonson et al. (23). The dashed curves were obtained by fitting the data to the theory of Jacob and Burke (21).

distribution for satellites n=2-6 which start at about -.5 to -.2 and increase with increasing photon energy. One notes that the ß-values decrease with increasing principal quantum number n due to the changing contributions of the subshells. An analysis based on Jacob and Burke's (21) calculation shows that the higher the principal quantum number n the smaller is the ns-contribution to the satellite state. The calculated curves shown fit quite well the experimental data which shows that the influence of higher l-contributions (l>2) is small. With increasing photon energy the ß values increase, converging to ß=2, as expected.

## DOUBLE PHOTOIONIZATION AT INTERMEDIATE ENERGIES

The ratio of double to single photoionization is a key parameter for characterizing electron correlations. From the late 70's until present, it has been measured (27) and calculated (28) extensively from threshold to about 12 keV and both calculations and measurements have been most challenging. Difficulties in the exact description of the two-electron continuum have led to various approximate treatments; perturbative approaches, such as many-body perturbation theory (MBPT), the Born approximation, or a distorted-wave Born-type approximation. From threshold to about 280 eV, various MBPT calculations have been carried out, differing in their choice of basis set and in their methods used to estimate higher order effects. In the early 80's, Carter and Kelly's (29) MBPT calculation correctly described the data (27). In the early 90's renewed interest in this problem resulted in work in the intermediate energy region and at high energies. At intermediate energies, between 280 and 1210 eV, the questions that arose were: a) how does the interplay of electron correlations, in both the initial and final states, affect the behavior of the ratio? b) what is the relative importance of the basis set and explicit consideration of higher-order correlation corrections. Berrah et al. (30) carried out measurements that tested the most recent theories of Pan and Kelly (31) and Hino (32) in this largely unexplored energy region.

The measurements were conducted at BESSY using monochromatic light from the high-energy toroidal grating monochromator (HE-TGM-1). The experimental technique was similar to the one used previously (33). He ions produced in the interaction region were analyzed by a 4.5-cm-long TOF spectrometer adapted to enable the measurement of both $He^+$ and $He^{++}$ within the 208 ns spacing of electron bunches in the ring. Several experimental effects can result in an inaccurate determination of the ratio of double-to-single ionization of He (33). In this experiment, special emphasis was placed on suppressing possible higher-order and stray light contributions to the ion signal produced by the monochromator (34). Both of these effects produce spurious $He^+$ signals, and therefore result in a lower ratio than one would measure if the source were "clean" of these two effects.

Our measurements (30) (circles) are shown in Fig. 3 along with the scaled data of Bartlett et al.(triangles) (35) and earlier measurements at threshold (squares) (27). They are compared with a few calculations; curve 'a' is a MBPT in the velocity form by Carter and Kelly (29) including both ground state correlation (GSC) and final state correlation (FSC), curve 'b' is a calculation by Hino (32) using the acceleration form of the dipole operator in the lowest-order MBPT, curve 'c'is a semiempirical calculation of Samson et al. (36), curves 'd,e' are MBPT of Pan and Kelly (31) in the velocity and length form respectively. They extended the previous calculations (29) and found that the inclusion of certain higher-order correlation corrections was significant. Curves 'f,g' are calculations in the velocity and acceleration form (respectively) by Meyer and Greene (37) using the R-matrix method as an alternative approach. The data

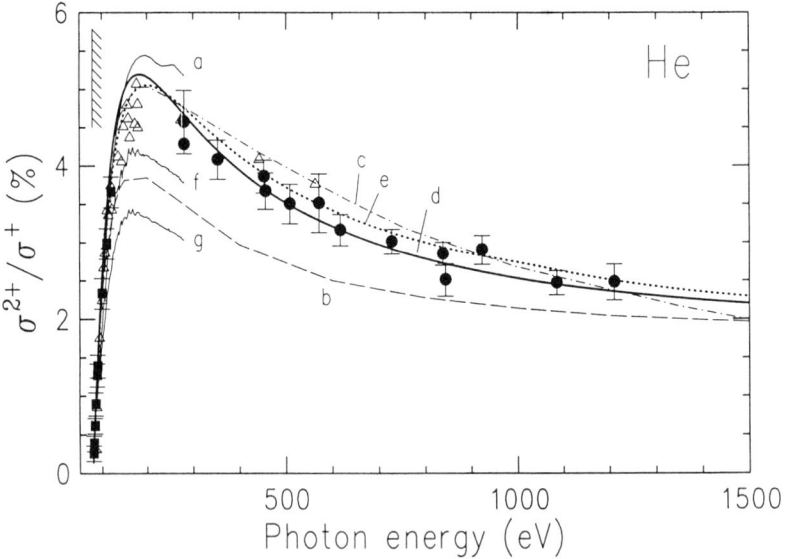

FIG. 3 Ratio of double-to-single photoionization as a function of photon energy. The symbols represent experimental data while the lines represent calculations. (see details in text).

show reasonable agreement with Samson's model (36) which conjectured that there should be a proportionality between producing a doubly charged ion by photon impact on a neutral atom and electron impact on a singly charged ion, although ground state correlations are different in the two cases. The data show excellent agreement with both the length and velocity form of Pan and Kelly (31). If indeed the main difference between Hino's MBPT calculation (32) and Pan and Kelly's MBPT calculation (31) is the inclusion of higher-order effects, both in GSC and FSC, and the use of a different basis set (important since the choice of a pertinent basis set enables the implicit inclusion of higher-order effects) we are led to conclude that, at intermediate energies, these higher order effects are very important for a good description of the present data. There seems to be a discrepancy between the data and Meyer and Greene's (37) results, and this discrepancy could be attributed to the uncertainty in their calculations, and also perhaps in the data. New measurements are underway to determine precisely the peak of the ratio since Manson and McGuire pointed out its importance (38) and also to reduce the error bars in our measurements in order to test unambiguously the various calculations.

## DOUBLE PHOTOIONIZATION AT HIGH ENERGIES

In the high photon energy limit, from 2- to 12- keV, measurements of the ratio using ion TOF spectrometry (33), conducted at BNL, have been reported by Levin et al. (39) to be 1.5 (±0.2)% and are shown in Fig. 4 (squares). This value is consistent with the data by Bartlett et al. (inverted triangles) (40) and with different, and in part conflicting, theoretical predictions of asymptotic value of 1.6-1.7% (28,41). It was only recently that an understanding of the relative importance of the different processes in the asymptotic limit was achieved. Dalgarno and Sadeghpour (42) explained how in this limit only ground-state-correlations need to be considered when using the acceleration gauge and that final-state correlation is not essential.

Above 2.5 keV, Compton scattering starts to play a role in the formation of $He^+$ (43) and above 4.6 keV that of $He^{++}$. At about 12 keV it dominates the ionization process. Our measurements, which collect all the ions formed at a particular photon energy, do not discriminate between the two processes. However, calculations of the ratio of double to single *photoionization* by Andersson and Burgdörfer (43) that have taken into account Compton scattering (curve 'h' of Fig. 4) agree very well with the data. In fact, we would expect the measured ratio at 12 keV to be produced mainly by Compton ionization because the ratio of incoherent scattering to the photoelectric effect is about a factor of ten at this energy (44). New calculations by Andersson and Burgdörfer (45) have evaluated with higher accuracy the asymptotic value of the ratio for Compton scattering and found it to be of the order of 0.8%. This value agrees well with an independent calculation by Suric et al. (46) but disagrees with calculations by Hino et al. (47) (curve 'i' in Fig. 4) who find an asymptotic ratio for Compton scattering of about 1.6%.

Recently, a new measurement by Spielberger et al. (48) differentiated between photoionization and Compton ionization. They have given i) a ratio of 1.72 (±0.12)% for photoionization at about 7.0 (+2.1,-1.6) keV (diamond 'ph' in Fig. 4) which agrees quite well with the photoionization calculations ('d'&'e') (31) and ii) a ratio of 1.22 (±0.06)% at 8.8 (+1.5,-1.65) keV (circle 'cp') which agrees with our data (39) and with calculations (43).

Experimental and theoretical work are continuing in this area in an effort to understand the similarity and differences between photoionization and Compton ionization and to establish the asymptotic limit in both cases.

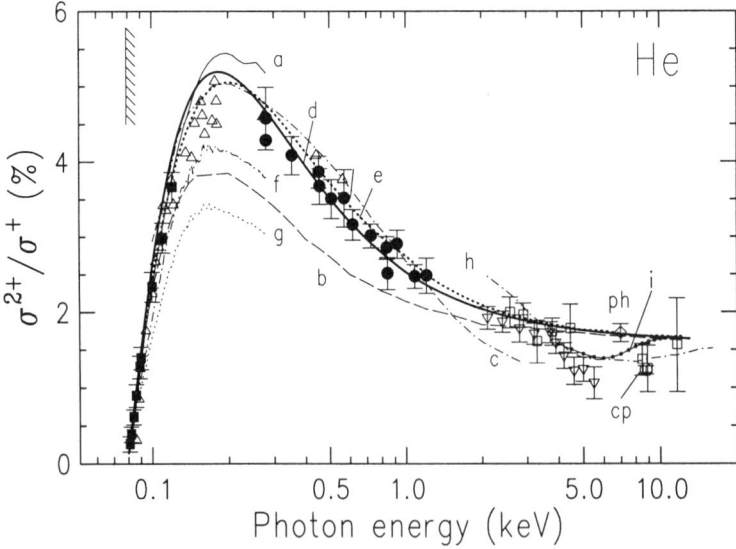

Fig. 4. Ratio of double to single ionization as a function of photon energy (see text for details).

## ACKNOWLEDGEMENTS

The angular distribution of the He satellites was carried out in collaboration with Wehlitz et al.(26), the intermediate energy measurements with Berrah et al. (30) and the high energy measurements with J. Levin et al. (33,39). This work was supported in part by the U.S. DOE, Office of Basic Energy Science, Division of Chemical Science under contract No. DE-FG02-92ER14299.

## REFERENCES

1. G. H. Wannier, Phys. Rev. **90**, 817 (1953).
2. A. R. P. Rau, Phys. Rev. A **4**, 207 (1971).
3. H. Klar and W. Schlecht, J. Phys. B **9**, 1699 (1976).
4. C. H. Greene and A. R. P. Rau, Phys. Rev. Lett. **48**, 533 (1982) and J. Phys. B **16**, 99 (1983).
5. H. Kossmann, V. Schmidt, and T. Andersen, Phys. Rev. Lett. **60**, 1266 (1988).
6. P. Lablanquie, K. Ito, P. Morin, I. Nenner, and J. H. D. Eland, Z. Phys. D **16**, 77 (1990).
7. R. I. Hall, A. G. McConkey, L. Avaldi, K. Ellis, M. A. MacDonald, G. Dawber and G. C. King, J. Phys. B: At. Mol. Opt. Phys. **25**, 1195 (1992).
8. G. Dawber, R. I. Hall, A. G. McConkey, M. A. MacDonald and G. C. King J. Phys. B: At. Mol. Opt. Phys. **27**, L341 (1994).
9. G. Dawber, L. Avaldi, A. G. McConkey, H. Rojas, M. A. MacDonald and G. C. King, J. Phys. B: At. Mol. Opt. Phys. **28**, L271 (1995).
10. A. Huetz, P. Selles, D. Waymel, and J. Mazeau, J. Phys. B: At. Mol. Opt. Phys.**24**, 1917 (1991)
11. A. Kazansky and V. N. Ostrovski, J. Phys. B: At. Mol. Opt. Phys. **27**, 447 (1994).
12. F. Maulbetsch and J. S. Briggs, Phys. Rev. Lett. **68**, 2004 (1994).
13. P. Lablanquie, J. Mazeau, L. Andric, P. Selles, and A. Huetz, Phys. Rev. Lett. **74,** 2192 (1995).
14. F. Heiser, U. Hergenhahn, J. Viefhaus, K. Wieliczek and U. Becker, J. Electron Spectrosc. Relat. Phenom. **60**, 337 (1992).
15. R. I. Hall, L. Avaldi, G. Dawber, M. Zubek, K. ellis, and G. C. King, J. Phys. B: At. Mol. Opt. Phys. **24**, 115 (1991).
16. M. O. Krause, and F. Wuilleumier J. Phys. B: At. Mol. Opt. Phys. **5**, L143 (1972).
17. V. Schmidt H. Derenbach and R. Malutzki J. Phys. B: At. Mol. Opt. Phys. **15**, L523 (1982).
18. J. M. Bizau, F. Wuilleumier, P. Dhez, D. L. Ederer, T. N. Chang, S. Krummacher and V. Schmidt, Phys. Rev. Lett. **48**, 588 (1982).
19. D. W. Lindle, T. Ferrett, U. Becker, P. H. Kobrin, C. M. Truesdale, H. G. Kerkhoff, and D. A Shirley, Phys. Rev. A **31**, 714 (1985).
20. M. Zubek, G. Dawber, R. I. Hall, L. Avaldi, K. Ellis and G. C.

King J. Phys. B: At. Mol. Opt. Phys. 24, L337 (1991).
21. V. L. Jacob and P. G. Burke J. Phys. B: At. Mol. Opt. Phys. 5, L67 (1972).
22. K. A. Berrington, P. G. Burke, W. C. Fon, K. T. Taylor J. Phys. B: At. Mol. Opt. Phys. 14, L603 (1980).
23. S. Salomonson, S. L. Carter and H. P. Kelly, Phys. Rev. A 39, 5111 (1989).
24. I. Sanchez and F. Martin, Phys. Rev. A 45, 4468 (1992).
25. C.H. Green Phys. Rev. Lett. 44, 869 (1980).
26. R. Wehlitz, B. Langer, N. Berrah, S. B. Whitfield, J. Viefhaus and U. Becker, J. Phys. B: At. Mol. Opt. Phys. 26, L783 (1993).
27. H. Kossmann, V. Schmidt, and T. Andersen, Phys. Rev. Lett. 60, 1266 (1988), and references therein.
28. T. Äberg, Phys. Rev. A 2, 1726 (1970), and references therein.
29. S. L. Carter and H. P. Kelly, Phys. Rev. A 24, 170 (1981).
30. N. Berrah, F. Heiser, R. Wehlitz, J. Levin, S. B. Whitfield, I. A. Sellin, and U. Becker, Phys. Rev. A 48, R1733 (1993).
31. C. Pan and H. P. Kelly, Phys. Rev. A (in press).
32. K. Hino, Phys. Rev. A 47, 4845 (1993).
33. J. C. Levin, D. W. Lindle, N. Keller, R. D. Miller, Y. Azuma, N. Berrah, H. G. Berry and I. A. Sellin, Phys. Rev. Lett. 67, 968 (1991).
34. H. Kossmann, O. Schwarzkopf, B. Kämmerling, W. Braun, and V. Schmidt, J. Phys. B: At. Mol. Opt. Phys. 22, L411 (1989).
35. R. J. Bartlett, P. J. Walsh, Z. X. He, Y. Chung, E.-M. Lee, and J. A. R. Samson, Phys. Rev. A 46, 5574 (1992).
36. J. A. R. Samson, R. J. Bartlett, and Z. X. He, Phys. Rev. A 46, 7277 (1992).
37. K. W. Meyer and C. H. Greene, Phys. Rev. a 50, R1 (1994).
38. S. T. Manson and J. H. McGuire, Phys. Rev. A 51, 400 (1995).
39. J. C. Levin, I. A. Sellin, B. M. Johnson, D. W. Lindle, R. D. Miller, N. Berrah, Y. Azuma and H. G. Berry and D.-H. Lee, Phys. Rev. Lett. 47, R16 (1993).
40. R. J. Bartlett. M. Sagurton, J. A. R. Samson and Z. X. He, J. Opt. Soc. Am 1994 (to be published).
41. F. W. Byron and C. J. Joachain, Phys. Rev. 164, 1 (1967.
42. A. Dalgarno and H. R. Sadeghpour, Phys. Rev. A 46, R3591 (1992).
43. L. R. Andersson and J. Burgdörfer, Phys. Rev. Lett. 71, 50 (1993); J. A. R. Samson, C. H. Greene, and R. J. Bartlett, *ibid.* 71, 201 (1993).
44. J. H. McGuire, N. Berrah, R. J. Bartlett, J. A. R. Samson, J. A. Tanis, C. L. Cocke, and A. S. Schlachter, J. Phys. B: At. Mol. Opt. Phys. 28, 913 (1995).
45. L. R. Andersson and J. Burgdörfer, Phys. Rev. A 50, R2810 (1994)
46. T. Suric, K. Pisk, B. A. Logan, and R. H. Pratt, Phys. Rev. Lett. 73, 790 (1994).
47. K. Hino, P. Bergstrom and J. Macek, Phys. Rev. Lett. 72, 1620 (1994).
48. L. Spielberger et al. Phys. Rev. Lett. 74, 4615 (1995).

# Signatures of Strong Correlations in Photoionization

Chris H. Greene

*Department of Physics and JILA, University of Colorado, Boulder, CO 80309-0440*

> Recent theoretical developments have led to an unprecedented capability to describe complex photoionization processes. In many systems that exhibit nonperturbative electron correlations, theory can reproduce or predict experimental spectra, essentially to spectroscopic accuracy. The aim of this review is to show how R-matrix techniques can determine the parameters of multichannel quantum defect theory. These parameters, in turn, determine the experimental observables that can be measured either with or without an external field present. I also summarize some relatively recent developments, such as the development of R-matrix methods to describe some two-electron ejection processes.

## INTRODUCTION

The early years of photoionization theory were devoted to the determination of the average cross section for liberating an electron from a bound orbital into a continuum state of the atom. In hydrogen, the cross section decreases smoothly and monotonically as the energy increases above the ionization threshold. Photoionization research in the 1960's demonstrated that hydrogen is anomalously simple. All nonhydrogenic atoms and ions show structured variations in the energy dependence of their photoionization cross sections.

One type of energy dependence arises already at the independent particle level: so-called Cooper minima [1] in which the radial matrix element between the oscillatory initial state and final state radial wavefunctions exhibits a cancellation at certain energies (the energies of the minima). Shape resonances can also cause a strong energy dependence of cross sections. Qualitatively, shape resonances arise already within the independent electron approximation; nevertheless improved treatments such as the random phase approximation are often needed in order to obtain theoretical resonance positions and widths that agree with experiment. Shape resonance decay widths are often (but not always) comparable to the energy of the resonance above its appropriate ionization threshold. Such widths run from tens of eV (for valence $d$- or $f$-subshell resonances in heavy elements) to meV (for negative ion shape resonances).

Autoionizing Feshbach resonances typically produce much stronger variations in the photoionization cross section as a function of energy. In this review paper, I discuss the insight into nonperturbative electron correlation physics that has been gleaned from the study of autoionizing resonances. Recent progress toward a theoretical description of two-electron escape is also described. The use of synchrotron radiation sources to photoionize atoms, negative ions, and molecular species produces spectra of amazing complexity. Laser spectroscopy of these species yields even richer spectra owing to the higher resolution of a laser; spectroscopy with lasers remains limited, however, to photons that are far less energetic than those which emerge from synchrotron sources.

Most spectra analyzed prior to 1970 involved smooth continua with isolated resonances superimposed. Such isolated resonances are reasonably-well described as a Breit-Wigner-type Lorentzian enhancement that rises far above the background continuum. If the continuum excitation becomes comparable to the resonance excitation strength, however, the resonant pathway to ionization can interfere appreciably with the continuum, which can generate an asymmetric Beutler-Fano-type lineshape or in some cases a window resonance. When a truly isolated resonance interacts with a smooth continuum, the dynamics involves only one observable time scale: the resonance decay lifetime $\tau$. A more interesting situation arises when several resonances are present, particularly when they overlap and result in a nontrivial spectral distribution of intensity versus energy. A single Rydberg series of resonances embedded in a simple continuum can be viewed as a sequence of isolated resonances. For this situation, the positions of autoionizing levels obey the usual Rydberg formula $E_n = E_{\text{th}} - \frac{1}{2}(n - \mu)^{-2}$, where $\mu$ is the real part of the quantum defect for this series. This type of spectrum now exhibits *two* characteristic time scales: the Rydberg resonance lifetime $\tau \propto \nu^3$, where $\nu$ is the effective quantum number $\nu = n - \mu$, and the Rydberg orbital period $\tau_n$ which also increases in proportion to $\nu^3$.

When a second Rydberg series is attached to a higher-lying ionization threshold, a new level of complexity arises. This situation leads to far more complicated spectra that show, at first glance anyway, an irregular appearance. In fact the classical counterpart of such a system (if the system can be meaningfully said to posess a classical counterpart) is frequently chaotic. In one limit, where the two energetically-closed channels have negligible interaction, two uncoupled autoionizing Rydberg series occur, each with its separate characteristic orbital time and autoionization lifetime. More typically the two series of bound levels interact, in which case a new time scale enters the equation, namely the average time required for an electron in one bound Rydberg channel to scatter into the other closed channel. This new time scale controls a great deal of the appearance of the resonances that are excited by different processes such as an electron-ion collision or ionization by photon impact. Quantum mechanical interferences can arise in this situation owing to the increasing number of different excitation and decay pathways. Among these interference effects are "zero-width states": states embedded in the ionization continuum whose decay pathways destructively interfere, leading to long (or even potentially infinite) autoionization lifetimes.

With the development of Seaton's multichannel quantum defect theory (MQDT) starting in the late 1960's [2], the nature of the resonances became quite well understood. The MQDT framework unifies the theoretical description of the resonance physics, regardless of the strength of the channel interactions. The resonance pattern is controlled by a set of a set of quantum mechanical amplitudes: these are often referred to as short-range MQDT scattering parameters because they can be expressed as a generalized type of $S$-matrix in a scattering-theory description of the interchannel interactions. Early applications of this theoretical MQDT description analyzed any given spectrum semiempirically; the MQDT parameters were fitted so as to reproduce the observed data. The reason this approach was used was because theoretical methods had not yet been devised to directly calculate the scattering parameters from first principles. Despite the semiempirical nature of these fits, the resulting fitted parameters permit a thorough interpretation of the spectrum; for instance, one obtains the probability for scattering from one (closed or open) Rydberg channel into another channel. The MQDT parameters also have predictive power: e.g., in some cases the parameters extracted from a fit of *total* cross sections were used to predict *partial* photoionization cross sections, photoelectron angular distributions, and the photoelectron spin polarization. These fits provided the first opportunity to characterize the nature of quite complex spectral data, and were a welcome improvement over theoretical techniques that existed in the 1960's and early 1970's. On the other hand, reliance on fitting alone is clearly not adequate; in order to fit the spectra with a manageable number of parameters, restrictive assumptions had to be imposed such as neglect of their energy dependence [3].

The success of the aforementioned fitting scheme led to a perception in the atomic physics community that this theoretical description could *only* be used semiempirically. In fact, in numerous atomic (and molecular) systems, the number of channels $N$ is so large that semiempirical fits can no longer uniquely determine the $N(N+1)/2$ unknown MQDT scattering parameters. Early efforts in the middle 1970's by Lee and Fano [4] showed that R-matrix methods could be used to determine the MQDT scattering parameters by a direct *ab initio* calculation. Several years later, the relativistic random-phase approximation was developed by Johnson, Lee, Cheng, and coworkers [5] to calculate the MQDT parameters to high accuracy for closed-shell species near their lowest ionization thresholds. The MQDT parameters from those calculations remain some of the most accurate available to this date for rare gas atoms and their isoelectronic sequences.

## APPLICATIONS TO TWO-ELECTRON SYSTEMS

Applications to systems like the alkaline earth atoms, whose electron correlations are notoriously strong, began in earnest a few years later. An efficient numerical scheme was developed that succeeded far beyond any plausible expectations. This scheme was based on: (A) The use of a model potential that describes the interaction of each valence electron with the doubly-charged ionic core, and is optimized to give accurate energy levels of the *alkali-like ion*. (B) Solution of the full two-electron Schrödinger equation inside a finite reaction volume, where each electron interacts with the doubly-charged ionic core and with the other valence electron; this step was achieved by using through an efficient version of the eigenchannel R-matrix method that was reformulated [6] and applied to numerous two-electron systems by O'Mahony, Greene, Kim, Aymar, Luc-Koenig, and others [7], [8], [9]. (C) A match of the calculated solution and its normal derivative at the reaction surface to a linear combination of Coulomb wavefunctions in each channel; the linear coefficients give the full matrix of short-range MQDT scattering parameters and photoabsorption amplitudes.

Photoionization calculations have now been performed for a large number of atomic species, utilizing the methodology just described. In many cases experimental measurements had been performed before the theoretical calculations were published, while in others the theory came first and constituted a prediction of the spectrum. Fig.1 shows an example of the latter, namely a prediction by Kim and Greene [10]. The predicted photoionization cross section of atomic calcium in its $4s^2$ ground state is shown versus the final state energy, for photon energies ranging from approximately 9 eV to 14 eV. This energy range exhibits a rich complexity that arises because of the large number of interacting Rydberg series present that converge to different ionization thresholds ($Ca^+(5s)$ - $Ca^+(6s)$) and overlap in energy. In classical language, one might say that the dynamics is chaotic and complicated by the large number of incommensurate time scales; but the spectrum of Fig.1 is fully quantum mechanical and could not be described by classical physics. The Fig.1 spectrum has not yet been tested against any experiments. Such an experimental test remains desirable, because at these relatively high energies the calculations are increasingly difficult.

Photoionization of excited state atoms is amenable to these theoretical techniques, although the experimental investigation of such processes is more difficult than for ground states. An example studied in recent years is laser photoionization of the barium $5d6p\,^3D_1^o$ level. This barium excited state was prepared using a different laser that pumped ground state barium atoms up to that level. Figure 2 shows experimental and theoretical spectra in the absence of any external field; the plots are shown as a mirror image in order to permit an easier judgement of the level of agreement. The good correspondence between theory and experiment is evidently not fortuitous, as it is hard to imagine how such complex spectra could agree accidentally. The description of such a complicated pattern of channel interactions by a nearly *ab initio* calculation seemed like an implausible theoretical task just a decade ago. Now such calculations in this energy range have become routine.

The spectra of Fig.2 are complicated even when no external fields are present. They become still more complex when the same basic experiment is conducted in the presence of a static, homogeneous electric field. A similar comparison between theory and experiment is shown in Fig.3. The electric field has a generally weak effect on the short range physics, but additional

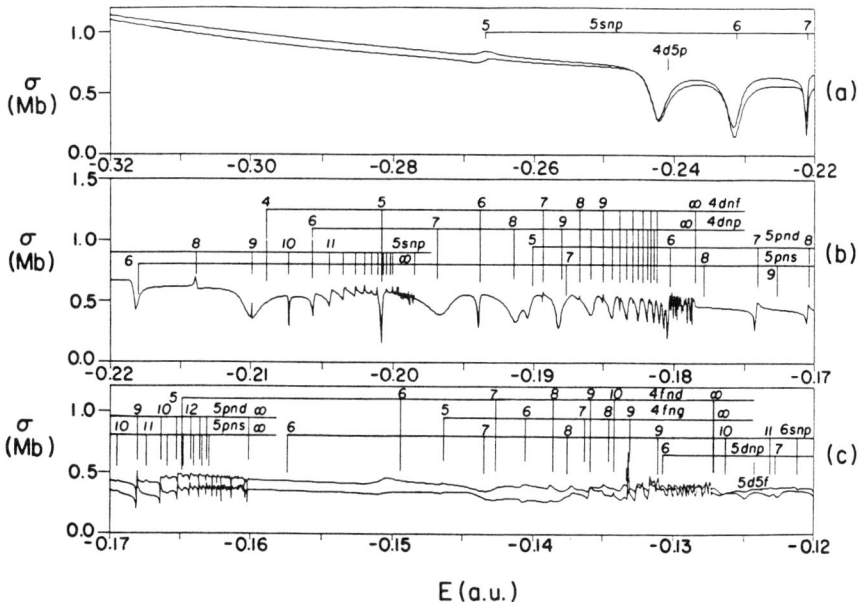

**FIG. 1.** The photoionization cross section of the ground state of the calcium atom is shown as a function of final state energy (in a.u.) relative to the double ionization threshold. The spectrum shown was predicted by Kim and Greene[10], but it has not yet been tested experimentally.

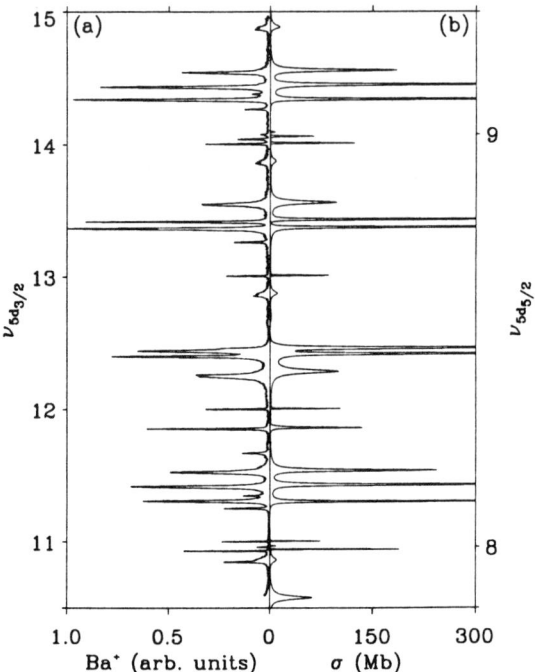

**FIG. 2.** (a) The experimental photoionization spectrum of the Ba state is shown as a function of final state energy. The vertical axis represents an energy variable which is the effective quantum number relative to the level of Ba+. (b) Calculated photoionization spectrum of the same level, including an energy convolution over the laser bandwidth. From Armstrong and Greene [12].

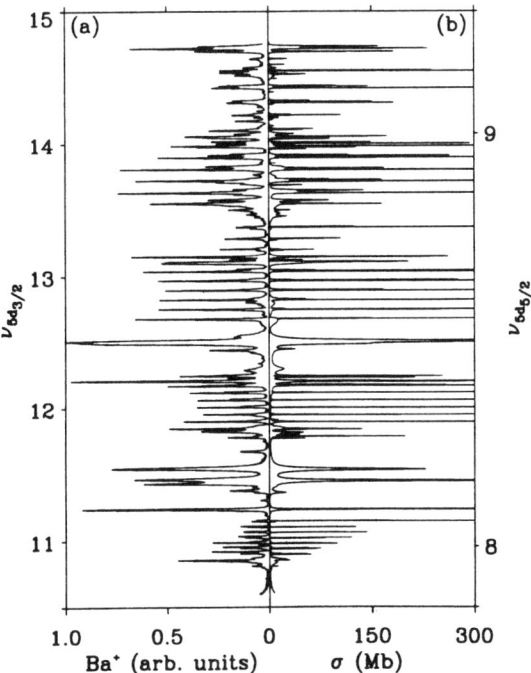

**FIG. 3.** (a) The experimental photoionization spectrum of the Ba $5d6p\ ^3D_1^o$ state in the presence of an applied electric field $F = 4kV/cm$ is shown as a function of final state energy. As in Fig.2, the vertical axis is an energy variable which is the effective quantum number $\nu_{5d_{5/2}}$ relative to the $5d_{5/2}$ level of $Ba^+$. (b) Calculated photoionization spectrum of the same initial level in the applied electric field. From Armstrong and Greene [12].

features arise in the spectrum as hydrogenic manifolds of states expand as the field increases. The weakness of the electric field compared to the atomic channel interactions permits a Harmin-Fano-type frame transformation [13] to be carried out. This theoretical technique amounts to a method by which the field dependent spectrum can be expressed in terms of field free scattering parameters for atomic barium, and hydrogenic Stark properties.

As laser techniques have improved, atomic photoionization experiments have begun to probe higher and higher final state energies that could previously be reached only using a synchrotron radiation photon. An example is the study of very high doubly-excited states of Sr by [16] and of Ba by [15]. Both of these experiments use several laser photons to excite the alkaline earth atom sequentially up to final states that lie approximately 0.1 a.u. in energy below the threshold energy for two-electron escape. Another feature common to both experiments is their use of a Stark switching method to prepare Rydberg states in which the outermost electron has very high orbital angular

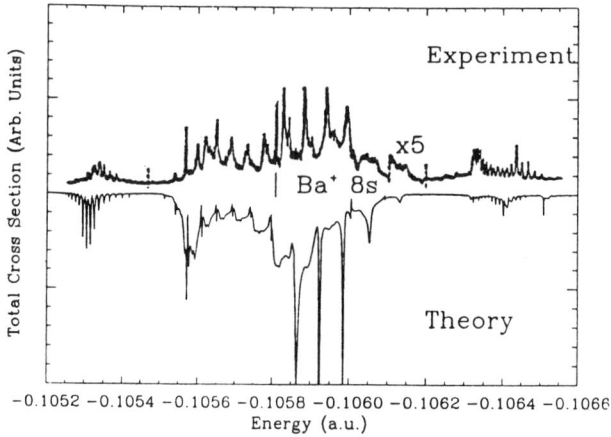

FIG. 4. Comparison of theoretical and experimental (Camus et al. [15]) photoionization spectra of Ba, produced by a multistep isolated core excitation scheme using lasers and a Stark switching method. From Wood and Greene[14].

momenta. For instance, in the experiment of Camus et al. [15]. this Stark switching technique is first used to produce Ba $6sn\ell$ levels with $\ell \approx 6 - 10$, and $n \approx 12 - 16$. These nonpenetrating Rydberg states have extremely small interactions with the $Ba^+$ core, and can be described by hydrogenic models quite adequately. On the other hand, the next step excites the inner ($6s$) electron to energies near the $Ba^+(8s)$ state. This is a two-photon variant of the isolated-core-excitation (ICE) method, for which the final states show a rich variety of nonperturbatively-coupled channels.

One usually thinks of two strongly correlated electrons as a pair that overlaps the same region of space, but in this case the outermost two electrons of Ba do not overlap but are strongly correlated nevertheless. A comparison of the theoretical and experimental spectra for this system is shown in Fig.4. This experimental scheme produces a final state whose symmetry is not straightforward to identify, and which probes the channel interactions of several different final states.

The preceding examples show that two-electron systems can be described remarkably well by current theoretical tools, at least for energies below the double ionization threshold. Above this threshold energy, the description of two-electron escape is less straightforward. Two applications adaptations of the eigenchannel R-matrix method to handle such processes have been made thus far. One of these is a treatment of helium double photoionization [17], which appears to produce sensible results, although disagreements between several experiments and theoretical studies make it difficult to definitively assess the accuracy of the calculations. Also the velocity and acceleration gauge results differ by about 20%, which is a discrepancy that needs to be reduced before we can have strong confidence in the method. Some improvements in this method have been developed and applied to model Hamiltonians [18], including the development of a frame transformation technique that transforms pseudostate ionization amplitudes into physical amplitudes. The improvements have produced accurate cross sections for electron impact ionization, but they have not yet been tested for double photoionization processes.

## MANY-ELECTRON ATOMS

Calculations of resonant photoionization spectra have been carried out for years by a number of groups, including those of Kelly, Burke, LeDourneuf, and Starace, to name a few. While these groups have used sophisticated configuration interaction wavefunctions to describe short-range electron correlations, most of them have not made use of the full power of multichannel quantum defect theory. For instance, one powerful capability of MQDT is to calculate the short-range scattering parameters using an R-matrix code that ignores all spin-orbit and relativistic effects; the spin-orbit physics is readily incorporated through the use of a recoupling frame transformation that is carried out before the final step of determining the photoionization cross section in the MQDT calculation.

F. Robicheaux developed a new set of computer programs that extend this basic theoretical approach to atoms or ions having an arbitrary number of open-shell electrons. Some key improvements to the original formulation were implemented more recently by Robicheaux and G. Miecznik [19,20]. The basic eigenchannel R-matrix approach followed for many-electrons resembles the two-electron method in principle, but in practice a number of additional issues must be faced in order to get accurate wavefunctions and observables. First and foremost is the fact that to treat photoionization of an $N$-electron atom, the $(N-1)$ electron "target" states must now be represented as a multiconfiguration expansion. For the alkaline earth atoms, each target state could be represented by a single radial orbital, with essentially no uncertainty in its calculated energy. Largely for this reason, the choice of the radial basis set to be used in constructing a variational basis set inside the R-matrix reaction volume requires far more attention in a many-electron calculation. Nevertheless, the use of model potentials, Löwdin-type natural orbitals, and multiconfiguration Hartree-Fock methods (MCHF) can all be used with varying degrees of ease and reliability to construct a basis set expansion that converges reasonably well. Systems in the periodic table studied thus far using this set of computer programs include the fluorine group, the oxygen group, the carbon group, in addition to Ne, Al, Sc, and Ti. For transition metal atoms the number of $d^n$ configurations grows explosively, making it desirable to find basis sets that converge more rapidly than the independent electron basis set. Reliable results have been obtained for the lighter transition metal elements near the lower ionization thresholds.

Fig.5 shows an example of Sc photoionization in the wavelength range 185

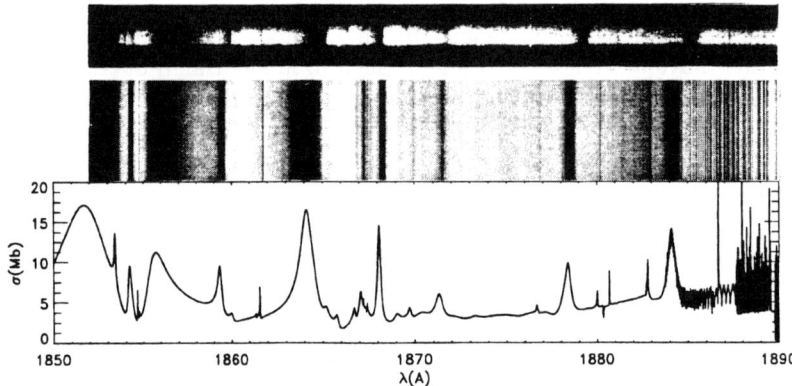

**FIG. 5.** Photograph(top) of an experimental plate measured by Garton et al. [21] that displays Sc photoabsorption versus photon wavelength. Simulation (middle) of a "theoretical plate" based on the calculated theoretical photoionization cross section in the length gauge, which is given at the bottom. From Robicheaux and Greene [22].

nm to 189 nm. This rich spectrum was measured long ago by Garton et al. [21], but it had proven difficult to understand from a theoretical viewpoint. The eigenchannel R-matrix calculation of Robicheaux and Greene [22], combined with a recoupling (fine-structure) frame transformation, succeeds reasonably well in identifying the key spectral features. As no densitometer tracing of the experimental photograph was published by Garton [21], it is not easy to compare intensities. Garton [21] sidestepped that problem by converting the theoretical spectrum into a "mock photographic plate" that can be compared directly to the experimental plate. The resulting comparison in Fig.5 suggests that spectroscopically useful information is indeed obtained by this approach to atomic photoionization calculations.

In summary, the state of the art of atomic photoionization calculations has advanced dramatically in the past decade. Systems of unprecedented complexity can now be treated theoretically, often to an accuracy competitive with experimental accuracy. A number of improvements remain desirable, e.g., to describe excitations that reach much higher in energy or in atoms like the lanthanides that can have more open-shell electrons. Improvements in treating spin-orbit effects remain desirable, as the currently implemented frame transformation theory is inadequate to describe these effects in atoms much heavier than Kr.

This work was supported in part by the National Science Foundation.

# REFERENCES

1. U. Fano and J. W. Cooper, Rev. Mod. Phys. **40**, 441 (1968). See also A. F. Starace, Handbuch der Physik, Vol.31 (S. Flügge and W. Mehlhorn, eds.). Springer-Verlag, Berlin and New York, 1982.
2. M. J. Seaton, Rep. Prog. Phys. **46**, 97 (1983), and references therein.
3. C. M. Lee and K. T. Lu, Phys. Rev. A **8**, 1241 (1973); M. Aymar, Phys. Rep. **110**, 163 (1984); J. A. Armstrong, J. J. Wynne, and P. Esherick, J. Opt. Soc. Am. **69**, 211 (1979).
4. U. Fano and C. M. Lee, Phys. Rev. Lett. **31**, 1573 (1973); C. M. Lee, Phys. Rev. A **10**, 584 (1974).
5. C. M. Lee and W. R. Johnson, Phys. Rev. A **22**, 979 (1980); W. R. Johnson, K. T. Cheng, K. N. Huang, and M. Le Dourneuf, *ibid.* **22**, 989 (1980).
6. C. H. Greene, Phys. Rev. A **8**, 2209 (1983). See also an independent derivation of the variational principle presented by H. Le Rouzo and G. Raseev, Phys. Rev. A **29**, 1214 (1984).
7. P. F. O'Mahony and C. H. Greene, Phys. Rev. A **31**, 250 (1985).
8. C. H. Greene and L. Kim, Phys. Rev. A **38**, 5953 (1988).
9. M. Aymar, E. Luc-Koenig, and S. Watanabe, J. Phys. B **20**, 4235 (1987); M. Aymar, J. Phys. B **20**, 6507 (1987). C. H. Greene and M. Aymar, Phys. Rev. A **44**, 6271 (1991).
10. L. Kim and C. H. Greene, Phys. Rev. A **38**, 2361 (1988).
11. L. Kim and C. H. Greene, Phys. Rev. A **36**, 4272 (1987).
12. D. J. Armstrong and C H. Greene, Phys. Rev. A **50**, 4956 (1994); see also D. J. Armstrong, C. H. Greene, R. P. Wood, and J. Cooper, Phys. Rev. Lett. **70**, 2379 (1993).
13. D. A. Harmin, Phys. Rev. A **26**, 2656 (1982); U. Fano, Phys. Rev. A **24**, 619 (1981).
14. R. P. Wood and C. H. Greene, Phys. Rev. A **49**, 1029 (1994).
15. P. Camus, S. Cohen, L. Pruvost, and A. Bolovinos, Phys. Rev. A **48**, R9 (1993).
16. U. Eichmann, V. Lange, and W. Sandner, Phys. Rev. Lett. **68**, 21 (1992).
17. K. W. Meyer and C. H. Greene, Phys. Rev. A **50**, R3573 (1994).
18. K. W. Meyer, C. H. Greene, and I. Bray, Phys. Rev. A **52**, 1334 (1995).
19. G. Miecznik, C. H. Greene, and F. Robicheaux, Phys. Rev. A **51**, 513 (1995), and references therein.
20. G. Miecznik and C. H. Greene, to be published.
21. W. R. S. Garton, E. M. Reeves, F. S. Tomkins, and B. Ercoli, Proc. R. Soc. London Ser. A **333**, 1 (1973).
22. F. Robicheaux and C. H. Greene, Phys. Rev. A **48**, 4429 (1993).

# ENERGY AND ANGULAR RESOLVED STUDIES OF DOUBLE PHOTOIONIZATION OF HELIUM AND RARE GASES

A. Huetz, L. Andric, A. Jean, P. Lablanquie[†], P. Selles
and J. Mazeau

*Laboratoire de Dynamique Moléculaire et Atomique, CNRS et Université Pierre et Marie Curie, 4, place Jussieu, T12-E5 75252 Paris Cedex 05, FRANCE*
[†]*LURE, Bat 209D, Université Paris Sud, 91405 Orsay, FRANCE*

**Abstract.** Experiments on double photoionization, where the two outgoing electrons are selected in angle and energy before being detected in coincidence, have started in 1991 on xenon [1] and krypton [2]. During the past four years they have been extended to lighter rare gases including helium, and a general understanding of the observed angular patterns has started to emerge. In the present paper we report on such recent measurements in helium, neon and argon, obtained with a multidetection apparatus and in the so called "coplanar" geometry. We try to analyse the physical effects underlying the shapes of the various angular patterns which are presented.

## I. INTRODUCTION

Since the multiple ionization of atoms and molecules by synchrotron radiation or other sources of light is under study, the exact behaviour of the electrons which are produced by such processes has long been inaccessible. This was true for the so called *direct* process responsible for helium double photoionization, which has been extensively investigated by measuring the $He^{++}/He^+$ ion ratio (see reference [3]). This was also the case for the *indirect* mechanism in which relaxation of an intermediate singly charged state produces multiply charged ions and Auger electrons. In the latter case and considering double photoionization, the commonly used description is a two step picture and experimental studies can be mainly divided into two classes in which either the photoelectron or the Auger electron is measured but where the behaviour of the two electrons as a pair is ignored.

In the present paper we report on measurements relying on two electrostatic electron analysers which select the energies and angles of each electron, yielding the so called triple differential cross section (TDCS): $\sigma^{(3)} = d^3\sigma / dE_1 d\Omega_1 d\Omega_2$.

Sections III to VI concern the direct process whereas section VII is devoted to the indirect mechanism in the specific case of valence Auger electrons.

## II. EXPERIMENTAL SET-UP

The experimental set-up is presented on figure 1, and the upper inset is a schematic view of the system seen from the top. Briefly the main component is a toroidal analyser, the design of which is close to that first adopted by Leckey and

Riley [4]. The photon beam issued from the synchrotron beam lines (SU6 undulator line and SA31 bending magnet line of Super Aco (Orsay, France) have been successively used in the following experiments) penetrates the toroidal analyser through holes in the outer toroid and is crossed with the gas beam on the axis of cylindrical symmetry. The cut of the analyser in the plane perpendicular to the photon beam is reported on figure 1, showing that the exit slit and optics are partly closed (see the right part of the figure) to hold the inner toroid. The detection plane for this analyser is the $(\hat{\varepsilon}, \hat{k})$ plane defined by the main axis of polarization $\hat{\varepsilon}$ and the photon beam direction $\hat{k}$. In the following the label "electron 2" is used for the electron detected on the position sensitive detector located at the bottom of the toroidal analyser. The emission angle $\theta_2$ for this electron is defined in trigonometrical convention with respect to the direction of $\hat{\varepsilon}$. This angle is determined by the position of electron 2 on the position sensitive detector, over a large range of 260 degrees. There are three forbidden sectors due to mechanical contraints, namely $85° \leq \theta_2 \leq 95°$ and $265° \leq \theta_2 < 275°$ (forward and backward directions of the photon beam) and $320° \leq \theta_2 \leq 360°$, $0° \leq \theta_2 \leq 40°$ (closed parts of exit slit and optics including edges effects).

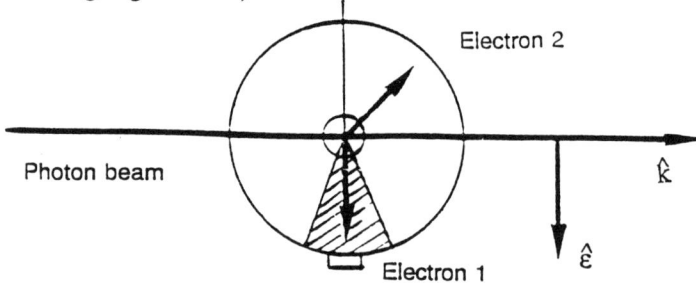

Figure 1

Experimental set-up

Electron 1 with wave vector $\hat{k}_1$ is measured also in the $(\hat{\epsilon}, \hat{k})$ plane but at a fixed angle $\theta_1=(\hat{\epsilon}, \hat{k}_1)=0$, by means of two concentric tubes which take advantage of the dead angle of the toroidal analyser and allow this electron to fly out of the toroidal analyser to be detected by a small hemispherical analyser (see figure 1). The energy resolutions $\Delta E_1$ and $\Delta E_2$ (FWHM) for the two electrons are determined by the pass energies of the two analysers, and the energies $E_1$ and $E_2$ can be varied independently. It should be stressed that in the present arrangement all vectors $\hat{\epsilon}$, $\hat{k}$, $\hat{k}_1$ and $\hat{k}_2$ belong to the same plane, in contrast with perpendicular plane experiments [1, 5 to 11]. Such geometry (to be called coplanar in the following) makes the TDCS measurements insensitive to circularly components of the light for equal sharing ($E_1 = E_2$) as well as unequal sharing ($E_1 \neq E_2$) conditions [12, 13]. In addition the linear component of the light which is orthogonal to the $(\hat{\epsilon}, \hat{k})$ plane does not contribute either in some specific cases, such as helium for instance [14, 15].

Finally we outline that the experimental conditions described above hold for all results presented in this report, and differ from those of references [14] and [15] in two aspects: (i) the range of accessible $\theta_2$ angles has been extended; (ii) a different technique has been used for the position sensitive detector, replacing the previous discrete anode (angular stepping 10°) by a resistive anode encoder which allows better accuracy and angular resolution (5°) when measuring $\theta_2$.

## III. HELIUM

It has been established previously [14,15,16] that the TDCS in the coplanar geometry is given by:

$$\sigma^{(3)} = |a_g(E_1,E_2,\theta_{12})(\cos\theta_1+\cos\theta_2) + a_u(E_1,E_2,\theta_{12})(\cos\theta_1-\cos\theta_2)|^2 \quad (1)$$

where the amplitudes $a_g$ and $a_u$ are respectively symmetric and antisymmetric in $E_1 \leftrightarrow E_2$ and where $\theta_{12}$ is the mutual angle between the two electrons. The above expression does not rely on any approximation and shows up very simply from the body fixed frame approach introduced in reference [16]. The main conclusions which have been drawn from the measurements at $E = E_1 + E_2 = 18.6$ eV and 4 eV, reported in reference [15], can be summarized as follows. At $E = 4$ eV the second term in (1) becomes negligible, and the first one independent of the $E_1/E_2$ ratio. This is in contrast with observations at $E = 18.6$ eV and higher energies [7] and indicates that a threshold regime exists for $0<E<4$ eV, where the TDCS becomes insensitive to $E_1/E_2$ at least on a relative scale. Such prediction has been subsequently confirmed by lower energy measurements [11] for $0.6 \leq E \leq 2$ eV. It is worth noting that this remarkable property was predictible from the original work of Wannier [17], as mentioned in reference [15] and illustrated in detail by the recent calculations of reference [18].

We have undertaken a new series of equal and unequal energy sharing measurements in the $2 \leq E \leq 20$ eV energy range, and with the experimental improvements reported in section II. As an example these new and more extended measurements at E = 18.6 eV and E = 4 eV are reported on figure 2 (circles), in comparison with previous ones [15](squares). The overall agreement is rather good, except for two points at the limits of the previous angular range which may have been affected by edges effects in previous experimental conditions. For equal sharing equation (1) becomes:

$$\sigma^{(3)} = |a_g(E/2,E/2,\theta_{12})|^2 (\cos\theta_1+\cos\theta_2)^2 \quad (2)$$

with $\theta_1=0°$, $\theta_{12}=\theta_2$ under the present experimental conditions, and the new measurements allow an accurate determination of the angular correlation function:

$$C(E,\theta_{12}) = |a_g(E/2,E/2,\theta_{12})|^2 \quad (3)$$

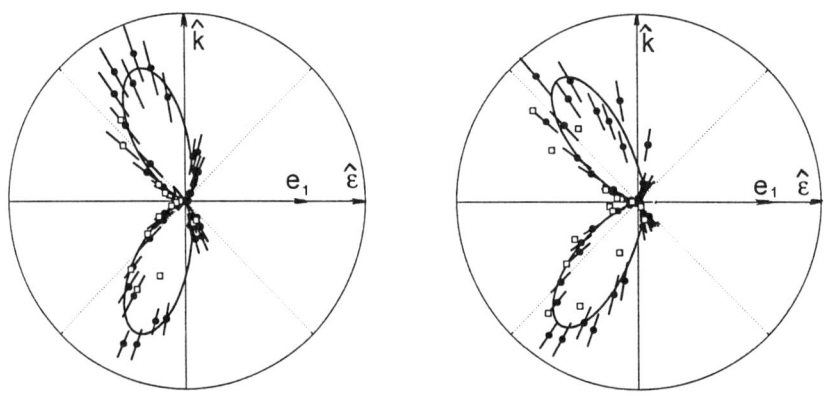

Figure 2 : TDCS for helium in polar coordinates. The photon beam is along $\hat{k}$ and electron 1 is emitted along $\hat{\epsilon}$, to the right of the figures. Same conditions hold for all TDCS reported below. Left: $E_1 = E_2 = 9.3$ eV; Right: $E_1 = E_2 = 2$ eV. Full lines: fitted curves (see text).

A polynomial form of $\cos\theta_{12}$ has been used and fitted to the results of figure 2 (full line). The associated C functions are reported on figure 3, and confirm the main features already observed from previous measurements [14,15]. The FWHM widths ($\theta_{1/2} = 106\pm14°$ at E = 18.6 eV and $\theta_{1/2} = 75\pm5°$ at E = 4 eV) are slightly larger than those obtained from the restricted angular range [14] but they also decrease slowly with energy. As already observed [15] spectacular dynamical effects occur in the wing of the C function (for example $C(18.6, 60°) = 9 \cdot 10^{-3}$; $C(4,60°) = 6 \cdot 10^{-4}$) and clearly indicates that the system becomes more correlated with decreasing E.

These new extended and more accurate measurements on helium will allow in the near future a detailed comparison with existing theories [18,19,20], but this is out of the scope of the present report.

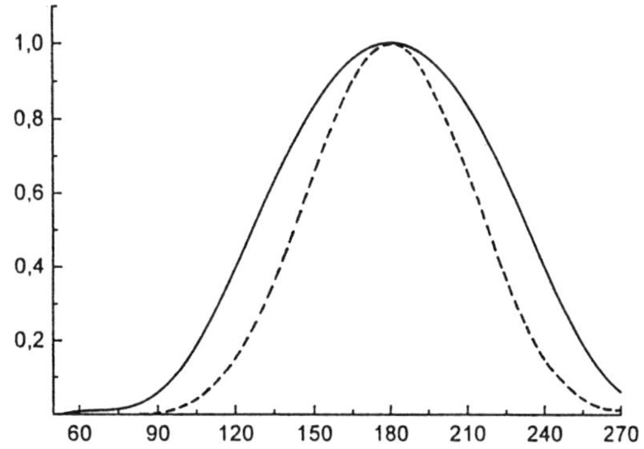

Figure 3

Angular correlation function $C(E,\theta_{12})$ from fitted curves of figure 1. Full line: E=18.6 eV; Dashed line: E = 4 eV.

## IV. INITIAL STATE EFFECT

The direct double photoionization also exists for atoms heavier than helium. For light atoms such as neon it can be easily distinguished from indirect (two-step) processes which give well defined energies for the photoelectrons as well as for the Auger electrons. Thus the energies $E_1$ and $E_2$ of the two detected electrons can be chosen to avoid interference with indirect double photoionization. When the $2s^22p^4$ $^1S^e$ final state of $Ne^{++}$ is selected the conservation rules impose the $^1P^o$ symmetry to the pair of outgoing electrons, and from this point of view the observed patterns should be similar to those for helium. Remarkably this is not the case, as shown on figure 4 at equal sharing ($E_1 = E_2 = 6.2$ eV): two additional lobes are clearly visible for angles around $\theta_{12} = 65°$, with a node of the TDCS around $\theta_{12} = 90°$. Such phenomenon has also been observed for $Xe^{++}$ $5s^25p^4$ $^1S^e$ [21] and $Ne^{++}$ $2s^22p^4$ $^1S^e$ at higher photon energy [8], and a physical explanation was first proposed by Kazansky and Ostrovsky [22, 23, 24]. Considering the two active electrons the initial state is a $p^2$ $^1S^e$ (because the remaining core is $p^4$ $^1S^e$) which contains a $\cos\theta_{12}$ factor due to the properties of spherical harmonics. Thus a node at $\theta_{12} = 90°$ exists in the initial state in contrast with the $1s^2$ $^1S^e$ initial state of helium, in a simple model where initial state correlations are neglected. In the wave packet formalism developed by Kazansky and Ostrovsky this node at $\theta_{12} = 90°$ survives (although it can be appreciably shifted) during the double escape process, and leads to the "butterfly" shape observed on figure 4. These calculations [24] are in fair agreement with the $Ne^{++}$ $2p^4$ $^1S^e$ data of Schaphorst et al [8] at $E_1 = E_2 = 20.3$ eV, and a comparison with the data of figure 4 at much lower energies is under progress.

Another approach to account for this initial state effect consist in using equation (2) and splitting the $C(E,\theta_{12})$ general function into a product of two terms [8,25]. The first one is connected to the initial state structure and accounts for the fact that from a $p^2$ initial state two continua, namely $\varepsilon s \varepsilon p$ and $\varepsilon d \varepsilon p$, are accessible in the independent particule picture. The second factor corrects this picture by including electron correlations, and is expected to be similar to the $C(E,\theta_{12})$ angular correlation function of helium. This approach has proved to reproduce quite well all $Ne^{++}$ $2p^4$ $^1S^e$ data (including those of figure 4) but the fitted intensities and phases associated to $\varepsilon s \varepsilon p$ and $\varepsilon d \varepsilon p$ continua disagree with Hartree Fock calculations for these quantities [25]. Thus the physical meaning of the parameters introduced in this approach is still open to question.

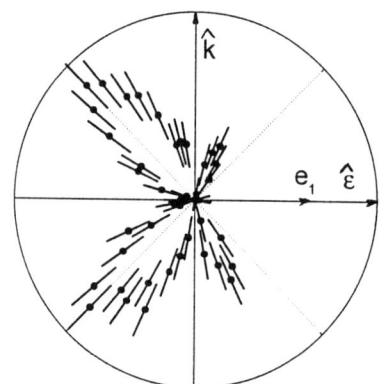

Figure 4

TDCS for $Ne^{++}$ $2p^4$ $^1S^e$ final state and:
$E_1 = E_2 = 6.2$ eV.

## V. ENERGY SHARING EFFECT

It has been shown in reference [15] and for helium that when the total energy E increases above 4 eV the threshold regime is replaced by another regime in which the TDCS tends to be maximum at $\theta_{12} = 180°$ for unequal sharing, whereas it is still zero there for equal sharing. This property is due to an increasing $a_u$ amplitude in equation (1) and becomes most spectacular at $E = 52.9$ eV [7]. Systematic studies of the direct process in neon and argon performed with our apparatus have shown that this is a quite general law: when a node exists at $\theta_{12}=180°$ for equal sharing due to symmetry properties, it is filled for unequal sharing and often becomes a maximum. An example is given on figure 5 for the $Ar^{++}$ $3s^23p^4$ $^1D^e$ final state where the energies have been chosen again to avoid interference with indirect double photoionization. In such a direct process:

$$\gamma + Ar\,(^1S^e) \rightarrow Ar^{++}\,(3s^23p^4\,^1D^e) + e + e$$

three different symmetries, namely $^1P^o$, $^1D^o$, $^1F^o$ are allowed for the electron pair. An experiment similar to that of figure 5a), but in the perpendicular plane, has been reported by Krässig et al [5]. The node at $\theta_{12} = 180°$ is clearly visible on figure 5a), and stems from the fact that all $^1P^o$, $^1D^o$, $^1F^o$ symmetries are of "unfavoured" type [26]. Expression (1) can be generalized to include $^1D^o$ and $^1F^o$ partial waves [16,27], thus introducing $a^P_{g,u}$, $a^D_{g,u}$ and $a^F_{g,u}$ amplitudes. All three partial waves have a node at $\theta_{12} = 180°$ when $E_1 = E_2$, leading to the node observed on figure 5a) (although it is not exactly zero due to experimental resolution). In contrast the pattern shown on figure 5b) for very asymmetric conditions ($E_1 = 1$ eV, $E_2 = 9.9$ eV) exhibits a maximum at $\theta_{12} = 180°$, implying a dominant contribution of $a^P_u$, $a^D_u$ and $a^F_u$ amplitudes. Between the two extreme cases of figure 5 the pattern is expected to evolve continously depending on intensity ratios and phase differences between the various amplitudes.

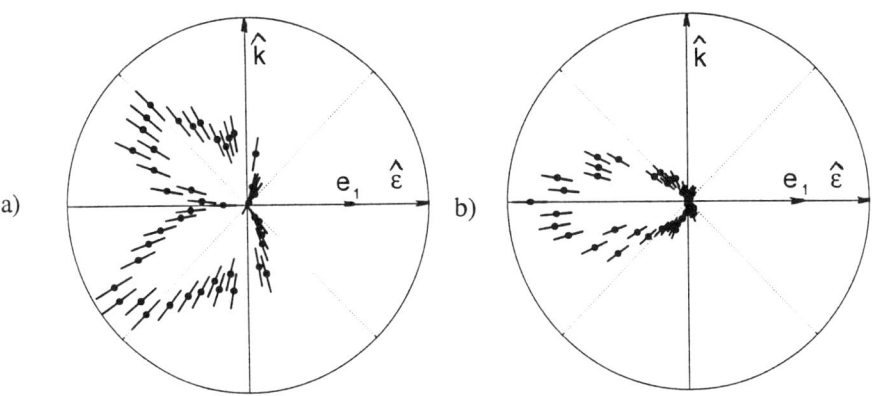

Figure 5  TDCS for $Ar^{++}$ $3p^4$ $^1D^e$ final state and: a) $E_1 = E_2 = 9.9$ eV; b) $E_1 = 1$ eV, $E_2 = 9.9$ eV

## VI.  SYMMETRY EFFECT

So far we have been dealing with direct double photoionization for which the allowed symmetries for the outgoing pair were all of unfavoured type, which could be easily recognized by the typical node at $\theta_{12} = 180°$ in the conditions of equal sharing. An interesting question in the systematic studies which have been undertaken on rare gases is the following: can direct double photoionization exist and be fully favoured by symmetry ? To answer this question we have investigated higher double photoionization limits of rare gases, in which inner valence shells are concerned. An example is shown on figure 6, corresponding to the direct process:

$$\gamma + Ar \rightarrow Ar^{++} \ 3s \ 3p^5 \ (^1P^o) + e + e$$

and at equal sharing $E_1 = E_2 = 3$ eV. This pattern is the first example of direct double photoionization where a maximum appears at $\theta_{12} = 180°$ for equal sharing. The physical explanation is quite simple, and only relies on symmetry considerations. As the final $Ar^{++}$ state is $^1P^o$, conservation rules give $^1S^e$, $^1P^e$ and $^1D^e$ as allowed symmetries for the pair. As shown in references [26] and [28] only two symmetries, namely $^3S^e$ and $^1P^e$, have the remarkable property to be fully antisymmetric with respect to radial interchange $r_1 \leftrightarrow r_2$, and as a consequence they can only give rise to "u" type amplitudes, which cancel down to zero at equal sharing $E_1 = E_2$. Thus the $^1P^e$ symmetry cannot be present in the conditions of figure 6, where only the "favoured" type [26,28] $^1S^e$ and $^1D^e$ symmetries participate. Taking the very simple $^1S^e$ contribution only one amplitude, $a_g^S$ ($E_1$, $E_2, \theta_{12}$), participates in the TDCS expression, and its propension to be maximum at $\theta_{12} = 180°$ due to electron electron repulsion is not anihilated by a multiplying factor such as the $(\cos\theta_1 + \cos\theta_2)$ term in equation (2) for the unfavoured $^1P^o$ symmetry. For the $^1D^e$ symmetry it can easily be shown that two amplitudes contribute in the conditions of figure 6, and that the multiplying "kinematical factors" have an antinode at $\theta_{12} = 180°$ [29]. Therefore the $^1D^e$ partial wave should be even more peaked at $\theta_{12} = 180°$ than the $^1S^e$, due to the conjunction of dynamical effects (through the amplitudes) and kinematical effects (through the multiplying factors).

These considerations are fully confirmed when looking at figure 6, where the TDCS is spectacularly peaked at $\theta_{12} = 180°$, i.e. when the two electrons fly back to back with equal energies. Similar observations have been obtained at different energies ($E_1 = E_2 = 1.5, 5.7$ and $12$ eV [30]). It is worth noting that the "symmetry favoured" double photoionization discussed in this section is closely linked to the original work of Wannier [17] who considered only the $^1S^e$ symmetry, for reason of simplicity. So far such a simplification was not realistic, as "unfavoured" symmetries were always present in (e,2e) and previous ($\gamma$,2e) studies. For the first time figure 6 well illustrates the "Wannier picture", which states that the two equal energy electrons tend to fly into opposite directions.

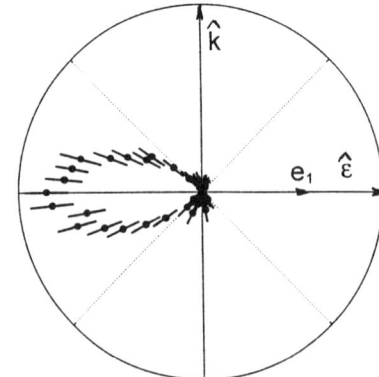

Figure 6

TDCS for $Ar^{++}$ $3s\,3p^5\,^1P^o$ final state and :
$E_1 = E_2 = 3$ eV

## VII. INDIRECT DOUBLE PHOTOIONIZATION

We now come to a physical mechanism also leading to doubly charged ions but which differs drastically from the direct process. TDCS for this so called indirect process have first been obtained in xenon [1,9] for $4d_{5/2}$ inner shell ionization, and in argon for valence double photoionization [5]. The xenon patterns have been analysed by means of a parametrized theory [31] based on the standard two step model, which considers the photoelectron emission and the subsequent Auger decay as completely independent events.

Considering the argon case the indirect process:

$$\gamma + Ar \rightarrow Ar^+ (.... 3s\, 3p^5\, (^1P^o)\, 4s\, ^2P^o ) + e_p$$
$$\downarrow$$
$$Ar^{++} (..3p^4\, ^1D^e) + e_A$$

has been proposed [5] to interpret the intense production of $E_A = 6.2$ eV electrons in the spectra. Recent high resolution spectra taken with our apparatus reveal that the natural width of this Auger electron is $\Gamma = 30 \pm 15$ meV. A TDCS measurement for the above process with conditions $E_1 = E_2 = 6.2$ eV is presented on figure 7, with resolutions $\Delta E_1 = 100$ meV, $\Delta E_2 = 80$ meV and $\Delta E_\gamma = 250$ meV. The shape of this pattern is similar to that obtained in the perpendicular plane by Krässig et al [5] which is not surprising as in both experiments electron 1 was aligned with the main axis of polarization, and a cylindrical symmetry around this line exists in the case of fully polarized light. It is typical of indirect patterns that the TDCS has two maxima, one at $\theta_{12} = 180°$ and the other at $\theta_{12} = 0°$ (see reference [32]). This rule is fulfilled on figure 7, where the increase of the TDCS on the right part of the figure allows to imagine a maximum at $\theta_{12} = 0°$, which unfortunately cannot be measured (dead angle of the toroidal analyser).

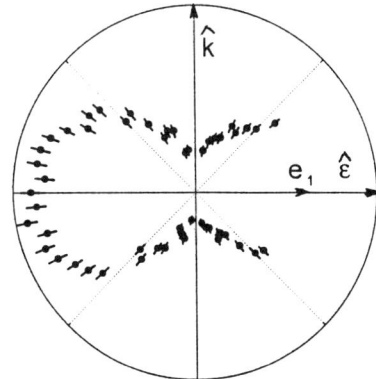

Figure 7

TDCS for indirect photoionization leading to the $Ar^{++}\, 3p^4\, ^1D^e$ final state and:
$E_1 = E_2 = 6.2$ eV

We wish to focus here onto a very interesting question, which has first been raised by Vegh and Macek [33] and further developed by the same authors [34]. This question can be simply introduced in this report from the preceding sections in which symmetry considerations have been extensively discussed for the direct process. The conservation rules used above to establish the symmetry of the outgoing pair rely on the final symmetry of the doubly charged ion. If one treats the $(e_p, e_A)$ couple of electrons as a pair and forgets about the intermediate $Ar^+$ state, the same rules must necessarily apply. In the above case this means that the pair $(e_p, e_A)$ is produced in $^1P^o, {}^1D^o, {}^1F^o$ symmetries, exactly as in section V and because of the $^1D^e$ final symmetry of $Ar^{++}$. Going one step further this means that a node at $\theta_{12} = 180°$ should be observed on figure 7, at equal sharing $E_1 = E_2 = 6.2$ eV. However a maximum is observed instead, leading to a contradiction.

In fact this contradiction is removed when looking at the TDCS expression proposed by Vegh and Macek [34] in which the resonant direct and exchange amplitudes are written $a_r(E_1, E_2) / (E_1 - E_A + i\Gamma/2)$ and $a_r(E_2, E_1) / (E_2 - E_A + i\Gamma/2)$ and therefore vary much more rapidly than the amplitudes for the direct process. In other words an equal sharing experiment in the case of indirect double photoionization is an experiment where the constraint $E_1 = E_2 = E_A$ is realized exactly, or at least where the $E_1 - E_A$ and $E_2 - E_A$ differences are small compared to $\Gamma$ so that the direct and exchange terms are nearly equal. Then they can cancel each other at $\theta_{12} = 180°$ according to symmetry properties. But in the case of figure 7 the above conditions are not satisfied, as the experimental energy resolutions are four or five times larger than $\Gamma$.

It follows from the preceding arguments that Auger processes with shorter lifetime and larger $\Gamma$ are better candidates to observe the interference effects predicted by Vegh and Macek. To confirm this statement we have studied another indirect process in neon:

$$\gamma + Ne \rightarrow Ne^+ (.. 2s2p^5 \, (^3P^o) \, 3p \, {}^2S^e) + e_p$$
$$\downarrow$$
$$Ne^{++} \, 2s^2 2p^4 \, {}^1D^e + e_A$$

for which the analysis of spectra gives $E_A = 13.2$ eV and $\Gamma = 155 \pm 10$ meV. The TDCS for equal sharing $E_1 = E_2 = 13.2$ eV has been measured for two different sets of resolutions, which are labelled PR (poor resolutions: $\Delta E_1 = 500$ meV, $\Delta E_2 = 400$ meV, $\Delta E_\gamma = 250$ meV) and HR (high resolutions: $\Delta E_1 = 100$ meV, $\Delta E_2 = 80$ meV, $\Delta E_\gamma = 250$ meV). Results with HR exhibit a marked dip at $\theta_{12} = 180°$, as shown on figure 8, whereas with PR this dip clearly fills in. These measurements confirm the predictions made by Vegh and Macek and will be presented elsewhere in more detail. It should be emphasized that another experimental confirmation has also been reported in $4d_{5/2}$ photoionization of xenon in fixed angle ($\theta_{12} = 180°$) experiments [35] and that similar coherence effects in autoionization have been observed in electron impact experiments [36].

The physical effect in indirect double photoionization described above has been presented in natural continuity with the properties of the TDCS for the direct process where the two electrons are indistinguishable. It can also be understood from the point of view of the standard two step model. If I is the double ionization potential the energy conservation imposes $E_\gamma = I + E_p + E_A$. For almost any photon energy (satisfying $|E_\gamma - I - 2E_A| \gg \Gamma$) then $|E_p - E_A| \gg \Gamma$ and the two electrons issued from the indirect mechanism are distinguishable: the first one, issued from the first step is labelled "photoelectron", while the second is labelled "Auger electron". In such case the exchange amplitude in the formula of Vegh and Macek is negligible and the two electrons can be well identified. On the other hand when the photon energy is tuned to satisfy $|E_\gamma - I - 2E_A| \ll \Gamma$ the two electrons become indistinguishable, as they are in the direct process. Then the standard two step model breaks down and strong interference between direct and exchange resonant amplitudes affect the angular patterns at the TDCS level. For doubly charged ions in singlet state and even parity these interferences are destructive at $\theta_{12} = 180°$, and lead to a node there as for the direct process for the same final state of the ion.

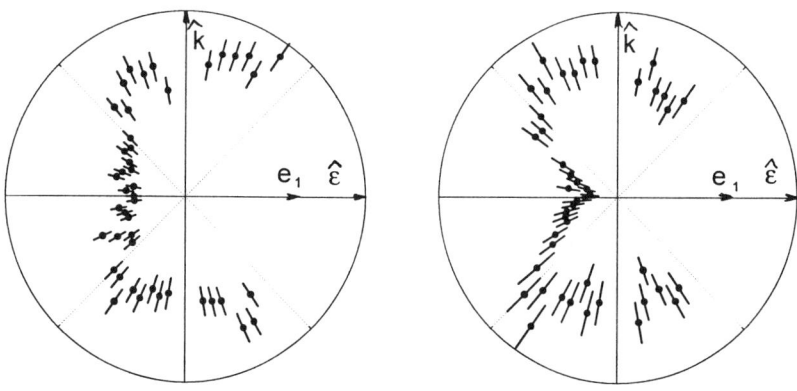

Figure 8  TDCS for indirect double photoionization leading to the Ne$^{++}$ 2p$^4$ $^1$D$^e$ final state and $E_1 = E_2 = 13.2$ eV. Left: poor resolutions; right: high resolutions.

## VIII.  SUMMARY

In this report we have tried to review briefly the physical mechanisms which are at work in the double photoionization of rare gases and which determine the shapes of the observed TDCS. Apart from the very specific case of helium at equal sharing for which the shape was predicted before being observed (see figure 2 of reference [16]) most of the features described above have received physical explanation after being measured. The investigation of angular behaviour of photoelectrons issued from single photoionization (asymmetry parameter measurements) has covered decades and has been most helpful for a deeper understanding of this process. TDCS studies for double photoionization open an even wider field, where experiments are already appealing for more elaborated and extended theoretical methods. It is predictible that these experiments will grow rapidly in the near future, to cover still unexplored energy ranges and other targets than rare gases.

## ACKNOWLEDGEMENTS

We are grateful to the staff of Super Aco (LURE) for technical support all along the experiments shown here. We also deeply acknowledge B. Krässig, S. Schaphorst, O. Schwarzkopf, and V. Schmidt from Freiburg, A. K. Kazansky and V.N. Ostrovsky from St Petersburg for communication of their results prior to publication and for permanent exchanges and stimulating discussions since the begining of this work.

## REFERENCES

1. Kammerling B., and Schmidt V., Phys. Rev. Lett. **67** (1991) 848
2. Mazeau J., Selles P., Waymel D., and Huetz A., Phys. Rev. Lett. **67** (1991) 820
3. Mc Guire J.H., Berrah N., Bartlett R.J., Samson J.A.R., Tanis J.A., Cocke C.L., and Schlachter A.S., J. Phys. B **28** (1995) 913
4. Leckey R.C.G., and Riley J.D., Appl. Surf. Sci. **22/23** (1985) 196
5. Krässig B., Schwarzkopf O., and Schmidt V., J. Phys. B **26** (1993) 2589
6. Schwarzkopf O., Krässig B., Elmiger J., and Schmidt V., Phys. Rev. Lett. **70** (1993) 3008
7. Schwarzkopf O., Krässig B., Schmidt V., Maulbetsch F., and Briggs J.S., J. Phys. B.**27** (1994) L347
8. Schaphorst S.J., Krässig B., Schwarzkopf O., Scherer N., and Schmidt V., J. Phys. B **28** (1995) L233
9. Kammerling B., and Schmidt V., J. Phys. B. **26** (1993) 1141
10. Schwarzkopf O., Krässig B., and Schmidt V., J. de Physique IV **3** (1993) 169
11. Dawber G., Avaldi L., Mc Conkey A. G., Rojas H., Mac Donald M.A., and King G.C., J. Phys. B. **28** (1995) L271
12. Berakdar J., and Klar H., Phys. Rev. Lett. **69** (1992) 1175
13. Berakdar J., Klar H., Huetz A., and Selles P., J. Phys. B. **26** (1993) 1463
14. Huetz A., Lablanquie P., Andric L., Selles P., and Mazeau J., J. Phys. B. **27** (1994) L13
15. Lablanquie P., Mazeau J., Andric L., Selles P., and Huetz A., Phys. Rev. Lett. **74** (1995) 2192
16. Huetz A., Selles P., Waymel D., and Mazeau J., J. Phys. B. **24** (1991) 1917
17. Wannier G.H., Phys. Rev. **90** (1953) 817
18. Kazansky A.K., and Ostrovsky V.N., Phys. Rev.A **51** (1995) 3698
19. Maulbetsch F., and Briggs J.S., J. Phys. B. **27** (1994) 4095
20. Pont M. and Shakeshaft R., Phys.Rev.A **51** (1995) R2671

21. Waymel D., Andric L., Mazeau J., Selles P. and Huetz A., J. Phys. B. **26** (1993) L123
22. Kazansky A.K. and Ostrovsky V.N., Phys. Rev. A. **51** (1995) 3712
23. Kazansky A.K. and Ostrovsky V.N., J. Phys. B. **28** (1995) 1453
24. Kazansky A.K. and Ostrovsky V.N., Phys. Rev. A., in press
25. Schaphorst S.J., Krässig B., Schwarzkopf O., Scherer N., Schmidt V., Lablanquie P., Andric L., Mazeau J. and Huetz A., submitted to Journal of Electron Spectroscopy.
26. Greene C.H. and Rau A.R.P., J. Phys. B. **16** (1983) 99
27. Krässig O., Universität Freiburg, Ph.D Thesis, (1994)
28. Stauffer A.D., Phys. Lett. **91A** (1982) 114
29. Selles P., Mazeau J. and Huetz A., J. Phys. B. **20** (1987) 5183
30. Mazeau J., Lablanquie P., Andric L, Selles P. and Huetz A., to be published
31. Kabachnik N.M., J. Phys. B. **25** (1992) L289
32. Schmidt V., Proc; 10th VUV Int. Conf., Paris (Saclay:SDEM) (1992)
33. Vegh L., Becker R.L. and Macek J.H., abstract, 15th X-ray Int.Conf., Knoxville Tennessee, (1990)
34. Vegh L. and Macek J.H., Phys. Rev. A **50**, (1994) 4031
35. Schwarzkopf O., Universität Freiburg, Ph.D Thesis, (1995)
36. De Gouw J.A., Van Eck J., Van Der Aart S. and Heideman H.G.M., J. de Physique IV **3** (1993) 207

# ELECTRON-ATOM COLLISIONS

Electron Collisions with Atoms in Excited States .............. 155
    C.C. LIN
Recent Progress in Polarized-Electron Scattering
from Atoms and Molecules ...................................... 163
    S. MAYER
The Role of Lasers in Electron-Atom Collision Physics ....... 173
    M. STANDAGE
Laser-Assisted Electron-Atom Collisions ....................... 189
    B. WALLBANK

# Electron Collisions with Atoms in Excited States

## Chun C. Lin

*Department of Physics, University of Wisconsin, Madison, Wisconsin 53706*

**Abstract.** Cross sections for electron excitation of helium metastable levels to higher levels have been measured. When the metastable atoms are generated by a hollow cathode discharge, the measurements are limited to incident electron energies below 16 eV. An alternate target source utilizes the charge-exchange reactions between $He^+$ ions and alkali atoms which yield predominately He atoms in the metastable levels. With this target source cross section measurements have been made with electron energies up to 300 eV. Recently laser-cooled atom traps have been used to measure electron scattering cross sections and electron-impact ionization cross sections. This technique may be well suited for studying electron collisions with excited atoms.

Electron excitation of atoms out of excited levels is an important basic process in a variety of natural and laboratory phenomena. However, most of the experiments on electron excitation reported in the literature dealt with excitation out of the ground level. Only in recent years have systematic studies of excitation out of excited states have been successfully made.

## ELECTRON EXCITATION OUT OF THE METASTABLE LEVELS OF HELIUM

Of great fundamental importance is the excitation out of the $2^1S$ and $2^3S$ levels of He. In a series of experiments initiated by Rall *et al.* [1]-[3] the metastable atoms are generated in a hollow cathode discharge. The He atomic beam emerging from the discharge contains $He(2^3S)$ metastable atoms at a concentration of about 3 parts in $10^5$. The ratio of the $He(2^1S)$ atoms to $He(2^3S)$ atoms varies from 1:3 to 1:16 depending on the discharge conditions. The metastable atom number density is typically $3\times10^9/cm^3$. An electron beam at an energy below 16 eV crosses the atomic beam perpendicularly and excites the metastable atoms to the higher levels. The intensities of the radiation from the various levels are measured to obtain the excitation cross sections. Since the electron energy is below the threshold of excitation out of the ground level, the observed emission is entirely due to electron collisions with the metastable atoms. When the relative abundance of the He($2^1$S) to the He($2^3$S) is as small as 1:10, only the He($2^3$S) metastables play a significant role in the excitation into the higher triplet levels because there are much fewer He($2^1$S) atoms and because a spin-changing excitation ($2^1$S → $n^3$L) is less favorable than the corresponding spin-conserving excitation ($2^3$S → $n^3$L). On the other hand, both kinds of metastables may contribute comparably to the excitation into the singlet levels even when the triplet metastables are present in a much higher concentration than the singlet metastables.

Interference filters are used to spectrally isolate the emission from a particular excited level of interest, and the intensity of the transition is measured by a photomultiplier tube (PMT). Because of the extremely low emission intensity, great care must be taken to keep any scattered light from reaching the optical detection system. Determination of the absolute cross sections requires a knowledge of the concentrations of the metastables. This is accomplished by measuring the laser-induced fluorescence (LIF) resulting from the $2^3$S→$3^3$P and $2^1$S→$4^1$P absorption. Details of the apparatus and the experimental procedure have been published [3].

Cross sections for electron excitation out of the metastable levels of He into the various higher singlet and triplet levels in the energy range from threshold to 16 eV have been reported in Refs. [1] and [2]. Of special interest is that the excitation cross section from the $2^3$S into the $3^3$P level is smaller than the cross section for excitation into the $3^3$S and the $3^3$D levels [1], in contrast to excitation out of the ground level in which excitation corresponding to a dipole-allowed transition is more favorable. Recently we have measured the cross section for excitation from the $2^3$S into the $2^3$P level which is found to be more than 50 times larger than the cross section from the $2^3$S into $3^3$P at 10 eV. The extraordinary large cross section for the $2^3$S→$2^3$P excitation may be attributed to the very large dipole matrix element connecting the two levels.

The very low metastable concentration within the He-atom beam emerging from the hollow cathode limits our experiments to electron energies below 16 eV, because if the incident energy exceeds the threshold of ground-level excitation, the observed emission from the $n^1$L or $n^3$L level would be overwhelmingly due to excitation out of the ground level. To overcome this limitation we must have a metastable target source in which the metastable number density is at least comparable to the ground-level atom density. This is accomplished by using charge-exchange reactions to produce the metastable atoms, i. e.,

$$He^+ + Cs \rightarrow He^* + Cs^+. \tag{1}$$

The He atoms formed in this reaction are mostly in the metastable levels because of the near-resonance energy relation. A new apparatus has been built utilizing this method for producing the metastable atoms. Figure 1 shows a schematic diagram of

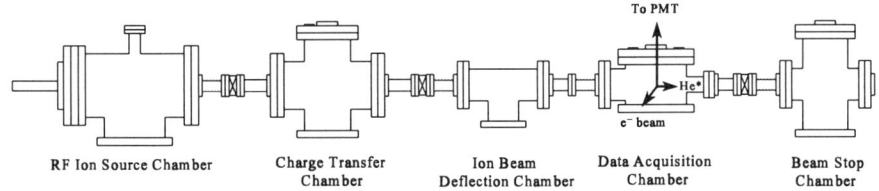

**FIGURE 1.** Schematic diagram of fast-beam metastable atom apparatus.

this apparatus. The He$^+$ ion beam is generated by a radio-frequency discharge. After being accelerated to about 1.6 keV, the ion beam enters the charge transfer chamber which contains Cs vapor. In the third chamber the He$^+$ ions are deflected away and the remaining metastable He beam moves onto the next chamber where collision with an intersecting electron beam takes place. The last chamber serves as a beam stop where the neutral beam flux is monitored via secondary electron ejection. Formation of the excited atoms by electron impact is again monitored optically. For absolute measurements of the cross sections a thermal detector is used to obtain the absolute neutral flux, and the profiles of the metastable atom beam and of the electron beam are determined as well.

Preliminary results of the cross sections for excitation out of the $2^3$S metastable level into the $3^3$D level are shown in Figure 2.

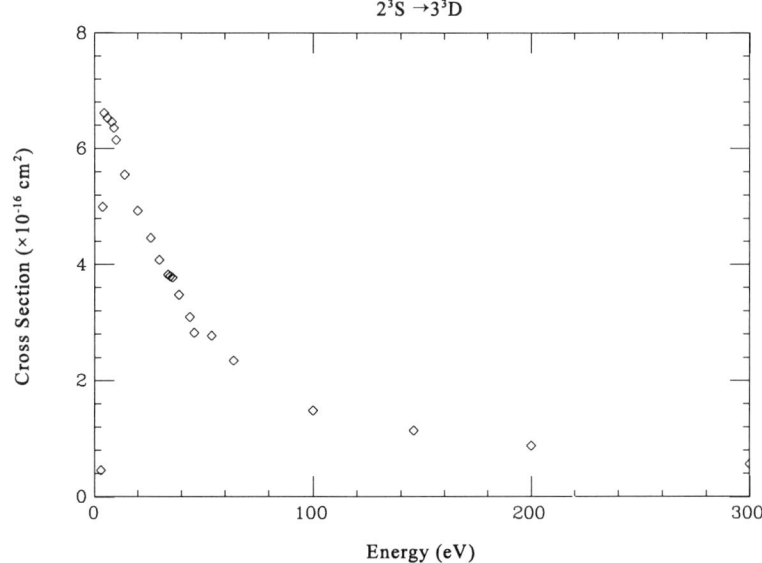

**FIGURE 2.** Excitation function for electron excitation of $2^3$S metastable helium into the $3^3$D level.

# USE OF ATOM TRAPS FOR MEASURING ELECTRON COLLISION CROSS SECTIONS

Recent development of the use of laser-cooled atoms in a magneto-optical trap for measuring electron scattering cross section has provided a potentially powerful means of studying electron excitation out of excited levels [4]. Imagine an electron beam passing through very cold Rb atoms inside a magneto-optical trap. An electron-atom collision imparts a recoil velocity to the Rb atom large enough that the atom is lost from the trap except for collisions with nearly zero momentum transfer. Thus the total electron scattering cross section can be determined by measuring the percentage loss of trapped atoms due to the electron beam. In the experiment reported in Ref. [4], there are about $10^6$ trapped Rb atoms within a volume of about $10^{-4}$ cm$^3$ at a temperature of 0.1 mK. The trapping region which is dictated by the size of the laser beams, however, is much larger, about 1 cm in diameter. The principle of the experiment is as follows. The trap originally containing $N_1$ atoms is turned off at time $t_1$ as shown schematically in Fig. 3. This is followed by a pulsed electron beam of duration T of the order of a few ms. After a delay time $t_d$ the trap is turned back on recapturing all the atoms within the trapping region. The number of trapped atoms is now decreased to $N_2$ because the atoms with large enough recoil velocity to exit the trapped region within the delay time $t_d$ are not recaptured. When $t_d$ is so large that even the scattered atoms with very small momentum transfer have enough time to escape the trap region, $N_1 - N_2$ accounts for all the scattered atoms. However, when the trap is turned off, collisions of the trapped atoms with the ambient warm atoms may also result in a loss even in the absence of the electron beam. Thus the decrease in the trapped atoms is

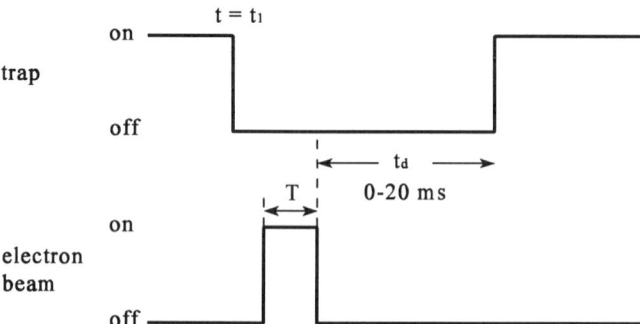

**FIGURE 3.** Timing sequence for the experiment using atom traps for measuring electron collision cross sections.

measured with and without the electron beam so that the difference between the two measurements is entirely due to electron collision. The experiment is repeated using increasingly larger values of $t_d$ until the loss rate due to electron collision reaches an asymptotic value. Since the trapped atoms are within a volume much smaller than the size of the electron beam, it is necessary to determine the current density of the electron beam at the location of the trapped atoms. Combining the current density with the fractional loss rate due to the electron beam gives the total scattering cross section which includes both elastic and inelastic collisions. The full experimental procedure has been described in Ref. [4]. The total scattering cross section for Rb has been measured by this method for electron energy from 7 to 500 eV [4], and the results are in good agreement with those of Visconti, Slevin, and Rubin [5] obtained by the atomic beam recoil method and with those of Parikh *et al.* by the electron beam transmission technique [6].

The use of atom traps to measure electron collision cross sections has several advantages. Since the cross section depends only on the percentage loss rate and the current density, it is not necessary to make absolute measurement for the target density which is difficult for alkali atoms. Thus the relative values of $N_1$ and $N_2$ are simply obtained by monitoring the 5p→5s fluorescence resulting form the absorption of the resonant radiation from the cooling laser, and the absolute values are not needed. The atomic beam recoil technique developed by Bederson and his co-workers [7] has the same advantage of requiring relative rather than absolute target number density. However, in crossed beam experiments the atoms with small-angle scattering are usually difficult to distinguish from those in the incident beam. The use of a trapped-atom target in place of an atomic beam allows us to fully account for the small-angle scattering in the total cross section. The construction of magneto-optical traps has been streamlined in recent years, and the procedure for determining the cross sections is straightforward.

As an illustration of the versatility of the trapped-atom target, suppose we set $t_d = 0$ in Fig. 3 for the electron collision experiment. When the laser is turned back on immediately after the electron-beam pulse, most the scattered atoms are recaptured except those with very large recoil velocity so that they can exit the trap region within the duration of the electron-beam pulse T. Loss of this kind can be reduced by decreasing T and extrapolating the results to the limiting case of T = 0. However, if an atom is ionized by electron impact, the resulting ion would not be affected by the cooling laser (which is tuned to the 5s→5p absorption of the neutral atom) and therefore would escape the trap. Furthermore, the relative number of Rb atoms in the trap is determined by the Rb (5p→5s) fluorescence, thus the Rb$^+$ ions would not contribute to this signal even if they stay in the trapped region. The fractional loss of the trapped atoms, $(N_1-N_2)/N_1$, can then be utilized to determine the ionization cross section. In practice instead of measuring directly $N_1-N_2$ which is very small, we measure the growth curve of the number of atoms (N) in the trap as we start loading the trap beginning with N=0 at t=0, in the manner described in Ref.

[4]. The time-derivative of N depends on the loading rate and the total loss rate, whereas the time constant in the exponential growth of N(t) depends only on the total loss rate as illustrated in Eqs. (2) and (3) of Ref. [4]. By analyzing the growth curve obtained without the pulsed electron beam, we determine the loss rate due to collisions with the warm ambient atoms. The experiment is repeated with the pulsed electron beam, and the resulting growth curve yields the combined loss rate due to both collisions with the warm atoms and electron-impact ionization. This allows us to isolate the loss rate due to electron-impact ionization alone. We emphasize that in order to extract the time constant from the growth curve, we need only the shape of the N(t) function but not the absolute magnitude. This is an important advantage since we determine the time constant of the growth curve by simply measuring, as a function of time, the intensity of the 5p→5s fluorescence resulting from the absorption of the resonant radiation from the cooling laser which is proportional to N(t). The time constants from the growth curves (with and without the electron beam) give the fractional loss rate due to the electron beam ($\Gamma_e$) similar to the procedure employed in Ref. [4]. The ionization cross section is equal to $e\,\Gamma_e/J$ where e is the magnitude of electron charge and J is the current density of the electron beam at the volume of the trapped atoms which is much smaller than the size of the electron beam. The method for measuring this current density has been described in Ref. [4].

We have carried out the experiment outlined in the preceding paragraph. At an electron energy of 500 eV we obtain a cross section of $2.4\times10^{-16}$ cm$^2$. In our experiment the "loss" includes the formation of all kinds of ions, i.e., $Rb^+$, $Rb^{++}$, $Rb^{3+}$, .... Therefore the cross sections we determine are the "total ionization cross sections" which is the sum of the cross sections for producing $Rb^+$, $Rb^{++}$, ... by electron impact. A full account of the experiment and comparison with the results of previous works will be reported in a forthcoming paper.

It is interesting to compare our method of using trapped-atom target with the crossed beam method for measuring the electron-impact ionization cross section. The latter requires a knowledge of the number of target atoms reaching the detector per unit time, the electron beam current, the overlap of the atomic and electron beams in the collision region, and the ion current produced by electron-impact ionization [8]. Measurement of the absolute number of ground-state neutral atoms is generally difficult. Furthermore the atomic beam does not have a uniform density, thus a measurement of the spatial distribution of the atoms in the beam is needed. The use of trapped-atom target avoids a major part of these complications. As remarked earlier, our method does not require absolute number of target atoms. The volume of the target (diameter ~ 0.5 mm) is so small that within the overlap region between the electron beam and the trapped atoms, the current density is a constant. The data reduction is very simple as the ionization cross section is simply equal to the electron charge times the fractional loss rate divided by the current density.

So far we have used atom traps for measuring cross sections for electron collisions with atoms in the ground states. The nature of the magneto-optical traps makes them well suited as target sources for excited atoms. Works are underway to use atom traps for electron collision experiments with excited atoms.

## ACKNOWLEDGMENTS

This work is supported by the U.S. National Science Foundation and the U.S. Air Force Office of Scientific Research.

## REFERENCES

1. Rall, D.L.A., Sharpton, F.A., Schulman, M.B., Anderson, L.W., Lawler, J.E., and Lin, C.C., Phys. Rev. Lett. **62**, 2253-2256 (1989).
2. Lockwood, R.B., Sharpton, F.A., Anderson L.W., and Lin, C.C., Phys. Lett. A **166,** 357-360 (1992).
3. Lockwood, R.B., Anderson, L.W., and Lin, C.C., Z Phys. D **24**, 155-160 (1992).
4. Schappe, R.S., Feng, P., Anderson, L.W., Lin, C.C., and Walker, T., Europhys. Lett. **29**, 439-444 (1995).
5. Visconti, P.J., Slevin, J.A., and Rubin, K., Phys. Rev. A **3**, 1310-1317 (1971).
6. Parikh, S.P., Kauppila, W.E., Kwan, C.K., Lukaszew, R.A., Przybyla, D., Stein, T.S., and Zhou, S., Phys. Rev. A **47**, 1535-1538 (1993).
7. Rubin, K., Perel, J., and Benderson, B., Phys. Rev. **117**, 151-158 (1960).
8. See, for example, deHeer, F.J. and Inokuti, M. in *Electron Impact Ionization*, edited by Märk, T.D. and Dunn, G.H. (Springer-Verlag, New York, 1985). p.242-243.

# Recent Progress in Polarized-Electron Scattering from Atoms and Molecules

Stefan Mayer

*Physikalisches Institut, Universität Münster,*
*Wilhelm-Klemm-Str. 10, D-48149 Münster, Germany*

---

A selection of the results which have recently been obtained at the Universität Münster by scattering experiments with polarized electrons is presented. Experimental data are given for inelastic scattering from argon and mercury atoms and are compared to calculations. The accuracy of two electron polarimeters, the Mott detector and the helium polarimeter, is discussed. The paper concludes with a presentation of recent results obtained by electron scattering from chiral molecules. The results of a recent transmission experiment on the attenuation of polarized electron beams in the vapor of chiral molecules confirm the existence of electron optic dichroism.

---

## INTRODUCTION

Spin polarized electrons have proved to be a versatile tool for the investigation of electron-atom and electron-molecule collisions since many years. By using polarized electrons as projectiles and/or measuring the polarization of the scattered electrons collision processes can be resolved into their various reaction channels and the different interaction mechanisms can be disentangled. It is thus possible to study the weak spin-dependent interactions between electrons and atoms which are usually masked by the much stronger Coulomb force. In special cases one can even make "complete" scattering experiments which yield the maximum possible information on the collision studied [1].

Due to the enormous progress which has recently been achieved in the development of efficient polarized-electron sources and accurate polarimeters it is nowadays possible to explore also complicated processes like inelastic scattering. The first part of this paper is focused on inelastic processes, which are difficult to examine because of the large number of possible reaction channels and because of low cross sections.

Furthermore it has now become possible to study the extremely small polarization effects which are expected to occur in electron scattering from unoriented chiral molecules. This will be shown in the last part of this paper after a short intermezzo about recent progress in electron polarimetry.

© 1995 American Institute of Physics

# INELASTIC ELECTRON SCATTERING FROM ATOMS

Spin dependent effects in elastic electron scattering from spinless atoms can only be produced by spin-orbit interaction of the projectile electrons (Mott scattering) [1]. The scattering process is described by two complex amplitudes which can be completely determined by a set of scattering experiments with polarized electrons.

The situation is much more complicated when inelastic processes are considered, even if the atoms do neither have spin nor orbital angular momentum in the ground state. This can be understood as follows: The atoms may gain angular momentum by the excitation, and different magnetic sublevels of the selected excited state can be populated, thus increasing the number of scattering amplitudes which are necessary to describe the process. In general, the scattering amplitudes cannot be completely determined by observing only the scattered electrons because some of the information about the scattering process is locked up in the excited atoms which are not observed. Furthermore it is no longer Mott scattering alone which produces spin dependent effects in inelastic scattering. It has been shown that also spin-orbit coupling of the bound electrons in combination with electron exchange can result in polarization effects if the excited state possesses orbital angular momentum (see e. g. [1] and references therein). This mechanism is sometimes called "fine-structure effect" because it results in typical polarization effects if the fine structure of the excited atoms is resolved in the experiment.

A well-known method of obtaining information about spin-dependent electron scattering is the measurement of the asymmetry function $S_A$ which describes an asymmetry

$$A_{LR} = \frac{L-R}{L+R} = S_A P \tag{1}$$

of the intensities $L$ and $R$ scattered to the left and right, respectively (see fig. 1). Such an asymmetry can occur if the incoming beam has a polarization

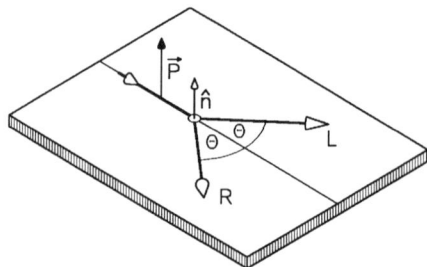

**FIG. 1.** Scattering geometry for the measurement of the asymmetry function $S_A$

component $P$ perpendicular to the scattering plane. The asymmetry function $S_A$ depends on the scattering angle $\Theta$ and the electron energy.

Fig. 2 shows some of the results which have been obtained recently for inelastic electron scattering from argon atoms. The experimental values are compared with calculations from Refs. [2] and [3]. The following conclusions can de drawn from the figure:

i) With increasing electron energy $S_A$ becomes smaller for the $^3P_1$ and particularly the $^1P_1$ excitation, as can be seen from the measured points and the theory curves as well. This indicates that electron exchange, which is less probable at higher energies, plays an important role in the generation of spin effects in the excitation of argon.

ii) It can be seen that the experimental data approximately obey the rule

$$S_A(^1P_1) = -S_A(^3P_1) . \qquad (2)$$

This equation can be derived within the approximations that the excited states are described in an LS- or intermediate coupling scheme and that the spin-orbit interaction of the projectile electrons is negligible [1]. The validity of eq. (2) for argon is a strong hint that spin effects in inelastic electron scattering

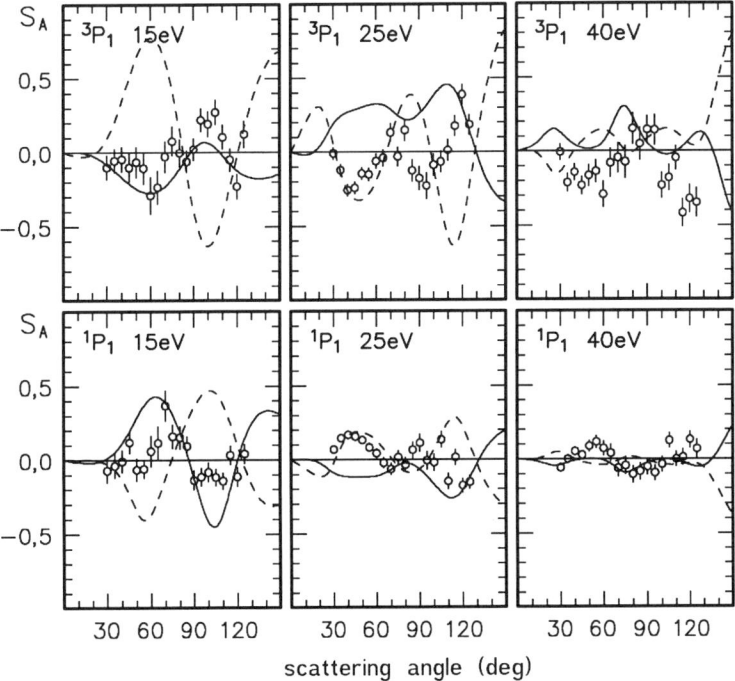

**FIG. 2.** Asymmetry function $S_A$ for inelastic electron scattering from argon atoms. Experimental values: o [4], calculations: ——— [2] and - - - [3].

from argon are predominantely produced by the aforementioned fine-structure effect.

iii) There is a strong disagreement between both sets of theoretical data and between theory and experiment. This disagreement is not yet understood.

Asymmetry measurements were also made for krypton and xenon. Whereas for krypton similar conclusions can be drawn as for argon, the xenon data show that Mott scattering can no longer be neglected and that the intermediate coupling scheme is not a good approximation for the heavy xenon atom. The total work on noble gas targets will be published soon [4].

More information about the spin-dependent scattering can be obtained by analyzing the polarization of the scattered electrons. Such experiments are difficult because one has to deal with low counting rates due to the low efficiency of common electron polarimeters and due to the small cross sections for inelastic scattering. Fig. 3 shows a sketch of the apparatus which has been used for polarization measurements of electrons scattered inelastically

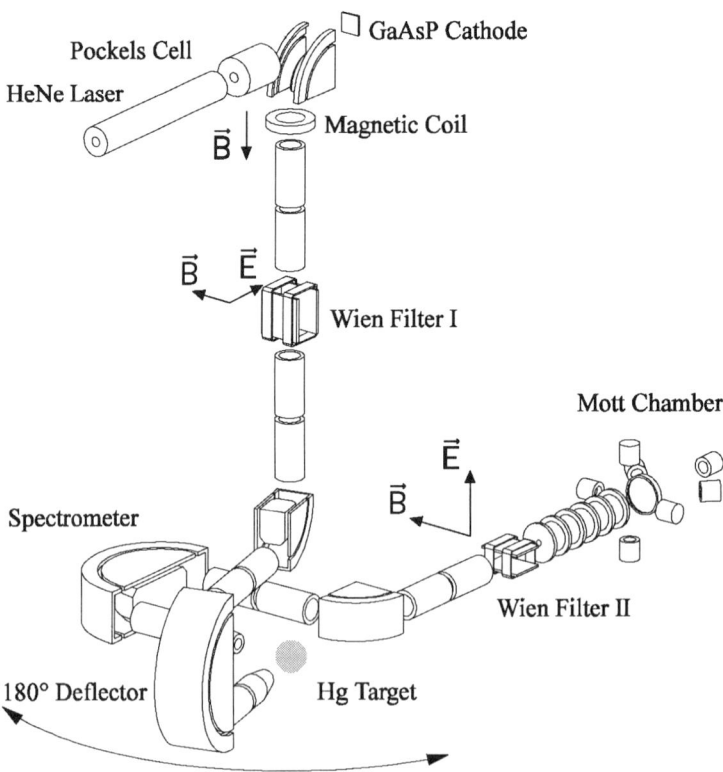

**FIG. 3.** Schematic view of the apparatus for polarization analysis of electrons scattered inelastically from mercury atoms

from mercury atoms [5,6]. It incorporates a polarized-electron photoemission source, spin manipulators for the incident and scattered beam, and a Mott electron polarimeter. The 180°-deflector, through which the incident beam is focused onto the atoms, can be swiveled around the mercury target beam, whereas the spectrometer for the energy selection of the scattered electrons is fixed in space.

In order to get the maximum information which can be obtained from the scattered electrons eight observables have to be measured. These observables are the scattering cross section for unpolarized electrons and seven polarization parameters, the so-called "generalized $STU$-parameters", which give the relation between an arbitrary incident polarization and the polarization of the scattered beam. A theoretical treatment of this situation was given e. g. by Bartschat and Madison [8]. A complete set of polarization parameters has recently been measured for singlet and triplet excitation of mercury atoms. A small selection of the experimental data is shown in fig. 4. The physical meaning of the parameters $T_Z$ and $U_{XZ}$ can be visualized as follows: If the polarization of the incident beam is parallel to the beam direction, i. e. longitudinal, $T_Z$ and $U_{XZ}$ describe the rotation of the polarization vector in the scattering plane. The rotation angle $\alpha$ is given by

$$\tan \alpha = \frac{U_{XZ}}{T_Z}. \tag{3}$$

Additionally, the polarization may be rotated out of the scattering plane and

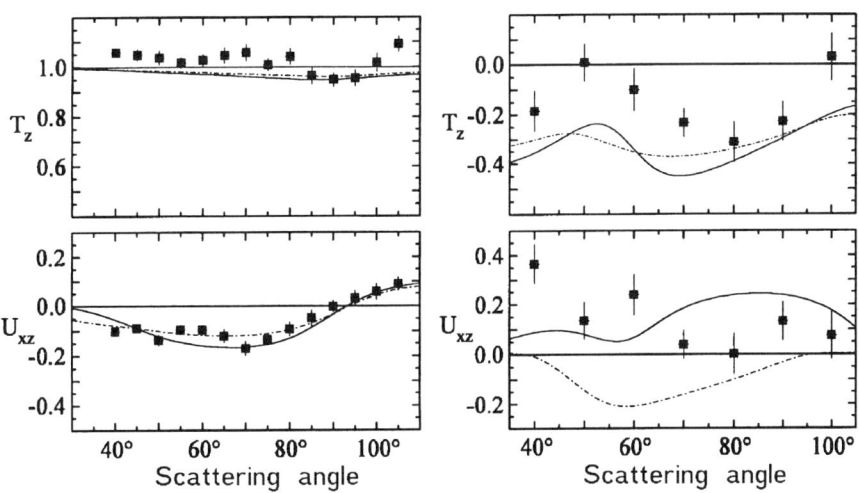

**FIG. 4.** Polarization parameters $T_Z$ and $U_{XZ}$ for the excitation of the Hg states $6\ ^1P_1$ (left) and $6\ ^3P_2$ (right) at 40 eV collision energy. The error bars give the statistical uncertainty of the measured points. Calculations: - - -, fits through numerical data from [7]; ———, data from [8].

its absolute value may change. This behaviour is described by further parameters which cannot be explained here.

## ELECTRON POLARIMETRY

It is clear that the accuracy of the electron polarimeter plays an essential role in experiments like the measurement of $STU$-parameters. In most experiments described in the literature polarization analysis is done by Mott scattering from heavy targets. An opinion found even in the newest literature is that the accuracy of this method is fundamentally limited to ~5%. This is of course not true! In one of our laboratories we use a Mott detector which has been experimentally calibrated with an accuracy of 0.3% and which is capable of detecting a polarization as low as $10^{-3}$. The details of our work on Mott polarimetry, which has already been published some years ago [9–11] and which clears up with the aforementioned prejudice, cannot be explained here. However, it should be mentioned that the accurate experimental calibration of a Mott detector is not an easy task and that in some existing applications it is even impossible to perform a calibration experiment without a considerable modification of the apparatus.

An approach which was thought to overcome the calibration problem of electron polarimeters is the so-called helium polarimeter [12] which can, in principle, be calibrated by theory. The helium polarimeter works as follows: The electron beam to be analyzed excites helium atoms into a triplet state. Since this is possible by exchange scattering only, the electron polarization is transferred to the atoms. The total angular momentum of the atoms becomes then oriented by spin orbit coupling. As a consequence, the radiation, which is emitted when the excited states decay, is polarized, and the light polarization can be taken as a measure for the electron polarization. The relation between the light polarization and the electron polarization, i. e. the analyzing power of the polarimeter, can be calculated for the threshold energy of the considered excitation.

In one of our projects [13] such a helium polarimeter was compared to a calibrated Mott detector. The relative accuracy of the helium polarimeter was found to be limited to ~2% by the uncertainties of the light polarization measurement. Furthermore it turned out that the analyzing power of the helium polarimeter shows an unexpected strong dependence on the electron energy especially near the threshold energy so that the theoretical calibration only applies at energies very close to the threshold and for electron beams of a narrow energy width. Unfortunately, the efficiency of the helium polarimeter is very low at the threshold, which makes it impossible to calibrate the polarimeter by theory within an accuracy much better than 10%.

However, a helium polarimeter is a good choice when an in-beam polarization monitor of moderate accuracy is required since the helium target does not much affect the intensity and shape of the electron beam.

# ELECTRON SCATTERING FROM CHIRAL MOLECULES

Chiral molecules, i. e. molecules which can be distinguished from their mirror image, are found in nature mainly in one handedness. For instance, nature produces only right-handed sugars and predominantly left-handed amino acids. When the violation of parity in nuclear $\beta$-decay had been dicovered in the mid-fifties, the idea came up that the longitudinal polarization of $\beta$-particles may be responsable for the origin of the molecular handedness in nature. Vester and Ulbricht [14] suggested in 1957 that destruction of biomolecules by polarized $\beta$-particles or by their circularly polarized bremsstrahlung could have tipped the balance between right- and left-handed molecules in an early stage of the Earth's evolution so that, as a consequence of some type of biological amplification, only one handedness survived.

Radiolysis experiments which have been made in the past, e. g. with polarized electrons from $\beta$-sources, in order to test this hypothesis showed rather inconclusive results. Therefore it seemed to be worthwhile to investigate the spin dependent interaction between electrons and chiral molecules under the simplest possible conditons. Farago [15] proposed a transmission experiment in order to study the spin dependent attenuation of electron beams in the vapor of chiral molecules (see fig. 5). He postulated from symmetry arguments an asymmetry

$$A = \frac{I(P) - I(-P)}{I(P) + I(-P)} \quad (4)$$

in the transmitted intensities $I(P)$ and $I(-P)$ of longitudinally polarized electrons of polarization $P$, an effect which he christened "electron optic dichroism" in analogy to circular dichroism in light optics.

Figure 6 shows the results of such a transmission experiment which we made during the last few months. The target density in the gas cell was adjusted to attenuate the incident beam by a factor of $\sim 10$ and the electron polarization $P$ was measured to be $0.4 \pm 0.01$. No transmission asymmetry

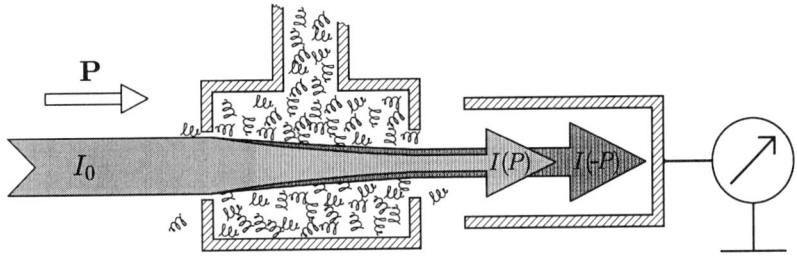

**FIG. 5.** Principle of the transmission experiment for the detection of electron optic dichroism

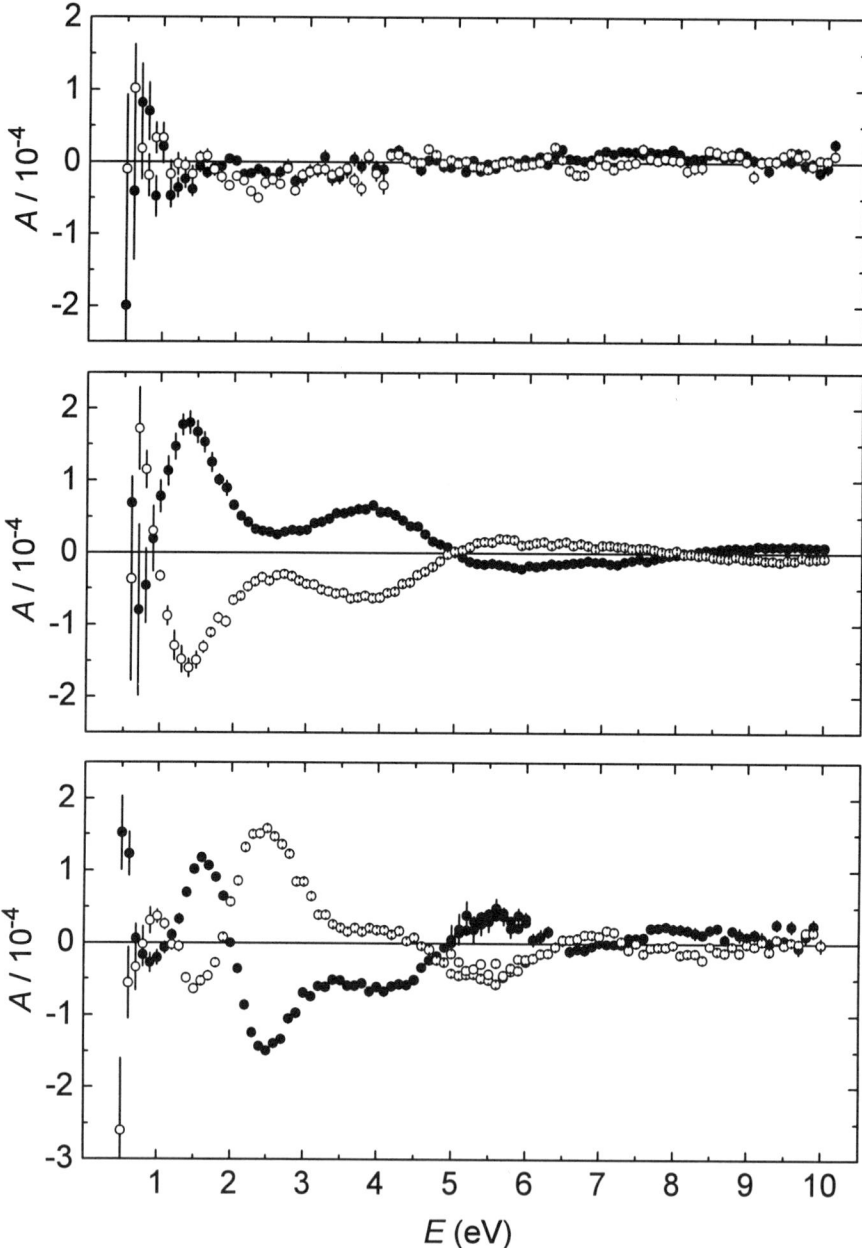

**FIG. 6.** Transmission asymmetry $A$ measured for the L-enantiomer (○) and the D-enantiomer (●) of camphor (upper diagram), bromocamphor (middle), and Yb(hfc)$_3$ (lower diagram) vs electron energy $E$. The error bars indicate the statistical uncertainty.

was found for camphor, thus contradicting a result obtained within a much lower accuracy by Campbell and Farago [16]. The negative result for camphor has also been confirmed by Gay [17]. However, asymmetries clearly above the detection limit were measured for Yb(hfc)$_3$ [18] and bromocamphor [19]. The measured asymmetries show opposite signs for L- and D-enantiomers, as is required by symmetry. Pronounced maxima and minima of the asymmetry $A$ can be seen at several values of the electron energy where resonances have also been detected in a different experiment [20]. The values measured for Yb(hfc)$_3$, a camphorlike molecule containing an Ytterbium atom, are the first experimental proof of electron optic dichroism. Our results are in good agreement with calculations [21] which predict asymmetries $\geq 10^{-4}$ only for molecules containing heavy atoms.

Although the principle of the experiment is simple the difficulties in measuring asymmetries below $10^{-4}$ even at energies below 1 eV should not be underestimated. In order to make such measurements feasible the stability of our polarized-electron beam, which was obtained from a GaAs photoemission source, had to be considerably improved. Among the necessary improvements the most important ones were thorough adjustment and temperature stabilization of the source's optical system and the installation of a feedback-controlled magnetic field compensation system. Another problem which had to be solved was condensation of target material inside the electron source. Since bromocamphor and Yb(hfc)$_3$ must be heated in order to reach a sufficient vapor pressure a special target chamber was constructed which can be heated up to 200 °C by halogen lamps. The chamber contains differentially pumped electron optics and electron deflectors that prevent the target molecules, which leave the gas cell through the electron entrance aperture, from contaminating the electron source.

## ACKNOWLEDGMENTS

The author wishes to thank Prof. J. Kessler and Prof. G. F. Hanne for their valuable support and all members of the Münster Polarized-Electron Group for fruitful cooperation. Stimulating discussions with Prof. K. Blum are gratefully acknowledged. This work has been supported by the Deutsche Forschungsgemeinschaft in the Sonderforschungsbereich 216.

## REFERENCES

1. Kessler, J., *Adv. At. Mol. Opt. Phys.* **27**, 81 (1991).
2. Bartschat, K., and Madison, D. H., *J. Phys. B* **20**, 5839 (1987), (erratum *J. Phys. B.* **25**, 1361 (1992)).
3. Zuo T., McEachran, R. P., and Stauffer, A. D., *J. Phys. B* **24**, 2853 (1991).
4. Dümmler, M., Hanne, G. F., and Kessler, J., *J. Phys. B.*, in press.

5. Müller, H., and Kessler, J., *J. Phys. B.* **27**, 5933 (1994), (corrigendum *J. Phys. B.* **28**, 911 (1995)).
6. Klose, M., and Kessler, J., to be published.
7. Srivastava, R., Zuo, T., McEachran, R. P., and Stauffer, A. D., *J. Phys. B.* **25**, 2409 (1992).
8. Bartschat, K., and Madison, D. H., *J. Phys. B.* **21**, 2621 (1988).
9. Gellrich, A., Jost, K., and Kessler, J., *Rev. Sci. Instr.* **61**, 3399 (1990).
10. Gellrich, A., and Kessler, J., *Phys. Rev. A* **43**, 204 (1991).
11. Mayer, S., Fischer, T., Blaschke, W., and Kessler, J., *Rev. Sci. Instr.* **64**, 952 (1993).
12. Gay, T. J., *J. Phys. B* **16**, L553 (1983).
13. Fischer, T., and Kessler, J., *Rev. Sci. Instr.*, in press.
14. Ulbricht, T. L. V., and Vester, F., *Tetrahedron* **18**, 629 (1962).
15. Farago, P. S., *J. Phys. B* **14**, L743 (1981).
16. Campbell, D. M., and Farago, P. S., *J. Phys. B* **20**, 5133 (1987).
17. Gay, T. J., (private communication); in *Proceedings of the Symposium on Electron Physics*, Edinburgh, 1995 (unpublished).
18. Mayer, S., and Kessler, J., *Phys. Rev. Lett.* **74**, 4803 (1995).
19. Nolting, C., and Kessler, J., to be published.
20. Nolting, C., *Thesis* Universität Münster (1993).
21. Fandreyer, R., Thompson, D., and Blum, K., *J. Phys. B* **23**, 3031 (1990).

# The Role of Lasers in Electron-Atom Collision Physics

## Max Standage

*Laser Atomic Physics Laboratory, School of Science, Griffith University, Brisbane, Queeensland 4111, Australia*

**Abstract.** A review is presented on the application of lasers to the field of electron-atom collisions. Developments in the theoretical treatment of the interactions between laser light and atoms are reviewed with respect to both spectroscopic processes and optically induced mechanical forces. A review of experimental developments in superelastic scattering, stepwise electron/laser excitation, laser assisted collisions, excited state (e,2e) momentum spectroscopy, optical traps and photon deflection techniques is also presented.

## INTRODUCTION

The introduction of tunable dye lasers to the field of atomic physics in the early 1970's brought about rapid advances in theory and experiment in the field of spectroscopy and quantum optics. At the same time, the application of lasers to atomic collision physics led to the introduction of a number of new techniques in this field. The well-known properties of lasers; their tunability, narrow spectral bandwidth, optical power and polarization have all been employed in the development of new methods for investigating atomic collision processes. Some of these methods have opened up new areas of investigation in atomic collision physics, or provided alternative experimental methods in existing areas.

The first application of lasers in electron-atom collision physics was the pioneering work of Hertel and coworkers, who investigated $e^-$-Na atom collisions. Hertel's group developed the electron superelastic scattering method in the early 1970's, publishing a series of papers[1-3] on further developments of the technique. The superelastic scattering technique has subsequently been extensively used by a number of experimental groups to measure atomic collision parameters for several different target atoms. In 1977, Bederson and coworkers[4] introduced a photon recoil method to investigate low energy scattering of electrons by laser excited Na atoms travelling in a highly collimated atomic beam. The possibility of using the optical force between a laser beam and atoms to measure collisionally induced alignment and orientation parameters has been discussed by Summy et al 1994[5]. Since the late 1970's, stepwise excitation techniques have been developed by several groups, notably C C Lin and coworkers[6-9] and MacGillivray & Standage and coworkers[10-18]. The latter group have been successful in performing the first stepwise electron-photon coincidence experiments[14,15,18]. In 1976, Andrick and Langhans[19] reported the first experiment on laser assisted collision processes involving single photon free-free transitions, otherwise known as inverse Bremsstrahlung processes. In 1977, Weingartshofer et al[20] reported the first observations of multiphoton free-free transitions. More recently, Mason & Newell[21], Weingartshofer et al[22,23] and Wallbank et al[24-27] have performed experiments in which laser-assisted elastic and inelastic electron-atom collisions have been observed.

In 1990, Weigold and coworkers reported[28] on the first experiment in which laser excited state electron momentum spectroscopy, or (e,2e) spectroscopy was carried out on optically pumped Na atoms. In 1995, C C Lin and coworkers have reported[29] on the first electron-atom collision experiment in which a magneto-optical trap has been used. In this work an electron gun was used to excite Rb atoms held in a magneto-optical trap. A very important application of lasers in the field of atomic collision physics has been in the production of spin polarized electrons using a GaAs cathode illuminated by circularly polarized infrared laser radiation. This technology has revolutionized the study of spin effects in atomic scattering physics and recent advances in this field have been reviewed by Kessler and others[30].

## ATOM-LIGHT INTERACTIONS

An essential requirement for the successful application of lasers to atomic collision processes is a good understanding of the interactions between laser light and the target atoms. With conventional light sources, such interactions are described in terms of a weak excitation process in which only one photon interacts with an atom during the lifetime of the excited state and it is then possible to treat the interaction perturbatively. However, there are many cases where the weak excitation approach is not applicable and other theoretical treatments must be used. The simplest approach is based on rate equations, which make use of the Einstein A and B coefficients to describe optical interactions. The limitation with rate equations is that they only deal with populations, or in terms of density matrix theory, diagonal density matrix elements and ignore coherences, or off-diagonal density matrix elements. Several non-perturbative techniques are now available, such as semiclassical density matrix and Heisenberg operator methods that provide an essentially complete description of the excitation of atoms by laser light and include a full treatment of optically pumped populations and coherences. The semiclassical density matrix equations are obtained from the density operator equation of motion[31]:

$$i\hbar \dot{\rho} = [H, \rho] + relaxation\ terms \qquad (1)$$

where H is the semiclassical Hamiltonian and the relaxation terms are added in an ad hoc manner. Farrell et al 1988[32] have generalised the atomic operator method introduced for two-state systems[33,34] in which the relaxation terms arise naturally from the quantum treatment of the optical field. This is a fully quantum-electrodynamic theory for the atomic operator whose time evolution is governed by the Heisenberg equation of motion:

$$i\hbar \dot{\sigma} = [\sigma, H] \qquad (2)$$

where the Hamiltonian is fully quantum mechanical and the operator elements are given by $\sigma_{ij} = |i\rangle\langle j|$, where $|i\rangle$ and $|j\rangle$ are states of the atom.

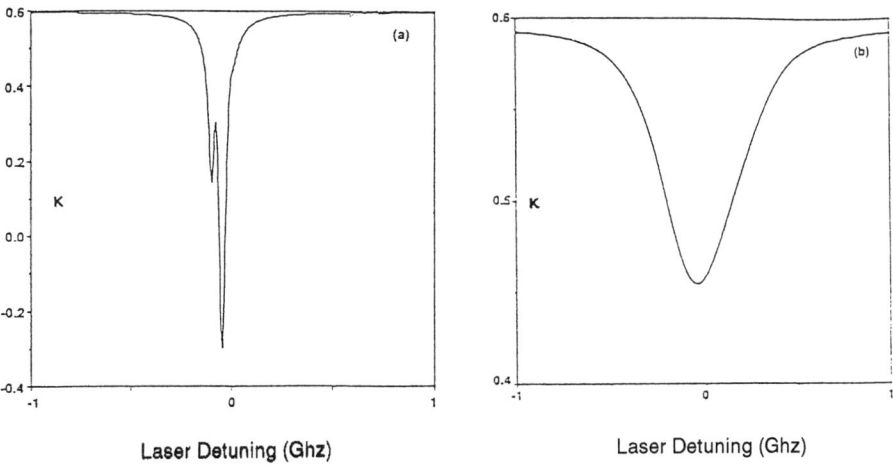

**FIGURE 1.** K calculated as a function of laser detuning for a uniform intensity laser beam. (a) Intensity of 0.2mWmm$^{-2}$ and zero Doppler width; (b) intensity of 50mWmm$^{-2}$ and a Doppler width of 300 MHz. The transit time of the atoms through the laser beam was assumed to be 2µs.

It is perhaps not fully appreciated that computational treatments using such techniques can include arbitrarily complicated atomic structure in the target atoms and all relevant experimental conditions, such as the intensity and detuning of the laser light. It is important to appreciate how sensitive the atomic populations and coherences formed in laser-atom interactions can be to various experimental parameters, such as the tuning and intensity of the laser light, the interaction time of the atom as it passes through the laser beam, and the Doppler width of the atomic beam. All these parameters can have a significant effect on the analysis of the collision parameters in laser based collision experiments. By way of illustration, Figure 1 show the results of a quantum electrodynamical calculation of the line polarization for the fluorescent light scattered perpendicular to the direction of a single mode laser beam, which was performed by Farrell et al[35] for the $3^2S_{1/2}(F=2) \rightarrow 3^2P_{3/2}(F)$ transitions of Na. For such a transition, the line polarization can be shown to be equal to an optical pumping parameter K which has a critical role in superelastic scattering experiments. The calculation used the Heisenberg equation of motion method. The atoms were assumed to be travelling in a beam perpendicular to the direction of the laser beam. The large variations and the structure of the weak intensity curve illustrates the effect of the resolved hyperfine structure at low laser intensity and points to the problem in determining K under such conditions. The results illustrate the importance of carefully considering the effect of such parameters in laser based atomic collision experiments.

Another important aspect of light-atom interactions is the optical force experienced by atoms subjected to laser light. The optical force can arise through one of two mechanisms. The spontaneous emission optical force is caused by absorption of a laser photon with a consequent momentum kick in the direction of the laser beam. If the subsequent emission is spontaneous, then provided it is not in the direction of the laser beam, a net transfer of momentum will occur. On average, the spontaneous emission fluorescence is approximately isotropic, so that after a number of absorptions n have occurred, the atom will have received a momentum which is n times the momentum of a single photon, hf/c. For sodium, one photon recoil produces a change in velocity of approximately $0.1 ms^{-1}$. In a travelling wave laser field, the expectation value of the deflection force, <F(t)> is given by

$$< F(t) > = -i\hbar k \sum_{eg} [\rho_{ge}(t) - \rho_{eg}(t)]\Omega_{eg} / 2 \qquad (3)$$

where k is the wave vector of the laser light, $\Omega_{eg}$ is the Rabi frequency associated with the transition between the states |g> and |e>, and $\rho_{eg}(t)$ and $\rho_{ge}(t)$ are slowly varying off-diagonal density matrix components of the atom that represent the optical coherences. It has been shown, Summy et al[5,36] that the momentum transferred to the atom by this force during a laser-atom interaction in the time interval t=0 to t, can be written as

$$< P(t) > = \sum_{mn} \alpha_{mn}(t) \rho_{mn}(0) \qquad (4)$$

where $\rho_{mn}(0)$ are the density matrix elements for the atom as it enters the laser beam. It can be seen that the transferred momentum depends in principle on the quantum state of the atom as it enters the laser beam. The terms $\alpha_{mn}(t)$ represent the optical pumping of the atom and are calculated using the Heisenberg operator method mentioned previously. The second mechanism makes use of stimulated emission processes in a standing wave laser field. It has been shown[5,36] that the deflection produced in this case is also dependent on the initial quantum state of the atom as it enters the laser beam.

# APPLICATION OF LASER TECHNIQUES TO ATOMIC COLLISION EXPERIMENTS

In this section a survey is presented of atomic collision techniques in which lasers are used. Experiments involving spin polarized beams produced using laser techniques are not included unless some other laser technique is also involved in the experiment.

## Superelastic Scattering Experiments

Since the early 1970's, considerable interest has been focussed on the development of experimental techniques that enable scattering processes to be completely characterised. Such experiments were first realised in the development of two methods, the superelastic scattering technique and the electron-photon coincidence technique of Kleinpoppen and coworkers[37-39]. Although both techniques appear at first sight to be quite different, they are in fact closely related. In the coincidence method, Figure 2(a), the atom is excited by the colliding electron and the subsequently emitted fluorescence photon is detected in coincidence with the electron inelastically scattered at some angle. For an S-P transition, measurement of the Stokes parameters for the coincidently detected photons yields the alignment and orientation of the atom in its excited state. The superelastic scattering method is illustrated in Figure 2(b). The laser optically pumps the atom into a well defined state with a particular alignment and orientation. The electron gains kinetic energy and is superelastically scattered. Unlike the coincidence technique, the superelastic method is not time resolved and requires only the measurement of the superelastic differential cross section as a function of laser polarization, which leads to the determination of the same information as obtained in a coincidence experiment. The relationship between the techniques is revealed by inspection of Figure 2, which shows that the two methods are the time inverse of each other. The cw nature of the superelastic technique means that it has a much higher data collection rate than the equivalent coincidence experiment. As well, the energy resolution of the laser is much greater than that experimentally realisable for an electron beam, so that atomic structure can be resolved and the depolarization effects of the atomic structure can be minimized. The most obvious limitation of the superelastic technique is the lack of suitable tunable lasers in the uv and vuv region of the spectrum, which limits the applicability of the technique and its inability to access metastable states. As might be expected, the theoretical form of the signals for both the

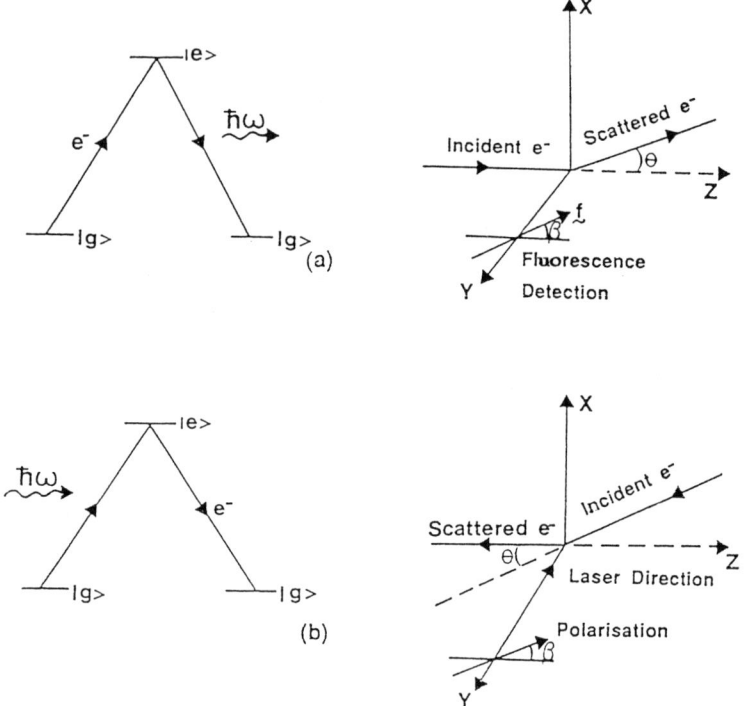

**FIGURE 2.** Schematic energy level diagram of the collision process and the geometry for (a) electron-photon coincidence method; (b) superelastic scattering method.

coincidence and superelastic scattering techniques are very similar. For the coincidence technique, the coincidence signal has the form[40]

$$I = \sum_{mn} \rho^e_{mn} L^F_{mn} \qquad (5)$$

where I is the intensity of the coincidently detected fluorescence signal and $\rho^e_{mn}$ represents the density matrix elements of the atomic state excited in the inelastic collision process. $L^F_{nm}$ is the fluorescence light monitoring operator. The polarization state of the fluorescence is fully represented by the Stokes parameters, $P^c_1 = (I_0 - I_{90})/I$, $P^c_2 = (I_{45} - I_{135})/I$ and $P^c_3 = (I_{RHC} - I_{LHC})/I$, where I is the total intensity.

The superelastic scattering signal has the form[40]

$$S = \sum_{mn} \rho^L_{mn} L^D_{nm} \tag{6}$$

where S is the differential cross section for the superelastically scattered electrons and $\rho^L_{mn}$ represents the density matrix elements for the laser excited atomic state. $L^D_{nm}$ is a de-excitation operator which represents the electron induced superelastic scattering process. As first pointed out by Macek & Hertel[41], application of the principle of microreversibility, which states that the matrix elements of the electron de-excitation process are the complex conjugates of the electron excitation process, gives[40]

$$S = \sum_{mn} \rho^L_{mn} \rho^e_{mn} \tag{7}$$

For consistency, if the quantization axis for the coincidence technique is chosen as the direction of the incident electron, then the quantization axis for the superelastic technique is chosen as the direction antiparallel to that of the scattered electron. Farrell et al 1989[42] introduced the concept of pseudo Stokes parameters, where $P^s_1 = (S_0 - S_{90})/S$, $P^s_2 = (S_{45} - S_{135})/S$ and $P^s_3 = (S_{RHC} - S_{LHC})/S$. The terms $S_\beta$ represent the superelastic differential cross section for the laser radiation linearly polarized at angle $\beta$ to the quantization axis, or for circularly polarized light.

A comparison of the coincidence and superelastic scattering signals can be illustrated by considering the scattering parameters associated with the inelastic excitation of an S-P atomic transition. In the collision frame, assuming positive reflection symmetry about the scattering plane is preserved, three independent parameters, $\lambda, \cos\chi, \sin\phi$ are required to fully represent the collision process. The scattering parameters are expressed in terms of the density matrix elements associated with a P(L=1) state as:

$$\lambda = \rho_{00}/(\rho_{00} + 2\rho_{11}), \cos\chi = \text{Re}(\rho_{10})/(\rho_{00}\rho_{11})^{\frac{1}{2}}, \sin\phi = \text{Im}(\rho_{10})/(\rho_{00}\rho_{11})^{\frac{1}{2}}$$

where the well known symmetry relations[43] for density matrix elements have been applied. Under the condition of pure P state excitation, with no spin-orbit interactions, the Stokes parameters are related to the atomic collision parameters by:

$$P_1 = 2\lambda - 1, P_2 = -2[\lambda(1-\lambda)]^{\frac{1}{2}} \cos\chi, P_3 = -2[\lambda(1-\lambda)]^{\frac{1}{2}} \sin\phi$$. The effect of spectroscopic structure, such as fine and hyperfine structure is generally to cause a depolarization of both coincidence and superelastic signals. In the case of coincidence signals, it has been shown[40] that $P_1^c = KP_1$, $P_2^c = KP_2$, $P_3^c = K'P_3$, where the factors K and K' account for the depolarization due to the spectroscopic structure. For the Na $3^2P$ state, K=0.141 and K'=0.557. This should be compared to cases where such effects are absent, such as for the He $2^1P$ state, where K=K'=1. It can be seen in the case of the Na $3^2P$ state, that considerable depolarization occurs due to the spectroscopic structure. In

the case of superelastic signals for the Na $3^2P$ state, the value of K and K' is determined by the optical pumping caused by the laser intensity and detuning and the Doppler width associated with the atomic beam. As discussed above, the sensitivity of K to changing experimental parameters is a striking feature of these results. Calculations show that K' is comparatively less sensitive to such experimental parameters. Typical experimental values for a superelastic scattering experiment would be $K \cong 0.40$ *and* $K' \cong 1$. Comprehensive theoretical and experimental studies of optical pumping effects in the Na $3^2P$-$3^2S$ transition have been reported [35,44].

The initial superelastic scattering experiments were performed by Hertel and Stoll in 1974[1] on e⁻-Na atom collisions over the energy range 5.1eV to 22.1eV and with the laser beam incident in the scattering plane. Later measurements by this group [2,3] included measurements performed with the laser beam incident from out of the plane.

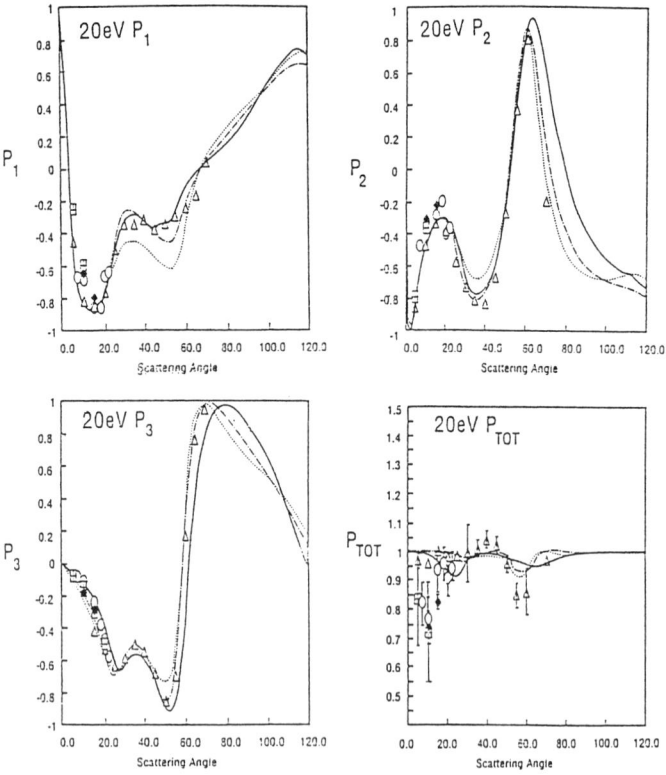

**Figure 3.** Stokes parameter data and theoretical calculations for 20eV incident energy as a function of scattering angle. —— (DWBA2), — — — (CCO), – – – (CC), Ⓒ (present), Δ (Teubner et al 1989[68]), □ (Hertel et al, see[3]), ◆ Farrell et al[42].

Other superelastic scattering measurements on sodium have been performed by Scholten et al 1988 [45] Farrell et al 1989[42] and Sang et al 1994[46] using an experimental geometry in which the laser beam was perpendicular to the scattering plane. Figure 3 shows data from superelastic scattering experiments for e⁻-Na collisions, obtained by several experimental groups together with several theoretical calculations for an incident energy of 20eV. Overall, the agreement between theory and experiment is very good, except for the parameter $P_{TOT}=(P_1^2+P_2^2+P_3^2)^{1/2}$, where there is some disagreement between experimental data and theoretical calculations, with a clear indication that at some electron scattering angles, $P_{TOT}$ is less than unity. $P_{TOT}$ provides a measure of the coherent nature of the electron collision process and a departure of $P_{TOT}$ from unity indicates that the process is not fully coherent. Sang et al 1994[46] have speculated that the observed loss of coherence could be due to the unresolved singlet and triplet channels for e⁻-Na scattering. They have introduced an expression for $P_{TOT}$ in terms of spin resolved differential cross sections and Stokes parameters such that

$$(P_{TOT})^2 = 1 + \frac{3}{8}\frac{\sigma^s \sigma^T}{\sigma^2}(S-1) \tag{8}$$

where the quantity S is given by $S = P_1^S P_1^T + P_2^S P_2^T + P_3^S P_3^T$. Theoretical calculations[46] show that the S parameter is the the most sensitive parameter for investigating loss of coherence in collision processes involving multiple spin scattering channels.

Superelastic scattering experiments involving spin polarized electron beams have been performed by Hanne et al 1982[47], McClelland and Kelly 1985[48], McClelland et al 1985, 1986a,b, 1987, 1989[49-53], Hedgemann et al 1991 [54] and Scholten et al 1991[55].

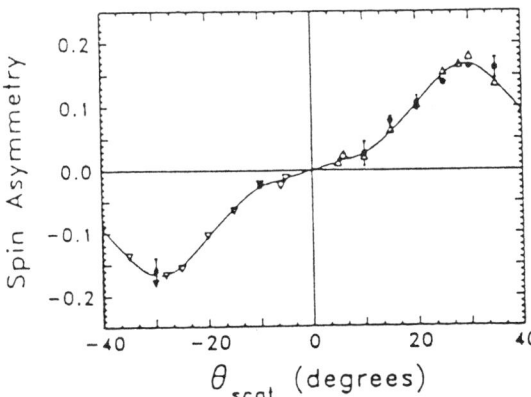

**FIGURE 4.** Spin asymmetry data for superelastically scattered 10eV spin polarized electrons from unpolarized sodium atoms in the $3^2P_{3/2}$ state. From [30].

The application of superelastic scattering measurements involving a spin polarized electron beam is well illustrated by studies of the "fine structure" effect. The fine structure effect is where the fine structure splitting of a target atom gives rise to significant spin polarization effects even if there is no spin-orbit interaction for the scattered electrons with the target atom. Evidence of the fine structure effect was first found by Hanne et al 1982[47] by measuring the polarization of 20eV electrons superelastically scattered from the $3^2P_{1/2}$ state of Na. McClelland et al[49,50] performed similar, but more comprehensive experiments in which the target atoms were excited to the $3^2P_{3/2}(F=3)$ hyperfine state. Figure 4 shows the measured spin asymmetry of 10eV polarized electrons superelastically scattered from unpolarized sodium atoms in the $3P_{3/2}(F=3)$ state. It should be noted that for a light atoms such as sodium, the spin-orbit interaction of the scattered electrons with the target is negligible, so that the spin asymmetry is attributable to the fine-structure effect.

Superelastic scattering experiments have also been performed by Register et al 1978, 1983 [56,57] on barium using a single mode laser to isolate the I=0 nuclear spin isotopes. In the course of these experiments, an unexpected asymmetry was noted in the superelastic scattering signal at small electron scattering angles. It was subsequently established that the cause of this anomaly was due to the finite volume of the interaction region and the finite size of the electron detector acceptance solid angles. Recently, Li et al 1994[59] have reported a superelastic scattering experiment on laser excited Yb. Yb is a heavy atom (Z=70) with substantial spin-orbit interaction which gives rise to strong intercombination lines. In this experiment, the $6^3P_1$-$6^1S_0$ (555.6nm) intercombination line was excited by circularly polarized laser light and measurements of the angular momentum $L_\perp$ were made as a function of electron scattering angle for 20eV and 40eV incident electron energy. At 20eV, and for the range of scattering angles $0^0$-$50^0$, $L_\perp$ showed little evidence of structure and reached a maximum value of about 0.5. The 40eV data showed some evidence of reaching a peak value of about 0.6 at $20°$. This behaviour is radically different to that found in light atoms for $^1P$-$^1S$ transitions, where $L_\perp$ strongly varies with scattering angle and can reach maximum values of almost unity. Li et al suggest that this behaviour may be due to the important role that electron exchange processes play in the excitation of the Yb $6^3P_1$ state.

## Stepwise Excitation

Two basic types of stepwise excitation schemes have been used in experiments to date. In a Type I scheme, electron excitation from the ground state to a first excited state is followed by laser excitation to a higher lying state. The intensity and polarization of the light emitted as the atom decays enables information about the initial electron excitation process. In a Type II scheme, the reverse sequence of excitation occurs with laser excitation from the ground state to a first excited state being followed by electron excitation to a higher lying excited state. In Type I stepwise excitation experiments, the intensity of the fluorescence emitted from the step emitted from the stepwise excited atom is given by[13]

$$I = \sum_{m'n'mn} F_{n'm'} A_{m'n'mn} \rho^e_{mn} \qquad (9)$$

where the term A represents the laser excitation process, $\rho^e_{mn}$ are matrix elements of the electron excitation density operator and $F_{n'm'}$ are matrix elements of the fluorescence emission operator which are given by $F_{n'm'} = \sum_q <n'|\underline{f}.\underline{P}|q><q|\underline{f}^*.\underline{P}|m'>$ where **f** is the polarisation vector of the the optical analyser, **P** is the electric dipole moment operator and |m'> and |q> are respectively substates of the upper excited and final states. In some cases, weak optical excitation applies and A then has an analytic form[13]. In Type II stepwise excitation schemes, it can be shown that the fluorescent intensity from the upper excited state is given by

$$I = \sum_{m'n'mn} F_{n'm'} O^e_{mn} f_{m'm} f^*_{n'n} \qquad (10)$$

where $O^e_{mn}$ are the matrix elements of the optical excitation operator for the laser excited step, and $f_{m'm}$ are amplitudes representing electron excitation between the lower and upper excited states. Another approach in using Type II excitation schemes is to detect the inelastically scattered electrons and measure their differential cross section as a function of laser polarization in a manner analogous to superelastic scattering experiments. In this case it can be shown that the differential cross section is given by

$$\sigma = \sum_{mnn'} O^e_{mn} f_{n'm} f^*_{n'n} \qquad (11)$$

Only one experiment has been reported using this latter technique[60].

Type I excitation schemes have been used to investigate electron collisions with neon, calcium, mercury and helium atoms. An important aspect of the Type I excitation technique is its use to investigate processes in which metastable atoms are collisionally excited. Absolute cross section measurements have been made using stepwise techniques on the neon metastable states of $1s_5(^3P_2)$ and $1s_3(^3P_0)$ as well as the states $1s_2(^1P_1)$ and $1s_4(^3P_1)$[6-9]. Investigations on calcium have been performed and produced absolute total cross sections for the $4^3P_{0,1,2}$ states. The near threshold cross section of the helium $2^1S$ has also been investigated[61]. Type I experiments have been used to investigate e⁻-Hg atom collisions in which the $6^1P_1$ state is excited. Relative total and partial total cross sections have been measured for the $6^1P_1$ state[10,11]. The spectral resolution offered by a combination of single mode laser excitation and atomic beam techniques enables the hyperfine structure to be resolved in the laser excited step enabling a direct experimental test to be made of the Percival-Seaton hypothesis[10]. Stepwise electron-photon coincidence experiments have also been performed to obtain the first alignment and orientation data for e⁻-Hg($6^1P_1$) collisions[14,15,18]. In these experiments, it was possible to measure data for selected isotopic species of the target Hg atoms. The additional experimental parameter provided by the laser polarization enables a complete analysis to be made of the atomic scattering parameters, including the spin-flip cross section, using only photon detection perpendicular to the scattering plane. Measurements of the partial total cross sections for the $6^3P_2$ metastable state of mercury near threshold have been performed using both c.w.[62] and pulsed laser[63] stepwise techniques. The use of stepwise techniques in this case provided information that can not be accessed using non-laser techniques. It has been proposed[14] that electron-photon coincidence techniques could be applied to obtain alignment and orientation data for collision processes involving metastable atoms.

Several Type II scheme experiments have been reported for sodium and barium target atoms. The first such experiment measured differential cross sections for the $3^2P$-$4^2S$ and $3^2P$-$3^2D$ transitions of sodium[1]. In a subsequent experiment, alignment parameters were obtained for these transitions[60]. The method used in this experiment was the same as that used in a superelastic scattering experiment except that inelastic, rather than superelastic differential cross sections were monitored as a function of laser polarization. Such experiments do require a sufficiently high electron analyser resolution that only the inelastic channel of interest is monitored. Other experiments have been performed to measure the excited state total cross sections for the 3P-3D transitions of sodium[64]. Differential cross sections for inelastic scattering from excited states of barium have also been performed[56].

## Laser-assisted collisions

A laser-assisted collision process is one in which the laser photon can be thought of as having the role of a third body in a collision involving an atom, and electron and a photon. Initial experiments performed by Andrick and Langhans 1976[19] Weingartschofer et al 1977[20] involved elastic electron collisions with Ar atoms in which the main interaction took place between the free electron and the laser field. More recently, Mason and Newell[21] and Wallbank et al[24-27] have performed experiments on He, Ne and Ar in which the simultaneous electron-photon excitation (SEPE) of metastable atomic states has been observed. The first experiments were performed by Mason & Newell[21] and were conducted near threshold so that neither the electron nor the photon by itself had sufficient energy to excite the transition. They observed an enhanced excitation cross section for the He $2^3S$ state below the excitation threshold in the presence of radiation from a 400W $CO_2$ laser. Wallbank et al[65] have performed differential cross section measurements for the inelastic scattering of electrons from the $2^1P_1$ state of helium in the presence of an intense pulsed $10^8$ Wcm$^{-2}$

laser field produced by a $CO_2$ laser. These experiments were performed with incident energies in the range 36-70eV and for scattering angles from $13^0$ to $31^0$. There are some indications that the theoretical treatment based on the Kroll & Watson treatment[66-67], which neglects the laser-atom interactions, may not adequately account for the experimental results. The Kroll & Watson treatment predicts that the laser assisted cross section is the same as the cross section in the absence of the laser, except that it is shifted in energy by a multiple of the photon energy and its magnitude is reduced by a multiplicative factor which depends on the laser intensity, wavelength, polarization, and the electron energy.

## Excited state (e,2e) spectroscopy

Electron momentum (e,2e) spectroscopy has been used extensively to investigate the electronic structure of atoms and molecules in the ground state. Weigold and coworkers performed the first excited state (e,2e) experiment in 1990[28] in which sodium target atoms were optically pumped to the $3^2P_{3/2}(F=3, M_F=3)$ hyperfine state. Non-coplanar symmetric (e,2e) measurements were made and the measured and calculated momentum profiles showed excellent agreement for the 3p state of Na.

## The application of atom optics and optical traps to collision studies.

The optical forces that underpin the application of atom optical techniques have already been discussed above. The pioneering work of Bederson and colleagues[4] was discussed along with the work of Summy et al[5,36] which suggests that new techniques can be developed that provide an alternative to the superelastic and coincidence techniques, and are of particular relevance to the investigation of collisionally excited metastable states. Very recently, the first measurements of collision cross sections have been reported[29] in which optically trapped atoms have been used as the target. Rb vapour was trapped in a magneto-optical trap which incorporated an electron gun. Rb atoms were trapped in a small cloud of approximately 0.5mm diameter. The experiment was operated in pulsed mode. The trap magnetic fields are turned off in less than 0.5ms and the laser frequency is shifted so that the trap is turned off. A pulse of electron gun beam current is fired at the target atoms for a duration of between 0.8 and 4ms, after which the atoms are allowed to travel ballistically for up to 18ms before the trap is turned on again and recaptured atoms return to the center of the trap. The loss rate from the trap depends both on the electron beam causing collisions that eject atoms from the trap and collisions due to the background gas. Measurements of the risetime of the trap fluorescence transient as the trap fills with and without the presence of the electron beam enables the total scattering cross section to be determined.

# CONCLUSION

In this paper, attention has been drawn to the range of experimental techniques now available in which lasers can be applied to the investigation of collision processes. Superelastic scattering techniques have a great capacity for rapid data acquisition, but are limited in application because of the relatively limited spectral range combined with sufficient optical power available with current laser technology. The introduction of tunable Ti:sapphire lasers and diode lasers has extended this range into the near infrared. The use of two or more lasers to access a greater range of atomic transitions and provides a means of probing the detailed dynamics of collision processes that involve Rydberg states. Pulsed lasers can access a greater spectral range and it should be possible to develop superelastic techniques that utilise pulsed lasers. Stepwise excitation techniques offer an alternative to existing methods for investigating collision processes involving vacuum ultra-violet transitions in atomic targets. These techniques also offer a means of investigating processes in which metastable states are excited and for which new partial total cross section data has already been obtained. It should be possible to extend stepwise coincidence techniques to the measurement of atomic alignment and orientation parameters for metastable states. Stepwise excitation techniques also provide a way of investigating collision processes involving excited state-excited state transitions.

Since the early work involving elastic collisions, the field of laser-assisted collisions has advanced significantly in recent years with very interesting developments occurring in experiments in which the effect of the presence of a powerful laser field on inelastic total and differential cross sections has been clearly demonstrated. The most recent work suggests that the theory of laser-assisted collision processes may need further development.

The introduction of excited state (e,2e) momentum spectroscopy has demonstrated a very interesting area of development for this field, which it can be anticipated will lead to further experiments utilizing laser techniques.

The first applications of laser cooling techniques to atomic collision physics are just starting to appear in the literature. These are based on the use of optical traps and the measurement of the change in loss rates of trapped atoms when they are subjected to electronic collisions. Optical manipulation of atomic beams to measure total cross sections was first demonstrated almost two decades ago. Recent work suggests that such techniques can be further refined to yield alignment and orientation parameters. It can be expected that as the field of atom optics further develops, the ability to control atoms with laser beams will lead to new developments in atomic collision physics.

Finally, it should be noted that extensive progress has been made on the theoretical treatment of atom-laser interactions, with quantum electrodynamical methods now available, which allow the full energy level structure of atoms to be included along with all relevant experimental parameters such as laser detuning, laser beam profile, atomic beam Doppler width and the flight time of atoms through the laser beam.

## REFERENCES

1. Hertel, I.V.and Stoll, W.J., *Phys.B:At.Mol.Phys.***7**, pp. 570 (1974).
2. *Adv.At.Mol.Phys.*, **13**, pp. 113 (and other references therein) (1977).
3. Andersen, N., Gallagher, J.W. and Hertel, I. V., *Phys. Rep.*, **165**, pp. 1 (and other references therein) (1988).
4. Bhaskar, N. D., Jaduszliwer, B. and Bederson, B. *Phys.Rev.Lett.*, **38**, pp. 14, (1977).
5. Summy, G. S., Lohmann, B., MacGillivray, W. R. and Standage, M. C., Z.*Physik D.*, **30**, pp. 155, (1994).
6. Phillips, M. H., Anderson, L. W., Lin, C. C. and Miers, R. E., *Phys. Lett .A*, **82**, pp. 404, (1981).
7. Miers, R. E., Gastineau, J. E., Phillips, M. H., Anderson, L. W. and Lin, C. C., *Phys. Rev.A*, **25**, pp. 1185, (1982).
8. Phelps, J. O., Phillips, M. H., Anderson, L. W. and Lin, C. C. *J.Phys.B.At.Mol.Phys.*, **16**, pp. 3825, (1982).
9. Phillips, M. H., Anderson, L. W. and Lin, C. C., *Phys. Rev.A.*, **32**, pp. 2117, (1985).
10. McLucas, C. W., MacGillivray, W. R. and Standage, M. C., *Phys. Rev. Lett.*, **48**, pp. 88, (1982).
11. McLucas, C. W., Wehr, H. J. E., MacGillivray, W. R. and Standage, M. C., *J.Phys.B:At.Mol.Phys.*, **15**, pp. 1883, (1982).
12. Webb, C. J., MacGillivray, W. R. and Standage, M. C., *J.Phys.B:At.Mol.Phys.*, **17**, pp. 2577, (1984).
13. MacGillivray, W. R. and Standage, M. C., *Phys.Rep.*,**168**, pp. 1, (1988).
14. Murray, A. J., Webb, C. J., MacGillivray, W. R. and Standage, M. C., *Phys.Rev.Lett.*, **62**, pp. 411, (1989).
15. Murray, A. J., MacGillivray, W. R. and Standage, M. C., *J.Phys.B:At.Mol.Opt.Phys.*, **23**, pp. 3373 (1990).
16. Murray. A. J., MacGillivray, W. R. and Standage, M. C., *Phys.Rev.A*, **44**, pp. 3162, (1991).
17. *J.Mod.Opt*, .**38**, pp.961, (1991).
18. Murray, A. J., Pascual, R., MacGillivray, W. R. and Standage, M. C., *J.Phys.B:At.Mol.Opt.Phys.*, **25**, pp. 1915, (1992).
19. Andrick, D. and Langhans, L., *J.Phys.B:At.Mol.Phys.*, **9**, pp. L459, (1976).
20. Weingartshofer, A., Holmes, J. K., Caudle, G., Clarke, E. M. and Kruger, H., *Phys.Rev.Lett.*, **39**, pp. 269, (1977).
21. Mason, N. J. and Newell, W. R., *J.Phys.B:At.Mol.Opt.Phys.*, **22**, pp. 777 (1989).
22. Weingartshofer, A., Clarke, E. M., Holmes, J. K., and Jung, C. *Phys.Rev.A.*, **19**, pp.2371, (1979).
23. Weingartshofer, A., Holmes, J. K., Sabbagh, J. and Chin, S. L., *J.Phys.B:At.Mol.Phys.*, **16**, pp. 1805, (1983).
24. Wallbank, B., Connors, V. W., Holmes, J. K. and Weingartshofer, A., *J.Phys.B:At.Mol.Phys.*, **20**, pp. L833, (1987).
25. Wallbank, B., Holmes, J. K. and Weingartshofer, A., *J.Phys.B:At.Mol.Phys.*, **20**, pp. 6121, (1987).
26. Wallbank, B., Holmes, J. K., MacIsaac, S. C. and Weingartshofer, A., *J.Phys.B:At.Mol.Opt.Phys.*, **25**, pp. 1265, (1992).
27. Wallbank, B. and Holmes, J. K., *J.Phys.B:At.Mol.Opt.Phys.*, .**27**, pp. 1221, (1992).
28. Zheng, Y., McCarthy, I. E., Weigold, E. and Zhang, D., *Phys. Rev.Lett.*, **64**, pp. 1358, (1990).
29. Schappe, R. S., Feng, P., Anderson, L. W., Lin, C. C. and Walker, T., *EuroPhys.Lett.*, **29**, pp. 439, (1995).

30. Kessler, J., *Adv.At.Mol.& Opt.Phys.*, **27**, pp. 81, (1991).
31. Allen, L. and Eberly, J. H., *Optical Resonance and Two-Level Atoms.*, New York, Wiley,1975, pp. 975.
32. Farrell, P. M., MacGillivray, W. R. and Standage, M. C., *Phys. Rev.A*, **37**, pp. 4240, (1988).
33. Ackerhalt, J. R., Knight, P. L. and Eberly, J. H., *Phys.Rev. Lett*, **30**, pp. 456, (1973).
34. Ackerhalt, J. R. and Eberly, J. H., *Phys. Rev.D.*, **10**, pp. 3350, (1974).
35. Farrell, P. M., MacGillivray, W. R. and Standage, M. C., *Phys. Rev.A.*, **44**, pp. 1828, (1991).
36. Summy, G. S., MacGillivray, W. R. and Standage, M. C., *Submitted to J.Phys.B:At.Mol.Opt.Phys.*, (1995).
37. Eminyan, M., MacAdam, K. B., Slevin, J. and Kleinpoppen, H. *J., Phys.B*, **7** 1519, (1974).
38. Standage, M. C. and Kleinpoppen, H., *Phys.Rev.Lett.*, **36,** 577, (1976).
39. Slevin, J., *Rep.Prog.Phys.*, **47,** 461, (1984).
40. MacGillivray, W. R. and Standage, M. C., *Comm.At.Mol.Phys.*, **26,** 179, (1991).
41. Macek, J. and Hertel, I. V., *J.Phys.B:At.Mol.Phys.*, **7,** 2173, (1974).
42. Farrell, P. M., Webb, C. J., MacGillivray, W. R. and Standage, M. C., *J.Phys.B:At.Mol.Opt.Phys.*, **22,** L527, (1989).
43. Blum, K., *Density Matrix Theory and Applications,* Plenum, New York.
44. Meng, X-K., MacGillivray, W. R. and Standage, M. C., *Phys. Rev.A* ,**45**, pp 1767, (1992).
45. Scholten, R. E., Anderson, T. and Teubner, P. J. O., *J.Phys.B: At.Mol.Opt.Phys.*, **21,** L473, (1988).
46. Sang, R. T., Farrell, P. M., Madison, D. H., MacGillivray, W. R. and Standage, M. C., *J.Phys.B:At.Mol.Opt.Phys,.***27**, pp. 1187, (1994).
47. Hanne, G. F., Szmytkowski ,Cz. and Van der Wiel, M., *J.Phys.B:*, **15,** L109, (1982).
48. McClelland, J. J. and Kelly, M. H., *Phys. Rev.A*, **31**, pp. 3704, (1985).
49. McClelland, J. J., Kelly, M. H. and Celotta, R. J,. *Phys.Lett.*, **55**, pp. 688, (1985).
50. *Phys.Rev.Lett,.***56**, pp. 1362,( 1986a).
51. *Phys.Rev.Lett,.***56** , pp. 2771, (1986b).
52. *J.Phys.B:At.Mol.Phys,* **20,** L385, (1987).
53. *Phys. Rev.A* , **40** , pp. 2321, (1989)
54. Hedgemann, T., Oberste-Vorth, M., Vogts, R. and Hanne, G. F., *Phys.Rev. Lett.,* **66**, pp. 2968, (1991).
55. Scholten, R. E., Lorentz, S. R., McClelland, J. J., Kelly, M. H. and Celotta, R. J., *J.Phys.B:At.Mol.Phys.*, **24,** L653, (1991).
56. Register, D. F., Trajmar, S., Jensen, S. W. and Poe, R. T., *Phys. Rev. Lett.*, **41**, pp. 749, (1978).
57. Register, D. F., Trajmar, S., Csanak, G., Jensen, S. W., Fineman, M. A. and Poe, R. T., *Phys.Rev.A*, **28,** pp 151, (1983).
58. Zetner, P. W., Trajmar, S., Csanak, G. and Clark, R. E. H., *Phys. Rev.A*, **39,** pp. 6022, (1989).
59. Li, Y., Wang, S., Zetner, P. W. and Trajmar, S., *J.Phys.B: At. Mol. Phys.*, **24,** L653, (1994).
60. Hermann, H. W., Hertel, I. V., Reiland, W., Stamatovic, A. and Stoll, W., *J.Phys.B.*, **10**, pp. 251, (1977).
61. Zetner, P. W., Westerweld, W. B., King, G. C. and McConkey, J. W., *J.Phys.B.*, **16**, pp 4205, (1986).
62. Webb, C. J., MacGillivray, W. R. and Standage, M. C., *J.Phys.B.*, **18,** L259, (1985).
63. Hanne, G. F., Nickich, V. and Sohn, M., *J.Phys.B.*, **18,** pp. 2037, (1985).
64. Stumpf, B. and Gallagher, A., *Phys.Rev.A.*, **32**, pp. 3344, (1985).
65. Wallbank, B., Holmes, J. K. and Weingartshofer, A., *Can.J. Phys.*, **71**, pp. 326, (1993).
66. Kroll, N. and Watson, K., *Phys.Rev.A.*, **8**, pp 804, (1973).
67. Mittleman, M., *J.Phys.B:At.Mol.Opt.Phys.*, **26**, pp. 2709, (1993).
68. Teubner, P. J. O., Scholten, R. E. and Shen, G. F., Proc.Int.Symp. *Correlation and Polarization in Electronic and Atomic Collisions*, pp. 45-50. (1989).

# Laser-Assisted Electron-Atom Collisions

### Barry Wallbank

*Department of Physics, St. Francis Xavier University,
Antigonish, Nova Scotia, B2G 2W5, Canada*

**Abstract.** Recent experimental measurements of laser-assisted elastic scattering of electrons from atoms in the presence of a carbon dioxide laser are reported. These data are discussed in terms of the Kroll-Watson approximation which is normally applied to such measurements. Large discrepancies between the predictions of such a theoretical treatment and the experimental data are observed.

## INTRODUCTION

Free-free transitions experienced by electrons scattered in the presence of a $CO_2$ laser field by an atomic target remaining in its ground state were observed for the first time almost 20 years ago with a CW laser [1] and a higher power pulsed TEA laser [2]. More recently, the observation of laser-assisted electron impact excitation, usually dubbed Simultaneous Electron-Photon Excitation (SEPE), has also been reported using a CW $CO_2$ laser [3], a pulsed $CO_2$ laser [4] and a pulsed Nd-YAG laser [5]. The focus of the last Progress Report for laser-assisted electron-atom collisions which was presented at ICPEAC XVII [6] was the theoretical treatment of the above processes. A simple theoretical model in the low frequency regime was shown to be in reasonable agreement, at least qualitatively, with the experimental data reported. The purpose of the present report is to discuss the experimental data obtained since that time. These new data raise several questions concerning the adequacy of the previously applied low frequency theoretical models.

For the theoretical treatment of laser-assisted elastic collisions, the first non-perturbative approaches to the interactions with the laser field were treatments of potential scattering [7,8]. The case of electron-atom collisions is, of course, more difficult to treat because of the internal structure of the target. However, by neglecting the laser-atom interaction, a simple expression for the scattering cross section in the presence of the laser in terms of the field-free scattering cross-section may be deduced [9,10] which should be valid at low frequencies. It is this treatment, usually called the Kroll-Watson approximation (KWA), that has been most widely applied to the experimental data. For fast incident electrons and laser field strengths which are small compared to the atomic unit of field, laser-assisted collisions have been described by treating the laser-electron interaction to all orders, while the laser-target and electron-target interactions are treated by perturbation theory [11,12]. In this way the effects of dressing of the target states by the laser field on fast electron-atom collisions may be examined. For slow electrons, where, for example, exchange scattering may be important, little has been reported with a realistic treatment of target dressing.

Experimentally, a significant amount of data for low energy elastic scattering through large scattering angles in the presence of a $CO_2$ laser has been reported [13] which has been shown to be in reasonable agreement with the KWA. Recently, however, our laboratory has obtained data for low energy scattering from argon and helium which are in disagreement with the predictions of the KWA. The driving force behind these experiments was our unsuccessful attempt to examine the low-lying resonances of argon in the presence of a laser, as we had already successfully reported measurements of the helium $^2S$ resonance [14]. In the argon experiments we observed a large background signal, under conditions for which the KWA predicts near-zero signals, that prevented us from observing the $^2P$ resonances. A larger than expected backround had also been reported previously [15] for similar experiments at lower $CO_2$ laser intensities. The experimental data which disagree so strongly with the theoretical predictions are the subject of the present report.

## EXPERIMENTAL

The experimental arrangements for laser-assisted electron atom collisions that have been used over the ~20 years of work in the field have all employed the same basic instrumentation: a conventional electron spectrometer and some means to bring into the scattering region a focussed laser beam. As these slightly differing instruments have been described in some detail [13] only a brief description of the present instrument used in our laboratory will be given.

In the electron spectrometer electrons, emitted from a sharp tungsten filament, are focussed into a 127° cylindrical deflector which produces an electron beam of narrow energy spread (~25meV) which is then accelerated and focussed to produce a beam of the required incident energy to collide with an atomic beam formed by a pulsed supersonic beam valve. The electrons scattered from the atomic beam through a particular angle are then decelerated and focussed into a double hemisperical sector deflector in the 'S' configuration for energy selection and finally detected with an electron multiplier after reacceleration. The spectrometer in which both the electron gun and the electron analyzer are rotatable is housed in a stainless steel vacuum

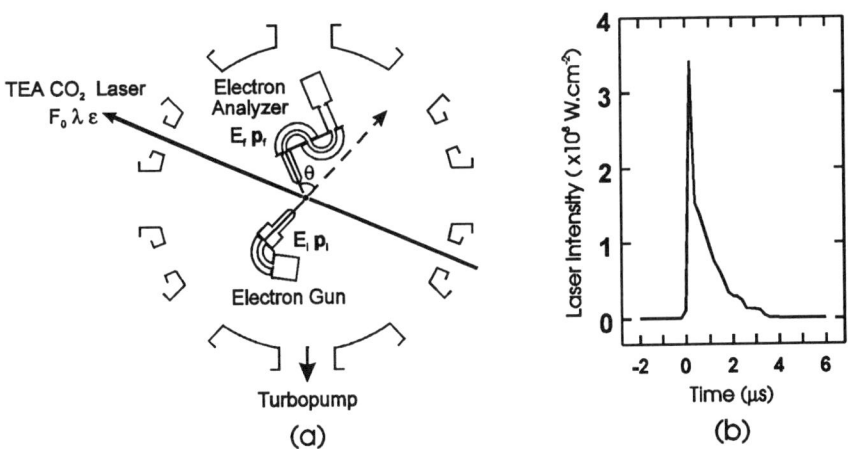

**FIGURE 1.** (a) Schematic diagram of the electron spectrometer and (b) a typical $CO_2$ laser pulse with the estimated laser intensity in the scattering region of the spectrometer.

chamber pumped by a 500 $ls^{-1}$ turbomolecular pump and is shielded from the earth's magnetic field. Radiation from a pulsed $CO_2$ TEA laser operating in a multilongitudinal mode optical configuration enters and exits the vacuum chamber through sodium chloride windows fitted at the Brewster angle and is focussed into the target region of the spectrometer. Detected electrons are recorded after amplification using a homebuilt counter [16] which also simultaneously digitizes and stores a small fraction of each laser pulse reflected from a sodium chloride optical element and detected with a photon drag detector. A schematic diagram of the present instrument is shown in Figure 1 together with the detected profile of a typical laser pulse. Further details of our experimental arrangement have been given elsewhere [17].

## RESULTS AND DISCUSSION

As mentioned earlier, the usual theoretical method used to discuss data of the type to be presented here is the Kroll-Watson approximation the main result from which is

$$\frac{d\sigma_{FF}}{d\Omega}(n) = \frac{p_f}{p_i} J_n^2(\Gamma) \frac{d\sigma_{EL}}{d\Omega} \qquad (1)$$

where $\Gamma^2 = 1.944 \times 10^{-12} \lambda^4 F_0 E_i \ (\varepsilon \cdot (\mathbf{p}_i - \mathbf{p}_f) / 2p_i)^2$ with the laser wavelength ($\lambda$)

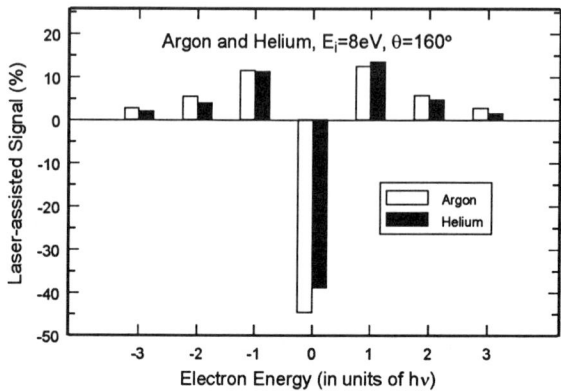

**FIGURE 2.** Electron spectra resulting from elastically scattering 8eV electrons from argon and helium through 160° in the presence of a carbon dioxide laser.

in μm, the laser intensity ($F_0$) in W cm$^{-2}$, the incident electron energy ($E_i$) in eV, the laser polarization is $\varepsilon$ and the incident and final electron momenta are $\mathbf{p_i}$ and $\mathbf{p}$ respectively. Thus, the free-free cross-section for an n-photon process ($d\sigma_{FF}(n)/d\Omega$) is expressed in terms of the field-free elastic cross-section ($d\sigma_{EL}/d\Omega$), the interaction of the electron with the laser field being expressed through the Bessel function of the first kind and order n. The model used to describe the laser defines the form of the function to describe the laser-electron interaction, the Bessel function being the consequence of assuming a single mode, homogeneous, purely coherent, linearly polarized, laser field.

Examples of free-free electron spectra are shown in Figure 2. The change in electron signal due to the laser, as a percentage of the field-free elastic scattering signal, is plotted against the scattered electron energy expressed as the difference in energy from the incident electron energy in units of the laser photon energy. These spectra were obtained for 8eV incident electrons scattering through 160° from two atomic targets, argon and helium, and are the result of summing the data from 5000 laser pulses at each electron energy. For both atoms, the laser polarization was parallel to the change in electron momenta ($\mathbf{p_i} - \mathbf{p_f}$). As stated earlier, for backward scattering conditions, the experimental data are in good agreement with equation (1). This appears to be the case for the data in Figure 2 as the spectra for the two atoms are in good agreement which implies neglect of the laser-atom interaction, inherent in the KWA, is valid for these scattering conditions at least. It is also apparent that there is a redistribution of the field-free elastically scattered signal over up to 3-photon processes with a corresponding loss in the 0-photon signal in the presence of the laser. This "sum-rule" has been discussed extensively previously [18].

In contrast to Figure 2, data obtained under near forward scattering conditions with the laser polarization parallel to the incoming electrons is presented in Figure 3

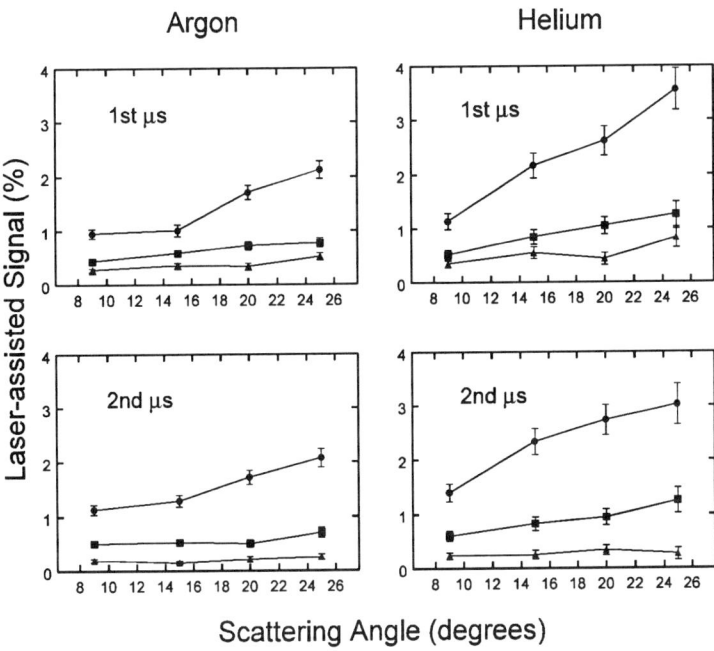

**FIGURE 3.** The relative laser-assisted signals expressed as percentages of the field-free elastic scattering signals for 10.5eV electrons from argon and helium as a function of scattering angle for the first two microseconds of the laser pulse (1-photon: circles; 2-photon: squares; 3-photon: triangles).

as the relative laser-assisted signal expressed as a percentage of the measured field-free elastic scattering signal for the 1-, 2- and 3-photon processes as a function of scattering angle in the range of 9° to 25°. For each atomic target, data obtained during the first two microseconds of the laser pulse are presented and the displayed points are averages of the observed photon absorption and emission signals. For both atoms, the 1-photon signal is ~1% at 9°, and increases to ~3% for argon and ~2% for helium at 25°. The intensities of the multiphoton processes are somewhat less but are measurable for both atoms. However, using equation (1), the KWA predicts near-zero free-free cross-sections, particularly for higher order processes. For example, for a laser intensity of $1.5 \times 10^8$ W.cm$^{-2}$ and 10eV electrons scattering through 10°, equation (1) predicts laser-assisted cross-sections as percentages of the field-free cross-section for the 1-, 2- and 3-photon processes to be $\sim 10^{-2}$, $\sim 10^{-6}$ and $\sim 10^{-11}$, respectively. For the multiphoton processes the observed signals are orders of magnitude greater than those predicted for this scattering geometry. We have also obtained data that show this unexpected free-free cross-section decreases with increasing electron energy [17,19,20]. These data may account for the problems experienced in detecting the argon resonances (at ~11eV) as compared to the helium

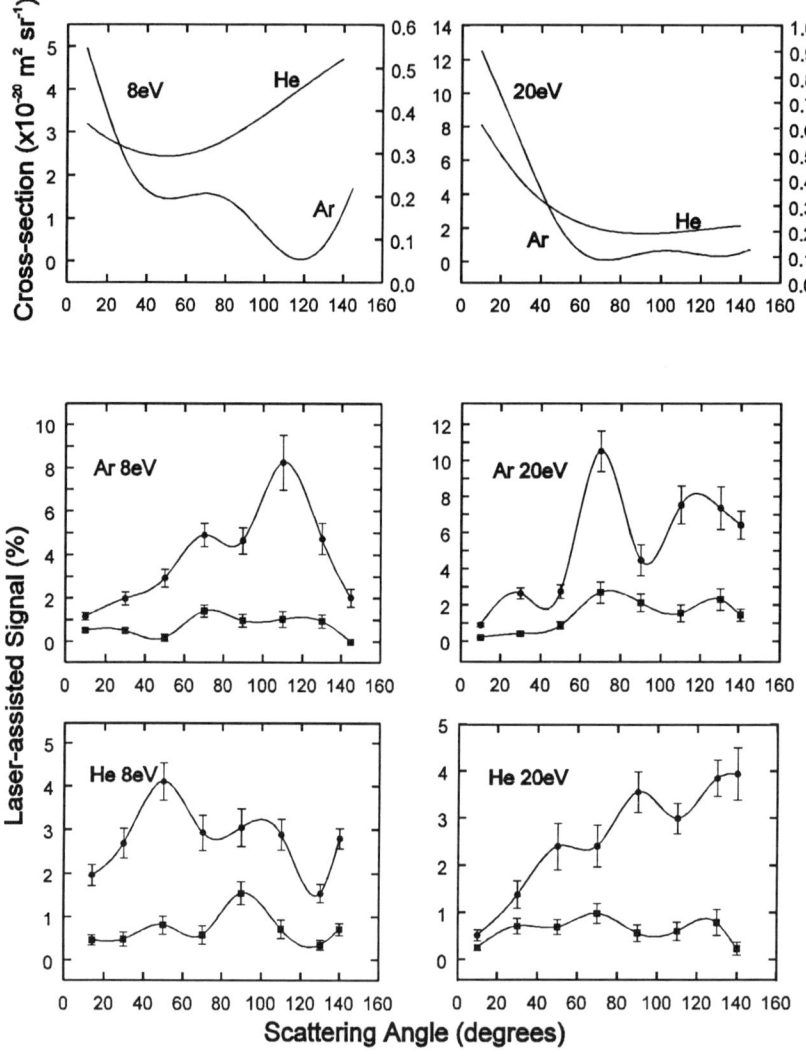

**FIGURE 4.** The relative laser-assisted signals as a function of scattering angle for scattering from argon and helium at 8eV and 20eV (1-photon: circles; 2-photon: squares). Also shown (top panels) are the corresponding field-free elastic scattering cross-sections. (In each of these two diagrams the left-hand scale is for argon and the right-hand scale for helium.)

resonance (at~18eV) due to a large non-resonant background. Such resonance experiments are performed at near-zero scattering angle to minimize the non-resonant background while maximizing the resonant signals. It appears from these data that the electron energy is more critical to this increased free-free cross-section than the change in atomic target.

It is possible with our experimental system to examine free-free transitions under conditions for which the KWA predicts zero cross-sections. We have performed such experiments over a large range of scattering angles with the laser polarization in the scattering plane but always perpendicular to the change in electron momenta for both helium and argon targets. Examples of the results of these measurements are presented in Figure 4 and also for comparison are the field-free elastic cross-sections for scattering from helium and argon [21]. The 1-photon free-free data for argon show a greater range of signals than do those for helium. For example, at 8eV, the argon data vary from ~1% to ~8% while for helium the range is ~1% to ~4%. It is also interesting that the maximum relative laser-assisted signals for argon occur at 110° for 8eV electrons and 70° for 20eV electrons. These are the angles for which the field-free cross-sections are minima at these energies. The 2-photon signals follow approximately the behaviour of the 1-photon signals. Although not presented here, we have obtained similar data at other electron energies.

The magnitude of the free-free signals is very surprising since the KWA predicts zero cross-sections with this beam geometry. Also, at least in the case of argon, the observed free-free signal does not appear to scale with the elastic cross-section since the maximum relative laser-assisted signal occurs for the minimum field-free cross-section. This is very different behaviour to that predicted by the KWA under any conditions.

Examples of data have been presented which are in drastic disagreement with the KWA, long accepted as an adequate theoretical treatment for electron-atom scattering in the presence of a $CO_2$ laser. There has been discussion of the role of the laser-atom interaction and, at small scattering angles, this interaction may dominate that of the laser-electron at sufficiently high laser intensities [11,12]. However, under the conditions of our experiment, it has been shown that such a contribution to the free-free cross-section is negligibly small [22,23]. An intriguing model has been proposed [24] which requires the collective and coherent scattering of the electrons by a quasi-macroscopic laser-induced dipole potential formed from the individual polarization potentials of the atoms in the gas beam. The results of this calculation are free-free cross-sections which are much more in line with those observed than are the ones for single-atom scattering using the KWA. However, as such a model requires quite high target gas pressures, it will have to be tested by, at least, repeating the experiments at lower atomic densities than have normally been used. The phenomenon of exchange scattering, which becomes more important in field-free electron scattering as the incident electron energy decreases, has not been investigated in any great detail for the case of laser-assisted scattering. Such theoretical studies may have a bearing on the data presented here.

# PROSPECTS

Although free-free transitions have been studied experimentally for almost 20 years, recent data indicate that there are still mechanisms to be studied and understood. As yet, no experimental results have been reported for low incident electron energies (<1eV) such that the electron energy and photon energy are becoming comparable and where the KWA would also be expected to become invalid. Although this presents experimental difficulties, we have already initiated such a series of experiments and have been able to make measurements for electron energies down to ~0.4eV. We have so far concentrated on experiments for which the free-free cross-section, according to the KWA, is large but also intend obtaining data under conditions similar to those disussed in this paper. These low energy experiments may be relevant to those which have examined the above-threshold photodetachment from negative ions with extremely high intensity lasers [25,26].

This paper has been concerned only with elastic scattering but there remains much to be done in the area of inelastic scattering in the presence of a laser. Experiments in which differential cross-sections are measured would be particularly useful as such data are finer probes of theoretical models than total cross-sections. As yet only one such series of experiments has been reported and these data were somewhat preliminary [27].

A fascinating prospect is the examination of ionization processes in the presence of a laser field. Although there have been some calculations for laser-assisted e-2e processes [6], such experiments are probably prohibitively difficult. However, relatively simple experiments which measure the total ion yield caused by electrons in the presence of a laser would also be of value. One can also envision laser-assisted ionization experiments in which the ejcted electrons are detected but employ a UV lamp or metastable atom beam as the primary ionizing source. These latter two experimental studies would have the benefit of preparing the electron in a well-defined state and would be investigating the interaction of this electron with the laser field in the presence of the Coulomb potential of the ion. Such measurements would be important for fundamental collision theory and in applied areas such as laser heating of plasmas.

## ACKNOWLEDGEMENTS

This work was financially supported by the Natural Sciences and Engineering Research Council of Canada and the St. Francis Xavier University Council for Research. I am indebted to Mr. James K. Holmes for his invaluable contributions to the experiments reported here and to Professor Antonio Weingartshofer for his continued interest in the work of our laboratory.

# REFERENCES

1. Andrick, D., and Langhans, L., *J. Physics B: At. Mol. Phys.* **9,** L459–L461 (1976).
2. Weingartshofer, A., Holmes, J.K., Caudle, G., Clarke, E.M., and Krüger, H., *Phys. Rev. Lett.* **39,** 269–270 (1977).
3. Mason, N.J., and Newell, W.R., *J. Phys. B: At. Mol. Phys.* **20,** L323–L325 (1987).
4. Wallbank, B., Holmes, J.K., LeBlanc, L., and Weingartshofer, A., *Z. Phys. D* **10,** 467–472 (1988).
5. Luan, S., Hippler, R., and Lutz, H.O., *J. Phys. B: At. Mol. Opt. Phys.* **24,** 3241–3249 (1991).
6. Maquet, A., *Electronic and Atomic Collisions,* Eds. W.R. MacGillivray, I.E. McCarthy and M.C. Standage, Bristol, UK: Hilger, 1992, pp. 327–336.
7. Bunkin, F.V., and Fedorov, M.V., *Sov. Phys. JETP* **22,** 844–847 (1966).
8. Kroll, N.M., and Watson, K.M., *Phys. Rev. A* **8,** 804–809 (1973).
9. Krüger, H., and Jung, C., *Phys. Rev. A* **17,** 1706–1712 (1978).
10. Mittleman, M.H., *Phys. Rev. A* **20,** 1965–1971 (1979).
11. Byron, F.W.Jr., and Joachain, C.J., *J. Phys. B: At. Mol. Phys.* **17,** L295–L301 (1984).
12. Byron, F.W.Jr., Francken, P., and Joachain, C.J., *J. Phys. B: At. Mol. Phys.* **20,** 5487–5503 (1987).
13. See for example, Mason, N.J., *Rep. Prog. Phys.* **56,** 1275–1346 (1993).
14. Wallbank, B., Holmes, J.K., MacIsaac, S.C., and Weingartshofer, A., *J. Phys. B: At. Mol. Opt. Phys.* **25,** 1265–1277 (1992).
15. Bader, H., *J. Phys. B: At. Mol. Phys.* **19,** 2177–2188 (1986).
16. Van Audenhove, G.L., Holmes. J.K., Wallbank, B., and Weingartshofer, A., *J. Phys. E: Sci Instrum.* **18,** 476–478 (1985).
17. Wallbank, B., and Holmes, J.K., *J. Phys. B: At. Mol. Opt. Phys.* **27,** 1221–1231 (1994).
18. Weingartshofer, A., and Jung, C., "Multiphoton Free-Free Transitions," in *Multiphoton Ionization of Atoms,* S.L. Chin and P. Lambropoulos, Eds., Toronto: Academic Press, 1984, pp. 155–187.
19. Wallbank, B., and Holmes, J.K., *Phys. Rev A* **48,** R2515–R2518 (1993).
20. Wallbank, B., and Holmes, J.K., *J. Phys. B: At. Mol. Opt. Phys.* **27,** 5405–5418 (1994).
21. Bitsch, A., *Diplomarbeit,* University of Kaiserslautern, Germany, (1972).
22. Rabadán, I., Méndez, L., and Dickinson, A.S., *J. Phys. B: At. Mol. Opt. Phys.* **27,** L535–L541 (1994).
23. Geltman, S., *Phys. Rev. A* **51,** R34–R37 (1995).
24. Varró, S., and Ehlotzky, F., *Phys. Rev. A* To be published (1995).
25. Davidson, M.D., Muller, H.G., and van Linden van den Heuvell, H.B., *Phys. Rev. Lett.* **67,** 1712–1715 (1991).
26. Stapelfeldt, H., Balling, P., Brink, C., and Haugen, H.K., *Phys. Rev. Lett.* **67,** 1731–1734 (1991).
27. Wallbank, B., and Holmes, J.K., *Can. J. Phys.* **71,** 326–333 (1993).

# ELECTRON-MOLECULE COLLISIONS
## – THEORY –

Electron Collisions with Oriented Molecules .................... 201
    C.M. FULLERTON
Low-Energy Electron Scattering from Polyatomic Molecules:
Recent Theoretical Results ...................................... 211
    F.A. GIANTURCO
Electronic Excitation in Electron-Molecule Scattering
using the R-Matrix Method ...................................... 233
    J. TENNYSON

# Electron Collisions with Oriented Molecules

## Christina M Fullerton[1]

*Institut für Theoretische Physik I, Universität Münster, Wilhelm - Klemm Str. 9, D - 48149 Münster, Germany*

> This report investigates electron scattering from oriented molecules. The spin - exchange effects are studied for low - energy electron collisions with both unoriented and oriented oxygen molecules. Electronic orientation and alignment parameters are defined and investigated in order to obtain a deeper insight into the dynamics of the collision. A brief discussion follows on electron scattering from oriented molecules.

## INTRODUCTION

Over the past few decades, there has been much progress in the understanding of electron collisions with randomly oriented molecules and our attention is now turning to collisions with molecules whose internuclear axis is fixed in space, for example in the case of molecules adsorbed on a surface. In most of this report, we consider electron collisions with free oriented molecules, in order to get a feeling about the effects we can expect. In the last section, I shall describe recent work on molecules at surfaces.

We define an oriented molecule as one whose internuclear axis is fixed in space and consider now why the study of oriented molecules is of interest to both theoretical and experimental physicists.

There has been great experimental progress in the production of beams of oriented molecules. It is now possible to produce a beam of oriented molecules through the preparation in the presence of an external field [1], [2]. Theoretical and experimental results for differential cross sections in elastic electron scattering show a strong dependence of the DCS on the orientation of the molecule [3], [4], [5], [6].

Electron scattering from adsorbed molecules is of increasing interest. One effect of the surface on the collision is a purely geometry one - the surface fixes the orientation of the molecule, and it is of interest to compare DCS for isolated, oriented molecules with those obtained for the adsorbed molecule in order to consider the effect of the surface on the collision.

By studying oriented molecules, we can define electronic orientation and alignment parameters which offer a deeper insight into the dynamics of the

---

[1]Permanent address: The Department of Applied Mathematics, Queens University, University Road, Belfast, BT7 1NN, N. Ireland

© 1995 American Institute of Physics

collision. It is also possible to investigate the symmetries which contribute to the scattering amplitudes for particular molecular orientations - e.g. when the molecule lies along the laboratory Z - axis for elastic electron scattering with oxygen, only the $\Sigma$ symmetries contribute to the scattering process and in particular when the scattering angle is 90° only the $\Sigma_g$ symmetries contribute [5].

## THE STUDY OF SPIN - EXCHANGE EFFECTS IN ELECTRON COLLISIONS WITH ORIENTED MOLECULES

We consider now the study of the spin - polarization of an electron beam, firstly for collisions with randomly oriented molecules, and then we look at collisions with oriented molecules in an effort to explain the results obtained for randomly oriented molecules.

In general, changes in the spin - polarization of an electron beam can be caused by electron - exchange, by explicit spin - dependent interactions such as spin - orbit, or by a combination of both of these effects. This change in the polarization is known as **depolarization** and is calculated as the polarization fraction P'/P , where P(P') is the initial (final) polarization of the electron beam. For collisions with light targets such as $O_2$, we can neglect the spin - orbit interaction and we expect the exchange effects to be responsible for changes in the spin - polarization of the electron beam.

Experiments have been carried out over the past few years, in which the spin - polarization of the electron beam before and after scattering is measured, enabling exchange effects to be observed [7], [8]. They observed the interesting result that in elastic collisions of transversely polarized electrons with the open - shell molecules $O_2$ and NO, the spin - exchange effects were significantly smaller than for the atomic cases Na and Hg.

In figure 1 we present theoretical calculations of the polarization fractions for elastic collisions at 10 eV with randomly oriented $O_2$ molecules. The solid line are the theoretical results of [10], the linespoints are of [9] and the experimental results are those of [8]. Our results show good agreement with experiment and confirm the experimental observations that there is very little deviation in the polarization fraction from unity.

For light molecules, such as $O_2$, the polarization fraction can be shown [9], to be related to the spin - flip cross section as follows:

$$\frac{P'_y}{P_y} = (1 - 2\frac{w_{SF}}{\sigma}) \qquad (1)$$

where $w_{SF}$ is the spin - flip cross section and $\sigma$ is the differential cross section

We define the spin - flip cross section as:

$$w_{SF} = \frac{4}{27}|F^{(\frac{1}{2})} - F^{(\frac{3}{2})}|^2 \qquad (2)$$

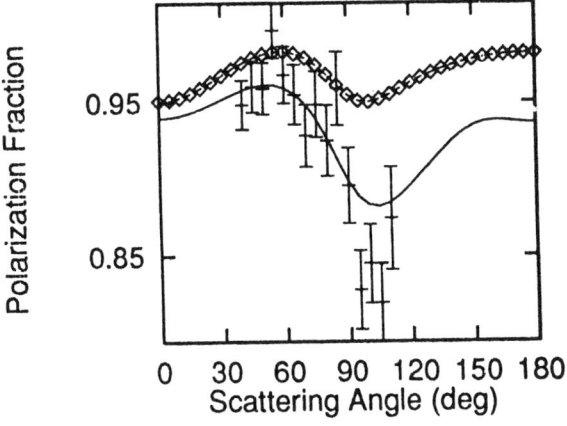

FIG. 1. Polarization Fractions at 10 eV

where $F^{(S)}$ is the spin - irreducible scattering amplitude for a total spin S. In our most recent calculations [10], these scattering amplitudes were calculated using T - matrices [11] for ten fixed - nuclei symmetries and using the vibrationally averaged resonant $^2\Pi_u$ and $^4\Sigma_u$ states [12]. Our earlier results were obtained using T - matrices at fixed internuclear separations for all symmetries [9]. In the neighbourhood of a resonance, there can be a large change in the spin - flip cross section, as there may be considerable differences between the doublet and quartet scattering amplitudes. Thus from equation (1) we can expect significant deviations in the polarization fraction from unity. Secondly, if the differential cross sections become very small, then there may be large depolarization. Theoretical calculations for the polarization fraction for elastic electron scattering from $O_2$ has also been carried out using the Schwinger multichannel method [13].

But why should there be such a difference in the behaviour of the polarization fractions for atoms and molecules? One would expect that there should also be significant deviations due to resonances and other interference effects in electron scattering with $O_2$ and NO. In order to discuss this problem, we consider now the case when the target is an oriented $O_2$ molecule. We have investigated the polarization fractions for electron scattering from an oxygen molecule with various orientations [5] and find that there are indeed large deviations in the polarization fraction. We consider, as an example, in figure 2 the case when the molecule lies in the scattering plane along the lab Z axis, described by Euler angles ($\alpha = 0°, \beta = 0°$) and for a scattering energy of 10 eV. We see that at $\theta = 90°$, there is considerable depolarization especially in comparison with the unoriented case, represented here by the dashed line. When we examine the DCS for this particular orientation, we see that the DCS becomes extremely small in the region of depolarization and we found similar results for other orientations. However since the scattered intensity is

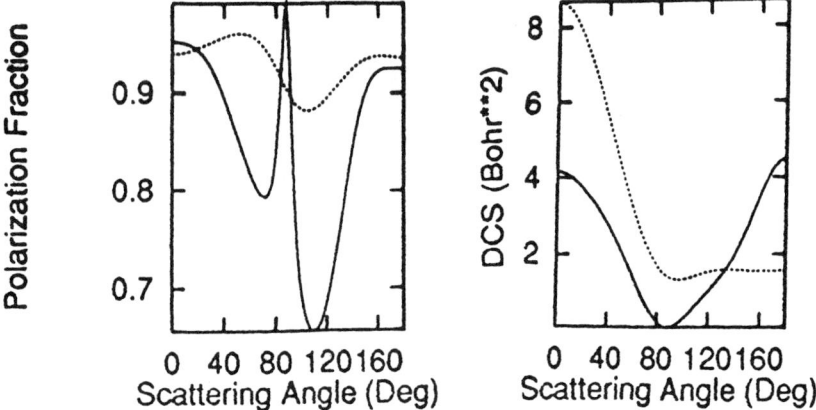

FIG. 2. Polarization fractions and DCS for an oriented $O_2$ molecule (solid line) with orientation $\alpha = 0°, \beta = 0°$ and for an unoriented molecule (dashed line)

very low, these effects wil not be observed when the average over all molecular orientations is taken and so their contribution to the unoriented calculation is very small.

## ELECTRONIC ORIENTATION AND ALIGNMENT PARAMETERS

We consider now scattering from CO, in particular the excitation of CO from its ground state $X^1\Sigma^+$ to the $a^3\Pi$ state. When discussing the excitation to states with $\mid \Lambda_j \mid \neq 0$, we must consider the phenomenon of $\Lambda$ - doubling - i.e. when the $\mid \Lambda_j \mid$ term splits into two states $\mid +\Lambda_j \rangle$ and $\mid -\Lambda_j \rangle$, where the different sign indicates the opposite direction of rotation of the electronic charge cloud around the internuclear axis. We consider $\Lambda$ to be the component of the orbital angular momentum along the internuclear axis for spinless molecules. The corresponding energetic splitting is normally negligible compared with other internal or collision energies, and the relevant distinction between the two states is not their energetic splitting but the difference in the direction of the orbitals relative to the molecular rotation plane [14]. It has been observed in chemical reactions, inelastic scattering and surface scattering that it is possible to preferentially populate one of these $\mid \pm\Lambda_j \rangle$ states. We have investigated the orientational dependence of these orbitals on the scattering process.

Since $\mid +\Lambda_j \rangle$ and $\mid -\Lambda_j \rangle$ are degenerate, we represent the final molecular state $\mid \Psi_j \rangle$ as the coherent superposition:

$$\mid \Psi_j \rangle = [f(\Lambda_j) \mid \Lambda_j \rangle + f(-\Lambda_j) \mid \Lambda_j \rangle] \quad (3)$$

where the corresponding scattering amplitudes are given by $f(\Lambda_j) = f(\Lambda_j \underline{k_j}; \Sigma^{\pm}\underline{k_i}; \underline{n})$. Equation (3) is characterized by the parameters $\mid f(\Lambda_j) \mid$,

$| f(-\Lambda_j) |$ and $\chi$, the relative phase. We now choose a new set of observables to completely characterize $| \Psi_j \rangle$, and these are:

(i) the differential cross section $\sigma$:

$$\sigma = | f(\Lambda_j) |^2 + | f(-\Lambda_j) |^2 \qquad (4)$$

(ii) the expectation value of the z - component (molecular system) of the electronic orbital angular momentum $L_z$:

$$\langle L_z \rangle = \frac{| \Lambda_j | \{| f(\Lambda_j) |^2 - | f(-\Lambda_j) |^2\}}{\sigma} \qquad (5)$$

(iii) the expectation value of the reflection operator in the molecular xz plane:

$$\langle \sigma_{xz} \rangle = \frac{(-1)^{\Lambda_j} 2 | f(\Lambda_j) || f(-\Lambda_j) | \cos \chi}{\sigma} \qquad (6)$$

(iv) and the expectation value:

$$\langle \sigma_{xz} L_z \rangle = \frac{i | \Lambda_j | 2 | f(\Lambda_j) || f(-\Lambda_j) | \sin \chi}{\sigma} \qquad (7)$$

Usually $\langle L_z \rangle$ corresponds to the rotation of the electronic charge cloud around the internuclear axis. It measures the amount of angular momentum transfer to the molecules during the collision. Similar to atomic physics we shall call $\langle L_z \rangle$ the **electronic orientation**. It characterizes the difference in the population of the $| +\Lambda_j \rangle$ and $| -\Lambda_j \rangle$ states, as can be seen from equation (5).

It can be shown that $\langle \sigma_{xz} \rangle$ and $\langle \sigma_{xz} L_z \rangle$ characterize the shape and spatial orientation of the charge cloud in the following way

$$| \Psi |^2 = \sigma | \Psi_{|\Lambda_j|} |^2 [1 + \langle \sigma_{xz} \rangle \cos(2 | \Lambda_j | \phi) + \frac{i}{|\Lambda_j|} \langle \sigma_{xz} L_z \rangle \sin(2 | \Lambda_j | \phi)] \qquad (8)$$

where we have followed the notation of [15]. We see from equation (8) how the shape of the charge cloud changes as we rotate the molecule around the internuclear axis. This change is determined by $\langle \sigma_{xz} \rangle$ and $\langle \sigma_{xz} L_z \rangle$. In atomic physics, the shape and spatial orientation are characterized by the alignment parameters. For this reason we call $\langle \sigma_{xz} \rangle$ and $\langle \sigma_{xz} L_z \rangle$ the **electronic alignment** parameters. In chemical physics, the parameter $\langle \sigma_{xz} \rangle$ has been discussed [14].

In order to depict the directional properties of the electronic charge density, we use linear combinations of the $| +\Lambda_j \rangle$ and $| -\Lambda_j \rangle$ states as basis states. For example, for the Π states, we use the following orthogonal combinations [6]:

$$| \Pi_x \rangle = -\frac{1}{\sqrt{2}}[| \Lambda = 1 \rangle - | \Lambda = -1 \rangle] \qquad (9)$$

and

$$| \Pi_y \rangle = \frac{i}{\sqrt{2}}[| \Lambda = 1 \rangle + | \Lambda = -1 \rangle] \qquad (10)$$

where we have chosen a notation in close analogy to the atomic states. These orbitals have the same shape and can be transformed into each other by a rotation of 90° around the internuclear axis. A simple interpretation can then be given to $\langle \sigma_{xz} \rangle$:

$$\langle \sigma_{xz} \rangle = \frac{| f(\Pi_x) |^2 - | f(\Pi_y) |^2}{\sigma} \qquad (11)$$

where for example f($\Pi_x$) is the scattering amplitude for excitation to the $| \Pi_x \rangle$ state and we see that $\langle \sigma_{xz} \rangle$ characterizes the difference in population of these $| \Pi_x \rangle$ and $| \Pi_y \rangle$ states. If the molecule has spin, or if spin - orbit effects are taken into account, the parameters $d\sigma$, $\langle L_z \rangle$, $\langle \sigma_{xz} \rangle$ and $\langle \sigma_{xz} L_z \rangle$ must be averaged over the spins.

We have investigated the electronic orientation and alignment parameters for the excitation of CO from its ground state $X^1\Sigma^+$ to the $a^3\Pi$ state, at various orientations in order to extract more information on the dynamics of the collision. These parameters were calculated using T - matrices calculated by [16]. Figure 3 shows the results obtained for the differential cross sections at 10 eV.

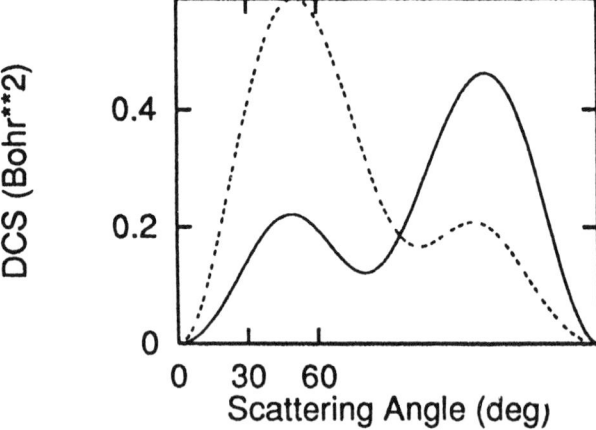

**FIG. 3.** DCS at 10 eV for electron scattering from an oriented CO molecule. The solid line are the results for the molecule with orientation $\alpha = 0°, \beta = 0°$, i.e. along the Z axis and the dashed line are for the orientation $\alpha = 0°, \beta = 180°$

The molecule is oriented along the lab Z axis, with orientation described by $\alpha = 0°, \beta = 0°$ (solid line) and then rotated by 180° around the Y axis (dashed line). In the former case we find that for lower scattering angles the DCS are much smaller than for the larger angles. However when we rotate the molecule around the Y axis, we find the reverse situation - the DCS are much larger for lower scattering angles.

We now consider the molecule to be oriented along the Y - axis, described by $\alpha = 90°, \beta = 90°$. In figure 4 we show the results obtained at 10 eV for the electronic orientation parameter $\langle L_z \rangle$. $\langle L_z \rangle$ is zero at $\theta = 0°$ and $\theta = 180°$ due to symmetry requirements. If we consider the results at 10 eV as an example, for scattering angles larger than about $\theta = 80°$, we see that the state $| \Lambda_j = -1 \rangle$ is preferentially populated. It should also be noted that $\langle L_z \rangle$ is energy dependent.

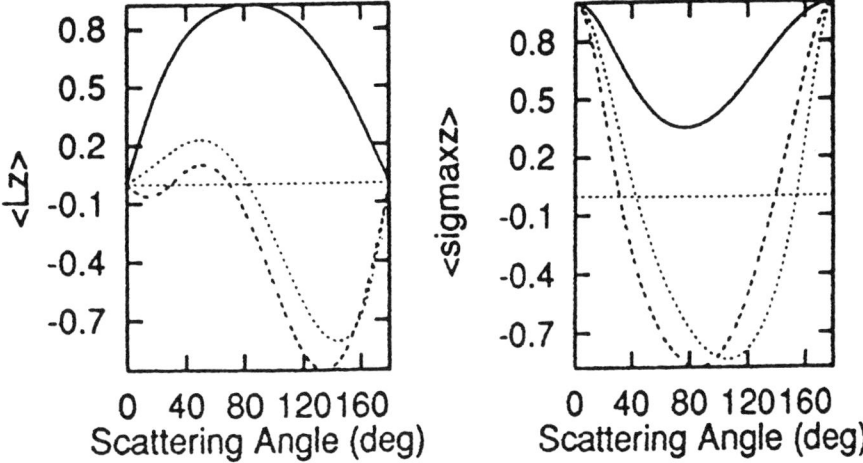

**FIG. 4.** Results obtained for the electronic orientation parameter $\langle L_z \rangle$ and the electronic alignment parameter $\langle \sigma_{xz} \rangle$ for the molecular orientation $\alpha = 90°, \beta = 90°$ for energies of 7 (solid line), 10 (dotted line) and 12 eV (dashed line)

We look next at the electronic alignment parameter $\langle \sigma_{xz} \rangle$ in figure 4 for the same orientation $\alpha = 90°, \beta = 90°$. Again we see that this parameter is dependent on the energy and we can use equation (17) to interpret our results, where we have changed to the basis states $| \Pi_x \rangle$ and $| \Pi_y \rangle$. At all energies, when $\theta = 0°$ and $\theta = 180°$, only the $| \Pi_x \rangle$ state is populated, due to symmetry requirements. At the higher energies, 10 and 12 eV, this state becomes less populated, as the scattering angle increases, until at $\theta \approx 90°$, the $| \Pi_y \rangle$ state is almost fully populated. At $\theta = 180°$, the $| \Pi_x \rangle$ state is again fully occupied.

## ELECTRON SCATTERING FROM ADSORBED MOLECULES

One method of obtaining oriented molecules is by adsorption on to a surface. Therefore the DCS obtained in this scattering situation differ considerably from those obtained for randomly oriented gas phase molecules.

The surface effects the collision in several ways - the first is a purely geometric effect where the surface fixes the orientation of the molecule. Secondly, the introduction of an attractive **image potential** modifies both the resonance lifetime and position and multiple scattering can occur - the electron can experience multiple scattering with the surface or with other adsorbates.

Resonances for physisorbed molecules were first observed in 1981 [18], [19] and decreases in both the resonance energy and lifetimes compared with the free molecule were reported. Since then, there have been considerable experimental studies carried out in this area, and I would refer the reader to recent reviews [20], [21].

Work is in progress to solve this scattering problem using R - matrix theory. For full details of this theory see [22] and [23]. Configuration space is divided into two regions. In the inner region, electron exchange and correlation effects are dominant and in the outer region these effects are negligible. Since the physical situations in each region are different, a different method of solving the collision problem is adopted in these regions.

The surface potential $V_S(\mathbf{r})$ is represented by a shifted classical image potential outside the surface [24] and smoothly joins a constant potential inside the surface:

$$V_S(\mathbf{r}) = \begin{cases} -|V_0| & z \leq 0 \\ -\frac{1}{4(z+z_0)} & z \geq 0 \end{cases}$$

where z is the distance of the electron above the surface. This potential is expanded in spherical harmonics about the molecular centre of mass and is included in the Hamiltonian to give:

$$H_{N+1} = H_N - \frac{1}{2}\nabla^2_{N+1} + V^{mol}(\mathbf{r}_{N+1}) + V_S(\mathbf{r}_{N+1}) \tag{12}$$

where $H_N$ is the fixed - nuclei Hamiltonian for the target, modified by the interaction with the surface, $V^{mol}$ is the electron - molecule interaction potential, $-\frac{1}{2}\nabla^2_{N+1}$ is the k.e. operator of the scattering electron and $V_S(\mathbf{r}_{N+1})$ is the electron - surface interaction potential.

Thus the calculations in **both** the inner and outer region are modified by the surface potential. One effect of this potential is the breaking of the molecular symmetry, allowing molecular target states and continuum states with different $\lambda$ symmetries to couple and therefore produce a much larger Hamiltonian matrix. Work is ongoing in this area.

Another theoretical approach has been adopted known as the Coupled Angular Modes method (CAM) [25]. As in the R - matrix method, configuration space is divided into two regions and in the outer region the electronic charge density is small. The effect of the inner region is represented by a boundary condition on the electronic wavefunction on the surface of the sphere. The electron - surface interaction potential is modelled as before using the step - potential.

Using this method [26], it was found that for adsorbed CO, as the molecule approached the surface, the resonance energy decreased. This was in agreement with the experimental results [27] who found a decrease in the resonance energy under adsorption. The isolated CO molecule has a $^2\Sigma$ resonance with an energy of 19.4 eV [28]. The image potential lowered this to 18 eV. This method has also been used in the study of adsorbed $N_2$ [29] and in dynamic studies, where the nuclear motion during the electron - molecule collision is taken into account [30].

Theoretical studies have also been conducted [31] in which multiple scattering effects are included using the Korringa - Kohn - Rostoker method. This is an extremely interesting new area of electron - molecule scattering theory which can be seen to yield some extremely interesting physical effects.

## CONCLUSION

We have seen how important it is to study electron collisions with oriented molecules to give us a deeper insight into what actually happens during the collision. By studying oriented molecules, we can explain the difference in the spin - exchange effects observed both experimentally and theoretically for the atomic and molecular cases.

It is possible to characterize the molecular charge cloud distribution completely using electronic orientation and alignment parameters. These parameters provide information on the rotation of the charge cloud around the internuclear axis and on the shape and spatial orientation of the orbitals. We have investigated how these parameters change during the collision.

Finally, one way of producing oriented molecules is to adsorb them on a surface. There is much interest now in this area of electron collisions with adsorbed molecules; in order to study the effects of the surface on the collision, we must compare the situation with collisions from isolated oriented molecules.

## ACKNOWLEDGEMENTS

This work was performed in collaboration with G Woeste and C J Noble of CCL Laboratory Daresbury, D G Thompson, P G Burke and K Higgins of Queens University, Belfast, K Blum and R - P Nordbeck of University of Münster, Germany and R Srivastava of Roorkee, India. This work has been supported by a grant from the EC - HCM programme under contract number ERB CHRX CT920013.

## REFERENCES

1. Kaesdorf S., Schönhense G. and Heinzmann U., *Phys. Rev. Lett.* **54**, 885 (1985).
2. Volkmer M., Meier C., Mihill A., Fink M. and Böwering N., *Phys. Rev. Lett.* **68**, 2289 (1992).

3. Mihill A. and Fink M., *Z. Phys. D.* **14**, 77 (1989)
4. Kohl D. A. and Shipley E. J., *Z. Phys. D.* **24**, 33 (1992).
5. Nordbeck R. P., Fullerton C. M., Woeste G., Thompson D. G. and Blum K., *J. Phys. B: At. Mol. Opt. Phys* **27**, 5375 (1994).
6. Woeste G., Fullerton C. M., Blum K. and Thompson D. G., *J. Phys. B: At. Mol. Opt. Phys* **27**, 2625 (1994).
7. Hegemann T., Oberste - Vorth M., Vogts R. and Hanne G. F.,*Phys. Rev. Lett.* **66**, 2968 (1991).
8. Hegemann T., *Doctoral Thesis* Universität Münster, (1993).
9. Fullerton C. M., Woeste G., Thompson D. G., Blum K. and Noble C. J., *J. Phys. B: At. Mol. Opt. Phys* **27**, 185 (1994).
10. Fullerton C. M., Woeste G. and Higgins K., *J. Phys. B: At. Mol. Opt. Phys*, submitted (1995)
11. Noble C. J. and Burke P. G., *Phys. Rev. Lett.* **68**, 2011 (1992).
12. Woeste G., Noble C. J., Higgins K., Burke P. G., Brunger M. J., Teubner P. J. O and Middleton A. G. *J. Phys. B: At. Mol. Opt. Phys*, submitted (1995).
13. da Paixão F. J., Lima M. A. P. and McKoy V., *Phys. Rev. Lett.* **68**, 1698 (1992).
14. Andresen P. and Rothe B. W.,*J. Chem. Phys* **82** 3634 (1985)
15. Green S. and Zare R. N., *Chem. Phys.* **7**, 62 (1975).
16. Morgan L. A. and J. Tennyson J.,*J. Phys. B: At. Mol. Opt. Phys* **26**, 2429 (1993)
17. Andersen N., Gallagher J. and Hertel I., *Physics Reports* **165**, 1 (1988)
18. Sanche L. and Michaud M., *Phys. Rev. Lett.* **47**, 1008 (1981).
19. Demuth J. E., Schmeisser D. and Avouris P., *Phys. Rev. Lett.* **47**, 1166 (1981).
20. Sanche L., *J. Phys. B: At. Mol. Opt. Phys* **23**, 1597 (1990)
21. Palmer R. E. and Rous P. J., *Rev. Mod. Phys.* **64**, 383 (1992)
22. Burke P. G., Mackey I. and Shimamura I., *J. Phys. B: At. Mol. Opt. Phys* **10**, 2497 (1977)
23. Gillan C. J., Nagy O., Burke P. G., Morgan L. A. and Noble C. J., *J. Phys. B: At. Mol. Opt. Phys* **20**, 4585 (1987)
24. Jennings P. J., Jones R. O. and Weinert M., *Phys. Rev. B* **37** (1988)
25. Teillet - Billy D. and Gauyacq J. P., *Surf. Sci.* **239**, 343 (1990)
26. Teillet - Billy D. and Gauyacq J. P., *Nucl. Inst. Meth. B* **58**, 393 (1991)
27. Jones T. S., Ashton M. R. and Richardson N. U., *J. Chem. Phys.* **90**, 7564 (1989)
28. Tronc M., Azria R. and le Coat Y., *J. Phys. B: At. Mol. Opt. Phys* **13**, 2327 (1980)
29. Teillet - Billy D., Djamo V. and Gauyacq J. P., *Surf. Sci.* **269/270**, 425 (1992)
30. Gauyacq J. P., Djamo V. and Teillet - Billy D., *Electron Collisions with Molecules, Clusters and Surfaces* New York: Plenum Press, 1994, 217 - 226
31. Rous P. J., *Surf. Sci.* **260**, 361 (1992)

# Low-Energy Electron Scattering From Polyatomic Molecules: Recent Theoretical Results

### Franco A. Gianturco

Department of Chemistry, The University of Rome, Città Universitaria, 00185 Rome, Italy.

**Abstract** - A review is given on the recent progress in the development of theoretical models for treating the scattering of slow electrons from polyatomic molecules. Several different theoretical methods have been intensively used in recent years thanks to the rapidly increasing power and availability of high performance computers. They have provided us for the first time with a rather detailed picture of the basic physics involved in the interaction and dynamics of electron collisions with complex targets.

## INTRODUCTION

Low-energy electron scattering from gaseous molecules of ever increasing complexity has already been studied for more than half a century [1], both from the experimental and the theoretical points of view, because of its relevance in many areas of fundamental and applied investigations. There are in fact several, and apparently different, topics for which a knowledge of the various possible processes activated by collisions of electrons with molecular gaseous targets is of fundamental importance for the global understanding of the more macroscopic phenomena. Suffice it to mention here the modelling of discharges in low-temperature gases, the behaviour of excitation and dissociation pathways in reactive gases employed in plasma processing and the effect of electron cooling after solar ionisation in the interstellar media to demonstrate the broad range of applicability of electron-molecule scattering data and the corresponding importance of their theoretical and computational description.

This short review will be concerned with a more limited range of phenomena, i.e. with scattering cross sections at low collision energies ($E_{coll} \leq 50$ eV) and mostly those which have been measured and computed in the last couple of years or so. This is because a similar report was already presented at the Aarhus meeting in 1993 [2]. We will briefly summarize below the various theoretical methods that have already come a long way in describing at the quantitative level the behaviour of the experimental cross sections and will also try to show how important has been the refining and development of articulate 'parameter-free', ab initio models of electron-molecule interaction in order to gain a fuller understanding of the molecular processes underlying the scattering events for targets of increasing complexity.

# CURRENT THEORETICAL METHODS

We will chiefly consider here the elastic and inelastic processes which occur with electrons and positrons as projectiles and without formation of new 'reactants' from the original molecular target states:

$$e^{\pm} + M(v_i, j_i | n_o) \rightarrow M(v_f, j_f | n_o) + e^{\pm} \qquad (1)$$

where the non linear polyatomic target is considered to be initially prepared in a given rotovibrational state $|j_i v_i\rangle$ and to be at most excited in a final $|j_f v_f\rangle$ state. Its electronic state, $|n_o\rangle$, is assumed to be left unchanged during collisions. This usually means that the scattered electron velocities, classically speaking, are smaller or at most of the same order, of those of the target electrons which play an active role in the collisions. It also means that, for positron projectiles, only scattering below the thresholds of positronium formation can be considered. We also assume that all relativistic effects can be neglected and restrict our treatment to molecules made up of light atoms and ions. At the end of this review we will also mention some results of computations that relax the limitation of the $|n_o\rangle$ state being unaltered.

The Schrödinger equation describing the collision of the electron with the N-electron target with M atomic nuclei is given by the familiar expression

$$H_{N+1} \Psi_\alpha(r, X) = E \Psi_\alpha(r, X) \qquad (2)$$

where E is the total energy of the (N+1) electron system and of the M nuclei and the Hamiltonian is given, in atomic units, by

$$H_{N+1} \hat{T} + \hat{V} + \hat{H}_{target} \qquad (3)$$

where $\hat{T}$ is the kinetic energy operator for the incident electron, $\hat{V}$ is its interaction energy with the bound particles of the target

$$\hat{V} = \sum_{y=1}^{N} |r - x_i|^{-1} + \sum_{y=1}^{M} Z_y |r - R_y|^{-1} \qquad (4)$$

$\hat{H}_{target}$ is the electronic Hamiltonian for the target electrons and r is the position of the continuum electron. The symbol X represents here the collective positions of the bound electron coordinates ($x_i$ i=1,N) and of the bound molecular nuclei $R_y$ (y=1,M). Usually its asymptotic eigenstates

$$\hat{H}_{target} \phi_\beta(x) = \varepsilon_\beta \phi_\beta(x) \qquad (5)$$

are employed to expand the total wavefunction of eq.(2) over the unknown coefficients for the continuum electron.

In order to first circumvent the additional problems created by the nuclear relative motion, it has been usually expedient to start by solving the scattering problem within the familiar separation of variables associated to the approximation described as the Fixed-Nuclei Approximation (FNA) [3]

$$\Psi_\alpha(r,X) = \sum_\beta A\{\Psi_\beta^\alpha(r)\phi_\beta(X)\} \quad (6)$$

where $A$ is the antisymmetrisation operator and $\varepsilon_\beta$ is the electronic eigenvalue for the $\beta$ asymptotic target state. The insertion of expansion [6] into eq.(2) leads, after multiplying it on the left by the conjugate of one of the representative target eigenstates, to a set of coupled partial integro-differential equations (PIDE's) given by

$$\left\{\frac{1}{2}\nabla^2 + (E - \varepsilon_\beta)\right\}\Psi_\beta^\alpha(r) = \sum_{\beta'}\int V_{\beta\beta'}^\alpha(r, r')\Psi_\beta(r')d^3r' \quad (7)$$

which holds for a given initial state of the target $|\alpha\rangle$ and where the kernel of the integral operator for the potential is in general the sum of diagonal (local) and non-diagonal (non-local) terms and runs over all the asymptotic states which are in principle necessary to have the sum of eq.(6) converge to an acceptable representation of the (N+1) electron system in its initial $|\alpha\rangle$ state. Its explicit expression obviously depends on the form of the target states used in the expansion of eq.(6). The above expansion is usually referred to as the close coupling (CC) expansion and was first introduced for atomic targets [1]. It was further developed in its modern form by generalising the Hartree-Fock (HF) equations for the bound states to the atomic continuum states [4].

Even when the nuclear dynamical coupling is separated out, the eq.s(7) constitute still a formidable problem because of the complicated nature of the $V_{\beta\beta'}^\alpha$ coupling potentials and of the complex interplay during the scattering process between the motion of the colliding electron ant that of the bound electrons. For solving the CC scattering equations a few theoretical methods have gained a certain amount of popularity in recent years and have been employed to treat collisions of electrons and positrons with polyatomic molecules. There are essentially two aspects of the problem which need to be considered:
(i) which theoretical method to apply to solve the CC scattering problem;
(ii) how to describe at best, within the chosen method, the coupling matrix elements on the r.h.s. of eq.(7). The above two aspects are naturally interrelated but are not necessarily fully dependent on each other. This means that each selection of the scattering framework within which polyatomic targets have been described can be matched with different levels of sophistication in treating the microscopic interaction forces represented by the above coupling matrix elements. In the following we will therefore briefly outline the most popular methods in which the scattering problems have been actually cast and then discuss the various ways in which the corresponding interactions have been treated within that method.

## Solving The Scattering Equations

Not many methods have been actually applied to polyatomic targets in the sense of being used to solve in an <u>ab initio</u> fashion the scattering equations and produce the $\underline{S}$ or $\underline{T}$-matrix as a result of the computational effort. We will list below the variational approaches, the R-matrix methods, the CC single center expansions as the ones applied thus far to polyatomic molecules.

### (1) The Kohn Variational Method

The complex Kohn variational method (CKVM) is an algebraic variational technique which, over the last few years, has been developed into an effective approach for studying both heavy-particle reactive collisions [5] and electron scattering problems for increasingly more complicated molecular targets [6]. Although variational methods based on the Kato identity [7], such as the CKVM [8], have been used successfully in electron-atom scattering problems they were only recently applied to electron scattering from molecular targets because of two fundamental difficulties: (i) the occurrence of anomalous singularities in the reactance or in the $\underline{K}$-matrices and (ii) the computational difficulties associated with the evaluation of multi-center integrals involving continuum functions.

The first problem can be easily solved by formulating the variational problem with physical, outgoing wave complex boundary conditions, a fact first noticed in nuclear physics [9] and later introduced in molecular scattering when treating reactive collisions [10].

The second problem has been practically approached by the judicious use of separable approximation in the handling of exchange integrals, thus transferring the major part of the computational effort to the construction and manipulation of bound-bound matrix elements [11]. An extensive review of the method has been recently given [12] and will not be repeated here. Suffice it only to say that the method is now providing a flexible computational tool, albeit one computationally intensive, for treating elastic and inelastic scattering from polyatomic targets and which also includes electronic excitation processes. We will discuss below some specific examples.

### (2) The Schwinger Multichannel Variational Method

When one decides to implement this approach for electron-polyatomic molecule scattering processes, then the (N+1) electron wavefunctions of the full system are expanded in a basis of discrete and/or analytic continuum functions and the coefficients are determined by finding the extremum of the Schwinger variational expression for the scattering amplitude. In the SMCV method the correct boundary condition is automatically incorporated through the use of the Green's function and therefore it can employ basis functions with arbitrary boundary conditions [13,14]. Such a feature therefore enables one to use an $L^2$ basis for scattering calculations and provides the principal motivation for applying this method to atomic and molecular scattering [15]. In fact, when Gaussian-type functions are used for the

$L^2$ basis centered over the various molecular nuclei and plane waves are employed to treat initially the incoming electron, then all the two-electron repulsion integrals can be calculated analytically. Furthermore, the integrals involving the Green's functions that are usually considered an important bottleneck of the computational aspects of the method have been efficiently and accurately computed using various computational devices [16]. The method has been recently reviewed [14,15] and its application to polyatomic targets also discussed extensively in a very recent volume [17]. We will therefore not go into its details any further but will discuss some of its most recent applications to large polyatomics in the following Section.

## (3) The R-Matrix Method for Polyatomics

Although the computational applications of R-matrix theory to electronic collisions with atomic and diatomic targets have been carried out quite extensively from the 1970's [18-20], it has found practical applications to polyatomic molecules only in very recent years [21,22]. The method has been extensively reviewed in recent books [20,22] and it will not be repeated here in detail. The essence of the R-matrix idea, in the context of electron collisions from polyatomic systems, is to treat the full (N+1)-electron problem and the M-nuclei problem within a confined region of space defined by a finite sphere of fixed radius. One will therefore enforce special boundary conditions which discretize the spectrum of the full electronic Hamiltonian for each nuclear configuration. Those discrete states are then used to expand the full scattering wavefunction inside the sphere and the result of the expansion is then matched to the correct asymptotic form on the surface of the chosen sphere.

Since no quantum chemistry code is designed to perform the solution of the (N+1) electron problem on a finite region, key computational modifications need to be introduced, a task pursued independently by the Belfast's group led by P.G. Burke [23] and by the Bonn group in his recent implementations [21, 22, 24]. Briefly, the latter results were attained by (i) confining the electron-repulsion integrals and the nuclear attraction integrals to the R-matrix sphere by using a multipole expansion to approximate an effective potential outside the sphere and (ii) by taking into account correlation and polarisation effects only for selected, energetically low-lying (N+1) electron CI roots (corresponding to the poles of the R-matrix) which have nonvanishing amplitudes on the R-matrix sphere. Higher lying roots are then approximated by a simple Hartree-Fock static-exchange calculation.

## (4) The Close-Coupling Single-Center expansions

In the treatment of electron (and positron) scattering from polyatomic molecules the earliest of the computational methods employed approaches the search for the needed continuum functions similarly to that used for electron scattering from atomic targets [25]. This means that the three-dimensional scattering functions are expanded in a set of symmetry-adapted angular functions and that the corresponding coefficients are represented on a numerical grid centered in the most convenient Body-Fixed frame of reference for the (N+1) electron problem [26]. In this approach, any arbitrary three-dimensional function $F^{p\mu}(r,\vartheta,\varphi)$ is expanded as

$$F^{p\mu}(r,\vartheta,\varphi) = \sum_{lh} r^{-1} f^{p\mu}_{lh}(r) X^{p\mu}_{lh}(\vartheta,\varphi) \qquad (8)$$

where the function $F^{p\mu}$ transforms as the $\mu$th element of the pth irreducible representation (IR) of the point group of the molecule in question. The functions $X^{p\mu}_{lh}$ are generalized harmonic functions which are eigenfunctions of the $\hat{L}^2$ operator and are given by symmetry-adapted linear combinations of the familiar spherical harmonics $Y_{lm}(\vartheta,\varphi)$ in the form

$$X^{p\mu}_{lh}(\vartheta,\varphi) = \sum_m b^{p\mu}_{lhm} Y_{lm}(\vartheta,\varphi) \qquad (9)$$

Further details about the computation of the above coefficients have been given elsewhere [27,28] and will not be repeated here.

Once all the three-dimensional functions have been expanded using the above relations, the full scattering equations are block-diagonalized for each I.R. of the FN molecular target and reduced to a set of ordinary Integro-differential equations (IDE's) describing the Close-coupling (CC) expansion of the (N+1) electron wavefunction

$$\left\{ \frac{1}{2} \frac{d^2}{dr^2} + (E - \varepsilon_\alpha) \right\} \Psi_{lh\alpha}(r) = \sum_{l'h'\beta} V_{lh\alpha, l'h'\beta}(r, r') \Psi_{l'h'\beta}(r') dr' \qquad (10)$$

where the potential terms on the r.h.s. describe both the local and non-local interaction of the scattering electron with the molecular target. In the case of positron projectiles the non-local interactions due to antisymmetrisation of the whole electronic wavefunction are not present.

The CC-SCE approach has been extensively employed by our group in Rome and results for polyatomic targets have also been recently obtained from the Y. Itikawa's group in Japan [29-31].

## Treating The Interaction Forces

Low-energy electron scattering is acutely sensitive to all three constituents of the electron-molecule interaction potential

$$\hat{V}_{int}(r,R) = \hat{V}_{st}(r,R) + \hat{V}_{ex}(r,R) + \hat{V}_{cp}(r,R) \qquad (11)$$

which are, first the electrostatic term arising from Coulomb interactions between the projectile and the bound electrons and nuclei of the target, second the exchange term arising from anti-symmetrisation requirement on the (N+1) particle electronic wavefunction of the system and finally the correlation-polarisation potential arising at short-range from bound-free many-body effects and at long-range from induced polarisation effects. All three terms are non-spherical and, strictly speaking, both the last two are non-local and dependent on the scattering energy of the impinging electron.

When only the first two contributions are considered, then the problem of calculating the full potential of eq.(11) is somewhat simplified and is similar to solving the quantum chemistry Hartree-Fock (HF) equations for the N bound electrons while adding the extra electron under the influence of the undistorted HF field from the target electrons. It is usually called the Static-Exchange approximation and, when computed following the above procedure, is called the Exact-Static-Exchange (ESE) approximation. All the methods for solving the scattering equations discussed before can be used at this level of treatment and often produce reasonable agreement with experiments. However, it is mandatory to correctly include correlation-polarisation effects to treat resonances and the very low-energy (threshold) behaviour of the elastic and inelastic cross sections, integral and differential.

The way of doing this within the language of the expansion procedures outlined before is by including an extensive set of additional configurations which will correlate the motion of the impinging electron with those of the bound electrons. In principle, the total wavefunction can be described to any desired level of accuracy with a trial function where the above additional configurations are increased in number till convergence is reached for any selected scattering observable. In practice, however, the open-channel expansion is usually truncated rather severely and the target states themselves are largely approximate and therefore the additional configurations play a limited role. These two facts, as often discussed in presenting applications to complex targets [2], can markedly affect the quality of the computed cross sections and it becomes very important to balance the inclusion of correlation corrections in the N-electron and the (N+1) electron systems in order to progress uniformly towards improved scattering quantities. The latter procedure, therefore, has developed into a rather refined art based on previous experience and strongly dependent on the particular coding of the method employed. In general, however, such procedures are all very slowly convergent and very seldom have full scattering calculations on polyatomic molecules been carried out with progressive extension of the CI expansion to methodically test convergence factors. This is obviously due to the computational efforts needed to complete well balanced electron scattering calculations off polyatomic molecules and we will show below some of the most recent results along those lines of ab inito expansions which are usually extended as much as possible.

An alternative approach has been directed to the development of model potentials for treating both positron and electron scattering from polyatomic targets of increased complexity. Within this context, model potentials are usually meant to be those constructed from first principles without the inclusion of any parametric dependence or of any procedure of "tuning" some scattering quantity to experimental observables. From the results which have been obtained in the last couple of years, it appears that such model treatments have indeed great potentials for handling realistic situations and, most important, to be able to go beyond the FNA simplification and to analyse energy loss data from electron-spectroscopy experiments [32].

## (1) The DFT Treatment Of Correlation Forces

Even when static and exchange interactions are accounted for exactly, one is still left with the all-important problem of including target polarisation effects. This task, as mentioned before, can be performed in a number of ways. The most straightforward approach is to include more states in the wavefunction expansion. These additional states can either be eigenstates of $\hat{H}_{target}$ or can be chosen to be additional pseudostates whose inclusion attempts to reproduce as accurately as possible the static and low-frequency dynamic polarisability of the target system [20]. A second possible approach is to include the appropriate optical potential. Such a potential can either be obtained by purely ab initio methods [33] or can be constructed by using a local density approximation and developing a Density Functional Theory (DFT) formalism to include correlation and polarisation effects [34-37]. The latter approach has been very successful in the last few years in producing quantitative accord between computed and measured dynamical observables while keeping the computational effort within reasonable limits [28]. Briefly, the Correlation-Polarisation effective potential, $V_{cp}$, contains a short-range correlation contribution which is smoothly connected to the long-range polarisation potential given by second-order perturbation expansions. The short-range part, the $V_c$ potential, is obtained as a real, local quantity by defining an average correlation energy of a single particle either within the formalism of the Kohn and Sham variational theorem [38] for a free-electron-gas [35,36] or by using a numerical procedure which starts from the HF target orbitals and includes gradient corrections to the evaluation of $V_c$ [37]. In both cases the correlation effect is introduced as a functional of the target electronic density [27]. The long-range part, the polarisation potential $V_p$, is obtained by first constructing a model polarisation potential which asymptotically agrees with the potential obtained from the static polarisability of the molecular target. It corresponds to including the first term in the second-order perturbation expansion of the $V_p$. This part of the full model potential can be obtained by assuming a single-center origin for the $V_p$ or by partitioning the static polarisability to different centers

$$V_p(r) = -\sum_{j=1}^{Q} \frac{\alpha_j}{2(r-R_j)^4} \qquad (12)$$

where the individual atomic polarisabilities are estimated using simple transferability criteria subject to the constraint that the total target polarisability is reproduced

$$\alpha_T = \sum_{j=1}^{Q} \alpha_j \qquad (13)$$

where Q may or may not coincide with the number of atoms in the molecular target. In general, the form of eq.(12) does not match exactly, at short distances, the form of $V_c$ from the DFT formulation at any given value of the electron-molecule distance. To select an appropriate matching radius, $r_{match}$, one can first expand both $V_p$ and $V_c$ on a symmetry-adapted set of angular functions centered on the c.o.m. of the target [28]. Then one can find the radial region where the two l=0 radial coefficients intersect and choose the smaller of the crossing radii as the $r_{match}$ selected for all the components of the two contributions to $V_{cp}$ [38]. The above prescription also provides an interesting distinction for the case of positron scattering, since in most systems the final results for low-energy integral cross sections are very sensitive to the $r_{match}$ values and indicate that they should be different when positron scattering is considered as opposed to electron scattering [39,40].

## (2) Modelling Exchange and Correlation Forces

A further level of approximation when one intends to analyse polyatomic targets has been the replacement of exact treatments of exchange with approximate local forms which allow for an easier numerical solution of the CC equations (7). One possibility is given by semiclassical approximations to the exchange interaction (SME) whereby the local velocity of the impinging electron is modified by the average local momentum of the bound electrons in each target MO [41]. Another popular alternative for polyatomic targets has been to use the extension, due to Hara [42], to scattering problems of the familiar Slater average exchange potential for bound states [43]. It was extended to polyatomic targets by Salvini and Thompson [44] and more recently by the Itikawa's group in Japan [29]. One often finds that the energy dependence of the Hara-free-electron-gas-exchange (HFEGE) potential is rather weak over moderate ranges of energy and therefore a further simplification is attained by using the above potential, computed at some mean energy value, for the whole range of energies of interest [38].

The above method has been recently implemented within the SCE approach and found to give reasonable results in connection with the additional free-electron-gas (FEG) modelling of the $V_{cp}$ potential [29,30,31].

## (3) Modelling The Full Interaction

As the size of the molecules increases, and if one is interested in extensive computational tests over several nuclear geometries and for a variety of excitation processes, one needs to further simplify the treatment of interaction forces, while still treating the scattering problem within one of the methods outlined before.

In the last couple of years two distinct approaches have been tried for polyatomic targets; the formulation of a pseudo potential (PP) method and the quasiclassical model R-matrix for dissociative attachment (DA). We will briefly discuss both methods below.

For low-energy collision processes only the valence electrons are often important to mediate the electronuclear interaction with the impinging projectile, while the core electrons simply become an unwanted burden that increases the computational time. One way of eliminating the core electrons is therefore to use some sort of pseudo potential (PP) model which should have the property of rendering the valence wavefunctions much smoother, thereby allowing their expansion onto smaller basis sets of either $L^2$ or sine-cosine functions [45]. In the case of electron scattering, the smooth and energy-independent PP which go under the name of norm-conserving PP were obtained from all-electron calculations performed within the LDA scheme [46-48]. These methods produce valence pseudo-wavefunctions which are identical to the true wavefunctions beyond some core radius $r_c$ and never require any orthogonalisation to the core states. Such quantities can include relativistic effects and can be fitted with familiar analytic functions so that the corresponding matrix elements describing the interaction between bound and continuum electrons can be easily evaluated [49].

These model potentials have been recently applied, within the SMCV method discussed before, to a few polyatomic targets of increasing size (e.g. $Si_2H_6$ and $CH_2O$) and have been found to yield reasonable agreement both with experiments and with previous calculations [50]. Further applications, at the Static-Exchange level, to other multielectron targets also seem to confirm the good general promises of this method [51], although no extensive inclusion of CI treatment for correlation-polarisation effects has been attempted thus far with this approach.

Another method for treating specific dynamical processes which occur between electrons and polyatomic targets has been implemented by the Nebraska group [52] for calculations of dissociative attachment cross sections in systems like $CH_3Cl$, $C_2H_5Cl$ and several other monochloroalkanes [53,54]. The basic theoretical model [55] employs the R-matrix approach to incorporate the resonant character of the scattering process and the long-range (dipole and polarisation) interactions between the impinging electron and the molecular target. The coupling between vibrational channels in the inner region is then included semiclassically and using empirical parameters dictated by the features of the experiments. In spite of its empirical nature, the above treatment is beginning to shed new light on the coupling sequence during the scattering dynamics of DA processes in large molecules, a type of experiment that could not be yet reproduced with fully ab initio methods for such complicated targets.

Finally, a rather useful computational tool for the study of shape resonances in polyatomic targets has been recently introduced by developing angular adiabatic basis function for the expansion of eq.(9). The idea is based on the observation that centrifugal effects control the penetration of low-energy particles into atoms and molecules. While in atoms the approximate spherical symmetry of the electrostatic field $V(r)$ allows it to be algebraically added to the centrifugal field that yields an effective potential for each partial wave, the large anisotropy of a molecular system make such simple addition not possible. However, since the anisotropy of the inner well of a molecular system only affects the few partial waves that penetrate into that region, one could try to replace the spherical basis of the atom with a new molecular basis which is radially adapted, i.e. with the adiabatic labelling of the allowed partial wave coming from the diagonalisation of the full potential [56,57]

$$\left\{ \sum_{l'h'} V^{p\mu}_{lh,l'h'}(r) + \frac{\hat{l}^2}{2r^2} \right\} \phi^{p\mu}_{lh}(r;\hat{r}) = \varepsilon^{p\mu}_{lh}(r) \phi^{p\mu}_{lh}(r;\hat{r}) \qquad (14)$$

for each value of the electron-molecule distance r. The new angular functions coincide with one of the symmetry-adapted radially diabatic functions of eq.(9)

whenever $l^2/2r^2 \gg$ potential coupling i.e.: 1) for very small r and 2) for very large r. The further expansion of eq.(8) on the new complete basis

$$F^{p\mu}(r,\vartheta,\varphi) = \sum_{lh} r^{-1} g_{lh}^{p\mu}(r) \phi_{lh}^{p\mu}(r|\vartheta,\varphi) \qquad (15)$$

reduces the relevant Schrödinger equation to a set of coupled radial equations, where the couplings further result from the dependence of the angular functions on r, and are accordingly confined only to that range of r and l values over which this dependence is appreciable. In other words, the whole effect of the anisotropy is thus compacted in an adaptive way into the coupling of a few adiabatic channels over a limited range of radial values. Such a method therefore allows one to visualize the microscopic dynamics, at the level of the relevant molecular space for the scattered electron, of some of the most studied, low-lying shape resonances of temporary molecular anions [58].

As an example we show in Figure 1 the behaviour of the adiabatic potential curves computed [38] using a model HFEGE exchange potential and a FEG modelling of the $V_{cp}$ potential as discussed before. The system examined was $SF_6$ and two specific shape resonances are examined: the $t_{2g}$ resonance (left) and the $a_{1g}$ resonance (right). The eigenphase sum for the former resonance is shown in Figure 2, where the CC calculations using 64 angular functions including l values up to 30 (solid line) are compared with the same calculations which use 10 adiabatic channels only and a procedure of piecewise diabatic (PD) representation to perform the scattering calculations [38]. We clearly see that the behaviour around the resonance position, and its width, are well described by the smaller, radially adapted expansion of eq.(15). This clearly indicates that such a representation of the full anisotropy given by a model, local interaction in molecular systems can provide an efficient, and pictorially useful, description on interchannel couplings at resonances involving one electron state only.

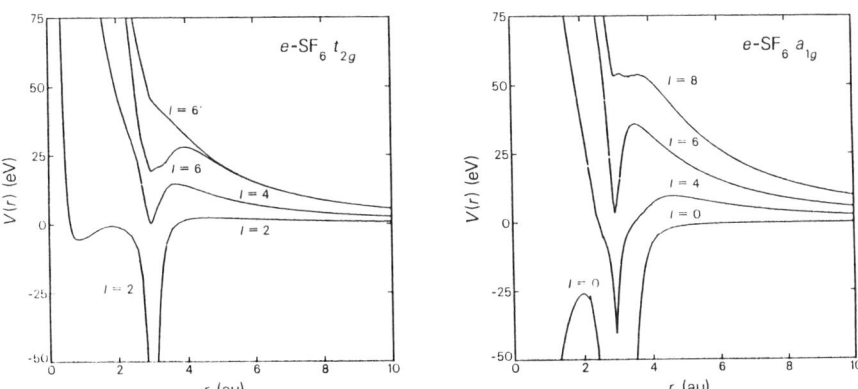

**FIGURE 1.** Computed PD representation fo the radially adaptive potentials in electron-$SF_6$ scattering fro two molecular IR's where shape resonances are observed. Left: $t_{2g}$ resonance. Right: $a_{1g}$ resonance (from ref. [38]).

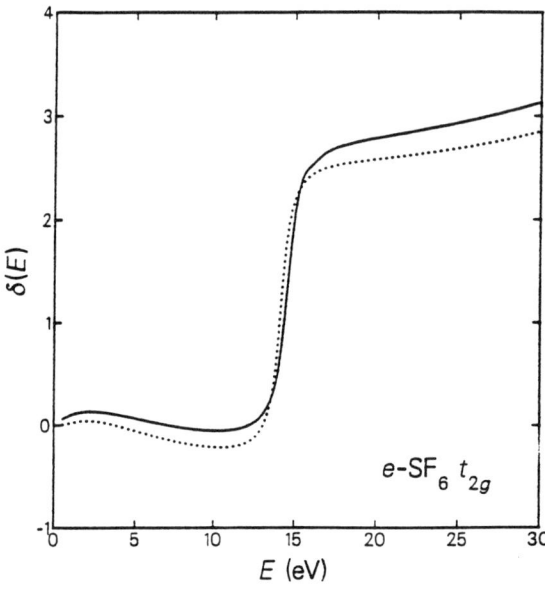

**FIGURE 2.** Computed eigenphase sum around resonance for the $t_{2g}$ IR. Solid line: full calculations with 64 channels. Dotted line: PD channels with only 10 contributions (from ref.[38]).

## RECENT COMPUTATIONAL RESULTS

As mentioned at the beginning, the last few years have witnessed a marked increase in the application of fairly sophisticated scattering methods to the computation of dynamical quantities connected with either elastic or inelastic collisions of low-energy electrons with polyatomic targets. We will try to summarize below some of the results from such calculations and to see what can be learned in general from the use of the various methods. Generally speaking, the expansion techniques which further employ a form of model interaction for either part of or for all of the contributions to eq.(11) have been the most prolific in producing new results and in extending the range of applicability to increasingly more complex molecular targets. Some of the methods which still keep a fully ab initio approach to the correlation problem, however, have also been extended beyond the traditional test case of the methane molecule, although very little in terms of convergence tests appears to be carried out within any of such applications.

### Calculations At Fixed Geometries

These calculations are usually the first ones for which theoretical methods are tested, since one can carry out the computation in the molecular frame and obtain directly the elastic, rotationally-summed, integral cross sections for each contributing IR [26]. The existence of accurate experimental data producing absolute differential cross sections is a further stimulus for improving the methodology of the full calculations. We give below examples for some polyatomic systems for which several theoretical methods have been tested and therefore give us a feeling on the relative accuracy that such calculations can achieve today.

The $CF_4$ molecule has been analysed by various experimental [59,60] and theoretical groups [61-63] because of its importance in several areas of fundamental research and directly applied physics. Thus, one could already draw some conclusions from the results on the importance of polarisation effects on angular distributions.

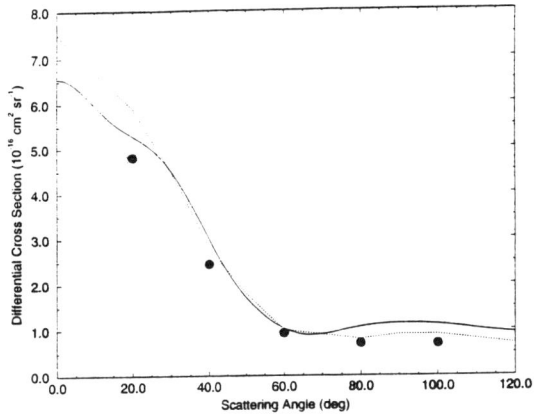

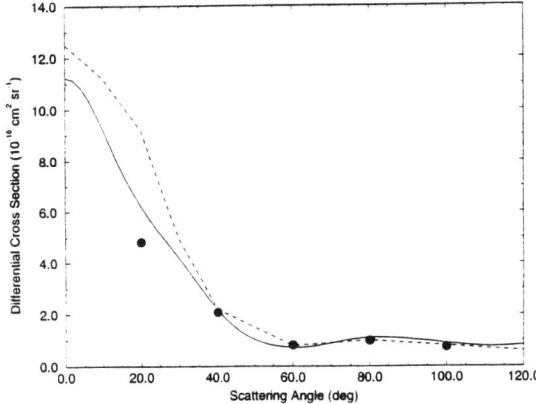

**FIGURE 3.** Elastic differential cross sections at 10 eV (top) and 15 eV (bottom) for electron-$CF_4$ scattering. Solid dots: experiments from [59]. Solid line: ESE calculations from ref.[65]. Dashed line: ESE calculations from ref. [61].

In Figure 3 we show the angular distributions measured and computed at 10 eV (top) and 15 eV (bottom) for the elastic scattering of electrons. The filled circles are the experimental data of ref. [59] and the solid line reports the Exact-Static-Exchange (ESE) calculations using the CC-SCE approach described before [63,65]. The dashed curve show similar ESE calculations using the MCSV approach [61]. We see there that the agreement with experiments is fairly good: in the range of angular distributions sampled by the experiments both methods appear to produce good results and the additional effect of polarisation forces possibly plays, at those angles, a rather minor role.

When one goes to lower energies, however, one finds that polarisation effects become important as shown by the comparison between ESE and ESEP calculations carried out by us using the DFT modelling of the $V_{cp}$ interaction [65]. As further example of such a comparison is reported in Figure 4, where the calculations of ref. [64] are compared with experiments.

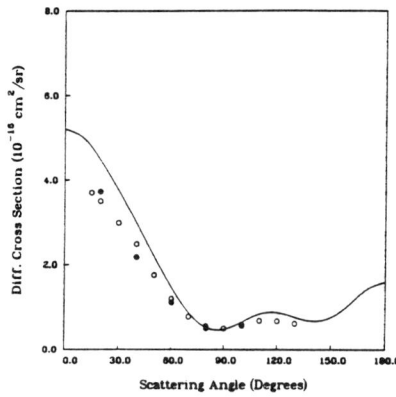

**FIGURE 4.** Computed and measured angular distribution for $e^-$-$CF_4$ scattering at 8 eV. Experiments (•) from ref. [59]. Experiments (o) from ref. [60]. ESEP calculations (—) from ref. [64].

One clearly sees there that the DCS behaviour at small angles requires the inclusion of polarisation forces, even if we see that further refinements of the employed CI expansion may be needed to improve the agreement with experiments.

Another system for which comparisons between various methods has been possible is the polar molecule $H_2S$, a target that has been analysed several times, both experimentally and theoretically, in the last few years.

The latest experimental cross sections reported absolute DCS values and obtained also elastic total cross sections and momentum transfer cross sections [66], thus providing interesting ground for comparing different theoretical methods over a fairly broad range of collision energies.

A global comparison of methods is shown in Figure 5 for the total cross sections. One sees there that the model potential using a fairly extended basis set [68] agrees very well with experiments and with the more computationally demanding CKVM calculations of ref. [70] given by a solid line. On the whole, however, theoretical methods are providing here rather good, nearly quantitative accord with the very accurate experiments. Such a gratifying agreement is also visible when angular distributions are further compared with experiments. An example at three different collision energies is shown in Figure 6, where the experiments from ref. [66] are compared with two different model potentials which employ the CC-SCE approach in carrying our the calculations [68, 71]. The agreement of the computed quantities with measurements is rather good, in keeping with the accord shown by the CKVM calculations of ref. [66].

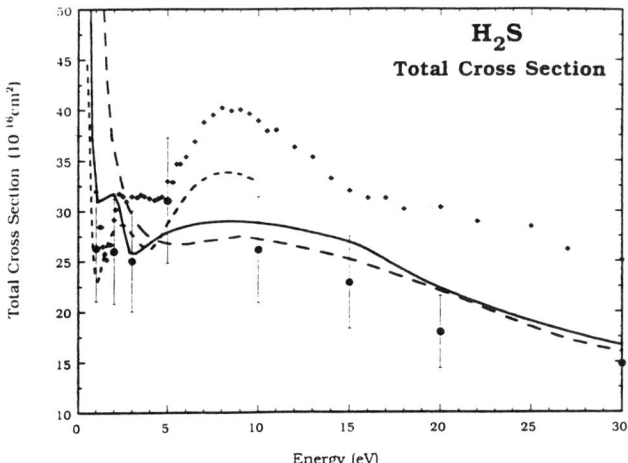

**FIGURE 5.** Total cross section for elastic scattering by $H_2S$. Elastic measurements (•) from ref. [66]. Total cross section measurements (◊) from ref. [67]. Calculations with CC-SCE model potentials: long dashes from ref. [68] and short dashes from ref. [69]. Calculations using the CKV method: solid line from ref. [70].

In other words, we have in this case a set of very accurate experimental data which are well reproduced by at least three different calculations and also by a fourth additional set of calculations [72] where exact exchange was used instead of the HFEGE employed by ref.s [68, 71]. It all indicates that the experimental and theoretical findings for polyatomic targets are coming consistently more and more together in producing the same overall picture for integral and differential cross sections. A recent confirmation of the quality of the CC-SCE approach for elastic cross sections has been given for the acetylene molecule [73].

## Beyond The FN Approximation

Much of the theoretical literature on electron-molecule scattering has focussed on the development of ab initio methods and of model potentials basically for solving the electronic fixed-nuclei (FN) problem, thereby implementing this approach within the framework of the adiabatic nuclei approximation [77]. In many cases, in fact, meaningful cross sections can be extracted solely from fixed-nuclei calculations while in other cases the FN solution can provide the necessary input for treating the fuller situation which includes the nuclear dynamics [75]. On the other hand, there are situations where the AN approximation breaks down and a different problem should be tackled. We shall give below the most recent example of a polyatomic molecule for which calculations beyond the AN approach have been attempted.

Recent reviews of electron scattering at energies around the Ramsauer-Townsend minimum [76] and around the 8 eV shape resonance [77] have been published on the $CH_4$ molecule using the FN approximation and therefore comparing only elastic cross sections with experiments. In treating only those cross sections the above

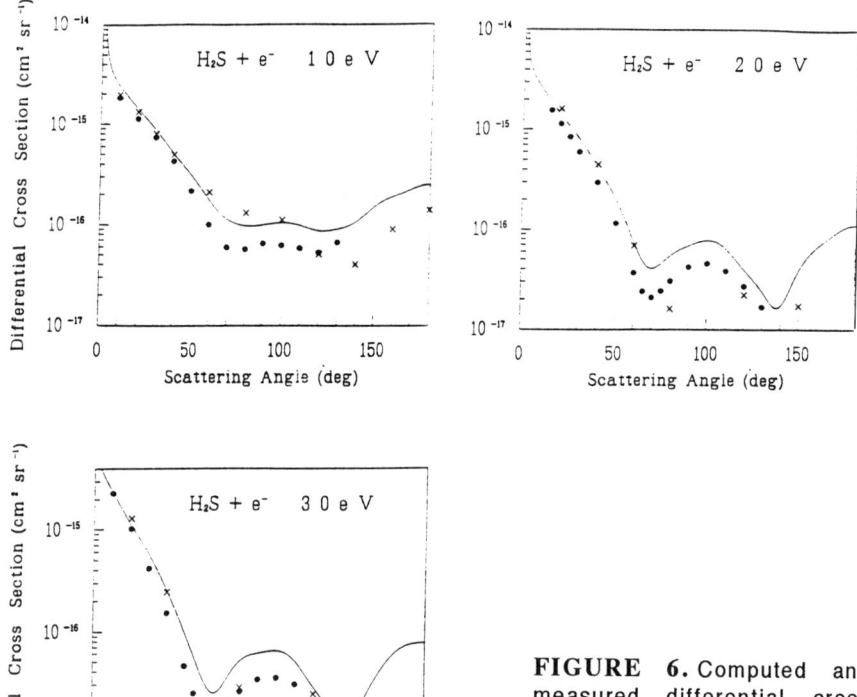

**FIGURE 6.** Computed and measured differential cross sections (elastic and rotationally summed) at three different collision energies. Experiments (•) from ref. [6]. Model calculations: solid line from ref. [71] and crosses from ref. [68].

calculations were able to employ a simple description of the nuclear dynamics, and hence concentrate on modelling the $e^-$-$CH_4$ interaction potential. The simplest way to extend such approaches to vibrationally inelastic scattering is to work within the AN approximation and therefore repeat elastic calculations at a set of nuclear geometries in order to construct the AN scattering wavefunctions for each vibrationally inelastic final channel. When applied to $CH_4$, however, corrections should be made for non-adiabatic effects since the AN approach is known to fail at low energy and to give bad descriptions of threshold scattering [78].

One way to introduce non-adiabatic corrections to a computed AN wavefunction is to substitute it into the r.h.s. of an exact Lippmann-Schwinger equation, this leading to the "off-shell" scattering amplitude [79,80,81]. This method has recently been applied to vibrationally inelastic scattering of electrons from $CH_4$ to calculate integral cross sections for excitation of the methane stretching modes at energies up to 2 eV [82] and using the CKV method discussed before.

More extensive calculations were also carried out using the CC-SCE treatment with ESE interaction and with correlation-polarisation forces described by the DFT method [83].

The conventional AN K-matrix is given by the following quadrature over nuclear normal coordinate for a given IR

$$K_{l'h'v'}^{lhv} = -\int \chi_v(Q) K_{lh,l'h'}^{FN} \chi_{v'} U_{lh,l'h'}^{\alpha}(Q) dQ \qquad (16)$$

and it assumes the nuclear and electronic motions to be adiabatically separable, a condition usually satisfied when $E \approx E_v > \varepsilon_v$, the latter being the vibrational energy spacing. One can replace the $K^{FN}$ matrix by the 'off-shell' $K_{hl,l'h'}^{OFF}(Q)$

$$K_{lh,l'h'}^{OFF}(Q) = -2\sqrt{k_v} \int_o^a j_l(k_v,r) U_{lh,l'h'}^{\alpha}(r,Q) r dr \qquad (17)$$

where

$$U_{lh,l'h'}^{\alpha}(r;Q) = \sum_{l''h''} V_{l''h''}^{hl}(r;Q) \Psi_{lh}^{\alpha}(r;Q) + \sum_{l''h''} \int_o^a W_{lh,l'h'}^{\alpha}(r,r';Q) \Psi_{lh}^{\alpha}(r;Q) r' dr' \qquad (18)$$

where a refers to any value of r beyond which the potential can be neglected and the $\Psi_{lh}^{\alpha}$ are the radial continuum functions of eq.(10). The improvement given by the formulation of $K^{OFF}$ can be recognized if one notices that it describes the asymptotic behaviour of the following functions

$$\psi_{lh,l'h'}^{off,v,v'}(r) = \sqrt{k_{v'}} r \, j_l(k_v r) \delta_{lh,l'h'} \delta_{vv'} +$$
$$+ 2\int_o^a G_{lhv}(r,r') \int \chi_{v'}(Q) U_{lh,l'h'}(r';Q) \chi_v(Q dQ dr') \qquad (19)$$

in which

$$G_{lhv}(r,r') = k_v rr' \, j_l(k_v,r_<) \eta_l(k_{v'},r_<) \qquad (20)$$

where $r_<$ and $r_>$ refer to the lesser and greater of r and r'. The eq. (19) is obtained, as mentioned before, by substituting the AN wavefunction into the r.h.s. of the llh> component of the Lippmann-Schwinger equations (LSE) which has as exact solutions the BF-vibrational close-coupling (BF-VCC) wavefunctions. The off-shell K matrix therefore includes non-adiabatic corrections and higher-order approximations from it could be obtained by substituting the $\psi_{lh,l'h'}^{off,vv}(r)$ back into the LSE and continuing in an iterative fashion to converge to the exact BF-VCC solutions [83]. The calculations carried out for CH4 indicate a threshold peak appearing in the inelastic cross sections due to the v4 mode excitation in the target molecule and a broader peak at higher collision energies as due to the $(v_1 + v_3)$ modes excitations [83]. Since the $(v_3 + v_4)$ are the infrared active modes, then the threshold peak could be viewed as brought about by the coupling of the scattered electron with the geometry dependence of the induced dipole moment, i.e. by the socalled direct mechanism for molecular excitation suggested by the experiments

[84]. The similar calculations by the CKVM [81] found a peak in the $(v_1 + v_3)$ cross section which is approximately the same as that given by the CC-SCE calculations [83]. The latter theoretical treatment also suggests that in the region of the shape resonance (~ 8.0 eV) the inelastic contributions from infrared active modes are comparable to those from the infrared active modes.

The total cross sections computed with the above method are shown in Figure 7, where they are compared with experiments and indeed exhibit very good accord with the total cross sections. As expected, the vibrational excitation processes are most effective in the energy region where resonant scattering is known to dominate.

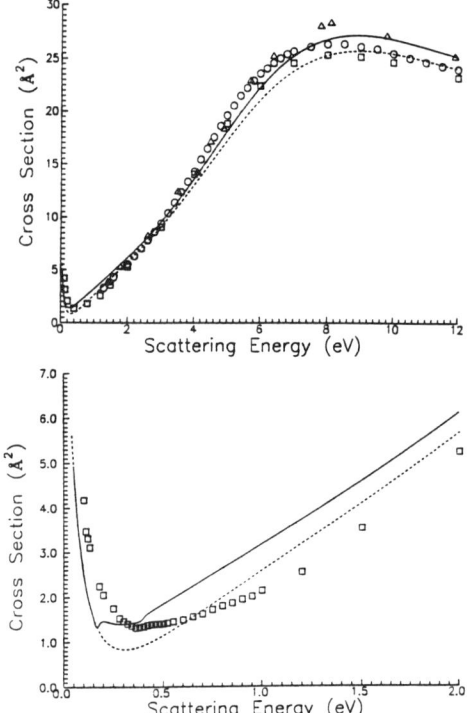

**FIGURE 7.** Computed and measured total cross sections for electron-$CH_4$ scattering. Solid line (—) shows the elastic + inelastic cross sections computed via the 'off-shell' K matrix of ref. [83]. The dashed line shows the corresponding vibrationally elastic part of the cross sections. The experiments are (top): squares from ref. [85], circles from ref. [86] and triangles from ref. [87]. At the bottom the Ramsauer minimum region is shown, reporting calculations and experiments.

## Further Excitation Processes

As mentioned in the previous sections, a great deal of the interest in electron-molecule scatering processes stems from the need to known as reliably as possible the energy dependence and the relative importance of excitation cross sections leading to a wide variety of final products of the electron-molecule 'reactions'

$$e^- + M(n_i | v_i, j_i) \to e^- + M(n_i | v_f, j_f) \quad (21a)$$

$$\to e^- + M(n_f | v_f, j_f) \quad (21b)$$

$$\rightarrow e^- + M_1(n_f|v_f,j_f) \\ + M_2(n_f|v_f,j_f)+\ldots \quad (21c)$$

where $|n\rangle$ labels the electronic states, $|v_j\rangle$ the rotovibrational states and $M_1$, $M_2$... correspond to molecular dissociation products in excited electronic states.

For polyatomic molecules, only very recently results have appeared on the application of the CKVM to the calculations of dissociative excitation cross sections of such systems and comparisons have been carried out between calculations and measurements involving $H_2O$ and $CH_4$ [88,89]. In particular, close-coupling CKVM calculations for electron impact excitation of water [88] into electronically dissociative states, in the energy range from 8 to 30 eV, have coupled five channels including $^1A_1$ ground electronic state and excited singlet and triplet states formed by promoting the occupied $3a_1$ and $1b_1$ electrons to the lowest unoccupied $4a_1$ orbital. These states have theoretical, vertical excitation energies in the range from 8 to 12 eV and are known to be dissociative. The calculations have confirmed in this system the existence of Feshbach resonances related to the $A_1$ and $B_1$ excited states responsible for dissociative electron attachment to water molecules at rather high collision energies [88].

Similar calculations have also been carried out for the $CH_4$ molecule [89] and for the $NF_3$ molecule [90]. In the case of the methane target the electronically dissociative states are found to be of Rydberg nature and were classified by analogy with the isoelectronic Ne atom as $2p\rightarrow 3s$, $2p\rightarrow 3p$ transitions, i.e. single excitations of the valence 2p electron into the n=3 Rydberg manifold, for which optical selection rules suggest large excitation probabilities to states analogues to the $^1P$ atomic odd-parity states. On the other hand, the CKVM calculations in the energy range 11-60 eV indicate a more complex dynamics with large contributions from triplet excitation processes and from even-parity ($2p\rightarrow 3p$) states [89]. The real status of the convergence level of such extended calculations was, however, rather uncertain.

It is further interesting to note that we are now beginning to have computational results for electronic excitation of polyatomic targets that are already at a very high level of sophistication and complexity and that indicate clearly the need to attack such problem with the widest possible variety of experimental data in order to really isolate the microscopic event which is being measured.

## CONCLUSIONS

The previous, very brief, description of the most recent results from various computational models which attempt to treat the full scale of the possible outcomes from electron scattering off polyatomic targets has indicated already that this field of research is still very active and full of questions which have been answered only in part by the current methodologies.

It is certainly true that the variety of experimental findings, and the high quality of the quantities which are being measured by the best experiments, are putting to test the current capabilities of the computational methods and are asking them to produce an ever increasing range of inelastic cross sections and of angular distributions from such inelastic processes.

The computational approaches, on the other hand, are attempting to treat (more or less at the same time) the following three main aspects of the problem:

(i) to select the most effective dynamical model for obtaining multichannel, CC **K**-matrices that are dependent on all the available electronuclear variables;
(ii) to treat the electronuclear coupling terms as accurately as possible within the full multidimensional space;
(iii) to advance into areas of calculations where the above quantities are needed for molecular targets of increasing complexity and size.

This is certainly a very tall order for the theoretical-computational community, but from the present analysis of the current literature the feeling is that several groups are already doing a great deal of work to meet such challenges.

## ACKNOWLEDGEMENTS

The many extensive and profitable discussions with all the temporary members of the fluctuating working group in Rome are gratefully acknowledged: with Nico Sanna, Juan-Antonio Rodriguez, Stuart Althorpe, Paola Paioletti, Bob Lucchese, Thierry Stoecklin, Frank Schneider and Charles Gillan. I am also grateful to those who kindly answered my request of reprints and preprints of their work to help me prepare the present review. Finally, the financial support of the EU-HCM network programme and of the Italian National Research Council is also acknowledged.

## REFERENCES

1. e.g. see: Massey, H.S.W. and Mohr, C.B.O., Proc. Roy. Soc. (London) **A136**, 289 (1953).
2. Mc Curdy, C.W., in *AIP Conference Proceedings n.259*, Andersen T., Fastrup B., Folkmann F., Knudsen H. and Andersen N. Eds., AIP Press (New York, 1993).
3. e.g. see: Lane, N.F., Rev. Mod. Phys. **52**, 29 (1980).
4. Seaton, M.J., Phyl. Trans. Roy. Soc. (London) **A245**, 469 (1953).
5. Zhang, J.Z.H. and Miller, W.H., J. Chem. Phys. **91**, 1528 (1989).
6. Mc Curdy, C.W. and Rescigno, T.N., Phys. Rev. **A39**, 4487 (1989).
7. Kato, T., Prog. Teor. Phys. **6**, 394 (1951).
8. Kohn, W., Phys. Rev. **74**, 1763 (1948).
9. Mito, Y. and Kamimura, M., Prog. Theo. Phys. **56**, 583 (1976).
10. Miller, W.H. and Jansen op de Haar, B.M.D.D., J. Chem. Phys. **86**, 6213 (1987).
11. Rescigno, T.N. and Schneider, B.I., Phys. Rev. **A37**, 1044 (1988).
12. Rescigno, T.N., Lensfield III, B.H. and Mc Curdy, C.W., in *Modern Electronic Structure Theory*, D. Yarkony Ed., World Scientific (Singapore, 1994).
13. Schwinger, J., Phys. Rev. **56**, 750 (1947).
14. Watson, D.K., Adv. At. Mol. Phys. **25**, 221 (1988).
15. Lucchese, R.R., Takatsuka, K. and Mc Koy, V., Phys. Rep. **131**, 147 (1986).
16. Winstead, C. and Mc Koy, V. in *Modern Electronic Structure Theory*, D. Yarkony Ed., World Scientific (Singapore, 1994).
17. Huo, W.M., in *Computational Methods for Electron-Molecule Collisions*, W.H. Huo and F.A. Gianturco Eds., Plenum Publ. Co. (New York, 1995).
18. Burke, P.G. and Seaton, M.J., Methods Comput. Phys. **10**, 1 (1971).
19. Burke, P.G. and Robb, W.D., Adv. At. Mol. Phys. **11**, 143 (1975)..

20. Burke, P.G. and Berrington, K.A. *R-matrix Theory of Atomic and Molecular Processes*, IOP Publishing (Bristol, 1993).
21. Nestmann, B.M., Pfingst, K. and Peyerimhoff, S.D., J. Phys. B **27**, 2297 (1994).
22. Pfingst, K., Nestmann, B.M. and Peyerimhoff, S.D. in *Computational Methods for Electron-Molecule Collisions*, Huo W.H. and Gianturco F.A. Eds., Plenum Publ. Co. (New York, 1995).
23. Burke, P.G. and Gillan, C.J., private communication (1995).
24. Nestmann, B.M., Nesbet, R.K. and Peyerimhoff, S.D., J. Phys. B **24**, 513 (1991).
25. Gianturco, F.A. and Thompson, D.G., J. Phys. B **9**, L383 (1976).
26. Gianturco, F.A. and Jain, A.K., Rep. Progr. Phys. **143**, 347 (1986).
27. Gianturco, F.A., Lucchese, R.R., Sanna, N. and Talamo, A. in *Electron Collisions with Molecules, Clusters and Surfaces*, Ehrhardt H. and Morgan L.A. Ed.s, Plenum Publ. Co. (New York, 1994).
28. Gianturco, F.A., Thompson, D.G. and Jain, A.K. in *Computational Methods for Electron-Molecule Collisions*, Huo W.M. and Gianturco F.A. Ed.s, Plenum Publ. Co. (New York, 1995).
29. Okamoto, Y., Onda, K. and Itikawa, Y., J. Phys. B **23**, 2405S (1993).
30. Mishimura, T. and Itikawa, Y., J. Phys. B **27**, 2309 (1994).
31. Okamoto, Y., Onda, K. and Itikawa, Y., Chem. Phys. Lett. **203**, 61 (1993).
32. Dunning, F.B., J. Phys. B **28**, 1645 (1995).
33. Klonover, A. and Kaldor, U., J. Phys. B **11**, 1623 (1978).
34. Perdew, J.P. and Zinger, A., Phys. Rev. A**23**, 5048 (1981).
35. Padial, N.T. and Norcross, D.W., A**29**, 1742 (1984).
36. Gianturco, F.A., Jain, A.K. and Pantano, L.C., J. Phys. B **20**, 571 (1987).
37. Gianturco, F.A., Rodriguez-Ruiz, J.A. and Jain, A.K., Phys. Rev. A **48**, 4321 (1993).
38. Lucchese, R.R. and Gianturco, F.A., Int. Rev. Phys. Chem., **xx**, xxx (1995).
39. Gianturco, F.A. and De Fazio, D.D., Phys. Rev. **50**, 4819 (1994).
40. Gianturco, F.A., Paioletti, P. and Rodriguez-Ruiz, J.A., Z. Phys. D **xx**, xxx (1995).
41. Gianturco, F.A. and Scialla, S., J. Phys. B **20**, 3171 (1987).
42. Hara, S., J. Phys. Soc. Jpn., **22**, 710 (1967).
43. Hara, S., J. Phys. Soc. Jpn., **27**, 1262 (1969).
44. Salvini, S. and Thompson, D.G., . Phys. B **14**, 3797 (1981).
45 Phillips, J.C. And Kleinman, L., Phys. Rev. **116**, 287 (1959).
46. Northrup, J.E., Ihm, J. and Cohen, M.L., Phys. Rev. Lett. **47**, 1910 (1981).
47. Saito, S. and Oshiyama, A., Rev. Lett. **66**, 2367 (1991).
48. Hamann, D.R., Schlüter, M. and Chiang, C., Phys. Rev. Lett. **43**, 1494 (1979).
49. Bachelet, G.B., Hamann, D.R. and Schlüter, M., Phys. Rev. B **26**, 4199 (1982).
50. Bettega, M.H.F., Ferreira, L.G. and Lima, M.A.P., Phys. Rev. A**47**, 1111 (1993).
51. Natalense, A.P.P., Bettega, M.H.F., Ferreira, L.G. and Lima, M.A.P., Phys. Rev. Lett. (submitted).
52. Pearl, D.M., Burrow, P.D., Fabrikant, I.I. and Gallup, G.A., J. Chem. Phys. **102**, 2737 (1995).
53. Pearl, D.M. and Burrow, P.D., J. Chem. Phys. **101**, 2940 (1994).
54. Pearl, D.M., Burrow, P.D., Nash, J.J., Morrison, H. and Jordan, K.D., J. Am. Chem. Soc. **115**, 9876 (1993).
55. Fabrikant, I.I., J. Phys. B **24**, 2213 (1991).
56. Battaglia, F. and Gianturco, F.A., Europhys. Lett. **10**, 117 (1989).
57. Gianturco, F.A., J. Molec. Struct. **254**, 99 (1992).
58. Le Dorneuf, M., Lan, V.K. and Launay, J.M., J. Phys. B **15**, L685 (1982).
59. Mann, A. and Linder, F., J. Phys. B **25**, 533 (1992).
60. Boesten, L., Tanaka, H., Kobayashi, A.K., Dillon, M.A. and Kimura, M., J. Phys. B **25**, 1607 (1992).
61. Winstead, C., Sun, Q. and Mc Koy, V., J. Chem. Phys. **98**, 1105 (1993).
62. Natalense, A.P.P., Bettega, M.H.F., Ferreira, L.G. and Lima, M.A.P., Phys. Rev. A **52**, R1 (1995).

63. Gianturco, F.A., Lucchese, R.R. and Sanna, N., J. Chem. Phys. **100**, 6464 (1994).
64. Huo, W.M. and Sheehy, J.A., (to be published).
65. Gianturco, F.A., Lucchese, R.R. and Sanna, N., (to be published).
66. Gulley, R.J., Brunger, M.J. and Buckman, S.J., J. Phys. B **26**, 2913 (1993).
67. Sinytkowski, Cz. and Maciag, K., Chem. Phys. Lett. **129**, 321 (1986).
68. Gianturco, F.A., J. Phys. B **24**, 4627 (1991).
69. Jain, A. and Thompson, D.G., J. Phys. B **17**, 443 (1983).
70. Lengsfield, B.H., Rescigno, T.N. and Mc Curdy, C.W., private communication as quoted in (66).
71. I. Itikawa, private communication.
72. Thompson, D.G. and Greer, R.A., J. Phys. B **27**, 3533 (1994).
73. Gianturco, F.A. and Stoecklin, T., J. Phys. B **27**, 5903 (1994).
74. Chase, D.M., Phys. Rev. **104**, 838 (1956).
75. Hazi, A.U., Rescigno, T.N. and Kurilla, M., Phys. Rev. **A2**, 1089 (1981).
76. Gianturco, F.A., Rodriguez-Ruiz, J.A. and Sanna, N., J. Phys. B **28**, 1287 (1995).
77. Gianturco, F.A., Rodriguez-Ruiz, J.A. and Sanna, N., Phys. Rev. **A52**, 1257 (1995).
78. Morrison, M.A., J. Phys. B **19**, L707 (1986).
79. Shugard, M. and Hazi, A., Phys. Rev. **A12**, 1985 419757.
80. Morrison, M.A., J. Phys. B **19**, L707 (1986).
81. Morrison, M.A., Abdolsalami, M. and Elza, B.K., Phys. Rev. **A43**, 3440 (1991).
82. Rescigno, T.N., Mc Curdy, C.W., Orel, A.E. and Lengsfield III, B.H., in *Electron Collisions with Molecules, Clusters and Surfaces*, Ehrhardt H. and Morgan L.A. Ed.s, Plenum Publ. Co. (New, York, 1993) pg.1.
83. Althorpe, S.C. and Gianturco, F.A., J. Phys. B **28**, xxx (1995)
84. e.g. see: Lunt, S.L., Randell, J., Ziesel, J.P., Mrotzeck, G. and Field, D., J. Phys. B**27**, 1407 (1994)
85. Lohmann, N. and Buckmann, S.J., J. Phys. B **19**, 2565 (1986).
86. Jones, R.K., J. Chem. Phys. **82**, 5424 (1985).
87. Dababneh, M.S., Hsieh, Y.-F., Kauppila, W.E., Kwan, C.K., Smith, S.J., Stein, T.S. and Uddin, M.N., Phys. Rev. **A3**, 1207 (1988).
88. Gil, T.J., Rescigno, T.N., McCurdy, C.W. and Lengsfield III, B.H., Phys. Rev. **49**, 2642 (1994).
89. Gil, T.J., Lengsfield III, B.H., McCurdy, C.W. and Rescigno, T.N.,Phys. Rev. **49**, 2551 (1994).
90. Rescigno, T.N., Phys. Rev. **A52**, 325 (1995).

# Electronic Excitation in Electron Molecule Scattering using the R-Matrix Method

Jonathan Tennyson

*Department of Physics and Astronomy,*
*University College London,*
*Gower Street, London WC1E 6BT, UK*

Recent progress made by the UK R-matrix collaboration is reviewed. Particular attention is paid to electronic excitation calculations with have increased significantly in their scope and sophistication. Calculations on electron impact electronic excitation of $N_2$, $O_2$ and CO are used as examples. Examples of other processes studied in recent calculations are also given. Future prospects are discussed.

## INTRODUCTION

The R-matrix method was originally developed for nuclear physics but has proved itself outstandingly successful for the treatment of electron collisions with neutral and ionized atoms (1). Indeed this method was the workhorse of the recently completed international Opacity Project which studied all atomic processes of importance for stellar atmospheres (2).

Attempts to perform molecular R-matrix calculations are more than two decades old. For the past decade workers at a number of UK institutions (Daresbury Laboratory, Queen's University Belfast, Royal Holloway College and University College London) have jointly maintained and developed a suite of programs for treating collisions between electrons (positrons) and diatomic molecules. This work has been supported by UK Collaborative Computational Project 2 ('Continuum States of Atoms and Molecules'). An outline of the UK R-matrix code as well an introduction to R-matrix theory is given in an extensive section on R-matrix methods in a recent book on 'Computational Methods in Electron-Molecule Collisions' (3).

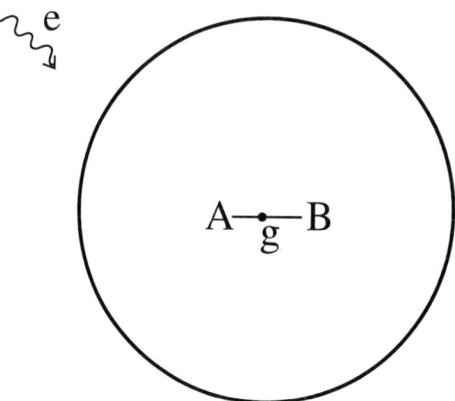

**FIG. 1.** R-matrix sphere for molecule AB. The sphere is centered on the center of gravity of the molecule, g.

Electronic excitation cross sections are needed for a variety of applications and their calculation places more emphasis on the accurate representation of target electronic wavefunctions than the study of other processes. This is because while experience has shown that target wavefunctions derived using the Hartree-Fock or Self Consistent Field (SCF) approximation often give excellent results for elastic scattering, this is not so for electronic excitation. In this case it is usually necessary to have an accurate representation of several target states and the energy gap between them. This can only be achieved by moving beyond the Hartree-Fock approximation and introducing configuration interaction (CI) into the target wavefunctions. How this is done within the R-matrix method, the limitations it gives for target representations and sample results for systems such as $N_2$, $O_2$ and CO are the major topic of this progress report.

## THE R-MATRIX METHOD

The R-matrix method divides electron configuration space into two regions, see Fig. 1. The spherical inner region, typically of radius 10–20 $a_o$, is centered on the center of gravity of the molecule and is assumed to entirely contain the $N$-electron wavefunction of the target molecule. In this region it is necessary to consider the full, multicentered interactions between the $N + 1$ electrons in the system including exchange. In the outer region it is assumed that a separate scattering electron can be identified and that exchange interactions can be neglected.

The accuracy of an R-matrix calculation is therefore crucially dependent on the representation of the problem in the inner region. In this region the scattering wavefunction is written

$$\psi_k = \sum_{ij} A\phi_i(x_1 \ldots x_N)u_{ij}(x_{N+1})a_{ijk} + \sum_n \chi_n(x_1 \ldots x_{N+1})b_{nk}, \qquad (1)$$

where the sum over $i$ runs over target states and $\phi_i$ are target wavefunctions. The $u_{ij}(x)$ are a partial wave expansion about $g$ of numerical continuum orbitals which represent the scattering electron inside the R-matrix sphere. This electron is anti-symmetrized to the target electrons by the operator $A$. The $\chi_n$ are two-centre quadratically integrable functions constructed from the target occupied and virtual molecular orbitals, which are, in our present programs, expanded as a linear combination of Slater Type Orbitals (STOs).

For studies of electronic excitation, the target wavefunctions are themselves expanded as a linear combination of configurations, $\eta_m(x_1 \ldots x_N)$

$$\phi_i = \sum_m \eta_m(x_1 \ldots x_N)c_{im}. \qquad (2)$$

Put crudely, the size of this expansion determines the quality of the target wavefunction.

Quantum chemists routinely perform electronic structure calculations on diatomics comprising C, N and/or O using a million of or so configurations in their calculations, eg (4). Scattering calculations, even with special coding to take advantage of the structure of the problem (5), have usually struggled to use more than a hundred or so target configurations. Furthermore, at least in the R-matrix method, it is necessary to represent all target states included in the calculation using a single set of molecular orbitals.

The problem can be seen in terms of the R-matrix method by looking at eq (1). The number of configurations that have to be generated in a scattering calculation is the number of target configurations *times* the number of continuum orbitals summed over all target states. Although calculations are sometimes presented in terms of total numbers of configurations generated, this measure is somewhat method dependent. It is therefore easier and safer to make comparisons between methods in terms of target representations. The quality of target representations discussed below are similar to or better than those used by workers employing other state-of-the-art scattering methods (eg (6,7)).

It is clear that in the longer term further improvements in target representations will be required. To this end considerable efficiencies have been achieved in the atomic R-matrix code by, amongst other things, removing explicit calculation of the target wavefunction from the scattering calculation (8,9). Similar developments for the UK molecular R-matrix codes are being investigated (10).

## CALCULATIONS

### Few electron targets

R-matrix electronic excitation calculations have been performed for a number of diatomic targets. I will here only consider those which used accurate target wavefunctions. Early calculations on electronic excitation of $H_2$ were performed by Branchett et al (11–13). These calculations were performed for the lowest 6 or 7 electronic states which were represented within by a full CI within the given orbital set. With a full CI representation the problem of balance between $N$ electron and $N+1$ electron systems is largely overcome. However such a representation is only possible for very simple targets.

Branchett et al only performed calculations at a single $H_2$ geometry corresponding to the equilibrium separation. This meant that their calculations were unable to probe either the detailed and complicated vibrational structure observed in the 10-12 eV resonance region (14) or the possibility of electron impact dissociation occurring via electronically excited states. New calculations are currently being performed which will address these problems (15).

Calculations have also been performed for electron impact excitation of $HeH^+$ (16) and $He_2^+$ (17). These calculations were for targets with only 2 or 3 electrons respectively and relatively few excited the states; they therefore suffer from few of the complications encountered with the heavier systems discussed below.

### Electronic excitation of $N_2$

In contrast to the experimental situation, there has been relatively little theoretical work on the electron impact electronic excitation of molecular nitrogen. This, in part, is because despite the importance of this system, its triple bond makes it a particularly difficult system to treat reliably at a low level of electronic structure theory.

Electronic excitation calculations have been performed by Huo et al (18) using the Schwinger Variational method who used excited target states represented by moving a single electron, and Gillan et al (19) using the R-matrix method who used an expansion of 4 target states represented by between 5 and 13 configurations. Neither of these calculations gave totally satisfactory agreement with experiment.

Besides the limitations of their target wavefunctions, it has become apparent that the calculations of Gillan et al suffered from a further technical problem. This was caused by the fact that in the original implementation in the R-matrix code, the phases of the individual coefficients in the target CI wavefunction, eq (2), could differ from the phases contained implicitly in the $N+1$ electron wavefunction of eq (1). This problem was identified by Orel et al (21), and is relatively easy to fix.

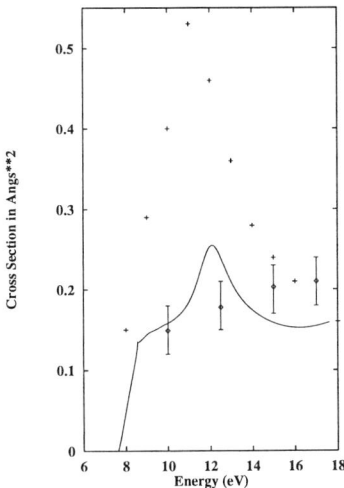

**FIG. 2.** Electron impact excitation of $N_2$: cross sections for $X^1\Sigma_g \to A^3\Sigma_u$. Curve: R-matrix calculations of Gillan et al (20); crosses: measurements of Borst (22); points: measurements of Cartwright et al (23).

For a variety of reasons therefore, Gillan et al (20) have recently completed a comprehensive new calculation of electron impact electronic excitation of $N_2$. This calculation, which was still confined to the equilibrium internuclear separation, considered the lowest 8 electronic states of $N_2$. These states were represented by a full CI within an active space of the $N_2$ valence orbitals. These orbitals comprised the occupied $2\sigma_g$, $3\sigma_g$, $2\sigma_u$ and $1\pi_u$ orbitals, and the $3\sigma_u$ and $1\pi_g$ virtual orbitals. This representation leads to between 68 and 164 configurations for the target states concerned. It is worth noting that Gillan et al found considerable sensitivity to how exactly they defined the orbitals they used.

These new calculations give significantly improved agreement with the available experimental data, see Figs. 2 and 3, as well as making predictions for a number of excitation processes for which no measurements are available. In particular the calculations of the cross sections for the optically forbidden $X^1\Sigma_g \to B^3\Pi_g$ transition show excellent agreement with experiment, see Fig 3. This is in marked contrast to the earlier R-matrix calculations.

Gillan et al (20) calculated differential cross sections for the various electronic excitation processes for the first time. There is limited experimental data available for comparisons with these and the agreement at this stage can only be described as fair; Fig. 4 gives a representative example.

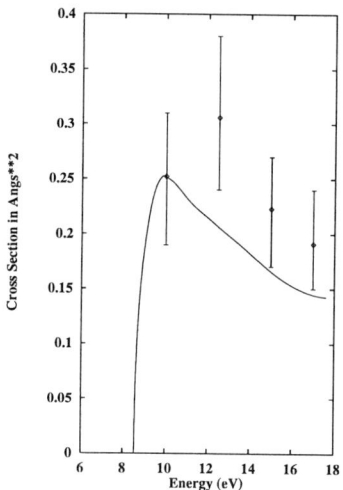

**FIG. 3.** Electron impact excitation of $N_2$: cross sections for $X^1\Sigma_g \to B^3\Pi_g$. Curve: R-matrix calculations of Gillan *et al* (20); points: measurements of Cartwright *et al* (23).

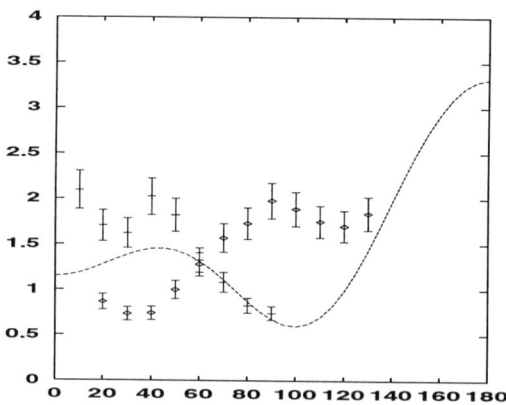

**FIG. 4.** Electron impact excitation of $N_2$: 15 eV differential cross sections for $X^1\Sigma_g \to A^3\Sigma_u$ in $10^{-18}$ cm$^2$ per steradian. Curve: R-matrix calculations of Gillan *et al* (20); crosses: measurements of Brunger and Teubner (24); points: measurements of Trajmar *et al* (25).

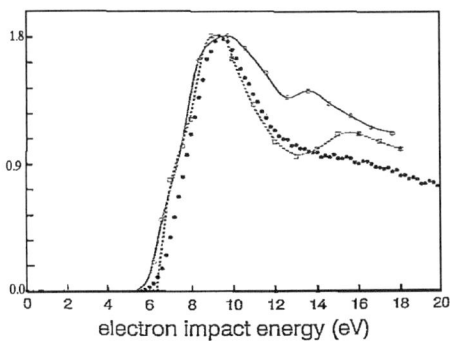

**FIG. 5.** Electron impact cross sections for excitation the low-lying metastable states of CO in $10^{-16}$ cm$^2$. Dashed curve: R-matrix calculations of Morgan and Tennyson (5); points: measurements of Le Clair et al (27); solid curve: measurements of Furlong and Newell (26). The experimental curves have been normalized to the peak of the calculated cross section.

## Electronic excitation of carbon monoxide

There have been a number of new experimental investigations of low energy electron CO excitation cross sections (26–28). Prior to these Morgan and Tennyson (5) performed R-matrix calculations for the lowest 8 electronic states of CO. These calculations were performed using a smaller complete active space than the most recent $N_2$ calculations described above which comprised $5\sigma$ and $1\pi$ occupied and the $6\sigma$ and $2\pi$ virtual orbitals. This model leads to up to 40 configurations per target state.

Morgan and Tennyson's calculations identified six resonances that had not been seen previously in more limited theoretical calculations. The positions and widths of these resonances were studied as a function of CO bondlength but no nuclear motion calculations were attempted. Instead Morgan and Tennyson concentrated on electronic excitation. In particular their predictions gave excellent agreement with subsequent experiments in the $a^3\Pi$ threshold region, see figs 5 and 6.

Previous theory (29) and earlier lower resolution experiment failed to give the pronounced peak near threshold. The differences between the results above 10 eV, where metastable states higher than the $a^3\Pi$ state become open, appears understood. The increased cross section measured by Furlong and Newell (26) is probably due to cascade effects in their experiment; whereas the structure in the R-matrix calculations is caused by a broad $^2\Pi$ (pseudo?) resonance at about 15 eV.

The recent experiments by Gibson et al have measured differential cross

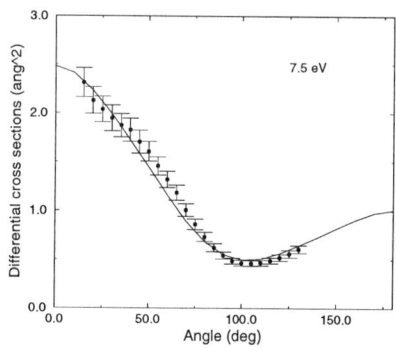

**FIG. 6.** Elastic scattering from CO: differential cross sections at 7.5 eV. Curve: R-matrix calculations of Morgan (unpublished); points: measurements of Gibson *et al* (28).

sections for a variety of processes, both elastic and inelastic, up to 10 eV. Detailed comparisons with the calculations of Morgan and Tennyson (5) and earlier R-matrix calculations by Morgan (30), which concentrated on vibrational excitation in the region of the well known $^2\Pi$ shape resonance, are still in progress.

Fig. 6 compares R-matrix calculations and experiment for an elastic differential cross section calculated from the T-matrices of Morgan and Tennyson at 7.5 eV. This is above the threshold of the $a^3\Pi$ state of CO. In contrast to previous calculations (31), the R-matrix results give excellent agreement with the new measurements.

The R-matrix T-matrices of Morgan and Tennyson (5) have also been used to predict coherence effects in electron impact excitation of the $a^3\Pi$ state of CO (32). This work found that in both forward and backward directions there was a strong preference for exciting the $\Lambda = +1$ component of the $\Lambda$-doublet.

### Electronic excitation of oxygen

The electronic excitation of $O_2$ is particularly interesting because of the presence of two lowing electronic excited states. Electron impact electronic excitation of this molecule has been the subject of a series of R-matrix calculations, of increasing sophistication, by Noble, Burke and co-workers (33–39). These calculations use up to 9 electronic states of the $O_2$ target and a target representation similar in spirit and size to those discussed for $N_2$ and CO above.

The first of these calculations to include the so-called '6 eV' electronic

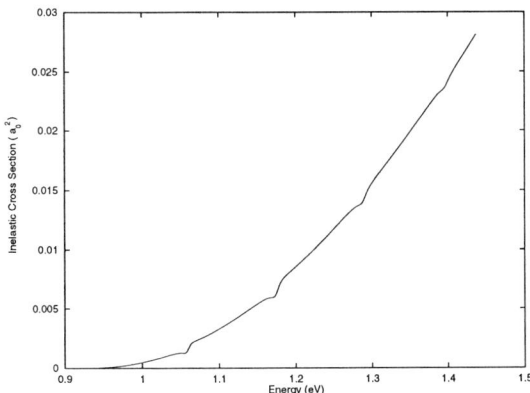

**FIG. 7.** Electron impact excitation of $O_2$: R-matrix calculations of Higgins et al (39) of the cross sections for $X^3\Sigma_g^-(v=0) \to a^1\Delta_u(v=0)$ in $a_o^2$. 7.5 eV which is above the $a^3\Pi$ threshold.

states of $O_2$ was performed for a single fixed $O_2$ bondlength. This led to the discovery, subsequently confirmed by experiment (40) of a new $^2\Pi_u$, resonance at about 8 eV. These calculations were extended to a range of geometries by Higgins et al (37).

Nuclear motion has been considered both for rotation (35) and vibration (39). The most recent calculation (39) has, for the first time, given a full vibronic treatment of the electronic excitation problem. These calculations used the non-adiabatic 'double R-matrix' formalism developed by Schneider et al (41) and applied previously to the study of vibrational excitation near a resonance (30,42).

The latest calculation concentrated on the low energy region, below 2 eV. It showed for both elastic and vibrationally inelastic collisions, the dominant effect was that the cross sections displayed series of sharp spikes very similar to those which had previously been observed experimentally (43). Calculations on the curve of well known low-lying $^2\Pi_g$ resonance showed that these spikes are close to, but do not coincide exactly with, vibrational states of $O_2^-$.

Vibrationally resolved electronic excitation cross sections, such as the one shown in Fig. 7, are somewhat different. These display a series of small dips which coincide with the spikes in the electronically elastic cross sections discussed above. The T-matrices from the $O_2$ calculations have in turn been used to provide differential cross sections (38) for a variety of processes, see Fig 8 for an example. The most recent differential cross sections are currently being compared with recent experimental work on the same problem.

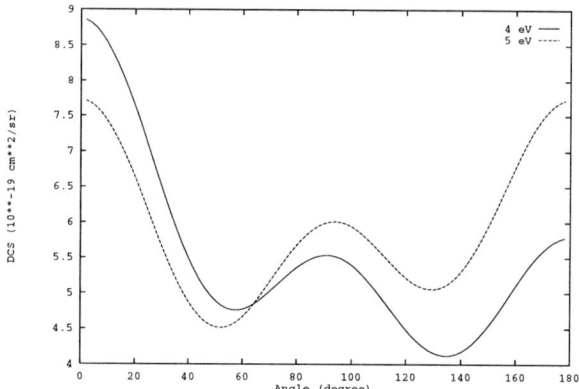

**FIG. 8.** Calculated differential cross sections for the $O_2$ electronic transition $X^3\Sigma_g^- \rightarrow a^1\Delta_g$ at 4 eV (solid curve) and 5 eV (dashed curve), see Middleton *et al* (38).

### Other processes

The discussion above has deliberately focused on electronic excitation calculation as this has been an area of broad advance in recent years. However the UK R-matrix code is a flexible package and has been to study a variety of processes. These include transitions of Rydberg molecules (44) and their associated transition intensities (45,46).

More recently the non-adiabatic approach of Schneider *et al* (41), mentioned above, has been generalized to allow the treatment of dissociative recombination. The particular problem focused on is the low energy dissociative recombination of $HeH^+$ (47). This process, whose magnitude as determined experimentally remains controversial, cannot proceed via conventional mechanisms, which involve curve crossing. However a detailed discussion of this problem is beyond the scope of this report.

### CONCLUSIONS

The R-matrix method has proved itself both flexible and durable over many years (1). However the discussion above has confined itself entirely to diatomic molecules. Electron impact electronic excitation is clearly an important process in molecules with more than two atoms as well.

An R-matrix code which treats polyatomic targets has been developed in Bonn (48). This code has a number of important differences from the R-matrix theory outlined here, particularly when it comes to representing CI target wavefunctions within a scattering calculation. However the Bonn code has yet to be applied to calculations which explicitly include more than one target electronic state. Development of a UK polyatomic R-matrix code, like the Bonn programs, based on the use of Gaussian, rather than mixed Slater and numerical, representation of the orbitals is nearing completion.

## ACKNOWLEDGEMENTS

I would like to thank all the people involved in the UK R-matrix collaboration and particularly those people, Charles Gillan, Katrina Higgins, Lesley Morgan, Cliff Noble, Darian Stibbe and Georg Wöste, who helped me prepare figures for this article. The collaboration has received support from the UK Engineering and Physical Science Research Council, and the European Community Human Capital and Mobility Network 'Electron and photon interactions with atoms, ions and molecules'.

## REFERENCES

1. P.G. Burke and K.A. Berrington, *Atomic and Molecular Processes: an R-Matrix Approach*, Bristol, England: IOP Press, 1993.
2. The Opacity Project Team, *The Opacity Project Volume 1*, Bristol, England: IOP Press, 1995.
3. *Computational methods for Electron-molecule collisions*, W. Huo and F.A. Gianturco (Eds.), New York: Plenum, 1995.
4. H. Partridge, S.R. Langhoff and C.W. Bauschlicher Jr, J. Chem. Phys., **93**, 7179 (1990).
5. L.A. Morgan and J. Tennyson, *J. Phys. B: At. Mol. Opt. Phys.*, **26**, 2429 (1993).
6. A.E. Orel, T.N. Rescigno and B.H. Lengsfield III, Phys. Rev. A, **42**, 5292 (1990).
7. T.N. Rescigno, Phys. Rev. A, **50**, 1382 (1994).
8. P.G. Burke, V.M. Burke and K.M. Dunseath, J. Phys. B: At. Mol. Opt. Phys., **27**, 5341 (1994).
9. V.M. Burke and C.J. Noble, Computer Phys. Comms., **84**, 19 (1994).
10. J. Tennyson and C.J. Noble, work in progress.
11. S.E. Branchett and J. Tennyson, Phys. Rev. Letts., **64**, 2889 (1990).
12. S.E. Branchett, J. Tennyson and L.A. Morgan, J. Phys. B: At. Mol. Opt. Phys., **23**, 4625 (1990).
13. S.E. Branchett, J. Tennyson and L.A. Morgan, J. Phys. B: At. Mol. Opt. Phys., **24**, 3479 (1991).
14. T.E. Sharp, Atomic data, **2**, 119 (1971).
15. D.T. Stibbe and J. Tennyson, work in progress.
16. B.K. Sarpal, J. Tennyson and L.A. Morgan, in *Dissociative recombination: theory, experiments and applications*, B.R. Rowe, L.B.A. Mitchell and A. Canosa (Eds.), NATO ASI series B, **313**, New York: Plenum, 1993.
17. B.M. McLaughlin, C.J. Gillan, P.G. Burke and J.S. Dahler, Phys. Rev. A, **47**, 1967 (1993).
18. W.M. Huo, C.A. Weatherford and T.L. Gibson, in Abstracts of 18th ICPEAC, A. Dalgarno *et al*, New York: Plenum, 1989, p 294.

19. C.J. Gillan, C.J. Noble and P.G. Burke, J. Phys. B: At. Mol. Opt. Phys., **23**, L407 (1990).
20. C.J. Gillan, J. Tennyson, B.M. McLaughlin and P.G. Burke, J. Phys. B: At. Mol. Opt. Phys., (to be submitted).
21. A.E. Orel, T.N. Rescigno and B.H. Lengsfield III, Phys. Rev. A, **44**, 4330 (1991).
22. W.L. Borst, Phys. Rev. A, **3**, 648 (1972).
23. D.C. Cartwright, A. Chutjian, S. Trajmar and W. Williams, Phys. Rev. A, **16**, 1013 and 1041 (1977).
24. M.J. Brunger and P.J. Teubner, Phys. Rev. A, **41**, 1413 (1990).
25. S. Trajmar, D.F. Register and A. Chutjian, Phys. Rep., **97**, 219 (1983).
26. J.M. Furlong and W.R. Newell, J. Phys. B: At. Mol. Opt. Phys., (submitted).
27. L.R. Le Clair, M.D. Brown and J.W. McConkey, Chem. Phys. **189**, 769 (1994).
28. J.C. Gibson, S.J. Buckman and M.J. Brunger, to be published.
29. Q. Sun, C. Winstead and V. McKoy, Phys. Rev. A, **46**, 6987 (1992).
30. L.A. Morgan, J. Phys. B: At. Mol. Opt. Phys., **24**, 4649 (1991).
31. A. Jain and D.W. Norcross, Phys. Rev. A, **45**, 1644 (1992).
32. A. Dellen, K. Blum and L.A. Morgan, J. Phys. B: At. Mol. Opt. Phys., **28**, 1067 (1995).
33. C.J. Noble and P.G. Burke, J. Phys. B: At. Mol. Opt. Phys., **19**, L35 (1986).
34. C.J. Noble and P.G. Burke, Phys. Rev. Lett., **68**, 2011 (1992).
35. R.P. Nordbeck, K. Blum, C.J. Noble and P.G. Burke, J. Phys. B: At. Mol. Opt. Phys., **26**, 3611 (1993)
36. C.M. Fullerton, G. Wöste, D.G. Thompson, K. Blum and C.J. Noble, J. Phys. B: At. Mol. Opt. Phys., **27**, 185 (1994).
37. K. Higgins, C.J. Noble and P.G. Burke, J. Phys. B: At. Mol. Opt. Phys., **27**, 4585 (1994).
38. A.G. Middleton, M.J. Brunger, P.J.O. Teubner, M.W.B. Anderson, C.J. Noble, G. Wöste, K. Blum, P.G. Burke, and C. Fullerton, J. Phys. B: At. Mol. Opt. Phys., **27**, 4057 (1994).
39. K. Higgins, C.J. Gillan, C.J. Noble and P.G. Burke, J. Phys. B: At. Mol. Opt. Phys., (in press).
40. A.G. Middleton, P.J.O. Teubner and M.J. Brunger, Phys. Rev. Lett., **69**, 2495 (1992).
41. B.I. Schneider, M. Le Dourneuf and P.G. Burke, J. Phys. B: At. Mol. Phys., **12**, L365 (1979).
42. L.A. Morgan, J. Phys. B: At. Mol. Opt. Phys., **19**, L439 (1986).
43. D. Field, G. Mrotzek, D.W. Knight, S. Lunt and J.P. Zeisel, J. Phys. B: At. Mol. Opt. Phys., **21**, 171 (1991).
44. B.K. Sarpal, S.E. Branchett, J. Tennyson and L.A. Morgan, J. Phys. B: At. Mol. Opt. Phys., **24**, 3685 (1991).
45. B.K. Sarpal and J. Tennyson, J. Phys. B: At. Mol. Opt. Phys., **25**, L49 (1992).
46. S.E. Branchett and J. Tennyson, J. Phys. B: At. Mol. Opt. Phys., **25**, 2017 (1992).
47. B.K. Sarpal, J. Tennyson and L.A. Morgan, J. Phys. B: At. Mol. Phys., **27**, 5943 (1994).
48. K. Pfingst, B.M. Nestmann and S.D. Peyerimhoff, J. Phys. B: At. Mol. Phys., **27**, 2283 (1994).

# ELECTRON-MOLECULE COLLISIONS
## – EXPERIMENT –

High Resolution Studies of the Dissociative Attachment
of Low Energy Electrons to State-Selected Sodium Dimers ....... 247
    K. BERGMANN, M. Keil, M. Külz,
    A. Kortyna, D. Weyh, and W. Meyer

Electron Scattering from Vibrationally Excited Molecules ........ 257
    P.D. BURROW

Studies of Low Energy Electron Collisions at
Sub-meV Resolution .................................................. 267
    H. HOTOP, D. Klar, J. Kreil,
    M.-W. Ruf, A. Schramm, and J.M. Weber

Absolute Cross Section Measurements for Electron Collisions
with Polyatomic Molecules of Plasma Chemistry .................. 279
    H. TANAKA and L. Boesten

# High Resolution Studies of the Dissociative Attachment of Low Energy Electrons to State-Selected Sodium Dimers

K.Bergmann[*], M.Keil[*], M.Külz[*], A.Kortyna[*], D.Weyh[#] and W.Meyer[#]

*) Fachbereich Physik der Universität, 67653 Kaiserslautern, Germany
#) Fachbereich Chemie der Universität, 67653 Kaiserslautern, Germany

**Abstract.** We present data on the vibrational level dependence of negative ion formation by dissociative attachment (DA) to $Na_2(v)$, together with preliminary theoretical results for vibrational enhancement derived from the relevant potential energy curves. We show that, for sodium molecules and our electron energy distribution, the DA rate increases with v" by more than three orders of magnitude. Above a critical value, here identified as $v"_c = 12$, the DA rate is rather insensitive to v". Preliminary results obtained with a newly developed photoelectron source show that the rate for DA to levels $v" \geq 12$ decreases rapidly with the electron energy $E_{el}$ in the range $0 \leq E_{el} < 10$ meV.

## INTRODUCTION

Molecular beam collision experiments with laser state selection were introduced some 15 years ago /1,2/. At that time, depletion of thermally populated levels in the vibrational ground state by optical pumping with lasers was used to label specific levels and to accurately measure state-to-state differential cross sections for heavy particle collisions. In the meantime new techniques have emerged which advance our ability to fully control the molecular level population /3/. In particular, we now can move population from thermally populated levels to unpopulated ones, including high lying vibrational states, very efficiently and selectively /4 - 6/. This ability is crucial to the attempt to study collision dynamics of highly vibrationally excited molecules under molecular beam conditions, since such levels are thermally not populated. Here, we report first results using new techniques for the investigation of the vibrational dependence of the process of dissociative attachment (DA) of low energy electrons.

It was also some 15 years ago that the high sensitivity of the DA rate to the vibrational excitation of a target molecule was experimentally clearly documented /7/. The underlying mechanism was subsequently theoretically analyzed /8,9/. It was recognized that the incoming low energy free electron can be captured by the molecule into a resonance state. Often this resonance is essentially repulsive (see middle one of the dashed curves in Fig.1) and the nuclei begin to seperate. Eventually the electron may be emitted, leaving the neutral molecule in a vibrationally excited level. If the negative ion of one of the fragments is stable, the resonance curve crosses the neutral curve at some distance $R_s$. The electron will remain bound to an atom for $R > R_s$. The DA rate is then given by the capture rate into the resonance curve and by the probability of the system to reach internuclear distances $R > R_s$. This survival probability is determined by the coupling of the resonance state to the continuum which can be characterized by the width $\Gamma(R)$ of the resonance. Typically, the capture of the electron by molecules in their vibrational ground state v" = 0 occurs at distances R significantly smaller than $R_s$, resulting in a relatively small survival probability. When the molecule is vibrationally excited capture is possible at larger R and the survival probability may increase dramatically with v" /7 - 10/ until the outer turning point of the vibrational motion is near $R_s$ for a level $v"_c$. For even higher

© 1995 American Institute of Physics

vibrational excitation than this critical level v"$_c$, the DA rate is expected to change only relatively little with v".

The first observation of Na$^-$ resulting from DA to vibrationally excited Na$_2$ molecules was reported by Ziesel et.al. /11/. These authors observed atomic negative ions using an effusive molecule beam. They infer that their signal derives from the very small thermal population in very high vibrational levels. Here, we report about progress towards very detailed experimental and theoretical characterisation of the vibrational dependence of the DA rate for Na$_2$ molecules.

## Theoretical potential curves

In contrast to the well studied case of H$_2$, the lowest molecular negative ion state with the electronic structure $^2\Sigma_u^-(1\sigma_g^2 1\sigma_u)$ is stable for Na$_2^-$ (see Fig. 1) and it is not involved in the DA process. Previous theoretical data on higher Na$_2^-$ states is scarce /12/.

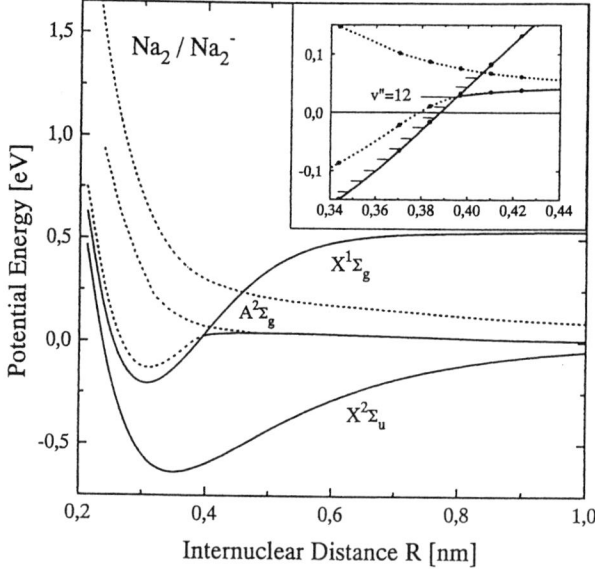

**Figure 1.** Potential energy curves for the neutral Na$_2$ molecule ($^1\Sigma$ - curve) and the molecular negative ion ($^2\Sigma$ - curves) as obtained from multi-reference configuration interaction (MR-CI) calculations /13/. The uppermost dashed curve results from a Feshbach projection while the lowest dashed curve is obtained when the 2$\sigma_g$ orbital is not constrained. The middle one of the dashed curve is obtained when the compact shape of the 2$\sigma_g$ orbital is preserved by constraining the relative population of Feshbach and shape-type electronic structures, respectively. Independent stabilization-type calculations provide evidence that this latter curve is close to the actual potential curve of the resonance.

The electronic structure responsible for the relevant negative ion resonance is $^2\Sigma_g^+(1\sigma_g 1\sigma_u^2)$. This Feshbach-type structure has a strong interaction with the electronic continuum, i.e. the resonance state acquires a significant contribution from the shape-type structure $1\sigma_g^2 2\sigma_g$

which amounts to 30 - 40% at shorter internuclear seperations. By constraining the relative population of these two structures during orbital optimizations, a compact shape of the $2\sigma_g$ orbital could be preserved also in the resonance region. The resonance potential curves resulting from subsequent MR-CI calculations are given in Fig. 1. The bound state $Na_2(X)$ and $Na_2^-$ (A) cross between the turning point of the vibrational states v"= 12 and v"= 13 /13/.

Our experiment provides, for the first time, results for the variation of the DA rate with v" for vibrational excitation below *and* far above v"$_c$. We will discuss various experimental arrangements for the study of DA with high electron energy resolution ( $\Delta E$ < 10 meV) and for molecules in a wide range of single (v", j")-levels. Such data allow stringent tests of the results from the ongoing theoretical effort to determine the energy of the resonance E(R) and its width $\Gamma$(R) and to model the DA process.

## EXPERIMENTAL

The experiment involves a supersonic beam (see Fig.2), lasers for state preparation and detection, a conventional indirectly heated surface emission electron source or a photoelectron source and a Wiley-McLaren type time-of-flight spectrometer /14/ for $Na^-$ detection.

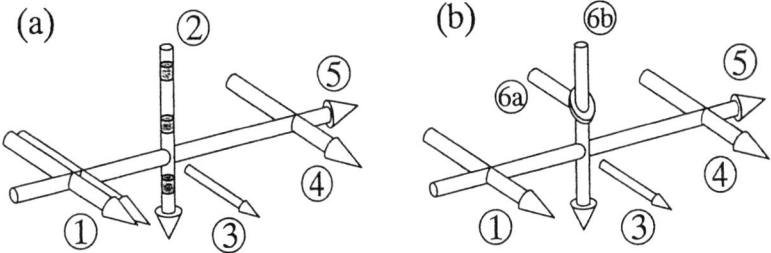

**Figure 2.** Schematics of the experimental arrangements, with the molecular beam (5), the laser(s) for vibrational state preparation (1), the thermal electron beam (2) or the lasers for the photoelectron source (6), the probe laser for testing the vibrational excitation (4) and the ion extraction system for time-of-flight analysis (3). Only one laser is used for state preparation when the Franck-Condon-Pumping technique used (see b). Two lasers are needed for the implementation of the STIRAP method (see a).

## Electron Sources

We employ two different types of electron sources. Most of the data is taken with a pulsed version of the electron beam source described in /15/. Electrons are thermally emitted from a low-temperature barium oxide cathode. They are extracted from the space charge region near the cathode by 250 ns long pulses at a repetition rate of 100 kHz. The electrons are then accelerated towards the scattering region by means of a set of three mesh-covered apertures. A magnetic guiding field of 100 Gauss is applied. The peak electron current of this source is 40 μA. The width of the electron energy distribution is approximately 0.7 eV full width at half maximum, as measured by observing $SF_6^-$ formation /16/. The width is primarily due to the thermal distribution of the emitted electrons. This distribution has been shifted energetically so as to maximize the $Na^- / Na_2$ (v" = 0) DA rate. We present also preliminary results obtained with a high resolution photoelectron source. The development of this source was inspired by earlier work of Hotop and coworkers /16 - 18/ on two step photoionisation of metastable Ar* atoms. We use the abundant Na atoms in the molecular beam as a source of photoelectrons.

The atoms are first excited to the 3p level (see inset of Fig.3) with radiation from a cw laser and then ionized by the light from a frequency doubled mode-locked Titanium-Sapphire laser (< 2 ps pulse width). The laser operates at a repetition rate of 80 MHz which is equivalent to continuous operation for our purpose. With an avarage power of up to 400 mW of blue light we achieve a photoelectron current of up to 1 nA and an energy resolution of ≤ 8 meV. The transform limited bandwidth of the radiation corresponds to an energy spread of about 1 meV. In the present set-up the observed energy spread is dominated by space charge broadening and by stray electric fields /18/. However, since the spread of the electron energy is smaller than the vibrational spacing in $Na_2$, the resolution is - at present - considered adequate. Fig. 3 shows also the variation of the photoelectron current.

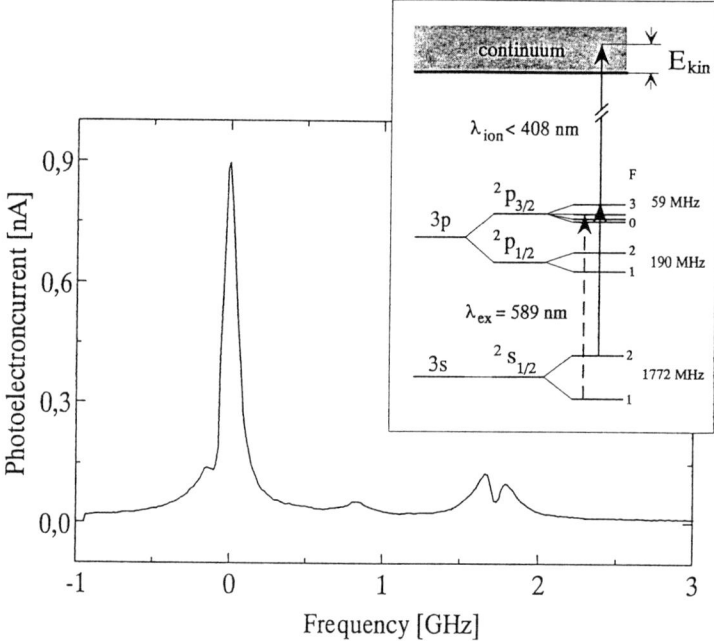

**Figure 3.** Level scheme for the two step photoionisation of the Na atoms (inset). The dependence of the photoelectron current on the frequency of the cw laser for the first excitation step, measured relative to the F = 2 ' F' = 3 transition frequency, is also shown.

The shape of the peaks can be understood by considering the hyperfine structure and optical pumping. Optical pumping can be detrimental because it renders the spatial overlap of the radiation fields for the first and second excitation step sensitive to the power of the cw laser beam /19/. In fact, the small peak in the middle occurs when the pump rate is equal for both hyperfine levels of the 3s - state and the detrimental effect of optical pumping is reduced. Soon we will employ a dual frequency laser to pump both hyperfine levels simultaneously /20/.

## Vibrational level preparation

One technique for vibrational state preparation, the Franck-Condon-Pumping (FCP) method, relies on spontaneous emission /21/. A judiciously chosen level in the electronically excited state $A^1\Sigma_u^+(v', j'= 10)$ is excited from the thermally populated level $X^1\Sigma_g^+(v'' = 0, j'' = 9)$ by a single mode laser which crosses the molecular beam at a right angle. The choice of v' determines the vibrational distribution in the $X^1\Sigma_g^+$ - state established by spontoneous emission (see Fig.4). This method involves only one laser in the preparation step, is easy to implement but does not provide the desired high state selectivity. However, the vibrational level distribution is precisely known and can be varied. The molecules encounter the low energy electrons further downstream. The population distribution is monitored even further downstream by a probe laser. The vibrational excitation is preserved along the flight path since the homonuclear molecules in their electronic ground state do not radiate.

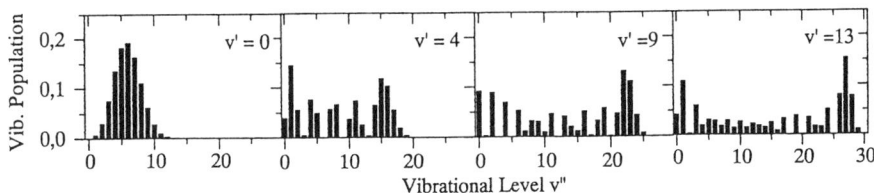

**Figure 4.** Vibrational population distribution resulting from spontaneous emission after excitation of the vibrational level v' in the $A^1\Sigma$ - electronic state.

The probe laser is used to confirm the full depletion of the level (v" = 0, j" = 9) after the preparation step. In fact, the interaction time between the pump laser and the Na$_2$ molecules is two orders of magnitude longer than the 12.5 ns electronic life time of the $A^1\Sigma_u^+$- state. Those molecules returning to the level (v" = 0, j"= 9) can undergo of the order of 100 pump-decay cycles. Eventually, the population in this level is depleted.

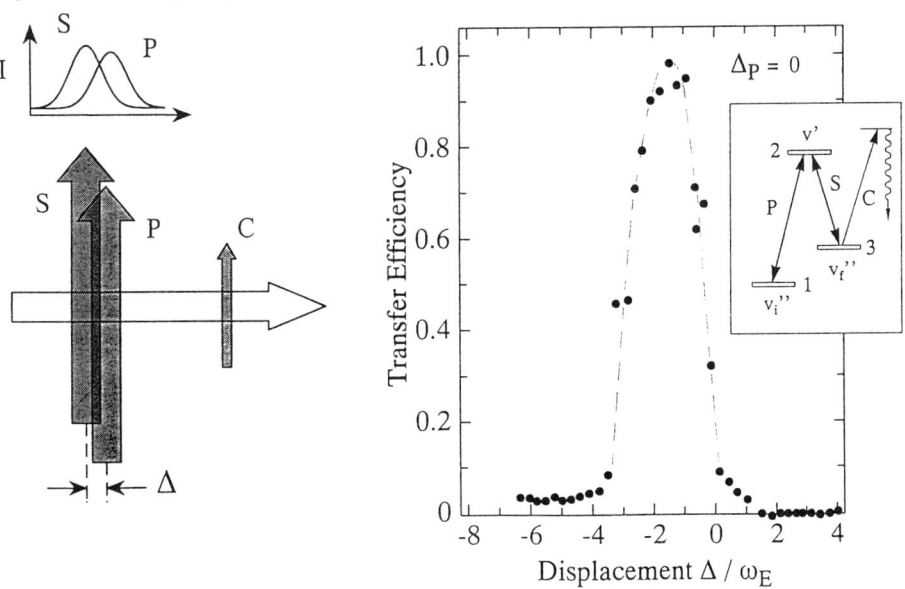

**Figure 5.** Geometric arrangement of the Stokes laser (S) and the pump laser (P) which cross the axis of a molecular beam at right angle (left hand side). The efficieny of the population transfer is monitored by a probe laser (C). The variation of the experimentally determined transfer efficiency from v"$_i$ = 0 to the v"$_f$ with the spatial separation $\Delta$, measured in units of the laser beam waist $\omega_E$, of the axis of the laser beams is shown at the right hand side (adapted from ref.4). Here, the pump laser was tuned to it's one-photon resonance, $\Delta_P = 0$.

Another technique, coherent population transfer (STIRAP) /4, 22/, addresses the main shortcoming of the FCP method. The experimentally more involved STIRAP technique appears to be an optimal technique for efficient and selective transfer of population to high vibrational levels. It requires two laser fields, a pump field which couples the thermally populated level to an intermediate level in the electronically excited state and a Stokes field, which couples the intermediate level to the final one. These two fields have to be arranged such that the molecules encounter the Stokes field *first* and the pump field *last*. However, the two fields need to have a suitable spatial overlap, see Fig. 5. Both lasers may be tuned to their respective resonance with the electronic transitions or they may be detuned by an equal amount from this resonance. The two photon resonance between the laser frequencies and the $v"_i = 0$ ' $v$'' $v"_f$ transition needs to be maintained. If properly implemented STIRAP will transfer all of the initial level population into a single level (see Fig.5), which is ($v"_f$, $j" = 9$) in the present case.

## RESULTS

We now present experimental data taken with (i) the thermal emission electron source and the FCP technique for vibrational state preparation, (ii) the same electron source and the highly selective STIRAP method, and (iii) the photoelectron source combined with the FCP technique.

### Data obtained with the Franck-Condon-Pumping method

A set of data, obtained with the FCP technique and the thermal electron source is shown in Fig. 6 together with results from a model calculation. The enhancement is determined from the comparison of the negative ion signal with and without vibrational excitation, using the known

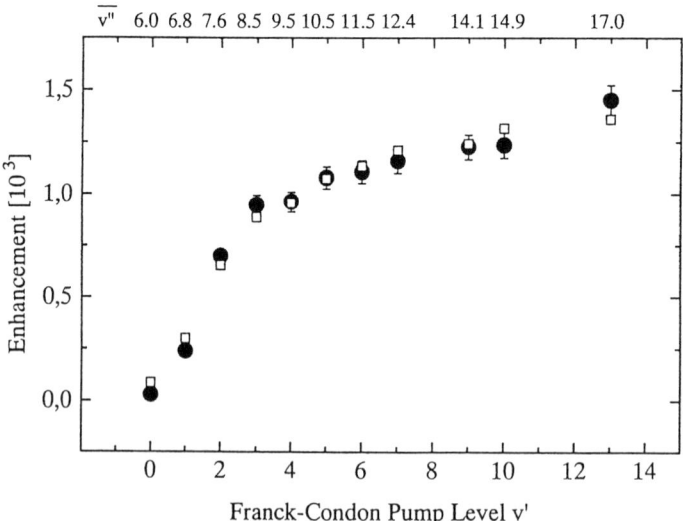

**Figure 6**. Vibrational enhancement of the DA rate as obtained experimentally using the FCP technique (full circles) and from the fit of an assumed functional form of k(v") (open squares). The lower scale shows the vibrational level of the electronically excited state, which determines the vibrational distribution in the ground electronic state. The upper scale identifies the mean vibrational excitation, which can also be used to lable the population distribution.

252

fraction of molecules excited out of the v" = 0 level (8 %) and assuming that the attachment rate is not sensitive to the rotational level j" /12/. Interpretation of these results in terms of the vibrational level dependence of the DA rate relies on model assumptions. Here we assume an exponential increase of the DA rate with v" up to a critical value $v"_c$ where the DA rate reaches a maximum value. For levels $v" > v"_c$ we assume the DA rate to be independent of v". Thus, the rate is given by

$$k(v') = \Sigma_{v"} \beta_{FCP}^{v'}(v") k(v") \qquad (1)$$

where $\beta_{FCP}^{v'}(v")$ is the known relative population established by spontaneous emission and normalized such that $\Sigma_{v"} \beta_{FCP}^{v'}(v") = \beta_{th}(v" = 0, j" = 9)$, i.e. the sum of the molecules transferred to levels v" > 0 equals the thermal population in the pumped level. The rate is assumed to vary as $k(v") = k_0 \exp(\alpha v")$ for $v" < v"_c$ and $k(v") = k_{max}$ for $v" > v"_c$. When two of the three parameters $k_{max}/k_0$, $\alpha$, $v_c$ are chosen the third one is fixed as well. For various choices of $v"_c$ the value of $k_{max}/k_0$ is derived from a least squares fit of eq.(1) to the experimental data. Fig. 6 shows the result for $v"_c = 12$, yielding $k_{max}/k_0 = 1.9*10^3$. A more detailed sensitivity analysis shows that the critical vibrational level can be confined to the range $10 \leq v"_c \leq 14$, in satisfying agreement with the results from quantum chemistry calculations (Fig.1) which shows the crossing of the neutral and negative ion potential curve near v" = 12.

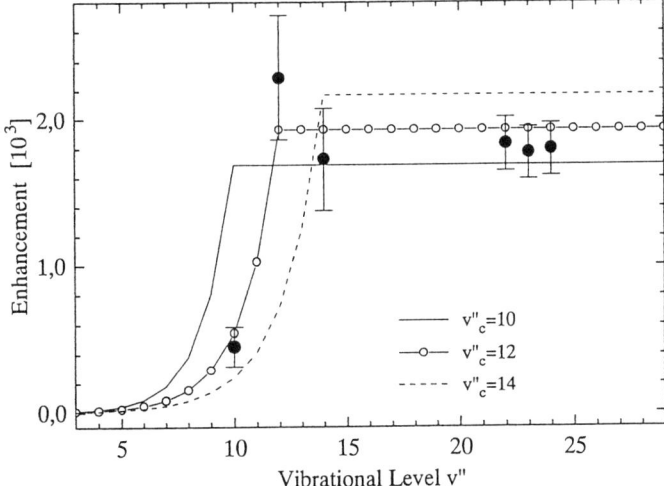

**Figure 7.** Vibrational enhancement of the DA rate as obtained experimentally using the STIRAP method (full circles) and from the fit of eq.(1) to the FCP data with various choices of the critical vibrational level $v"_c$.

## Data obtained with the STIRAP method

Figure 7 shows the results obtained with the STIRAP method, which provides data for the enhancement of the DA rate for *individual* levels v", together with the variation of k(v") as determined from a fit to the FCP data (Fig.6). Clearly, the result for v" = 10 is not compatible with $v"_c = 10$, while the data for v" = 22, 23 and 24 is not compatible with $v"_c = 14$. The data shown in Fig. 7 provide strong evidence that the DA rate is (nearly) independent of v" for vibrational levels above $v"_c$. The data also confirms that the maximum enhancement is $k_{max}/k_0 = 1.9*10^3$. These results finally suggest a maximum of the DA rate for $v" = v_c$. However, further experiments must provide data with smaller error bars before a conclusive statement can be made.

## Theoretical enhancement factors

Using traditional resonance theory within the local approximation, as previously applied to, e.g., $H_2$ /8/ and $Li_2$ /9/, we have derived the vibrational enhancement factors shown in Fig. 8. As in the experiment, an electron energy distribution with 0.7 eV FWHM was positioned to maximize the v"= 0 DA rate, resulting in a peak energy of 0.40 eV. The results are preliminary in so far as the resonance width was simply related to the difference potential by Wigner's

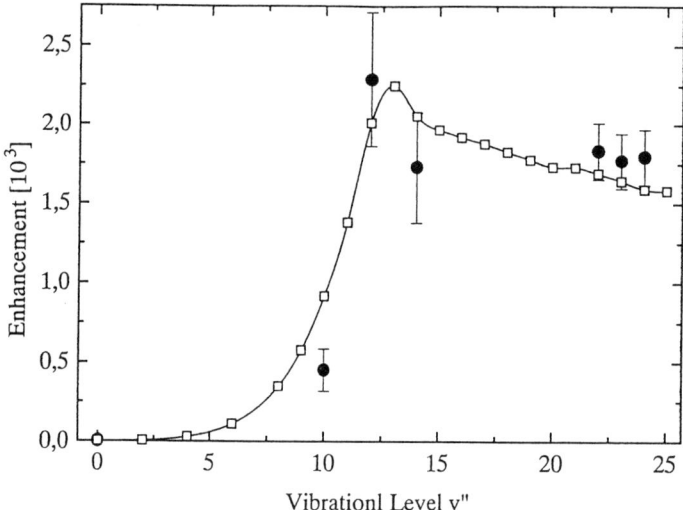

**Figure 8.** Theoretical enhancement factor of the DA rate. The experimental values measured with the STIRAP-method (solid circles) from Fig. 7 are also shown.

threshold law, i.e. $\Gamma(R) = \Gamma_0 (2m(V(A)-V(X)))^{1/2}$ with $\Gamma_0 = 0.029$ au. Furthermore, the resonance potential was slightly shifted to match the calculated crossing point accurately. Rather satisfying overall agreement with the measurement is observed. For v" > 12. the apparent width as taken from the dependence of the DA rate on the electron energy is ~ 20 meV.

## Data obtained with the photoelectron source

Preliminary data, obtained with the photoelectron source in combination with the FCP technique (see Fig. 2 (b) ), are shown in Fig. 9. The level v' = 2 has been chosen in order to maximize the population near v" = 12. Below the ionisation threshold, Rydberg levels of the sodium atom are excited. Weakly bound Rydberg electrons may transfer to the $Na_2$ molecules /23/. When attachment to molecules occurs near the crossing in the level v"$_c$ the negative ion receives very little kinetic energy. Because of the low relative velocity of the Na* atoms and the $Na_2$ molecules, copropagating in the supersonic beam, the resulting positive and negative atomic ions can only separate if they are formed at a large distance. Thus, except for very high Rydberg states, the charge transfer process will not lead to a free negative ion and a positively charged atom /24/. This explains the rapid decrease of the negative ion signal with decreasing energy of the photons. Above threshold, the photoelectron current is essentially independent of the photon energy over at least the first 50 meV

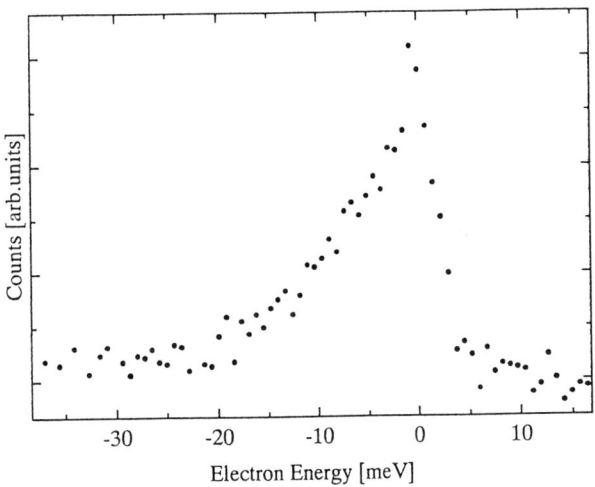

**Figure 9.** The Na⁻ - signal from vibrationally excited Na$_2$ molecules (FCP method with v' = 2) is observed as the frequency of the ionizing pulsed laser, yielding the photoelectrons, is scanned across the ionisation threshold of the Na atoms (see Fig.3). At present, the uncertainty of the energy E of the photoelectrons is about < 2 meV. Negative energies correspond to electrons in Rydberg levels. Very high lying Rydberg levels may be ionized by stray electric fields. The maximum count rate is in the order of 50 per second.

The rapid decrease of the negative ion signal with increasing energy of the free electrons shows that only very low energy electrons are attached to the sodium dimers which are vibrationally excited to levels near v" ≥ 12. At the present signal-to-noise ration there is no detectable Na⁻ - signal for electron energies above 10 meV.

## SUMMARY AND OUTLOOK

We have provided data on the vibrational level dependence of negative ion formation by dissociative attachment (DA). We show that, for Na$_2$ and our electron energy distribtion, the DA rate increases with v" by more than three orders of magnitude. This is substantially larger than predicted for Li$_2$ /9/. Above a critical value, here identified as v"$_c$ = 12, the DA rate is rather insensitive to v". First results obtained with a newly developed photoelectron source show that the rate for DA to levels v" ≥ 12 decreases rapidly with the electron energy $E_{el}$ in the range $0 \leq E_{el} < 10$ meV.

In future experiments we will combine the high resolution photoelectron source with the efficient STIRAP method for selective vibrational excitation. The combination of these two elements should eventually allow the experimental characterization of the energy E(R) and the width Γ(R) of the negative ion resonance, as a function of the internuclear distance R of the atoms in the molecule. Eventually, we expect to be able to even identify non-local effects in the dynamics of the process /25/.

## ACKNOWLEDGMENTS

We thank R.Setzkorn, M.Rudorf, B.Schellhaaß, J.Hauck, Th.Kolling, and P.Münch for their contributions to the development of this experiment. We thank in particular H.Hotop for stimulating discussions on the subject of negative ion formation and on photoelectron sources. This work is supported by the Deutsche Forschungsgemeinschaft and by the „Stiftung für Innovation Rheinland-Pfalz".

## REFERENCES

/1/ K.Bergmann, U.Hefter and J.Witt *Electronic and Atomic Collisions*, N.Oda and K.Takayanaki eds. North Holland publ. comp. 1980, p.523

/2/ K.Bergmann, U.Hefter, and J.Witt , *J.Chem.Phys*. **72**, 4777 (1980)

/3/ *Molecular Dynamics and Spectroscopy by Stimulated Emission Pumping*, H.L.Dai and R.W.Field, eds., World Scientific, Singapore (1995)

/4/ U.Gaubatz, P.Rudecki, S.Schiemann, and K.Bergmann, *J.Chem.Phys.* **92**, 5363 (1990)

/5/ S.Schiemann, A.Kuhn, S.Steuerwald, and K.Bergmann, *Phys.Rev.Lett.* **71**, 3637 (1990)

/6/ B.W.Shore, J.Martin, M.P.Fewell, and K.Bergmann, *Phys.Rev.* A **52**, 566 (1995)
J.Martin, B.W.Shore, and K.Bergmann, ibid. p.583

/7/ M.Allan and S.F.Wong, *Phys.Rev.Lett* **41**, 1791 (1978);
M.Allan and S.F.Wong, *J.Chem.Phys.* **74**, 1687 (1981)

/8/ J.M.Wadehra and J.N.Bardsley *Phys.Rev.Lett* **41**, 1795 (1978);
J.M.Bardsley and J.M.Wadehra *Phys.Rev A* **20**, 1398 (1979); A.P.Hickman *Phys.Rev.* A **43**, 3495 (1991),
I.I.Fabrikant in *Dissociative Recombination*, B.R.Rowe ed., Plenum Press, N.Y. 1993

/9/ J.M.Wadehra *Phys.Rev.A.* **41**, 3607 (1990)

/10/ I.Cadez, R.I.Hall, M.Landau, F.Pichou, and C.Schermann, *J.Phys.* B **21**, 3271 (1988);
D.Popovic, I.Cadez, M.Landau, C.Schermann, and R.I.Hall, *Meas.Sci.Tech.* **1**, 1040 (1990)

/11/ J.P.Ziesel, D.Teillet-Billy, and L.Bouby, *Chem.Phys.Lett* **123**, 371 (1986)

/12/ K.K.Sunil and K.D.Jordan, *Chem.Phys.Lett* **104**, 343 (1984)
H.Partridge, D.A.Dixon, S.P.Walch, C.W.Bauschlicher Jr., and J.L.Gole, *J.Chem.Phys.* **79**, 1859 (1983)

/13/ M.Külz, M.Keil, A.Kortyna, B.Schellhaaß, J.Hauck, K.Bergmann, D.Weyh, and W.Meyer, *Dissociative attachment of low energy electrons to state selected diatomic molecules*, *Phys.Rev. A*, submitted (1995)

/14/ W.C.Wiley and I.H.McLaren, *Rev.Sci.Instr.* **26**, 1150 (1955)

/15/ R.E.Collins, B.B.Aubrey, P.N.Eisner, and R.J.Celotta *Rev.Sci.Instr.* **41**, 1403 (1970)

/16/ D.Klar, M.W.Ruf and H.Hotop, *Chem.Phys.Lett.* **189**, 488 (1992)

/17/ D.Klar, M.-W.Ruf, and H.Hotop, *Aust.J.Phys.* **45**, 263 (1992)
D.Klar, M.-W.Ruf, and H.Hotop, *Meas.Sci.Technol.* **5**, 1248 (1994)

/18/ H.Hotop, D.Klar, J.Kreil, M.-W.Ruf, A.Schramm, and J.M.Weber, contribution to this volume

/19/ U.Hefter and K.Bergmann *Spectroscopic detection methods* in: *Atomic and Molecular Beam Methods* G.Scoles ed., Oxford University Press (1988)

/20/ I.V.Hertel and A.S.Stamatovic, *IEEE J.QuantumElectron.* **QE-11**, 210 (1975)

/21/ K.Bergmann *State selection via optical methods* in: *Atomic and Molecular Beam Methods* G.Scoles ed., Oxford University Press (1988)

/22/ K.Bergmann and B. W. Shore, "*Coherent Population Transfer*" , chap. 9 of ref. 3

/23/ F.B.Dunning, *J.Phys.* B **28**, 1645 (1995)

/24/ K.Harth, M.Raab, J.Ganz, A.Siegel, M.-W.Ruf, and H.Hotop, *Opt.Commun.* **54**, 343 (1985)
B.G.Zollar, C.W.Walter, F.Lu, C.B.Johnson, K.A..Smith, and F.B.Dunning , *J.Chem.Phys.* **84**, 5589 (1986)

/25/ D.E.Atems and L.M.Wadehra, *Phys.Rev.* A **42**, 5201 (1990)
W.Domcke, *Phys.Rep.* **208**, 97 (1991)

# Electron Scattering from Vibrationally Excited Molecules

Paul D. Burrow

Department of Physics and Astronomy
University of Nebraska
Lincoln, Nebraska, USA 68588
pburrow@unlinfo.unl.edu

**Abstract.** This Progress Report discusses a number of experiments, both old and new, in which low energy electrons are scattered from vibrationally excited molecules. The effects of the excited species on the thresholds for electronic excitation and ionization are described, as well as their influence on the formation and decay of temporary negative ion states.

## INTRODUCTION

Despite the fact that low energy electron scattering studies with vibrationally excited targets produced by non-thermal means were carried out almost 30 years ago [1], the field remains a difficult one for experimentalists, with considerable challenges in producing adequate densities of target molecules in well-characterized levels. The goal of producing selectively pumped vibrational levels in an arbitrary molecule remains to be achieved in the future. The majority of workers in this area have employed either thermal or plasma discharge methods to produce excited species, although laser techniques ultimately will play the dominant role.

As this field has been recently, but briefly, reviewed by Mason *et al.* [2] at a Satellite Meeting associated with the XVIIIth ICPEAC, this presentation will touch, for the most part, on work not included in that paper. I will do some violence to the meaning of "Progress Report" by using a few rather older references to illustrate the concepts. I will also omit discussions of work likely to be described in other Reports at the present meeting.

# ELECTRONIC EXCITATION NEAR THRESHOLD

The most obvious effects arising from scattering from vibrationally excited molecules appear because of the shift of the threshold energies to lower values and the expansion of the Franck-Condon region to portions of the potential surfaces not previously accessible. In some cases, shifts due to the latter effect may well exceed the former.

Because of the ease with which vibrationally excited $N_2$ is produced in discharges, threshold changes in electronic excitation are readily observed. Figure 1 illustrates a spectrum [3] in which the current of scattered electrons, the "trapped electron current", whose final energy is between zero and a few tens of meVs, is plotted as a function of electron impact energy. In the lower panel, measured at room temperature, the major peaks reflect excitation to the vibrational levels of the $B^3\Pi_g$ state, just at threshold, with a smaller contribution from the $A^3\Sigma_u^+$ state below 7.2 eV. The solid line shows how well the data can be fit with a synthetic spectrum constructed using an experimentally determined instrument function,

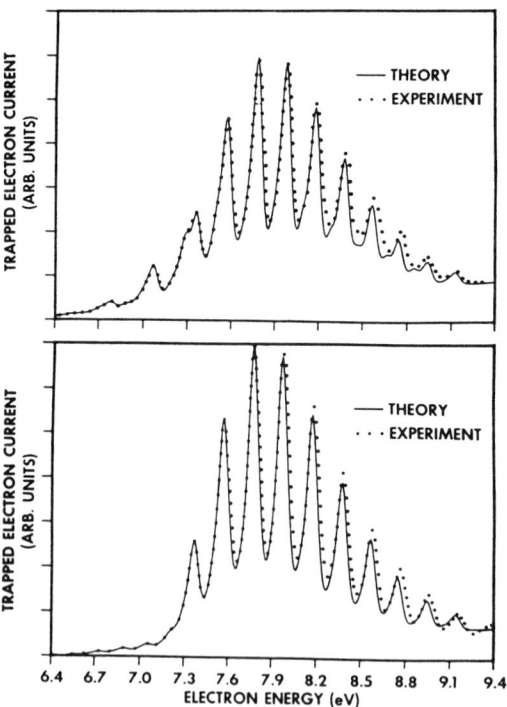

**Figure 1.** The current of slow electrons following excitation of the A and B states of $N_2$ as a function of energy. Lower panel, excitation from the ground vibrational state. Upper panel, excitation from molecules at a vibrational temperature of 1930K.

scaling the size of the peaks according to the known Franck-Condon factors, and positioning them according to the known energies.

The upper panel of Figure 1 shows how the spectrum changes in the presence of vibrationally excited molecules. Both panels are drawn to the same vertical scale. The curve marked "theory" was fit to the data using only a single parameter, namely the vibrational temperature, along with the known Franck-Condon factors. In this example the temperature was 1930K, giving fractional vibrational populations of 0.824, 0.145, 0.026 and 0.004 for $v = 0-3$ respectively. The shift of the threshold for production of the B state is readily apparent, along with changes in the intensities. The purpose of this work was to examine whether low energy electron techniques could be used to determine the vibrational temperature, and in the extremely favorable case of $N_2$, this appeared to be the case. A similar study was carried out concurrently using electrostatic analyzers with higher energy resolution and even higher vibrational temperatures by Huetz *et al.* [4].

## IONIZATION NEAR THRESHOLD

Similar shifts in the onset for production of positive ions would also be anticipated from the discussion above. $N_2$ provides an interesting example in that a superficial examination suggests that in the presence of vibrationally hot molecules, no shift would occur. The ground state of $N_2^+$ has a potential curve which is very similar in shape, vibrational spacing and equilibrium distance to that of the ground state of the neutral molecule. Consequently the Franck-Condon factors peak very sharply between the *same* vibrational levels of the neutral and positive ion. In fact, they tend to decrease roughly one order of magnitude for each increment of $\Delta v$, where $\Delta v = v - v'$. As a result, very little shift in the onset might be expected.

In unpublished work by J.A. Michejda and myself [5], we examined this question, using the techniques described earlier [3] to determine the vibrational temperature. Figure 2 shows a plot of the positive ion current near threshold for the room temperature gas and gas traversing a microwave discharge and having a vibrational temperature near 2000K. Also shown are some model fits to the data. Clearly there is a significant shift to lower energies. An ionization model incorporating only direct excitation to the $X^2\Sigma_g^+$ ground state of the ion is completely incapable of explaining these data. By including indirect processes such as the autoionization of Rydberg levels of $N_2$ converging to the ground ionic state and to the first excited $A^2\Pi_u$ state of $N_2^+$, a consistent measure of the contribution of each of the processes was obtained. This was carried out by modeling these processes and fitting to our data at several different vibrational temperatures. In contrast to a superficial guess, the cross sections for ionization from $v = 1$ and 2 at their thresholds are the same size as that from $\upsilon = 0$. Table 1 reports these cross section ratios over two narrow energy regimes above threshold. The key result is that they are large enough to give rise to a clear shift in the onset for ionization.

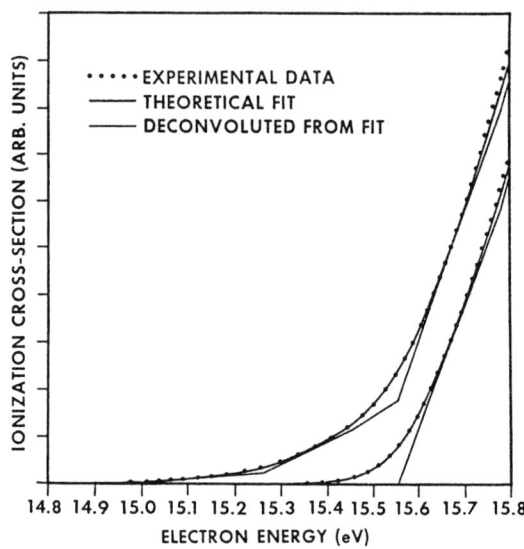

**Figure 2.** The positive ion current near threshold as a function of electron energy in $N_2$ at room temperature, on the right, and vibrationally excited $N_2$ at a temperature near 2000K, on the left.

**Table 1.** Ratios of the ionization cross sections near threshold from $N_2$ ($\upsilon = 1$ and 2) to that from $N_2$ ($\upsilon = 0$).

| Ratio | 0-300 meV above Threshold | 300-600 meV above Threshold |
|---|---|---|
| $N_2(\upsilon = 1)/N_2(\upsilon = 0)$ | $1.0 \pm 0.2$ | $0.8 \pm 0.3$ |
| $N_2(\upsilon = 2)/N_2(\upsilon = 0)$ | $1.2 \pm 0.3$ | $0.9 \pm 0.4$ |

In a study of vibrationally excited $N_2$ molecules formed by decomposition of ammonia on platinum, Foner and Hudson [6] observed shifted onsets for production of $N_2^+$ and also invoked the autoionization mechanism. In very recent work by Kuelz et al. [7], the vibrational dependence of the ionization cross section of $Na_2$ was studied using laser production of the excited levels.

# VIBRATIONAL EFFECTS AND TEMPORARY NEGATIVE ION STATES

The presence of vibrationally excited molecules can produce much stronger effects, proportionally, in electron scattering processes that involve temporary negative ion formation. Indeed, many of the examples in the previous review of this subject by Mason *et al.* [2] deal with this area. Because the total electron scattering cross sections of most molecules at low energies tend to be dominated by the presence of one or more temporary negative ion states, vibrationally excited levels can produce changes which appear more dramatic because the shifts may well be a significant proportion of the incident electron energy. In contrast to the case of excitation to the B state of $N_2$ mentioned earlier, the magnitudes of the changes are not generally given by simple Franck-Condon factors, because the lifetimes of the anion states are not long enough to display true vibrational levels.

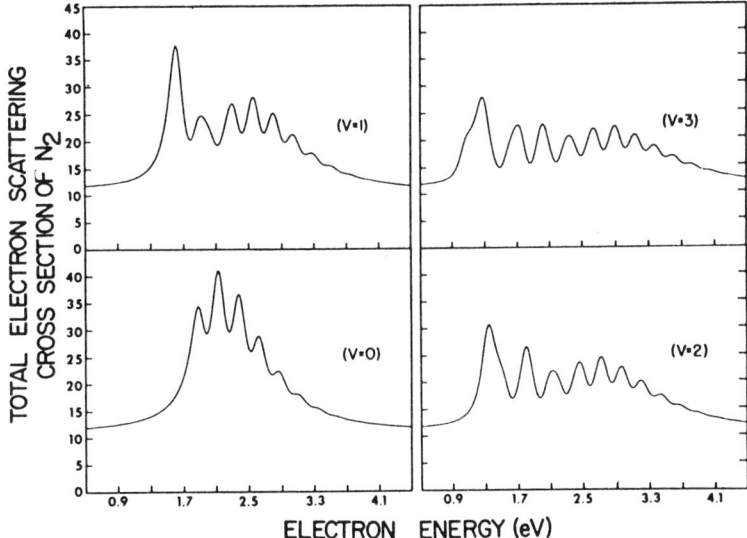

**Figure 3**. Total scattering cross section for $N_2$ in $v = 0,1,2$ and 3 as a function of energy as computed by Dubé and Herzenberg.

These effects are readily illustrated in $N_2$, and results from a more complex molecule will be shown below. As is well-known, the $N_2$ scattering cross section is dominated by the quasi-vibrational peaks of a temporary negative ion state located near 2 eV. Figure 3 shows the total scattering cross sections for molecules in the $v = 0 - 3$ vibrational levels computed by Dubé and Herzenberg using the boomerang model, as quoted in the paper by Michejda *et al.*[3]. The calculations were only for the resonant portion of the scattering, and a constant value has been added to represent the non-resonant part. The curves show dramatic shifts of the vibrational structure to lower impact energies. These features can easily be seen in vibrationally hot $N_2$ even with an unmonochromatized beam of slow electrons transmitted through the gas [8].

In molecules containing temporary negative ion states of short lifetime, which do not display anionic vibrational structure, one expects to see shifts of the resonance peak appearing in the total cross section toward lower energies, along with a broadening of the peak. This was observed by Ferch et al.[9] in $CO_2$. Figure 4 shows the total cross section at two different temperatures, 250K (solid line) and 520K (dashed line), along with triangular points from Buckman et al.[10] taken at 573K. The enhancement in the region below the resonance is attributed to the dipole scattering from molecules populating the bending mode. The cross section in the $^2\Pi_u$ resonance region, peaking near 4 eV, is shifted slightly lower in energy and enhanced at the higher temperature.

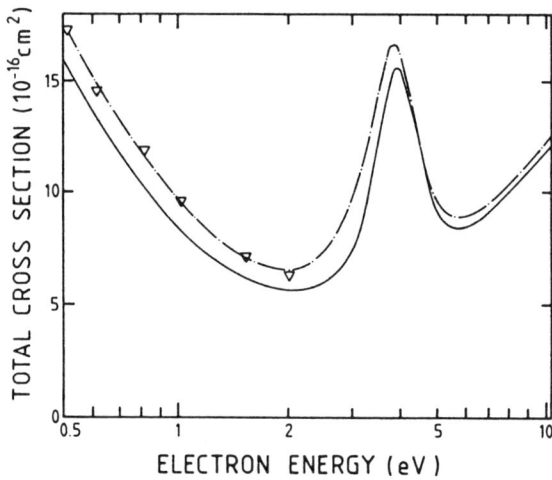

Figure 4. Total cross sections for electron scattering from $CO_2$ at 250K, solid line, and 520K, dashed line. The triangles show the results of Buckman et al. [10]. From Ferch et al. [9].

The effect in the total cross section at this rather modest temperature is considerably less dramatic than seen in the earlier $N_2$ work. However, when the cross section, as shown in Figure 5, is broken into its components reflecting scattering from molecules in the vibrational ground state (dashed line) and from molecules in excited levels of the bending vibration, the effect is seen to be quite substantial, considering the small energy of the bending mode, 82.8 meV. The cross section is shifted roughly 0.3 eV lower in energy and increased in magnitude. For this partitioning, a number of assumptions were made, among them that all members of the bending series have the same cross section. The effect however should be dominated by the first excited level.

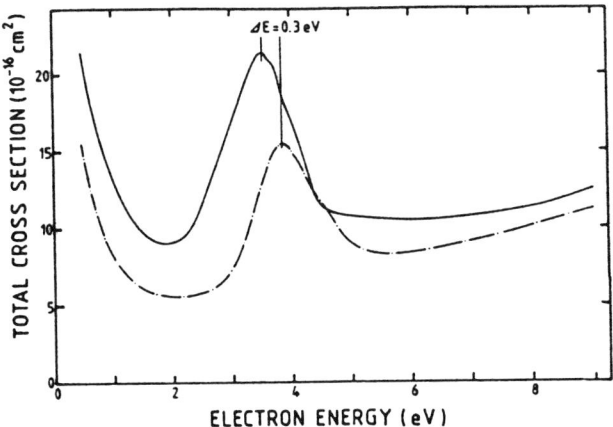

**Figure 5.** Total cross sections for scattering from $CO_2$ in its ground vibrational level, dashed curve, and from molecules in excited states of the bending vibration, solid curve.

Although detailed calculations were not been made, Ferch *et al.* point out that this behavior is consistent with known characteristics of the anion potential surface, in particular, the breaking of the orbital degeneracy as the molecule departs from linearity and the rapid decline in energy of one anionic component as the bending angle moves away from 180°.

In more complex targets, particularly those with high symmetry, studies using vibrationally hot molecules could provide a way to examine portions of the anionic potential surface that are not accessible from levels populated at room temperature. In the ideal case, mode selective excitation using lasers would be desirable, although little work combining such methods with high resolution electron beams has been carried out. A small step in this direction was taken by Stricklett and Burrow[11]. In this work, the effect of internal energy on the cross section for electron scattering from $SF_6$ was examined. The objective was to enhance the small differences in the total cross sections for laser-excited and non-excited molecules, particularly the energy dependence of these changes. No attempt was made to put these changes on an absolute scale as carried out in the work of Buckman *et al.* and Ferch *et al.*

In the apparatus, a triply crossed beam geometry was employed as shown in Figure 6. A free jet of $SF_6$ was irradiated with infrared light from a $CO_2$ laser, and a magnetically collimated electron beam probed the gas downstream from the laser. The positions of the laser and the jet orifice were both independently adjustable with respect to that of the electron beam. The output of the laser was chopped and the change in the electron current transmitted through the gas jet was recorded synchronously.

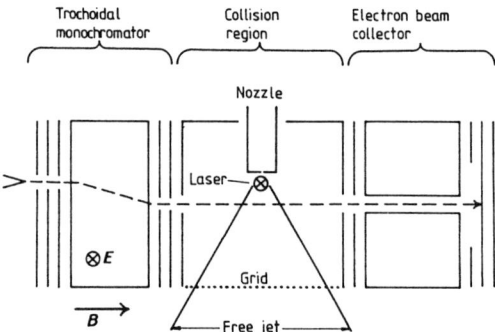

Figure 6. Schematic drawing of the crossed electron, laser and gas beam apparatus. The electron trajectory is shown by the broken line.

Laser excitation took place just outside the aperature through which the gas was expanding, in a regime in which many gas collisions take place. Although the $\nu_3$ mode was pumped, rapid V-V transfer effectively couples this mode to the vibrational manifold, producing a thermal distribution of levels. The modulation scheme therefore allows us to observe at each impact energy the difference between scattering in the hot and cold gas, without having to subtract two absolutely measured cross sections.

Curve (a) of Figure 7 shows the results using laser pumping in the high density region just outside the jet orifice. The data are displayed such that the positive direction (upward) corresponds to an *increase* in the total scattering cross section with the laser light on. As the laser is moved away from the orifice, the signal drops rapidly. Curve (b) shows the signal acquired when the laser illuminates the jet in a region in which molecule-molecule collisions are greatly reduced. Although the signal is much smaller, the profiles of the structures are the same as in curve (a). Curve (c) shows the total scattering cross section as measured by Kennerly *et al.*[12]; note the suppressed zero. The vertical dashed lines locate the vertical attachment energies of three temporary negative ion states of $SF_6$.

Curves (a) and (b) display a strong enhancement at energies below 1 eV. A number of mechanisms may contribute to this, such as superelastic scattering and elastic scattering from vibrational levels with instantaneous dipole moments. The lowest lying negative ion state, $^2A_{1g}$, may also contribute. Above 1 eV, the signal displays regions of enhancement and depletion. These structures appear to be associated with the two $^2T$ resonances; that for the $^2T_{1u}$ resonance shows a depletion at the resonance center at 7.05 eV and an enhancement in scattering which peaks approximately 2 eV lower, a substantial shift of the anionic potential surface now accessible from the excited molecules. This effect is similar to that observed in $CO_2$ by Ferch *et al.* Calculations of the anionic potential surfaces along the normal mode coordinates would be most useful in interpreting these measurements.

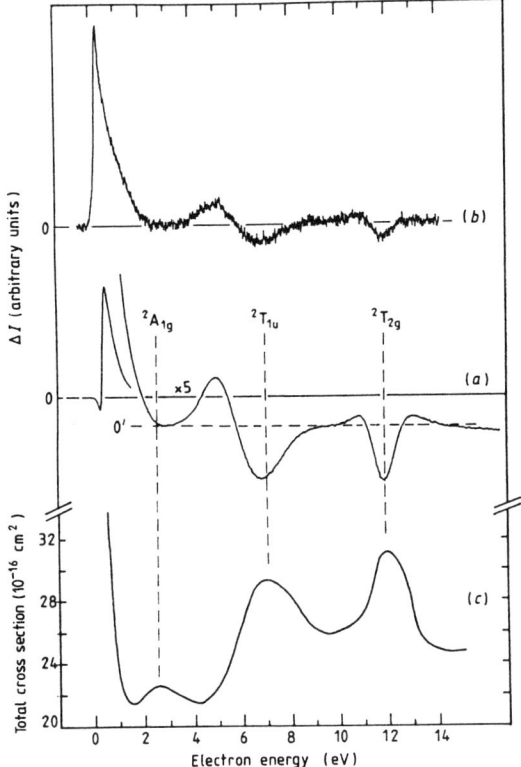

**Figure 7.** The upper two curves show the ac component of the electron current transmitted through a jet of $SF_6$ pumped with light from a modulated $CO_2$ laser as a function of electron energy. The bottom curve shows the total scattering cross section of $SF_6$ as measured by Kennerly et al.[12].

Space permits only a brief mention of measurements of the individual processes that make up the total scattering cross section from vibrationally excited molecules. Superelastic scattering from the lower vibrational levels populated by thermal means has been frequently seen. Measurements from highly excited levels are rare, although superelastic scattering from $N_2(v = 5)$, produced by collisional energy transfer from $Rb^*$, has been observed[1]. Scattering studies from vibrationally excited $CO_2$, described previously by Johnstone et al.[2], are continuing[13,14].

By far the most dramatic enhancement in an individual process due to vibration takes place in the dissociative attachment (DA) channel. The extreme dependence on vibrational level in $H_2$, for example, along with the favorably peaked cross section shape, has made DA a suitable diagnostic for $H_2^v$[15]. Progress, both theoretical and experimental, has been made in understanding the effect of vibrational temperature on DA in more complex molecules such as $CH_3Cl$[16], although the calculations are still restricted to the use of the diatomic approximation. Problems encountered with thermal decomposition of such molecules on the oven walls[17] reinforce the desirability of direct laser pumping into vibrational levels as a means of production.

## ACKNOWLEDGMENTS

This work was supported by the National Science Foundation.

## REFERENCES

1. See, for example, Burrow, P.D., and Davidovits, D., Phys. Rev. Letters **21**, 1789 (1968).
2. Mason, N.J., Johnstone, W.M., and Akther, P., *Electron Collisions with Molecules, Clusters, and Surfaces* (Ed. by H. Ehrhardt and L.A. Morgan, Plenum Press, New York, 1994, pp. 47-62
3. Michejda, J.A., Dubé, L. and Burrow, P.D., J. Appl. Phys. **52**, 3121 (1981).
4. Huetz, A., Gresteau, F., Hall, R.I., and Mazeau, J., J. Chem. Phys. **72**, 5297 (1980).
5. Michejda, J.A., and Burrow, P.D., unpublished. Also, Michejda, J.A., Ph.D. Thesis (Yale University, 1977).
6. Foner, S.N., and Hudson, R.L., J. Chem. Phys. **80**, 518 (1984).
7. Kuelz, M., Kortyna, A., Keil, M., Schellhaass, B., and Bergmann, K., Z. Phys. D **33**, 109 (1995).
8. Michejda, J.A., and Burrow, P.D., J. Appl. Phys. **47**, 2780 (1976).
9. Ferch, J., Masche, C., Raith, W., and Wieman, L., Phys. Rev. A **40**, 5407 (1989).
10. Buckman, S.J., Elford, M.T., and Newman, D.S., J. Phys. B **20**, 5175 (1987).
11. Stricklett, K.L., and Burrow, P.D., J. Phys. B **24**, L149 (1991).
12. Kennerly, R.E., Bonham, R.A., and McMillan, M., J. Chem. Phys. **70**, 2039 (1979).
13. Johnstone, W.M., Mason, N.J., and Newell, W.R., J. Phys. B **26**, L147 (1993).
14. Johnstone, W.M., Akther, P., and Newell, W.R., J. Phys. B **28**, 743 (1995).
15. Hall, R.I., Cadez, I., Landau, M., Pichou, F., and C. Schermann, Phys. Rev. Letters **60**, 337 (1988).
16. Pearl, D.M., Burrow, P.D., Fabrikant, I.I., and Gallup, G.A., J. Chem. Phys. **102**, 2737 (1995).
17. Pearl, D.M., and Burrow, P.D., Chem. Phys. Letters **206**, 483 (1993).

# Studies of low energy electron collisions at sub-meV resolution

H. Hotop, D. Klar, J. Kreil, M.-W. Ruf, A. Schramm and J. M. Weber

Fachbereich Physik, Univ. Kaiserslautern, D-67653 Kaiserslautern, FRG

**Abstract.** Recent developments and important aspects for achieving very high resolution in low energy electron collisions are summarized. An atomic beam laser photoelectron source with sub-meV resolution and its application to studies of electron attachment to molecules in the electron energy range 0- 200 meV is described.

## INTRODUCTION

The discovery of the sharp $He^-(2^2S_{1/2})$ scattering resonance in 1963 [1] stands at the beginning of a renewed interest in electron collision phenomena, and many groups have subsequently developed experimental techniques with the aim of improving the effective energy resolution and thereby providing detailed data which - in conjunction with theoretical calculations - lead to a deeper understanding of the elementary processes. Twenty years ago typical energy widths were in the range (20 - 50) meV [2,3], and Frank Read [4] concluded an important paper on broadening effects in electron scattering experiments by stating: "Although very much has already been learned from low energy electron impact spectrocopy, it is clear that some aspects of this spectroscopy will only become truly fruitful when better energy resolutions are available. This is particularly true in cases of very narrow resonances and cusps, and also in the case of vibrational and rotational excitation of molecules". In spite of several advances [5-21] which we have witnessed over the last two decades, this statement is still relevant today. In this report we briefly summarize some of these achievements, and we describe in some detail a laser photoelectron source with sub-meV energy resolution, which our group has developed [19,20], and its application to studies of electron attachment to molecules in the electron energy range 0 - 200 meV [19,22,23]. The cross sections $\sigma_e(E)$ for these processes (which are important for gaseous dielectrics, see for example [24])

$$e^-(E) + XY \xrightarrow{\sigma_e(E)} XY^-(X^- + Y) \qquad (1)$$

can now be studied with sufficient resolution and down to sufficiently low energies to test theoretical threshold laws [25,26]; moreover, cusp structure at vibrationally inelastic scattering thresholds can be investigated with unprecedented clarity [19,23]. Using the measured velocity-dependent cross sections $\sigma_e(v)$ one can calculate dependable rate coefficients $k_e(T_e)$ for reaction (1),

$$k_e(T_e) = \int \sigma_e(v) \, v \, f_v(v; T_e) \, dv \qquad (2)$$

as relevant for, e.g., a Maxwellian electron ensemble with the electron temperature $T_e$ for a given gas temperature $T_g$ which describes the population of the internal degrees of freedom of the molecule under study [27]. For the first time, the free-electron model [28,29] for Rydberg electron attachment reactions

$$A^*(nl) + XY \xrightarrow{k_{nl}} A^+ + XY^- (X^- + Y) \qquad (3)$$

can be tested in a reliable way by comparing - at sufficiently high principal quantum numbers n - measured rate coefficients $k_{nl}$ with those calculated from the cross section $\sigma_e(v)$ for free electron attachment [22,29]

$$k_{nl} = \int \sigma_e(v) \, v \, f_{nl}(v) \, dv \qquad (4)$$

where $f_{nl}(v)$ represents the normalized velocity distribution of the Rydberg electron.

## EXPERIMENTAL ASPECTS AND DEVELOPMENTS

The factors which determine the resolution in an electron collision experiment (as expressed by the effective energy width $\Delta E$ (FWHM)) have been discussed by several authors (see e.g. [4,5,7,8,10,14,20]): 1) Energy widths $\Delta E_S$ of the electron source and $\Delta E_A$ of the electron analyzer used to detect scattered electrons; 2) energy width $\Delta E_F$ associated with the presence of residual electric fields in the reaction volume; 3) energy width $\Delta E_D$ introduced by kinematic effects due to the motion of the target atoms or molecules (Doppler broadening).

Two approaches have mainly been used to realize electron sources with high resolution: 1a) thermionic electron emitter (hot filament) in conjunction with an electrostatic condenser; 1b) production of photoelectrons from an atomic target. In the "conventional" approach 1a) the electrostatic monochromator selects a narrow band from the broad thermionic distribution. At a given resolution $\Delta E$ a space-charge limited, monochromatized electron current $I \sim (\Delta E)^{5/2}$ can be expected in essential agreement with observations [5,8], see Fig. 1. Ibach and coworkers [17,18] have recently optimized such electron spectrometers for use in electron-surface scattering under ultrahigh vacuum conditions and achieved resolutions around 1 meV (as measured for the elastic peak in specular reflection [17]): Their monochromator includes a special input and output lens and two successive toroidal condensers of novel design, which permit to reduce angular aberrations and to compensate space-charge induced aberrations [18]. The performance of such an electron source is illustrated in Fig. 1 (dash-dotted curve). It remains to be seen whether this impressive performance can also be realized in studies of gaseous targets. In energy analysis of scattered electrons space charge is normally not a problem and the resolution $\Delta E_A$ of an optimized analyzer should therefore match or surpass the figure achieved for the monochromator as demonstrated by Ibach et al. [17,18]. Using an optimized single stage hemispherical condenser with multichannel detection in conjunction with a VUV resonance lamp of narrow bandwidth Baltzer et al. [30] have achieved resolutions down to about 2 meV in the analysis of VUV photoelectrons.

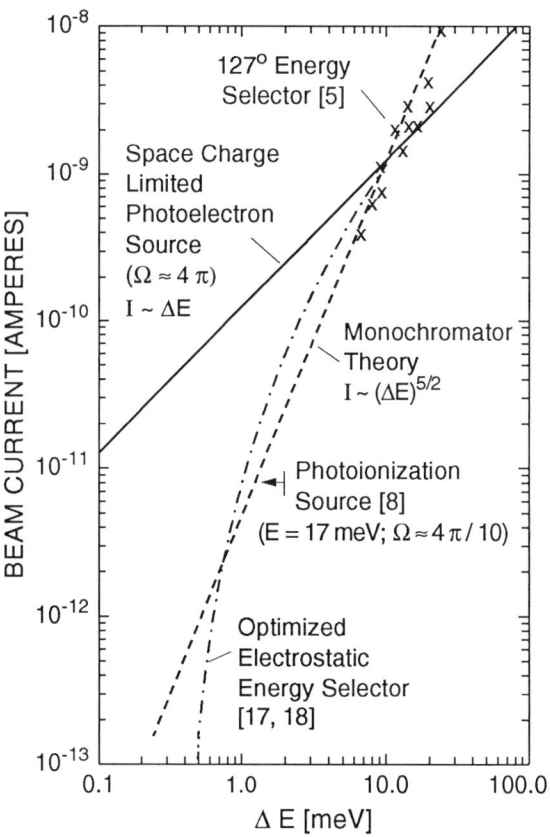

**FIGURE 1.** Performance of high resolution electron sources

Electron sources based on photoionization of atoms near threshold hold a lot of promise as indicated by the full line in Fig. 1 which represents the space-charged limited photoelectron current which is in principle available at a given resolution level $\Delta E$. Note that photoelectron sources are expected to deliver higher currents than optimized conventional sources for resolutions < 10 meV. Gallagher and coworkers [8] were the first to give a detailed description of a laser photoelectron source. It is based on photoionization of metastable $Ba^*(6s5d\ ^1D_2)$ atoms by an intracavity Helium-Cadmium laser (325 nm); photoelectrons with 17 meV initial energy are created and currents up to $8 \cdot 10^{-12}$ A - corresponding to the use of about 10% of the photoelectrons through extraction with a weak electric field - were available in an accelerated focussed beam for scattering from a nozzle target beam of atoms or molecules. The resolution of the electron source was estimated to be < 2 meV [8], and beautiful results were obtained for the $He^-(2^2S_{1/2})$ resonance at an overall resolution of about 5 meV [9]. Subsequently, Chutjian et al. and Field et al. have employed VUV photoionization of rare gas atoms to produce photoelectrons at typical energy widths around 5 meV for use in electron attachment [14] and electron scattering [15] experiments. In their case the resolution was limited by the bandwidth of the monochromatized radiation and, in part, by the presence of weak electric fields. Baltzer et al. [21] have recently studied rotationally inelastic and superelastic forward scattering of electrons from dipolar molecules such as HF by energy analyzing photoelectrons which were created by NeI-VUV photoionization of Ar and which lost or gained energy according to $\Delta J = \pm 1$ rotational transitions in subsequent electron-molecule collisions; an overall resolution around 2.5 meV was achieved by careful reduction of residual electric fields in the photoionization and scattering chamber [21,30].

Recently our group has developed a laser photoelectron source with sub-meV energy bandwidth which is based on resonant two photon ionization of metastable $Ar^*(4s\ ^3P_2)$ atoms in a collimated beam [19,20]. The principle of the method and the experimental setup, as relevant for laser photoelectron attachment to molecules or clusters in a supersonic target beam [31], are illustrated in Fig. 2. Metastable $Ar^*(4s\ ^3P_2)$ atoms, originating from a differentially-pumped dc discharge source, interact with two transverse cw lasers in the region where they cross a skimmed, differentially-pumped supersonic target beam. An infrared single mode laser ($\lambda = 811.75$ nm), frequency-stabilized to the $Ar^*(4s\ ^3P_2 - 4p\ ^3D_3)$ transition by saturation spectroscopy in an auxiliary argon discharge, produces a substantial quasi-stationary population of $Ar^*(4p\ ^3D_3)$ atoms which are further excited by a blue tunable cw dye laser (Stilbene 3, 430 - 478 nm) to $Ar^{**}(nl)$ Rydberg states ($\lambda > 462$ nm) or to the

$Ar^+(^2P_{3/2}) + e^-(E)$ continuum. Negative ions, produced in electron attachment reactions of molecules (or clusters) in the target beam with the $Ar^{**}(nl)$ Rydberg atoms (subsequently labelled by negative electron energies E < 0) or by the photoelectrons (E > 0), are mass-analyzed with a quadrupole mass spectrometer and detected by an off-axis dual channel plate electron multiplier, as described elsewhere [32]. In our first studies with this method a stationary gas target was admitted to the reaction chamber (temperature $T_g$ = 300 K), and the experiment was pulsed at a high repetition rate (140 kHz); excitation/ionization and attachment processes occur for time intervals of ≈ 3 µs, and - after a short delay - negative ions due to attachment processes are extracted by application of a pulsed electric field while the excitation laser is switched off acousto-optically. With the skimmed target beam cw measurements are possible: tests showed that $SF_6^-$ ions from $e^- + SF_6$ and $Ar^{**}(nl) + SF_6$ attachment reactions are efficiently detected without application of a positive bias on the extraction electrode E (see Fig. 2). Field penetration from the ion optics through the hole in plate E suffices to extract ions without introducing significant electric fields in the reaction region which is magnetically shielded (1 - 2 µT). All the electrodes of the reaction chamber are coated with colloidal graphite to ensure nearly homogeneous surface potentials and voltage ripple is reduced by RC circuits.

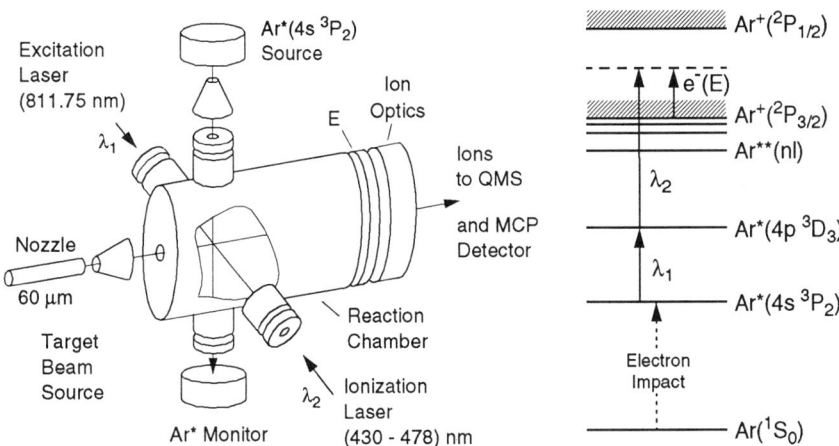

**FIGURE 2.** Principle of laser photoelectron source and schematic drawing of the experimental setup relevant for laser photoelectron attachment to molecules or clusters

In order to achieve sub-meV resolution all broadening factors have to be carefully assessed. Space charge effects can be simply estimated from the density of the $Ar^+$ photoions; assuming for simplicity spherical symmetry of the ion cloud with N ions distributed uniformly within a radius R, one calculates a potential drop within the sphere of $\Delta\varphi = Ne/8\pi\varepsilon_0 R$ and a maximum electric field (located on the surface of the sphere) of $F_R = Ne/4\pi\varepsilon_0 R^2$. Our metastable $Ar^*$ beam laser photoelectron source yields a current of about 1 pA, when an intracavity photoionization laser (2 W) is used; the corresponding ion density amounts to about 10 ions in a sphere with 1 mm diameter resulting in $\Delta\varphi \approx 14\ \mu V$ and $F_R \approx 56\ \mu V/mm$ [20].

Residual electric fields associated with potential variations on surrounding surfaces and with ac voltages (ripple) are hard to estimate, and therefore one has to develop a suitable diagnostic method to quantitatively measure and remove the residual fields to negligible levels. It is well known that highly excited Rydberg atoms are sensitive probes of electric fields, see e. g. [33,34]. The Stark broadening/shifting of Rydberg levels, the field-induced occurrence of optically forbidden states and the reduction of the ionization energy can be used to diagnose the local electric field. In sub-meV resolution experiments one has to reliably probe and establish fields $< 1$ mV/mm. In recent experiments with the target beam setup (Fig. 2) we found it an efficient and rather precise procedure to determine the onset of field ionization relative to the field-free ionization limit (E = 0 in Fig. 3), known to within $\pm\ 0.6\ \mu eV$ from extrapolation of unperturbed Rydberg series. The corresponding energy shift $S_F$, induced by the residual (or an externally applied) electric field F, is classically given by $S_F = 76\ \mu eV \times (F[V/m])^{1/2}$. Exciting $Ar^*(4p\ ^3D_3)$ atoms with a cw single mode laser in known externally applied fields (F > 0.02 V/m) we found shifts $S_F$ in agreement with the classical formula. For illustration Fig. 3 presents two data sets for the energy range $- 175 < E < 75\ \mu eV$, obtained with optimally compensated residual field (lower part) and with an additional external field $F_z = 0.1$ V/m applied along the direction of the target beam. The vertical lines indicate the respective onsets of field ionization, corresponding to shifts of $S_F = 8\ \mu eV$ and $S_F = 24\ \mu eV$. From $S_F = 8\ \mu eV$ a residual stray electric field of $F_S \approx 0.01$ V/m is deduced. At such a low field meaningful free electron attachment data down to very low energies (about 10 µeV) may be extracted. Note that under the experimental conditions of Fig. 3 (ionizing laser power about 30 mW, laser diameter about 1 mm) the effects of ion space charge are negligible. We mention that the electric field due to the motional Stark effect acting on the highly excited Rydberg atoms amounts to values around $10^{-3}$ V/m (B $\leq 2\ \mu T$, v(Ar)=600 m/s) and is therefore negligible compared with the value of the stray electric field.

In our published work [19,20,22] we used a multimode ionization laser with a bandwidth of 150 µeV or 50 µeV; in these studies the residual electric field was determined from a comparison of measured threshold electron attachment data with calculated yields taking into account the influence of the (fitted) electric field on the photoelectron motion prior to attachment. This approach can yield realistic values for the residual field down to values of about 50 µV/mm (for 50 µeV photon bandwidth) [20,22], but the use of a single mode ionization laser for determining $F_S$ from measured shifts $S_F$ is much more precise and reliable as demonstrated by data such as those in Fig. 3. Our present laser photoelectron source works with a photoionization laser which can be readily changed between single mode operation for ultrahigh resolution studies and intracavity broadband operation by extending the laser cavity through the reaction region and thereby increasing the available laser power by a factor of about 100.

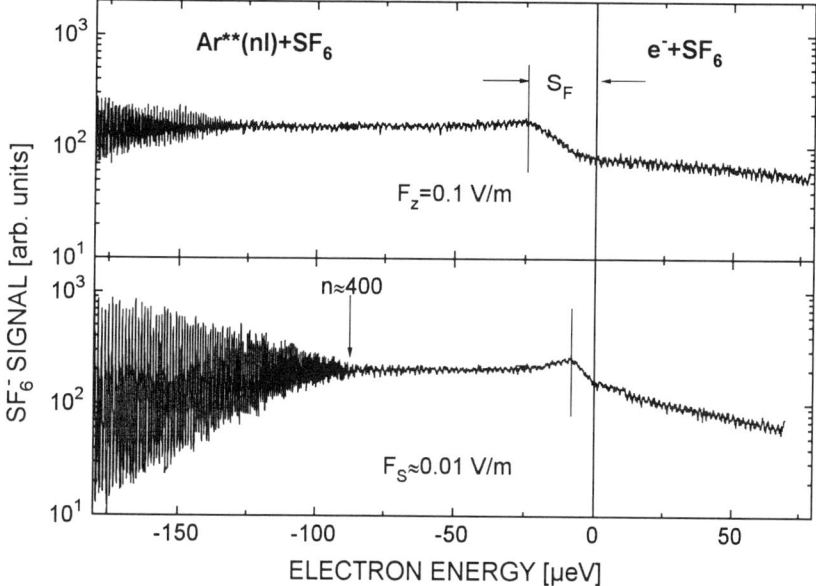

**FIGURE 3.** Weak electric fields probed by electron attachment near threshold. Data taken with single mode ionization laser.

At sub-meV resolution kinematic energy shifts (Doppler broadening) [4,8] can easily play an important role. In our photoelectron attachment experiment two effects contribute: i) the spread in laboratory electron energies $\Delta E_D(Ar)$ due to the motion of the electron emitting argon atoms $\Delta E_D(Ar) = m < \vec{v}_{CM} \vec{v}_{Ar} >_{1/2}$ (m = electron mass; $< >_{1/2}$ denotes the full width at half maximum of the Doppler spread function; the recoil energy $(m/m_{Ar})E_{Ar}$, i. e. the laboratory energy of an electron with zero center-of-mass velocity $v_{CM}$, amounts to 0.9 µeV). ii) the spread in center-of-mass collision energies for the $e^- + XY$ system

(XY = target molecule) $\Delta E_D(XY) = m <\bar{v}_e \cdot \bar{v}_{XY}>_{1/2}$ ($v_e$ = lab electron velocity). For the molecules discussed in this report (XY = $SF_6$, $CCl_4$, $CH_3I$) the dominant contribution is due to the first term i) and amounts to $\Delta E_D(Ar) \approx 0.06$ meV$(E[meV])^{1/2}$ [20].

## LASER PHOTOELECTRON ATTACHMENT TO MOLECULES AT SUB-meV RESOLUTION

Using a static gas target in conjunction with pulsed electron production/ attachment and ion extraction we have so far studied laser photoelectron attachment (LPA) to seven molecules ($SF_6$ [19,20,22], HI [22], $CCl_4$, $CFCl_3$, $C_2F_3Cl_3$ [23], $CH_3I$ and $CH_2Br_2$) in the range 0-170 meV at resolutions below 1 meV. Using a skimmed supersonic target beam we have studied XY = $SF_6$, $CH_3I$, $CH_2Br_2$, $CFCl_3$ and also - over the limited range of 0-7 meV - the formation of $(SF_6)_2^-$ and $(SF_6)_3^-$ cluster ions [31]. The important molecule $SF_6$ has been investigated in most detail and over the largest energy range (10 μeV - 250 meV); the published results for E = 0.3-170 meV [19,20] were extended to lower energies by data taken with a single mode laser over the range 0-14 meV and to higher energies by photoionizing laser-excited $Kr^*(5p\ ^3D_3)$ atoms. Fig. 4 summarizes all these data for $SF_6^-$ formation (lifetime > 100 μs). The absolute cross section scale was obtained by normalization to the T = 300 K (equal electron and gas temperature) rate coefficient $k_e(T = 300\ K) = 2.27(9) \cdot 10^{-7}$ cm$^3$ s$^{-1}$, measured by Petrovic and Crompton [35] with the Cavalleri swarm technique. At low energies (E < 1 meV) the measured LPA cross section approaches the $E^{-1/2}$ behaviour expected for s-wave attachment [19,25,26] and agrees well with the theoretical cross section $\sigma_c$ for s-wave electron capture in the $1/r^4$ polarization potential [19,26]. The rate coefficients $k_{nl}$ for Rydberg electron attachment calculated from the LPA cross sections with formula (4) [22,29] are in good agreement with the values measured by Dunning and coworkers [29] who find $k_{nl} = (4 \pm 1) \cdot 10^{-7}$ cm$^3$ s$^{-1}$ for n > 50. For energies up to $E(v_1) = 95.4$ meV, the threshold for excitation of one quantum of the symmetrical stretch vibration, the LPA cross section is found to be well described by the simple analytical formula [19,20,22]

$$\sigma_e(E) = (\sigma_0 / E)[1 - \exp(-\beta\ E^{1/2})] \qquad (5)$$

with $\sigma_0 = 7130 \cdot 10^{-20}$ m$^2$, β = 0.405, E in meV.

The parameter β in eq. (5) characterizes the deviation of the cross section from the limiting $E^{-1/2}$ threshold behaviour. At the onset for $v_1$ vibrational excitation a sharp downward cusp due to channel coupling is observed [19] in essential agreement with the theoretical prediction of Gauyacq and Herzenberg [36]. At $E \approx 191$ meV, the $(2v_1)$ threshold, an analogous feature is observed, see also [36].

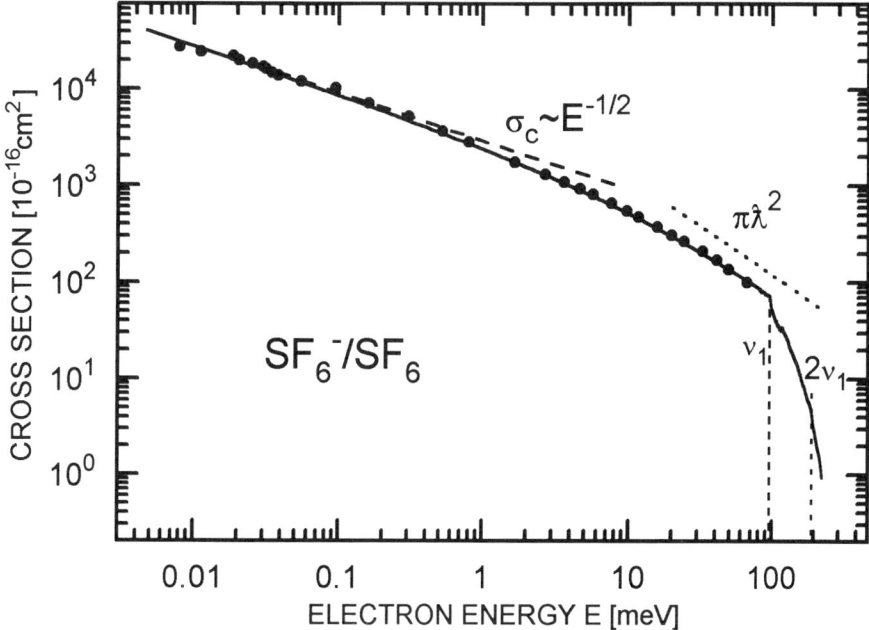

**FIGURE 4.** Laser photoelectron attachment cross section for $SF_6^-$ formation from $SF_6$ molecules ($T_g$ = 300 K). The smooth line is the fit cross section eq. (5) for E < 90 meV. For a comparison with previous data see [19,22,29] and text.

These threshold structures are more clearly seen in plots of the energy dependent rate coefficient $k_e(E) = \sigma_e(E) \, v$ ($v = (2E/m)^{1/2}$), as illustrated for dissociative attachment (DA) to $CCl_4$ molecules [23] in Fig. 5a: at $E(v_1) = 56.9$ meV and $E(2v_1)$ clear downward cusps appear due to coupling of the DA channel with the symmetrical stretch vibration which - as for $SF_6$ - plays an important role in the s-wave attachment process. Vibrations with other symmetries can also induce threshold structure and may exhibit different appearance [19,23]. Below the $v_1$ threshold the DA cross section data for $CCl_4$, again normalized to swarm results [37], are well described by formula (5) with $\sigma_0 = 11160 \cdot 10^{-20}$ m$^2$ and β = 0.588 [23]. Good agreement is found between the Rydberg rate coefficients $k_{nl}$ calculated from the LPA cross section with the recent results of Dunning's group [29]. We note that for both $SF_6$ and $CCl_4$ our LPA data predict a substantial decrease of the attachment rate coefficients $k_e(T_e)$ with rising electron temperature $T_e$ for a Maxwellian electron ensemble at the fixed gas temperatures $T_g$ = 300 K in satisfactory agreement with recent swarm data [27,38].

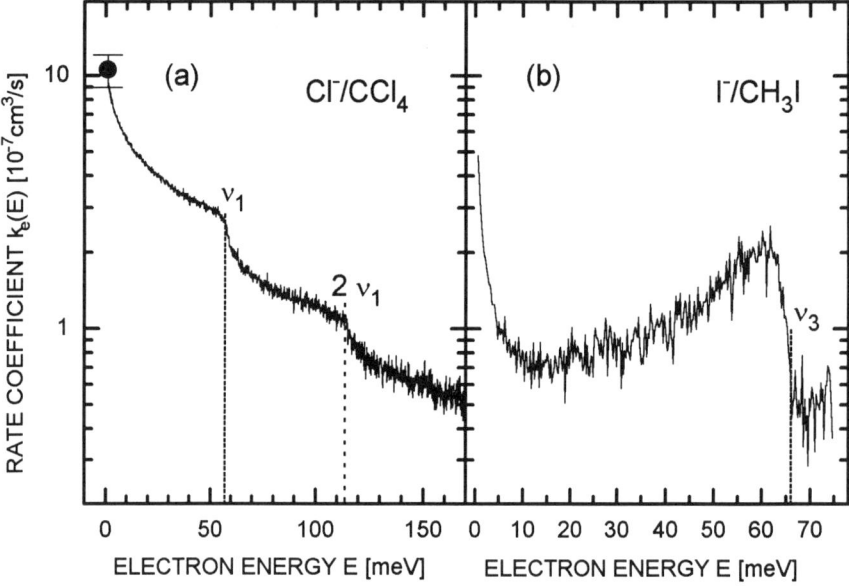

**FIGURE 5.** Energy dependent rate coefficients $k_e(E)$ for dissociative electron attachment to the molecules $CCl_4$ (a) and $CH_3I$ (b). The absolute scale is established by normalization to thermal rate coefficients ($T_e = T_g$ = 300 K) [37,38,40]. The filled circle with error bar in (a) represents the rate coefficient for $Cl^-$ formation in collisions of $CCl_4$ with high n Rydberg atoms (n > 70) [29].

In the DA cross section for $CH_3I$ molecules ($I^-$ formation) we detect a remarkably strong downward threshold cusp associated with the opening of the $CH_3I(v_3)$ threshold ($E(v_3)$ = 66.1 meV). A corresponding feature is present in the DA data of Alajajian et al. (resolution about 6.5 meV) [39] who attributed the intermediate maximum in the DA cross section near E = 57 meV to a second crossing of the $CH_3I$ and $CH_3I^-$ potential surfaces. Fig. 5b presents our energy dependent rate coefficients $k_e(E)$ as normalized to the thermal swarm value $k_e(T = 300 \text{ K}) = 9.5(5) \cdot 10^{-8}$ cm$^3$ s$^{-1}$ [40]. The width of the threshold structure suggests broadening due to rotational transitions which are known to be strong for this dipolar molecule [41] (dipole moment 1.62 D).

## PERSPECTIVES

Laser photoelectron sources have for the first time allowed studies of low energy electron collisions with gaseous targets at sub-meV resolution. An essential feature of such a source is the reliable diagnosis of residual electric fields arising from inhomogeneous surface potentials, space charge effects and voltage ripple. It will be attractive to use such a source with some modifications to investigate electron scattering processes from skimmed supersonic target beams and surfaces with (sub) meV resolution. Higher currents than with our $Ar^*$ photoionization scheme can be realized by photoionizing laser-excited alkali atoms [20]. We are in the process of replacing the $Ar^*(4s)$ beam by a $K(4s)$ beam with the aim to study photoelectron attachment to clusters. Photoionization of $Na^*(3p)$ atoms with a frequency-doubled modelocked titanium-sapphire laser is being exploited by Bergmann's group [42] to investigate DA to vibrationally excited $Na_2$ molecules.

## Acknowledgements

We gratefully acknowledge support of our work by the Deutsche Forschungsgemeinschaft and by the Stiftung Rheinland-Pfalz für Innovation.

## REFERENCES

1. Schulz, G.J., Phys. Rev. Lett. **10**, 104 (1963)
2. Schulz, G.J., Rev. Mod. Phys. **45**, 378 (1973)
3. Linder, F., Endeavour **XXXIII**, 124 (1974)
4. Read, F. H., J. Phys. **B8**, 1034 (1975)
5. Rohr, K., Linder F., J. Phys. **B9**, 2521 (1976)
6. Chang, E. S., Wong, S. F., Phys. Rev. Lett. **38**, 1327 (1977)
7. Brunt, J. N. H., Read, F. H., King, G. C., J. Phys. **E10**, 134 (1977); Read, F. H., Phys. Scripta **27**, 103 (1983)
8. Gallagher, A., York, G., Rev. Sci. Instrum. **45**, 662 (1974); van Brunt, R. J., Gallagher, A., in Electronic and Atomic Collisions (Watel, G., Ed.), pp. 129-142, North Holland, Amsterdam, 1978
9. Kennerly, R. E., van Brunt, R. J., Gallagher, A., Phys. Rev. **A23**, 2430 (1981)
10. Jung, K., Antoni, Th., Müller, R., Kochem, K. H., Ehrhardt, H., J. Phys. **B15**, 3535 (1982) Ehrhardt, H., in Molecular Processes in Space (Watanabe, T., Shimamura, I., Shimizu, M., Itikawa, Y., Eds.), pp. 41-64, Plenum Publ. Corp., New York, 1990
11. Ferch, J., Raith, W., Schröder, K., J. Phys. **B13**, 1481 (1980)
12. Kennerly, R. E., Phys. Rev. **A21**, 1876 (1980)
13. Buckman, S. J., Lohmann, B., J. Phys. **B19**, 2547 (1986)

14. Chutjian, A., Alajajian, S. H., Phys. Rev. **A31**, 2885 (1985); Chutjian A., in Physics of Electronic and Atomic Collisions (MacGillivray W. R., McCarthy, I. E., Standage, M. C., Eds.), pp. 127-138, Adam Hilger, Bristol, 1992
15. Field, D., Ziesel, J. P., Guyon, P. M., Govers, T. R., J. Phys. **B17**, 4565 (1984); Field D., Knight, D. W., Mrotzek, G., Randell, J., Lunt, S. L., Ozenne, J. B., Ziesel, J. P., Meas. Sci. Technol. **2**, 757 (1991); Ziesel, J. P., Randell, J., Field, D., Lunt, S. L., Mrotzek, G., Martin, P., J. Phys. **B26**, 527 (1993)
16. Allan, M., in Electronic and Atomic Collisions (Gilbody, H. B., Newell, W. R., Read, F. H., Smith, A. C. H., Eds.), pp. 93-104, Elsevier Science Publ. B. V., 1988; Allan, M., J. Electron Spectrosc. Rel. Phen. **48**, 219 (1989); Allan, M., J. Phys. **B25**, 1559 (1992)
17. Kisters, G., Chen, J. G., Lehwald, S., Ibach, H., Surf. Sci. **245**, 65 (1991); Ibach, H., Balden, M., Bruchmann, D., Lehwald, S., Surf. Sci. **269/270**, 94 (1992)
18. Ibach, H., J. Electron Spectrosc. Rel. Phen. **64/65**, 819 (1993); Ibach, H., Electron Energy Loss Spectrometers, Springer, Berlin-Heidelberg, 1991
19. Klar, D., Ruf, M.-W., Hotop, H., Chem. Phys. Lett. 189, 448 (1992); Aust. J. Phys. **45**, 263 (1992)
20. Klar, D., Ruf, M.-W., Hotop, H., Meas. Sci. Technol. **5**, 1248 (1994)
21. Baltzer, P., Lundqvist, M., Edvardsson, D., Wannberg, B., Karlsson, L. (to be publ.); Baltzer, P., Karlsson, L., private communication
22. Klar, D., Mirbach, B., Korsch, H. J., Ruf, M.-W., Hotop, H., Z. Phys. **D31**, 235 (1994)
23. Klar, D., Ruf, M.-W., Hotop, H., Z. Phys. D (to be publ.), Klar, D., Dissertation, Univ. Kaiserslautern (1993)
24. Christophorou, L. G. (Ed.), Electron-molecule interactions and their applications, Vol. 1 and 2 (Academic Press, New York, 1984)
25. Bethe, H. A., Phys. Rev. **47**, 747 (1935); Wigner, E. P., Phys. Rev. **73**, 1002 (1948)
26. Vogt, E., Wannier, G. H., Phys. Rev. **95**, 1190 (1954)
27. Smith, D., Spanel, P., Adv. At. Mol. Opt. Phys. **32**, 307 (1994); Spanel, P., Smith, D., Int. J. Mass. Spectrom. Ion Proc. **129**, 193 (1993)
28. Fermi, E., Nuovo Cimento **11**, 157 (1934); Matsuzawa, M., in Rydberg States of Atoms and Molecules (Stebbings, R. F., Dunning, F. B., Eds.), pp. 267-314, Cambridge Univ. Press, New York, 1983
29. Dunning, F. B., J. Phys. **B28**, 1645 (1995)
30. Baltzer, P., Wannberg, B., Göthe, M. C., Rev. Sci. Instrum. **62**, 643 (1991)
    Baltzer, P., Karlsson, L., Lundqvist, M., Wannberg, B., Rev. Sci. Instrum. **64**, 2179 (1993)
31. Kreil, J., Weber, J. M., Klar, D., Schramm, A., Ruf, M.-W., Hotop, H., Symp. Atomic, Cluster and Surface Physics 1994, Contributions pp. 339-342 (Märk, T. D., Schrittwieser, R., Smith, D., Eds.) Maria Alm, Austria, 20-26 March 1994; Kreil, J., Dissertation, Univ. Kaiserslautern (1995)
32. Kraft, T., Ruf, M.-W., Hotop, H., Z. Phys. **D14**, 179 (1989)
33. Neukammer, J., Rinneberg, H., Vietzke, K., König, A., Hieronymus, H., Kohl, M., Grabka, H. J., Wunner, G., Phys. Rev. Lett. **59**, 2947 (1987)
34. Frey, M. T., Ling, X., Lindsay, B. G., Smith, K. A., Dunning, F. B., Rev. Sci. Instrum. **64**, 3649 (1993)
35. Petrovic, Z. Lj., Crompton, R. W., J. Phys. **B18**, 2777 (1985)
36. Gauyacq, J. P., Herzenberg, A., J. Phys. **B17**, 1155 (1984)
37. Orient, O. J., Chutjian, A., Crompton, R. W., Cheung, B., Phys. Rev. **A39**, 4494 (1989)
38. Shimamori, H., Tatsumi, Y., Ogawa, Y., Sunagawa, T., J. Chem. Phys. **97**, 6335 (1992)
39. Alajajian, S. H., Bernius, M. T., Chutjian, A., J. Phys. **B21**, 4021 (1988)
40. Shimamori, H., Tatsumi, Y., Ogawa, Y., Sunagawa, T., Chem. Phys. Lett. **194**, 223 (1992)
41. Ling, X., Smith, K. A., Dunning, F. B., Phys. Rev. **A47**, R1 (1993)
42. Bergmann, K., private communication; see also contribution in this volume

# Absolute Cross Section Measurements for Electron Collisions with Polyatomic Molecules of Plasma Chemistry

H. Tanaka and L. Boesten

*Dept. of Physic, Sophia University, Chiyoda-ku 7-1, Tokyo 102, Japan*

**Abstract.** Modelling of low temperature plasma requires a detailed knowledge of the electron collisional cross sections for the initial reactions as well as for the formation of reactants. After a short review of the main experimental methods of their determination, i.e. transmission, swarm techniques, and beam methods, recent measurements of the of electron - polyatomic molecule collision cross sections are summarized for the major gases encountered in chemical plasma processing.

## INTRODUCTION

As the series of past ICPEAC publications shows, electron and atom/molecule collisions constitute an ever lasting central theme in the long history of atomic and molecular physics. In the previous conference in Aarhus, Burke [1] gave a detailed review of the theory on electron, atom, ion and molecule collisions, while the results of a the satellite meeting in London on electron - molecule, cluster, and surface interactions [2] was published only last year. Since the early days of quantum physics progress in this field was sustained by the pure scientific interest, but there also was and continues a significant undercurrent of stimuli arising from a wide range of applications.

The same can be said about the growing interest in low energy electron-polyatomic collision studies [3,4,5]. One example is the attempt to grasp the physics of low-temperature, low-pressure reactive plasma reactions which are now widely used in the fabrication of super LSI chips and amorphous solar cells etc., at the atomic-molecular level [6]. Such a chemical plasma generally is produced by glow discharge in "process gases" at a pressure of a few torr. With an ionization degree of approximately $10^{-4}$, most of the gas consists of neutral molecules. The electrons within the bulk plasma have a high average temperature of $1 \sim 10$ eV, and the plasma is in a non-equilibrium state, since the neutral molecules or ions remain at room temperature. Collision processes within the plasma can be classified as follows: (1) Primary reactions with electrons as the main carrier (electrons accelerated by the electric fields collide with molecules and excite or ionize these molecules, or produce ionization or dissociation and thus generate radicals). (2) Secondary reactions by these radicals (collisions between radicals, or radicals and molecules leading to recombination or dissociation) and (3) Surface reactions (deposition, etching, or rearrangement of the surface structure).

We will review the present state of cross sections for electrons colliding with polyatomic molecules as used in plasma processing. These collisions form the initial trigger for the long chain of chemical and physical processes evolving in reactive plasmas.

## CROSS SECTION DATA FOR ELECTRON - POLYATOMIC COLLISIONS

The probability that an electron colliding with a molecule and after exciting state "s" is scattered into a unit solid angle at a scattering angle $\theta$ with respect to the incident direction is called the "differential cross section" ( DCS, $\sigma_s(\theta)$ ) of this particular process, the integral of $\sigma_s(\theta)$ over all deflection angles is called the "integrated cross section" ($q_s$) for this process, and their sum over all particular processes is called the "total cross section" $Q = \Sigma\ q_s$.

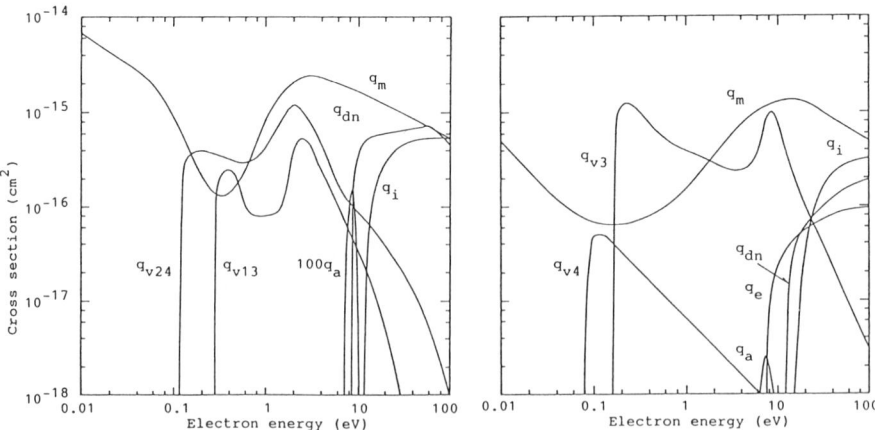

**Figure 1.** A typical "set" of electron collision cross sections for $SiH_4$ (left) and $CF_4$ (right). The symbols are explained in the text.

Figure 1 gives a typical example [7] of a graphical compilation of the cross sections for the representative plasma molecules $SiH_4$ and $CF_4$. Collisions of electrons in the energy range included are extremely important for the determination of the macroscopic behavior of plasmas ("plasma modelling"). The momentum transfer cross section $q_m$, obtained by multiplying $\{1 - \cos(\theta)\}$ into $\sigma_s(\theta)$ and integrating over all angles, shows the Ramsauer-Townsend minimum in the low energy region and a wide maximum at ~ 2.3 eV ($SiH_4$) and ~ 10 eV ($CF_4$). The particular cross sections $q_s$ (s = v: vibration; s = a: attachment; s = e: electronic excitations; s = d: dissociation; s = i: ionization) increase sharply near their threshold and depend strongly impact energy of the electron and the type of molecule involved. A review of these electron collision cross sections can be found in vol. 33 of *Advances in Atomic, Molecular and Optical Physics* [8].

## REPRESENTATIVE MEASUREMENT METHODS FOR ELECTRON COLLISIONS

To determine particular cross sections over a wide range of energies, several experimental methods like swarm experiments, cross beam methods, and transmission experiments, must be combined. They differ in the energy region covered and in the ease of obtaining absolute cross sectional values.

### Electron swarm experiments [9]

In this method, one measures transport properties like drift velocity and diffusion coefficients of electron swarms in carrier gases moving under the influence of an electric field together with the excitation-, ionization-, attachment- and other rates as a function the ratio of electric field strength and gas density, and converts the set of these coefficients into cross sections by use of the Boltzmann equation. This method is the most effective way to determination of absolute cross sections below 1 eV where beam methods become difficult. By admixing small amounts of $SiH_4$ or $CF_4$ into Ar gas and working near the Ramsauer Townsend minimum of Ar, Kurachi et al. [7] were able to determine the vibrational cross section of these gases from the transport parameters of the mixture. However, at higher energies other experimental methods must be relied upon to obtain a breakdown of the DCS into the various particular processes.

### Transmission methods

Here an electron beam is transmitted through a cell containing the gas of interest and the total cross section is derived from the intensity loss of the beam on passage through the cell. All electrons deflected by elastic collisions and electrons from all non-elastic collisions give an upper limit to the cross section. Good quality data can be obtained over a wide energy. However care is needed in the correction for the effective cell length (some gas leaks from the containment apertures) and forward scattering. Like swarm experiments, no subdivision of the total DCS into particular processes is possible. Figure 2 shows the total cross section of $CF_4$. While the measurements of Jones [10] and Sueoka [11] agree well, the total cross section of Mann [12] and Boesten et al. [13], determined by extrapolation of the elastic DCS and subsequent integration, differ in the whole energy region above 5 eV from one another and from the total cross sections. The reason for the former may be the extrapolation method, for the latter the exclusion of vibrational cross sections from the integration. Several theoretical calculations [14,15,16] have been performed, but so far, do not yet reproduce the experimental data. Trajmar [17] has reviewed the literature of total cross section measurements up to 1993.

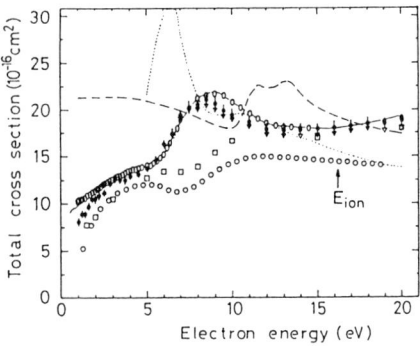

**Figure 2.** Total cross sections for electron $CF_4$ collisions. Solid and open circles (upper) are the total cross sections of Jones and Sueoka, the lower circles and squares those of Mann et al. and Boesten et al. Other lines represent theory.

## Beam methods

An electron beam of known energy $E_0$ is crossed at right angles with a narrow beam of target molecules and the angular distribution of the scattered electrons at fixed impact energy, or the energy-distribution at a fixed scattering angle θ and a fixed energy loss (excitation function for process "s") is measured. Because these are differential measurements for particular processes, they provide much more information than the previous two methods. Disadvantages are the rather difficult control of a low energy electron beam and the contradictory requirements for high energy resolution and high beam intensity. Direct absolute determination of the DCS is difficult because too many experimental parameters have to be controlled like the density and shape of the target gas beam, the determination of the collision volume and view-cone etc. The very careful check on these parameters as performed by Bromberg [18] at high impact energies for their absolute cross section determinations become extremely difficult at low energies. Fortunately, it is known [19] that the shape of the gas beam is rather invariant against a change in the gas as long as approximately equal Knudson numbers are obtained. This allows the use of a reference gas (r) with known DCS like He [20] by repeating the He-measurements under identical experimental conditions as used for the target gas. The unknown DCS of the gas (x) can then be determined from their pressure ratios as

$$\sigma_x(\theta) = (P_r / P_x)(I_x / I_r) \sigma_r(\theta)$$

where I is the observed intensity and P the pressure. The procedure is known as the "relative flow method" [21,22] and has an accuracy of about 15-30%. With this method we have obtained the absolute DCS of many gases in the energy range from 1.5 to 100 eV and scattering angles from $20^0$ to $130^0$ with a resolution of 30-40 meV. Table 1 gives a schematic view of the series of molecules investigated. Of course,

Table 1. Molecules investigated in the author's laboratory

| | | | | | | | | |
|---|---|---|---|---|---|---|---|---|
| | | | | | | | | $C_3H_4$ |
| | | | | | | | | $C_3H_6$ |
| $C_3F_8$ | $C_2F_6$ | $CF_4$ | $CHF_3$ | $CH_2F_2$ | $CH_3F$ | $CH_4$ | $C_2H_6$ | $C_3H_8$ |
| | | | | | | $SiH_4$ | $Si_2H_6$ | |
| | | | | | | $GeH_4$ | | |

rotational DCS cannot be resolved at that resolution, and the "elastic" cross sections obtained correspond rather to "vibrationally elastic DCS". For some molecules even the elastic DCS may contain contributions from the lowest vibrational modes of the molecule.

## ELASTIC CROSS SECTIONS: ENERGY- AND ANGULAR DEPENDENCE

Elastic scattering occurs in all collisions and generally its cross sections are much larger than those of inelastic scattering. It is therefore much easier to obtain their absolute cross sections for example with the relative flow method just described. Fortunately, as long as the transmission rate of selectors and lens systems is guaranteed, the elastic cross sections can be measured simultaneously with the inelastic cross section and thus can serve as calibration standard for the absolute determination of the inelastic DCS. The cross sections determined so far in this fashion in our laboratory can be classified as follows.

## $XH_4$ molecules (X = C, Si, Ge)

The most fundamental molecule of the hydrocarbon series, $CH_4$, has been extensively investigated theoretically and experimentally. Some recent publications include the experiments of Mapstone et al. [23] and Lunt et al. [24], and a theoretical paper by Gianturco et al. [25]. In contrast to this, measurements of the similarly structured $SiH_4$ [26] and $GeH_4$ [27] are quite scarce. The main features of the DCS of these molecules can be summarized as follows: The size of the DCS increases with the geometrical size of the mole-cule, that is the C-H, Si-H, and Ge-H bond lengths of 1.09, 1.48, and 1.53 Å. The broad maximum in $Q_T$, $q_o$, and $q_m$ of $CH_4$ at about 7.5 eV [28] shifts towards lower energies ($SiH_4$: 2.3 eV [26], $GeH_4$: ~ 5 eV [27]). Theoretical calculations of these molecules belonging to symmetry group $T_d$ show that resonance enhancements for $CH_4$ and $SiH_4$ occur in the contributions of irreducible species $T_2$, and for $CF_4$ in $T_2$, and possibly $A_1$ [15], and especially that the contribution of the angular moment with $\ell = 2$ is large. Figure 3 shows the observed DCS of $CH_4$ [28], $SiH_4$ [26], and $GeH_4$ [27] as a function of impact energy and scattering angle. At low energies a trend towards a d-wave ($\ell = 2$) is clearly recognizable but this broad feature appears in nearly all polyatomic molecules. Because of the presence of strong direct scattering in the elastic DCS, it nearly impossible to discern resonant scattering in these DCS. These are much easier to detect in the vibrational scattering described below.

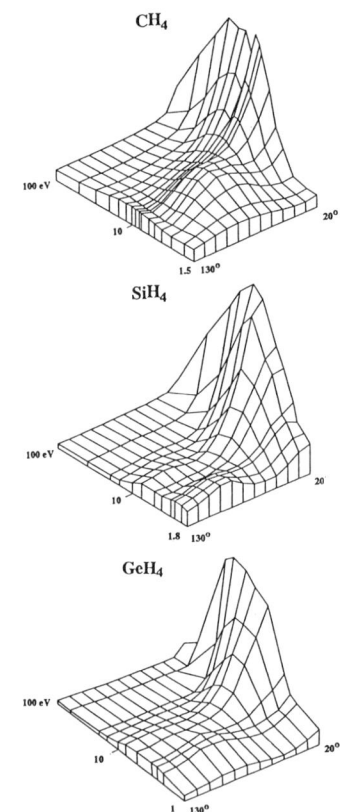

**Figure 3.** Elastic DCS (linear scale) of three $XH_4$ molecules

## $C_nY_{2n+2}$ (Y=H, F; n = 1 ~ 3) and $NY_3$

Because the C-H bond length in the saturated carbons ($CH_4$, $C_2H_6$, $C_3H_8$) remains essentially constant, the size of the molecule is determined solely by the length of the C-C coupling. In turn, this is reflected in $q_0$ as can be seen in Figure 4 [29]. The broad maximum of all three hydrocarbons occurs at the same position of ~ 7.5 eV. The integrated cross section $q_0$ of this figure was determined by extrapolating the differential DCS $\sigma(\theta)$ measurements in the angular range $20^0 \sim 130^0$ with the help of phase shift fitting. Taking $C_2H_6$ [30] as an example, the peak structure at 7.5 eV shows an angular distribution which reflects a dominant shape resonance of angular component $\ell = 3$ (f-wave). This also shows up in the Complex-Kohn calculations of Sun et al. [31].

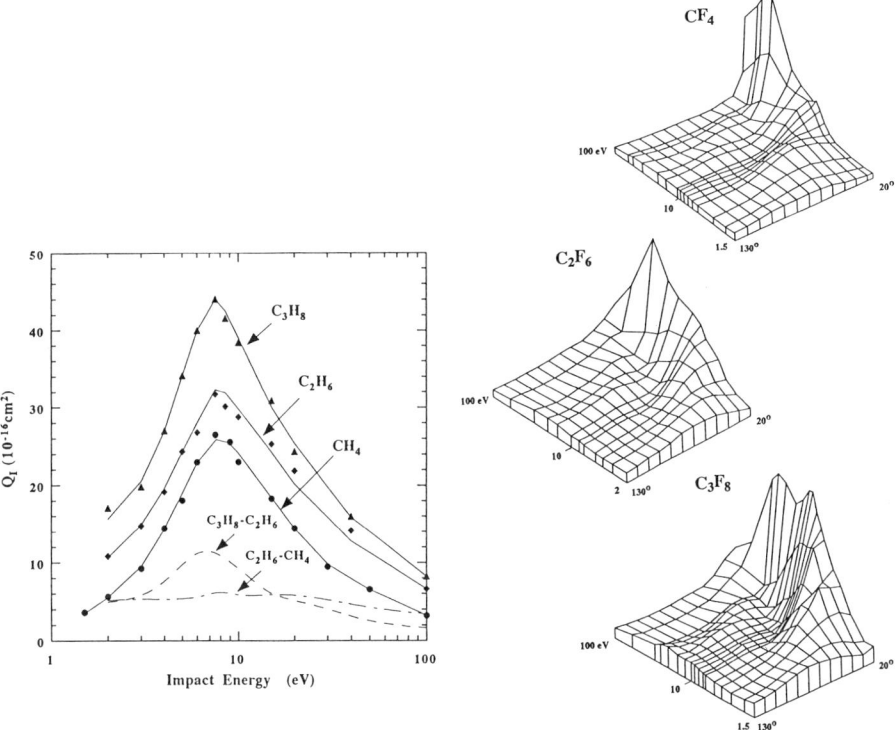

**Figure 4.** Integrated elastic cross sections of three hydrocarbons.

**Figure 5.** The same as Figure 3 for $C_nY_{2n+2}$ (Y=H, F; n = 1 ~ 3)

Next we discuss molecules in which the H of the hydrocarbons has been replaced by F-atoms. Figure 5 shows the observed elastic DCS of $CF_4$, $C_2F_6$ [32], and $C_3F_8$. These DCS show much stronger "undulations" than their hydrocarbon equivalents. One reason may be that larger $\ell$-components are required because $CF_4$ is larger than $CH_4$ (C-F bond length 1.31 Å). The low energy, low angle DCS of C-F molecules tend to be lower, and in forward scattering at high energies, there clearly appear small periodic structures which are due to interference of the scattering

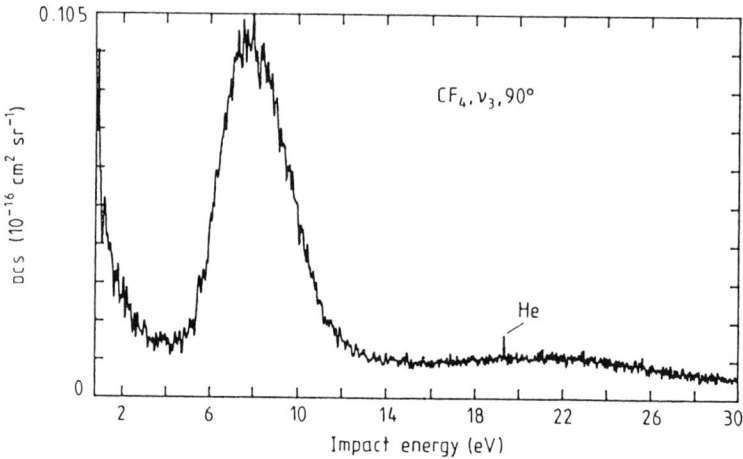

**Figure 6.** Excitation function of the stretching vibrational mode of $CF_4$.

waves from individual atoms. As already mentioned, theory indicates that the integrated cross section $q_0$ of $CF_4$ (see Figure 2) contains many more resonances than that of $CH_4$. However, the irreducible symmetry species contributing to these resonances cannot be clearly discerned in the elastic angular distributions. So far no calculations have been performed for $C_2F_6$ and $C_3F_8$.

For the elastic DCS of $NH_3$ [33] and $NF_3$ [34] a similar trend as in the C-H and C-F systems exists: the resonance of $NH_3$ occurs at ~ 7 eV while that of $NF_3$ peaks at a lower ~ 3 eV. Of much interest are also comparisons with the double bond molecules $C_2H_4$ [35], $C_3H_4$ [36], and $C_3H_6$ [37] (propylene), triple bond molecules $C_2H_2$ [38] and $C_3H_4$ [36], "long" molecules, and the highly symmetric molecules $XY_4$ and $C_3H_6$ [37].

## INELASTIC COLLISIONAL CROSS SECTIONS

### Vibrational excitation and resonances

$CF_4$ possesses 4 fundamental vibrational modes ($v_1 = 113$, $v_2 = 54$, $v_3 = 159$, $v_4 = 78$ meV). As shown in Figure 1, $q_{13}$ ($v_1+v_3$) and $q_{24}$ depend strongly on the impact energy of the electron, and over a certain energy region even surpass the elastic cross section. They constitute an important loss mechanisms for the low energy electrons of plasma gases. In Figure 6 [13] we present the vibrational excitation function for $v_{13}$ of $CF_4$ at $90^0$. The trace shows a sharp increase towards low energies, while the peaks near ~ 8 eV and ~ 21 eV indicate the presence of $t_2$ shape resonances. As stated previously, resonances are generally difficult to observe in elastic scattering where strong direct scattering is dominant, but in vibrational excitations where the DCS of direct scattering themselves are small, the enhancements due to resonances appear rather clearly. Generally, the resonances of polyatomic molecules possess wide smooth shapes without the detailed structures as common in scattering from diatomic molecules. Because of the limitations of energy resolution, not all vibrational modes can be distinguished and one has to resort to wave-decomposition techniques to obtain the desired information. Allan [40] gives some examples of low energy resonances, the saturated linear molecule n-propane ($C_3H_8$), the strained cyclic molecules cyclopropane ($C_3H_6$) and the poly-cyclic molecule [1.1.1] propellane ($C_5H_6$), the double bonded molecules ethylene ($C_2H_4$) and allene ($C_3H_4$). In view of the limitations of the Koopmann theorem - when combined with HF/6-31* level molecular calculations - he tries to identify the resonant state directly from the selectivity of the dominant vibrations excited and the angular distribution of the scattered electron. He concludes that the relatively narrow resonance at 2 ~ 3 eV of single bonded molecules arises from a trapping of the electron in the cyclic bond, and that the wide resonance at 7 ~ 10 eV is due to a capture of the electron in the $\sigma^*$ antibonding orbital of the C-H bond.

We, too, have performed systematic studies on the molecules listed in table 1. Like Wong and Schulz [40] in their analysis of their $C_6H_6$ data we have used symmetry arguments and the angular correlation theory of Andrick and Read [41]. Just as Allan, we assume that the irreducible species of the quasi-stable resonance state which captures the electron, is the same as that of the LUMO of the neutral molecule. However intentionally we have reverted to a Gaussian-92 calculations with minimal basis sets. Vibrational excitation is possible only when the transition probability for vibration $<\chi_f|H'|\chi_i>$ expressed in terms of direct products $\Gamma_f \times \Gamma' \times \Gamma_i$ is the totally symmetric. Here it has been assumed that the symmetry species of $H'$ is determined by that of the electron distribution in the quasi-stable resonant state

**Table 2.** The symmetry species involved in the resonance of $C_2H_6$

| State | Representation | |
|---|---|---|
| | Case A | Case B |
| Excited state $\psi_r$ | $a_{2u}$ | $e_u$ |
| Final vibrational state $\psi_f$ Modes | $a_{1g}$ $\nu_1, \nu_2, \nu_3$ | $a_{1g}$  $\nu_1, \nu_2, \nu_3$   or   $e_g$  $\nu_{10}, \nu_{11}, \nu_{12}$ |
| Electron wave $l$ | $a_{2u}$ 1, 3 | $e_u$  1, 3   or   $a_{1u}, a_{2u}, e_u$  1, 3, 5 |

(Hellman-Feynman theorem). One simply has to insert the symmetry species of the initial and final vibrational states for $\chi_i$ and $\chi_f$. One example of such an analysis for $C_2H_6$ [42] is given in table 2. Using only the vibrational modes allowed according to this symmetry analysis we have made a wave decomposition of the energy loss spectrum near 7.5 eV. As shown in Figure 7, vibrational modes $v_{10}$, $v_{11}$, and $v_{12}$ are needed to recover the shape of the loss spectrum and thus it was possible to derive the symmetry species of the resonant state as $e_u$. The angular distribution of the electron scattered by way of this resonant state and leaving the molecule in an excited vibrational state is then compared with the angular moment component as calculated from the angular correlation theory (see table 3). The result of these calculations can be stated as follows: (1) the common resonance at ~ 2 eV appearing in double and triple bond hydrocarbons can be explained systematically by a temporary trapping of the electron (shape resonance) in the $\pi^*$ orbital of the C-C bond, and the high energy resonance (~8 eV) by a trapping in the $\sigma^*$ orbital of the C-H bond, see Figure 8a. (2) In the single bonded molecules with C-H or C-F bonds, the resonance at ~7.5 eV is due to the trapping of the electron in the $\sigma^*$ orbital of the C-H or C-F bond. (3) The low energy resonance seen in single bonded cyclic C-H or C-F molecules like $C_3H_6$ or $C_5H_4$ can be explained by trapping of the electron in a pseudo $\pi^*$ orbital formed by the C-H or C-F bonds.

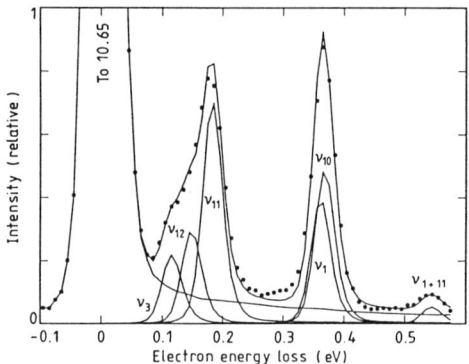

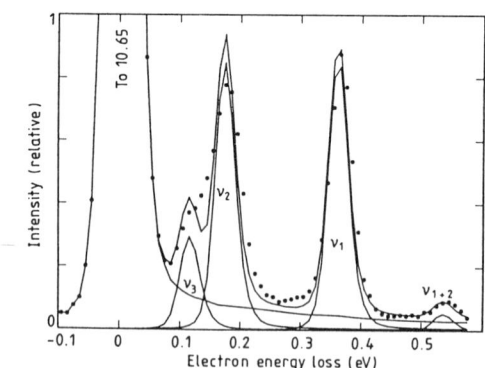

**Figure 7.** Decomposition of the $C_2H_6$ vibrational mode. Top: including all modes belonging to species $e_u$. Bottom: only modes of $a_{2u}$.

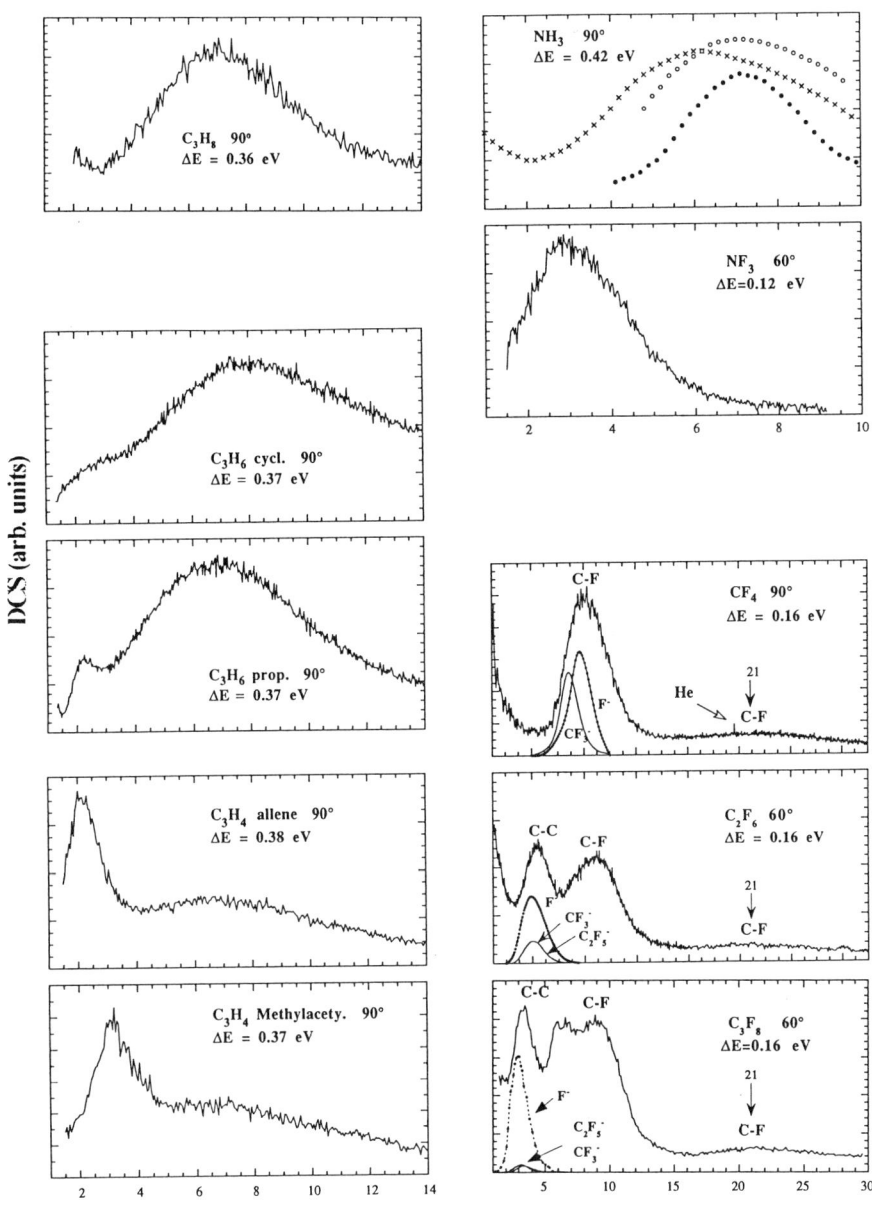

**Figure 8.** Vibrational excitatonal excitation functions for some hydrocarbons (a) and (b) fluorine molecules.

## Dissociative attachment and ionization

Ionization is an important aspect in plasma physics as it helps to sustain the glow discharge and the activated positive ions generated have a high reactivity and perform an important role in the plasma. Becker [43] has given a review of the ionization of halogen-containing molecules. The $q_a$ of Figure 1 is the cross section for the unstable negative ion originating in a dissociative attachment process. Its energy dependence shows resonant behavior. The same can be seen in Figure 9b, and was recently investigated by Le Coat et al. [44] with the help of an energy analysis of the dissociation products of $CF_4$ and a detailed measurement of the angular distribution. Since negative ions accumulate between the electrodes even a minuscule amount in a plasma discharge can influence the structure of its fields considerably. Figure 9 shows the TOF mass spectrum analysis of the negative ion products from electron collisions for $CF_4$ [45]. Similar results have been obtained for $SiH_4$ [46].

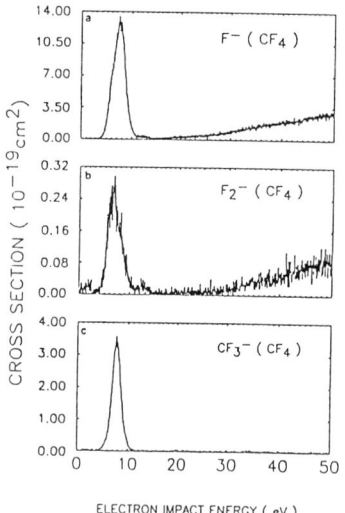
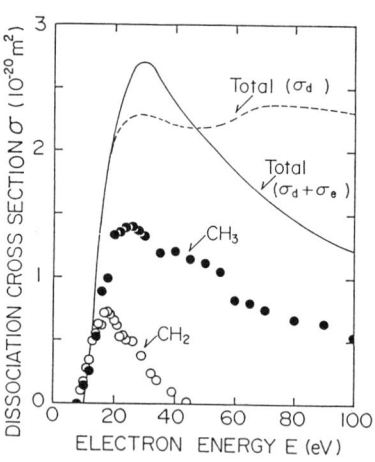

**Figure 9.** Dissociative attachment of $CF_4$     **Figure 10.** Partial dissociation of $CH_4$

## Electronic excitation and dissociation

It is generally held that the electronic excitation states of $CH_4$, $SiH_4$, and $CF_4$ are repulsive states. That implies that in any plasma containing these molecules, either chemically active neutral molecules or radicals will be formed with high efficiency. Among these are also neutral, non-emitting radicals (radicals in their ground state) like $SiH_3$, $SiH_2$, $CH_3$, and $CF_3$, all with rather important roles in plasma physics. However, it is very difficult to measure their concentration in an operating plasma because of their non-emitting nature. So far the cross sections of the neutral radicals were determined in the following way. Parent gas was sealed into a collision chamber and the drop in pressure when irradiating with electrons was observed. The only measurements done were the total cross sections for $SiH_4$ [47], $CH_4$ [48], and $CF_4$ [49]. Recently Nakano et al. [50], and Nakano and Sugai [51] succeeded in the absolute determination of the cross section of the neutral radicals by a threshold mass spectrum analysis of the fragments $CH_3$, $CH_2$ and $CH$ from $CH_4$ and of $CF_3$, $CF_2$, and $CF$ from $CF_4$, see Figure 10. Let us shortly introduce their setup using the $CH_3$ radical as example. The ionization energy of $CH_3$ by electron impact is 9.8 eV while that of $CH_4$ is 14.3 eV. They extract $CH_3$ and $CH_4$ from the plasma, ionize once more by electron impact, and observe the ions produced in a quadrupole mass spectrometer. At the low impact energies only $CH_3^+$ ions are extracted; at higher electron energies $CH_3^+$ from $CH_4$ will dominate. By observing the $CH_3^+$ ions while glow discharge is turned on and off, it is thus possible to determine the concentration of $CH_3$ radicals. Inspired by this experiment, Winstead et al. [52,53] performed SMC calculations on the electronic states of $CH_4$, while Gil et al. [54] used the complex Kohn methods for the same.

## CONCLUSION

We have given a short review of the present state of cross section measurements for electron-polyatomic molecule collisions taking our examples from the electron polyatomic collision mechanisms occurring in the reactive chemical plasmas. In the past 10 years, much progress has been made in quality as well as in quantity in the development of data on the basic collision processes required for modelling, measurement, and control of plasmas. Especially, experimentalists are grateful that finally theorist have begun to get actively involved with more difficult polyatomic molecules. It would be of great help for laser physics, radiation physics, astrophysics, environmental sciences, and nuclear fusion studies, if a set of similar data for non-plasma molecules could be established. Especially, since it is difficult to determine the cross sections over a wide energy range in a single laboratory, it will be mandatory, to increase the reliability of such data by compilations, evaluation and supplementation from other sources.

# ACKNOWLEDGMENT

We want to thank Dr. M. A. Dillon and M. Kimura of Argonne National Laboratory for many discussions on this subject over many years.

# REFERENCES

1. Burke P. G. XVIII ICPEAC (Aarhus, Denmark 1993), AIP Conference Proceedings **295** (AIP Press, Am. Inst. Phys. New York 1993, p. 26-47
2. Ehrhardt H. and Morgan L.A., ed. *Electron Collisions with Molecules, Clusters, and Surfaces* (Plenum Press N.Y. and London 1994)
3. Allan M., ref. 2, p.1005-118
4. Rescigno T. N., ref. 2, p 1-13
5. McCurdy C. W., ref. 1, p. 361-370
6. Itatani R., ed. *Report on Control of Reactive Plasmas (1989-1991 FY)*. Dept. of Engineering, Kyoto Univ. Grant-in-Aid for Scientific Research on Priority Areas, Ministry of Education, Science and Culture, March 1992
7. Kurachi M. and Nakamura Y., *Proc. 13th. Symp. Ion-Sources and Ion-assisted Technology*, Kyoto 1990, ed. Takagi, T. p. 205
8. M. Inokuti, *Advances in Atomic, Molecular, and Optical Physics*, vol. 33 (Academic Press, Boston, 1994)
9. Crompton R. W., ref. 8, p. 47-148
10. Jones R. K., *J. Chem. Phys.* **84** 813-819(1986)
11. Sueoka O., Mori S., and Hamada A., *J. Phys. B* **27** 1453-1456 (1994)
12. Mann A. and Linder F., *J. Phys. B* **25** 533-543 (1992)
13. Boesten L., Tanaka H., Kobayashi A., Dillon M. A., and Kimura M., *J. Phys. B* **25** 1607-1620 (1992)
14. Huo W. M., *Phys. Rev. A.* **38** 3303-3309 (1988)
15. Winstead C., Sun Q., and McKoy V., *J. Chem. Phys.* **98** 1105-1109 (1993)
16. Gianturco F. A., Lucchese R. R., andSanna N., *J. Chem. Phys.* **100** 6464-6471 (1994)
17. Trajmar S. and McConkey J. W., ref. 8, p. 63-96
18. Bromberg J. P., *J. Chem. Phys.* **50** 3906-3921 (1969)
19. Olander D. R. and Kruger V., *J. Appl. Phys.* **41** 2769-2776 (1970)
20. Boesten L. and Tanaka H., *At. Data Nucl. Data Tables* **52** 25-29 (1992)
21. Srivastava S. K., Chutjian A., and Trajmar S., *J. Chem. Phys.* **63** 2659-2665 (1975)
22. Nickel J. C., Zenter P. Z., Shen G., and Trajmar S., *J. Phys. E* **22** 730-738 (1989)
23. Mapstone B. and Newell W. R., *J. Phys. B* **27** 5761-5772 (1994)
24. Lunt S. L., Randell J., Ziesel J. P., Mrotzek G., and Field D., *J. Phys. B* **27** 1407-1422 (1994)
25. Gianturco F. A., Rodrigues-Ruiz J. A., and Sanna N., *J. Phys. B* **28** 1287-1300 (1995)
26. Tanaka H., Boesten L., Sato H., Kimura M., Dillon M. A., and Spence D., *J. Phys. B* **23** 577-588 (1990)
27. Dillon M. A., Boesten L., Tanaka H., Kimura M., and Sato H., *J. Phys. B* **26** 3147-3158 (1993)

28. Boesten L. and Tanaka H., *J. Phys. B* **24** 821-832 (1991)
29. Boesten L., Dillon M. A., H. Tanaka, M. Kimura, andSato H., *J. Phys. B* **27** 1845-1855 (1994)
30. Tanaka H., Boesten L., Matsunaga D., and Kudo T., *J. Phys. B* **21** 1255-1263 (1988)
31. Weiguo Sun, McCurdy C. W., Lengsfield III B. H., *J. Chem. Phys.* **97** 5480-5488 (1992)
32. Takagi T., Boesten L., Tanaka H., and Dillon M. A., *J. Phys. B* **27** 5389-5404 (1994)
33. Alle D. T., Gulley R. J., Buckman S. J., and Brunger M. J., *J. Phys. B* **25** 1533-1542 (1992)
34. Shinhara T, Tachibana Y. Yuri M., Tanaka H, and Boesten L., ICPEAC 1995
35. Mapstone B. and Newell W. R., *J. Phys. B* **25** 491-506 (1992)
36. Nakano Y, Yuri M. Boesten L., and Tanaka H, ICPEAC 1995
37. Boesten L, Takagi T., Tanaka H., Kimura M., Sato H. and Dillon M. A., ICPEAC XVIII (Aarhus, Denmark 1993), Abstracts of contributed papers, p. 264
38. Khakoo M. A., Jayaweera T., Wang S., and Trajmar S., *J. Phys. B* **26** 4845-4860 (1993)
39. Allan M., ref. 2, p. 105-118
40. Wong S. F. and Schulz G. J., *Phys. Rev. Lett.* **24** 1429-1432 (1975)
41. Andrick D. and Read F. H., *J. Phys. B* **4** 389-396 (1971)
42. Boesten L., Tanaka H., Kubo M., Sato H., Kimura M., and Dillon M. A., *J. Phys. B* **23** 1905-1913 (1990)
43. Becker K. H., ref. 2, p. 127-140
44. Le Coat Y., Ziesel J. P., and Guillotin J. P., *J. Phys. B* **27** 965-979 (1994)
45. Iga I., Rao M. V. V. S., Srivastava S. K., and Nagueira J. C., *Z. Phys. D* **24** 111-115 (1992)
46. Haarland P., *J. Chem. Phys.* **93** 4066-4072 (1990)
47. Perrin J., Schmitt J. P. M., de Rosny G., Drevillon B., Huc J., and Lloret A., *Chem. Phys.* **73** 383-394 (1982)
48. Winters H. F., *J. Chem. Phys.* **63** 3462-3466 (1975)
49. Winters H. F. and Inokuti H., *Phys. Rev. A* **25** 1420-1429 (1982)
50. Nakano T., Toyoda H., and Sugai H., *Jap. J. Appl. Phys.* **30** 2912-2915 (1991)
51. Nakano T. and Susgai H., *Jap. J. Appl. Phys.* **31** 2919-2924 (1992)
52. Winstead C., Sun Q., McKoy V., Lino J. L. S., andLima M. P. A., *J. Chem. Phys.* **98** 2132-2137 (1993)
53. Winstead C., Pritchard H. P., McKoy V., *J. Chem. Phys.* **101** 338-342 (1994)
54. Gil T. J., Lengsfield B. H., McCurdy C. W., and Rescigno T. N., *Phys. Rev. A* **49** 2551-2560 (1994)

# ELECTRON-ION COLLISIONS

Experimental Studies of Electron Impact Dissociation
of Molecular Ions ................................................... 297
    N. DJURIĆ, Y.S. Chung, B. Wallbank, and G.H. Dunn

New Mechanisms for Dissociative Recombination ................... 307
    S.L. GUBERMAN

Electron-Ion Collisions in Storage Rings ........................... 317
    A. MÜLLER

Dissociative Recombination of Molecular Ions
in a Cooler Ring ................................................... 329
    T. TANABE, I. Katayama, H. Kamegaya,
    K. Chida, T. Watanabe, Y. Arakaki, M. Yoshizawa,
    M. Saito, Y. Haruyama, K. Hosono, K. Hatanaka,
    T. Honma, K. Noda, S. Ohtani, and H. Takagi

# Experimental Studies of Electron Impact Dissociation of Molecular Ions

Nada Djurić[*], Yang-Soo Chung, Barry Wallbank[**]
and Gordon H. Dunn

*JILA, National Institute of Standards and Technology
and the University of Colorado, Boulder, Colorado 80309-0440*

**Abstract.** Recent progress on experimental studies of electron impact dissociation of molecular ions is presented. A technique especially configured to detect and measure "light particles" is described. Preliminary results of absolute cross sections for dissociation of $CD_2^+$ and $Cl_2^+$ are presented.

## INTRODUCTION

Attention is focused here on electron impact dissociation of molecular ions leading to at least one ionic fragment as well as neutral fragments. Thus, the processes of interest may be represented,

$$e + AB^+ \quad \begin{array}{l} \rightarrow A^+ + B^+ + 2e \\ \rightarrow A^+ + B + e \\ \rightarrow A^+ + B^- \end{array} \quad , \qquad (1)$$

and are referred to respectively as dissociative ionization (DI), dissociative excitation (DE), and ion pair formation. Despite the fact that early studies [1,2,3,4] gave interesting results, there has been little progress in the field during the past 30 years. A dramatic dependence of the cross section for dissociative excitation (DE) on the vibrational level of $H_2^+$ was illustrated in the early studies, and at low energies the measured DE cross section is dominated by transitions from the populated higher vibrational levels, with possible contributions from autoionizing states.

---

[*] Permanent address: Institut of Physics, 11000 Beograd, Yugoslavia

[**] Permanent address: Physics Department, St Francis Xavier University, Antigonish, N. S. Canada

There is renewed interest in experimental DE studies. The importance of reactions of molecular ions in plasma generators for etching and for deposition [5], and the prominence of molecular ions in the edge plasmas of fusion devices [6] has stimulated interest in such investigations.

Experiments on DE are made inherently difficult by the fact that dissociation products are born with broad kinetic energy and angular distributions. This leads to problems of collecting and quantitatively measuring a dispersed and separated product signal. Simple kinematic considerations (ignoring the minute momentum associated with the electron) show that for a molecular ion of mass $M = (m_1 + m_2)$, laboratory energy $E = MV^2/2$, and released kinetic energy $\epsilon_d$, a product particle of mass $m_1$ will have an energy $\epsilon_1$ in the center of mass system and maximum and minimum laboratory energies of $\mathscr{E}_1^+$ and $\mathscr{E}_1^-$ given by

$$\epsilon_1 = \frac{m_2}{M}\epsilon_d = \frac{1}{2}m_1 v_1^2$$

$$\mathscr{E}_1^+ = \frac{1}{2}m_1[V + v_1]^2 = \left[\sqrt{\frac{m_1 E}{M}} + \sqrt{\frac{m_2 \epsilon_d}{M}}\right]^2 \quad (2)$$

$$\mathscr{E}_1^- = \frac{1}{2}m_1[V - v_1]^2 = \left[\sqrt{\frac{m_1 E}{M}} - \sqrt{\frac{m_2 \epsilon_d}{M}}\right]^2.$$

Similarly, the product particle will diverge at a maximum angle in the laboratory when dissociating in the center of mass system at an angle of 90° with respect to **V**. This leads to a maximum laboratory angle of

$$\tan \theta_{max} = \left(\frac{m_2 \, \epsilon_d}{m_1 \, E}\right)^{\frac{1}{2}}. \quad (3)$$

A numerical example serves to emphasize the extreme conditions encountered. Clearly, the worst situations arise when one seeks to detect a light fragment, e.g., $H^+$ from the dissociation of $CH_4^+$. Suppose for the example's sake that the target ion has 7 keV energy and that the released kinetic energy is $\epsilon_d = 10$ eV. Then $\epsilon_1 = 9.4$ eV, $\mathscr{E}_1^+ \approx 581$ eV, $\mathscr{E}_1^- \approx 316$ eV, and $\tan \theta_m \approx 0.14$. The particle diverges a whopping 1.4 cm for every 10 cm traveled, and the energy distribution is 60% of the mean energy of the product particles!

Another issue alluded to in connection with earlier studies is the presence of vibrationally exited states in the primary beam. Molecular ions produced in an ion source may be vibrationally hot and are not readily cooled by conventional techniques. Serious efforts have been made by others [7] to obtain target ions in low vibrational states or in the ground state, but the techniques were not broadly applied.

# OUTLOOK AND PERSPECTIVE

Dramatic progress in DE studies with vibrationally relaxed molecular ions can be expected as broader use is made of heavy-ion storage rings for such studies. With these devices, one can obviate both the kinematic and internal states problems. Some work has already been performed [8,9] and more can be expected given the already-mentioned renewed interest. Though the storage ring methodology should get past most of the problems now encountered with measurement of DE, there are some cautions that should be noted. First, since strong dipole fields are used and since velocities of the ions are high, it is possible to field ionize high-Rydberg state atoms produced in dissociative recombination. The recombination process then gets detected as DE and may appear at energies that are energetically impossible for DE. Second, the space charge potential of the electron beam in the "electron cooler" can be high and can trap a number of "stationary" ions produced on background gas. These ions are then an additional target for producing DE of the ion beam. If the cooler pulsing is slow, such ion target densities can become large enough to affect the experimental results. These effects can be detected by multiscaling techniques and can in principle be obviated by pulsing the electrons fast enough.

In this paper we discuss a technique that we judge will be complementary to storage ring methods, since the method focuses on detecting and measuring light dissociation-product ions from a heavy target. It is speculated that the detectors at rings will not soon be configured for these light species. Furthermore, although experiments on vibrationally cold ions are the most appealing in terms of simplicity and hope for comparing with theory, for the applications mentioned above and also for a comparison with results with cold ions, it is desirable to have measurements on vibrationally hot ions.

It is instructive to review some of the results obtained in the past, and more recently with a storage ring, in order to emphasize the continuing need for alternative methods of study. The simplest of all molecules, $H_2^+$, serves as the paradigm for this.

Figure 1 presents results for $H_2^+$ ions showing cross section versus collision energy as measured [1,2,3] for vibrationally hot ions populated in formation approximately according to the Franck-Condon principle [10]. Also shown are results [8] for $HD^+$ ions when the ions are vibrationally cold after storage and relaxation in the TSR storage ring in Heidelberg. The experiments on hot ions are seen to agree with one another very well and with the theory [4] that includes the effects of the ion vibrations. The experimental results for the cold ions are dramatically different, highlighting the importance of the effects of

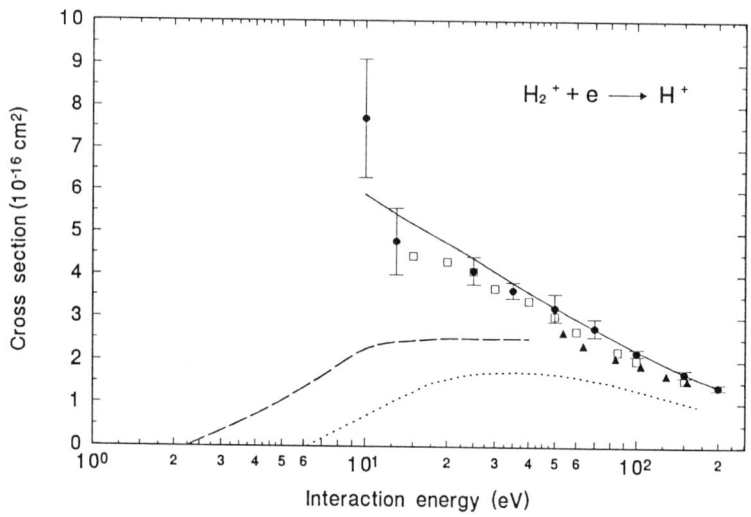

**FIGURE 1.** Cross section for the DE of $H_2^+$: ●, Ref.[1]; □, Ref.[2]; ▲, Ref.[3]; ---, normalized relative measurements Ref.[8]. Theoretical estimates, Ref.[4]: ... for v=0; — for Franck-Condon distribution (see text, Ref. [10]).

vibrations in the dissociation process. The latter experiment also differs significantly from the theory [4]. It can be speculated that this is due primarily to the neglect in the Peek theory, a first Born calculation, of the effects of capture and subsequent autoionization. This effect could be particularly important at low energies. Also, it is to be noted that the Heidelberg results [8] were put on an absolute scale by normalization to dissociative recombination results. Later reporting of the results [9] apparently uses different normalization and shows different values in absolute magnitude, but as these are not published as yet, they are not shown here. Not shown in Fig. 1 are results to be published by Yousif and Mitchell [11] giving cross sections for "hot" ions down to interaction energies of 10 meV. At energies of overlap, these results are in good agreement with those of Refs. [1-3].

These results demonstrate the importance of making measurements both with and without vibrational excitation of the target ions.

## EXPERIMENTAL APPROACH

Methods previously employed to study DE [1,2,3,7] typically have not been configured to detect a product ion if it is much lighter than the parent and thus carries away most of the kinetic energy of dissociation. These "single-pass" experiments including both crossed and merged beams configurations have been limited primarily to the study of homonuclear ions [1,2,3,7,12], with a few exceptions that include measurements on $CO^+$ [13], $H_3O^+$ ($D_3O^+$) [14], $CO_2^+$ [15], and $CD_4^+$ [16] where "heavy" dissociation products were detected, e.g., C or O from $CO^+$, $OH^+$ from $H_3O^+$, or $C^+$ from $CD_4^+$. The experiments are, of course, subject to the same limitations and difficulties as other charged particle beams experiments: low target densities and hence low signal rates accompanied by high backgrounds and space charge and pressure modulation of backgrounds leading to false signals. A major source of absolute uncertainties for these measurements is the evaluation of the fraction of signal fragment ions collected at the detector because of the kinematic divergence and broadening. Despite the difficulties, data of high quality have been obtained, and for $H_2^+$, experiments at different laboratories are in good agreement.

As already discussed, because of their energy and angular spread, dissociation fragments may escape detection, reducing the apparent cross section. A large number of different fragments with a variety of masses is formed even from a simple hydrocarbon like $CH_4^+$. Since every dissociation process is characterized by unique energy and angular distributions, it is almost impossible to have a relatively simple apparatus that will provide 100% collection of each fragment ion.

We set out to configure an apparatus that would be suited specifically to detect and measure light product ions. This was done in recognition of the inherent difficulties that have limited previous studies to detection of "heavy" product ions, and we speculated [17] that it would be difficult and some time in the future before detectors for the light products would be installed on storage rings. Complementary measurements on heavy fragment ions will be made at Oak Ridge National Laboratory (ORNL).

The JILA crossed beams apparatus [18,19] has been modified as illustrated in Fig. 2. Briefly, an ion beam is extracted from a Colutron ion source [20], accelerated through up to 8 kV potential and mass analyzed by a sector magnet. The ions are transported through a beam line containing optics and differential pumping to the collision chamber ($10^{-7}$ Pa). Upon entering this chamber the ion beam is crossed at right angles by a magnetically confined (0.006 T) electron beam [21].

The electron beam is chopped at about 1000 Hz, and the detector is gated to record counts both with electrons on and electrons off, so that background counts can be separated. A scanning slit probe is located in the center of the collision volume and can be rotated to measure spatial profiles of either the electron or the ion beam.

After colliding with the electrons, the ions (parent and fragment) enter a cylindrically symmetric lens system designed to collect, accelerate, and transport the fragment ions of interest into the analyzer chamber. The parent ion beam in large measure retains its initial collimation, while other product ions fly to sundry places. The analyzer chamber consists of two 45° electrostatic analyzers. The

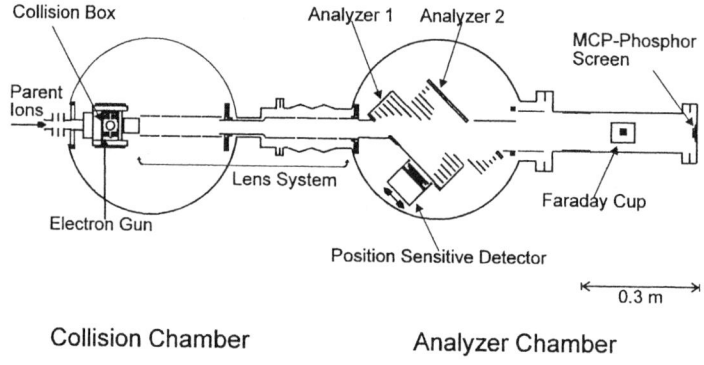

**FIGURE 2.** Crossed electron-ion beams interaction and fragment ions analysis aparatus

first analyzer separates the investigated fragment ions from other fragment ions and from the primary beam. The primary ions are slightly deflected, and the second 45° analyzer is used to redirect them out of the analyzer chamber toward the Faraday cup, located in the electrically isolated small chamber. This ancillary chamber may serve as a large ion collector when necessary, but also contains an in-line removable Faraday cup. The signal dissociation products proceed to the position sensitive detector (PSD) with sensitive area of diameter 40 mm mounted on a linear motion feedthrough with linear travel of 50 mm.

Background on the detector arises from the dissociation of the parent ions in collisions with residual gas and strongly depends on which fragmentation channel is favored in that dissociation and, of course, upon background pressure. The detector is broad and relatively "open", because of the wide energy and angular spread of the signal "beam". This makes it vulnerable to backgrounds from photons produced by ions hitting any surface in the analyzer chamber and traveling back directly or via reflections to the detector. This source of background is most insidious, since it is susceptible to modulation by the space charge potential of the electron beam, and hence, it is phased with the electrons, giving rise to a false signal. In order to minimize these effects the parent ion beam is buried deep in the auxiliary chamber as described above.

Extensive ion trajectory modeling using SIMION [22] was carried out to design and investigate the capability of the apparatus to collect fragment ions under investigation. The model calculations assumed break-up of simple hydrocarbon ions ($CH_4^+$, $CH_3^+$, $CH_2^+$, $CH^+$, and corresponding deuterated species), traveling at a fixed incident velocity (corresponding to 7-8 keV) into the light fragments ($H^+$, $H_2^+$, $H_3^+$ or deuterated species) and unspecified other fragments. The excess kinetic energy in our simulation was chosen to be not more than 10 eV per fragment ion. The relevant broad ranges of transverse and longitudinal energies were inserted in the modeling, and conditions determined whereby the desired particles would be fully collected. Both the transport and analyzer parameters vary as the desired signal ion changes.

Although the fragment ions start with approximately the same laboratory velocity as the parent ions, the arrival times of the various fragment species are quite different because of the accelerations and decelerations in the transport system. Thus, species separation using time of flight techniques (TOF) is possible with a short electron pulse as the time marker. For this mode of taking data, short electron pulses, 200 ns wide and 5 ns rise time, will be used. The duty factor in this mode is estimated to be only about 10%, and so far no effort has been made to put it into practice.

Measurements are currently in progress using the apparatus described with the ion targets $CD_4^+$, $CD_3^+$, $CD_2^+$, and $CD^+$. Preliminary results have been obtained for producing $D_2^+$ from $CD_2^+$ and are presented below. Measurements have also been made and are discussed below for dissociation of $Cl_2^+$ using only a single 45° analyzer and a particle multiplier after the collision region [19].

## RESULTS

In the edge plasma of fusion devices, the production of simple hydrocarbons follows the interaction and collision of hydrogen and its ions with the carbon limiters and the associated edge gases. There is an almost total lack of data for DE of light hydrocarbon ions, and one of the aims of our work is to obtain relevant data.

Figure 3 displays preliminary results obtained by using the technique described in this paper for the process

$$e + CD_2^+ \rightarrow D_2^+ + \text{products} \qquad (4)$$

where the term "products" may represent C, C$^*$, or C$^-$. The data were obtained with ion currents between 5 and 20 nA, and electron currents between 10 and 250 μA, yielding signal to background count rates that ranged between 0.002 at the lowest energy to 0.066 at the highest energy.

The preliminary character of the data is emphasized, noting that a previously measured detector efficiency has been used and a number of systematic checks have not been performed. Thus, for example, it is not clear whether the dip in the cross section between 30 and 40 eV is real.

The minimum energy [23] for obtaining ground-state carbon and $H_2^+$ is about 8 eV. For obtaining C$^-$ and $H_2^+$ the minimum energy is about 7.2 eV. One can only say that the data are consistent with either of these thresholds and cannot say anything meaningful about the level of vibrational state populations of the target $CD_2^+$.

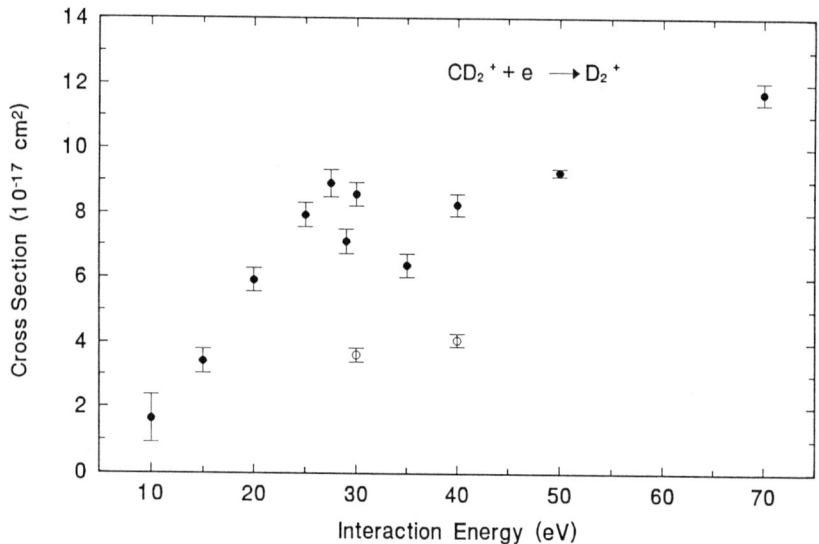

**FIGURE 3.** Cross section versus interaction energy for the process of Eq. 4

Chlorine molecules were chosen for investigation because of their importance in the modeling of plasmas used in etching processes. Clearly, the fragment ions share equally the excess kinetic energy, and one would expect less energy and angular spread of the fragments than for the cases discussed above.

The preliminary data are shown in Fig. 4 for cross section versus collision energy for the process,

$$e + Cl_2^+ \rightarrow Cl^+ + \text{products}$$
$$\rightarrow Cl_2^{2+} + 2e \qquad (5)$$

The cross section is finite even at 1 eV, well below the dissociation energy of the molecular ion ($D_o$=3.95 eV [24]). One may only speculate on explanations for the low-energy data. Presence of vibrationally excited molecular ions is clearly a reality when the ions are made in a discharge ion source. If vibrationally hot ions are present, electron capture followed by autoionization could contribute to the cross section at such low energies. Finally, one could consider ion pair formation ($Cl^+ + Cl^-$). There has been no previous measurement of the cross section for simultaneous production of $Cl^+$ and $Cl^-$. There are results, however, for the production of $H^+$ and $H^-$ from $H_2^+$ [25], and the measured cross section is an order of magnitude smaller than that for DE at the same energies.

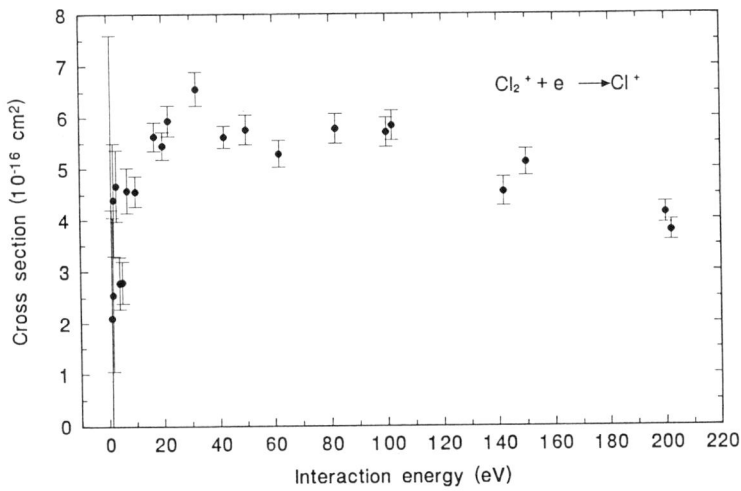

**FIGURE 4.** Cross section versus interaction energy for the process of Eq. 5

## CONCLUSIONS

In this brief report we have discussed progress on a technique to measure DE resulting in light fragment ions. The technique is complementary to that of the ion storage rings in that it provides for the study of vibrationally hot ions as encountered in the environments of a gas discharge. Preliminary measurements on $CD_2^+$ and $Cl_2^+$ have been presented, illustrating the technique.

The obvious opportunities for future DE studies are in using the storage ring technique, though some issues remain to be addressed. Although it has not been emphasized earlier, it should be noted here that there is a distinct dearth of theoretical results for DE of molecular ions.

## ACKNOWLEDGMENTS

This work was supported in part by the Office of Fusion Energy of the Department of Energy, Contract No. DE-A105-86ER53237 with the National Institute of Standards and Technology. B.W. acknowledges financial support from the JILA Visiting Fellows Program.

## REFERENCES

1. Dunn, G. H., and Van Zyl, B., *Phys. Rev.* **154**, 40–50 (1967).
2. Dance, D. F., Harrison, M. F. A., Rundel, R. D., and Smith. A. C. H., *Proc. Phys. Soc.* **92**, 577–88 (1967).

3. Peart, B., and Dolder, K. T., *J. Phys. B (London)* **6**, 1496–05 (1971); **5**, 1554–58 (1972).
4. Peek, J. M., *Phys. Rev.* **154**, 52–56 (1967); Peek, J. M., and Green, T. A., *Phys. Rev.* **183**, 202–212 (1969).
5. Flamm, D. L., "Introduction to plasma chemistry," in *Plasma Etching - An Introduction* (Eds: Manos, D., amd Flamm, D. L.): Boston, Academic Press, 1989, ch. 2, pp 91–183.
6. Dunn, G. H., *Nucl. Fus. Supl. At. Plasma Mater. Interaction Data Fusion* **2**, 25–39 (1972)
7. Hus, H., Yousif, F., Noren, C., Sen, A., and Mitchell, J. B. A., *Phys. Rev. Lett.* **60**, 1006–09 (1988)
8. Forck, P., Grieser, M., Habs, D., Lampert, A., Repnow, R., Schwalm, D., Wolf, A., and D. Zajfman, *Nucl. Instr. and Meth. B* **79**, 273–75 (1993).
9. Forck, P., PhD Thesis, Ruprecht-Karls-Universität, Heidelberg, 1994.
10. In fact, the vibrational distribution of the ions has been shown to be slightly different than given by the Franck-Condon principle, but the difference is slight enough that one can generically speak in these terms. See von Busch, F., and Dunn, G. H., *Phys. Rev. A* **5**, 1726–43 (1972).
11. Yousif, F. B., and Mitchell, J. B. A., *Zeitschrift für Physik D*, to be published (1995).
12. Van Zyl, B., and Dunn, G. H., *Phys. Rev.* **163**, 43–45 (1967).
13. Mitchell, J. B. A., and Hus, H., *J. Phys. B (London)* **18**, 547–55 (1985).
14. Schulz, P. A., Gregory, D. C., Meyer, F. W., and Phaneuf, R. A., *J. Chem. Phys.* **85**, 3386–94 (1986).
15. Müller, A., Salzborn, E., Frodl, R., Becker, R., and Klein, H., *J. Phys. B (London)* **13**, L221–23 (1980).
16. Gregory, D. C., and Tawara, H., "Dissociation and Ionization of $CD_4^+$ by electron imact," *Abstracts of Contributed Papers* (Eds: Dalgarno, A., Freund, R. S., Lubell, M. S., and Lucatorto, T. B.), XVI ICPEAC, New York, 1989, p. 352.
17. This speculation may have been in error. Private communication with L. Andersen, July 1995, indicates that such detectors are essentially ready to go at the ASTRID storage ring. Nevertheless, there remains the desirability of obtaining results for vibrationally hot ions.
18. Rogers, W. T., Stefani, G., Camillioni, R., Dunn, G. H., Msezane, A. Z., and Henry, R. J. W., *Phys. Rev. A* **25**, 737–48 (1982).
19. Djurić, N., Bell, E. W., Daniel, E., and Dunn, G. H., *Phys. Rev. A* **46**, 270–74 (1992).
20. Menzinger, M., and Wåhlin, L., *Rev. Sci. Instrum.* **40**, 102–05 (1969).
21. Taylor, P. O., Dolder, K. T., Kauppila, W. E., and Dunn, G. H., *Rev. Sci. Instrum.* **45**, 538–44 (1974).
22. Dahl, D. A., and Delmore, J. E., SIMION, Version 4.02, Technical Report, Idaho National Engineering Laboratory, EG&G Idaho Inc., Idaho Falls, ID 83415.
23. Minimum energy for obtaining C and $H_2^+$. Schuette, G. F., and Gentry, W. R., *J. Chem. Phys.* **78**, 1777–85 (1983).
24. Huber, K. P., and Herzberg, G., *Molecular Spectra and Molecular Structure IV. Constants of Diatomic Molecules*, New York: Van Nostrand, 1979, p. 150.
25. Peart, B., and Dolder, K. T., *J. Phys. B* **8**, 1570–73 (1975).

# New Mechanisms for Dissociative Recombination

Steven L. Guberman

*Institute for Scientific Research, 33 Bedford St., Suite 19A, Lexington, MA 02173, USA*

> Three important mechanisms for the dissociative recombination of an electron with a molecular ion are discussed and each is illustrated with the results of recent calculations. The mechanisms include direct capture by derivative coupling for ion and neutral curves which do not cross, capture by the second order electronic coupling to the intermediate vibrationally excited Rydberg states for a crossing of ion and neutral curves, and dissociative recombination through intermediate bound excited core Rydberg states.

## INTRODUCTION

Much of our understanding of dissociative recombination (DR) has been based upon the pioneering papers of Bates (1) and Bardsley (2,3), who introduced the direct and indirect DR mechanisms respectively. In recent years there have been considerable advances in both the theoretical and the laboratory methods that are used to study DR. These advances have uncovered new mechanisms which supplement the earlier mechanisms. In the next section the highlights of the Bates and Bardsley mechanisms are reviewed. This is followed by discussions of DR in the case where there is no crossing of ion and dissociative states, of the second order electronic coupling, and finally of recombination through a bound excited core Rydberg state.

## DIRECT AND INDIRECT DISSOCIATIVE RECOMBINATION

The mechanism for direct dissociative recombination was first described over 40 years ago by Sir David Bates. (1) In direct dissociative recombination,

$$AB^+ + e^- \rightarrow AB^* \rightarrow A + B, \tag{1}$$

the reactant ion $AB^+$ captures an electron, $e^-$, into an electronically excited state of the neutral molecule, $AB^*$, which dissociates into the products, $A$ and $B$. If $AB^+$ is a diatomic, $A$ and $B$ can be ground or excited state atoms. The mechanism for Eq. (1) is illustrated in Fig. (1) where an electron with energy $\epsilon$ is captured directly into a dissociative state. After capture, the molecule may emit an electron leaving the ion in the ground or a vibrationally

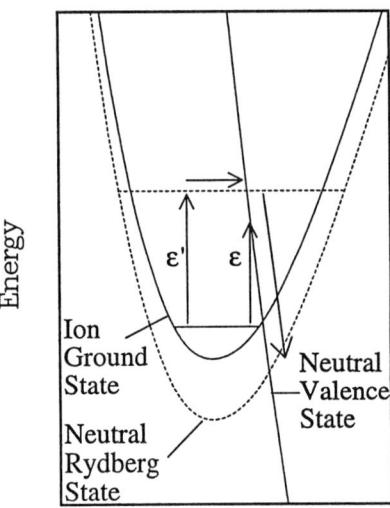

FIG. 1. The direct and vibronic indirect dissociative recombination mechanisms.

excited state. On the other hand, if the potential curve is steep and if the autoionization lifetime is longer than the dissociative lifetime, DR is likely. Once the fragments separate beyond the ion potential curve, autoionization is no longer possible. The dissociative states are diabatic states which cross through an infinite number of Rydberg states in order to cross the ion. The matrix element that couples the reactants to the products is over the total Hamiltonian, $H^{Total} = H + T$, where H is the electronic Hamiltonian and T is the nuclear kinetic energy. Because the states are diabatic and do not change character with R, the contribution from T (see Eq. (4)) is negligible compared to that from H. The matrix element over H is given by

$$V_{v,d}^{k,\ell} = \langle \Psi_d X_d \mid H \mid \Psi^{k,\ell} X_v \rangle \qquad (2)$$

where $\Psi_d$ and $X_d$ are the electronic and vibrational continuum wave functions, respectively, for the dissociative state. $\Psi^{k,\ell}$ is the electronic wave function for both the ion and the free electron with wave number $k$ and angular momentum, $\ell$. $X_v$ is the bound ion vibrational wave function. The direct cross section for DR is approximately proportional to the square of $V_{v,d}^{k,\ell}$. (2,4)

In the indirect recombination mechanism, first proposed by Bardsley (3), the electron is captured into a Rydberg state of the neutral molecule which is predissociated by the repulsive state of the direct mechanism. The indirect mechanism is illustrated in Fig. (1) for capture of an electron with energy $\epsilon'$. The coupling matrix element which drives the capture into the Rydberg vibrational level is again given by a matrix element of the total Hamiltonian, $H^{Total} = H + T$. The matrix element over H vanishes and we are left with

$$T^{n^*,k}_{v,v',\ell,\ell'} = \langle \Psi^{n^*,\ell'} X_v \mid T \mid \Psi^{k,\ell} X_{v'} \rangle \qquad (3)$$

where the wave functions are the same as in Eq. (2) except that now $\Psi^{n^*,\ell'}$ is the electronic wave function of the Rydberg level with effective principal quantum number, $n^*$. The set of vibrational wave functions for the Rydberg levels is taken to be the same as the set for the ion. The matrix element over the electronic Hamiltonian is zero since the Rydberg states and the free electron states, both with the same ion core, comprise the same spectrum of states that diagonalize the electronic Hamiltonian. This matrix element falls off as $(1/n^*)^{3/2}$ and favors transitions between ion and Rydberg vibrational levels that differ by a single quantum. Bardsley (3) showed that the indirect DR cross section took a Breit-Wigner form in which the resonances contribute peaks to the DR cross section. In this approach, indirect recombination is treated separately from and not allowed to interfere with direct recombination. In the Multichannel Quantum Defect Theory (MQDT) approach to DR introduced by Giusti, (4) this interference is included and it has been shown (4,5) that the cross section near the resonance energy may take the form of a peak (constructive interference), a dip (destructive interference), or a peak on one side of the resonance center and a dip on the other side. The shape is determined from the relative magnitudes of the direct recombination matrix element (Eq. (2)) and the matrix elements for the interaction of the Rydberg level with the electron ion continuum (Eq. (3)) and with the dissociative state. (5,6) The resonances take the same form as that predicted by Fano for resonances seen in atomic photoionization. (7)

## DISSOCIATIVE RECOMBINATION WITHOUT A CURVE CROSSING

If we shift the repulsive curve in Fig. (1) to smaller R so that it does not cross the ion at all, we would expect the DR cross section to be small because of the small vibrational wave function overlap that enters the matrix element in Eq. (2). Molecules that exhibit this lack of a crossing include the rare gas hydrides and $H_3^+$. Potential curves calculated (8,9) for HeH are shown in Fig. (2). These curves have been calculated with large Gaussian basis sets and multireference configuration interaction (CI) wave functions. In this system, the dissociative curve equivalent to the repulsive curve of Fig. (1) is the ground state which, however, does not cross the ion. The only remaining dissociative routes are the A and C Rydberg levels which have asymptotes below the v=0 level of the ion. (We neglect the $B^2\Pi$ state because its rotational couplings to the electron-ion state are small.) Because these Rydberg states have the same core as the ion ground state, they too do not cross the ion. Despite this theoretical evidence that might lead one to the conclusion that the HeH DR rate constant may be small, recent experimental results for both $^3HeH^+$ (10,11) and $^4HeH^+$ (12,13) indicate that the cross sections are not negligible.

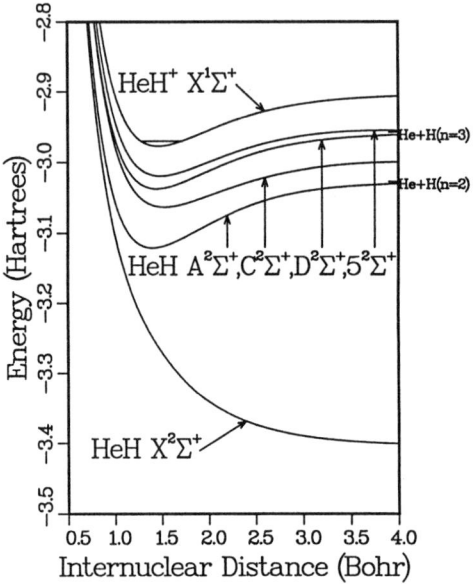

**FIG. 2.** The calculated potential curves for HeH.

Indeed, a nonnegligible cross section can occur in the absence of a crossing if a new mechanism is introduced in which the electron capture into the dissociative state is driven by the nuclear kinetic energy operator, (8) the same operator that drives the Bardsley indirect capture mechanism. Because this matrix element is nonzero between electron-ion and Rydberg states with the same core, it can drive direct DR along the Rydberg A and C states shown in Fig. (2). In order to calculate DR cross sections and rates, we have included these matrix elements in a revised version of the first order MQDT of Giusti. (4)

For the cross section calculations, we must determine matrix elements over the nuclear kinetic energy operator,

$$T = \frac{-\hbar^2}{2\mu} \frac{1}{R^2} \frac{\partial}{\partial R} R^2 \frac{\partial}{\partial R}, \qquad (4)$$

where $R$ is the internuclear distance, $\hbar$ is Planck's constant, and $\mu$ is the reduced mass. The matrix element over T reduces to

$$T^{k,\ell}_{v,d} = \rho^{\frac{1}{2}} \langle \Psi_d X_d \mid T \mid \Psi^\ell X_v \rangle = \rho^{\frac{1}{2}} \frac{-\hbar^2}{2\mu} \langle X_d \mid B^\ell_d(R) + 2A^\ell_d(R) \frac{\partial}{\partial R} \mid X_v \rangle, \quad (5)$$

where

$$A^\ell_d(R) = \langle \Psi_d \mid \frac{\partial}{\partial R} \mid \Psi^\ell \rangle, \qquad (6)$$

$$B^\ell_d(R) = \langle \Psi_d \mid \frac{\partial^2}{\partial R^2} \mid \Psi^\ell \rangle, \qquad (7)$$

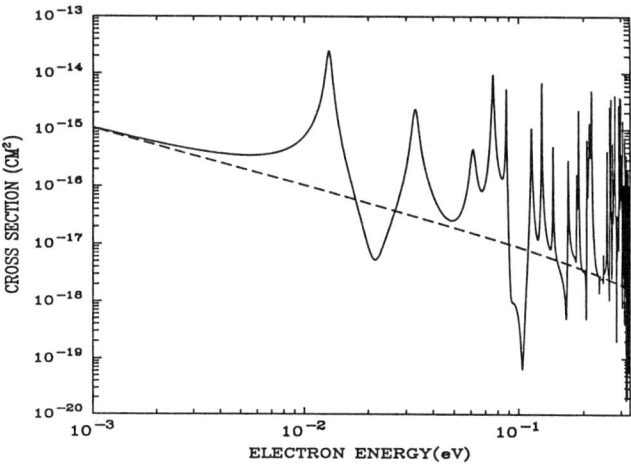

**FIG. 3.** Calculated cross sections for DR of $^4HeH^+$.

and $\rho$ is a density of states which converts the matrix elements to those appropriate for an electron-ion continuum. (8,9) The $k$ superscript has been dropped on the right side of Eqs. (5-7) because the matrix elements are calculated for zero energy electrons. Matrix elements of $A_d^\ell(R)$ between the neutral HeH states have already been reported in the literature (14) as have the $\langle \frac{\partial \Psi_d}{\partial R} | \frac{\partial \Psi^\ell}{\partial R} \rangle$ matrix elements. (14) The $B_d^\ell(R)$ matrix elements can be derived from these matrix elements by using

$$B_d^\ell(R) = \frac{\partial A_d^\ell}{\partial R} - \langle \frac{\partial \Psi_d}{\partial R} | \frac{\partial \Psi^\ell}{\partial R} \rangle. \tag{8}$$

Denoting the Rydberg orbitals by $(n, \ell)$, the A and C Rydberg states have an electron in the (2,0) and (2,1) Rydberg orbitals respectively. The published couplings (14) show that the strongest interaction is for the capture of an $\ell = 0$ electron into the C state, i.e. a $\Delta \ell = 1$ transition. Furthermore, because the C state is the closest dissociative state to the ion, it has the largest nuclear wave function overlap contribution to the matrix elements in Eq. (5).

We have calculated DR cross sections for $^4HeH^+$ and these are shown in Fig. (3). For $^4HeH^+$, ten bound vibrational levels were included in the first order MQDT calculations. The direct cross section is shown by the dashed line. The direct DR cross sections are small compared to those for many molecules that have favorable diabatic curve crossings. The direct rate at 300K derived from this direct cross section is $4.2x10^{-10} cm^3/sec$. Denoting the resonances by $(n, v, \ell)$, the lowest energy resonance structure near 0.01eV is due to (3,4,1), i.e. it is the v=4 level of the $5^2\Sigma^+$ state perturbed by the interference between direct and indirect recombination. The large vibrational spacing in this low reduced mass system allows Rydberg states with both low n and low v to lie above the v=0 level of the ion. While the matrix element in Eq. (3) might be expected to be small because it is a $\Delta v = 4$ transition, this is offset by the low value of n. Moving to higher energies, the (6,1,1) resonance appears with a peak near 0.03eV. The lifetime for predissociation of the resonances can be estimated from the matrix element, (5) $T_{v,d}^{k,\ell}$ in Eq. (5)

$$\tau_{n,v}^{pre} = \frac{\hbar}{\frac{2\pi\rho}{(n-\mu_0)^3} \mid T_{v,d}^{k,\ell} \mid^2}. \tag{9}$$

where $\mu_0$ is the value of the quantum defect at the R value corresponding to the minimum of the ion well.

The lifetimes for predissociation of the (3,4,1) level by the X, A, and C states are $1.0x10^{-9}$, $0.95x10^{-9}$, and $2.6x10^{-13}sec$ respectively. Clearly, the C state is the dominant dissociative route for v=4. In fact, the C state is also the dominant dissociative route for every Rydberg vibrational level. These results are confirmed by the full calculated DR rates at 300K which are $7.7x10^{-12}$, $3.0x10^{-11}$, and $6.6x10^{-9}cm^3/sec$ along the X, A, and C states respectively. Indeed, the plot of the total cross section in Fig. (3) is nearly identical with the cross section for the C state only. The time for a vibration in the $v = 4$ level is given approximately by (15)

$$\tau_v^{vib} = \frac{1}{c(\omega_e - \omega_e x_e - 2v\omega_e x_e)} \tag{10}$$

where c is the speed of light, $\omega$ is the fundamental frequency, and $\omega_e x_e$ is the anharmonicity for $^4HeH$. Taking the Rydberg state to have the same calculated spectroscopic constants as those for the ion, the lifetime of a vibration is found to be approximately $1.8x10^{-14}sec$. After capture into the (3,4,1) level, the molecule undergoes about 14 vibrations before falling apart. The autoionization lifetime, $\tau_{n,v}^{auto}$, can be calculated approximately from (16)

$$\tau_{n,v}^{auto} = \frac{\hbar}{\frac{2\pi}{(n-\mu_0)^3} \mid \langle v_0 \mid \mu(R) \mid v \rangle \mid^2}. \tag{11}$$

For $v = 4$, the autoionization lifetime is $3x10^{-11}sec$. Therefore, once excited, the (3,4,1) state will almost entirely dissociate.

For $^3HeH$ and $^4HeH$ the direct DR cross sections are quite similar with those for $^3HeH$ slightly larger than those for $^4HeH$. However, for the lighter species, the calculated full rate coefficient at room temperature is $2.6x10^{-8}cm^3/sec$. (8) For $^3HeH$, the energy difference between vibrational levels is greater than in $^4HeH$ and a (4,2,1) falls above threshold but is below threshold in $^4HeH$. The (4,2,1) level plays a major role in enhancing the $^3HeH$ DR cross section and rate coefficient over that for $^4HeH$.

## THE SECOND ORDER ELECTRONIC COUPLING

The indirect mechanism of Bardsley (3) is driven by the Born-Oppenheimer breakdown coupling of Eq. (3). However, it is also possible to have capture into a Rydberg state by an electronic coupling. The first order coupling, in which the electronic Hamiltonian appears once, vanishes between an electron-ion state and a Rydberg state with the same ion core. However, if the Hamiltonian acts twice, using the dissociative state as an intermediate, the electronic coupling can be nonzero. A mechanism involving this type of electronic coupling was first discussed by O'Malley (17) and was used in preliminary calculations on $H_2^+$ DR by Hickman (18). It was first included in a full MQDT calculation in 1991 (5). The coupling matrix element takes the form

$$\int\int X_v(R)\langle\Psi_v^{k,\ell}|H(R)|\Psi_d\rangle F_d(R_<)G_d(R_>)\langle\Psi_d|H(R')|\Psi_v^{n,\ell}\rangle X_{v'}(R')dRdR'$$

(12)

where $F_d$ and $G_d$ are the regular and irregular vibrational wave functions in the dissociative state. An important difference between this mechanism for indirect recombination and the Born-Oppenheimer breakdown mechanism is that we are no longer restricted to the $\Delta v = 1$ propensity rule. A large $\Delta v$ is possible as long as the initial and final bound vibrational wave functions have a large overlap with the continuum vibrational wave functions of the dissociative state.

The second order mechanism arises from the perturbation expansion of the K matrix in the MQDT approach. (4,5) We have found that the second order mechanism is often more important in some systems than the Born-Oppenheimer breakdown mechanism. In an analysis (5) of the $v = 1$ contributions to the DR cross section for $O_2^+$ leading to $O(^1S)$ products, we found that the shape of the resonances was due to the interaction between the vibronic and the electronic indirect mechanisms and that the electronic indirect mechanism was dominant. For DR of $O_2^+$ along the $^1\Delta_u$ route, (19) the cross section near threshold is depressed below the direct cross section by five orders of magnitude due to the electronic indirect mechanism alone and by one order of magnitude due to the vibronic indirect mechanism. Interference between the two processes leads to a full cross section which is three orders of magnitude below the direct cross section. The effect is due to an n=3, v=13 resonance lying 0.02eV below threshold. This level has a wing which interferes destructively with direct recombination above threshold.

The second order coupling is now included in all our calculations involving diabatic electronic couplings. It is clear that in most molecules it must be included in order to calculate accurate cross sections. In molecules having large capture widths, it may be necessary to go beyond second order. Calculations including the full higher order K matrix have been reported by Takagi. (20)

In the case of DR proceeding entirely by derivative coupling, as in the

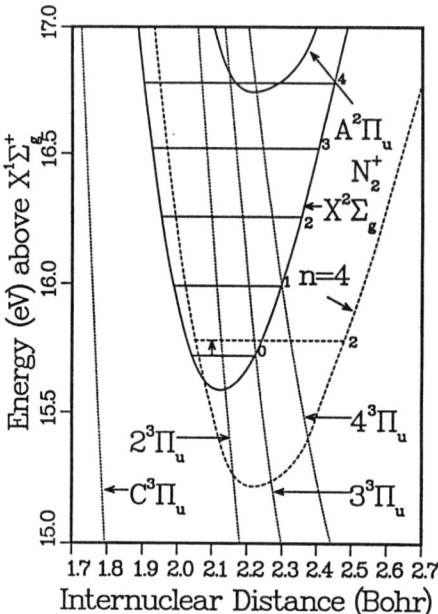

**FIG. 4.** Indirect recombination through the n=4 $^3\Pi_u$ Rydberg state of $N_2$ with the excited $A^2\Pi_u$ core. The small vertical arrow indicates capture from the v=0 ion ground state into the (4,2,0) $^3\Pi_u$ state with the $A^2\Pi_u$ core.

previous section, an equivalent second order derivative coupling also can be constructed. In such a coupling for HeH, the initial electron-ion state would be coupled to a Rydberg state by the C state acting as an intermediate. These couplings are all vibronic as opposed to the above electronic couplings. This coupling has not been included in the HeH MQDT calculations to date and will be the subject of future calculations.

## EXCITED ION CORE BOUND RYDBERG STATES

Indirect DR through a bound excited core Rydberg state proceeds as shown in Fig. (4) for $N_2^+$. Because the intermediate Rydberg state does not have the ground state of the ion as the core, it differs from the free electron-ion system in the reactant channel of Eq. (1) by a double excitation, i.e. by moving two electrons to previously unoccupied orbitals. Therefore, the electronic Hamiltonian can drive the capture. The capture matrix element is

$$V_{v,v',\ell,\ell'}^{n^*,k} = \langle \Psi^{n^*,\ell} X_v \mid H \mid \Psi^{k,\ell'} X_{v'} \rangle. \tag{13}$$

This process has been described qualitatively in the literature for DR of $N_2^+$ (21) and $CD^+$, (22) but there have been no reported prior calculations. (Note that in the diabatic treatments of $H_2^+$ DR, all the dissociative routes (23) are excited core Rydberg states but they are all repulsive.) An important qualitative feature of this recombination mechanism is that it is possible to have capture into the $v = 0$ level of excited core states. Such a capture can have large Franck-Condon factors and large capture matrix elements. This type of capture is impossible in the vibronic indirect mechanism since all the $v = 0$ Rydberg levels with the same core as the ion are at lower energies than the ion v=0 level.

We have calculated (25) the effect of intermediate excited core Rydberg levels on the DR of the v=0 level of $N_2^+$. The relevant potential curves are shown in Fig. (4). The dominant dissociative channels (shown as dotted lines) are the four lowest valence $^3\Pi_u$ states with $2\,^3\Pi_u$ the dominant route for v=0. (24) The calculation includes 18 vibrational levels for the ion ground state and Rydberg states and the four dissociative routes. We have included Rydberg states having the $A^2\Pi_u$ excited core. The inclusion of this state substantially increases the amount of required molecular data and the amount of computer time needed to complete the calculation of cross sections and rates. Since we must treat the A state as well as we treat the X ion core, 18 vibrational levels are also included for the A state and its Rydberg states. In total, the electronic K matrix needed in the MQDT approach includes 36 bound vibrational levels and the four dissociative $^3\Pi_u$ states. In addition, in order to build the K matrix, we need electronic widths for capture of the free electron by the A state into each of the four dissociative states. These widths also give us the predissociation widths of the intermediate A core Rydberg states. We have treated capture of an $\ell = 0, \sigma$ free electron by the $A^2\Pi_u$ ion. Finally, we need the width matrix element of Eq. (13), where the right side describes a "free" electron in the field of the X core and the left side describes a Rydberg state with the A core. The A core lies $0.11a_o$ to larger R than the X core and the minima are separated by only $1.14eV$. The v=0 level of the calculated $n = 3\,^3\Pi_u$ Rydberg state with the A core lies $2.3eV$ below the $v = 0$ level of the $X^2\Sigma_g^+$ state. The first calculated A core Rydberg state above threshold is the $n = 3, v = 11$ state near $0.03eV$ followed by the $n = 4, v = 2$ state near $0.06eV$. The latter level is shown in Fig. (4). The effect of the excited core Rydberg levels is to increase the DR rate by about 10% over the rate calculated with only the four dissociative routes and the ground state ion core Rydberg levels. Note that capture into the $n = 4, v = 2$ level shown in Fig. (4) allows for dissociation along the $4\,^3\Pi_u$ channel which has only small Franck-Condon factors for direct capture from v=0. In this manner, indirect DR can strongly affect the quantum yield of the atomic products. Further details on these calculations will be published separately. (25)

## ACKNOWLEDGEMENTS

This work is supported by NASA grants NAGW 2832 and 1404 and by NSF grant ATM-9503224. The computations were done at the National Center for Atmospheric Research and at the Pittsburgh Supercomputer Center which are both supported by NSF.

## REFERENCES

1. D. R. Bates, Phys. Rev. A **78**, 492 (1950).
2. J. N. Bardsley, J. Phys. B **1**, 349(1968).
3. J. N. Bardsley, J. Phys. B **1**, 365 (1968).
4. A. Giusti, J. Phys. B **13**, 3867 (1980).
5. S. L. Guberman and A. Giusti-Suzor, J. Chem. Phys. **95**, 2602 (1991).
6. S. L. Guberman, "Recent Theoretical Developments in Dissociative Recombination," in *Atomic Collisions: A Symposium in Honor of Christopher Bottcher (1945-1993)* (American Institute of Physics Press, New York, 1995).
7. U. Fano, Phys. Rev. **17**, 93 (1978).
8. S. L. Guberman, Phys. Rev. A **49**, R4277 (1994).
9. S. L. Guberman, "The dissociative recombination of $^4HeH^+$", in preparation.
10. G. Sundstrom, S. Datz, J. R. Mowat, S. Mannervik, L. Brostrom, M. Carlson, H. Danared and M. Larsson, Phys. Rev. A **50**, R2806 (1994).
11. J. R. Mowat, H. Danared, G. Sundstrom, M. Carlson, L. H. Andersen, L. Vejby-Christensen, M. af Ugglas, and M. Larsson, Phys. Rev. Lett. **74**, 50 (1995).
12. F. B. Yousif and J. B. A. Mitchell, Phys. Rev. A **40**, 4318 (1989).
13. F. B. Yousif, J. B . A. Mitchell, M. Rogelstad, A. Le Paddelec, A. Canosa and M. I. Chibisov, Phys. Rev. A **49**, 4610 (1994).
14. I. D. Petsalakis, G. Theodorakopoulos, C. A. Nicolaides and R. J. Buenker, J. Phys. B **20**, 5959 (1987).
15. G. Herzberg, *Molecular Spectra and Molecular Structure, I. Spectra of Diatomic Molecules* (Van Nostrand Reinhold Co., New York, 1950), p.98.
16. H. Nakamura, Int. Rev. Phys. Chem. **10**, 123 (1991).
17. T. F. O'Malley, J. Phys. B. **14**, 1229 (1981).
18. A. P. Hickman, J. Phys. B **20**, 2091 (1987).
19. S. L. Guberman, "The dissociative recombination of $O_2^+$ along the $^1\Delta_u$ route", in preparation.
20. H. Takagi, "Theoretical Problems in the Dissociative Recombination of $H_2^+ + e^-$", in *Dissociative Recombination: Theory, Experiment and Applications*, ed. by B. R. Rowe, J. B. A. Mitchell, and A. Canosa (Plenum Press, New York, 1993), p.75.
21. S. L. Guberman, "Ab Initio Studies of Dissociative Recombination", in *Dissociative Recombination: Theory, Experiment and Applications*, ed. by J. B. A. Mitchell and S. L. Guberman (World Scientific, Singapore, 1989), p.45.
22. P. Forck, C. Broude, M. Grieser, D. Habs, J. Kenntner, J. Liebmann, R. Repnow, D. Schwalm, A. Wolf, Z. Amitay, and D. Zajfman, Phys. Rev. Lett. **72**, 2002 (1994).
23. S. L. Guberman, J. Chem. Phys. **78**, 1404 (1983).
24. S. L. Guberman, Geophys. Res. Lett. **18**, 1051 (1991).
25. S. L. Guberman, "The dissociative recombination of the v=0 level of $N_2^+$", in preparation.

# Electron-Ion Collisions in Storage Rings

### Alfred Müller

*Institut für Strahlenphysik, Universität Stuttgart,*
*D-70550 Stuttgart, Federal Republic of Germany*

---

Recent progress in electron-impact ionization and recombination of atomic ions in merged-beams experiments is reviewed. The use of heavy-ion storage rings and the method of adiabatic magnetic expansion of the electron-cooler beam have provided the technical basis for measurements of cross sections and rates with unprecedented quality regarding both statistics and energy resolution. Electron-ion resonances showing widths as small as 0.025 eV have been observed. Thus the data provide extremely detailed information about collision processes and atomic structure of ions ranging from $He^+$ to $U^{89+}$.

---

## INTRODUCTION

Heavy-ion storage rings with integrated electron cooling devices provide new excellent possibilities to study electron-ion interactions. Ion beams of high brightness are merged with cold dense electron beams providing low collision-energy spreads and long interaction path lengths for measurements of collision cross sections and rates of a quality that has not been accessible before, especially not with highly charged ions. By the technical development enormous progress was facilitated in a field that has traditionally been suffering from low counting rates, problems with metastable contaminations of parent ion beams, low signal-to-background ratios, difficult signal recovery, troublesome determination of beam overlaps and the apparent incompatibility of high electron density and low electron energy spread.

Among all possible electron-ion interactions in storage rings [1] the present paper particularly addresses recombination and ionization (including detachment) of atomic ions in collisions with free electrons. The electron cooling of ions, the X-ray spectroscopy of very highly charged ions recombining with cooler electrons, the population of specific states by electron-ion collisions, the laser-stimulated electron-ion recombination, and the interaction of electrons with molecular ions are not subjects of this progress report.

With the pioneering experiments at Oak Ridge [2] and at London, Ontario, [3] combining ion accelerator technology with electron-beam target facilities, a new era began in electron-ion collision studies little over a decade ago. Since then, new single-pass merged-beams experiments have been performed at the

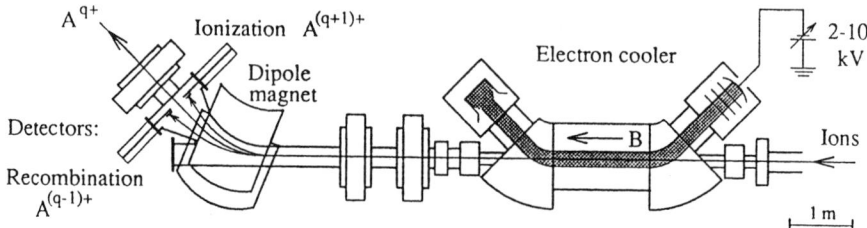

**FIG. 1.** Scheme of cross section measurements at the TSR. The circulating $A^{q+}$ ion beam enters from the right and is merged with the electron cooler beam which is guided by a magnetic field B. The dipole magnet separates the ion charge states. Ionized and recombined product ions are detected in separate detectors. Neutrals are counted behind the straight-through port of the magnet. The experimental arrangements are similar in all of the rings under discussion.

university of Aarhus [4], at GSI in Darmstadt [5] and at the Indiana cooler ring [6]. In addition, the storage rings TSR in Heidelberg, ESR in Darmstadt, CRYRING in Stockholm, TARN in Tokyo, and ASTRID in Aarhus are now providing excellent opportunities to study interactions of electrons with atomic or molecular ions [1]. A wide range of collision energies can be accessed with atomic ions from $H^-$ to $U^{92+}$. The low temperatures in magnetically expanded electron beams [7] provide unprecedented energy resolution and allow to obtain useful information on collisions at center-of-mass energies even below 1 meV.

## EXPERIMENTS

The processes of interest in the present context are electron-ion recombination:

$$e + A^{q+} \rightarrow A^{(q-1)+} + photons \qquad (1)$$

and electron-impact ionization of ions:

$$e + A^{q+} \rightarrow A^{(q+1)+} + 2e \qquad (2)$$

In the experiments the ions which have changed their charge state are detected behind the first dipole magnet downbeam from the interaction region (see Fig.1). The normalized counting rate of these ions is measured as a function of the center-of-mass energy. Absolute rates are determined from

$$\alpha(E_{Rel}) = \frac{R_{exp} \gamma^2 v_i q e}{I_i \ell_{eff} n_e \varepsilon} \qquad (3)$$

The number of charge-changing collisions per unit time $R$ is determined from the counting rates $R_{exp} = R\varepsilon$ of appropriate particle detectors. Detection efficiencies $\varepsilon$ are usually close to 1. Eq.(3) holds because the density $n_e$ of the electrons is uniform in space with negligible variations across the magnetically guided electron beam throughout the whole interaction region. Further, $v_i$ is the ion velocity, $q$ the ion charge state, $e$ the charge of an electron, and $\ell_{eff}$ the length of the interaction path. In single-pass experiments the total electrical ion current $I_i$ is measured by collecting the parent ions in a Faraday cup. In storage ring experiments the current of the circulating ion beam is determined from the magnetic-field signal of a ferrite ring detector surrounding the ion beam. An additional factor $\gamma^2 = [1 - (v_i/c)^2]^{-1}$ accounts for the relativistic transformation between the laboratory and the center-of-mass frames. The rate $\alpha$ is a convolution of the recombination cross section $\sigma$ with the experimental distribution $f(v, v_{Rel})$ of the electron-ion center-of-mass velocity $v$ centered at $v_{Rel}$. Apparent cross sections can be determined from $\sigma = \alpha/v_{Rel}$.

Since photons or electrons are not detected the observation of net recombination and net ionization comprises a number of mechanisms all of which produce ions $A^{(q-1)+}$ or $A^{(q+1)+}$, respectively. Beyond their characteristic energy dependences these mechanisms are not experimentally distinguished. Net ionization for example can result from a direct knock-off process, but also from multi-step processes such as inner-shell excitation and subsequent autoionization (EA), or from resonant excitation (resonant dielectronic capture of the incident electron) and subsequent emission of two electrons in a double autoionization process (REDA). Similarly, recombination can be the result of different processes. Examples are the direct radiative recombination (RR), where a photon carries away the excess energy liberated by the binding of an initially free electron, and the two-step process of dielectronic recombination (DR), where -in a first step- the incident electron is captured into a bound state while a core electron is excited (resonant dielectronic capture of the incident electron as in REDA) and where -in a second step- the intermediate multiply excited state stabilizes the new charge state by photon emission. There are other possibilities for an electron and an ion to recombine (or to be caused to recombine) which are not addressed here.

## RESULTS

The following atomic ions have been studied at storage rings
(a) with respect to energy-dependences of recombination:
$He^+$ [8], $C^{4+}$ [9], $C^{5+}$ [10], $C^{6+}$ [11], $O^{7+}$ [12], $Ne^{7+}$ [13], $Ne^{10+}$ [13], $Si^{11+}$ [14], $S^{15+}$ [15], $Cl^{6+}$ [16], $Cl^{14+}$ [14], $Ar^{13+}$ [13], $Ar^{15+}$ [13], $Fe^{15+}$ [16], $Cu^{26+}$ [17], $Se^{23+}$ [16], $Se^{25+}$ [18], $Au^{76+}$ [19,20], $Bi^{80+}$ [21], $U^{89+}$ [20,21];
(b) with respect to energy-dependences of ionization (and detachment from negative ions):
$D^-$ [22], $Si^{11+}$ [14], $Cl^{6+}$ [16], $Cl^{14+}$ [14], $Fe^{15+}$ [23], $Se^{23+}$ [16].

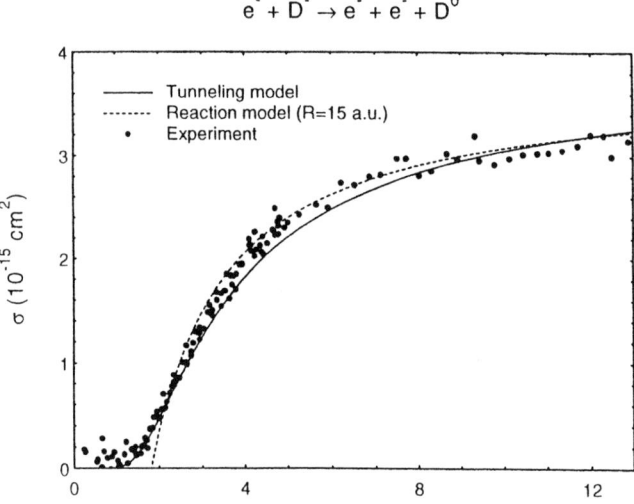

**FIG. 2.** Experimental and theoretical electron-detachment cross sections of $D^-$ ions [22].

The ion species listed above have not all been studied with equally precise techniques nor are the energy ranges of the different investigations comparable. Additional ions have been investigated in laser-induced recombination measurements, however, extended energy dependences are only available for the ions given above. In the following, several representative new results of storage-ring experiments with atomic ions are discussed.

The lightest ions studied with respect to recombination and electron removal are $He^+$ and $D^-$. For the low atomic numbers correlation effects between the electrons active in the collisions are of particular importance. Interest in the detachment of $D^-$ by electron impact arises for example from the peculiar situation in the detachment threshold region where two slow outgoing electrons interact in the vicinity of a neutral residual ion. This scenario differs from the one described by the Wannier threshold law [24] where the heavy residual particle is a positive ion instead of an atom. The differences in the correlated movement of the slow outgoing electrons in the two cases result in different threshold behaviours. With a measurement of the detachment cross section at low energies, new insight into the Coulomb three-body problem can be expected when very precise data for energies barely above threshold become available. At ASTRID the detachment of $D^-$ ions by electron impact was studied [22] in a center-of-mass energy range from 0 to 20 eV. A part of the experimental results is displayed in Fig.2. Although the measured data are in very reasonable overall agreement with model calculations, the preci-

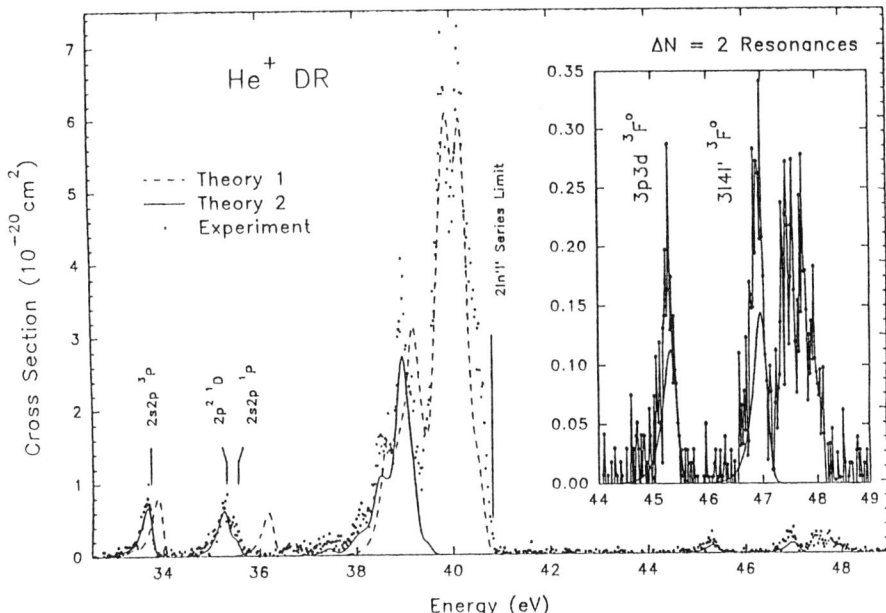

**FIG. 3.** The $\Delta n = 1$ and $\Delta n = 2$ dielectronic recombination resonances of He [8]. The dashed line (theory 1) is by Badnell and the full line (theory 2) is by Lindroth. Only theory 2 includes electron correlation "to all orders".

sion at very low energies is not sufficient yet for a rigorous test of theory. An important result of the Aarhus experiment was to falsify a previous observation of the formation of $H^{--}$ resonances [25] in collisions of $H^-$ ions with electrons which had stirred excitement in the literature for nearly 25 years.

The second system ideal for studies of electron correlations is the $e+He^+$ complex. Measurements for dielectronic recombination were carried out at CRYRING [8] subsequent to two previous measurements with $He^+$, one at TARN [26] and one at the Indiana cooler ring [6], both, however, with considerably lower energy resolution and lower statistical significance. The dielectronic resonances investigated with H-like ions are of the type $e + (1s) \to (2\ell n\ell')$ with $n = 2, 3, ..., \infty$ and $e + (1s) \to (3\ell n\ell')$ with $n = 3, ..., \infty$. Experimental results for both groups of resonances are shown in Fig.3 for $He^+$ ions. The cross sections for the $\Delta n = 2$ resonances (K $\to$ M) are about a factor 20 smaller than those for $\Delta n = 1$ (K $\to$ L) transitions. In the $(2\ell, n\ell')$ series several specified groups of states are experimentally resolved. The experimental $n$-distribution is cut off by field ionization already above $n_{max} = 5$. It was shown that only the inclusion of correlation to all orders in theoretical calculations yields satisfactory agreement with the experimental results. The agreement of theory and experiment is remarkable, though some discrepancies in the energies, the sizes and the shapes of resonances still remain to be explained.

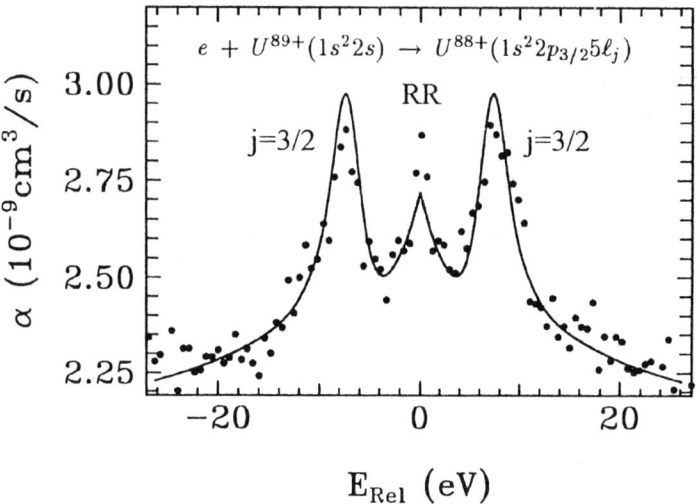

**FIG. 4.** Recombination rates of $U^{89+}$ ions measured at the ESR [20,21]. The solid line is a calculation of the rate based on theory for dielectronic [29] and radiative [30] recombination modelled to the specific experimental conditions. The meaning of positive and negative energies is explained in the text.

As the parent ion charge state is increased correlation effects are overwhelmed by the Coulomb forces. The lowest excitation energies increase rapidly along isoelectronic sequences with increasing atomic numbers. Energy splittings between electronic levels in general become very big for the heaviest few-electron ions. Since the low energy spread in the electron-ion collision experiments at storage rings does not explicitly depend on the ion charge state, measurements with highly charged ions are attractive because they provide very detailed experimental spectra. Many resonances and other cross section features can be resolved for highly charged ions which would never be seen in ions of low charge state. An example is the fine structure of the first excited $^2P$ terms in the lithium isoelectronic sequence. The energy splitting increases from $4.2 \cdot 10^{-5}$ eV [27] for the Li atom to 4.178 keV for the Li-like $U^{89+}$ ion [28]. Similar effects occur in other isoelectronic sequences.

Presently, the ESR of GSI is the only facility where comfortable intensities of the heaviest few-electron ions are available for merged-beams experiments. First experiments on recombination of such ions have been carried out with $Au^{76+}$, $Bi^{80+}$ and $U^{89+}$ [19–21]. Fig.4 shows the result of a measurement with Li-like uranium at the ESR [20,21]. Recombination rates are displayed at positive relative energies $E_{Rel}$ when the electrons are faster than the ions and at negative energies when the electrons are slower than the ions. Of course, both measurements are equivalent and ought to give the same results. At zero center-of-mass energy, radiative recombination dominates. Then at about 10 eV a dielectronic resonance is observed which is associated with the process

$$U^{89+}(1s^2 2s) + e \to U^{88+}(1s^2 2p_{3/2}\, 5\ell_{3/2})^{**} \to U^{88+} + photons. \qquad (4)$$

The resonant state is characterized by the total angular momentum quantum number $j = 3/2$ of the Rydberg electron and $j = 3/2$ for the excited $2p$ electron. This feature is one of the strongest dielectronic recombination resonances ever observed. The solid curve was calculated by modeling a theoretical calculation of DR [29] and of RR [30] to the special conditions in this experiment. For the calculation beam temperatures typical for the first ESR measurements were used and an angle between the beams of about 0.3° was invoked to get reasonable agreement with these and other measurements carried out with $U^{89+}$ ions in the same experimental run [20,21]. Improved experimental conditions are expected to be available at the ESR after a reconstruction period in 1995.

Apart from the clarification of the Z-dependence of electron-ion collision cross sections and rates another new and important aspect is coming up with the capability to study the highest-Z few-electron systems. Relativistic effects in the wave functions and on transition probabilities can be tested and, moreover, a new precision tool to study quantum electrodynamic contributions to atomic binding energies may evolve from the present studies of DR. Lamb shifts in the heaviest one-electron systems are up to several hundreds of eV. Considering that energy spreads well below 0.1 eV are possible for dielectronic recombination experiments, even with 60 GeV uranium ions, the potential for a high-precision spectroscopy by DR provides a fascinating new prospect for the field [20].

Besides lithium-like ions also sodium-like ions have recently been studied in great detail in both the ionization and the recombination channels. The wide energy ranges investigated at the TSR cover transitions of L- and M-shell electrons in $Cl^{6+}$, $Fe^{15+}$ and $Se^{23+}$ ions [16]. As an example, Fig. 5 shows measured and calculated DR rates for $Fe^{15+}$ ions in the $E_{cm}$ energy range 0 to 40 eV which spans $3s \to 3p, 3d, 4\ell$ transitions in the ion core. In the measurement the technique of adiabatic transverse expansion of the electron beam was employed. The magnetic guiding field $B$ was lowered by a factor of 7.4 along the beam axis. Thus, the transverse temperature of the expanded beam could be reduced to $T_\perp = (15\pm3)$ meV/k, where k is Boltzmann's constant and also a lower parallel temperature $T_\parallel = (0.15\pm0.05)$ meV/k was obtained. This resulted in an excellent resolution of resonances $e+3s \to 3p_{1/2}n\ell$ and $e+3s \to 3p_{3/2}n\ell$ with series limits at 34.0 eV and 36.6 eV, respectively. Within the energy range presented in Fig. 5 there are also clear signatures of $3s \to 3d7\ell$ resonances. Some contributions corresponding to transitions $e+3s \to 4\ell4\ell'$ are also present. The figure includes theoretical calculations (dotted line). Apart from discrepancies at the series limits theory and experiment are in excellent agreement.

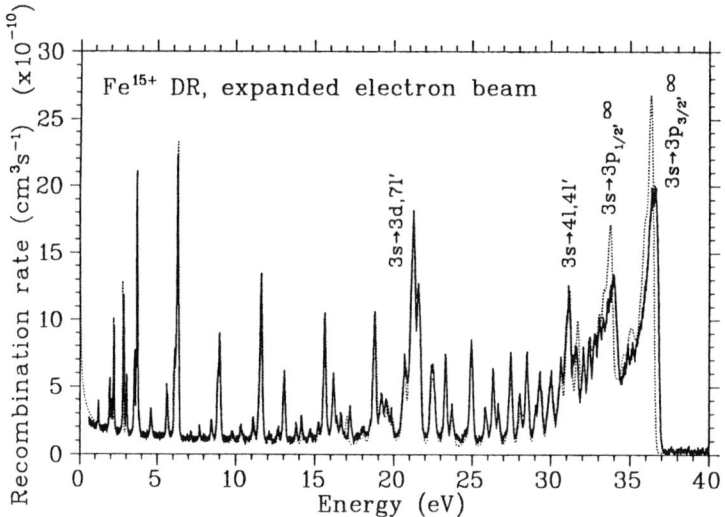

**FIG. 5.** Recombination rate of $Fe^{15+}$ ions [16]. The dotted line is the theoretical prediction convoluted with the experimental energy distribution function.

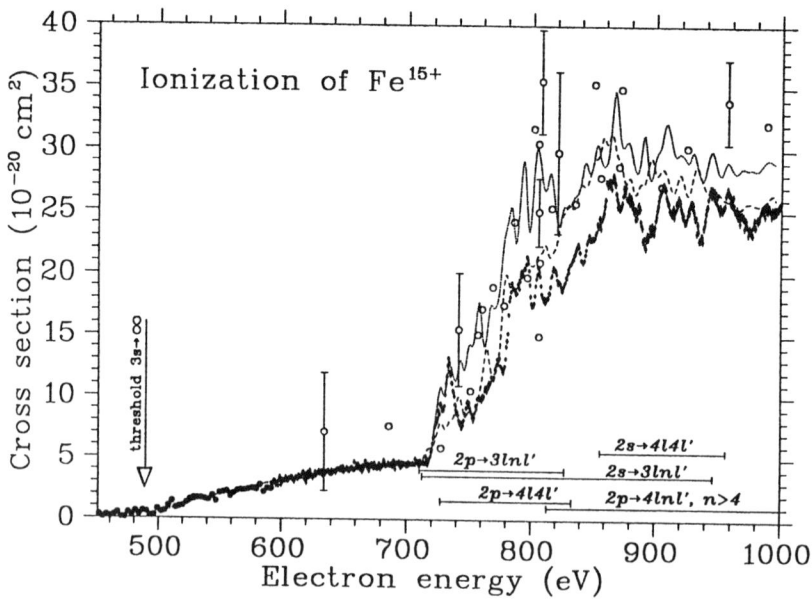

**FIG. 6.** Ionization cross sections of $Fe^{15+}$ ions. The solid dots are from a storage ring measurement [23], the open circles are from a crossed-beams experiment [31]. Theoretical calculations by Chen et al. [32] (dashed line) and Badnell and Pindzola [23] (solid line) are indicated. The ground-state ionization energy (489 eV) is marked by an arrow. Ranges of intermediate excited states are displayed by horizontal bars. The EA threshold is 710 eV.

Strong indirect contributions to ionization of sodium–like $Fe^{15+}$, with REDA even dominating the ionization cross section at certain energies, had been estimated 15 years ago by LaGattuta and Hahn [33]. It was not before the advent of heavy ion storage rings, however, that detailed measurements for this ion could be performed. The Heidelberg storage ring TSR was used to measure electron–impact ionization of $^{56}Fe^{15+}$ ions stored at a beam energy of 350 MeV [23]. By variation of the cathode voltage the electron–ion impact energy was set between 450 and 1030 eV. This range covers the ground–state ionization threshold and the energies corresponding to the most important EA and REDA cross section contributions.

The experimental results are shown in Fig. 6 by the solid dots. Within total error bars they agree with previous data of Gregory et al. [31] (open circles). They clearly reveal the rich structure caused by EA steps and REDA peaks in the cross section function. Obviously, the indirect ionization mechanisms dominate the ionization cross section of $Fe^{15+}$ ions at energies above 800 eV. Fig. 6 includes the results of two detailed calculations by Chen et al. [32] (dashed line) and by Badnell and Pindzola [23] (solid line).

Apart from studies of resonances and other characteristic cross section features in the channels of ionization and recombination the merged beams installations at accelerators facilitate access to very low center-of-mass energies. This was employed for the experiments with $D^-$ and $U^{89+}$ discussed above. Limitations are only set by the longitudinal and the transverse velocity spreads in the electron beam. With the excellent cooler beams presently available, meaningful measurements become possible at relative energies even below 1 meV. At such low energies electron-ion recombination reveals new unexpected features. In several experiments at zero center-of-mass energy $E_{cm}$ recombination rates were observed exceeding the expectations for radiative recombination by large factors [10]. Apparently, there are additional recombination mechanisms which are especially important at very low energies. The most surprising results so far were obtained with the electron target facility at GSI [34,35] with $U^{28+}$ ions. The observed recombination rate at $E_{cm} = 0\,eV$ exceeds the number expected for radiative recombination by more than a factor of 20.

## OUTLOOK

The field of electron-ion collision studies has reached a level of maturity where both, in recombination and ionization, total cross sections are investigated with excellent precision. Particularly with very highly charged ions, however, the new opportunities could not really be fully exploited yet. Considerable progress and new insight are expected in the near future for very heavy few-electron systems. Moreover, technical development is continued to provide further improvements in the energy resolution of storage ring experiments by employing even colder electron beams. Thus, even more detailed experimental information will become available. The improvement of energy resolution is also important for the spectroscopic aspect of DR with envisaged studies of QED effects in very highly charged few-electron ions. Measurements with these ions in wide ranges of energies will be facilitated by new electron-beam targets in storage rings additional to the existing electron cooling devices. Further progress will require measurements of partial cross sections differential e.g. in energy and angle of emitted photons and electrons. Special efforts have to be (and are being) undertaken to obtain a better understanding of the rate enhancement of recombination at very low energies. And last not least, the effects of external electric fields in the collision region on DR will be a subject of future research with multiply charged ions in storage rings.

## ACKNOWLEDGEMENTS

The author wants to express his gratitude for fruitful scientific interaction and collaboration with W. Spies, J. Linkemann, O. Uwira, A. Frank, T. Cramer, J. Kenntner, A. Wolf, D. Habs, M. Grieser, D. Schwalm, R. Becker, M. Kleinod, C. Kozhuharov, P. H. Mokler, N. Angert, F. Bosch, M. Steck, P. Spädtke, S. Schennach, B. Franzke, M. S. Pindzola, N. R. Badnell, P. Zimmerer, N. Grün, W. Scheid, K. J. Reed, and many others who have been involved in the work presented in this review. Support by the Gesellschaft für Schwerionenforschung (GSI), Darmstadt, by the Max-Planck-Institut für Kernphysik, Heidelberg, and by the German Ministry of Education, Science, Research and Technology (BMBF) is gratefully acknowledged.

## REFERENCES

1. A. Müller, Nucl. Instrum. Meth. B **87**, 34 (1994)
2. P. F. Dittner, S. Datz, in *Recombination of Atomic Ions*, editors: W. G. Graham et al., NATO ASI Series B: Physics Vol. 296, Plenum, New York, 1992, p. 133
3. D. Auerbach et al., J. Phys. B **10**, 3797 (1977); and J. B. A. Mitchell et al., Phys. Rev. Lett. **50**, 335 (1983)
4. L. H. Andersen, in *The Physics of Electronic and Atomic collisions*, editors: T. Andersen et al., AIP Press, New York, 1993, p. 432
5. S. Schennach et al., Z. Phys. D **30**, 291 (1994)
6. R. R. Haar et al., Phys. Rev. A **47**, R3472 (1993)
7. H. Danared et al., Phys. Rev. Lett. **72**, 3775 (1994)
8. D. R. DeWitt et al., J. Phys. B **28**, L147 (1995); and D. R. DeWitt et al., Phys. Rev. A **50**, 1257 (1994)
9. G. Kilgus et al., Phys. Rev. A **47**, 4859 (1993)
10. A. Wolf et al., in *Physics of Highly-Charged Ions*, editors: E. Salzborn et al., Suppl. Z. Phys. D **21**, 69 (1991)

11. O. Uwira et al., to be published
12. G. Kilgus et al., Phys. Rev. Lett. **64**, 737 (1990)
13. D. R. DeWitt et al., Phys. Rev. A, to be published; and Nucl. Instr. and Meth. B (1995), in print; see also: S. Asp et al., Annual Report 1994, Manne Siegbahn Laboratory, Stockholm University
14. J. Kenntner et al., in *Physics of Highly Charged Ions*, editors: F. Aumayr et al., Nucl. Instr. and Meth. B **98**, (1995), in print
15. A. Wolf, report MPI H-V15-1992 (unpublished)
16. J. Linkemann et al., in *Physics of Highly Charged Ions*, editors: F. Aumayr et al., Nucl. Instr. and Meth. B **98**, 1995, in print
17. G. Kilgus et al., Phys. Rev. **A46**, 5730 (1992)
18. A. Lampert et al., in *Atomic Physics of Highly-Charged Ions*, AIP Conference Proceedings 274, editors: P. Richard, et al., American Institute of Physics, New York, 1993, p. 537
19. W. Spies et al., Phys. Rev. Lett. **69**, 2768 (1992)
20. W. Spies et al., in *Physics of Highly Charged Ions*, editors: F. Aumayr et al., Nucl. Instr. and Meth. B **98**, 1995, in print
21. W. Spies, Thesis, Giessen 1995, unpublished
22. L. H. Andersen et al., Phys. Rev. Lett. **74**, 892 (1995)
23. J. Linkemann et al., Phys. Rev. Lett. **74**, 4173 (1995)
24. G. H. Wannier, Phys. Rev. **90**, 817 (1953)
25. D. S. Walton et al., J. Phys. B **3**, L148 (1970); J. Phys. B **4**, 1343 (1970); and B. Peart, K. T. Dolder, J. Phys. B **6**, 1497 (1973)
26. T. Tanabe et al., Phys. Rev. A **45**, 276 (1992)
27. S. Bashkin and J. O. Stoner,jr., *Atomic Energy Levels & Grotrian Diagrams 1*, North Holland, Amsterdam, Oxford, and American Elsevier, New York, 1975
28. Y.-K. Kim et al., Phys. Rev. A **44**, 148 (1991)
29. M. S. Pindzola, private communication
30. M. Stobbe, Ann. Physik **7**, 661 (1930)
31. D. C. Gregory et al., Phys. Rev. A **35**, 3256 (1987)
32. M. H. Chen et al., Phys. Rev. Lett. **64**, 1350 (1990)
33. K. J. LaGattuta, Y. Hahn, Phys. Rev. A **24**, 2273 (1981)
34. S. Schennach et al., in *Physics of Highly-Charged Ions*, editors: E. Salzborn et al., Suppl. Z. Phys. D **21**, 205 (1991)
35. O. Uwira et al., in *Physics of Highly Charged Ions*, editors: F. Aumayr et al., Nucl. Instr. and Meth. B **98**, 1995, in print

# Dissociative Recombination of Molecular Ions in a Cooler Ring

T.Tanabe,[1] I.Katayama,[1] H.Kamegaya,[1] K.Chida,[1] T.Watanabe,[1]
Y.Arakaki,[1] M.Yoshizawa[1] M.Saito,[2] Y.Haruyama,[2] K.Hosono,[3]
K.Hatanaka,[3] T.Honma,[4] K.Noda,[5] S.Ohtani[6] and H.Takagi[7]

[1] *Institute for Nuclear Study, University of Tokyo, Tanashi, Tokyo 188, Japan*
[2] *Kyoto Prefectural University, Kyoto 606, Japan*
[3] *Research Center for Nuclear Physics, Osaka University, Ibaraki 567, Japan*
[4] *Cyclotron and Radioisotope Center, Tohoku University, Sendai 980, Japan*
[5] *National Institute of Radiological Sciences, Anagawa, Chiba 260, Japan*
[6] *University of Electro-Communications, Chofu, Tokyo 182, Japan*
[7] *Physics Laboratory, School of Medicine, Kitasato University, Sagamihara 228, Japan*

**Abstract.** The dissociative recombination of $HD^+$ was studied with an adiabatically expanded low-temperature electron beam in the cooler-ring TARN II. Measurements were performed over a wide energy range from 0 to 40 eV in the center-of-mass system. The spectrum at low electron energies of less than 1 eV shows structures even in the energy region less than 10 meV. The experimental results agree well with theoretical calculations based on a multichannel quantum-defect theory including rotational motions. The dissociative recombination of $^4HeH^+$ with the expanded electron beam is also described.

## INTRODUCTION

Studies of dissociative recombinations (DR) are now rapidly progressing since the onset of experiments using the storage rings: TARN II in Tokyo, TSR in Heidelberg, CRYRING in Stockholm and recently ASTRID in Aarhus. Three experiments published in 1993 for the DR on $^4HeH^+$[1], $HD^+$[2] and $H_3^+$[3] revealed the existence of high energy resonances and they were just the beginning of experiments using cooler ring Afterwards, many experiments have been reported on $H_2^+$[4], $D_2^+$[5], $H_3^+$[6], $^3HeH^+$[7,8], $^4HeH^+$[9] and $CD^+$[10]. A lot of elaborate experiments using the storage rings have also been presented at the Third International Symposium on Dissociative Recombination recently held in Ein Gedi, Israel [11].

We carried out the first DR experiment [12] using the storage ring in 1990, which was aimed at beam diagnostics of the electron cooling device in the cooler ring TARN II at the Institute for Nuclear Study, University of Tokyo. In 1991, the first experiment for DR physics was also made on $H_3^+$ and later on $HeH^+$ with TARN II [1]. The main advantages of the storage-ring technique for the study of the DR process, compared with single-pass experiments, are : 1) the luminosity is higher by a factor of $10^2$ or more; 2) the electron-beam quality is good and the ion-beam quality can be improved if the electron cooling works; 3) the relative velocity between electron and ion beams can be changed over a wide range, especially in the region where electron velocity is smaller than the ion velocity, since the ion velocity is high; 4) background is low due to the good vacuum and the high ion-velocity; and 5) vibrationally excited ions can be quenched if they can be stored longer than their spontaneous emission lifetime. We have been performing DR experiments on light molecular ions with TARN II. The electron cooler had been an ordinary type in which the electron-beam size at the cooling section was the same as that at the cathode. In this system electrons have a temperature of about 100 meV ($=\Delta E_0$), corresponding to a cathode temperature of about 1200 K. At the cooling section, the longitudinal electron temperature, as seen from a co-moving system with the electron beam, is given by $\Delta E=(\Delta E_0)^2/4E$, where E is the electron energy. This value is quite small compared with $\Delta E_0$, because E is usually on the order of keV, while the electron temperature in the transverse direction remains the same as $\Delta E_0$. The energy resolution is thus limited by the transverse temperature. On the other hand, an old idea to reduce the transverse temperature by expanding the electron beam has successfully been realized at the Stockholm CRYRING [13]. Recently, our cooler was also modified to such a new type, and the transverse electron temperature has been greatly reduced. In the paper we first describe a DR experiment on $HD^+$ [14] using an adiabatically expanded low-temperature beam. Then, the DR experiment on $^4HeH^+$ also with the cold electron beam is discussed.

## EXPERIMENTAL SETUP

### On the $HD^+$ Measurements

An investigation was performed using TARN II and its associated electron cooler [12]. $HD^+$ ions were produced in a pulsed Penning ion source and accelerated to either 10 or 12 MeV in the injector cyclotron. Slight $H_3^+$ ion impurities were completely removed by the different resonance condition between $H_3^+$ and $HD^+$ in the cyclotron. The beam was then multiturn injected into the storage ring. The ion beam merged with an electron beam with a current of 70-140 mA. The 1/e-lifetime of the ion beam was about 5 s at an average vacuum pressure of $1\times10^{-10}$ mbar. The stored ions were cooled to a diameter of 1 mm within a time of less than 1 s. The electron-beam diameter was expanded from 1.4 cm to 5.2 cm in a gradually decreasing field from 4.8 kG to 0.35 kG. After being enlarged by a factor of about 14 in cross-sectional area, the electron beam enters into the cooling solenoid. Since the transverse electron temperature is proportional to the solenoid field, we can simply expect a temperature on the order of 7 meV, assuming an initial temperature of about 100 meV. The neutral atoms produced in the electron cooler were detected by a solid-state detector (SSD) installed in the vacuum extension of the cooler straight section outside the dipole magnet. The neutral products from the DR on HD$^+$ are identified based on the energies deposited in the SSD. The absolute cross sections were hard to determine, because the circulating ion current was too low to measure with a DC current transformer.

The DR experiments were performed by measuring the rate of H+D production as a function of the electron acceleration voltage. Before measurements, ions were stored for 6 s which is more than sufficient to allow for both longitudinal and transverse cooling of the ion beam and deexcitation of the vibrationally excited states. After this period the electron acceleration voltage was stepped up (or down) to a measuring voltage and switched back and forth between the measuring voltages and the cooling voltage at a rate of 50 Hz in order to avoid an energy shift of the ions due to the drag force of the electrons.

## On the $^4$HeH$^+$ Measurements

The $^4$HeH$^+$ ions were produced from a natural He and H$_2$ gas mixture in the Penning ion source and accelerated to 11.5 MeV in the injector cyclotron, and then injected into the storage ring. The ion beam merged with the adiabatically expanded electron beam with a current of 80 mA. The 1/e-lifetime of the beam in the ring was about 1.5 s. The DR measurements were started after storage of the beam for 5 s, which is long enough for the phase-space cooling and also vibrational cooling of the ion beam [15]. Other experimental conditions are the same as those for the HD$^+$ DR experiments.

## DISSOCIATIVE RECOMBINATION OF HD$^+$

The DR of the H$_2^+$ ion (and its isotopes) is a most fundamental and important collision process in plasmas. Extensive experiments [16-18] and calculations [19-24] on H$_2^+$ DR have been performed in recent years. For low-energy electrons (less than 1 eV), besides a mechanism called the direct process, where the initial state directly couples with the lowest doubly excited state of the H$_2^{**}$: $(2p\sigma_u)^2$ $^1\Sigma_g^+$ state, there is also an indirect DR process, represented by e+H$_2^+\rightarrow$H$_2^R\rightarrow$H$_2^{**}((2p\sigma_u)^2$ $^1\Sigma_g^+)\rightarrow$H($1s$)+H($nl$). In this case, electrons are initially captured into rot-vibrationally excited Rydberg states of the neutral molecule (H$_2^R$). These states then decay via crossing to the $(2p\sigma_u)^2$ $^1\Sigma_g^+$ state, followed by dissociation.

Low-energy DR has been studied by single-pass merged-beam measurements. The problem in single-pass experiments is that the original population of the vibrationally excited levels in the $H_2^+$ beam can not be completely removed, and that the beam populations are unknown. The situation is the same even for the storage ring experiments [4] where ions can be stored for a long time. Although the cross section below 1 eV has been measured with ions in low vibrational states in the single-pass experiment [18], the statistics of the cross section were not sufficiently good. Accordingly, the details concerning the structures of the cross sections at low-energy are still not completely clear. On the other hand, a high-energy resonance at around 10 eV has recently been found in experiments using the Heidelberg storage ring with the $HD^+$ ion [2]. In this case, $HD^+$ is expected to be vibrationally cooled within a time shorter than 1 s, because of spontaneous emission [25]. However, the structure of the low-energy peak below 0.3 eV has not yet been studied using the storage-ring technique.

Theoretically, multichannel quantum-defect theory (MQDT) predicts a complex resonance structure in the low-energy cross sections of $H_2^+$ and its isotopes, mostly in the form of dips, by taking into account the interference between direct and indirect recombination processes as well as overlapping effects between closely lying resonances [19,21,22]. The rotational effects on the DR shift the energies of the dips; also, the widely spread rotational distribution causes the structure of the DR cross sections to smear out [23,24]. No decisive comparison between theory and experiment has yet been presented.

Figure 1 (a) shows the rate of neutral products emitted in DR as a function of the electron energy. As expected from ref. 2, double DR peaks at around 10 and 15 eV were observed for electron velocities both slower and faster than the ion beam velocity. These double peaks result from the DR through doubly excited Rydberg states, converging to the $2p\sigma_u$ and $2p\pi_u$ states. Figure 1 (b) shows a closeup view of the low-energy peak, which was obtained by fine scanning the electron-acceleration voltage. As can be seen from the figure, there are many stair-like structures which appear almost symmetrically with respect to the zero c.m. energy. The structure can be recognized even in the neighborhood of 5 meV c.m. energy, which reflects a temperature decrease to below 10 meV. The relative cross sections over the entire energy range are shown in Fig. 2. The results are compared with the theoretical calculations.

The appropriate DR mechanism to be considered depends on whether the collision energies are higher or lower than 1 eV. In the lower energy region, significant effects are caused by the rotational motion and the higher order dynamical effect (off-the-energy-shell effect) induced by the electronic configuration interaction (CI). On the other hand, in the higher energy region, many highly excited states both of electronic and nuclear relative motions are important. We now focus on the lower energy region. The adopted method for the present calculation is based on the MQDT with the rotational motion [23], including the off-the-energy-shell contribution of CI [26]. The only dissociative state considered is the lowest resonance state, $(2p\sigma_u)^2\ ^1\Sigma_g^+$. The adopted vibrational and

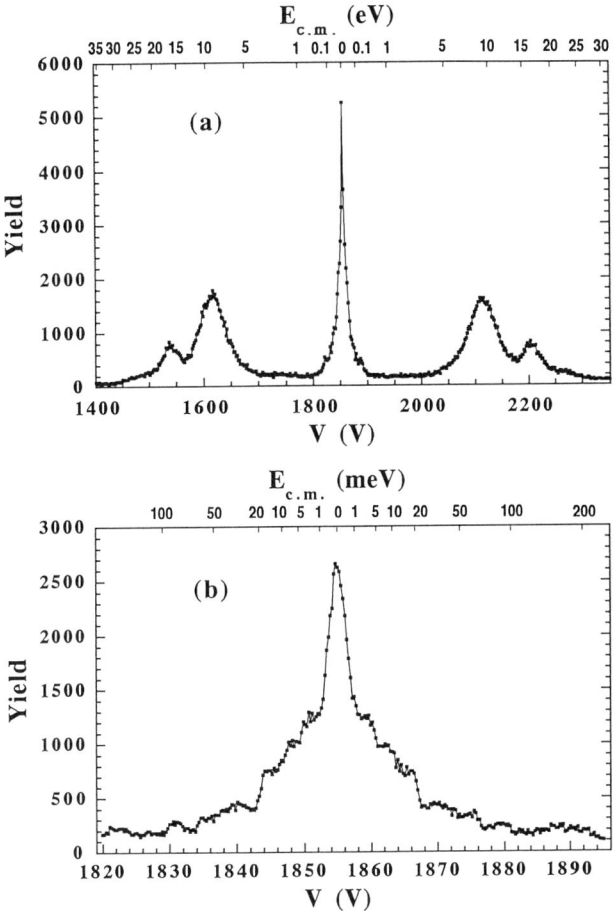

**FIGURE 1.** (a) Yields of neutral H+D atoms formed in the DR of HD$^+$ as a function of the electron acceleration voltage. The c.m. energy scale is also shown. (b) Closeup view of the low-energy DR peak [14].

rotational states are the lowest 10 and 15 states, respectively. In order to compare the calculation with the experiment, we must convolute the calculated cross section with the rotational distribution of the target ion and the energy resolution of the experiment. Since the stored time as 6 s is not sufficient to quench the rotational state [2], we assume a Maxwell distribution with a temperature of 800 K for the distribution of the rotational states. At this temperature about 90% of the population is distributed among states with a rotational quantum number of less than 8. For the experimental energy resolution, we assume a finite energy width in the transverse direction and neglect that in the longitudinal direction. The solid curve in Fig. 2 shows the result of convolution assuming a Maxwell distribution with $kT_{e\perp}=5$

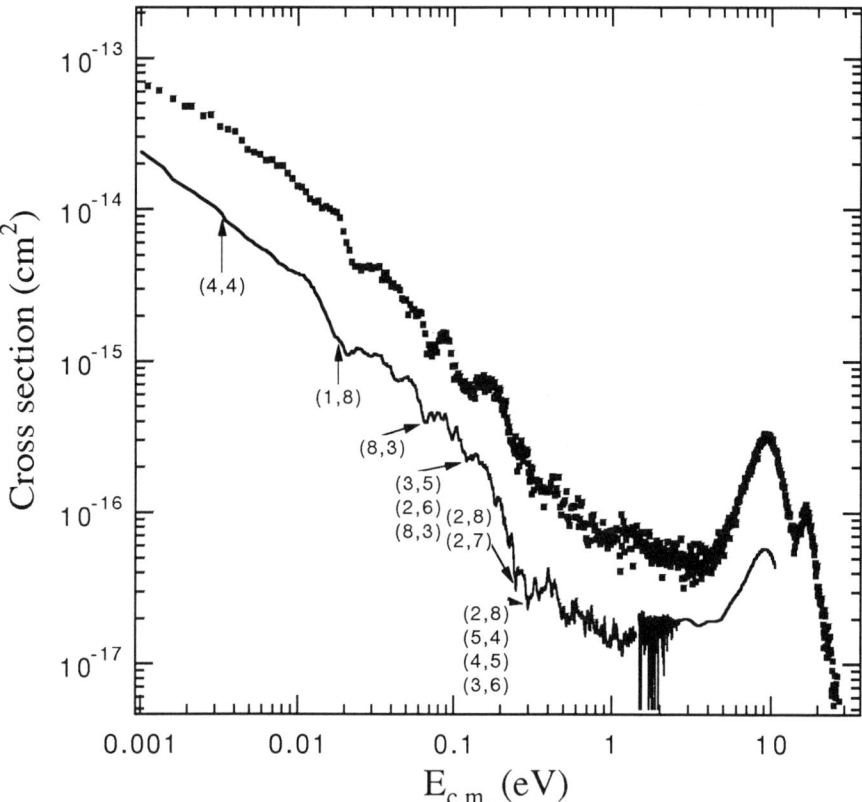

**FIGURE 2.** Relative experimental DR cross sections (dots) of HD$^+$ as a function of the c.m. energy. The curve represents the theoretical cross sections (absolute value). The experimental cross sections are arbitrarily scaled in magnitude to facilitate comparison with theory. The numbers in parentheses represent (v, n), where v and n are the vibrational and principal quantum numbers of the Rydberg states, respectively.

meV, where $T_{e\perp}$ is the electron temperature in the transverse direction. At this resolution, the global structure of the cross section remains almost unchanged, even if the rotational temperature changes from 800 K to 200 K. The calculated results reproduce the global energy dependence of the experiment: the large shoulder structures at 0.02 eV and 0.2 eV and the gradient. A good general agreement is observed over the entire energy region. Some outstanding structures are assigned to resonance states, which have the vibrational (v) and principal (n) quantum numbers indicated in Fig. 2. The convolution with the rotational state considerably smears the structure. The off-the-energy-shell effect also tends to reduce the large structures. Moreover, this effect strongly enlarges the cross section for energies lower than 0.2 eV, which can be seen in the shoulder at around 0.2 eV. Both the rotational motion and the off-the-energy-shell effect are indispensable to reproduce the experimental result.

For the energy range higher than 1.5 eV, major Rydberg series converging to the lowest excited states of HD$^+$ (2p$\sigma_u$) were taken into account, which are $^1\Sigma_g$, $^1\Sigma_u$, $^1\Pi_g$, $^3\Pi_g$ and $^3\Pi_u$. Cross sections were calculated using these 5 Rydberg manifolds where the adiabatic CI strengths were taken after the results by Tennyson and Noble [27], and neglecting the rotational motion and the contribution from the off-the-energy-shell term. The agreement with the experiment is nice although the theory seems to underestimate still around the top of the peak.

The above results were calculated assuming the initial vibrational quantum number (v=0) for the ion. We also calculated the cross sections for v=1. However, the results do not agree with the experiments. This supports the view of the vibrational cooling of stored ions.

In summary, the detailed shape of the cross sections for the HD$^+$ DR with slow electrons was revealed using an ultracold electron beam and a vibrationally and phase-space cooled ion beam. MQDT calculations including the rotational motions and off-the-energy-shell effects are in good agreement with the experimental results.

# DISSOCIATIVE RECOMBINATION OF $^4$HeH$^+$

For the DR process of HeH[+], a big feature is the absence of the potential curve crossings between neutral dissociating molecular states and the ground state of HeH[+] ion. These crossings are normally believed to be the main source of the DR process. Despite such a situation the experimental results show appreciable cross sections for low-energy DR. Relating to this phenomenon, it has been proposed by a single-pass experiment that DR at low energies is due to the presence of a metastable triplet component of the ion beam which has many potential energy curve crossings suitable for DR [28]. However, it is not self-evident that the low-energy peak studied with the storage ring technique and reported previously [1] is caused by such a metastable state, since the types of ion sources and the experimental conditions are different between the works [1] and [28]. In order to understand the mechanism of DR, it is very important to clarify whether the DR spectra originate from ground state component or metastable one. If the initial ion is in the ground state, DR is only possible by the non-adiabatic coupling between the ionizing state and lower neutral states. The storage ring experiments have been performed on this subject. These results suggest that the low energy component observed at the storage ring experiments originates from the $X^1\Sigma^+$ ground state ions [7,9]. Recently, a fine structure of the low energy peak has been found for the $^3$HeH[+] DR [8], which was compared with the theoretical calculations based on the nonadiabatic coupling [29]. The strength of the theoretical rate is comparable to the experiment. However, the details of the structure are far from agreement with the experiments. Further investigations on the fine structure are required experimentally, especially on the isotope dependence, as well as development in theory.

Figure 3 shows the rate of neutral products emitted in DR as a function of the electron energy. As can be seen from the figure, there are peaks which appear almost symmetrically with respect to the zero c.m. energy. Figure 4 shows the relative rate coefficients as a function of the center-of-mass energy. The structure is similar to that on $^3$HeH$^+$ DR [8]. However, there are some definite differences from the $^3$HeH$^+$ DR. First, the energies of the peaks at around 15 and 210 meV for $^4$HeH$^+$ are lower than the corresponding ones at around 40 and 250 meV for $^3$HeH$^+$. Second, the peak-height ratio of 15 meV to 210 meV is about 3.5, which is about twice as big as the corresponding value of $^3$HeH$^+$. Third, a small peak between them appears only weakly for $^4$HeH$^+$. The position of resonances and accordingly their interference effects depend on the locations of rot-vibrational levels of the neutral Rydberg states which strongly depend on the isotopic mass of the ground state ions. Therefore, these differences seem to come from the isotope effect rather than the different experimental conditions. The global similarity between these two storage ring experiments and the disagreements with the single-pass experiment [28] suggest that the original ion state is the $X^1\Sigma^+$ ground state.

In summary, the relative rate of the low energy component for the $^4$HeH$^+$ DR was measured with the cold electron beam. The energies and relative intensities of the peaks for the rate are clearly different from those on $^3$HeH$^+$ DR, although the structures are globally similar each other. This seems to be due to an isotope effect.

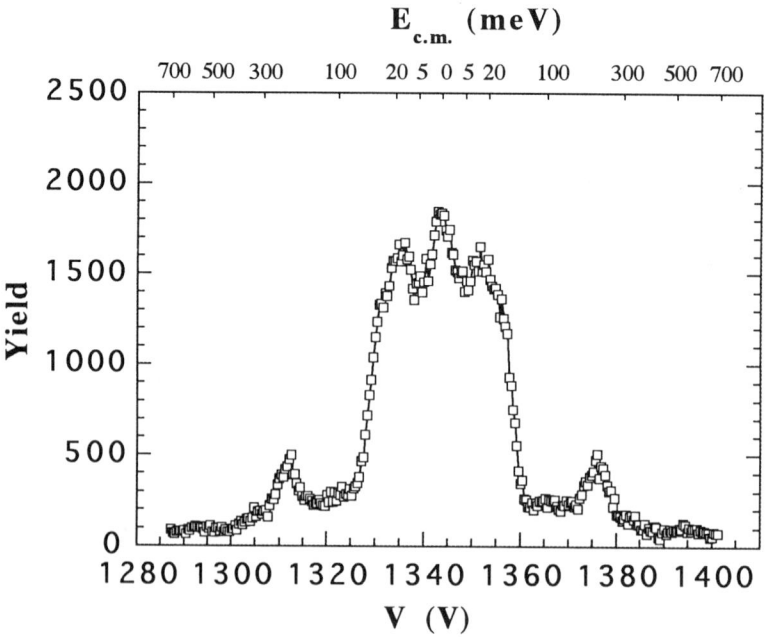

**FIGURE 3.** Yield of neutral $^4$He+H atoms formed in the DR of $^4$HeH$^+$ as a function of the electron acceleration voltage. The c.m. energy scale is also shown.

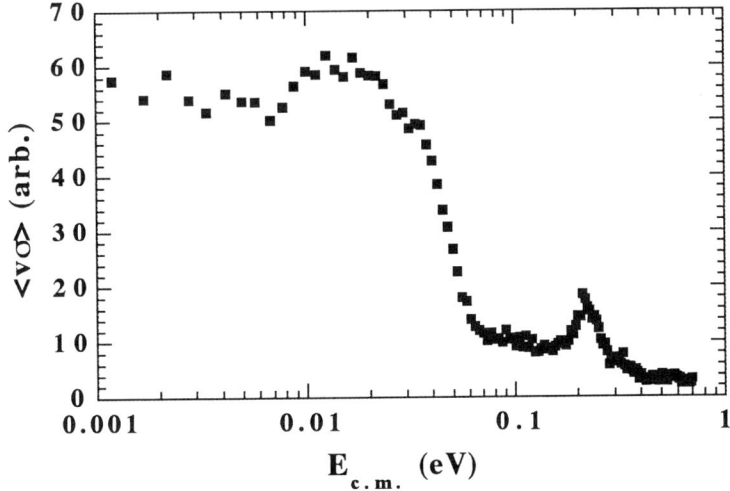

**FIGURE 4.** Relative rate coefficients on the $^4$HeH$^+$ DR as a function of c.m. energy.

## ACKNOWLEDGMENTS

We thank the cyclotron staff for their helpful cooperation. This work was performed under Grants-in-Aid for Scientific Research (A) of the Ministry of Education, Science and Culture.

## REFERENCES

1. Tanabe,T., Katayama,I., Inoue,N., Chida,K., Arakaki,Y., Watanabe,T., Yoshizawa,M., Ohtani,S., and Noda,K., *Phys. Rev. Lett.* **70**, 422-425 (1993).
2. Forck,P., Grieser,M., Habs,D., Lampert,A., Repnow,R., Schwalm,D., Wolf,A., and Zajfman,D., *Phys. Rev. Lett.* **70**, 426-429 (1993).
3. Larsson,M., Danared,H., Mowat,J.R., Sigray,P., Sundström,G., Broström,L., Filevich,A., Källberg,A., Mannervik,S., Rensfelt, K.-G., and Datz, S., *Phys. Rev. Lett.* **70**, 430-433 (1993).
4. Larsson,M., Broström,L., Carlson,M., Danared,H., Datz, S., Mannervik,S., and Sundström,G., *Physica Scripta* **51**, 354-358 (1995).
5. Larsson,M., Carlson,M., Danared,H., Broström,L., Mannervik,S., and Sundström,G., *J. Phys. B: At. Mol. Opt. Phys.* **27**, 1397-1406 (1994).
6. Datz, S., Sundström,G., Biedermann,Ch., Broström,L., Danared,H., Mannervik,S., Mowat,J.R., and Larsson,M., *Phys. Rev. Lett.* **74**, 896-899 (1995).
7. Sundström,G., Datz,S., Mowat,J.R., Mannervik,S., Broström,L., Carlson,M., Danared,H., and Larsson, M., *Phys. Rev. A* **50**, R2806-R2809 (1994).
8. Mowat,J.R., Danared,H., Sundström,G., Carlson,M., Andersen,L.H., Vejby-Christensen,L., af Ugglas,M., and Larsson, M., *Phys. Rev. Lett.* **74**, 50-53 (1995).

9. Tanabe,T., Katayama,I., Inoue,N., Chida,K., Arakaki,Y., Watanabe,T., Yoshizawa,M., Saito, M., Haruyama,Y., Hosono, K., Honma, T., Noda,K., Ohtani,S., and Takagi, H., *Phys. Rev. A* **49**, R1531-R1534 (1994).
10. Fork,P., Broude,C., Grieser,M., Habs, D., Kenntner, J., Liebmann,J., Repnow,R., Schwalm,D., Wolf,A., Amitay,Z., and Zajfman,D., *Phys. Rev. Lett.* **72**, 2002-2005 (1994).
11. *Dissociative Recombiantion: Theory, Experiment and Application* :World Scientific, Singapore, 1996.
12. Tanabe,T., Noda,K., Honma,T., Kodaira,M., Chida,K., Watanabe,T., Noda,A., Watanabe,S., Mizobuchi,A., Yoshizawa,M., Katayama,T., and Muto,H., *Nucl. Instrum. Methods Phys. Res., Sect. A* **307**, 7-25 (1991).
13. Danared,H., Andler,G., Bagge,L., Herrlander,C.J., Hilke,J., Jeansson,J., Källberg,A., Nilsson,A., Paál,A., Rensfelt,K.-G., Rosengård,U., Starker,J., and af Ugglas,M., *Phys. Rev. Lett.* **72**, 3775-3778 (1994).
14. Tanabe,T., Katayama,I., Kamegaya,H., Chida,K., Arakaki,Y., Watanabe,T., Yoshizawa,M., Saito, M., Haruyama,Y., Hosono, K., Hatanaka,K., Honma, T., Noda,K., Ohtani,S., and Takagi, H., *Phys. Rev. Lett.* (in press).
15. Datz,S., and Larsson,M., *Physica Scripta* **46**, 343-347 (1992).
16. Auerbach,D., Cacak,R., Caudano,R., Gaily,T.D., Keyser,C.J., McGowan,J.Wm., Mitchell,J.B.A., and Wilk,S.F.J., *J. Phys. B: At. Mol. Phys.* **10**, 3797- 3820 (1977).
17. Hus,H., Yousif,F., Noren,C., Sen,A., and Mitchell,J.B.A., *Phys. Rev. Lett.* **60**, 1006-1009 (1988).
18. Van der Donk,P., Yousif,F.B., Mitchell,J.B.A., and Hickman,A.P., *Phys. Rev. Lett.* **67**, 42- 45 (1991).
19. Giusti-Suzor,A., Bardsley,J.N., and Derkits,C., *Phys. Rev. A* **28**, 682- 691 (1983).
20. Hickman,A.P. *J. Phys. B: At. Mol. Phys.* **20**, 2091- 2099 (1987).
21. Nakashima,K., Takagi,H., and Nakamura,H., *J.Chem. Phys.* **86**, 726- 737 (1987).
22. Schneider,I.F., Dulieu,O., and Giusti-Suzor,A., *J. Phys. B: At. Mol. Opt. Phys.* **24**, L289-L297 (1991).
23. Takagi,H., *J. Phys. B: At. Mol. Opt. Phys.* **26**, 4815-4832 (1993).
24. Takagi,H., "Basic Ploblems in Dissociative Recombiantion of Diatomic Molecules," in *The Physics of Electronic and Atomic Collisions, XVIII International Conference, Aarhus, 1993*, edited by T.Andersen, B.Fastrup, F.Folkmann, H.Knudsen and N.Andersen, AIP Conf. Proc., No.295 (AIP, New York, 1993), pp. 442-451.
25. Colbourn,E.A., and Bunker,P.R., *J. Mol. Spectrosc.* **63**, 155-163 (1976).
26. Takagi,H., Hara,S., and Sato,H., "Off-the-energy-shell Effects in Dissociative Recombiantion of $H_2^+$+e, II," in *Abstracts of Contributed Papers, XVIII Intenational Conference on the Physics of Electronic and Atomic Collisions (ICPEAC), Aarhus, Denmark, 1993* (unpublished), p. 339.
27. Tennyson,J., and Noble, C.J., *J.Phys. B : At. Mol. Phys.* **18**, 155-165 (1985).
28. Yousif,F.B., Mitchell,J.B.A., Rogelstad,M., LePaddelec,A.,Canosa,A., and Chibisov,M.I., *Phys. Rev. A* **49**, 4610-4615 (1994).
29. Guberman,Steven L., *Phys. Rev. A* **49**, R4277-R4280 (1994).

# ELECTRON COLLISIONS
## – IONIZATION –

Ionization of Atoms as a Three-Body Process ...................... 341
    S. JONES and D.H. Madison
Basis Spline Method for $e^- + H$ Collisions ......................... 347
    S.V. PASSOVETS, J.H. Macek, and S.Yu Ovchinnikov
(e, 2e) Momentum Spectroscopy of Thin Films .................... 357
    P. STORER, Y.Q. Cai, S.A. Canney, R. Caprari,
    S.A.C. Clark, A.S. Kheifets, I.E. McCarthy,
    S. Utteridge, M. Vos, and E. Weigold

# Ionization of Atoms as a Three-Body Process

## S. Jones and D. H. Madison

*Physics Department, University of Missouri-Rolla, Rolla, MO 65401*

> We report the results of a preliminary calculation for electron-hydrogen ionization that uses a final-state wave function that satisfies the proper boundary condition for this three-body problem. In our calculation, we approximate the exact final-state wave function by a wave function that is asymptotically correct even if *only one* of the escaping electrons is far from the nucleus [E. O. Alt and A. M. Mukhamedzhanov, Phys. Rev. A **47**, 2004 (1993)]. The regions of validity for this wave function thus *enclose* the region that contributes to scattering. This is an improvement over Redmond's asymptotic form, which is valid only if *both* electrons are far from the nucleus.

## INTRODUCTION

In just the last few years, remarkable progress has been made in the theoretical treatment of atomic three-body processes. Much of this progress is due to the widespread availability of powerful yet relatively inexpensive workstations easily capable of performing numerical calculations that were prohibitive just a decade ago.

Our understanding of three-body processes has improved considerably through a spirited interplay between experiment and two distinct yet complementary theoretical approaches. In the perturbative approach (*e.g.*, distorted-wave methods), one seeks to include exactly only "strong" interactions (*e.g.*, long-range contributions from Coulomb interactions and static electron-atom interactions), thereby leaving only "weak" interactions to be treated perturbatively. Non-perturbative methods, on the other hand, in principle make no assumption about the relative importance of interactions. Nevertheless, in practice these methods rely on perturbation theory as a tool. Similarly, non-perturbative methods have been instrumental in guiding perturbation theory, particularly when the relative importance of various interactions is poorly understood.

One of the most important developments of the last few years is the emergence of the *convergent* close-coupling method, which takes the close-coupling method to completeness if convergence can be obtained for the observable of interest. For electron-hydrogen ionization, convergence in the cross section has been achieved for incident energies as low as 54.4 eV in asymmetric kinematics [1].

© 1995 American Institute of Physics

A similar improvement may be in the future for perturbative methods. Recently, Alt and Mukhamedzhanov [2], hereafter referred to as AM, derived the proper boundary condition for three charged particles in the continuum. The AM wave function is asymptotically correct if *any* interparticle separation is large. This means that the AM wave function has the correct asymptotic form for three continuum particles not only when all interparticle separations are large, but also when any two particles are arbitrarily close, provided the third particle is far away from the center of mass of the other two. As a result, the regions of validity for the AM wave function *enclose* the scattering region (the closed and finite region that contributes to scattering). In contrast, Redmond's form is valid only when *all* interparticle separations are large. The region of validity for Redmond's form is *disjoint* to the scattering region. A perturbation series containing the AM wave function includes all long-range Coulomb interactions explicitly, and therefore could conceivably converge rapidly even for fairly low energies.

In this report, we present the results of a preliminary calculation for electron-impact ionization of hydrogen using the AM wave function as an approximation for the exact final-state wave function. Atomic units are used throughout this paper and unit vectors are denoted by a "hat", *e.g.*, $\hat{\mathbf{r}} = \mathbf{r}/r$.

## THEORY

In this report, we consider only asymmetric collisions where the contribution from electron exchange is small. The TDCS (triply-differential cross section) neglecting exchange is given by

$$\frac{d^3\sigma}{d\Omega_a d\Omega_b dE_b} = \frac{k_a k_b}{k_i}|f|^2, \tag{1}$$

where $f$ is the direct amplitude which is given by

$$f = -(2\pi)^{-5/2}\left\langle \Psi_f^-(\mathbf{r}_a, \mathbf{r}_b) \left| -\frac{1}{r_a} + \frac{1}{r_{ab}} \right| e^{i\mathbf{k}_i \cdot \mathbf{r}_a}\psi_i(\mathbf{r}_b) \right\rangle. \tag{2}$$

Here $\psi_i$ is the wave function for the hydrogen atom, $\mathbf{r}_{ab} = \mathbf{r}_a - \mathbf{r}_b$, and $\mathbf{k}_i$, $\mathbf{k}_a$, and $\mathbf{k}_b$ are the wave vectors for the incident, faster final-state, and slower final-state electrons respectively.

To obtain the exact final-state wave function, $\Psi_f$, we would need to solve

$$(H - E)\Psi_f = 0, \tag{3}$$

where

$$H = -\frac{1}{2}\nabla_{\mathbf{r}_a}^2 - \frac{1}{2}\nabla_{\mathbf{r}_b}^2 - \frac{1}{r_a} - \frac{1}{r_b} + \frac{1}{r_{ab}} \tag{4}$$

is the Hamiltonian and E is the energy. Since there is no known solution for the three-body Hamiltonian, one must approximate $\Psi_f$. In this work, we have used the AM wave function to approximate $\Psi_f$. The AM wave function satisfies the proper three-body boundary condition for $\Psi_f$ and is given by

$$\Psi^-_{f,AM} = e^{i(\mathbf{k}_a\cdot\mathbf{r}_a+\mathbf{k}_b\cdot\mathbf{r}_b)} C(-1/k'_a, \mathbf{k}'_a, \mathbf{r}_a) C(-1/k'_b, \mathbf{k}'_b, \mathbf{r}_b) C(\mu/k'_{ab}, \mathbf{k}'_{ab}, \mathbf{r}_{ab}). \tag{5}$$

Here

$$C(\alpha, \mathbf{k}, \mathbf{r}) = \Gamma(1 - i\alpha) e^{-\pi\alpha/2} F(i\alpha, 1; -i(kr + \mathbf{k}\cdot\mathbf{r})) \tag{6}$$

are the ordinary Coulombic distortion factors with $\Gamma$ the gamma function and $F$ the confluent hypergeometric function. The new property of the AM wave function is its dependence on *local* wavevectors. These are given by

$$\begin{aligned}
\mathbf{k}'_a &= \mathbf{k}_a + \mathbf{K}\left(\mu/k_{ba}, \mathbf{k}_{ba}, \mathbf{r}_b\right), \\
\mathbf{k}'_b &= \mathbf{k}_b + \mathbf{K}\left(\mu/k_{ab}, \mathbf{k}_{ab}, \mathbf{r}_a\right), \\
\mathbf{k}'_{ab} &= \mathbf{k}_{ab} + \mathbf{K}\left(-1/k_a, \mathbf{k}_a, \boldsymbol{\rho}\right) - \mathbf{K}\left(-1/k_b, \mathbf{k}_b, \boldsymbol{\rho}\right).
\end{aligned} \tag{7}$$

In the above, $\mathbf{k}_{ab} = \mu(\mathbf{k}_a - \mathbf{k}_b)$ with the reduced mass of two electrons denoted by $\mu = 1/2$, $\boldsymbol{\rho} = \mu(\mathbf{r}_a + \mathbf{r}_b)$ is the coordinate for the center of mass of the two electrons, and

$$\mathbf{K}(\eta, \mathbf{k}, \mathbf{r}) = \frac{r}{R}\left[\frac{F(1+i\eta, 2; -i(kr+\mathbf{k}\cdot\mathbf{r}))}{-iF(i\eta, 1; -i(kr+\mathbf{k}\cdot\mathbf{r}))}\right](\hat{\mathbf{k}} + \hat{\mathbf{r}}) \tag{8}$$

are the local wavevector modifications with $R = r_a + r_b + r_{ab}$. Note that $\mathbf{K}$ is a *complex* six-dimensional vector function in coordinate space. Since only the real part is needed to satisfy the boundary condition, we have neglected the imaginary part of $\mathbf{K}$ in the present work.

## RESULTS

To calculate the scattering amplitude we used six-dimensional numerical quadrature. We estimate our numerical uncertainty to be 10%. In Fig. 1 we compare our results (denoted AM) for coplanar TDCS with relative experimental data from Schlemmer et al. [5] and the CCC (convergent close-coupling) [1], BBK (Brauner, Briggs, and Klar) [3], and 3DWBA (three-body distorted-wave Born approximation) [4] theories. We use the same normalization of the experimental data that was use in Ref. [1]. This normalization gave the best overall agreement between the CCC results and experiment for the four available cases at 54.4-eV incident energy ($\theta_a = 4$, 10, 16 and 23 degrees). The high-energy approximations BBK and 3DWBA use final-state wave functions satisfying Redmond's form. At this relatively low energy, none of the theories are in quantitative agreement with the experimental data, although all show more or less qualitative agreement. Obtaining accurate TDCS for intermediate and low energies has been a major problem for theoreticians

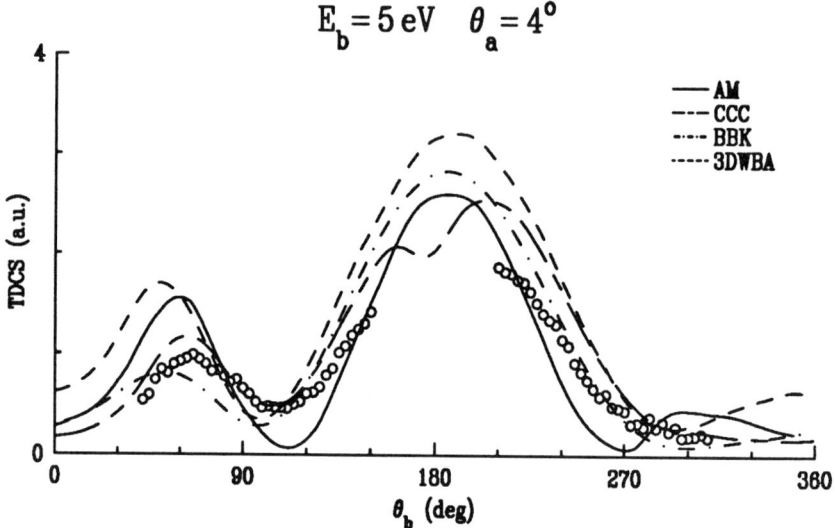

**FIG. 1.** Triply-differential cross section for 54.4-eV electron-impact ionization of hydrogen. The circles are relative experimental data from Schlemmer et al. [5]. The angle $\theta_a$ of the faster final-state electron is measured counter-clockwise from the forward beam direction while the angle $\theta_b$ of the slower electron is taken clockwise.

for some time now and we are encouraged by the fact that these preliminary results are in qualitative agreement with experiment and the best currently available theories. One of the major debates in the field has been whether or not it is important to get the asymptotic form of the wave function correct. One argument is that since the asymptotic region does not contribute to the T-matrix integrals, the form of the wave function in this region does not matter. On the other hand, the form of the wavefunction in the asymptotic region effects the behavior in the close region. Consequently, it is reasonable to expect that the close-in behavior will be better represented by a wavefunction that is asymptotically correct than one which is not. In any case, this asymptotic form is needed if one wishes to obtain accurate results working from a final-state representation, since a variational principle requires an asymptotic form that *encloses* the scattering region. The present approach can be viewed as a first step in this direction.

## CONCLUSION

We presented the results of a preliminary calculation for electron-hydrogen ionization that uses a final-state wave function satisfying the proper three-body boundary condition. Numerically, we have found it difficult to accurately include the fairly long-range dipole contribution, and this is the main source of uncertainty in the present results. Our goal is to be able to routinely obtain results numerically accurate to 5%. Although these first results are encouraging, further calculations are needed in order to draw any definite conclusions. Calculations at lower energies are underway and will provide a more rigorous test of the present approach.

## ACKNOWLEDGMENTS

Helpful discussions with V. Kravtsov and J. L. Peacher are gratefully acknowledged. This work was supported by the National Science Foundation.

## REFERENCES

1. I. Bray, D. A. Konovalov, I. E. McCarthy, and A. T. Stelbovics, Phys. Rev. A **50**, R2818 (1994).
2. E. O. Alt and A. M. Mukhamedzhanov, Phys. Rev. A **47**, 2004 (1993).
3. M. Brauner, J. S. Briggs, and H. Klar, J. Phys. B **22**, 2265 (1989).
4. S. Jones, D. H. Madison, A. Franz, and P. L. Altick, Phys. Rev. A **48**, R22 (1993).
5. P. Schlemmer, T. Rösel, K. Jung, and H. Ehrhardt, J. Phys. B **22**, 2179 (1989).

# Basis Spline Method for $e^- + H$ Collisions

## S. V. Passovets

*Ioffe Physical Technical Institute, St. Petersburg, Russia*

## J. H. Macek and S. Yu Ovchinnikov*

*Department of Physics and Astronomy, University of Tennessee, Knoxville, TN 37996-1501,
and
Oak Ridge National Laboratory†, Post Office Box 2009
Oak Ridge, TN 37831*

---

We study a particular utilization of the basis-spline collocation method (BSCM) for the *ab initio* computations of process of the fragmentation of an atom into two electrons and a positive ion by the electron impact on atomic hydrogen. Computed ionization cross sections and spin asymmetry agree within 10% with measurements.

---

The fragmentation of an atom or molecule into two charged fragments is well understood on the basis of conventional theory, for example on the basis of R-matrix theory, frame transformations, and the associated multi-channel-quantum-defect (MQDT) method [1]. Extension of these methods to treat fragmentation into three or more charged fragments has proved particularly difficult. The central feature of such fragmentation, namely the Wannier threshold law [2–9] has not emerged from conventional representations. Assumed three-body final state wave functions appear to have limited or unknown regions of applicability. This remark applies to both the Rau-Peterkop functions, which emphasize the collective motion of two electrons that underlies the Wannier threshold law, and the Brauner, Briggs, and Klar function [10] which emphasizes the role of two-body interactions acting nearly independently. For ion-atom interactions the hidden crossing theory [11] of atomic transitions has been applied in the threshold region to compute absolute cross sections over an extended energy range. This theory can be applied to electron-atom interactions to investigate the asymptotic expansion for which the Wannier power law represents the first term. In addition, *ab initio* calculations of the cross section are not limited to the near threshold region.

The hidden crossing theory has its genesis in the work of Landau [12] on transitions when the motion is quasi-classical. Then adiabatic energy eigenvalues $\varepsilon_n(R)$ play a central role. This role was placed on a much broader

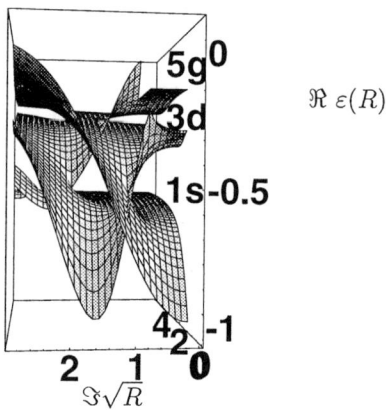

**FIG. 1.** Riemann surface constructed by plotting $\Re\varepsilon(R)$ vs. $R^{1/2}$.

footing by Demkov's [13] remark that the individual eigenvalues $\varepsilon_n(R)$ represent a single function $\varepsilon(R)$ on a multisheeted Riemann surface considered as a function of the complex variable $R$. The coordinate $R$ is usually taken as the distance between two nuclei in an ion-atom collision. For real values of $R$ the function $\varepsilon(R)$ takes on different values $\varepsilon_n(R)$ depending upon which sheet $n$ of the Riemann surface $R$ is located. This Riemann surface is constructed as in Fig.(1) by plotting $\Re\varepsilon(R)$ vs $\sqrt{R}$ for $H_2^+$.

Note that the different sheets join at branch points for complex $R$. Also note that the values of this function for $\Im R = 0$ are just the normal adiabatic energy eigenvalues. Solov'ev [11] further developed the theory in order to treat fragmentation processes. This theory is quite successful for computing total cross sections for ionization of hydrogen by proton impact for energies in the 5 – 25 keV range. Our own work has focused upon adapting the method for electron impact and computing energy and angular distributions of electrons. The key quantity in the hidden crossing theory is seen to be the function $\varepsilon(R)$. To find $S$-matrix element it is necessary to compute this quantity for arbitrary complex $R$ to locate the branch points where different sheets of the Riemann surface are connected. This has been done for the $H_2^+$ diatomic molecular system and the results used to compute ionization cross sections [14].

For ionization of atomic hydrogen by electron impact, it is necessary to use the adiabatic approximation in the hyperspherical representation [15]. The hyperradius $R$ and hyperangle $\alpha$ are defined in terms of the coordinates $\mathbf{r}_1$ and $\mathbf{r}_2$ as

$$R = \sqrt{r_1^2 + r_2^2}, \qquad \tan\alpha = r_2/r_1 \tag{1}$$

In this representation the Schrödinger equation takes the form

$$H = -\frac{1}{2}\frac{d^2}{dR^2} + \frac{\Lambda^2}{2R^2} + RC(\Omega), \tag{2}$$

where

$$\Lambda^2 = -\frac{d^2}{d\alpha^2} + \frac{\hat{\ell}_1^2}{\sin^2\alpha} + \frac{\hat{\ell}_2}{\cos^2\alpha^2} - \frac{1}{4}, \quad (3)$$

and where $C(\Omega)/R$ is the potential energy of the two electrons written in terms of a set of hyperangles $\Omega$, and $\hat{\ell}_i$ is the angular momentum operator of electron i. Unlike the situation for $H_2^+$, the adiabatic "Hamiltonian" $\Lambda^2 + 2RC(\Omega)$ in Eq.(2) is not separable in any known coordinate system. Accordingly, it is necessary to solve the partial differential equation directly using appropriate numerical techniques. We have used the basis spline programs developed by Bottcher and co-workers [16] to find the energy eigenvalues at complex R.

The hyperspherical adiabatic basis functions are eigenfunctions of the operator $H_{\text{eff}} = \Lambda^2 + 2RC(\Omega)$. This operator is diagonalized by writing the eigenfunction in terms of Euler angles $\omega_1, \omega_2$, and $\omega_3$ following Zhou and Lin [17]:

$$\Phi(R;\Omega) = \sum_{I=0}^{L} f_I(R;\alpha,\theta)\mathcal{D}_{|I|,M}^{(L)}(\omega_1,\omega_2,\omega_3), \quad (4)$$

where $I$ runs from 0 to $L$ for states with parity $P = (-1)^L$, and from 1 to $L$ for states with parity $(-1)^{L+1}$. The reader is referred to Ref. [17] for the definition of the rotation functions. The operator $H_{\text{eff}}$ then takes the matrix form, represented by the symbol $h$ with elements

$$h_{I,I} = -\frac{\partial^2}{\partial\alpha^2} - \frac{4}{\sin^2 2\alpha}\left[\frac{\partial^2}{\partial\theta^2} + \cot\theta\frac{\partial}{\partial\theta} + \frac{I^2}{\sin^2\theta}\right] - \frac{2I^2 - L(L+1)}{\cos^2\alpha} + 2RC(\alpha,\theta),$$

$$C(\alpha,\theta) = -\frac{Z}{\sin\alpha} - \frac{Z}{\cos\alpha} + \frac{1}{\sqrt{1-\sin 2\alpha\cos\theta}},$$

$$h_{0,1} = -\frac{\sqrt{2L(L+1)}}{\cos^2\alpha}\left(\frac{\partial}{\partial\theta} + \cot\theta\right), \quad \text{for} \quad I = 0,$$

$$h_{I,I+1} = -\frac{\sqrt{(L+I+1)(L-I)}}{\cos^2\alpha}\left(\frac{\partial}{\partial\theta} + (I+1)\cot\theta\right), \quad \text{for} \quad I = 1,2\ldots L,$$

$$h_{1,0} = \frac{\sqrt{2L(L+1)}}{\cos^2\alpha}\frac{\partial}{\partial\theta}, \quad \text{for} \quad I = 1,$$

$$h_{I,I-1} = \frac{\sqrt{(L-I+1)(L+I)}}{\cos^2\alpha}\left(\frac{\partial}{\partial\theta} - (I-1)\cot\theta\right), \quad \text{for} \quad I = 2, 3 \ldots L. \quad (5)$$

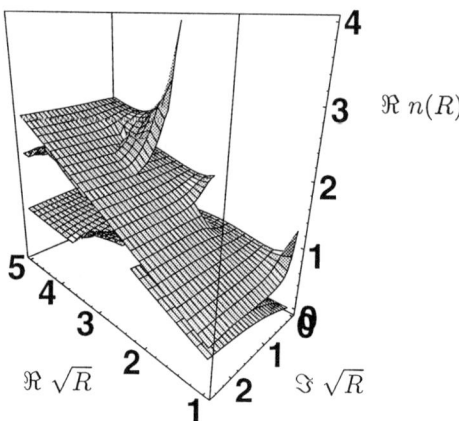

**FIG. 2.** Plot of the real part of $n(R) = 1/[-\varepsilon(R)]^{1/2}$ vs. $R^{1/2}$ for two electrons in the field of a proton.

The coupled partial differential equations $hf = \varepsilon(R)R^2 f$ are then solved approximately using the basis-spline method. This method employs polynomials of order N on an interval in the range of a coordinate $\alpha$, say. The polynomials centered on different intervals of the range and their $N - 2$'th order derivatives are required to match at collocation points where the functions overlap. This gives a non-diagonal representation of derivatives with respect to $\alpha$ and $\theta$ and a diagonal representation of $C(\alpha, \theta)$. The programs of Bottcher and co-workers [16] are used to compute the first derivative matrices. The Hamiltonian matrix in Eq.(5) is then constructed from these elementary matrices and diagonalized using the LAPACK library to find $\varepsilon_n(R)R^2$ for complex $R$. We employ 7'th order polynomials on a lattice of 9 uniform intervals in the variable $0 \leq \theta \leq \pi$ and 11'th order polynomials on a lattice of 23 uniform intervals in $0 \leq \alpha \leq \pi/2$ to set up the Hamiltonian matrix. For $L = 0$ and 1 the diagonalization programs run sufficiently fast on a Sparc 5 workstation to get accurate results in an acceptable amount of time. For $L > 1$, it is necessary to use a parallel machine to achieve the desired performance goals. The different branches of the function $\varepsilon(R)$ are joined at branch points to form the Riemann surface. The adiabatic energy $\varepsilon(R)$ is then use to construct a Riemann surface on which the function $\varepsilon(R)$ is single-valued.

To analyze the Riemann surface for $\varepsilon(R)$ we plot the function $n(R) = 1/\sqrt{-\varepsilon(R)}$ vs $R$ in Fig.(2). In this plot one sees a series of branch points extending to infinite distance. These are the top-of-barrier branch points that correspond to correlated electron motion such that the electrons are localized in the Wannier configuration $\mathbf{r}_1 = -\mathbf{r}_2$. Notice the broad, flat, sloping region extending to infinite $R$. In this region, the motion is that of bound harmonic oscillators in both the $\alpha$ and $\theta_{12}$ coordinates. The corresponding energy eigenvalues $\varepsilon(R)$ have the harmonic oscillator structure given by the asymptotic expansion

$$\varepsilon_{\text{asy}}(R) = C_0/R + C_1/R^{3/2} + C_2/R^2 + C_3/R^{5/2} + \ldots \quad (6)$$

and where

$$C_0 = -\frac{4Z-1}{\sqrt{2}}, \quad C_1 = -(2n_\theta + I + 1)2^{-1/4} - i(2n_\alpha + 1)2^{-5/4}\sqrt{12Z-1}, \quad (7)$$

where $I$ is the projection of the electron angular momentum $\mathbf{L}$ onto the vector $(\mathbf{r}_1 - \mathbf{r}_2)$, $n_\theta$ and $n_\alpha$ are harmonic oscillator quantum numbers, and the higher order terms represent anharmonic corrections. We consider that the initial 1s channel is connected with the final ionization state by the flat harmonic oscillator region seen in Fig.(2). We find that for $L = 0$ and 1 there is such a connection, but that for higher $L$ the first top-of-barrier branch point connects the harmonic oscillator region only with excited initial states. For $L = 2$ and $S = 0$ the 1s adiabatic hyperspherical energy eigenvalue is degenerate at $R = 0$ with a hyperspherical energy eigenvalue that connects to an excited $n = 2$ state of the hydrogen atom. In this case, there occurs the analogue of united-atom rotational coupling in proton-hydrogen collisions so that there is a transition from the 1s channel to a channel with a top-of-barrier branch point. Ionization occurs via a transition at $R \approx 0$ followed by propagation to infinite distances through the harmonic oscillator region. The $S$-matrix element in the hidden crossing theory involves the action integral

$$\exp\left[i\int_{R_0}^\infty [K(R) - K_0(R)]dR\right] \quad (8)$$

Here $K(R) = \sqrt{2[E - \varepsilon(R)]}$, $K_0(R) = \sqrt{2(E + C_0/R)}$. This integral is evaluated along a curve that starts at the turning point $R_0$ on the real axis, circles the first top-of-barrier branch point and goes to infinity through the harmonic oscillator region. At some large value of $R$ called $R_Q$ the exact function $\varepsilon(R)$ is replaced by its asymptotic value $\varepsilon_{\text{asy}}(R)$ of Eq.(6). Since the asymptotic expression has no branch points, the integral from $R_Q$ to $\infty$ can be taken along a path that returns to $R_0$ and goes to infinity along the real axis. The integration path is shown in Fig.(3)

The $S$-matrix element is then written as the product

$$S = \exp\left[i\int_{R_0}^{R_Q}[K(R) - K_{\text{asy}}(R)]dR\right]$$
$$\times \exp\left[i\int_{R_0}^\infty [K_{\text{asy}}(R) - K_0(R)]dR\right] \varphi(R_E, \alpha, \theta_{12}) \quad (9)$$

where $K_{\text{asy}}(R) = \sqrt{2[E - \varepsilon_{\text{asy}}(R)]}$, $\varphi$ is the top-of-barrier wave function, and the angles $\alpha$, $\theta_{12}$ now refer to the angular coordinates of the six-dimensional wave vector $K_\infty = \sqrt{2E}$. The radius $R_E$ equals $4C_0/E$. The ionization cross section is obtained by summing over all appropriate paths, and integrating

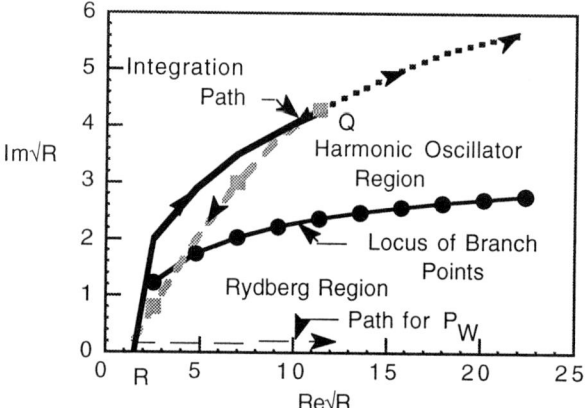

**FIG. 3.** Plot showing the integration path in the complex $R$-plane to determine the ionization cross section. The locus of branch points separates the Rydberg region from the harmonic oscillator region. The integration path (solid path) for the ionization channel starts at $R_i$ on the real axis and goes around the first Top-of-Barrier branch point to infinite distances through the harmonic oscillator region. At $R_Q$ the exact $\varepsilon(R)$ is replaced by the first few terms in its asymptotic expansion $\varepsilon_{\text{asy}}(R)$ (dotted path). Then the integration path may be distorted to return to $R_0$ (dashed path) and then to infinity along the real axis. Integration of the asymptotic expression along this latter path determines the function $P_W(E)$.

the squared magnitude over $\alpha$ and $\theta_{12}$. For the ionization cross section for a given spin we obtain

$$\frac{d\sigma^S}{dE} = \frac{\pi}{2E+1}\sum_L (2L+1)P_L^S(E)P_W^{(L,S)}(E), \tag{10}$$

where the computation of $P_L^S(E)$ is described below and the "Wannier" function $P_W^{(L,S)}(E)$ is given by

$$P_W^{(L,S)}(E) = \left|\exp\left[i\int_{R_0}^{\infty}[K_{\text{asy}}(R) - K_0(R)]dR\right]\right|^2$$
$$\times \int_0^\pi \int_0^{\pi/2} |\varphi(R_E,\alpha,\theta_{12})|^2 d\alpha \sin\theta_{12} d\theta_{12}$$
$$\approx \text{const} E^{(2n_\alpha+1)\zeta_W}, \qquad E \to 0. \tag{11}$$

where $\zeta_W$ is the Wannier index equal to 1.1269 for ionization of neutral atoms by electron impact. For total angular momenta $L$ equal to 0 and 1, computations of the hyperspherical $\varepsilon(E)$ for complex $R$ shows that the initial state connects to the harmonic oscillator region at a top-of-barrier branch point. In this circumstance there are always two paths around the branch point which contribute. In the semiclassical sense, one path $c_1$ represents ionization as the

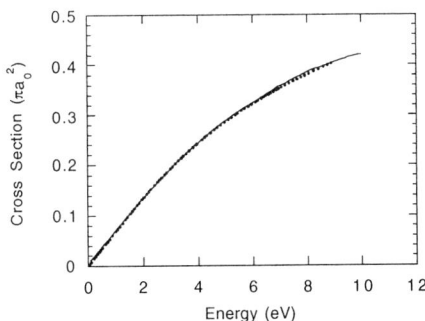

**FIG. 4.** Cross section for the ionization of atomic hydrogen by electron impact. The solid curve is a calculation based on the hidden crossing theory, and the dots are the experimental data of Ref.(18). The theory has been multiplied by 0.93 to agree with the measurements at 9 eV in order to show the agreement between the measured and computed energy variation of the cross section.

incident electron approaches the atom and one path $c_2$ represents ionization as it leaves. $P_L^S(E)$ in then given by

$$P_L^S(E) = \left| \exp\left[ i \int_{c_1}^{R_Q} K(R) dR \right] + \exp\left[ i \int_{c_2}^{R_Q} K(R) dR \right] \right|^2. \tag{12}$$

For $L$ greater than 1, the initial state and harmonic oscillator region are not connected by a top-of-barrier branch point. The initial $^1D$ state is, however, indirectly connected. At $R = 0$ the initial $(\ell_1, \ell_2) = (0, 2)$ $^1D$ channel mixes with the $(1, 1)$ channel since the eigenvalues are degenerate [19]. The $(1, 1)$ state does have a top-of-barrier branch point so that the harmonic oscillator region is reached by a transition near $R = 0$ followed by a transition to the harmonic oscillator region. Letting $p$ denote the transition probability for the $(0, 2)$ state to transfer to the $(1, 1)$ state near $R = 0$ we have for $P_L^S(E)$ in this case the result

$$P_L^S(E) = p \left| \left[ i \int_{R_t}^{R_Q} K(R) dR \right] \right|^2 \tag{13}$$

where $R_t$ is the classical turning point for the $(1, 1)$ channel. With the exception of the $^3F$ state, no other states contribute significantly to the ionization cross section in the hidden crossing theory. Our results for the total ionization cross section are compared with the experimental results of McGowan and Clarke [20] in Fig.(4).

The cross sections agree well in shape, but the theory exceeds experiment by 7%. In order to show the agreement in shape we have multiplied the theory by a factor of 0.93 to normalize to the experimental results at 9 eV. There

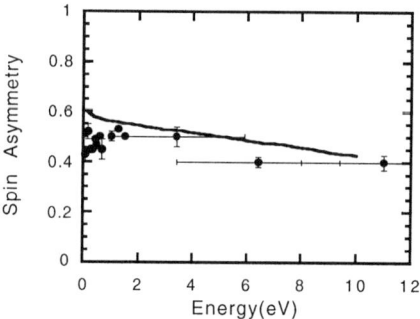

**FIG. 5.** Comparison of the measured spin asymmetry (points with error bars) of Ref.(21) for electron impact on atomic hydrogen with the hidden crossing theory (solid curve).

are some uncertainties in the theory, mainly related dynamical phases and the coupling at $R = 0$ for the $^1D$ and $^3F$ contributions, but the comparison shows that the hidden crossing theory gives a good description of the ionization of hydrogen atoms by electron impact.

The spin asymmetry defined as

$$A = \frac{\sigma^{(1)} - \sigma^{(3)}}{\sigma^{(1)} + 3\sigma^{(3)}} \quad (14)$$

is compared with the measurements of Lubell [21,22] and co-workers in Fig.(5). The agreement is good. It is noteworthy that only the $^3P$ and $^3F$ contribute to the triplet cross section. The good agreement between theory and experiment supports particularly the conclusion that the $L = 0, 1$ and 2 partial waves contribute to the singlet cross section but that only $L = 1$ and 3 contribute to the triplet. Since the $^1D$ and $^3F$ contributions are essential for both the total cross section and the spin asymmetry our results show that the correlations at $R = 0$, emphasized by Cavagnero [19], play an important role in ionization.

## ACKNOWLEDGMENTS

Support for collaboration with the Ioffe Physical Technical Institute, St. Petersburg, Russia is provided by the National Science Foundation under grant no. PHY-9213953. This research is sponsored by the Division of Chemical Sciences, U. S. Department of Energy, under Contract No. DE-AC05-84OR21400. Thanks are also due to Jack Wells for teaching us about basis spline techniques and for his invaluable assistance in using basis spline codes.

## REFERENCES

* Permanent Address: Ioffe Physical Technical Institute, St. Petersburg, Russia.
† Managed by Lockheed-Martin Corporation for the U. S. Department of Energy.

1. U. Fano, *Rep. Prog. Phys.* **46**, 97 (1983).
2. G. H. Wannier, *Phys. Rev.* **90**, 817 (1953).
3. A. R. P. Rau, *Phys. Rev. A* **4**, 207 (1971).
4. R. Peterkop, *J. Phys. B* **4**, 513 (1971).
5. H. Klar and W. Schlecht, *J. Phys. B* **9**, 1699 (1976).
6. James M. Feagin, *J. Phys. B* **17**, 2433 (1984).
7. D. S. Crothers, *J. Phys. B* **19**, 463 (1986).
8. C. Bottcher, *Adv. in At. and Mol. Phys.*, **25**, 303 (1989).
9. A. K. Kazansky and V. N. Ostrovsky, *J. Phys. B* **25**, 2121 (1992).
10. M. Brauner, J. S. Briggs, and H. Klar, *J. Phys. B* **22**, 2265 (1989).
11. E. A. Solov'ev, Usp. Fiz. Nauk. **157**, 437 (1989) [Sov. Phys. Usp. **32**, 228 (1989)].
12. L.D.Landau and E.M.Lifshitz. *Quantum Mechanics:Non-Relativistic Theory*, 2nd. ed., (Pergamon Press, Oxford, England 1965).
13. Yu. N. Demkov, Proceedings of invited talks of the V ICPEAC, Leningrad, USSR, July (1967)(Published by Joint Institute for Laboratory Astrophysics, Boulder, Colorado,1968), p 186.
14. M. Pieksma and S. Y. Ovchinnikov, *J. Phys. B* **24**, 2699 (1991).
15. J. H. Macek *J. Phys. B* **1**, 1 (1968).
16. J. Wells, V. E. Oberacker, A. S. Umar, C. Bottcher, M. R. Strayer, J. S. Wu, and G. Plunien, *Phys. Rev. A*, 6296 (1992).
17. Y. Zhou and C. D. Lin, J. Phys. B **27**, 5065 (1994).
18. J. Macek and S. Yu Ovchinnikov *Phys. Rev. A* **50**, 468 (1994).
19. M. Cavagnero, *Phys. Rev. A* **30**, 1169 (1984).
20. J. W. McGowan and E. M. Clarke, *Phys. Rev.* **167**, 43 (1968)
21. M. J. Alguard, V. W. Hughes, M. S. Lubell and P. F. Wainwright, *Phy. Rev. Lett.* **39**, 334 (1977).
22. M. Lubell, *Phys Rev. A* **47**, R2450 (1993).

# (e,2e) Momentum Spectroscopy of Thin Films

P. Storer[†], Y.Q. Cai[‡], S.A. Canney, R. Caprari[♭], S.A.C. Clark[♮] A.S. Kheifets, I.E. McCarthy, S. Utteridge, M. Vos and E. Weigold[♮]

*Electronic Structure of Materials Centre, Faculty of Science and Engineering, The Flinders University of South Australia, Adelaide 5001, Australia*

**Abstract.** Recent developments in (e,2e) momentum spectroscopy of thin films have resulted in the study of a diverse range of solid targets. These studies have revealed the electronic structure of solids in much more detail than has been previously available using this technique. A summary of the developments which have led up to this is presented here. Some details of a spectrometer that represents the state of the art are given. Recent results from this spectrometer are discussed.

## INTRODUCTION

The (e,2e) reaction has been used for many years as a probe of target electronic structure. In particular the reaction is an ideal way to measure target electron momenta, when suitable kinematic conditions are chosen. A schematic of the kinematics is shown in Fig. 1(a), with the incident electron represented by the vector $\mathbf{p}_o$. Using conservation of energy and momentum it is straightforward to calculate the binding energy and the momentum of the target electron before the knockout collision. For an incident electron energy of $E_o$ and scattered and ejected electrons of energy $E_s$ and $E_e$ (momenta $\mathbf{p}_s$ and $\mathbf{p}_e$) respectively, the binding energy $\varepsilon$ of the target electron before knockout is

$$\varepsilon = E_o - E_s - E_e, \qquad (1)$$

and the recoil momentum $\mathbf{q}$, which in the plane wave approximation is equal and opposite to the momentum of the target electron before the collision, is

$$\mathbf{q} = \mathbf{p}_o - \mathbf{p}_s - \mathbf{p}_e. \qquad (2)$$

The momentum transfer for this reaction is defined as

$$K = |\mathbf{p}_o - \mathbf{p}_s|. \qquad (3)$$

Momentum transfer above about $6a_0^{-1}$ is generally required for momentum spectroscopy. Throughout this paper atomic units for momentum ($a_0^{-1}$) will be used. Energies will be in units of electron volts (eV).

The first (e,2e) measurements were on solid targets, which were thin Formvar films [CH2(OCH3)2]. These early measurements by Amaldi *et al* [1] demonstrated the feasibility of measuring target structure using (e,2e). By measuring the relative cross sections for core and valence electrons under two different kinematic conditions, which corresponded to low and high target electron momenta, they were able to show that carbon core state electrons have a much wider momentum distribution than Formvar valence electrons. However, no valence states were resolved in these measurements, as they were taken with an energy resolution of about 150 eV. In solid targets there is considerable interest in valence states since these are strongly influenced by the local bonding structure and the degree of order, and this provides a window into the potential technological application of the material.

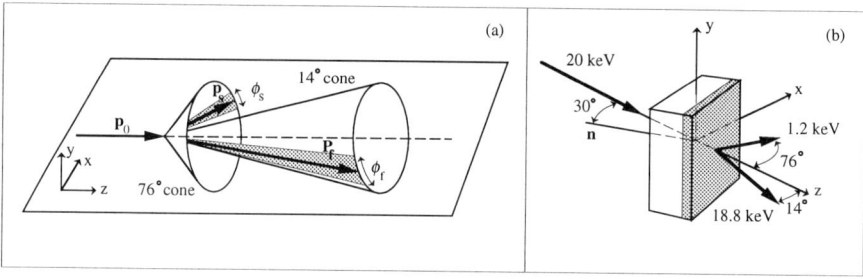

FIG. 1. (a) Kinematics for a transmission (e,2e) experiment. A range of outgoing azimuthal angles are highlighted to show how different target electron momenta can be measured. (b) An illustration of the surface that is studied in a transmission (e,2e) measurement on a thin film. The depth that is probed depends on the energy of the electrons and the angle of the target. This depth is dominated by the energy of the slowest electron that is detected.

Until recently, poor energy resolution has proven to be a major impediment to solid state (e,2e) measurements. Ritter *et al* were the first to resolve different valence states in a solid target [2]. They studied evaporated carbon films and were able to resolve two valence bands. At zero target electron momentum the binding energy of the two bands was 9eV and 23eV. Although they were able to show that the inner valence band dispersed to lower binding energy at higher target electron momentum, they were not able to reveal much more detail since their energy resolution was about 6 eV. In addition to poor energy resolution this measurement had a very low count rate, taking some months for the full measurement.

The difficulty in obtaining good energy and momentum resolution and reasonable counting statistics is common to solid state and gas phase (e,2e) measurements, but for solids the problem is particularly pernicious because of the high density of solid targets. Gas phase (e,2e) studies are typically performed at an incident electron beam energy of about 1keV, and this energy allows sufficiently high momentum transfer that the reaction mechanism is well understood. This means that momentum spectroscopic information can be easily extracted from cross section measurements.

To probe close to zero target electron momentum in (e,2e) on solids the incident beam energy must be significantly higher than that used for gas measurements. This is to allow transmission of the incident electrons through the high density target, minimising multiple scattering. Transmission allows the incoming and outgoing electrons to be coplanar, which is necessary to obtain a cross section for target electrons with zero momentum as illustrated in Fig. 1. To obtain sufficient transmission with even the thinnest targets (ie. avoiding severe multiple scattering problems) requires incident beam energies of greater than 10keV.

Experiments have been proposed and attempted using diffracted beams in reflection from a solid surface. Recently Iacobucci et al have reported some (e,2e) measurements at 300eV incident beam energy in a grazing angle reflection geometry [3]. These measurements from a graphite surface are the first successful grazing incidence measurements to be reported and they establish the feasibility of this technique. The reaction in these measurements occurs between an incident beam electron that is specularly reflected from the graphite surface and the target electron. In these measurements the lowest momentum studied was $0.65a_0^{-1}$ which is just past halfway to the first Brillouin zone boundary in graphite. This experiment shows that grazing incidence (e,2e) is an area which holds promise of some interesting results, if the data aquisition rate can be improved. A higher incident beam energy will be needed if the reaction mechanism is to be easily modelled.

## (e,2e) SPECTROMETERS

In designing an (e,2e) spectrometer for solids it is important to allow for parallel data aquisition of a wide range of the phase space of interest. This has been achieved by using PSD's and electrostatic analysers. By using 1D parallel detection Lower et al [4] and Hayes et al [5] were able to significantly improve on the 4 month data aquisition time of Ritter et al, who used single channel detectors [6]. These 1D multichannel spectrometers were able to obtain reasonable statistics within weeks rather than months, in both cases with reasonable energy and momentum resolution.

Hayes et al used an incident beam energy of 7.5keV, which is about the lower limit for a practical solid state (e,2e) study using transmission, because of multiple scattering in the target. Lower et al operated at 10keV incident energy, once again near the lower limit. At these energies the short mean free paths for the incoming and outgoing electrons mean that even with the thinnest carbon films it is still only 10%-20% likely that a given (e,2e) reaction will occur without one or more further scatterings of the incoming or outgoing electrons.

Clearly multiple scattering favours operation at higher energies. The spectrometer described by Storer et al [7] operates at an incident energy of 20keV incident energy. Two additional features of this spectrometer that differ significantly from earlier spectrometers are the use of 2D position sensitive detection and the use of an asymmetric geometry. Using asymmetric geometry with 18.8keV and 1.2keV outgoing energies means that it is significantly easier to obtain good energy resolution. This is because the high voltage supply for the incident electron beam can also be used for the 18.8keV electron analyser. A smaller (and hence more stable) supply can be used to provide an offset between these two energies. In this way drifts or ripple in the high voltage are easily compensated.

A further advantage is obtained from using asymmetric geometry. As the momentum transfer is decreased the (e,2e) cross section increases significantly, roughly proportional to $K^{3/2}$. As long as the momentum transfer is high enough the measurement is still able to be directly related to the target structure [9]. A standard test of this is to measure the momentum profiles of the valence states of argon. The results of such a measurement in the high energy asymmetric spectrometer are shown in Fig. 2. These results show excellent agreement with the theoretical argon momentum profiles (folded with the experimental momentum resolution) under the plane wave impulse approximation. In addition the excellent momentum resolution is indicated by the strength of the minimum in the $3p^{-1}$ cross section, which is theoretically zero at zero momentum.

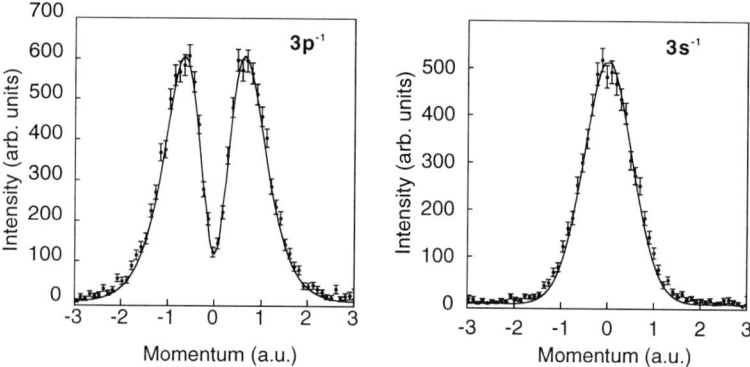

**FIG. 2.** The momentum profiles of the argon $3p^{-1}$ state at 15.7 eV and the main $3s^{-1}$ state at 29.3 eV, from a gas phase measurement with incident energy of 20 keV, compared to a PWIA calculation. The PWIA calculation includes the effect of the finite experimental acceptance angles on the momentum resolution using a standard folding technique. Both sets of data are normalised to theory, using a single normalisation factor.

Recently an electron monochromator has been developed to produce the incident beam for the spectrometer. This is a standard hemispherical deflector style monochromator which is floated at -20kV and produces an incident beam of 100nA into a spot of 150$\mu$m diameter. Preliminary results using this monochromator show that it significantly improves the coincidence energy resolution. This is determined by measuring the energy width of the 1s core state in carbon (Fig. 3). The momentum profile of this state is well described by the momentum profile of the 1s state of atomic carbon and it shows no dispersion in energy [10]. Measurements of the core state of an annealed amorphous carbon target show an energy linewidth of 1.4eV full width at half maximum (FWHM). This gives an estimated energy resolution of under 1eV for the spectrometer, after deconvoluting the linewidth of the carbon core state, compared to about 2eV before monochromation of the incident beam. By lowering the pass energies of the two outgoing electron analysers (presently set at 100 and 200eV for the scattered and ejected electrons, respectively) it should be possible to obtain resolution of around 0.5eV FWHM.

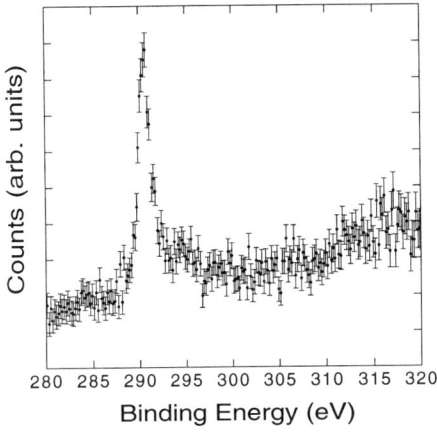

**FIG. 3.** The $1s^{-1}$ carbon core state measured for a 4.5nm thick evaporated carbon target, to determine the coincidence energy resolution and the absolute binding energy scale.

## TARGET PREPARATION

Considerable effort has been made in the area of target preparation, and this is an important part of any solid state (e,2e) spectrometer. In photoelectron spectroscopy it is relatively straightforward to produce a bulk crystalline target with a clean crystalline surface by processes of cleaving, sputtering and annealing targets under UHV conditions. For transmission (e,2e) spectroscopy it is not possible to produce suitable targets by cleaving under UHV conditions without further processing, such as plasma etching. Suitable techniques of target preparation are plasma etching, evaporation and annealing. Using electrochemical and chemical etching it is possible to thin targets to a preliminary stage, suitable for further thinning *in situ* using plasma etching. Sputtering may be useful in the early stages of target preparation, but is unlikely to be useful at the final stages because the thickness of damaged layers in the target is of the order of the final thickness that is required.

The asymmetric spectrometer described earlier has target preparation facilities in two vacuum chambers connected in series. Targets can be transferred between these chambers under vacuum. The chamber furthest from the main spectrometer chamber has a dual function. It serves both as a "dirty" etch chamber, and as a chamber where etch gas pressures of the order of Torr range can be used. The buffer provided by the intermediate chamber (low $10^{-9}$Torr) allows relatively fast transfers into the UHV main chamber (operating in the low $10^{-10}$Torr range). The intermediate chamber has an Auger/LEED system and an annealing stage. Targets can be prepared in the intermediate chamber by evaporation onto one surface of a thin free standing film.

Coincidence counts have been obtained from thin crystals of silicon, prepared by a process of chemical etching followed by plasma etching. At this stage a count rate of up to 20 counts/minute has been obtained, but there have been difficulties due to charging of the target. At these count rates reasonable statistics could be obtained within a month or so. With some refinements in the target preparation technique much higher count rates should be achievable.

Evaporation of materials onto an amorphous carbon substrate has proven to be a convenient way of producing amorphous targets suitable for transmission (e,2e) studies. Materials studied in this way include aluminium, silicon, germanium, copper and $C_{60}$. In the case of the silicon target it was possible to produce SiC targets by subsequent annealing of the C/Si layers.

The thickness of the overlayer required for an (e2e) study is determined by the mean free path of the slower outgoing electron. For silicon the elastic mean free path is 1.5nm and the inelastic mean free path is 2.2nm for a 1.2keV electron. This means that the signal from the carbon backing is attenuated by about a factor of 250 by a 3.5nm Si overlayer [11]. Note that the effective thickness of this overlayer is 5nm since the slow electron leaves the target at about 45° to the surface. The maximum thickness of any overlayer is limited by the electron mean free paths in the material used, and decreases with increasing atomic weight. Successful studies of Ge overlayers show that it is practical to study at least the first row of transition elements [12].

## RESULTS

By far the easiest material to produce as a thin film for transmission is e-C and it is the most studied material in solid (e,2e) spectrometers. The reason for this is that it is a low atomic weight material which easily forms thin films. The low atomic weight means that carbon has a relatively low electron scattering cross section which means the problem of multiple scattering is minimised. It can be thinned by in situ oxygen/argon plasma etching down to 2nm thickness when supported on an electron microscope grid. In addition, carbon has the advantage of being a relatively stable material that can endure the relatively long periods required to acquire data in an (e,2e) measurement without significant contamination. By studying this material earlier researchers were able to obtain reasonable results from spectrometers operating at lower than optimal energies.

Carbon is a particularly interesting element because it presents itself in a diverse range of allotropes, including graphite, diamond as crystalline structures, moelcular fullerenes and amorphous carbon in the form of soot, glassy carbon, evaporated carbon and tetrahedral amorphous carbon. Not surprisingly the amorphous forms of carbon have all been characterised as having a localised bonding structure that falls somewhere between the extremes of diamond and graphite, with tetrahedral and trigonal bonding patterns respectively. Of particular technological interest are the diamond-like forms of carbon, since it is possible that they can be applied in the areas of electronics and protective coatings.

In a recent study of diamond-like amorphous carbon it has been demonstrated that there is a form of diamond-like amorphous carbon known as tetrahedral amorphous carbon (ta-C) which is relatively stable and has an electronic structure that is more diamond-like than graphitic [13]. This material forms a graphitic surface layer after high temperature annealing ($\approx 900°C$), but the graphitic layer can be removed by oxygen/argon plasma etching. Measurements of the spectral momentum density of ta-C before and after annealing are shown in Fig. 4. In this figure the experimental data is compared to angularly averaged theoretical calculations of crystalline graphite and diamond band structure. The experimental data has been corrected for multiple scattering by deconvolution of a multiple plasmon energy loss function.

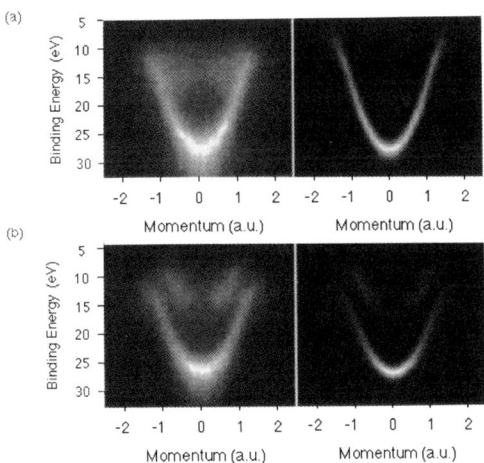

**FIG. 4.** (e,2e) data from the ta-C sample after Ar/O plasma etching (a), and after annealing (b). These data sets have been deconvoluted to reduce the contribution of multiple scattering effects and are compared to LMTO calculations which have been spherically averaged and folded with experimental resolution functions.

Electron momentum spectroscopy of graphite has shown good agreement between measurement and the band structure and momentum distributions predicted by theory [8]. Graphite is a layered material with van der Waals bonding between layers. The bonding within layers is represented by planar trigonal $\sigma$ bonds, and delocalized $\pi$ bonds formed between the $p_z$ orbitals of each carbon atom. In directions in the plane of bonding all the energy bands show the characteristic free electron dispersion curve expected for electrons in a solid. In the direction perpendicular to the plane there is almost no dispersion. This is as expected since the electrons are not free to move in this direction and hence they show an atomic-like binding energy level.

Fig. 4(b) is to be compared with the angular average of the graphite band structure. In the theoretical plot the $\pi$ band strength shows up above the inner valence band dispersion curve. This $\pi$ band strength can be clearly seen in the annealed diamond-like carbon data, showing that the electronic structure is graphitic. When the annealed diamond-like carbon is etched in an argon/oxygen plasma, the graphitic surface layer is removed. This can be seen in Fig. 4(a), where the $\pi$ intensity is no longer present. In this case the dispersion of the peaks is closer to the diamond band structure than to graphite [13].

A material which has been studied as an overlayer on an e-C substrate is amorphous silicon. The nature of the band structure of amorphous silicon (a-Si) was discussed many years ago by Ziman [14]. He observed that the width of an a-Si band would be greatest in the regions of the greatest dispersion. It is in these regions that the differences between the dispersion in different crystal directions are greatest, so a spherical average of the silicon band structure will yield the greatest width. In contrast the width of the a-Si band structure was predicted to be narrowest at zero momentum, since the almost flat nature of the band structure at zero momentum means that an angular average will not widen the band. This prediction that the band structure would be sharp in the energy direction at zero momentum is not verified by the results of the (e,2e) experimental study on a-Si [11]. It is not likely that this is due to poor resolution of the spectrometer, since this was independently measured to be better than 2eV, while the energy width of the a-Si band at zero momentum is about 3.5eV. However, as far as the dispersion is concerned, the results show a general agreement between the measured a-Si spectral momentum density and an angular average of crystalline silicon band struture.

When an overlayer of Si on e-C was annealled at 900°C the (e,2e) results showed that an SiC structure had formed [15]. This material gives a clear illustration of basic solid state theory since it has a band gap in the valence region. The band gap appears because the size of the SiC Brillouin zone is about half the size of the silicon Brillouin zone. As the valence band disperses through the zone edge the periodicity of the electron wavefunction matches that of the Si and C atomic separation, which lowers or raises the binding energy depending on the density concentration either at or between the atomic cores. The random orientation does not destroy the gap in the case of polycrystalline

SiC because the differences between different directions are sufficiently small. Fig. 5 shows the (e,2e) results from C, Si and SiC. The SiC band structure lies between that of C and Si and the band gap is clearly visible.

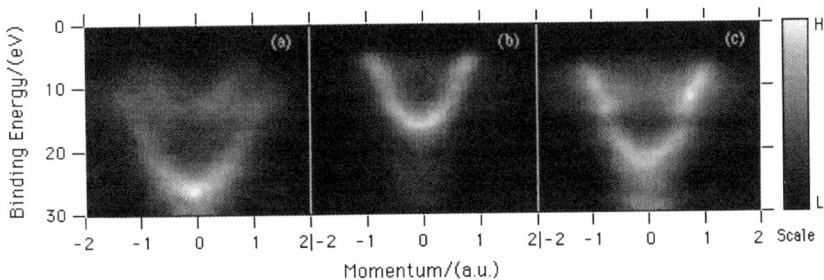

**FIG. 5.** (e,2e) data from an annealed e-C target, an amorphous Si overlayer, and from the SiC sample after annealing the C/Si film. The data have been deconvoluted to reduce the contribution of multiple scattering effects.

## FUTURE DIRECTIONS

It is clear from the present research that (e,2e) on solids has not yet reached its full potential. Some improvements can be made in the area of theory, data analysis and spectrometer design. It is likely that these will all work together to advance the present state of our knowledge of the electronic stucture of solids. In the area of theory it is important that correlation effects are included in the calculation of the (e,2e) cross section. In addition theory could be developed for an amorphous solid, rather than for the angular average of a crystal.

In the area of data analysis it is important that multiple scattering effects are accurately dealt with. It is reasonable to expect that this can be modelled accurately with a Monte Carlo technique, since the elastic and inelastic cross sections are available. This would allow a more accurate assessment of the contribution of satellite structure or correlation effects to the measured spectra. At present it is not possible to unambiguously separate satellite structure from multiple scattering effects. Monte Carlo modelling is currently being developed at Flinders University.

In the area of spectrometer design there are improvements available by increasing the incident beam energy and the energy and momentum resolution. Higher energies will mean that in some cases target preparation will become easier, while in other cases it will become possible. There are two possible approaches to making use of a higher incident beam energy in an (e,2e) spectrometer. On one hand the momentum transfer could be held at around $10a_0^{-1}$, while on the other hand the momentum transfer could be increased.

With the fixed momentum transfer approach the reaction cross section remains constant, so there is no need to increase the incident beam current. This is advantageous when trying to maintain good energy resolution. However there will still be significant multiple scattering of the slower outgoing electron. In the other approach the momentum transfer could be increased. This will require an increase in the incident beam current because of the decrease in (e,2e) cross section, but the effect of multiple scattering will be reduced. In addition the measurement will become more sensitive to the bulk of the target. In such a machine the option of increasing surface sensitivity would still be available by rotating the target so that the slower electron leaves at a shallow angle to the surface.

(e,2e) momentum spectroscopy of thin films has reached a point where a diverse range of targets have been studied in some detail. Even so there is still a considerable amount of research to be done and it is likely that this technique will be making significant and interesting contributions for some years to come.

**REFERENCES**

† Present address: MCI Ltd., 40 Maple Ave, Forestville, SA 5035, Australia.
‡ Present address: UVSOR, Institute for Molecular Science, Myodaiji, Okazaki 444, Japan.
* Present address: Defence Science and Technology Organisation, PO Box 1500, Salisbury SA 5108, Australia.
? Present address: PO Box 12, Buffalo Narrows, Saskatchewan 50M2TO, Canada.
℘ Present address: Research School of Physical Sciences and Engineering, Institute of Advanced Studies, Australian National University, Canberra, ACT 0200, Australia.

1. Amaldi, U. jr., Egidi, A., Marconero, R., and Pizzella, G., Rev. Sci. Instr., **40**, 1001-1005 (1969).
2. Ritter, A.L., Dennison, J.R., and Jones, R., Phys. Rev. Lett., **53**, 2054-2057 (1984).
3. Iacobucci, S., Marassi, L., Camilloni, R., Nannarone, S., and Stefani, G. Phys. Rev. B, **51**, 10252-10259 (1995).
4. Lower, J., Bharathi, S.M., Chen, Y., Nygaard, K.J., and Weigold, E., Surf. Sci., **251/252** 213-218 (1991).
5. Hayes, P., Williams, J., and Flexman, J., Phys. Rev. B, **43**, 1928-1936 (1991).
6. Ritter, A.L., Dennison, J.R., and Dunn, J., Rev. Sci. Instr., **55**, 1280-1289 (1984).
7. Storer, P., Caprari, R.S., Clark, S.A.C., Vos, M., and Weigold, E., Rev. Sci. Instrum. **65**, 2214-2226 (1994).
8. Vos, M., Storer, P., Canney, S.A., Kheifets, A.S., McCarthy, I.E., and Weigold, E., Phys. Rev. B **50**, 5635-5644 (1994).
9. Avaldi, L., Camilloni, R., Fainelli, E., and Stefani, G., J. Phys. B: At. Mol. Phys., **20**, 4163-4169 (1987).
10. Caprari, R.S., Clark, S.A., McCarthy, I.E., Storer, P., Vos, M., and Weigold, E., Phys. Rev. B **50**, 12078-12083 (1994).
11. Vos, M., Storer, P., Cai, Y.Q., Kheifets, A.S., McCarthy, I.E., and Weigold, E., J. Phys.: Condens. Matter **7**, 279-288 (1995).
12. Cai, Y.Q., Storer, P., Kheifets, A.S., McCarthy, I.E., and Weigold, E., Surface Science, *in press*, 1995.
13. Storer, P., Cai, Y.Q., Canney, S.A., Clark, S.A.C., Kheifets, A.S., McCarthy, I.E., Utteridge, S., Vos, M., and Weigold, E., submitted to J. Appl.Phys., (1995).
14. Ziman, J.M., J. Phys. C: Solid State Phys., **4**, 3129-3134 (1971).
15. Cai, Y.Q., Vos, M., Storer, P., Kheifets, A.S., McCarthy, I.E., and Weigold, E., Phys. Rev. B **51**, 3449-3457 (1995).

# POSITRON AND MUON COLLISIONS

Why Positron Physics is Fun .................................................. 369
    R.J. DRACHMAN
Collisions of Positrons and Positronium
with Atoms and Molecules .................................................. 385
    G. LARICCHIA
Overview of the Present Theoretical Status
of Positron-Atom Collisions ................................................ 397
    H.R.J. WALTERS, A.A. Kernoghan, and M.T. McAlinden
Spin Relaxation Phenomena in Muon Collision
Processes in Gases ............................................................ 413
    D.G. FLEMING, M. Senba, J.J. Pan, and D.J. Arseneau

# Why Positron Physics is Fun

Richard J. Drachman

*Laboratory for Astronomy and Solar Physics*
*Goddard Space Flight Center, Greenbelt, Maryland 20771*

**Abstract.** In this Review I will describe some properties of the positron ("antimatter") in interaction with ordinary matter at low energies, in order to explain why positron physics inspires such devotion and enthusiasm in its practitioners. Positron scattering is much like electron scattering, may involve unusual bound states and resonances, often satisfies simple dispersion relations, and can usually be analyzed without invoking the Pauli principle. The experimental production of low-energy positron beams has been an unusual cross-disciplinary challenge, and, finally, the positron dies in a high-energy burst of photons which can carry diagnostic information about atoms, molecules, and astrophysical environments.

## INTRODUCTION

For years there has been a small but loyal group of physicists and chemists, both theoretical and experimental, who have found pleasure and enlightenment in studying the positron. This particle is, of course, just the antiparticle of the electron, and as such should have properties identical to those of the electron, with the notable exception that the sign of its electric charge is opposite. In an antimatter world there would not be anything different to see: antihydrogen atoms would look just like our familiar hydrogen atoms. But here on earth, where protons are positive and electrons are negative, the intrusion of a single positron can cause a variety of intriguing phenomena and can pose interesting theoretical questions. In addition, positron annihilation can be a tool for the understanding of practical systems: angular-correlation experiments probe the Fermi surface of solids, PET scans of the human brain localize thought processes, and annihilation radiation from outer space carries information about otherwise inaccessible parts of the Universe.

The surprise is, however, that not every atomic physicist is aware of the wide array of amusing and interesting things that can be done with the positron. I hope, in this talk, to communicate a sense of excitement about positron physics to those who are not yet working in this field while at the same time reminding myself and my positron colleagues of the reasons why we are.

The outline of the Review is as follows: First, I will present a summary of the unique features, both theoretical and experimental, of the positron and its interactions with atomic systems. This will not be exhaustive in any sense, but I hope it will give a feeling for the flavor of positron physics. In subsequent sections I will discuss in greater detail the physics and astrophysics of positron scattering and annihilation, emphasizing my personal favorites. I think you will see that, although much work and ingenuity has gone into this field, many exciting things still remain to be done.

It must be emphasized at the outset that annihilation, although spectacular and unique, is very improbable when compared with the usual atomic processes. For this reason, it is possible to treat the low-energy positron as an ordinary, non-relativistic, positively-charged particle with the same mass as the electron; later, the annihilation can be included as a very weak perturbation. Thus, we can write the non-relativistic Hamiltonian for a positron interacting with an atom as the sum of the following three terms:

$$H_A = -\sum_{i=1}^{Z} \left[ \nabla_i^2 + \frac{2Z}{r_i} \right] + \sum_{i<j}^{Z} \frac{2}{|\vec{r}_i - \vec{r}_j|},$$

$$H_+ = -\nabla_x^2,$$

$$H' = \frac{2Z}{x} - \sum_{i=1}^{Z} \frac{2}{|\vec{x} - \vec{r}_i|}.$$

(1)

(Here $\vec{r}_i$ is the coordinate of the $i^{th}$ electron and $\vec{x}$ is the positron coordinate, all measured from the position of the nucleus of charge Z, assumed fixed.) Please notice that this Hamiltonian is practically the same as the usual one for electron scattering from an atom, except for the change in the sign of $H'$. This "minor" difference has several major implications!

## UNIQUE THEORETICAL FEATURES

First and most obvious, the positron feels a net short-range repulsion as it approaches and penetrates the atom (assumed to be unperturbed), while an electron would feel an attraction. This makes it unlikely that any bound state of a positron and a neutral atom could exist. (But the first attempt at calculating a scattering cross-section is usually the Born approximation, and in that approximation the overall sign of the perturbing potential is irrelevant.) In second order, however, the atom is polarized by the incoming particle, and an effective potential is induced that falls off like $1/x^4$ and is attractive for both electrons and positrons. This simple discussion is at the root of one of the features of positron scattering: the first- and second-order effective potentials are of opposite sign and tend to cancel, so one must compute quite carefully to avoid inaccuracy. In addition, there is a sort of Ramsauer effect, since the effects of the two opposite potentials cancel (for the S-wave) at a certain low energy. This makes physical sense; the long-range attractive potential is most important for low-energy positrons, while the short-range one dominates for the higher-energy particles that penetrate closer to the nucleus. The first really good calculation [1] in this energy region used the Kohn variational principle with a correlated trial function of Hylleraas type to get S-wave phase shifts for $e^+$-H elastic scattering:

$$\Psi(\vec{r}, \vec{x}) = \chi(\vec{x})\phi(r) + e^{-\alpha r - \beta x} \sum_{ijk} C_{ijk} x^i r^j |\vec{x}-\vec{r}|^k \qquad (2)$$

All the features mentioned above appear in this pioneering calculation, shown in Fig.1, and later calculations have improved the accuracy only slightly. In fact, this now classic problem has been used as a benchmark to test newer calculation techniques.

Secondly, the fact that H' has an attractive term makes it possible to form positronium whenever energetically possible. (Positronium is the "isotope" of hydrogen in which the proton is replaced by a positron; because of the effect of reduced mass its spectrum is hydrogenic with all energies reduced by 1/2.) Above an energy of 1/2 Rydberg (6.8 eV), just beyond the range of Fig.1, this re-arrangement begins to occur in hydrogen. It is interesting and difficult to calculate the cross-section for this process, chiefly for kinematic reasons; it is even harder for systems with more than one electron.

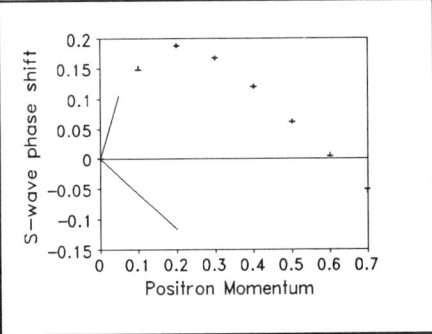

FIGURE 1. Positron-hydrogen S-wave phase shift in the elastic region. Vertical markers are from [1], and horizontal markers are the best modern values. The negative-slope line is the result omitting polarizability.

The third special property of the Hamiltonian of Eq.(1) is that the positron is distinguishable from the electrons, and hence the wave function does not have to be antisymmetrized with respect to it. This brings about some computational simplifications, and it makes the dispersion theory practical for positron scattering [2].

## UNIQUE EXPERIMENTAL FEATURES

There are two sources of annihilation radiation that can be observed, depending on whether positronium formation is energetically forbidden or allowed. By observing two properties of the radiation one can learn much about the target atom and its environment: the annihilation cross-section (or lifetime) and the angular correlation of the two annihilation gamma rays (or the line shape of one). For energies below threshold annihilation may occur when the positron contacts an atomic electron, and above threshold positronium formation and decay becomes the dominant source of annihilation radiation. The Ps atom (to give it an appropriate chemical symbol) has two independent modes of annihilation: ortho-positronium (o-Ps), the triplet spin state, decays into three photons and is "long-lived" (mean life $1.4 \times 10^{-7}$ seconds), while para-positronium (p-Ps), the singlet spin state, decays into two photons with a mean life of $1.24 \times 10^{-10}$ seconds.

For years there were only two possible kinds of experiments involving positrons interacting with atoms or molecules, both of which depended directly on the observation of annihilation photons. Both used radioactive positron sources, whose energy was, of course, not controllable. The first [3] was a simple time measurement. A positron would be allowed to enter a chamber in which high-density gas was confined. The entrance of the positron started a

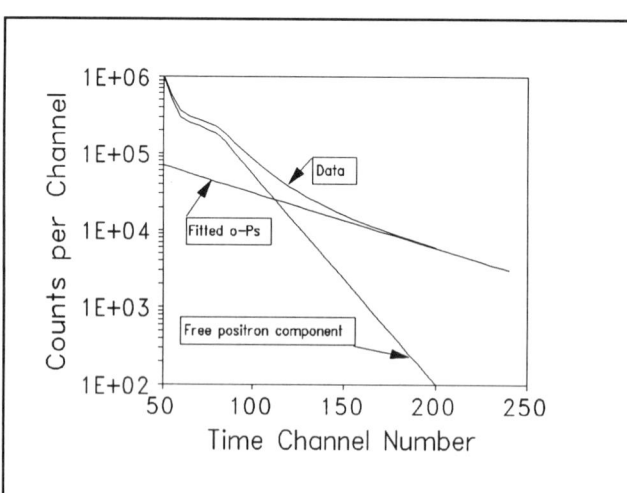

**FIGURE 2.** Typical data for argon with channel width = 1.92 ns. (Re-drawn from Ref.[3]).

clock, and after some time annihilation occurred, and the clock was stopped by the detection of one of the annihilation gamma rays. The distribution of time intervals was recorded as shown in Fig.2. There is clearly some relation between the time of annihilation and the energy the positron had when it was annihilated, since it slows down continuously after entering the medium. Qualitative features of the curve (the "shoulder" and "plateau") can be identified and they can be translated, at least qualitatively, into information about the positron-atom thermalization cross-section. (But more quantitative understanding of the processes requires solution of an equation that describes diffusion in energy.) At later times the spectrum becomes a superposition of two exponentials; one measures the lifetime of o-Ps formed in the gas and the other measures the annihilation cross-section of free positrons in the gas at thermal energy.

The second type is a correlation measurement between the two gamma rays emerging from the annihilation region. They are approximately 180° apart because of momentum conservation, but small angular deviations occur due to the motion of the positron-electron pair at the instant of annihilation. If the positron is completely thermalized, then the angular correlation essentially measures the Fourier transform (with respect to the momentum components perpendicular to the line of sight) of the two-particle wave function when the positron and electron are in contact. It is more difficult to measure the shape of the annihilation line (due to motion in the line of sight), but this has also been done. I will come back to this later, in connection with the observation of astrophysical annihilation radiation.

Much ingenuity went into these two types of measurements, but even with the superposition of an electric field (to increase the equilibrium energy of the positrons) it was not possible to make all the detailed measurements on positrons that were routine for electrons. What was needed was a good low-energy controllable positron beam. For electrons that was easy. An early attempt to make low energy positrons by thermalizing a positron beam from an accelerator in a foil resulted in a surprising outcome. Although most of the incident positrons did annihilate in the foil, a reasonable number were re-emitted with a narrow energy distribution around a mean energy of about 1 eV. I remember the excitement this discovery caused among the experimenters and also among the solid-state theorists who tried to make sense of this unexpected phenomenon. Eventually the effect was explained as due to potentials existing near the surface of the solid and coming from the electron distribution inside, combined with the image charge induced by the positron. A variety of materials and surface coatings were tried, and there are now standard sources of low-energy positron beams. Experience has gradually increased the efficiency and the flux of positrons in these sources, while "brightness-enhancement" has made narrow, parallel, well resolved beams available at controllable energy.

With such sources available, a growing number of real scattering experiments were carried out. At first these were total cross-section measurements, done in poor geometry, so it was necessary to make corrections for a wide cone of forward scattering. These have now been followed by differential cross-section determinations as well as measurements of Ps formation and ionization; these will be touched on later. To a large extent, positron-atom scattering has become just an ordinary part of experimental atomic physics.

The interesting point about these developments is, however, that positron scattering on the hydrogen atom, the process best known theoretically, is not at all easy to investigate experimentally. For years helium was the target of choice for the positron experimentalists, while it was still difficult to calculate anything really accurately. Gradually, theory and experiment have approached each other in both these elementary systems, as I will show later.

Finally, there is the challenge posed by the special bound states and resonances that have been predicted by theory. The $Ps^-$ ion, consisting of two electrons and a positron, is a test case for theoretical three-body computations; it has been observed, but its binding energy has not been measured, and its lifetime was measured only roughly. A tour de force would be to produce and observe the $Ps^+$ ion, its antiparticle, but even the present high-brightness beams would have difficulty focusing two positrons simultaneously onto a small enough spot. Similarly, the molecule $Ps_2$ would be interesting to produce, but difficult. I will discuss positronium hydride [PsH] later.

## ASTROPHYSICAL ANNIHILATION

About 15 years ago, the first reliable balloon-borne observation of gamma rays from the direction of the Galactic Center [GC] was made [4]. They showed unequivocally that positron-electron annihilation radiation was being received from that direction. The radiation was weak, with a flux of only about $10^{-3}$ photons/cm$^2$-sec, but if the source is actually at or near the GC it is quite powerful. Since that historic balloon flight many subsequent observations have been made, and the results found have been interesting, puzzling, and controversial. As a NASA employee and a positron enthusiast I have followed and contributed to the development of this new astronomical window as much as possible.

The GC region is particularly interesting for astronomy. It is the center of mass of the whole Galaxy, with a considerable local concentration of matter, but it cannot be seen in visible light because of the large amount of dust that lies in our line of sight. Any radiation that does manage to get through the obscuration is, therefore, valuable. There are indications (mainly from microwave and infra-red observations) that there may be a black hole at the GC, although if it exists it would not be as massive and active as those that seem to lie in the center of some other galaxies. What information (in addition to the total annihilation rate at the source) can we hope to obtain? While trying to answer this question we have re-examined all the basic processes involving positrons and the most abundant atoms, molecules, and ions. For that reason, the study of astrophysical annihilation is an excellent outline for much general positron physics.

The most straightforward way (in principle) to study the physics of an inaccessible region like the GC (or in solar flares, which also sometimes emit annihilation radiation) would be to set up an experimental simulation of the expected conditions there, inject positrons, and observe the radiation emitted. With the notable exception of the beautiful experiments of Brown *et al* [5] this has not been done; the astronomically low density of scatterers is not easy to simulate in the laboratory. (In Ref.[5] this problem was partially overcome by confining the positron swarm with a magnetic field.) So for the most part it has been necessary to use theory to obtain the basic cross-sections and other needed properties of the positronic systems and then to apply them to the astrophysical environment.

It is generally assumed in the analysis of Galactic positron annihilation [6] that high-energy positrons are injected into a neutral or partly-ionized cloud consisting mainly of atomic and/or molecular hydrogen. The source of the positrons is presumably some energetic object, perhaps the accretion disk of a black hole. (As an atomic physicist I have always left the specification of the source up to the "real" astrophysicists, although I will mention some recent speculations later.) Our problem is to calculate or measure every possibly relevant process involving positrons, free electrons, and hydrogen (atoms, molecules, or negative ions). There is also some helium, the other main constituent of the universe, but most analyses omit this. In molecular clouds there is also a significant amount of dust; from my point of view this just complicates things, but realistically, it must be taken into account too [7]. The idea is to follow the positron as it decreases in energy until it finally ends its life by annihilating. The question is exactly what the positron was doing at the moment of its annihilation because that determines the properties of the radiation that eventually reaches us. In Fig. 3 I show what the observations look like. This is a stylized re-drawing of the actual situation; the errors in the individual points were quite large. Here, a least-squares fit to the radiation spectrum is shown, while the background is dotted. In astrophysics you will often see continuum radiation like this fitted to a power-law spectrum by convention; in this case it seems to fit well. The two-photon annihilation line can be seen; it is narrow (about 3 keV FWHM, most of which is instrumental) and is quite accurately centered at the laboratory energy of 511 keV. (This is important; the small Doppler shift shows that the source of radiation is at rest relative to the earth.) To the left of the peak is the 3-photon contribution from o-Ps. We try to reproduce the properties of the observed spectrum by

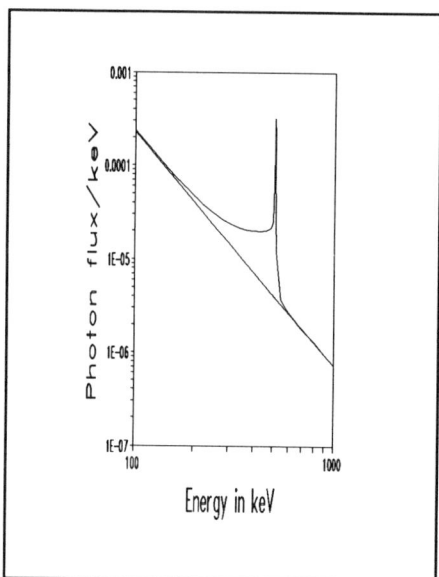

**FIGURE 3.** Spectrum of annihilation radiation from the GC. (Redrawn from Ref.[4]).

constructing a scenario describing the life history of the positrons, using the most accurate information available on cross-sections and annihilation parameters. We can use our imagination in this effort, but the final energy and time dependence of the radiation must be in reasonable agreement with experiment.

For example, one might assume that the annihilation region consisted of neutral atomic hydrogen at low temperature. Then the arriving positrons would slow down, mainly by ionization and excitation, until they reached about 200 eV; above this energy the best available cross-sections for Ps formation are small, and annihilation in flight is totally negligible. Below this energy the Ps formation process competes more and more strongly, until at about 30 eV there is equal probability that the next collision will be inelastic scattering or Ps formation. This slowing-down process was treated by a Monte Carlo technique in Ref.[6], but it is also fairly well described using a continuous slowing-down approximation [8]. In this approximation one half of the initial positrons have formed Ps (and rapidly annihilated) by the time they have reached an energy of 39 eV.

What Doppler shift of the photon energy does this correspond to? Since its binding energy is 6.8 eV, Ps formed at this point has a translational energy of 32 eV. The velocity of the Ps atom for this energy is $\frac{v}{c} = \sqrt{\frac{E}{mc^2}}$, so the maximum shift in the photon energy is $\Delta = \pm\sqrt{Emc^2}$, corresponding to a full width of the line of 8 keV. All this has been examined in more detail in an earlier review [9], giving a line width of 6.75 keV after all the appropriate kinematics is folded in; this agrees pretty well with the results of Ref.[6]. At this stage only very rough values of the various cross-sections need to be known, because this predicted line width is at least twice that observed.

But all the work discussed above used highly imaginative estimates for all the cross-sections, since nothing was really known except for the elastic region. One of the most exciting recent developments is the gradual appearance of reports on the measurement of various inelastic cross-sections in atomic hydrogen: Ps formation [10], impact ionization [11], and the total

[12]. At the moment several different determinations of the impact ionization cross-section are significantly in disagreement, they are not quite what was used in the earlier analysis, and it will eventually be of astrophysical interest to know how things turn out.

The next part of the scenario is quite interesting. Depending on exactly how probable inelastic processes are as compared with Ps formation, a certain fraction of the positrons can make their last collision and fall below the Ps threshold at 6.8 eV. (Using the rough cross-sections which were available at the time, mostly taken to be the same as those for electrons, it was found [6] that 5% fall into this category.) Once that happens they are subject to only two possible processes: elastic scattering and annihilation in flight with the atomic electron. The first of these has a large cross-section, of order $\pi a_o^2$, and it dominates the second which is of order $\alpha^4 \pi a_o^2 \frac{c}{v}$. So these positrons lose energy very slowly and finally annihilate, giving a second line width component, about 1.3 keV, determined mostly by the momentum distribution of the atomic electron. (Some details of this will be given later.) But please note that essentially all of these annihilations give rise to two photons, while only 1/4 of the Ps annihilations do. This 5% component contributes 20% to the line intensity, and it narrows the effective width considerably.

The final point to be made concerns the apparent time variation of the annihilation radiation. If it is real it implies a time variation and compact structure of the source and a fairly high density of the medium. Suppose the source turns on and off in a short time; there is some evidence that just that happens with the source 1E 1740.7-2942, familiarly called "The Great Annihilator". It was observed [13], only once, to increase in intensity and maximum energy for about one day, and it is thought to be a transient source of positron-electron pairs. If these pairs enter what seems to be a molecular cloud [14] there could be an interesting time history of the radiation line shape. It might initially be broad, due to the Ps produced during slowing down, and subsequently become narrow as the remnant positrons fall below the Ps threshold to annihilate in flight. All this is practically pure speculation!

This is only one of the possible scenarios that one might consider. In particular, if an accurate measurement of the ratio of 3$\gamma$ to 2$\gamma$ intensity can be made then one will know how much Ps is being formed. Different processes give different amounts of Ps. For example, low-energy annihilation in H gives practically no Ps. If there is an appreciable density of free electrons present, then either direct annihilation (giving only 2$\gamma$ annihilation) or radiative recombination to Ps (giving a 3:1 ratio) can occur. Which process dominates depends on the temperature. Then there are the two more exotic processes, $e^+ + H^- \rightarrow Ps(NL) + H$ and $e^+ + H_2 \rightarrow PsH + H^+$, which may be significant; these will be discussed below.

# SOME PROCESSES AND HOW TO CALCULATE THEM

The last section was intended to persuade you that positrons have some real significance in the Universe outside the laboratory. I will now come back to a description of several special positron processes and how they can be treated theoretically.

## Angular Correlation, Line Width, and Annihilation Rate

If we know the wave function $\Psi(\vec{r}_1, \vec{r}_2, \vec{x})$ describing scattering of a positron $(\vec{x})$ from a two-electron atom or ion, all three of these observables can be computed from one basic expression which is

$$\Lambda_N(\vec{q}) = \left| \int d^3x\, d^3r_1\, \Psi(\vec{r}_1, \vec{x}, \vec{x})\, \phi_N^*(\vec{r}_1)\, e^{i\vec{q}\cdot\vec{x}} \right|^2 \quad (3)$$

Here $\Lambda_N(\vec{q})$ is proportional to the annihilation cross-section, N is the quantum state of the one-electron system left after annihilation, and $\vec{q}$ is the total momentum of the two annihilation gamma rays. In this simple, physically reasonable formula we have set $\vec{r}_2 = \vec{x}$ on the assumption that annihilation occurs at contact; this is an approximation, valid because $\lambda_c$ is much smaller than $a_0$. A well-known constant multiplies this expression[1].

Usually, we are not interested in such a complete expression for the annihilation. In fact, there has not yet been a measurement of the final state of the remaining electron(s). Making use of the completeness relation for the final-state eigenfunctions

$$\sum_N \phi_N(\vec{r})\, \phi_N^*(\vec{r}') = \delta(\vec{r}-\vec{r}') \quad (4)$$

we can write an expression for the probability of emitting two photons of total momentum $\vec{q}$, without regard for the final state:

$$\Lambda(\vec{q}) = \int d^3r \left| \int d^3x\, \Psi(\vec{r}, \vec{x}, \vec{x},)\, e^{i\vec{q}\cdot\vec{x}} \right|^2 \quad (5)$$

The usual observable, as I mentioned before, is the distribution of momentum either along the line of sight or perpendicular to it, so we must integrate over

---

[1] I had trouble convincing my colleague Joe Sucher that this actually works.

the other two momentum components. If the positron is at rest when it annihilates the result is

$$P(q_z) = \int_{q_z}^{\infty} q\,dq\,\Lambda(q^2) \tag{6}$$

As an example let us use the simple, approximate (unnormalized) helium wave function, $e^{-Z(r_1+r_2)}$, with $Z=27/16$, and assume a plane wave for the positron. Then we find the following expression for the one-dimensional distribution, normalized to 1 for $q_z=0$:

$$P(q_z) = \left(1 + \frac{q_z^2}{Z^2}\right)^{-3} \tag{7}$$

It is quite easy to work out both the angular correlation of the two gamma rays and the Doppler width of the gamma ray line.[2] The FWHM of the angular correlation for helium is 12.56 mrad in this approximation, and the FWHM of the line is 3.2 keV. These overestimate by about 35%; this can be understood qualitatively by noticing that the Coulomb repulsion of the nucleus prevents the positron from sampling the higher momentum parts of the wave function, which lie near the nucleus. In the same approximation for hydrogen, we get 7.44 mrad and 1.9 keV for the same two quantities, overestimating in this case by about 45%. The line width obtained from a variational scattering treatment [15] for hydrogen, $\Gamma_{FWHM}=1.3\,\text{keV}$, is the value I used in the astrophysical scenario above to describe the contribution of those positrons which thermalized and died in collision with neutral hydrogen atoms.

The total annihilation cross-section is then obtained by integrating Eq.(5) over $d^3q$ and applying the resulting $\delta(\vec{x}-\vec{x}')$ to produce the following simple formula for the so-called effective electron number:

$$Z_{eff} = 2\iint d^3x\,d^3r\,|\Psi(\vec{r},\vec{x},\vec{x})|^2 \tag{8}$$

(The factor of 2 really comes from the two ways in which the positron can annihilate, with either electron. For multi-electron atoms it is also necessary to consider whether the positron is in a singlet state with each of the electrons.) Then the cross-section is $\sigma_a = Z_{eff}\alpha^4\pi a_o^2 \frac{c}{v}$. The definition of $Z_{eff}$ is such that for a non-interacting, plane-wave positron it is equal to the number of electrons Z (for systems with zero spin or when averaged over

---

[2] Although I got the latter wrong by a factor of 2 for hydrogen the first time!

positron polarizations.) The attractive polarization potential acting on the positron makes this quantity larger than Z in almost all cases; for hydrogen at zero energy [16] it is 8.868 instead of 1, and for helium the theory [17] gives 4.01 instead of 2 while a good experimental value is 3.94±.02 at an average energy slightly above zero [3].

## Positronium Hydride and Its Formation

PsH is one of the best known exotic systems from a theoretical point of view. The binding energy of its single particle-stable state is known [18] to be 1.0597 eV, and its lifetime is very close to four times that of p-Ps; these facts suggest that the system looks like a symmetrized, loosely bound combination of ground-state Ps and H atoms. But from a molecular viewpoint PsH looks like an $H_2$ molecule, one of whose protons has been replaced by a positron. For a long time I believed and said frequently that it would not be possible to make PsH by simply bombarding $H_2$ with positrons, since the difference between the two masses implied that the positron would not be able to transfer enough energy to the target proton to eject it. In that case, it would be necessary to do something different, if we wanted to make some PsH in the laboratory.

An obvious possibility would be radiative capture (recombination?) of very low-energy $e^+$ and $H^-$. It is clear, however, that the charge-exchange reaction producing Ps and H would dominate, since it is non-radiative and has an atomic-sized cross-section. In fact, it began to be more interesting to investigate [19] the Ps-formation process, which might even contribute to astrophysical annihilation. I tried to interest someone in doing a merging-beam experiment, so that the relative velocity of the two ions would be low in order to take advantage of the $v^{-2}$ dependence of the cross-section coming from the exothermic nature of the process and the Coulomb attraction. I even tried to persuade people interested in making antihydrogen by collision of $e^+$ and $\bar{H}$ that $H^-$ would be an excellent stand-in for $\bar{H}$ and would be easier to produce. But until recently, when Schrader and some of his colleagues became interested, I had no luck. I am looking forward to their results in the future. For astrophysics, the total cross-section is of interest. Perhaps of equal or greater interest is a measurement or reliable calculation of the branching ratio, at low incident energy, of the formation of Ps in each of the states that are energetically accessible. Even at zero energy the 1s, 2s, and 2p states of Ps can be formed; a different outgoing kinetic energy of the Ps yields a different Doppler width of the subsequent annihilation lines. This process is of quite high interest at present.

In fact, however, I was too pessimistic about collisional formation of PsH, since I have difficulty thinking like a molecule. Although direct collision with one of the nuclei would indeed be ineffective, resonant states might exist. That is, a positron-$H_2$ excited state might be formed and subsequently separate into PsH and $H^+$. This should be an interesting theoretical study, and a proposal to make such a measurement has appeared [20]. PsH itself has been produced in methane and detected by energy-balance considerations [21], but so far none of its detailed properties (lifetime, angular correlation, final-state excitation) has been investigated.

## Resonances

I once wrote that "no-one likes a cross-section that is too smooth", and a referee refused to let me publish that phrase; he felt that some people might disagree. There is little doubt, however, that bumps in a cross-section are interesting and that they draw extra attention. It is especially amusing when a target system that is fairly simple should have some sort of intricate scattering structure. I think it is still true that no experimental evidence for any resonances in $e^+$-atom scattering has been found, but there is quite a lot of theoretical interest and activity in this subject.

The earliest discussion of this subject is probably the paper by Mittleman [22], who suggested that the infinite series of resonances predicted to lie below the $n=2$ threshold in $e^-$-H scattering, due to the long-range effects of the s-p degeneracy, should have counterparts in the $e^+$ case. The doubts expressed about this did not depend on the sign of charge itself, but the existence of the Ps-formation threshold below the $n=2$ threshold produced a two-channel scattering problem that worried some of us [23]. Two definitive papers [24] showed that the resonances did exist and established their positions with some accuracy, at least in the non-relativistic approximation. (The Lamb shift breaks the degeneracy, so only a finite number of resonances remain.) Since then, there have been too many accurate computations, extending to higher thresholds of both H and Ps, to be included here [25]. The bottom line is that these resonances are reliably predicted to exist. They are quite narrow, however, and will be difficult to find experimentally.

Another kind of resonance that certainly exists concerns the PsH system discussed above. In addition to the particle-stable ground state there is an infinite series of Rydberg states, constructed out of $[H^- e^+]$ configurations; these look approximately like ordinary hydrogenic excited states. They are, however, unstable against breakup into Ps+H, so they represent resonances in that open channel. It is possible to estimate the shift in the energies of these unstable states and to represent them in terms of complex quantum defects. I would like to see experimental verification of these states, but the experiments are difficult.

Besides these known resonances there are also some questionable ones that have been reported from time to time. Some of them are predicted to occur at energies where the usual formation mechanisms are unlikely to work. A study [26] of one of these using the complex-rotation method [25] failed to verify its existence. Another most interesting broad s-wave resonance [27] seemed to appear above the ionization threshold, but its position varied significantly in different approximations. It is almost certainly spurious and due to incomplete representation of certain open channels [28].

The moral is that one must be very careful when searching for resonances that the physics of the problem is properly represented; otherwise almost anything can turn up.

## CONCLUSIONS

If I have done my job right no formal conclusions should really be necessary. You will have seen that the positron is a unique probe of ordinary matter, that its annihilation carries a kind of information about the target that is not easy to get in other ways. In astrophysics, the annihilation radiation penetrates obscured regions of space and brings us coded news about exotic places; our task as atomic physicists is to learn to break the code. But above all I find the study of exotic bits of matter like the positron (but also the muon) is its own reward; it seems to me to be among the most fascinating parts of physics.

## REFERENCES

1. C. Schwartz, Phys. Rev. **124**, 1468 (1961).
2. A. Tip, J. Phys. B. **10**, L11 (1977); W. E. Kauppila, T. S. Stein, J. H. Smart, M. S. Dababneh, Y. K. Ho, J. P. Downing, and V. Pol, Phys. Rev. A **24**, 725 (1981).
3. G. R. Heyland, M. Charlton, T. C. Griffith, and G. L. Wright, Can. J. Phys. **60**, 503 (1982).
4. M. Leventhal, C. J. MacCallum, and P. D. Stang, Astrophys. J. **225**, L11 (1978).
5. B. L. Brown, M. Leventhal, A. P. Mills, Jr., and D. W. Gidley, Phys. Rev. Lett. **53**, 2347 (1984); B. L. Brown and M. Leventhal, Phys. Rev. Lett. **57**, 1651 (1986); B. L. Brown, M. Leventhal, and A. P. Mills, Jr., Phys. Rev. A **33**, 2281 (1986).
6. R. W. Bussard, R. Ramaty, and R. J. Drachman, Astrophys. J. **228**, 928 (1979).
7. W. H. Zurek, Astrophys. J. **289**, 603 (1985).
8. F. W. Stecker, *Cosmic Gamma Rays*, Washington, D. C., NASA, 1971, p146.

9. R. J. Drachman,"Low-Temperature Positron Annihilation," in *Positron-Electron Pairs in Astrophysics*, edited by M. L. Burns, A. K. Harding, and R. Ramaty, New York, American Institute of Physics, 1983, pp. 242-252.
10. W. Sperber, D. Becker, K. G. Lynn, W. Raith, A. Schwab, G. Sinapius, G. Spicher, and W. Weber, Phys. Rev. Lett. **68**, 3690 (1992).
11. G. Spicher, B. Olsson, W. Raith, G. Sinapius, and W. Sperber, Phys. Rev. Lett. **64**, 1019 (1990); G. O. Jones, M. Charlton, J. Slevin, G. Laricchia, A. Köver, M. R. Poulsen, and S. Nic Chormaic, J. Phys. B **26**, L483 (1993).
12. M. Weber, A. Hofmann, W. Raith, W. Sperber, F. Jacobsen, and K. G. Lynn, Hyperfine Interactions **89**, 221 (1994).
13. L. Bouchet *et al*, Astrophys. J **383**, L45 (1991).
14. J. Bally and M. Leventhal, Nature **353**, 234 (1991); I. F. Mirabel, M. Morris, J. Wink, J. Paul, and B. Cordier, Astron. Astrophys. **251**, L43 (1991).
15. J. W. Humberston and J. B. G. Wallace, J. Phys. B **2**, 1278 (1972).
16. S. K. Houston and R. J. Drachman (unpublished), quoted in A. K. Bhatia, R. J. Drachman, and A. Temkin, Phys. Rev. A **9**, 223 (1974).
17. R. I. Campeanu and J. W. Humberston, J. Phys. B **8**, L244 (1975).
18. Y. K. Ho, Phys. Rev. A **34**, 609 (1986).
19. J. C. Straton and R. J. Drachman, Phys. Rev. A **44**, 7335 (1991).
20. D. M. Schrader, G. Laricchia, and T. N. Horsky, Hyperfine Interactions **89**, 355 (1994)
21. D. M. Schrader, F. M. Jacobsen, N.-P. Frandsen, and U. Mikkelsen, Phys. Rev. Lett. **69**, 57 (1992),
22. M. H. Mittleman, Phys. Rev. **152**, 76 (1966).
23. R. J. Drachman, Phys. Rev. A **12**, 340 (1975).
24. G. D. Doolen, J. Nuttall, and C. J. Wherry, Phys. Rev. Lett. **40**, 313 (1978); L. T. Choo, M. C. Crocker, and J. Nuttall, J. Phys. B **11**, 1313 (1978).
25. Y. K. Ho, Phys. Lett. **120A**, 348 (1984).
26. A. K. Bhatia and R. J. Drachman, Phys. Rev. A **42**, 5117 (1990).
27. K. Higgins and P. G. Burke, J. Phys. B **24**, L343 (1991); T. T. Gien, *ibid* **27**, L25 (1994).
28. A. A. Kernoghan, M. T. McAlinden, and H. R. J. Walters, J. Phys. B **27**, L543 (1994); *ibid*, L211 (1994).

# Collisions of Positrons and Positronium with Atoms and Molecules

## G. Laricchia

Department of Physics and Astronomy, University College London, Gower Street, London WC1E 6BT, UK

**Abstract.** Experimental investigations of positron collisions with atoms and molecules have advanced to near-threshold and differential studies of selected scattering channels, positronium (Ps) formation and direct ionization in particular. Coupling effects among the various channels are also topical. Progress with the production of Ps beams and their application to the first direct studies of Ps collisions with atomic and molecular targets are also reviewed.

## INTRODUCTION

Studies of atomic collisions involving particles and antiparticles projectiles can serve to highlight mass and charge effects and aid the development of accurate theoretical descriptions [1]. Experimental investigations with positron ($e^+$) projectiles have advanced from the study of total cross-sections for room-temperature atoms and molecules [2], to detailed investigations of their interaction with targets such as hydrogen [3],[4] and alkali atoms [5] and of specific scattering channels, including studies differential in the energy and/or angle of the scattered projectile or the emitted particle [6]. In recent years, beams of Ps atoms have also become available [7] and the first direct measurements of total cross-sections of Ps scattering from simple atomic and molecular systems have been performed [8],[9].

In this progress report, some of the new results in these areas will be reviewed. A more detailed overview may be gained by consulting the proceedings of the satellite workshop of this conference [10].

## POSITRONIUM FORMATION

At energies below ~100eV, an important process in the scattering of positrons from gases is electron capture which results in the formation of positronium [11]. At its peak, the Ps formation cross-section ($Q_{Ps}$) accounts for ~50% of the total cross-section ($Q_t$). The peak in $Q_{Ps}$ is expected to occur at projectile velocities comparable to the target electron orbital velocity and experimental results are in broad agreement with this expectation. At higher energies, $Q_{Ps}$ is found experimentally to decrease approximately as $E^{-2.5}$ [12],[13]

© 1995 American Institute of Physics

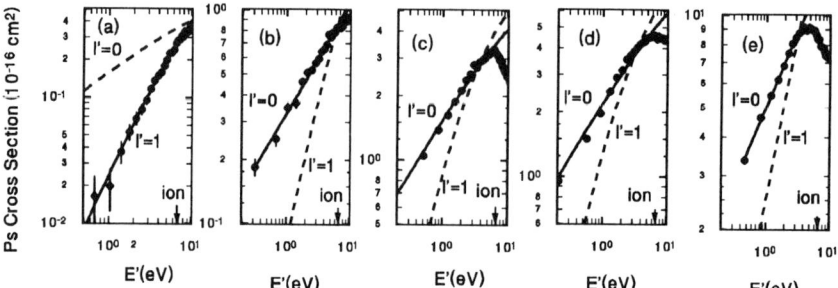

**Figure 1.** $Q_{Ps}$ vs $e^+$ energy (E') above threshold for (a) He, (b) Ne, (c) Ar, (d) Kr and (e) Xe. The lines correspond to $Q_{Ps} \propto (E')^{l'+0.5}$. The magnitude of the dashed curves is arbitrary. From [15].

with most theories [14] predicting powers in the range (3-5). Contributions from Ps(n>1) to the measured $Q_{Ps}$ might, at least partially, be responsible for some of the discrepancies. The study of its energy dependence close to its threshold ($E_{Ps}$) can yield information on the dominant angular momentum of the outgoing Ps and on the behaviour (near-$E_{Ps}$) of the cross-sections for other open channels.

The results of a recent near-threshold investigation for the inert atoms [15] are shown in Figure 1, where E' represents the projectile energy above $E_{Ps}$. This study has found that, in this energy region and at least from 1eV above threshold, $Q_{Ps}$ is dominantly p-wave in character for He and s-wave in all other cases, in qualitative agreement with measurements of the efficiency for collimated Ps production from He, Ar and Xe [16] and theoretical calculations [17]. A more explicit corroboration of these results has been recently found in the case of He by accurate variational calculations, according to which, the s-wave contribution is ~18% of the measured $Q_{Ps}$ at ~1eV above threshold [18].

The observed target dependence of the angular momentum of the outgoing Ps might be related to the strength of the (repulsive) interaction of the $e^+$ with the undistorted target which appears to result in progressively broader angular distributions of the Ps formed from targets of increasing atomic number. The latter hypothesis [19] is in accordance with recent studies on the degree of collimation of Ps formed from $H_2$ [9],[20] and the near-threshold importance of p and d partial waves in $Q_{Ps}$ for H [21],[22].

Upper and lower limits on the magnitude of $Q_{Ps}$ have recently been set for K and Na [23]. Interest in the alkali atoms arises from the fact that Ps formation from these systems is exothermic and their relatively simple atomic structure is amenable to theoretical approximations. Results for K are shown in Figure 2 where comparison with theory [24] suggests that the maximum at ~6eV arises from the formation of Ps(n=2), the latter perhaps accounting for up to 80% of $Q_{Ps}$ at this energy. In contrast, the monotonic increase of $Q_{Ps}$ for Na at low energies is thought to be due to a preponderance of Ps(n=1) [23].

Recently, the first (relative) measurements of $dQ_{Ps}/d\Omega$ for Ar, using crossed beams, have been reported [25]. The results at 75eV incident energy are shown in Figure 3 where they are compared to calculations [17] and to the angular distribution of secondary e⁻ following e⁺ impact ionization ($dQ_i/d\Omega$) [25]. Here, $dQ_{Ps}/d\Omega$ can be seen to be considerably more forward peaked that $dQ_i/d\Omega$ but, whilst good agreement is found with theory at this energy when allowance is made for the experimental angular resolution, discrepancies exist at 30eV where theory [17] has predicted a peak at around 15°.

## POSITRON IMPACT IONIZATION

At intermediate energies and starting from about (20-40)eV above threshold ($E_i$), direct single ionization is the dominant inelastic process in e⁺ - atom scattering. For the inert atoms, its cross-section ($Q_i^+$) peaks at ~100 eV above threshold and there, as discussed below, it exceeds the corresponding e⁻ cross-section by an amount which decreases with increasing atomic number of the target. The excess of $Q_i^+$ by e⁺ over e⁻ impact is thought to arise from the attractive Coulomb interaction between the projectile and the ionized e⁻. If so, the decreasing importance of this interaction with Z could be a result of the correspondingly larger impact parameters which might be involved in e⁺ impact ionization of targets with progressively higher atomic numbers. At low incident energies, $Q_i^+$ by e⁺ impact is found to be smaller than e⁻ impact, probably due to the importance of Ps formation in this energy range.

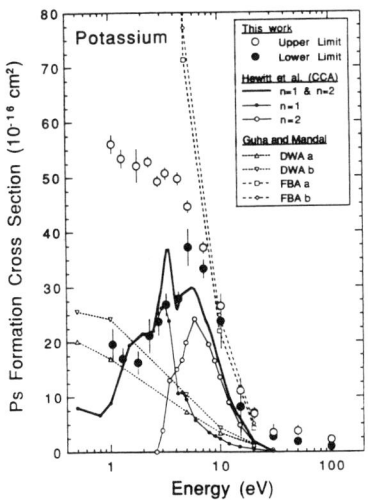
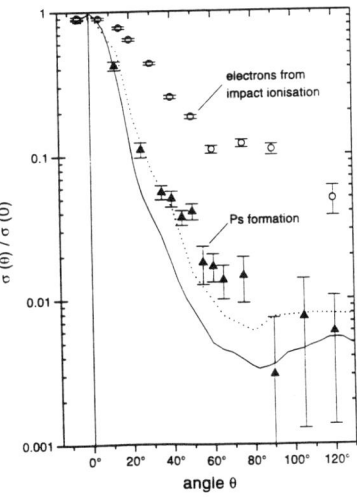

**Figure 2.** Comparison of $Q_{Ps}$ measurements for e⁺ - K scattering with theoretical calculations. From [23].

**Figure 3.** Relative $dQ_{Ps}/d\Omega$ and $dQ_i/d\Omega$ for Ar at 75eV [25]. The dashed line is theory [17] convoluted with the experimental angular resolution [25].

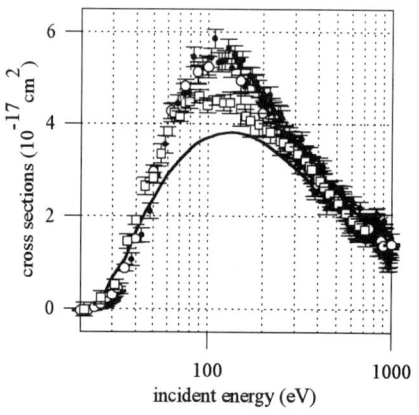
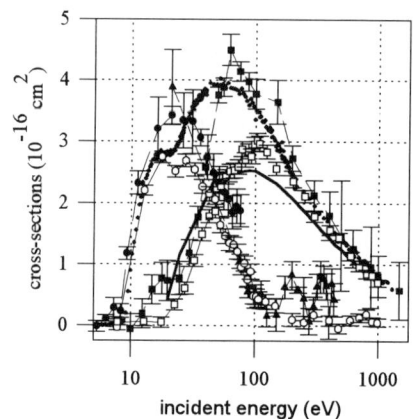

**Figure 4.** Single ionization cross-section for $e^+$ - He scattering: filled circles [27], hollow circles [28], squares [29]. The solid line is the single ionization cross-section for electron impact [32].

**Figure 5.** Cross-sections in $e^+$-Ar scattering. $Q_i^t$, dots [13]; $Q_i^+$, hollow [27] and filled [28] squares; $Q_{Ps}$, filled circles [33], filled triangles [34]; $(Q_i^t - Q_i^+) \sim Q_{Ps}$, hollow circles [13]. The solid line is $Q_i^+$ for $e^-$ [32].

New results are available for the total ionization cross-section ($Q_i^t$, comprising all ion producing processes) for all the inert atoms [13],[26] and the direct single ionization cross-section ($Q_i^+$) for He, Ar, Kr and $H_2$ [27]. In the case of He and $H_2$, reasonable agreement exists among various experimental determinations, as illustrated for He in Figure 4 [27]. At intermediate energies, good agreement is found between the data of Moxom et al [27] and that of Knudsen et al [28], with the results of Fromme et al [29] being ~18% lower around the peak. The latter, however, are in good agreement with distorted wave calculations [30] whilst a CTMC calculation [31] predict a higher peak but at a lower energy. Also shown in the figure is $Q_i^+$ for $e^-$ impact [32] for comparison. In the case of higher-Z targets, considerable discrepancies exist. As an example, the results for Ar are shown in Figure 5. Consistency is found in the energy dependence of $Q_i^t$ of Laricchia et al [13] and $Q_i^+$ of Moxom et al [27] at energies where $Q_i^+$ is expected to be the major contributor to $Q_i^t$. Major discrepancies are found in the case of $Q_i^+$ between the results of Moxom et al [27] and Knudsen et al [28], the latter being up to ~70% higher around the peak. Disagreements are also found between the low energy results for $Q_{Ps}$ of Fornari et al [33] and $(Q_i^t - Q_i^+) \sim Q_{Ps}$ [13] and, at higher energies, between the latter and those of Diana et al [34] whose results exhibit an oscillatory structure. The results for $Q_i^+$ for $e^-$ impact [32] are also shown for comparison. The absolute scale to the measurements of Knudsen et al [28], Laricchia et al [13] and Moxom et al [27] has been assigned by assuming convergence of the direct ionization cross-sections by $e^-$ and $e^+$ impact at the highest energies investigated.

Progress is also being made in the investigation of the near-threshold behaviour of $Q_i^+$ by $e^+$ impact. The measurements of Ashley et al [35] for $Q_i^+$ in

He ($E_i$=24.59eV) are shown in Figure 6 where they are compared to previous determinations [28],[29],[36]. Contrary to previous surmises [28],[36], the near-threshold *energy dependence* of $Q_i^+$, as well as its magnitude, has been found to be significantly different from that by e$^-$ impact. In the range of excess energies (1.3≤E'≤7)eV, the results of Ashley et al [35] can be fitted to a power law $Q_i^+ \propto$ (E')$^n$ with n=(2.1±0.1). These results are in qualitative agreement with extensions of the Wannier threshold theory to e$^+$ impact [37] which predict n=2.65 up to E'~3eV in contrast to the e$^-$ impact case which yields n=1.127. In the case of e$^+$ impact, the quantitative discrepancy between experimental and theoretical results might be due to the different E' ranges considered and, indeed, there is some evidence that closer to threshold the experimentally determined exponent increases. Further measurements in this energy region are in progress and should help to clarify the situation.

New results have also been obtained in the investigation of multiple ionization. Cross-sections for double ($Q_i^{2+}$) and triple ionization ($Q_i^{3+}$) of Ar, Kr and Xe have been measured relative to $Q_i^+$ [38]. Figure 7 shows the ratio of $Q_i^{2+}/Q_i^+$ for Kr by both e$^-$ [39] and e$^+$ [38] impact. The lines represent theoretical estimates for inner shell contributions to double ionization [38]. Here, the difference between the two projectiles is thought to arise from the different accelerations which the particles undergo in the Coulomb field of the nucleus.

Finally, ionization and fragmentation of organic molecules by e$^+$ impact have been observed at energies below the Ps formation threshold and have been attributed to energy transfer from e$^+$ annihilation [40]. Positron attachment to large

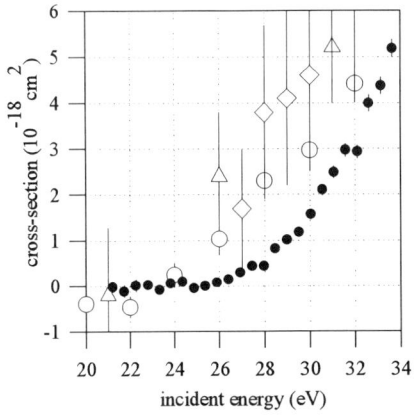
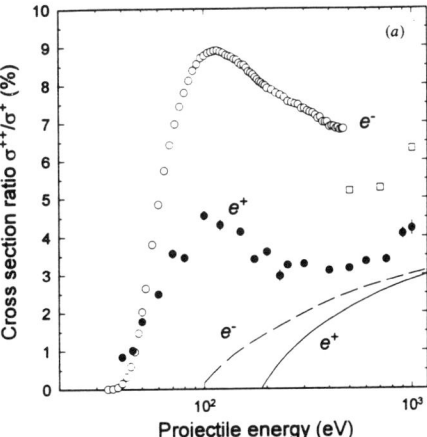

**Figure 6.** Near-threshold single ionization cross-section for e$^+$ - He scattering: filled circled [35], hollow circles [28], triangles [29], diamonds [36]. From [35].

**Figure 7.** $Q_i^{2+}/Q_i^+$ cross-section ratio for Kr by e$^+$ [38] and e$^-$ [39] impact. Curves represent theoretical estimates for inner shell contributions to double ionization [38].

molecules has been observed by using trapped clouds of thermal $e^+$ [41] and studies of the formation of compounds comprising a Ps atom have begun [42]. Differential ionization cross-section measurements of improved sensitivity are in progress at UCL to investigate further the phenomenon known as electron-capture-to-the-continuum (ECC) in which the ejected $e^-$ and the scattered projectile move with a small relative velocity [43]. This process is now expected to give rise to smaller structures in the differential cross-sections for ionization by $e^+$ impact in comparison to those observed by the impact of protons and heavier ions [44]. The dissimilarity is thought to arise from the projectile masses which result in the light $e^+$ being deflected from larger impact parameters with corresponding weakening of $e^+$-$e^-$ correlations and distribution of ECC events over a wide range of angles.

## CHANNEL COUPLING EFFECTS

At the last conference in this series, preliminary measurements of $Q_t$ for $e^+$ - H were reviewed [45]. Detailed measurements are now available [4] which indicate (see Figure 8) the cross-sections for $e^-$ and $e^+$ impact to be merged within 5% between 31 and 302eV. The observed merging, although supported by theory [46], is noteworthy since large differences are expected in corresponding partial cross-sections over this energy range. It has been suggested that strong coupling

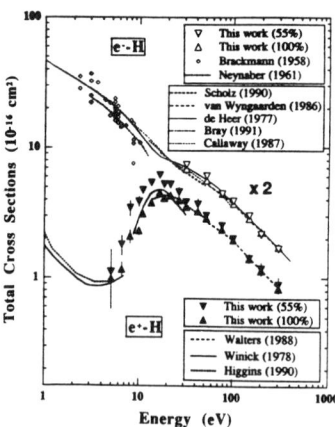
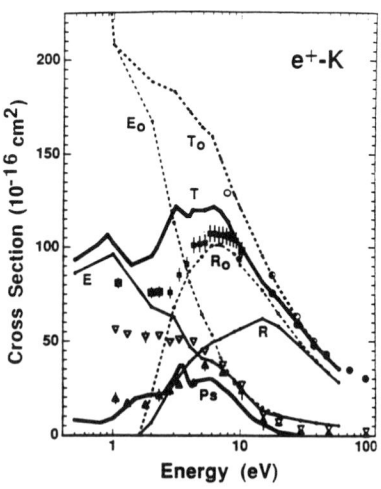

**Figure 8.** Total cross-sections of atomic hydrogen by positron and electron impact. The points are measurements [4] and are compared to various theories as indicated in the captions. From [4].

**Figure 9.** Cross-sections for $e^+$-K scattering. $Q_t$: squares and filled circles; $Q_{Ps}$: triangles [48]; total (T), elastic (E), resonance excitation (R) and Ps formation (Ps) cross-section calculations; subscript o indicates neglect of Ps formation [49]. From [48].

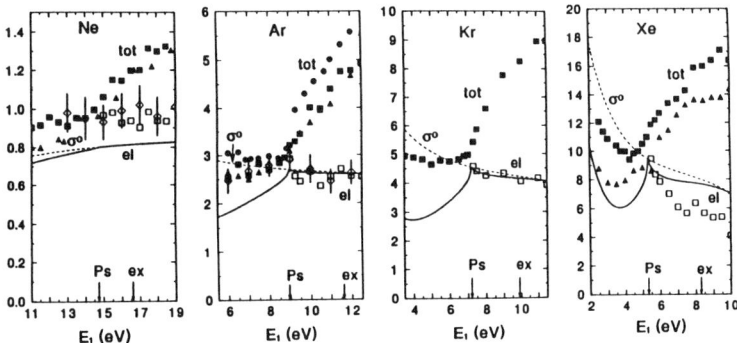

**Figure 10.** Total (tot) and elastic scattering (el) cross-sections for $e^+$ + (noble-gas atom) near the Ps formation threshold as a function of incident energy. The points are experimental results, the solid and broken curves are theoretical predictions for $Q_{el}$ with and without allowance for Ps formation. From [51].

among the various channels is responsible for the almost identical $Q_t$ for the two projectiles [4]. Similar effects are noted in the case of He and $H_2$ [47].

Interference effects between channels have also been observed in the case of Na and K [48]. The case for K is illustrated in Figure 9 where the general shape of the experimental results is reproduced by theory [49] only after allowing for Ps formation, indicating that the latter does not merely constitute an additive scattering channel.

Possible interference effects between elastic scattering and Ps formation have also been examined for the inert atoms [15]. Wigner cusps may be expected to occur in the elastic scattering cross-section ($Q_{el}$) at $E_{Ps}$ if $Q_{Ps}$ starts at threshold with an infinite slope, i.e. if the partial s-wave dominates $Q_{Ps}$. As discussed above, $Q_{Ps}$ has been found, near-threshold, to be predominantly p-wave for He [15] explaining the smooth variation, within experimental resolution, of $Q_{el}$ across $E_{Ps}$ [50]. In the case of the heavier inert atoms, however, the dominant s partial wave contribution to $Q_{Ps}$ suggests that cusps in $Q_{el}$ should be expected. Semi-empirical estimates of the magnitude of these effects, based on new absolute values of $Q_{Ps}$ [13], have been made using R-matrix and threshold theories [51]. Results are shown in Figure 10 and await experimental verification via a direct measurement of $Q_{el}$.

It has been suggested [48] that some collision systems involving $e^+$ allow a better opportunity to study channel-coupling effects which may be also important, although possibly more subtle, in many other collision systems. In the measurements of the differential elastic cross-section of Ar and Kr versus projectile energy at fixed angles, pronounced decreases were found at intermediate energies which were attributed to possible interference effects with ionization [52]. Preliminary measurements [53] of the single ionization cross-section for Ar at 60° suggest that this indeed remains an important channel over the relevant energy region. The absence, however, of similar structures in corresponding measurements of differential elastic cross-sections with e⁻ projectiles, raises the question as to whether these effects can be unique to $e^+$-atom interactions [54].

## POSITRONIUM BEAMS AND SCATTERING

Collimated monoenergetic Ps atoms may be produced by neutralizing a $e^+$ beam either in a gaseous target or, at higher energies, by partial transmission through a C-film [7]. The characteristics of the Ps beam thus produced depend on the neutralizer employed and on the $e^+$ beam itself. In the case of gaseous neutralizers, the production efficiency is directly proportional to $dQ_{Ps}/d\Omega$ integrated over a suitably narrow angular range. The total cross-section of Ps scattering from the neutralizer sets an upper limit on its pressure. The energy resolution of the Ps beam depends on the $e^+$ beam energy spread, the relative production of Ps in its various quantum states [55],[16] and its production simultaneous to other scattering processes [56]. A time-of-flight method [57] has been developed to characterize the Ps beam with respect to incident energy and, to some extent, quantum states [16] and to measure $Q_t$ for Ps scattering from atomic and molecular targets [8],[9].

The total cross-section for Ps-Ar scattering, as shown in Figure 11, has been measured in the range 16-95eV incident energy [8]. After a rapid rise from a value of approximately $9 \times 10^{-20} m^2$ to $15 \times 10^{-20} m^2$ in the 16-30eV range, the

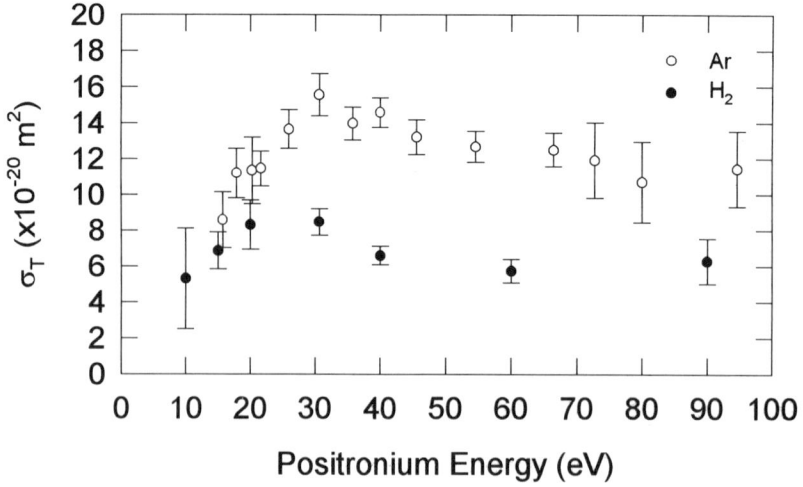

**Figure 11.** Total cross-sections for Ps scattering from Ar, hollow symbols [8] and $H_2$, filled symbols [9].

cross-section is seen to slowly decline to $\sim 12 \times 10^{-20} m^2$ at higher energies. These measurements represent the first direct study of atomic interactions of energetic Ps. Comparison with theory [58] indicates that the rise in $Q_t$ at intermediate energies is the result of increasing inelastic scattering, with projectile ionization being the dominant channel according to some predictions. It is expected that these studies will enhance our understanding of atomic collision in general as well as aiding the description of Ps scattering from more complex systems, including atoms on surfaces [59].

Studies on the production efficiency of Ps beams from gaseous targets had concluded that He resulted in a higher degree of collimation than other inert atoms [16]. This finding was contrary to expectations based on considerations of the magnitude of the ionization energy of target [60] but might be understood, again, in terms of the comparatively weak (repulsive) static interaction between the $e^+$ and the He atom. For the purpose of Ps beam production, the magnitude of $Q_{Ps}$ (as well as its angular dependence) is, of course, important and, in practice, Ar is a better neutralizer than He, $Q_{Ps}(Ar)$ being $\sim 10 \times Q_{Ps}(He)$ at their maximum. So, it is perhaps not surprising that $H_2$, which combines a low-Z with a fairly large $Q_{Ps}$ ($\sim 0.75 \times Q_{Ps}(Ar)$), turns out to be as much as three times as efficient as Ar [9]. Additionally, the dependence of the production efficiency with neutralizer pressure suggested a Ps-$H_2$ total cross-section smaller than that from Ar. Preliminary results for the former [9], also shown in Figure 11, have confirmed this hypothesis.

## CONCLUSIONS

New results in the field of $e^+$ and Ps scattering from atoms and molecules have been reviewed. In the case of $e^+$ scattering, progress has been made particularly in the study of Ps formation and ionization including near-threshold and differential cross-sections. Direct measurements of the total cross-section of Ps from atomic and molecular targets have also been performed, providing a basis for the understanding of atomic interactions of energetic Ps. Some of the results have been here qualitatively interpreted by comparison to corresponding collisions involving other projectiles but it is hoped that more rigorous descriptions, aided by increasing experimental sophistication, will be achieved in the not-too-distant future.

## ACKNOWLEDGMENTS

I wish to thank all my co-workers and those who have provided (p)re-prints of their work for this progress report. I am also grateful to the Nuffield Foundation for the award of a Science Research Fellowship, The Royal Society and the Engineering and Physical Sciences Research Council for supporting the $e^+$ research at UCL.

# REFERENCES

1. eg Schultz DR, Olson RE, Reinhold CO, J Phys B**25**, 4601 (1991); Knudsen H and Reading JF, Phys Rep **212**, 107 (1992)
2. eg Griffith TC, Hyp Int **89**, 3 (1994)
3. Jones GO et al, J Phys B**26**, L483 (1993); Weber M et al, Hyp Int **89**, 221 (1994)
4. Zhou S et al, Phys Rev Lett **72**, 144 (1994)
5. Kauppila WE et al, J Phys B**26**, L557 (1994)
6. eg Kover A, Laricchia G and Charlton M, Hyp Int **89**, 129 (1994); Schmitt A et al, Hyp Int **89**, 133 (1994)
7. eg Laricchia G "Positronium Beams and Surfaces" in *Proceedings of the International School of Physics <<Enrico Fermi>> CXXV Course* (Societa' Italiana di Fisica; Italy) 1995 in press
8. Zafar N et al, Phys Rev Letts, submitted (1995)
9. Garner A and Laricchia G, Can Journ Phys submitted (1995)
10. McEachran RP and Stauffer AD eds, *Proceedings of "Positron Workshop"* Can Jour Phys (1995)
11. eg Charlton M and Laricchia G, J Phys B**23**, 1047 (1990)
12. Overton N, Mills RJ and Coleman PG, J Phys B**26**, 3951 (1993)
13. Laricchia G, Moxom J and Hunter D, in preparation (1995)
14. Mandal P et al, J Phys B**12**, 2913 (1979); Khan P and Gosh AS, Phys Rev A**28**, 2181 (1983); Deb NC et al, Phys Rev A**36**, 1082 (1987); Schultz DR and Olson RE, Phys Rev A**38**, 1861 (1988); Tripathi S, Sinha C and Sil NC, Phys Rev A**39**, 2924 (1988); Bransden BH, Joachain CH and McCann JF, J Phys B**25**, 4965 (1992); McAlinden MT and Walters HRJ, Hyp Int **73**, 65 (1992); Fojon OA et al, Physica Scripta **51**,204 (1995)
15. Moxom J et al, Phys Rev A**50**, 3129 (1994)
16. Laricchia G et al, J Phys B**20**, L99 (1987); Zafar N et al, J Phys B**24**, 4461 (1991); Laricchia G et al, Hyp Int **73**, 133 (1992); Laricchia G (1987) unpublished results.
17. McAlinden MT and Walters HRJ, Hyp Int **89**, 407 (1994)
18. Van Reeth P and Humberston JW, J Phys B (1995) submitted
19. Laricchia G, Nucl Inst Meth (1995) in press
20. Tang S and Surko C, Phys Rev A**47**, 743 (1993)
21. Brown CJ and Humberston JW, J Phys B**18**, L401 (1985)
22. McAlinden MT, Kernogham AA and Walters HRJ, Hyp Int **89**, 161 (1994)
23. Zhou S et al, Phys Rev Letts **73**, 236 (1994)
24. Hewitt RN, Noble CJ and Bransden BH, J Phys B**26**, 3661 (1993)
25. Falke T, Raith W, Weber M and Wesskamp U, J Phys B (1995) in press
26. Moxom J, Laricchia G and Charlton M, J Phys B**28**, 1331 (1995)
27. Moxom J, Ashley P and Laricchia G, Can Jour Phys (1995) submitted
28. Knudsen H, Brun-Nielsen L, Charlton M and Poulsen MR, J Phys B**23**, 3995 (1990)
29. Fromme D, Kruse G, Raith W and Sinapius G, Phys Rev Lett **57**, 3031 (1986)
30. Campeanu RI, McEachran RP and Stauffer AD, J Phys B**20**, 1635 (1987)
31. Schultz DR and Olson RE, Phys Rev A**38**, 1866 (1988)
32. Krishnakumar E and Srivastava SK, J Phys B**21**, 1055 (1988)
33. Fornari L, Diana LM and Coleman PG, Phys Rev Letts **51**, 2276 (1983)
34. Diana LM et al, in *Positron Annihilation*, ed RM Singru and PC Jain, Singapore:World Scientific, 1985 pp428
35. Ashley P, Moxom J and Laricchia G, in preparation (1995)
36. Sueoka O, Jin B and Hamada A, Appl Surf Sci **85**, 59 (1995)
37. Klar H, J Phys B**14**, 4165 (1981); Rost M, Phys Rev A**49** (1995)
38. Helms S et al, J Phys B**28**, 1095 (1995)
39. Syage JA, Phys Rev A**46**, 5666 (1992); Schram BL et al, Physica **32**, 185 (1966)

40. Xu J et al, Phys Rev A**49**, R3151 (1994)
41. Iwata K et al, Phys Rev A**51**, 473 (1995)
42. Schrader DM et al, Phys Rev Letts **69**, 57 (1992); Can Jour Phys (1995) submitted
43. Brauner M and Briggs JS, J Phys B**19**, L325 (1986)
44. Schultz DR and Reinhold CR, J Phys B**19**, L9 (1990); Moxom J et al, J Phys B25, L613 (1992); Beradkar J and Klar H, J Phys B**26**, 3891 (1993); Kover A, Laricchia G and Charlton M, J Phys B**26**, L575 (1993) and J Phys B**27**, 2409 (1994); Sparrow RA and Olson RE, J Phys B**27**, 2647 (1994)
45. Charlton M and Laricchia G, "Progress in positron scattering" in *Proceedings of XVIII Int Conf on the Physics of Electronic and Atomic Collisions*, New York : AIP Press, pp445
46. Walters HRJ, J Phys B**21**, 1893 (1988); van Wyngaarden WL and Walters HRJ, J Phys B**19**, 929 (1986)
47. eg Kauppila W and Stein TS, Adv At Mol Opt Phys **26**, 1 (1990)
48. Kauppila W et al, J Phys B**26**, L557 (1994)
49. Hewitt RN, Noble CJ and Bransden BH, J Phys B**26**, 3661 (1993)
50. Coleman PG et al, J Phys B**25**, L585 (1993); Moxom J, Laricchia G and Charlton M, J Phys B**26**, L367 (1993)
51. Meyerhof WE et al, Can Jour Phys, submitted (1995)
52. Dou L et al, Phys Rev Letts **68**, 2913 (1992); Phys Rev A**46**, R5327 (1992)
53. Finch R et al, Can Journ Phys, submitted (1995)
54. Cvejanovic D and Crowe A, J Phys B**27**, L723 (1994)
55. Laricchia G et al, Phys Letts A**109**, 97 (1985)
56. Laricchia G and Moxom J, Phys Lett A**174**, 255 (1993)
57. Laricchia G et al, J Phys E**21**, 886 (1988)
58. see Charlton M and Laricchia G, Comm At Mol Phys **26**, 253 (1991) for a compilation; Peach G (1993) and Walters HRJ (1994) Private Communications
59. Canter KF in *Positron Scattering in Gases*, ed JW Humberston and MRC McDowell, New York: Plenum, 1984 pp219; Weber MH et al, Phys Rev Letts 61, 2542 (1988)
60. eg Brown BL, *Positron studies of Solids, Surfaces and Atoms*, World Scientific: Singapore, 1986, pp160

# Overview of the Present Theoretical Status of Positron-Atom Collisions

H.R.J.Walters*, Ann A.Kernoghan* and Mary T.McAlinden[+]

*Department of Applied Mathematics and Theoretical Physics, The Queen's University of Belfast, Belfast BT7 1NN, UK
[+]Department of Applied Mathematics and Theoretical Physics, University of Cambridge, Cambridge CB3 9EW, UK, and Gonville and Caius College, Cambridge CB2 1TA, UK

**Abstract.** Recent advances in the theoretical treatment of positron - atom scattering are reviewed with particular emphasis on coupled-state methods. Resonances appearing above the ionization threshold in positron - atomic hydrogen collisions are shown to be artefacts of simple approximations. Reliable results can now be obtained for all the main transitions in positron scattering off ground state atomic hydrogen. Calculations now exist for all the alkali metals from Li to Cs. These show a pronounced growth in excited state positronium formation on going up the sequence but with a corresponding collapse in the Ps(1s) formation cross section. Despite a wealth of experimental data, coupled-state calculations on the noble gases which include positronium formation are rather sparse.

## INTRODUCTION

When a positron collides with, for example, a hydrogen atom the following processes can take place :

$$
\begin{aligned}
e^+ + H(1s) &\longrightarrow e^+ + H(1s) \quad &\text{Elastic scattering} \\
&\longrightarrow e^+ + H(nl \neq 1s) \quad &\text{Atom excitation} \\
&\longrightarrow e^+ + e^- + p \quad &\text{Ionization} \\
&\longrightarrow Ps(nl) + p \quad &\text{Positronium formation}
\end{aligned}
$$

With more complicated atoms other reactions are possible, for example, $Ps^-$ formation, transfer ionization (ie, Ps formation and ionization together) etc.

From a theoretical viewpoint, what really makes positron - atom scattering really different from electron - atom scattering is the existence of the positronium formation channel. Although a rearrangement channel also exists in electron - atom scattering, namely, electron exchange, there is only one natural centre in this problem – the atomic nucleus. By contrast positronium formation is a two - centre problem, the first centre being the nucleus, the second the positronium centre of mass. It is this two - centre nature which makes the theoretical treatment difficult. The charge conjugate of the inverse reaction to positronium formation, ie,

$$\text{Ps}(nl) + \bar{p} \longrightarrow e^+ + \bar{H}(n'l'),$$

where $\bar{p}$ denotes an antiproton, is relevant to the production of anihydrogen[1], $\bar{H}$, and useful technical information on this process may be obtained as a by-product of $e^+$ – H calculations[2-4].

## THEORETICAL METHODS

Space permits us to mention here only the more sophisticated approximations which have been of recent interest.

At low energies there are accurate variational calculations of positron scattering by atomic hydrogen[5-17], helium[18-23] and lithium[24,25]. These results are invaluable in "pinning down" other approximations.

The hyperspherical coordinate approach has been used for positron scattering off atomic hydrogen by Archer et al[26], who looked only at S-wave scattering below the H(n=4) threshold, by Zhou and Lin[27], who have examined S, P, D and F partial waves below the H(n=2) threshold, and by Igarashi and Toshima[28], who have studied scattering up to 30eV and calculated partial waves with total orbital angular momentum J = 0 to 6. The virtue of the hyperspherical method is that it treats the positron and electron impartially, although it seems to become less tractable with increasing angular momentum[28].

However, in our opinion, the greatest advance in recent years has come from the application of coupled-state methods, to which we shall now restrict the discussion. In a coupled-state approximation the system wave function, for positron scattering by atomic hydrogen for example, is expanded as

$$\Psi = \sum_n F_n(\mathbf{r}_p)\psi_n(\mathbf{r}_e) + \sum_m G_m(\mathbf{R})\phi_m(\mathbf{t}) \qquad (1)$$

Here $\mathbf{r}_p$ ($\mathbf{r}_e$) is the position vector of the positron (atomic electron) relative to the nucleus and $\mathbf{R} \equiv (\mathbf{r}_p + \mathbf{r}_e)/2$ and $\mathbf{t} \equiv \mathbf{r}_p - \mathbf{r}_e$ denote respectively the position vector of the positronium centre of mass relative to the nucleus and the positronium internal coordinate. The first sum is over states $\psi_n$ of the hydrogen atom, the second over states $\phi_m$ of positronium. These states may be either eigenstates or pseudostates. It should be noted that were the sets $\psi_n$ and $\phi_m$ complete then either

$$\sum_n F_n(\mathbf{r}_p)\psi_n(\mathbf{r}_e) \quad \text{or} \quad \sum_m G_m(\mathbf{R})\phi_m(\mathbf{t}) \tag{2}$$

alone would give an exact expansion of $\Psi$. Coupled equations for the partial wave components of the $F_n$ and $G_m$ are obtained in the usual way by substituting the partial wave form of (1) into the Schrödinger equation and projecting with $\psi_n$ and $\phi_m$. Two main options then exist, whether to solve these equations in coordinate space or to convert to momentum space. The coordinate space approach has been used by Higgins and Burke[29,30] and Kernoghan, McAlinden and Walters[31-39] who have both adopted the R-matrix procedure, and by Gien and Liu[40-45] who have employed the Harris - Nesbet algebraic variational method. The R-matrix technique is good at low and intermediate energies where positronium formation is important and is efficient at giving a detailed picture of cross sections over the energy range. In the momentum space formulation coupled integral equations are obtained for the T-matrix elements[73] in which the driving terms are first Born amplitudes. As a result the first Born approximation is guaranteed and so this approach has advantages at high energies. The momentum space formulation for positron-atom scattering was pioneered by the Calcutta group[46-59] and is also the basis of the calculations by Hewitt, Noble and Bransden[60-66] and by Mitroy, Stelbovics and co-workers[67-75].

## POSITRON SCATTERING BY ATOMIC HYDROGEN

### Resonances Above the Ionization Threshold

While testing their computer programs using one of the simplest approximations of all, coupled-static, Higgins and Burke[29] discovered a totally unexpected, but very prominent, resonance in the S-wave $e^+$ - H(1s) cross section for Ps(1s) formation, this is shown in figure 1. The resonance is located at 2.62 Ryd (35.6eV), well above the ionization threshold, and has a width of 0.31 Ryd (4.2eV). Coupled-static is the name given to the two-state approximation

$$\Psi = F(\mathbf{r}_p)\psi_{1s}(\mathbf{r}_e) + G(\mathbf{R})\phi_{1s}(\mathbf{t}), \tag{3}$$

where the functions $F$ and $G$ satisfy the coupled equations

$$\begin{aligned}(\nabla_p^2 + k^2)F(\mathbf{r}_p) &= 2V_{1s,1s}(\mathbf{r}_p)F(\mathbf{r}_p) + 2\int K(\mathbf{r}_p,\mathbf{R})G(\mathbf{R})d\mathbf{R} \\ (\nabla_R^2 + p^2)G(\mathbf{R}) &= 4\int K(\mathbf{r}_p,\mathbf{R})F(\mathbf{r}_p)d\mathbf{r}_p\end{aligned} \tag{4}$$

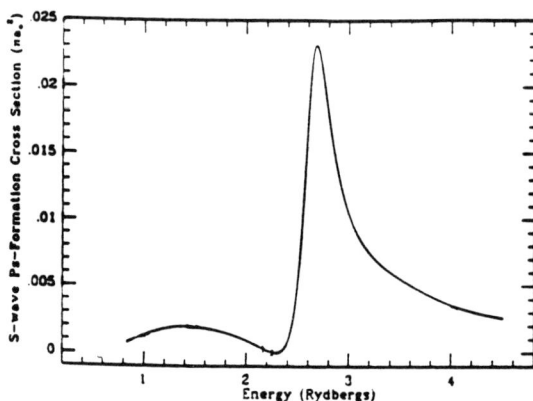

**FIGURE 1.** S-wave coupled-static cross section of Higgins and Burke[29] for Ps(1s) formation.

The direct potential in the atom channel, $V_{1s,1s}$, is the static potential which for positrons is repulsive; there is no direct potential in the positronium channel. The resonance can therefore only be trapped through the action of the non-diagonal positronium formation kernel $K$. For this reason Higgins and Burke termed it a "coupled-channel shape resonance". This type of mechanism is not known in electron-atom scattering and its discovery in positron-atom scattering caused much excitement. Later, resonances above the ionization threshold were also seen in coupled-static calculations on Ar, Kr and Xe by McAlinden and Walters[31,32], while experiments on Ar and Kr targets[76-78] revealed structure which, it was thought, might originate from such resonances. However, further surprises awaited in store. When the coupled-static approximation was upgraded to the 6-state Ps(1s,2s,2p) + H(1s,2s,2p) approximation[33,36,56,70,72], not only were two more resonances seen in the S-wave cross sections, near 15 and 18eV (figure 2)¶ , but six more resonances were identified in the P-, D- and F-waves[70]. In the 6-state approximation the Higgins-Burke S-wave resonance continues to appear prominently but its position has moved to 3.75 Ryd (51.0eV) and its width has narrowed to 0.21 Ryd (2.9eV) (figure 2), not a good sign for a real physical resonance.

An important physical component missing from the aforementioned approximations is break-up, ie, ionization. What effect would this have on the new-type resonances ? In the coupled-state approximation ionization can be represented by employing atom and/or positronium pseudostates ($\overline{\psi}_n$ and $\overline{\phi}_m$)

---

¶ Mitroy[68] had earlier seen the 15eV resonance in the Ps(1s,2s) + H(1s,2s) approximation.

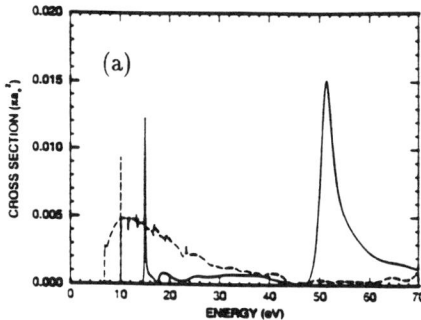
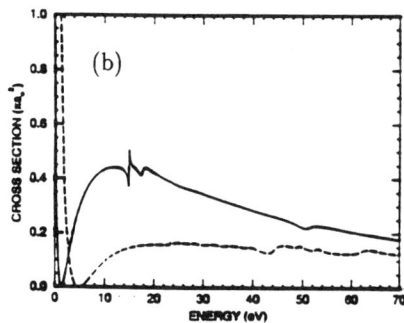

**FIGURE 2.** S-wave cross sections for positron scattering by H(1s)[36] : (a) Ps(1s) formation; (b) elastic scattering. Solid curve, 6-state approximation; dash curve 18-state approximation.

in the expansion (1). Pseudostates are normally constructed by diagonalizing the atom/positronium Hamiltonian in some basis :

$$\left\langle \overline{\psi}_n \left| H_A \right| \overline{\psi}_m \right\rangle = \overline{\epsilon}_n \delta_{nm} \qquad \left\langle \overline{\phi}_n \left| H_{Ps} \right| \overline{\phi}_m \right\rangle = \overline{E}_n \delta_{nm} \qquad (5)$$

But, pseudostates have their foibles – they can give rise to false threshold and resonance structures. However, it has been found in electron scattering that, if enough pseudostates can be used, pseudostructures become small and unimportant[79,80].

Kernoghan et al[36] have made calculations in an 18-state Ps(1s,2s,$\overline{3s},\overline{4s}$,2p, $\overline{3p},\overline{4p},\overline{3d},\overline{4d}$) + H(1s,2s,$\overline{3s},\overline{4s}$,2p,$\overline{3p},\overline{4p},\overline{3d},\overline{4d}$) approximation, where a bar denotes a pseudostate. The 18-state results are compared with the 6-state cross sections in figure 2. The 18-state numbers display a "lumpy" pseudoresonant behaviour but there is no obvious evidence of the three pronounced S-wave resonances of the 6-state calculation. It would appear that the coupling to the ionization channels has removed these resonances. As further evidence, the Higgins-Burke resonance was not seen in the hyperspherical calculations of Igarashi and Toshima[28] either. The conclusion is that the resonances above the ionization threshold are artefacts of simple approximations[§]. Why then do they appear ? According to Kernoghan et al[36], they are an attempt by the simpler approximations to compensate, in a rather perverse way, for the missing ionization channels.

This is all in accord with a theorem proved by Simon[81,82] in the 1970s. This states that resonances cannot exist at energies above the threshold for complete disintegration of a many particle system experiencing only Coulomb

---

§ Even over-simple pseudostate approximations[30,36].

forces. In electron scattering there seemed to be a contradiction to this result. In an experiment[83-85] on electron impact ionization of H$^-$ two resonances were found just above the threshold for complete disintegration, ie, e$^-$ + H$^-$ ⟶ 3e$^-$ + p. Reasons were then advanced for believing that Simon's theorem was not perfectly general. However, recently both theory[86] and experiment[87] have disproved the existence of these resonances. Simon's theorem therefore remains intact.

As predicted by Mittleman[88], real Feshbach resonances, similar to those seen in electron scattering, do exist near excitation thresholds of the atom and the positronium[26,33,36,38,45,72,75]. Much work has been done on resonances by Y.K.Ho[89,90].

## Pseudostate Calculations

At low energies pseudostate calculations have now become so refined that they are probably better than the variational results[74]. The largest two-centre pseudostate calculation over an extended energy range is the 18-state Ps(1s,2s,$\overline{3s},\overline{4s}$,2p,$\overline{3p},\overline{4p},\overline{3d},\overline{4d}$) + H(1s,2s,$\overline{3s},\overline{4s}$,2p,$\overline{3p},\overline{4p},\overline{3d},\overline{4d}$) approximation of Kernoghan et al[33,36,38]. Figure 3 shows the results in this approximation for elastic scattering, positronium formation and the total cross section. Small amplitude pseudostructure has been removed from these cross sections by smoothing. Comparison is made with the work of Bray and Stelbovics[91], the intermediate energy R-matrix approximation of Higgins et al[92], and the multipseudostate close-coupling approximation of Walters[93]. All of these are single-centre pseudostate approximations, ie,

$$\Psi = \sum_n F_n(\mathbf{r}_p)\overline{\psi}_n(\mathbf{r}_e) \qquad (6)$$

Also shown are the moment T-matrix calculations of Winick and Reinhardt[94,95]. The degree of agreement between all of the theoretical approximations for elastic scattering is very encouraging indeed. The 18-state results for positronium formation are in fairly good, but not perfect, accord with the measurements of Weber et al[96], the calculations indicate that positronium formation is mainly into the 1s state. For the total cross section there is more discord between the theories but, since only the 18-state approximation models positronium formation explicitly, it is probably to be preferred. Except near 30 and 50eV, the 18-state total cross section is consistent with the upper and lower bound measurements of Zhou et al[97].

Figure 4 shows the 18-state ionization cross section. Here the pseudostructure has not been removed. This cross section has been calculated according to the ansatz

$$\sigma_{ion} = \sum_n (1 - a_n)(\sigma_H(n) + \sigma_{Ps}(n)) \qquad (7)$$

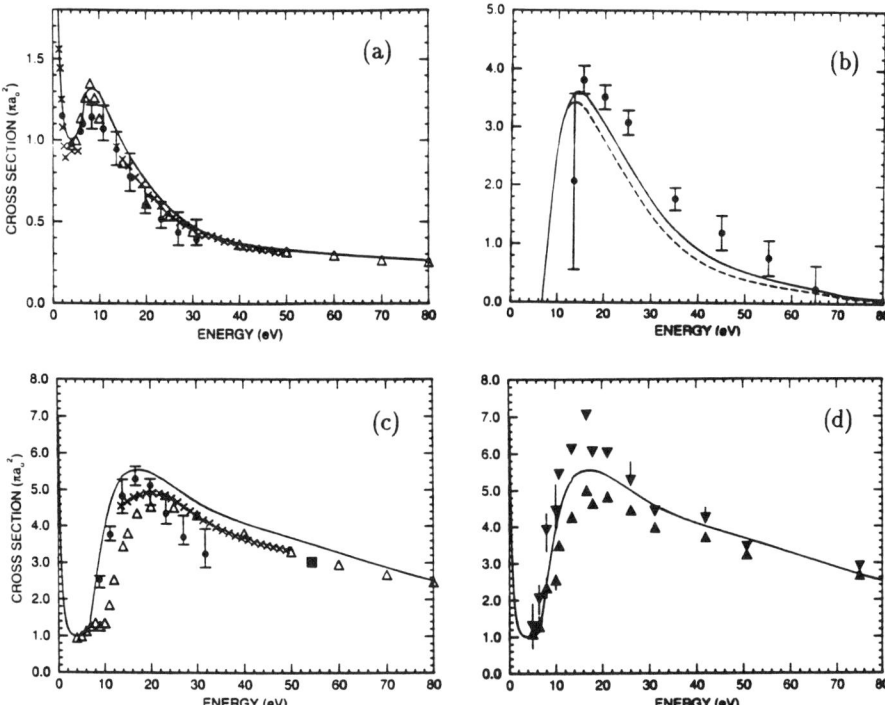

**FIGURE 3.** 18-state cross sections of Kernoghan et al[38] : (a) elastic scattering; (b) Ps formation; (c) and (d) total cross section. In (a) and (c) comparison is made with the theories of Bray and Stelbovics[91] (triangles), Higgins et al[92] (crosses), Walters[93] (squares) and Winick and Reinhardt[94,95] (circles). In (b) the solid curve is total Ps formation, the dash curve Ps(1s) formation; experimental data are from Weber et al[96]. In (d) triangles up (down) are the lower (upper) bound measurements of Zhou et al[97].

where the sum is over the pseudostates $n = \overline{3s},\ \overline{4s},\ \overline{3p},\ \overline{4p},\ \overline{3d},\ \overline{4d}$ and $\sigma_H(n)(\sigma_{Ps}(n))$ is the cross section for exciting the n pseudostate of atomic hydrogen (positronium). The factors $(1 - a_n)$ give the fraction of the pseudostate lying in the continuum and are easily evaluated[38]. There is very pleasing agreement with the measurements of Jones et al[98], but not with the data of Weber et al[96] (not shown) which lie somewhat higher, the reader is directed to Kernoghan et al[38] for a discussion. Also shown in figure 4 is the contribution

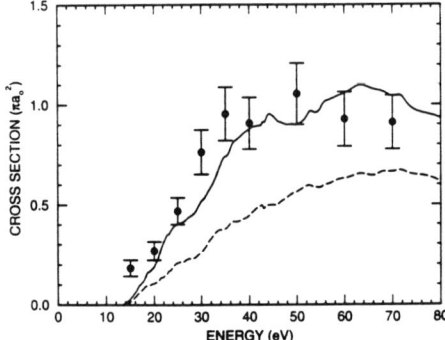

**FIGURE 4.** Ionization cross section of Kernoghan et al[38]. Solid curve, full cross section (7); dash curve, contribution from atom pseudostates alone (8); experimental data from Jones et al[98].

from the atom pseudostates alone, ie,

$$\sum_n (1 - a_n)\sigma_H(n) \qquad (8)$$

It is interesting to note that, in the given energy range, this accounts for little more than half of the cross section, ionization is clearly represented by both the atom and positronium pseudostates.

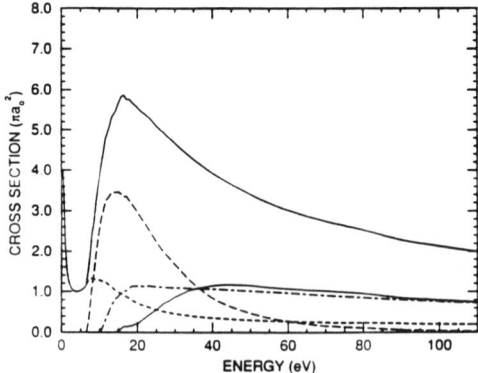

**FIGURE 5.** Composition of the total cross section in the 33-state approximation of Kernoghan et al[99]. Curves : solid, total cross section; long-dash, total Ps formation; short-dash, elastic; dash-dot, H(2p) excitation; dotted, ionization.

Because of the presence of too much pseudostructure, Kernoghan et al[38] felt it unwise to quote any other cross sections. Very recently, however, Kernoghan et al[99] have made a new 33-state calculation in which the pseudostructure is very small. This new calculation largely confirms the results of the 18-state approximation and now gives reliable cross sections for all the main transitions. The total cross section, and the main components that make it up, in this new approximation, are shown in figure 5.

## POSITRON SCATTERING BY THE ALKALI METALS

Unlike atomic hydrogen, the low ionization potentials of the alkali metals mean that positronium can be formed in the 1s state at any impact energy of the incident positron on the ground state atom. As the incident energy tends to zero the Ps(1s) formation cross section tends to infinity as $1/k_0$, where $k_0$ is the momentum of the incident positron[24,25,33]. Alkali cross sections are typically over an order of magnitude bigger than those for atomic hydrogen. Except near zero energy, where it becomes infinite, the positronium formation cross section, while being large, is not such a dominant component of the total cross section as in the case of atomic hydrogen. Neither is ionization as significant a process. Rather the total cross section is dominated by elastic scattering at low energies¶ and by the resonance transition at high energies.

To date theoretical approximations have treated the alkalis as single electron atoms with a frozen core. Pioneering coupled-state work on positron - alkali scattering has been done by the Calcutta group[49,54,100–102], using both one- and two-centre approximations, and by the York University group[103–111] using one-centre approximations. More recent calculations using two-centre approximations have been made by Hewitt, Noble and Bransden[63,65,66] and by Kernoghan, McAlinden and Walters[33,37,39]. To the best of our knowledge the only variational calculations are those of Watts and Humberston on Li[24,25].

On the experimental front, there are measurements of the total cross section and total positronium formation cross section for Na, K and Rb from the Wayne State group at Detroit[112–116].

A very interesting discovery made by Hewitt et al[65] was the pronounced growth in excited state positronium formation on going from positron scattering by Li and Na to positron scattering by K. Figure 6 illustrates the situation using the more extended calculations of McAlinden et al[39] in the Ps(1s,2s,2p,3s,3p,3d) + K(4s,4p,5s,5p,3d) approximation. These calculations are close to the lower bound measurements of Zhou et al[114]. Although confirming the general result of Hewitt et al, about the importance of excited state positronium formation, the calculations of McAlinden et al give a different

---

¶ Except near zero energy.

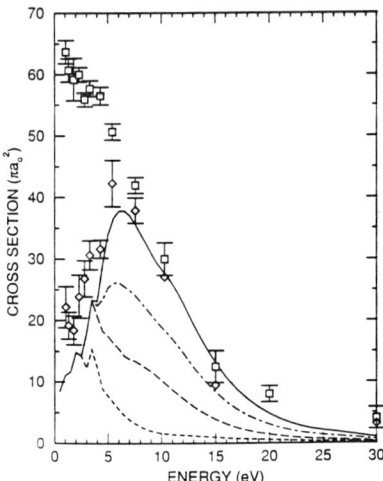

**FIGURE 6.** Ps formation in K. Curves, as calculated by McAlinden et al[39] : short dash, Ps(1s); long dash, Ps(n=1+2); dash-dot; Ps(n=1+2+3); solid; total Ps formation assuming $n^3$ scaling for $n \geq 4$. Squares (diamonds) are upper (lower) bound measurements of Zhou et al[114].

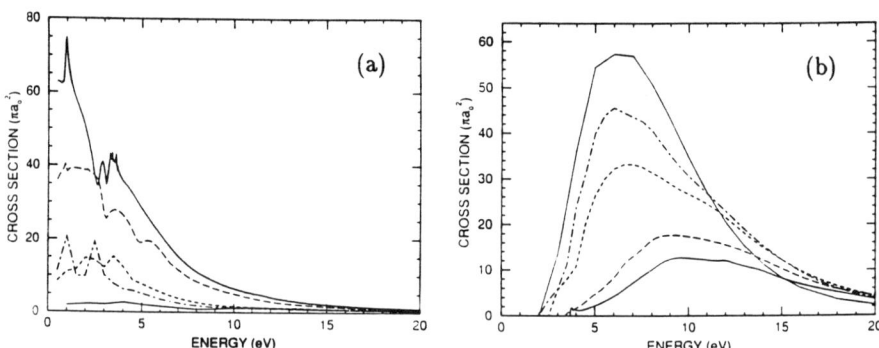

**FIGURE 7.** Calculated[117] Ps formation in Li, Na, K, Rb, Cs : (a) Ps(1s); (b) excited Ps. Curves : solid, Li; long-dash, Na; Short-dash, K; dash-dot, Rb, dotted, Cs.

distribution over the excited states. Hewitt et al used the simpler approximation Ps(1s,2s,2p) + K(4s,4p,5s,5p).

Coupled-state calculations now exist for the full set of alkali metals Li to Cs[117]. Figure 7 shows the trend in cross sections for Ps(1s) formation and for excited Ps formation in the energy range 0.5 to 20eV. Here we see a collapse in the Ps(1s) cross section¶ on going from Li to Cs, with a corresponding rise in excited state positronium formation. It should, however, be emphasised that the results of figure 7 come from calculations using only atom and positronium eigenstates, the effect of including pseudostates has not been examined and could perhaps change the picture. So far, at least, the eigenstate calculations seem to be in reasonable agreement with the available measurements, in Na, K and Rb, of total cross section and total positronium production.

## POSITRON SCATTERING BY THE NOBLE GASES

Unlike atomic hydrogen and the quasi one-electron alkali metals, the noble gases provide a more difficult challenge to theory. Yet there is a wealth of experimental data on these systems (see references 118-120) : on total cross sections; on positronium formation; on single and double ionization; on differential ionization; on differential elastic scattering, and now on differential positronium formation[121]. Needless to say, theoretical effort has been concentrated on the simplest of the noble gases, namely He. Amongst the most recent works on He we would highlight the variational calculations of Van Reeth and Humberston[22,23] and the coupled-state calculations of Hewitt et al[62]. Although there has also been significant work on the heavier noble gases, for example, the polarized orbital calculations of the York University group[122-125], theories which explicitly allow for positronium formation in the heavier targets are somewhat sparse and crude: for Ne, Ar, Kr and Xe there are (truncated) coupled-static calculations by McAlinden and Walters[31,32], while for Ne and Ar there are first-order Born and distorted-wave calculations by Gillespie and Thompson[126]; some results have also been obtained using the classical-trajectory Monte Carlo method[127,128].

Potentially the heavier noble gases are the most interesting, with the possibility of seeing shell effects in the positronium formation[31,32] and the intriguing question of whether, like the alkalis, excited state positronium formation grows with the size of the target. Clearly, a proper treatment of the noble gases must be the next step in the theoretical programme.

## CONCLUSIONS

It is now clear that resonances seen above the ionization threshold in simple coupled-state calculations of $e^+$ – H scattering are artefacts of low order approximations. With the latest 33-state results[99] we believe that the main cross sections for positrons incident on ground state atomic hydrogen are now known to a high degree of accuracy. Here, theory awaits experiment.

---

¶ Except near zero energy where it becomes infinite.

For the alkali metals coupled eigenstate approximations have given reasonable agreement with total cross section and total Ps measurements. Theory has also predicted a large growth in excited state Ps formation on progressing from Li to Cs. This needs to be tested experimentally, and if possible the distribution over excited states measured. Pseudostate calculations are also needed to check on the eigenstate results.

More high level theoretical effort needs to be applied to the noble gases.

## REFERENCES

1. Charlton M., Eades J., Horváth D., Hughes R.J. and Zimmermann C., Phys. Repts. **241** 65 (1994).
2. Humberston J.W., Charlton M., Jacobsen F.M. and Deutch B.I., J. Phys. B **20** L25 (1987).
3. Mitroy J. and Stelbovics A.T., Phys. Rev. Lett. **72** 3495 (1994).
4. Mitroy J. and Stelbovics A.T., J. Phys. B **27** L79 (1994).
5. Schwartz C., Phys. Rev. **124** 1468 (1961).
6. Armstead R.L., Phys. Rev. **171** 91 (1968).
7. Bhatia A.K., Temkin A., Drachman R.J. and Eisereke H., Phys. Rev. A **3** 1328 (1971).
8. Stein J. and Sternlicht R., Phys. Rev. A **6** 2165 (1972).
9. Humberston J.W. and Wallace J.B., J. Phys. B **5** 1138 (1972).
10. Bhatia A.K., Temkin A. and Eisereke H., Phys. Rev. A **9** 219 (1974).
11. Register D. and Poe R., Phys. Lett. **51A** 431 (1975).
12. Humberston J.W., Canadian J. Physics. **60** 591 (1982).
13. Humberston J.W., J. Phys. B **17** 2353 (1984).
14. Brown C.J. and Humberston J.W., J. Phys. B **18** L401 (1985).
15. Roy U. and Mandal P., J. Phys. B **23** L55 (1990).
16. Roy U. and Mandal P., Phys. Rev. A **48** 223 (1993).
17. Roy U. and Mandal P., Phys. Rev. A **48** 2952 (1993).
18. Humberston J.W., J. Phys. B **6** L305 (1973).
19. Campeanu R.I. and Humberston J.W., J. Phys. B **8** L244 (1975).
20. Campeanu R.I. and Humberston J.W., J. Phys. B **10** L153 (1977).
21. Humberston J.W., Adv. At. Mol. Phys. **15** 101 (1979).
22. Van Reeth P. and Humberston J.W., J. Phys. B **28** L23 (1995).
23. Van Reeth P. and Humberston J.W., J. Phys. B, to be published.
24. Watts M.S.T. and Humberston J.W., J. Phys. B **25** L491 (1992).
25. Humberston J.W. and Watts M.S.T., Hyperfine Interactions **89** 47 (1994).
26. Archer B.J., Parker G.A. and Pack R.T., Phys. Rev. A **41** 1303 (1990).
27. Zhou Y. and Lin C.D., J. Phys. B **27** 5065 (1994).
28. Igarashi A. and Toshima N., Phys. Rev. A **50** 232 (1994).
29. Higgins K. and Burke P.G., J. Phys. B **24** L343 (1991).
30. Higgins K. and Burke P.G., J. Phys. B **26** 4269 (1993).

31. McAlinden M.T., PhD Thesis, The Queen's University of Belfast (1992).
32. McAlinden M.T. and Walters H.R.J., Hyperfine Interactions **73** 65 (1992).
33. McAlinden M.T., Kernoghan A.A. and Walters H.R.J., Hyperfine Interactions **89** 161 (1994).
34. McAlinden M.T. and Walters H.R.J., Hyperfine Interactions **89** 407 (1994).
35. Kernoghan A.A., McAlinden M.T. and Walters H.R.J., J. Phys. B **27** L211 (1994).
36. Kernoghan A.A., McAlinden M.T. and Walters H.R.J., J. Phys. B **27** L543 (1994).
37. Kernoghan A.A., McAlinden M.T. and Walters H.R.J., J. Phys. B **27** L625 (1994).
38. Kernoghan A.A., McAlinden M.T. and Walters H.R.J., J. Phys. B **28** 1079 (1995).
39. McAlinden M.T., Kernoghan A.A. and Walters H.R.J., J. Phys. B, to be published.
40. Liu G. and Gien T.T., Phys. Rev. A **46** 3918 (1992).
41. Gien T.T. and Liu G., Phys. Rev. A **48** 3386 (1993).
42. Gien T.T., J. Phys. B **27** L25 (1994).
43. Gien T.T. and Liu G., J. Phys. B **27** L179 (1994).
44. Liu G. and Gien T.T., Phys. Rev. A **49** 5157 (1994).
45. Gien T.T., J. Phys. B **28** L313 (1995).
46. Mandal P., Ghosh A.S. and Sil N.C., J. Phys. B **8** 2377 (1975).
47. Basu D., Banerji G. and Ghosh A.S., Phys. Rev. A **13** 1381 (1976).
48. Mandal P., Basu D. and Ghosh A.S., J. Phys. B **9** 2633 (1976).
49. Guha S. and Ghosh A.S., Phys. Rev. A **23** 743 (1981).
50. Ghosh A.S., Sil N.C. and Mandal P., Phys. Repts. **87** 313 (1982).
51. Basu M., Mukherjee M. and Ghosh A.S., J. Phys. B **22** 2195 (1989).
52. Mukherjee M., Basu M. and Ghosh A.S., J. Phys. B **23** 757 (1990).
53. Basu M., Mukherjee M. and Ghosh A.S., J. Phys. B **23** 2641 (1990).
54. Basu M. and Ghosh A.S., Phys. Rev. A **43** 4746 (1991).
55. Mukherjee M. and Basu M., Z.Phys.D **22** 2195 (1991).
56. Sarkar N.K., Mukherjee M., Basu M. and Ghosh A.S., J. Phys. B **26** L427 (1993).
57. Sarkar N.K., Basu M. and Ghosh A.S., J. Phys. B **26** L799 (1993).
58. Sarkar N.K., Jha L.K. and Ghosh A.S., Ind.J. Phys. B **67** 189 (1993).
59. Sarkar N.K. and Ghosh A.S., J. Phys. B **27** 759 (1994).
60. Hewitt R.N., Noble C.J. and Bransden B.H., J. Phys. B **23** 4185 (1990).
61. Hewitt R.N., Noble C.J. and Bransden B.H., J. Phys. B **24** L635 (1991).
62. Hewitt R.N., Noble C.J. and Bransden B.H., J. Phys. B **25** 557 (1992).
63. Hewitt R.N., Noble C.J. and Bransden B.H., J. Phys. B **25** 2683 (1992).
64. Bransden B.H., Noble C.J. and Hewitt R.N., J. Phys. B **26** 2487 (1993).
65. Hewitt R.N., Noble C.J. and Bransden B.H., J. Phys. B **26** 3661 (1993).

66. Hewitt R.N., Noble C.J. and Bransden B.H., Hyperfine Interactions **89** 195 (1994).
67. Mitroy J., Aust. J. Phys. **46** 751 (1993).
68. Mitroy J., J. Phys. B **26** L625 (1993).
69. Mitroy J., J. Phys. B **26** 4861 (1993).
70. Mitroy J. and Stelbovics A.T., J. Phys. B **27** L55 (1994).
71. Mitroy J. and von Geramb H.V., J. Phys. B **27** L427 (1994).
72. Mitroy J. and Stelbovics A.T., J. Phys. B **27** 3257 (1994).
73. Mitroy J. and Stelbovics A.T., Phys. Rev. Lett. **72** 3495 (1994).
74. Mitroy J., Berge L. and Stelbovics A.T., Phys. Rev. Lett. **73** 2966 (1994).
75. Mitroy J. and Ratnavelu K., J. Phys. B **28** 287 (1995).
76. Kauppila W.E. and Stein T.S., Hyperfine Interactions **73** 87 (1992).
77. Dou L., Kauppila W.E., Kwan C.K., Przybyla D., Smith S.J. and Stein T.S., Phys. Rev. A **46** R5327 (1992).
78. Dou L., Kauppila W.E., Kwan C.K. and Stein T.S., Phys. Rev. Lett. **68** 2913 (1992).
79. Scholz T.T., J. Phys. B **24** 2127 (1991).
80. Bray I. and Stelbovics A.T. Phys. Rev. Lett. **69** 53 (1992).
81. Simon B., Math. Ann. **207** 133 (1974).
82. Simon B., Int. J. Quantum Chem. **14** 529 (1978).
83. Walton D.S., Peart B. and Dolder K., J. Phys. B **3** L148 (1970).
84. Walton D.S., Peart B. and Dolder K., J. Phys. B **4** 1343 (1971).
85. Peart B. and Dolder K., J. Phys. B. **6** 1497 (1973).
86. Robicheaux F., Wood R.P. and Greene C.H., Phys. Rev. A **49** 1866 (1994).
87. Andersen L.H., Mathur D., Schmidt H.T. and Vejby-Christensen L., Phys. Rev. Lett. **74** 892 (1995).
88. Mittleman M.H., Phys. Rev. **152** 76 (1966).
89. Ho Y.K., Phys. Repts. **99** 1 (1983).
90. Ho Y.K., Hyperfine Interactions **73** 109 (1992).
91. Bray I. and Stelbovics A.T., Phys. Rev. A **49** R2224 (1994).
92. Higgins K., Burke P.G. and Walters H.R.J., J. Phys. B **23** 1345 (1990).
93. Walters H.R.J., J. Phys. B **21** 1893 (1988).
94. Winick J.R. and Reinhardt W.P., Phys. Rev. A **18** 910 (1978).
95. Winick J.R. and Reinhardt W.P., Phys. Rev. A **18** 925 (1978).
96. Weber M., Hofmann A., Raith W., Sperber W., Jacobsen F. and Lynn K.G., Hyperfine Interactions **89** 221 (1994).
97. Zhou S., Kauppila W.E., Kwan C.K. and Stein T.S., Phys. Rev. Lett. **72** 1443 (1994).
98. Jones G.O., Charlton M., Sleviin J., Laricchia G., Kövér Á., Poulsen M.R. and Nic Chormaic S., J. Phys. B **26** L483 (1993).
99. Kernoghan A.A., Robinson D., McAlinden M.T. and Walters H.R.J., to be published.

100. Khan P., Dutta S. and Ghosh A.S., J. Phys. B **20** 2927 (1987).
101. Sarkar K.P., Basu M. and Ghosh A.S., J. Phys. B **21** 1649 (1988).
102. Sarkar K.P. and Ghosh A.S., J. Phys. B **22** 105 (1989).
103. Ward S.J., Horbatsch M., McEachran R.P. and Stauffer A.D., in "Atomic Physics with Positrons" ed. Humberston J.W. and Armour E.A.G. (New York, Plenum) p265 (1988).
104. Ward S.J., Horbatsch M., McEachran R.P. and Stauffer A.D., J. Phys. B **21** L611 (1988).
105. Ward S.J., Horbatsch M., McEachran R.P. and Stauffer A.D., J. Phys. B **22** 1845 (1989).
106. Ward S.J., Horbatsch M., McEachran R.P. and Stauffer A.D., J. Phys. B **22** 3763 (1989).
107. Ward S.J., Horbatsch M., McEachran R.P. and Stauffer A.D., Nucl. Instrum. Methods B **42** 472 (1989).
108. McEachran R.P., Horbatsch M. and Stauffer A.D., J. Phys. B **24** 1107 (1991).
109. De Vries K.M., Bartschat K., McEachran R.P. and Stauffer A.D., J. Phys. B **25** L653 (1992).
110. Bartschat K., McEachran R.P. and Stauffer A.D., Hyperfine Interactions **73** 99 (1992).
111. Bartschat K., De Vries K.M., McEachran R.P. and Stauffer A.D., Hyperfine Interactions **89** 57 (1994).
112. Kwan C.K., Kauppila W.E., Lukaszew R.A., Parikh S.P., Stein T.S., Wan Y.J. and Dababneh M.S., Phys. Rev. A **44** 1620 (1991).
113. Parikh S.P., Kauppila W.E., Kwan C.K., Lukaszew R.A., Przybyla D., Stein T.S. and Zhou S., Phys. Rev. A **47** 1535 (1993).
114. Zhou S., Parikh S.P., Kauppila W.E., Kwan C.K., Lin D., Surdutovich A. and Stein T.S., Phys. Rev. Lett. **73** 236 (1994).
115. Kauppila W.E., Kwan C.K., Stein T.S. and Zhou S., J. Phys. B **27** L551 (1994).
116. Kauppila W.E. and Stein T.S., to be published.
117. Kernoghan A.A., McAlinden M.T. and Walters H.R.J., to be published.
118. Kauppila W.E. and Stein T.S., Adv. At. Mol. Phys. **26** 1 (1989).
119. Charlton M. and Laricchia G., J. Phys. B **23** 1045 (1990).
120. Charlton M. and Laricchia G. in "The Physics of Electronic and Atomic Collisions, 18$^{th}$ International Conference, Aarhus, Denmark" ed. Andersen et al. (AIP Conference Proceedings 295) p455 (1993).
121. Falke T., Raith W., Weber M. and Wesskamp U., J. Phys. B, to be published.
122. McEachran R.P., Morgan D.L., Ryman A.G. and Stauffer A.D., J. Phys. B **10** 663 (1977).
123. McEachran R.P., Ryman A.G. and Stauffer A.D., J. Phys. B **11** 551 (1978).

124. McEachran R.P., Ryman A.G. and Stauffer A.D., J. Phys. B **12** 1031 (1979).
125. McEachran R.P., Stauffer A.D. and Campbell L.E.M., J. Phys. B **13** 1281 (1980).
126. Gillespie E.S. and Thompson D.G., J. Phys. B **10** 3543 (1977).
127. Schultz D.R., Reinhold C.O. and Olson R.E., Phys. Rev. A **40** 4947 (1989).
128. Sparrow R.A. and Olson R.E., J. Phys. B **27** 2647 (1994).

# SPIN RELAXATION PHENOMENA IN MUON COLLISION PROCESSES IN GASES

Donald G. Fleming, Masayoshi Senba, James J. Pan and Donald J. Arseneau

*TRIUMF and Department of Chemistry, University of British Columbia, Vancouver, B.C., Canada, V6T 1Y6*

**Abstract.** The Muonium atom (Mu = $\mu^+e^-$) can simply be regarded as an ultralight isotope of the H atom, with 1/9 the mass. Since the $\mu^+$ is produced 100% polarized, and is coupled to the $e^-$ spin by the hyperfine interaction, Mu is an ideal probe of spin-relaxation processes, exemplified by (*inter*molecular) spin-exchange collisions with paramagnetic species. For Mu-substituted free radicals (e.g., $MuC_2H_4$), the (*intra*molecular) spin-rotation interaction provides an additional spin relaxation mechanism; one which is difficult to study by ESR in gases due to the very broad lines. Current experiments from TRIUMF on the relaxation rates of Mu with Cs, $O_2$, NO, and $C_2H_4$ are discussed, along with recent studies of He$\mu$ + Rb spin exchange from LAMPF using *negative* muons.

## I. INTRODUCTION

The muon, like its much lighter leptonic cousin, the electron, comes in two charge states, $\mu^+$ and $\mu^-$, and undergoes a variety of interactions in matter. However, whereas the interactions of the $\mu^-$ are those of a heavy electron ($m_\mu/m_e \approx 200$) the *positive* muon behaves much more like a light proton, a tendency which has been exploited in wide-ranging studies utilizing the $\mu$SR technique [1]. This report is primarily concerned with recent developments in the area of spin-relaxation collisions of positive muons in gases, though some related recent studies with negative muons, notably the He$\mu$ + Rb repolarization [2], is also of interest. Most of the recent work on the interactions of the $\mu^+$ in gases has been carried out at TRIUMF (in Vancouver, Canada), but related gas-phase studies are also underway at PSI (near Zürich, Switzerland) [3] and at the "BOOM" facility at KEK (Tsukuba, Japan) [4].

Although the ionization and charge-exchange interactions of the charged muon ($\mu^+$) in its slowing-down processes in gases are of interest in their own right [5-8], the present article focusses on the interactions of the neutral species, muonium (Mu=$\mu^+e^-$). With a mass only 1/9 that of the more familiar H atom (but still 200 times the electron mass), Mu can simply be regarded as an ultralight isotope of hydrogen. It was first identified and characterized in Ar gas by Hughes and collaborators [9], and today the study of Mu reactivity in gases, both in the realm of spin-exchange collisions [7, 10–14] and in chemical reaction dynamics, [3, 4, 15, 16], is well established. Though chemical reactions are not the topic here, Mu addition to unsaturated bonds to form Mu-radicals (e.g., $MuC_2H_4$) [16, 17] is relevant. It is the *spin relaxation* of these free radical species of the Mu atom itself that provide a rich "laboratory" for the study of spin physics by $\mu$SR and form the topic of the present treatise.

## II. $\mu$SR, MUONIUM AND MU-RADICALS

### A. Muonium Formation and the Basic Technique

The $\mu^+$ can be produced essentially monoenergetic and 100% spin polarized in the parity-violating decay of positive pions, $\pi^+ \to \mu^+ \nu_\mu$. The subject of $\mu$SR (Muon Spin Relaxation, Rotation,...) [1] is inherently the study of the time evolution of this polarization, due to interactions of the muon spin with its environment, manifest by the spatial asymmetry of decay positrons ($\mu^+ \to e^+ \nu_e \bar{\nu}_\mu$). These are emitted preferentially along the muon spin, giving rise to a "$\mu$SR time histogram" of accumulated $e^+$ decay events having the form

$$N(t) = N_0 \exp(-t/\tau_\mu) \left[ 1 + \sum_i S_i(t) \right] + B \qquad (1)$$

where $N_0$ is a normalization, $\tau_\mu$ the muon lifetime (2.2 $\mu$s) and $S_i(t)$ represents the "signal" of interest, corresponding to muons in the $i$th environment.

There are two basic (time-differential, TD) geometries in $\mu$SR: transverse field (TF, $T_2$ processes) and longitudinal field (LF, $T_1$ processes), in which the incident muon spin is either perpendicular (TF) or parallel (LF), to the applied magnetic field [18, 19]. The incident $\mu^+$ triggers a thin plastic scintillator and its decay positron triggers an array of scintillators (a "counter telescope") positioned at a fixed angle to the muon spin. Signals from the counters start and stop a high frequency ($\sim$ 1 GHz) clock, respectively, such that each good recorded event is incremented in computer memory, giving rise to the form of Eqn. (1). The nature of the spin relaxation, which is the focus of the present report, is contained in the signal $S(t)$. The data acquisition system ensures there is only *one* muon in the system at a time, so it can be matched with a single decay positron. This is an important feature of the $\mu$SR technique, obviating concerns about "probe-probe" interactions which can often plague the corresponding H-atom experiments, including some aspects of magnetic resonance.

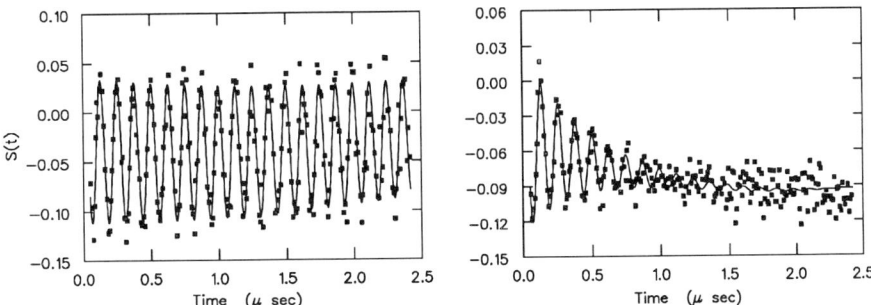

**FIG. 1.** The $\mu$SR signals $S(t)$ at 556 K for Mu precession in a TF of 8 G in (left) pure $N_2$ at 1500 torr; and (right) with an added $10 \times 10^{15}$ molec cm$^{-3}$ of Cs (21 ppm). The solid line is a fit of Eqn. (2) to the data. The pronounced increase in relaxation rate $\lambda_{Mu}$ is due to Mu + Cs spin-exchange collisions. From Ref. 11.

After the $\mu^+$ stops in a gas, the observed muon polarization can be distributed between three principal environments: diamagnetic, with polarization $P_D$, as muonium, with polarization $P_{Mu}$, or as a muonium substituted free radical, with polarization $P_R$. In addition, there will frequently be a "lost" fraction, $P_L$, which is strongly density and environment dependent such that $\sum P_i = 1$. In an inert moderating gases, like $N_2$, free radical formation is not expected and the $\mu$SR signal in a (weak) TF has the form

$$S_T(t) = A_D \cos(\omega_D t + \phi_D) + A_{Mu} \exp(-\lambda_{Mu} t) \cos(\omega_{Mu} t - \phi_{Mu}) \qquad (2)$$

where the parameters $A_D$, $A_{Mu}$, $\phi_D$, $\phi_{Mu}$, $\omega_D$, $\omega_{Mu}$ have the meaning of initial amplitude, phase and larmor frequency for muons in diamagnetic environments and as Mu, respectively ($\omega_{Mu} = 103 \omega_D$), and $\lambda_{Mu}$ is a relaxation rate due to the interaction of Mu with its environment ($\lambda_{Mu} = 1/T_2$). In a LF there is a similar form, but without the oscillatory factor.

Example precession signals are shown in Fig. 1 (left), for $\mu^+$ stopping in $N_2$ gas. In almost all gases, which having ionization potentials less than that of Mu itself (13.6 eV), most incident $\mu^+$ form Mu as a as indicated by the large initial signal amplitude, $A_{Mu}$, in Fig. 1, giving about 85% Mu formation in $N_2$. (Exceptions are He and Ne, where almost all $\mu^+$ thermalize in diamagnetic environments, as HeMu$^+$ and NeMu$^+$ [6].) The solid lines in Fig. 1 are fits of Eqn. (2) to the data. In the presence of a reactive species (Cs), there is a noticeable increase in relaxation rate (Fig.1, right) due in this case to Mu + Cs spin exchange [11].

### B. The $\mu^+$-$e^-$ Muonium Hyperfine Interaction

The time-dependence of the muon polarization is the key to interpreting observations of Mu, particularly to its spin relaxation. For the isolated Mu

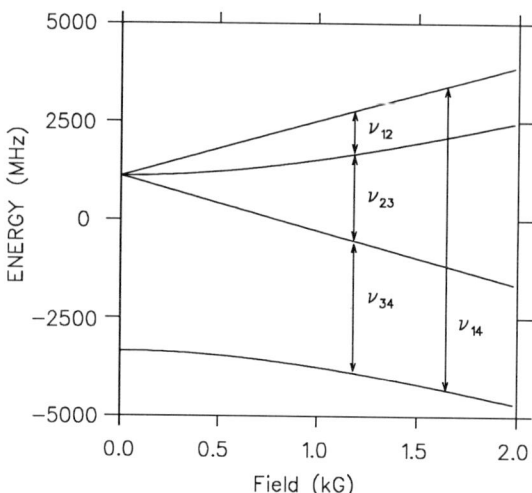

FIG. 2. The Breit–Rabi diagram for the isotropic $\mu^+$-$e^-$ magnetic hyperfine interaction of Eqn. (3). The allowed transition frequencies in a transverse magnetic field are indicated. In a weak TF characteristic of many μSR experiments in gases ($\lesssim 10\,\text{G}$), the observed precession is due to the essentially degenerate frequencies $\nu_{12}$ and $\nu_{23}$.

atom, the total Breit–Rabi Hamiltonian describing the isotropic hyperfine interaction and the separate Zeeman interactions of the muon and electron spins with an external magnetic field is given by [12,17,19,20]

$$H = \hbar[\omega_e S_e \cdot B - \omega_\mu I_\mu \cdot B + \omega_0 S_e \cdot I_\mu] \quad (3)$$

where $\omega_e = 2\pi\nu_e$ and $\omega_\mu$ are the Larmor precession frequencies of electron ($\omega_e/2\pi = 2.8$ MHz/G) and muon ($\omega_\mu/2\pi = 13.5$ kHz/G) and $\omega_0$ is the muonium hyperfine coupling constant in zero field ($\omega_0/2\pi = 4463.3$ MHz). The eigenstates of Eqn. (3) are plotted in Fig. 2, in frequency units, as a function of magnetic field up to 2 kG. In low field, they are given by the hyperfine coupled states $|F,M\rangle$, whereas in high fields, the uncoupled product spin projections are good quantum numbers.

Recalling that the muon beam is 100% polarized and choosing a quantization axis along the beam direction ($|\alpha_\mu\rangle$), muonium forms in two spin states with equal probability; $|A\rangle = |\alpha_\mu \alpha_e\rangle$ and $|B\rangle = |\alpha_\mu \beta_e\rangle$, since the electrons in the moderator are normally unpolarized. In a (weak) longitudinal or zero magnetic field, $|A\rangle = |1,1\rangle$ is a stationary eigenstate but $|B\rangle$ is a superposition of $|1,0\rangle$ and $|0,0\rangle$ eigenstates, oscillating at the hyperfine frequency. Thus on a time scale of order $1/\omega_0 \sim 0.04$ ns, this half of the muon ensemble appears effectively depolarized, given the ≈1 ns time resolution in a typical experiment. In a TF environment, with the quantization axis along the field ($z$) direction, the $|A\rangle$ and $|B\rangle$ states are superpositions of the four Zeeman eigenstates, with the four allowed transitions $\nu_{12}$, $\nu_{23}$, $\nu_{14}$, and $\nu_{43}$, indicated in Fig. 2.

In a TF, the muon polarization in (free) Mu is given by the expectation value $\langle \sigma_\mu^+ \rangle$ [13,19], leading to the expression

$$P_T^{Mu}(t) = \tfrac{1}{4}[(1+\delta)(e^{i\omega_{12}t} + e^{-i\omega_{34}t}) + (1-\delta)(e^{i\omega_{23}t} + e^{i\omega_{14}t})] \quad (4)$$

where $\omega_{ij} = \omega_i - \omega_j = 2\pi\nu_{ij}$ (see Fig. 2) with $\delta = X/\sqrt{1+X^2}$ and $X = B/B_0$ ($B_0 = 1585$ G is the contact field of the muon at the electron). In magnetic fields $< 200$ G, both $\nu_{12}$ and $\nu_{23}$ are experimentally observable, but $\nu_{14}$ and $\nu_{34}$ are comparable to $\nu_0$ and are generally not observed. The frequencies $\nu_{12}$ and $\nu_{23}$ become equal in fields $< 10$ G, such that only a single frequency, $\omega_{Mu}$, is observed. In this limit, with $\delta = 0$, Eqn. (4) reduces to the simple result of coherently precessing "triplet" Mu (state $|A\rangle$),

$$P_{Mu}(t) = \tfrac{1}{2} e^{i\omega_{Mu} t} \tag{5}$$

which is contained in Eqn. (2). The initial amplitude though, $A_{Mu}$, is always empirically determined.

### C. $\mu$SR and Muonium-Substituted Radicals in Gases

In an inert moderator (e.g., $N_2$), the initial (100%) muon polarization is distributed between $P_{Mu}$ and $P_D$, but in unsaturated gases (e.g., $C_2H_4$), the Mu atom undergoes addition reactions giving rise to muonium-substituted free-radicals, such as Mu-ethyl (MuC$_2$H$_4$), with polarization $P_R$, according to the kinetic scheme [3, 16, 17],

$$\text{Mu} + \text{C}_2\text{H}_4 \underset{k_d}{\overset{k_a}{\rightleftharpoons}} \text{MuC}_2\text{H}_4^* \xrightarrow[\text{M}]{k_s} \text{MuC}_2\text{H}_4 \tag{6}$$

where $k_a$, $k_d$ and $k_s$ are the corresponding rate constants for Mu addition, dissociation and stabilization (by moderator 'M'), respectively. The time evolution of the muon polarization in each species, $P_{Mu}$, $P_R^*$ and $P_R$ depends on the magnitude of these rate constants and on the densities involved [17]. With increasing density of $C_2H_4$, in the high pressure limit, $P_{Mu}$ quickly becomes unobservable on the (2.2 $\mu$s) $\mu$SR time scale and all the polarization is transferred to the (stabilized) MuC$_2$H$_4$ radical.

Eqn. (7) gives the hamiltonian for a gas-phase (isotropic) Mu-radical in a magnetic field,

$$H/h = \nu_e S_e \cdot B - \nu_\mu I_\mu \cdot B - \sum \nu_p I_p \cdot B + A_\mu S_e \cdot I_\mu + \sum A_p S_e \cdot I_p \tag{7}$$

where $\nu_i = \gamma_i B$ are the Larmor frequencies for electron, muon and proton and $A_\mu$ and $A_p$ are the muon and proton isotropic hyperfine couplings, respectively, and the summations are over all the protons in the absence of nuclear moments. In moderately high fields, $\sim 1$ kG, Eqn. (7) is dominated by just *two* transitions: $\nu_{12}$ and $\nu_{34}$ (see Fig. 2). This "ENDOR" limit is the basis for the detection of Mu-radicals, which can be seen from Eqn. (4) as well: for a typical Mu-radical $\omega_0^R \ll \omega_0^{Mu}$ (e.g., for MuC$_2$H$_4$, $\nu_0^R = 330$ MHz vs. $\nu_0^{Mu} = 4463$ MHz), the field parameter $\delta \to 1$ in even modest fields, leaving just $\nu_{12}$ and $\nu_{34}$ [20–22].

In the time domain, the TD TF-μSR signal $S(t)$ for a Mu-radical has a form similar to Eqn. (2):

$$S_{TF}(t) = \sum_j [A_{R_j} e^{-\lambda_{R_j} t} \cos(\omega_{R_j} t + \phi_{R_j})] + A_D e^{-\lambda_D t} \cos(\omega_D t + \phi_D) \qquad (8)$$

where $\lambda_{R_j}$ is the ($T_2$) relaxation rate for a given radical frequency $\omega_{R_j}$ having amplitude $A_{R_j}$ and initial phase $\phi_{R_j}$. In practice, only one (or two) radical frequencies contribute to the sum.

In an LF, the μSR signal for a Mu-radical can also be written in a form like Eqn. 8, but again without the oscillatory term. At certain applied longitudinal magnetic fields, a pair of nearly degenerate muon and proton energy levels with different nuclear spin orientations are mixed by the hyperfine interaction, resulting in a "resonance" (an avoided level crossing, ALCR), where the muon spin evolution changes dramatically [20, 23, 24]. The prime interest in this report, however, is with measurements of muon-radical relaxation rates *off-resonance*, which have a simple exponential form, with $\lambda_R = 1/T_1$.

### D. Basic Concepts in μSR Spin Relaxation

As well established in magnetic resonance and other phenomena, e.g., optical pumping [2], spin relaxation *per sé* involves a change of (electron) spin state resulting from a specific collision process. In the present context, this is most easily visualized in the case of Mu colliding with a *paramagnetic* species (e.g., Cs), represented by

$$\text{Mu}(|\alpha_\mu \alpha_e\rangle) + \text{Cs}(|\beta_{e'}\rangle) \to \text{Mu}(|\alpha_\mu \beta_{e'}\rangle) + \text{Cs}(|\alpha_e\rangle) \qquad (9)$$

where Mu in state $|A\rangle$ has been changed to Mu in state $|B\rangle$ by an electron spin flip ("exchange"), but the $|B\rangle$ state is not an eigenstate of the hyperfine interaction [Eqn. (3)] so Mu is rapidly depolarized on a time scale of $1/\omega_0 \approx 0.04$ ns. This is the situation illustrated for Mu + Cs in Fig. 1 in a weak TF. The relaxation rates $\lambda_{Mu}$ are related to the SF cross sections as described below.

A SF collision of this nature involves the *intermolecular* potential $V_{int}(r)$, and the exchange operator $\mathbf{P}_{12} = 1 + \boldsymbol{\sigma}_1 \cdot \boldsymbol{\sigma}_2$, for $S = 1/2$ collision partners, (or $\mathbf{P}_{12} + \mathbf{P}_{13}$ for $S = 1$ $O_2$) and gives maximum effect at short to intermediate ranges where the *difference* in scattering potentials is largest. This can be thought of as the physical change of (indistinguishable) electrons, which can be of "like" ($\alpha_e \alpha_{e'}$) or "unlike" ($\alpha_e \beta_{e'}$) spins, but only the latter (SF) exchange is normally measurable. The situation is physically different but quantum mechanically similar in the case of Mu-radicals, where the spin Hamiltonian (Eqn. 7) offers additional possibilities for spin relaxation. In the case of a typical radical like MuC$_2$H$_4\cdot$, there is *no* paramagnetic partner and thus *no intermolecular SF*. However, because of collisions perturbing the spin-coupled system in the radical, there is an *intra*molecular spin relaxation, due, e.g., to the electron spin rotation (SR) interaction, $\mathbf{S}\cdot\mathbf{C_S}\cdot\mathbf{J}$, where $C_S$ is an appropriate (electron) SR coupling constant and $J$ is the rotational angular momentum of the molecule. The unpaired electron spin in the radical is flipped by collisions and this can be qualitatively viewed as a change of some equivalent (free radical) state from $|A\rangle \rightarrow |B\rangle$, effecting a depolarization of the muon spin.

## III. MUONIUM SPIN EXCHANGE COLLISIONS

### A. Theory: Slow Spin Exchange

Of principal interest here is the regime of "slow" spin exchange where the spin flip rate is slow compared to the muonium hyperfine frequency. In Senba's stochastic treatment of spin exchange [13, 14], for a Poisson process, the statistically averaged muon polarization observed at time $t$ is

$$P_{\lambda_{SF}}(t) = \sum_{n=0}^{\infty} e^{-\lambda_{SF} t} \lambda_{SF}^n \int_0^{t_2} dt_1 \int_0^{t_3} dt_2 \cdots \int_0^{t_n} dt_{n-1} \int_0^t dt_n$$
$$G(t - t_n) G(t_n - t_{n-1}) \cdots G(t_2 - t_1) G(t_1 - t_0) \quad (10)$$

where $G(t)$ is the time evolution function of the muon spin in muonium.

### B. Transverse Field (TF)

In a TF, the function $G(t)$ is just Eqn. (4) exhibiting the four allowed hyperfine transitions (Fig. 2), which can be rewritten

$$G_T(t) = \frac{1}{4} e^{i\omega_M t} \left[ (1-\delta)\left(e^{i(\omega_0 + \Omega)t} + e^{i\Omega t}\right) + (1+\delta)\left(e^{-i(\omega_0 + \Omega)t} + e^{-i\Omega t}\right) \right] \quad (11)$$

where $\omega_M = (\omega_e - \omega_\mu)/2$ is the Larmor precession frequency of "triplet" muonium, given by $\omega_M/2\pi = 1.39$ MHz/G, and $\Omega$ is the split (or "beat") frequency, given by $\Omega = (\omega_{23} - \omega_{12})/2 = \omega_0(\sqrt{1+x^2} - 1)/2$, with again $\delta = X/\sqrt{X^2+1}$.

If the applied transverse field is low ($B < 10$ G), both $\delta$ and $\Omega$ are $\approx 0$, and, from Eqn. (10), the muon polarization averaged over $\omega_0$ is

$$P_T(t) = \tfrac{1}{2} e^{i\omega_M t} e^{-\lambda_{SF} t/2} \quad (12)$$

It is important to note, in comparison with Eqn. (2), that the observed muonium depolarization rate is *half* of the spin flip rate

$$\lambda_T = \lambda_{Mu} = \tfrac{1}{2}\lambda_{SF} \quad (13)$$

The result in Eqn. (13) is valid for $S=1/2$ collision partners (Cs, NO). For spin partners of general spin $S$ (e.g., for $S=1$ $O_2$), it is convenient to introduce a spin statistical factor $f$, which emerges from consideration of the angular momentum states involved [12], such that in a weak TF

$$\lambda_T = \lambda_{Mu} = f \lambda_{SF} . \qquad (14)$$

(For $S=1/2$, $f=1/2$; for $S=1$, $f=16/27$; for $S=3/2$, $f=15/24$.) These factors are important in extracting the correct SF cross section, $\sigma_{SF}$ for comparison with theoretical calculations. The Mu + Cs reaction is compared below in this context with the equivalent reactions for both Mu and He$\mu$ + Rb spin exchange [2].

If the applied field is $> 10$ G, but much smaller than $B_0$ so that $\delta \ll 1$, one observes two frequencies, and the $\mu$SR polarization is given by

$$P_T(t) = \tfrac{1}{4} \left[ e^{i(\omega_M + \Omega)t} + e^{i(\omega_M - \Omega)t} \right] e^{-3\lambda_{SF} t/4} \qquad (15)$$

with the observed depolarization rate now

$$\lambda_T = \tfrac{3}{4} \lambda_{SF} \qquad (16)$$

The physical basis for Eqns. (15) and (16) can be found in Eqn. (4), where it can be seen that only one quarter of the amplitude will survive each spin flip at a given (coherent) frequency. Therefore, the portion lost is $1 - (1/4) = 3/4$, corresponding to the depolarization rate $(3/4)\lambda_{SF}$ (and $1 - (1/2) = 1/2$ in weak TF's).

A chemical reaction that places the muon in a sufficiently different magnetic environment will destroy the precession coherence completely, yielding a depolarization rate that is independent of field. This feature can be used to distinguish spin exchange interactions ($\lambda_{SF}$) from chemical reactions ($\lambda_c$) in a TF, which would otherwise just contribute to the total observed relaxation rate. In practice, $\lambda_c$ is often $\ll \lambda_{SF}$ and the ratio of Eqn. (16) to Eqn. (13) = 3:2 provides a distinguishing characteristic for a SF collision. This ratio can be called the "3/2-effect", which is independent of the spin of the collision partner, and has been demonstrated recently in the cases of Mu + Cs [11] and Mu + $O_2$ [14] and with surprising indications in Mu + CO (in progress).

### C. Longitudinal Field (LF)

In a longitudinal field (LF, applied along the $z$ axis), the (initial) muon polarization in Mu is given by the expectation value $\langle \sigma_\mu^z \rangle$, leading to an expression for the time evolution of the muon spin, averaged over the hyperfine modulation (for $S=1/2$) of

$$P_L(t) = \frac{1+2X^2}{2(1+X^2)} \exp(-\lambda_L t) \quad (17)$$

$$\lambda_L = \tfrac{1}{2}\lambda_{\text{SF}} 1/1+X^2 \quad (18)$$

where $X = B/B_0$ has the same meaning as above. Note that as $X \to 0$, the LF result of Eqn. (18) and the (weak) TF result of Eqn. (13) are identical, as they should be. Moreover, from Eqn. (18), as $X \to \infty$, contributions to the measured relaxation rate from SF go to zero, giving no relaxation at all. In the presence of a competing *chemical* reaction rate, $\lambda_c$, the total relaxation rate is given by

$$\lambda_L = \tfrac{1}{2}\lambda_{\text{SF}} 1/1+X^2 + \lambda_c \quad (19)$$

even when, as for a diamagnetic product, the chemical reaction gives no spin relaxation. If $\lambda_L$ is plotted against $1/(1+X^2)$, the intercept is just $\lambda_c$ and the slope gives $\lambda_{\text{SF}}$. An example can be seen in Ref. 10 for the Mu+NO reaction.

### D. Experimental: Mu+Cs, Mu+O$_2$ and Mu+NO

The SF cross sections, $\sigma_{\text{SF}}$, have been measured for Mu+Cs [11] and Mu+NO [10] (both $S=1/2$), and for Mu+O$_2$ ($S=1$) [14,26]. Also measured, in the case of Mu+NO, was the rate for the chemical addition reaction forming (diamagnetic) MuNO* (which remains in a low pressure, "termolecular" regime). The experimental, thermally averaged, SF cross section can be found from the weak TF result of Eqn. (14), by the relationship

$$\lambda_{\text{SF}} = \lambda_T / f = n\,\overline{v}\,\overline{\sigma}_{\text{SF}} \quad (20)$$

where $n$ is the number density of the reactant (e.g., Cs, O$_2$, NO) and $\overline{v} = \sqrt{8k_B T/\pi\mu}$ is the mean thermal velocity. The cross section $\overline{\sigma}_{\text{SF}}$ can be compared with the theoretical result, from a sum over partial waves. Generally the agreement with theory is poor.

The relaxation rate for the Mu+Cs SF reaction was measured in TF over the temperature range $\sim$ 540–640 K in a nickel-plated steel reaction vessel using a Cs "boiler" and a titration technique to determine the Cs concentration [11]. The time histogram $N(t)$ was fit to Eqn. (2) for low fields, and at intermediate fields by the two-frequency equivalent based on Eqn. (15). A 3/2-effect was clearly seen establishing that the collision is due soley to an electron spin-flip. Experimentally, $\sigma_{\text{SF}}(T)$ is determined by measuring $\lambda$ at several values of $n$ at a given temperature and fitting to Eqn. (20). This is illustrated for Mu+Cs in Fig. 3 for data taken at 565 K. The slope gives the thermal SF cross section, $\sigma_{\text{SF}}(T) = 41.8 \pm 1.2 \times 10^{-16}$ cm$^2$ at 565 K. Results over the temperature range $\sim$ 540–640 K are plotted below. A straight average of these data gives $\sigma_{\text{SF}} = 39.7 \pm 6.0 \times 10^{-16}$ cm$^2$

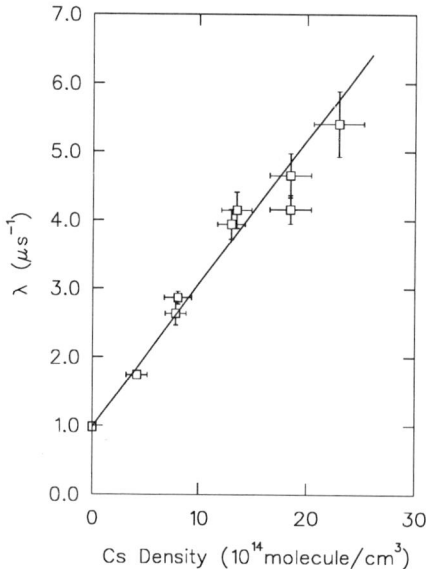

**FIG. 3.** Spin relaxation rate $\lambda$ vs. Cs number density at 566 K and an $N_2$ moderator pressure 2 atm. The vertical error bars are due to counting statistics; the horizontals are due to uncertainty in Cs density. The spin flip cross section, $\sigma_{SF}(T)$, is obtained from the slope and Eqn. (20) and is equal to $(41.8 \pm 1.2) \times 10^{-16}$ cm$^2$. The non-zero intercept is due to field inhomogeneities. The quoted error ($\pm 5\%$) is statistical only. From Ref. 11.

The $Mu + O_2$ reaction was measured in an Al vessel in weak TF's over the temperature range $\sim$90–500 K at pressures up to a few atm [26], and at room temperature up to 60 atm [14]. Both weak and intermediate TF's and the fitting functions given above were used, in comparison with $Mu + N_2O$ in 10 atm $N_2$ in order to check the aforementioned 3/2-effect. For the $Mu + N_2O$ reaction there should be no spin exchange possible at these conditions. Figure 4 shows the observed relaxation rates for $Mu + O_2$ (solid circles) as a function of field, along with those in pure $N_2$ (open circles). The solid line denotes the predicted $\lambda$ at intermediate fields based on the low field value at 4.5 G (dotted line). The $Mu + N_2O$ reaction is a purely chemical reaction at these conditions and hence shows no such field dependence, with the two relaxation rates at 4.5 G and 60 G (triangles) agreeing within the uncertainty of the measurement.

For Mu + NO, competing chemical addition and spin exchange

$$Mu + NO \underset{k_d}{\overset{k_a}{\rightleftharpoons}} MuNO^* \overset{k_s}{\underset{M}{\longrightarrow}} MuNO \qquad (21)$$

$$Mu(\uparrow) + NO(\downarrow) \longrightarrow Mu(\downarrow) + NO(\uparrow) \qquad (22)$$

have been measured by LF-$\mu$SR at room temperature at fields up to 15 kG, and with up to 60 atm of $N_2$ moderator. Since MuNO is *diamagnetic* it does not relax in a LF, leaving just the spin exchange depolarization mechanism. A plot of the LF $\mu$SR signal, $S_L(t)$, in a field of 7.6 kG is shown in Fig. 5. In this case, spin exchange is largely "quenched". From a plot of $\lambda_L$ vs. applied magnetic field, fitting Eqn. (19) to the data for different NO concentrations and moderator pressures, the spin exchange ($\lambda_{SF}$) and chemical reaction ($\lambda_c$)

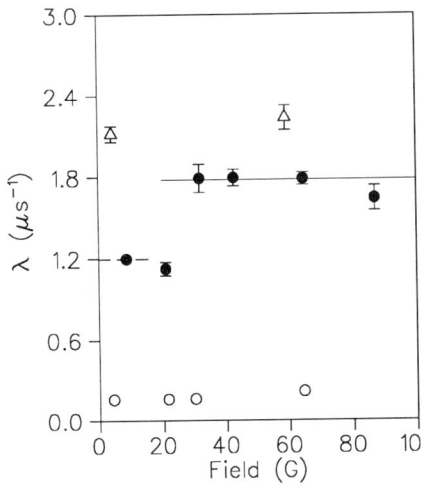

FIG. 4. Spin relaxation at low and intermediate transverse fields for Mu + $O_2$ (solid circles) in comparison with Mu + $N_2O$ (open triangles). A "3/2-effect" is seen for Mu + $O_2$ for small $O_2$ concentrations since only the spin exchange interaction relaxes the muon spin in this case. The solid line denotes the relaxation rate, $\lambda_{obs}$, predicted from the low-field value at 4.5 G (dotted line). In the Mu + $N_2O$ chemical reaction, at these pressures, there should be no spin exchange, so no field dependence. The open circles are for pure $N_2$ in the same field range.

relaxation rates can be obtained [10]. The chemical reaction rate is due to the formation of unstable MuNO* and hence is a strong function of moderator pressure. Comparison with H-atom results indicates a marked isotope effect in the unimolecular dissociation channel, $k_d$. In contrast, no (moderator) pressure dependence was observed in the spin exchange rates, the average value of which gives for the SF cross section, $\sigma_{SF}(T) = 8.0 \pm 0.3 \times 10^{-16}$ cm$^2$, in good agreement with an earlier value quoted in Ref. 12.

### E. He$\mu$ + Rb and Isotope Effects in Spin Exchange

As noted earlier, when *negative* muons are slowed down in a gas they are captured into "muonic orbits", at kinetic energies $\sim 10$ keV, which, in the case of He, is accompanied by the Auger loss of two electrons, resulting in a positive ion. Since the $\mu^-$ mass (as for the $\mu^+$) is 200 times the electron mass, the 1s muonic orbit is close to the nucleus [2] so that this ion looks just like a heavy proton, with a single positive charge. (Related studies have used *antiprotons* and other "exotic particles" captured in He [27]). It can then undergo charge exchange with added dopants (Xe, CH$_4$) in exact analogy with the $\mu^+$, forming the *neutral* species, He$\mu$ (He$^{++}\mu^- e^-$) [2, 28], which, can be viewed as an isotopic analogue of muonium, effectively looking just like a *heavy* H atom ($m_{He\mu}/m_H \approx 4$). This is an amazing 35 times heavier than the lightest H isotope, Mu. In the $\mu$SR realm, the principal difference between Mu and He$\mu$ is that the muon polarization in the ground state of the He$\mu$ atom is essentially zero, whereas, for Mu, in a weak magnetic field, it is 50%. In an elegant recent experiment, Barton et al. [2] have succeeded in *repolarizing* the $\mu^-$ in He$\mu$ by spin exchange collisions with optically-pumped (polarized) Rb,

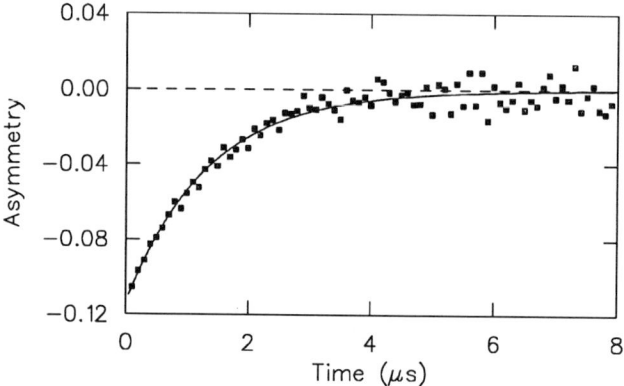

**FIG. 5.** The LF μSR signal for the Mu + NO reaction at ~ 300 K obtained with $3.5 \times 10^{16}$ molec cm$^{-3}$ of NO and 20 atm of N$_2$ moderator in a field of 7.6 kG. The solid line is a fit of the LF Signal $S_L(t)$ to the data, giving the relaxation rate $\lambda_L = 0.6$ μs$^{-1}$.

providing a measure of the SF cross section for the process

$$\text{Rb}(\uparrow) + \text{He}\mu(\downarrow) \longrightarrow \text{Rb}(\downarrow) + \text{He}\mu(\uparrow) . \tag{23}$$

The experiment measured the muon asymmetry from decay electrons in the manner outlined above for the $\mu^+$. A related set of experiments, using positive muons, measured the equivalent repolarization of muonium. (The *repolarization* curves for all species look much like Fig. 5 but asymptotic instead to maximum polarization.) From global fits to both the He$\mu$ and Mu + Rb repolarization data, under the (apparent) assumption that the cross section $\sigma_{\text{SF}}(T)$ is the *same* for all three isotopic species He$\mu$, H and Mu + Rb spin exchange, Barton et al. obtained the result $\sigma_{\text{SF}}(T) = 136 \pm 30 \times 10^{-16}$ cm$^2$ at about 475 K. (Separate errors are given in Ref. 2, but these we have added in quadrature.) This is quadruple the average value quoted above for the Mu + Cs reaction, $\sigma_{\text{SF}}(T) = 39.7 \pm 6.0 \times 10^{-16}$ cm$^2$, over somewhat higher temperatures. Why?

Before addressing this question, first reconsider the Mu + Cs results and, in particular, their comparison with theory. Results for $\sigma_{\text{SF}}(T)$ at different temperatures are plotted in Fig. 6 with the theoretical calculations of Dalgarno and Rudge [29] and Cole and Olson [30] for H + Cs. (To our knowledge, there are no direct experimental measurements.) The thermal spin-flip cross section $\sigma_{\text{SF}}(T)$ is found from the experimental relaxation rates and is related to the *theoretically calculated* quantity, $\sigma_{\text{SF}}(E)$ by

$$\sigma_{\text{SF}}(T) = \frac{1}{(k_B T)^2} \int_0^\infty \sigma_{\text{SF}}(E) \, E \, e^{-E/k_B T} \, dE \tag{24}$$

with

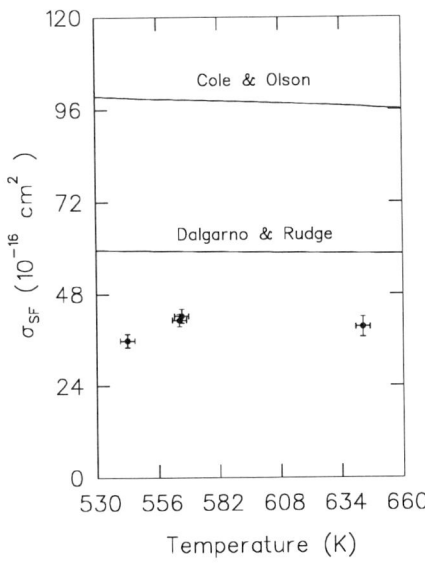

**FIG. 6.** Spin flip cross sections $\sigma_{SF}(T)$ for Mu–Cs. Solid lines are based on calculations for H–Cs from the work of Cole and Olson (top) [30] and Dalgarno and Rudge (middle) [29], as discussed in the text. The vertical errors on the experimental points are one standard deviation for the fits of Eqn. (20) to the data. Systematic errors are 10%. The possible decrease in the experimental $\sigma_{SF}(T)$ at the lowest temperature is seen also in related SE studies for the Mu + $O_2$ reaction [26].

$$\sigma_{SF}(E) = \frac{\pi}{k^2} \sum_{l=0}^{\infty} (2l+1) \sin^2(\Delta_l) \qquad (25)$$

where $k = \mu v/\hbar$ is the incident wave number, and $\Delta_l = \delta_l^3 - \delta_l^1$ is the *difference* in phase shifts for scattering from the interaction potentials for triplet and singlet states (for $S = 1/2$ collision partners).

At present, there are no theoretical calculations available for Mu–Cs spin-exchange collisions. The spin-flip cross sections for H–Cs calculated by Dalgarno and Rudge have used a straight-line trajectory method [29]. Those of Cole and Olson used a pseudopotential molecular-structure method to calculate the interaction potentials and a full quantal S-matrix technique, as well as a straight-line trajectory method to determine the spin-flip cross sections [30]. The total cross sections shown in Fig. 6 were calculated using Eqn. (24) and the parameters given by Dalgarno and Ridge in Ref. 29 and estimated from the reaction rate curve given by Cole and Olson in Ref. 30. Our measured Mu–Cs thermal spin-flip cross sections are significantly *lower* than the *calculated* H–Cs values: two thirds of the value of Dalgarno and Rudge, and less than half of the presumably more accurate value of Cole and Olson, indicating a *significant* isotopic effect. From comparisons with related Alkalai–Alkalai calculations, indications are that the H–A spin-flip cross sections are also underestimated so that the magnitude of this isotope effect could be greater still. The same trend has also been seen in the experimental results for Mu–$O_2$ *vs.* H–$O_2$ and Mu–NO *vs.* H–NO spin exchange, where the ratio of $\sigma_{SF}^{Mu} : \sigma_{SF}^{H}$ is about 1:3 in both cases [12,26].

For Mu + Cs spin exchange, we can estimate $\sigma_{SF}^{Mu}(T)$ from straight-line trajectory methods [29,30]. Since Mu has essentially the same IP as H,

$$\left[\sigma_{SF}^{Mu}(E)\right]^{1/2} = A - B\ln(\mu_H/\mu_{Mu}) - B\ln(E) \tag{26}$$

where $A$ and $B$ are constants defined by $\left[\sigma_{SF}^{H}(E)\right]^{1/2}$ and the $\mu$'s are reduced masses. Eqn. (26) with Dalgarno and Rudge's parameters for H–Cs give $\sigma_{SF}^{Mu}(T) \approx 52 \times 10^{-16}$ cm$^2$ at 560 K. Though this is within 20% of the experimental value at this temperature, $41.8 \times 10^{-16}$ cm$^2$, the calculations may in fact underestimate the SF cross section. Similarly, Cole and Olson's method would give $\sigma_{SF}^{Mu}(T) \approx 85 \times 10^{-16}$ cm$^2$ at the same temperature, in disagreement with experiment. From these comparisons it seems that the calculations do *not* account for the isotope effect seen in Fig. 6, though a trend to a decreased spin-flip cross section for Mu is indicated.

In Eqn. (25), the difference in phase shift, $\Delta_l = \delta_l^3 - \delta_l^1$, will disappear at large separations, so that there is a maximum number ($l_{max}$) of partial waves contributing to the cross section. Since Mu is lighter than the H atom by a factor of 9, $l_{max}$ for Mu is about three times less than for H at a given energy. The measured isotope effect is likely due to the fewer partial waves for Mu scattering [12,26], as well as different resonances, but accurate calculations are called for.

Now consider again the Mu+Cs and He$\mu$+Rb SF cross sections, $\sigma_{SF}^{He\mu+Rb} = 136 \pm 30 \times 10^{-16}$ cm$^2$ vs. $\sigma_{SF}^{Mu+Cs} = 40 \pm 6 \times 10^{-16}$ cm$^2$. In the light of the above discussion, the enhancement for the much heavier He$\mu$ isotope seems reasonable, with perhaps a factor of six increase in $l_{max}$ expected, consistent with the aforementioned trend in the measured SF cross sections for Mu(H)+NO and O$_2$. In short, we *do* expect an isotope effect in thermal SF cross sections at typical thermal energies, $\sim$ 100–1000 K. (At much higher energies, where an RPA approximation to Eqn. (25) may be appropriate, the SF cross sections would be mass and temperature independent.) The puzzling aspect in the treatment of Barton et al. [2] is their seeming assumption that there should be *no* isotope effect in spin exchange. It may be that the experimental techniques are more different than appears at first glance; for example, the polarized Rb target used in Ref. 2 could perhaps account for a factor of two. It can also be noted that the He$\mu$+Rb SF cross sections were meausured at temperatures about 100 K lower than the Mu+Cs data (though this only tends to exacerbate the difference).

## IV. SPIN RELAXATION OF MUONIUM RADICALS IN GASES

### A. Model for spin relaxation

The general theory of (NMR) spin relaxation in gases includes both *inter-* and *intra*molecular relaxation mechanisms. However, at low densities, intermolecular spin relaxation is of consequence only for atomic radicals [31] or for collisions with other paramagnetic species (e.g., in spin exchange [3, 11, 26]). For MuC$_2$H$_4$ and related molecules, with many degrees of freedom ($\langle J \rangle \approx 15$), it is *intra*molecular spin relaxation which dominates. There is only *one* such free radical present at a given time, which is therefore studied in an ultra-dilute limit. This situation would be impossible in the corresponding ESR studies, which is one of the prime advantages of the $\mu$SR technique in testing theories of spin relaxation. Although intermolecular spin coupling does not directly cause spin relaxation, collisions play a vital role by perturbing the spin-coupled system, causing primarily a reorientation of angular momentum ($|J, M\rangle$).

Spin relaxation in such a multilevel system as MuC$_2$H$_4$ can be described by a superposition of relaxation terms corresponding to multiple contributions from (off-diagonal) transitions between pairs of energy levels $|i\rangle \to |j\rangle$. Generally one relaxation time (the longest) will dominate, consistent with both experimental observation of a single relaxation and recent theoretical calculations [32], and hence the total relaxation rate is expected to have the classic form of Eqn. (27),

$$\lambda_L = 1/T_1 = \sum W_{ij} J(\omega_{ij}) = \sum W_{ij} 2\tau_c / (1 + (\omega_{ij}\tau_c)^2) \qquad (27)$$

familiar from magnetic resonance, where $W_{ij}$ is due to specific contributions from off-diagonal matrix elements of particular local field components ($W_{ij} = \langle j|H_{\text{int}}|i\rangle^2$) and $J(\omega_{ij})$ is the usual spectral density response of transition frequencies. The quantities $W_{ij}$ and $\omega_{ij}$ must be calculated or found from specific models, whereas $\tau_c$, the motional correlation time, is determined by collisions through the intermoleclar potential. In the kinetic-molecular theory of spin relaxation [12,31,32], $\tau_c$ involves a collison integral of the density operator, which implies it is proportional to the time between collisons. Our model approach seeks to identify the allowed transitions and their contributions to the muon spin relaxation via various matrix elements.

We may expect relaxation of the muon spin to be caused by fluctuations in the *isotropic* hyperfine interaction ($\langle A_\mu(\mathbf{I}\cdot\mathbf{S})\rangle$), in the *anisotropic* hyperfine interaction ($\langle \delta A(\mathbf{I_+S_- + I_-S_+})\rangle$, or $\langle \delta A \mathbf{S_z I_+}\rangle$ or $\delta A \langle \mathbf{I_+S_+}\rangle$) and in the *spin-rotation* interactions, both of the electron ($C_S \langle \mathbf{S}\cdot\mathbf{J}\rangle$) and the muon ($C_I \langle \mathbf{I}\cdot\mathbf{J}\rangle$). (These are in fact tensor couplings.) In our model a Mu-radical is treated as a psuedo Mu-atom, with reduced hyperfine coupling ($\nu_0(\text{R}) \ll \nu_0(\text{Mu})$), for which the solutions of the isotropic muonium Breit-Rabi hamiltonian of Eqn. (7) (with terms in $I_p = 0$) corresponding to Fig. 2 are well known. From the energy levels of Fig. 2, these fluctuating fields will give rise to "spin flip" transitions relaxing the electron ($\Delta M_e = \pm 1$), which in turn are coupled to the muon, to direct muon-electron "flip-flop" ($\Delta M_{\mu,e} = 0$) or "flip-flip" ($\Delta M_{\mu,e} = 2$), and to direct muon spin flip ($\Delta M_\mu = \pm 1$) with strengths $W_{ij}$, denoted subsequently by $\Delta_E^2$, $\Delta_{ME}^2$, and $\Delta_M^2$, respectively.

In a *longitudinal* field, for $T_1$ gas phase relaxation rates, this model gives rise to expression with several terms having different field-weightings and spectral densities, [33], but, for the magnetic fields of interest here ($B \gtrsim 0.5\,\text{kG}$), this can be written accurately as

$$\lambda_L = \frac{1}{T_1} = \frac{\Delta_E^2}{1+X^2}\frac{2\tau_c}{1+(\omega_e\tau_c)^2} + \frac{\Delta_{ME}^2\, 2\tau_c}{1+(\omega_e\tau_c)^2} + \Delta_M^2[J(\omega_{12})+J(\omega_{34})] \quad (28)$$

Here $\tau_c$ (essentially the time between collisions), is assumed to be the same for all terms, while the parameters $\Delta_E$, $\Delta_{ME}$ and $\Delta_M$ represent the aforementioned local field fluctuations for a given radical.

In a *transverse* field, electron ($T_1$) relaxation contributes directly as well to muon ($T_2$) relaxation, and hence

$$\lambda_T = \frac{1}{T_2} = (\Delta_E^2 + \Delta_{ME}^2)\frac{\tau_c}{1+(\omega_e\tau_c)^2} + \Delta_S^2\tau_c \quad (29)$$

where it is anticipated that the $\Delta_M$ term in Eqn. (28) makes a negligible contribution to the $T_2$ rate. In addition to off-diagonal ($T_1$) contributions to the muon $T_2$ relaxation rates, Eqn. (29) assumes a (diagonal) "secular" term, $\Delta_S$, which also includes inhomogeneous line broadening. The $T_2$ relaxation rates were obtained either from the widths of Fourier transform peaks or directly from the time histograms, from fitting the data to the $\mu$SR signal of Eqn. 8.

## B. Results and Interpretation of the Parameters

Typical of a variety of polyatomic free radicals is MuC$_2$H$_4$ formed by reaction (6). Both LF and TF relaxation rates were measured over a range of pressures from $\sim$1-15 atm and for applied fields in the range $\sim$0.5-35 kG. Since the same parameters should fit both the $T_1$ and $T_2$ data (excluding the $\Delta_S$ parameter in the $T_2$ relaxations), a *global* fit of *all* the data to Eqns. (28) and (29) was carried out. Fig. 7 shows the LF data at pressures of 1, 3.8 and 11.9 atm *vs.* applied longitudinal field with their fits. Note the log scale! The quality of fit is indicative of that seen for other radicals as well. The parameters from the fits for isotopically-substituted Mu-ethyl radicals as well as for the Mu-t-butyl radical are listed in Table I, along with the values of the isotropic hfc's, $A_\mu$, for each radical. A similar quality of fit is seen for the $T_2$ data which is actually better described by a slight ($\pm 5\%$) parameter variation (the * parameters in Table I). It is satisfying to note that the present model also accounts for the $T_2$ data of Roduner and Garner for MuC$_2$H$_4$ at much lower fields and much higher pressures [34].

**The $\Delta_E$ Parameter:** The spin rotation (SR) interaction has the form $\mathbf{J}\cdot\mathbf{C}\cdot\mathbf{S}$, where $C$ is a coupling tensor that depends on the moment of inertia $I_r$ of the rotating molecule and on the anisotropy of the *electron g* factor, $\delta g_e$. This

**TABLE I:** Isotropic hfc ($A_\mu$) and Relaxation Parameters for $MuC_2H_4$, $MuC_2D_4$, $Mu^{13}C_2H_4$ and $MuC_4H_8$.

| Parameter | $MuC_2H_4$ | $MuC_2D_4$ | $Mu^{13}C_2H_4$ | $MuC_4H_8$ |
|---|---|---|---|---|
| $A_\mu$ (MHz) | 330 | 340 | 331 | 290 |
| $\Delta_E$ ($\mu s^{-1}$) | $2124 \pm 20$ | $1389 \pm 38$ | $2474 \pm 55$ | $379 \pm 34$ |
| $\Delta_{ME}$ ($\mu s^{-1}$) | $328 \pm 2$ | $339 \pm 28$ | $320 \pm 10$ | $156 \pm 7$ |
| $\Delta_M$ ($\mu s^{-1}$) | $23 \pm 2$ | $30 \pm 4$ | $50 \pm 11$ | $84 \pm 5$ |
| $\Delta_S$ ($\mu s^{-1}$) | $441 \pm 9$ | $859 \pm 47$ | | |
| $\tau_c$ (1 atm)(ps) | $65 \pm 1$ | $93 \pm 2$ | $54 \pm 9$ | $128 \pm 19$ |
| $\Delta_E^*$ ($\mu s^{-1}$) | $2254 \pm 53$ | $1381 \pm 58$ | | |
| $\Delta_S^*$ ($\mu s^{-1}$) | $408 \pm 16$ | $845 \pm 58$ | | |
| $\tau_c^*$(1 atm)(ps) | $74 \pm 1$ | $93 \pm 1$ | | |

interaction leads to spin relaxation, because both **J** and **C** are modulated by molecular collisions in the gas and hence are time dependent. Relaxation due to the SR interaction is mainly expected to contribute to the $\Delta_E$ term, having the form [31, 34, 35],

$$\frac{1}{T_1(SR)} = \frac{2}{3}\langle J(J+1)\rangle C_S^2 \cdot J(\omega_{ij}) = \frac{4k_B T\, I_r C_S^2}{3\hbar^2} J(\omega_{ij}) \propto \Delta_E^2 J(\omega_{ij}) \quad (30)$$

It is well known from the "Curl formula" in ESR studies [35, 36] that the magnitude of the electron SR coupling constant for a polyatomic radical can be written in the form,

$$C_S \approx -(\hbar^2/I_r) \times \delta g_e \quad (31)$$

Calculations give $C_S = 360\,\mu s^{-1}$ and hence, from Eqn. (30), one finds

$$\Delta_E^2 = \frac{4 I k_B T}{3\hbar^2} \times C_S^2 = 7.1 \times 10^{18}\,s^{-2}$$

or $\Delta_E \approx 2700\,\mu s^{-1}$. The calculated value of $2700\,\mu s^{-1}$ is certainly in acceptably good agreement with the experimentally determined value for $\Delta_E$ in Table I, supporting the expected importance of the SR interaction in relaxing the muon in $MuC_2H_4$ (as $X \to 0$). It is important to note from Eqns. (30) and (31) that the parameter $\Delta_E$ should *decrease* with increasing moment of inertia, $I_r$, as $1/\sqrt{I_r}$, a trend which is clearly seen in the values in Table I. It is worth remarking that a SR interaction corresponding to $\Delta_E \sim 3000\,\mu s^{-1}$, would render ESR linewidths far too broad ($\gtrsim 200\,G$) to be easily seen in the gas phase at normal pressures, a fact which has compromised ESR studies in gases for many years, exemplified by one of the few such results reported long ago by Schafsma and Kivelson for $(CF_3)_2NO$ [36].

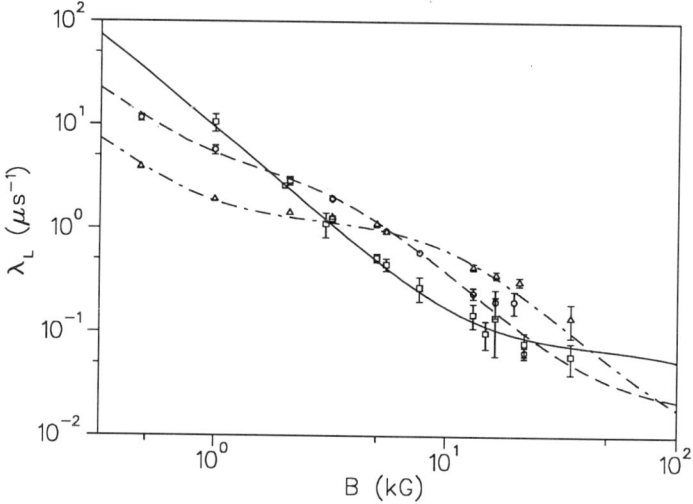

**FIG. 7.** The LF $T_1$ spin relaxation rates for the MuC$_2$H$_4$ radical vs. applied LF, for pressures of 1 (□), 3.8 (○), and 11.9 (△) atm pure ethene. Note the log scale! The solid curves are *global fits* of the data from Eqns. (28) and (29) and are typical of fits to other radicals, examples of which are given in Table I.

**The $\Delta_{ME}$ Parameter:** Dipolar coupling between the muon and the electron contains the terms of the anisotropic $\mu - e$ hfc and hence can be considered as being responsible for that part of the $\Delta_{ME}$ term which is due to hyperfine anisotropy. Direct muon spin relaxation due to dipole–dipole coupling is of the form

$$\frac{1}{T_{1d}} = \frac{A(J) \cdot \gamma_I^2 \gamma_S^2 \hbar^2}{r^6} J(\omega_{ij}) \propto \Delta_{ME}^2 J(\omega_{ij}) \qquad (32)$$

where $A(J)$ is a numerical factor that depends on the rotational state $(J)$ of the target molecule and $\gamma_I$ and $\gamma_S$ are the gyromagnetic ratios of the ($\beta$) muon and electron, respectively, while $r$ is the muon–electron distance in the radical. For large $J$, here assumed to be some single (average-large) $\langle J \rangle$ that is changing orientation by collisions, $A(J)$ is ~1 [31].

From the form of Eqn. (32) and utilizing data for the ethyl radical, we obtain $\Delta_{ME} \sim 180\,\mu\text{s}^{-1}$, about a factor of two lower than the experimentally determined value given in Table I. Qualitatively similar results could be expected for other alkyl radicals as well. Although contributions from hyperfine anisotropy are surely important, the fitted parameters are appreciably larger than this interaction alone would suggest, indicating that modulation of the anisotropic part of the SR interaction and/or modulation of the *isotropic* muon hfc by collisions also contributes. The latter contribution is likely most important for the ($\beta$-hfc) Mu-t-butyl radical judging from a recent (liquid phase)

ESR study of the corresponding H-radical [37].

**The $\Delta_M$ Parameter:** In the present model, the $\Delta_M$ interaction parameter is due to the same types of matrix elements that give rise to $\Delta_E$ but involving instead the operator $\mathbf{I}_\pm$, which gives rise to *direct* muon spin flips. Assuming, in parallel with the situation for $\Delta_E$, that $\Delta_M$ derives largely from terms in the *nuclear* spin-rotation interaction, $C_I \langle \mathbf{I} \cdot \mathbf{J} \rangle$, and hence with matrix elements of the form $\langle \mathbf{I}_\pm \mathbf{J}_z \rangle$, from the form of Eqn. (30) but for $C_I$, we would expect $\Delta_M \sim 12\,\mu s^{-1}$. This is in reasonable agreement with the experimental value of $23\,\mu s^{-1}$ in Table I, when the possibility of an appreciable systematic error ($\approx 30\%$) for these slow relaxation rates is considered. By these same arguments, a similar value could be expected for $\Delta_M$ for the other alkyl radicals in Table I, not inconsistent with the data. It may also be that the size of the $\Delta_M$ parameter is at least partly due to the "spreading width" from the (anisotropic, $\Delta M = 2$) Level-Crossing Resonance [24]. It is noted that possible contributions from $\mu-e$ hypefine anisotropy would be of the same order as those giving rise to $\Delta_{ME}$ and hence would be too large to account for the size of the $\Delta_M$ parameter.

## REFERENCES

1. Proc. of the 6'th Int'l Conference on $\mu$SR, *"Muon Spin Rotation, Relaxation and Resonance"*, Maui, Hawaii, June 1993, Hyp. Int., **87**, 1994; and references therein to previous conferences in this series.
2. A.S. Barton et al., Phys. Rev. Letts., **70**, 758 (1993).
3. H. Dilger, E. Roduner and D.G. Fleming et al., Hyp. Int., **87**, 899 (1994); and J. Phys. Chem, *submitted*.
4. T. Sugai, M. Sakamoto, A. Matsushita, K. Nishiyama, K. Nagamine and T. Kondow, J. Chem. Phys., **101**, 209 (1994).
5. M. Senba, J. Phys. B: At. Mol. Opt. Phys. **23** 1545 (1990); D.G. Fleming and M. Senba, *Atomic Physics with Positrons*, J. Humberston and E. Armour, eds., Plenum Press, London, 1987, P.343.
6. D.J. Arseneau, D.G. Fleming and M. Senba et al., Can. J. Chem., **66**, 2018 (1988).
7. D.G. Fleming and M. Senba, in *Recent Advances in Meson Science*, T. Yamazaki, K. Nakai and K. Nagamine, eds., (1992), p. 219.
8. D.G. Fleming, in *Proc. of XII'th ICPEAC*, Gatlinburg, North Holland Press, 1981, Pg. 297.
9. V.W. Hughes, D.W. McColm, K. Ziock and R. Prepost, Phys. Rev. **1A**, 595 (1970); R.M. Mobley, J.J. Amato and V.W. Hughes et al., J. Chem. Phys. **47**, 3074 (1967).
10. J.J. Pan, A.C. Gonzalez, and M. Senba et al., J. Phys. Chem., *submitted*.
11. J.J. Pan, M. Senba and D.J. Arseneau et al., Phys. Rev. **48A**, 1218 (1993).
12. R.E. Turner, R.F. Snider and D.G. Fleming, Phys. Rev. **A41**, 1505, (1990).

13. M. Senba, Phys. Rev. **50A**, 214 (1994); M. Senba, J.Phys. **B26**, 3215, (1993).
14. M. Senba, J.J. Pan and S. Baer et al., Hyp. Int., **87**, 965 (1994).
15. R. Snooks, D.J. Arseneau and D.G. Fleming et al., **102**, 4860 (1995).
16. D.M. Garner, D.G. Fleming, D.J. Arseneau, et al., J. Chem. Phys., **93** 1732 (1990).
17. R.J. Duchovic. A.F. Wagner, R.E. Turner, D.M. Garner and D.G. Fleming, J .Chem. Phys. **94** 2794 (1991).
18. R.H. Heffner and D.G. Fleming, Physics Today, December 1984
19. D.G. Fleming, D.M. Garner and J.H. Brewer et al., Adv. in Chem. **175**, 279 (1979); J.H. Brewer and K.M. Crowe, Ann. Rev. Nucl. Sci., **28**, 239 (1978).
20. E. Roduner, Chem. Soc. Revs.(London), 1993; E. Roduner, *Lecture Notes in Chemistry*, Springer Verlag, **49** (1988).
21. P.W. Percival, R.F. Kiefl and D.G. Fleming et al., Chem. Phys. Letts, **163**, 241 (1989)
22. M. Senba, D.J. Arseneau and J.J. Pan et al., J. Radio. and Anal. Chem., **190**, 493 (1995).
23. R.F. Kiefl and S.R. Kreitzman, in *Perspectives in Meson Science*, Elsevier, 1992, pg. 265; S.R. Kreitzman, Chem. Phys., **152**, 353 (1991).
24. S. R. Kreitzman, Chem. Phys., **192**, 189 (1995).
25. C.J. Jameson, A.K. Jameson, N.C. Smith, J.K. Wang and T. Zia, J. Phys. Chem., **95**,1092 (1991); C. Lemaire and R.L. Armstrong, J. Chem. Phys., **81**, 1626 (1984).
26. M. Senba, D.G. Fleming, D.J. Arseneau, et al., Phys. Rev. **A39**, 3871 (1989)
27. R.S. Hayano, M. Iwasaki and T. Yamazaki, in *Perspectives in Meson Science*, Elsevier, 1992, pg. 417.
28. H. Orth, Hyp. Int., **19**, 829 (1984); P.A. Souder, Phys. Rev. **22A** 33, 1980.
29. A. Dalgarno and M.R.H. Rudge, Proc. R. Soc. **A 286**, 519 (1965).
30. H.R. Cole and R.E. Olson, Phys. Rev. **A31**, 2137 (1985).
31. F.M. Chen and R.F. Snider, J. Chem. Phys., **48**, 3185 (1968); *ibid*, **46**, 3937 (1967).
32. R.E. Turner and R.F. Snider, Phys. Rev. **50A**, 4743 (1994).
33. J.J. Pan, D.G. Fleming and M. Senba et al., Hyp. Int., **87**, 865 (1994); D.G. Fleming, R.F. Kiefl and S.F. J. Cox et al., Hyp. Int., **65**, 767 (1990).
34. E. Roduner and D.M. Garner, Hyp. Int. **32**, 733 (1986).
35. W. Weltner in *Magnetic Atoms and Molecules*, Van Nostrand Press, 1982.
36. T.J. Schaafsma and D. Kivelson, J. Chem. Phys., **49**, 5235 (1968).
37. G.H. Goudsmit, F. Jent and H. Paul, Zeitschrift. für Phys. Chem., **180**, 51 (1993).

# ION-ATOM COLLISIONS
## – THEORY –

Time Dependent Description of Inner Shell Excitation
and Transfer in Ion-Atom Collisions .................................... 435
    B. FRICKE, P. Kürpick, and W.-D. Sepp
Molecular Treatment of Ion-Atom Collisions
at Intermediate Energies ................................................. 445
    C. HAREL, H. Jouin, B. Pons,
    L.F. Errea, L. Mendez, and A. Riera
Time-Dependent, Lattice Approach to Atomic Collisions ......... 455
    D. R. SCHULTZ
Theory of Low Energy Ion-Atom Collisions ........................ 471
    E.A. SOLOV'EV

# Time dependent description of inner shell excitation and transfer in ion-atom collisions

B. Fricke, P. Kürpick[1], W.–D. Sepp

*Fachbereich Physik, Universität Kassel, D-34109 Kassel*

---

In order to achieve an ab initio description of electronic inner shell excitation and transfer processes in many electron heavy ion–atom collision systems the time dependent Dirac equation has to be solved. In recent years we have developed a program solving the time–dependency of the electrons in the Dirac-Fock-Slater (DFS) approximation using a relativistic many–electron (LCAO–MO) molecular basis. Thereby the nuclei move on classical trajectories. This basis set ansatz leads to the wellknown close coupling equations which are solved for all initially occupied electronic levels. The actual experimental question which should be answered is calculated within the framework of inclusive probabilities which reinstates the many–particle aspect of the collision system.

As examples of the progress which has been achieved during the last years several heavy–ion atom collision systems are presented and compared with experimental highly differential cross-sections. The improvements for the impact parameter dependent excitation and transfer processes are significant compared to available theories used so far. It is explained which improvement is due to the relativistic treatment resp. the many–particle aspect taken into account through the inclusive probability formalism. In the last part it is shown, that even more complicated ion–solid target collision systems can be described within our method.

---

## INTRODUCTION

Nearly 30 years ago Fano and Lichten [1] came up with the idea that inner shell excitation and transfer during the collision can simply be described by rotational coupling of the two quasi–molecular levels $2p\sigma$ and $2p\pi$ at very small internuclear distances. This strong coupling between the adiabatic molecular levels is due to the fact that the motion of the electrons in such a colliding system is described by the time-dependent Schrödinger– or Dirac–equation where the operator $\frac{d}{dt}$ is responsible for the coupling. This process has been described nearly 20 years ago in a theoretical paper by Taulbjerg and Briggs

---

[1]Present address: Kansas State University, Department of Physics

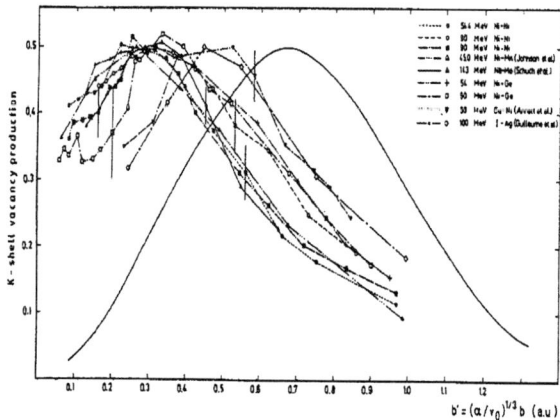

**FIG. 1.** Ion–solid target measurements versus Taulbjerg–Briggs scaling law (for details see Ref. [7])

[2]. They derived a scaling law for the impact parameter dependent probability of a charge transfer from an empty atomic L-shell to the K-shell of the target atom. Following this early theoretical paper an impressive number of experimental results on inner shell charge transfer has been published up to now [3–10]. Two main results showed up: for the gas target collision experiments the Taulbjerg–Briggs theory seemed to describe reasonably well the position of the maximum of the adiabatic peak whereas its maximum was shifted to almost twice the experimental value when compared to solid target measurements as shown in Fig. 1 taken from the paper of Schuch et al. [7]. The authors therefore concluded that an unknown solid–state effect should be the reason for this discrepancy. The apparent mismatch in the position of the maximum as well as the unknown absolute value of the charge transfer cross section was unexplained until recently.

In this progress report we present results showing that both the absolute position and value of the experimental P(b) curves can be explained within our method therefore laying to rest this old discrepancy. In addition we are able to give a first ab initio description of the target thickness dependence of charge transfer cross sections for ion–solid collision systems. In the next chapter we briefly describe the theoretical method and in the last chapter we present results for various collision systems and mainly focus on ion–solid target systems.

## GENERAL DESCRIPTION OF THE METHOD

In order to describe the time dependent behavior of N electrons in the collision system of the incoming ion plus the target atom or ion we solve the time dependent Dirac-Fock-Slater equation

$$\left(\underbrace{\hat{t} + \hat{V}^N(t) + \hat{V}^C(t) + \hat{V}^{Ex}_\alpha(t)}_{\hat{h}^{TDDFS}(t)} - i\hbar\frac{\partial}{\partial t}\right)\psi_i(t) = 0; \quad i = 1, \cdots, N \qquad (1)$$

which is obtained from the variation of the functional

$$I = \int_{-\infty}^{+\infty} dt < \Psi \mid H - i\hbar\frac{\partial}{\partial t} \mid \Psi >; \quad \Psi = det(\psi_1, \cdots, \psi_N).$$

$\hat{V}^N(t)$ is the electron–nuclear potential, $\hat{V}^C(t)$ is the Coulomb potential and $\hat{V}^{Ex}_\alpha(t)$ is the Slater exchange potential.

In our ansatz the time dependent wavefunction $\psi_i(\vec{r}, t)$ is expanded in a set of self consistent single–particle molecular wavefunctions

$$\psi_i(\vec{r}, t) = \sum_j^M a_{ij}(t)\phi_j(\vec{r}; \vec{R}(t)). \qquad (2)$$

Using this molecular basis set [11] in eqn1 leads to a set of equivalent first order differential equations

$$\dot{a}_{ij} = \sum_j -a_{im} < \phi_j \mid \frac{\partial}{\partial t} \mid \phi_m > e^{-\frac{i}{\hbar}\int(\varepsilon_m - \varepsilon_j)dt'} \qquad (3)$$

for the time dependent single–particle amplitudes $a_{ij}$. To calculate many-particle probabilities in the scheme of inclusive probabilities [13,14] we need to evaluate the single–particle density matrix and therefore have to solve the coupled equation (3) for all the $N$ initial conditions

$$\lim_{t\to-\infty}\left(\psi_i(t) - \psi_i^0(t)\right) = 0 \quad ; i = 1, \cdots, N.$$

$\psi_i^0(t)$ is the initially occupied atomic orbital to which the molecular orbital $\psi_i(t)$ tends to for $t = -\infty$.

The molecular basis functions $\phi_j$ are chosen to be eigenfunctions of the time independent Dirac–Fock–Slater equation [11]

$$\left(\underbrace{\hat{t} + \hat{V}^N(\vec{R}) + \hat{V}^C(\vec{R}) + \hat{V}^{Ex}_\alpha(\vec{R})}_{\hat{h}^{MO}(\vec{R})}\right)\phi_j(\vec{r}; \vec{R}) = \varepsilon_j(\vec{R})\phi_j(\vec{r}; \vec{R}). \qquad (4)$$

The solution of equation (4) is achieved by expanding the molecular wavefunctions $\phi_j$ in numerically given atomic four-component Dirac spinors

$$\phi_j(\vec{r}; \vec{R}) = \sum_{\nu=1}^S c_{j\nu}(\vec{R})\xi_\nu(\vec{r}; \vec{R})$$

which can be located anywhere on the internuclear axis. The static Dirac–Fock–Slater equation (4) is therefore reduced to the secular equation

$$\mathbf{h}^{MO}\mathbf{c} = \varepsilon \mathbf{S}\mathbf{c} \tag{5}$$

which has to be solved selfconsistently for a large number of internuclear distances. Typically we solve equation (5) at about 100 internuclear distances to get both accurate eigenvalues and matrix elements along the whole trajectory.

The molecular dynamic coupling matrix element

$$<\phi_l \mid \frac{\partial}{\partial t} \mid \phi_m> = \dot{R} <\phi_l \mid \frac{\partial}{\partial R} \mid \phi_m> - \frac{i\dot{\theta}}{\hbar} <\phi_l \mid \hat{j}_y^{CM} \mid \phi_m>$$

is reduced by means of the atomic basis set representation of the molecular orbitals to four matrix elements involving only one– and two–center atomic matrix elements [12].

After solving the matrix equation (3) from $-\infty \leq t \leq +\infty$ the formalism of inclusive probabilities allows to calculate many particle probabilities from the time dependent single–particle amplitudes $a_{ij}$ [13,14]

The classical trajectories of the nuclei can either be deduced by using a model potential like the Coulomb potential or the screened Bohr potential or by the fully self consistent interatomic potential which we gain from solving equation (5).

## RESULTS AND DISCUSSION

### Gas target collision systems

We will not discuss here the "simple" systems like p on H because their successfull treatment is mainly achieved by using purely atomic basis functions as an ansatz in the time dependent Schrödinger equation which has extensively been done in the last decade (see for example ref. [15]).

A fairly simple system where our molecular basis set description is appropriate is $S^{15+}$ on Ar where one initial 1s vacancy in the $S^{15+}$ is partially transfered during the collision to the 1s shell of Ar. Fig.2 shows our result for this impact parameter dependent vacancy transfer probability as full line at various impact energies versus both experimental results and other theoretical calculations [16]. Both the experimental oscillatory structure which results from interferences between the incoming an outgoing channel and the absolute height are reproduced by our calculations.

Collision systems involving two holes in the initial channel are actually more complicated to handle as we have to evaluate the manifold of different finally possible many–hole states within the framework of inclusive probabilities and add those up that contribute to the experimental measurement. This has been done by us in two different ways as is shown as an example for the collision system 16 MeV $S^{16+}$ on Ar.

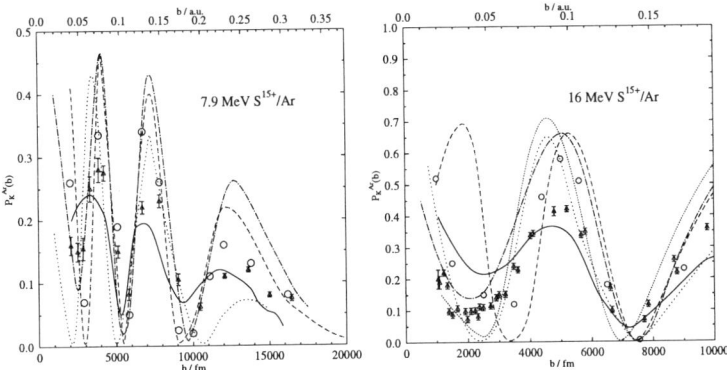

**FIG. 2.** Ar–K vacancy transfer probability for $S^{15+}$ on Ar at 7.9 and 16 MeV. The symbols used are explained in Ref. [16]

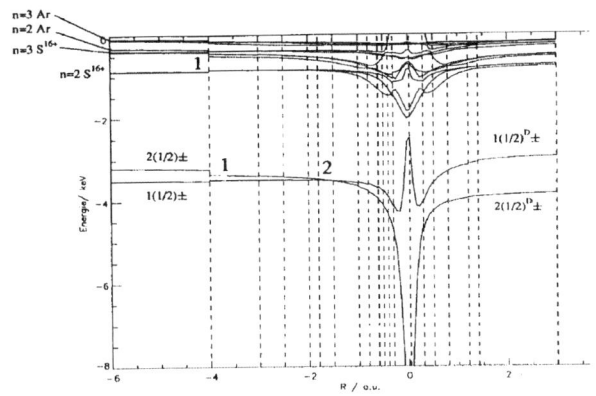

**FIG. 3.** TDHF–correlation diagram for the system 16 MeV $S^{16+}$ on Ar.

Results of close coupling calculation [17] using matrix elements and eigenvalues from static DFS calculations are presented in the left part of Fig.4 while a result of a full TDHF–calculation [18] is shown in the right part. The three many–vacancy probabilities shown are $P_{SS}$, the probability to have two vacancies in the S-1s shell, $P_{AA}$, the probability to have two vacancies in the Ar-1s shell and $P_{SA}$, the probability to have both one vacancy in the S-1s shell and one in the Ar-1s shell. The later TDHF–calculation fully includes the dynamical screening of all electrons therefore leading to a strong asymmetry in the matrix elements and eigenvalues between the incoming and outgoing part of the trajectory as can be seen in Fig.3.

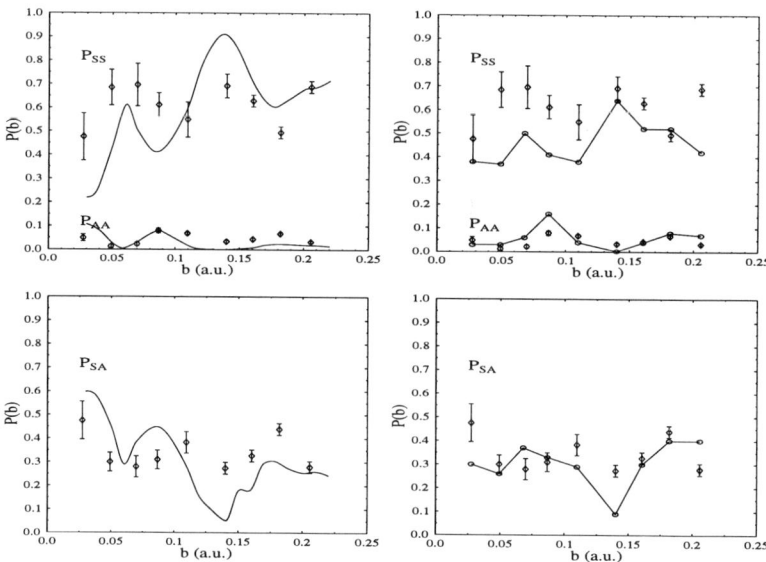

**FIG. 4.** TDHF- and SCF-results for $P_{SS}$, $P_{AA}$ and $P_{SA}$.

### Solid target collision systems

As stated in the introduction heavy–ion solid target measurements unlike the extensively studied ion–gas target systems showed systematic deviations from available theories. These deviations were interpreted as to reveal a fundamental difference between collisions with gas and those with solid targets. In this progress report we present recent results showing that solid target collision systems can be fully understood within the framework of our ab initio method. From our analysis we developed the following procedure:

Firstly, large impact parameter calculations allow us to determine how the L– and M–shells of initially low ionised projectiles are stripped within a few atomic layers of the solid target and therefore define the mean charge state of the ion and the approximate distribution of vacancies among the projectile L–shell. Utilising this information to define our projectile initial state we secondly do calculations at small impact parameters which give very good agreement with experimental results on K–vacancy charge transfer probability. As an example Fig.5 shows the probability for Ca–K, Ar–K vacancy probability and the sum of both for the collision system 40.6 MeV $Ar^{12+}$ on Ca–solid target [9,12].

Additionally we are able to understand target thickness dependencies within our ansatz as shown for the collision system 50 MeV Cu on Ni–solid target. Fig.6 shows the correlation diagram for Cu–Ni. The original measurements by Annett et al. [6] were done with $Cu^{9+}$ as incoming projectile therefore having neither initial M– or L–vacancies. The measurements of the final charge state

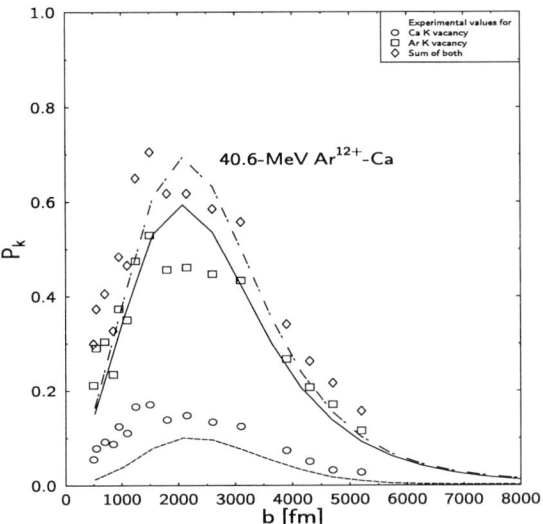

**FIG. 5.** $P_K$ vacancy probability for Ca–K, Ar–K and the sum of both for the system 40.6 MeV $Ar^{12+}$ on Ca.

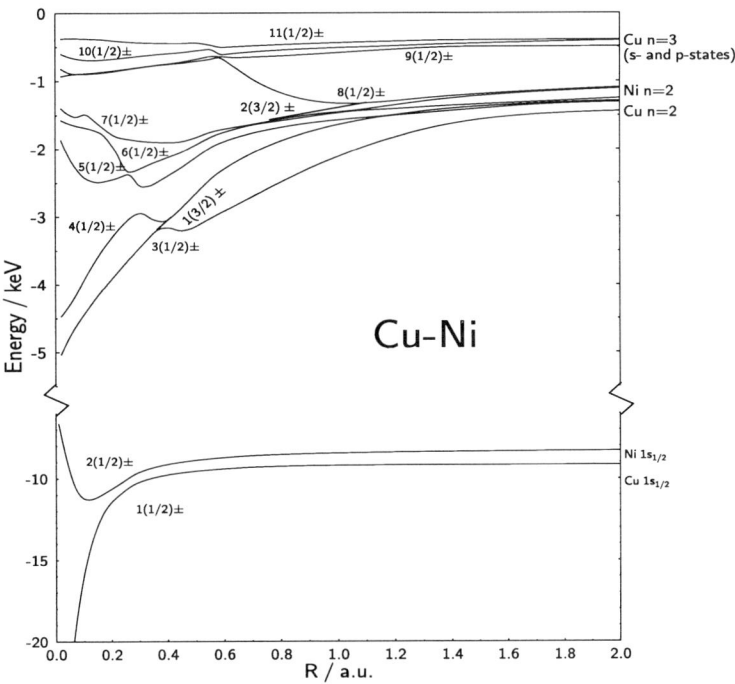

**FIG. 6.** Cu–Ni correlation diagram

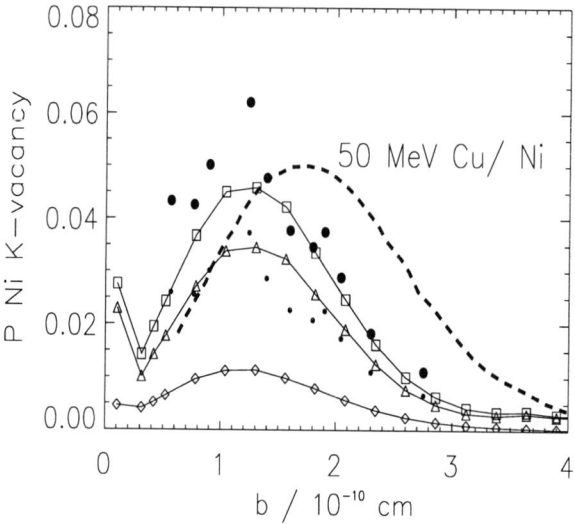

**FIG. 7.** $P_{Ni-K}$ vacancy probability for 50 MeV Cu on Ni.

of the projectile showed a broad spreading up to $Cu^{22+}$. We therefore calculated the Ni–K vacancy probability for each of the possible final Cu charge states and weighted these partial probabilities with the experimentally determined weights. In Fig.7 the triangles show the weighted contribution of the M-shell holes to the Ni–K vacancy probability. As in very thin targets M-shell vacancies are more likely to be created as L-shell vacancies this results should be compared to the small black circles which show the experimental low target thickness results. The squares show our total Ni–K vacancy probability therefore including both the contribution of M- and L-shell vacancies. It agrees very well with the thick–target measurement drawn as big black circles.

The former stated argument that we can split up the individual contributions of the L-shell manifold to the K-shell vacancy production is shown in Fig.8 for the collision system 108 MeV Br on Ni–solid target. The partial probabilities from the molecular $3(1/2)\pm$, $4(1/2)\pm$ and $1(3/2)\pm$ are presented while the total K-shell vacancy production probability is drawn as full line. The diamonds show experimental results from Schuch et al. [8]. The broken line was gained with the Taulbjerg–Briggs scaling law. The plot clearly identifies the contribution of the molecular $1(3/2)\pm$ levels as the non relativistic $2p\pi$–$2p\sigma$ counterpart but makes it clear that the contribution of the other levels among the L-shell manifold is not negligible.

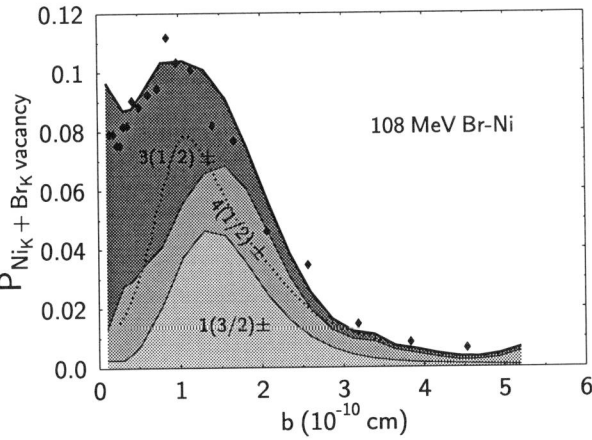

**FIG. 8.** Partial contributions of the molecular levels to the total K-vacancy production probability for 108 MeV Br on Ni.

## CONCLUSION

In this progress report we presented ab initio time dependent relativistic many particle calculations for the inner shell charge transfer in heavy ion atom collision systems. As a summary we highlight the following important points:

- The non-relativistic $2p\pi$–$2p\sigma$ scaling law of Taulbjerg and Briggs keeps its full validity only in the small Z region and is focussed on the soley contribution of the $2p\pi$ vacancies via rotational coupling.

- The many particle behavior of the collision system which is taken into account in our calculation both for the
  - energy eigenvalues and
  - matrix elements

and the evaluation of experimental questions within the framework of inclusive probabilities allows to describe both ion–gas and ion–solid target systems with high accuracy.

- Finally the inclusion of relativistic effects within our ansatz which mainly influence both the
  - energy eigenvalues and therefore the phases and the
  - matrix elements

also contribute to a quantitative description of ion–atom collision systems especially for heavy collision systems.

Using this well tested method we have now undertaken time dependent relativistic calculations for the inner shell charge transfer in highly stripped super heavy collision systems as $U^{91+}$ on Pb which have not been measured up to now.

## REFERENCES

1. U. Fano, W. Lichten, *Phys. Rev. Lett.* **14** (1965) 627
2. K. Taulbjerg, J. S. Briggs, *J. Phys. B.* **8** (1975) 1895
   K. Taulbjerg, J. S. Briggs, J. Vaaben, *J. Phys. B.* **9** (1976) 1351
3. B. M. Johnson, K. W. Jones, W. Brandt, F. C. Jundt, G. Guillaume, T. H. Kruse, *Phys. Rev. A* **19** (1977) 81
4. I. Tserruya, H. Schmidt–Böcking, R. Schuch, *Phys. Rev. A* **18** (1978) 2482
5. R. Schuch, G. Nolte, *Z. Phys. A.* **293** (1979) 91
6. C. H. Annett, B. Curnutte, C. L. Cocke, *Phys. Rev. A* **19** (1979) 1038
7. R. Schuch, G. Nolte, H. Schmidt–Böcking, *Phys. Rev. A* **22** (1980) 1447
8. R. Schuch, R. Hoffmann, K. Müller, E. Pflanz, H. Schmidt–Böcking and H. J. Specht, *Z. Phys. A.* **316** (1984) 5
9. T. Kambara, R. Schuch, Y. Awaya, T. Mizogawa, H. Kumagai, Y. Kanai, H. Shibata, K. Shima, *Z. Phys D.* **22** (1992) 451
10. M. Jäger, M. Schulz, T. Kandler, A. Warczak, H. Bräuning, A. Demian, M. Damrau, A. El–Sadek, K. Freitag, J. Ullrich, H. Schmidt–Böcking, *7. HCI–Conference Proceedings, Vienna (Austria) (to be published in Nucl. Inst. Meth. B.)* (1995)
11. W.–D. Sepp, D. Kolb, W. Sengler, H. Hartung, B. Fricke, *Phys. Rev. A* **33** (1986) 3679
12. P. Kürpick, W.–D. Sepp, B. Fricke, *Phys. Rev. A* **51** (1995) 369 3
13. P. Kürpick, H. J. Lüdde, W.–D. Sepp, B. Fricke, *Z. Phys D.* **25** (1992) 17
14. P. Kürpick, H. J. Lüdde, *Comp. Phys. Comm.* **75** (1993) 127
15. W. Fritsch, C. D. Lin, *Phys. Rep.* **202** (1991) 1
16. P. Kürpick, W.–D. Sepp, B. Fricke, *J. Phys. B.* **25** (1992) 5431
17. P. Kürpick, W.–D. Sepp, B. Fricke, *J. Phys. B.* **24** (1991) L139
18. P. Kürpick, W.–D. Sepp, B. Fricke, *Nucl. Instr. Meth. B.* **94** (1994) 183

# Molecular Treatment of Ion-Atom Collisions at Intermediate Energies.

C. Harel*, H. Jouin*, B. Pons*, L.F.Errea†, L.Mendez† and A.Riera†

*Centre de Physique Théorique et de Modélisation, Laboratoire des Collisions Atomiques, Université de Bordeaux I, 351 cours de la Libération, 33405 Talence, France.

†Departamento de Química, CIX, Universidad Autónoma de Madrid, Cantoblanco, E-28049 Madrid, Spain

**Abstract.** We report an extension of the molecular treatment of Ion-Atom collisions at intermediate impact energies. It is first analysed how the ionization mechanism is taken into account through the population of bound excited states, leading mainly to a saturation of electron capture cross-sections. Finally, the introduction of molecular pseudostates in the expansion of the time dependant electronic wave function allows to explicitly take into account the ionization channel and to calculate with a good accuracy ionization, excitation and electron capture cross sections in a large range of impact velocities ($0.1 < v < 3$ a.u.).

## INTRODUCTION

In recent papers [1,2], we have shown that the close coupling method using a molecular expansion modified by a common translation factor (CTF/MF method) is a very powerful method for calculating electron capture cross sections at low energies, but also for obtaining electron loss and excitation cross sections in a large range of impact velocities. However, in the CTF/MF approach, a systematic overestimation of capture cross sections is observed at high impact velocities. In that case, at small internuclear distances, the total wave function represented by the CTF/MF method allows to describe all the features of dynamics including the ionization process whereas no explicit representation of the ionization channel was introduced in the expansion [2]. For example, in the benchmark case $He^{2+}$-H(1s), we have shown that the overestimation of capture cross-sections at high impact velocities is due to the trapping of the ionizing flux onto the higher excited states included in the CTF/MF expansion. In the present work, first, this phenomenon is analysed in terms of the properties of the CTF/MF orbitals. In a second step, an explicit treatment of the ionization process is proposed.

© 1995 American Institute of Physics

# THE MOLECULAR APPROACH

## Theory

We restrict the formulation to the case with one active electron (the application to any number of electrons can be found in [2]) and we start from a semi-classical treatment (see [2] and references therein); assuming rectilinear trajectories for the motion of nuclei **R**=ρ+**v**t (**R** being the distance between both nuclei, **v** the constant impact velocity and ρ the impact parameter), the wave function which describes the electron Φ(r,t) fulfils the time dependant Schrödinger equation:

$$i\frac{\partial}{\partial t}\Phi(\mathbf{r},t) = H_{el}\Phi(\mathbf{r},t) \tag{1}$$

where **r** stands for electron position and $H_{el}$ is the Born-Oppenheimer electronic Hamiltonian. In a modified molecular-close coupling treatment the equation (1) is solved by expanding the electronic wave function Φ(r,t) in terms of a finite set of One Electron Diatom Molecule orbitals $\varphi_j(\mathbf{r},R)$ eigenstates of the Hamiltonian $H_{el}$ at fixed internuclear distance R with energy $E_j$ ( as usually, the OEDM are labelled according to their atomic unit limit nlm).

$$\Phi(\mathbf{r},t) = \exp[iU(\mathbf{r},t)] \sum_{j=1}^{N} a_j(t)\, \varphi_j(\mathbf{r},R)\, \exp\left[-i\int E_j(R)dt\right] \tag{2}$$

with
$$\varphi_j(\mathbf{r},R) = \Lambda_j(\xi,R) M_j(\eta,R) \Omega_j(\Phi) \tag{3}$$

where η=($r_A$-$r_B$)/R, ξ=($r_A$+$r_B$)/R and Φ are prolate spheroïdal coordinates and the phase term exp(iU(r,t)) is a common electron translation factor (CTF), as proposed in [3]. The function U is written as follows [4]:

$$U(\mathbf{r},t) = f(\mathbf{r},R)\mathbf{v}\cdot\mathbf{r} - \frac{1}{2}f^2(\mathbf{r},t)\mathbf{v}^2 t \tag{4}$$

with
$$f(\mathbf{r},R) = \frac{1}{2}[g_\alpha(\eta) + d] \tag{5}$$

$$g_\alpha(\eta) = \alpha^{\alpha/2} \frac{\eta}{(\alpha-1+\eta^2)^{\alpha/2}}$$

where d=1-2p (pR being the distance from nucleus A to the origin of the electronic coordinates), and α is the parameter allowing to vary the form of the switching function. The form (5) choosen for the switching function f is well adapted to the use in expansion (2) of OEDM orbitals [5]. The present calculations have been performed using for $\alpha$ a value $\alpha_0$ such that the collisional results remain unchanged when varying $\alpha$ around $\alpha_0$ ( $\alpha_0 = 1.25$).

Substituting the expansion (2) in equation (1) leads to a new system of first order linear differential coupled equations where the impact parameter and velocity dependence of all the coupling matrix elements can be factored out as for the usual dynamical couplings. Then, the system of coupled equations is integrated using a version of the program [6] modified to include the new coupling terms arising from the CTF treatment. Finally, total cross-sections corresponding to the various processes are determinated from the final transition amplitudes $|a_j(t_{max})|^2$ calculated in this way [2].

## Treatment without pseudostates

We have performed calculations for various collisional systems ($I^{Z+}+H$, with $Z=4,5,6$) in the impact velocity range $0.1 a.u.<v<2.5.u.a.$ to verify that the analysis made in [2] is general. As shown in figure 1, our total capture cross sections are systematically close to the 'electron loss' (total capture plus ionization) cross

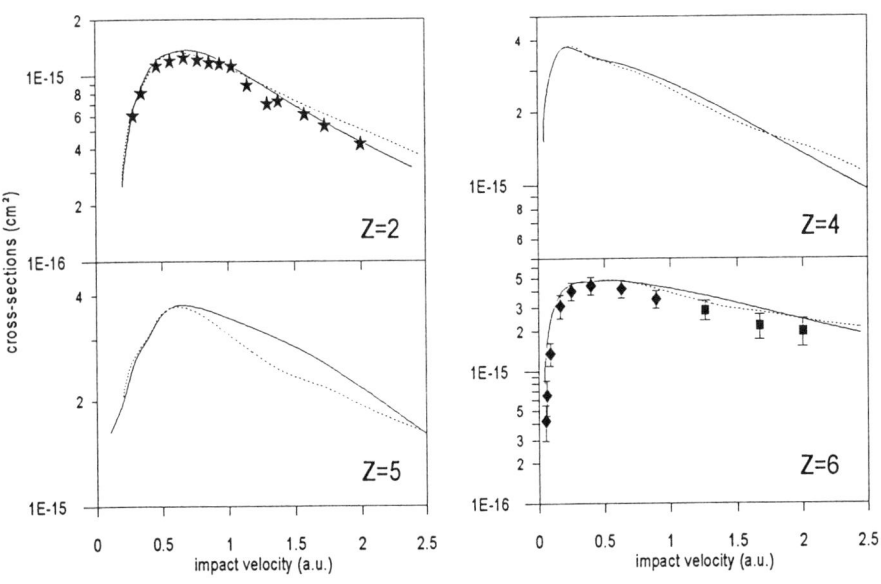

**FIGURE 1.** Total capture cross sections of present work $\sigma$ (cm$^2$) for the collision $I^{Z+}+H$ as a function of the impact velocity (———), compared with electron loss cross sections: (---), theoretical calculations [7]; experiments: (★) [8] and [9], (♦) and (■) [10].

sections which have been measured (whenever available) and of the Plane Wave Translation Factor/Atomic Orbitals (PWTF/AO) calculations of Toshima [7]. Moreover, our results for the excitation cross sections (not shown) agree reasonably well with the other theoretical results when available ([11] and [12] for $He^{2+}$-H and $C^{6+}$-H respectively). Hence, in all the multicharged systems investigated in the present work, the CTF/MF close coupling approach is able to describe globally the total loss of flux from the entry channel including the ionizing one. Starting from basis sets corresponding in each case to the treatment of the very dominant processes, there is mainly an overestimation of cross sections corresponding to capture and, to a lesser extent, to excitation due to the trapping of the ionizing flux onto the bound states. In more extensive calculations, adding in the expansion (2) molecular orbitals corresponding to higher excited states, one observes a lowering of cross sections corresponding to lower states due to a transfer of the ionizing flux to the molecular states associated to the higher multiplets introduced in the expansion (this promotion of flux from the entry channel up to the continuum through more and more excited molecular states has been described in [1] as a ladder mechanism).

As an illustration, in the $He^{2+}$-H case, 'histories' of collision of figure 2 allow to understand how the CTF/MF expansion manages to describe the ionization process: for a characteristic collisional trajectory ($\rho$=2.2a.u., v=1.6a.u.) we have

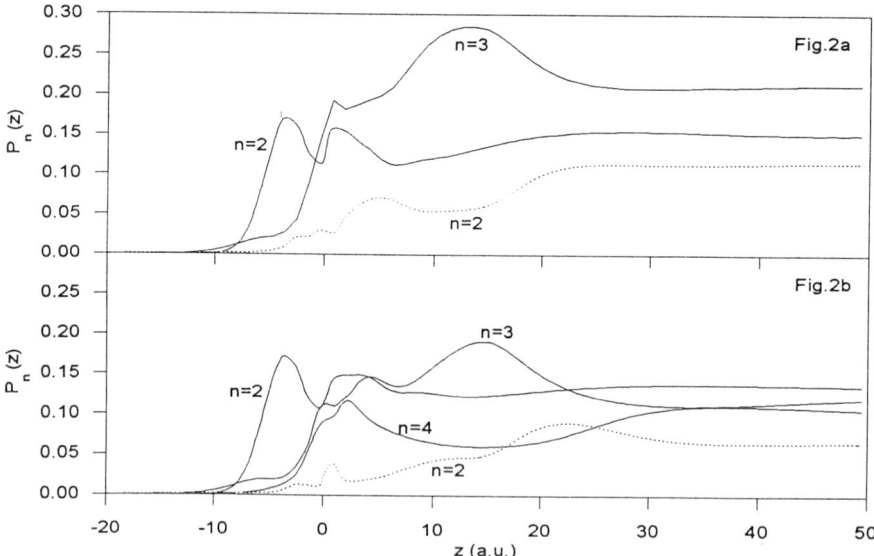

**FIGURE 2.** Population $P_n(z) = \sum_{lm} |a_{nlm}(z)|^2$ of the excited states as a function of z= v t for the collisional trajectory (v=1.6au, $\rho$=2.2au). $He^{2+}$-H(1s)→$He^+$(n)+$H^+$: full lines; $He^{2+}$-H(1s)→$He^{2+}$+H(n): dashed lines. a) 14-states calculation, b) 23-states calculation.

plotted populations of various channels obtained in a 14-states (fig 2a) and a 23-states (fig 2b) calculations. The transfer of flux from lower to upper states appears clearly on figure 2b: transitions to the states (correlated with the $He^+(n=4)$ multiplet) added in the 23-states-calculation strongly lower the final population of the n=3 capture channel and, to a lesser extent, of excitation channel obtained in the 14-state calculation (the global flux leaving the entry channel remaining quite constant). Note that the population of the dominant n=2 capture channel is very similar in both calculations; the population of this channel is already stabilized in the 14-states calculation. Two distinct ranges of values of z are involved in the promotion of the ionizing flux: around z=0, first, in the 23-states-calculation the population of the additional states lowers the flux absorbed by the states corresponding to the multiplet n=3 (compare figures 2a and 2b); at larger z, in the second part of the collision, a significant fraction of the flux accumulated in the n=3 capture channel is promoted on the higher channel (around z=20.a.u.). Also, one can observe a non negligible depopulation of the excitation channel around z=30.a.u.. A more detailed analysis of the role played by the various states corresponding to a same multiplet on the transfer of flux can be drawn from figure 3. Obviously, in the case of σ substates, the promotion of the ionizing flux populates around z=0 the more compact OEDM's n l m (with l=0,1), whereas the mechanism at large z is fully due to the population of the basis state 6 5 0. A similar analysis made on π state populations shows that these states are dominantly populated around z=0 by rotational transitions. The promotion of flux at large z is fully due to transitions at pseudo-crossings between molecular orbitals of σ character having a maximum electronic density between both nuclei; only the n n-1 0 type orbital presents this property for a given multiplet [13]. Both mechanisms of transfer of the ionizing flux from lower to upper multiplets can be schematically represented as on figure 4c. We could tentatively say that the present analysis is related with the work of Pieksma and Ovchinikov[14]. The mechanism around z=0 is equivalent to transitions through the series $S_{lm}$ associated by these authors to direct ionization at small internuclear distances, whereas the large z

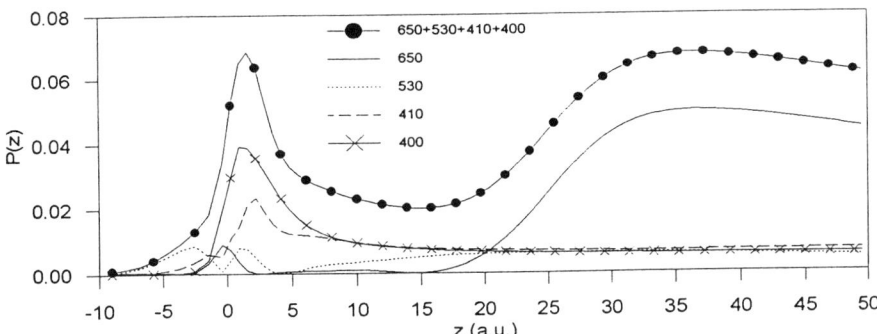

**FIGURE 3.** Populations of the σ states corresponding to the multiplet n=4 in a 23-states calculation for v=1.6 and ρ=2.2a.u. as a function of z=vt ($He^{2+}$-H(1s) collision).

ladder mechanism is the same as the mechanism involving successive transitions through the super-series T promoting the electronic flux up to the continuum.

An important information on the properties of the travelling molecular orbitals included in expansion (2) is available on figures 4.a,b. The basic idea is to look at the energies of these orbitals ($\varphi_n(r,R) \exp[iU(r,t)]$) in inertial frames moving with

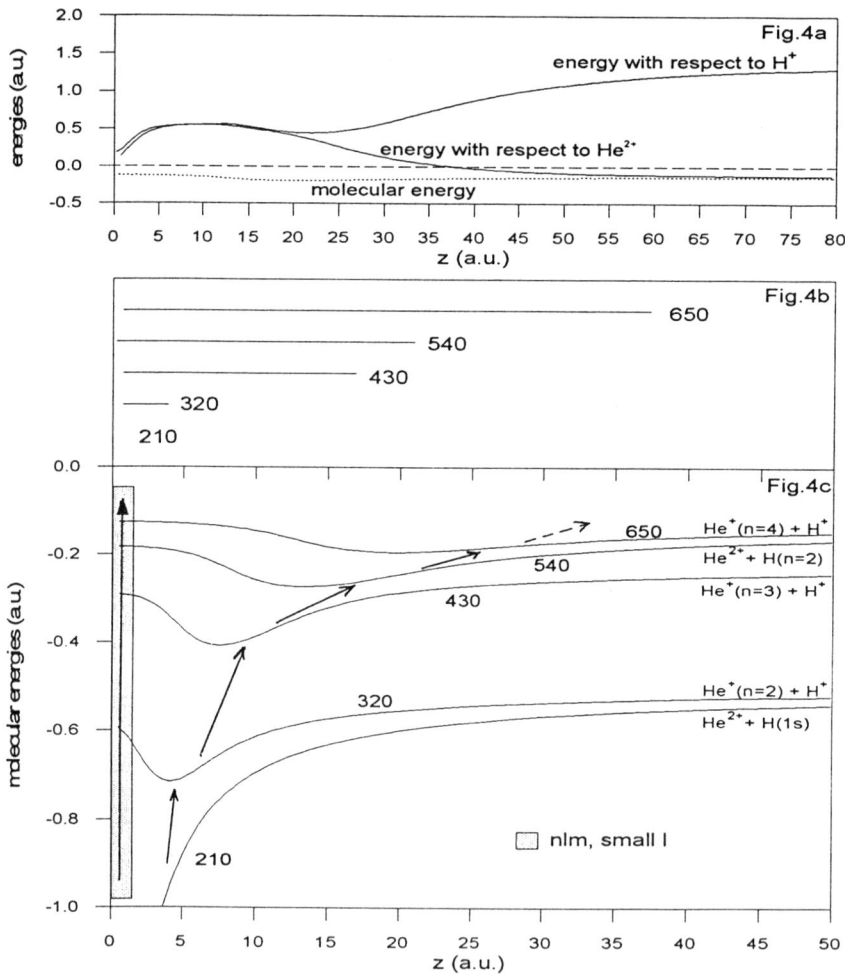

**FIGURE 4. a)** Energies of the OEDM 6 5 0 of $He^{2+}$-H as a function of z=vt.: with respect to $H^+$ or $He^{2+}$ (see text), full lines; molecular energy, dashed line. **b)** the continuous line marks the range of z values for which both energies of the travelling OEDM respect with $He^{2+}$ and $H^+$ are positive. **c)** schematic representation of both processes of desaturation (ionization) from the molecular energy diagram of $He^{2+}$-H. All the figures are drawn for the collisionnal trajectory (v=1.6au, ρ=2.2au)

the target or the projectile. These quantities which depend on the velocity and the impact parameter are plotted on figure 4a for the orbital 6 5 0 of (HeH)$^{2+}$as a function of z. In that case, they are both positive up to a maximum value of z, noted $z_{max}$, of the order of 37a.u.; at larger z, as the orbital 6 5 0 localizes on He$^+$(n=4), the energy with respect He$^{2+}$ tends to a negative limit, whereas the energy with respect to H remains positive. On the figure 4b, we have reported the values of $z_{max}$ for the n n-1 0 orbitals of figure 4c: clearly, in the range where transitions corresponding to the ladder climber mechanism occur the corresponding travelling molecular orbitals present positive energies with respect to both nuclei. Hence, although these orbitals tend to represent an unbound electron in these regions of transition, finally, at the end of the 'relay race' [2] the flux accumulated on the higher states is trapped on the corresponding capture channel (for z>$z_{max}$). This behaviour has been also found by Bandarage and Parson [15] in CTMC calculations: even at very large times, an important part of the ionizing trajectories studied by these authors is associated with bound molecular states. In addition, the pseudo-states introduced by Winter and Lin [16] in their triple-center atomic approach have similar properties as the travelling molecular states when they account for the ionization mechanism: electronic density centered between both nuclei and positive energies respect to both atomic centers. The previous analysis doesn't apply to the compact states responsible for the desaturation mechanism at small z: the concept of travelling orbitals has no signification in the molecular region.

Finally, the representation of the ionization process by the CTF/MF method is a positive point. However, the trapping effect of the ionizing flux onto the bound states is at the origin of the well known failure of this method to calculate the capture cross sections at intermediate energies. Addition of pseudostates in the expansion (2) does allow to improve the completeness of the basis and to separate both components of the electron-loss cross section.

## Treatment with pseudostates

As explained in [17] the representation of the ionization channel has been performed by addition in expansion (2) of pseudostates having a maximum electronic density between both nuclei. They are built as the OEDM orbitals except for the η-part which is replaced by an R-independent expansion M(η). This form offers some flexibility in the representation of the electronic density corresponding to ionization. Varying quantum numbers and charges affects only the ξ-part of the pseudostates, and, hence, allows to change the spatial extension of the electronic density. As a benchmark case, we have chosen to apply the present approach to the collision He$^+$(1s)+H$^+$ where the capture process is very selective and populates dominantly the fundamental state of H. A basis set of 14 OEDM is sufficient for calculating with a good accuracy the various transitions responsible for the loss of

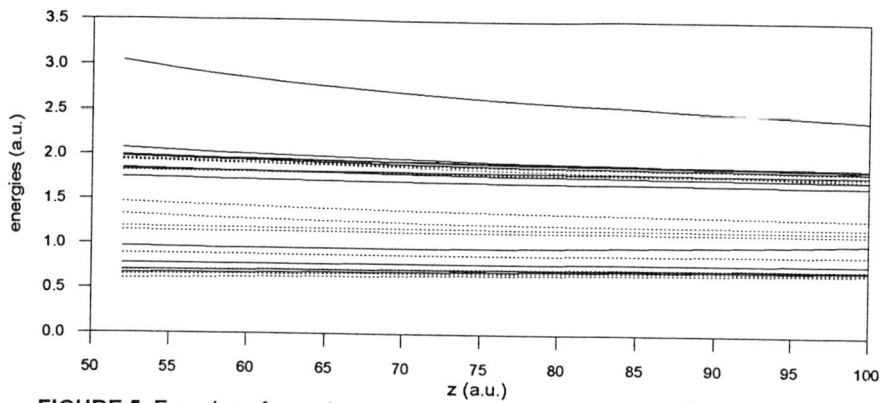

**FIGURE 5.** Energies of pseudostates at large z: with respect to $H^+$, full lines; with respect to $He^{2+}$, dashed lines

flux by the entry channel; 12 pseudostates (9 of sigma character and 3 of pi character), built so as to cover the region between the nuclei, have been added to model the ionization process. After orthogonalisation of the pseudostates to bound states, we have calculated as in the previous section their energies in fixed nuclei referentials. As shown on figure 5, all the corresponding energies are positive. Then, the ionization cross section is calculated from the pseudostate populations and is generally found in good agreement with the experiments. The results are given on figure 6a,b. Note the correct decrease of the capture cross section to the ground state at high velocities obtained for the first time using a molecular treatment. The effect due to the inclusion of pseudostates is clearly seen on figure 7: around z=0 one can observe an important population of these states; at

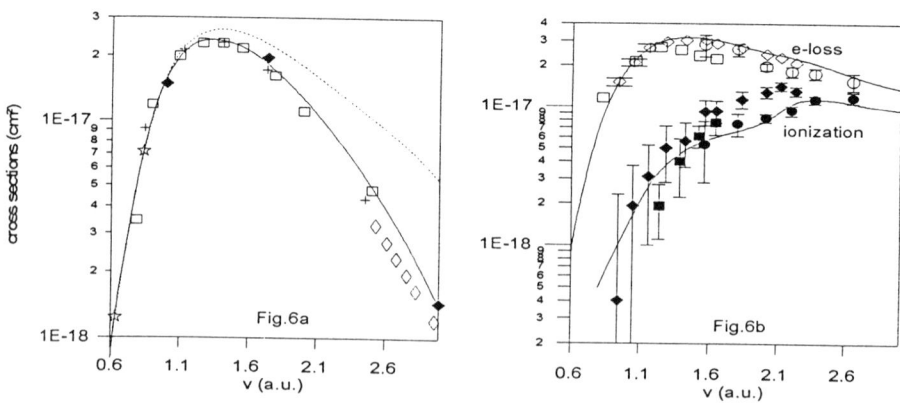

**FIGURE 6.** $H^+ + He^+(1s)$ collision; **a)** Cross-section of capture to the ground state as a function of the impact velocity: Theory, present calculations, 14 OEDM + 12 pseudostates, full line; 14 OEDM , dashed line; ☆ [18]; +[19];◆[20]; ✧ [21]; □[22]. **b)** electron loss and ionization cross sections: Present theory, full lines. Experiments: (✧,◆) [23]; (□,■) [24]; (○,●) [25].

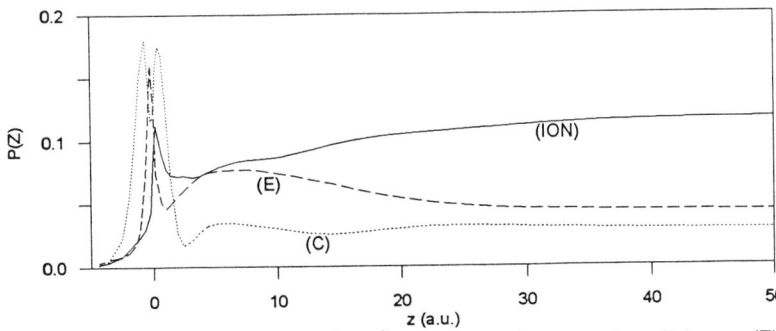

FIGURE 7. 'History' of the collision $H^+$-$He^+$(1s) for v=3.a.u. and ρ=0.4a.u.: (E), excitation; (C) capture; (ION), ionization.

large z, the populations of capture and excitation channels decrease significantly. Clearly, the pseudostates allow to diminuate the trapping effect.

Information on the ionization mechanisms is included in the temporal evolution of the total wave function. Only a careful dynamical study of the completeness of the basis set used in expansion (2) can determine the reliability of its representation; this total wave function is, at large distances, sensitive to variations of the basis of pseudostates, whereas the collisional results are stable. This is also true for the PWTF/AO calculations for ionization (see for example [26], [27]) in which different definitions of pseudostates are used. Hence, to date, the strong experimental controversy [28, 29] on the so called 'saddle point' mechanism cannot be settled. To improve the completeness of our molecular expansion, more flexible forms of pseudostates are then required; that work, using gaussian forms in place of the OEDM ones, is in progress.

## CONCLUSION

A new light on the understanding of the CTF/MF method has been given in the present work; we have clearly identified that the excited molecular states account for the ionizing flux in two steps: i) around z=0 through the population of the more compact states (of small l value) and ii) at large z through the successive population of the molecular states which present a maximum electronic density between both nuclei (n n-1 0 type states). Moreover, at intermediate impact energies, the travelling molecular orbitals involved in the latter mechanism represent an unbound electron with respect to both nuclei at times where transitions occur. Finally, the CTF/MF expansion modified by inclusion of pseudostates has allowed to calculate the three basic processes occurring in one electron ion-atom collisions at intermediate impact energies. The method is competitive and offers an alternative to the PWTF/AO one in that range of impact energies.

# ACKNOWLEDGEMENTS

This work has been partially supported by the DGICYT project No. PB93-288-C02, The 'Action Intégrée Franco-Espagnole' No 96B.

# REFERENCES

1. Errea L.F., Harel C., Jouin H., Maidagan J.M., Mendez L., Pons B. and Riera A., *Phys.Rev.A* **46**, 5617 (1992).
2. Errea L.F., Harel C., Jouin H., Mendez L., Pons B. and Riera A., *J.Phys.B* **27**, 3603 (1994).
3. Schneiderman S.B. and Russek A., *Phys.Rev.* 181, 311 (1969).
4. Harel C. and Jouin.H., *J.Phys.B* 24, 3219 (1990).
5. Power J.D., *Phil.Soc.Trans.Roy.Soc.* 274, 663 (1973).
6. Gaussorgues C., Piacentini R.D. and Salin A., *Comput.Physics Comm.* 10, 224 (1975).
7. Toshima N., *Phys.Rev.A* 50, 3940-47 (1994).
8. Shah M.B. and Gilbody H.B., *J.PhysB* 11, 121 (1978).
9. Shah M.B. and Gilbody H.B., *J.PhysB* 14, 2361 (1981).
10. Phaneuf R.A., Janev R.K. and Pindzola M.S., *Atomic Data for Fusion*, vol 5 (1987).
11. Fritsch W, Shingal R.and Lin C.D., *Phys Rev.A* 44, 5686 (1991).
12. Fritsch W, Gayet R., Gilbody M.B. Olson R.E. and Shartner K., International Nuclear Data Commitee, *IAEA Nuclear Data Session* 253/N2, 29 (1992).
13. Harel C. and Salin A.,in *Invited Papers of XV ICPEAC*, Brighton 1987 , Elsevier Science Publishers, 631-42 (1988).
14. Pieksma M. and Ovchinikov S.Y., *J.Phys.B* 24, 2699 (1991).
15. Bandarage G. and Parson R., *Phys.Rev.A* 41, 5878 (1990).
16. Winter T.G. and Lin C.D., *Phys.Rev.A* 29, 3071 (1984).
17. Errea L.F., Harel C., Jouin H., Mendez L., Pons B. and Riera A., *Phys.Rev.A* (to be published)
18. Winter T.G., Hatton G.J. and Lane N.F., *Phys.Rev.A* 22, 930 (1980).
19. Winter T.G., *Phys.Rev.A* 25, 697 (1982).
20. Bransden B.H., Noble C.J. and Chandler J., *J.Phys.B* 16, 4191 (1983).
21. Gayet R., *private communication*
22. Fritsch W. and Lin C.D., *J.Phys.B* 15, 1255 (1982).
23. Rinn K., Melchert F. and Salzborn E., *J.Phys.B* 19, 3717 (1986).
24. Peart B., Rinn K. and Dolder K., *J.Phys.B* 16, 1461 (1983).
25. Watts K., Dunn K.F. and Gilbody H.B., *J.Phys.B* 19, L355 (1986).
26. Shingal R. and Lin C.D., *J.Phys.B* 22, 1445 (1989).
27. Winter T.G., *Phys.Rev.A* 37, 4656 (1988).
28. Bernardi G. and Meckbach W., *Phys.Rev.A* 51, 1709 (1995).
29. Irby V.D., *Phys.Rev.A* 51, 1713 (1995).

# Time-Dependent, Lattice Approach to Atomic Collisions

## David R. Schultz

*Physics Division, Oak Ridge National Laboratory*
*Oak Ridge, Tennessee 37831-6373*

---

Recent progress in developing and applying methods of direct numerical solution of atomic collision problems is described. Various forms of the three-body problem are used to illustrate these techniques. Specifically, the process of ionization in proton-, antiproton-, and electron-impact of atomic hydrogen is considered in applications ranging in computational intensity from collisions simulated in two spatial dimensions to treatment of the three-dimensional, fully correlated two-electron Schrödinger equation. These examples demonstrate the utility and feasibility of treating strongly interacting atomic systems through time-dependent, lattice approaches.

---

### INTRODUCTION

It would not be unusual for a paper describing methods of treating atomic collisions on a numerical lattice to begin with a sentence such as "With the advent of contemporary supercomputing resources, direct solution of the time-dependent Schrödinger equation for collisions of ions or electrons with atoms has finally become a feasible undertaking." Indeed, a number of works have appeared in which pilot or proof-of-principle calculations have been carried out demonstrating that multidimensional wavefunctions and operators may be discretized and the collision dynamics faithfully described (see e.g. Figure 1). But it has also been a goal of atomic physicists to bring these techniques to bear on the very accurate treatment of fundamental atomic collision problems on a more routine basis.

What do we mean by routine and why in particular should we seek the application of calculations performed on a numerical grid?

The goal of making lattice approaches part of the atomic physicist's tool kit alongside perturbation theory, more traditional close coupling schemes, and classical scattering theory, requires that individual calculations be accomplished in times, say, shorter than a day. Only then could results which must be repeated for many impact parameters or angular momenta, or to prove convergence with lattice size or grid spacing be said to be routine. The objective is thus techniques which can significantly impact physical interpretation of phenomena, and to do that often requires computing probabilities or cross sections over a range of collision energies or geometries. It is this

© 1995 American Institute of Physics

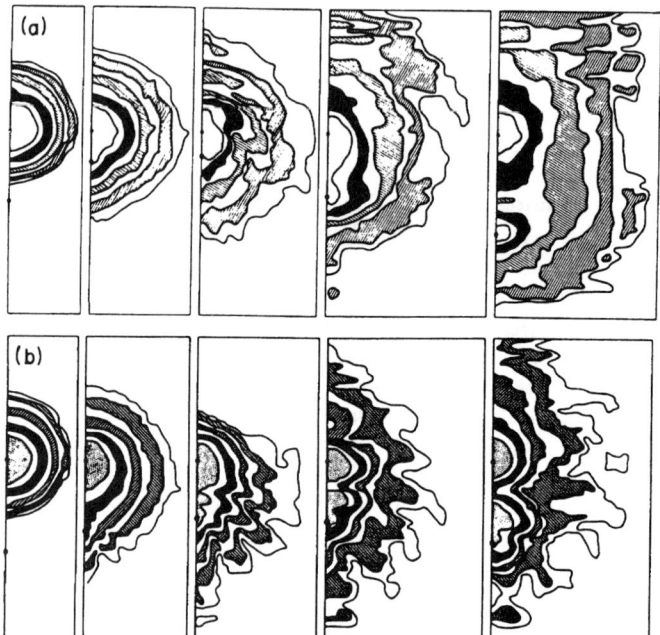

FIG. 1. Contour density plots illustrating the time evolution of the electronic wavefunction for (a) $H^+$ and (b) $C^{6+}$ with atomic hydrogen for several time slices throughout the collision. These calculations [1] were performed in 1982 using the fastest available supercomputers, but still covered only a very small collision volume and relied on the approximation of axial decoupling to reduce the problem to two spatial dimensions.

desire for physical insight that also coincides with one of the advantages of the method. That is, the lattice approach lends itself to a very ready visualization of the collision process in time. Furthermore, many lattice methods rely on flexible basis functions so that a very wide variety of phenomena can be represented. They also seek to solve as directly as possible the equations governing the interaction (e.g. the Schrödinger or Hartree-Fock equations) for even strongly interacting systems, and therefore provide a powerful nonperturbative treatment for inherently time-dependent problems.

Here we report on our recent progress in making such approaches routine and in deriving insight from them regarding fundamental atomic collisions. In particular, we describe our treatment of three, three-body problems, namely ionization in proton-, antiproton-, and electron-impact of atomic hydrogen.

## MODEL H$^+$ + H COLLISIONS

Recently, there has been great activity in developing theories of ionization in ion-atom collisions especially in the low- to intermediate-energy regime, motivated in large part by new experimental measurements. The most elementary quantum mechanical theory of ionization, the first Born approximation, has been perhaps remarkably successful but simply treats the electron as being ejected into the continuum of the residual target ion through a perturbative interaction with the projectile. Evidence that this theory is inadequate to describe the full ejected electron spectrum has been found for some time.

A number of works (see e.g. [2–6]) in the past decade have therefore emphasized the need to treat the electron ejected in ionization as moving in the combined field of both the target and projectile ions. Such theories include the classical trajectory Monte Carlo (CTMC) method [6,7], the continuum-distorted-wave–eikonal-initial-state (CDW-EIS) approximation [2,5] and the strong-potential-Born (DSPB) approximation [8]. These theories treat the electron as being ejected under the combined influence of both the target and projectile ions. To varying degrees each has been successful in describing the ejected electron spectrum for a rather wide range of collision energies. For example, at intermediate collision energies, the CTMC method provided a good description of the formation and resulting symmetry of the "electron-capture-to-the-continuum" peak, the magnitude and shift of the binary peak resulting from two-center effects, and the contribution to the spectrum coming from "saddle-point" electrons.

In order to provide a means of examining the electronic probability density ejected into the two-center continuum without the need for approximations which theory and experiment have shown to be inadequate, we have sought to directly solve the Schrödinger equation for protons colliding with atomic hydrogen on a numerical lattice. Hopefully in this context, the most clear picture of the physics underlying ion-atom collisions, and in particular ionization, can be obtained.

Earlier approaches based on this type of approach were developed to directly solve the time-dependent Schrödinger [1,9,10] equation or the time-dependent Hartree-Fock [11,12] equations. Both finite difference and finite element methods have been used for discretizing the wavefunction and the actions of operators. Despite the high performance of modern supercomputers, these works indicated that the numerical study of even a simple three-body system, such as H$^+$ + H, remained a difficult task. Due to the computational limits imposed by storing and operating on a very large multidimensional matrix representing the wavefunction, the earlier studies reduced the dimensionality of the problem by considering only zero impact parameter, imposing axial symmetry, or using a rotating frame [13] which would mimic the effect of a non-zero impact parameter.

In order to explore the behavior of the ionization process over a wide range of collision energies and impact parameters, we have recently [14] considered

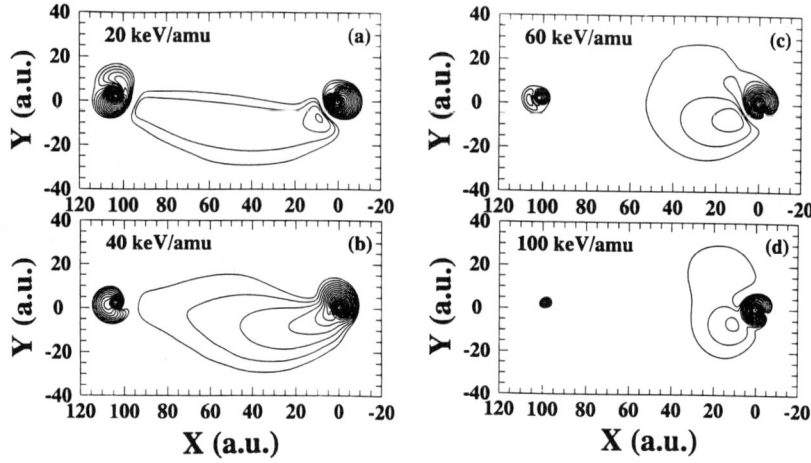

FIG. 2. The electronic probability density in 2D model collisions of a proton with atomic hydrogen displayed at a constant impact parameter (2 a.u.) for several energies long after the collision with the target.

collisions in two-dimensional (2D) Cartesian space. In addition, to ameliorate the difficulties associated with trying to represent the singular Coulomb potential on a numerical lattice of finite spacing, we introduce a model interaction based on the inclusion of a so-called softcore parameter to remove the singularity. We choose a value of the softcore parameter which yields the same binding energy as in the three-dimensional (3D) hydrogen atom, and also possesses radial and momentum distributions closely resembling those found in three dimensions. Thus, the dynamics of a collision in 3D are reasonably well reproduced. By using a smaller collision volume, we have demonstrated that this is the case by performing full 3D calculations as checks. The results of these simulations yield insight into the collision dynamics leading to such features as saddle point ionization, and are described in some detail below.

We solve the time-dependent Schrödinger equation in this model space using lattice techniques to obtain a discrete representation of the wavefunction, i.e. $\psi(x,y) \to \psi(x_i, y_j) \equiv \psi_{i,j}$, and all coordinate-space operators on a two-dimensional Cartesian mesh. Local operators such as potentials simply become diagonal matrices composed of their values at the lattice points, i.e. $V(x,y) \to V(x_i, x_j)\delta_{i,i'}\delta_{j,j'} \equiv V_{i,j}$. Derivative operators, such as the kinetic energy, have lattice representations in terms of matrices, i.e. $\partial/\partial x \to D_{i,j}^{(x)}$. We have implemented our solutions using both a low-order three-point finite-difference method [15], resulting in banded derivative matrices, and high-order methods such as Fourier and basis-spline collocation [16], which give full-matrix representations for the derivatives.

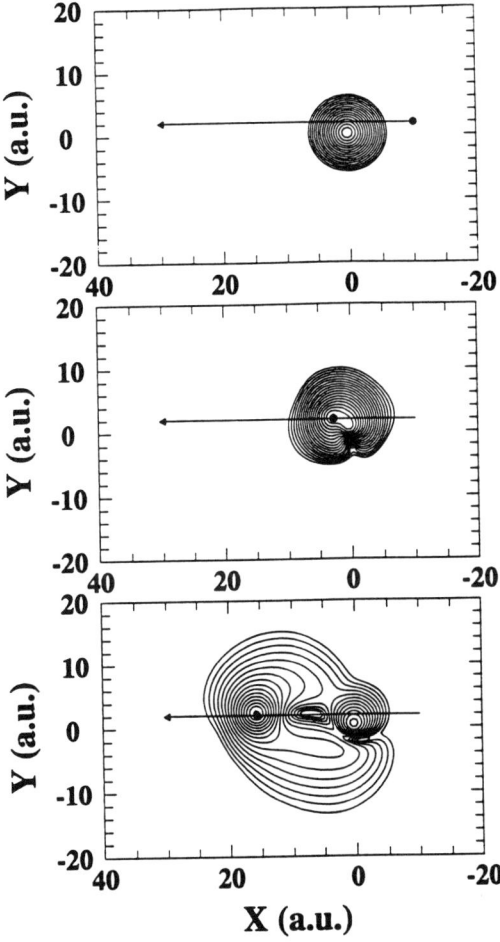

**FIG. 3.** The electronic probability density in a 2D model collision of a proton with atomic hydrogen at 10 keV and an impact parameter of 2 a.u. (continued in Figure 4).

Figure 2 illustrates the late collision time topology of the electronic density for a fixed impact parameter (2 a.u.) for several energies (20, 40, 60, 100 keV) obtained from these simulations. Clearly seen is the transition from saddle-point-dominated ionization at the lower energies to direct, target-centered ionization for higher energies. Also, the decrease of density on the projectile illustrates the rapid falloff of charge exchange with increasing impact energy. Also, by monitoring the overlap of the time-evolved wavefunction with a basis of target and projectile states, we have computed elastic, excitation, capture, and ionization probabilities in these collisions. Work is under way to derive the same information and visualizations in full 3D calculations.

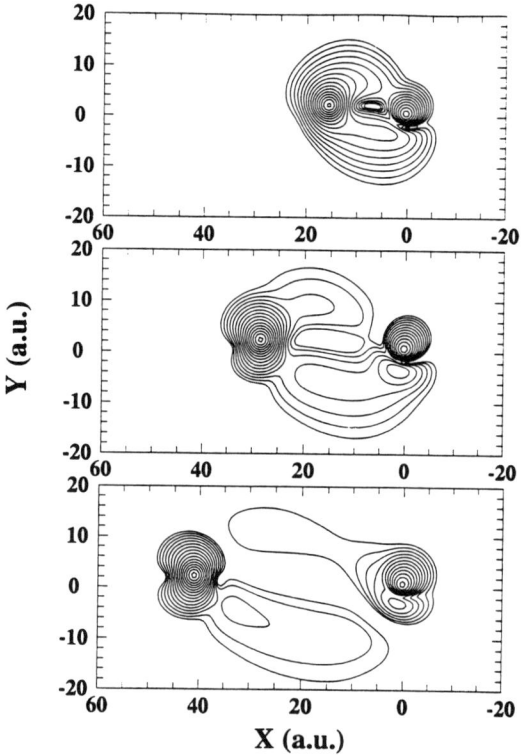

**FIG. 4.** The electronic probability density in a 2D model collision of a proton with atomic hydrogen at 10 keV and an impact parameter of 2 a.u. The time slices show the evolution of a zero-density island centered about the saddle or equiforce point between the two protons (continued from Figure 3).

Another typical result of the 2D time propagations is the case shown in Figures 3 and 4, which we describe in some detail due to the significance of the result to discussion of the production of saddle point ionization. The projectile is introduced 10 a.u. to the right of the target and moves on a straight line for 140 a.u., here with an impact energy of 10 keV and impact parameter of 2 a.u. The first segment in Figure 3 shows the initial unperturbed state of the 2D hydrogen atom, while successive frames show the evolution of the density through an intermediate $H_2^+$ quasi-molecule temporarily formed in the collision to that present as the protons separate. Obvious in these frames is the large transfer of density to the projectile and what will later in Figure 4 be seen to be ejection to the continuum.

The collision is followed to larger times in Figure 4, and is fairly typical of what we find for collision energies between about 6 and 15 keV for impact parameters in the range 0.4 to 2.6 a.u. In these frames, one sees the formation of a "zero density island" between the projectile and target in the location one would expect at first thought to see a significant density associated with the equiforce or saddle point of the projectile-target potential. This result is in qualitative agreement with a very recent description [17] for this low energy collision range, based on the so-called theory of hidden crossings.

## 3D $P^-$ + H COLLISIONS

Motivated especially by recent experimental measurements [18] of the total cross section for ionization of atomic hydrogen by antiproton-impact, we have applied our 2D and 3D lattice approaches for proton-hydrogen scattering to this new system. A typical result of the 2D time-evolution of the electronic probability density is shown in Figure 5 for a collision energy of 10 keV. The most striking feature is the development of a large region surrounding the antiproton from which electronic density is repelled. This feature is the analog of the electron-capture-to-the-continuum cusp found in positive particle impact, and has been termed the "anti-cusp."

By computing the overlap of the time-evolved wavefunction with various target states as the collision progresses, we can determine the probability that the target will be ionized. Figure 6 shows these probabilities as a function of time for a full 3D collision at 1 keV for three different impact parameters. This figure shows that the probability of ionization rises sharply at a point in time coinciding with the antiproton passing the distance of closest approach to the target nucleus. Owing to the unitarity of the channel probabilities, the elastic and excitation probability falls in complement to the rise in ionization. At long times the probabilities stabilize and may be integrated to yield total channel cross sections. This figure also shows the impact parameter (b) dependence of the channel probabilites. For example, at b=0.3 a.u., the ionization probability is almost ninety percent, but drops to just over fifteen percent by b=1.8 a.u.

By time-evolving the initial H(1s) wavefunction in the presence of the impinging antiproton for four impact energies with approximately 10 to 15 impact parameters per energy, we have obtained preliminary total ionization cross sections between 1 and 500 keV. These are displayed in Figure 7 along with the experimental measurements of Knudsen *et al.* [18] and various other theoretical approaches. The antiproton-hydrogen system is of particular interest in that it is the simplest atomic collision, involving only three parti-

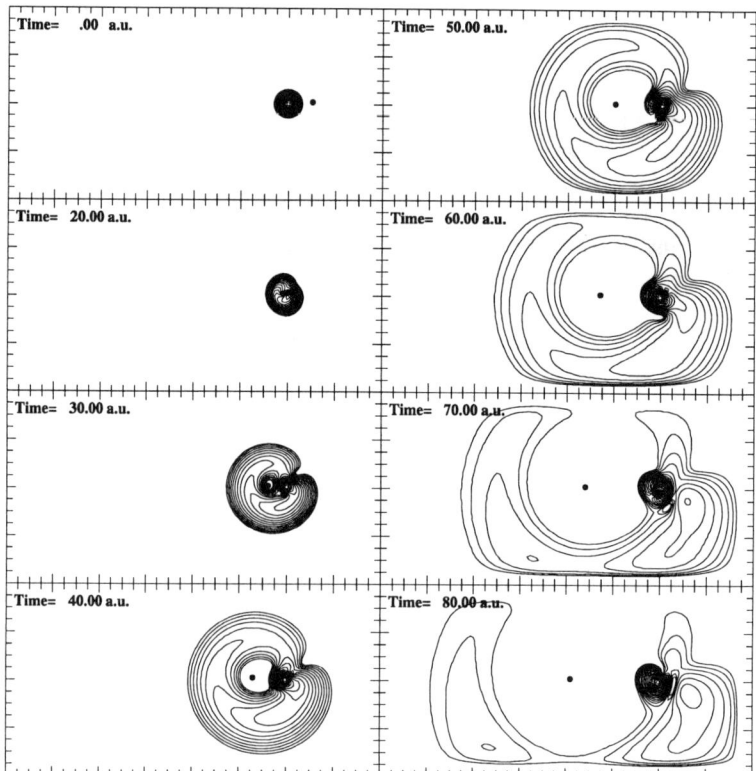

**FIG. 5.** The electronic probability density in 2D model collisions of an antiproton with atomic hydrogen at 10 keV and an impact parameter of 1 a.u. The time slices show the formation of an anti-cusp centered on the projectile, the analog of the electron-capture-to-the-continuum peak in proton-atom collisions.

cles, and therefore provides a fundamental test of theory. In particular, it is simpler than electron- or proton-impact of atomic hydrogen since neither a two-electron wavefunction nor the charge transfer channel need be treated. The strong coupling to the continuum also leads to the result that the ionization cross section does not drop at low collision energy. In Figure 7 we show the low energy limits of the ionization cross section for this system imposed by the Fermi-Teller (FT) model [19] and our recent hidden crossings (HC) model [20]. Also shown are the close coupling (CC) results of Toshima [21] and our classical trajectory Monte Carlo (CTMC) and continuum-distorted-wave–eikonal-initial-state (CDW-EIS) calculations. The 3D lattice (s3D) calculations reproduce very well the behavior of the cross section over a wide range of collision energies.

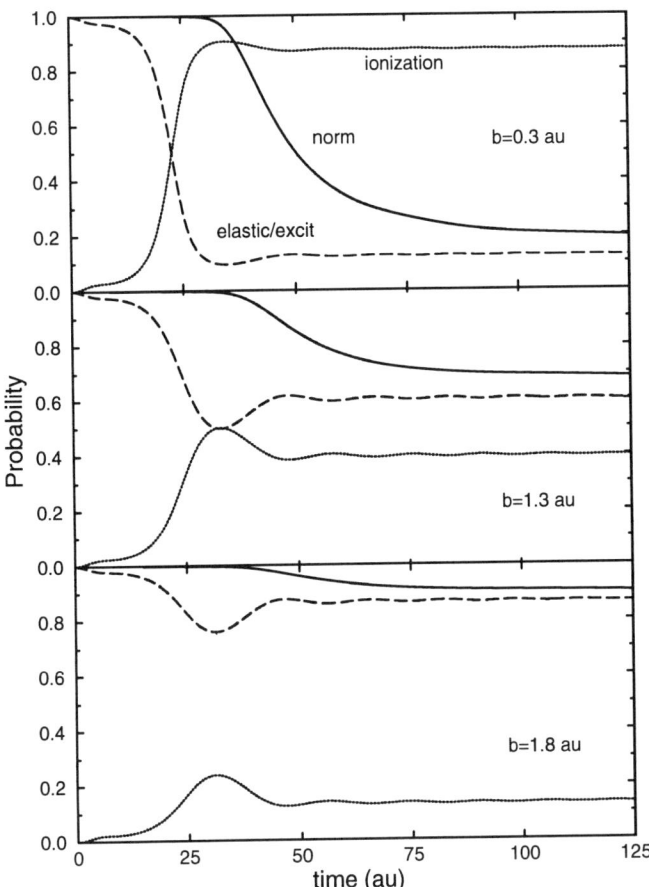

**FIG. 6.** The probability for various processes as a function of collision time for three impact parameters in full 3-dimensional calculations for antiprotons colliding with atomic hydrogen at 1 keV. The wavefunction is absorbed at the edges of the numerical lattice and thus the norm as depicted drops when there is a large amount of density ejected to the continuum. The dashed curve gives the sum of elastic and excitation to low-lying states and the dotted curve shows the ionization probability.

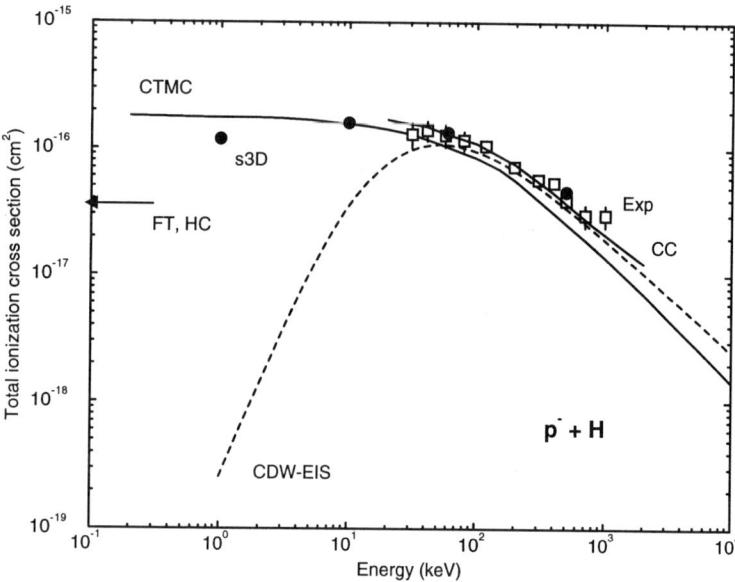

**FIG. 7.** The total cross section for ionization in antiproton collisions with atomic hydrogen. The experimental points are from Knudsen *et al.* [18]. The theoretical methods denoted by CTMC, CDW-EIS, *etc.* are described in the text. The present 3-dimensional lattice calculations are given by the filled circles (s3D).

## FULLY CORRELATED TWO-ELECTRON SYSTEMS

Another type of fundamental atomic three-body system is that composed of a heavy particle and two electrons. Rather than possessing the difficulties associated with describing an electron moving in a two-center field as in the $p^+, p^- + H$ problems discussed above, complete description of this system involves accurately treating the electron-electron interaction. A sketch of how such fully correlated two-electron systems could be treated on a numerical lattice was given by Bottcher [22] several years ago at ICPEAC XVI. Since then, we have completed the development of this technique [23], have applied it to a number of basic problems [23,24], and are investigating several others. These include structure calculations of the ground, singly, and doubly excited states of He, $H^-$, and two-, or pseudo-two-electron ions, the time-dependent description of autoionization, electron impact excitation and ionization of H, and the photoexcitation and photoionization of He and Be. All of these systems are treated with the same underlying technique, the direct solution of the fully correlated two-electron Schrödinger equation on a numerical lattice.

Key to efficiently implementing this method is the ability to describe the two electrons in a reduced number of coordinates and through the solution of only a small number of coupled partial differential equations. In this approach, the wavefunction takes the form [23]

$$\Psi(\hat{r}_1, \hat{r}_2) = \sum_{l=\varpi}^{L} \psi_l(r_1, r_2, \vartheta) \mathcal{Y}_{l,L+\varpi-l}^{LM}(\hat{r}_1, \hat{r}_2), \qquad (1)$$

where $\vec{r}_1$ and $\vec{r}_2$ are vectors locating the two electrons, $\vartheta$ is the angle between $\vec{r}_1$ and $\vec{r}_2$, $L$ and $M$ are the total orbital angular momentum and its projection, $\varpi$ is the parity, and $\mathcal{Y}$ denotes a coupled spherical harmonic. For example, for a $^1S$ state such as the ground state of helium, the sum contains only one term. Thus, the simplification of the problem is the reduction of the formally infinite sum over coupled spherical harmonics to a remarkably few terms, and of the full six-dimensional space (e.g. $r_1, \theta_1, \phi_1, r_2, \theta_2, \phi_2$) to three dimensions $(r_1, r_2, \vartheta)$.

By using the variational procedure, relatively straightforward angular momentum algebra allows the Schrödinger equation to be replaced by a set of coupled equations involving the expansion coefficients $\psi_l(r_1, r_2, \vartheta)$,

$$(h_1 + h_2 + h_\vartheta + \frac{1}{r_{12}} - E)\psi_l + \sum_{l'=\varpi}^{L} (\mathcal{U}_{l'l}^{(1)} + \mathcal{U}_{l'l}^{(2)})\psi_{l'} = 0 \qquad (2)$$

where

$$h_p = -\frac{1}{2}\frac{\partial^2}{\partial r_p^2} - \frac{Z}{r_p} + \frac{l_p(l_p+1)}{2r_p^2}, \qquad (3)$$

$$h_\vartheta = (\frac{1}{r_1^2} + \frac{1}{r_2^2})(-\frac{1}{2}\frac{1}{sin\vartheta}\frac{\partial}{\partial \vartheta}sin\vartheta\frac{\partial}{\partial \vartheta}) \qquad (4)$$

$$\mathcal{U}_{l'l}^{(p)} = -\frac{\mathcal{Z}_{l'l}^{(p)}(\vartheta)}{\mathcal{Z}_{l'l}^{(0)}(\vartheta)}\frac{1}{r_p^2 sin\vartheta}\frac{\partial}{\partial \vartheta} \qquad (5)$$

where $p = 1, 2$, $r_{12} = (r_1^2 + r_2^2 - 2r_1r_2cos\vartheta)^{\frac{1}{2}}$, and $Z$ is the charge of the nucleus. The so-called $\mathcal{Z}$-coefficients, $\mathcal{Z}^{(0)}$, $\mathcal{Z}^{(1)}$, and $\mathcal{Z}^{(2)}$, are functions of $\vartheta$ and are given in Reference [23] as matrix elements of various angular momentum operators connecting coupled spherical harmonics. We represent the wavefunction and the action of the various operators utilizing the basis-spline collocation method.

In this case, the basis-splines, which incorporate the boundary conditions required, are used to expand the coefficients $\psi_l$ in a product, $u_i(r_1), u_j(r_2), w_k(\vartheta)$, i.e.

$$\psi_l(r_1, r_2, \vartheta) = \sum_{i,j,k} u_i(r_1) u_j(r_2) w_k(\vartheta). \qquad (6)$$

**TABLE 1.** The table lists the computed expectation values of total energy along with accurate reference values for states of He, H$^-$, and Be. For the doubly excited state, the decay width is reported as the imaginary part of the energy. All quantities are given in atomic units.

| State | $\langle E \rangle_{reference}$ | $\langle E \rangle$ |
|---|---|---|
| He$(1s^2)\,^1S$ | $-2.903\,72$ | $-2.903\,2$ |
| He$(1s2s)\,^1S$ | $-2.145\,97$ | $-2.144\,8$ |
| He$(1s2p)\,^1P$ | $-2.123\,84$ | $-2.123\,4$ |
| He$(1s3d)\,^1D$ | $-2.055\,63$ | $-2.052$ |
| He$(2s^2)\,^1S$ | $-0.778\,81 - i0.002\,28$ | $-0.787 - i0.002\,4$ |
| H$^-\,^1S$ | $-0.527\,75$ | $-0.526\,5$ |
| Be$(1s^22s^2)\,^1S$ | $-1.003$ | $-1.011$ |

Then the principle of collocation is applied in which we require the residual to vanish at a set of points, called collocation points, intermediate to the knots. Since we use a set of knots which is not uniformly spaced in order to have a good density of points where the wavefunction has its greatest density, the resulting lattice representation of the Hamiltonian is not Hermitian. Therefore, we compute in addition to the wavefunction, $\Psi$, its adjoint, $\Phi$, so the vectors $\Psi$ and $\Phi$ are biorthogonal, and norms, expectation values, and other operators will be properly defined.

By utilizing methods of partial eigensolution [22,23] (since the rank of the Hamiltonian is too large for practical diagonalization), we can begin with a trial wavefunction, such as the product of hydrogenic $1s$ orbitals, and iteratively approach the true ground state of helium. The accuracy of this approach is ultimately limited by the number of basis-splines that we can incorporate into the expansion of the wavefunction, but with a reasonably small number (e.g. 78, fifth-order basis splines on a mesh extending from 0 to 6 a.u.), we have been able to obtain an accuracy of about 0.02 percent with about 1500 iterations of the damped relaxation. A summary of our results for the ground and several low-lying excited states of He is given in Table 1. In addition, we show trial results for the ground state of H$^-$ and Be, the latter calculation utilizing a pseudo-potential to simulate the presence of the core electrons. Finally, the table gives our result for the energy and decay width of a doubly excited state of He, determined by spectral analysis of the time-dependent overlap of the initial state with the time-evolved state. These tests have helped us to validate the theoretical and numerical techniques which we have developed to treat the fully correlated two-electron problem.

Also, plots of the probability density obtained using the lattice wavefunction can give insight into the effects of correlation, for the structure of helium, or more importantly for the description of time-dependent processes. Figure 8 shows, for example, that for the ground state, the electrons are most likely to be found at equal radial distances, since the density is mostly peaked along the line $r_1 = r_2$, while the greatest probability is for the electrons to have an angle between their position vectors near $\pi$, owing to their mutual repulsion. In addition, for small values of $\vartheta$, the density peaks on either side of the line $r_1 = r_2$, indicating that at small angular separations, one or the other electron is pushed inward (outward) from the other due to the repulsion. All these simple observations are clearly in accord with our intuitive picture of the consequences of electron-electron interaction in helium.

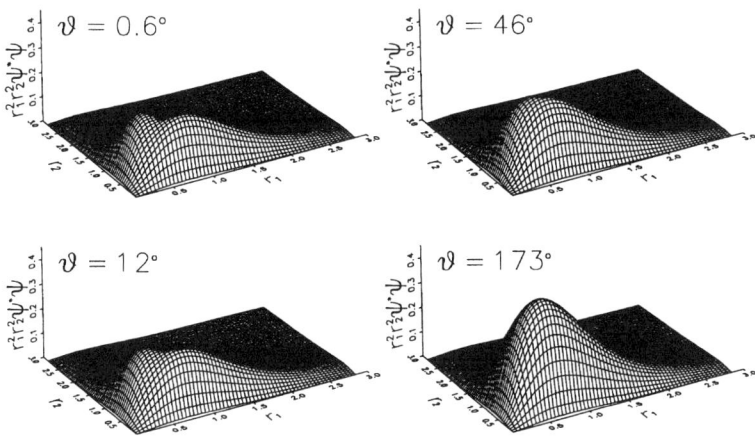

**FIG. 8.** The probability density in atomic units for the ground state of helium computed using the expansion coefficient $\psi_0(r_1, r_2, \vartheta)$ for several values of $\vartheta$.

We have also used the present lattice approach to follow up much earlier related models which used restricted dimensions, model wavefunctions, and less sophisticated numerical procedures [25] regarding the electron impact of atomic hydrogen. In this case, we represent the incoming electron by a wavepacket of a given angular momentum and time-propagate the total wavefunction through the collision. Figure 9 shows the electronic probability density at several time steps for a collision energy of 27.2 eV and total angular momentum of L=0. The left column is for the singlet configuration and the right column for the triplet. The axes are the radial coordinate of either electron (i.e. $r_1$ and $r_2$) and the results are plotted for an interparticle separation of $\vartheta = \pi$. One can see the development of a ridge of density in the continuum for the singlet case (along the line $r_1 = r_2$) and the exclusion from this ridge for the triplet case. By monitoring the overlap of the time evolved wavefunction with a basis of target states, elastic, excitation, and ionization probabilities can be computed. In Table 2 we compare probabilities computed in this way with results of the second-order distored-wave Born approximation regarding elastic scattering, finding generally very good agreement. We will follow up this preliminary calculation with computations on a finer and larger numerical grid and extend it to larger L so as to obtain convergence of

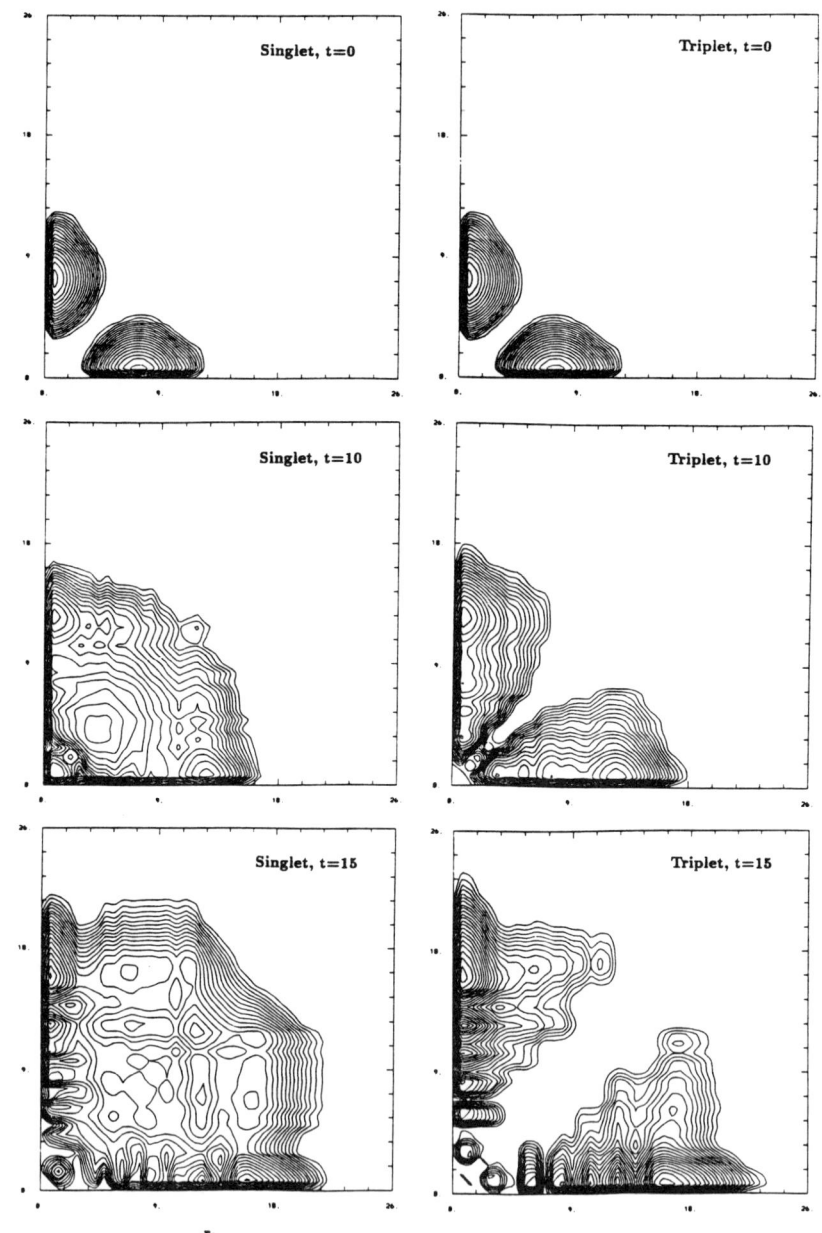

**FIG. 9.** The electronic probability density computed as the square of the coefficient $\psi_l(r_1, r_2, \vartheta)$ (Eq. 6) for 27.2 eV electron-impact of atomic hydrogen. The total angular momentum is zero, $\vartheta \approx \pi$, and the two columns show the result for either singlet or triplet symmetry. Times are given in atomic units, the projectile electron being represented by a wavepacket initially centered at 7.5 a.u. from the target center.

the total cross sections.

TABLE 2. The absolute value of the transition probability as a function of total angular momentum for 27.2 eV electron impact of atomic hydrogen. Listed are the results for elastic scattering. Preliminary calculations using the fully correlated two-electron lattice approach are labeled "lattice" and are compared with results of the second-order distorted-wave Born approximation ("DWB2").

| | e + H(1s) → e + H(1s) | | | |
|---|---|---|---|---|
| L | Singlet (lattice) | Singlet (DWB2) | Triplet (lattice) | Triplet (DWB2) |
| 0 | 0.58 | 0.596 | 0.96 | 0.980 |
| 1 | 0.12 | 0.148 | 0.41 | 0.424 |

Work is also in progress to treat the photo-excitation and ionization of two-electron atoms and ions as well as electron impact in a electromagnetic field. We also note that other groups have been pursuing somewhat similar lattice approaches for computing scattering phase shifts [26,27], wavepacket scattering from hydrogen (L=0) [28], and a model atom approach (2-dimensional) for autoionization [29,24], a comprehensive critique of which is beyond the scope of this progress report.

## CONCLUSIONS

It has been our recent goal to develop and apply time-dependent, lattice approaches to the description of atomic collisions. These methods possess the advantage of providing a great deal of insight through the easy visualization of the collision process in time. In addition, they provide a powerful method of studying strongly interacting systems by directly solving in a non-perturbative way the relevant equations of motion (e.g. the Schrödinger or Hartree-Fock equations). Lattice techniques are now being routinely applied to study atomic collisions and are indeed becoming a part of the standard technology employed by the atomic physicist.

## ACKNOWLEDGEMENTS

The research reported on here has been supported by the U.S. Department of Energy, Office of Basic Energy Sciences through Contract Number DE-AC05-84OR21400 managed by Lockheed Martin Energy Systems, Inc. The author wishes to also acknowledge the collaboration of the following colleagues who have been partners in various aspects of the works reported on here: Chris Bottcher, Don Madison, Gavin Buffington, Jerry Peacher, Mitch Pindzola, Panos Gavras, Jack Wells, Carlos Reinhold and Predrag Krstic.

# REFERENCES

1. C. Bottcher, Phys. Rev. Lett. **48**, 85 (1982).
2. D.S.F. Crothers and J.F. McCann, J. Phys. B **16**, 3229 (1983).
3. R.E. Olson, Phys. Rev. A **33**, 4397 (1986).
4. N. Stolterfoht, D. Schneider, J. Tanis, H. Altervogt, A. Salin, P.D. Fainstein, R. Rivarola, J.P. Grandin, J.N. Scheurer, S. Andraimonje, D. Bertault, and J.F. Chemin, Europhys. Lett. **4**, 899 (1987).
5. P.D. Fainstein, V.H. Ponce, and R.D. Rivarola, J. Phys. B **21**, 287 (1988).
6. C.O. Reinhold and R.E. Olson, Phys. Rev. A **39**, 3861 (1989).
7. R.E. Olson and A. Salop, Phys. Rev. A **16**, 531 (1977).
8. J.H. Macek, J. Phys. B **24**, 5121 (1991) and references therein.
9. V. Maruhn-Rezwani, N. Grün, and W. Scheid, Phys. Rev. Lett. **43**, 512 (1979).
10. N. Grün, A. Mühlhans, and W. Scheid, J. Phys. B **15** 4043 (1982).
11. K.C. Kulander, D.K.R. Sandhya, and S.E. Koonin, Phys. Rev. A **25**, 2968 (1982).
12. J.D. Garcia, Nucl. Instr. and Methods in Phys. Res. A **240**, 552 (1985).
13. N.H. Kwong, K.J. Schaudt, and J.D. Garcia, Comput. Phys. Commun. **63**, 171 (1991).
14. P. Gavras, M.S. Pindzola, D.R. Schultz, and J.C. Wells, Phys. Rev. A (1995), to be published.
15. M.S. Pindzola and C. Bottcher, Laser Phys. **3**, 748 (1993); M.W. Pindzola, T.W. Gorczyca, and C. Bottcher, Phys. Rev. A **47**, 4982 (1993).
16. J.C. Wells, V.E. Oberacker, M.R. Strayer, A.S. Umar, Int. J. Mod. Phys. C **6**, 143 (1995); A.S. Umar, J.-S. Wu, M.R. Strayer, and C. Bottcher, J. Comp. Phys. **93**, 426 (1991).
17. S.Yu. Ovchinnikov and J.H. Macek, Phys. Rev. Lett. (1995) to be published.
18. H. Knudsen, U. Mikkelsen, K. Paludan, K. Kirsebom, S.P. Moller, E. Uggerhoj, J. Slevin, M. Charlton, and E. Morenzoni, Phys. Rev. Lett. **74**, 4627 (1995).
19. E. Fermi and E. Teller, Phys. Rev. **72**, 399 (1947).
20. P.S. Krstic, D.R. Schultz, C.O. Reinhold, and J.C. Wells (1995).
21. N. Toshima, Phys. Lett. A **175**, 133 (1993).
22. C. Bottcher and M.R. Strayer, in *Proceedings of the XVI International Conference on the Physics of Electronic and Atomic Collisions*, edited by A. Dalgarno, R.S. Freud, P.M. Koch, M.S. Lubell, and T.B. Lucatorto, AIP Conf. Proc. No. 205 (AIP, New York, 1989), p. 658.
23. C. Bottcher, D.R. Schultz, and D.H. Madison, Phys. Rev. A 49, 1714 (1994).
24. D.R. Schultz, C. Bottcher, D.H. Madison, J.L. Peacher, G. Buffington, M.S. Pindzola, T.W. Gorczyca, P. Gavras, and D.C. Griffin, Phys. Rev. A **50**, 1348 (1994).
25. C. Bottcher, J. Phys. B **14**, L349 (1981); C. Bottcher, J. Phys. B **15**, L463 (1982); C. Bottcher, Adv. At. Mol. Phys. **20**, 241 (1985).
26. J. Botero and J. Shertzer, Phys. Rev. A **46**, R1155 (1992); J. Shertzer and J. Botero, Phys. Rev. A **49**, 3673 (1994).
27. Y.D. Wang and J. Callaway, Phys. Rev. A **48**, 2058 (1993); Y.D. Wang and J. Callaway, Phys. Rev. A **50**, 2327 (1994).
28. L. Zhang, J. Feagin, V. Engel, and A. Nakano, Phys. Rev. A **49**, 3457 (1994).
29. S.L Haan, R. Grobe, and J.H. Eberly, Phys. Rev. A **50**, 378 (1994).

# Theory of Low Energy Ion-Atom Collisions

Eugene A. Solov'ev [1]

*The Macedonian Academy of Sciences and Arts, Skopje 91000, Macedonia*

> The advanced adiabatic approach to slow ion-atom collisions provides a self-consistent unified description of all inelastic processes within the same conceptual framework. The basic physical aspects of this approach are herewith reviewed, as well as its recent developments. Possible further extensions of the method are also discussed.

## INTRODUCTION

During the last fifteen years an advanced adiabatic approach has been developed and widely used for calculation of inelastic processes in low energy ion-atom collisions (see e.g. [1-16]) as well as in exotic reactions involving antiproton and muon [17-22]. It is based on the semiclassical approximation for the nuclear motion employing a modified adiabatic basis which is compatible with the physical boundary conditions [23,24] and on the theory of hidden crossings [1,2,25], which provides a complete description of nonadiabatic couplings. This theory can compete with direct numerical calculations with respect to precision and, particularly, to required computer facilities. It also provides important insight in the mechanisms of the studied processes and on the characteristic internuclear distances at which they occur. Below we briefly review the main aspects of the theory.

In the centre of mass (c.m.) frame of reference the electronic transitions are described by the Schrödinger equation (atomic units are used throughout this work)

$$\left[-\frac{1}{2M}\Delta_{\mathbf{R}} + \frac{Z_1 Z_2}{R} + H(\mathbf{R})\right]\Psi(\mathbf{R},\mathbf{r}) = \mathcal{E}\Psi(\mathbf{R},\mathbf{r}) \tag{1}$$

where $\mathbf{R}$ is internuclear distance, $Z_1, Z_2$ are charges and $M$ is the reduced mass of nuclei, $\mathbf{r}$ is radius-vector of electron referred to c.m. of nuclei, $H(\mathbf{R})$ is the electronic part of total Hamiltonian. In the adiabatic representation the wave function $\Psi(\mathbf{R},\mathbf{r})$ is sought in the form

---

[1] Permanent address: Department of Theoretical Physics, St-Petersburg University, St-Petersburg 198904, Russia

© 1995 American Institute of Physics

$$\Psi(\mathbf{R}, \mathbf{r}) = \sum_{L,p} \frac{F_p^{(L)}(R)}{R} P_L(\theta) \varphi_p(\mathbf{r}, R) \quad (2)$$

where $P_L(\theta)$ are Legendre polynomials (angular part of the nuclear wave function), $\varphi_p(\mathbf{r}, R)$ are the adiabatic wave functions, and $E_p(R)$ are the adiabatic potential curves

$$H(R)\varphi_p(\mathbf{r}, R) = E_p(R)\varphi_p(\mathbf{r}, R),$$

which depend on R as an external parameter occurring in the Hamiltonian $H(R)$. After substitution Eq.(2) into Eq.(1), the Schrödinger equation takes the form of a system of equations for the radial nuclear wave functions

$$\left[ -\frac{1}{2M}\frac{d^2}{dR^2} + \frac{L(L+1)}{2MR^2} + \frac{Z_1 Z_2}{R} + E_p(R) - \mathcal{E} \right] F_p^{(L)} = \sum W_{pp'}^{LL'}(R) F_{p'}^{(L')}, \quad (3)$$

where $W_{pp'}^{LL'}$ is the operator of the non-adiabatic coupling.

In the adiabatic representation the boundary condition is formulated as follows. In the limit $R \to \infty$, the adiabatic potential curves $E_p(R)$ reduce to energy levels of isolated atoms, whereas $\varphi_p(\mathbf{r}, R)$ transform into related atomic wave functions $\varphi_p^{(a)}$. If q is the set of quantum numbers of the initial atomic state, the asymptotic form of the nuclear wave function is

$$F_p^{(L)}(R) = \begin{cases} (-1)^{L+1} e^{-iK_q R} + S_{qq}^{(L)} e^{iK_q R}, & \text{p=q}, \\ S_{pq}^{(L)} e^{iK_p R}, & \text{p} \neq \text{q}, \end{cases} \quad (4)$$

where $K_p = \sqrt{2M(\mathcal{E} - E_p(\infty))}$ and $S_{pq}^{(L)}$ are the S-matrix elements. In terms of $S_{pq}^{(L)}$ the cross section reads [26]

$$\sigma_{qq} = \frac{\pi}{K_q^2} \sum_{L=0}^{\infty} (2L+1)|1 - S_{qq}^{(L)}|^2, \quad (5)$$

for elastic scattering and

$$\sigma_{pq} = \frac{\pi}{K_q^2} \sum_{L=0}^{\infty} (2L+1)|S_{pq}^{(L)}|^2, \quad (6)$$

for an inelastic transition.

If the impact energy is far enough from threshold, one can utilize the classical approximation for the nuclear motion. In this case, a large number of partial waves contribute to the cross sections. Then, using the relation $(L + \frac{1}{2}) = K_q b$, the sum in Eqs.(5,6) can be replaced by an integral over the impact parameter $b$,

$$\frac{\pi}{K_q^2} \sum_{L=0}^{\infty} (2L+1) \longrightarrow 2\pi \int_0^{\infty} b\, db, \qquad (7)$$

whereas $S_{pq}^{(L)}$ is replaced by a probability of the inelastic transition as a function of impact parameter, $P_{pq}(b)$.

## AVOIDED CROSSINGS AND BRANCH POINTS OF ADIABATIC POTENTIAL CURVES

In the adiabatic approximation, inelastic transitions occur in the regions of closest approach of the potential curves. According to Neumann-Wigner theorem [27], the exact crossing of two adiabatic potential curves is an exception, so that the most frequent case is that of an avoided crossing. The avoided crossing of two curves $E_1(R)$ and $E_2(R)$ reflects their exact crossing at the complex value $R_c$ near the real axis of $R$. Degeneracy of the energy levels at $R = R_c$ has an important special feature. In the complex $R$-plane, the Hamiltonian $H(R)$ is no longer self-adjoint so that, when two eigenvalues merge $[E_1(R_c) = E_2(R_c) \equiv E_c]$, it is reduced not to a diagonal form but rather to a Jordan form [28]

$$H(R_c) = \begin{pmatrix} E_c & 1 \\ 0 & E_c \end{pmatrix}.$$

In the vicinity of $R_c$ the values of $E_1(R)$ and $E_2(R)$ can be found using perturbation theory with respect to a small parameter $\Delta R = R - R_c$. In the standard case of a linear perturbation

$$U(R) = \Delta R \begin{pmatrix} U_{11} & U_{12} \\ U_{21} & U_{22} \end{pmatrix} \quad (U_{ij} = const),$$

the first approximation yields

$$E_{1,2}(R) = E_c \pm (U_{21}\Delta R)^{1/2}, \qquad (8)$$

i.e., instead of the usual linear dependence on the small parameter $\Delta R$, we have now a square-root dependence. The square-root branch point combines the two adiabatic potential curves into a single analytic function, so that when the point $R_c$ is circled once, the sign in front of the radical in Eq.(8) is reversed: the first adiabatic potential curve transforms into the second one, and vice versa. Obviously, the same property is exhibited also by the corresponding adiabatic wave functions.

## PROBABILITY OF NON-ADIABATIC TRANSITIONS

The technique for calculation of the probability of a transition in the region of an avoided crossing utilizes the smallness of the quantity $1/M$ [22]. In this approximation, the right hand side of Eq.(3) is a small correction of order $M^{-\frac{1}{2}}$. Therefore in the limit $1/M \to 0$ the system of equations (3) is decoupled and its solution has a semiclassical type of asymptotics, namely,

$$F_p^{(as.)}(R) = \mathcal{P}_p(R)^{1/2} \exp\left(\pm i \int^R \mathcal{P}_p(R')dR'\right), \qquad (9)$$

where

$$\mathcal{P}_p(R) = \sqrt{2M\left[\mathcal{E} - \frac{L(L+1)}{2MR^2} - \frac{Z_1 Z_2}{R} - E_p(R)\right]}$$

is the radial momentum of nuclei. To calculate an exponentially small transition probability one should continue the asymptotics (9) into the complex $R$-plane. It turns out that this asymptotics is valid everywhere except the small region $\Omega$ around a complex branch point $R_c$ where the operator of non-adiabatic coupling $W_{pp'}^{LL'}$ has a singularity. Inside $\Omega$ it is necessary to find a solvable system (called the *comparison system*) of equations which takes into account the singularity at the point $R_c$ explicitly. The amplitude of the transition probability is obtained by matching, at the boundary of the region $\Omega$, the solution of the comparison system to the asymptotics given by Eq.(9) and satisfying the boundary conditions of Eq.(4). The final expression for the transition probability reads [22]

$$|S_{pq}^{(L)}|^2 = e^{-2\Delta_{pq}}, \qquad (10)$$

where

$$\Delta_{pq} = \left|\text{Im} \oint_{C_{pq}} \mathcal{P}(R)\, dR\right| = \left|\text{Im} \int_{\text{Re } R_c}^{R_c} \left[\mathcal{P}_p(R) - \mathcal{P}_q(R)\right] dR\right| \qquad (10a)$$

is the Stueckelberg parameter [34], and $C_{pq}$ is a contour in the complex $R$-plane enclosing the branch point $R_c$. If $\mathcal{E} \gg E_p(R)$, the classical approximation for nuclear motion can be employed. Treating $E_p(R)$ as a small quantity the momentum $\mathcal{P}_p(R)$ can be approximated by the expression, $\mathcal{P}_p(R) = MV(R,b) + E_p(R)/V(R,b)$, where $V(R,b)$ is the radial nuclear velocity. Equation (10a) then takes the form of the Massey parameter

$$\Delta_{pq} = \left| \text{Im} \oint_{C_{pq}} E(R) \frac{dR}{V(R,b)} \right| = \left| \text{Im} \int_{Re R_c}^{R_c} [E_p(R) - E_q(R)] \frac{dR}{V(R,b)} \right|. \quad (10b)$$

For the spectrum of ionized electrons the adiabatic theory has been developed in [29] where the energy distribution of ionized electrons was found in the form

$$W(E) = \frac{1}{2\pi v} \left| \frac{dR_q(E)}{dE} C(E) \exp\left[-\frac{2}{v} \int_{E_q^0}^{E} R_q(E') dE'\right] \right|$$

where $v$ is the impact velocity, $E_q^0$ is the energy of initial atomic state, $R_q(E)$ is the inverse function to $E_q(R)$ and $C(E)$ is a quantity defined in [29] which has the meaning of density of states in the continuous spectrum. The function $R_q(E)$ is complex at $E > 0$; it is treated as an analytic continuation from the discrete spectrum to the region of positive electron energies.

## HIDDEN CROSSINGS

In the past, the use of adiabatic approximation has been restricted to processes taking place via narrow avoided crossings caused by the underbarrier resonance interaction of the adiabatic states located on different nuclei. Recently the theory has been advanced by acknowledging the various new mechanisms of non-adiabatic transitions via so-called "hidden crossings" [1,2,25]. The hidden crossings play an important role because only by taking them into account can a complete set of non-adiabatic couplings be constructed.

In the one-electron-two-Coulomb-center problem there are four types of hidden crossings which are denoted as S-, T-, P- and Q- hidden crossings.

i) *S-superseries of hidden crossings*
Each of the S-superseries consists of an infinite set of branch points connecting pairwise the states $(N,l,m)$ and $(N+1,l,m)$ consecutively for all $N$ ($Nlm$ are the united atom spherical quantum numbers). They are distributed in a small domain of the complex $R$-plane according to the relation $\text{Re} R_{N,lm}^S > \text{Re} R_{N+1,lm}^S$ and for $N \to \infty$ the limit point of the series is [14,30]

$$R^{(S)}_{lm} = \frac{(l+\frac{1}{2})^2}{Z} \exp\left[i\frac{\pi(m+1)}{(2l+1)}\right].$$

The S-superseries are physically associated with the so called "superpromotion" of diabatic potential curves to the continuous spectrum [1,25,30]. At this place the centrifugal barrier of the united atom is formed and the topology of adiabatic wave function changes from the two-center geometry of a quasimolecule to the one-center geometry of a united atom.

ii) *T-superseries of hidden crossings*

The T-superseries exist in the symmetric case, $Z_1 = Z_2$. They consists of branch points which connect pairwise the states $(N, l, m)$ and $(N+2, l+2, m)$. The position of T branch points can be estimated by the expression [6]

$$R_c^{(T\pm)} = (n_1 + n_2 + m + 1)[6n_2 + 3m + 3 + (6 \mp 2)i],$$

where $n_1, n_2, m$ denote the parabolic quantum numbers of the electron in the separated atom limit and $(\pm)$ indicate the parity of the state as $z \to -z$ (the $(\pm)$ parity $W$ is related to the $(g, u)$ parity $I$ by $I = We^{im\pi}$). The appearance of the T-superseries reflects the passing of the $E(R)$ over the top of the potential barrier separating the Coulomb potential wells. This leads to a qualitative modification of adiabatic states: if $R < \mathrm{Re} R_c^{(T\pm)}$, an electron moves in a shared potential well of two center and its wave function is essentially quasimolecular. If $R > \mathrm{Re} R_c^{(T\pm)}$, the regions of classically allowed motion of an electron near the nucleus are separated from one another by a barrier and the wave function can be represented approximately by superposition (symmetric or antisymmetric) of the wave functions of two isolated atoms $(Z_1 e)$ and $(Z_2 e)$.

iii) *P-superseries of hidden crossings*

In the case $Z_1$ close to $Z_2$, the quasimolecule has approximate $(g, u)$ symmetry and so-called P-superseries of hidden crossings arise. They are related to the branch points which sew together the two adiabatic potential curves $(N, l, m)$ and $(N+1, l+1, m)$ uniformly along a line perpendicular to the real $R$ axis . In the limit $R \to \infty$ these pairs of states transform into hydrogen-like states which are located at different centers and have an identical set of parabolic quantum numbers $(n_1, n_2, m)$. The P-superseries are caused by the Rosen-Zener-Demkov coupling [31,32]. They are related to a breaking of the approximate $(g, u)$ symmetry. To the left of the P-superseries $(R < \mathrm{Re} R^{(P)})$, one can ignore the resonance defect in comparison with the underbarrier interaction. The situation is qualitatively close to the symmetric case $Z_1 = Z_2$, i.e. the wave functions have an approximate $(g, u)$ symmetry. To the right of this series $(R > \mathrm{Re} R^{(P)})$, the resonance defect dominates; the approximate $(g, u)$ symmetry is lost and the adiabatic state is localized at one of the nuclei.

iv) *Q-superseries of hidden crossings*

The Q-superseries are formed from the quasimolecular T- and P- superseries in the course of the increase of the discrepancy between the charges $Z_1$ and $Z_2$. They consist of branch points which connect pairwise the states $(N, l, m)$ and $(N + 1, l + 1, m)$ and they have the same origin as the T-superseries.

The occurence of hidden crossings is closely related to the case of what is called *limiting motion* (or unstable periodic trajectories) in the classical description of an electron state. The semiclassical theory of hidden crossings has been formulated and applied to S-superseries in [30]. In [33] it was elaborated for all other cases in two-Coulomb-center problem.

## DYNAMICAL ADIABATIC BASIS COMPATIBLE WITH BOUNDARY CONDITIONS

The calculations of the conventional adiabatic basis sets are usually based on the assumption that the nuclei are fixed, so that the basis sets are not matched to the physical boundary conditions in the limit $R \to \infty$. The lack of matching is manifested in different ways, depending on whether the motion of nuclei is treated classically or quantum-mechanically. In the quantum-mechanical approach, the adiabatic basis suffers from the fact that, instead of the reduced mass, it contains the electron mass, whereas the defect in the classical approach is that the basis does not contain the electron translational factor. The problem of the boundary conditions under the quantum treatment of nucler motion has been considered in [22]. Here we concentrate on the classical approach.

In [21] the method of nonstationary scaling of the length has been proposed in order to reduce the problem of the electron translational factor to the determination of so-called dynamical adiabatic states. Thus, in the case of a collisional system consisting of one electron and two bare nuclei $A$ and $B$ travelling along straight-line classical trajectories, the time-dependent Schrödinger equation to be solved is

$$\left(-\frac{1}{2}\nabla_r^2 - \frac{Z_1}{|\mathbf{r} + \alpha\mathbf{R}|} - \frac{Z_2}{|\mathbf{r} - \beta\mathbf{R}|}\right)\Psi(\mathbf{r},t) = i\frac{\partial}{\partial t}\Psi(\mathbf{r},t), \qquad (11)$$

where $\mathbf{R} = \mathbf{R}_2 - \mathbf{R}_1 = (vt, \rho, 0)$ is the vector connecting the nuclei $A$ and $B$, $v$ is the relative collision velocity, $\rho$ is the impact parameter and the point on the internuclear axis defining the origin of the reference frame is determined by the parameters $\alpha$ and $\beta$ ($\mathbf{R}_1 = -\alpha\mathbf{R}$, $\mathbf{R}_2 = \beta\mathbf{R}$, $\alpha + \beta = 1$). The initial condition for $t \to -\infty$ requires that $\Psi$ takes the form of a product of the initial atomic wave function $\Phi_\gamma^{(a)}(\mathbf{r}_j)$ located at one of the two centers ($j = A, B$) with the Galilean translational factor which takes into account the motion of the nuclei

$$\lim_{t \to -\infty} \Psi(\mathbf{r},t) = \Phi_\gamma^{(a)}(\mathbf{r}_j)\exp[i(\mathbf{v}_j \cdot \mathbf{r}_j - \frac{1}{2}v_j^2 t - E_\gamma^{(a)}t)], \qquad (12)$$

where $r_j = r - R_j$ and $v_j$ is the velocity of the $j$th nucleus. We introduce now the nonstationary scaling of the length by dividing electronic coordinates $(x, y, z)$ by the internuclear separation $R(t)$ and subsequently make the transformation to the rotating coordinate system $(q_1, q_2, q_3)$ with the $q_1$ axis directed along the internuclear axis

$$q_1 = \frac{1}{R(t)}[x\cos\varphi(t) + y\sin\varphi(t)], \quad q_2 = \frac{1}{R(t)}[-x\sin\varphi(t) + y\cos\varphi(t)], \quad q_3 = \frac{z}{R(t)},$$

where $\varphi(t) = \arctan(\rho/vt)$ is the polar angle of $\mathbf{R}(t)$ in the scattering $(x, y)$ plane. We also represent the wave function in the form

$$\Psi(\mathbf{r}, t) = R^{-3/2} \exp\left(i\frac{r^2}{2R}\frac{dR}{dt}\right) f(\mathbf{q}, t) \tag{13}$$

and introduce a new time like variable ($\omega = \rho v$)

$$\tau(t) = \int_0^t \frac{dt'}{R^2(t')} = \omega^{-1} \arctan(vt/\rho). \tag{14}$$

The factor $R^{-3/2}$ in (13) ensures the normalization and the exponent is a generalized translational factor. The variation of $t$ from $-\infty$ to $+\infty$ corresponds to variation of $\tau$ from $-\pi/(2\omega)$ to $+\pi/(2\omega)$. Substituting (14) into (11) we obtain the modified Schrödinger equation in $\mathbf{q}$ space

$$\mathcal{H}f = \left[-\frac{1}{2}\nabla_\mathbf{q}^2 - \frac{Z_1 R}{|\mathbf{q} + \alpha\hat{\mathbf{q}}_1|} + \frac{Z_2 R}{|\mathbf{q} - \beta\hat{\mathbf{q}}_1|} + \omega L_3 + \frac{1}{2}\omega^2 q^2\right] f = i\frac{\partial f}{\partial \tau}, \tag{15}$$

where $\mathcal{H}$ is the effective Hamiltonian, $\hat{\mathbf{q}}_1$ is the unit vector along the $q_1$ axis and $L_3$ is the operator of the projection of the electronic angular momentum on the direction perpendicular to the scattering plane.

For slow collisions we look for the solution of Eq.(15) by expanding $f(\mathbf{q}, \tau)$ in terms of eigenfunctions $\Phi_\gamma(\mathbf{q}, \tau)$ of the effective instantaneous Hamiltonian

$$\mathcal{H}(\tau)\Phi_\gamma(\mathbf{q}, \tau) = \mathcal{E}(\tau)\Phi_\gamma(\mathbf{q}, \tau). \tag{16}$$

In the new representation both centers are at rest and the modified basis is compatible with asymptotic boundary conditions. The correct translational factor is obtained automatically from the exponential factor in (13), when transforming back to the original wave function $\Psi$, i.e.

$$\exp\left(i\frac{r^2}{2R}\frac{dR}{dt}\right) = \exp\left(i\frac{|\mathbf{r}_j + \mathbf{R}_j|^2}{2R}\frac{dR}{dt}\right) = \exp[i(\mathbf{v}_j \cdot \mathbf{r}_j - \frac{1}{2}v_j^2 t)]|_{Rr_j^{-1} \to \infty}.$$

We call the complete set of functions, $\Phi_\gamma$, dynamical adiabatic states and the eigenvalues, $\mathcal{E}_\gamma$, dynamical potential energy curves, since in addition to the internuclear separation they also depend on $\omega = \rho v$. The use of dynamical basis solves another well known problem of the standard adiabatic basis, namely, the dependence of the transition probability on the selected coordinate system: the center–of– mass system, the laboratory reference system, or the reference system with a center at one of the nuclei. All these reference systems are distinguished, in terms of the new variables, by a shift of a constant vector along the $q_1$ axis. It is easy to show that the transition probability is invariant with respect to this transformation.

Contrary to the simpler, two-Coulomb-center problem, the variables in the eigenvalue problem (16) cannot be separated. However in the framework of the adiabatic approach, we can consider the additional dynamical potential as a perturbation. This perturbation has a very important property which allows us to reduce this problem to the two- Coulomb-center problem with respect to the calculation of the Massey parameter. It is easy to show, that the first correction to eigenvalues caused by both terms of the dynamical interaction in the Hamiltonian $\mathcal{H}$ is proportional $v^2$. The following relation holds between the Massey parameter $\Delta^{(0)}$ in the standard basis and the Massey parameter $\Delta$ in the dynamical adiabatic basis, $\Delta = \Delta^{(0)} + v^2 \Delta^{(1)} + O(v^3)$, or between the related probabilities $P = P^{(0)}(1 + O(v))$ (see Eqs.(10,10b)). Therefore, the probabilities, $P$ and $P^{(0)}$, coincide in the region of validity of the adiabatic theory. This represents a rigorous mathematical justification of the exact asymptotic character of the results based on the standard adiabatic states.

## ISOTOPE EFFECTS IN THE ADIABATIC APPROXIMATION

In reactions involving different isotopes of the same chemical element, the exact symmetry with respect to mutual transposition of the nuclei breaks down because their masses are different ($M_A \neq M_B$). This gives rise to significant effects not encountered in the case of symmetric quasimolecules. They are associated entirely with the motion of the nuclei so that the standard adiabatic approach must be modified to incorporate them. Such a modification of adiabatic collision theory has been done in [35]. Following [35], let us introduce the following coordinates

$$\mathbf{q} = \frac{1}{2}(\sqrt{\mu_A} + \sqrt{\mu_B})\mathbf{R}_C - \frac{1}{2}(\sqrt{\mu_A}\mathbf{R}_A + \sqrt{\mu_B}\mathbf{R}_B)$$

$$\mathbf{Q} = (\sqrt{\mu_A} - \sqrt{\mu_B})\mathbf{R}_C - (\sqrt{\mu_A}\mathbf{R}_A - \sqrt{\mu_B}\mathbf{R}_B)$$

$$\mathbf{R}_{c.m.} = \frac{\mathbf{R}_C + M_A \mathbf{R}_A + M_B \mathbf{R}_B}{1 + M_A + M_B} \qquad (17)$$

where $\mu_A = M_A/(1+M_A)$ and $\mu_B = M_B/(1+M_B)$ are the reduced masses of atoms AC and BC ($M_C = m = 1$), $\mathbf{R}_A, \mathbf{R}_B$ and $\mathbf{R}_C$ are the radius-vectors of the first and second nucleus in the electron and the lab frame of reference, respectively. After separation of the center of mass motion the exact three-body Hamiltonian takes the form

$$\mathcal{H} = -\frac{1}{4}(1+\sqrt{\mu_A\mu_B})\Delta_\mathbf{q} - (1-\sqrt{\mu_A\mu_B})\Delta_\mathbf{Q} - \frac{\sqrt{\mu_A}Z_1}{|\mathbf{q}+\frac{1}{2}\mathbf{Q}|}$$

$$-\frac{\sqrt{\mu_B}Z_2}{|\mathbf{q}-\frac{1}{2}\mathbf{Q}|} + \frac{\sqrt{\mu_A\mu_B}Z_1Z_2}{|\frac{1}{2}(\sqrt{\mu_A}+\sqrt{\mu_B})\mathbf{Q}+(\sqrt{\mu_B}-\sqrt{\mu_A})\mathbf{q}|} \,. \quad (18)$$

If we ignore temporarily the dependence of the internuclear interaction on $\mathbf{q}$ (last term in Eq.(18)), we can see that the Hamiltonian (18) is exactly the same as the Hamiltonian of an effective three-body system composed of a "light particle" with a mass $m^* = (\mu_A\mu_B)^{-\frac{1}{2}} \approx 1$ and two "heavy particles" with identical masses $M^* = (1-\sqrt{\mu_A\mu_B})^{-1} \gg 1$ whereas $\mathbf{q}$ and $\mathbf{Q}$ play the role of the Jacobian coordinates of such an effective system. Therefore, if $M_A \neq M_B$ and $Z_1 = Z_2 \equiv Z$, the transformation of Eq.(17) converts the asymmetry of the masses of heavy particles into the asymmetry of charges: $Z_1^{eff.} = \sqrt{\mu_A}Z$ and $Z_2^{eff.} = \sqrt{\mu_B}Z$. After adiabatic separation of the motion of the "heavy particles" with respect to the coordinates $\mathbf{Q}$, we have the problem of two Coulomb centers with "isotopically" different charges. This leads to the appearance of the Rosen-Zener-Demkov coupling of adiabatic potential curves [31,32] (P-superseries of hidden crossings [2]).

The dependence of the internuclear interaction on $\mathbf{q}$ in Eq.(18) gives rise to an additional asymmetry which is of the order $(M^*Q^2)^{-1}$. It can be taken into account in the framework of perturbation theory. However for non-adiabatic transitions via P-superseries of hidden crossings this correction can be omitted.

## AUTOMATED PROGRAM-PACKAGE ARSENY

In the adiabatic approach, inelastic transitions are decomposed into a sequence of individual two-level transitions via hidden crossings whose probabilities are determined by (10). In addition, rotational coupling at large and small internuclear distances has to be included as described in [36,37]. According to this scheme, a program package ARSENY has been developed in [38]. The program package ARSENY is based on a program from [1] for exact calculation of adiabatic potential curves of two Coulomb center problem in complex $R$-plane and designed for use as a black-box routine which only requires as input parameters: i) the charges of the nuclei, ii) the list of energies of the colliding particles and iii) the basis size. As a first step, it searches for all branch points and calculates the corresponding Stueckelberg parameter (10a) and probability (10) as function of $L$ for the entire set of nonadiabatic transitions. Then the $S$ matrix is calculated as a product of elementary $S$ matrices for the individual transitions induced by the separated branch points. Finally, the cross sections are calculated as sum over $L$ (Eqs.(5,6)). This yields simultaneously all partial cross sections for arbitrary initial and final states in a given molecular-orbital basis set.

## ACKNOWLEDGEMENTS

I am indebted to staff of the Isaac Newton Institute for Mathematical Sciences for the warm hospitality during my stay at the University of Cambridge.

## REFERENCES

1. Solov'ev E.A., Sov. Phys.- JETP **54**, 893 (1981).
2. Ovchinnikov S.Yu. and Solov'ev E.A., Sov. Phys.- JETP **63**, 538 (1986).
3. Grozdanov T. and Solov'ev E.A., Phys. A **38**, 4333 (1988).
4. Ovchinnikov S.Yu. and Solov'ev E.A., Comments At. Mol. Phys. **22**, 69 (1988).
5. Grozdanov T. and Solov'ev E.A., Phys. Rev. A **42**, 2703 (1990).
6. Solov'ev E.A., Phys. Rev. A **42**, 1331 (1990).
7. Ovchinnikov S.Yu., Phys. Rev. A **42**, 3865 (1990).
8. Pieksma M. and Ovchinnikov S.Yu., J. Phys. B **24**, 2699 (1991).
9. Janev R.K. and Kristic P.S., Phys. Rev. A **44**, R1453 (1991).
10. Janev R.K. and Kristic P.S., Phys. Rev. A **46**, 5554 (1992).
11. Kristic P.S., Radmilovic M. and Janev R.K., At. Plasma-Mater.Int.Data.Fusion **3**, 113 (1992).
12. Kristic P.S. and Janev R.K., Phys. Rev. A **47**, 3894 (1993).
13. Richter.K and Solov'ev E.A., Phys. Rev. A **48**, 432 (1993).
14. Janev R.K., Ivanovski G. and Solov'ev E.A., Phys. Rev. A **49**, R645 (1994).
15. Macek J.H. and Ovchinnikov S.Yu., Phys. Rev. A **50**, 486 (1994).
16. Macek J.H., Ovchinnikov S.Yu. and Pasovets S.V., Phys. Rev. Let. **74**, 4631 (1995).
17. Richter.K, Rost M.J., Thurwachter R., Briggs J.S., Wintgen D. and Solov'ev E.A., Phys. Rev. Let. **66**, 149 (1991).
18. Richter.K, Rost M.J., Briggs J.S., Wintgen D. and Solov'ev E.A., J. Phys. B **25**, 347 (1992).
19. Gusev V.V., Ponomarev L.I. and Solov'ev E.A., Muon Catalized Fusion **7**, 594 (1993).
20. Gzaplinski W., Gula A., Kravtsov A., Mikhailov A. and Popov N., Phys. Rev. A **50**, 518 (1994).
21. Gzaplinski W., Gula A., Kravtsov A., Mikhailov A. and Popov N., Phys. Rev. A **50**, 525 (1994).
22. Janev R.K., Solov'ev E.A. and Jakimovski D., J. Phys. B (1995) (to be published).
23. Solov'ev E.A., Sov. J. Theor. Math. Phys. **28**, 757 (1976).

24. Solov'ev E.A. and Vinitsky S.I., J. Phys. B **18**, L557 (1985).
25. Solov'ev E.A., Sov. Phys.- Usp. **32**, 228 (1989).
26. Landau L.D. and Lifshitz E.M., *Quantum Mechanics: Non- Relativistic Theory*, 2nd ed., (Pergamon Press, Oxford 1965).
27. Neumann J. and Wigner E., Phys. Z. **30**, 467 (1929).
28. Stoll R.R. and Wong E.T., *Linear Algebra*, (Academic Press, New York 1968).
29. Solov'ev E.A., Sov. Phys.- JETP **43**, 453 (1976).
30. Solov'ev E.A., Sov. Phys.- JETP **63**, 678 (1986).
31. Rosen N. and Zener C., Phys. Rev. **40**, 502 (1932).
32. Demkov Yu.N., Sov. Phys.- JETP **18**, 138 (1964).
33. Abramov D.I., Ovchinnikov S.Yu. and Solov'ev E.A., Phys. Rev. A **42**, 6366 (1990).
34. Stueckelberg E.C.C., Helv. Phys. Acta **5**, 369 (1932).
35. Solov'ev E.A., Sov. J. Nucl. Phys. **45** (1986).
36. Grozdanov T. and Solov'ev E.A., J. Phys. B **15**, 3871 (1982).
37. Grozdanov T. and Solov'ev E.A., Phys. Rev. A **44**, 5605 (1991).
38. Solov'ev E.A. in *Workshop on "Hidden crossings in Ion - Collisions and in Other Nonadiabatic Transitions"*, (Harvard Smithsonian Center for Astrophysics, Cambrige 1991).

# ION-ATOM COLLISIONS
# – EXPERIMENT –

Non-Statistically Populated Autoionizing Levels
of Li-like Carbon: Hidden Crossings .................................. 485
  E.F. DEVENEY, H.F. Krause, N.L. Jones,
  J.M. Sanders, C.R. Vane, W. Wu, S.Datz,
  M. Breinig, D.Desai, S.Yu Ovchinnikov,
  Q.C. Kessel, and S.M. Shafroth

Cold Target Recoil Ion Momentum Spectroscopy .................. 495
  R. DÖRNER, V. Mergel, L. Spielberger,
  O. Jagutzki, M. Unverzagt, W. Schmitt,
  J. Ullrich, R. Moshammer, H. Khemliche,
  M. Prior, R.E. Olson, L. Zhaoyuan, W. Wu,
  C.L. Cocke, and H. Schmidt-Böcking

Electron-Electron and Electron-Nuclear Interactions
in Dressed Ion-Atom Collisions ....................................... 505
  R.D. DUBOIS

Projectile Electron Excitation and Loss in Ion-Atom Collisions .. 515
  E.C. MONTENEGRO, W.E. Meyerhof,
  J.H. McGuire, and C.L. Cocke

Radiative and Resonant Electron Capture
Studies for High-Z Projectiles ........................................ 525
  Th. STÖHLKER

# Non-Statistically Populated Autoionizing Levels of Li-like Carbon: Hidden-Crossings

### E. F. Deveney*, H.F. Krause, N.L. Jones, J.M. Sanders**, C.R. Vane, W.Wu and S. Datz.
*Physics Division, Oak Ridge National Laboratory, Oak Ridge, TN 37831-6377 USA*

### M. Breinig, D. Desai and S.Y. Ovchinnikov
*Department of Physics, The University of Tennessee, Knoxville, TN 37996-1200 USA and Oak Ridge National Laboratory, Oak Ridge, TN 37831-6377 USA*

### Q.C. Kessel
*Department of Physics, The University of Connecticut, Storrs, CT 06269-3046 USA*

### S.M. Shafroth
*Department of Physics and Astronomy, The University of North Carolina, Chapel Hill NC 27599-3255 USA*

**Abstract.** The intensities of the Auger-electron lines from autoionizing (AI) states of Li-like ($1s2s2l$) configurations excited in ion-atom collisions vary as functions of the collision parameters such as, for example, the collision velocity. A statistical population of the three-electron levels is at best incomplete and underscores the intricate dynamical development of the electronic states. We compare several experimental studies to calculations using 'hidden-crossing' techniques to explore some of the details of these Auger-electron intensity variation phenomena. Our investigations show promising results suggesting that Auger-electron intensity variations can be used to probe collision dynamics.

## INTRODUCTION

It is instructive to first discuss the autoionizing levels of the Li-like configurations which will be the subjects of investigation. There are four distinct levels of $(1s2s2l)$ configurations which autoionize; $(1s2s^2)\,^2S$, $^2P_-$ and $^2P_+$ both from $(1s2s2p)$ [1,2], and $^4P$ from $(1s2s2p)$ with all spins aligned. The three doublets decay to the $(\{1s^2\}^1S + e_c l_c)\,^2L$ ground state configuration where $e_c$ represents the energy of a continuum electron with angular momentum $l_c$ ($l_c = L$). The quartet will also decay via an Auger-electron transition, however the decay is spin forbidden and the state is therefore metastable. Our investigations involve the measurements of Auger-electrons from these levels that were excited from $(1s^2 2s)^2S$ during ion-atom collisions and how and why the intensities vary as functions of different collision parameters.

© 1995 American Institute of Physics

Ziem et al. [3] measured the intensities of Auger-electrons from $Li^{**}(1s2s2l)$ targets excited by protons with impact energies from 22.5 to 500 KeV using the crossed beam method. The Auger-electrons from $^2S$, $^2P_-$ and $^2P_+$ states formed from a single K-shell electron excitation all exhibited intensity variations as a function of the proton energy. In particular, at 22.5 KeV, the $^2P_-/^2P_+$ intensity ratio was 1:1 while at 400 KeV the ratio had changed to 20:1. This they note is in sharp contrast to the idea of a simple statistical excitation of levels independent of the collision energy. Deveney et al. [2] measured and identified Auger-electrons from the series of $(1s2snl)^2L$ AI states with $n = 2,3,....\infty$ from Li-like carbon projectiles excited by $He$ targets in 1 MeV/amu collisions and note that a simple statistical excitation of the levels is insufficient to explain the relative intensities of the Auger-electron peaks. D.H. Lee [4] observed strong energy dependent variations of the same Auger-electron lines emitted from $O^{5+}$ and $F^{6+}$ projectiles excited by $He$ and $H_2$ targets for total collision energies between 4 and 33 MeV. The same variations are evident in Stolterfoht's work [5,6].

There has not been, to the best of our knowledge, an unambiguous explanation of all of the before mentioned Auger-electron intensity variation phenomena. In one of our present studies the $^2S$, $^2P_-$ and $^2P_+$ Auger-electron peak intensities from $C^{3+}$ projectiles excited from $He, Ne, Ar, Kr$ and $Xe$ targets were measured. Rich intensity variations of the Auger-electron peaks were observed as a function of increasing Z. To determine what effect the target recoil charge state has on the Auger-electron intensity distribution, a coincidence experiment was done using the $He$ system where the Auger-electrons were measured with the specific recoil charge state produced in the collision event. Again strong intensity variations were observed. A 'quasi' one-electron model was considered in which the transition probabilities for a single $C^{3+}$ K-shell electron dynamically developing in two distinct 'outward' collision fields (corresponding to the different potential fields from $C^{3+}$ with either the $He^{1+}$ or $He^{2+}$) gives good agreement with the experimental observations. The transition probabilities are calculated using 'hidden-crossings' [7-11]. If this picture is correct, the Auger-electron intensities and variations provide a window into the collision dynamics. In a final study to further explore this picture, we compare calculations to experimental results for the intensity variation of the $^2S$ Auger-electron as function of collision energy.

## Auger-electron Intensities as a Function Noble Gas Targets

There are five spectra in figure 1 showing electrons measured from the collision systems 4.8 MeV (4 a.u. velocity) $C^{3+} + He, Ne, Ar, Kr$ and $Xe$ [12]. This experiment was performed at the ORNL EN Tandem accelerator facility under the same conditions described in [2]. Briefly, electrons were collected from a differentially pumped gas-cell region at 10° in the laboratory frame into a double-pass parallel-plate spectrometer [13] without any deceleration. Table 1 is given along side of figure 1 to identify the states and energies (in the frame of a $C^{3+}$ emitter ) of the Auger-electrons observed [2]. In

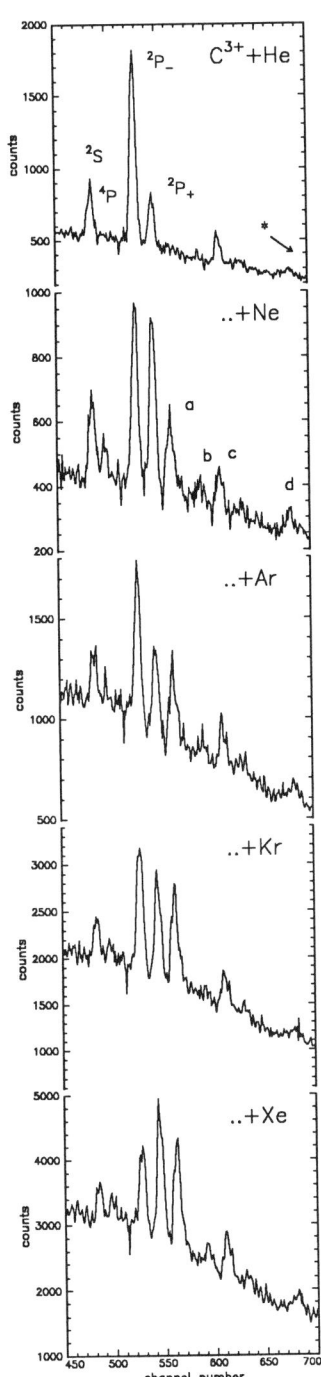

**Table 1.** The Auger-electron energies from doubly excited $C^{3+**}$ ($1s2snl$) autoionizing levels, $|d\rangle$, with respect to the $(C^{4+}\{1s^2\}^1S + e_cl_c)^2L$ ground state ($l = l_c = L$) and other AI configurations ($^4P$ and a-d) observed in figure 1 [2].

| line | $|d\rangle$ | Auger-$e^-$ energy (eV) |
|---|---|---|
| $^2S$ | $(1s\{2s^2\}^1S)^2S$ | 227.5 |
| $^4P$ | $(1s2s2p)^4P$ | 229.9 |
| $^2P_-$ | $(1s2s2p)^2P_-$ | 235.9 |
| $^2P_+$ | $(1s2s2p)^2P_+$ | 239.3 |
| * (not in view) | $(1s2s\{n\geq 3\}l)^2L$ | ≥271 |

| a | $(1s\{2p^2\}^1D)^2D$ | 242.0 |
|---|---|---|
| b | $(1s\{2p^2\}^1S)^2S$ | 248.0 |
| c | $(1s2s2p^2)^3D$ (transfer and excitation) | 252.6 |
| d | $(2s\{2s \text{ or } 2p\})$ | 265 |

**Figure 1.** Electrons measured at 10° in the laboratory frame from 4.8 MeV $C^{3+}$ + $He, Ne, Ar, Kr$ and $Xe$ collisions.

comparing the spectra it should be noted that the only parameter changed from run-to-run was the target gas.

There are intensity variations for almost all of the Auger features. The $^2P_-/^2P_+$ intensity ratio changes from 4:1 in $He$, to 1:1 in $Ne$, to 2:1 in $Ar$, to 1:1 in $Kr$ and finally to 2:3 in $Xe$. We note the absence of the $^4P$ Auger-peak. This is attributed to it's long life time, $10^{-9} - 10^{-8}$ seconds [14], which means it decays outside of the effective target length seen by the electron spectrometer [2]. In measurements of zero degree electrons from a very long effective target length (~20 cm) by Lee [4] the $^4P$ is one of the dominant Auger-electron features. An important conclusion that we draw from one of Lee's studies using retarding fields to separate in energy electrons born in different parts of the target length is that the $^2S$, $^2P_-$ and $^2P_+$ Auger-electron intensity variations are not target length dependent.

## Auger-electron and Recoil Charge State Coincidences

To investigate whether or not the particular charge state of the recoil had any effect on the Auger-electron intensity distributions we measured the $^2S$, $^2P_-$ and $^2P_+$ Auger-electron peaks in coincidence with either the $He^{1+}$ or the $He^{2+}$ recoil charge state in $C^{3+} + He$ (4 a.u. velocity) collisions. A Univ. of TN spectrometer at ORNL was used for this measurement [15]. Here, ~4 pA $C^{3+}$ beam was steered through a 1 cm target gas cell maintained at 4 mTorr. Recoils were extracted at 90° and down stream at 0±10° electrons where directed into a single pass parallel plate spectrometer and onto a two-dimensional position-sensitive-detector. A PC collected the position and coincidence information which was partitioned into three separate 64 by 64 arrays; two corresponding to those electrons in coincidence with either the $He^{1+}$ or $He^{2+}$ and a third for all other non- and random-coincidences. The collection time for this data set was 84 hours. The total number of electrons in coincidence with $He^{1+}$ was approximately equal to the number in coincidence with $He^{2+}$. An amount due to randoms was estimated and subtracted from each of the coincidence arrays and the data was binned at the cost of a slight loss in energy resolution. At this stage, it must be noted, the data analysis is preliminary and error analysis has not been completed.

Figures 2a and 2b are the preliminary results of the coincidence measurements. The solid line represents electrons measured in coincidence with $He^{1+}$ and the dashed line represents electrons measure in coincidence with $He^{2+}$ in figures 2a and 2b. Fig. 2a shows the data before a binary-encounter electron (BEE) background subtraction. Fig. 2b is the data following a BEE background subtraction. The line-shape of the BEE distribution was approximated from our calculations using a CDW-EIS (continuum distorted wave - eikonal initial state) code by Reinhold [16]. The x-axis is energy in the emitter, or projectile, frame. There is good overall agreement with the energies of the Auger-electrons from table 1.

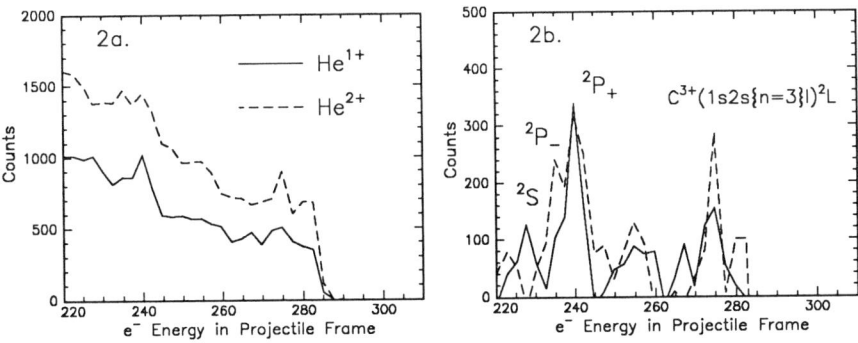

**Figures 2a and 2b.** Emitted-electron distributions measured in coincidence with $He^{1+}$ and $He^{2+}$. In fig. 2b, a binary-encounter background has been subtracted.

There appears to be significant Auger-electron intensity variation for the $^2S$, $^2P_-$ and $^2P_+$ Auger-electron peaks when measured in coincidence with either $He^{1+}$ or $He^{2+}$. The $^2P_-/^2P_+$ intensity ratio changes drastically and the $(1s2s^2)^2S$ Auger-electron disappears in the $He^{2+}$ spectrum in conjunction with the increase in the number of Auger-electrons from the $(1s2s3l)**$ state (the peak at 273 eV). It is this later aspect of the intensity variations that we will consider more deeply.

## Model and Calculations for the Intensity Variations

The coincidence measurements suggests that there is an interaction characterized by the distinct post-collision fields produced by either the $He^{1+}$ or the $He^{2+}$ acting on the outward $C^{3+**}$ that plays a role in determining the final Auger-electron intensity distribution. Having selected small impact parameters (the final AI configurations, $1s2snl$, can be associated with $1s \to nl$ excitations at small impact parameters of ~1/6 a.u. for carbon) on the inward part of the collision a carbon K-shell electron will develop in the combined field of it's own nucleus and an unscreened $He$. If the $He$ target is ionized quickly on the inward part of the collision by impulsive Coulomb interaction with the projectile (binary encounters), the K-shell electron will develop on the outward part of the collision in a post-collision field that depends on the ionization state of the $He$. The outward potential curves for $C^{3+**}$ with either the $He^{1+}$ or the $He^{2+}$ are very different as are the inelastic excitations and population probabilities amongst the quasi-molecular states calculated using hidden-crossings. In the limits of the far separated atoms (ions), it is the expansion of the differently populated outward quasi-molecular levels for either the $C^{3+}$ and $He^{1+}$ or $C^{3+}$ and $He^{2+}$ systems onto the final AI atomic basis states that can lead to some of the Auger-electron intensity variations.

Hidden-crossing theory and techniques offer a means to calculate nonadiabatic or inelastic radially-coupled transitions [7-11]. In this theory, transition probabilities rely only on the evaluation of the Massey-parameter around branch points connecting different potential surfaces that are constructed from time independent eigen-value solutions to the Schroedinger equation for complex internuclear distance. The probability for a transition from an initial state, $i$, to a final state, $j$, is given by;

$$P_{i-j} = \exp\left\{-2|\Delta_{i-j}|/v_0\right\} \qquad (1)$$

where $\Delta_{i-j}$ is the Massey parameter and $v_0$ is the collision velocity. Equation (1) is expected to give good results for $\Delta_{i-j} > v_0$. Krstic and Janev [10] used hidden-crossings to calculate $1s \rightarrow 2s$ excitations in $He^{2+} + H(1s)$ collisions up to 1.7 a.u. of internuclear velocity, which is larger than the $H(1s)$ electron velocity, with good results. It should be noted that hidden-crossing techniques have thus far only been developed for and used in one-electron systems and ours is the first extension of such techniques to a complex 'quasi' one-electron system. The potential curves for complex internuclear distances needed to evaluate the Massey parameter are numerical solutions to the three-body (two-center) instantaneous eigen-value problem. A cut of the potential surfaces with zero component of complex internuclear distance is shown in figure 3.

The first potential curve in fig. 3 takes the ion-atom collision event from an internuclear distance of minus infinity (shown starting at -5) to 0 a.u. for the inward part of the collision. The effective nuclear charges experienced by the carbon K-shell electron developing from $-\infty$ to the united atom limit are $Z^* = 5.7$ and $Z^* = 5.7 + 2$ respectively (using Slater's screening rules [17]). There is a strong radial coupling of the $1s\sigma$ to the $2p\sigma$ molecular orbital which shows up as a branch point at $0.4 + i0.3$ a.u. in the complex plane. For this coupling, the Massey parameter is 5.2 which is greater than $v_0$ (4.0 a.u.). This T-series coupling is characteristic of saddle-point excitation in which the electronic probability amplitudes build up on the unstable equilibrium point or saddle of the two nuclear coulomb centers. Electrons promoted to the $2p\sigma$ may also experience a rotational coupling to the $2p\pi$ that can be estimated from a model by Demkov [18]. At plus infinity, a $2p\sigma$ molecular orbital collapses into an atomic orbital resulting in a $C^{3+**}(1s2s^2)^2S$ three-electron level. Similarly, the $2p\pi$ and $2s\sigma$ electrons yield the $C^{3+**}(1s2s2p)^2P_-$ and $C^{3+**}(1s2s2p)^2P_+$ levels respectively. K-shell electrons promoted into principle quantum states, $n > 2$, develop into $C^{3+**}(1s2s\{n \geq 3\}l)^2L$ levels in the separated ion limit.

Also shown in fig. 3 are the outward potential curves calculated for the cases $C^{3+} + He^{1+}$ (top) and $C^{3+} + He^{2+}$ (bottom). At an internuclear distance of zero $Z^* = 4.8 + 1$ and $Z^* = 4.8 + 2$ are used as the united atom effective nuclear charges for the $He^{1+}$ and $He^{2+}$ cases respectively. The 4.8 is the effective charge for the $2l$

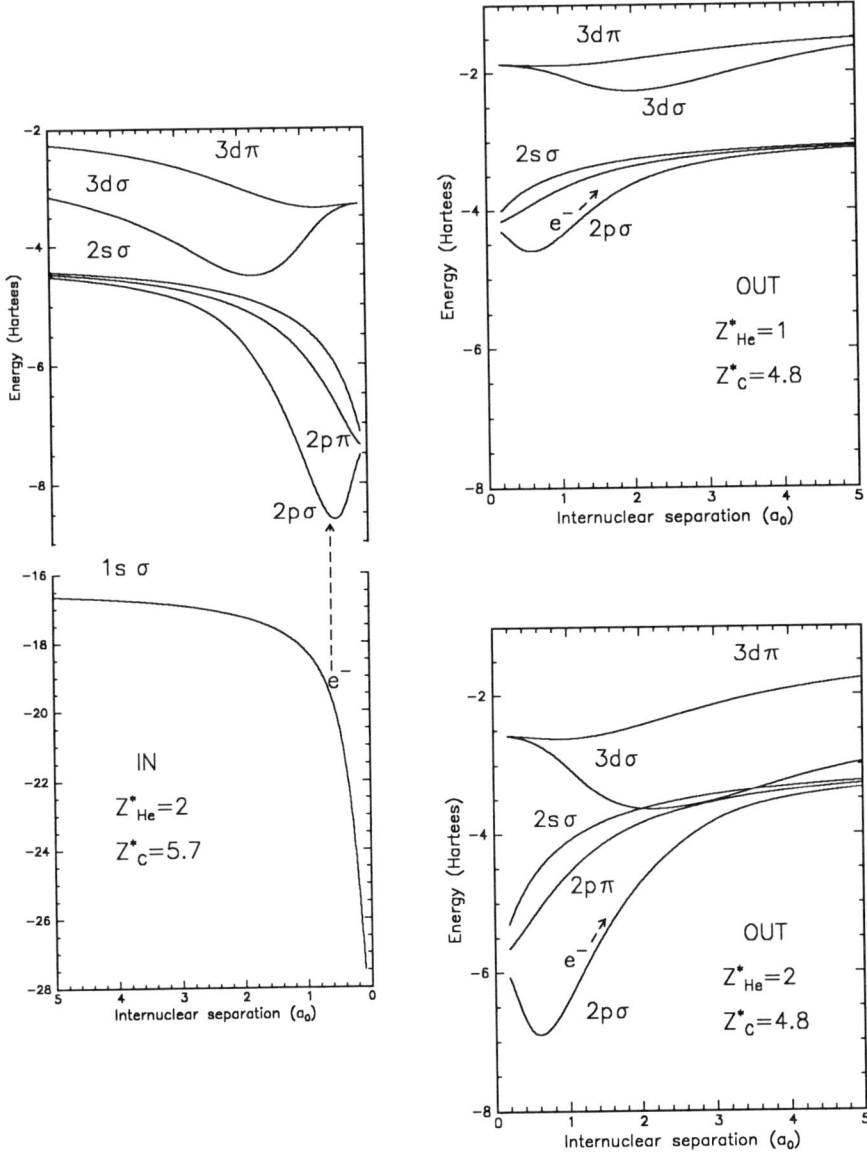

**Figure 3.** A single K-shell electron on an 'inward' and two distinct, corresponding to $He^{1+}$ or $He^{2+}$, 'outward' potential curves.

$C^{3+}(1s2s2l)$ electron screened by a $1s$ and $2s$ electron. In both of the outward potential curves there is a branch point connecting the 2pσ level to the 3dσ level by a T-series transition at 2-4 a.u. of internuclear distance. The Massey-parameters evaluated for the 2pσ to 3dσ transitions in the $He^{1+}$ system is 1.4 while in the $He^{2+}$ system it is 0.12. Under these conditions the absolute probabilities might not be expected to be exact but the relative ratio of probabilities for the two cases should be good. Using equ. 1, the 2pσ →3dσ transition probability in the $C^{3+} + He^{2+}$ system is found to be 10 times larger than it is in the $C^{3+} + He^{1+}$ system. In other words, the 2pσ population (connected to the $C^{3+**}(1s2s^2)^2S$ in the separated atomic basis) is essentially dumped into the 3dσ (connected to the $C^{3+**}(1s2s\{n=3\}l)^2L$ population by a factor of 10 more often in $He^{2+}$ than it is for $He^{1+}$. This is evident in fig. 2b; the Auger-electron intensity from $C^{3+**}(1s2s^2)^2S$ in $He^{2+}$ has nearly vanished while the intensity of Auger-electrons from $C^{3+**}(1s2s\{n=3\}l)^2L$ at roughly 273 eV has increased.

## Energy Variation of the $^2S$ Auger-electron Intensity

To investigate the promotion of the 2pσ into the $n=3$ level more closely, previous singles data from the $C^{3+} + He$ system at 4.8, 6.0, 8.0, 12.0 and 16 MeV total collision energies was analyzed for energy dependence of the $^2S$ and the $C^{3+**}(1s2s\{n=3\}l)^2L$ Auger-electron intensities with 227.5 and 273 eV Auger-electron energies respectively. The autoionizing level $C^{3+**}(1s2s\{n=3\}l)^2L$ is associated with a K-shell electron promotion into the lowest lying $n=3$ level, the 3dσ, without further excitation. Figure 4a is the energy dependent Auger-electron spectra taken under the same experimental conditions as in [2]. As a function of the collision internuclear velocity, the ratios of the number of $n=3$ Auger-electrons at 273 eV divided by the number of $^2S$ Auger-electrons are plotted (open circles) in fig. 4b.

Using equ. (1) the probabilities $P_{i-f}$, for 2pσ - 3dσ promotion and the number of $n=3$ electrons, and $1 - P_{i-f}$, electrons that stay in the 2pσ and the number of $^2S$ electrons, are computed as a function of the internuclear velocity and for each of the two distinct outward potential fields. The ratio $P_{i-f}/(1-P_{i-f})$, is given for each case; $C^{3+}$ with $He^{1+}$ and with $He^{2+}$ (diamond and star symbols) respectively in fig. 4c. To compare with the experimental data, the calculated ratios were averaged (the singles data does not differentiate the charge state of the recoil and we have determined that the number of $He^{1+}$ is roughly equivalent to the number of $He^{2+}$) and scaled down by a factor of 8.5. In fig. 4b, the comparison is made between theory, dashed curve, and experiment, open circles. There is good overall agreement between theory and experiment. Scaling the computed values down is reasonable because additional transitions out of the 3dσ and into the n≥4 levels are not taken into account.

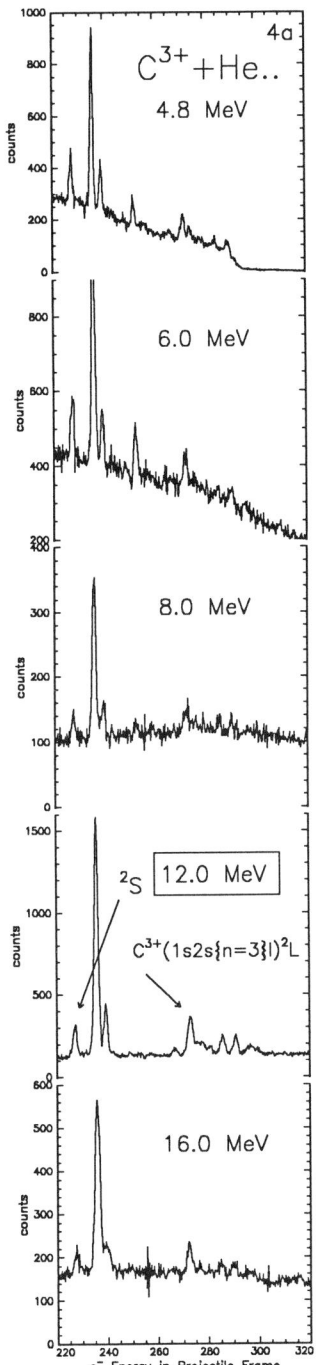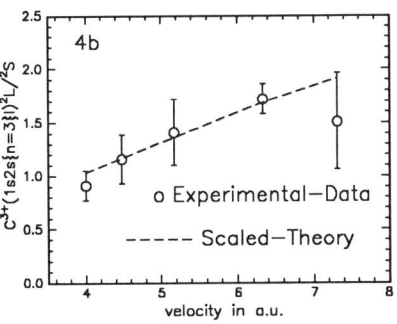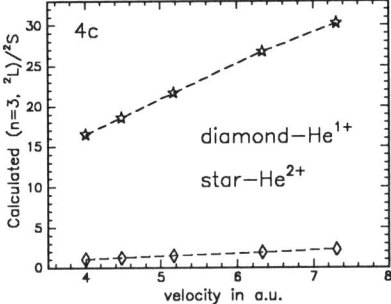

**Figures 4a, b and c.** Energy dependent study of the $C^{3+**}(1s2s\{n=3\}l)^2L$ and $C^{3+**}(1s2s^2)^2S$ Auger-electron peaks.

**Fig. 4a.** Electrons at 10° measured from the $C^{3+}+He$ system at 4.8, 6.0, 8.0, 12.0 and 16.0 MeV total collision energies. In the 12 MeV data, the $(n=3,^2L)$ and the $^2S$ are pointed out.

**Fig. 4b.** Comparison of the experimental ratios of the measured intensities for $(n=3,^2L)/^2S$ with our theoretical results, from fig. 4c, after averaging and scaling by 1/8.5.

**Fig. 4c.** The calculated ratios of $(n=3,^2L)/^2S$ for $He^{1+}$ recoils (diamonds) and for $He^{2+}$ recoils (stars).

## Conclusions

Having measured the Auger-electron intensity variations from $C^{3+**}(1s2snl)$ levels excited in various $C^{3+}(1s^22s)^1S + atom$ collisions, we also explored the possibility that the variations can be used as a window into the detail of some of the intricate dynamical developments of electrons during collision events. A 'quasi' one-electron approximation in which a single active K-shell electron develops in different outward potential fields and whose transitions can be calculated using 'hidden-crossings' techniques shows promise for the $^2S$ Auger-electron variation in the $He$ coincidence data and for the data with $He$ as a function of the internuclear velocity. One of the strongest intensity variation features observed remains unexplained; the $^2P_-/^2P_+$ intensity ratio. One possible explanation is the mixing of quasi-molecular levels on the outward part of the collision from Stark and Zeeman interactions connecting $\Delta l \pm 1$ and $\sigma \to \pi$ levels together respectively. In this picture, the $^2S$, $^2P_-$ and $^2P_+$ share the population of electrons from $n = 2$ (the 2p$\sigma$, 2p$\pi$ and 2s$\sigma$ levels) in a complicated manner depending on the internuclear velocity, the charge state of the recoiled target and the Auger-electron launch angle. Exciting experimental and theoretical work remain for these questions.

## Acknowledgments

We thank P.S. Krstic for many helpful suggestions and discussions. We also thank D.H. Lee for his help and a copy of his Ph.D. thesis. Research for the ORNL participants was sponsored by the U.S. Department of Energy, Office of Basic Energy Sciences, Division of Chemical Sciences, under contract No. DE-AC05-84OR21400 with Lockheed Martin Energy Systems, Inc. Research for E.F.D., J.M.S., and W.W. was supported in part by an appointment to the ORNL Postdoctoral Research Associates Program administered by Oak Ridge Institute for Science and Education and ORNL.

---

\* Present address; Dept. of Physics, The University of Connecticut, Storrs, CT 06269 USA.
\*\* Present address; Dept. of Physics, The University of South Alabama, Mobile, AL 36688 USA.

## References

[1] Cowan, R.D., *The Theory of Atomic Structure and Spectra*, Cal., Univ. of Cal. Press, 1981.
[2] Deveney, E.F. et al., Phys. Rev. A, **48**, 2926-2933, 1993.
[3] Ziem, P., et al. J. Phys. B: Atom. Molec. Phys. **13**, 2071-2081, 1980.
[4] Lee, D.H., Ph.D thesis, Kansas State University, 1990.
[5] Stolterfoht, N., Physics Reports, **146**(6), 315-424, 1987.
[6] Stolterfoht, N., et al., Phys. Rev. A, **48**, 2986-2994, 1993.
[7] Solov'ev, E.A., Sov. Phys. Usp. **32** (3), 228-250, 1989.
[8] Grozdanov, T.P. and Solov'ev, E.A., Phys. Rev. A, **42**, 2703-2718, 1990.
[9] Ovchinnikov, S.Y., Phys. Rev. A, **42**, 3865-3877, 1990.
[10] Krstic, P.S. and Janev, R.K., Phys. Rev. A, **47**, 3894-3912, 1993.
[11] Pieksma, M., Ph.D. thesis, Utrecht University, 1993.
[12] Deveney, E.F. et al., Abstracts papers of the 18th ICPEAC Vol. 2. 499, 1993.
[13] Bechthold, U., Masters thesis, The University of Frankfurt, 1991.
[14] Davis, B.F. and Chung, K.T., Phys. Rev. A, **39**, 3942-3955, 1989.
[15] Desai, D., Ph.D. thesis, The University of Tennessee, 1996.
[16] Reinhold, C., CDW-EIS program, private communication
[17] Slater, J.C., Phys. Rev., **36**, 57, 1930.
[18] Demkov, Y.N., Kunasz, C.V. and Ostrovskii, V.N., Phys. Rev. A, **18**, 2097-2105, 1978.

# Cold Target Recoil Ion Momentum Spectroscopy

R. Dörner[†‡1], V. Mergel[†], L. Spielberger[†], O. Jagutzki[†],
M. Unverzagt[†], W. Schmitt[†], J. Ullrich[*], R. Moshammer[†*]
H. Khemliche[#], M. Prior[#], R.E. Olson[°], L. Zhaoyuan[□],
W. Wu[‡], C.L. Cocke[‡] and H. Schmidt-Böcking[†]

[†] *Institut für Kernphysik, Universität Frankfurt
D60486 Frankfurt, Germany*
[*] *GSI, D64291 Darmstadt, Germany*
[#] *Lawrence Berkeley National Laboratory, Berkeley, CA 94720*
[°] *University of Missouri, Rolla*
[□] *Lanzhou University, PR China*
[‡] *Kansas State University, Manhattan, Kansas 66506*

---

The experimental technique of Cold Target Recoil Ion Momentum Spectroscopy (COLTRIMS) is described. It allows a three dimensional imaging of momentum space of the recoiling ion for all ionizing atomic reaction with $4\pi$ solid angle for momentum measurement. The resolution presently achieved is $\pm 0.035$ a.u.. Depending on the collision system this corresponds to a resolution in projectile energy loss of down to $\Delta E/E = 10^{-9}$ and a scattering angle resolution of down to $10^{-9}$ rad for fast heavy ion collisions. We discuss the experimental technique and some recent results on dynamics of recoil ion production for electron capture, target ionization and projectile electron loss.

---

## INTRODUCTION

For a detailed understanding of charged particle or photon induced electronic transition processes in atoms, ions or molecules experimental techniques are desirable which provide highly differential cross sections, i.e. approaching the ideal 'complete' experiment. The dynamics of reactions such as single or multiple target ionization, projectile ionization, single and multiple electron capture or transfer ionization can be unveiled by fully determining the final state momentum distribution of the reaction products. In this paper we illustrate the new experimental approach of Cold Target Recoil Ion Momentum Spectroscopy (COLTRIMS), which allows a three dimensional imaging of the final state momentum space of the recoiling target ion for any ionizing reaction. This technique combines $4\pi$ solid angle detection with a high momentum resolution (<0.1 atomic units (a.u.)). The high detection efficiency

---
[1] E-mail: Doerner@ikf007.ikf.physik.uni-frankfurt.de

© 1995 American Institute of Physics

makes COLTRIMS ideally suited for multicoincidence experiments. For example the momentum distribution of other reaction products such as one or more electrons, the emerging projectile (scattering angle) or photons can be easily measured in coincidence to the recoil ion. Thus, it paves the way towards experiments complete in momentum space for all ionizing atomic processes. We briefly describe the experimental technique and then discuss the kinematics of recoil ion production by heavy charged particle impact along with some recent results. We do not discuss the case of photoionization, where COLTRIMS has recently also been successfully applied [1,2].

## EXPERIMENTAL TECHNIQUE

The momenta of ions emerging from most atomic reactions is in the range of a few a.u., which is about the width of thermal momentum distribution at room temperature. Therefore the experimental key to enable such measurements is a preparation of an internally cold gas target. First experiments used extended warm or cooled gas cells [3–5] or warm effusive jets [6–9]. The highest resolution was achieved in recent experiments by using precooled supersonic gas jets [10–12].

The He gas is precooled by a cryogenic cold head to 10-30K and than expands through a 30$\mu m$ nozzle. Ne and other gases can be used at a higher temperature, to prevent freezing and forming of clusters. The inner part of this supersonic gas jet passes through a skimmer of 0.3mm diameter about 1 cm above the nozzle into the scattering chamber. The internal momentum spread of the target in the direction of the gas jet is determined by the parameters of the expansion and is typically below 0.1 a.u.. The supersonic expansion gives the He atoms an offset velocity of about $v_{jet} = 2$ a.u. at a nozzle temperature of 30 K. In both directions perpendicular to the direction of the jet the momentum spread is typically 0.03 to 0.07 a.u.; this is determined by $v_{jet}$ and the skimmer diameter and its distance to the nozzle.

The gas jet is intersected with the ion, electron or photon beam about 2 cm above the skimmer. Recoil ions created at the intersection region are accelerated by a weak homogeneous electric field (0.3-20 V/cm, depending on the range of momenta expected in the experiment). After passing a field-free drift region they are detected by a two-dimensional position-sensitive channel-plate detector with wedge-and-strip readout. From the time of flight of the ion, measured by a coincidence with the projectile, an electron, photon or a pulsed beam, the charge state is determined and the momentum in field direction is calculated. From the position information on the channel-plate detector and the time-of-flight one obtains the two momentum components perpendicular to the field. One example of this type of spectrometer is shown in figure 1.

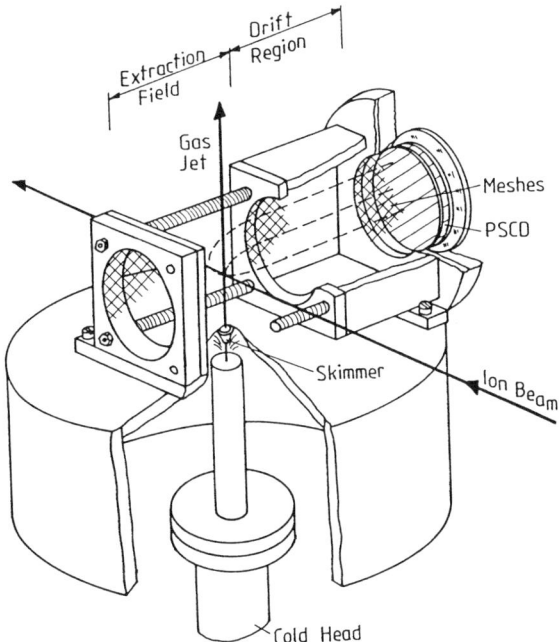

**FIG. 1.** Example of a gas jet and a spectrometer for COLTRIMS (from ref. 9). The gas nozzle is mounted on a cryogenic cold head. The ions created at the interaction point of gas jet and ion beam are projected by a 0.3V/cm homogeneous electrical field onto a position-sensitive channel-plate detector. The ion time of flight is measured by a coincidence with the projectile or any other reaction product.

In the field direction the momentum measurement becomes insensitive to the different starting points of the ions within the jet due to the time focussing properties of the configuration of the field and drift regions. In the two directions perpendicular to the field the uncertainty of the starting point of the ion limits the momentum resolution of the spectrometer. To circumvent this problem an additional electrostatic lens can be implemented in the field region of the spectrometer which focuses the ions starting from different positions within the jet to the same position on the channel plate.

## KINEMATICS OF RECOIL ION PRODUCTION FOR FAST HEAVY PARTICLE IMPACT

For heavy ion collisions the energy loss or gain of the projectile is typically small compared to the total energy of the projectile. Also the scattering angles are in the range of only a few mrad. In this case the momentum components of the recoil ion longitudinal to the beam ($p_{\|_{rec}}$) and perpendicular to the beam ($p_{\perp_{rec}}$) are fully decoupled and carry different information about the collision process.

$p_{\perp rec}$ results from the internuclear repulsion (which makes it sensitive to the impact parameter of the collision) and the transverse momentum of the emitted electrons. Thus in a pure capture collision, where no electron is emitted to the continuum, $p_{\perp rec}$ is exactly equal to the transverse projectile momentum given by the projectile scattering angle. For reactions where one or more electrons are emitted to the continuum the transverse momentum of these electrons couples partly to the recoil ion and partly to the projectile transverse motion. The details of this transverse momentum exchange depend strongly on the collision system. For fast collisions the contribution of the electron to $p_{\perp rec}$ is typically in the range of the Compton profile of the initial state of the emitted electron. Thus larger $p_{\perp rec}$ can be related to the impact parameter. [3,13,4,14,5,8,15,9–11,16]. For fast ion collisions a measurement of projectile transverse momentum becomes very inaccurate due to the large initial longitudinal projectile momentum. However $p_{\perp rec}$ can be measured even for these cases with high accuracy. A resolution of ±0.1 a.u. in the recoil ion transverse momentum, which is typically achieved, corresponds for example in a 1 GeV/u U on He collision to a projectile scattering angle resolution of $10^{-9}$ rad.

The longitudinal momentum of the recoil ion for a complex reaction involving multiple target ionization, projectile ionization (loss), electron capture and excitation, can be calculated from energy and momentum conservation to be: [17]

$$p_{\|rec} = p_{\|rec}^{capture} + p_{\|rec}^{ionisation} + p_{\|rec}^{loss} + p_{\|rec}^{excitation} \quad (1)$$

$$p_{\|rec}^{capture} = -\frac{n_c v_{pro}}{2} + \frac{Q_c}{v_{pro}}$$

$$p_{\|rec}^{ionisation} = \sum_{k=1}^{n_i} \frac{E_{bind}^k + E_{cont}^k}{v_{pro}} - p_{\|}^{e_k}$$

$$p_{\|rec}^{loss} = \sum_{j=1}^{n_l} \frac{E_{bind}^j + E_{cont}^j}{v_{pro}}$$

$$p_{\|rec}^{excitation} = \frac{E_{exc}}{v_{pro}}$$

Atomic units are used throughout this paper. $n_c$, $n_i$ and $n_l$ is the number of captured, ejected target and projectile electrons. $Q_c$ is the differences in binding energy in the initial and final state summed over all captured electrons (exothermic reactions leading to $Q_c > 0$), $E_{bind}$ and $E_{cont}$ are the binding and continuum energies of the target and projectile electron in their parent rest frame and $p_{\|}^{e_k}$ is the longitudinal momentum of target electron k in the final state. $E_{exc}$ is the sum of the excitation energies of target and projectile (if not already counted in $Q_c$). In the following section we will discuss equation 1 for ionization, electron capture and electron loss along with some examples.

## EXPERIMENTAL RESULTS AND DISCUSSION

We first consider the case of a bare projectile, where no projectile electron loss or projectile excitation is possible ($p_{\parallel rec}^{loss} = 0$). For simplicity we consider one electron processes only. In this case the recoil ion longitudinal distribution has two distinct parts:

$$p_{\parallel rec} < p_{\parallel rec}^{cusp} \tag{2}$$

$$p_{\parallel rec} \geq p_{\parallel rec}^{cusp} \tag{3}$$

The momentum $p_{\parallel rec}^{cusp} = E_{bind}^0/v_{pro} - v_{pro}/2$ results from an electron being captured to an continuum state of the projectile [18]. $E_{bind}^0$ is the initial state binding energy of the active electron. Part (2) results from electron capture reactions only, it shows discrete lines related to the quantized Q-value for capture to the different projectile final states. Part (3) emerges from electron emission to the continuum and shows in general no discrete structure.

Figure 2 shows the $p_{\parallel rec}$ distribution of He$^+$ ions created by 15 keV p impact. For this collision system all reaction channels are endothermic. The capture to the projectile ground state, which requires the smallest energy transfer, is by far the dominant channel. The capture to the continuum is located at $p_{\parallel rec}^{cusp} = 0.77$ a.u.. The ionization part of the spectrum which is shown as a solid line in the inset has been measured separately by detecting all emitted electrons in coincidence with the recoil ion. The peak at 2.06 a.u. results from a two electron process where one electron is captured to H(n=1) and simultaneously the target is excited to $He^+(n=2)$. The resolution of ±0.035 a.u. achieved in this experiment is limited by the internal momentum distribution of the gas jet. An electrostatic lens has been used to compensate for the different starting points of the ions within the 1 mm jet diameter. The resolution correspond to ± 0.7 eV of energy loss of the projectile. As can be seen from eq. 1 this resolution in energy gain depends, only in second order, on the energy spread of the incoming beam. Thus high resolution energy gain spectroscopy becomes possible even for fast heavy ion beams [17,7,11,12]. The recoil ion momentum resolution of ±0.035 a.u. for example corresponds to $\Delta E/E = \pm 0.7 \times 10^{-9}$ for 1GeV/u U on He collisions.

The momentum distribution for ions resulting from ionization strongly depends on the collision system. For low velocities the ions are emitted forward (see inset in figure 2). The main contribution of this forward shift is the term $E_{bind}/v_{pro}$ which reflects the decrease of longitudinal momentum of the projectile which supplies energy to the electron necessary to escape into the continuum. For fast collisions and small perturbation position and shape of the recoil ion longitudinal momentum distribution changes significantly. For 0.5 MeV p on He it is centered close to zero with a width approximately given by the initial state Compton profile [16,19] (see fig. 3).

For these fast collisions the $E_{bind}^k/v_{pro}$ becomes small compared to the typical continuum momenta of the electron. In this perturbative regime the elec-

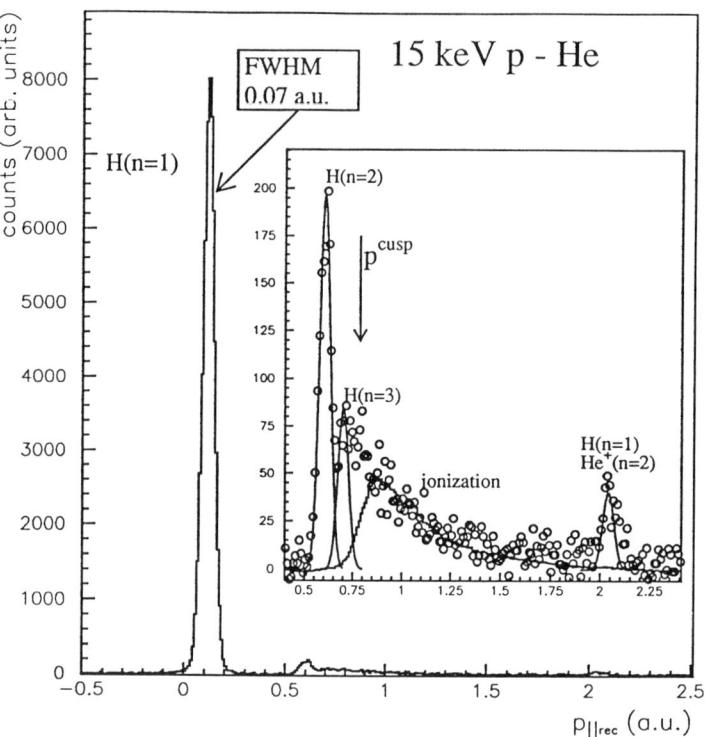

FIG. 2. Longitudinal momentum distribution of He$^+$ ions from 15 keV proton impact. The dominant peak is due to capture to the projectile ground state. The arrow indicates the position of the capture to the projectile continuum. The full line right of the arrow shows the momentum distribution for ionization only, which has been measured separately by detecting an electron in coincidence with the recoil ion. The momentum resolution is $\pm 0.035 a.u.$, equivalent to an energy gain of $\pm 0.7 eV$ and an recoil ion energy of $\pm 4.5 \mu eV$.

tron is ejected by a short interaction with the projectile leaving the recoil ion with its initial state momentum distribution behind. For fast highly charged ion impact, the recoil-ion emission pattern changes again. In this case the long range force of the highly charged projectile pushes the recoil-ions backward [12,18].

Let us now consider reactions where the projectile is ionized. Projectile ionization can be either a product of an interaction of the target nucleus with the projectile electron (ne), or of one of the target electrons with the projectile electron (ee) [20–23]. In both cases the projectile experiences a backward momentum transfer of at least $E^j_{bind}/v_{pro}$ due to the energy required to overcome the projectile binding energy $E^j_{bind}$. This momentum is compensated by the active agent in the collision, the target electron (ee) or the target recoil nucleus (ne). In the second case this leads to forward emission of the recoil ion.

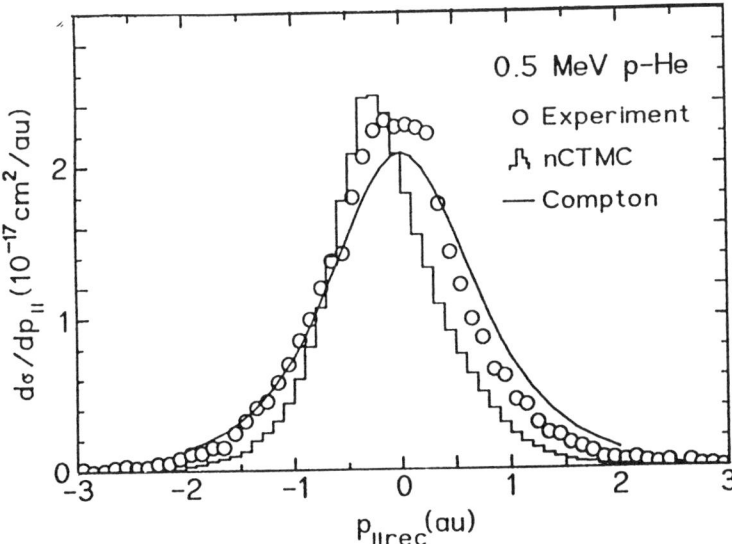

FIG. 3. Longitudinal momentum distribution of $He^+$ ions from the reaction 0.5 MeV $p + He \to p + He^+ + e^-$ (from ref. 16). The full line shows the Compton profile of the He atom, the histogram the result of a Classical Trajectory Monte Carlo calculation with two electrons (nCTMC).

For the case of an (ee) interaction the recoil nucleus is only a spectator to the process and gets only a little momentum. $((E^k_{cont})/v_{pro} - p^{ek}_{\|} \approx -E^j_{bind}/v_{pro}$, notation as in eq. 1) Thus, detecting the recoil-ion momentum has allowed, for the first time, experimental separation of the (ne) and (ee) process [10,9,15].

Figure 4 shows the momentum distribution of $He^+$ ions from simultaneous target and projectile ionization for 1MeV $He^+$ on He collisions (from ref [10]). The maximum close to zero momentum results from the (ee) interaction. The contribution from the (ne) interaction is shifted forward and to larger transverse momentum transfer, indicating that the cross section for this process peaks at smaller impact parameters than the (ee) contribution [24].

## CONCLUSION

We have demonstrated how processes like target ionization, electron capture and projectile electron loss are characterized by very different momentum transfer to the recoil ion. The new experimental technique of COLTRIMS allows to measure this quantity with a unique combination of high resolution

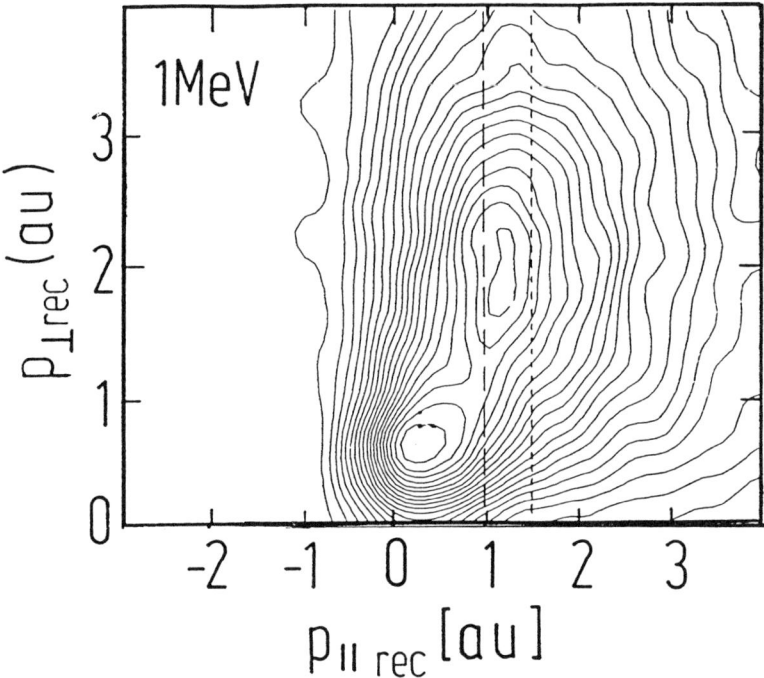

FIG. 4. Longitudinal (to the ion beam) and transverse momentum distribution of $He^+$ recoil ions from simultaneous target and projectile ionization. The peak close to zero at 1 MeV is due to the $(ee)$ interaction, the second peak is due to the $(ne)$ interaction (from ref (8)).

in momentum space and $4\pi$ solid angle. It can easily be combined with detection devices for other reaction products allowing for complete experiments in momentum space for ion, electron, or photon atom and molecule collisions. Since the primary ion beam is not affected by the recoil ion detection, this technique is ideally suited for implementation in storage rings.

## I. ACKNOWLEGMENT

The work was financially supported by DFG, BMFT and DOE grant 82ER53128 (KSU) and DOE contract No. DE-AC03-76SF00098 (LBNL). One of us (R.D.) was supported was supported by the Feodor Lynen Progam of the Alexander von Humboldt Stiftung. We also acknowledge financial support from Max Planck Forschungspreis of the Humboldt foundation. We are thankful to our colleagues W.E. Meyerhof, E. Montenegro, Y.D. Wang, V.D. Rodriguez, C.D. Lin, and U. Buck for helpful discussions.

## REFERENCES

1. L. Spielberger, O. Jagutzki, R. Dörner, J. Ullrich, U. Meyer, V. Mergel, M. Unverzagt, M. Damrau, T. Vogt, I. Ali, Kh. Khayyat, D. Bahr, H.G. Schmidt, R. Frahm, and H. Schmidt-Böcking. *Phys. Rev. Lett.*, 74:4615, 1995.
2. T. Vogt. *Diploma Thesis, University Frankfurt 1995, to be published.*
3. J. Ullrich, R.E. Olson, R. Dörner, V. Dangendorf, S. Kelbch, H. Berg, and H. Schmidt-Böcking. *J. Phys*, B22:627, 1989.
4. R. Dörner, J. Ullrich, H. Schmidt-Böcking, and R.E. Olson. *Phys. Rev. Lett*, 63:147, 1989.
5. S. Lencinas, J. Ullrich, R. Dörner, R.E. Olson, W. Wolff, L. Spielberger, S. Hagmann, M. Horbatsch, C.L. Cocke, and H. Schmidt-Böcking. *J. Phys*, B27:287, 1994.
6. V. Frohne, S. Cheng, R. Ali, M. Raphaelian, C.L. Cocke, and R.E. Olson. *Phys. Rev. Lett*, 71:696, 1993.
7. R. Ali, V. Frohne, C.L. Cocke, M. Stöckli, S. Cheng, and M.L.A. Raphaelian. *Phys. Rev. Lett*, 69:2491, 1992.
8. W. Wu, J.P. Giese, Z. Chen, R. Ali, C.L. Cocke, P. Richard, and M. Stöckli. *Phys. Rev.*, A50:502, 1994.
9. W. Wu, R. Ali, C.L. Cocke, V. Frohne, J.P. Giese, B. Walch, K.L. Wong, R. Dörner, V. Mergel, H. Schmidt-Böcking, and W.E. Meyerhof. *Phys. Rev. Lett*, 72:3170, 1994.
10. R. Dörner, V. Mergel, R. Ali, U. Buck, C.L. Cocke, K. Froschauer, O. Jagutzki, S. Lencinas, W.E. Meyerhof, S. Nüttgens, R.E. Olson, H. Schmidt-Böcking, L. Spielberger, K. Tökesi, J. Ullrich, M. Unverzagt, and W. Wu. *Phys. Rev. Lett*, 72:3166, 1994.
11. V. Mergel, R. Dörner, J. Ullrich, O. Jagutzki, S. Lencinas, S. Nüttgens, L. Spielberger, M. Unverzagt, C.L. Cocke, R.E. Olson, M. Schulz, U. Buck, E. Zanger, W. Theisinger, M. Isser, S. Geis, and H. Schmidt-Böcking. *Phys. Rev. Lett*, 74:2200, 1995.
12. R. Moshammer, J. Ullrich, M. Unverzagt, V. Schmidt, P. Jardin, R.E. Olson, R. Mann, R. Dörner, V. Mergel, U. Buck, and H. Schmidt-Böcking. *Phys. Rev. Lett*, 73:3371, 1994.
13. R.E. Olson, J. Ullrich, and H. Schmidt-Böcking. *Phys. Rev.*, A39:5572, 1989.
14. A. Gensmantel, J. Ullrich, R. Dörner, R.E. Olson, K. Ullmann, E. Forberich, S. Lencinas, and H. Schmidt-Böcking. *Phys. Rev.*, A45:4572, 1992.
15. W. Wu, K.L. Wong, E.C. Montenegro, R. Ali, C.Y Chen, C.L Cocke, R. Dörner, V. Frohne, J.P Giese, V. Mergel, W.E. Meyerhof, M. Raphaelian, H. Schmidt-Böcking, and B. Walch. *Phys. Rev.*, A, 1995. submitted.
16. R. Dörner, V. Mergel, L. Zhaoyuan, J. Ullrich, L. Spielberger, R.E. Olson, and H. Schmidt-Böcking. *J. Phys*, B28:435, 1995.
17. R. Dörner, J. Ullrich, O. Jagutzki, S. Lencinas, A. Gensmantel, and H. Schmidt-Böcking. In W.R. MacGillivray, I.E. McCarthy, and M.C. Standage, editors, *Electronic and Atomic Collisions, Invited Papers of the ICPEAC XVII*, page 351. Adam Hilger, 1991.

18. Y.D. Wang, V.D. Rodriguez, and C.D. Lin. *Phys. Rev.*, page accepted for publication, 1995.
19. V.D. Rodriguez, Y.D. Wang, and C.D. Lin. *J. Phys*, accepted for publication as Letter.
20. D.R. Bates and G. Griffin. *Proc. Phys. Soc. London*, A67:663, 1954.
21. D.R. Bates and G. Griffin. *Proc. Phys. Soc. London*, A68:90, 1955.
22. E.C. Montenegro, W.S. Melo, W.E. Meyerhof, and A.G. dePinho. *Phys. Rev. Lett*, 69:3033, 1992.
23. H.-P. Hülskötter, B. Feinberg, W.E. Meyerhof, A. Belkacem, J.R. Alonso, L. Blumenfeld, E.A. Dillard, H. Gould, G.F. Krebs, M.A. McMahan, M.E. Rhoades-Brown, B.S. Rude, J. Schweppe, D.W. Spooner, K. Street, P. Thieberger, and H.E. Wegner. *Phys. Rev.*, 44:1712, 1991.
24. E.C. Montenegro and W.E. Meyerhof. *Phys. Rev.*, A46:5506, 1992.

# Electron-Electron and Electron-Nuclear Interactions in Dressed Ion-atom Collisions

## R.D. DuBois

Pacific Northwest Laboratory
P.O. Box 999
Richland, WA 99337 USA

**Abstract.** Electron-electron and electron-nuclear interactions leading to excitation and ionization of the collision partners in dressed ion-atom collisions are discussed. After discussing various signatures of these processes which can be used to isolate and identify them, examples of experimental and thoeretical investigations of electron-electron and electron-nuclear processes are given. It is shown how these studies are generating an improved understanding of dressed ion-atom collisions.

Interactions between heavy atomic particles are one of the fundamental processes of nature. At high impact energies where target ionization induced by fully stripped ion impact dominates, these processes which are reasonably well understood. However, for a wide range of intermediate energies the interactions involve partially stripped projectiles and our understanding is considerably poorer because many interaction channels are possible. For example, when both charge centers initially possess loosely bound electrons, ionization of either or both the projectile and target is possible via interactions where the electron(s) bound to one center interact with the screened nuclear charge or with the electron(s) of the other center. This is illustrated below for atomic hydrogen impact on helium.

$$
\begin{aligned}
H + He &\rightarrow H + He^* & (a) \\
&\rightarrow H + He^+ + e_T & (b)
\end{aligned}
\quad \Big\} \; e_T - n_P
$$

$$
\begin{aligned}
&\rightarrow H^* + He & (c) \\
&\rightarrow H^+ + He + e_P & (d)
\end{aligned}
\quad \Big\} \; e_P - n_T
$$

$$
\begin{aligned}
&\rightarrow H^* + He^* & (e) \\
&\rightarrow H^* + He^+ + e_T & (f) \\
&\rightarrow H^+ + He^* + e_P & (g) \\
&\rightarrow H^+ + He^+ + e_P + e_T & (h)
\end{aligned}
\quad \Bigg\} \; e_P - e_T
$$

© 1995 American Institute of Physics

Here *e-n* and *e-e* signify electron-screened nuclear and electron-electron interactions respectively, with the subscripts standing for target and projectile. In *e-n* processes the bound "projectile" electrons remain strongly attached to the nucleus which merely alters the Coulomb field by partially screening the nuclear charge. In *e-e* processes the electrons bound to the two charge centers directly interact and the nuclei serve only to provide initial distributions of electron momenta. The overall interaction can only be described after all of these channels are modeled and combined. In order to evaluate our understanding of these processes, detailed information about the individual electron-nuclear and electron-electron processes is required.

At this point, the nomenclature must be clarified. One of the first theoretical treatments of electron-nuclear and electron-electron ionization mechanisms, (1) referred to these processes as "single-" and "double-transitions". Renewed interest at various times has resulted in other names being attached to these same processes. The "electron-nuclear" interaction, or "screening" role, has also been referred to as a "monoelectronic" process or as the "singly inelastic" channel. The "electron-electron" interaction has been referred to as a "two-center dielectronic" interaction, as the "doubly inelastic" channel, as "simultaneous-" or "mutual-ionization" or as "antiscreening". For consistancy, the terms "electron-nuclear" and "electron-electron" interactions, designated by *e-n* and *e-e*, will be used here.

Distinguishing features associated with *e-n* and *e-e* processes permit identifying and isolating them in certain cases. First of all, coincidence measurements between various interaction products can be used. For example, projectile ion-target ion coincidences isolate channel (h) above; electron-projectile ion coincidences select channels (d + h) which can further be separated by using information about the electron momenta or impact energy thresholds for these processes. Other methods include investigating cross section scaling, momentum transfer, and impact energy thresholds. These can be illustrated by examining a first Born description of the DDCS for *e-n* and the *e-e* target ionization processes induced by incident particles having a single occupied electron subshell. (2)

$$d^2\sigma(\varepsilon,\Theta)_{e-n} \approx \int_{K_m}^{K_M} \left\{ \left| Z_P - N_P F_{o',o'} \right|^2 \right\} \left| F_{o,n} \right|^2 \frac{dK}{K^3}$$

$$d^2\sigma(\varepsilon,\Theta)_{e-e} \approx \int_{K'_m}^{K'_M} N_P \left\{ 1 - \left| F_{o',o'} \right|^2 \right\} \left| F_{o,n} \right|^2 \frac{dK}{K^3}$$

(1)

In these formulae, the terms in brackets describe how the incident particle's nuclear and electronic charges combine to produce an effective charge seen by the target electron, $\varepsilon$, $\Theta$ are the laboratory energy and emission angle of the ejected target electron, $\hbar K$ is the momentum transfer, and $F(K)$ is the single electron form factor for the initial state. By definition, $1 \geq F(K) \geq 0$ for $0 \leq K < \infty$ which corresponds to large and small impact parameter collisions.

Note that the lower and upper limits of integration are different for the *e-n* and *e-e* processes. This implies different threshold energies for initiating *e-n* and *e-e* processes and different maximum and average momentum that can be transferred during the collision. In *e-e* processes, both collision partners are excited/ionized; this requires a larger threshold energy than do *e-n* processes where only one electron is excited/ionized. This threshold energy is related to the impact energy, $E_P$, by $\Delta E = \frac{m}{M} E_P$, $m$ and $M$ being the electron and projectile masses. Because the *e-n* process involves interactions with a massive nucleus, the maximum velocity that can be aquired by the target electron is approximately twice the incoming projectile velocity. In contrast, for the *e-e* process the interaction is between two electrons which means that the maximum velocity is approximately that of the incoming projectile. These are "fuzzy" limits since the electrons have distributions of momenta. Keep in mind that the recoil ion momenta will demonstrate these same features; the only difference being several orders of magnitude in the velocities of the recoil ions and the ionized electrons.

The bracketed terms in the equations indicate that the cross sections for *e-e* interactions scale linearly with the number of electrons of the collision partner whereas the *e-n* interactions scale quadratically with the screened nuclear charge of the collision partner. Also the combination of the different limits and the variation of $F(K)$ with $K$ leads to different momentum transfer for the *e-n* and the *e-e* processes, i.e., to small momentum transfer in *e-e* processes and larger momentum transfer in *e-n* processes. Finally, the relative importance of the *e-e* and *e-n* processes is seen to depend on $K$, $N_P$ and $Z_P$. For example, the relative importance of *e-e* processes is largest for neutral particle impact (when $N_P = Z_P$) and for small $K$; for increasing $Z_P$, it rapidly decreases in importance. (2)

Let us now examine the progress that has been made in obtaining information about *e-n* and *e-e* processes. I wish to emphasize that the examples and references chosen are rather arbitrary. It is not meant to imply that they are the only efforts in this field. Space prohibits me from referencing many studies that have made equally important contributions to increasing our understanding of interactions between dressed particles.

Fifty years ago Bates and Griffing(1) outlined the basic theoretical procedures for modeling *e-n* and *e-e* interactions between two dressed particles and used the plane wave Born approximation to calculate total cross sections for H - H collisions. More than a decade later, Bell and collaborators (3) extended these studies to dressed hydrogen and helium impact on helium. These studies provide the basic framework for performing plane-wave Born calculations of *e-n* and *e-e* interactions and total cross sections for the individual processes in simple systems.

A major step was made in the early 80's when Manson and Toburen (4) used the first Born approximation to calculate doubly differential electron emission cross sections for the individual $e$-$n$ and $e$-$e$ channels in $He^+$ - He collisions. Similar calculations were performed later for other systems.(5,6) Another significant advance was made more recently when Montenegro et al. (7) calculated the contribution of second-order $e$-$n$ processes to mutual ionization of the collision partners in $He^+$ - He collisions. These second-order processes are indistinguishable from first-order $e$-$e$ processes if only coincidence charge state information is available. Following up on the idea that second-order $e$-$n$ processes are important, DuBois et al. (8) recently calculated the influence of second-order $e$-$n$ processes on the doubly differential target electron emission. CTMC calculations of the recoil ion momenta in first-order $e$-$n$ and $e$-$e$ processes (9) also provide valuable information in advancing out understanding of dressed ion-atom collisions.

Somewhat simultaneous with these efforts, first- and second-order contributions leading to ionization of the projectile were also being calculated. (10-13) In these studies, a primary concern is to achieve the proper shape and intensity of the electron loss peak because $e$-$e$ processes enhance the low-energy region of the electron emission spectra which, when transformed to the laboratory frame, results in an enhanced electron loss peak intensity. In addition, $e$-$n$ and $e$-$e$ processes leading to projectile ionization can be viewed as elastic and inelastic scattering of a projectile electron from the target. (14,15) Thus, unshifted and shifted electron loss peak components correspond to $e$-$e$ and $e$-$n$ processes. These components have different intensities and widths which influences the overall peak shape. Unfortunately, calculated $e$-$n$ and $e$-$e$ cross sections must first be added in order to compare with experimental data; thus detailed information, as can be obtained from studies of the target ionization channel, is often lacking.

On the experimental side, early experiments only provided total cross section information about the sum of all the $e$-$n$ and $e$-$e$ ionization channels. However, with the advent of coincidence measurements, more detailed tests became possible. For example, in 1979 Horsdal Pedersen and Larsen (16) used projectile ion-target ion coincidence techniques to isolate and measure total cross sections for the $e$-$e$ channel in energetic H-atom collisions. Since then data have been added for H, He, and $He^+$ impact on He. (7, 17-19) These studies provide the necessary experimental information for testing theoretical predictions of the $e$-$e$ channel in simple collision systems. In addition, some of these studies also investigated the $e$-$n$ channel.

Experimental (dotted curves) and theoretical results (solid and dashed curves) are compared in Fig. 1. The important findings are that $e$-$e$ processes are dominant target ionization mechanisms for energetic atom impact but are far less important for ion impact. Quantitatively, at the higher impact energies the first Born approximation (solid curves) does a reasonable to excellent job in describing both $e$-$n$ and $e$-e interactions. For He atom impact some of the discrepancies between experiment and theory may be associated with metastable contamination of the beam. Discrepancies for the e-e channel for He⁺ impact will be discussed shortly. At lower impact energies, large discrepancies are seen-particularly for the *e-e* channels where theory severely underestimates the measured cross sections. Also note that the measured cross sections remain large well below the expected thresholds for the *e-e* process (approximately 70, 90 and 150 keV/u for H, He and He⁺ impact).

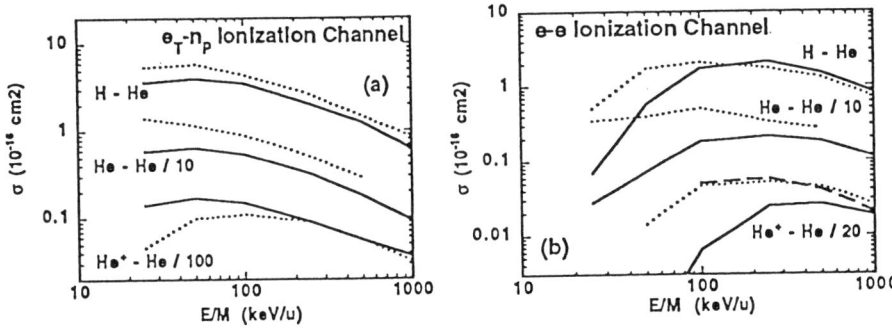

Figure 1. Total cross sections for *e-n* and *e-e* processes for H, He, and He⁺ impact on He. The theoretical and experimental data are from references (3,7,17-19).

By including second-order independent *e-n* processes leading to mutual ionization of the collision partners in the theoretical treatment, Montenegro et al. (7) accounted for discrepancies for He⁺ impact as shown by the dashed curve in Fig. 1b. Presumably the addition of second-order *e-n* processes will also improve the agreement between experiment and theory for atomic hydrogen and helium impact.

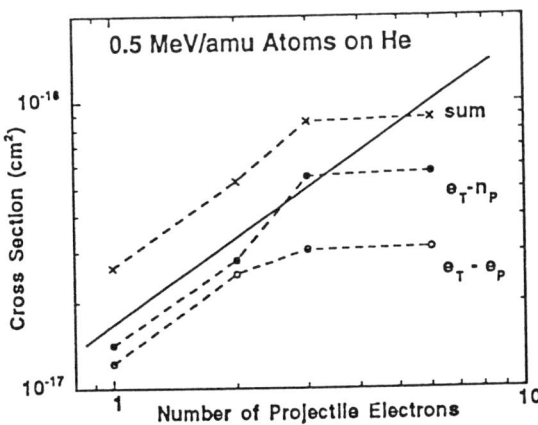

FIGURE 2. Total cross sections for *e-e e-n* processes and their sum for fast atom impact on helium. The solid curve illustrates a linear $N_P$ dependence. Data from ref. (8,17-20).

Examing how the total cross sections scale for different projectiles, specifically looking for a linear dependence with the number of active projectile electrons, is another means of identifying whether the *e-e* channel is important. However, as shown in Fig. 2, a linear cross section dependence on the number of projectile electrons is not sufficient for identifying *e-e* interaction processes. Note that for $N_P < 3$ both the *e-e* and the *e-n* processes, as well as their sum, demonstrate reasonable approximations to a linear dependence on $N_P$. Using only a linear scaling criteria, DuBois and Manson (2) incorrectly concluded that *e-e* processes dominate the target ionization process, which obviously is incorrect. An additional problem with this method occurs when the projectile has loosely, and tightly, bound electrons for then the number of participating electrons is not clearly defined, which may account for the carbon *e-e* data not falling on the linear curve.

Differential electron emission studies have also added to our understanding of *e-n* and *e-e* processes. It is interesting to note that in the very first experimental differential study for light dressed ion impact, Wilson and Toburen (15) recognized the fact that the *e-e* and *e-n* processes lead to shifted and unshifted electron loss peak components and were able to identify these contributions. As shown in Fig. 3, they observed an electron loss peak centered at the projectile velocity and a broadening of the peak on the low-energy side that shifted toward lower emission energies with increasing laboratory emission angles. By fitting gaussian line shapes to the peak, they extracted cross sections for the two processes. Their data implies that the relative importance of *e-e* processes increases for larger laboratory emission angles and for higher impact energies.

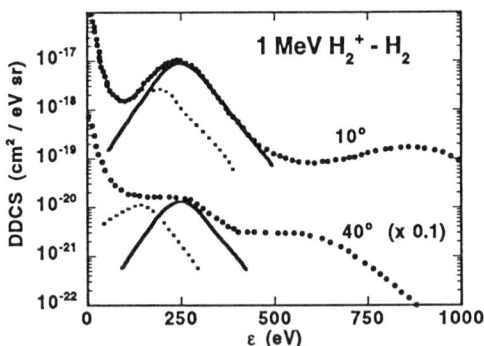

FIGURE 3. Differential electron emission for $H_2^+ - H_2$ collisions. The solid and dotted curves fitted to the data are attributed to *e-n* and *e-e* processes. From ref. (15).

Although no further attempts have been made to experimentally separate the electron loss peak into its two components, in a later study Rudd et al. (21) also noted that the electron loss peak was shifted toward lower energies. However, their interest was in lower emission energies where they observed that the cross sections increased with decreasing emission energy. They attributed this increase to two-center dielectronic processes; this constitutes the first example of dielectronic processes contributing to ionization of the target.

During the 1980's several attempts to theoretically describe the differential electron emission resulting from two-center dielectronic interactions were made. (4,10,11) At the time, no detailed experimental information relating to dielectronic processes was available, so comparisons between experiment and theory tended to concentrate on electron loss peak intensities, positions and widths. Unfortunately, conflicting results were found and attempts to clarify the inconsistancies and to provide additional information were hampered by experimental difficulties in isolating the electron loss peak.

To overcome this problem, DuBois and Manson (5,22) measured coincidences between post-collision projectile ions and electrons. This provided the first differential cross section information about the sum of ionization channels (d) and (h). Similar, but more extensive, studies were performed later at the University of Frankfurt. (23-25) Recently, carbon particle impact has been investigated and information about higher-order $e$-$e$ processes has been obtained.(26) Experimental results for H and He$^+$ impact on helium and theoretical predictions for the various first- and second-order $e$-$e$ and $e$-$n$ processes, are shown in Fig. 4.

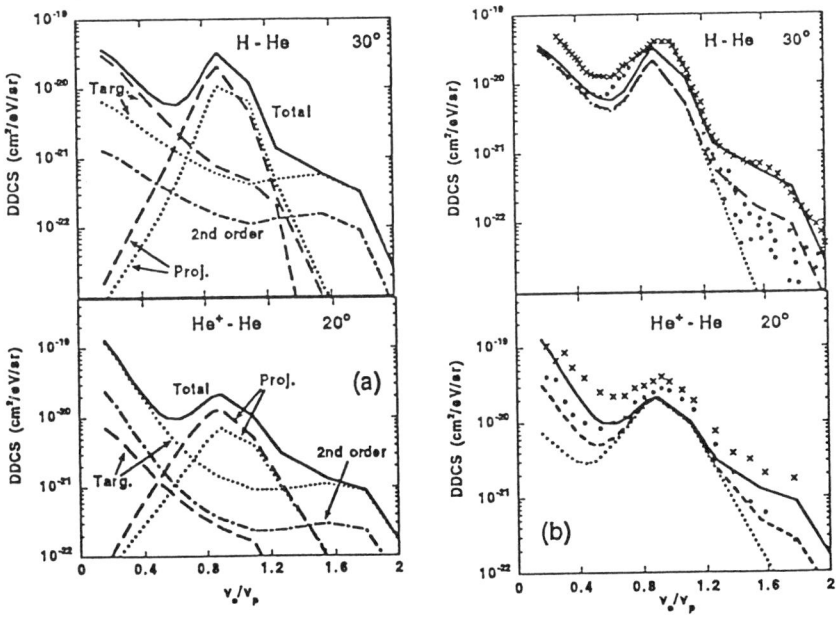

Figure 4. Theoretical and experiment cross sections for ionization in H and He$^+$ – He collisions. Part a shows theoretical predictions for the various ionization channels. Part b compares appropriate sums of these channels with experimental data. Theory: ref. (5,6,8); exp. ref. (5,23).

Part a shows the Born predictions for the first-order *e-e* and *e-n* processes leading to target and projectile ionization.(5,6) In part a, the dotted curves are for *e-n* precesses, the dashed curves for *e-e* processes, the chain curves for independent second-order *e-n* processes leading to ionization of the target, (8) and the solid curves are the sum of all processes. Note how the relative importance of the *e-e* process varies for target and projectile ionization for H and He$^+$ impact. This reflects the importance of *e-e* and *e-n* processes for neutral and charged projectile impact. Also note that the second-order *e-n* process approximately mimics the target ionization *e-n* process but is smaller by roughly a factor of 3. Because the target ionization *e-n* cross sections scale with the screened nuclear charge, second-order *e-n* processes can be major contributors to the target component of the mutual ionization channel for ion impact but should play minor roles for neutral atom impact. Of particular importance is the fact that, in both cases, second-order *e-n* processes are the only means of producing energetic ($v_e \gg v_p$) electrons when the target and projectile are both ionized in the collision.

In Fig. 4b, the experimental cross sections for the total doubly differential electron emission (x) and for electrons measured in coincidence with an ionized projectile ion (o) are shown. Note the great improvement between the experimental coincidence data and theory in the low-energy region for He$^+$ impact and in the high-energy region for both H and He$^+$ impact when second-order processes are included in the theoretical treatment (dotted and dashed curves).

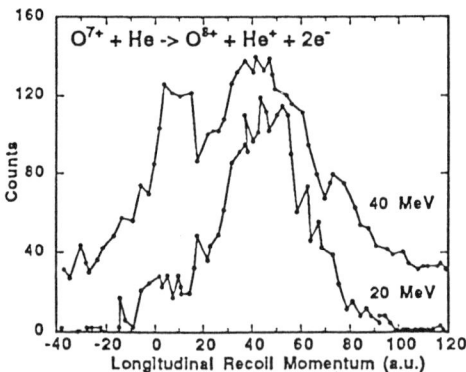

Figure 5. Longitudinal momentum aquired by the target in $O^{7+}$ + He → $O^{8+}$ + He$^+$ collisions From ref. (27).

Another method of separating *e-e*, *e-n*, and second order *e-n* processes is to study the recoil ion momenta. In 1994 groups at the University of Frankfurt (9) and Kansas State University (27) measured the momentum transferred to recoil ions in dressed ion-atom collisions. An example of these first data are shown in Fig. 5. Note that when both the target and the projectile are ionized in the collision the longitudinal component of the recoil ion momentum has two components: a broad peak indicating large momentum transfer to the target and a sharper peak centered at very small momentum transfer. The sharp peak represents the *e-e* component of the ionization and the broad peak is the second-order *e-n* component. Since this case involves a highly charged projectile having tightly bound electrons, $Z_P \gg N_P$. Thus, the target *e-n* cross section is large which leads to a large second-order *e-n* process.

Examples thus far have provided little or no information about excitation processes. However, a very nice example of *e-e* and *e-n* processes contributing to the excitation of the projectile exists. Zouros et al. (28) used zero-degree Auger spectroscopy to study K-shell excitation cross sections for $F^{6+}$ - $H_2$ collisions. Some of these data are shown in Fig. 6 along with PWBA and IA (impulse approximations) for the *e-n* and *e-e* excitation channels. The threshold for *e-e* excitation of the K-shell is indicated by the arrow. Note that in the vicinity of this threshold, the cross sections clearly demonstrate the onset of a new excitation channel-which theory verifies as the *e-e* channel. Also note that since the $^4P$ state requires an electron exchange process, it can only be produced via *e-e* interactions.

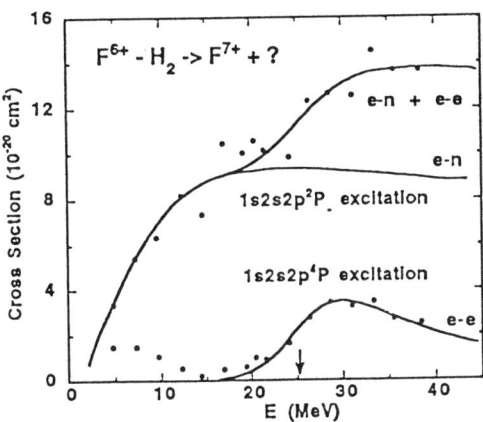

FIGURE. 6. 1s2s2P excitation in the $F^{6+}$ + $H_2$ → $F^{7+}$ + ? channel. Calculations (solid curves) for *e-e* *e-n* processes and their sum are compared with experiment. From ref. (28).

To summarize, the examples discussed have demonstrated how our understanding of dressed ion-atom collisions has advanced over the years. They also have demonstrated that careful interrogation of the differential details of the interaction products may be required to avoid misinterpreting the data. Considering the important roles that interactions between dressed particles play in a wide variety of fields, improving and supplementing this differential information will remain an important atomic physics problem for some time.

## ACKNOWLEDGEMENTS

This work was funded by the U.S. Department of Energy, Office of Health and Environmental Research under contract DE-AC06-76RLO-1830.

## REFERENCES

1. Bates, D. R. and Griffing, G. W., Proc Phys Soc A **68**, 90-6 (1955).
2. DuBois, R. D. and Manson, S. T., Nucl Inst and Meth B **86**, 161-4 (1994).
3. Bell, K. L., Dose, V. and Kingston, A. E., J Phys B **2**, 831-8 (1969) and **3**, 129-36 (1970).
4. Manson, S. T. and Toburen, L. H., Phys Rev Lett **46**, 529-32 (1981).
5. DuBois, R. D. and Manson, S. T., Phys Rev A **42**, 1222-30 (1990).
6. Manson, S. T. and DuBois, R. D., Phys Rev A **46**, R6773-6 (1992).
7. Montenegro, E. C., Melo, W. S., Meyerhof, W. E. and de Pinho, A. G., Phys Rev Lett **69**, 3033-6 (1992) and Phys Rev A **48**, 4259-66 (1993).
8. DuBois, R. D., Stolterfoht, N. and Schiwietz, G., submitted to Phys Rev A (1995).
9. Dörner, R, Mergel, V., Ali, R., Buck, U., Cocke, C. L., Froschauer, K., Jagutzki, O., Lencinas, S., Meyerhof, W. E., Nuttgens, S., Olson, R. E., Schmidt-Böcking, H., Spielberger, L., Tokesi, K., Ullrich, J., Unverzagt, M. and Wu, W., Phys Rev Lett **72**, 3166-9 (1994).
10. Hartley, H. M. and Walters, H. R. J., J Phys B **20**, 3811-30 (1987).
11. Jakubassa, D. H., J Phys B **13**, 2099-2108 (1979).
12. Jakubassa-Amundsen, D. H., Z Phys D **22**, 701-11 (1992).
13. Wang, J., Reinhold, C. O. and Burgdörfer, J., Phys Rev A **45**, 4507-18 (1992).
14. Burch, D., Wieman, H. and Ingalls, W. B., Phys Rev Lett **30**, 823-6 (1973).
15. Wilson, W. E. and Toburen, L. H., Phys Rev A **7**, 1535-44 (1973).
16. Horsdal Pedersen, E. and Larsen, L., J Phys B **24**, 4099-4112 (1979).
17. DuBois, R. D. and Kövèr, À., Phys Rev A **40**, 3605-12 (1989).
18. DuBois, R. D., private communication (1995).
19. DuBois, R. D., Phys Rev **39** 4440-50 (1989).
20. Sanders, J. M., Datz, S., DuBois, R. D. and Manson, S. T., contributed paper at this conference (1995).
21. Rudd, M. E., Risley, J. S., Fryar, J. and Rolfes, R. G., Phys Rev A **21**, 506-14 (1980).
22. DuBois, R. D. and Manson, S. T., Phys Rev Lett **57**, 1130-2 (1986).
23. Heil, O., DuBois, R. D., Maier, R., Kuzel, M. and Groeneveld, K-O., Z Phys D **21**, 235-9 (1991), Nucl Inst and Meth B **56/57**, 282-4 (1991), Phys Rev A **45**, 2850-8 (1992).
24. Kuzel, M., DuBois, R. D., Maier, R., Heil, O., Jakubassa-Amundsen, D. H., Lucus, M. W. and Groeneveld, K-O., J Phys B **27**, 1993-2008 (1994).
25. Trabold, H., Sigaud, G. M., Jakubassa-Amundsen, D., H., Kuzel, M., Heil, O. and Groeneveld, K-O., Phys Rev A **46**, 1270-8 (1992).
26. DuBois, R. D., Phys Rev A **50**, 364-70 (1994).
27. Wu, W., Wong, K. L., Ali, R., Chen, C. Y., Cocke, C. L., Frohne, V., Giese, J. P., Raphaelian, M., Walch, B., Dorner, R., Mergel, V., Schmidt-Böcking, H. and Meyerhof, W. E., Phys Rev Lett **72**, 3170-3 (1994).
28. Zouros, T. J. M., Lee, D. H. and Richard, P., Phys Rev Lett **62**, 2261-4 (1989).

# Projectile Electron Excitation and Loss in Ion-Atom Collisions

E.C.Montenegro
*Departamento de Física, Pontifícia Universidade Católica do Rio de Janeiro*
*Caixa Postal 38071, Rio de Janeiro, RJ 22452, Brazil*

W.E.Meyerhof
*Department of Physics, Stanford University, Stanford, CA 94305, USA*

J.H. McGuire
*Department of Physics, Tulane University, New Orleans, LA 70117, USA*

C.L. Cocke
*Department of Physics, Kansas State University, Manhattan, KS 66506, USA*

## Abstract

The contribution from the nucleus-electron and the electron-electron interactions to the electron loss process by ions passing through neutral targets is reviewed with particular emphasis on recent theoretical concepts and experimental techniques which have been developed in this field. One important conclusion of these studies, namely the connection between the electron-loss process and electron-ion ionization phenomena, is also discussed.

## Introduction

The doping of a Ge matrix by Li ions through ion implantation, the presence of Hydrogen lines in the Aurora spectrum, or the injection of neutral atoms in a fusion device are examples of three apparently disconnected situations which depend strongly on the balance among capture, loss, ionization and excitation in collisions between dressed ions and atoms [1]-[3]. Atomic collisions involving energetic multielectron ions and neutral atoms usually result in multielectronic excitation of both collision partners. The simplest way to describe this complex system as a whole is through the independent-electron approximation (IEA), where the probability distribution for a particular process is calculated in terms of the single-electron probabilities of the various competing channels [4]. As a consequence of coupling of probabilities, a detailed study of each collision channel is needed in order to understand the system as a whole. Within this scenario, the loss channel is particularly important because of its role in determining the charge state of the projectile, a parameter which affects practically all collision channels. In this work, we report some recent advances obtained in the understanding of electron loss in the intermediate-to-high velocity regime.

## Total Cross Section for Electron Loss

If a swift, dressed ion collides with a neutral atom, the eventual loss of a projectile electron is due to the joint action of the target nucleus and of the target electrons on it. In order to organize the concepts involved in this combined action, it is convenient to separate the interaction between the projectile and the target electrons into two modes, characterized by the state in which the target is left after being disturbed by this two-center electron-electron interaction.

If the target is left in the ground state, its electron cloud does not change during the collision process and the electron-electron (e-e) action can be coherently added to that of the target nucleus (n-e). Under these conditions, the loss process can be pictured as resulting from the ionizing action of a screened target nucleus impinging on the projectile. This is called the *screening mode*. On the other hand, if the target is left in any available excited state due to the e-e interaction with the active projectile electron, the loss process is said to be in the *antiscreening mode*. In this way, the antiscreening mode is related solely to the e-e interaction while the screening mode is related to both n-e and e-e interactions, although in the screening mode the final target state is selected to be the ground state [5]. In the case of light targets, such as $H_2$ or He, these two modes contribute almost equally to the total loss cross section of tightly bound electrons, a result which is not obvious from a first sight. Figure 1 shows the electron loss cross section for $O^{7+}$ on $H_2$ and He [6]. A clear indication of the existence of a onset for the antiscreening mode around 1.5 MeV/amu as well as the importance of the contribution from this mode for high velocity projectiles, principally for the $H_2$ target, can be seen from this figure.

The theoretical treatment of two-center processes in atom-atom collisions was given originally by Bates and Griffing about forty years ago using the plane wave Born approximation (PWBA) in a study of the H + H system. This set

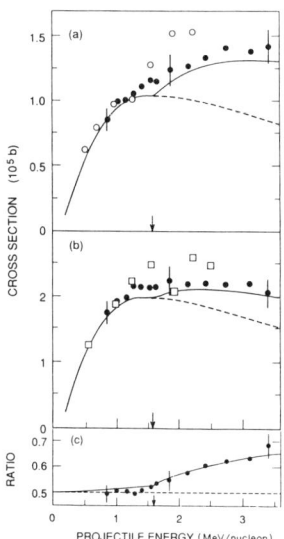

Figure 1: *Electron loss cross section for $O^{7+}$ projectiles colliding with (a)$H_2$ and (b) He targets as a function of the projectile energy. The dashed curves give the contribution from the screening mode above the antiscreening threshold. The solid curve is the PWBA cross section. (c) Ratio of $H_2$ to He target cross sections.[From [6].]*

the basis of the methodology which has been followed later by many authors in an effort to generalize the use of the PWBA approach to multielectron atoms (see [5] for a recent review). The description of electron loss within the PWBA framework has been successfully carried out for one-electron projectiles impinging on light targets. For multielectron projectiles or heavier targets, however, either a different point of view or more sophisticated theories are necessary to properly describe the electron loss process.

## Impact Parameter Dependence for Electron Loss

Figure 2 shows the one-electron loss cross section for $C^{3+}$ on $H_2$ and He. Although the energy region scanned in Fig.2 is around the antiscreening threshold, it can be seen from this figure that the agreement between theory and experiment is not as good as in the $O^{7+}$ case, shown in Fig. 1. This disagreement appears to be more pronounced at lower energies, where the capture channel begins to compete with the electron loss channel. In fact, the transfer-loss process becomes competitive with single electron loss at low velocities, causing the premature decrease of the experimental single-electron loss cross section, if compared with the PWBA theory, as the projectile energy decreases.

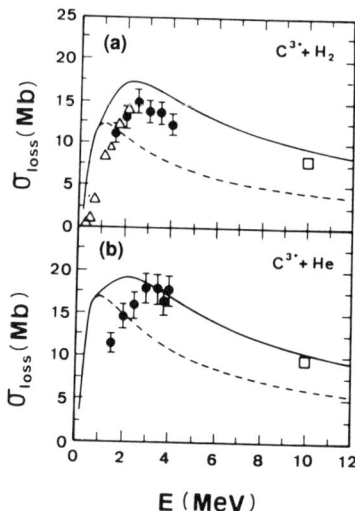

Figure 2: *One-electron loss cross section for $C^{3+}$ projectiles colliding with $H_2$ and He targets as a function of the projectile energy. The dashed curves give the contribution from the screening mode above the antiscreening threshold. The solid curve is the PWBA cross section. [From [7].]*

As stated earlier, the simplest way to treat multi-electron processes, such as transfer-loss, is through the IEA. The IEA, however, is formulated in the impact-parameter space and requires knowledge of the probability distributions for the various competing channels. Only recently, the electron loss process has been studied within the semiclassical approximation (SCA) [8]-[13], allowing the probability distribution to be obtained for this process. These studies result in a useful geometrical picture of the screening and antiscreening modes, which can be illustrated with the aid of Figs. 3a and 3c [12, 13]. In the projectile frame, the effect of the joint action of the target nucleus and of the target electrons can be simulated by a superposition of two kinds of point-like particles: the target nucleus and the volume element $\mid \Phi(\xi) \mid^2$ of the electron cloud. These two effective projectiles follow a straight-line trajectory and impact parameters, $b$ and $b_{eff}$, as well scattering amplitudes, $a_{ne}$ and $a_{ee}$, respectively, are associated with each of them. In the screening mode, the amplitudes for the nucleus-electron ($a_{ne}$) and electron-electron($a_{ee}$) interactions add coherently, interfering destructively when inducing a transition from an initial state $\mid s >$ to a final state $\mid f >$ of the active projectile electron. The effect of this interference is clearly shown in Fig. 3b through the node appearing in the probability distribution $bP(b)$ for the 1s-2s transition in H+H collisions.

Because of the interference effect, the screening distribution is narrower than the corresponding distribution for a bare nucleus.

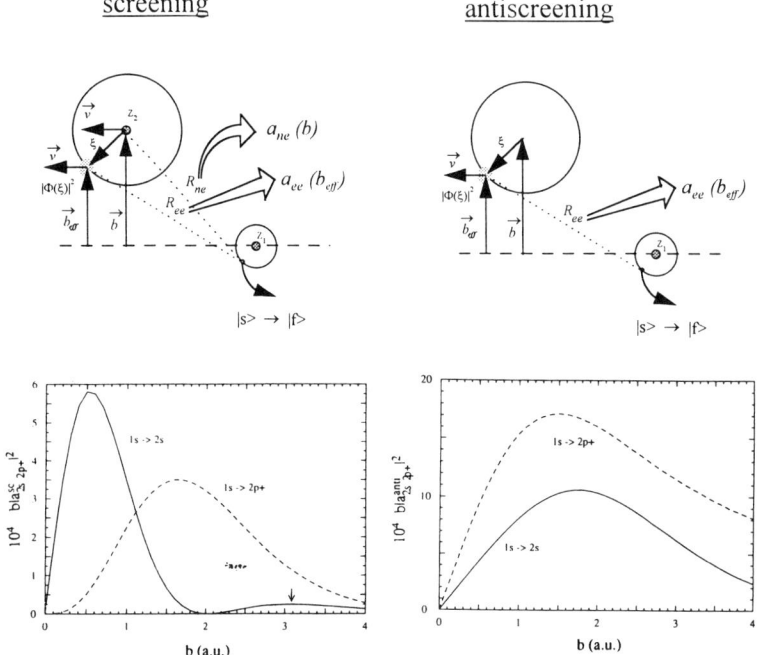

Figure 3: *Impact parameter description of the loss process.(a) Sketch of the screening contribution for electron loss in the impact parameter picture; (b) screening probability distribution, bP(b), for 1s-2s (full curve)and 1s-2p+ (dashed curve) excitation in H+H collisions at a projectile velocity of 5.0 a.u. [12]; (c) and (d): same as (a) and (b) for the antiscreening contribution.*

For the antiscreening case (Fig. 3c), the amplitudes $a_{ee}$ corresponding to each element of the electron cloud $\mid \Phi(\xi) \mid^2$ add incoherently [12, 13]. Since there is no true point-like effective projectile in this case, the probability distribution for antiscreening is significantly broader than that of the screening mode, a result which can be clearly seen comparing Figs. 3b and 3d.

We are aware of only two attempts to observe experimentally the probability distribution for electron loss. Montenegro et al.[14] used a finely collimated $Li^{2+}$ and $C^{5+}$ beams to measure the scattering-angle distribution of projectiles having electron loss in $H_2$ and He targets. Although a good agreement with the SCA was obtained for the distribution corresponding to the total loss, the experiment could not discriminate between the screening and antiscreening distributions. More recently, Wu et al. [15] used the newly developed recoil ion momentum spectroscopy (RIMS)[16]-[19] to obtain the probability distributions for the loss-ionization and antiscreening processes in 59 MeV $O^{7+}$ + He collisions (Fig. 4). The shape of the probability distribution for the loss-ionization process is similar to that corresponding to the screening mode. The experiment clearly shows the broader distribution arising from the electron-electron interaction (antiscreening) if compared to that related to nucleus-electron interaction (transfer-loss).

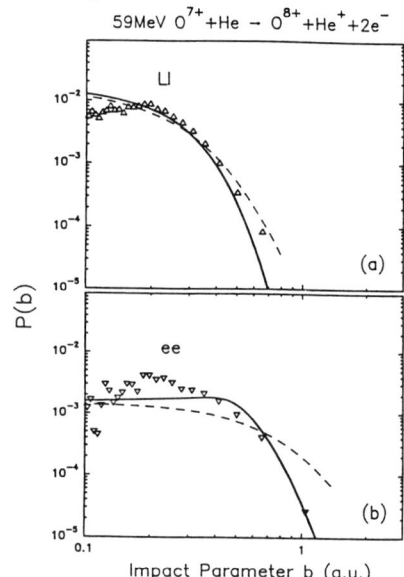

Figure 4: *Probability distributions as a function of the impact parameter for $O^{7+}$ on He. (a) loss-ionization; (b) antiscreening. The full and dashed curves correspond to screened and not screened coulomb deflection functions, respectively. The deflection functions are used to transform the transverse momentum of the recoil ion into impact parameter. [From [15].]*

## Antiscreening and Electron-Ion Collisions

The idea that the antiscreening process can be viewed as a collision between a "free" and a bound electron, which underlies the model pictured in Fig. 3c, was first introduced by Anholt [20]. However, instead of considering the collision in the configuration space, Anholt conceived the antiscreening mode as a collision between a "beam" of electrons (originating from the target) and the projectile ion. In this picture, the antiscreening process should have an abrupt threshold, similar to the electron ionization process, if the translational kinetic energy of the target cloud in the projectile frame is equal to the binding energy of the active projectile electron. A few years later, Zouros et al. [21] showed the presence of a smooth threshold in projectile excitation, observing the $1s2s2p^4P$ Auger decay in $F^{6+} + H_2$ collisions. Montenegro et al. [22] used the coincidence between the projectile and the recoil ion to isolate the antiscreening process in $He^+ + H_2$ and He electron-loss collisions at high energies, but the behavior around the threshold was obscured by the presence of the loss-ionization process. Recently, Wu et al. [17] and Dörner et al. [18] were able to isolate completely the antiscreening from the loss-ionization process using the RIMS technique. Figure 5 shows the antiscreening cross section as a function of the projectile energy for $O^{7+}$ and $F^{8+}$+ He collisions [23]. The presence of a threshold is quite clear. Also, the agreement between the experimental data, obtained in a ion-atom collision, and the theory of ionization by electron impact is impressive. Montenegro and Zouros [24] have derived this important connection between ion-atom and electron-ion collisions using the PWBA framework and assuming that only the continuous target states are effective for the antiscreening process.

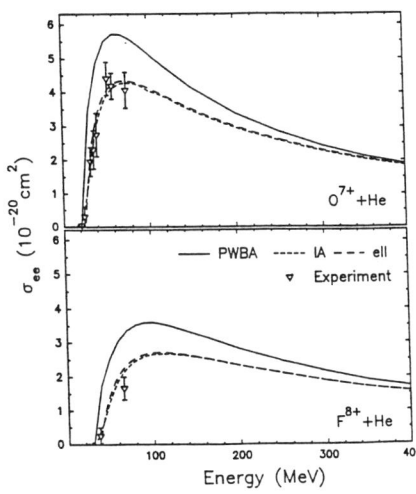

Figure 5: *Cross section for the antiscreening mode as a function of the projectile energy for (a) $O^{7+}$ on He, (b) $F^{8+}$ on He. The full curve is the extended PWBA calculation [25]. The short-dashed curve is the impulse approximation calculation [24]. The long-dashed curve is calculated for electron impact ionization [26]. Experiment: Ref. [23].*

The good understanding of the relative contributions from the screening and antiscreening modes to projectile electron loss, obtained for light targets, is not so clear if heavier targets are considered. When the target atomic number ($Z$) increases, the screening cross section does not increase as $Z^2$, as expected from first-order theories, but tends to saturate for elements heavier than Ar. This tendency, which appears in more elaborate theories such as time-dependent coupled-states calculations [27], is clearly verified in total cross section measurements of $He^+$ loss in several gases, performed by Sant'Anna et al. around the antiscreening threshold [28](see also Ref. [29] for $C^{3+}$ projectile loss in various gases). However, the role of the antiscreening mode in these cases is not well understood. The PWBA calculations based on Ref. [25] seem to give too high cross sections [27]. On the other hand, looking from the electron-impact point of view, which works well in the He case as Fig. 5 shows, there is the question which electrons of a many-electron target can be considered as "free". Experiments which are able to separate the e-e interaction from the e-n interaction would be particularly useful to help answering these kind of questions. Such attempts are under way in the case of projectile excitation, through the observation of the Auger decay of $^4P$ states induced in Li-like ions impinging in multi-electron gases [30].

## Conclusions

In this short paper we show that our understanding of the electron loss in the intermediate-to-high velocity regime has improved considerably in recent years. This progress is due in part to the development of new theoretical concepts based on simple models and in part to the appearance of powerful experimental techniques to study this process. Although our present stage of knowledge is satisfactory for light targets, this is not true for the heavy ones where more sophisticated calculations are needed to account for deviation from the perturbative regime which occurs in former cases.

This work was supported in part by the CNPq (Brazil), by National Science Foundation Grants Nos. PHY-9019293 and INT-9101057 and by the Division of Chemical Sciences, Office of Energy Research, U.S. Department of Energy.

# References

[1] Schiwietz, G. and P.L. Grande, Nucl. Inst. Meth. **B69** (1992) 10.

[2] Van Zyl,B., XVIII ICPEAC, AIP Conf. Proc. 295, edts. T. Andersen, B. Fastrup, F. Folkmann, H. Knudsen and N. Andersen (1993) p.684.

[3] Melchert,F., XVIII ICPEAC, AIP Conf. Proc. 295, edts. T. Andersen, B. Fastrup, F. Folkmann, H. Knudsen and N. Andersen (1993) p.574.

[4] McGuire, J.H., Adv. At. Mol. Opt. Phys. **29** (1992) 217.

[5] Montenegro, E.C., Meyerhof, W.E. and McGuire, J.H., Adv. At. Mol. Opt. Phys. **34** (1994) 249.

[6] Hülskötter, H.P., Feinberg, B., Meyerhof, W.E., Belkacem, A., Alonso, J.R., Blumenfeld, L., Dillard, E.D., Gould, H., Guardala, N., Krebs, N., McMahan, M.A., Rhoades-Brown, M.E., Rude, B., Schweppe, J. Spooner, D.W., Street, K., Thieberger, P. and Wegner, H.E., Phys. Rev. **A44** (1991) 1712.

[7] Montenegro, E.C., Sigaud, G.M., and Meyerhof, W.E., Phys. Rev. **A45** (1992) 1575.

[8] Montenegro, E.C. and Meyerhof, W.E, Phys. Rev. **A44** (1991) 7229.

[9] Montenegro, E.C. and Meyerhof, W.E, Phys. Rev. **A46** (1992) 5506.

[10] Ricz, S., Sulik, B., Stolterfoht, N. and Kadar,J., Phys. Rev. **A48** (1993) 1930.

[11] Kabachnik, N.B., J. Phys. **B26** (1993) 3803.

[12] Wang, J., McGuire, J.H., and Montenegro, E.C., Phys. Rev. **A51** (1995) 504.

[13] Montenegro, E.C. and Meyerhof, W.E., "Two-Center Effects in Ion-Atom Collisions", AIP Conf. Proc. ***, edts. T.J. Gay and A.F. Starace (1995) p.***.

[14] Montenegro, E.C., Belkacem, A., Spooner, D.W, Meyerhof, W.E. and Shah, M.B., Phys. Rev. **A47** (1993) 1045.

[15] Wu, W., Wong, K.L, Montenegro, E.C., Ali, R., Chen, C.Y., Cocke, C.L., Dörner, R., Frohne, V., Giese, J.P., Mergel, V., Meyerhof, W.E., Raphaelian, M., Schimidt Böking, H. and Walch, B. (to be published)

[16] Ali, R., Frohne, V., Cocke, C.L., Stökli, M., Cheng, S. and Raphaelian, M.L.A., Phys. Rev. Lett. **69** (1992) 2491.

[17] Wu, W., Wong, K.L., Ali, R., Chen, C.Y., Cocke, C.L., Frohne, V., Giese, J.P., Raphaelian, M., Walch, B., Dörner,R., Mergel, V., Schmidt-Böking, H. and Meyerhof, W.E., Phys. Rev. Lett. **72** (1994) 3170.

[18] Dörner, R., Mergel, V., Ali, R., Buck, U., Cocke, C.L., Froschauer, K., Jagutzki, O., Lencinas, S., Meyerhof, W.E., Nüttgens, S., Olson, R.E., Schmidt-Böking, H., Spielberger, L., Tökesi, K., Ullrich, j., Unverzagt, M. and Wu, W., Phys. Rev. Lett. **72** (1994) 3166.

[19] Moshammer, R., Ullrich, J., Unverzagt, M., Schmidt, W., Jardin, P., Olson, R.E., Mann, R., Dörner, R., Mergel, V., Buck, U., and Schmidt-Böking, H., Phys. Rev. Lett. **73** (1994) 3371.

[20] Anholt, R., Phys. Lett. **114A** (1986) 126.

[21] Zouros, T.J.M., Lee, T.H., and Richard, P., Phys. Rev. Lett. **62** (1989) 2261.

[22] Montenegro, E.C., Melo, W.S., Meyerhof, W.E. and de Pinho, A.G., Phys. Rev. Lett. **69** (1992) 3033.

[23] W.Wu, PH.D Thesis, Kansas State University, 1994.

[24] Montenegro, E.C. and Zouros, T.J.M., Phys. Rev. **A50** (1994) 3186.

[25] Montenegro, E.C. and Meyerhof, W.E, Phys. Rev. **A43** (1991) 2289.

[26] Moores, D.L., Golden, L.B. and Sampson, D.H, J.Phys. **B13** (1980) 385.

[27] Grande, P., Sigaud, G.M., Montenegro, E.C. and Schiwietz, (to be published)

[28] Sant'Anna, M.M., Melo, W.S., Santos, A.C.F., Sigaud, G.M. and Montenegro, E.C., Nucl. Instr. Meth. Phys. Res. **B99** (1995) 46.

[29] Melo, W.S.,SantÁnna, M.M., Santos, A.C.F. Santos, G.M. Sigaud and E.C. Montenegro, this conference.

[30] Richard, P., Toth, G., Montenegro, E.C., Zouros, T.J.M., Hagmann, S., Grabbe, S.R. and Bhalla, C.P., this conference.

# Radiative and Resonant Electron Capture Studies for High-Z Projectiles

Th. Stöhlker

Gesellschaft für Schwerionenforschung, GSI, D-64220 Darmstadt, Germany

## ABSTRACT

The experimental studies of Radiative (REC) as well as of Resonant Electron Capture (RTE) into highly charged high-Z ions are reviewed. For REC, total and subshell differential cross sections are presented. In particular, the data of an L subshell resolved angular distribution measurement performed for He-like uranium are discussed. The comparison with exact relativistic calculations reveals that such investigations are very sensitive to the final state wave functions and provide unique information on magnetic interactions occuring in relativistic ion-atom collisions. For Resonant Capture the results of a subshell as well as angular differential measurement of KLL-RTE in $U^{90+} \to C$ collisions are given. The comparison with a full relativistic treatment underlines the importance of the Breit interaction for such high-Z few electron systems. Moreover, the data exhibit a strong anisotropy for $K\alpha_1$ transitions associated with the $KL_{1/2}L_{3/2}$ intermediate states.

## INTRODUCTION

In Radiative Electron Capture (REC) the transfer of a loosely bound 'quasifree' target electron into a bound state of the projectile is accompanied by the emission of a photon. Within the impulse approximation it can be treated as the time reversed photoionization process [1,2] and constitutes one of the most important projectile charge-exchange channels in fast encounters of high-Z ions and low-Z target atoms. Up to now most of the experimental work has been dedicated to total electron pick-up and K-shell capture cross-sections related to REC into bare and H-like ions [3-9]. By means of the available experimental systematics a universal scaling law for K-REC cross-sections could be established [9] based on the simple non-relativistic dipole-approximation [10]. This scaling law is in addition supported by rigorous relativistic calculations and can be explained in terms of an approximate cancellation among relativistic, retardation, and multipole effects [11]. This phenomenon occurs only for REC transitions into bound projectile s-states and only with respect to absolute cross section values. In particular, this fortuitous agreement between the dipole approximation and the exact relativistic treatment [2] also does not arise for angular-differential cross sections. In general, the angular-differential cross-sections are extremely sensitive to details of the atomic wave functions and give, therefore, detailed insight into the atomic structure of high-Z ions [2]. In the case of REC into projectile s-states such information is of particular interest as it reveals the influence of the electron spin in a unique way [2,12].

Similar to REC, Resonant Transfer and Excitation can be considered as a quasifree electron capture process [13,14]. Here, a loosely bound target electron is captured via resonant excitation of a projectile electron, forming in general a doubly excited state which decays by the emission of characteristic x-rays. This process is closely related to the time reversed Auger effect (Dielectronic Recombination) which refers to an initially free instead of a quasifree electron. RTE has been studied in great detail for light and medium ions up to Xe [15,16]. However, for high-Z projectiles only one RTE experiment was reported where the resonance profiles were scanned via total charge exchange measurements [14]. In high-Z ions, the large fine structure splitting causes a splitting of the KLL resonance into three well separated resonance groups, where the subshell resonance cross sections are in particular sensitive to the Breit interaction between the electrons. Compared to total charge-exchange measurements, more detailed information is accessible when the resonant capture is measured by a $K\alpha_1/K\alpha_2$ resolved detection of the emitted x-rays. Here, angular distributions are in particular of relevance as they are sensitive to the RTE population amplitudes for the various intermediate states. Such experiments may provide information on possible interferences between the REC and RTE processes. Evidence for such interferences have been found for highly-charged uranium ions in a recent DR experiment performed at the Super-EBIT [17].

In section 2 all available electron pick up cross sections for bare high-$Z$ ions related to REC are presented and compared with the dipole approximation as well as with correct relativistic calculations. In section 3 the L-subshell resolved angular differential data for REC into He-like uranium are shown. By comparing with rigorous relativistic calculations the relevance of the final state wave function is outlined. In section 4 a KLL-RTE experiment is discussed, performed for $U^{90+} \rightarrow C$ collisions. The data obtained are compared with complete relativistic calculations. Here, the importance of angular differential RTE measurements is stressed. Finally, in section 5, a short summary of the presented results is given.

## TOTAL CHARGE EXCHANGE CROSS SECTIONS RELATED TO REC

In the case of K-REC the available data cover now almost the entire range of projectile charges. It was shown [9] that the various experimental data fall closely together on one common curve if plotted as a function of the 'adiabaticity parameter' $\eta$, where $\eta$ is defined by

$$\eta = v^2/Z^2 \simeq 40.31 \times \frac{E_{kin}(MeV/u)}{Z^2}, \quad (1)$$

with $v$ denoting the velocity, $E_{kin}$ the kinetic energy, and $Z$ the atomic charge of the projectile. It is important to note that for the $\eta$-definition of Eq. (1) the relation $E_{kin} = \frac{1}{2}Mv^2$ was applied as if the projectile speed was non-relativistic ($M$ is the mass of the projectile). In Fig. 1 all measured total cross section data for REC into bare projectiles, normalized to the number of available target electrons, are plotted versus the $\eta$-parameter (lower x-axis) as defined by Eq. 1. The data cover a projectile Z regime ranging from Z=54 up to uranium (for references see [9]). In addition the data are compared with the predictions of the non-relativistic dipole approximation [10] (solid line in Fig. 1). From the figure, a good agreement between the experimental results and the simple theory applied can be deduced. This must be attributed to the fact that for fast bare ions, the main contribution to the total REC cross section arises from REC into projectile s-states ($\eta > 1$) and in particular from K-REC. The exact relativistic cross sections for various Z values do not follow the simple $\eta$-scaling rule. However, it has been shown that for low-$\eta$ values ($\eta < 1$) these cross sections also fall on one common curve independent of Z whereas for higher $\eta$-values the theoretical data diverge slightly. Here, within the overall experimental accuracy, the available data are not sensitive to this slight cross section variation. Therefore only the relativistic exact results for Z=80 are plotted in Fig. 1 (dashed line) as it is close to the atomic number of the projectile systems used at the largest $\eta$-values. For this particular case of Z=80 the upper x-axis gives the cross section values as a function of projectile energy. As can be seen from the figure the relativistic prediction deviates at large $\eta$-values from the general scaling of the dipole-approximation. In fact, the results of the latter approach underestimate slightly the experimental cross section data at large $\eta$-values ($\eta > 3$ measured for Au and U ions at energies greater than 580 MeV/u).

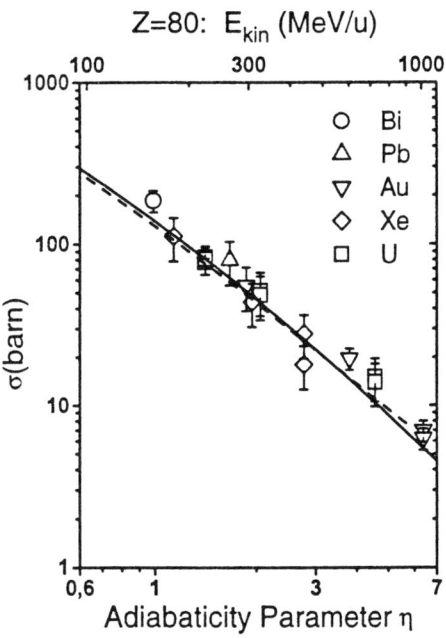

Figure 1: Total electron-capture cross sections per target electron measured for bare ions ($Z \geq 54$) in collisions with light gaseous and solid targets. The results are plotted as a function of the $\eta$-parameter as defined by Eq. (1) (lower x-axis).

## ANGULAR DIFFERENTIAL REC INVESTIGATIONS

Until recently only one dedicated angular distribution study of the REC process was available [18]. There it was found that for 197 MeV/u Xe $\rightarrow$ Be collisions the K-REC angular pattern closely follows a $\sin^2 \theta_{LAB}$ distribution in accordance with the non-relativistic approach, where $\theta_{LAB}$ denotes the laboratory observation angle. In order to elucidate in more detail the REC capture mechanism we conducted a first dedicated angular differential study of the L-REC process in 89 MeV/u $U^{90+}$ $\rightarrow$ C collisions

[12]. Here, the large L-subshell splitting between the $j = 1/2$ and $j = 3/2$ levels of about 4 keV and the moderate Doppler broadening in the experiment allowed us to separate REC into the different $j$-sublevels and to investigate the associated angular distributions.

For the experiment, performed at the Fragment Separator (FRS) at GSI in Darmstadt, the reaction target area was surrounded by five solid state Ge(i) detectors mounted at observation angles of 30°, 45°, 90°, 135° and 150° with respect to the beam axis. At 30° and 150° specially designed, granular detectors were installed which consist of seven equidistant, parallel strips allowing us to cover observation angles $\theta$ of $27° \leq \theta \leq 33°$ and of $146° \leq \theta \leq 154°$, respectively. The x-ray emission registered by each detector was recorded event by event in coincidence with one-electron capture into the projectile. For the latter purpose, the ion beam was magnetically analysed behind the target and the different outgoing charge states were detected by scintillator counters.

In Fig. 2 one REC x-ray spectrum is shown, measured at 135° observation angle in coincidence with electron pickup by the He-like uranium projectiles. The REC transition lines appear broadened by the Compton-profile of the carbon target atoms and, in addition, by the Doppler effect. However, the various REC contributions due to capture into the L-, M-, N-... shell of the projectile are well resolved. In particular, the L-REC line is split into the two subshell components – j = 3/2 and j = 1/2.

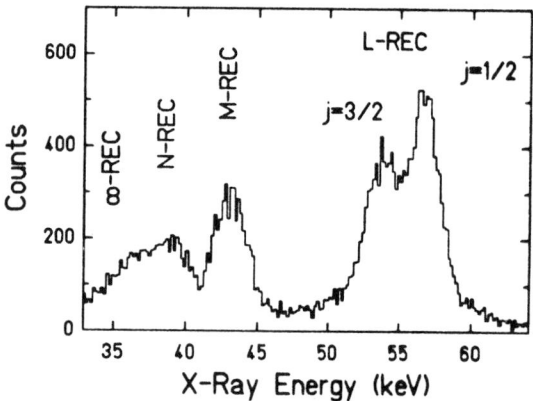

Figure 2: Charge-state-coincident REC spectrum measured at an observation angle of 135° for 89 MeV/u $U^{90+} \rightarrow$ C collisions.

In Fig. 3 the obtained cross section ratios for radiative capture into the $j = 3/2$ state and into the $j = 1/2$ levels are normalized to the sum of both distributions and given as a function of the observation angle (solid points in Fig. 3). These ratios $\mathcal{R}$ are essentially not affected by possible systematic uncertainties. The solid and dashed lines in Fig. 3 show the corresponding theoretical predictions, based on the rigorous treatment of the REC process [2]. The main feature of the experimental data is a very strong forward/backward asymmetry as predicted by the exact relativistic theory. Also the results of the non-relativistic dipole approximation for $\mathcal{R}(j=1/2)$ are shown which considers also lowest order retardation effects (dotted line in Fig. 3) [19]. Obviously, this approach fails completely in describing the experimental results.

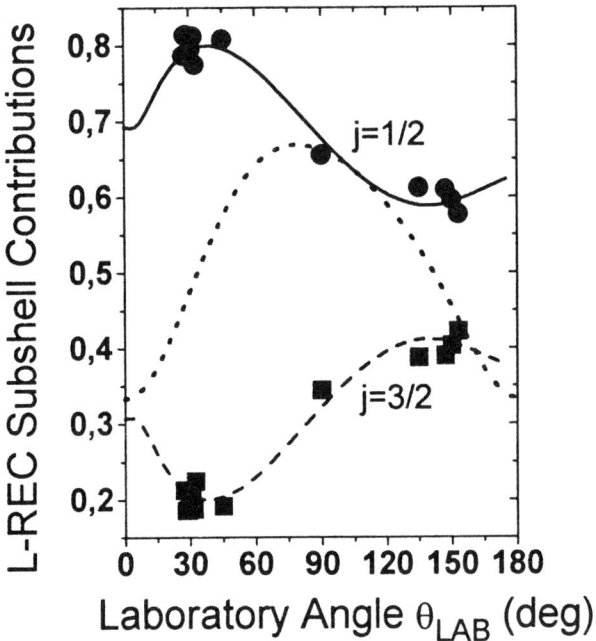

Figure 3: Measured differential L-REC cross sections for capture into the j = 1/2 and j = 3/2 sublevels in He-like uranium, normalized to the sum of both contributions.

In Fig. 4 the measured subshell-resolved L-REC angular-distributions are plotted. In order to compare the experimental results with the exact calculations, all measured data points were multiplied by one common factor of 0.65 which is still within the total normalization uncertainty. The predictions of the relativistic theory are given in the figure by the solid lines for the $j = 1/2$ levels and the $j = 3/2$ state. The data for the $j = 1/2$-levels show a considerable bending of the angular distribution into the forward direction, whereas the distribution for capture into the $j = 3/2$-state exhibits a slight enhancement at backward angles. This confirms in particular the prediction of the relativistic theory that the Lorentz transformation to the laboratory system is not sufficient to cancel for the $2p_{3/2}$ distribution the bending towards backward angles caused by retardation. Also, the calculated individual angular-distributions for the $2s_{1/2}$ and the $2p_{1/2}$ states are presented in Fig. 4 (the dotted and dashed lines, respectively). The $2s_{1/2}$ distribution deviates markedly from a $\sin^2\theta$ form. This distribution determines essentially the radiation pattern for capture into the $j = 1/2$ states at forward angles and the distribution for the $2p_{1/2}$ level follows closely the form for REC into the $2p_{3/2}$ state. The most remarkable aspect of the calculated distribution for the $2s_{1/2}$ state is that it predicts non-vanishing values at 180° and in particular at 0° observation angle. As has been pointed out in Ref. [2,20], angular momentum conservation requires that cross sections in the forward or backward directions can only be attributed to spin-flip transitions which are not considered in a non-relativistic theory. For K-REC into bare

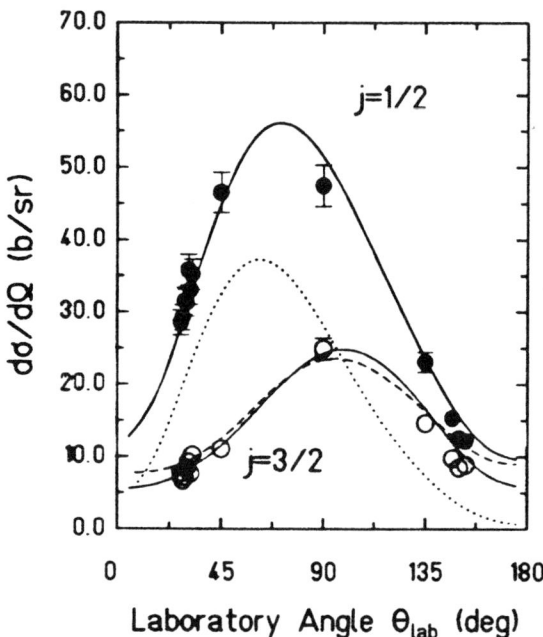

Figure 4: Experimental angular distribution for REC into the $j_{1/2}$ states (solid points) and the $2p_{3/2}$ level (full circles) in comparison with exact relativistic calculations (solid lines). In addition, the theoretical results for capture into the $2s_{1/2}$ (dotted line) and the $2p_{1/2}$ (dashed line) states are shown separately.

high-Z ions ($Z > 54$) at high energies ($E_{KIN} \approx 300$ MeV/u) this effect is predicted to be much more pronounced [20]. In this case, i.e. capture into the $1s_{1/2}$ ground state, the spin-flip contributions can be measured directly by experiments. There is no other case known either by theory or from experiment, in which spin-flip effects mediated by magnetic interactions can be identified so unambiguously in a relativistic atomic collision. This is depicted in Fig. 5, where the K-REC angular distribution for 295 MeV/u $U^{92+} \rightarrow N$ is plotted (dotted line: $\sin^2 \theta_{LAB}$ form; solid line: relativistic correct calculations; shaded area: spin-flip contribution). In the lower part the ratio of spin-flip to total K-REC is displayed separately. Due to the chosen observation angle the only experiment performed up to now for high-Z ions was not sensitive to this effect (solid point in Fig. 5) [9]. However, it demonstrates that the ESR storage ring provides well defined experimental conditions for such challenging investigations.

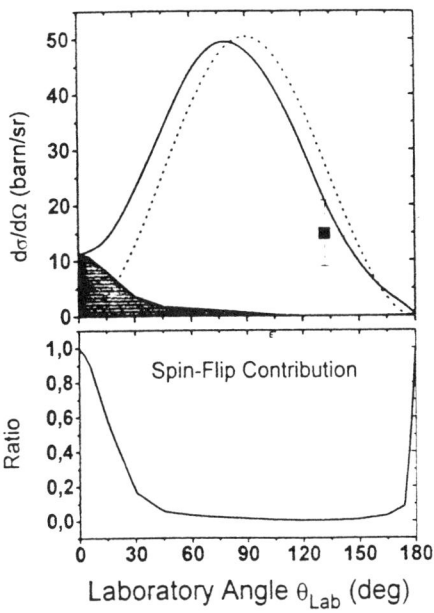

Figure 5: K-REC angular distribution for 295 MeV/u $U^{92+} \to N$ collisions; dotted line: $\sin^2 \theta_{LAB}$ form; solid line: relativistic correct calculations [2]; shaded area: spin-flip contribution. In the lower part the ratio of the spin-flip contribution to the total REC distribution is displayed separately.

## RESONANT ELECTRON CAPTURE

Using the same experimental set-up as for the L-REC studies the resonant KLL-RTE process was also investigated for $U^{90+} \to C$ collisions by scanning the beam energies in 20 steps from 105 to 140 MeV/u [21]. For He-like ions the $K\alpha$ emission along with electron capture into the projectile is an unambiguous signature of the resonant capture mechanism, i.e. $U^{90+}(1s^2) + e^{--} \to U^{89+}(1s2lj2l'j') \to U^{89+}(1s^22l''j'') + K\alpha$. Due to the closed K-shell of incident He-like ions neither the non-radiative nor the radiative capture events into excited projectile states can produce $K\alpha$ transitions.

Figure 6 shows the Doppler corrected, charge state coincident spectra for the three KLL-RTE resonance maxima located at the beam energies of 116 MeV/u for the $KL_{1/2}L_{1/2}$-RTE, of 124 MeV/u for the $KL_{1/2}L_{3/2}$-RTE, and of 132 MeV/u for the $KL_{3/2}L_{3/2}$-RTE, respectively. The spectra were generated by adding up the individual spectra of the seven segments of the 30° detector. At this observation angle the characteristic $K\alpha$ transitions are best resolved. Also at this angle, which is close to the transformed 54.7° center-of-mass angle, the $K\alpha$ lines are almost not affected by angular distributions. The $K\alpha_2$ and $K\alpha_1$ lines (shaded area) are fixed for all collision velocities at the center-of mass energies of 96 and 100 keV in contrast to the REC transitions into the projectile L- and M-shell (hatched areas). The main feature of the presented spectra is the drastic variation of the relative $K\alpha_2$, $K\alpha_1$ intensities as a function of beam energy. It is

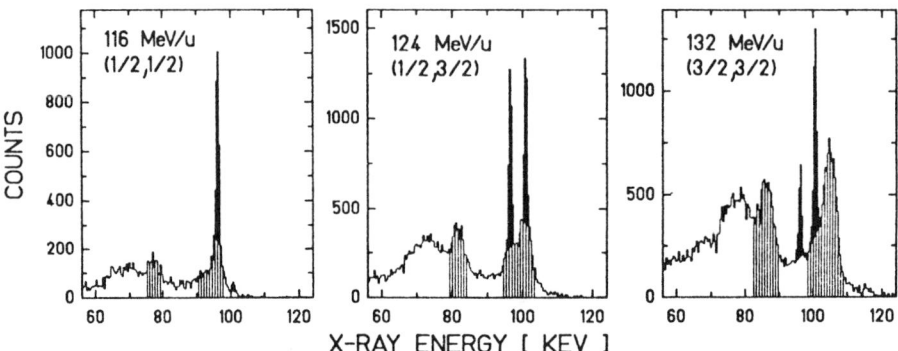

Figure 6: Coincident CM KLL-RTE x-ray spectra measured at 30° observation angle close to the three resonance maxima.

important to note that all intra-L subshell transitions are in general slower by two orders of magnitude than the K$\alpha$ transitions. Consequently, the subshell population cross sections for the various subshell resonances can be deduced directly from the K$\alpha_2$, K$\alpha_1$ intensities. For the data analysis, all x-ray intensities were normalized to the M-REC intensity in the spectrum by using theoretical M-REC cross sections [2]. As discussed in the previous section for the case of L-REC, the latter was found to deliver an appropriate description of the REC process. Similarly, the measured M-REC angular distributions are in excellent agreement with the applied rigorous relativistic approach [22].

| Resonance | Experiment | Theory | Experiment | Theory |
|---|---|---|---|---|
| | K$\alpha_1$ | | K$\alpha_2$ | |
| KL$_{1/2}$L$_{1/2}$ | – | – | 19.1±0.4 | 16.5(10.4) |
| KL$_{1/2}$L$_{3/2}$ | 7.3±0.2 | 7.0(7.1) | 5.2±0.3 | 5.2(4.7) |
| KL$_{3/2}$L$_{3/2}$ | 3.1±0.2 | 3.3(2.9) | – | – |

Table 1: Comparsion between theoretical [23] and experimental K$\alpha_1$ and K$\alpha_2$ cross section strengths [21] for the KLL-RTE subshell resonances in He-like uranium (see text). All values given are in kb·eV/sr.

The cross section strengths of the subshell resonances obtained from the K$\alpha_1$ and the K$\alpha_2$ intensities are summarized in table 1. In addition the results are compared with theoretical resonance strengths [23]. For theoretical predictions the Breit term of the Auger matrix elements was considered in one case and neglected in the other (see number in parentheses). Good agreement is found for the relativistically correct calculations including the Breit interaction. Moreover, the experimental data reveal that an isotropic K$\alpha_1$ emission can in general not be assumed. This is depicted in Fig. 7 where the measured K$\alpha_1$/ K$\alpha_2$ intensities are given for the KL$_{1/2}$L$_{3/2}$ resonance as a function of the laboratory observation angle. This ratio varies by up to a factor of two. For comparison, also plotted in the figure is the intensity ratio of 1.36 which was derived from the theoretical RTE cross sections by assuming an isotropic emission pattern. The observed strong variation of the relative $K\alpha$ intensities can be assumed to be essentially caused by the radiation pattern of the $K\alpha_1$ transitions, since the $K\alpha_2$ line contains only $j = 1/2$ to $j = 1/2$ transitions which are almost isotropic in the CM system. From this it can be deduced that there is a strong alignment of the $j = 3/2$ states produced by RTE.

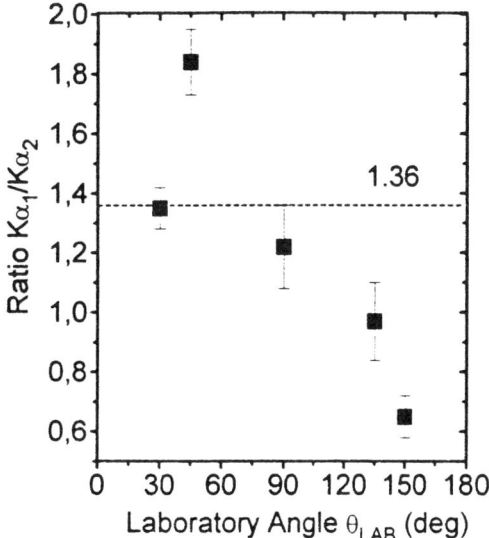

Figure 7: Measured angular distribution of the $K\alpha_1/K\alpha_2$ intensity ratio for the $KL_{1/2}L_{3/2}$ resonance group [21].

## SUMMARY

In conclusion, the limits of applicability of the non-relativistic dipole-approximation have been outlined by comparing this simple theory with total charge exchange cross sections as well as with subshell resolved L-REC angular distributions. The comparison with the differential data reveals a complete breakdown of the dipole-approximation in describing the subtleties of the REC process, and the good agreement found between the non-relativistic approach and the measured total cross sections must be regarded as fortuitous. For the angular differential results the applied rigorous relativistic calculations are shown to be most appropriate. The latter comparison exhibits that such studies are very sensitive to the final state wave function and they provide a novel access for the investigation of the magnetic interaction in relativistic ion-atom collisions. Moreover, the results of a dedicated KLL measurement for He-like uranium are given which were obtained by applying an x-ray/particle coincidence technique. Here, the $K\alpha$ production cross sections are found to be in excellent agreement with complete relativistic calculations including the Breit interaction. Also, a pronounced non-isotropic angular distribution for the $K\alpha_1/K\alpha_2$ intensity ratio has been found at the $KL_{1/2}L_{3/2}$ resonance. This indicates a strong alignment of the $p_{3/2}$ electrons in the intermediate state formed by RTE. With respect to this effect no theoretical information is presently available.

## ACKNOWLEDGEMENT

This work was done in collaboration with T. Kandler, C. Kozhuharov, P.H. Mokler, P. Rymuza, Z. Stachura, and A. Warczak. The fruitful cooperation with Hans Geissel and C. Scheidenberger is gratefully acknowledged. The author would like to thank J. Eichler (HMI Berlin), A. Ichihara (JAERI, Japan) and T. Shirai (JAERI, Japan) for the close collaboration and stimulating discussions.

## REFERENCES

[1] P. Kienle et al., 1099 (1973).

[2] A. Ichihara, T. Shirai, and J. Eichler, Phys. Rev. A 49, 1975 (1994).

[3] H.W. Schnopper et al., 898 (1972).

[4] H. Tawara, P. Richard, K. Kawatsura, Phys. Rev. A 26, 1975 (1982).

[5] S. Andriamonje et al., Phys. Rev. Lett. 59, 2271 (1987).

[6] Th. Stöhlker et al., Z. Phys. D 23, 121 (1992).

[7] L.C. Tribedi et al., Phys. Rev. A 49, 374 (1994).

[8] C.R. Vane et al., Phys. Rev. A 49, 1847 (1994).

[9] Th. Stöhlker et al., Phys. Rev. A 51 2098 (1995).

[10] M. Stobbe, Ann. Phys. 7, 661 (1930).

[11] A. Ron et al., Phys. Rev. A 50, 1312 (1994).

[12] Th. Stöhlker et al., Phys. Rev. Lett. 73 3520 (1994).

[13] P.H. Mokler, S. Reusch, Z. Phys. D 8, 393 (1988).

[14] W.G. Graham et al., Phys. Rev. Lett. 65, 2773 (1990).

[15] W.G. Graham, W. Fritsch, Y. Hahn, J.A. Tanis (Eds.), Recombination of Atomic Ions, Plenum Press, New York, 1992.

[16] S. Andriamonje et al., Phys. Lett. A 154, 194 (1994).

[17] D. A. Knapp et al., Phys. Rev. Lett. 74, 54 (1990).

[18] R. Anholt, et al., 53, 234 (1984).

[19] K.I. Hino and T. Watanabe, Phys. Rev. A. 36, 581 (1987).

[20] J. Eichler, A. Ichihara, and T. Shirai, Phys. Rev. A 51, 3027 (1995).

[21] T. Kandler et al., Phys. Lett. A accepted (1995).

[22] T. Kandler et al., submitted to Z. Phys. D (1995).

[23] M. Zimmermann, private communication and P. Zimmerer, N. Grün, W. Scheid, Phys. Lett. A 148, 457 (1990).

# ION-ATOM COLLISIONS
# – CHARGE TRANSFER –

Electron Spectroscopy of Rydberg States
Produced in Capture Processes .................................... 537
    A. BORDENAVE-MONTESQUIEU and P. Moretto-Capelle
Gathering and Evaluation of State Selective Capture
Cross Sections for Plasma Diagnostics ........................... 547
    R. HOEKSTRA
Double Electron Capture: Complex Amplitudes
from Auger Anisotropy Measurements ............................ 557
    M.H. PRIOR and H. Khemliche

# Electron Spectroscopy of Rydberg States Produced in Capture Processes

## A. Bordenave-Montesquieu and P. Moretto-Capelle

*Laboratoire CAR, IRSAMC, URA CNRS 770, Université Paul Sabatier*
*118, Route de Narbonne, 31062 Toulouse Cédex, France*

**Abstract.** Double capture processes populate many terms of a Rydberg series. Various kinds of information can be got from high resolution electron spectroscopy. We mainly focus on the data which have some connection with the population mechanisms. Recent results which concern the L-distribution inside each n-manifold, the n-dependence of the capture cross section and the ATR process are discussed.

## INTRODUCTION

Two-electron transfer on (N,n) doubly excited states of multicharged ions which interact with gas targets is generally studied by electron spectroscopy because these states are often purely autoionizing [1]. Here N and n designate the principal quantum numbers of the inner and Rydberg captured electrons respectively. At low velocity, this reaction is often considered to be rather selective in n and N. However in almost all cases several terms of a (N,n) Rydberg series, and sometimes terms of several series, are more or less discernible in the electron spectra [1-5]. In this report the double capture processes which involve ions with charge $q \leq 10$ and two-electron targets will be considered at low collision velocity ($v < 1$ au). In this velocity range, they generally occur near crossings of potential curves within a quasimolecule (for exothermic reactions). In these conditions N = 2, 3 and 4 Rydberg series may be populated.

The structure of the electron spectra does not always allow a clear identification of the different terms of a Rydberg series. Indeed, when using filled or partially filled K-shell incident ions, an overlap of the autoionizing partial decays of the doubly excited states into the available ionization continua occurs, which often precludes a separation of the various n-terms. However, if high enough ion charges and high resolution electron spectroscopy are used, an interesting spectroscopic work on Coster-Kronig transitions remains possible. For example, in the $Ne^{9+}(1s)$ - Ne collision, the recent study of the n > 8 part of the $Ne^{7+}(1s2lnl')$ series around the $Ne^{8+}(1s2s\ ^3S)$ ionization threshold, has allowed us to gain informations on several spectroscopic problems [7]: the sudden rise of the Auger yield at the opening of a new ionization threshold, the fine structure of $Ne^{8+}(1s2p\ ^3P_J)$ and the spin core conserving transitions $Ne^{7+}(1s2p\ ^3P)nl \rightarrow Ne^{8+}(1s2s\ ^3S) + e^-$. In contrast, a double capture by bare ions allows a straightforward identification of many terms of the Rydberg series. This is because the hydrogenic ionization thresholds of the residual ion formed after autoionization are degenerated [1]. Indeed, even for a system like $N^{7+}$-He, where a strong selectivity of the double capture on the first term of the $N^{5+}(3,n)$ series is observed, many n values have been recently identified, up to n = 16 [8].

When the target ionization potential is decreased (or the ion charge increased), the capture shifts towards higher terms of a given (N,n) series or towards the next (N+1,n) series. It may happen that the first terms of the (N,n) series are lacking, the higher terms being more strongly excited, as observed when the (3,n) series is populated in $N^{7+}$- $H_2$ [1] or $Ne^{8+}$- $H_2$ [2,3,9]. This shift is qualitatively understood [2] within the framework of the Extended Classical overBarrier model (ECB) of Niehaus [10]. It strictly predicts that n ≈ N symmetrical configurations are produced. This was only through the width of the reaction window introduced by Niehaus, that a population of the higher terms of a Rydberg series was "understood" [2]. This "prediction" contradicts the build-in assumptions of the ECB model [11]. Stolterfoht was the first to focus on the true *primary mechanisms* which explain an excitation of n > N asymmetric configurations; they were thought to be governed by the electron-electron interaction [12,13]. A population of asymmetric terms of a Rydberg series may then occur at small internuclear distances R, typically below R = 10 au, through a direct interaction of the exit Rydberg channel with the entrance one, the so-called Correlated Double Capture processes (CDC); it implies a simultaneous jump of two electrons near the crossing of each n-term with the entrance channel [12]. An alternative two-step process [14] involves an intermediate single capture channel (the so-called Correlated Transfer Excitation or CTE). These primary processes were later named "autoexcitation" by the author since they have some similarity with the Configuration Interaction (CI) which explains the autoionization [13]. More recently, an indirect population of the upper terms of the Rydberg series *through a secondary post-collisional mechanism*, the so-called ATR process (Auto Transfer to Rydberg states), has been proposed to explain the surprisingly high radiative stabilization of some multiply excited configurations [15,16]. It will be considered more throroughly in this report to understand some features of the electron spectra. Therefore, it appears that the electron spectroscopy of the Rydberg series may help to gain informations on the population mechanisms.

By far, up to now, most of the experimental and theoretical studies have focused on the n = N and n = N+1 terms to elucidate the capture processes. We will not consider these results here; instead we will focus on the n > N part of the Rydberg series, populated when bare ions interact with two-electron targets.

# L-DISTRIBUTIONS ALONG THE (N,n) SERIES (n < 10)

The L-distributions within the n-manifolds give informations on the secondary mechanisms which explain the redistribution of the initial population (Stark mixing [17,18]). For the N = 2 series, high angular momentum states were observed to be favoured (n = 6, 7) in the reaction [6]: $O^{6+}(1s^2)$ + He $\rightarrow$ $O^{4+}(1s^2 2pnl)$ $\rightarrow$ $O^{5+}(1s^2\,^1S)$ + e⁻. More recently a peaking of the L-distribution on the high angular momenta states along the $N^{5+}(3,n)$ series was reported at 70 keV [8], in the reaction: $N^{7+}$ + He $\rightarrow$ $N^{5+}(3,n)$ $\rightarrow$ $N^{6+}(n' = 2)$ + e⁻. The collision energy dependence is shown in figure 1. The L-distribution of the n = 3 term [19,20] is related to specific population mechanisms [17]; an intense excitation of the $^1F°$ and $^1G$ states is always seen. The n = 4 term shows several lines peaked on L = 3-5 states. Then, at a given collision energy, the L-distribution is found to concentrate on the high angular momentum states when n ≥ 4 (Fig. 1a: compare n = 4 and 5 lines). This was thought [8] to be a consequence of a post-collisional Stark mixing of the Rydberg states by the Coulomb field of the target ion, as shown in [21]. This interpretation was recently found to explain the quoted observations in $O^{6+}(1s^2)$ + He [22]; these data were initially understood as resulting from an angular momentum exchange between the electrons captured in a primary process [6]. When the collision energy increases, the high angular momentum states are favoured whatever the n value is (Fig. 1). This was already noted for the n = 3 term [17,20] but imperfectly reproduced by theory [17]. The observed velocity

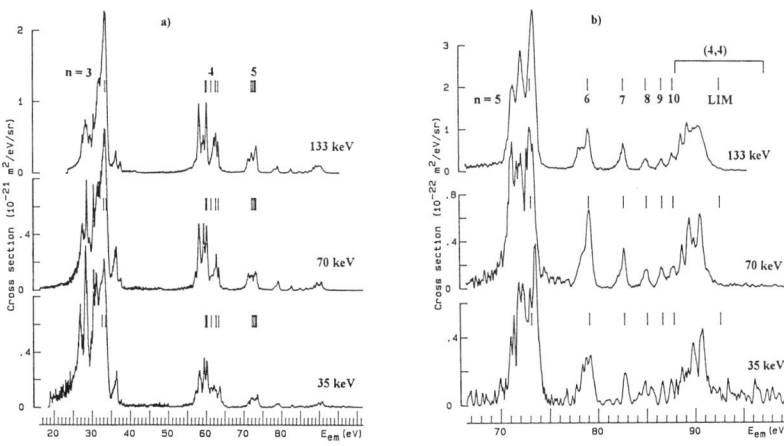

**FIGURE 1.** a): (3,n) series produced in $N^{7+}$- He at 10° and various collision energies. Energy resolution = 0.35 %. Vertical bars indicate the calculated positions of L = 3, 4 (n = 3), L = 3-5 (n = 4) and L = 4, 5 (n = 5) high angular momentum states [23]. b): enlarged view of the n ≥ 5 part of the spectra; as in a) but the positions of the high L states are deduced from a Rydberg formula using a zero quantum defect [8]. The positions of pure (4,4) states are also indicated. The (3,n) series limit is noted « LIM ». Note in figure 1b the perturbation of the Rydberg series at 35 keV, above n = 7 (see text).

dependence does not match the Stark mixing mechanism which becomes stronger at low velocity [22]. Therefore another mechanism which has a velocity dependence opposed to the one of the Stark effect, is certainly superimposed. For example it could happen that the population of the magnetic sublevels becomes more statistical at high energy.

## n-DEPENDENCE OF THE CAPTURE CROSS SECTION (n < 10)

The N series populated at a given collision velocity depends on the ion charge. The N = 2 one is mainly produced with $q \leq 6$ ions [2,5,24]; N = 3 and 4 series are observed with higher charges, up to q = 10 [1,5]. In this section we will focus on the n < 10 part of the N = 3 one. A comparison will be made with the data which concern the N = 2 and 4 series. The region of overlap of (4,4) and (3,n) states (n > 10) will be discussed in the next section.

A few years ago it has been emphasized that the measured cross section evolves as $n^{-m}$ along the Rydberg series, where m is a constant [9,25]. The fluorescence yield of the Rydberg states may affect the electron spectra and thus the measured n-dependence. The radiative rates are roughly given by the transition of the inner electron. They are equal to about $A(3s) = 1.5 \ 10^{10} \ s^{-1} / 6.3 \ 10^{10} \ s^{-1}$, $A(3p) = 4.5 \ 10^{11} \ s^{-1} / 1.9 \ 10^{12} \ s^{-1}$ and $A(3d) = 1.5 \ 10^{11} \ s^{-1} / 6.4 \ 10^{11} s^{-1}$ for $N^{6+}(3l)$ and $Ne^{9+}(3l)$ ions respectively [26]. In the case of neon these values begin to be almost equal to the calculated autoionization rates for (3,12) states [27]. In the present n-range the fluorescence yields do not too seriously affect the electron emission cross section along the (3,n) series; so, the measured cross sections may be considered as capture cross sections. A more accurate evaluation of the L-dependent fluorescence yields implies a detailed knowledge of the L-distribution inside each n-manifold; presently, only the results discussed in the preceding section are available on this topic. Some attempts to calculate the fluorescence yields with an arbitrary L-distribution inside each n-manifold have been reported [5,24].

From the analysis of our data as well as of those published by other authors, we can now draw some general tendencies:

i) *The capture cross section for the first excited term of the series does not generally follow the same n-dependence* than that of the upper ones. This indicates that the population mechanisms are different for the two n groups. On the other hand, the simple n-dependence of the capture cross section ($\sigma \approx n^{-m}$) measured for the medium terms of the series is probably the signature that the same underlying mechanism explains the population of all these terms. This property applies to almost all the systems investigated till now. In $C^{6+}$- He it has been found that the n = 3 term (N = 2 series) obeys a different mechanism than the n > 3 ones [28].

ii) From the data obtained at 10 qkeV it appears, as will be shown in the following, that *the m power evolves regularly when changing the collisional system*. For many years the m = 3 law found in the N = 2 series, with q = 6 ions, was considered as a rule [24,25,29]. However it was already known that the capture cross section along the N = 3 series, using q = 7 ions, does not follow the same m law [9]; the same observation was made more recently by other authors for q = 10 ions and the N = 4 series [5]. Then, the situation becomes confusing and the search for an underlying consistency among all the experimental results becomes necessary. The result of our analysis is presented in Table 1. Collisional systems are classified against the energy defect $\Delta E$ of the reaction of capture into the N = 3 series. Three $\Delta E$ values are given in Table 1 for three final states: the first and last (3,4) states and the (3,10) states respectively. Positive and negative values correspond to exothermic and endothermic processes respectively. The m power appears to decrease regularly when the exothermicity increases. This corresponds to a shift of the active crossing regions from large to short internuclear distances when $\Delta E$ increases. This may be verified with the help of the crossing radius $R_c$ with the entrance channel (possible CDC mechanisms) which is given by $R_c = (2(q-2)/\Delta E$.

Let us consider now the (3,n) data. Two borderline situations are found at the left and the right of the table:

- at the left, crossings with the entrance channel (CDC) or with the n = 5 single capture channel (CTE) in $N^{7+}$- He occur at very large R and the $R_c$ value strongly depends on n. These crossings are very diabatic and hence improbable. The double capture into (3,n) states becomes almost completely endothermic with $C^{6+}$- He. The similarity of the m values in $C^{6+}$- He (no crossings) and $N^{7+}$- He seems to indicate that unlocalized Demkov-type transitions [30] between almost parallel potential curves play a significant role (see also [25]).

TABLE 1. n-dependence of the double capture cross sections at 10 qkeV. Our results and some of other authors have been fitted with a $\sigma \approx n^m$ law to deduce the m value. The table gives our results unless otherwise stated. The collisional systems are approximately classified according to the energy defect $\Delta E$ of the reaction. When necessary, another n value associated with the $\Delta E$ value is specified within parentheses for the (2,n) and (4,n) series.

| Rydberg series | (3,n) | | | | | | (2,n) | | (4,n) |
|---|---|---|---|---|---|---|---|---|---|
| ion | $C^{6+}$ | $N^{7+}$ | $C^{6+}$ | $O^{8+}$ | $N^{7+}$ | $Ne^{10+}$ | $C^{6+}$ | $O^{6+}$ | $Ne^{10+}$ |
| target | He | He | $H_2$ | He | $H_2$ | He | He | He | He |
| $\Delta E$(ua) (n=4) | 0.04 | 1.1 | 1.1 | 2.4 | 2.2 | 5.4 | 5 | 5 (6) | 2 (5) |
|  | -0.2 | 0.8 | 0.9 | 2.1 | 1.9 | 5 |  |  | 1.7 |
| (n=10) | -0.8 | 0.002 | 0.3 | 0.9 | 1.1 | 3.2 | 1.7 | 4.8 | 0.6 |
| m ($\sigma \approx n^m$) | 6.1[a] | 7.4 | 6.9[a] | 6.5 | 4 | 3[a]/3 | 3.4[a] | 3[a]/3[b] | 7[a] |

[a] $C^{6+}$- He: [24]; $C^{6+}$- $H_2$: [31]; $Ne^{10+}$- He:[5]; $O^{8+}$- He: [25]
[b] our result; in [32] the n = 6 line intensity was found to be affected by a transmission problem of the detector

- at the right, the largest energy defect is found for $Ne^{10+}$- He. In this case all the terms of the series cross the entrance channel in a restricted R range; so, the couplings between the (3,n) exit channels and the (assumed) entrance channel become almost independent of R. This situation has been analyzed in terms of Landau-Zener crossings [33] even if the crossings are evidently not well separated [5]; the measured $n^{-3}$ dependence of the cross section is thought to result from the normalization factor of the Rydberg atomic wave function [5,34]. However as this law is observed when the transitions occur at small R in the quasimolecule this explanation is not necessarily valid.

The interest of the classification proposed in Table 1 appears when the data obtained for different (N,n) series are compared. In this respect the $Ne^{10+}$- He system appears to be very interesting. First, the m = 3 law found in the (3,n) series is about the same as the one found for the N = 2 series with q = 6 ions (Table 1); this is because the crossing situation is the same in all three cases. Secondly, both N = 3 and N = 4 series are populated, and different m values have been measured in the two cases [5]. Following our previous discussion, this is because two different R ranges (and hence different mechanisms) are involved. Let us consider this situation in more detail. Within the CDC process the N = 4 series is populated near crossings at rather large $R_c$ values (8 - 27 au); this situation looks like the one encountered with $C^{6+}$- $H_2$ for the N = 3 series (7 - 29 au), and indeed about the same m law is obtained. The same remark holds with a CTE process which certainly involves the n = 5 and 4 single capture channels respectively.

The electron-electron interaction is generally thought to explain the population of n > N terms [12,13]. From Table 1 it appears that there is no univocal relation between this mechanism and the m = 3 power law, as first thought [5,25]. The m = 3 value corresponds to a particular situation where electron transfer occurs within a restricted R-range. It has been reproduced by calculations which explain the population of the (2,n) series in $C^{6+}$- He by a CTE process [28], but no clear explanation arises from these calculations on the origin of this m value [29]. From the present analysis, which covers most of the collisional situations found in double capture processes, the m value appears to be roughly limited within the m = 3 to 8 range. Therefore it can be predicted that unmeasured systems must also obey the classification scheme proposed in Table 1.

It would be interesting to see how the $n^{-m}$ law, i.e. the electron transfer mechanism, behaves with the collision energy [17]. The only measurement in $N^{7+}$- He does not show any variation of m: m ≈ 7.3 ± 1.0 at 35 keV and 133 keV. Clearly, other systems must be investigated.

Finally, it must be noted that an unexplained perturbation of the Rydberg series is observed on the 35 keV spectrum (Fig. 1b) above n = 7. An identification of the n > 7 lines becomes difficult. This is not an artefact because it was already observed in a high statistics, lower resolution, spectrum (Fig. 8 in [34]).

# POPULATION OF THE n > 10 PART OF THE (3,n) SERIES, THE ATR MODEL

In the n > 10 range, no special behaviour is expected for the cross section for capture into the upper terms of the (2,n) series; the m power law reported above should also apply here. This is no longer true for the (3,n) series because it is perturbed by the first term of the (4,n) one; this was recently illustrated by our data obtained in $N^{7+}$- He anf $Ne^{10+}$- He (Fig. 1b and 2 respectively and [8]). In this region an unexpected large radiative decay of doubly excited states was first observed in $O^{7+}$(1s) - Ar whereas the double capture was believed to happen into the autoionizing single-configuration (4,4) states [15]. Many other similar situations have been recently found in various collisional systems [35,36]. It was first recognized by Roncin and coll. [15] that the important radiative stabilization comes from a transfer of population from the pure autoionizing (4,4) states into the adjacent (3,n) $^1$L Rydberg states which may have significant fluorescence yields if n and L are high enough. The post-collisional ATR mechanism was proposed by Bachau and Roncin [16,34] to understand this population transfer. The basic idea is that the (4,4) states can cross part of the (3,n) series at small internuclear distances in the quasimolecule. This is because the Rydberg electron does not shield the nuclear charge when $R < R_{max}$ with $R_{max} = 3n^2/2Z'$, where Z' is the core charge seen by the Rydberg electron. Let us consider that a (4,4) state is resonant with (3,$n_0$) states in the isolated atom and that this pure (4,4) state is populated by a primary mechanism at a small enough R value. Then, it can transfer part of its population in the receding part of the collision near crossings with the (3,n) series (n < $n_0$). When R > $R_{max}$ the potential curves become parallel [34] and a static CI develops between the (4,4) and several single-configuration (3,n) states (n ≈ $n_0$). Finally, after the collision, the initial population of the pure (4,4) state will be observed to be distributed among all these multi-configuration (3,n) and (4,4) states. The static CI alone was put forward by other authors as the main source of electron stabilization [37-39].

What can be learned from the electron spectroscopy of the (3,n) series near the n = 3 limit? High resolution electron spectra have been measured in $N^{7+}$- He (Fig. 1b) and $Ne^{10+}$- He (Fig. 2). They clearly show that the (3,n) Rydberg series is perturbed by the (4,4) states. An enhanced population of the n = 10 to 16 terms is found and no signal is detected above n = 16 [40]. The mechanisms discussed previously for the n < 10 part of the series only gives a smooth contribution in this n-range; they cannot explain the observed enhanced population. It is due to an initial population of the (4,4) states followed by a transfer of population induced by a dielectronic interaction through the ATR process. The contribution of the mechanism characterized by the $n^{-m}$ law increases when m decreases. So, the worse collisional system to see the ATR contribution is $Ne^{10+}$- He [8], but this is also in this case that the initial (4,4) population is a priori the highest... Nevertheless, even for this system, a clear enhancement of the (3,n) line intensities is observed in the region where the (4,4) states lie (Fig. 2). In a first attempt to inter

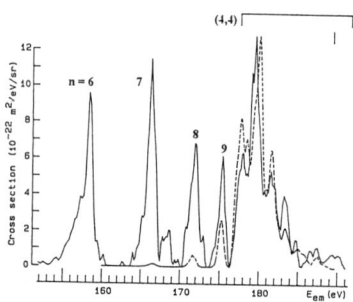

**FIGURE 2.** Ne$^{10+}$- He collision at 100 keV, 10°. Energy resolution = 0.35 %. Comparison of the measured (heavy line) and calculated (dotted line, normalized to experiment; from H. Bachau, private communication) electron spectra, below the N = 3 threshold. The calculation incorporates the ATR process in the whole R range (see text and [27,41]).

pret these results, only the dynamic part of the ATR model has been used ([8]: a contribution of single-configuration states was assumed to be observed in the electron spectra). More recently, a complete calculation which takes into account the ATR mechanism in the whole R range has been published [27,41]. Without specific collisional calculations, an arbitrary L-distribution inside the (4,4) manifold was assumed (the same population for all states [41]). Clearly, the dynamic part of the ATR process contributes to the population of the n = 7 to 10 states (Fig. 2) whereas the static CI at large R appears to be the main source of population of the large n values, say the n > 10 ones, in agreement with [39]. The absence of signal observed above about n = 16 is explained both by the dynamic part of the ATR process (n < $n_0$ states are first populated efficiently at small R) and by the radiative decay of the higher n ≈ $n_0$ states populated in the static part of the ATR process at large R. The autoexcitation process alone has recently been proposed to explain in this system the population of large n values [42], and hence to be at the origin of the large radiative stabilization of doubly excited states; this conclusion is at variance with the present interpretation and with the available calculation [22,27,41]. The systems which appear at the right of Table 1 are particular in the sense that the initial condition, a population of pure (4,4) states at small R, in order that the dynamic part of the ATR process may develop, is not straightforward. In N$^{7+}$-He, the dynamic ATR can only contribute through a (4,4) population arising from the incoming part of the collision since it probably occurs through Demkov-type transitions at large internuclear distances. In C$^{6+}$-He no enhancement of the population of the (3,n) series is observed [24] because the energy gap between the (4,4) states and the entrance channel is too large (Table 1) to make a population of the (4,4) states by Demkov-type transitions probable.

Finally, it must be recalled that, within the static CI, the interference between the (3,n) and (4,4) components of multi-configuration (3,n) and (4,4) states can induce large abnormal radiative probabilities for the hybrid states [39,43,44], and thus affect the electron spectra.

The features reported above are general and apply to all the electron spectra, whatever the cross sections or velocities are (see other examples in [34]). These perturbations may be observed with other ions than bare nuclei but the analysis of the electron spectra becomes more difficult in other cases (see for example the $Ne^{8+}$- $H_2$ spectrum in [45]).

Finally it is worth noting that the perturbation of the Rydberg series of helium observed in photoabsorption experiments [46] looks like the ones reported here even if the underlying processes are different (no post-collisional effect). For example the (5,n) series has been observed to be perturbed by the (6,6) states.

## ACKNOWLEDGEMENTS

The authors are indebted to H. Bachau and H. Jouin for a careful reading of the manuscript and for providing us with data.

## REFERENCES

1. Bordenave-Montesquieu, A., Benoit-Cattin, P., Gleizes, A., Marrakchi, A.I., Dousson, S., and Hitz, D., *J. Phys. B* **17**, L127-131 (1984)
2. Mack, M., *Nuclear Instr. Methods B* **23**, 74-85 (1987)
3. Meyer, F.W., Griffin, D.C., Havener, C.C., Huq, M.S., Phaneuf, R.A., Swenson, J.K., and Stolterfoht, N., *Proceedings of the 15$^{th}$ Int. Conf. on the Physics of Electronic and Atomic Collisions* (Brighton 1987), ed. Gilbody, H.B., et al, North-Holland, 1988, pp. 673-683
4. Sakaue, H.A., Awaya, Y., Danjo, A., Kambara, T., Kanai, Y., Nabeshima, T., Nakamura, N., Ohtani, S., Suzuki, H., Takayanagi, T., Wakiya, K., Yamada, I., and Yoshino, M., *J. Phys. B* **24**, 3787-3795 (1991)
5. Fremont, F., Merabet, H., Chesnel, J.Y., Husson, X., Lepoutre, A., Lecler, D., and Rieger, G., *Phys. Rev. A* **50**, 3117-3123 (1994)
6. Meyer, F.W., Griffin, D.C., Havener, C.C., Huq, M.S., Phaneuf, R.A., Swenson, J.K., and Stolterfoht, N., *Phys. Rev. Letters* **60**, 1821-1824 (1988)
7. Moretto-Capelle, P., Bordenave-Montesquieu, A., González, A., and Benhenni, M., *J. Phys. B* **27**, L317-323 (1994)
8. Bordenave-Montesquieu, A., Moretto-Capelle, P., González, A., Benhenni, M.,Bachau; H., and Sánchez, I., *J. Phys. B* **27**, 4243-4261 (1994)
9. Bordenave-Montesquieu, A., Benoit-Cattin, P., Boudjema, M., and Gleizes, A., *Proceedings of the 15$^{th}$ Int. Conf. on the Physics of Electronic and Atomic Collisions, book of Invited Papers,* (Brighton 1987), ed. Gilbody, H.B., et al, North-Holland, 1988, pp.643-653
10. Niehaus, A., *J. Phys. B* **19**, 2925-2937 (1986)
11. Barat, M., and Roncin, P., *J. Phys. B* **25**, 2205-2243 (1992)
12. Stolterfoht, N., Havener, C.C., Phaneuf, R.A., Swenson, J.K.,Shafroth, S.M., and Meyer, F.W., *Phys. Rev. Letters* **57**, 74-77 (1986)
13. Stolterfoht, N., *Physica Scripta* **2**,192-204 (1990); idem **T 46**, 22-33 (1993)
14. Winter, H., Mack, M., Hoekstra, R., Niehaus, A, and de Heer, F.J., *Phys. Rev. Letters* **57**, 957 (1987)
15. Roncin, P., Gaboriaud, M.N., and Barat, M., *Europhysics Letters* **16**, 551-556 (1991)

16. Bachau, H., Roncin, P., and Harel, C., *J. Phys. B* **25**, L109-115 (1992)
17. Harel, C., and Jouin, H., *J. Phys. B* **25**, 221-237 (1992)
18. Burgdörfer, J., *Phys. Rev. A* **24**, 1756-1767 (1981)
19. Bordenave-Montesquieu, A., Benoit-Cattin, P., Boudjema, M., Gleizes, A., and Bachau, H., *J. Phys. B* **20**, L695-L703 (1987)
20. Moretto-Capelle, P., *thesis*, Université Paul Sabatier, Toulouse (1990)
21. Kazansky, A.K., *J. Phys. B* **26**, 3863-3869 (1993)
22. Kazansky, A.K., and Roncin, P., *J. Phys. B* **27**, 5537-5550 (1994)
23. Bachau, H., Martín, F., Riera, A., and Yañez, M., *Atomic Data and Nuclear Data Tables* **48**, 167-212 (1991)
24. Stolterfoht, N., Sommer, K., Swenson, J.K., Havener, C.C., and Meyer, F.W., *Phys. Rev. A* **42**, 5396-5405 (1990)
25. Stolterfoht, N., Sommer, K., Griffin, D.C., Havener, C.C., Huq, M.S., Phaneuf, R.A., Swenson, J.K., and Meyer, F.W., *Nucl. Instr. Methods B* **40/41**, 28-32 (1989)
26. Bethe, H.A., and Salpeter, E.E., *Quantum Mechanics of One- and Two-Electrons Atoms*, New-York: Plenum Press, 1957
27. Sánchez, I., and Bachau, H., *J. Phys. B* **28**, 795-806 (1995) and private communication
28. Harel, C., Jouin, H., and Pons, B., in *Proceedings of the 6$^{th}$ Int. Conf. on Highly charged Ions* (Manhattan 1992) 179-182 (1993)
29. Harel, C., Jouin, H., and Pons, B., *J. Phys. B* **24** L425-L430 (1991)
30. Demkov, Yu.N., *Sov. Phys. - JETP* **18**, 138-142 (1964)
31. Mack, M., Nijland, J.H., Straten, P.v.d., Niehaus, A., and Morgenstern, R., *Phys. Rev. A* **39**, 3846-3854 (1989)
32. Bordenave-Montesquieu, A., Benoit-Cattin, P., Boudjema, M., and Gleizes, A., *Proceedings of the 15$^{th}$ Int. Conf. on the Physics of Electronic and Atomic Collisions* (Brighton 1987), ed. Gilbody, H.B., et al, North-Holland, 1988, p. 551
33. Landau, C.D., *Phys. Sov.* **2**, 46 (1932) and Zener, C., *Proc. Roy. Soc. A* **137**, 696 (1932)
34. Roncin, P., Gaboriaud, M.N., and Barat, M., Bordenave-Montesquieu, A., Moretto-Capelle, P., Benhenni, M., Bachau; H., and Harel, C., *J. Phys. B* **26**, 4181-4199 (1993)
35. Martin, S., Bernard, J., Denis, J., Désesquelles, J., Li Chen, and Ouerdane, Y., *Phys. Rev. A* **50**, 2322-2326 (1994)
36. Martin, S., Bernard, J., Denis, J., Désesquelles, J., and Li Chen, to be publshed in *Nucl. Instr. Methods B* (1995)
37. Vaeck, N., van der Hart, H.W., and Hansen, J.E., *Proceedings of the 18$^{th}$ Int. Conf. on the Physics of Electronic and Atomic Collisions, Book of Invited Papers*, (Aarhus 1993), ed. Andersen, T., et al, AIP vol. 295, 1993, pp. 794-802
38. Vaeck, N., van der Hart, H.W., and Hansen, J.E., *J. Phys. B* **27**, 3489-3514 (1994)
39. van der Hart, H.W., and Hansen, J.E., *J. Phys. B* **27**, L395-L400 (1994)
40. Moretto-Capelle, P., Oza, D.H., Benoit-Cattin, P., Bordenave-Montesquieu, A., Boudjema, M., Gleizes, A., Dousson, S., and Hitz, D., *J. Phys. B* **22**, 271-286 (1989)
41. Bachau, H., and Sánchez, I., to be published in *Nucl. Instr. Methods B* **98**, 78-80 (1995)
42. Merabet, H., Frémont, F., Chesnel, J.Y., Cremer, G., Husson, X., Lecler, D., Lepoutre, A., Rieger, G., and Stolterfoht, N., to be published in *Nucl. Instr. Methods B* (1995)
43. Draeger, M., Handke, G., Ihra, W., and Friedrich, H., *Phys. Rev. A* **50**, 3793-3808 (1994)
44. Tang, J.Z., Watanabe, S., Matsuzawa, M., and Lin, C.D., *Phys. Rev. Lett.* **69**, 1633-1635 (1992)
45. Boudjema, M., Cornille, M., Dubau, J., Moretto-Capelle, P., Bordenave-Montesquieu, A., Benoit-Cattin, P., and Gleizes, A., *J. Phys. B* **24**, 1713-1737 (1991)
46. Domke, M., Xue, C., Puschmann, A., Mandel, T., Hudson, E., Shirley, D.A., Kaindl, G., Greene, C.H., Sandeghpour, H.R., and Petersen, H., *Phys. Rev. Letters* **66**, 1306-1309 (1992)

# Gathering and Evaluation of State Selective Capture Cross Sections for Plasma Diagnostics

Ronnie Hoekstra

*KVI, Atomic Physics, Zernikelaan 25,*
*9747 AA Groningen, The Netherlands*
and
*JET, Joint Undertaking,*
*Abingdon, Oxon OX14 3EA, U.K.*

---

For charge exchange spectroscopy at large tokamaks accurate knowledge is needed of the underlying electron capture processes. Recent experimental advances in determining state selective cross sections for electron capture by multiply charged ions are discussed. Special attention is paid to the possibly crucial role of metastable atoms as electron donors and to the first photon spectra obtained at eV/amu collision energies.

---

## INTRODUCTION

Over the last 10 years Charge eXchange Spectroscopy (CXS) has evolved to a standard diagnostics of fusion plasmas [1–4]. At the larger tokamaks, such as the Joint European Torus (JET) in particular, CXS based on visible light emission subsequent to electron capture from the neutral heating beams is used to determine local parameters of the fusion plasma. Plasma quantities such as the ion temperature, plasma rotation velocity and impurity densities are extracted routinely from the CXS signals. For obtaining reliable information on these fusion-plasma quantities it is of great importance to know accurately the cross sections for the basic charge transfer processes. Electron capture by a multiply charged plasma ion, $A^{q+}$, from a neutral particle, B, is schematically given by:

$$A^{q+} + B \longrightarrow A^{(q-1)+}(nl) + B^+$$
$$\longrightarrow A^{(q-1)+}(n'l') + h\nu + B^+. \quad (1)$$

The electron is mainly captured into an excited state, $nl$, of the ion, $A^{(q-1)+}(nl)$. Subsequent to the electron capture processes the excited $A^{(q-1)+}(nl)$ particles decay via photon emission. The photon emission spectrum can be considered and exploited as the "fingerprint" of the primary $nl$-state selective electron capture processes between plasma ions and neutrals.

© 1995 American Institute of Physics

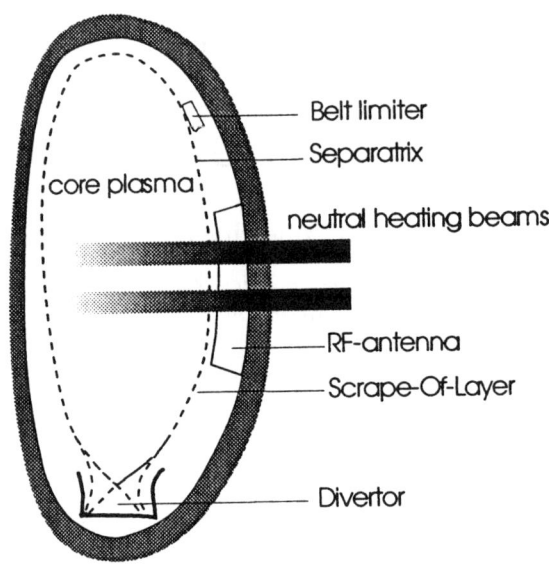

**FIG. 1.** Schematic intersection of the JET torus. The broken curves indicate magnetic field lines.

In the following, first the processes and collision systems of importance for CXS will be identified. Since neutrals are needed as electron donors, CXS on tokamaks such as JET is restricted to the regions where the neutral heating beams penetrate through the plasma and to the outer plasma regions near the walls, i.e., the so called scrape-of-layer and the divertor. Thereafter the status of the cross sectional databases for processes relevant for the neutral beam and the divertor regime will be discussed. It will be shown that recent experimental and theoretical work has filled in many gaps in the databases but that much more work is asked for.

## CROSS SECTIONS NEEDED FOR CXS

Figure 1 shows schematically an intersection of the JET torus. Indicated are the neutral sources, the neutral heating beams (power between 5 and 20 MW), the scrape-of-layer and the divertor. Since there are distinct differences between the collision processes that are of interest for neutral beam based CXS spectroscopy and the ones for CXS of the cold outside regions of the fusion plasma, they will be treated seperately.

### Neutral heating beams

Since 1991 not only deuterium beams but also He-beams are in operation on the JET machine. For JET with its carbon and beryllium coated walls, in addition to $He^{2+}$ ions the most important impurities are carbon and beryllium and to a lesser extent oxygen.

The energies of the neutral heating beams are typically in the energy range of 30 to 60 keVamu$^{-1}$. The energy range in which cross sections for charge transfer reactions between plasma impurities and neutral beam atoms have to be known is larger, approximately 10 - 80 keVamu$^{-1}$. This is due to the fact that the energy of the plasma ions can be considerable. In the plasma core ion temperatures up to 30 keV are observed. At these temperatures of several keV the impurity ions become fully stripped. One may be inclined to assume that in first approximation the plasma temperature is negligible with respect to the beam energy and so that knowing the relevant capture cross sections at the energy of the heating beam is sufficient. This may however lead to large errors in the determination of plasma parameters, such as the ion temperature ( [6] and references therein). So it is really important to have accurate knowledge of the fundamental cross sections over a wide energy range encompassing the neutral beam energy.

Because of the fiber optic transport of the photon signals to spectrometers located outside the JET vault, the neutral beam based CXS employs light emitted in the visible spectral range. This limits the number of accessible transitions. For example for $He^+$ ions produced by electron capture by $He^{2+}$ plasma ions, the HeII($n = 4 \longrightarrow n = 3$) transition at 468.6 nm has to be used. In general due to the resonant nature of electron capture processes of type (1), the cross sections for populating the high-$n$ levels radiating in the visible spectral range are only a small fraction, $\ll 1\%$, of the total electron capture cross sections. This severely complicates calculations, especially for He as an electron donor [5], and experiments to determine the relevant cross sections with absolute errors on the 20% level.

Since a fraction of the heating beams, in particular the He beams, may be in an excited state also cross sections have to be known for capture from metastable He. In one of the next sections it will be pointed out that a small fraction of metastables is already of importance due to the larger cross sections for populating the high-$n$ levels.

## Divertor and Scrape-Of-Layer

In the divertor and scrape-of-layer of the tokamak the ion temperatures are much lower from about 1 keVamu$^{-1}$ in the scrape-of-layer near the seperatrix (the last closed magnetic field line) down to energies as low as a few eVamu$^{-1}$. Therefore the emphasis is shifted from charge transfer collisions with fully stripped ions to only partially stripped ions. Once again the species to be considered are He, C, Be and O. Furthermore experiments with gas puffing into the divertor, to enhance radiative plasma cooling near the walls, may introduce other ions in the plasma. This means that databases have to be extended beyond the intrinsic impurities. At JET for example $N_2$ has recently been used.

The divertor region is not only studied by visible light but also by a vacuum-ultra-violet spectroscopy. This allows also for the detection of line emission originating from the states resonantly populated by electron capture. It will be indicated below that at these low collision energies the visible light emission from high-$n$ levels will be fully dominated by capture from metastables. So accurate cross sections are needed both for capture from ground state and metastable neutral hydrogen and helium.

Finally, although not yet operational, a Li-beam for diagnostics of the scrape-of-layer is installed near the top of the torus. Li-beam diagnostics is succesfully used at TEXTOR (Jülich, Germany) [7]. At JET the Li beam energy will be 8 to 9 keVamu$^{-1}$. At such keVamu$^{-1}$ energies electron capture from Li by multiply charged ions has been studied extensively [6,8]. The data available seem sufficient for the time being.

## CROSS SECTION DATA

The status of the database for state selective electron capture cross sections will be discussed for two different regimes of collision energies. The high energy regime, energies above several keVamu$^{-1}$, is relevant for the neutral beam based CXS, while the low energy regime deals with the processes taking place in the divertor and scrape-of-layer. It is noteworthy that both energy regimes are not directly accessible with highly charged ion beams extracted from ECR ion sources which typically deliver beams with energies in the range of 1 to 10 keVamu$^{-1}$. To access the high energy regime we have built at the KVI a small RF post accelerator, which is operational since 1992. With this compact accelerator we have beams available with energies up to some 75 keVamu$^{-1}$. First experiments were performed on the He$^{2+}$ - He system [9].

To enter the eVamu$^{-1}$ energy range we have designed an experimental set-up in which the ECR ion beams can be decelerated. First test measurements seem very promising. This will be discussed in more detail in the low energy subsection.

### High energy regime

Experiments [10–12] and theory [13,14] for collisions on hydrogen, cq. deuterium form a solid and reliable basis for CXS based on the injection of neutral deuterium beams [4]. In contrast to the database for hydrogen the one for collisions on helium shows many gaps. Actually only very recently the first data have been published for He$^{2+}$ ions colliding on He [9,15]. Figure 2 summarizes these results and depicts the recommended data for CXS use. At high energies the recommended curve follows the theoretical CDWA results [16] down to about 200 keVamu$^{-1}$, thereafter the CTMC results are followed to approximately 100 keVamu$^{-1}$. From thereon a smooth interpolation is made

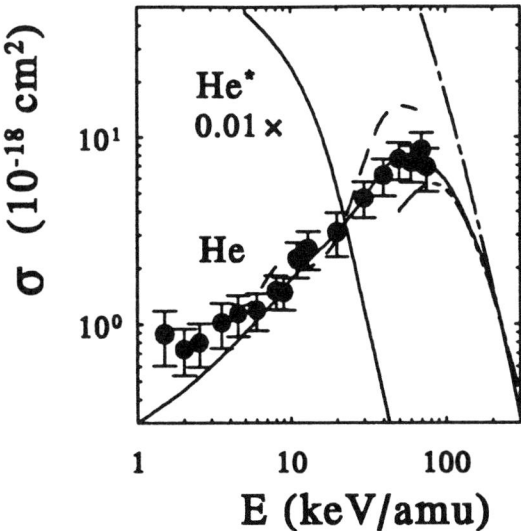

**FIG. 2.** Recommended (full curve) and summary of $He^+(n = 4)$ electron capture cross sections for $He^{2+}$ - He collisions. Experiment and CTMC (double chain) [9], AO (broken curve) [15] and CDWA (chain) [16]. The thin full line indicates the same cross section (multiplied by 0.01) for capture from metastable He.

to the experimental data, which are available below 75 keVamu$^{-1}$. At energies below 8 keVamu$^{-1}$ the recommended curve lies below the experimental data, because of a cascade contribution to the experimental data [9]. This is the only multiply charged ion - He system studied in sufficient detail by both theory and experiment to allow for its usage as a reliable tool for CXS diagnostics.

For carbon and beryllium ions data is urgently needed. For carbon there exists some data from atomic orbital (AO) calculations [5] and from our experiments [17]. Although the work is still in progress it is not to be expected that in the near future a database for $C^{6+}$ colliding on He will be available of a quality similar to the $He^{2+}$ - He one, especially not for capture into high-$n$ levels such as $n = 8$. This may not be that critical at the moment, because of a larger uncertainty arising from the unknown fraction of metastables present in the He beam. Due to the reduced binding energy of the electron, metastables populate more easily the high-$n$ levels radiating in the visible. This can be seen from figure 2 for $He^{2+}$ ions. For example at 40 keVamu$^{-1}$ the $He^+(n = 4)$ cross section for capture from metastable He exceeds the one for capture from ground state helium by a factor of 8. This implies that for example a metastable beam fraction of 2.5% will enhance the CXS signals by not less than 20% Therefore one of the main objectives of the CXS group at JET is to find a way to determine the fraction of metastables. The fraction is thought to be on the few percent level but still below 7% [18]. For $C^{6+}$ ions, preliminary results of CTMC calculations by Olson (University of Missouri, Rolla, USA) indicate that even at a relatively high energy of 50 keVamu$^{-1}$ the ratio for capture into C$^{5+}$($n = 8$) from metastables and ground state is almost 500! Note that in this case a metastable fraction of not more than 0.2% already doubles the CXS signals.

The intensity of the C$^{5+}$($n = 8 \longrightarrow n = 7$) line emission in C$^{6+}$ - He collisions is therefore practically completely determined by capture from the metastable fraction of the He beam, at least as soon as this fraction is $\gg 0.5\%$. The C$^{5+}$($n = 8 \longrightarrow n = 7$) line emission therefore seems a good candidate to be used as a tool to actually determine the fraction of metastables present in the helium beams. The so determined fraction may than be used to correct the results obtained from the HeII($n = 4 \longrightarrow n = 3$) line. However, before achieving a routinely and reliable use of this procedure the corresponding cross sections have to be known better. Theory for metastable He can not be tested directly because it is not feasible to produce a dense metastable He target. The quality of the theoretical predictions can be assessed by benchmarking these theories for systems with comparable electron binding energies (for example C$^{6+}$ on Li, Na and laser excited Na [6]) and by further experiments on C$^{6+}$ - He collisions.

**Low energy regime**

At low energies capture from metastables is even orders of magnitude more dominant over capture from ground state neutrals than at the neutral beam energies of tens keVamu$^{-1}$ [17]. Visible light emission from high-$n$ levels will therefore be indicative for metastable hydrogen and helium in the scrape-of-layer and the divertor. Accurate data are still lacking. For the dominant capture processes some theoretical results are available. The database is actually far from complete and often the different theoretical predictions are contradictive. To test some of the theoretical methods and to get a feeling for the reliability of the calculations we have constructed an experimental set-up in which collisions at eV/amu energies can be studied. The collisions take place inside an RF octopole trapping the ion beam in radial direction, cf figure 1. By floating the octopole on a potential V$_{trap}$, the energy at which collisions take place is given by:

$$\frac{E}{m} = \frac{q}{m}(V_s - V_{trap}) \qquad (2)$$

with V$_s$ the potential of the ion source and $q$ and $m$ the charge state and mass of the ion.

Radial trapping of the ions is based on the principle of radio frequency (RF) multipole beam guiding [19]. Supplying to the multipole an RF-voltage, V$_{RF}$, with a phase difference of 180° between neighboring poles and with sufficiently high frequency (MHz range) the effective potential experienced by the ions, confining their radial motion, is given by [19]:

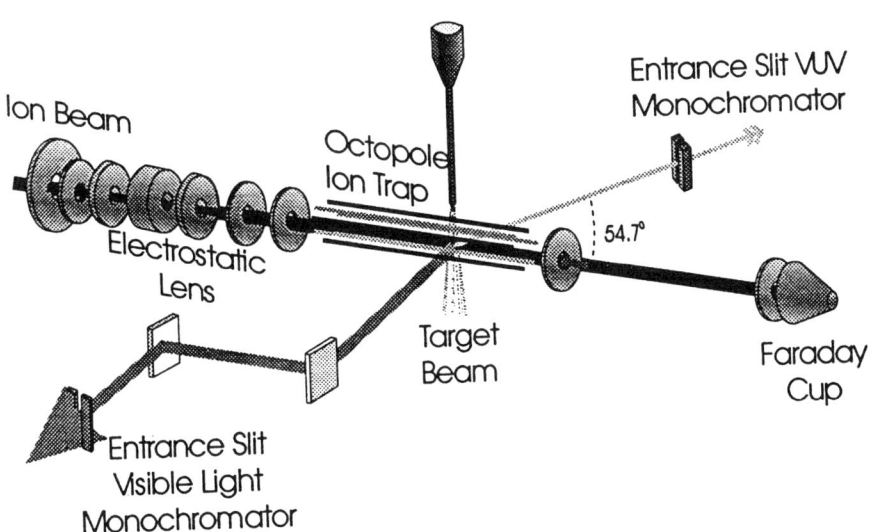

**FIG. 3.** Schematic representation of our set up for low energy experiments.

$$V_{eff} = \frac{(NqV_{RF})^2}{4m(\omega a)^2} \left(\frac{r}{a}\right)^{2(N-1)} \qquad (3)$$

with $2N$ the number of poles, $q$ and $m$ the charge state and mass of the ion, $\omega$ and $V_{RF}$ the frequency and amplitude of the RF-voltage and $a$ the inner radius of the multipole. Our trap consists of a set of 8 rods with a diameter of 1 mm and a length of 100 mm, which are inscribed on a circle of 5.5 mm. Just in front of the trap there is a set of electrostatic lens elements to focus the ion beam into the trap. The power supplies for the electrostatic lens elements and the octopole are connected in series with the voltage of the source. This is necessary to cancel changes in the ECR power supply. The configuration of our octopole has been chosen such that a set of two monochromators can view the collision region simultaneously. One of the monochromators is sensitive in the visible (300 - 650 nm) and the other one is a grazing incidence VUV-monochromator (10 - 80 nm) see e.g. ref. [11]. The geometry is optimized such that the monochromators have a maximum view of the collision region through the trap.

The capability of the guiding system was checked with a 3 kV $O^{6+}$ beam. The beam could be guided succesfully through the octopole at energies down to about 20 V without beam loss which shows that the octopole provides a potential well that is sufficiently steep and stable to trap the ions in radial direction. The first photon emission spectra at very low energies have been obtained from $C^{4+}$ colliding on $H_2$. For this collision system capture into $C^{4+}$ predominantly takes place into the n=3 level of $C^{3+}$. Our VUV- monochromator can view all three $3l \longrightarrow 2l$ emission lines simultaneuosly (see figure 4). It can clearly be seen from these spectra that in this range of collision

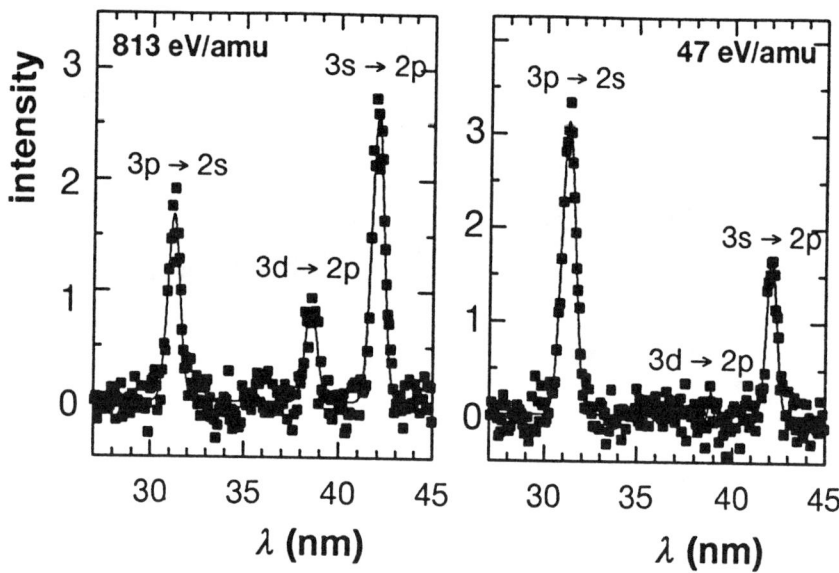

**FIG. 4.** VUV photon emission spectra resulting from electron from $H_2$ by $C^{4+}$ at 813 and 47 eVamu$^{-1}$.

energies the population of the different $3l$ states changes dramatically.

With this set up we hope to gather in the near future the data relevant for CXS of the colder regions of a tokamak plasma.

## CONCLUSION

Charge exchange spectroscopy at large tokamaks requires an accurate knowledge of the underlying electron capture processes. Recent theoretical and experimental work on state selective cross sections for electron capture by multiply charged ions has produced a reliable data base for neutral hydrogen beam based CXS. For neutral helium beams CXS is on less solid ground, firstly due to a lack of data for bare C, Be and O ions and secondly because of the unknown fraction of metastable helium present in the beam. Especially for impurity ions other than $He^{2+}$ capture from metastable He* may blend the visible light spectra due to the huge cross sections for capture from metastable helium. If the cross section for capture from metastables exceeds the one for capture from ground state helium by more than 2 orders of magnitude the signals may be used benificially, namely to determine the fraction of metastables.

A set up has been built to perform collision experiments at eVamu$^{-1}$ energies, which will allow us to study the processes relevant for the scrape-of-layer and divertor of large tokamaks such as JET.

## ACKNOWLEDGMENT

The author gratefully acknowledges the contributions of H.P. Summers and M. von Hellermann (JET, Abingdon, UK), F.W. Bliek, H.O. Folkerts, R. Morgenstern, J.P.M. Beijers and S. Schippers (KVI, Groningen, The Netherlands), R.E. Olson (University of Missouri, Rolla, USA) and W. Fritsch (Hahn-Meitner Institute, Berlin, Germany) to the work described. The KVI based research is part of the research program of the Stichting voor Fundamenteel Onderzoek der Materie (FOM), which is financially supported by the Nederlandse Organisatie voor Wetenschappelijk Onderzoek (NWO). It also receives support from EURATOM via an article 14 contract between JET Joint Undertaking and KVI.

## REFERENCES

1. R. C. Isler, Plasma Phys. Contr. Fusion **36**, 171 (1994).
2. E. Wolfrum, F. Aumayr, D. Wutte, HP. Winter, E. Hintz, D. Rusbüldt and R. P. Schorn, Rev. Sci. Instrum. **64**, 2285 (1993).
3. A. Boileau, M. von Hellermann, L. D. Horton and H. P. Summers, Plasma Phys. Contr. Fusion **31**, 779 (1989).
4. M. von Hellermann, W. Mandl, H. P. Summers, A. Boileau, R. Hoekstra, F. J. de Heer and G. J. Frieling, Plasma Phys. Contr. Fusion **33**, 1805 (1991).
5. W. Fritsch, Nucl. Instrum. Meth. B. **98**, 246 (1995).
6. R. Hoekstra, Comm. At. Mol. Phys. **30**, 361 (1995).
7. R. P. Schorn, E. Wolfrum, F. Aumayr, E. Hintz, D. Rusbüldt and HP. Winter, Nucl. Fusion **32**, 352 (1992).
8. J. Schweinzer, D. Wutte and HP. Winter, J. Phys. B: At. Mol. Opt. Phys. **27**, 137 (1994)
9. H. O. Folkerts, F. W. Bliek, L. Meng, R. E. Olson, R. Morgenstern, M. von Hellermann, H. P. Summers and R. Hoekstra, J. Phys. B: At. Mol. Opt. Phys. **27**, 3475 (1994)
10. R. Hoekstra, D. Ćirić, F. J. de Heer and R. Morgenstern, Phys. Scr. T. **28**, 81 (1989)
11. R. Hoekstra, F. J. de Heer and R. Morgenstern, J. Phys. B: At. Mol. Phys. **24**, 4025 (1991).
12. G.J. Frieling, R. Hoekstra, E. Smulders, W. J. Dickson, A. N. Zinoviev, S. J. Kuppens and F. J. de Heer, J. Phys. B: At. Mol. Phys. **25**, 1245 (1992).
13. W. Fritsch, J. Physique. Coll. **50**, 87 (1989).
14. R. E. Olson and D. R. Schultz, Phys. Scr. T. **28**, 71 (1989).
15. W. Fritsch, J. Phys. B: At. Mol. Opt. Phys. **27**, 3461 (1994).
16. R. Gayet, D. Belkić and A. Salin, private communication.
17. F. W. Bliek, H.O. Folkerts, R. Morgenstern, R. Hoekstra, L. Meng, R. E. Olson, W. Fritsch, M. von Hellermann and H. P. Summers, Nucl. Instrum. Meth. B. **98**, 195 (1995).
18. H. P. Summers, M. von Hellermann, P. Breger, J. Frieling, L. D. Horton, R. König, W. Mandl, H. Morsi, R. Wolf, F. J. de Heer and R. Hoekstra, *Atomic Processes in Plasmas* AIP Conf. Proceedings **257**, 111 (1992).

19. D. Gerlich, Adv. Chem. Phys. **LXXXII**, 1 (1992)

# Double electron capture: complex amplitudes from Auger anisotropy measurements

M. H. Prior
and
Hocine Khemliche

University of California, Lawrence Berkeley National Laboratory

Chemical Sciences Division

Berkeley, California

U.S.A. 94720

We report the scattering angle dependence of the complex population amplitudes of magnetic substates formed in doubly excited $C^{3+}$ ions in double electron transfer from He onto $C^{5+}$ projectiles. These results are derived from analysis of the anisotropic distribution of Auger electron/scattered ion coincidence events. Relative velocities in the range 0.2-0.4 au. have been studied. The results are the most complete experimental description of the double electron capture process, and provide insight into the impact parameter dependence of the orientation of the double excited electron wave function and the angular momentum transferred from the nuclear to the electron motion.

## 1. Introduction

A slow collision of a multiply charged ion projectile with a neutral target nearly always results in electron capture into an excited state on the projectile product ion. Characteristics of the photon or electron emission (energy, angular distribution, polarization) from the decay of these states are essential indicators of their nature, and these combined with measurement of the momentum transfer in the collision provide deep insight into the dynamics of the capture process. For the case of single electron capture, the excited state decays by photon emission and recent studies (see e.g. [1,2]) have measured the distribution of scattered projectiles in coincidence with the polarization or angular distribution of emitted photons from which details of the orientation and angular momentum transferred to the excited state can be inferred. Important closely related studies of single electron capture from laser excited states have also been carried out ([3] is a recent example). In the case of multiple electron capture, the excited final projectile states generally lie in the ionization continuum, and hence usually decay by

© 1995 American Institute of Physics

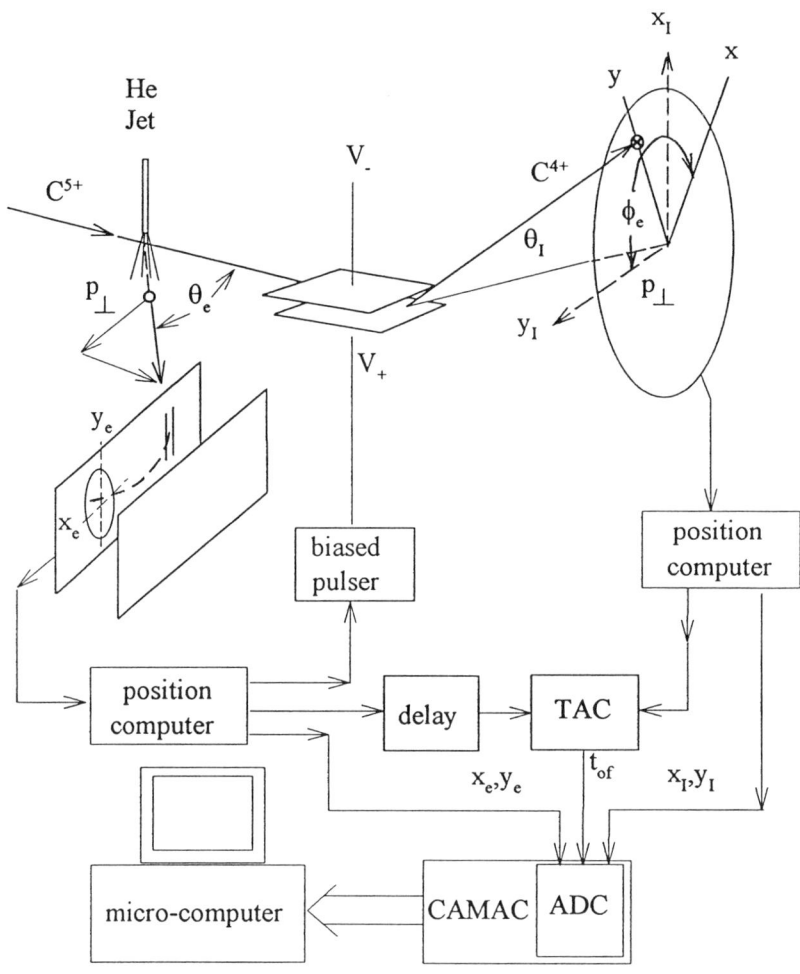

**Figure 1.** Scheme of the experiment. The electron spectrometer views the intersection of the $C^{5+}$ beam and the He jet at angles $\theta_e$ in the range $\approx 35°$ to $105°$. $C^{4+}$ products are selected and steered to the projectile detector by a parallel plate deflection field. The azimuthal angle of emission, $\varphi_e$, of the Auger electron, defined with respect to the projectile scattering frame, follows from a simple transformation of each event from fixed laboratory $x_l, y_l$ -axes. Five parameters are recorded for each event, the electron $x_e$, $y_e$, the time-of-flight difference between electron and ion, tof, and the impact position of the projectile $x_l$, $y_l$. The deflection voltage is pulsed to steer a $C^{4+}$ ion to the detector only when an electron is received. This ion sorting protects the projectile detector from the large flux of $C^{4+}$ ions produced by single capture.

emission of one or more electrons by the Auger process. In this work we report new studies of Auger emission anisotropy in coincidence with the projectile scattering in a double electron capture process. The results provide descriptions of the product ion state at nearly the same level of detail derivable from photon/ion coincidence studies of single electron capture, while accessing more complex doubly excited states.

We have studied the process:

$$C^{5+}(1s) + He(1s^2) \rightarrow C^{3+}(1s2l2l') + He^{2+},$$

at collision velocities in the range v=0.2-0.4 au., by coincident measurement of the Auger electron emitted in fast ($\approx 10^{-14}$ sec) decay:

$$C^{3+}(1s2l2l') \rightarrow C^{4+}(1s^2) + e^-,$$

and the scattering of the $C^{4+}$ ($1s^2$) final projectile state. The total spin of the captured electrons is preserved and hence only levels where the 2l2l' excited configuration has spin zero are formed; there are two such states with L=0, ($1s2s^2$ and $1s2p^2$ $^2$S), one with L=1 ($1s2s2p$ $^2$P), and one with L=2 ($1s2p^2$ $^2$D). These levels are split by a few to several eV, and hence are resolvable with a relatively simple electron spectrometer (this splitting is caused, in large part, by the variation in the nuclear screening by the 1s electron for the different 1s2l2l' levels; this is why a H-like rather than a bare projectile was chosen). Since the initial and final states of the projectile and target are S states, all information regarding the angular character of the $C^{3+}$ (1s2l2l') level populated in the double capture is carried in the anisotropy of the Auger emission.

## 2. Experimental Method

Our approach has been described recently [4]; here we summarize the method and report new results at v=0.22 au (15keV). To determine the relative phases of the $a_M$, the amplitudes of the final |L,M> substates, the Auger anisotropy must be measured with respect to the collision plane; thus we observe the 1s2l2l' Auger spectrum [5] emitted into a small solid angle at polar angle $\theta_e$ with respect to the $C^{5+}$ beam direction (z-axis) in coincidence with position sensitive detection of the scattered $C^{4+}$ final state ions. Figure 1 shows a schematic of the experimental arrangement. The $C^{5+}$ beam was produced by the LBNL Electron Cyclotron Resonance ion source; it was collimated by two 1.5x1.5 mm apertures separated by 1.2m before entering a chamber containing a He jet target and electron spectrometer. The Auger electrons from the 1s2l2l' levels have

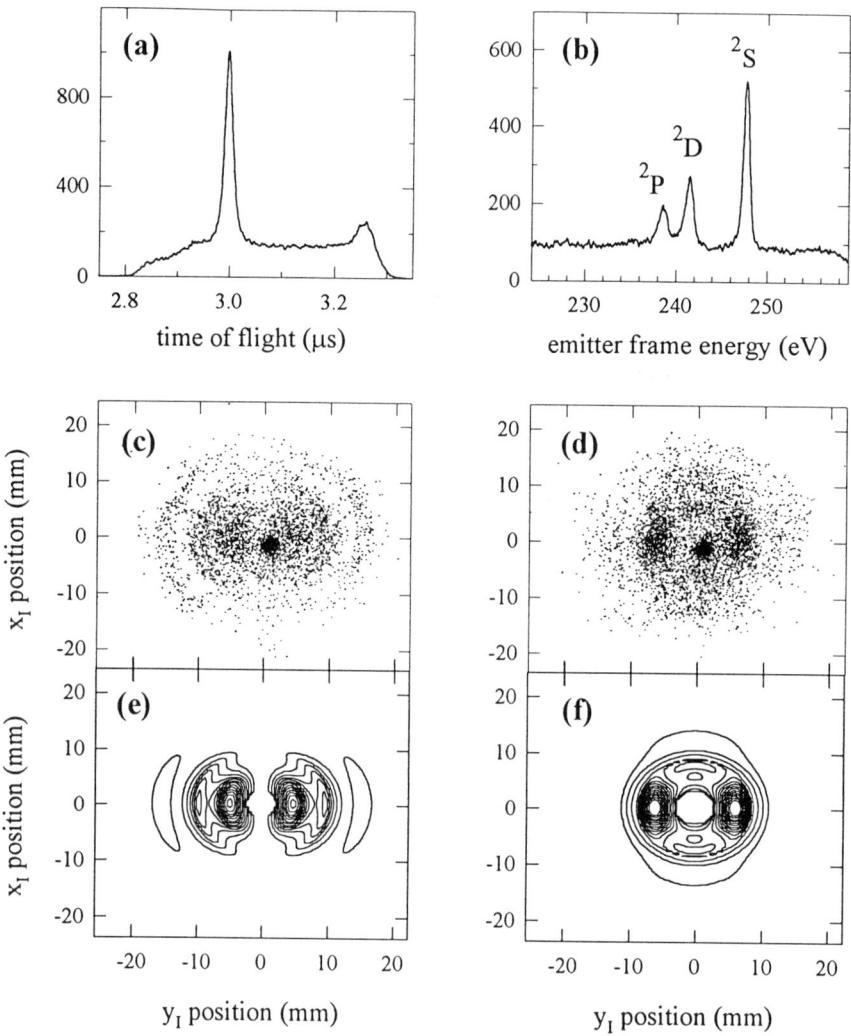

**Figure 2.** Example of data obtained at electron polar angle $\theta_e=90°$. (a) the TAC spectrum; the peak is located near the beginning of the deflection voltage pulse that steers the $C^{4+}$ product ions onto the projectile detector. (b) the Auger electron spectrum for events within the deflection pulse. (c) the projectile detector pattern for events gated on the TAC peak and the 1s2p $^2$P Auger line. (d) as in (c) but for events in the 2p$^2$ $^2$D Auger line. The small intense spots near the center of the patterns in (c) and (d) are random events included under the TAC peak, primarily from $C^{4+}$ products of single capture; these spots are offset from the center of the double capture patterns by recoil from the Auger emission and differences in the Q values for double vs. single capture. (d) and (f) are contour plots representing the patterns in (c) and (e) respectively (after removal of the random events). These were generated from the results of the analysis of all data (see text).

energies near 240 eV (see [5] and references therein), and are decelerated to ≈70 eV before entering the spectrometer; this provides a resolution ≈0.7 eV and spreads the lines across the electron detector. Typical beam currents used were ≈ 25 pA yielding net coincidence rates in the range 15-30 per minute. An ion sorting scheme deflects the $C^{4+}$ product ion onto the projectile detector only when an electron signal is received from the spectrometer; this unburdens the projectile detector from the otherwise overwhelming flux of $C^{4+}$ ions produced by single capture.

The variation of the Auger emission with electron azimuthal angle, $\varphi_e$, (for fixed selected polar angles $\theta_e$) and the projectile differential scattering are contained in the intensity patterns on the projectile detector for events coincident with each Auger line. Figure 2 shows an example of these patterns for $C^{5+}$ impact at 15 keV (v=0.22 au) for events coincident with the Auger lines from the $2s2p\ ^2P$ and $2p^2\ ^2D$ levels observed at $\theta_e = 90°$ (emitter frame). Extraction of the complex population amplitudes and their scattering angle dependences follows from analysis of these patterns taken at selected electron polar angles.

## 3. Analysis and Results

As indicated in Fig. 1, we use a right handed coordinate system with z-axis along the beam direction and y-axis in the collision plane along the direction of the scattered ion's transverse momentum. Conservation of positive reflection symmetery with respect to the scattering plane requires that $a_M = a_{-M}$. For decay of a P state to an S state (the case here), the coincidence rate, $S_P$, for ions scattered into a polar angle range $\Delta\theta_I$ at angle $\theta_I$ and emission of electrons into small solid angle $\Delta\Omega_e$ at angles $\theta_e$, $\varphi_e$ is:

$$S_P(\theta_I, \theta_e, \varphi_e) = K \cdot \frac{d\sigma_P}{d\theta_I} \cdot \frac{3}{4\pi} \cdot \left[ |a_1|^2 \sin^2\theta_e (1 - \cos 2\varphi_e) + |a_0|^2 \cos^2\theta_e \right.$$
$$\left. - |a_1| \cdot |a_0|\ 2^{1/2} \sin 2\theta_e \sin\varphi_e \sin\Delta\beta_{01} \right].$$

Where the amplitudes for the M=1,0 substates have magnitudes $|a_{1,0}|$ and relative phase $\Delta\beta_{01} = \beta_0 - \beta_1$, and $d\sigma_P/d\theta_I$ is the differential cross section for creation of the P state; all of these being functions of $\theta_I$. K is a factor including $\Delta\Omega_e$, $\Delta\theta_I$, the beam flux, target thickness and detector efficiencies. With inclusion of the normalization, $\Sigma|a_M|^2 = 1$, the P state anisotropy at $\theta_I$ depends upon two parameters, i.e., one of the $|a_M|$, and $\Delta\beta_{01}$. The more complex expression for $S_D$, the coincidence rate from the $1s2p^2\ ^2D$ state, depends upon four parameters.

At selected angles $\theta_e$, the anisotropy depends upon a subset of the $|a_M|$ and $\Delta\beta_{MM'}$; for example, at $\theta_e = 90°$, one has:

$$S_P = K \cdot \frac{d\sigma_P}{d\theta_I} \cdot \frac{3}{4\pi} \cdot |a_1|^2 (1-\cos 2\phi_e)$$

and,

$$S_D = K \cdot \frac{d\sigma_D}{d\theta_I} \cdot \frac{5}{16\pi} \cdot \left[ 6|a_2|^2 + |a_0|^2 - 2 \cdot 6^{1/2} |a_2||a_0| \cos 2\phi_e \cos\Delta\beta_{02} \right].$$

These are the forms used in analyzing the data and constructing the contour plots shown in Fig 2.

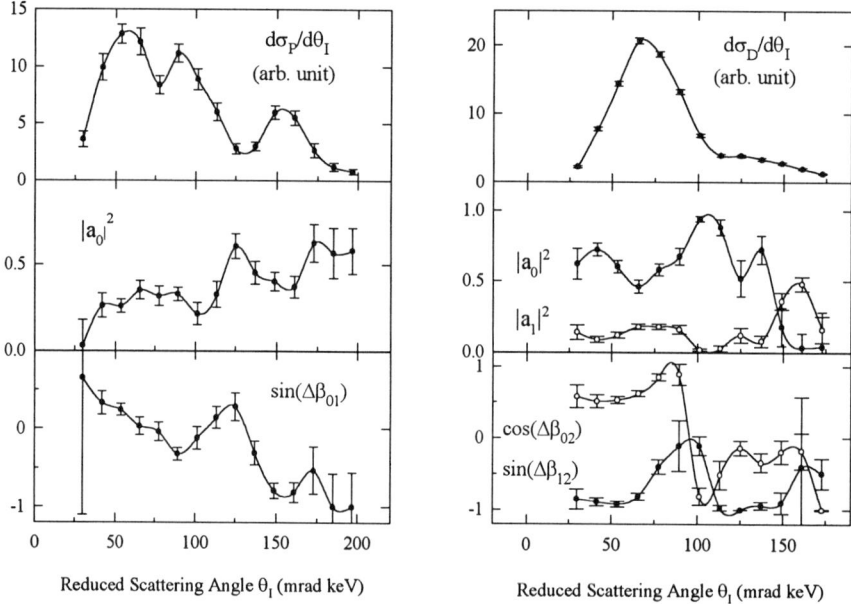

**Figure 3.** Differential cross sections, amplitudes and phase results for capture to 1s2s2p $^2$P (left) and 1s2p$^2$ $^2$D (right) at v=0.22 au (15keV). The lines are guides only.

We have collected data for $S_P$ and $S_D$ at electron polar angles $\theta_e=45°$, 54.7° and 90°, from which we extract, by fitting the $\phi_e$ dependences for separate regions of $\theta_I$, the $|a_M|$ and (for the P state) $\sin\Delta\beta_{01}$, and (for the D state) $\sin\Delta\beta_{12}$, and $\cos\Delta\beta_{02}$. We also determine relative angular scattering cross-sections (that is the $d\sigma_{P,D}/d\theta_I$ in arbitrary units). Figure 3 summarizes these results.

## 4. Discussion

An overall picture of the energetics and locations of relevant quasi-molecular crossing radii is given by the approximate diabatic energy level curves shown in Fig. 4. To avoid confusion only one each of the $C^{3+}$ 1s2l2l' and $C^{4+}$ 2l double and single capture levels are shown. Note that in the region between $\approx 2$ and $\approx 20$ au internuclear separation, the $C^{3+}$ 2l2l' and the $C^{4+}$ 2l levels have multiple crossings and that all of the 2l2l' and 2l levels intersect the initial $C^{5+}$+He level in the region 1.8-2.8 au.

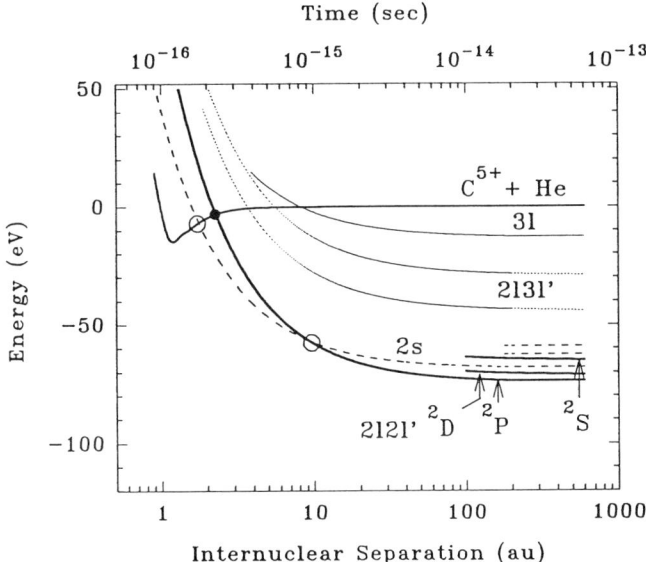

**Figure 4.** Approximate adiabatic energy levels relevant to double and single electron capture in $C^{5+}$+ He collisions. Single step double capture to the 1s2l2l' levels of $C^{3+}$ occurs at crossing radii near 2.1 au, as e.g. shown by the solid circle. A two step mechanism involving an intemediate $C^{4+}$ 2l level is indicated by the circled crossings. The time scale for the collision (v=.22 au) is at the top of the figure. Arrows are placed at the location of the mean lifetime for the Auger decay of the 1s2l2l' states studied in this work.

The oscillations in $d\sigma_p/d\theta_I$ vs $\theta_I$, seen in the data of fig 2 (c) and in fig 3, could result from interference between direct double capture, marked with the solid circle in Fig. 4, and the two step process with a capture into a $C^{4+}$ n=2 level followed by a second capture into the final $C^{3+}$1s(2s2p) $^2$P level at the circled crossings in the figure. A simple straight line trajectory calculation, utilizing

Coulomb potentials to calculate both phase integrals and deflection functions for this interference process, yields oscillations of approximately the same period (≈70 keV mrad) and near the same angles as seen in our results. Another mechanism possibly contributing to peaks in the angular scattering is analogous to that observed in double capture from He by slow $C^{4+}$ [6], i.e., interference between the direct double capture to the final state on the "way in" and elastic scattering from the repulsive wall of the initial state potential followed by capture on the "way out".

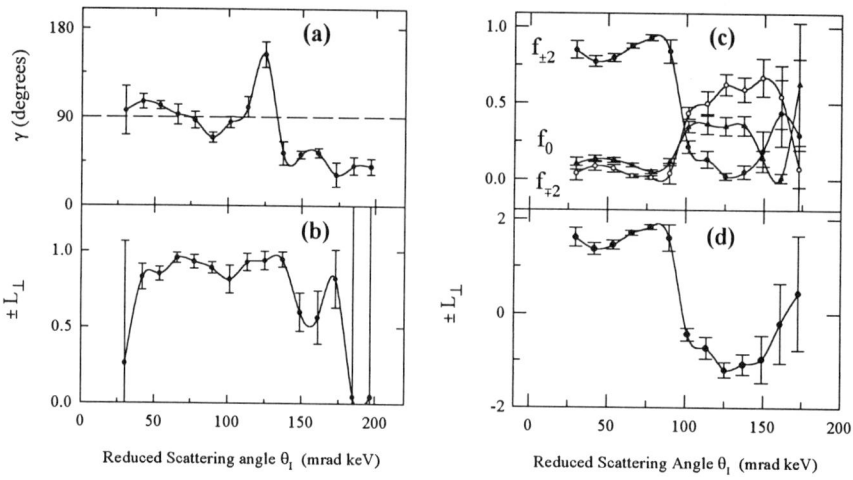

**Figure 5.** Scattering angle dependence, at v=0.22 au, of (a) the alignment angle, $\gamma$, and (b) $\pm L_\perp$ for the $1s2s2p\ ^2P$ final state and (c) the population fractions for the |LM> substates in the natural frame and (d) $\pm L_\perp$ for the $1s2p^2\ ^2D$ final state. Lines are guides only.

The results in Fig. 3 form a near complete description of the outcome of a double electron capture collision in the $C^{5+}$, He system. Indeed, from the measured parameters one can construct pictures of the angular part of the doubly excited wavefunction for the P state for any scattering angle within the measured range. These are summarized, in part, by the alignment angle, $\gamma$, in the scattering plane, between the major axis of the P state charge cloud and the z axis. Another quantity of interest is the component of angular momentum, $L_\perp$, perpendicular to the scattering plane transferred from the nuclear motion to the excited final state. Recently theoretical [7-10] and experimental studies of single electron capture [2,11] have shown the usefullness of a propensity rule which favors population of states with M = -L in the natural frame ($z_n$ axis normal to the scattering plane, $x_n$ axis along the initial relative velocity, reached by a 90° rotation of our laboratory system about the $y = y_n$ axis). Our measurements are not sensitive to the orientation of the normal to the scattering plane and hence do not determine the sign of $L_\perp$. Formally this is because the trigonometric functions, e.g. $\sin\Delta\beta_{01}$, determined by our measurements leave ambiguity in the relative phases, e.g. $\Delta\beta_{01}=\sin^{-1}(\sin\Delta\beta_{01})$, or $\pi + \sin^{-1}(\sin\Delta\beta_{01})$.

Figures 5(a) and 5(b) show $\gamma$, and $\pm L_\perp$ for the P state. Although there is a point at large $\gamma$ ($\approx 160°$) the general trend is a slow fall from near $90°$ to $\approx 40°$ from the smallest to the largest scattering angles. The rotational coupling interaction accounts for the inability of the captured electrons to follow the rotation of the internuclear axis during the collision. Radial couplings alone produce M=0 final states, that is $\gamma=180°$ (since the internuclear axis rotates one half turn during the collision). $\pm L_\perp$ is near unity over most of the range where $d\sigma_P/d\theta_I$ is large, however, there is a significant dip in the region of the outer most peak in $d\sigma_P/d\theta_I$. Overall this behavior of the P state is similar to that observed at 25keV (v=.29 au) [4]. These results are consistent with the single capture propensity rule, as might be expected for capture to a nominal 2s2p configuration, where only the 2p electron contributes to $L_\perp$.

From the D state results we have calculated $\pm L_\perp$ and the population fractions $f_M=|a_M|^2$ in the natural frame (reflection symmetry requires $f_{\pm 1}=0$). These are shown in Fig. 5(c) and (d). In contrast to our observations at 25keV [4], here we observe a sign change in $L_\perp$ which occurs near the scattering angle ($\approx 100$mrad keV) marking the onset of the weak outer peak in $d\sigma_D/d\theta_I$. In the region of the maximum in $d\sigma_D/d\theta_I$, $\pm L_\perp$ is near 2; this suggest applicability of the propensity rule to the $2p^2$ electron configuration. Since the substate $|D,M = -2\rangle$ is well described as the product of 2p orbitals each with m=-1, its strong population would follow from application of the single capture rule to each electron separately. The rapid sign change in $L_\perp$ near 100 mrad keV may reflect the onset of a different capture mechanism from that present in the region of the maximum in $d\sigma_D/d\theta_I$.

A detailed experimental description of the final state amplitudes and their scattering angle dependence can be an important element in achieving understanding of the dynamics of transient two electron motion. This will grow from further experimental measurement, but cannot be complete without comparison to, and guidance from, theoretical studies which report the same level of detail. Calculations of substate populations integrated over all impact parameters have been made for $C^{5+}$ + He [12] and closely related $B^{4+}$ + He [12,13] systems, and there have been related measurements [5] preceeding this work. Since those calculations include computaton of impact parameter dependent complex amplitudes, perhaps we can look forward to further comparisons with, e.g. relative phase information derived from them or similar new theoretical work. At the same time we encourage the reporting of complex amplitudes and their impact parameter dependence, determined in the course of calculations of total or state selective cross sections, even if the related experiments have not yet been carried out.

## Acknowledgments

We gratefully thank C. Lyneis and Z. Xie of the LBNL Nuclear Science Division for valuable assistance and D. Schneider of the Lawrence Livermore National Laboratory for the loan of important portions of the experimental setup. This work received support from the Director, Office of Energy Research, Office of Basic Energy Sciences, U.S. Department of Energy under Contract No. DE-AC03-76SF00098.

## References

[1] R. Hoekstra, M. G. Sauraud, F. J. de Heer, and R. Morgenstern, *J. Phys. C* **50**, 387 (1989).
[2] P. Roncin, C. Adjouri, N. Andersen, M. Barat, A. Dubois, M.N. Gaboriaud, J. P. Hansen, S.E. Nielsen and S.Z. Szilagy, *J. Phys. B* **27**, 3079 (1994).
[3] J.W. Thomsen, N. Andersen, D. Dowek, J.C. Houver, M.O. Larsson, J.H.V. Lauritsen, U Müller, J.O.P. Pedersen, J. Salgado and A. Svensson, *J. Phys B* **28**, L93 (1995).
[4] H. Khemliche, M.H. Prior and D. Schneider, *Phys. Rev. Lett.* **74**, 5013 (1995).
[5] M.H. Prior, R.A. Holt, D. Schneider, K.L. Randall and R. Hutton, Phys. Rev. A **48**, 1964 (1993).
[6] A. Bárány, H. Danared, H. Cederquist, P. Hvelplund, H. Knudsen, J. O. K. Pedersen, C.L. Cocke, L.N. Tunnell, W. Waggoner and J. P. Giese, *J.Phys B* **19**, L427 (1986).
[7] J. P. Hansen, L. Kochbach, A. Dubois, and S.E. Nielsen, *Phys. Rev. Lett.* **64**, 2491 (1990).
[8] M.F.V. Lundsgaard and C.D. Lin, *J. Phys B* **25**, L429 (1992).
[9] N. Toshima and C.D. Lin, Phys. Rev. A **47**, 4831 (1993).
[10] M. Gargaud, M.C. Bachus-Montabonel, R. McCarroll and T. Grozdanov, *J. Phys B* **27**, 4675, (1994).
[11] P. Roncin, C. Adjouri, M.N. Gaboriaud, L. Guillemot, M. Barat and N. Andersen, *Phys. Rev. Lett.* **65**, 3261 (1990).
[12] J.P. Hansen and K. Taulbjerg, *Phys. Rev. A* **47**, 2987 (1993).
[13] W. Fritsch and C.D. Lin, *Phys. Rev. A* **45**, 6411 (1992).

# ION-MOLECULE COLLISIONS

Coupled Wavepackets Study of Ion-Molecule Collisions ........... 569
    F. AGUILLON
Direct Ionization in Diatomic and Triatomic
Quasimolecules ....................................................... 579
    Y.S. GORDEEV and G.N. Ogurtsov
Wavefunction Overlap Effects in
Collisional Excitation of Molecules ................................. 587
    D. MATHUR and M. Krishnamurthy

# Coupled Wavepackets Study of Ion-Molecule Collisions

## F. Aguillon

*Laboratoire des Collisions Atomiques et Moléculaires, URA du CNRS n°281, Bât. 351, Université Paris XI, 91405 Orsay cedex, France.*

**Abstract.** A semiclassical coupled wavepacket method is used to study non adiabatic dynamics of ion molecule collisions. In this method, some of the nuclear degrees of freedom are treated classically. Simultaneously, the quantal degrees of freedom are handled numerically, resulting in a coupled wavepacket description. Applications to the study of dissociative charge transfer (e.g. $He^++H_2 \rightarrow He+H+H^+$) and non adiabatic reaction ($Ar^++H_2 \rightarrow ArH^++H$) are presented.

## INTRODUCTION

Compared to collisions between neutral species, ion-molecule collisions present very often three peculiar features. First, the interaction range is larger. Second, non adiabatic effects are more likely to play an important role in such collisions. Third, the collision energy range of interest often reaches higher energies. The conjunction of these features makes the theoretical treatment of such processes delicate. On one hand, the high collision energy prevents from using an exact quantum mechanical treatment. On the other hand, classical mechanics is unable to account for non adiabatic effects. Then approximate methods have been developed : the trajectory surface hopping method[1] handles classically the nuclear motion and mimics the non adiabatic behaviour of the actual system in the vicinity of the transition zone such as potential energy surface (PES) crossings. The infinite order sudden (IOS) approximation[2] treats quantally the nuclear motion, but reduces the dimensionality of the problem by keeping fixed the Jacobi angle. Finally the semi-classical method[3] used here shares the nuclear degrees of freedom in two groups : one is treated classically, thus reducing the quantal dimensionality of the problem ; the quantal treatment of the other nuclear degrees of freedom makes the description of non adiabatic transitions exact. In the first implementations of this semi-classical method, the wavefunction was expanded over a (ro)vibronic basis set. Such an expansion becomes very computer time consuming when the number of vibronic states becomes large ; it even becomes untractable when one of the electronic states involved in the process is a dissociative state and/or when a molecule dissociates. To circumvent these drawbacks, a numerical description of the wavefunction has been introduced.[4] This numerical description has made it possible to investigate charge exchange processes involving dissociation[4-6] or not.[7,8] It has also been applied to the study of non adiabatic reactive

© 1995 American Institute of Physics

collisions.[9,10] In the following will be presented the general equations, some results obtained in the study of DCE processes and in the study of reactive processes.

## GENERAL EQUATIONS

The total wavefunction $\Xi$ of an ion+molecule system depends on the nuclear coordinates denoted collectively by $\vec{\mathcal{R}}$ and on the electronic coordinates denoted collectively by $\vec{r}$. It is expanded over an electronic basis set $\zeta_j$ according to

$$\Xi(\vec{\mathcal{R}},\vec{r}) = \sum_j \Phi_j(\vec{\mathcal{R}}) \zeta_j(\vec{r};\vec{\mathcal{R}}) \tag{1}$$

If the electronic basis set is diabatic,[11] i.e. the $\zeta_j$ functions vary smoothly with the nuclear coordinates, the time dependent Schrödinger equation may be written :

$$i\frac{\partial \Phi_j(\vec{\mathcal{R}})}{\partial t} = T_{\vec{\mathcal{R}}} \Phi_j(\vec{\mathcal{R}}) + \sum_k V_{jk}(\vec{\mathcal{R}}) \Phi_k(\vec{\mathcal{R}}) \tag{2}$$

where $T_{\vec{\mathcal{R}}}$ is the kinetic energy operator of the nuclei and $V_{jk}$ is the matrix element of the electronic hamiltonian $H_{el}$ (which includes the nuclear repulsion) :

$$V_{jk}(\vec{\mathcal{R}}) = \int \zeta_j^*(\vec{r};\vec{\mathcal{R}}) H_{el} \zeta_k(\vec{r};\vec{\mathcal{R}}) d\vec{r} \tag{3}$$

Even in the simplest case of a triatomic system, the set of coupled equations (2) is generally not tractable. To overcome this difficulty, one can use the semiclassical approximation,[3] which consists in handling classically one part of the nuclear degrees of freedom. As an example, a triatomic system described in Jacobi coordinates by the two Jacobi vector $\vec{R}$ and $\vec{r}$ can be treated in the semiclassical approximation by using a classical description of $\vec{R}$ and a quantal description of $\vec{r}$. The quantal part of the system obeys the equation :

$$i\frac{\partial \psi_j(\vec{r})}{\partial t} = T_{\vec{r}} \psi_j(\vec{r}) + \sum_k V_{jk}(\vec{r},\vec{R}(t)) \psi_k(\vec{r}) \tag{4}$$

while the time evolution of $\vec{R}$ is governed by the classical hamiltonian :

$$H_{cl} = T_{\vec{R}} + \sum_{jk} \int \psi_j^*(\vec{r}) V_{jk}(\vec{r},\vec{R}) \psi_k(\vec{r}) d\vec{r} \tag{5}$$

The set of equations (4) is very close to the exact quantal equations (2) except that i) its dimensionality is reduced and ii) the potential matrix elements depend on time through the classical trajectory $\vec{R}(t)$. In the coupled wavepacket method, it is solved numerically by representing the wavefunctions $\psi_j$ over a spatial grid. The evaluation of the hamiltonian operator and the time integration are performed numerically.

## DISSOCIATIVE CHARGE EXCHANGE

The main advantage of the wavepacket description is that it can handle very efficiently the continuum states. It has been applied to the study of dissociative charge exchange (DCE) such as[6]

$$He^+ + H_2 \rightarrow He + H + H^+ \tag{6}$$

At collision energy above 1eV, the collision time is much shorter than the rotational period of $H_2$. In this situation, one can neglect the kinetic energy operator $T_{\hat{r}}$ associated with the rotation of $H_2$.[12] Within this frozen rotor approximation, equation (4) reduces to its radial component :

$$i\frac{\partial \psi_j(r)}{\partial t} = -\frac{1}{2m}\frac{\partial^2 \psi_j(r)}{\partial r^2} + \sum_k V_{jk}\left(r,\hat{r},\vec{R}(t)\right)\psi_k(r) \tag{7}$$

where m is the reduced He-$H_2$ mass. This equation is solved for each initial orientation $\hat{r}$ and for all impact parameters. Using the PES of McLaughlin and Thomson[13a] fitted by Aguado et al,[13b] DCE cross sections have been calculated in the energy range 2-10eV. They show good agreement with the most recent experimental data.[14]

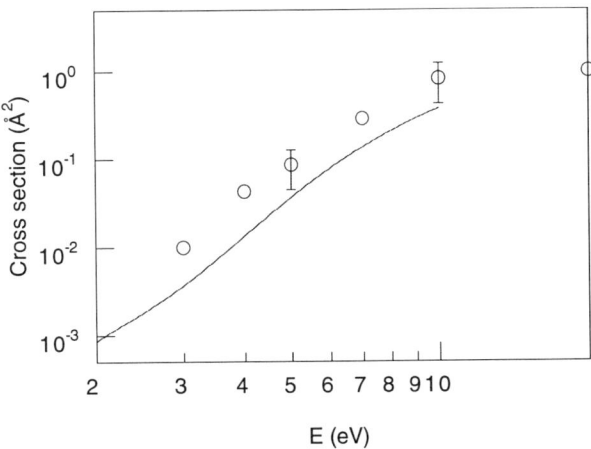

**FIGURE 1.** DCE cross section for the process (7). The full line shows the semiclassical coupled wavepacket results. The circles are the experimental results of ref.14.

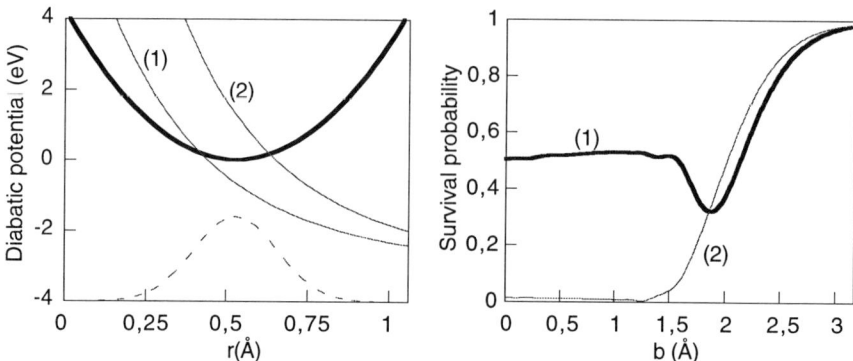

**FIGURE 2.** Left : the full lines show the model diabatic potential energy curves for the $A^+ + BC$ system. The bound state is associated with the $A^+ + BC$ state. Two different repulsive curves labelled (1) and (2) associated with the $A+B+C^+$ state have been investigated. The dashed line shows the initial probability density in the $A^+ + BC$ state. Right : survival probability of the BC molecule in its bound states as a function of the impact parameter for the repulsive potential curves (1) and (2). At small impact parameter, the system (1) is partially preserved against dissociation by the strength of the coupling.

One strong feature of the semiclassical coupled wavepacket method is its ability to treat exactly the non adiabatic effects. It has thus been possible to investigate the DCE dynamics in a wide range of conditions on a model system[4]

$$A^+ + BC \rightarrow A + B + C^+ \qquad (8)$$

The model diabatic PES for this system are displayed in fig.2. The coupling between the diabatic states is assumed to vary as $e^{-R}$. One of the most interesting results is that a strong coupling between the initial and the dissociative states can actually inhibit the dissociation by trapping the system in the upper adiabatic potential energy curve in the vicinity of an avoided crossing. This of course assumes that the system does not have time enough to dissociate before that the coupling becomes large, i.e. the collision energy is not too low. The efficiency of this trapping phenomenon depends on the way the system shares between the two adiabatic curves in the strong coupling region. If the collision energy is such that the onset time of the coupling is shorter than a vibrational period then the system cannot move during the branching of this interaction. It will be trapped or not depending on its initial position with respect to the crossing point (fig.2).

This simple picture also provides valuable insights in the role of vibrational excitation on the charge exchange probability. It is illustrated here on a model case of collision between a diatomic molecule and a diatomic ion[5]

$$AB(v) + CD^+(v') \rightarrow A^+ + B + CD \qquad (9)$$

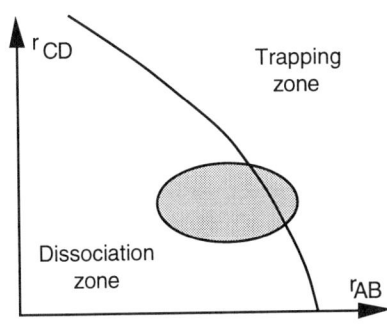 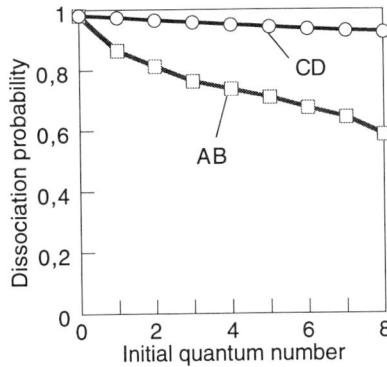

**FIGURE 3.** Influence of vibrational excitation on the dissociative charge exchange probability in reaction (9). Left : the grey zone shows the shape of an initial v=0 v'=0 wavepacket. The line splits the space in two regions : the part of the wavepacket initially located at small $r_{AB}$ and $r_{CD}$ values tends to dissociate, the rest of the wavepacket is trapped in an adiabatic bound state at strong coupling. As seen on the right frame, a vibrational excitation of AB efficiently inhibits the dissociation, since it tends to increase the part of the initial wavepacket which lies in the trapping zone. Vibrational excitation of CD is less efficient, since the crossing line is roughly parallel to the $r_{CD}$ axis.

In this calculation, the relative motion is handled classically. The two vibrational coordinates $r_{AB}$ and $r_{CD}$ are treated quantally. The frozen rotor approximation is made for both AB and CD molecules. The event of the vibrational trapping depends in this 2D calculation on the initial position of the wavepacket with respect to the crossing line between the bound and the dissociative PES : vibrational excitation of AB and CD can inhibit or enhance the dissociation (fig.3). Note on this figure that an initial excitation of the AB molecule tends to lower its dissociation probability, which may seem surprising at first sight.

## NON ADIABATIC REACTIVE COLLISIONS

Hyperspherical coordinates are needed to describe semi classically the reactive collisions. Modified Smith-Witten type coordinates[15] have been use to study the non adiabatic reaction[10]

$$Ar^+(J) + H_2 \rightarrow ArH^+ + H \qquad (10)$$

which competes with four other processes

$$\begin{aligned} Ar^+(J) + H_2 &\rightarrow Ar + H_2^+ \\ Ar^+(J) + H_2 &\rightarrow Ar^+(J') + H_2 \\ Ar^+(J) + H_2 &\rightarrow Ar + H + H^+ \\ Ar^+(J) + H_2 &\rightarrow Ar^+(J') + H + H \end{aligned} \qquad (11)$$

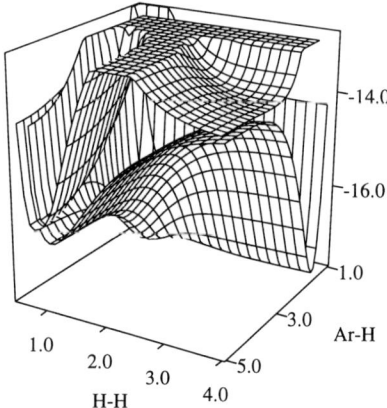

**FIGURE 4.** Ground and first excited adiabatic PES for the ArH$_2^+$ system in collinear geometry. Distances are in Ångströms and energy in eV.

In the coplanar approximation,[16] the three Euler angle which define the orientation of the triatomic triangle are not coupled neither with the hyperspherical angles $\theta$ and $\varphi$, which give the shape of the triatomic triangle, neither with the hyperspherical radius $\rho$ which determines its size. The semi-classical approximation consists in treating classically the $\rho$ motion, and quantally the $\theta$ and $\varphi$ motions. The semi-classical coupled wavepackets $\psi_j(\theta,\varphi)$ evolve according to[15,16]

$$i\frac{\partial \psi_j}{\partial t} = H_q \psi_j = -\frac{2}{\mu \rho^2}\left(\frac{\partial^2 \psi_j}{\partial \theta^2} + \frac{1}{\sin^2\theta}\left(\frac{\partial^2 \psi_j}{\partial \varphi^2} - \frac{\mathcal{J}^2}{4} + \mathcal{J}\cos\theta\frac{\partial \psi_j}{\partial \varphi}\right)\right) + V(\rho,\theta,\varphi) \quad (12)$$

where $\mathcal{J}$ is the total angular momentum and $\mu$ the hyperspherical reduced mass which depends on the mass $m_{Ar}$ and $m_H$ of the argon and hydrogen atoms :

$$\mu = \sqrt{\frac{m_{Ar}m_H^2}{m_{Ar} + 2m_H}} \quad (13)$$

The time evolution of the hyperspherical radius is governed by the hamiltonian :

$$H_{cl} = \frac{p_\rho^2}{2\mu} + \sum_{jk}\langle \psi_j | H_q | \psi_k \rangle \quad (14)$$

Eight spin-orbit dependent diabatic PES have been determined using the Diatomics in Molecule (DIM) method.[17] The main characteristics of these PES are that
i) the reactants and the products are not coupled together ;
ii) the ground adiabatic is reactive. Adiabatic excited states are not reactive ;

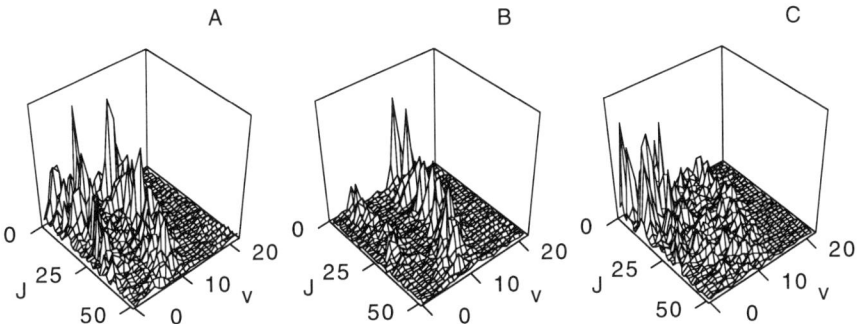

**FIGURE 5.** Rovibrational population distribution of the ArH$^+$ molecular ions produced by a collision at $E_{coll}$=3eV between Ar$^+$(J=3/2 M$_J$=1/2) and an H$_2$ molecule. A : total population. B : population formed through direct mechanism. C : population formed through knock-out mechanism.

iii) except in the entrance valley, the ground adiabatic electronic state is energetically far from the excited states (fig.4).
Dynamical calculations at zero total angular momentum show that non adiabatic transitions between ground and excited states occur in the entrance valley. Thereafter the reaction itself can be considered as taking place on the ground adiabatic PES only.[1] The scrutinization of the time evolution of the ground adiabatic wavepaket $\phi_0$, defined by

$$\phi_0(\theta,\varphi) = \sum_j \langle \xi_0 | \zeta_j \rangle \psi_j(\theta,\varphi) \qquad (15)$$

where $\xi_0$ is the ground adiabatic electronic state, allows one to identify two different regimes. At low collision energy (typically $E_{coll}$<1eV), the molecular H$_2$ axis first aligns along the intermolecular Ar-H$_2$ axis, because the (ArH$_2$)$^+$ ground state geometry is collinear. Then, at short intermolecular distance, the intermediate H$_a$ atom is jammed between the Ar atom and the H$_b$ atom. It is then ejected, and the molecule ArH$_b^+$ is formed. At highest energy (typically above 3eV), the collision motion is faster and the H$_2$ molecule does not have time enough to align along the intermolecular Ar-H$_2$ axis. In that case, the analysis of the time evolution of the wavepackets show that two mechanisms lead to the formation of the ArH$^+$ molecular ion. In the first so-called knock-out mechanism,[18] the Ar atom first hits the H$_a$ atom, transfer most of the collision energy to this atom and forms a low internal energy ArH$_b^+$ molecule. Conversely, in the so-called direct mechanism the Ar atom first hits the H$_a$ atom to form directly a high internal energy ArH$_a^+$ molecular ion (fig.5).

---

[1] The calculation is performed using the eight diabatic PES throughout the collision.

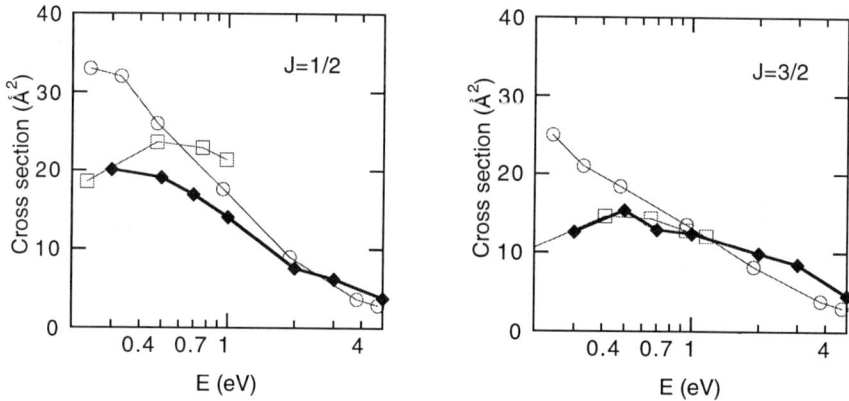

**FIGURE 6.** Total cross section for the reaction $Ar^+(J)+H_2 \rightarrow ArH^+ + H$. Circles : experimental measurements of Ng et al[19]. Squares : theoretical results of Baer et al[20]. Diamonds : present work.

Finally, a preliminary determination of the fine structure dependent reaction cross section has been made in the collision energy range 0.3eV to 5eV (fig.6). For this cross section calculation the classical trajectory has been computed on the ground adiabatic PES, since this surface is the only one which leads to reaction. It shows a nice agreement with the experimental data[19] and IOS calculations[20], except at low collision energy. The main reason for this discrepancy observed at low energy is probably the DIM PES used in this calculation, which do not properly account for polarisation effects : indeed IOS calculations[20] based upon the same PES exhibit the same feature at low energy.

## CONCLUSION

The semi-classical coupled wavepacket method presented here is a powerful technique to describe ion-molecule collisions. Its two main strong features are that i) it can exactly describe the non adiabatic effects and ii) it efficiently provides an accurate description of the wavefunctions, whatever is the number of rovibronic states involved in the process. Its main limitation arises from the semi-classical approximation, which assumes that some degrees of freedom have a classical behaviour. This approximation is likely to break down at low collision energy.

Besides its ability to provide numerically accurate results as those shown on the $He^+ + H_2$ system or on the $Ar^+ + H_2$ system, this method presents the non negligible advantage to get insight into the non adiabatic dynamics. Indeed the display of the time dependence of the coupled wavepacket, both diabatic and adiabatic, provides a direct view of the quantal collision dynamics which sometimes retrieves the simplest classical models needed to understand the phenomena, and simultaneously demonstrates their validity and show their limitations.

# ACKNOWLEDGEMENTS

Part of this work was supported by the European Union "Human Capital and Mobility" Program through the Structure and Reactivity of Molecular Ions Network under contract number CHRX-CT93-0150. Some of the calculations have been carried out at the " Institut de Développement et des Ressources en Informatique Scientifique" (IDRIS).

I am indebted to my collaborators in the mentioned works : J. P. Gauyacq for the work on DCE, G. Billing, N. Marković and V. Zenevich for the study of reactive collisions. Finally it is a pleasure to thank M. Sizun and V. Sidis for their important and essential contribution to the work presented here.

# REFERENCES

1. S. Chapman in *State Selected and State-to-State Ion-Molecule Reaction Dynamics*, Part 2: Theory, edited by M. Baer and C. Y. Ng, Adv. Chem. Phys. LXXXII (New York: Wiley) pp 424-483.
2. M. Baer in *State Selected and State-to-State Ion-Molecule Reaction Dynamics*, Part 2: Theory, edited by M. Baer and C. Y. Ng, Adv. Chem. Phys. LXXXII (New York: Wiley) pp 187-241.
3. G. D. Billing, Computer Phys. Rept. **1**, 239 (1984).
4. J. P. Gauyacq and V. Sidis, Europhys. Lett. **10**, 225 (1989); F. Aguillon, V. Sidis and J. P. Gauyacq, J. Chem. Phys. **95**, 1020 (1991).
5. F. Aguillon, V. Sidis and J. P. Gauyacq, Chem. Phys. **171**, 363 (1993).
6. F. Aguillon, Chem. Phys. Lett. **222**, 69 (1994).
7. M. Sizun and F. Aguillon, Chem. Phys. **177**, 157 (1993).
8. F. Aguillon, V. Sidis and J. P. Gauyacq, Mol. Phys. **81**, 169 (1994).
9. N. Marković and G. D. Billing, Chem. Phys. **191**, 247 (1995).
10. F. Aguillon, M. Sizun, V. Sidis, G. D. Billing, and N. Marković, to be published.
11. V. Sidis, J. Phys. Chem. **93**, 8128 (1989).
12. M. Sizun, D. Grimbert, and V. Sidis, Chem. Phys. Lett. **195**, 412 (1992).
13. (a) D. R. McLaughlin and D. L. Thomson, J. Chem. Phys. **70**, 2748 (1979); (b) A. Aguado, C.Suárez, and M. Paniagua, J. Chem. Phys. **98**, 309 (1993).
14. O. Lehner, P. Reinig, and F. Linder, "Energy and angular distribution of the product ions from dissociative charge transfer in $He^+ + H_2$ collisions", presented at the 10th European Conference on Dynamics of Molecular Collisions, Salamanca, Spain, August 28-September 2, 1994.
15. J. T. Muckerman, R. D. Gilbert, and G. D. Billing, J. Chem. Phys. **88**, 4779 (1988).
16. N. Marković and G. D. Billing, Chem. Phys. **173**, 385 (1993).
17. P. J. Kuntz and A. C. Roach, J. Chem. Soc. Faradey Trans. II, **68**, 259 (1972).
18. M. Sizun, G. Parlant, and E. A. Gislason, Chem. Phys. Lett. **139**, 1 (1987).
19. C.-L. Liao, R. Xu, S. Noubarkhsh, G. D. Flesh, M. Baer, and C. Y. Ng, J. Chem. Phys. **93**, 4832 (1990).
20. M. Baer, C.-L. Liao, R. Xu, S. Noubarkhsh, G. D. Flesh, M. Baer, C. Y. Ng, and D. Neuhauser, J. Chem. Phys. **93**, 4845 (1990).

# Direct Ionization in Diatomic and Triatomic Quasimolecules

## Yu.S. Gordeev, G.N. Ogurtsov

*A.F. Ioffe Physico-Technical Institute, 194021 St.-Petersburg, Russia*

---

Recent experimental results on direct ionization in slow ion-atom and ion-molecule collisions are discussed on the basis of a quasimolecular approximation. It is shown that the parameters of the quasimolecules formed in slow collisions can be determined from analysis of experimental data on the energy spectra of the ejected electrons. The data obtained on parameters characterizing the analytical features of the energy surfaces of quasimolecules consisting of two and more atoms are presented.

---

### INTRODUCTION

In recent years, considerable interest has been attracted to the study of direct ionization in slow atomic collisions. This study was stimulated by experimental [1] and theoretical [2-4] findings in which a very simple relation was established between differential cross section for electron ejection and parameters of a quasimolecule formed during the collision:

$$\frac{d\sigma}{dE} = A(E) \exp\left(\frac{-\alpha(E)}{v}\right) \quad (1)$$

$$A(E) = \frac{4\pi |R(E)|^2 Im R(E)}{\alpha(E)} \quad (2)$$

$$\alpha(E) = 2 \int_{E_o}^{E} Im R(E) dE \quad (3)$$

where E is the ejected electron energy, v is the projectile velocity, $E_o$ is the energy of an ionized quasimolecular state in the limit of united atom, R(E) is a reciprocal function of E(R) which depends also on the effective charge of a core and on the electron angular momentum.

Two features of Eq.(1) are of particular interest. First, the pre-exponential factor does not depend on the projectile velocity, this dependence is contained only in the exponent. Second, the strong exponential dependence of the cross section on the quasimolecular level energy $E_o$ suggests the possibility that well-separated energy ranges exist in the ejected electron energy

© 1995 American Institute of Physics

spectra which correspond to only one quasimolecular orbital contribution to direct ionization. These features facilitate the analysis of experimental data and show a promising perspective to develop a new method for quantitative spectroscopy of quasimolecules, in particular, for determination of such parameters as, the energy of the level coupled with continuum, the effective charge of a core, and the real and imaginary parts of internuclear distance as functions of electron energy. It is important that determination of most parameters can be made using only relative (not absolute) values of cross sections which can be measured very accurately. Proposals for procedures to determine quasimolecular parameters directly from experimental spectra has been put forward in [5-7]. Some data on parameters of simple diatomic and triatomic quasimolecules have been reported in short communications [8,9] , a thorough analysis of the system $H - He$ is given in our recent paper [10].

In this report, we will try to summarize our results on the parameters of quasimolecules consisting of two, three and four atoms, to discuss the features of the extracted parameters and the possibility of extending the theoretical relations derived for the system $Z_1 e Z_2$ to more complicated systems with many electrons and several atoms. In the theory [3] non-adiabatic transitions in quasimolecules are associated with existence of different series of branch points of the function E(R) situated in different ranges of internuclear distances where the structure of potential surfaces changes crucially. In this report, we will deal only with S-series of the branch points situated in the region of transformation of quasimolecular wave functions to those of the separated atoms. All the experiments have been made using our experimental apparatus with ion accelerator, mass-analyzer and electron spectrometer — cylindrical mirror with entrance angle $54.5^o$ and energy resolution 0.63% [10]. Ion energy range was 2-15 keV.

## PROCEDURE FOR DETERMINATION OF PARAMETERS

Using the relation (1) the Massey parameter $\alpha(E)$ can be determined as follows [5-7]:

$$\alpha(E) = \left[\frac{1}{v_2} - \frac{1}{v_1}\right]^{-1} \ln \frac{\sigma'(v_1, E)}{\sigma'(v_2, E)} \quad (4)$$

where $\sigma' = \frac{d\sigma}{dE}$. So to determine $\alpha(E)$, it is necessary to take a ratio of differential cross sections measured at the same electron energy but at different projectile velocities. Presenting eq.(3) in the form $\alpha(E) = \alpha_o(E)(E - E_o)$ where $\alpha_o(E)$ is a smooth function of energy, $\alpha_o(0) = 2 Im R(0)$, and fitting this expression to eq.(4) one can determine the parameters $E_o, \alpha_o$ and $ImR(0)$ and then, from $ImR(0)$, effective charge $Z_{eff}$ of the core. The function $ImR(E)$ is determined from differentiation of the function $\alpha(E)$, and $ReR(E)$ is determined using eq.(2) in which the values of cross sections and the parameters

$\alpha(E)$ and ImR(E) found before are substituted.

It should be noted that the use of the theoretical relations given above is not straightforward. For this purpose, it is necessary to transform doubly differential cross sections measured in the laboratory coordinate system into singly differential cross sections in the center of mass system, so it is necessary to know the angular distribution of ejected electrons. Expressions for angular distributions have been derived in [10,11] using the united atom approximation. Influence of kinematics is characterized by the ratio $\epsilon = v_{CM}/k$ where $v_{CM}$ is the velocity of the mass center which is usually small, $k = \sqrt[2]{2E}$, electron velocity in the laboratory system. This influence can be considerable at low electron energies and high (non-adiabatic) projectile velocities. Fortunately, in our case ($\theta_{lab} = 54.5°$, close to the "magic" angle) the ratio between $\sigma'_{CM}$ and $\sigma''_{lab}$ differs from $4\pi$ by terms of the order $O(\epsilon^2)$ [10]. Another factor which should be taken into account is the "survival" factor , $1 - exp(-\frac{\alpha(E)}{v})$, which accounts for a loss of the flux. Contributions from kinematics and the "survival" factor compensate each other. As a result, alpha values determined by us from cross sections in the center of mass system and in the laboratory system practically coincide in most cases under study. In addition, we should note that, to increase accuracy of the extracted parameters, a large set of alpha values for many combinations ($v_1, v_2$) has been collected and then the regression and correlation analysis has been made to find a regression line which is to be fitted to the theoretical relations.

## RESULTS AND DISCUSSION

Figs. 1-4 show continuous parts of energy spectra of electrons measured for $H - He, H^+ - N_2, H_2^+ - He$ systems [10,12,13] and calculated for $H^+ - H$ system [14]. In all the cases the dominant mechanism for electron emission is direct ionization. A remarkable similarity of the spectra, their exponential behavior, imply that they can be described by functions of the same kind with analogous parameters. We begin our discussion with analysis of the data on diatomic quasimolecules formed in slow ion(atom)-atom collisions parameters of which can be compared with those of the united atom.

### Diatomic quasimolecules

The systems $H - He, H^+ - He$ have been studied to get information about parameters of these relatively simple systems accessible for theoretical analysis. Parameters of the quasimolecules extracted from experimental data on energy spectra are given in Table 1 together with analogous values for the united atom.

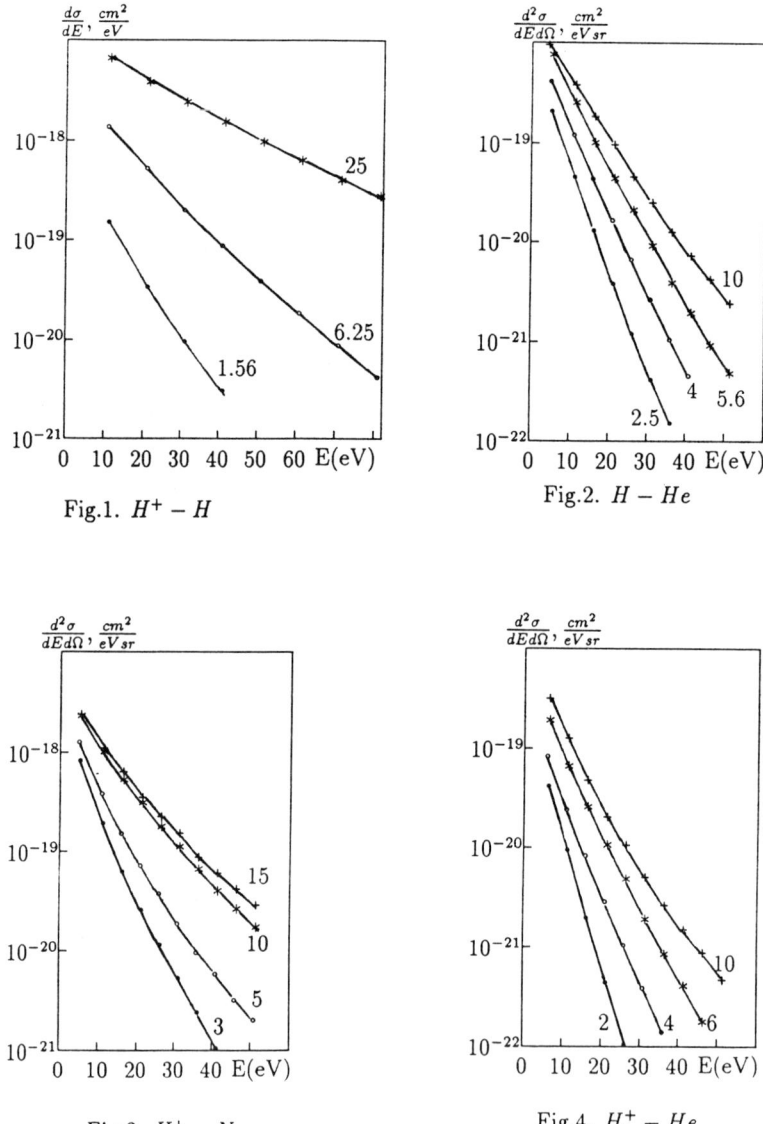

**FIGS. 1-4.** Continuous parts of energy distributions of electrons ejected in $H^+ - H$, $H - He$, $H^+ - N_2$ and $H_2^+ - He$ collisions. Numbers near curves correspond to the projectile energies in keV. Symbols are connected with solid lines.

TABLE 1. Parameters of diatomic quasimolecules.

| System | Orbital | $-E_o$ (eV) exp | $-E_o$ (eV) UA | $Z_{eff}$ exp | $Z_{eff}$ UA | ReR(0) (a.u.) | ImR(0) (a.u.) |
|---|---|---|---|---|---|---|---|
| $H - He$ | $2p\sigma$ | 3.4 | 3.54 | 1.1 | 1.02 | 1.7 | 1.5 |
| $H^+ - He$ | $2p\sigma$ | 13.5 | 13.42 | 2.5 | 2.0 | 0.65 | 0.8 |
| $H^+ - He$ | $3d\sigma$ | 7.0 | 6.05 | 2.0 | 2.0 | 2.9 | 1.75 |

As seen from the table, the parameters characterizing properties of the quasimolecules in the limit of united atom ($E_o$ and $Z_{eff}$) agree well with the analogous values for united atom taken from the spectroscopical tables. Difference in $Z_{eff}$ values for $H^+ - He$ system may indicate contribution of the states with two $2p\sigma$ electrons. The functions ImR(E) for the ionized $2p\sigma$ orbital in $H - He$ and $H^+ - He$ systems are given in Figs.5 and 6 in comparison with theoretical calculations [10]. In the case of $H^+ - He$ agreement of the data is quite satisfactory whereas in the case of $H - He$ the data differ considerably, especially at higher electron energies. This can be explained by the fact that the system $H - He$ is not Coulombic and the diabatic term responsible for direct ionization is "superpromoted" to the continuum more sharply than in the system $Z_1 e Z_2$ used in calculations [10]. From the above analysis we can conclude that our method of determination of quasimolecular parameters gives quite good results for the values relevant to the properties of the system in the limit of united atom. As for the dynamical parameters, ImR(E), Re(E), the screened Coulomb approximation is not applicable in all cases and more realistic models taking into account behavior of diabatic terms are needed to explain the results obtained.

## Triatomic and four-atomic quasimolecules

The possibility to determine parameters of quasimolecules at small internuclear distances discussed above encouraged us to apply the same procedure to energy spectra of electrons ejected via direct ionization in ion(atom) - molecule collisions. In this case many-atomic quasimolecules are formed in close approach of colliding particles. The parameters extracted from such analysis are given in Table 2.

TABLE 2. Parameters of many-atomic quasimolecules.

| System | Orbital | $-E_o(eV)$ | $Z_{eff}$ | ReR(0)(a.u.) | ImR(0)(a.u.) |
|---|---|---|---|---|---|
| $H - H_2$ | "$2p\sigma$" | 1.7 | 1.2 | 1.4 | 1.6 |
| $H^+ - H_2$ | "$2p\sigma$" | 5.5 | 2.2 | 0.8 | 0.95 |
| $H^+ - N_2$ | "$3d\sigma$" | 7.0 | 3.7 | 1.5 | 0.95 |
| $H^+ - CO$ | "$3d\sigma$" | 11.1 | 3.6 | 1.6 | 0.95 |
| $H_2^+ - H_2$ | "$2p\sigma$" | 5.0 | 2.0 | 0.9 | 1.0 |
| $H_2^+ - He$ | "$2p\sigma$" | 4.3 | 2.7 | 0.7 | 0.8 |

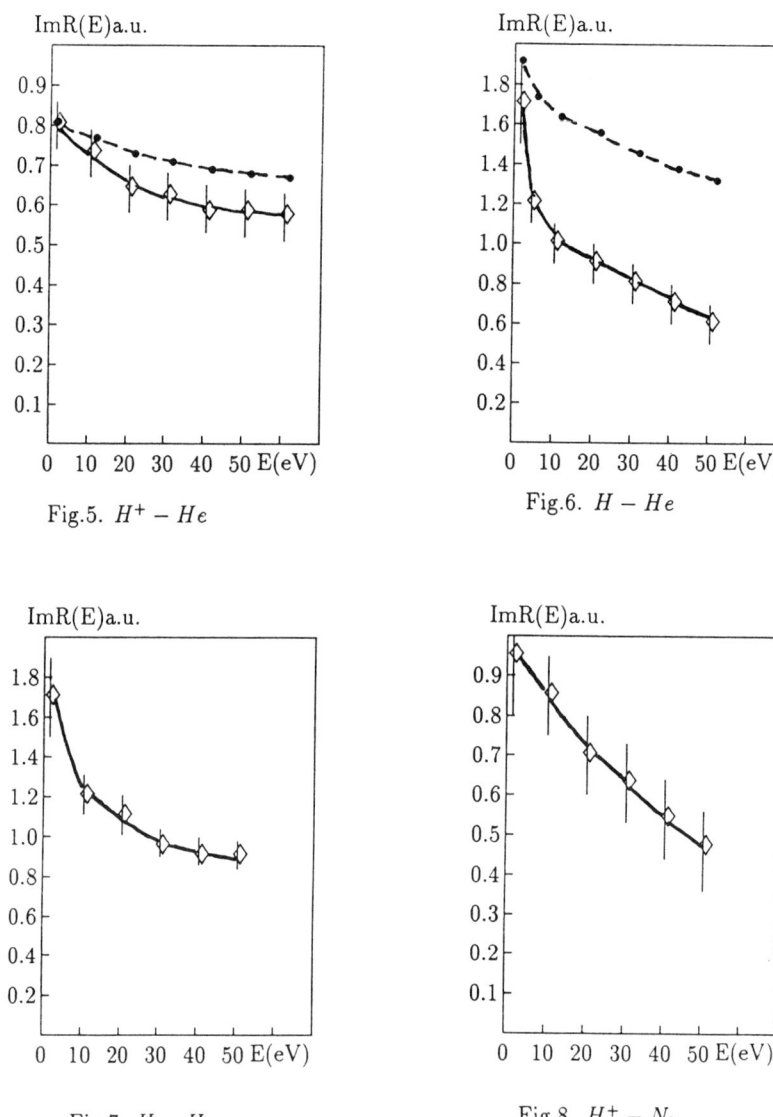

**FIGS. 5-8.** Imaginary parts of charateristic internuclear distances as functions of energy of electrons ejected in $H^+ - He$, $H - He$, $H - H_2$ and $H^+ - N_2$ collisions. Experimental data (diamonds) are connected by solid lines and the calculations (dots) are joined by dashed lines.

The symbols with quotation marks correspond to the orbitals to which energy surfaces converge in merging atoms of colliding molecules. Accuracy of the parameters given in Tables 1 and 2 is 10-20% . Comparison of the data given in the tables reveals the following features. The values $|E_o|$ for the systems "ion - homonuclear molecule" are considerably lower than those for isoelectronic diatomic quasimolecules (i.e. than energies of united atom levels). They are also considerably lower than energies of levels of a molecule formed in the projectile merging in one of atoms of the target molecule (e.g. $HeH$ for $H-H_2$ collisions). Moreover, the $|E_o|$ value for the system $H^+ - N_2$ ("ion - homonuclear molecule") is smaller than that for $H^+-CO$ (isoelectronic system "ion - heteronuclear molecule"). This fact has got a good physical explanation by "conical" promotion of potential surfaces in the systems of $C_{2v}$ symmetry group at small distances between the projectile and the mass center of the target molecule [15]. Other parameters of isoelectronic quasimolecules are very close. Since calculation of many-atomic quasimolecules in the complex plane is very difficult we have no theoretical data to compare with the parameters given in Table 2 as well as with the functions ImR(E) given in Figs.7 and 8. However analysis of the parameters implies a great importance of the region near the mass center of the target molecule where wave function of "active" electron changes crucially, and a significant contribution of trajectories passing through this center, especially those directed along the symmetry axis. This creates perspectives to reduce the number of degrees of freedom in theoretical description of ionization processes in ion-molecule collisions by using some kind of stationary phase approximation.

## CONCLUSIONS

1. The study of direct ionization in slow atomic and molecular collisions can give unique information on the parameters of quasimolecules formed during the collision. On the basis of this study a new method for quantitative spectroscopy of quasimolecules can be developed.

2. The range of applicability of the theoretical relations derived in a semiclassical version of the adiabatic approximation can be extended to the systems with non-Coulomb interaction although proper corrections should be made to take account of the specific features of the quasimolecular terms. This extension is caused by the one-electron character of direct ionization.

3. The results obtained imply that the most important range of R in direct ionization of triatomic quasimolecules is situated in the vicinity of the mass center of the target molecule (not in the vicinity of individual atoms).

The later conclusion will be checked by experiments using the coincidence technique "electron-fragment" which are under way in our laboratory. In these experiments, electron energy spectra will be studied as functions of the molecular target orientation.

## ACKNOWLEDGMENTS

This work was made possible in part by Grants No NU 7000,7300 from the International Science Foundation.

## REFERENCES

1. P.H. Woerlee, Yu.S. Gordeev, H. de Waard and F. Saris, J.Phys. B**14**, 527 (1981)
2. E.A. Soloviev, Zh.Eksp.Teor.Fiz. **81**, 1681 (1981)
3. S.Yu. Ovchinnikov and E.A. Soloviev, Zh.Eksp.Teor.Fiz. **90**, 921 (1986)
4. S.Yu. Ovchinnikov and E.A. Soloviev, Zh.Eksp.Teor.Fiz. **91**, 477 (1986)
5. A.N. Zinoviev, S.Yu. Ovchinnikov and Yu.S. Gordeev, Proc.12 Int. Conf. on the Physics of Electronic and Atomic Collisions, Gatlinburg 1981 (edited by S.Datz), Abstracts, p.900
6. G.N. Ogurtsov and M.G. Sargsyan, Proc.11 Int.Conf. on Atomic Physics, Paris 1988 (edited by G.Fabre, D.Delande), Abstracts, p.XI-68
7. G.N. Ogurtsov, Proc.22 Eur.Group of Atomic Spectroscopy, Uppsala 1990 (edited by A.Wannstrom), Abstracts, p.88
8. V.R. Asatryan, S.Yu. Ovchinnikov and A.P. Shergin, Proc.10 Int.Conf.on Atomic Physics, Tokyo 1986 (edited by H.Narumi, I.Shimamura), Abstracts, p.479
9. A.G. Kroupyshev and G.N. Ogurtsov, Proc.18 Int.Conf. on the Physics of Electronic and Atomic Collisions, Aarhus 1993 (edited by T.Andersen, B.Fastrup, F.Folkmann, H.Knudsen), Abstracts, p.463
10. G.N. Ogurtsov, A.G. Kroupyshev, M.G. Sargsyan, Yu.S. Gordeev and S.Yu. Ovchinnikov (submitted to Phys.Rev. A)
11. M. Pieksma and S.Yu. Ovchinnikov, J.Phys. B **27**, 4573 (1994)
12. G.N. Ogurtsov, A.G. Kroupyshev and Yu.S. Gordeev, Pis'ma Zh.Tech.Fiz. **19**, 23 (1993)
13. G.N. Ogurtsov, A.G. Kroupyshev and M.G. Sargsyan, Proc.10 Eur.Conf.on Dynamics of Molecular Collisions, Salamanca 1994 (edited by G.Delgado-Barrio), Abstracts, p.276
14. S.Yu. Ovchinnikov and E.A. Soloviev, *Electronic and Atomic Collisions*, eds. H.B. Gilbody, W.R. Newell, F.H. Read, A.C.H. Smith, (North-Holland, Amsterdam 1988) p.439
15. D. Dowek, D. Dhuicq, V. Sidis and M. Barat, Phys.Rev. A **26**, 746 (1982)

# Wavefunction Overlap Effects In Collisional Excitation of Molecules

## D. Mathur and M. Krishnamurthy[1]

*Tata Institute of Fundamental Research, Homi Bhabha Road, Bombay 400 005, India*

---

Energy loss experiments on electronic excitation of molecules in low-energy collisions with molecular ion projectiles have yielded new information on wave-function overlap effects and enabled the discovery of a new propensity rule which relates excitation cross sections to the quantal descriptions of the projectile-target system.

---

### INTRODUCTION

This report focuses attention on the use of high-resolution translational energy spectrometry (TES) to investigate wave-function overlap effects in collisional excitation of molecules. The information forthcoming from such studies is of importance in molecular dynamics, structure and spectroscopy; our particular interest is to investigate how structural considerations influence the dynamics of excitation processes in molecule-molecule interactions. Energy-loss studies of molecular excitation processes are also significant from the viewpoint of applications, particularly for developing insights into physico-chemical processes occurring in terrestrial and planetary atmospheres, in interstellar space and in diverse plasma processes and devices ranging from combustion chemistry and semiconductor processing to designs of new gas laser systems.

Excitation of simple diatomic molecules (like $H_2$, $N_2$, CO), triatomics ($CO_2$ and $CS_2$) and polyatomics (such as $NH_3$, $CH_4$, $CCl_4$) has been induced in collisions with projectiles such as $H_2^+$, $N_2^+$, $CO^+$, $O_2^+$, $CS_2^+$ and $CO_2^+$ and high-resolution energy loss spectra have been measured which reveal transitions to optically-allowed and dipole-forbidden states of the target molecules, as well as to ionic states. Relative oscillator strengths for transitions to individual states have been obtained in such measurements [1,2]. *A new propensity rule has been discovered which enables the "switching on" and "switching off" of excitation channels by altering the quantal description of the projectile ion.* Experiments have revealed that in all collisional situations where the projectile ion possessed $\Sigma$ symmetry, and the ground electronic state of the target molecule was also a $\Sigma$ state, collisional excitation cross sections were large.

---

[1] Present address: Joint Institute for Laboratory Astrophysics, University of Col orado, Boulder, CO 80309-0440, U.S.A.

When the projectile ions possessed Π symmetry, the corresponding excitation cross sections became at least an order of magnitude smaller.

Such observations have been interpreted in terms of wave-function overlap effects: in the case of "Σ-Σ collisions", there is a large overlap of the wavefunctions of the highest occupied molecular orbitals (HOMO's) of the collision partners, leading to large cross sections. In contrast, the low cross sections in "Π-Σ collisions" reflect the small overlap integrals between the projectile and target HOMO's. Calculated overlap integrals have been found to be in good accord with the measured cross section ratios [1].

## EXPERIMENTAL METHOD

In TES experiments, the measured quantity is the energy loss accompanying collision processes of the type:

$$AB^+ + CD \longrightarrow AB^+ + CD^\star - \Delta E$$
$$AB^+ + CD \longrightarrow AB^{+\star} + CD - \Delta E$$

where $AB^{+\star}$, $CD^\star$ indicate excited states of the fast projectile ion and "static" molecular target, respectively, and $\Delta E$ is the kinetic energy lost by the projectile ion in initiating excitation. Energy resolution comparable to, or even better than, that obtainable in electron energy loss spectrometry, is attainable in contemporary TES experiments [3]. Our energy loss measurements were made using a high-sensitivity, high-resolution, multi-sector ion translational energy spectrometer which has been described in detail elsewhere [4]; its most notable feature are large (55 cm radius) cylindrical electrostatic sectors used for pre-collision monochromation and post-collision energy analysis. Energy resolving powers in excess of $2 \times 10^4$ (in the laboratory frame) are attained; the angular resolution is better than 0.1° so that our energy loss spectra pertains to differential excitation cross sections at 0±0.1° scattering angle.

Our low-pressure ion source produces a mixture of ground and metastable state ions. By judicious choice of operating conditions we can verify elimination of projectile beam contamination by excited species by the non-observation of superelastic peaks (from collisional de-excitation) in measured translational energy spectra. Any remaining excited state component in the projectile beam can yield additional energy loss peaks due to projectile de-excitation accompanied by simultaneous target excitation; such peaks can be easily identified by simple energy considerations.

## WAVE-FUNCTION OVERLAP EFFECTS

### Excitation of diatomic molecules

We demonstrate wave-function overlap effects in collisional excitation of diatomic molecules with reference to 1.8-keV collisions of $H_2^+$, $N_2^+$ and $CO_2^+$

with CO. Fig.1 shows the energy loss envelope observed when CO molecules are excited by $H_2^+$ ions; the top portion of the figure indicates all the different electronic states that can be accessed in such an interaction with the expected energy loss values. Excitation to optically-allowed singlet states as well as dipole-forbidden triplet electronic states $a\ ^3\Pi$, $a'\ ^3\Sigma_u^+$, $A\ ^1\Pi$, $B\ ^1\Sigma^+$ and $G^1\Pi$ is observed along with ionizing transitions to $CO^+$ states $X\ ^2\Sigma^+$, $A\ ^2\Pi$, and $B^2\Sigma^+$. Transitions to ionic states have not been previously observed in TES spectra. Similar spectra were obtained with $N_2^+$ and $CO^+$ projectiles, both of

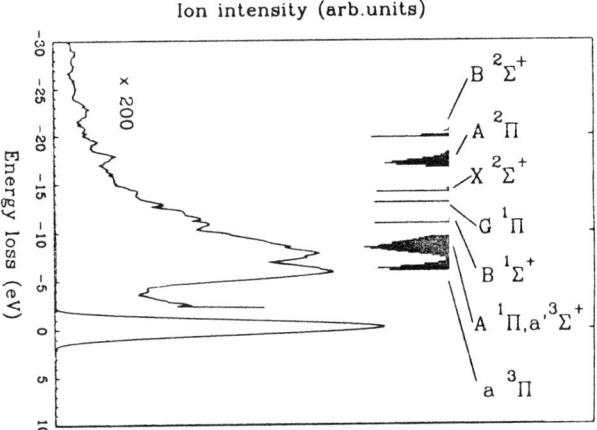

**FIG. 1.** Energy loss spectrum of $H_2^+$ ($^2\Sigma^+$) in collision with CO at 1.8 keV impact energy.

which also possess $^2\Sigma^+$ symmetry in their ground electronic state. In these cases additional structure is also produced because simultaneous excitation of both the interacting systems becomes possible:

$$N_2^+(CO^+) + CO \rightarrow N_2^{+\star}(CO^{+\star}) + CO^\star - \Delta E$$

where $\star$ indicates an excited electronic state with an overall endoergicity given by $\Delta E$. The resulting energy loss spectrum is smeared by the large number of close-lying states.

When energy loss spectra were measured with $CO_2^+$, $O_2^+$ or $CS_2^+$ projectile ions, each of which has a $^2\Pi_g$ ground electronic state, *no* energy loss structure was observed (see Fig.2). Several series of energy loss spectra were scanned with cyclic changes of projectile species ($N_2^+$, $CO_2^+$, $N_2^+$, $CO_2^+$,...). It was clearly observed that the energy loss structure observed in the energy loss region -8 eV to -12 eV in the case of $N_2^+$ projectiles almost totally "switches off" when the projectile ion is $CO_2^+$, to "switch on" when the projectile species again becomes $N_2^+$. On the basis of a large number of similar measurements with $O_2^+$ and $CS_2^+$ projectiles, the excitation cross section in the case of projectiles with $^2\Pi$ symmetry was estimated to be a factor of 10 smaller, compared

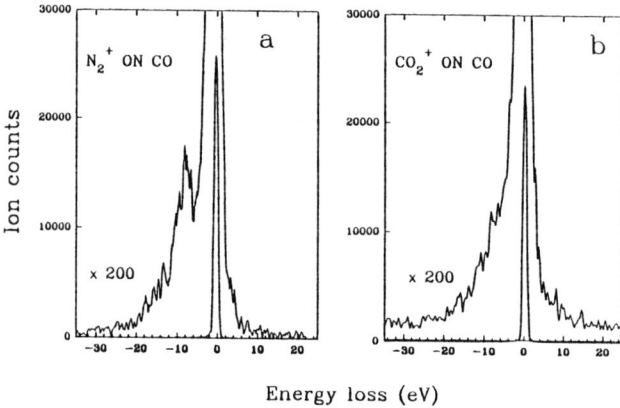

**FIG. 2.** Energy loss spectra of $N_2^+$ ($^2\Sigma^+$) and $CO_2^+$ ($^2\Pi$) in collision with CO at 1.8 keV impact energy.

to the excitation cross section obtained with projectiles possessing $^2\Sigma$ symmetry. Such dependence of excitation cross sections on the nature of the projectile cannot be attributed to the small differences in the velocity as the projectile changes from $N_2^+$ (or $H_2^+$ or $CO^+$) to $O_2^+$ (or $CO_2^+$ or $CS_2^+$). It is also not possible to make any correlation between the set of projectile ions which excite the target, and which have low target excitation cross sections, and dissociation/excitation energies of the projectiles [1]. The only link is the quantal description and the symmetry of the HOMO's of the projectile-target system. We put forward the following molecular orbital arguments in this respect.

The collisional interaction between the projectile and the target is a function of the overlap of the two corresponding HOMO's of the collision partners. The ratio of the square of the overlap integrals for two different collision partners is a measure of their relative cross sections. Calculation of overlap integrals for the HOMO's of molecules for *all* possible orientations of the projectile charge cloud relative to the target is a formidable task. However, by using symmetry arguments and choosing certain orientations that contribute maximum to the overall overlap integrals, a comparative study can be made. The necessary wavefunctions are obtained by *ab inito* computations, solving the all-electron hamiltonian by the linear combination of atomic orbitals (LCAO) approximation [1]. The overlap integral is then calculated by fixing the projectile-target orientation, and the relative excitation cross section is proportional to the ratio of the square of the overlap integrals. Fig. 3 depicts the ground-state wave-functions of the HOMO's of neutral CO, and of $H_2^+$ and $O_2^+$. In case of $^2\Sigma^+$ projectiles, the HOMO is a $\sigma_g$ orbital and due to favorable symmetries there is substantial overlap with the HOMO of the CO target molecule, which is also $\sigma$. Consequently, larger excitation cross sections are obtained.

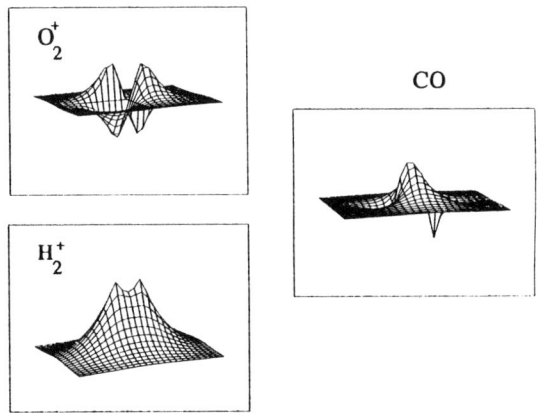

FIG. 3. Plots of the highest occupied molecular orbitals of $O_2^+$, $H_2^+$ and CO.

However, for $^2\Pi$ projectiles, the HOMO is a $\pi_g^*$ orbital and the overlap of this with the $\sigma_g$ orbital of CO is very poor, resulting in a significantly lower excitation cross section. By way of quantitative illustration, Table 1 shows results of calculations of the square of the overlap integrals obtained for $H_2^+$ + CO and $O_2^+$ +CO collinear interactions at two values of impact parameter (3Å and 4.5Å). The ratio of the square of the overlap integral of $H_2^+$ and $O_2^+$ with the CO target, which gives a relative measure of the cross sections,

$$R = [I(H_2^+ - CO)]^2/[I(O_2^+ - CO)]^2.$$

is also tabulated. The chosen values of impact parameter are in conformity with the high angular resolution of the apparatus which discriminates strongly in favour of collisional processes resulting from long-range interactions. The results indicate that suppression of excitation cross section by a factor of $\sim$20 is expected when $\Sigma$ projectiles are replaced by $\Pi$ projectiles. This is in good semi-quantitative conformity with our experimental findings. In the case of $O_2^+$ projectiles, when the internuclear axis of $O_2^+$ is oriented perpendicular to that of CO, R is about 4.7 at an impact parameter of 3Å. This is the most favorable orientation for $O_2^+$ to have large overlap with CO.

TABLE 1. Computed values of the ratios of overlap integrals, R, for different collision partners at two different impact parameters.

| Impact parameter | $|I(H_2^+\text{-}CO)|$ | $|I(O_2^+\text{-}CO)|$ | R |
|---|---|---|---|
| 3.0 | 0.176 | 0.038 | 21.4 |
| 4.5 | 0.038 | 0.008 | 22.6 |

## Excitation of non-planar molecules

To further explore the effect of wave-function overlaps, we have also studied non-planar molecules, $NH_3$, $CH_4$ and $CCl_4$ [1], so as to examine systems with different symmetries of HOMO's. We discuss here excitation of $NH_3$ as an illustrative example. The HOMO in the case of $NH_3$ is the $3a_1$ orbital, which essentially comprises the lone pair of electrons on the N-atom. The wave-function is similar in symmetry to an atomic p-orbital, with one lobe buried in the pyramidal structure formed by the three hydrogen atoms and the other lobe on the N-atom protruding out of the pyramid. The lobe in the pyramid in sterically enveloped by the 3 H-atoms. The HOMO can essentially be considered to possess $\sigma$ symmetry in any collisional process since one half of the lobe, which is in the pyramid, is sterically hindered by the 3 H-atoms and, consequently, contributes little to the overlap integral of any projectile-target system. This aspect of the structure of the molecular orbitals of ammonia make it a particularly interesting species in the study of wave-function overlap effects. When measurements are made with $\Sigma$ projectiles ($H_2^+$ and $N_2^+$), excitation of ammonia from the ground $^1A_1$ state to $\tilde{A}\ ^1A_2''$, $\tilde{B}\ ^1E''$, $\tilde{D}\ ^1A_2''$ and $\tilde{G}$ electronic states is observed along with transitions to ionic states $\tilde{X}\ ^2A'$ and $\tilde{A}\ ^2E'$ (Fig.4).

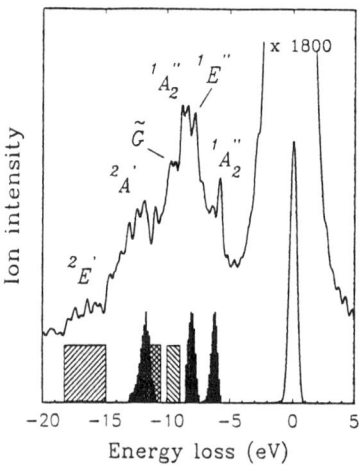

**FIG. 4.** Energy loss spectrum of $H_2^+$ colliding with $NH_3$.

However when experiments are carried out with projectiles of $^2\Pi$ symmetry ($O_2^+$ and $CO_2^+$), the energy loss features are "switched off". As before, differences in excitation cross sections are understood in terms of differences in the overlap of the HOMO's which are evaluated in two different orientations of the projectile-target system (see Fig.5). In this case the $C_\infty$ axis of $O_2^+$ is held parallel to the $C_{3v}$ axis of $NH_3$ at an impact parameter of b=3 Å and the $NH_3$ molecule is moved by varying c, the horizontal distance between the N-atom and the mid-point of the internuclear axis of $O_2^+$. The results of

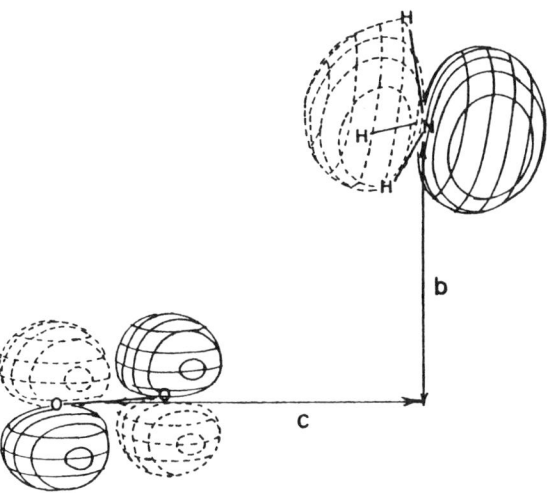

**FIG. 5.** Plot of the approach of $O_2^+$ towards $NH_3$, with the $C_\infty$ axis held parallel to the $C_{3v}$ axis at b Å and at a horizontal distance of c Å from the N-atom of $NH_3$.

overlap integral calculations for the $H_2^+$-$NH_3$ and $O_2^+$-$NH_3$ collision systems are tabulated in Table 2.

The ratio of the overlap integral averaged over different values of c for $H_2^+$-$NH_3$ and $O_2^+$-$NH_3$ is 3.2. We have considered another orientation of approach wherein the $C_2$ axis of $O_2^+$ is collinear with the $C_{3v}$ axis of $NH_3$, and the impact parameter b is varied while c=0. Table 2 also shows the magnitude of the overlap integrals obtained in the case of interactions of $H_2^+$ and $O_2^+$ with $NH_3$ for different values of impact parameter. The ratio of the overlap integral, $I$, averaged over the different impact parameters chosen for the case of $H_2^+$-$NH_3$ collisions and $O_2^+$-$NH_3$ collisions is 19.5.

It is tedious to calculate the R factor taking into account all the possible orientations. The chosen orientations critically evaluate the differences in the overlaps between the $\Sigma$ and $\Pi$ projectile wave-functions and the results of the calculation of the R factor shows a distinct and significant suppression in the excitation cross section obtained for $\Pi$ projectiles compared to that obtained for $\Sigma$ projectiles.

## Molecular symmetry or orbital symmetry ?

$CO_2$ and $CS_2$ are valence-isoelectronic triatomic molecules which provide particularly interesting targets for wave-function overlap studies because although the ground electronic state in each case has $\Sigma$ symmetry, with zero total orbital angular momentum, the valence electrons occupy a pair of degenerate anti-bonding $\pi^*$ orbitals. Intuitively, it may therefore be expected that

**TABLE 2.** Calculated values of the overlap integral ($I$) for $H_2^+ + NH_3$ and $O_2^+ + NH_3$ for different values of b and c (see Fig.4). Also shown are the squares of the overlap integrals summed over a range of values of impact parameters.

| b | c | $I(H_2^+\text{-}NH_3)$ | $I(O_2^+\text{-}NH_3)$ |
|---|---|---|---|
| 3 | -2 | 0.083 | 0.024 |
| 3 | -1 | 0.019 | 0.066 |
| 3 | 0 | 0.057 | 0.043 |
| 3 | 1 | 0.089 | 0.003 |
| 3 | 2 | 0.073 | 0.024 |
| $\sum_{c=-2}^{2} |I|^2$ | | 0.024 | 0.007 |
| 3 | 0 | 0.275 | 0.064 |
| 4 | 0 | 0.130 | 0.026 |
| 5 | 0 | 0.046 | 0.008 |
| 6 | 0 | 0.014 | 0.002 |
| $\sum_{b=3}^{6} |I|^2$ | | 0.0946 | 0.005 |

the overlap of the projectile-target HOMO's would be larger with projectiles having $\pi^*$ HOMO's and smaller with ions having bonding $\sigma$ orbitals occupied in the HOMO. Consequently, it might be expected that the excitation cross section for $CO_2$ and $CS_2$ might be larger in the case of $O_2^+$ and $CO_2^+$ projectiles than for $H_2^+$ and $N_2^+$ projectiles. However, the contrary is found to prevail in our experiments: we find that the excitation cross section is largest when $H_2^+$ projectiles are used (Fig.6). Peaks due to target excitation are weakly observable in collisions with $N_2^+$ projectiles. But in energy loss spectra obtained with $O_2^+$ and $CO_2^+$ projectiles, excitation is totally suppressed (Fig.7). Table 3 shows the square of the overlap integrals (I) obtained for different impact parameters b and c, the distances between the center of masses of the target and each projectile as defined in Fig.5. When the colliding systems are exactly one over the other (c=0), the intuitively-expected symmetry rules are perfectly valid and the overlap integrals of $CO_2$ and $CS_2$ with the two $\Sigma$ projectiles ($H_2^+$ and $N_2^+$) are zero. However, at non-zero values of c, the overlap integrals are much larger with $H_2^+$ projectiles than with any other colliding ion. The overlap integrals are only moderate in the case of collisions with $N_2^+$ and are distinctly smaller in the case of $O_2^+$ and $CO_2^+$ projectiles (Fig.7).

In summary, contrary to the expectations that the excitation cross section for $CO_2$ and $CS_2$ might be larger with $\Pi$ projectiles ($O_2^+$, $CO_2^+$) compared with $\Sigma$ projectiles ($H_2^+$, $N_2^+$), our measurements show that the excitation cross section is largest for $H_2^+$ projectiles. Concomitant wave-function overlap calculations yield results which are in accord with these observations.

It is clear that in order to explore the dependence of collisional excitation cross sections on the quantal properties of the collidants, it is not sufficient to

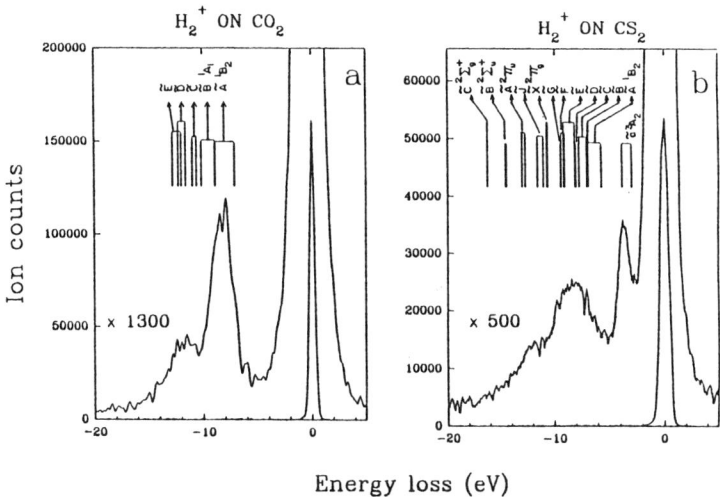

**FIG. 6.** Energy loss spectrum of $H_2^+$ colliding with $CO_2$ and $CS_2$.

**TABLE 3.** Overlap integrals for collisions on $CO_2$ and $CS_2$ with different projectiles.

| b (Å) | c (Å) | $I(H_2^+-CO_2)$ | $I(O_2^+-CO_2)$ | $I(N_2^+-CO_2)$ | $I(CO_2^+-CO_2)$ |
|---|---|---|---|---|---|
| 2 | 2 | 0.0332 | 0.0012 | 0.0493 | 0.00028 |
| 2 | 1 | 0.0423 | 0.0044 | 0.0432 | 0.00438 |
| 2 | 0 | 0.0 | 0.0324 | 0.0 | 0.02080 |
| 3 | 2 | 0.0163 | 0.0024 | 0.0054 | 0.0039 |
| 3 | 1 | 0.0162 | 0.0077 | 0.0011 | 0.0231 |
| 3 | 0 | 0.0 | 0.0271 | 0.0 | 0.0604 |
| 4 | 2 | 0.0030 | 0.0003 | 0.0 | 0.0022 |
| 4 | 1 | 0.0025 | 0.0018 | 0.0001 | 0.0088 |
| 4 | 0 | 0.0 | 0.0041 | 0.0 | 0.0162 |
| b (Å) | c (Å) | $I(H_2^+-CS_2)$ | $I(O_2^+-CS_2)$ | $I(N_2^+-CS_2)$ | $I(CO_2^+-CS_2)$ |
| 2 | 2 | 0.1065 | 0.0 | 0.1207 | 0.0004 |
| 2 | 1 | 0.0706 | 0.0001 | 0.0805 | 0.0009 |
| 2 | 0 | 0.0 | 0.0021 | 0.0 | 0.0020 |
| 3 | 2 | 0.0765 | 0.0025 | 0.0571 | 0.0056 |
| 3 | 1 | 0.0520 | 0.0087 | 0.0343 | 0.0181 |
| 3 | 0 | 0.0 | 0.0174 | 0.0 | 0.0210 |
| 4 | 2 | 0.0270 | 0.0026 | 0.0068 | 0.0004 |
| 4 | 1 | 0.0170 | 0.0089 | 0.0024 | 0.0009 |
| 4 | 0 | 0.0 | 0.0185 | 0.0 | 0.0020 |

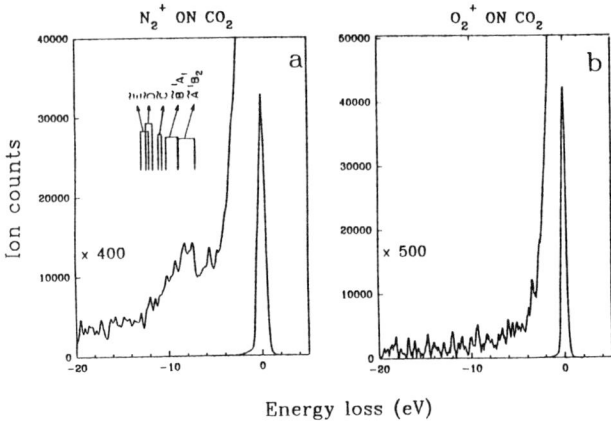

FIG. 7. Energy loss spectra of $N_2^+$ (a) and $O_2^+$ (b) colliding with $CO_2$.

consider the overall symmetry of the colliding partners: consideration of the nature of specific (outermost) molecular orbitals appears to be mandatory for wave-function overlap arguments to be applicable.

## ACKNOWLEDGMENTS

We gratefully acknowledge the skill and tenacity of U.T. Raheja in the design and fabrication of the translational energy spectrometer used in these studies.

## REFERENCES

1. M. Krishnamurthy, and D. Mathur, J. Phys. B. **27**, 1172 (1994); J. Phys. B. **27**, 3435 (1994); Chem. Phys. Lett. **231**, 127 (1994); J. Phys B **28**, L367 (1995).
2. M. Krishnamurthy, P. Gross and D. Mathur, Phys. Rev. A. **50** (1994) 2383.
3. N. Kobayashi, in *Electronic and Atomic Collisions*, ed. H.B. Gilbody, W.R. Newell, F.H. Read, and A.C.H. Smith (North-Holland, Amsterdam, 1988) p.333; M. Hamdan and A.G. Brenton, in *Physics of Ion Impact Phenomena*, ed. D. Mathur (Springer-Verlag, Berlin, 1991) Chapter 6.
4. M. Krishnamurthy, U. T. Raheja, and D. Mathur, Pramana-J. Phys. **41** (1993) 271.

# RYDBERG COLLISIONS

**Charge Transfer between Rydberg Atoms and
Polar Molecules or Clusters** .................................... 599
    C. DESFRANÇOIS,
    H. Abdoul-Carime, and J.P. Schermann

**Application of Coherent Rydberg States
for Collision Studies** .............................................. 609
    E. HORSDAL-PEDERSEN, J.C. Day,
    B. DePaola, T. Ehrenreich, S.B. Hansen, Y. Leontiev,
    K.B. MacAdam, and K.S. Mogensen

**Atomic Scattering from Oriented Rydberg Atoms** ............. 619
    J. WANG, J.H. McGuire, and R.E. Olson

# Charge Transfer Between Rydberg Atoms and Polar Molecules or Clusters

C. Desfrançois, H. Abdoul-Carime and J.P. Schermann

*Laboratoire de Physique des Lasers, URA CNRS, Institut Galilée,
Université Paris-Nord, F-93430 Villetaneuse, France*

**Abstract.** Electron transfer collisions between state-selected Rydberg atoms and polar closed-shell molecules or molecular clusters is a very efficient way to create ground-state dipole-bound molecular anions. This process can be selective with respect to the total electric dipole moment of the neutral parent and is well understood in terms of a ionic-covalent curve-crossing model. It is also a very gentle and reversible way to ionize polar molecules or clusters, with minimal change of the neutral geometry. The application of Rydberg charge exchange to the isomeric selection of small size-selected polar molecular clusters is discussed.

## INTRODUCTION

For many years, it has been theoretically known that a non-rotating closed-shell polar molecule can form a stable negative ion, provided that its permanent electric dipole moment is large enough [1]. Further numerical computations have shown that the extra electron is then very loosely bound essentially by the molecular dipole field, leading to a very diffuse outermost orbital [2]. However, the first experimental evidence for such dipole-bound negative ions only appeared some years ago as excited states of radical anions [3] or for a couple of molecules [4] or molecular dimers [5]. Recently, we have demonstrated the existence of stable ground-state dipole-bound anions for a set of closed-shell isolated molecules, whose dipole moments lie in the range 2.5 - 4D [6], and for several molecular clusters [7]. In both cases, the Rydberg electron transfer (RET) technique appeared as an efficient and selective way to create this new type of anion. Recent photodetachment studies [8] and semi-empirical [9] or *ab initio* [10] calculations have also brought new experimental and theoretical information on the nature of the binding of the excess electron in these systems.

In this paper, we summarize our present knowledge about ground-state dipole-bound negative ions. We first present a simple picture of what they are as compared to conventional covalent anions. In a second step we present our experimental results using RET, and interpret them within the framework of a simple curve-crossing model. We then purpose a new mass-spectrometry method which will allow one to perform both **mass and isomer** selection for some small neutral polar molecular clusters produced in a supersonic beam. Finally, we suggest some possible future propects concerning this special type of ion.

## WHAT ARE DIPOLE-BOUND ANIONS ?

At large distances r from a polar molecule (dipole moment $\mu$ and polarizability $\alpha$), an extra electron is attracted by the permanent dipole $-\mu \cos(\theta)/r^2$ plus charge induced-dipole $-\alpha/2r^4$ interaction potentials. At small distances, near the framework of a closed-shell molecule, in which all bonding orbitals are filled, the electron interaction with the core and valence electrons can be empirically expressed as a repulsive potential term $\exp(-r/r_c)$. One can thus construct a semi-empirical electron-molecule potential $V(r,\theta)$ which only depends on the molecular parameters $\mu$ and $\alpha$ plus the empirical parameter $r_c$ which determines the distance at which the extra electron begins to feel the repulsive interaction. Using Clary's rotationally adiabatic theory [11], it is possible to calculate the energy levels of the total anion Hamiltonian, which also includes the rotational operator of the neutral molecule and the relative angular-momentum operator of the excess electron about the molecule [9]. For symmetric top molecules whose dipole moments lie along the symmetry axis, it appears that all rotational energy levels of the neutral molecule are simply shifted to lower energies by the same amount corresponding to the excess electron binding energy or adiabatic electron affinity, EA. The EA is thus independent of the rotational state of the molecule. In this simple model, if the molecular parameters $\mu$ and $\alpha$ (and rotational constant, B) are known, EA values depend only upon the empirical parameter $r_c$. Figure 1 displays an example of such numerical calculations for the radial excess electron density and the rotationally adiabatic potential corresponding to an accurate electron affinity (e.g. EA = 3meV for $r_c = 3.5a_0$) for dipole-bound anions of acetone as experimentally determined as discussed in the next section.

One can clearly see from this figure that the potential well has a very short range and that the only bound state possesses a very low binding energy which corresponds to a very diffuse outermost orbital. The asymptotic behaviour of the radial electron density is of the form $N^2 e^{-2\sqrt{2EA}\,r}$, where N is a normalization constant, which is typical for distances larger than the classical turning point. One could then consider such dipole-bound states as the anionic dipolar counterpart of neutral coulombic Rydberg atoms. However, since the dipolar potential has a much shorter range than the coulombic potential, the radial electron density for a Rydberg state with the same electron binding energy (IP = EA) would be even more extended. At short distances, the assumption that the electron-molecule interaction is purely repulsive may not always be totally justified. Recent *ab initio* calculations [10] have shown that a fraction of the excess electron density is also located in close proximity to the molecule, especially near the electropositive

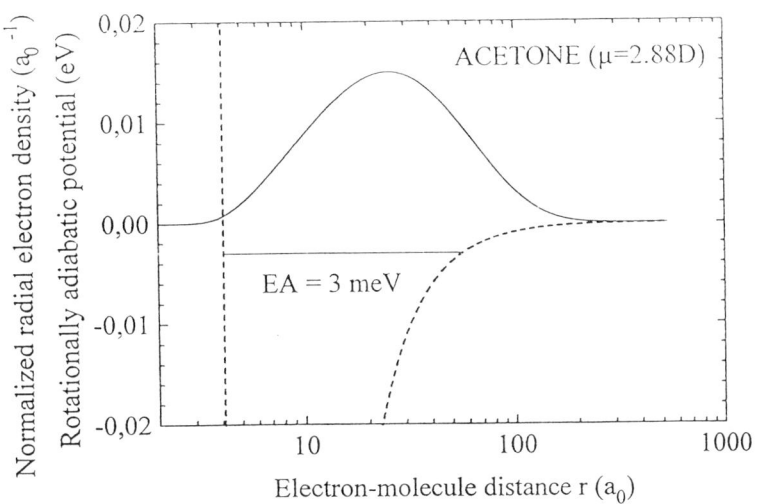

**FIGURE 1.** Normalized radial electron density and rotationally adiabatic potential, as a function of the electron-molecule distance r, for ground-state dipole-bound anions of acetone with an electron affinity of 3 meV corresponding to an empirical repulsion parameter $r_c$ of 3.5 $a_0$.

atoms. This fraction may be large for strongly polar molecules (as in the case of alkali-halide anions) or for aromatic molecules for which the lowest unoccupied molecular orbital (LUMO) is not too high in energy. It should however be negligible in the case of aliphatic molecules considered here for which the electron affinities are very low (EA << 0.1eV). Since EA values are much lower than intramolecular binding energies (few eV), we also assume that the geometry (and the dipole moment) of the neutral molecule is not greatly modified by electron attachment. This assumption is valid for covalently bound molecules with very low electron affinities, as it has been shown in recent photodetachment experiments [8] and *ab initio* calculations [10]. In the case of van der Waals or H-bonded molecular clusters, for which intermolecular binding energies are much weaker (few tens of meV), this may no longer be true as we shall discuss in the next sections.

In the first order picture presented above, dipole binding appears to be a very gentle method to ionize dipolar species, since the attached electron is very weakly bound resulting in very little change in the neutral geometries and internal energies. The vertical electron affinity, adiabatic electron affinity and electron binding energy (or vertical detachment energy) are then all equal and small. For simplicity, in this paper we refer to all three of these quantities as EA. This property of dipole binding makes this type of ionization technique very specific as compared to conventional methods in which the ionic geometries and internal energies are generally very different from that of the neutrals. Before discussing potential applications which can be drawn from these properties, we next present our experimental technique and typical results for the formation of dipole-bound anions.

## RYDBERG ELECTRON TRANSFER RESULTS

Our experimental set-up has been described previously in detail [12]. Briefly, it consists of a dual pulsed crossed beams / time-of-flight mass spectrometer apparatus. A beam of laser-excited nf (6 < n < 40) Rydberg xenon atoms crosses at right angle with a supersonic beam of molecules or molecular clusters seeded in a rare gas. Anions formed by charge transfer under single collision conditions are mass selected in a conventional time-of-flight mass spectrometer. An additional effusive thermal beam of $SF_6$ allows us to calibrate the number of Rydberg atoms produced as function of the principal quantum number n and the laser intensity. We can thus measure the relative n-dependences of the rate constants for negative ion formation. For weakly bound anions we can also perform static electric field detachment measurements, using a set of three wire mesh grids situated perpendicular to the ion path. The two external grids are grounded while the internal one is at a variable high voltage, so that this set-up does not change the energies of the ions. A fourth grid which can be biased with a repelling voltage allows us to detect either the neutrals or both neutrals plus ions. We thus measure the fraction of remaining anions as a function of the high electric field (up to 30 kV/cm) between the grids.

Such experimental results are displayed in Figures 2 and 3 for some molecules or small molecular clusters. The most striking feature in Fig. 2 is the sharply peaked shape observed for the formation rate constants as a function of the Rydberg quantum number n, even for large n values. This sharp n-dependence represents a very good signature of the formation of dipole-bound anions since in most Rydberg electron transfer experiments for conventional anion formation, these n-dependences are generally flat, monotonic or much broader [13]. In the case of acetonitrile molecules, we have also been able to obtain an estimate of the absolute value at the maximum rate constant (n = 13) which amounts to about $10^{-8} cm^3 / s$ and to verify that the $Xe^+$ ion production is equal to the negative ion production. The field detachment results of Fig. 3 confirm that electron binding energies are very low providing evidence that the created anions are, in fact, dipole-bound species. These experimental results can be simply interpreted within a curve-crossing model for the rate constants measurements and an electron tunneling model for field detachment measurements [9]. These two models only depend on the adiabatic electron affinity EA and the normalization constant N, which are solely determined by the unique empirical repulsion parameter, $r_c$ as described in the previous section. By fitting these models to experimental results, we can obtain two independent values for EAs which are found to be in good

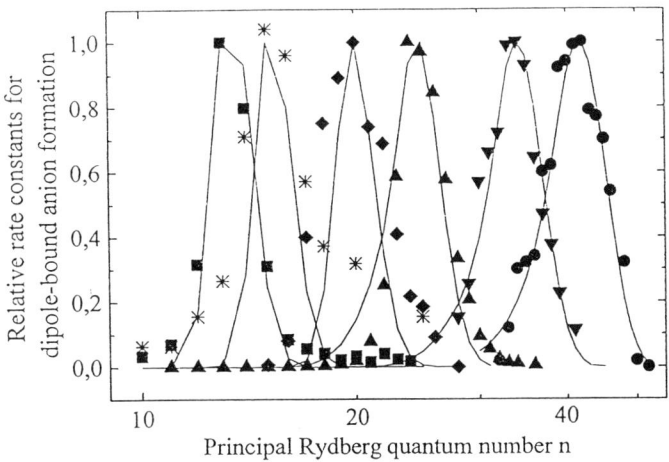

**FIGURE 2.** Experimental relative rate constants for anion formation in charge transfer collisions between nf Rydberg atoms and neutral polar molecules or clusters with the following symbols and dipole moments: water-ammonia dimers (squares; $\mu$ = 2.9 D), acetonitrile trimers (stars; $\mu$ = 3.6 D), methanol dimers (diamonds; $\mu$ = 3.0 D), acetone (up-triangles; $\mu$ = 2.88 D), pivaldehyde (down-triangles; $\mu$ = 2.66 D) and acetaldehyde molecules (circles; $\mu$ = 2.75 D). Full lines correspond to model calculations with respective EA values of 15, 11, 5.2, 3.0, 1.2 and 0.70 meV.

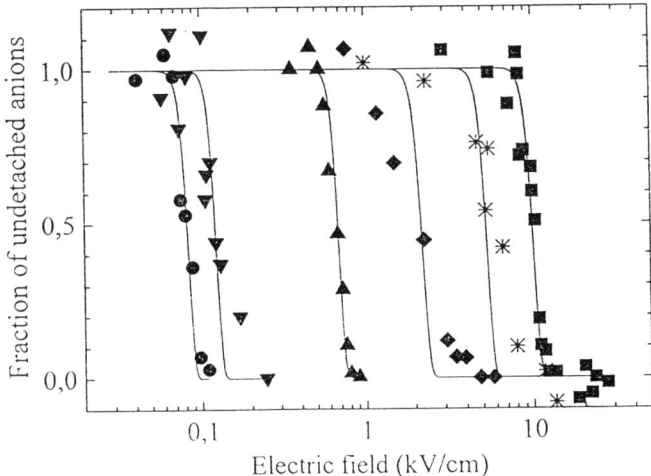

**FIGURE 3.** Experimental results for field-detachment experiments of the same species as in Fig. 2 (same symbols). Full lines are the results of tunneling model calculations with respective EA values of 15, 10, 5.2, 2.6, 0.80 and 0.60 meV, to be compared with the previous set of EA values of Fig. 2.

agreement (see captions of Fig. 2 and 3).

As expected, smaller molecular dipole moments generally correspond to lower EA values. From these results, it is difficult to assign a precise value for the critical dipole moment value above which dipole-bound anions are stable because for very low EA values (below .5 meV) these anions are very easily field-detached by low electric fields (less than 50 V/cm). From our experimental results, however, it is clear that this threshold value appears to lie just below our practical minimum value $\mu_0 = 2.5$ D [6], as inferred many years ago by Crawford [2] from numerical simulations. Anyhow, as discussed in the previous section, electron affinities not only depend upon the dipole moments but also upon the polarizabilities and the repulsion parameters $r_c$ which account for the finite size of neutrals. This is illustrated in Figure 4 where it can be seen that there is some correlation, but no precise relationship, between the relative dipole moment ($\mu - \mu_0$) and the Rydberg quantum number $n_{max}$ at which anion formation rate constants are maximum, whereas there is a very good relation between EA and $n_{max}$. This relation can be qualitatively understood as follows. During the collision, electron transfer occurs when the potential energy of the covalent pair (Rydberg atom + neutral molecule) is equal to that of the ionic pair (dipole-bound anion + positive Rydberg core). This corresponds to a critical value $R_c$ for the Rydberg-molecule distance which is reached either when the collision partners come close together or when they break apart. In order to have a maximum electron transfer probability at $R_c$, significant

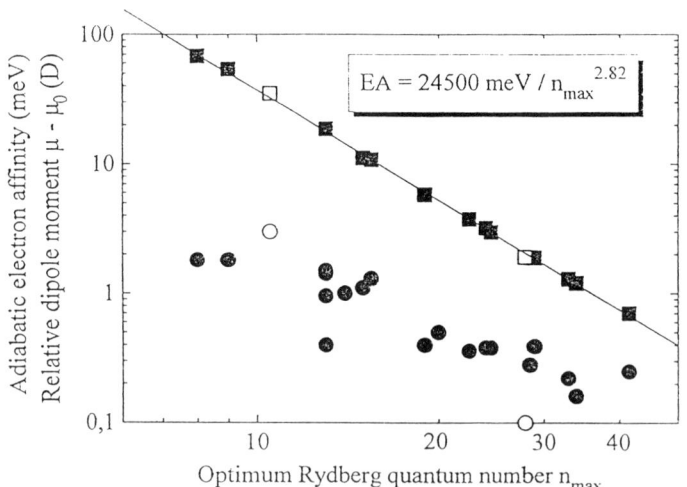

**FIGURE 4.** Electron affinities EA (squares) and relative dipole moments $\mu - \mu_0$ (circles for $\mu_0 = 2.5$ D) as a function of the principal Rydberg quantum numbers at which dipole-bound anion formation rate constants are maximum for several polar molecules or polar clusters. The straight line corresponds to the relation for EAs as given in the insert. Open squares and open circles are the relevant values for the two low-lying isomers of the water-acetonitrile dimer as described in the following section.

overlap must occur between the Rydberg orbital and the dipole-bound orbital. If the overlap is too small, the electron has almost no chance to jump from the Rydberg atom to the polar species. If it is maximum, the electron jumps twice so that in both cases anions are not formed.

An other interesting aspect of this problem is the rather surprising success of the ionic-covalent curve-crossing model used for interpreting our experimental data for unusually large Rydberg quantum numbers n up to more than 40. In this range of n-values, quasi-free electron models are generally thought to be more suitable since electron transfer occurs at very large distances between Rydberg atoms and molecules [14]. The observed energy resonance for production of $Ca^- + Ca^+$, in collisions between Rydberg Ca(nd) atoms and ground-state Ca atoms, for n =25 has been interpreted in such a manner [15]. In the present case, the electron affinities EA are always much smaller than the ionization potentials IP of Rydberg states at which rate constants are maximum. The resulting energy transfer IP-EA, between the translational nuclear motion and the electronic motion, has thus to be taken into account in a real three-body collision model [9]. Dipole-bound anion formation in collisions between Rydberg atoms and polar molecular species is thus a good illustration as well as a test of ionic-covalent curve-crossing models for charge transfer.

## CLUSTER ISOMER SELECTION

We now come to a posssible application of dipole-bound anion formation to the selection of isomers of either ionic or neutral small polar molecular clusters. In the previous sections, we have seen that dipole-binding is a very gentle way to ionize sufficiently polar species and that RET can be an efficient and selective production mechanism. For several clusters of polar molecules, we have also demonstrated that when dipole-bound anions are observed in RET experiments there is always at least one low-lying geometrical configuration of neutral parents for which the total dipole moment is high (typically larger than 3 D) [7]. For large clusters (more than a few monomer units), the number of isomers populated at an experimental temperature rapidly increases and the potential barriers (saddle points) which separate them generally decrease, so that the notion of distinct isomers becomes meaningless. The comparison between calculated minimum configurations, and their corresponding dipole moments, and experimental results are then less obvious and of lower interest. On the contrary, if a small cluster possesses two distinct geometrical isomers, with two different dipole moments, dipole-bound anion formation via RET should be able to non-destructively discriminate between the two isomers.

Among the small clusters studied, an interresting example is the water-acetonitrile dimer since it displays two, almost degenerate (within 3 meV), low-lying minima separated by a rather high potentiel barrier of 35 meV. The first minimum corresponds to a linear hydrogen bond and a high dipole moment of 5.5

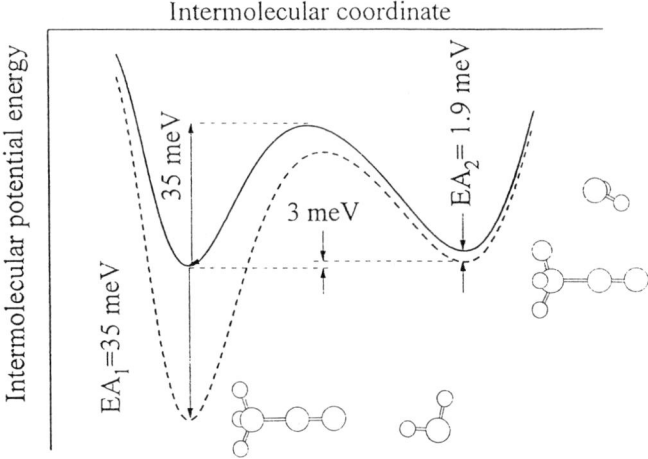

**FIGURE 5.** Schematic potential energy diagram of water-acetonitrile neutral dimers and dipole-bound anions. Neutral isomer energies and geometries are the result of theoretical calculations [7] and EA values are determined from present field detachment measurements (see text) and formation rate constants (see Fig. 6) results.

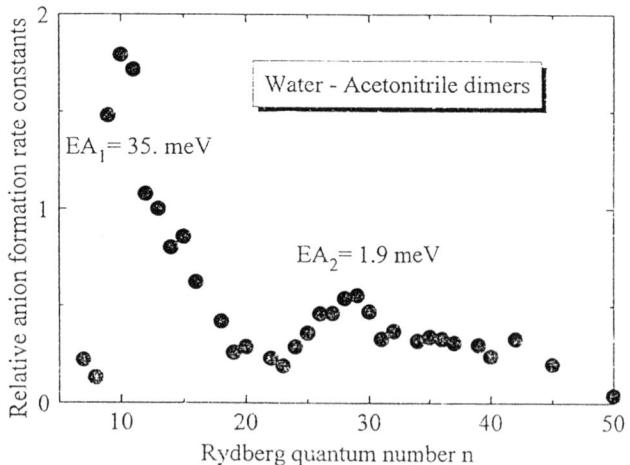

**FIGURE 6.** Experimental relative rate constants for dipole-bound anion formation in charge transfer collisions between nf Rydberg atoms and neutral water-acetonitrile dimers. The first peak at n = 10-11 corresponds to the high dipole isomer (5.5 D) and the second one at n around 28 to the low dipole isomer (2.6 D).

D, while in the second one the water molecule is bridged on the acetonitrile molecule (see Fig. 5) and its dipole moment is only 2.6 D [7]. As displayed in Figure 6, the rate constants for anion formation effectively show two distinct maxima for different principal Rydberg quantum number, $n_{max1} = 10-11$ and $n_{max2} \approx 28$. Field detachment experiments performed at $n_{max1}$ lead to a value of $EA_1 = 35$ meV and those at $n_{max2}$ give $EA_2 = 1.9$ meV. These values are in good agreement with the above relation between EA and $n_{max}$ and are also compatible with the above calculated total dipole moment values (see Fig. 4). Furthermore, they correspond to empirical repulsion parameters, in the above first order dipole-bound model, of $r_{c1} = 4.45 a_0$ and $r_{c2} = 3.32 a_0$, as compared to the value determined for isolated acetonitrile molecules $r_c = 3.38 a_0$ [9]. This repulsion parameter must be correlated to the distance between the origin of the dipole moment and the last atoms in the direction of the dipole moment, i.e. the methyl group in the present case (see Fig. 5). As it is expected from the isomer geometries, we effectively find $r_{c2} \approx r_c$ and $r_{c1} > r_c$. We therefore attribute the two peaks seen in the rate constants curve, at $n_{max1} = 10-11$ and $n_{max2} \approx 28$, to the formation of dipole-bound anions of the two calculated neutral isomers. RET then achieves isomeric selection of anions which can be further easily neutralized in order to obtain a beam of neutral dimers of only one selected isomer [16]. Further experiments and calculations are in progress in order to ascertain the above interpretation.

## FUTURE PROSPECTS

Since the early ideas of Fermi and Teller [1], dipole-bound negative ions have been of considerable theoretical and experimental interest. Here we attempt to demonstrate that these ions are interesting mainly because of their main property which is to retain the geometry of their neutral parents, even for weakly bound molecular complexes. This has to be further checked on different molecular systems. We are currently involved in experimental and theoretical studies on DNA and RNA base pairs [16], and on water dimers [17]. The first complexes can also have different polar isomers in the gas phase which are of theoretical and biological interest and which may be selected and determined with the above techniques. The case of the water dimer is interesting because the neutral complex has only one stable configuration with a rather low dipole moment of 2.6 D while we have some experimental and theoretical evidence that the dipole-bound anions correspond to different geometries with higher dipole moments. Unlike the water-ammonia dimer, the water dimer dipole-bound anions seem to be an exception to the above property of non-perturbative electron attachment. We have also studied nitromethane anions as an illustrative example of the case where both covalent and dipole-bound negative ions can be formed [18].

Throughout the results of this work, dipole-bound anion formation appears to be a very gentle, and easily reversible, way to ionize neutral polar molecules or complexes, which minimizes the amount of internal energy deposited into them. One can then imagine to perform the same experiments with neutrals as it is possible with ions, such as collisional cooling of internal and external degrees of freedom in ion traps, using dipole-bound ionization and subsequent photodetachment. Such applications, however, may require anion sources which are probably less selective but more intense than our present RET source, and will be still limited to polar species. The last aspect in which we are indeed interested is to experimentally investigate whether "polarizability-bound" anions can exists, i.e. if it is possible for a neutral closed-shell molecular species, with a low or null dipole moment but possessing a large polarizability, to form a loosely bound but stable negative ion. Using the above first-order model, with a zero dipole moment and a repulsion parameter $r_c = 4 a_0$ characteristic of cyclic molecules like cyclohexanone [6], we find that a polarizability of $\alpha \approx 20 - 25 \text{Å}^3$ is required for the existence of one electronic bound state. This is about twice the value of several substitued monocyclic aromatic molecules so one can hope that a stacking dimer of such molecules could meet the above requirement. However, such molecular species may be only very few.

## REFERENCES

[1] Fermi E. and Teller E., Phys. Rev. **72**, 399 (1947); Levy-Leblond J.M., Phys. Rev. **153**, 1 (1967); Turner J.E., Am. J. Phys. **45**, 758 (1977) and references therein.
[2] Crawford, O.H., Mol. Phys. **20**, 585 (1970); Garrett, W.R., J. Chem. Phys. **69**, 2621 (1978).
[3] Mullin A.S., Murray K.K., Schultz C.P. and Lineberger W.C., J. Phys. Chem. **97**, 10281 (1993); Brinkman E.A., Berger S., Marks J. and Brauman J.I., J. Chem. Phys. **99**, 7586 (1993).
[4] Stockdale J.A., Davis F.J., Compton R.N. and Klots C.E., J. Chem. Phys. **60**, 4279 (1974).
[5] Coe J.V., Lee G.H., Eaton J.G., Arnold S.T., Sarkas H.W., Bowen K.H., Ludewigt C., Haberland H. and Worsnop D.R., J. Chem. Phys. **92**, 3980 (1990).
[6] Desfrançois C., Abdoul-Carime H., Khelifa N. and Schermann J.P., Phys. Rev. Lett. **73**, 2436 (1994).
[7] Desfrançois C., Abdoul-Carime H., Khelifa N., Schermann J.P., Brenner V. and Millié P., J. Chem. Phys. **102**, 4952 (1995).
[8] Bowen, T- U- H2O2-
[9] Desfrançois C., Phys. Rev. A **51**, 3667 (1995).
[10] Oyler N.A. and Adamowicz L., J. Phys. Chem. **97**, 11122 (1993); ibid. Chem. Phys. Lett. **219**, 223 (1994).
[11] Clary D.C., J. Phys. Chem. **92**, 3173 (1988).
[12] Desfrançois C., Khelifa N., Lisfi A. and Schermann J.P., J. Chem. Phys. **96**, 5009 (1992).
[13] Harth K., Ruf M.W. and Hotop H., Z. Phys. D **14**, 149 and 179 (1989); Desfrançois C., Khelifa N., Schermann J.P., Kraft T., Ruf M.W. and Hotop H., Z. Phys. D **27**, 365 (1993).
[14] Matsusawa M., J. Phys. B **8**, 2114 (1975); Petitjean L., Gounand F. and Fournier P.R., Phys. Rev. A **30**, 736 (1984).
[15] Fabrikant I.I., Phys. Rev. A **48**, R3411 (1993) and references therein.
[16] Desfrançois C., Abdoul-Carime H., Schultz C.P. and Schermann J.P., submitted to Science.
[17] Bouteiller Y., Desfrançois C., Abdoul-Carime H. and Schermann J.P., in preparation.
[18] Carman H.S., Compton R.N., Hendricks J.H., Lyapustina S.A., Bowen K.H., Desfrançois C., Abdoul-Carime H. and Schermann J.P., in preparation.

# Application of Coherent Rydberg States for Collision Studies

E. Horsdal-Pedersen[*], J.C. Day[†], B. DePaola[#], T. Ehrenreich[*],
S.B. Hansen[*], Y. Leontiev[*&], K.B. MacAdam[†], and K.S. Mogensen[*]

[*]Institute of Physics and Astronomy, University of Aarhus, DK-8000 Aarhus C, Denmark
[†]Department of Physics and Astronomy, University of Kentucky, Lexington, KY 40506, USA
[#]Department of Physics, Kansas State University, Manhattan, KS 66506, USA
[&]Institute of Physics, University of St Petersburg, 198904, St Petersburg, Russia

**Abstract.** The cross section for charge transfer in fast collisions between Rydberg atoms in coherent elliptic states of $n=25$ and singly charged ions has been measured as a function of the eccentricity and the orientation of the initial state. Dramatic variations are seen. Theoretical results based on classical collision dynamics are in good qualitative agreement with the experimental data for initial conditions leading to relatively large cross sections. Quantal close-coupling results scaled from $n=4$ to $n=25$ are in better overall agreement with experiment.

## INTRODUCTION

In recent years, significant progress has been made in the production of semiclassical, stationary states of hydrogenic systems. Well-known examples of such states are the so-called circular states $|n,l,|m|\rangle = |n,n-1,n-1\rangle$ characterized by the quantum numbers $l=l_{max}=n-1$ and $|m|=m_{max}=l$, where $n$, $l$ and $m$ are the principal, the angular momentum and the magnetic quantum numbers, respectively [1]. Circular states have minimum quantum fluctuations consistent with the uncertainty relations, and for hydrogenic systems they share this property with an infinite set of states called Coherent Elliptic States (CES), which are characterized by $n$ and the expectation values of the angular momentum $\langle \mathbf{L} \rangle$ and the Runge-Lenz-Pauli vector $\langle \mathbf{A} \rangle$ [2]. The wave function in configuration space for a CES is concentrated near a classical elliptic orbit with the atomic nucleus at one of the focal points, $\langle \mathbf{L} \rangle$ is perpendicular to the plane of the orbit, and $\langle \mathbf{A} \rangle$ is directed from the nucleus towards the point of closest approach on the elliptic orbit (perihelion in classical planetary mechanics, but perikarpion or peripyrenion from the Greek words for nucleus would perhaps be more appropriate for atomic systems).

A CES may be produced by the adiabatic crossed fields method (ACFM) which involves resonant laser excitation of a Rydberg state in the presence of a weak magnetic field **B** and a somewhat stronger, orthogonal electric field $\mathbf{E}_S$ ($\omega_S \gg \omega_Z$, where $\omega_s = 3/2nE_s$ and $\omega_z = 1/2B$ are the Stark and Zeeman frequencies in a.u.) which is subsequently lowered slowly to a final value **E** such that $\omega_s \approx \omega_z$. The external fields control the vectors $\langle \mathbf{L} \rangle$ and $\langle \mathbf{A} \rangle$ or equivalently the eccentricity, $e$, and the spatial orientation of the CES [3]. The semiclassical character of the CES and the detailed experimental control of their size ($n$), shape ($e$), and orientation ($\langle \mathbf{L} \rangle$ and $\langle \mathbf{A} \rangle$) by tunable lasers and external fields open new possibilities for detailed studies of atomic dynamics in the semiclassical limit. The progress presented here on the experimental determination of relative cross sections for charge transfer into all final projectile states in ion-atom(CES) collisions at intermediate to high impact velocities, accomplished within the last 2½ years at the University of Aarhus, hold promise that this new field of research may prove very fruitful in the future. Here we show selected cross sections for 2.5keV $^{23}$Na$^+$ on CES of different orientations and eccentricities and we discuss experimental tests which illustrate how stable the target of CES is under particle bombardment and to what extend the experiment is capable of accommodating the full distribution of final states formed in charge transfer reactions.

## COHERENT ELLIPTIC STATES

With the technique used in the present investigations the orientation vectors $\langle \mathbf{L} \rangle$ and $\langle \mathbf{A} \rangle$ and the eccentricity $e$ are given by ($B=|\mathbf{B}|$, $E=|\mathbf{E}|$, and atomic units)

$$\langle \mathbf{L} \rangle = (n-1) \sqrt{1-e^2} \, \frac{\mathbf{B}}{B} \quad , \quad \langle \mathbf{A} \rangle = -(n-1) \, e \, \frac{\mathbf{E}}{E} \quad , \text{ and}$$

$$e = 1 / \sqrt{1 + \frac{\omega_z^2}{\omega_s^2}} = 1 / \sqrt{1 + \left( \frac{2.18 \, B[Gauss]}{3 \, n \, E[V/cm]} \right)^2}$$

The relative quantum fluctuations of **L** and **A** are small at $n=25$ and decrease like $n^{-1/2}$ as $n$ increases. Figure 1 shows a typical wavefunction for a CES.

When presenting reaction cross sections for CES it is convenient to use a *generalized* eccentricity $\varepsilon$ defined by

$$\varepsilon = \frac{\mathbf{v} \cdot \mathbf{v}_p}{|\mathbf{v} \cdot \mathbf{v}_p|} e \quad \text{with} \quad \mathbf{v}_p = \frac{1}{n} \sqrt{\frac{1+e}{1-e}} \, \frac{\langle \mathbf{L} \rangle \times \langle \mathbf{A} \rangle}{|\langle \mathbf{L} \rangle \times \langle \mathbf{A} \rangle|}$$

where **v** is the impact velocity and $\mathbf{v}_p$ is the orbital velocity at perihelion. As seen from its definition, $\varepsilon$ deviates from the *geometric* eccentricity $e$ only by the sign which is positive when the projection of $\mathbf{v}_p$ on **v** is in the direction of **v** and negative otherwise. Figure 2 illustrates the geometry of the present experimental situation for which **v**·**A**=0.

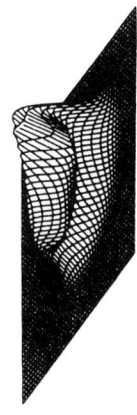

FIGURE 1. CES of eccentricity *e*=0.6.

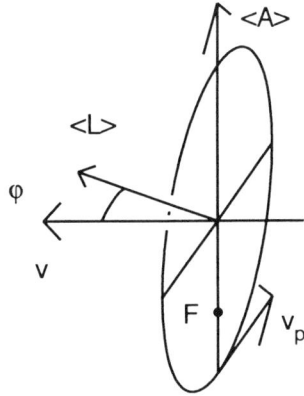

FIGURE 2. Orbit with $\varepsilon=-e$. Nucleus at F.

## EXPERIMENTAL ARRANGEMENT AND TESTS

The setup and procedures used were discussed earlier [4]. For convenience a schematic diagram of the apparatus is shown in Figure 3 and in the following we give a short description of the main features. Figure 4 shows a detailed diagram of the detection system for fast Rydberg atoms formed in charge transfer reactions with CES. This arrangement allows an experimental assessment to be made of the possible errors arising from incomplete detection of final charge transfer products.

### Production of CES

A thermal beam of Li atoms from an oven (Fig.3) is crossed by three collinear laser beams (from dye lasers pumped at 14Hz) which by resonant absorption excite the uppermost state of the *n*=25 Stark manifold according to the scheme 2s→2p→3d→$|n,k,m\rangle=|25,24,0\rangle$, where *k* is the polarization quantum number of the Stark state and the quantization axis is in the direction (vertical) of an external Stark field $\mathbf{E}_s$ formed by a stack of plates held at suitable potentials. As the excited Li atoms drift upwards from the point of laser excitation they first experience a gradually decreasing electric field and then a constant one which, together with an orthogonal magnetic field, defines the CES as discussed above.

The magnetic field is formed by an electromagnet with four poles placed symmetrically around the Stark plates and it is variable in size and direction within the horizontal plane. A set of vertical plates connected to a linear high-voltage ramp (slew rate 300V/μsec) fired when the Rydberg target has drifted into the space between the plates serve to field ionize the Rydberg target. The resulting spectrum of ions, a selective field ionization SFI spectrum, is detected by a secondary electron multiplier and a digital oscilloscope. The SFI spectrum is used to monitor the quality and density of the Rydberg target.

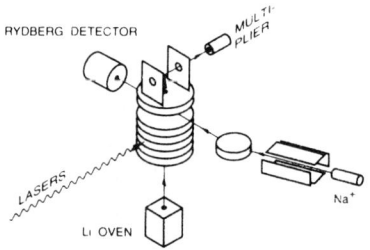

**FIGURE 3.** Experimental arrangement.

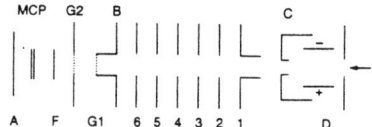

**FIGURE 4** Detection system for Rydbergs.

## Charge Transfer Cross Sections

As shown in Figure 3 an accelerated and mass analyzed beam of $^{23}Na^+$ ions is directed through the Rydberg target where Na atoms are formed as a result of charge transfer reactions from the CES. On theoretical [5] and experimental [6] grounds one expects that the binding energy of the charge transfer products is similar to the binding energy of the initial state. The products may therefore be selectively detected by an arrangement employing field ionization such as the one shown in Figure 4. Under standard operating conditions the electrodes 1-6 are at ground potential and there is a transverse electric field at D, strong enough to deflect $Na^+$ ions into the Faraday cup, C. Neutral Na atoms in highly excited states are field ionized immediately (12kV/cm) as they pass into the region between the grids G1 (at ground potential) and G2 (at −12kV). The ions produced are accelerated and impinge upon a 5μg/cm² C-foil F at a kinetic energy of 12keV+$T$, where $T$ is the kinetic energy of the neutral atoms before field ionization. At this energy a substantial fraction of the ions penetrate the foil and are detected by the microchannel plate and anode arrangement (MCP, A) while virtually all ground state atoms of kinetic energy $T$ formed by charge transfer from the background gas are stopped [7]. The charge transfer cross section $\sigma$ is given by

$$\sigma = K \cdot \frac{N}{Q \cdot S}$$

where $N$, $Q$, and $S$ are the number of particle counts, the beam charge, and the SFI-counts, respectively, obtained within a certain period of time, and $K$ is an unknown factor given by the detection efficiencies and the overlap between the fast Na$^+$ beam and the thermal target of Rydberg atoms.

## Purity of Initial States

The initial CES is very sensitive to external fields, and it has been suggested that deviation between theory and experiment could be due to $l$-change induced by the beam [8]. This was studied in detail by the comparison of SFI-spectra measured at beam currents above as well as below a typical current of a few nA. The shape of the SFI-spectra, which is very sensitive to changes of the initial-state quantum numbers, was found to be insensitive to currents of 2.5keV $^{23}$Na$^+$ ions ranging from zero to at least 15nA. Measurements of selected charge transfer cross sections as a function of beam current (Fig.5) also indicate that the Rydberg target is virtually unaffected by the presence of the ion beam. This is perhaps surprising in view of the large cross sections for transitions between degenerate $l$-states reported earlier [9], but may be caused by stabilizing Stark- and Zeeman shifts induced by the external DC-fields used in preparing the CES.

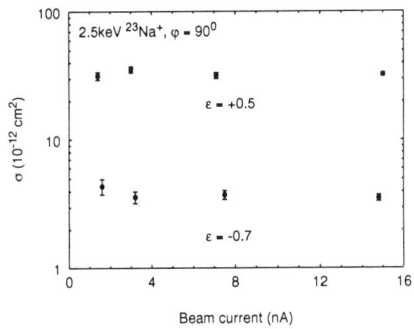

**FIGURE 5.** Cross section *vs* current.

## Distributions of Final States

With the above technique, some highly excited atoms may escape detection by the MCP (see Fig.4) because they are field ionized at D and subsequently deflected into C. Likewise, atoms in states of relatively strong binding energies may miss detection because they do not field ionize between G1 and G2.

The influence on the cross section from these two sources of error was investigated as follows. The electrodes labelled 1-6 and B (Fig.4) can act as a selective barrier for Na$^+$ ions while transmitting highly excited Na atoms and thus be a replacement for the transverse field at D. As illustrated in Figure 6 (for a 2.5keV beam approaching from the right) the conditions are the following. The barrier potential $V_B$ on B (Fig.4) must be larger than the acceleration potential and the ramp potentials on 1-6 (Fig.4) must lead to a *weak* axial electric field. If a Na atom stays neutral past a critical point $x_c$ it has enough kinetic energy to pass over the barrier even if it field-ionizes at a later time. The critical field $E_c$ at $x_c$ (43.3 or 270V/cm for the examples shown in Fig.6) determines an *upper* excitation limit on the final state distribution covered by the detector.

The *lower* limit on the final-state distribution was investigated by studying the dependence of the measured cross section on the electric field $E_{G12}$ between G1 and G2 (Fig.4). In practice, this was done by varying the potential on G2 while keeping G1,

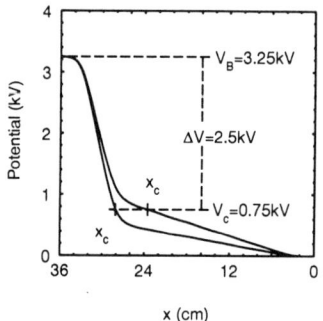

**FIGURE 6.** Ramp and barrier potentials.

B, and 1-6 at ground potential. Since the foil F is at the same potential as G2 this implies a varying ion impact energy $T_{foil}$, foil-transmission coefficient, and overall detection efficiency. The efficiency was found by sending the ion beam directly onto the detector and observing the response as a function of $V_{G2}$ ($T_{foil}=eV_{G2}+T$). After reduction for the varying efficiency a remaining variation of the measured cross section would be due to incomplete field ionization of the final state distribution.

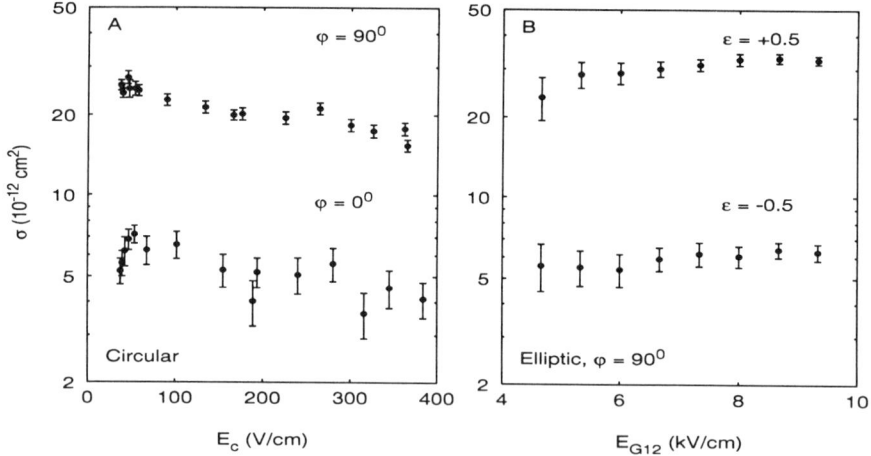

**FIGURE 7.** Cross section *vs* analyzing fields for 2.5keV $^{23}$Na$^+$ on CES of $n=25$.

Figure 7 shows selected experimental results. Extensive measurements of this type have revealed that under standard running conditions ($E_c$=180V/cm and $E_{G12}$=12keV), less than or about 20% of the part of the final-state distribution which is most easily field-ionized is cut off independent of eccentricity $e$, orientation $\varphi$, or collision speed $|\mathbf{v}|$ (Fig.7A) and no intensity is lost from the opposite part of the distribution (Fig.7B). Consequently, we conclude that the dependence of the charge transfer cross section on $e$, $\varphi$, and $|\mathbf{v}|$ is given correctly by the present experimental technique.

## RESULTS AND DISCUSSION

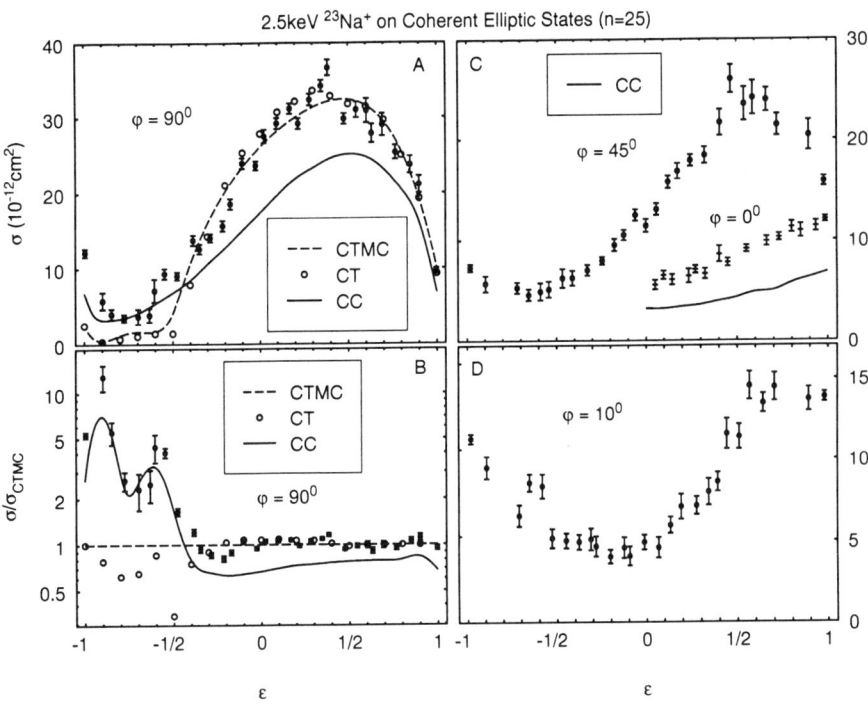

**FIGURE 8A, C, and D**: Charge transfer cross sections *vs* generalized eccentricity ε for different orientation angle φ. **B**: Ratios of cross sections. Theory: CTMC [8], CT [12], and scaled CC [10].

Figures 8 and 9 show selected experimental charge transfer cross sections for CES illustrating the dependence on eccentricity ε and orientation angle φ at a reduced impact velocity $v_{red}$ of 1.65 ($v_{red}=|\mathbf{v}|/v_e$, where $v_e=1/n$ a.u. is the orbital velocity of the initial state). The dependence on $v_{red}$ will be given soon in a forthcoming full publication. The figures also show results of *ab initio* theoretical calculations. The CC is a classically scaled Close-Coupling calculation for $n=4$ which includes as a basis set all atomic states up to $n=5$ on both centers [10] ($\sigma(n=25)=\sigma(n=4)\cdot(25/4)^4$), CTMC refers to Classical Trajectory Monte Carlo calculations based on classical dynamics and statistical sampling of all collision parameters including a microcanonical ensemble of classical elliptic orbits to represent the true initial CES [8, 11], and CT is a Classical Trajectory formulation with a single elliptic orbit representing the initial CES [12]. As mentioned earlier, the cross sections are determined only to within an unknown calibration constant. The constant depends on geometrical factors (beam overlap) and detection efficiencies (SFI- and fast Rydberg signals) which could not be determined experimentally. Instead, it was found by normalizing the experimental data to the CTMC results [8] for $\varphi=90°$ in the full $\varepsilon$-range from 0 to 1 for which theory and experiment show almost exactly the same $\varepsilon$-dependence (Figs.8A and 8B). It is not unusual that a detailed technique, like the present, prevents a precise determination of *absolute* values. *Relative* differential cross sections may often be normalized to absolute total cross sections by integration, but the present data are the only experimental results for CES. The best alternative is therefore to rely on normalization to theory.

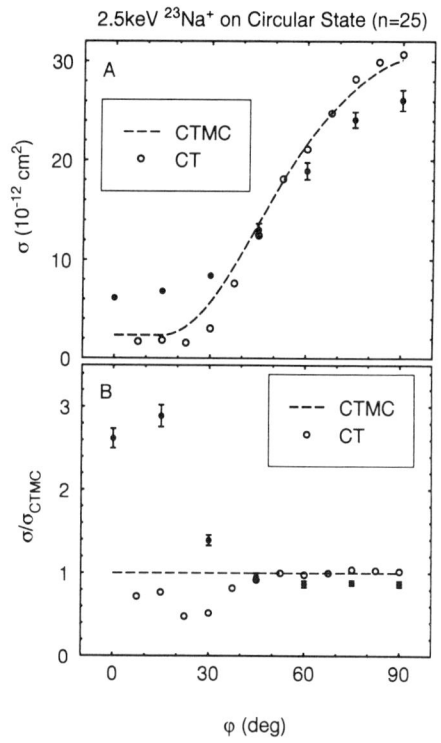

FIGURE 9. Charge transfer cross section $vs$ orientation angle $\varphi$. Theory: CT [12] and CTMC [11].

We wish to emphasize the following experimental trends. For ion impact normal to the plane of the CES ($\varphi=0°$) the cross section (Fig.8C) increases monotonously by a factor of 2.2 from a minimum at $\varepsilon=0$ to a maximum at $\varepsilon=1$. As $\varphi$ is increased from $0°$ to $90°$, which makes the impact velocity parallel to the plane of the CES while it is still perpendicular to the major axis of the ellipse, the minimum value of $\sigma$ decreases by a factor of 1.6 and the position of the minimum shifts to negative $\varepsilon$-values. The maximum increases by a factor of 2.8 for the same $\varphi$-variation and the position shifts gradually from $\varepsilon=1$ to $½$ but the interval $\Delta\varepsilon=\varepsilon_{max}-\varepsilon_{min}$ stays nearly constant. We also note that the dependence of $\sigma$ on $\varphi$ is particularly strong for small values of $\varphi$ (Figs.8C and 8D). The cross section for circular states (Fig.9) rises significantly as the direction of the impact velocity is varied from perpendicular ($\varphi=0°$) to parallel ($\varphi=90°$) relative to the plane of the orbit.

In connection with the normalization of the experimental data we already noticed that the CTMC describes the dependence of $\sigma$ on $\varepsilon$ very well (Fig.8) in the full range of ε-values from 1 to 0 and perhaps even down to ε=−0.4 when the impact velocity is parallel to the plane of the CES. This is almost exactly the range of ε-values for which the classical collision dynamics is dominated by just *one* strong interaction between the electron and the projectile [13]. The same trend is seen in Figure 9. The φ-dependence of the CTMC is in very good agreement with experiment in the region φ≥45° where the classical dynamics as before is dominated by only *one* strong interaction [12]. Since classical- and quantum mechanics lead to identical cross sections for two-body Coulomb interactions this is a further justification for the normalization of the experimental data as suggested, [8] and [11]. Further, the relatively broad distribution of final $n$-states with a long tail towards large $n$-values predicted by CTMC [11] for the case tested in Figure 7A is clearly at variance with the experimental findings.

Figures 8B and 9B show that there are also quite substantial discrepancies between experiment and classical theory. This happens in regions where the classical dynamics is dominated by *multiple* (more than one), strong electron-projectile or electron-target interactions. Extensive experimental tests (like the ones discussed previously in this report) show that the discrepancy is probably not due to systematic errors of the experiment. The case of ion impact parallel to the plane (φ=90°) of a circular state (ε=0) is common to Figures 8 and 9. It reveals a 15% discrepancy between the two sets of CTMC calculations ($\sigma=26\times10^{-12}$ cm$^2$ in [8] (Fig.8A) and $30\times10^{-12}$ cm$^2$ in [11] (Fig.9A)). This is beyond the computational statistical error or the difference due to slightly different reduced impact velocities (1.66 [8] and 1.65 [11]) and shows that the CTMC has interpretation- or systematic errors as well, originating perhaps from the classification of the final electronic states into discrete categories.

Figures 8 and 9 show that the CT- and CTMC results are in good agreement in absolute magnitude as well as in the dependence on ε and φ except in regions dominated by multiple, strong interactions and in this region a discrepancy between CT and experiment, somewhat larger than the one found for the CTMC, is seen. To the extent that the CTMC represents the outcome of the collisions if the dynamics were purely classical, one may conclude that in spite of the large values of the initial $n$ and $l$ quantum numbers (or in general $n$ and $\langle L \rangle$ for $e\neq0$) certain aspects of the collisions are still dominated by quantum phenomena.

The results of quantal close-coupling calculations [10] scaled according to the classical $n^4$-scaling rule from $n=4$ to $n=25$ are compared with experiment in Figures 8A, 8B, and 8C. For φ=0° (Fig.8C) the absolute magnitude is off by a factor of two, but since the absolute scale is not determined experimentally and in view of the extrapolation of theory, this is not so serious. The dependence on ε, however, is in good agreement with the experimental findings. This is true also at φ=90° (Figs.8A and 8B), and Figure 8B further shows that the (scaled) CC is in better overall agreement with experiment than the classical theories. Conclusions based on these observations are made difficult by the necessary scaling of the theoretical results.

## CONCLUSION

An experimental facility for the production and use of coherent elliptic Rydberg atoms has been built. The experimental arrangement offers unprecedented possibilities for controlling the initial conditions in a new class of ion-atom collisions presumably describable by classical physics in the limit as the principal quantum number $n$ and the angular momentum $\langle L \rangle$ of the initial state go towards infinity. As theoretical and experimental abilities develop within this new field, it is expected to expand our understanding of the correspondence between classical and quantum descriptions of atomic collision dynamics.

## ACKNOWLEDGMENTS

The authors are indebted to S Bradenbrink, M J Cavagnero, L J Dubé, D M Homan, M F V Lundsgaard, H O Lutz, K Taulbjerg, and J Wang for fruitful discussions and collaboration.

## REFERENCES

1. Hare J., Gross M., and Goy P., Phys. Rev. Lett. **61**, 1938 (1988).
2. Delande D. and Gay J.C., Europhys. Lett. **5**, 303 (1988) and Bommier A., Delande D., and Gay J.C., "Atoms in Strong Fields" Ed. C.A. Nicolaides *et al.*,Plenum Press, New York, (1990) p155.
3. Mogensen K.S., Day J.C., Ehrenreich T., Horsdal-Pedersen E., and Taulbjerg K., Phys. Rev. A **51**, 4038 (1995) and Day J C, Ehrenreich T, Hansen S B, Horsdal-Pedersen E, Mogensen K S, and Taulbjerg K, Phys. Rev. Lett., **72**, 1612 (1994).
4. Ehrenreich T., Day J.C., Hansen S.B., Horsdal-Pedersen E., MacAdam K.B., and Mogensen K.S., J. Phys. B, **27** L383 (1994), Hansen S.B., Ehrenreich E., Horsdal-Pedersen E., MacAdam K.B., and Dubé L.J., Phys. Rev. Lett., **71**, 1522 (1993), and Hansen S.B., Ehrenreich E., Horsdal-Pedersen E., and MacAdam K.B., in *Electronic and Atomic Collisions* eds. T Andersen, B Fastrup, F Folkmann, H Knudsen, and N Andersen (AIP Conference Proceedings 295, AIP Press), 1993, p. 828.
5. Shakeshaft R. and Spruch L., Rev. Mod. Phys. **51**, 369 (1979).
6. MacAdam K.B., in *Electronic and Atomic Collisions* eds. T Andersen, B Fastrup, F Folkmann, H Knudsen, and N Andersen (AIP Conference Proceedings 295, AIP Press), 1993, p. 183.
7. Hansen S.B., Gray L.G., Horsdal-Pedersen E., and MacAdam K.B., J. Phys. B, **24**, L315 (1991), and corrigendum, J. Phys. B, **24**, 4475 (1991).
8. Bradenbrink S., Reihl H., Wörmann Th., Roller-Lutz Z., and Lutz H.O., J. Phys. B, **27** L391 (1994).
9. Sun X. and MacAdam K.B., Phys. Rev. A **47**, 3913 (1993).
10. Lundsgaard M.F.V., Toshima N., Chen Z., and Lin C.D., J. Phys. B, **27**, L611 (1994).
11. Wang J and Olson R E, Phys. Rev. Lett. **72** 332 (1994).
12. Homan D.M., Cavagnero M.J., and Harmin D.A., Phys.Rev. A **50**, R1965 (1994).
13. Bradenbrink S., Reihl H., Roller-Lutz Z., and Lutz H.O., J. Phys. B, **28** L133 (1995).

# Atomic Scattering from Oriented Rydberg Atoms

Jianyi Wang[†], J. H. McGuire[†], and R. E. Olson[‡]

[†]*Department of Physics, Tulane University, New Orleans, LA 70118, USA*
[‡]*Department of Physics, University of Missouri-Rolla, Rolla, MO 65401, USA*

### Abstract

Highly-excited and oriented Rydberg atoms of varying angular momenta are being used to study atomic interactions. The "exotic" semiclassical properties of these atoms present an opportunity for fine control of dynamic pathways. We report initial application of oriented Rydberg atoms to collision studies of electron capture by protons and positrons at intermediate collision speeds. Several novel features have been revealed, such as the double-peak structures predicted in the angular scattering of projectiles under certain orientations. These structures are traced to the Thomas two-step capture mechanism.

## INTRODUCTION

Rydberg atoms have played an important role in modern physics since Bohr's model of the quantized atomic orbits. They continue to provide fertile ground for many areas of fundamental and applied studies including spectroscopy, scattering, astrophysics, nonlinear phenomena, and energy-related research [1]. Recent progress in experimental techniques for the production of Rydberg atoms has expanded the scope of selectivity previously limited by the use of tunable lasers alone. It is now possible to produce Rydberg atoms not only of a desired orientation, but of any angular momentum $\ell$ within the manifold of a given principal quantum number $n$ for $n \sim 25$ [2]. The orientational, localized, and coherent properties of these Rydberg atoms offer a good degree of pathway control. This may serve to revive interest in exploring physics in the semiclassical regime, and to gain new understanding and insights under conditions otherwise unavailable. As discussed below, use of oriented Rydberg atoms in collision studies in the past two years has already made a contribution in unlocking some puzzles in electron capture.

Rapid progress is being made since the first report two years ago of the experimental study using oriented circular Rydberg atoms [3]. Here we consider electron capture by protons and positrons from initially oriented Rydberg atoms. Our focus is on circular Rydberg atoms, atoms of the largest angular momentum $\ell = \ell_{max} = n - 1$ within a given $n$-manifold. We study capture in reactions

$$\text{H}^+ \text{ or } \text{e}^+ + \text{H}(n\ell m) \rightarrow \text{H}(n'\ell'm') \text{ or } \text{Ps}(n'\ell'm') + \text{H}^+ \qquad (1)$$

with $n = 25$, $\ell = n - 1 = 24$. The magnetic quantum number $m$ is varied between 0 and $\ell$ depending on orientation, although the end points will be studied in detail. The collision speed is chosen at $v_p/v_e \sim 1.5$ where $v_e = 1/n = 1/25$ is the electron orbital speed. The above parameters correspond to experimental conditions currently realizable. We shall first review the special properties of circular Rydberg atoms. These properties are essential in the interpretation of the electron capture mechanisms to be discussed below.

## CIRCULAR RYDBERG ATOMS

The classical Kepler orbit of an electron in a bound state of the Coulomb field consists of a family ellipses of a given eccentricity. The eccentricity, $\epsilon$, related to the angular momentum $l$ and binding energy $E = -1/2n^2$ by

$$\epsilon = \sqrt{1 - l^2/n^2}, \qquad (2)$$

defines the shape of the Kepler orbit, ranging from a straight line when $l \to l_{min} = 0, \epsilon \to 1$ to a circle when $l \to l_{max} = n, \epsilon \to 0$. By analogy, an atom in a quantum state of the maximum-allowed angular momentum within a given manifold $n$, $l = l_{max} = n - 1$, is referred to as the circular Rydberg atom.

Unlike Rydberg states of low angular momenta accessible by laser excitation, selective production of circular Rydberg states is impractical by optical excitation alone because of the many units of angular momenta and the range of frequencies needed. Several schemes based on a combination of optical excitation and external fields have overcome the barrier, rendering oriented circular Rydberg states accessible. They include methods [2] using adiabatic microwave transfer, crossed electric and magnetic fields, and circularly polarized microwave fields. Continuing experimental improvement is being made in this area as reviewed by E. Horsdal-Pedersen in this book.

The circular Rydberg atoms possess semiclassical properties with well localized spatial and angular distributions in both configuration and momentum space. One useful property, often taken for granted, is the minimum quantum fluctuation of the circular state. We recall that the relative uncertainties associated with the position $r$ and momentum $p$ of the electron in a circular orbit scale as $\Delta r / \langle r \rangle \simeq \Delta p / \langle p \rangle \simeq 1/\sqrt{n}$, $\Delta r \Delta p \simeq \hbar$ [4] such that the position and momentum can be specified very accurately for high Rydberg states ($n \gg 1$) without violating the Heisenberg uncertainty principle. The concept of a trajectory, fundamental in Newtonian mechanics, is then well defined for such circular states.

Shown in figure 1 is the density distribution of the circular state $n = 25$, $l = 24$ for two orientations $m = 24$ and $m = 0$. The distributions are sharp and reflect the minimum fluctuation noted earlier. Because of rotational symmetry, the actual three dimensional distribution under a rotation of $180^0$ around $z$ axis resemble a torus for $m = 24$ orientation, and a dumbbell for the $m = 0$ orientation. The thin torus lies approximately within a plane perpendicular to the quantization axis. Classical motion of the electron for $m = 24$ orientation is confined almost entirely within this plane. On the other hand, for the $m = 0$ orientation, the quantization axis lies approximately within the plane of the electron orbit.

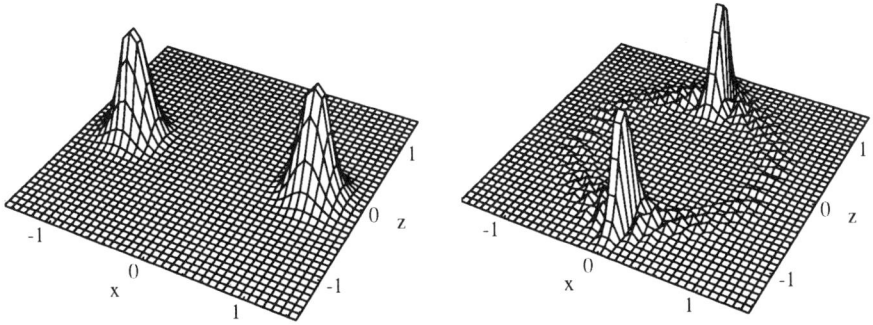

**Figure 1.** The quantum spatial density distribution of the circular Rydberg state $[n\ell = 25, 24]$ for two orientations $m = 24$ (left) and $m = 0$ (right). It is shown in the $x$-$z$ plane where $\hat{z}$ is defined as the quantization axis. The length is expressed in the scaled units of $n^2 a_0$ (or 625 a.u.).

The angular distribution is shown in figure 2. The directional (orientational) properties are clearly contrasted for the two orientations, again showing the concentration of the density in the plane perpendicular to the quantization axis for $m = 24$, as opposed to the plane parallel to the quantization axis for $m = 0$. We note that the

**Figure 2.** Polar plot of the angular distribution of the circular Rydberg state $[n\ell = 25, 24]$ for two orientations $m = 24$ and 0. Unlike the spatial density, the angular distribution is the same and equally applicable for both configuration and momentum space.

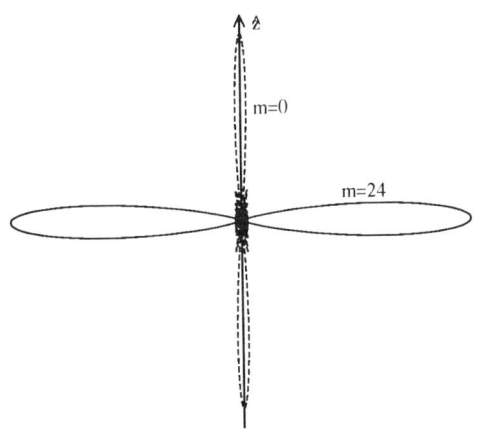

angular distribution shown in figure 2 is valid for both configuration and momentum space as given by the spherical harmonics $|Y_{\ell m}(\theta, \varphi)|^2$. It is seen from figure 2 that the electron in the $m = 24$ orientation has minimum parallel velocity component along the quantization axis, while the opposite is true for the $m = 0$ orientation which has maximum parallel velocity component. This is as expected from the orientations of the orbits. In the electron capture to be presented below, the quantization axis is usually chosen to be the projectile's incident direction. One may expect that, depending on the initial orientation, the parallel velocity component will be important in differentiating mechanisms in the velocity matching regime. As will be shown later, the capture mechanism changes from direct capture in the

$m = 0$ orientation to two-step capture in the $m = 24$ orientation.

## THEORETICAL CONSIDERATION

Among many important physical processes potentially capable of benefiting from the use of oriented Rydberg atoms, electron capture may signify an initial success in the resolution of two fundamental issues of capture theory. The first is the dominance of the double scattering capture mechanism proposed by Thomas [5] in 1927 in the context of classical mechanics but not shown unambiguously to exist in classical simulations [6] until recently [7]; and the second is whether the *dominance* is observable either classically or quantum mechanically, given the fact only a small fraction had been shown to come from double scattering nonrelativistically. Two important factors accentuate this resolution: controlling dynamical pathways via orientation; and enhanced semiclassical properties. The combined effect is that two-step scattering is necessary for certain orientations even at intermediate speeds. At the same time, there is sufficient interaction time for capture to occur so double scattering can be dominant. Even at high velocities, one may hope to diminish direct capture [8] because of the velocity dependence on the initial state angular momentum. The first Born term scales as $1/v^{12+2\ell+2\ell'}$ and may become negligible at moderate speeds compared to the second Born term $1/v^{11}$ if $\ell$ or $\ell'$ is sufficiently large, as theoretical state-selective studies have suggested [9].

Perturbative or nonperturbative models ranging from classical, semiclassical, to quantal methods may be applied and tested when studying scattering from Rydberg atoms [10]. In addition to the semiclassical properties stated earlier of the circular Rydberg atoms, it is known that capture from a Rydberg state populates predominantly a final high-$n$ Rydberg state of the projectile [11, 12]. Therefore, classical dynamics are expected to be valid throughout the collision. We use the classical trajectory Monte Carlo method [13] in this work to describe the collision process.

In this method [14], the initial conditions of the collision system are sampled randomly from a microcanonical subensemble from which $\ell, m$ are chosen by selecting the appropriate eccentricity and orientation angle of the Kepler orbit. The axis of quantization $\hat{z}$ is defined along the incoming projectile's direction, and the orientation angle refers to the angle between the angular momentum and $\hat{z}$. The system is then propagated according to its full three-body classical Hamiltonian. The exit channels are analyzed and capture events are recorded at the end of the evolution when the free particles are sufficiently far apart. Total and differential cross sections are determined by the number of capture events and the impact parameter range.

## INTEGRAL CROSS SECTIONS

First studies of electron capture from circular Rydberg atoms have focused on total capture cross sections as a function of the continuous orientation angle. They are displayed in figure 3 for protons and positrons at the reduced speed of $v^* = v_p/v_e$

= 1.65. The angular range $[0°, 90°]$ corresponds approximately to the magnetic

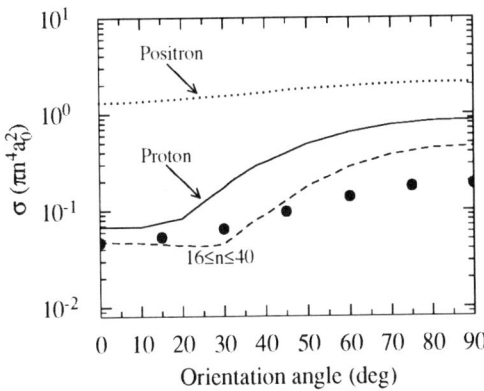

**Figure 3.** Total cross sections, in units of $\pi n^4 a_0^2$, for capture from H$^+$ and e$^+$ + H($nl$=25,24,-$m$=0-24) at the reduced speed $v^* = v_p/v_e = 1.65$ as a function of the orientation angle. Solid curve: results for protons; dotted curve: results for positrons; dashed curve: partial capture cross section into final states of $16 \leq n \leq 40$ for protons; full circle: relative expt. data of Ref. [3] normalized to the dashed curve at $0°$.

substates from $m = 24$ to $m = 0$. The cross section generally increases from $0°$ to $90°$ reflecting the importance of direct velocity matching [13]. In the $m = 0$ orientation, there is sufficient high-momentum component along $\hat{z}$ for direct capture. But when the parallel component of the electron orbital velocity is strongly suppressed due to orientation ($m = 24$), higher order interactions are required to mediate capture, resulting in smaller cross sections. The angular variation for protons is much stronger than for positrons due to the mass factor [14]. This factor is also attributed to the smaller cross sections for protons than for positrons.

Also shown in figure 3 is the experimental data [3] for Na$^+$ + Li where only final states in the range of $16 \leq n \leq 40$ are presumably detected by Stark ionization. It is to be compared to the results for H$^+$ + H (dashed curve in figure 3), since both mass and quantum defect factors are negligible. We find qualitative agreement between experiment and theory. Quantitatively, theory shows a ratio of $\sigma(90°)/\sigma(0°) \sim 12$, while experiment has a ratio of $\sim 4$. One possible cause of this discrepancy is due to the final state $nlm$ distribution [12]. Close-coupling calculations for $[n\ell = 5, 4]$ gives a ratio in agreement with experiment [15]. But it is unclear whether the ratio has converged for such low-$n$ and how the final states would map.

## DIFFERENTIAL SCATTERING

To elucidate the capture mechanisms, we show in figure 4 the differential capture cross section as a function of projectile scattering angle $\theta_p$ for $m = 24$ and $m = 0$ orientations. The overall features between proton impact and positron impact are similar. (Note the different $x$-axis scale, rad for positrons vs. mrad for protons.) Though there are differences in several aspects mostly due to the mass factor, the following observations about protons are equally applicable to positrons.

The behaviors of the cross sections for the two orientations exhibit differences that may be attributed to different pathways of capture. The cross section for $m = 0$ orientation peaks at zero scattering angle with very little structure. As we

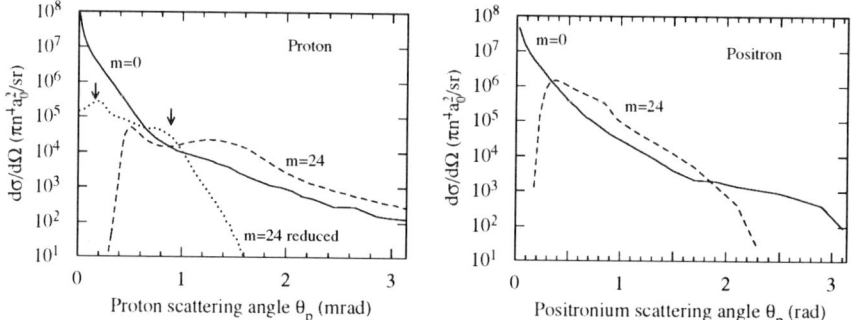

**Figure 4.** Differential cross section for capture from H$^+$ and e$^+$ + H($nl$=25,24, $m$=0,24) as a function of the proton (left) and positronium (right) scattering angle at the reduced speed $v_p/v_e = 1.5$. Results of full 3-body simulation for $m = 0$ (solid curves) and $m = 24$ (dashed curves); dotted curve: reduced 3-body simulation by switching off the internuclear interaction for $m = 24$ proton impact. The arrows mark the positions of the proton deflection angles in a binary collision with the electron using 2-body kinematics (see text).

remarked earlier, the electron has the maximum parallel velocity component along the projectile direction. As a result, capture can occur effectively by direct velocity matching in which the electron has sufficient velocity component along the proton direction to "smoothly glide" into the proton's moving frame. Soft interactions are involved and the proton is only marginally deflected.

The $m = 24$ orientation shows just the opposite behavior: It has negligible intensity in the small angle region, excluding direct velocity matching as a significant contributing mechanism. This may be understood if we recall that in this orientation the motion of the electron is confined almost entirely within the plane perpendicular to the beam direction, and the electron has little parallel velocity component to that of the projectile. Direct velocity matching is therefore strongly suppressed. In addition, instead of a single peak, there are two peaks in the $m = 24$ orientation at rather large angles. The explanation of the double peak structure has been found to be related the quasi-Thomas double scattering mechanism at finite collision speeds [7].

## TWO-STEP CAPTURE MECHANISMS

The formation of the two peaks may be explained by the required interaction between the proton and the electron to accelerate and deflect the electron in the forward direction. The basic kinetic argument is illustrated in figure 5 in terms of momentum transfers. Let us assume for simplicity that the orbital plane is exactly perpendicular to the $\hat{z}$ direction, so that the electron orbital velocity is perpendicular to $z$-axis as well. Conservation of momentum for the first collision between the proton and the electron yields

$$\theta_p = (v'_e \sin \theta_e \pm v_e)/(M_p v_p) \qquad (3)$$

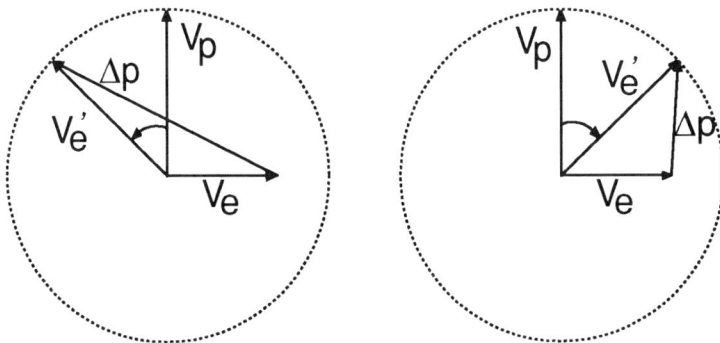

**Figure 5.** The momentum transfer diagram, in the plane formed by the projectile's ($v_p$) and the electron's ($v_e$) initial velocity vectors, showing two different momentum transfers ($\Delta p$) necessary to accelerate the electron to the speed of the proton ($v'_e \simeq v_p$) for $m = 24$ orientation. Initially the electron's motion is confined in a plane perpendicular to the incident direction. The electron can scatter either to the left or to the right of the projectile with the same scattering angle. The finite electron orbital speed renders left- and right-scattering non-equivalent, causing the projectile deflections (too small to be visible on this scale) given by Eq. (3).

where $\theta_p$ is the (small) proton scattering angle in radians ($M_p = 1836$ a.u. is the proton mass), and to zeroth order in $1/M_p$

$$v'_e = v_p \cos\theta_e + \sqrt{v_p^2 \cos^2\theta_e + v_e^2}. \tag{4}$$

Eq. (4) expresses the magnitude of the scattered electron velocity $v'_e$ in terms of the its scattering angle $\theta_e$. We note that because of the finite, non-negligible electron speed, there are two proton scattering angles depending on whether the electron scatters to the right or the left of the proton. The scattering angle of the electron $\theta_e$ may be determined by the capture requirement $v'_e \simeq v_p$. Combining eqs. (3) and (4), the values of $\theta_p$ in this case ($v^* = 1.5$) are predicted to be 0.16 mrad and 0.89 mrad, respectively. The two peak angles appearing in figure 4 (0.5 mrad and 1.2 mrad respectively) are obviously much larger than the predictions. It is attributed to the internuclear Coulomb interaction that influences the angular scattering due to the small laboratory projectile speed, which is only 0.06 a.u.

This effect is simulated in a reduced system by only switching off the nuclear-nuclear interaction. The results are shown in figure 4. The two peaks shift dramatically downward to their new positions 0.19 mrad and 0.72 mrad, respectively. The new values are in good agreement with the predictions of the two-body kinematics. It is worth pointing out that in the limiting case $v_p/v_e \gg 1$, the well-known binary encounter law $v'_e = 2v_p \cos\theta_e$ is recovered from eq. (4), and the two peaks converge into a single one – the Thomas peak at $\theta_p = (\sin 60°)/M_p$ for $v'_e = v_p$.

Two typical trajectories leading to capture from the two orientations are illustrated in figure 6. They show visually direct velocity matching vs. two-step scattering discussed in terms of momentum transfers above. Most capture events occur in a

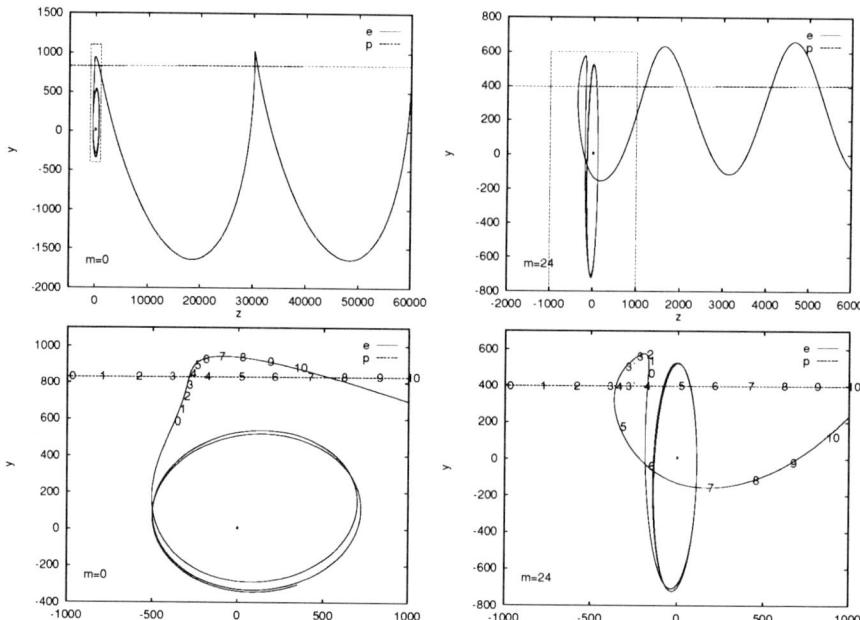

**Figure 6.** Two typical trajectories (top row) leading to capture from orientations $m = 0$ (upper left) and $m = 24$ (upper right), shown projected onto the collision plane in the laboratory frame (units of length in a.u.). Solid curves: electron path; dashed curves: proton path. The rectangular boxes are enlarged and shown in the bottom row with approximately correct aspect ratio. The numerals indicate simultaneous positions of the proton and the electron in the sequence of stroboscopic snapshots in time. The dot near the origin marks the target nucleus.

larger impact parameter range for $m = 0$ orientation than for $m = 24$ orientation. It may be readily seen that in the $m = 0$ orientation (figure 6, lower left) the electron is gradually pulled into a bound state of the proton. Here capture is most probable when the proton and the electron are on the same side of the target nucleus with the same sense of rotation. Once the electron enters the projectile field, it rarely comes back to the target field again. For the $m = 24$ orientation, however, the picture is entirely different. The electron has very little parallel velocity component and must be deflected and accelerated. This can be seen as a hard collision occurring between frames 3 and 4 near 3' (figure 6, lower right) where the electron loops around the projectile. It also picks up speed as evidenced by the increased distance between snapshots after the collision with the projectile. The electron comes back and scatters at the target nucleus before entering the proton field. It thus completes the two-step, quasi Thomas scattering process [7]. The corresponding processes have been termed 1-swap and 3-swap [16] counting the number of times the electron crosses the midplane between the projectile and the target.

## OTHER RESULTS AND DISCUSSIONS

Electron capture from oriented elliptic Rydberg states not discussed here has yielded fruitful results [17–19]. It has been shown that the velocity matching condition near perihelion is largely responsible for capture. But there are discrepancies between theory and experiment for very elliptic orbits having opposite sense of rotation between the projectile and the electron near perihelion. Quantum defects of these elliptic orbits may be important. Further studies in this region may yield useful core effects absent in a pure Coulomb field. Also, orientations of the orbital plane out of the beam axis may reveal further interesting features.

Other aspects of scattering from oriented Rydberg atoms such as ionization and state-changing collisions largely unexplored so far may prove just as successful in discovering new insights. For example, the parameter space maps [16] show reaction islands having irregular boundaries. It is interesting to see if the boundaries exhibit self-similar fractal patterns that may be useful for studying nonlinear instabilities.

The distribution of the final states shows peculiar properties that may be fundamentally interesting and practically useful. For example, large $\ell, m$ states are predicted to be populated [12]. These distributions may be useful as a unique way of producing high $\ell$ positronium through capture for spectroscopic and nonlinear studies. Another area of interest may lie in the large scattering asymmetry induced by Rydberg atoms if the quantization axis is chosen perpendicular to the incident beam [14]. One may deduce the effects of dipole forces which are seldomly observable in capture from ground state atoms.

## CONCLUSIONS

Studies on electron capture from oriented Rydberg atoms have shown many surprising results in just the last two years. In particular, the change from direct capture to two-step capture has been clearly demonstrated by varying the orientation at a given scaled collision velocity. This change may be directly tested experimentally in the angular scattering of the projectile. Interesting questions still to be answered include the final state $nlm$ distribution of the captured electron, high velocity scaling of the cross section, and possible quantum effects. Oriented Rydberg atoms may be used to control pathways and effective collision energies, and may yield comparable success when applied to a host of potential problems in the semiclassical regime such as laser-Rydberg atom interactions and nonlinear dynamics.

## ACKNOWLEDGMENTS

We would like to thank Drs. J. Burgdörfer, K. B. MacAdam, E. Horsdal-Pedersen, T. Ehrenreich, and M. Lundsgaard for helpful discussions, and Dr. K. Tökési for assistance in producing figure 6. Work supported in part by the Division of Chemical Sciences, Office of Basic Energy Sciences, Office of Energy Research, and Office of Fusion Energy, U.S. Department of Energy.

# References

1. R. F. Stebbings and F. B. Dunning, *Rydberg States of Atoms and Molecules* (Cambridge, New York, 1983); T. F. Gallagher, *Rydberg Atoms* (Cambridge, Cambridge, 1994).
2. D. Delande and J. C. Gay, *Europhys. Lett.* **5**, 303 (1988); J. Hare, M. Gross, and P. Goy, *Phys. Rev. Lett.* **61**, 1938 (1988); J. C. Day, T. Ehrenreich, S. B. Hansen, E. Horsdal-Pedersen, K. S. Mogensen, and K. Taulbjerg, *Phys. Rev. Lett.* **72**, 1612 (1994); C. H. Cheng, C. Y. Lee, and T. F. Gallagher, *Phys. Rev. Lett.* **73**, 3078 (1994); R. G. Hulet and D. Kleppner, *Phys. Rev. Lett.* **51**, 1430 (1983).
3. S. B. Hansen, T. Ehrenreich, E. Horsdal-Pedersen, and K. B. MacAdam, in *Electronic and Atomic Collisions*, ed. T. Andersen, B. Fastrup, F. Folkmann, H. Knudsen, and N. Andersen, (AIP Conf. Proc. **295**, Aarhus, Denmark, 1993) p. 828; S. B. Hansen, T. Ehrenreich, E. Horsdal-Pedersen, K. B. MacAdam, and L. J. Dubé, *Phys. Rev. Lett.* **71**, 1522 (1993).
4. H. A. Bethe and E. E. Salpeter, *Quantum Mechanics of One- and Two-Electron Atoms* (Springer, Berlin, 1957), pp. 17, 39-40.
5. L. H. Thomas, *Proc. R. Soc. (London) A* **114**, 561 (1927).
6. N. Toshima, *Phys. Rev. A* **45**, R2663 (1992); D. R. Schultz, C. O. Reinhold, and R. E. Olson, *Phys. Rev. A* **46**, 666 (1992).
7. J. Wang and R. E. Olson, *Phys. Rev. Lett.* **72**, 332 (1994).
8. R. Shakeshaft and L. Spruch, *Rev. Mod. Phys.* **51**, 369 (1979).
9. J. Burgdörfer and L. J. Dubé, *Phys. Rev. A* **31**, 634 (1985).
10. D. P. Dewangan and J. Eichler, *Phys. Rep.* **247**, 59 (1994).
11. K. B. MacAdam, in *Electronic and Atomic Collisions*, ed. T. Andersen, B. Fastrup, F. Folkmann, H. Knudsen, and N. Andersen, (AIP Conf. Proc. **295**, Aarhus, Denmark, 1993) p. 183.
12. J. Wang and R. E. Olson, *J. Phys. B* **26**, L817 (1993).
13. G. A. Kohring, A. E. Wetmore, and R. E. Olson, *Phys. Rev. A* **28**, 2526 (1983), and references therein.
14. J. Wang and R. E. Olson, *J. Phys. B* **27**, 3707 (1994); J. Wang, R. E. Olson, and J. H. McGuire, *Nucl. Instrum. Meth.* (1995).
15. M. F. V. Lundsgaard, Z. Chen, C. D. Lin, and N. Toshima, *Phys. Rev. A* **51**, 1347 (1995).
16. D. M. Homan, M. J. Cavagnero, and D. A. Harmin, *Phys. Rev. A* **50**, R1965 (1994); *Phys. Rev. A* **51**, 2075 (1995).
17. T. Ehrenreich, J. C. Day, S. B. Hansen, E. Horsdal-Pedersen, K. B. MacAdam, and K. S. Mogensen, *J. Phys. B* **27**, L383 (1994).
18. M. F. V. Lundsgaard, N. Toshima, Z. Chen, and C. D. Lin, *J. Phys. B* **27**, L611 (1994).
19. S. Bradenbrink, H. Reihl, Th. Wörmann, Z. Roller-Lutz, and H. O. Lutz, *J. Phys. B* **27**, L391 (1994); S. Bradenbrink, H. Reihl, Z. Roller-Lutz, and H. O. Lutz, *J. Phys. B* **28**, L133 (1995).

# COLLISIONS WITH SURFACES AND CLUSTERS

Interaction of Highly Charged Ions
with Metal and Insulator Surfaces .............................. 631
    F. AUMAYR
Inelastic Ion Surface Collisions .................................. 647
    V.A. ESAULOV, L. Guillemot, S. Lacombe,
    and V. Ngoc Tuan
Molecular Collisions on Large van der Waals Clusters ........ 657
    J.P. VISTICOT, J. Berlande, X. Biquard,
    M.A. Gaveau, A. Lallement, J.M. Mestdagh,
    and O. Sublemontier

# Interaction of Highly Charged Ions with Metal and Insulator Surfaces

## Friedrich Aumayr

*Institut für Allgemeine Physik, Technische Universität Wien,*
*Wiedner Hauptstraße 8-10, A-1040 Wien/Austria*

**Abstract.** Recent experimental investigations concerning the interaction of slow highly charged ions with clean metal and insulator surfaces are presented, with main emphasis on the ion-induced electron emission. At projectile impact energies close to the lower limit set by the projectile´s image charge attraction multiply excited "hollow atoms" are transiently formed which primarily decay by potential electron emission (PE). Essential differences for metal and insulator surfaces will be discussed by comparing results for polycrystalline Au - and LiF targets.

## 1. INTRODUCTION

The "hollow atom" is one of the latest and most exotic creations of atomic collision physics and represents a multiply excited (neutral) atom with most or all of its electrons in outer energy levels, while inner shells remain empty. Of course such a multiply inverted atom is a rather short - lived, rapidly changing complex with lifetimes of at most some ten femtoseconds and it is therefore rarely found in nature. "Hollow atoms" can be produced by colliding slow ($< 10^6$ m/s) highly charged ions (HCI) with metal surfaces. The conversion of a HCI into a "hollow atom" and its subsequent relaxation by emission of electrons and/or X-rays has been the subject of extensive experimental and theoretical studies in the past few years (for recent reviews see e.g. [1-5]).

In this review, we will first discuss an overall scenario of hollow atom formation and decay for metallic targets, since there already a rather consistent picture has emerged (chap. 2). Instead of a comprehensive (historical) review, selected experimental results will be presented to examplify the latest progress in this field (chap. 3) primarily referring to most recent work. In contrary to metallic targets much less is known about the interaction of HCI with an insulating surface. In the last part (chap. 4) we will therefore discuss the similarities and differences that have to be expected when bombarding insulators with HCI instead of metallic targets, and present some of the few experimental results so far available.

## 2. HOLLOW ATOMS

The current scenario for the interaction of slow HCI with metal surfaces has been derived from numerous experimental investigations of total electron yields [6-14], electron emission statistics [8, 11, 14-16] and fast Auger electron energy distributions [17-38], together with the analysis of scattered projectiles [39-46] and soft X-ray emission [31, 47-50]. It basically involves three stages (cf. fig. 1), i.e
(I)   the projectile's approach towards the surface
(II)  its close contact with the surface
(III) its penetration into the target bulk (or eventual backscattering into vaccum).

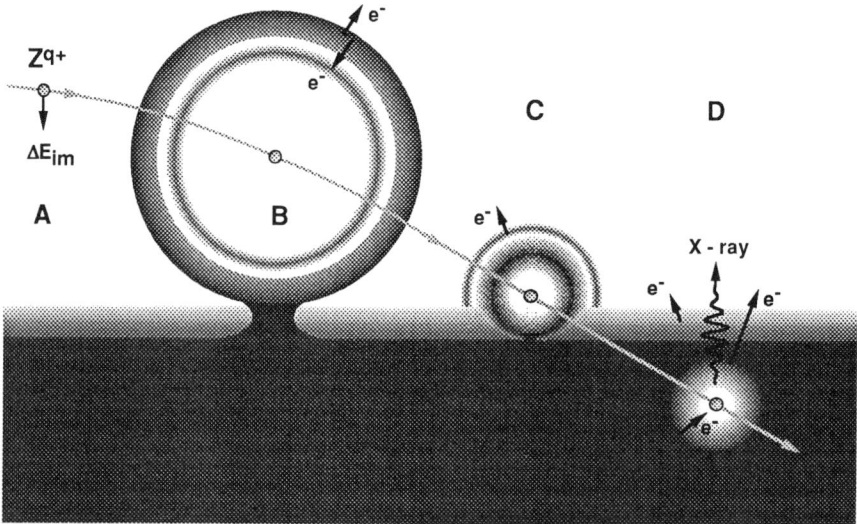

**FIGURE 1** Scenario of hollow atom formation and relaxation.
A ... an approaching HCI is accelerated towards the surface by its own image charge.
B ... at a critical distance electrons from the surface are captured resonantly into highly excited projectile states. A hollow atom is formed and decays via different electron emitting processes.
C ... screening of the projectile's core by metal electrons peels off remaining outer shell electrons.
D ... inner shell vacancies of the projectile are filled in close collisions with surface atoms or via Auger processes (cf. text).

The scenario's main features are based on the so-called "classical over-the-barrier model" (COB-model), which has been developed by Burgdörfer et al. [4, 51]. The approach of a slow HCI (charge state q; projectile velocity $v_p \ll v_F$, $v_F$ being the Fermi velocity of electrons inside the metal target) towards a metal surface (usually characterized in the jellium approximation by a conduction band with the work function $W_\Phi$ and Fermi energy $E_F$) causes a collective response of the metal electrons, which under the above assumptions at large distances can be described by a classical image potential. This image potential accelerates the HCI towards the metal surface (cf. fig. 1) and therefore poses a lower limit to the projectile impact velocity, corresponding to an upper limit for the available HCI - surface interaction time. In addition, the image interaction causes a shift of the projectile electron states and decreases the height of the electronic potential barrier between the HCI and the surface (cf. fig. 2a), which is formed by the projectile's potential, its image potential and the image potential of the particular electron to be captured. At a critical distance (atomic units)

$$d_c(q) \approx \sqrt{2q}/W_\Phi \tag{1}$$

classical overbarrier transitions of conduction band electrons from near the Fermi level into projectile states become possible (cf. figs. 1 and 2a). In general, resonant transitions (resonant neutralisation RN) will be favoured and therefore predominantly highly excited states of the projectile will be populated (cf. fig. 2b).

With further approach the projectile's levels are shifted upwards in energy due to image interaction (IS) and screening of the projectile charge by already captured electrons (SS), and somewhat lower projectile n-shells are now populated, either by a cascade of auto-ionizing transitions (auto-ionization AI) or because they now come into resonance with states at the Fermi edge. On the other hand, previously populated higher levels are emptied by resonant ionisation (RI) into empty states of the conduction band, auto-ionization and/or promotion above the vacuum level (fig. 2b). This interplay of electronic interactions continues during the projectile's approach towards the surface and gradually shifts the population to somewhat lower n-levels [1].

All the atomic states involved are populated and de-excited within a few femtoseconds only and can therefore not be considered as stationary states, as has already been pointed out by Burgdörfer [4]. Although a considerable number of electrons are emitted (mainly via AI and to a less extent via IS/SS promotion into vacuum) these losses are rapidly compensated by RN and the projectile is finally completely neutralized well above the surface with its electrons distributed among a number of excited levels, such forming the so-called "hollow atom" of highly transient nature (cf. fig. 1).

The image charge attraction limits the available interaction time above the surface. That is why the relaxation of the hollow atom to its neutral ground state (via the above described manifold of electronic interactions) is usually far from being complete when the projectile finally enters the region of appreciable surface electron density (stage II). Metal electrons almost immediately (timescale: inverse plasmon frequency) form a dynamic screening cloud around the ionic projectile core, which "peels off" (PO, [4, 5, 8]) all electrons with larger Rydberg radii than the respective screening distance $v_F/\omega_p$ ($\omega_p$ ... surface plasmon frequency).

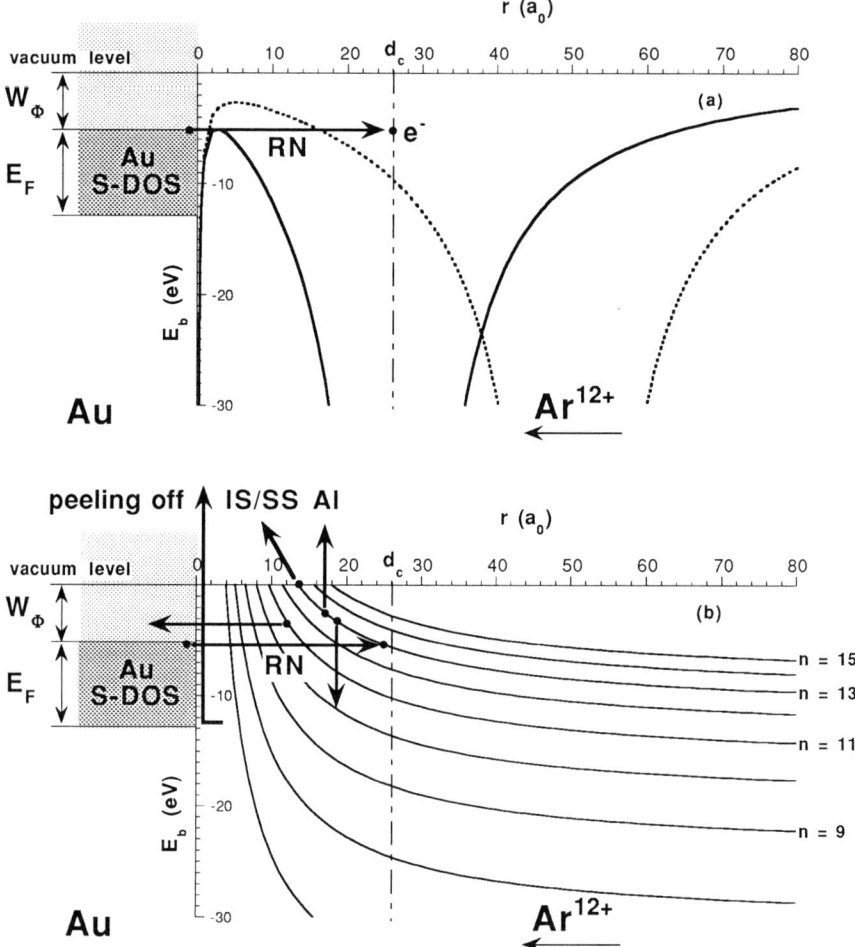

**FIGURE 2 (a)** Potential barrier between a metal surface (Au) and a HCI ($Ar^{12+}$) at a distance from the surface of about 50 a.u. (dotted curve) and 26 a.u. (full curve), respectively. In the second case the potential barrier has decreased below the Fermi level of Au, and electron capture becomes classically allowed. **(b)** States of a neutralizing HCI approaching a metal surface (cf. text).

Further relaxation of this now secondary "hollow atom" is governed by two competing processes (stage III). At low projectile velocity Auger transitions (Auger neutralization AN) fill the remaining L- or M shell holes of the projectile [52], while resonant vacancy transfer between the hollow projectile and target atomic levels via level crossing (Landau - Zener) and orbital promotion (Fano - Lichten) in close collisions dominate at higher impact velocities [53-55]. Filling of the so far surviving inner shell holes (including possible K shell holes) terminates the relaxation of the projectile and provides the majority of the observed fast (projectile) Auger electrons, in competition with X-ray emission.

Alternatively, the projectiles may become backscattered at any point on their trajectory at or inside the solid. For grazingly incident HCI, this backscattering can correspond to a "specular reflection" of the projectiles at the repulsive planar surface potential [40], in which case the projectile deexcitation can proceed on the outgoing trajectory [54].

## 3. EXPERIMENTAL RESULTS FOR METAL TARGETS

The formation and relaxation of a hollow atom as described in the previous chapter is accompanied by the emission of slow electrons, faster Auger electrons as well as characteristic X-rays. In addition, the ion - surface interaction determines the projectile's trajectory (incoming and outgoing) and its final charge state and kinetic energy. Appropriate experimental techniques can therefore be used to extract information on the underlying physical processes.

### 3.1 Electron Emission

The interaction of a HCI with a metal surface can produce a large number of electrons. Electron emitting mechanisms like AI, IS/SS promotion, PO, AN etc. are active during all stages (I-III) of the interaction [5]. The large majority of emitted electrons involves kinetic energies of a few eV only [17, 36] with a broad, unstructured energy distributions that cannot be assigned to specific electronic interactions. The filling of projectile inner shell holes leads to emission of energetic electrons (e.g. KLL Auger electrons) [17-38], and X-rays [31, 47-50] at characteristic energies. Fast Auger electrons emitted inside the solid either from the projectile or the target atoms may also produce slow secondary electrons [5, 12].

By using a very sensitive technique (electron emission statistics [8]) it has become possible [10] to measure total yields for emission of slow ($E_e \leq 50$ eV) electrons due to the impact of very slow ($v_p \leq 5.10^5$ m/s) highly charged ions ($Ar^{q+}$ ($q \leq 18$), $Xe^{q+}$ ($q \leq 51$) and $Th^{q+}$ ($q \leq 80$) provided by the Lawrence Livermore National Laboratory electron beam ion trap EBIT [56]) on clean polycrystalline gold. Huge electron yields of up to 280 electrons per incident ion have been observed (cf. fig. 3) which for a fixed nominal projectile impact energy increase more or less linearly with projectile charge state q. All measured total yields showed a gradual decrease with increasing impact velocity, with a levelling off towards an apparently velocity-independent part [9, 11]. Accompanying model calculations [9, 15] indicated that the PO mechanism delivers a contribution to the total electron yield which is nearly independent of the projectile's velocity (we remark that the IS/SS promotion - mechanism was also found to be rather insensitive to $v_p$ but always remained comparably unimportant), whereas the Al-contribution from above the surface decreases with increasing projectile velocity, because of the decreasing time available for Al cascades in front of the surface. For a more quantitative comparison between measured electron yields and corresponding calculated results, the fractions of electrons which actually can escape into vacuum and therefore be detected by an experimental setup have to be known for each electron emission mechanism. Preliminary calculations of such "escape fractions" are available for electrons emitted via above - surface AI and during PO as a function of electron energy and place of origin [57]. The importance of secondary emission processes from below the surface induced by fast Auger electrons is unsettled and currently a matter of debate [5, 12, 14].

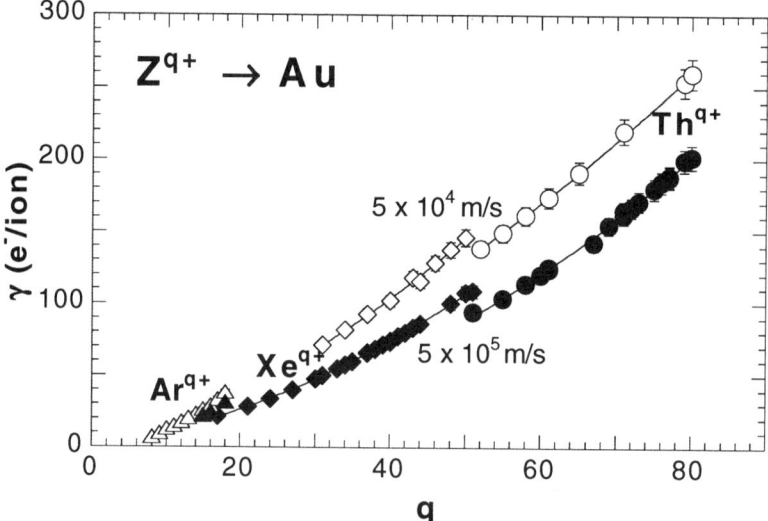

**FIGURE 3** Total electron yield $\gamma$ plotted vs. ion charge state q for impact of highly charged $Ar^{q+}$ q≤18 (triangles), $Xe^{q+}$ q≤ 51 (diamonds) and $Th^{q+}$ q≤ 80 (circles) ions on clean polycrystalline gold at impact velocities of $v_p \approx 5\times10^4$ m/s (open symbols) and $v_p \approx 5\times10^5$ m/s (full symbols), respectively (data from Ref. [10]).

Furthermore, from precise measurements of the velocity dependence of measured total slow electron yields at very low ($v_p < 3 \times 10^4$ m/s) projectile velocities, the ultimate low impact energy limit due to projectile image charge acceleration towards the metal surface could be demonstrated and quantitively evaluated [10]. The results of Aumayr et al. (e.g. impact energy limit of about 700 (±160) eV for $Th^{71+}$ ions on a Au surface) confirmed the validity of equ. (1) up to the so far highest accessible HCI charge states.

Recently it has been shown, that measuring the electron number statistics, apart from its use for deriving total electron emission yields, also contains additional information on the first phases of hollow atom formation and deexcitation [58].

A large amount of work [17-38] has been devoted to Auger electron spectroscopy, which provides experimental evidence for the last steps of the relaxation cascade in a hollow atom. The filling of (empty) K-shell or L-shell holes in the projectile can be identified from the respective characteristic high energy peaks in the fast electron energy spectra. Moreover, the spectra show profound structures, which, by means of Hartree-Fock atomic structure calculations, can be used to identify the hollow atom states present at the moment of Auger decay [21, 24]. Doppler shift measurements of Auger lines as a function of observation angle are consistent with electron emission from projectiles still travelling along their original direction of incidence. Because of this fact, in earlier work all these electrons have been attributed to phase I, but more recent experiments showed that only a small fraction of the inner shell vacancies (typically less than a few %) can be filled above the surface [3, 20, 30, 34, 35, 37], whereas the majority of the fast Auger electrons will be emitted only after the projectile has entered the metal. The generally minute above - surface contribution to the fast Auger electron peaks is limited by the image charge acceleration and can only be quantified in careful measurements for very low impact energy and/or grazing impact angle [3, 20, 30, 34, 35, 37]. Observations of characteristic target-Auger electron lines [27] definitely showed that at least some inner shell vacancies in the projectiles can survive sufficiently long to become transferred in close encounters to the target atoms. Recent experiments in connection with calculations [24, 25, 27, 52, 53, 55] showed that only at rather low impact velocities a substantial fraction of L-shell holes in the projectile below the surface are filled via Auger transitions (e.g. Auger neutralization AN). At high impact velocities vacancy transfer between the hollow projectile and target atomic levels via level crossing (Landau - Zener) and orbital promotion (Fano - Lichten) in close collisions provides a much faster (velocity proportional) filling of the L - shell.

## 3.2 X-Ray Emission

For heavier projectiles, recombination of inner shell vacancies can also proceed via characteristic X-ray photon emission. Detailed information can be obtained from highly resolved X-ray spectra produced in HCI-surface collisions [31, 47-50] complementary to the observation of fast Auger electron emission. The number of spectator electrons residing in higher principal quantum shells at the moment of the (now radiative) inner shell transition gives a rather direct evidence for the existence of "hollow" projectiles. In case of bare projectile ions the sequential emission of two $L_\alpha$ photons has been used to monitor the change in L-shell filling between these two emission events, by comparing the corresponding satellite spectra [48].

## 3.3 Ion Scattering

Ion scattering under grazing conditions provides a rather direct means of measuring projectile impact energy gain due to the image charge interaction [40]. Such measuments have to be performed on carefully prepared, flat and clean surfaces of single crystals. Data are available for $Ar^{q+}$ ($q \leq 6$) on Al(111) [40], $Xe^{q+}$ ($q \leq 12$) on Fe(110) [59], $Xe^{q+}$ ($q \leq 33$) on Al(111) [41], $I^{q+}$ ($q \leq 25$) and $Pb^{q+}$ ($q \leq 36$) on Au(110) [45] and $Xe^{q+}$ ($q \leq 15$) on LiF(100) [42]). Because of the image charge interaction, the HCI are accelerated on their incident path towards the surface plane (cf. fig.1) until their complete neutralization. This acceleration results in an increased effective angle of incidence for the projectiles. The image charge - induced energy gain can then be determined from the measured angular distributions of (specular) reflected neutralized projectiles. Experimental results are consistent with the predictions of the COB model.

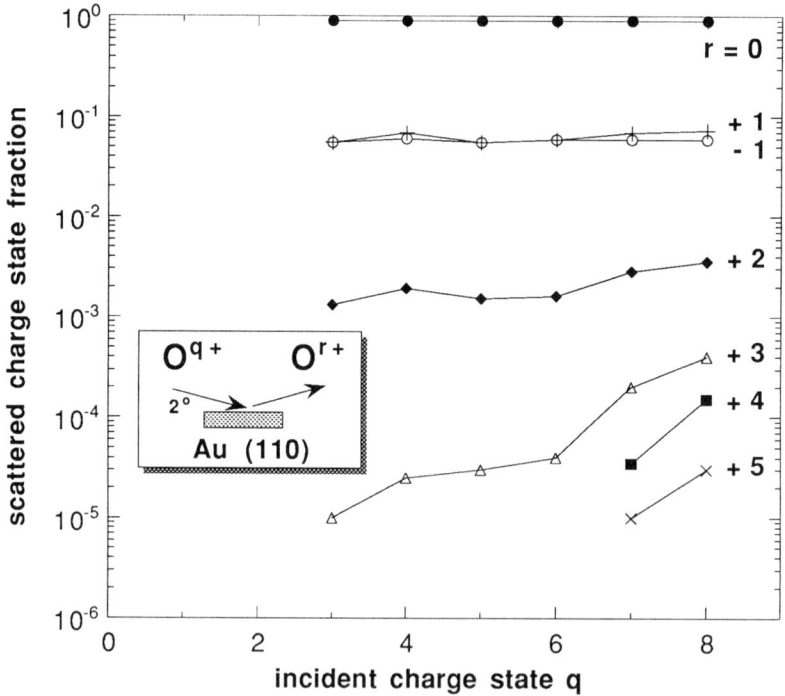

**FIGURE 4** Scattered projectile charge fractions vs. charge state of incident $O^{q+}$ projectiles (incidence angle 2°, impact energy 3.75 keV/amu, Au(110) single crystal target). The dominating charge state fractions (0, +1, -1) are almost independent of the initial charge state. (adapted from Ref. [44]).

Recently, experimental data on the charge state distributions of scattered projectiles for slow $O^{q+}$ ($q \leq 8$) ions surface channeled along a Au single crystal surface have become available [44], where surface penetration could be ruled out. Projectile trajectory calculations indicated that the projectiles spend less than 30 fs within 2 Å of the topmost Au surface layer. The measured charge state fractions of the reflected projectiles were found to be essentially independent of the incident charge state (cf. fig. 4). This remarkable result indicates that charge state equilibration occurs on a time scale of less than 100 fs. Corresponding calculations within the COB model have been performed, that treat the direct und efficient filling of inner shells by quasi-resonant charge transfer into the L shell of the projectile [54]. They showed, that the speed of neutralization is significantly enhanced by the upward shift of energy levels at the surface due to dynamical screening, which leads to almost complete relaxation of the projectiles within the interaction time found by the above experiment. Formation of negative oxygen ions (up to 6%) shows a kinematic resonance that has been explained by the Galilei shifted Fermi sphere (due to the high velocity component of the projectiles parallel to the surface) in conjunction with a shift (increase) of the electron affinity due to image interaction [44].

## 4. EXPERIMENTAL RESULTS FOR INSULATOR TARGETS

For bombardment of insulator surfaces with slow HCI one expects the above described scenario to change considerably, since electrons are not as abundantly available and much less mobile than in free electron metals. So far only a few experimental investigations involving insulator targets have been carried out, all of them with LiF surfaces. LiF has been chosen because at elevated temperatures it exhibits an ionic conductivity which is sufficient to avoid charging of the surface by impact of low intensity ion beams [60].

In fig. 5 the occupied, un-occupied and forbidden electron energy states of LiF are compared with those of a typical metal (e.g. Au). Whereas in metals the conduction band is divided at the Fermi level in electronic bands of occupied and unoccupied states, in LiF the $F^-(2p)$ electrons form a completely occupied valence band with binding energies of about 12 eV (bulk value; surface value can somewhat differ). This valence band exhibits a flat dispersion curve which results in large effective electron masses and a localisation of the 2p-electrons at the fluorine atoms in the lattice. A broad energy gap of about 14 eV (bulk value) extends well above the vacuum level to the first empty allowed states. Not only the electronic structure but also the dielectric response of LiF clearly differ from a perfect conductor. The classical over barrier condition has to be re-calculated by taking into account a dynamic image potential, which involves the (frequency dependent) permittivity $\varepsilon(\omega)$ of LiF between about $\varepsilon(0) = 8.6$ (static value) and $\varepsilon(\infty) = 1.9$ (optical value).

A COB treatment for finite ε has recently been presented [61]. The reduced image interaction lowers the potential barrier between the impinging projectile and the surface, so that the critical distance $d_c$ for over-the-barrier transitions is larger for LiF than for a perfect conductor of similar work function. However, the smaller work function of typical metals (4 - 6 eV) more than compensates for this effect. As a consequence, in order to capture electrons the HCI has to come much closer to the LiF - than to the metal surface (almost a factor of 2). Whether HCI projectiles can fully be neutralized until impact on the LiF surface is still an open question. Because of the highly localized valence band electrons it is probably difficult to extract rapidly several electrons from a specific surface region, especially at normal incidence conditions. However, complete neutralization (and in the case of oxygen projectiles very efficient negative ion formation [43]) has been found in grazing collisions with LiF single crystals [42], where electrons can be captured from different lattice sites. These measurements also reported projectile impact energy gains due to image acceleration, comparable to those of typical metals [42]. The results for LiF could be explained from a compensation of the smaller critical distance with the weaker image attraction.

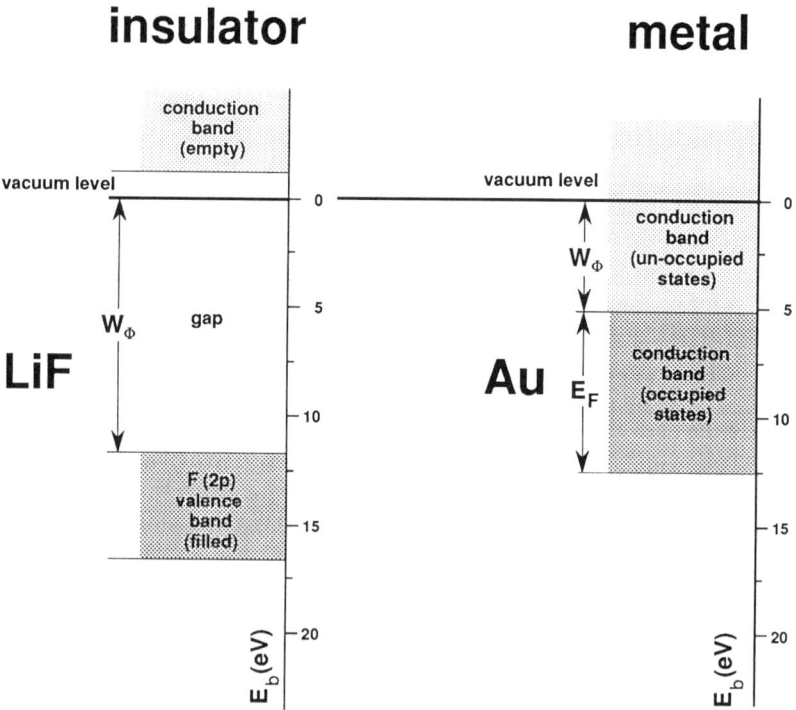

**FIGURE 5** Occupied, un-occupied and forbidden electron energy states (schematic) in a typical insulator (LiF) compared to a typical metal (Au). $W_\Phi$ ... work function, $E_F$ ... Fermi energy, $E_b$ ... binding energy.

Because the neutralization sequence starts closer to the surface less time is available for the autoionization processes until surface impact. This is consistent with measurements of total slow electron yields from polycrystalline LiF under slow MCI impact [13, 16], where for projectiles without inner shell vacancies (e.g. $N^{5+}$) smaller potential emission yields have been found than for a gold targets (cf. fig. 6.).

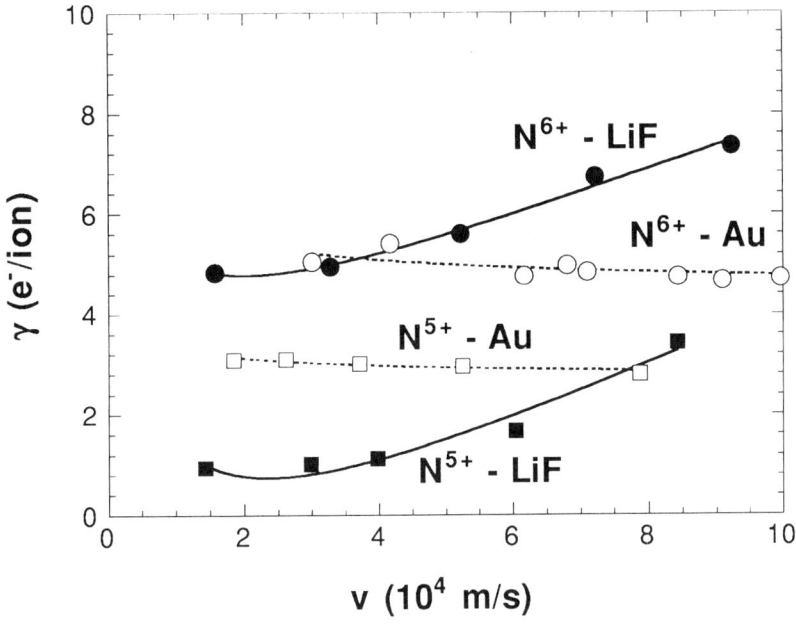

**FIGURE 6** Total slow electron yields γ for normally incident $N^{5+}$ (squares) and $N^{6+}$ (circles) projectiles on LiF (full symbols) and Au (dotted curves), respectively vs. projectile impact velocity (data from ref. [13]).

However, the changes in PE yields from $N^{5+}$ to $N^{6+}$ (cf. fig. 6) are considerably larger for LiF than for Au. This has been attributed [13] to the more efficient secondary electron emission induced by the fast ($E_e \approx 350$ eV) KLL-electron (from $N^{6+}$-projectile K-shell vacancy decay) inside LiF, which is mainly due to the larger inelastic mean free path in the insulator. Again there is a compensation of the less efficient "above-surface" electron production by an increased electron emission from below surface.

Pronounced differences in the dynamics of the above surface deexcitaion of HCI on metal and LiF surfaces have also been found in high resolution KLL Auger electron emission measurements [26]. In Fig. 7 we compare KLL spectra obtained with $N^{6+}$ projectiles on Si(100) and LiF(100) surfaces, respectively.

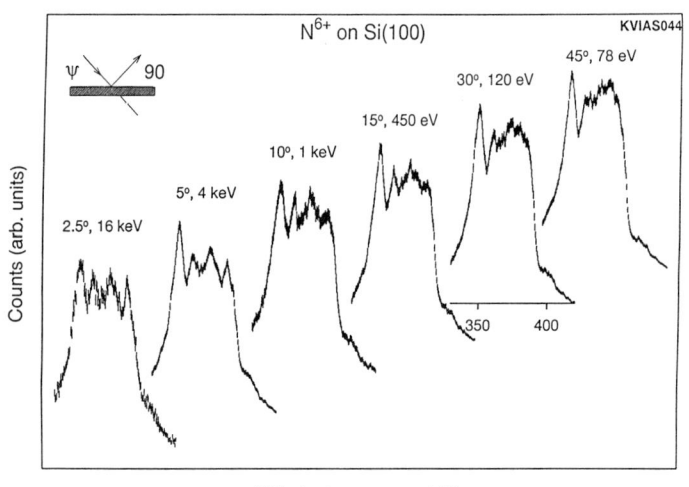

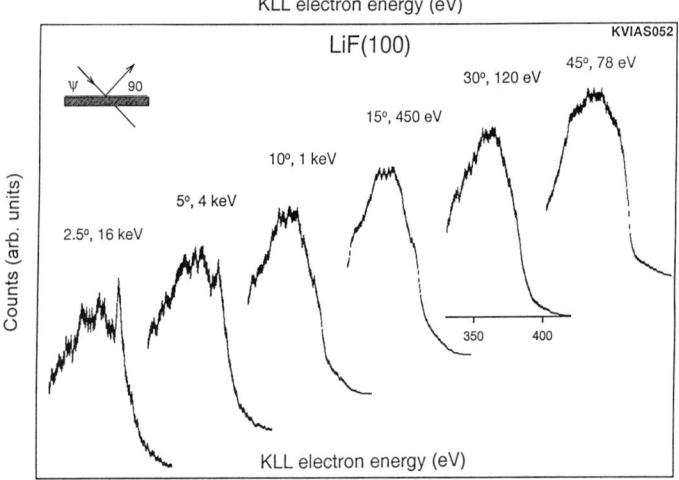

**FIGURE 7** KLL-Auger electron spectra obtained for $N^{6+}$ impact on Si(100) and LiF(100) surfaces, respectively. The ion energy and angle of incidence have been variied in order to keep the projectile's velocity component normal to the surface constant (from ref. [26]).

The prominent peak, which for the Si(100) target shows up on the low energy side of the KLL Auger electron distribution, is missing for the LiF(100) target, but for both targets a sharp peak appears with increasing collision energy at the high energy side. These peaks have previously been identified on the basis of atomic structure calculations to result from KLL decay of $N(1s2s^23l^4)$ and $N(1s2s^22p^4)$ configurations, respectively [21, 24]. The two peaks have been associated with two competing mechanisms filling the projectile L-shell. One of them proceeds via Auger cascades and starts as soon as the hollow atom is formed. The associated L-shell filling rate is small compared to typical KLL Auger rates and KLL decay will thus occur readily after a second L-electron has been captured ($N(1s2s^23l^4)$ configuration, low energy peak). However, as soon as the ion experiences close collisions with target atoms, the L-shell can be much more rapidly filled by direct transfer of target core electrons. This process is localized and the associated filling rate is proportional to the ion-target atom collision frequency. For sufficiently high projectile velocities the L-shell becomes completely filled before KLL decay takes place ($N(1s2s^22p^4)$ configuration, high energy peak). As seen from fig. 7 the latter mechanism is active for both Si and LiF, whereas the slow L-shell filling by Auger cascades is absent for LiF. This can in principle be explained [26] by assuming that no hollow atoms are formed above the surface, just as one would expect from LiF being an insulator (cf. above). The large binding energy and small mobility of the LiF valence band electrons blocks multiple electron capture from a single lattice site. Hollow atom formation is only possible by electron transfer from different sites, i.e. for high projectile velocities at grazing incidence. However, even if under such conditions a complete neutralization takes place, the $N(1s2s^23l^4)$ peak would not show up in the spectra, since a population shift towards lower n-levels inside the hollow atom mainly arises from the dynamic interplay between resonant neutralization, screening/image shift of projectile levels and losses to empty target states (cf. chapter 2). The large bandgap for LiF (12eV) merely blocks these processes and therefore a noticeable population of the L-shell can probably not be reached before the close collision regime is entered. This blocking of an important electron loss channel due to the energy gap of LiF is also responsible for the large fractions of negative oxygen ions found for ion scattering from a LiF surface [43].

Finally, we want to draw attention to a new type of sputtering ("potential sputtering") that has been observed when bombarding LiF surfaces with $Ar^{q+}$ ($q \leq 9$) ions [60]. The measured sputter yields are proportional to the potential energy carried by the Ar ions. Effective sputtering of predominantly neutrals [62] already takes place well below 100 eV impact energy and has been related to defect production (color centers) in LiF, following electron capture by the HCI. No such charge state dependent sputtering process can be observed for metal (Au) and semiconductor (Si) targets [63]. So far it is not clear whether this mechanism is only characteristic for alkali-halides or other insulating materials as well (which would have important practical applications in HCI - induced modification of materials).

## ACKNOWLEDGEMENTS

The author is grateful to HP.Winter (Wien) for a long and fruitful collaboration in this field and would also like to thank Dr. H. Kurz and Mssrs. M. Vana and C. Lemell for their assistance in parts of the work. Valuable discussions with A. Arnau, J. Burgdörfer, P.M. Echenique, W. Heiland, F. Meyer, R. Morgenstern, N. Stolterfoht, P. Varga and H. Winter (Berlin) are acknowledged. Work has been supported by Fonds zur Förderung der wissenschaftlichen Forschung and by the EU Human Capital and Mobility Program.

# REFERENCES

1. H. J. Andrä, et al., XVII ICPEAC, W. R. MacGillivray, I. E. McCarty, M. C. Standages, Eds., (IOP Conference Proceedings, Brisbane, 1991), p. 89.
2. P. Varga, HP. Winter, in *Particle Induced Electron Emission II* G. Höhler, Eds. (Springer, Heidelberg, 1992), Vol. 123, p. 149.
3. J. Das, R. Morgenstern, *Comments At.Mol.Phys.* **29**, 205 (1993).
4. J. Burgdörfer, in *Fundamental Processes and Applications of Atoms and Ions* C. D. Lin, Eds. (World Scientific, 1993).
5. F. Aumayr, HP. Winter, *Comments At.Mol.Phys.* **29**, 275 (1994).
6. M. Delaunay, M. Fehringer, R. Geller, D. Hitz, P. Varga, HP. Winter, *Phys.Rev.B* **35**, 4232 (1987).
7. J. W. McDonald, D. Schneider, M. W. Clark, D. Dewitt, *Phys.Rev.Lett.* **68**, 2297 (1992).
8. H. Kurz, K. Töglhofer, HP. Winter, F. Aumayr, R. Mann, *Phys.Rev.Lett.* **69**, 1140 (1992).
9. H. Kurz, F. Aumayr, C. Lemell, K. Töglhofer, HP. Winter, *Phys.Rev.A* **48**, 2182 (1993).
10. F. Aumayr, H. Kurz, D. Schneider, M. A. Briere, J. W. McDonald, C. E. Cunningham, HP. Winter, *Phys.Rev.Lett.* **71**, 1943 (1993).
11. H. Kurz, F. Aumayr, D. Schneider, M. A. Briere, J. W. McDonald, HP. Winter, *Phys.Rev.A* **49**, 4693 (1994).
12. I. G. Hughes, J. Burgdörfer, L. Folkerts, C. C. Havener, S. H. Overbury, M. T. Robinson, D. M. Zehner, P. A. Zeijlmans van Emmichhoven, F. W. Meyer, *Phys.Rev.Lett.* **71**, 291 (1993).
13. M. Vana, F. Aumayr, P. Varga, HP. Winter, *Europhys. Lett.* **29**, 55 - 60 (1995).
14. M. Vana, H. Kurz, HP. Winter, F. Aumayr, *Nucl.Instrum.Meth.Phys.Res.B* **100**, 402 (1995).
15. H. Kurz, F. Aumayr, C. Lemell, K. Töglhofer, HP. Winter, *Phys.Rev.A* **48**, 2192 (1993).
16. M. Vana, F. Aumayr, P. Varga, HP. Winter, *Nucl.Instrum.Meth.Phys.Res.B* **100**, 284 (1995).
17. M. Delaunay, M. Fehringer, R. Geller, P. Varga, HP. Winter, *Europhys.Lett.* **4**, 377 (1987).
18. S. T. de Zwart, A. G. Drentje, A. L. Boers, R. Morgenstern, *Surf.Sci.* **217**, 298 (1989).
19. L. Folkerts, R. Morgenstern, *Europhys.Lett.* **13**, 377 (1990).
20. J. Das, L. Folkerts, R. Morgenstern, *Phys.Rev.A* **45**, 4669 (1992).
21. S. Schippers, J. Limburg, J. Das, R. Hoekstra, R. Morgenstern, *Phys.Rev.A* **50**, 540 (1994).
22. J. Limburg, J. Das, S. Schippers, R. Hoekstra, R. Morgenstern, *Surf.Sci.* **313**, 355 (1994).
23. J. Limburg, J. Das, S. Schippers, R. Hoekstra, R. Morgenstern, *Phys.Rev.Lett.* **73**, 786 (1994).
24. J. Limburg, S. Schippers, I. Hughes, R. Hoekstra, R. Morgenstern, S. Hustedt, N. Hatke, W. Heiland, *Phys.Rev.A* **51**, 3873 (1995).
25. J. Limburg, S. Schippers, I. Hughes, R. Hoekstra, R. Morgenstern, S. Hustedt, N. Hatke, W. Heiland, *Nucl.Instrum.Meth.Phys.Res.B* **98**, 436 (1995).
26. J. Limburg, S. Schippers, R. Hoekstra, R. Morgenstern, H. Kurz, F. Aumayr, HP. Winter, *Phys.Rev.Lett.* **75**, (July 10, 1995).
27. S. Schippers, S. Hustedt, W. Heiland, R. Köhrbrück, J. Bleck-Neuhaus, J. Kemmler, D. Lecler, N. Stolterfoht, *Phys.Rev.A* **46**, 4003 (1992).
28. S. Schippers, S. Hustedt, W. Heiland, R. Köhrbrück, J. Bleck-Neuhaus, J. Kemmler, D. Lecler, N. Stolterfoht, *Nucl.Instrum.Methods B* **78**, 106 (1993).
29. R. Köhrbrück, N. Stolterfoht, S. Schippers, S. Hustedt, W. Heiland, D. Lecler, J. Kemmler, J. Bleck-Neuhaus, *Phys.Rev.A* **48**, 3731 (1993).
30. R. Köhrbrück, M. Grether, A. Spieler, N. Stolterfoht, R. Page, A. Saal, J. Bleck-Neuhaus, *Phys.Rev.A* **50**, 1429 (1994).
31. H. J. Andrä, et al., *Suppl. to Z.Phys.D* **21**, S135 (1991).
32. H. J. Andrä, A. Simionovici, T. Lamy, A. Brenac, A. Pesnelle, *Europhys.Lett.* **23**, 361 (1993).

33. P. A. Zeijlmans van Emmichoven, C. C. Havener, F. W. Meyer,
    *Phys.Rev.A* **43**, 1405 (1991).
34. F. W. Meyer, S. H. Overbury, C. C. Havener, P. A. Zeijlmans van Emmichhoven,
    D. M. Zehner, *Phys.Rev.Lett.* **67**, 723 (1991).
35. F. W. Meyer, S. H. Overbury, C. C. Havener, P. A. Zeijlmans van Emmichhoven,
    J. Burgdörfer, D. M. Zehner, *Phys.Rev.A* **44**, 7214 (1991).
36. P. A. Zeijlmans van Emmichoven, C. C. Havener, I.G. Hughes, D. M. Zehner,
    F. W. Meyer, *Phys.Rev.A* **47**, 3998 (1993).
37. F. W. Meyer, C. C. Havener, P. A. Zeijlmans van Emmichhoven,
    *Phys.Rev.A* **48**, 4476 (1993).
38. F. W. Meyer, L. Folkerts, I. G. Hughes, S. H. Overbury, D. M. Zehner,
    P. A. Zeijlmans van Emmichhoven, J. Burgdörfer, *Phys.Rev.A* **48**, 4479 (1993).
39. S. T. de Zwart, T. Fried, U. Jellen, A. L. Boers, A. G. Drentje, *J.Phys.B* **18**, L623 (1985).
40. H. Winter, *Europhys.Lett.* **18**, 207 (1992).
41. H. Winter, C. Auth, R. Schuch, E. Beebe, *Phys.Rev.Lett.* **71**, 1939 (1993).
42. C. Auth, T. Hecht, T. Igel, H. Winter, *Phys.Rev.Lett.* **74**, 5244 (1995).
43. C. Auth, A. G. Borisov, H. Winter, *to be published* , (1995).
44. L. Folkerts, S. Schippers, D. M. Zehner, F. W. Meyer, *Phys.Rev.Lett.* **74**, 2204 (1995)
    and corresponding erratum (in print).
45. F. W. Meyer, L. Folkerts, H. O. Folkerts, S. Schippers,
    *Nucl.Instrum.Meth.Phys.Res.B* **98**, 441 (1995).
46. I. Hughes, J. Limburg, R. Hoekstra, R. Morgenstern, S. Hustedt, N. Hatke, W. Heiland,
    *Nucl.Instrum.Meth.Phys.Res.B* **98**, 458 (1995).
47. J. P. Briand, L. de Billy, P. Charles, S. Essabaa, P. Briand, R. Geller, J. P. Desclaux,
    S. Bliman, C. Ristori, *Phys.Rev.Lett.* **65**, 159 (1990).
48. J. P. Briand, L. de Billy, P. Charles, S. Essabaa, P. Briand, R. Geller, J. P. Desclaux,
    S. Bliman, C. Ristori, *Phys.Rev.A* **43**, 565 (1991).
49. B. d´Etat, J. P. Briand, G. Ban, L. de Billy, J. P. Desclaux, P. Briand,
    *Phys.Rev.A* **48**, 1098 (1993).
50. D. Schneider, M. A. Briere, J. McDonald, J. Biersack, *Rad.Eff.Def.Solids* **127**, 113 (1993).
51. J. Burgdörfer, P. Lerner, F. W. Meyer, *Phys.Rev.A* **44**, 5647 (1991).
52. A. Arnau, P. A. Zeijlmans van Emmichoven, J. I. Juaristi, E. Zaremba,
    *Nucl.Instrum.Meth. Phys.Res.B* **100**, 297 (1995).
53. A. Arnau, R. Köhrbrück, M. Grether, A. Spieler, N. Stolterfoht,
    *Phys.Rev.A* **51**, R3399 (1995).
54. J. Burgdörfer, C. Reinhold, F. Meyer, *Nucl.Instrum.Meth.Phys.Res.B* **98**, 415 (1995).
55. N. Stolterfoht, A. Arnau, M. Grether, R. Köhrbrück, A. Spieler, R. Page, A. Saal,
    J. Thomaschewski, J. Bleck-Neuhaus, *Phys.Rev.A* , in print (1995).
56. D. Schneider, M. W. Clark, B. Penetrante, J. McDonald, D. DeWitt, J. N. Bardsley,
    *Phys.Rev.A* **44**, 3119 (1991).
57. C. Lemell, HP. Winter, F. Aumayr, J. Burgdörfer, C. Reinhold,
    *Nucl.Instrum.Meth.Phys.Res.B* (1995, in print).
58. M. Vana, F. Aumayr, C. Lemell, HP. Winter, *Intern.J.MassSpectr.IonProc.* (1995, in print)
59. H. Winter, 6th HCI, P. Richard, M. Stöckli, C. L. Cocke, C. D. Lins, Eds.,
    (AIP Conf. Proceedings, Manhattan, KA, 1992), Vol. 274, p. 583.
60. T. Neidhart, F. Pichler, F. Aumayr, HP. Winter, M. Schmid, P. Varga,
    *Phys.Rev.Lett.* **74**, 5280 (1995).
61. A. Bárány, C. J. Setterlind, *Nucl.Instrum.Meth.Phys.Res.B* **98**, 184 (1995).
62. T. Neidhart, F. Pichler, F. Aumayr, HP. Winter, M. Schmid, P. Varga,
    *Nucl.Instrum.Meth. Phys.Res.B* **98**, 465 (1995).
63. T. Neidhart, F. Pichler, F. Aumayr, HP. Winter, M. Schmid, P. Varga, 3S´95 Symp. on
    Surface Science, P. Varga and F. Aumayr, Eds., Kaprun, Salzburg, Austria, 1995), p. 74.

# Inelastic ion surface collisions

V.A.Esaulov, L.Guillemot, S.Lacombe and Vu Ngoc Tuan

*Laboratoire des Collisions Atomiques et Moléculaires, (Unité Assosciée au CNRS), bât. 351, Université de Paris Sud, 91405 Orsay, FRANCE*

**Abstract.** Results of an investigation of dynamics of excited state production in collisions of inert gas projectiles on Na, Mg, Al and Si surfaces are presented. Angle resolved measurements of charge fractions and electron spectra are reported. The efficiency of excited autoionising state production as a function of energy is determined. These results are compared to known gas phase collision results. A similarity in the general trends of excited state production, suggests that both in gas phase and in ion-surface collisions similar *primary* excitation mechanisms are operative. However *surface specific* electron capture, loss, deexcitation and core rearrangement processes lead to a strong modification in final state distributions and a dominant selective production of the lower lying excited states occurs. Modifications induced by oxygen adsorption are discussed.

## Introduction

This report presents the results of an investigation of the trends and dynamics of excitation processes in collisions of inert gas projectiles on Na, Mg, Al and Si surfaces. Some of these systems as well as others, have been experimentally studied in this laboratory and elsewhere [1-12,14-16]. In some previous studies of ion scattering on surfaces, ionisation and production of excited states that could not be explained in terms of the usual framework of resonant electron capture and loss processes, was ascribed to "binary" collisions of the projectile with surface atoms. The excitation processes were then arbitrarily discussed in terms of the molecular orbital (MO) electron promotion model [13], which has described successfullly outer shell excitation in gas phase collisions. Its use for describing such processes in ion-surface collisions has not, to our knowledge, been justified and no attempt was made to study systematically systems that have also been studied in gas phase collisions.

Here we shall discuss results of investigations of inelastic inert gas ion scattering on some simple targets like Na, Mg, Al and Si. The work at our laboratory includes studies of angle resolved ion scattering and electron spectroscopy [10,23]. The choice of these systems was dictated by the fact that in the gas phase, collisions of Na, Mg and Al ions with Ne atoms have been studied [17-22] and these results can thus serve for comparison with surface data. In the following we shall discuss ion and neutral scattering spectroscopy, charge fraction measurements and electron and photon spectroscopy. We shall try to delineate effects related to the binary violent collisions and compare the data to existing gas phase studies to analyse the modifications brought about by the neighbouring surface. Finally, we will consider the effect of adsorbate atoms.

© 1995 American Institute of Physics

## Results

Usually in low energy ion-surface scattering a large fraction of ions is neutralised by resonant and Auger neutralisation processes and any scattered ions are assumed to correspond to a surviving fraction. We performed measurements of angular distributions (N($\theta$)) of scattered ions and neutrals for Ne atoms and ions incident on Na, Mg, Al and Si surfaces [10, 14-16], from which we deduced scattered ion charge fractions defined as $Y^+$ = Ions/ (Ions+Neutrals). Here we mention a very important point, viz. scattered ions are observed independently of the charge state of the incident particle and the angular distributions and charge fractions are similar for incident Ne and $Ne^+$. We found that the ion fractions are quite large: for 2keV $Ne^+$ ions incident at 6 deg the total ion fractions integrated over scattering angles are of 35%, 14%, 10% and 5% for Na, Mg, Al and Si surfaces respectively. These studies clearly demonstrate the existence of ionisation in collisions of incident and surface atoms. It is most interesting to note that the general trend is that, for a given energy the charge fractions increase as one goes from Si to Na, i.e. as the collision becomes more "symmetric". We note that this is similar to the well known trend in gas-phase collisions, where the excitation cross-sections are large for "symmetric" systems [13].

The identification of some of the possible excitation channels is rendered possible by photon and electron spectroscopy measurements.

Typical electron spectra obtained for He and Ne ions incident at 6° to the target surfaces are shown in fig.1a. These consist of a continuous distribution on which are superimposed a series of peaks. The energy position and the shape of these structures in the electron spectrum, depend on the observation angle (fig.1) and are determined by ejection kinematics, related to emission occuring from a moving source [3,5,11]. For all targets there are two main large peaks due to the decay of Ne** $2p^4$ ($^3P$ and $^1D$) $3s^2$ states, whose center of mass energy is close to the 20.35eV and 23.55eV positions corresponding to the decay of the excited atom in free space. One generally also observes (fig.1a (curve 3)) a series of much smaller structures, the largest of which lies close to 30 eV. These are attributable to higher lying autoionising states of Ne and $Ne^+$ [11,12,14] with a $2p^3 nln'l'$ configuration. Structures due to target excitation are observed for all cases. Thus in case of Mg, the structures lying at about 35, 39.5 and 44 eV are due to excited states of $Mg^+$ and Mg [3,11,14]. These are situated atop of a broad structure, which is attributable to bulk LVV emission, involving transitions from the metal valence band [1-6,11,14]. Similarly, in the case of Na the peak at 26 eV is attributed to decay of excited sputtered Na atoms in the $2p^5 3s^2$ state and the shoulder below it, to bulk Na emission. An important point to be stressed is that for the Mg [3] and Al [9] targets the *electron spectrum from ion and neutral scattering are similar*, consistent with our charge fraction results.

The results for incident $He^+$ are summarised in fig.1a (curve 1). The spectrum for scattering on Na displays a peak due to decay of excited sputtered Na atoms in the $2p^5 3s^2$ state and a peak lying at higher energies which is due to autoionisation of the He** $2s^2$ state, with a rest frame energy of 33.2 eV. Electron energy spectra for He scattering on Mg and Al also show excitation of this He state. We were not able to observe its production for scattering on Si.

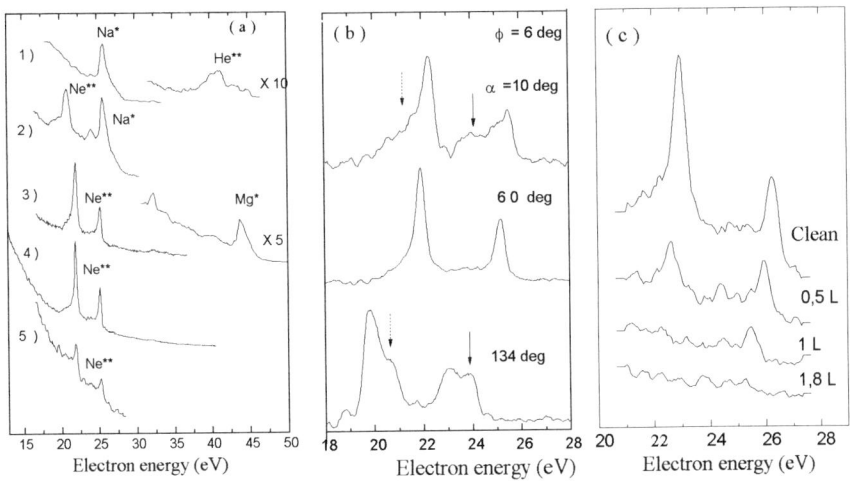

Fig.1. Energy spectra of electrons in the autoionising state region, produced in a) 2 kev $He^+$ incident on Na at 6° incidence and 45° collection; 2 keV $Ne^+$ collisions incident at 6° on 2) Na, 90° detection, 3) Mg 45° detection, 4) Al 45° detection and 5) Si 45° detection; b) Observation angle dependence of the line shapes of Ne** peaks produced in $Ne^+$ scattering on Al; c) Effect of oxygen adsorption on the Ne** excitation (Mg surface).

In the case of $Ar^+$ scattering we did not observe Ar** state production. The electron energy spectra display peaks due to target excitation [1,9,10] only.

The electron spectroscopy results clearly demonstrate the existence of electronic excitation of the projectile atoms. This was further confirmed in a collaboration with Drobnitch and Bandourin[14], where VUV photon spectra were measured for 15 keV $Ne^+$ scattering on Mg and Al surfaces for 10° incidence angles. Strong resonance line emission at 743.7nm and 735.9 nm was observed indicating that a substantial number of resonantly neutralised ions are scattered off the surface [ see also 5]. Additionally VUV lines are due to the $(2p^4)^3P3s/3p$, the $(2p^4)$ $^1D3s/3p$, the $2s2p^6$ excited states of $Ne^+$ and the $2s2p^5$ state of $Ne^{++*}$.

Our electron spectroscopy data show that, in all the cases considered our results follow the same pattern as in gas phase collisions. Indeed existing experimental data [17-22] for $Na^+$, $Mg^+$ and $Al^+$ scattering on He, Ne and Ar and $He^+$ scattering on Na vapours, shows that He and Ne are excited, while Ar excitation is extremely weak. Note that in all cases excitation of the target metal atom was observed. Gas phase data for Si were not found.

We next studied the energy dependence of excited state production and compared with data in gas phase and ion-solid collisions. The intensities of the

main autoionisation peaks were determined as a function of incident energy for 6°, 15° and 30° incidence angles. These are shown for the Mg target (fig.2.). The general trends apparent from our measurements are that :
a) the efficiency of autoionising state production generally increases with collision energy in the range of energies below 3keV,
b) the ratio of the intensities of peaks due to $^3P3s^2$ and the $^1D3s^2$ states *decreases with increasing* collision energy,
and c) the ratio of the intensity of the peak due to the $^1D3s^2$ state to that due to the $^3P3s^2$ state decreases as the incident angle increases.

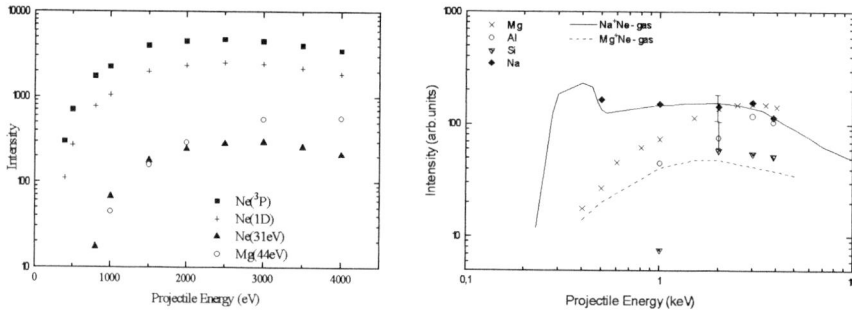

*Fig.2. a) Energy dependence of autoionising state production in Ne scattering on Mg for various incidence angles ($\phi$) and a detection angle of $\alpha = 45°$. b) Efficiency of autoionising state production in $Ne^+$ scattering on Na, Mg, Al and Si surfaces at 15 deg incidence. The lines represent the total cross section for Ne** autoionising state production in $Na^+$ and $Mg^+$ collisions [17,19] with a Ne target.*

Fig. 2b. shows a comparative plot of the intensities summed over the Ne $^3P3s^2$ and the $^1D3s^2$ states for the different targets for a 15° incidence angle. As may be seen, in the studied energy range, the intensity increases as we go from the Si to the Na target. This trend in the excited state production was also observed in measurements of ion fractions, as mentioned above.

These results are similar to what is observed in gas phase atomic collisions. Fig.2b also shows the cross section of summed Ne** excited state production measured for gas phase collisions of $Na^+$ [17] and $Mg^+$ [19] with Ne. As may be seen in gas phase collisions in case of the Na projectile, excitation processes are important at low energies. This is also the case for the Na solid target. In case of Mg both in the gas phase and for the solid target excitation processes become important at higher energies.

## Discussion

The similar general trends observed suggest that both in gas phase collisions and in ion-surface scattering, similar *primary excitation mechanisms* are operative [13,17]. In gas phase atom-atom collisions these are discussed in terms of the molecular orbital promotion model [13,17-22]. Excitation processes in e.g. all the cases considered here for Ne, involve the promotion of the 4fσ orbital, which is correlated to the outermost *2p orbital of Ne*. Excited state production results from one or two electron transitions from the $[...3d\pi^4 4f\sigma^2..]$ core molecular state. This explains the *strong excitation of e.g. Ne\*\** [9-11,17-20]. Note that experimental and theoretical studies in gas phase collisions of Na, Mg and Al scattering on Ne, indicate that the internuclear distances at which the 4fσ promotion occurs, are of the order of 1.45 atomic units ($a_0$) [17], 1.2$a_0$ [18,19] and 0.85$a_0$ [20] respectively. From a purely geometric point of view, this indicates that the excitation cross section will decrease from Na to Al. This corresponds to our observations for the solid target and the Si data follows this trend.

The similarities observed between the gas phase and solid target data, suggest that the ionisation and excitation processes of Ne, are also due to one and two electron transitions from the 4fσ MO following its promotion. In case of He they involve the 3dσ MO promotion [21,22]. Construction of an MO diagram for the Ar-Mg system on the other hand shows that in this case the 2p orbital of Mg correlates to the promoted 4fσ orbital and Ar should not be efficiently excited. A discussion of target excitation may be found in [16,17].

There do exist intriguing differences in the nature of excited states produced in gas phase and in collisions with surfaces, which point to the *role of surface-specific effects*. Thus in the Ne/Ne$^+$ surface collisions one observes: i) the strong excitation of the $^3P3s^2$ state, whereas such states are only weakly excited in gas phase collisions, in agreement with what is expected in the MO description for dissociation of the $3d\pi^4$ core into only the $^1D/^1S$ core atomic states and ii) the very weak excitation of higher lying states such as the 3s3p and $3p^2$ ones.

The nature of the excited states formed in the surface collision, involving valence orbitals is not clear, since it is not clear how one should consider transitions from the promoted orbital to higher lying orbitals. However, it should be remembered here, that the scattered particle starts from "inside" the surface, at distances of the order of an atomic unit from surface atoms. Therefore, because of image potential induced level shifts in most cases considered, excitation may actually lead to ionisation near the surface. One electron excitation will lead to e.g. Ne$^+$ production, while two electron excitation processes lead in particular in the production of Ne$^{++}$($^1$D) near the surface [14-16,26,27]. This turns out to be equivalent to considering, as proposed by Joyes [29], that for a collision involving a solid the 4fσ orbital promoted above the Fermi level limit leads to one or two electron loss. The formation of excited particles and ions is then a result of a series of electron capture, loss and deexcitation processes which occur as the particles fly away from the surface [26, 27]. The above noted difference, i.e. the very small magnitude of higher lying Ne\*\* excited states in case of solid targets results from the fact that their binding energies are small and most of these lie above the Fermi level of the solid. Electron capture to these states will therefore not occur. States like the Ne\*\*3s3p/$3p^2$ with intermediate binding energies of the order of 4 eV can only be formed in some cases and then only at large ion-surface distances. Their population will then be small, as may be seen from numerical simulations [28]. This explains the dominance of the more strongly bound (*circa* 7.2eV) $3s^2$ states.

Note that it seems reasonable to conserve the aspect of the gas phase interpretation, concerning the $^1D$ core production, since we are dealing with an inner ($3d\pi$), unpromoted orbital. As discussed in [27, 28], there exist several rearrangement processes, which must account for the dominant $^3P$ core production observed experimentally. A discussion of the experimental evidence in favour of the surface specific rearrangement effects will now be given, followed by that of the rearrangement processes.

As shown in fig.1b. for Ne scattering on Al the positions and the shapes of the peaks in the electron spectra change when the observation angle ($\alpha$) changes. The spectra are broad, with a low energy tail for small $\alpha$, thin down for larger $\alpha$, and for still larger values are very broad, with a high energy tail. A most interesting aspect of these spectra is the difference in the shape of the two Ne peaks. As pointed out by arrows in fig.1b, for a 135° observation angle the peak due to the $^1D3s^2$ state has a larger low energy shoulder than that due to the $^3P3s^2$ state.

We showed [11, 30], that the shape of the spectra can be well accounted for using the experimentally determined scattered ion energy and angular distribution, which extends to large angles and is characterised by large energy losses. This analysis shows that the larger low energy hump in the $^1D3s^2$ state peak is due to the greater intensity of excited atoms scattered into large angles. Thus these measurements indicate a very interesting point, i.e. that *the $^3P3s^2$ is formed preferentially close to the surface, i.e.* when the perpendicular velocity is smaller and the particles spend more time near the surface [11, 30].

Further indications of this behaviour were recently obtained in a collaboration with O.Grizzi et al, where the intensity of the Ne** states produced in glancing collisions with monocrystalline Al(111) sample were measured as a function of azimuthal incidence angle $\psi$ [26]. We found as shown in fig.3. that not only the intensities of the peaks *but also the ratio of the intensities* of the $^1D3s^2$ peak to that of the $^3P3s2$ peak change as a function of $\psi$.

In order to understand this behaviour, the scattered particle trajectories were analysed using a simple computer model based on the Marlowe code [25] of ion scattering as a series of binary collisions. The model was adapted to include excited state production [26] using a two state quasimolecular model of excitations. The swarm of outgoing trajectories for "inelastic" scattering was then analysed. The total number of these was taken to be representative of the intensity of excited state production. This is shown in fig .3. as a function of azimuthal angle. As may be seen the maximum Ne** intensity is found for $\phi=10°$. This direction corresponds to a most "open" direction, a situation where the screening of one atom by another is reduced. In order to model the effect of incident beam divergence, residual surface roughness and an inhomogenity in crystal face positioning as a function of azimuthal angle, calculations were performed for a 1.4° incident beam divergence.

Fig.4. shows the distribution of exit trajectories as a function of exit azimuth ($\psi$) and angle with respect to the surface plane ($\theta$) for several incidence azimuths. Dramatic changes in this distribution occur in a small range of variation of incidence azimuths around $\phi=10°$. Note that in all cases discussed here the total scattering angle and hence the *final collision energy is the same*. For $\phi=9°$, the maximum of the distribution lies for $\theta=8°$. This angle rapidly increases to 13° for $\phi=11°$. Then when $\phi$ increase further to e.g. 14°, we again find $\theta=8°$. The width of the distribution is also found to change. It was found that for $\phi=11°$ to 13°, there is a clear increase in the perpendicular velocity component, which thus explains the existence of the maximum in the $^1D3s^2$ state production observed for these angles in fig.3. in agreement with the core rearrangement picture.

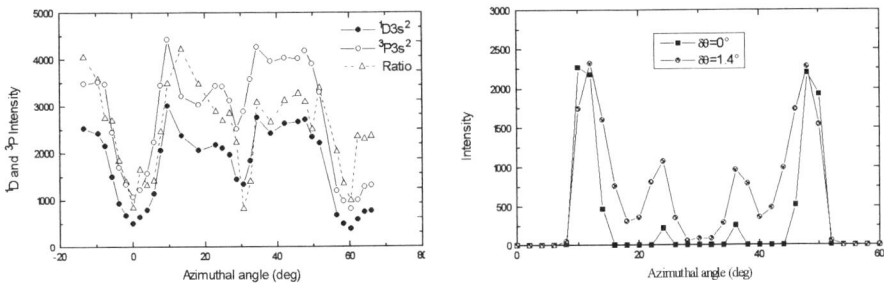

*Fig.3. Azimuthal dependence of intensities of the triplet and singlet core states of Ne and their ratio as a function of the azimuthal angle and computer simulations of the integral peak intensities for two incident beam divergences.*

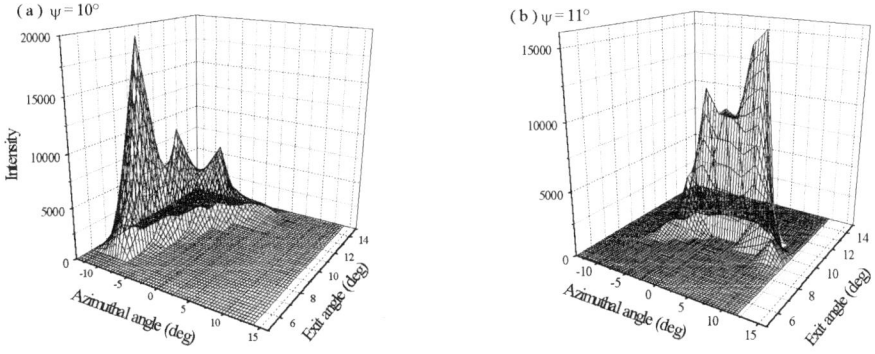

*Fig.4. Distribution of scattered excited Ne atoms as a function of the exit azimuhal angle and the angle with respect to the surface plane.*

The *quasimolecular rearrangement* process envisaged initially [3,10, 31] is an extension of the gas phase description [17, 20], where this occurs at a crossing of the $(3d\pi^4)$nl n'l' → $(3d\pi^3 4f\sigma)$ n'l' n"l" states. The $3d\pi^3 4f\sigma$ core state with an inner vacancy should be more repulsive than the $3d\pi^4$ core state [9,17]. In the case of a solid target we can envisage an Auger transition between these states, which would be active in the *outgoing* leg of the collision for distances larger than the crossing point of these states. It would involve electrons close to the Fermi level with energies comprised between zero (at the crossing point) to ones corresponding to the energy difference between these states at infinity. This process would only be active for the rather short time of the separation of the quasimolecule (circa $10^{-15}$ sec for 1 keV Ne).

An alternative to this are "atomic" rearrangement processes [14, 27, 28], which are active after the collision has occured as the excited particle flies away from the surface. In this case the effective distances are larger and the process would be the more effective the longer the particle spends near the surface, i.e. for small perpendicular velocities.

A *resonant autoionisation rearrangement* process was suggested recently by us [28]. We attribute rearrangement to ionisation into the continuum represented by : *metal* + $Ne^{++}$ *($^3P$)*, followed by electron capture processes. This ionisation is the *analogue of the autoionisation of excited states of open shell atoms in their free states*. In case of Ne, most $^1D$ core states lie *below* the $^3P$ $Ne^{++}$ continuum limit. However, when we consider an atom or ion near a metal surface, image charge effects, will shift levels above the Fermi level. Actually a level can be bound with respect to its parent continuum : *metal* + $Ne^{++}$ *($^1D$)*, *but be autoionising with respect to the continuum* : *metal* + $Ne^{++}$ *($^3P$)*. An excited atom or ion formed collisionally in the $^1D$ core configuration can thus autoionise into its parent *and* into the $^3P$ core related continuum. The resulting $Ne^{++}$ $^3P$ or $Ne^{+*}$ $^3Pnl$ can then capture electrons giving the final state distribution.

Note that a modification of the rearrangement process will occur if the workfunction changes since this will affect the range of atom surface distances during which resonant autoionisation can occur.

Additionally, autoionising states such as Ne** ($^1D$ nln'l') can decay into $Ne^{+*}$ ($^3P$ nl), a process observed by us in e.g., F⁻ ** ($^1D$ $3s^2$) autodetachment.

A *direct Auger rearrangement* process, between atomic excited states corresponding to the two cores may also be considered [14]. This process involves electrons from a band of an approximately 3.2eV width, corresponding to the ($^3P$-$^1D$), located at the Fermi level. This process may be envisaged for $Ne^{++}$, $Ne^{+*}$ and Ne** particles.

Finally we should also consider Auger capture processes into states with 2p vacancies such as the $2p^3 3s^2$ state seen at 31eV in the electron spectrum. The filling of this 2p vacancy will lead to $2p^4$ core states and if we consider a sudden filling this would occur in a statistical ratio of $9(^3P) : 5(^1D) : 1 (^1S)$.

In a "real" situation all of these processes should exist and we see that the dominant observation of the $^3P$ core states can thus be explained.

## Effects of adsorbates

We performed a study of the effects of initial stages of oxydation on the production of autoionising states in scattering on Mg, Al and Si. A characterisation of oxygen adsorption was performed, showing in particular a decrease of the workfunction ($\Phi$) of Mg and Al in the early stages of adsorption. For the Si target we did not observe a variation in the work function.

Fig.1c. shows the results of exposing an Mg surface to small doses of $O_2$. As may be seen a dramatic drop in intensity of the $3s^2$ states is observed. The most intriguing feature is the initial strong drop and disappearance of the peak due to the $^3P3s^2$ state. A similar effect was observed for the Al target. For He scattering the He** peak is also attenuated. In the case of Ar collisions the 44eV peak assigned to neutral Mg* $2p^53s^23p$ is the most affected at small coverages.

In the case of the Si target we also observe a decrease of Ne** peak intensities, but here they disappear at about the same rate.

It is interesting to note that the decrease of the intensities of the He, Ne ($^1D$) $3s^2$ and of the $Mg^{+*}$ states on the one hand and of that of the Ne ($^3P$) $3s^2$ and the Mg* states follows different trends. This led us to suggest that the initial drop in the Ne ($^3P$) $3s^2$ state intensity is due to a decrease of the core rearrangement rate. This effect is compatible with the prediction of the resonant autoionisation model [28] of core rearrangement, since in this model the decrease of $\Phi$ will lead to a stabilisation of $^1D$ core states against resonant autoionisation and hence lead to a decrease in rearrangement. In the Si case since $\Phi$ did not change, the rearrangement would not be affected in this model.

The general decrease of the intensity of all the peaks and also the above mentioned drop in charge fractions to screening of substrate Mg atoms by oxygen atoms. This is compatible with results of direct recoil measurements under grazing incidence conditions, which show a drop of Mg recoil and increase of O recoil ones [32]. Ne excitation is not observed in Ne-O collisions [33].

## Conclusions

We thus see that excited state production in inelastic "binary" collisions with surface atoms can be accounted for in terms of the gas phase electron promotion model insofar as the primary excitation mechanism is concerned and reasonable conclusions concerning the particle that will be dominantly excited can be drawn from this description. The final excited state distribution is however determined by surface specific electron capture, loss, deexcitation and core rearrangement processes, which lead to a dominant selective production of the lower lying excited states. Strong modifications are observed upon adsorption of oxygen. Part of these modifications are due to changes in electron transfer rates and rearrangement processes and some are due to modifications due to collisions on both substrate and target atoms.

## Acknowledgements

The authors are grateful to M.Maazouz for help in the final stages of experiments and to Yu.Bandourin, R.Baragiola, V.Drobnitch, O.Grizzi, N.Mandarino, E.Sanchez and F.Xu for interesting discussions.

## References

1. S.Valeri , Surf.Sci.Rep.17, (1993),85.
2. C.Benazeth, N.Benazeth, L.Viel Surf.Sci. 78, (1978),625.
3. G.Zampieri, F.Meier and R.Baragiola, Phys Rev A29, (1984),116.
4. T.E.Gallon and A.P.Nixon J.Phys: Condens.Mat , 4, (1992), 9761
5. S.V.Pepper and P.R.Aron Surf.Sci 169, (1986), 14.
6. O.Grizzi, M.Shi, H.Bu, J.W.Rabalais and R.A.Baragiola Phys Rev B41, (1990), 4789.
7. J.W.Rabalais, J.N.Chen and R.Kumar Phys Rev Let 55, (1985),1124.
8. R.Souda and M.Aono Nucl.Inst.Methods B15, (1986),114.
9. V.A.Esaulov, L.Guillemot and S.Lacombe, Nucl.Inst. Methods Phys. Res B90, (1993), 305
10. S.Lacombe, L.Guillemot, M.Huels, Vu Ngoc Tuan and V.A.Esaulov Surf.Sci.Lett 295, (1993), L1011.
11. L.Guillemot, M.Maazouz and V.A.Esaulov, 1995 submitted to J.Phys C
12. F.Xu, N.Mandarino, A.Oliva, P.Zoccoli, M.Camarca , A.Bonanno and R.Baragiola, Phys Rev A50, (1994), 4040
13. M.Barat and W.Lichten Phys Rev A6, (1972),211.
14. L.Guillemot, S.Lacombe, V.Esaulov, M.Maazouz, N.Mandarino, E.Sanchez, V.Drobnich, Yu.Bandurin and A.I.Daschenko, in preparation for Surf.Sci
15. S.Lacombe, V.Esaulov, L.Guillemot, M.Maazouz, N.Mandarino and E.Sanchez in preparation for Surf.Sci
16. S.Lacombe, V.Esaulov, L.Guillemot, M.Maazouz and E.Sanchez, 1995, sub. to Surf. Sci.
17. J.Ostgaard Olsen et al Phys Rev A 19,(1979), 1457
18. J.Fayeton, N.Anderson and M.Barat J.Phys B: Atom.Molec.Phys.9, (1976), L149.
19. J.Fayeton, Thesis, 1976, Universite de Paris Sud, Orsay, France
20. D.Doweck, Thesis, 1978, Universite de Paris Sud, Orsay, France
21. Vu Ngoc Tuan and J.Pommier, 8ème Colloque sur la Physique des Collisions Atomiques et Moléculaires, Louvain-la-Neuve, (1980)
22. C.Courbin-Gaussorgues, V.Vaaben and V.Sidis J.Phys.B16, 2817, (1983)
23. V.A.Esaulov, L.Guillemot, O.Grizzi, M.Huels and Vu Ngoc Tuan 1995, sub to RSI
24. J.C. Brenot . et al , Phys Rev A11, (1975), 1245
25. M.T.Robinson and I.M.Torrens Phys Rev B9, (1974), 5008
26. G.P.Prigliasco, E.Sanchez, O.Grizzi, V.A.Esaulov and Vu Ngoc Tuan, (1995) submitted to
27. V.Esaulov, L.Guillemot, S.Lacombe, and Vu Ngoc Tuan, Nucl.Inst.Meth. B100, (1995), 232
28. V.Esaulov, J.Phys C: Condens.Matter 6, (1994), L699
29. P.Joyes, J. de Physique 30, (1969) ,243
30. S.Lacombe, L.Guillemot, M.Huels, Vu Ngoc Tuan and V.A.Esaulov Izvestia Akademii Nauk, (Sov Phys Izvestia in Russian ), 58, (1994), 8
31. F. Xu, R.A. Baragiola, A.Bonanno, P.Zoccali, M.Camarca and A.Oliva, Phys Rev Let. 72, (1994), 4041
32. J.A.Shultz, M.H.Mintz, T.R.Schuler and J.W.Rabalais Surf. Sci. 146, (1984), 438
33. S.Boumsellek and V.A.Esaulov J.Phys.B : Atom.Molec.Phys 23, (1990), 1303

# Molecular Collisions on Large van der Waals Clusters

J.P. Visticot, J. Berlande, X. Biquard, M.A. Gaveau,
A. Lallement, J.M. Mestdagh, and O. Sublemontier

*Commissariat à l'Energie Atomique*
*Service des Photons, Atomes et Molécules*
*Centre d'Etudes de Saclay*
*91191 Gif sur Yvette Cedex - France*

Large van der Waals clusters are used as microreactors to study collision processes at the contact of a solvent. A barium atom and different molecules are deposited on the surface of an argon cluster through the pick-up technique. The location of the barium with respect to the cluster is determined through the spectroscopy. By analyzing the changes in the barium fluorescence when adding molecules on the cluster, the collision rate can be determined and consequently the relative mobility of the barium-molecule pair at the cluster surface.

## INTRODUCTION

Most of the studies in reaction dynamics have been performed in very diluted media in order to insure single collision conditions and to obtain a microscopic view of the chemical reactivity. Molecular beams have been largely used to also achieve well-defined initial conditions [1]. On the other hand, a large number of reactions do not occur in the gas phase but in a condensed medium or in an heterogeneous medium. To investigate the influence of a solvent on a chemical reaction is thus an important goal. A number of studies has used liquid or matrix phase to this effect. However, in an infinite medium like a liquid or a matrix, the diffusion of the reactants makes very difficult the isolation of a single pair of reactants, even with very dilute concentrations.

In our group, we have used an alternative way to investigate the effect of a condensed medium on the reactivity. This consists in using large van der Waals clusters as reaction cells of microscopic size [2]. The finite size of the cluster allows us to deposit a single pair of reactant on it and, thus, to investigate in this condensed medium the collisions of this pair of reactants. By varying the nature of the cluster, it is possible to look at various types of solvents and to consider different effects. In a first step, we have used rare gas clusters. In that case, the solvent is non reactive and does not perturb strongly the electronic properties of the reactants. This allows us to investigate specifically the mechanical effects due to the solvent on the dynamics.

© 1995 American Institute of Physics

Once a pair of reactants is deposited on a rare gas cluster, two questions arise. The first is about the location of the reactant, whether they stay at the surface of the cluster of whether they prefer to migrate in the interior. The second question concerns the relative diffusion of the reactants on the cluster. These two questions will have important consequences upon the reactivity. In the present work, we have addressed these two questions through the spectroscopy of a reactant deposited on the cluster. The location of a barium atom on an argon cluster is deduced from the comparison of the experimental excitation spectrum to predictions from molecular dynamics simulations. The variations of the intensity of fluorescence when adding molecules on the argon cluster is then analyzed in terms of collisions between the barium and the molecule at the cluster surface. Results concerning the reactivity of barium atoms on clusters are presented elsewhere [3]–[5].

## EXPERIMENT

The experiment can be decomposed in several parts. First the pure argon clusters are generated. Then barium atoms and molecules are deposited on the cluster. Finally, the laser induced fluorescence is performed. We will consider successively these points.

### Cluster generation

The argon clusters are generated from a supersonic molecular beam source of the Campargue type [6]. The beam is skimmed from a free jet expanded through a 0.2 mm nozzle and passes through a differentially pumped chamber before entering the main chamber. The argon backing pressure can be varied between 3 and 35 bars that corresponds to average cluster sizes between 300 and 4000. These average sizes have been determined by measuring the changes in velocity of the clusters when they pass through a chamber containing a defined buffer gas pressure [7].

### Pick-up

Once the homogeneous argon clusters have been formed, barium atoms and molecules ($CH_4$, $SF_6$, $O_2$, ...) can be deposited on the clusters by the pick-up technique. A barium tube is located at the entrance of the main chamber and is heated to yield a small pressure of barium. The argon clusters pass through the tube and by collision can trap barium atoms. Because of the large mass difference, there is almost no change in the direction of velocity of the clusters that continue towards the observation region in the center of the main chamber. The barium pressure is maintained low enough so that there is at the most one collision between a cluster and the barium vapor.

Consequently, the fraction of clusters containing more than one barium atom is negligible.

Concerning the molecules, they are also deposited by the pick-up technique. In this case, the cluster beam crosses an effusive beam of molecules generated in the differentially pumped chamber. The number of deposited molecules is varied by changing the effusive beam pressure. We have observed that we can deposit up to 15 molecules of $CH_4$ without perturbing the cluster beam by more than 10 %. As a collision process is a Poisson process, the number of molecules deposited on the cluster will follow a Poisson statistics. This means that the knowledge of the average number of molecules is sufficient to determine the fraction of clusters that contain 0, 1, 2 or more molecules [8]. This average number can be determined by the measurement of the relative fluxes of argon and molecules contained in the cluster beam and the knowledge of the average cluster size [3,9].

### Excitation

The beam of a CW dye laser is focused onto the entrance of an optical fiber. The output of the fiber is imaged onto the observation zone. The laser can be scanned between 535 and 560 nm in order to photoexcite the barium atoms trapped on the cluster in the vicinity of the Ba resonance line ($^1S_0 \rightarrow {^1P_1}$). The resulting fluorescence is collected onto the entrance slit of a monochromator followed by a photomultiplier. Two different types of spectra can be recorded. First, the fluorescence excitation spectrum is obtained by scanning the laser frequency while integrating the total undispersed fluorescence. Second, the emission spectrum is obtained for a fixed frequency of the laser and by scanning the frequency of the monochromator. The last experiments that have been performed are the kinetic measurements where the fluorescence intensity is recorded as a function of the average number of additional molecules picked up.

## SPECTROSCOPY OF THE BA-$(AR)_N$ CLUSTER

An example of excitation spectrum in displayed in the upper part of figure 1. The spectrum is obtained for a cluster size of about 300 and does not depend strongly upon the cluster size [10]. It exhibits two main bands located at each side of the unperturbed Ba($6s^2\,^1S_0 \rightarrow 6s6p^1P_1$) atomic line. In order to interpret the spectrum in term of barium location, molecular dynamics simulations have been performed. The calculation is based upon classical dynamics and pairwise additive potentials. Two calculations have been performed corresponding to the two extreme situations, with the barium at the surface of a 125-atoms argon cluster and with the barium in the middle of a 113-atoms argon cluster. In each case, the evolution of the barium on the ground state surface has been calculated by solving the equations for classical

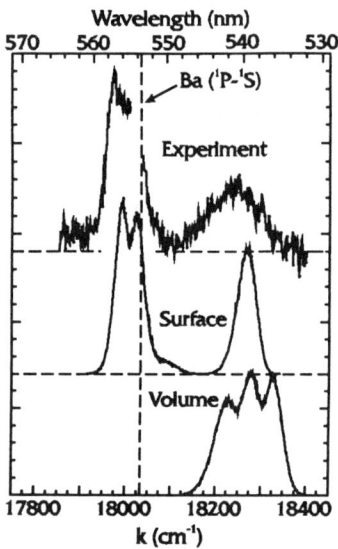

**FIG. 1.** The upper part is the $Ba(Ar)_n$ fluorescence excitation spectrum obtained for a cluster size of about 300. It is compared to the results of two molecular dynamics simulations. The middle part corresponds to a 125-argon cluster where the barium stays at the surface while the bottom part corresponds to a 113-argon cluster with the barium located inside

motion and at each time step the three instantaneous excitation frequencies have been obtained from the diagonalization of a 3x3 hamiltonian resulting from the multiplicity of the excited Ba P state. The final histogram of these frequencies directly yields the simulated excitation spectrum [10].

The results of the two simulations are also plotted in figure 1. There is a very good agreement between the experiment and the simulation with the barium at the surface of the argon cluster. Indeed, the particular shape of the excitation spectrum in two bands can be qualitatively described in term of surface location through a quasi diatomic picture by considering the argon cluster as a large rare gas atom. In that case, the excited P state of the barium atom of cylindrical symmetry is expected to be split in two surfaces corresponding to the two possible configurations of the P orbital with respect to the cluster surface, a "Σ" with the P orbital perpendicular to the surface and two degenerate or almost degenerate "Π" with the P orbital in the plane of the cluster. Moreover, the "Σ" surface is expected to be more repulsive due to the larger electron density at shorter barium-cluster distance in that case.

## MOLECULAR COLLISIONS ON THE CLUSTER

The emission spectrum corresponding to the excitation of the $Ba(Ar)_n$ cluster in the "Σ" blue band is displayed in figure 2. It presents two components.

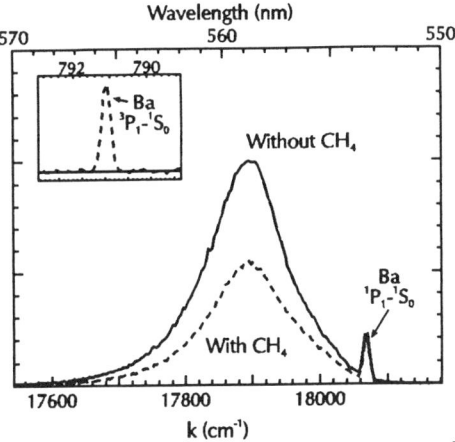

FIG. 2. The full line correspond to the emission spectrum of the Ba(Ar)$_n$ cluster when excited in the "$\Sigma$" blue component. The dashed line correspond to the same spectrum but when several CH$_4$ molecules have been added to the cluster. Note that, in this case, the fluorescence of the free Ba(6s6p $^3$P$_1$) state is observed as it is shown in the insert.

The first is a wide band in the red of the barium resonance line. This component is independent of the excitation frequency in the "$\Sigma$" or "$\Pi$" band. When the excitation frequency is high enough the emission of the free barium atomic resonance line is also observed. These two components have been interpreted by a competition between the relaxation and the desorption of the excited barium at the surface of the argon cluster [9]. The barium atomic line corresponds to the desorption while the red band corresponds to the emission of a barium atom that stays solvated on the cluster and is relaxed to the most stable "$\Pi$" state. This interpretation has been confirmed by the observation of the polarizations of the two emissions relative to the excitation laser, the relaxation from a "$\Sigma$" to a "$\Pi$" state implying an orbital rearrangement [11].

In figure 2 is also plotted the same emission spectrum when an average number of 1 CH$_4$ molecule has been added to the cluster. The intensity of the barium resonance line is almost unperturbed while that of the broad emission has decreased by a factor 2. Moreover, in the red a new line appears that is the intercombination Ba(6s6p$^3$P$_1 \rightarrow$6s$^2$ $^1$S$_0$) line. In figure 3, we have plotted for three different molecules (O$_2$, SF$_6$, and CH$_4$) the evolutions of the intensities of these three components as a function of the average number of molecules that have been added to the cluster. There are important differences between CH$_4$ and O$_2$. In the oxygen case, the intensities of the Ba($^1$P$_1 \rightarrow ^1$S$_0$) line and of the emission band decrease at approximately the same rate and no emission of the Ba($^3$P$_1 \rightarrow ^1$S$_0$) line is observed. On the other hand, in the methane case, the Ba($^1$P$_1 \rightarrow ^1$S$_0$) intensity decreases at a much lower rate than that of the emission band and the intensity of the Ba($^3$P$_1 \rightarrow ^1$S$_0$) line increases regularly. The SF$_6$ molecule appears as an intermediate case.

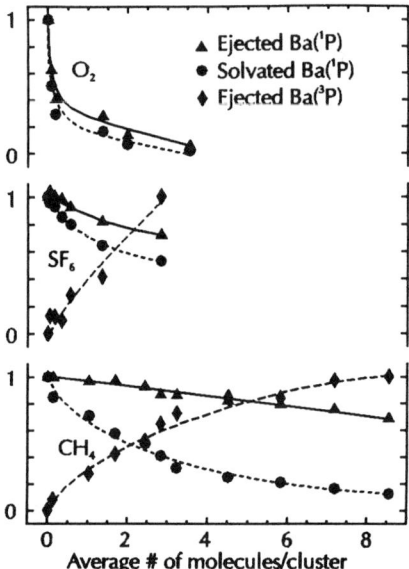

FIG. 3. Plots of the relative intensities of the three observed components (atomic Ba($^1P_1 \rightarrow {}^1S_0$) and Ba($^3P_1 \rightarrow {}^1S_0$) lines and broad emission) as a function of the average number of molecules picked up. Three molecules have been considered: $O_2$, $SF_6$, and $CH_4$. Notice that, for $O_2$, no emission from the triplet P state of barium is observed.

The origin of the two components allows to understand the differences between the different evolutions. First, the barium resonance line results from a photodesorption that occurs a short time after the photoexcitation. This time should be of the order of a few vibrations of the barium-cluster system and is short compared to any diffusion process on the cluster. On the other hand, the emission of the red band corresponds to a solvated barium. In that case, the time of interest is now the radiative lifetime of the excited barium that is 8 ns [12]. Consequently, the decrease of the intensity of the barium resonance line means that the barium must be in interaction with the molecule before the excitation and the additional decrease of the intensity of the red band will be interpreted as a quenching due to a collision between the barium and the molecule that occurs after the excitation [9]. This interpretation is consistent with the nature of the interactions that exist between the barium and the three molecules that have been used. In the case of $O_2$, the two intensities ($^1P_1$ and solvated) decrease at the same rate and the remaining intensity corresponds to the fraction of clusters that do not contain any molecule. This can be understood by a chemical reaction between the ground state barium and the oxygen molecule. Indeed, this reaction is exoergic by 0.6 eV and, because of the delay between the pick-up of the molecules and the laser photoexcitation that is of the order of 50 $\mu$s in our experimental conditions, this reaction can be complete. On the other hand, the methane molecule has no chemical

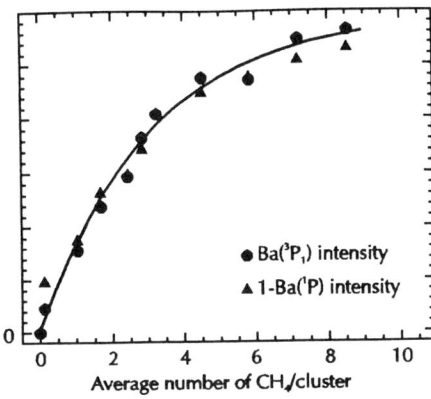

**FIG. 4.** Correlation between the decrease of the intensity of emission of the solvated barium and the increase of intensity of the Ba($^3P_1 \rightarrow{}^1S_0$) line in the case of the quenching by CH$_4$ molecules.

affinity with barium. The slow decrease on the intensity of the barium resonance line with the number of methane molecules correspond simply to the probability that a molecule be in contact with the barium and is related to the relative sizes of the molecule and the cluster surface [9]. The SF$_6$ molecule is intermediate between the oxygen and the methane. This can be interpreted by an equilibrium with the formation of a Ba-SF$_6$ complex at the surface of the cluster [9].

The difference between the decrease of the Ba resonance line and the solvated Ba intensities results from a collision of the excited barium and the molecule at the surface of the cluster. This is further confirmed by the apparition of the free Ba($^3P_1 \rightarrow{}^1S_0$) line. This product is formed through the quenching of the excited barium and the decrease of the solvated Ba corresponds exactly to the increase in intensity of the triplet line as it is shown in figure 4. The evolution of the intensity of the solvated barium emission with the number of molecules picked up has been modeled by a first order kinetic law in order to estimate the collision rate between the excited barium and the CH$_4$ molecule on the argon cluster [8,9]. This leads to a quenching rate of about $7.5\,10^7\,s^{-1}$ for a cluster size of the order of 400 argon atoms. This rate is large and corresponds to a relative mobility of the barium and the molecule at the surface of the argon cluster and not in the inside. The surface diffusion constant that can be deduced from these measured rates is of the order of $2\,10^{-6}\,cm^2.s^{-1}$. Molecular dynamics simulations of the diffusion of a ground state barium atom or a N$_2$O molecule at the surface of a 147 atoms argon cluster have been recently performed [13]. The values found are of the order of $4\,10^{-7}$ and $1.2\,10^{-5}\,cm^2.s^{-1}$ for Ba and N$_2$O, respectively, in good agreement with the present determination.

## CONCLUSION

We have shown that large van der Waals clusters can be considered as microscopic solvent where reactants can be deposited in a controlled manner. The spectroscopy coupled with molecular dynamics simulation appears as a very efficient way to determine the location of a chromophore with respect to the cluster. Once a pair of reactants has been deposited by pick-up at the surface of such a cluster, the mobility is sufficient to allow collisions within a short time (a few nanoseconds for an excited barium at the surface of a cluster of about 400 argon atoms). These properties together with the fact that the cluster is a reservoir of energy that can act as a thermostat, make these clusters very good candidates to study chemical reactions of non isolated systems at a microscopic level [3-5].

## REFERENCES

1. R. D. Levine and R. B. Bernstein, *Molecular Reaction Dynamics and Chemical Reactivity*, Oxford University Press, New York, 1987.
2. J. M. Mestdagh, A. J. Bell, J. Berlande, X. Biquard, M. A. Gaveau, A. Lallement, O. Sublemontier, and J. P. Visticot, in *Reactions Dynamics in Clusters and Condensed Phases*, edited by J. Jortner, R. D. Levine, and B. Pullman, volume 26 of *The Jerusalem Symposium on Quantum Chemistry and Biochemistry*, pp. 101-114, Kluwer, Dordrecht, 1994.
3. A. Lallement, J. M. Mestdagh, P. Meynadier, P. de Pujo, O. Sublemontier, J. P. Visticot, J. Berlande, X. Biquard, J. Cuvellier, and C. G. Hickman, J. Chem. Phys. **99**, 8705 (1993).
4. X. Biquard, O. Sublemontier, J. Berlande, M. A. Gaveau, J. M. Mestdagh, B. Schilling, and J. P. Visticot, J. Chimie Phys. **92**, 264 (1995).
5. X. Biquard, O. Sublemontier, J. Berlande, M. A. Gaveau, J. M. Mestdagh, and J. P. Visticot, to be published in J. Chem. Phys., 1995.
6. R. Campargue, J. Phys. Chem. **88**, 4466 (1984).
7. J. Cuvellier, P. Meynadier, P. de Pujo, O. Sublemontier, J. P. Visticot, J. Berlande, A. Lallement, and J. M. Mestdagh, Z. Phys. D **21**, 265 (1991).
8. A. Lallement, A. J. Bell, J. Berlande, J. Cuvellier, J. M. Mestdagh, P. Meynadier, O. Sublemontier, and J. P. Visticot, Chem. Phys. Lett. **204**, 440 (1993).
9. X. Biquard, O. Sublemontier, J. P. Visticot, J. M. Mestdagh, P. Meynadier, M. A. Gaveau, and J. Berlande, Z. Phys. D **30**, 45 (1994).
10. J. P. Visticot, P. de Pujo, J. M. Mestdagh, A. Lallement, J. Berlande, O. Sublemontier, P. Meynadier, and J. Cuvellier, J. Chem. Phys. **100**, 158 (1994).
11. B. Schilling, M. A. Gaveau, O. Sublemontier, J. M. Mestdagh, J. P. Visticot, X. Biquard, and J. Berlande, J. Chem. Phys. **101**, 5772 (1994).
12. F. Kelly and M. Mathur, Can. J. Phys. **55**, 83 (1977).
13. M. P. Gaigeot, P. Millié, and P. de Pujo, to be published, 1995.

# COLD ATOMS AND ION-ION COLLISIONS

Photoassociation in Ultracold Collisions: High Resolution
Spectroscopy from the Collision Continuum .................... 667
    P.D. LETT
Differential Scattering in Ion-Ion Collisions of He ............ 677
    S. KRÜDENER
Collision Processes Involving Negative Hydrogen Ions ........ 687
    D.B. USKOV

# Photoassociation in Ultracold Collisions: High Resolution Spectroscopy from the Collision Continuum

## P. D. Lett

National Institute of Standards and Technology
Physics A-167
Gaithersburg, MD 20899 USA

**Abstract**: Research into collisions at ultracold temperatures has revealed new physics and created new spectroscopic techniques. A brief review of ultracold collisions in general and an overview of photoassociation experiments with laser-cooled Na atoms are given here. These experiments point out some of the interesting and unique features of ultracold collisions and some of the possibilities for their application in high resolution spectroscopy and the manipulation of atomic collisions.

## INTRODUCTION

Collisions of ultracold, laser-cooled atoms fall into a whole new regime of atomic collision physics [1, 2, 3]. The collision durations can easily be longer than the excited state lifetimes of the collision partners and the collisions can be modified quite readily by light-induced forces. These collisions can be studied using a number of different approaches. Some of the techniques used thus far have led to new types of spectroscopy. In particular, these spectroscopies fall under the heading of photoassociation spectroscopy [4]. Photoassociation is the binding of two colliding atoms into an excited molecular state by the addition of a photon. The use of ultracold atoms in the photoassociation, first suggested in 1987 [5], gives the technique a resolution comparable to, or better than conventional spectroscopic techniques and allows the technique to access states that are difficult or impossible to reach from bound molecular states.

Ultracold temperatures are understood, at least within the laser cooling community, to encompass kinetic energies corresponding to ~1 mK or less. This is an energy range that is well below what can typically be studied in an atomic beam but corresponds to temperatures routinely produced by laser cooling techniques. 1 mK is also a temperature at which the lighter elements have moved into a regime where s-wave scattering begins to dominate. The centrifugal barriers for the higher angular momentum partial waves will exclude them unless there is a specific barrier-penetrating resonance. 1 mK is nominally the temperature where the mean kinetic energy of the colliding atoms equals the p-wave barrier height for Na-Na collisions and heavier elements will pass into this "quantum scattering regime" at slightly lower temperatures.

An ultracold collision has a number of properties that set it apart, in the way in which it must be treated theoretically, from those at higher temperatures.

© 1995 American Institute of Physics

First and most noticeably, the collision duration can be quite long. If we define a collision time to be the time during which the atom-atom interaction energy is the same size as, or larger than the thermal energy, then the collision durations can easily last for several tens of nanoseconds. The lifetime of the Na (3P) state, for comparison, is 16 ns. Dissipation in the form of spontaneous emission must therefore be considered in the theoretical description of any such collision process involving excited states [6, 7].

A second feature of ultracold collisions is that such slow atoms have large de Broglie wavelengths. It is not uncommon in experiments today to trap atoms in potential wells, formed by laser beams, that have dimensions on the order of an optical wavelength at temperatures on the order of only a few times the "recoil temperature" for the atom. This temperature, $T_R$, is the temperature at which the mean kinetic energy equals the kinetic energy of an atom after absorbing a single photon from at rest. Since the de Broglie wavelength is directly related to the momentum, such an atom will have a wavelength that is equal to that of the absorbed photon. As the atoms become this cold their de Broglie wavelengths will become comparable to the scale length of such trapping potentials as well as to the scale of the interatomic interaction potentials. At this point the behavior can no longer be described semiclassically.

Finally, external fields can easily cause energy level shifts in the atoms that are comparable to the mean thermal energies being discussed here. Magnetic field shifts and ac Stark shifts can be used to produce forces on the atoms that can easily confine them at these temperatures. This is important to many other atomic physics experiments in addition to collision studies and it is important to remember that these interactions can be a major perturbation on the atom and its motion.

## LASER COOLING AND LASER TRAPS

The development of efficient laser cooling techniques has allowed the investigation of ultracold collision phenomena. The slowing of atomic beams and the development of laser traps have been the most important advances in this regard. Beam sources can be used to study intra-beam collisions at energies that can approach the "ultracold" regime [8]. Laser-cooled atomic beams provide a bright source of atoms, compressed in velocity space and often at very low velocity [9]. Optical molasses [10] is the term given to a configuration of pairs of counter-propagating laser beams, tuned below but near to the atomic resonance, that damps the atomic motion. Optical molasses cools, but does not confine atoms. Laser traps, typically not deeper than around 1 K, can be made and then loaded either from beam sources or directly from the low velocity tail of the thermal distribution of atoms in a vapor cell. The most robust and popular of these traps is known as a Magneto-Optical Trap (MOT) [11]. This trap combines a quadrupole magnetic field with three pairs of counter-propagating laser beams to both cool and confine the atoms. The magnetic field has a zero in the center of the trap, between two coils carrying oppositely-directed current, and a linearly increasing field in each direction away from this point. The laser beams, in pairs with $\sigma^+$ - $\sigma^-$ polarizations, intersect at the field zero. The interaction of the circularly polarized light with the atoms in the magnetic field provides a restoring force toward the zero of the magnetic field. This trap is, typically, a few hundred microns in diameter and holds atoms at densities in the $10^{10}$ - $10^{12}$ cm$^{-3}$ range and typically at temperatures in the few tens of microkelvin to few millikelvin regime.

Other traps are based on the dipole force that an atom experiences in a gradient of an oscillating electric field, such as a focussed laser beam. The energy $W = -\mathbf{E} \cdot \mathbf{d}$ is lower in the highest-intensity portions of the field for a laser tuned below the atomic resonance, where the dipole moment $\mathbf{d}$ is in phase with the oscillating laser field $\mathbf{E}$. These traps are typically much smaller in volume; limited to a fraction of the focal spot of a laser beam that has typically a 100 µm diameter. Nonetheless, if a source of atoms that are cold enough such as a MOT is available, the trap can be quite effective and one such trap, the Far Off-Resonant Trap (FORT) has been used quite effectively in studying photoassociation collisions [12, 13]. Densities of $\sim 10^{12}$ cm$^{-3}$ have also been obtained in such a trap. Another variety of dipole-force trap, the two-focus dipole trap [14] is tuned much closer to resonance and must be alternated in time with a cooling field such as optical molasses, which also serves to load the trap.

## Na COLLISIONS

The first studies of a specific collision process under ultracold conditions focussed on a reaction related to associative ionization in Na [14]. At higher temperatures this reaction is indicated as follows:

$$Na(3P) + Na(3P) \rightarrow Na_2^+ + e^-. \tag{1}$$

It is now accepted that under ultracold conditions this is a poor description of the reaction and it is more properly referred to as photo-associative ionization:

$$Na(3S) + Na(3S) + h\nu \rightarrow Na_2^* \tag{2}$$

followed by

$$Na_2^* + h\nu \rightarrow Na_2^+ + e^- \tag{3}$$

or by

$$Na_2^* + h\nu \rightarrow Na_2^{**} \rightarrow Na_2^+ + e^-. \tag{4}$$

The first step above is the photoassociation of two colliding ground state Na atoms to an excited molecular state. Following that the molecule can either be ionized by another photon (which can be of the same color as the first in Na) or be promoted to a doubly excited molecular state and autoionize. Alternatively, the excited molecule produced in the first step can decay radiatively and no ion is produced.

The first experiments were carried out in the two-focus dipole trap mentioned above. This trap has two very different types of operating conditions. During the periods when the trap lasers are on the light is rather intense ($2 \times 10^5$ mW/cm$^2$) but also detuned some distance below resonance ($\Delta \sim 2.1$ GHz in the case shown below). On the other hand, during the periods when the trapping lasers are off to facilitate cooling, the conditions are those of the near resonant ($\Delta \sim 10$ MHz or one natural linewidth for Na) and relatively weak ($\sim 8$ mW/cm$^2$) light of the optical molasses. These two conditions lead to very different rates of ion production from the above reaction, even though the fraction of atoms in the excited state is very similar in the two cases.

First consider the low-intensity, small-detuning conditions present when the trap light is off. The atoms can be excited by the laser at very large internuclear separations, where the atom-atom interaction potential is small. As the atoms move closer together their interaction gets larger and they will shift out of resonance and decouple from the laser field at a rather large distance - being left to survive to small internuclear separation where the ionization reaction can occur without the possibility of being re-excited. The decoupling occurs at distances ~2000 $a_0$ and the atoms need to survive some tens of nanoseconds in the excited state if they are to produce a molecular ion. Consequently, most of these events end in either an S + P or an S + S collision, neither of which have sufficient energy to produce the ion that the P + P collision of Eq. (1) is capable of. (See Figure 1.)

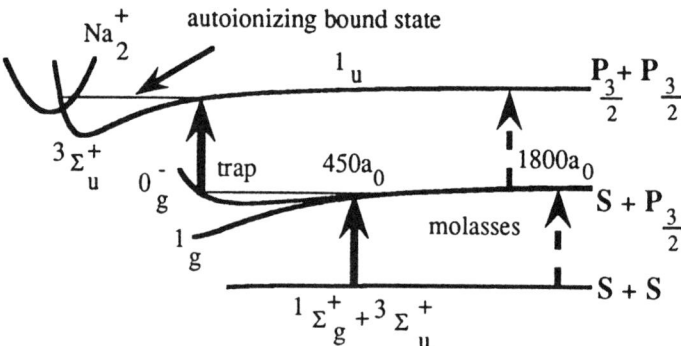

Figure 1. Schematic figure (not to scale) of the potentials and excitation mechanisms for associative ionization/photoassociative ionization in the two-focus trap of Ref [15].

During the trapping periods, however, where the detuning is larger and the laser intensity higher as well, the excitation takes place at a much smaller distance (~450 $a_0$). At this distance there are a number of molecular states that the atom pair can be excited to. They can then be further promoted to a doubly-excited molecular state that autoionizes. The process is enhanced by the increased slope of the interaction potential in the region of excitation, causing the atoms to

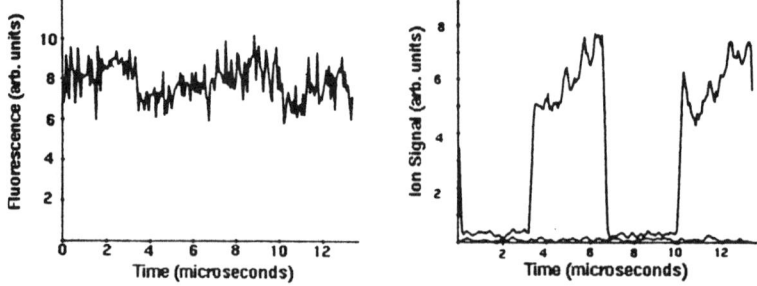

Figure 2. Time-resolved ion and fluorescence signals from the two-focus trap indicating much higher ionization rates during the trapping periods, in spite of a similar P-state population, as indicated by the fluorescence.

accelerate toward each other. The increased approach velocity, coupled with the decreased distance that the excited atoms have to travel, results in much more survival in the excited state and a higher ionization rate under these conditions [15, 16]. Under the conditions of Figure 2, where the fluorescence (and thus the P-state population) is almost constant, the ion signal is almost 100× larger during the trapping periods than during the cooling periods. Clearly it is very important to consider the intensity and detuning of any light present during ultracold collisions, as well as the possibility of spontaneous emission if excited states are involved.

## PHOTOASSOCIATION

The photoassociation process represented in Eq. (2) above is the essential part of a new type of spectrosopy [5], developed over the last 2-3 years. The photoassociation of ultracold atoms can be used to produce a high-resolution spectrum of the resulting dimer molecules. In particular, the vibrational states near to the dissociation limit of the molecule can be probed in great detail - a region of the spectrum that is difficult if not impossible to investigate with traditional spectroscopic methods. This technique uses the frequency selectivity of the photoassociation step (the excitation can only take place if there is an available molecular level at the particular detuning of the laser being used) followed by a non-frequency-selective detection scheme. This detection can be as simple as monitoring the fluorescence from (and hence the number of atoms in) the atom trap; as atoms are taken away to make molecules the fluorescence is reduced. Alternatively, the molecules formed in the photoassociation can be detected directly, for instance by ionizing them with a second photon, as indicated in Eqs. (3) and (4) above. If these detection processes are sufficiently broad in their excitation spectrum then the structure from the photoassociation step will dominate the spectrum.

Figure 3 shows the interatomic potential energy curves for the states involved in photoassociation spectroscopy. The ground state potentials are dominated by the attractive van der Waals potential, varying as $R^{-6}$. The first excited states of dimer molecules, dissociating to S + P state atoms, on the other hand, are dominated by the dipole interaction. The transition dipole moment induced by the light couples the atoms like two dipole antennae and produces a stronger interaction, varying as $R^{-3}$. These long range potentials can be either attractive or repulsive. Both the ground state and excited state potentials can have a large influence on the types of spectra observed. The relatively short range of the ground state interaction compared to the large de Broglie wavelength means that only atoms having very small relative angular momentum can penetrate the centrifugal barrier to a collision. In fact, as $T \rightarrow 0$ only s-wave ($l = 0$) collisions are allowed, although typically at the temperatures used in photoassociation experiments several partial waves will contribute to the signal.

Photoassociation spectroscopy, involving transitions from free atoms to bound molecular states, can be viewed as complimentary to traditional bound-bound molecular spectroscopy. Transitions favored in photoassociation tend to be to the outer turning points of vibrational states near to dissociation. In these states the vibrational motion of the nuclei near the outer turning point of the anharmonic potential is slow and looks very much like the motion of slowly colliding free atoms. Quantum mechanically, the transition probability is proportional to a

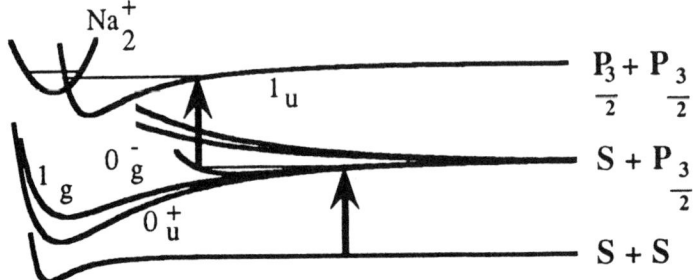

Figure 3 Sketch of interatomic potential energy curves involved in photoassociation spectroscopy in Na collisions.

Franck-Condon factor that depends on the overlap integral between wavefunctions of the ground and excited state wavefunctions. This is dominated by the contribution near the Condon point, where the quasimolecule is in resonance with the exciting light. Since the wavefunction of the excited state is largest near the outer turning point of the classical motion and has a long wavelength near this point, similar to the colliding free atoms, the excitation is localized in this region. This means that long range states and "purely long range" states, states whose inner turning point is at a larger distance than the outer turning points of most chemically bound states, are accessible via photoassociation. These states, the "purely long range" states in particular, are very difficult to access via bound-bound spectroscopy. Photoassociation spectroscopy has in fact resulted in the first reported spectra of these "purely long range" states in $Rb_2$ and $Na_2$ [12, 17, 18].

An example of the spectra that can be observed is shown in Figure 4. The spectra shown are taken by scanning the frequency of a probe laser that interacts with Na atoms confined in a MOT. The trapping lasers are alternated in time (~10

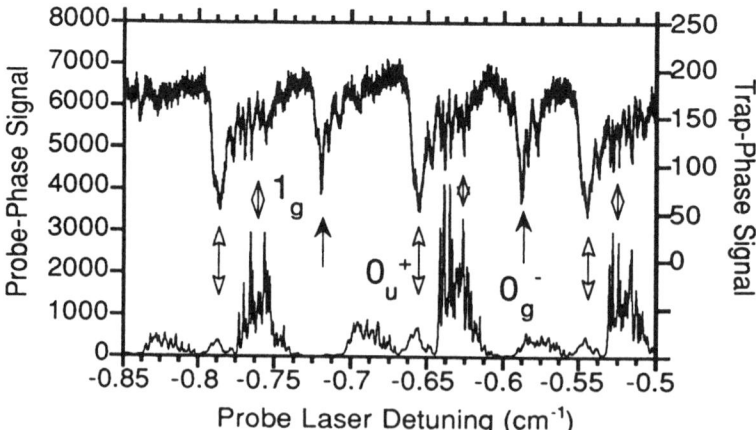

Figure 4. Spectra obtained from monitoring the probe ionization and trap loss in a photoassociation experiment in Na [17].

μs periods) with the probe laser to avoid any two-frequency processes that might involve a trap laser photon as well as a probe photon. Molecular ions from the processes (3) and (4) are collected and gated into two channels corresponding to whether they were created by the probe laser or by the (fixed-frequency) trapping lasers. The spectrum generated directly by the probe laser itself demonstrates the frequency selectivity of the method and the high signal-to-noise ratio obtainable. This probe spectrum favors the detection of states that ionize easily. The upper spectrum in the figure is obtained by simultaneously monitoring the loss of atoms from the trap via the ionization produced by the trapping lasers. In this spectrum one can observe, albeit with less signal-to-noise, additional states such as the "purely long range" $0_g^-$ states indicated in the figure. The $0_g^-$ potential in $Na_2$ has an inner turning point near 55 $a_0$ and is only 2 cm$^{-1}$ (60 GHz) deep at the potential minimum.

Spectra such as shown in Figure 4 can be extended over a much larger range, as indicated in Fig. 5. A number of states converge to dissociation limits of S + $P_{1/2}$ (discernable near 16956 cm$^{-1}$) and S + $P_{3/2}$ (near 16973 cm$^{-1}$). This spectrum of $Na_2$ is dominated by a vibrational series corresponding to the $1_g$ ($^1\Pi_g$) state that dissociates to S + $P_{3/2}$. The angular momentum coupling case changes from the Hund's case (c) ($1_g$ label) near dissociation, characterized by the molecular hyperfine structure seen in Figure 4, to Hund's case (a) ($^1\Pi_g$ label) at short range[19]. The spectrum in this region is characterized by a simple vibrational-rotational progression, although the transition becomes forbidden in this limit. An expanded view of the $1_g$ (v=48) vibrational line is shown in Figure 6. In spite of the "forbidden transition" (the natural linewidth in this region of the spectrum is approximately 70 KHz) the signal-to-noise is still quite impressive. A

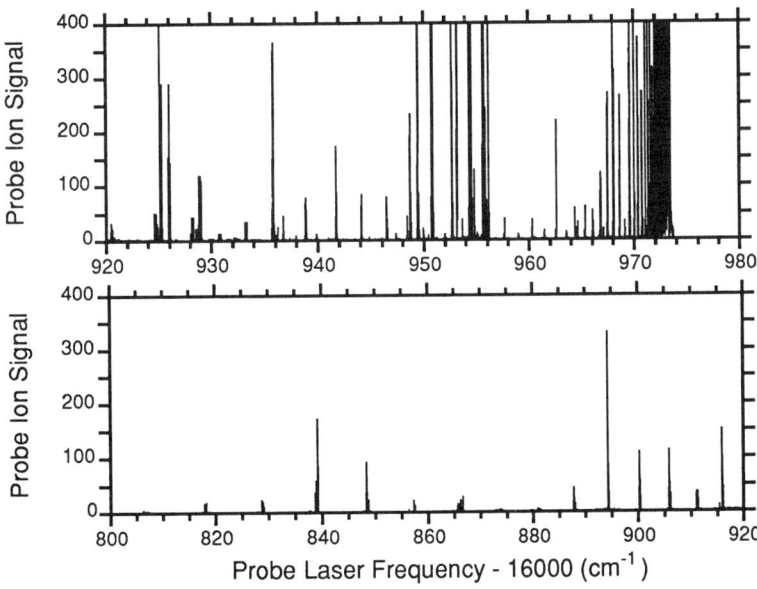

Figure 5. Extended photoassociation spectrum of $Na_2$ taken by monitoring the probe ionization.

further expansion of the J=2 peak of this line is shown in the inset. The remaining structure evident in the figure represents residual molecular hyperfine structure indicating the incomplete transition to Hund's case (a).

The asymmetric lineshape in Figure 6 is due to the kinetic energy spread of the colliding atoms. The laser linewidth and Doppler broadening in this experiment were both in the 1-2 MHz range. As the laser is tuned from blue to red across the line each molecular hyperfine component turns on with its appropriate Wigner threshold law behavior. The exponential tail on the red side of the line indicates the range of collision energies available to contribute to the transition. The ~13 MHz exponential tail indicates a temperature of approximately 600 μK as calculated from a detailed lineshape analysis [20].

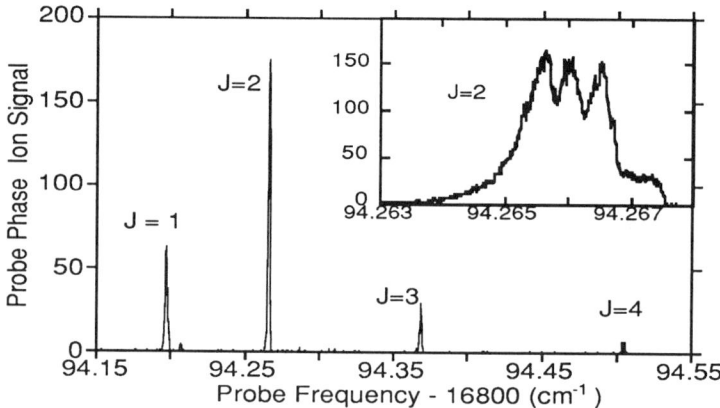

Figure 6  Expanded view of the spectrum near the $1_g$ (v=48) line.

A number of the alkali species have now been investigated by photoassociation spectroscopy. $Na_2$ has also been studied by Weiner, Bagnato, and coworkers [3, 21, 22]. $Rb_2$ has been extensively investigated by Heinzen and coworkers [12, 13] by looking at trap loss from a FORT. $Li_2$ spectroscopy has been reported by Hulet and coworkers [23] using fluorescence to monitor trap loss from a MOT. Gould, Stwalley and coworkers have also used this technique to report photoassociation spectra in $K_2$ [24]. The spectra discussed above have all been performed with a single color probe. There are, however, many multiple-resonance techniques that have been used in bound-bound spectroscopy and a number of these have been already adapted to photoassociation spectroscopy.

Two-color techniques have extended the ionization detection method to states that would not otherwise ionize with a single color, such as states of the $0_g^-$ potential in $Na_2$ [25]. In addition it allows ion detection to be used in species that do not allow 1-color two-photon ionization at all, such as Rb [26]. Two color techniques have allowed us to investigate structure in the ionization continuum, the ground state continuum, and the spectroscopy of doubly-excited states as well [25]. Two-color techniques can also be used to search for the uppermost bound states of the molecular ground state potentials [27], important in pinning down the ground state scattering length, which is relevant to the pursuit of Bose Condensation.

As a dramatic example of how easy it is to affect the result of an ultracold collision one can examine a spectrum of the rotational peaks in one vibrational line, such as that shown in Fig. 6, as a function of the probe laser intensity. At the (relatively) low intensities usually used for this one finds that only three partial waves (contributing to four rotational levels) make up the signal. At a hundred fold higher intensity approximately ten rotational states appear in the spectrum. The intensities of some of these "new" lines are larger than those of the lowest four rotational states. The explanation for the appearance of these higher rotational lines is that the tightly focussed probe laser beam forms a potential well which affects the colliding atoms. The additional kinetic energy gained by the atoms accelerating into this potental results in an effectively higher temperature, which accounts for the extra partial waves.

## SUMMARY

Ultracold collisions in general and photoassociation spectroscopy in particular are new fields that have benefited greatly from the advances in laser cooling and trapping over the past decade. An entirely new energy regime for collision studies has been opened up - one where quantum effects are commonplace and where one must be very careful before applying traditional theoretical approaches to the problem. Photoassociation can be generally considered a new form of high resolution spectroscopy that is complementary to existing bound-bound molecular spectroscopy - for instance, the lower panel of Figure 5 overlaps previous spectroscopy of the $1_g(^1\Pi_g)$ state in $Na_2$ [28], while the upper panel represents an extension of the spectra to cover levels with v = 53-95. On the other hand, such spectra also present new opportunities to investigate physics not previously accessible. Detection of the "purely long range" $0_g^-$ states, for example, or the use of such spectra near dissociation to extract a precision *atomic* lifetime from *molecular* spetroscopy [12, 17, 23, 29]. The possibilities that are presented for the investigation of phenomena involved in ultracold collisions is perhaps most clearly brought out by the recent observation of Bose-Einstein condensation by evaporative cooling of a Rb vapor [30]. Clearly there is much physics that is left to be understood in this energy regime.

## ACKNOWLEDGEMENTS

This work has been supported by the U. S. Office of Naval Research. I would like to thank L. Ratliff, M. Wagshul, S. Maleki, K. Jones, W. Phillips, S. Rolston, S. Bize, P. Julienne, C. Williams, and E. Tiesinga for their valuable contributions to this work.

## REFERENCES

1. P. Julienne, A. Smith and K. Burnett, Adv. At. Mol. Opt. Phys. **30**, 141 (1993).
2. T. Walker and P. Feng, Adv. At. Mol. Opt. Phys. **34**, (1994).
3. J. Weiner, Adv. At. Mol. Opt. Phys. to be published, (1995).
4. P. D. Lett, P. S. Julienne and W. D. Phillips, Annu. Rev. Phys. Chem. **46**, 423 (1995).
5. H. R. Thorsheim, J. Weiner and P. S. Julienne, Phys. Rev. Lett. **58**, 2420 (1987).
6. P. S. Julienne and F. H. Mies, J. Opt. Soc. Am. B **6**, 2257 (1989).
7. P. Julienne, in *Laser Manipulation of Atoms and Ions (Proceedings of the International School of Physics "Enrico Fermi", Course CXVIII)* W. P. E. Arimondo and F. Strumia, Eds.

(North Holland, Amsterdam, 1992).
8. J. Weiner, J. Opt. Soc. Am. B **6**, 2270 (1989).
9. See the article by K. A. H. van Leeuwen in these proceedings.
10. S. Chu, L. Hollberg, J. Bjorkholm, A. Cable and A. Ashkin, Phys. Rev. Lett. **55**, 48 (1985).
11. E. Raab, M. Prentiss, A. Cable, S. Chu and D. Pritchard, Phys. Rev. Lett. **59**, 2631 (1987).
12. R. A. Cline, Miller. J. D. and D. J. Heinzen, Phys. Rev. Lett. **73**, 632 (1994).
13. J. D. Miller, R. A. Cline and D. J. Heinzen, Phys. Rev. Lett. **71**, 2204 (1993).
14. P. L. Gould, et al., Phys. Rev. Lett. **60**, 788 (1988).
15. P. D. Lett, et al., Phys. Rev. Lett. **67**, 2139 (1991).
16. P. S. Julienne and R. Heather, Phys. Rev. Lett. **67**, 2135 (1991).
17. L. P. Ratliff, M. E. Wagshul, P. D. Lett, S. L. Rolston and W. D. Phillips, J. Chem. Phys. **101**, 2638 (1994).
18. P. van der Straten, private communication, (1995).
19. C. Williams and P. S. Julienne, J. Chem. Phys. **101**, 2634 (1994).
20. R. Napolitano, J. Weiner, C. J. Williams and P. S. Julienne, Phys. Rev. Lett. **73**, 1352 (1994).
21. V. Bagnato, L. Marcassa, C. Tsao, Y. Wang and J. Weiner, Phys. Rev. Lett. **70**, 3225 (1993).
22. L. Marcassa, et al., Phys. Rev. Lett. **73**, 1911 (1994).
23. W. McAlexander, et al., Phys. Rev. A **51**, R871 (1995).
24. H. Wang, P. Gould and W. Stwalley, Bull. Am. Phys. Soc. **40**, 1347 (1995).
25. P. D. Lett, L. P. Ratliff, M. E. Wagshul, S. L. Rolston and W. D. Phillips, in *Resonance Ionization Spectroscopy 1994* J. E. P. a. K. W. H.-J. Kluge, Eds. (American Institute of Physics, 1995), vol. AIP Conference Proceedings 329, pp. 289.
26. D. Leonhardt and J. Weiner, preprint (1995).
27. E. R. Abraham, W. I. McAlexander, C. A. Sackett and R. G. Hulet, Phys. Rev. Lett. **74**, 1315 (1995).
28. R. F. Barrow, J. Verges, C. Effantin, K. Hussein and J. D'Incan, Chem. Phys. Lett. **104**, 179 (1984).
29. W. J. Meath, J. Chem. Phys. **48**, 227 (1968).
30. M. H. Anderson, J. R. Ensher, M. R. Matthews, C. E. Wieman and E. A. Cornell, Science **269**, 198 (1995).

# Differential Scattering in Ion-Ion Collisions of He

## Stephan Krüdener

*Institut für Kernphysik, Strahlenzentrum der Universität Giessen, D-35392 Giessen, Germany*

**Abstract.** By means of an ion-ion crossed-beams experiment angular, differential cross sections have been measured for charge transfer in $He^{2+}$ - $He^+$ collisions at barycentric energies between 0.5 and 16.3 keV. The measurements show an oscillatory structure as predicted by theory. These oscillations can be interpreted in terms of the interference between scattering into gerade and ungerade molecular states, which arise due to the identity of the nuclear charges.

## INTRODUCTION

Charge changing collisions between ions belong to the fundamental processes in astrophysical and laboratory plasmas, including tokamak plasmas. For many years charge exchange in $He^{2+}$-$He^+$ collisions has been investigated both experimentally [1-3] and theoretically [4-7]:

$$He^{2+} + He^+ \rightarrow He^+ + He^{2+} \quad (1)$$

Reaction (1) represents a unique collision system since it is a one electron, charge symmetric system, which is comparatively easy to describe by theory. Therefore this system is an ideal testing ground for various theories. Another important feature of reaction (1) are large total cross sections due to the fact that charge exchange is dominated by resonant capture between the 1s ground states of the $He^+$ ions. Finally, it is the simplest ion-ion collision system with Coulomb repulsion, both in the incoming and outgoing channel, leading to remarkable angular scattering of the reaction products. In contrast to the theory [7], experiments could deliver, up to now, only total cross sections for reaction (1). Measurements of angular differential cross sections are therefore an excellent test for the various theoretical approaches, which may yield similar total cross sections by integrating different differential cross sections.

The Giessen ion-ion crossed-beams facility has been used to measure lab frame scattering distributions by position sensitive detection of the He$^+$ reaction products in coincidence with the He$^{2+}$ reaction products of reaction (1). When extracting differential cross sections from the measured scattering distributions, the primary ion beam profile has to be taken into account as well as the transformation between lab- and cm-system. Theoretical and experimental data can be compared in two ways. The easier way is to convolute the theoretical data with the apparatus function (primary ion beam profile) of the experiment. The disadvantage of this method is a loss of information about the abundant maxima and minima in the course of the differential cross section. Therefore it is more convenient to deconvolute the experimental data. This can be done by a Fourier method or an iterative method.

## EXPERIMENTAL TECHNIQUE

As a general approach we used the crossed-beams method. Two well collimated beams of He$^+$ and He$^{2+}$ ions with defined energy intersect in an ultrahigh vacuum of about $10^{-10}$ mbar. Since a $^4$He$^{2+}$ primary ion beam can be contaminated by H$_2^+$ ions, which have a virtually identical charge to mass ratio, for measurements of total cross sections the $^3$He isotope was used to produce $^3$He$^{2+}$ ions. For the measurements of the angular differential cross sections we have used the $^4$He isotope, because the uncertainty arises from the primary ion beam current measurement only. The normalization of the angular differential cross section was done with the measured values for the total cross section [3].
When the beams collide, charge exchange processes occur. Since the energy transferred between the He-ions is much smaller than the beam energy, the reaction products remain inside the beam until they are electrostaticly separated and detected by single particle detectors. The basic principles are simple but their realization is invariably complicated. The main difficulties stem from low ion beam densities. They are limited by space charge effects and the availability of intense, high brightness ion sources. Consequently the rate at which the required collisions are produced is typically in the region of a single event per second. Performing position sensitive measurements using a channelplate detector, enhance the count rate problem. In this case the small reaction rate is distributed over a position matrix of typical 100 * 100 channels. This leads to an average count rate of about $10^{-4}$ events per channel per second, which is masked by about four orders of magnitude higher background rates due to ion-residual gas reactions of the $10^{12}$ times more intense primary beam. Therefore it is necessary to discriminate signal from background events by means of a coincidence technique.

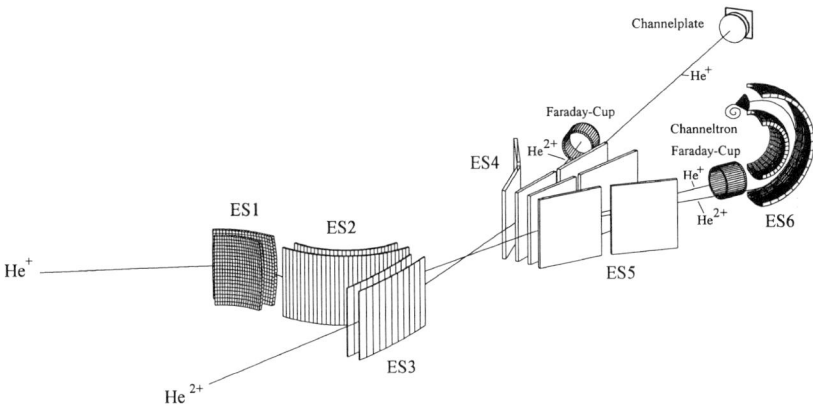

**FIGURE 1.** Schematic diagram of the charge-state analysis before and after the intersection. ES1-ES6: electrostatic analyzers

In order to reach lower collision energies, we have changed the interaction angle between both ion beams from 45°, of the setup described previously [3], to a smaller interaction angle of 17.5°. Figure 1 shows the detailed experimental setup of the modified interaction region. Both the $He^{2+}$ and the $He^+$ beam are collimated to about 1.5 mm · 1.5 mm and are cleaned shortly before intersection by electrostatic deflectors ES1-ES3 from particles in other charge states, resulting from charge-changing collisions in the residual gas on the path from the analyzing magnets, not shown in figure 1, to the interaction region. In the slow $He^+$ beam line (energies up to 20 keV) a two stage electrostatic system ES1, ES2 is used to make the ion beams intersect. This allows for a small interaction angle of 17.5°. Downstream from the intersection, the reaction products $He^+$ and $He^{2+}$ are separated from their parent ion beams. Aiming at a reduction of background, the doubly charged $He^{2+}$ ions pass through a two stage analyzer system ES5, ES6. In the fast beam line, the $He^+$ reaction products are separated from their parent $He^{2+}$ ion beam by the electrostatic deflector ES4, which is similar to ES5. The parent ion beams are recorded by biased Faraday cups whilst the reaction products are counted individually, either by a position sensitive channelplate detector or a channeltron-based single particle detector. The channelplate detector generates a position signal using a resistive anode and a timing signal for the coincidence condition.

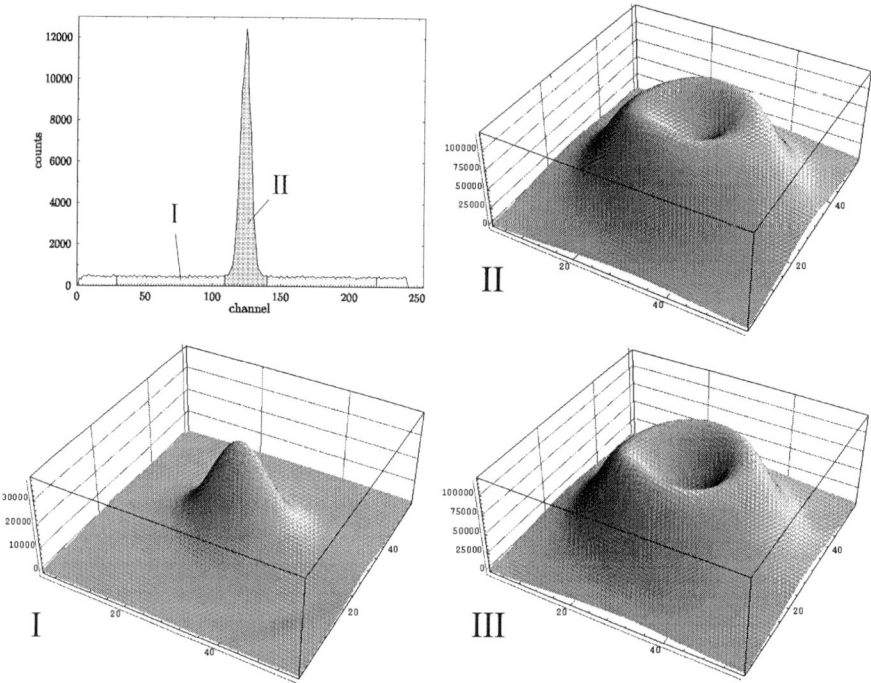

**FIGURE 2.** Time coincidence spectrum (top, left) of product ions $He^+$ and $He^{2+}$ from reaction (1) measured at $E_{cm} = 2.5$ keV. The events are stored with respect to their arrival times in matrix I (bottom left), which contains only background events due to ion-residual gas reactions, and matrix II (top, right), which contains the ion-ion signal and background events. Matrix III (bottom, right) shows the pure ion-ion signal. All scattering distributions are plotted with a linear z-scale.

Since in the charge exchange process (1), both reaction products are generated simultaneously at the interaction point and their flight times to the detectors are fixed, the corresponding output pulses of both detectors have a fixed time delay in the case of true signals, whereas there is no time correlation for background events. This leads to a coincidence spectrum shown in figure 2 (top, left). A sharp peak originated from ion-ion collisions sits on a flat background due to random coincidences. Both position and time information are recorded by a computer in list-mode. The accumulated data are sorted "off-line" with respect to their event time yielding different position matrices as indicated by figure 2. In this example all events registered in time section I are collected in the position matrix I, which represents the background distribution, whereas the events in section II are sorted into matrix II, which contains both the ion-ion events and background events due to ion-residual gas reactions. The position matrix of true ion-ion signals III can be derived by subtracting the normalized matrix I from II. It is obvious that the minimum in forward direction is deeper due to the missing background contribution. All position matrices, shown in figure 2, contain smoothed data in order to present the data more clearly.

To extract angular differential cross sections from the measured scattering distributions, as in the case of matrix III, it is necessary to know the primary ion beam profile. Therefore matrix I can be interpreted as an image of the beam shape since it contains mainly background events due to ion-residual gas reactions. In contrast to ion-ion reactions, where Coulomb-repulsion at low cm-energies plays an important role, fast ion-residual gas collisions produce reaction products, scattered by small scattering angles, forming a peak in the forward direction. In addition, the primary ion beam profile is measured on an absolute scale in one dimension by moving a horizontal slit across the collision plane. Together with the background scattering distribution of the $He^+$ ions produced by ion-residual gas collisions, which represents the relative beam profile, we can assess the density profile in two dimensions.

## Deconvolution of the Scattering Distributions

The finite angular width of the incident beam ($\Delta\alpha \approx 0.2°$) smears out structures that are finer than $\Delta\alpha$. When comparing the theoretical data with the experiment, the scattering distributions have to be deconvoluted with respect to the relative density profile of the parent ion beam. The following two methods are introduced to solve this problem.

### Fourier Method

This method makes use of the Fourier transformation of both the scattering distribution $S(\vec{r})$ and the resolution function, represented by the primary beam profile $P(\vec{r})$:

$$\tilde{S}(\vec{s}) = \frac{1}{2\pi} \int_{-\infty}^{\infty} S(\vec{r}) e^{-i\vec{s}\vec{r}} d\vec{r} \tag{2}$$

$$\tilde{P}(\vec{s}) = \frac{1}{2\pi} \int_{-\infty}^{\infty} P(\vec{r}) e^{-i\vec{s}\vec{r}} d\vec{r} \tag{3}$$

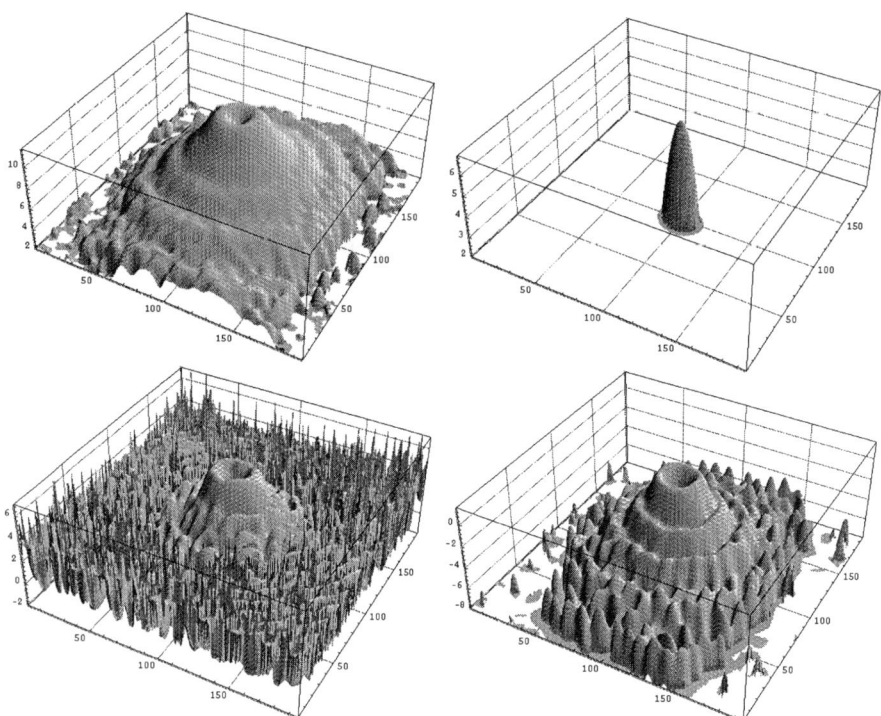

**FIGURE 3.** Three-dimensional view of the measured scattering distribution of He$^+$ ions produced in He$^{2+}$ - He$^+$ collisions at E$_{cm}$ = 2.5 keV (top, left). Due to the influence of the primary ion beam profile (top, right) the oscillatory structure is "smeared out". The results of two deconvolution methods are shown below: Fourier method (bottom, left), iterative method (bottom, right); all data are plotted with a natural logarithmic z-axis.

The deconvoluted scattering distribution $L(\vec{r})$ can then be obtained by the reverse Fourier transformation of the quotient $\widetilde{S}(\vec{s}) / \widetilde{P}(\vec{s})$ [8]:

$$L(\vec{r}) = \left(\frac{1}{2\pi}\right)^2 \int_{-\infty}^{\infty} \frac{\widetilde{S}(\vec{s})}{\widetilde{P}(\vec{s})} e^{i\vec{s}\vec{r}} d\vec{s} \tag{4}$$

Figure 3 shows the result of this procedure for a measurement at E$_{cm}$ = 2.5 keV. Beneath the filtered experimental data (top, left), which are smeared out by the primary ion beam profile (top, right), the figure contains the deconvoluted results (bottom, left). In comparison to figure 2 the measured scattering distribution is plotted on a larger x, y scale. Additionally, the z-axis is drawn in a natural logarithmic scale. It can be seen that the minimum in the forward direction is surrounded by symmetric structures with rapidly decreasing amplitudes. It is evident that only the structures at smaller scattering angles, especially the structure nearest to the central minimum, are now better resolved compared with the filtered experimental data (top, left). In the region of large scattering angles the Fourier method leads to strong numerical fluctuations, which can not be distinguished from the physical oscillations. As a result this method delivers only reliable results for smaller scattering angles.

## Iterative Method

The idea of this method is to "guess" a reasonable start value for the deconvoluted result $L°(\vec{r})$ and to vary $L^n(\vec{r})$ until the convolution of these data, $S^n(\vec{r})$, with respect to the primary beam $P(\vec{r})$, yields the original measurement $S(\vec{r})$. The method is based on a Bayesian deconvolution technique for one-dimensional spectra [9, 10, 11]. In this article, filtered experimental data are taken as the starting matrix $L°(\vec{r})$. The algorithm can therefore be written as

$$S^n(\vec{r}) = \int_{-\infty}^{\infty} L^n(\vec{r}') P(\vec{r} - \vec{r}')\, d\vec{r}' \tag{5}$$

$$L^{n+1}(\vec{r}) = L^n(\vec{r}) \int_{-\infty}^{\infty} \frac{S(\vec{r}')}{S^n(\vec{r}')} P(\vec{r} - \vec{r}')\, d\vec{r}' \tag{6}$$

Of course the matrices $S^n(\vec{r})$ and $L^n(\vec{r})$ have to be normalized after each iteration. The advantage of this procedure (it does not lead to strong numerical fluctuations) has to be compared with the disadvantage, that at least 100 iterations are necessary to get a reliable result. So far, in comparison to the Fourier method discussed above this method is more time-consuming. The result for the iterative procedure is also shown in figure 3 (bottom, right). It is clearly evident that the oscillating structures are indeed better resolved. In comparison to the Fourier method the absence of numerical fluctuations at the edge of the matrix is obvious. The only numerical fluctuations left, remain within the oscillating structures themselves. Since one has to integrate over the azimutal angle φ when extracting angular differential cross sections from those deconvoluted scattering distributions, the influence of these oscillations on the angular differential cross section is negligible. The areas of the matrix, where no counts occur, were shielded during the measurements by edges within the apparatus.

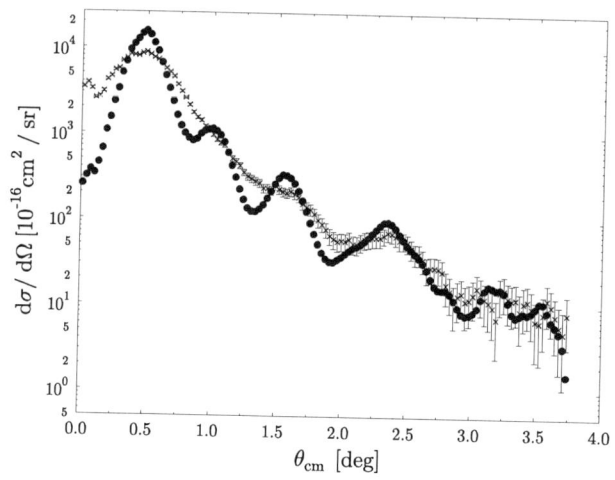

**FIGURE 4.** Comparison of the angular differential cross sections at a cm-energy of 2.5 keV extracted both from the raw data (×) and, the by means of the iterative method, deconvoluted data (•). The raw data are plotted together with their statistical error bars.

## RESULTS AND DISCUSSION

Integration over the azimutal angle φ of the deconvoluted (by means of the iterative method) data of figure 3 yields the angular differential cross section $d\sigma/d\Omega$. Figure 4 shows the result as a function of the cm-scattering angle at a cm-energy of 2.5 keV. In order to assess the effect of the primary beam profile the raw data and their error bars are plotted together with the deconvoluted data on a logarithmic scale for the y-axis. It can be seen that the shape of the angular differential cross sections, extracted from the deconvoluted data, shows an oscillatory structure. In contrast, the shape of the raw data is smeared out due to the influence of the primary ion beam profile. As expected, the influence of the primary ion beam profile becomes smaller with increasing scattering angles. The angular differential cross section of the deconvoluted data is showing five maxima up to cm-scattering angles of 4°. In addition to the minimum in the forward direction, another four minima are resolved experimentally. The amplitude of the

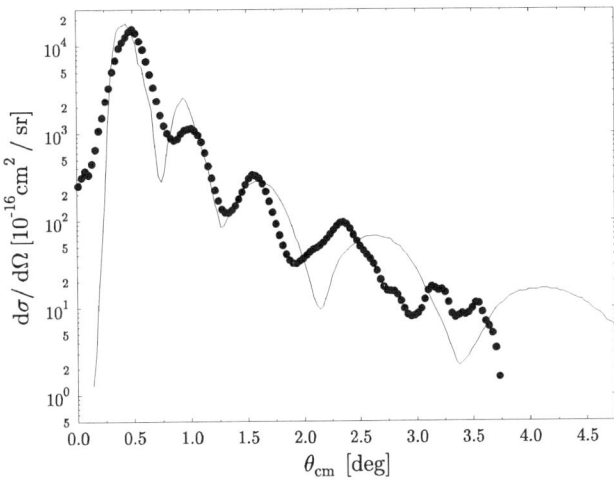

**FIGURE 5.** Comparison of the experimental (•) and theoretical data [7] ( — ) for the angular differential cross section in the $He^{2+}$-$He^+$ system at a cm-energy of 2.5 keV.

structure within the fifth maximum has to be compared with the statistical errors of the raw data and can therefore hardly be interpreted physically. Another feature of the measured angular differential cross section is the increasing period of the oscillation with increasing scattering angles. Figure 5 compares the experimental data with a semiclassical calculation by Forster et al. [7] using Coulomb trajectories. Within this theory, the oscillatory structure can be interpreted as an interference between scattering in the gerade and ungerade molecular states. Up to the third maximum there is a fair agreement between the experimental and theoretical data. For larger scattering angles, however, it seems that theory shows a longer period for the next structures, resulting in a kind of phase shift between experimental and theoretical curve. The calculation is restricted to ground state transitions. Although these transitions dominate the total cross section of reaction (1), it can be expected that the inclusion of a small flux going into excited states of the $He^+$ ion in the outgoing channel leads to a better agreement between the theoretical and experimental data [12, 13]. This concerns the position of maxima and minima as well as the amplitudes of the oscillations. Similar results, as presented in figure 4, have been obtained in the cm-energy range between 0.5 and 16.3 keV.

# CONCLUSION AND OUTLOOK

In recent years experiments investigating ion-ion collisions were restricted to measurements of total cross sections. The experimental determination of angular differential cross sections now opens a wide field for investigations, which allows the testing of various theoretical approaches on a more sophisticated level. Since differential cross sections are a more sensitive probe for the scattering process than the integral cross section, experiments can distinguish between theoretical calculations, which often give similar values for the total cross sections but disagree on the prediction of angular differential cross sections. This is a real challenge both for experimentalists and theoreticians.

# ACKNOWLEDGEMENT

The author would like to thank Prof. E. Salzborn for his support during many years which made the presented work possible. Special thanks to Dr. F. Melchert, Dr. K. Huber, Dipl. Phys. K. v. Diemar and Dipl. Phys. A. Pfeiffer for fruitful collaboration.

Financial support by Deutsche Forschungsgemeinschaft (DFG) is gratefully acknowledged.

# REFERENCES

1. Jognaux, A., Brouillard, F., Szücs, S., J. Phys. **B 11**, L669 (1978)
2. Peart, B., Dolder, K., J. Phys. **B 12**, 4155 (1979)
3. Melchert, F., Krüdener, S., Schulze, R., Petri, S., Pfaff, S., Salzborn, E., J. Phys. **B 28**, L355 (1995)
4. Dickinson, A.S., Hardie, D.J.W., J. Phys. **B 12**, 4147 (1979)
5. Mukherjee, S., Sil, N.C., J. Phys. **B 13**, 3421 (1980)
6. Lal, M., Srivastava, K., Tripathi, A.N., Phys. Rev. **A 26**, 305 (1982)
7. Forster, C., Shingal, R., Flower, D.R., Bransden, B.H., Dickinson, A.S., J. Phys. **B 21**, 3941 (1988)
8. Presnyakov, L.P., private communication
9. Kennett, T.J., Prestwich, W. V., Robertson, A., Nucl. Instr. and Meth. **151**, 285 (1978)
10. Kennett, T.J., Prestwich, W. V., Robertson, A., Nucl. Instr. and Meth. **151**, 293 (1978)
11. Kennett, T.J., Brewster, P.M., Prestwich, W. V., Robertson, A., Nucl. Instr. and Meth. **153**, 125 (1978)
12. Schuch, R., Ingwersen, H., Justiniano, E., Schmidt-Böcking, H., Schulz, M., Ziegler, F., J. Phys. **B 17**, 2319 (1984)
13. Schulz, M., Justiniano, E., Konrad, J., Schuch, R., Salin, A., J. Phys. **B 20**, 2057 (1987)

# Collision Processes Involving Negative Hydrogen Ions

### D. B. Uskov

*P.N.Lebedev Physical Institute, Leninsky Prospect
117924 Moscow, Russia*

---

The paper is devoted to the theoretical description of single and double electron rearrangement collisions involving hydrogen negative ions: $H^- + X^{q+}$, $H^- + H^-$. Calculated cross sections are compared with recent experimental data.

---

## INTRODUCTION

The recent interest in collisions involving negative hydrogen ions was mainly stimulated by the possible application of the energetic beams of $H^-$ ions in the technology of auxiliary heating of magnetically confined fusion plasmas. One of the promising methods of neutralization of $H^-$ ions necessary to obtain the beam of neutral hydrogen is based on the mechanism of electron detachment in collisions of H ions with multiply charged ions in the plasma neutralizer [1]. The following processes are quite important for the calculations of neutralization efficiencies and stability of $H^-$ beam

$$H^- + X^{q+} \longrightarrow H^0 + \ldots \qquad (\sigma_{-0})$$
$$H^- + X^{q+} \longrightarrow H^+ + \ldots \qquad (\sigma_{-+})$$
$$H^- + H^- \longrightarrow H^0 + H^- + e^-$$
$$H^- + H^- \longrightarrow H^0 + H^0 + 2e^-$$

Rearrangement collisions of $H^-$ ions with neutral atoms have been studied in detail during the last decades [2]. The experimental studies of collisions between negative and positive ions required development of crossed-beam and merged-beam techniques [1] which stimulated the development of reliable theoretical methods for calculations of various related cross sections in a wide range of collision energies and ion charge states. From the theoretical point of view negative hydrogen ions are "exotic" atomic species having very low ionization potential and the outer electron in many cases can be considered as being bound by a short-range forces. The theory presented in the present short report makes extensive use of the methods developed in the theory of multiphoton ionization by a strong electromagnetic field. Extension of these methods to the rearrangement collision problems allowed to perform calculations of various cross sections in very good agreement with recent experimental

© 1995 American Institute of Physics

data and to predict rather reliable scaling properties of some collision cross sections.

## ELECTRON DETACHMENT FROM $H^-$ IONS IN COLLISIONS WITH MULTIPLY CHARGED IONS

The presented calculations are based on the theoretical method in ion-atom collisions [3,4,5] developed as a generalization of the Keldysh [6] theory of multiphoton ionization. The following approximations constitute the basis of the theory. One-electron removal from $H^-$ occurs at large internuclear distances where the positive ion field can be described as a pure Coulomb field. The dynamic part of the problem in the final reaction channel has an exact solution if one expands the electron-projectile interaction over multipoles and retain the monopole and dipole terms. Thus the physical mechanism of electron removal from the $H^-$ ion i.e. the process of electron capture by a multiply charged ion and ionization of the $H^-$ electron into continuum, is described as under-barrier and overbarrier transition in the non-stationary potential created by $H^0$ atom and by the field of an ion moving in the vicinity of the $H^-$. In the formalism of quasistationary states the binding energy of the $H^-$ ion acquires an imaginary part related to the decay of the initial state during the collision, and the wave function is the wave packet of the Volkov- Keldysh states. In the stationary phase three-dimensional approximation [3,4] the imaginary part of the binding energy, unitary transition amplitudes and transition probabilities are expressed in terms of the reduced transition amplitudes given by

$$h(\mathbf{p}) = \int_{-\infty}^{+\infty} dt \ < \Phi_\mathbf{p}(\mathbf{r},t)|\mathbf{r}\mathbf{F}(t)|\Phi_0(\mathbf{r})e^{-i\epsilon t} > \qquad (1)$$

Here, $\mathbf{p}$ and $\mathbf{r}$ are the momentum of the particle and coordinate of the active electron, $\Phi_0$ is the unperturbed wave function of the the negative ion, and $\epsilon$ is the real part of the binding energy. The magnitude of the ion field

$$\mathbf{F}(t) = q \ \mathbf{R}(t)/R^3(t) \qquad (2)$$

depends on the positive ion charge q and on the time-dependent internuclear distance $\mathbf{R}(t)$. The Volkov-Keldysh states

$$\Phi_\mathbf{p}(\mathbf{r},t) = (2\pi)^{-3/2} \exp\left[i\mathbf{k}(t)\mathbf{r} - \frac{i}{2}\int_0^t d\tau \ k^2(\tau)\right] \qquad (3)$$

with

$$\mathbf{k}(t) = \mathbf{p} - \mathbf{A}(t) \qquad \mathbf{A}(t) = -\int_0^t d\tau \ \mathbf{F}(\tau)$$

satisfy the nonstationary Schrödinger equation

$$\left[i\frac{d}{dt} + \frac{1}{2}\Delta_r + \mathbf{rF}(t)\right] \Phi_\mathbf{p}(\mathbf{r}, t) = 0 \qquad (4)$$

and describe the motion of the unbound electron with definite values of momentum **k** in the time-dependent field (2). The total probability of electron removal from the negative ion as function of impact parameter is given by [3,4]

$$W = 1 - \exp\left[-\int d\mathbf{p} \, |h(\mathbf{p})|^2\right] . \qquad (5)$$

In the present approach the unperturbed negative ion wave function is obtained on the basis of a short-range one-electron potential approach. The two-electron correlation effects are important for calculation of the binding energy of the unperturbed $H^-$ ion. However, if the binding energy is known, the short-range potential model has been shown to be very effective for the analysis of bound-free process: electron detachment in a static field, photoionization, multiphoton ionization, etc [6-8]. Using the momentum representation it can be shown that the matrix element in Eq.(1) has its main contribution from the asymptotic region of $r$ where the wave function is defined by

$$\Phi_0(r) \approx \Phi_0^{as}(r) = \frac{B\gamma^{1/2}}{(2\pi)^{1/2}} \frac{e^{-\gamma r}}{r} \qquad (6)$$

where $\gamma/2 = -\epsilon = 0.0275$ a.u. $= 0.75$ eV and $B = 1.56$. The value of $B$ is obtained from electron-density calculations with Pekeris many-parameter two-electron wave function [9].

For many applications [10] the problem of cross section parametrization is more important than elaborate calculations of some particular cases. For electron removal from neutral hydrogen atom a scaling law was established and compared with experimental data cross section [10-12]. In the present case the approximation of the ion field by Eq.(2) is leading to the following scaling law [5] for the the electron removal cross section (charge transfer plus ionization)

$$\sigma = 2\pi \int_0^\infty W \, bdb = q \, Q(v^2/q) \qquad (7)$$

where $W$ is given by Eq.(5) and $v$ is the collision velocity. The numerically calculated function $Q$ in Eq.(7) can be approximated by the analytical approximate formula

$$\frac{\sigma}{q}[10^{-14}\text{cm}^2] = \begin{cases} 0.239 \ln(2.5/y + 0.8) & , \quad y \le 0.3 \\ (0.129/y) \ln(8.12y + 1.01) & , \quad y \ge 0.3 \end{cases} \qquad (8)$$

with $y = 0.25 v^2/q = E/100q$ and where $v$ is taken in atomic units $v_0 = 2.2 \, 10^8$ cm/s and $E$ in keV/amu. The discrepancy between the two

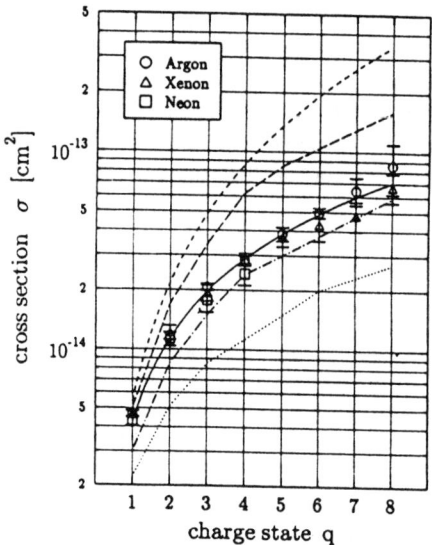

**FIG. 1.** Cross sections $\sigma_{-0}$ for $H^- + X^{q+} \to H^0 + \ldots$. Experimental data are from [15,16,17], solid curve — present theory, dashed curve - [13], long-dashed, dot-dashed and dotted curves represent CTMC calculations [15].

formulas is less than 1% in the region of $0.1 < y < 1$. The analysis [4] shows that the theory presented is valid for $0.02/q^2 < y < 1.5q$. At larger energies the dipole potential in Eq. (2) should be replaced by the Coulomb potential [4]. The result for $y > q$ can be expressed in the form

$$\frac{\sigma}{q}[10^{-14}\text{cm}^2] = \frac{1.29}{y} \ln\left\{\frac{8.12}{(1+0.577y/q)^{1/2}} + 1.01\right\} \quad (9)$$

which agrees with the ab initio Bethe-Born calculations [13,14] in the asymptotic region $y \gg q$ for ionization by bare ion projectiles. Here we observe small deviation from the scaling law (7) and (8) at high energies (see Fig 2 below). Cross section for electron detachment as a function of ion charge $q$ are shown in Fig.1 for center of mass energy $E_{cm} = 50$ keV.

The results of the classical trajectory Monte Carlo (CTMC) calculations shown in Fig.1 significantly depend on the choice of the binding potential and are roughly proportional to the square of the classical turning radius. Calculations with a short-range potential, giving the correct $H^-$ binding energy, are a factor of 3 smaller than the experimental values. This is due to the neglect of the tunneling mechanism as well as to the difference between the classical microcanonical distribution and the quantum Wigner distribution function, especially pronounced for short-range potentials. Classical mechanics, often giving good results for Coulomb interactions, should be replaced by quantum

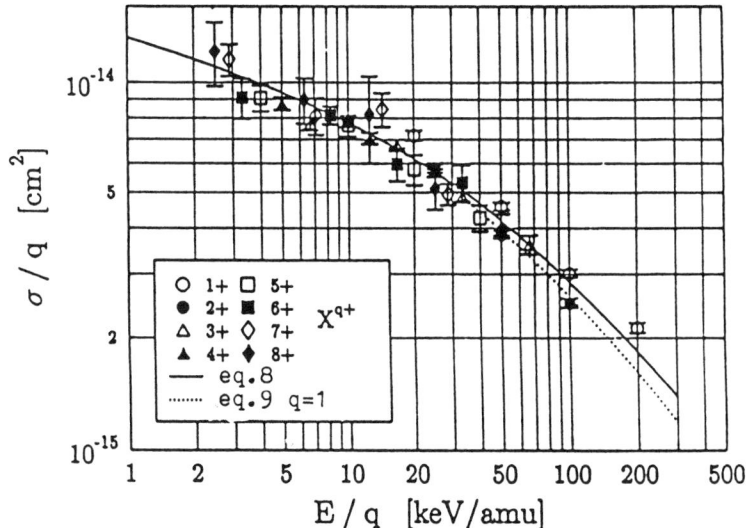

**FIG. 2.** Scaled cross section $\sigma_{-0}/q$ as a function of scaled energy $E/q$. Solid line - Eqs.(7) and (8). Experimental data are from [15,16,17]. Deviation from the scaling law (8) for $q = 1$ are demonstrated by the dotted curve, Eq.(9).

treatment when either underbarrier transitions are important or the WKB approximation is not applicable.

In Fig. 2 scaled cross sections $\sigma_{-0}/q$ are shown as a function of scaled energy $E/q$. The scaling law proposed is confirmed by experiment, allowing accurate predictions of cross sections for $H^-$ neutralization in collisions with multiply charged ions in a wide energy range.

## SINGLE AND DOUBLE ELECTRON DETACHMENT IN $H^- + H^-$ COLLISIONS

Electron detachment in slow $H^- + H^-$ collisions is defined by the physical mechanism of the promotion of the initial channel quasimolecular term by Coulomb interaction first into the continuum of $H_2^-$ quasimolecule at the internuclear distance $R \approx 36$ a.u [18] and then into the continuum of $H_2$ quasimolecule at $R \approx 18$ a.u. [19,20]. Single electron detachment is related to the decay of quasistationary state of electron bound to $H^0$ atom in the presence of the external Coulomb repulsive potential.

Two-electron detachment is defined by mechanisms which we call two-step transition and direct transition. The two-step mechanism corresponds to release of one electron and ionization of another one in subsequent collision of $H^-$ with $H^0$ [16]. The direct mechanism is a simultaneous transition of two electrons into continuum with the population of $H_2$ quasimolecular term. For this mechanism it is essential that the quasistationary state formed during $H^- + H^-$ collision is resonantly coupled to both one-electron continuum of $H_2^-$ and two-electron continuum of $H_2$ quasimolecules.

One- and two-electron detachment cross sections are defined as integrals of corresponding population probabilities in final reaction channels over the impact parameter. It is important to take into account the two-step mechanism as well as the direct mechanism. A method of calculation of one- and two- electron decay rates in the initial reaction channel is described below. For the intermediate $H^0 + H^-$ channel we employ the results of Bardsley and Cohen [21], Bieniek and Dalgarno [22]. The comparison of calculated cross sections with experimental results reveals the importance of various transition mechanisms of the double electron detachment for different collision energy ranges.

In the independent particle approximation, which describes the weakly bound electron state of negative ion by some model short-range potential $V(r)$, the Hamiltonian for two active electrons has the form

$$\hat{H} = \sum_{i=1,2} \left[ -\frac{1}{2}\Delta_{r_i} + V(|\mathbf{r}_i - \mathbf{R}_i|) \right] + \frac{1}{|\mathbf{r}_1 - \mathbf{r}_2|} \quad (10)$$

where $\mathbf{r}_1$ and $\mathbf{r}_2$ are the electron position-vectors; $\mathbf{R}_1$ and $\mathbf{R}_2$ are the position-vectors of the nuclei. The applicability of the independent-particle approximation has been discussed in connection with $H^-$ electron detachment by many authors [16,23].

For the one-electron detachment problem the Hamiltonian (10) can be reduced to

$$\hat{H}^I = \hat{H}_i^I + \frac{1}{|\mathbf{r} - \mathbf{R}|} = \hat{H}_f^I + V(r) = -\frac{1}{2}\Delta_r + V(r) + \frac{1}{|\mathbf{r} - \mathbf{R}|} \quad . \quad (11)$$

Here we introduced unperturbed initial and final channel Hamiltonians $\hat{H}_i^I$ and $\hat{H}_f^I$ respectively. By formal scattering theory [24,25] we define the quasistationary state width and corresponding transition rate as

$$\Gamma^I = 2\pi \int |<\phi_i|V|\phi_f>|^2 \, \delta(E_i^I - E_f^I) \, d\mathbf{f}$$
$$= \mp <\phi_i|V \, 2 \, \text{Im}\left[(E_i^I - \hat{H}_f^I \pm i\epsilon)^{-1}\right] V|\phi_i> \quad (12)$$

Expression (12) is similar to the Fermi Golden rule applied for a rearrangement transition with different initial and final channel Hamiltonians (see [25] for details).

Initial and final wavefunctions satisfy the stationary Schrödinger equation

$$\hat{H}_i^I|\phi_i> = -\frac{\kappa^2}{2}|\phi_i> \quad , \quad \hat{H}_f^I|\phi_f> = -\frac{\kappa^2}{2}|\phi_f> \quad . \quad (13)$$

The energy $E_i^I$ is equal to the real part of the complex energy of the quasistationary state and, neglecting the polarization term, is given by

$$E_i^I = -\frac{\kappa^2}{2} + \frac{1}{R} \approx <\phi_i|\hat{H}^I|\phi_i> \qquad (14)$$

To analyze the matrix element in Eq.(12) we rewrite it in the momentum representation. Using the Schrödinger equation (13) one obtains

$$<\phi_i|V|\phi_f> = -\int \frac{1}{2}(k^2 + \kappa^2) <\phi_i|\mathbf{k}><\mathbf{k}|\phi_f> d\mathbf{k} \quad . \qquad (15)$$

We note that in terms of the JWKB approximation $<\mathbf{k}|\phi_f>$ is a rapidly oscillating function in the region of $k \approx \kappa$ defined by the behavior of $<\phi_i|\mathbf{k}>$. In the complex $k$-plane, the function $<\mathbf{k}|\phi_f>$, with the energy $E_i^I = E_f^I$, given by (14), has a stationary phase point when $k^2 + \kappa^2 = 0$. In the vicinity of this point the other factor in the integrand of (15) is a smoothly varying function. For this reason we use an approximation

$$<\phi_i|V|\phi_f> = \left\{ \lim_{(k^2+\kappa^2)\to 0} \left[ -\frac{1}{2}(k^2+\kappa^2) <\phi_i|\mathbf{k}> \right] \right\} \int <\mathbf{k}|\phi_f> d\mathbf{k}$$
$$= -(2\pi\kappa)^{1/2} B\phi_f(r)|_{r=0} \qquad (16)$$

The constant $B$ in this equation is related to the asymptotic behavior of the unperturbed wavefunction $\phi_i$ Eq.(6). In the case of zero-range potentials, the factor $(k^2 + \kappa^2) <\phi_i|\mathbf{k}>$ is a constant equal to $\kappa^{1/2}/\pi$ and thereby formula (16) is exact.

Employing the Coulomb Green function [26,27] and Eq.(16) one can obtain the following expression for the width

$$\Gamma^I = \frac{4\pi\kappa^2 B}{\exp(2\pi/k) - 1} \left[ \left(\frac{dM}{dz}\right)^2 - \left(\frac{1}{4} - \frac{1}{2k^2R}\right) M^2 \right] \qquad (17)$$

with

$$M(z) = M_{-i/k, 1/2}(-2ikR)$$

where $M$ is the Whitthaker function [28]. The variable $k$ denotes the magnitude of electron momentum in the continuum and its value is defined by the resonant condition

$$\frac{k^2}{2} = \frac{1}{R} - \frac{\kappa^2}{2} \quad . \qquad (18)$$

The result (17) is the final form of transition rate used in present calculations. It is worth noting that the basic limiting cases - the result of Drukarev and Demkov for a homogeneous field [29], the result of Komarov and Solov'ev

[30] for Coulomb field and the approximate formula by Smirnov and Chibisov [18] - can be derived from Eq.(17) by means of the appropriate asymptotes of the $M$-functions. It is important to define the range of internuclear distances where the tunneling probability can be approximated by the result of the homogeneous field model. By requiring that the homogeneous field Green function [31] can be substituted for the Coulomb Green function in Eq. (13) one obtains two inequalities

$$R \gg 1 \quad \text{and} \quad R \ll 2/\kappa^2 . \tag{19}$$

Interestingly, these conditions are not enough to provide applicability of asymptotic results obtained in [18,29]. An additional requirement, which is not important for our approach, is that the quasiclassical exponential factor should be small, i.e. $R^2\kappa^3/3 \gg 1$. In fact, this considerably restricts the applicability of asymptotic results [18,29,30] for $H^- + H^-$ collision problem considered here.

For the two-electron tunneling probability quite similar to (12) we take

$$\Gamma^{II} = \mp <\psi_i|U\, 2\, \text{Im}\left[\left(E_i^{II} - \hat{H}_f^{II} \pm i\epsilon\right)^{-1}\right]U|\psi_i> \tag{20}$$

where

$$U(\mathbf{r}_1, \mathbf{r}_2) = V(|\mathbf{r}_1 - \mathbf{R}_1|) + V(|\mathbf{r}_2 - \mathbf{R}_2|) \tag{21}$$

$$E_i^{II} = \frac{1}{R} - \kappa^2 \tag{22}$$

$$\psi_i((\mathbf{r}_1, \mathbf{r}_2) = \phi_i(\mathbf{r}_1 - \mathbf{R}_1)\phi_i(\mathbf{r}_2 - \mathbf{R}_2) \tag{23}$$

and $\hat{H}_f^{II}$ is defined as

$$\hat{H}_f^{II} = -\sum_{i=1,2}\frac{1}{2}\Delta_{r_i} + \frac{1}{|\mathbf{r}_1 - \mathbf{r}_2|} . \tag{24}$$

The same approximation as (16) gives the resulting formula

$$\Gamma^{II} = 16\kappa^2 B^4 \int_0^{2(1/R-\kappa^2)^{1/2}} K_{\text{cm}}^2\, G(K_{\text{cm}})\, dK_{\text{cm}} \tag{25}$$

$$G(K_{\text{cm}}) = \left\{[\exp(\pi/k) - 1]\left(\kappa^2 + K_{\text{cm}}^2/4\right)^2\right\}^{-1}$$
$$\left[\left(\frac{d\mathcal{M}}{dz}\right)^2 - \left(\frac{1}{4} - \frac{1}{4k^2R}\right)\mathcal{M}^2\right] \tag{26}$$

with

$$\mathcal{M}(z) = M_{-i/2k,1/2}(-2ikR) \quad .$$

The variable $k$ in (26) is a function of $K_{cm}$ in (25), defined by

$$\frac{K_{cm}^2}{4} + k^2 = \frac{1}{R} - \kappa^2 \quad . \tag{27}$$

The quantities $K_{cm}$ and $k$ are two-electron center of mass momentum and relative electron momentum, respectively. Expressions (17) and (25,26) are used to calculate the cross sections for low-energy collisions.

The theory based on the description of the transition as the multichannel decay of quasistationary states of $H^- H^-$ and $H_2^-$ systems, formed during the collision, is valid up to the energies of about 5 keV/amu. For these energies the characteristic time of electron underbarrier motion is still less than the characteristic time of the potential variation and the nonstationary electron wavefunction at each instant could be approximated by a wavefunction of a quasistationary state. An extension of the theory for higher energies is based on the Keldysh nonstationary approach for rearrangement collision problems [3,4,5].

First consider the single-electron detachment. For small collision energies the transition probability is close to unity in the wide range of impact parameters contributing to the total cross section. With the energy increasing this range is diminishing roughly as the square root of the cross section. In terms of condition (11) that implies that the approximation of potential by the potential of non-stationary homogeneous field becomes appropriate for assessment of reaction mechanism. For non-stationary field the corresponding dynamic width $\hat{\Gamma}_{din}$ was defined in the framework of the Keldysh formalism [4].We introduce the dynamic tunneling factor for $H^- + H^-$ collision as the ratio of $\hat{\Gamma}_{din}$ and the stationary width $\hat{\Gamma}_{st}$ (calculated in the same dipole approximation as $\hat{\Gamma}_{din}$) and extend the stationary formula for $\Gamma^I$ as follows

$$\Gamma_{din}^I(\rho, v, t) = \Gamma^I(R) \, \Omega(\rho, v, t) \qquad \Omega(\rho, v, t) = \hat{\Gamma}_{din}(\rho, v, t)/\hat{\Gamma}_{st}(R) \tag{28}$$

where $\rho$ and $v$ are impact parameter and collision velocity respectively. Thereby we provide the stationary result for $v \to 0$ when $\Omega \to 1$. For higher energies the impact parameters contributing to the cross section is small enough so that $\Gamma^I/\hat{\Gamma}_{st} \approx 1$ and consequently $\Gamma_{din}^I \approx \hat{\Gamma}_{din}$. For the distances $R \ll 1$ and $R \gg 1$ formula (28) retains some features of the Coulomb interaction. Approximation (28) comprises a phenomenological element in a sense that the stationary width calculated for the Coulomb potential in the Hamiltonian (11) is corrected for the dynamic tunneling effect by means of Keldysh-type approximation retaining only monopole and dipole potential terms. In the high- energy limiting case the integral of $\hat{\Gamma}_{din}$ over the trajectory reduces to the ionization probability in the Born dipole approximation

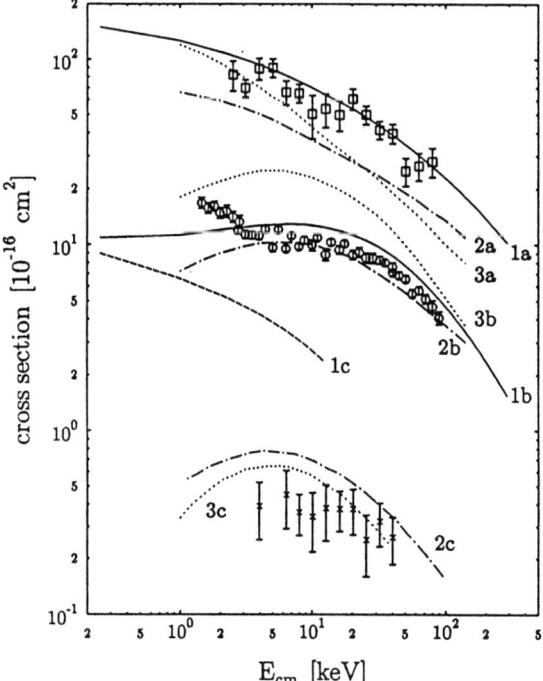

**FIG. 3.** Single and double detachment cross sections as functions of the center of mass collision energy: experiment [18], single and double detachment, respectively; curves 1a,1b – the theory, curve 1c – test calculations with $\Gamma^{II} = 0$; curves 2a,b and 3a,b – CTMC results [16]; curves 3c,2c – $H^- + H^- \to H^+ + H^0 + 3e$.

[32,33]. Asymptotically the cross section of single ionization in the $H^- + H^-$ collision approaches the cross section of $H^-$ ionization by electron or proton impact multiplied by a factor of 2. The dynamic corrections for the two-electron decay rate is introduced in the form:

$$\Gamma^{II}_{din}(\rho, v, t) = \Gamma^{II}(R)\Omega^2(\rho, v, t) \quad . \tag{29}$$

The analysis of formula (28) shows that the dynamic effects are quite small up to the energies of $E = 10$ keV/amu.

Experimental results [15,16] and calculated cross sections are presented in Fig. 3 as a function of the center of mass collision energy $E_{cm}$. For energies less than 50 eV a distortion of the straight-line heavy particle trajectory by the Coulomb interaction may become important. The agreement between experiment and theory for single ionization cross section is quite good. The theory tends to overestimate cross section for double ionization except for the low-energy region of $E_{cm} < 2.5$ keV. That could be due to the overestimation of the direct two-electron transition rate in the region of middle energies. Several experimental values for double ionization at low energies exhibit the trend to grow for lower energies while the theory is giving practically constant cross section. A reason of this effect is not quite well understood. At the limit of low energies the present theoretical approach gives a cross section asymptotically approaching the cross section of single electron detachment in $H^- + H^0$ collision. To demonstrate the importance of the double electron tunneling mechanism the test calculations with the rate $\Gamma^{II} = 0$ is shown in Fig.3 (curve 1c). The comparison indicates that the mechanism of simultaneous two-electron tunneling is the dominant one for the double ionization process at $E_{cm} > 10$ keV. The present results is the first theoretical study of this specific two-electron quantum process.

Two sets of calculations by CTMC calculations in Fig. 3 correspond to two different fitting potentials describing electron interaction with the hydrogen atom [15]. One of the first calculations [23,34], performed by CTMC and quantum JWKB method of Smirnov and Chibisov [18], for $H^- + \bar{p}$ ionizing collisions showed that CTMC results are about factor of 4 less than results of the JWKB tunneling approach. Some problems connected with the applicability of classical mechanics for collisions of $H^-$ ions with multicharged ions were also discussed in [5]. For one-electron processes, considered in these publications, general trends of CTMC data are in agreement with other theories and experimental results. For two-electron ionization in $H^- + H^-$ collision both CTMC data sets (in Fig. 3, curves 2b,3b) have maxima at some energies, which is not confirmed by experiment. The reason of this effect is that the classical mechanics does not describe the process of ionization in $H^- + H^0$ collision for low collision energies. This process occurs mainly via the quantum mechanism of the promotion of two initial quasimolecular terms into the continuum of $H_2$ quasimolecule [21,22]. From our test calculations (Fig 3 curve 1c) we found that this intermediate reaction channel is very important at low energies. The classical mechanic cross section for this intermediate channel decreases adiabatically at low energies, affecting the double ionization cross section . In general the theory, based on the non-stationary tunneling approach, is in a good agreement with the experiment. The new two-electron simultaneous tunneling process is introduced. The importance of this phenomena is confirmed by the experiment [16] and quantitatively described by the proposed theoretical method.

## DOUBLE ELECTRON REMOVAL FROM $H^-$ IN COLLISIONS WITH MULTIPLY CHARGED IONS

There are three physical mechanisms for the process of collisional double ionization: i) one of the atomic electrons after interaction with a projectile

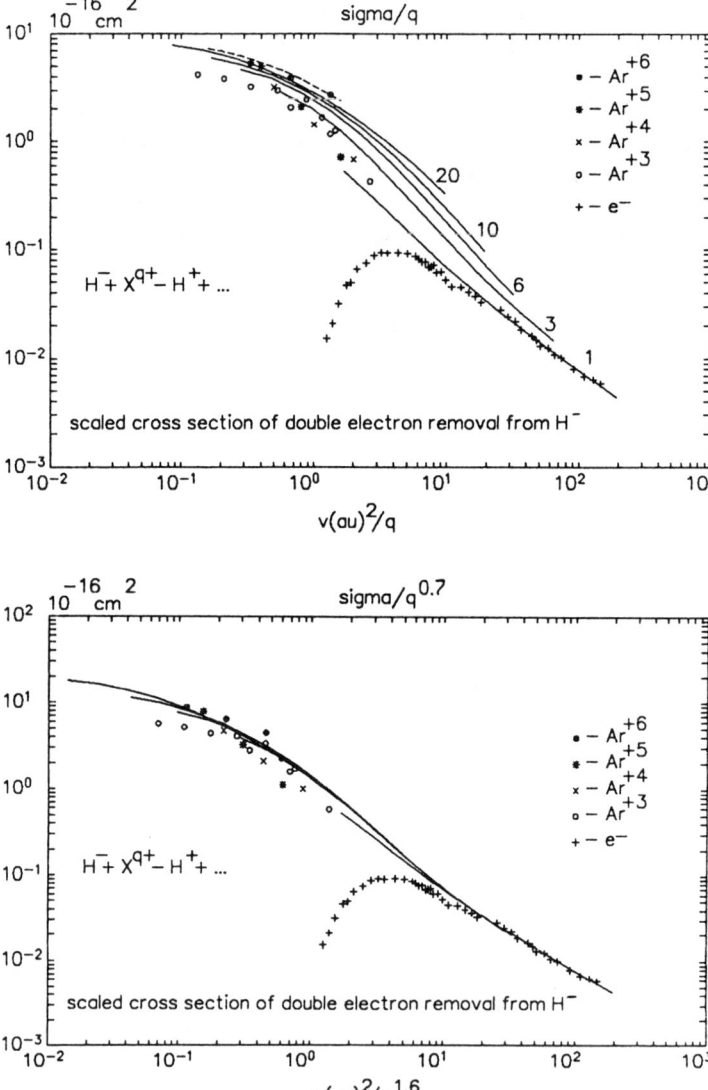

**FIG. 4.** Two scaling laws (top, **4a**; bottom, **4b**) for the cross section of double electron removal in collisions of $H^-$ with positive ions of charge $q$. The scaled cross section is plotted as a function of the scaled square collision velocity. Solid curves - present theory for $q = 1, 3, 6, 10, 20$, dashed curve is the scaled cross section for electron detachment from H by multiply charged ion impact [3], experimental data for collisions with ions are taken from [1,36], and those for double electron ionization by electron impact are from [35].

is ejected with high velocity and the other one due to the correlation in initial state undergoes a relaxation leading to additional ionization (Shakeoff), ii) primarily ejected electron collides with the second one resulting in double ionization (TS-1), iii) projectile interacts consequently with two electrons ionizing both (TS-2). The Shakeoff and TS-1 mechanisms are important at high relative velocities, while TS-2 is the dominant one for low energies. Since TS-1 and TS-2 imply that a large momentum is transferred to the atomic electrons these mechanisms are assessed on the basis of classical mechanics.

The Shakeoff process is defined by the region of distant collisions. The cross section is expressed in terms of the double photoionization of $H^-$. The present results and experimental data [1,35,36] are presented in Figs 4a,b. The theoretical calculations of the double electron detachment are analyzed from the point of view of scaling laws. Beside quite valuable physical information gained in the case of a favorable scaling, this procedure provides the possibility of a comprehensive comparison of theory and experiment when a cross section is a function of two parameters (collision velocity and charge of impinging ion). If the charge dependence is basically reduced to a convenient scaling then, the set of energy dependence curves for various charges should be close to a single universal "reduced" curve and experimental data are also expected to follow this curve.

The result of the application of the basic scaling law [5,10,33] is shown in Fig. 4a. The important effect revealed by Fig. 4a is the convergence of theoretical curves (for $q > 10$) to the universal curve which practically coincides with the scaled cross section for the process of electron removal from the neutral H [3,4]. This effect could be qualitatively explained using an independent electron approximation [15]. As demonstrated in Fig. 4a, this result holds only in the very limited range of energies and charges. For higher energies the independent electron model fails to reproduce even the order of magnitude of the cross section. The high energy experimental data is available only for electron impact double ionization [35], shown in Fig. 4a by the crosses (we remind that equivelocity electron and proton impact double electron ionization cross section converge asymptotically for large energies).

Another scaling law peculiar for this particular process is shown in Fig. 4b. The scaling procedure provides the best convergence to the theoretical curves for $q = 1 - 20$. Fig. 4b shows that the experimental data available allow to establish already the cross section behavior in both high and low energy regions. On the other hand, it would be quite important to obtain experimental results covering the intermediate energy region, where, at present, data are practically absent.

## ACKNOWLEDGEMENTS

This research was supported by grants ISF MK1000, MK1300 and ISTC grant 076-95.

# REFERENCES

1. F. Melchert, W. Debus, S. Kruedener, R. Schulze, R.E.Olson and E.Salzborn, Suppl. Z.Phys **D21**, S 249 (1991).
2. J.S. Risley in *Proc. XI ICPEAC*, eds. N. Oda, K. Takayanagi (Amsterdam, North Holland 1980) p. 619.
3. L.P. Presnyakov and D.B. Uskov, Sov.Phys. JETP **59**, 515 (1984).
4. L.P. Presnyakov and D.B. Uskov in *Proceedings of the P.N.Lebedev Physical Institute*, edited by I.I Sobelman (Nova Science Publishers, New York 1987) Vol 179, p. 137.
5. F. Melchert, M. Benner, S. Kruedener, R. Shulze, S. Meuser, K. Huber, E. Salzborn, D.B. Uskov and L.P. Presnyakov., Phys.Rev.Lett. **74**, 888 (1995).
6. L.V. Keldysh, Sov.Phys. JETP **20**, 1307 (1965).
7. A.I. Nikishov and V.I. Ritus, Sov.Phys. JETP **25**, 145 (1968).
8. P.A. Golovinsky and I.Yu. Kiyan, Usp. Fiz. Nauk. **160**, 97 (1990).
9. C.L. Pekeris,, Phys.Rev. **126**, 1470 (1972).
10. . R.K. Janev, L.P. Presnyakov V.P. Shevelko, *Physics of Highly Charged Ions* (Springer, Heidelberg 1985).
11. K.H. Berkner, W.G. Graham, R.V. Pyle, A.S. Schlahter, J.W.Stearns, and R.E.Olson, J.Phys. **B11**, 875 (1978).
12. H.B. Gibody, Phys.Scripta **24**, 712 (1981).); M.B. Shah and H.B. Gilbody, J.Phys. **B14**, 2361 (1981).
13. Y.K. Kim and M. Inokuti, Phys.Rev. **A4**, 665 (1971).
14. H.S.W. Massey, *Negative Ions* (Cambridge University Press, Cambridge 1976).
15. F. Melchert, W. Debus, M. Liehr, R.E. Olson and E. Salzborn, Europhys. Lett. **9**, 433 (1989).
16. R. Shulze, F. Melchert, M. Hagmann, S. Kruedener, J. Kruger, E. Salzborn, C.O. Reinhold, and R.E. Olson, J.Phys. **B24**, L7 (1991).
17. F. Melchert in *AIP Conference Proceedings 295, XVIII ICPEAC 1993 Denmark* (AIP Press, New York 1993) p.575.
18. B.M. Smirnov and M.I. Chibisov, Sov.Phys. JETP **22**, 585 (1966).
19. D.B. Uskov, A.D. Ulantsev and L.P. Presnyakov, *Book of Abstracts Aarhus Denmark 21-27 July 1993*, eds. T. Andersen, B. Fastrup, F. Folkmann and I.H. Knudsen, p.644.
20. M.I. Chibisov and Yu.N. Yavlinslii, Kurchatov Institute Preprint IAE-5738/6 (1994).
21. J.N. Bardsley and J.S. Cohen, J.Phys. **B11**, 3645 (1978).
22. R.J. Bieniek and A. Dalgarno, Astrophys. J. **228**, 635 (1979).
23. J.S. Cohen and G. Fiorentini, Phys.Rev. **A33**, 1590 (1986).
24. M.L. Goldberger and K.M. Watson, *Collision Theory* (New York, Wiley 1964).
25. A.M. Urnov and D.B. Uskov, J.Phys. **B26**, 268 (1993).
26. L.H. Hostler and R. Pratt, Phys.Rev.Lett **10**, 469 (1963).
27. L. Hostler, J.Math.Phys. **5**, 591 (1964).
28. M.Abramowits and I.A. Stegun, *Handbook of Mathematical Functions*, Applied mathematics Series 55 (National Bureau of Standards 1964).
29. Yu. N. Demkov and G.F. Drukarev, Sov. Phys. JETP **20**, 616 (1965).
30. I.V. Komarov and E.A. Solov'ev, Theor. Math. Phys. (USSR) **32**, 736 (1966).
31. V.Z. Slonim and F.I. Dalidchik, Sov.Phys. JETP **44**, 1081 (1976).

32. M.J. Seaton, Proc.Phys.Soc. **79**, 1105 (1962).
33. R.K. Janev and L.P. Presnyakov, J.Phys. B**13**, 4233 (1980).
34. G. Fiorentini and R. Tripiccone, Phys.Rev. A**23**, 737 (1983).
35. D.J. Yu, S. Rachafi, J. Jureta, P. Defrance, J.Phys. B**25**, 4593 (1992).
36. E. Salzborn, private communication (1995)

# NOVEL TECHNIQUES

Recent Results from the Super EBIT .............................. 705
    R.E. MARRS
Attaining Low Electron Temperatures in Electron Coolers ........ 721
    H. DANARED
Production of Bright, Cold Beams
for Atomic Collision Experiments .................................. 731
    K.A.H. VAN LEEUWEN, E.J.D. Vredenbregt, P.G.M. Sebel,
    J.P.J. Driessen, M.D. Hoogerland, and H.C.W. Beijerinck
"Complete" Measurement of
Molecular Coulomb-Explosions ...................................... 741
    U. Werner and H.O. LUTZ

# Recent Results from the Super EBIT

## R. E. Marrs

*Lawrence Livermore National Laboratory*
*Livermore, CA 94550*

**Abstract.** The Super EBIT device at LLNL can produce and trap any highly charged ion at rest in the laboratory, including bare $U^{92+}$ ions. Recently, the ionization cross sections for high-Z hydrogenlike ions have been measured for the first time, and measurements of the L-shell ionization cross sections for uranium ions are in progress. The two-electron contributions to the ground state energies of heliumlike ions have been directly measured using a novel technique, and spectra of 2s-2p transitions in highly ionized thorium and uranium have been used to test QED corrections to the energy levels of few electron high-Z ions. A new capability for the study of rare isotopes has been demonstrated. Ion cooling has been used to reduce the thermal broadening of x-ray emission lines to the point where natural line widths can be observed in some cases.

## INTRODUCTION

The super electron beam ion trap (Super EBIT) is a device for producing and trapping the few-electron ions of heavy elements for studies of atomic structure and electron-ion collisions. It is called "Super" because of the high (~200 keV) electron energy. Any highly charged ion can be produced in a Super EBIT, including bare $U^{92+}$ [1]. Such highly charged ions provide an opportunity to understand the contributions of relativity and QED in atomic structure and interactions. For example, QED energy corrections increase as $Z^4$, so they are most visible and most important at high atomic number.

In an EBIT, atomic physics information is derived from the x-ray emission spectra of trapped ions colliding with beam electrons. The ions are confined radially by the space charge potential of the electron beam. In the axial direction the ions are confined by bias voltages applied to three drift tubes. The electron beam is compressed to high density by the magnetic field from a split pair of superconducting coils; the magnetic field has little effect on the ion motion. X rays are observed at 90 degrees to the electron beam through the gap between the coils as indicated in Fig. 1. A more detailed description of the LLNL Super EBIT may be found elsewhere [2].

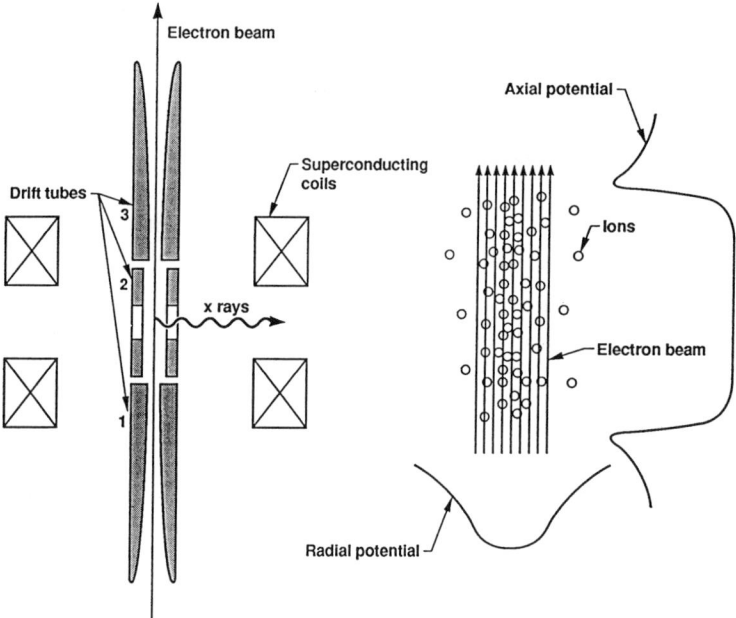

**FIGURE 1.** Key features of the LLNL Super EBIT. Left: Scale drawing of the trap electrodes and magnet coils. Right: Schematic enlargement of the trapping potentials. Ions are confined in drift tube number 2.

Recent Super EBIT experiments at LLNL span two orders of magnitude in x-ray energy and address several different areas of highly-charged-ion physics. Most of the recent results at high electron energy are summarized in what follows. The results of many other EBIT experiments at lower electron energy can be found in the literature.

## PRODUCTION OF BARE $U^{92+}$ IONS

Bare $U^{92+}$ is sometimes called the ultimate ion, since it is the highest ionization stage of the heaviest naturally occurring element, as well as the most difficult ion to produce. The threshold for producing $U^{92+}$ by electron impact ionization of the preceding hydrogenlike ionization stage is approximately 130 keV, within the range of a Super EBIT. Trapped bare uranium ions were first produced in the LLNL Super EBIT, operating at its maximum energy of 200 keV [1]. Unfortunately, at this energy, the radiative recombination cross section for destroying $U^{92+}$ is roughly 40 times larger than the ionization cross section for producing it, so only a small fraction of the trapped uranium ions can be in the fully stripped charge state. However, as discussed below, enough bare uranium is produced to use its abundance as a signal for the measurement of the ionization cross section that produces it. The other (few electron) ionization stages of uranium can be produced in sufficient abundance for high resolution spectroscopy of their characteristic x-ray lines. Table I lists the observed uranium ionization balance as derived from radiative recombination spectra such as the one shown in Fig. 2.

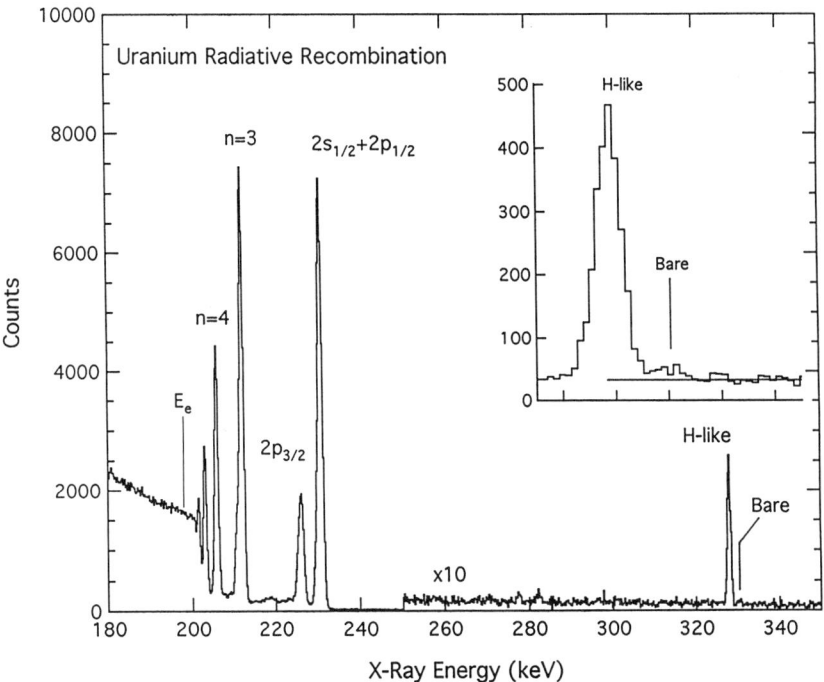

**FIGURE 2.** Uranium radiative recombination x-ray spectrum at 198-keV electron energy observed in a 40-cm$^3$ coaxial germanium detector. The inset shows the n=1 feature from a second (90-cm$^3$) detector.

## ELECTRON-ION COLLISION CROSS SECTIONS

Every x ray emitted by the ions trapped in an EBIT is associated with an electron-ion collision. The x-ray spectra, in combination with the ability to vary the electron energy that produces them, enable the study of almost every type of electron-ion collision process. Moreover, the cross section for one collision

**TABLE I.** The observed equilibrium ionization balance in the LLNL Super EBIT for uranium at approximately 200-keV electron beam energy.

| Ionization stage | Abundance (%) |
|---|---|
| Bare | 0.02% |
| H-like | 1.0% |
| He-like | 17% |
| Li-like | 34% |
| Be-like | 31% |
| B-like | 15% |
| C-like | 3% |

process, radiative recombination, is well known [3], so it can be used to normalize measurements of the cross sections for other processes such as ionization, excitation, or dielectronic recombination of the same target ions. This avoids the errors associated with an absolute cross section measurement. Almost all EBIT cross section measurements have been normalized to radiative recombination, which produces easily observable x-ray lines for every target ion.

Recent Super EBIT experiments at LLNL succeeded in making the first direct measurement of the ionization cross section for hydrogenlike uranium, as well as several other hydrogenlike high-Z ions. Measurements of ionization cross sections for the uranium L shell are in progress. Another electron-ion collision process, dielectronic recombination, has been studied for the first time for highly charged uranium ions with excitation of K-shell electrons.

## Ionization

The electron impact ionization cross sections for the tightly bound 1s electrons of high-Z elements are of interest as a test of relativistic interactions in a simple atomic system. Until now, the size of such cross sections could only be obtained indirectly from the stripping of accelerator beams [4]. Although the electron impact ionization of a hydrogenlike or heliumlike ion does not directly involve the emission of an x ray, we have developed an x-ray technique based on the condition of steady-state ionization balance in Super EBIT and used it to measure the ionization cross section of the hydrogenlike ions of several elements, including uranium [1]. In this technique the bare-to-hydrogenlike abundance ratio, $N_{Bare}/N_H$, is determined from the relative intensity of K-shell radiative recombination x-rays. The hydrogenlike ionization cross section is then obtained from the abundance ratio using the relation

$$\sigma^{ion}_{H \to Bare} = \frac{N_{Bare}}{N_H} (\sigma^{RR}_{Bare \to H}). \qquad (1)$$

A small correction for charge-exchange-recombination with neutral background gas is required to complete the measurement.

Cross section data for uranium at 198-keV electron energy are shown in Fig. 2. The x-ray spectrum consists of a series of peaks corresponding to radiative recombination into the open shells of the uranium target ions. The lines at approximately 330-keV are from K-shell capture by bare and hydrogenlike target ions. The electron beam energy spread is known from other measurements to be approximately 100 eV FWHM, so its contribution to the observed peak width is hidden by the 900-eV FWHM detector resolution. The intensity of the weak bare uranium radiative recombination line was obtained from a least-squares fit that fixed its width, position, and shape relative to the hydrogenlike radiative recombination line, and used a background level determined from the spectral region above the peak. Separate results were obtained from each of two detectors and averaged. For hydrogenlike uranium, our measured cross section is 50% larger than theoretical values available at the time [5], a discrepancy that has since been resolved by improvements to the theory[6,7].

Now that the K-shell ionization cross sections have been determined, we are attempting to apply the same technique to the measurement of ionization cross sections for the L shell of uranium. We determine the equilibrium ionization balance for a population of uranium ions in charge states from $U^{89+}$ (lithiumlike) to $U^{82+}$ (neonlike) from the intensity of radiative recombination x rays associated with each charge state. However, unlike radiative recombination into the K shell, for which the x-ray lines corresponding to bare and hydrogenlike target ions are completely resolved in a germanium detector, radiative recombination into the L shell of uranium produces incompletely resolved x-ray lines. This is illustrated in Fig. 3, which shows an x-ray spectrum obtained at 60-keV electron energy. The intensity of the radiative recombination x-ray lines associated with the different charge states can be determined from least squares fits to the n=2 j=3/2 and j=1/2 features using the known positions of the x-ray lines. This provides the ionization balance, which is then used to determine the ionization cross sections in a manner similar that used for the hydrogenlike ions. We have obtained data at 45, 60, and 75 keV electron energy.

## Dielectronic Recombination

In dielectronic recombination, an electron (from the Super EBIT beam) is captured by an ion with the simultaneous excitation of a bound electron, resulting in an excited intermediate state. Decay of the intermediate state by photon (x-ray) emission completes the recombination process. Dielectronic recombination is a resonant process because the incident electron energy must exactly match the energy required for excitation (with electron capture). The cross sections are usually very large, producing an easily observed x-ray signal. Resonances in which

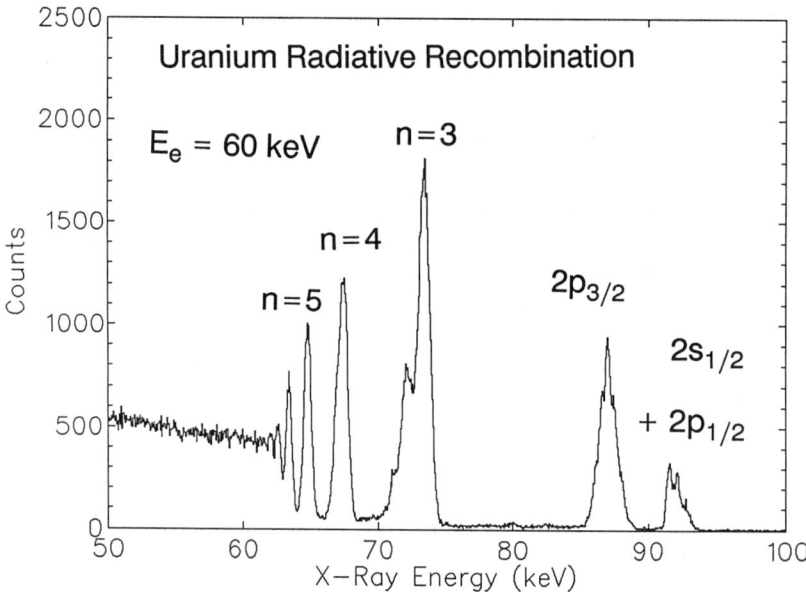

**FIGURE 3.** Uranium radiative recombination x-ray spectrum at 60-keV electron energy observed in a planar germanium detector. The structure in the n=2 j=3/2 and j=1/2 features corresponds to the different ionization stages of the target ions.

a K electron is excited to the L shell and an incident electron is captured into the L shell are called KLL resonances. The KLL resonances for several heliumlike target ions have been studied in previous EBIT experiments [8,9], but only Super EBIT has sufficient energy to excite the K electrons in the heaviest elements.

Recently, KLL dielectronic recombination was measured at the LLNL Super EBIT for heliumlike and adjacent ionization stages of uranium, for which the K-L transition energy is roughly 100 keV [10]. This experiment was done by scanning the electron beam energy through the resonance region and detecting dielectronic recombination photons in a germanium detector. The most remarkable result of this work is the unexpected discovery of interference between dielectronic and radiative recombination [10].

## QED IN HIGH-Z IONS

Spectroscopic studies of the energy levels of few-electron high-Z ions can be used to understand the effects of quantum electrodynamics (QED) and multielectron corrections in strong Coulomb fields. These effects appear as a difference between the physical energy and the Dirac-Coulomb energy of an atomic state. The one-electron (hydrogenlike) ions are the easiest to treat theoretically, and calculations of their energy levels are thought to be highly accurate even for very high Z. The heliumlike ions are the simplest multielectron system. However, in contrast to the hydrogenlike ions, there is no exact solution for the structure of high-Z heliumlike ions. The ions corresponding to the range of charge states from lithiumlike to neonlike challenge our understanding of the electron correlation terms that contribute to the energies of these multielectron ions. These issues are addressed by two recent precision x-ray experiments with Super EBIT. One experiment measures the high energy x rays emitted in radiative electron capture into the K shell of bare and hydrogenlike target ions. The other experiment measures the much lower energy n=2 to n=2 transitions in multielectron ions.

## Ground-State Energies In Heliumlike Ions

Heliumlike ions are more than just hydrogenlike ions with two independent electrons. What makes them different and interesting is the interaction between the two electrons. We have used a novel experimental approach that exploits radiative recombination transitions for a direct measurement of the two-electron contributions to the ground-state energy in heliumlike ions [11]. We measure the difference in the energy of radiative recombination x rays emitted in K-shell electron capture by stationary bare and hydrogenlike target ions. This difference is equal to the difference in the ionization potential between the hydrogenlike and heliumlike ion, which is exactly the two-electron contribution to the ground state energy of the heliumlike ion.

It should be noted that the one-electron contributions to the binding energy, such as the finite-nuclear-size correction and the one-electron self energy, cancel out in this type of experiment, which makes such measurements unique. The two-electron contribution to the energy of a heliumlike ion includes, in addition to the dominant Coulomb term, effects from electron correlation, the Breit interaction, screening of the Lamb shift, and higher order radiative corrections. Until now, the precision of available experimental data has not been sufficient to test the theoretical calculation of these various contributions.

The experimental setup at the LLNL Super EBIT is shown in Fig. 4. Energy calibration lines from a radioactive source are mixed with the radiative recombination x-ray lines from trapped ions and detected in a coaxial germanium detector. We have obtained data for six different elements: Ge(Z=32), Xe(Z=54), Dy(Z=66), W(Z=74), Os(Z=76), and Bi(Z=83). A typical spectrum from the xenon data is shown in Fig. 5. The roughly 100-eV energy spread of the Super EBIT electron beam is much less than the resolution of the x-ray detector and does not make a significant contribution to the observed peak width. Because of the rapid decrease in 1s ionization cross sections with increasing atomic number, the abundance of bare target ions is, at present, insufficient to extend these measurements beyond Z=83 (see Table I).

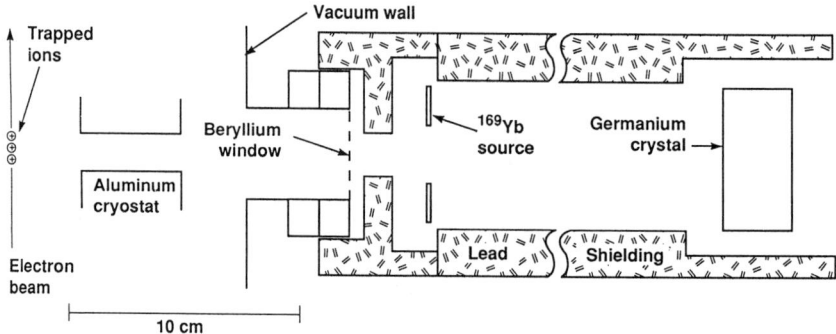

**FIGURE 4.** Experimental arrangement used for the binding energy measurements. The $^{169}$Yb source consists of several removable pieces attached to an annular holder. The detector was located 46 cm from the trapped ions.

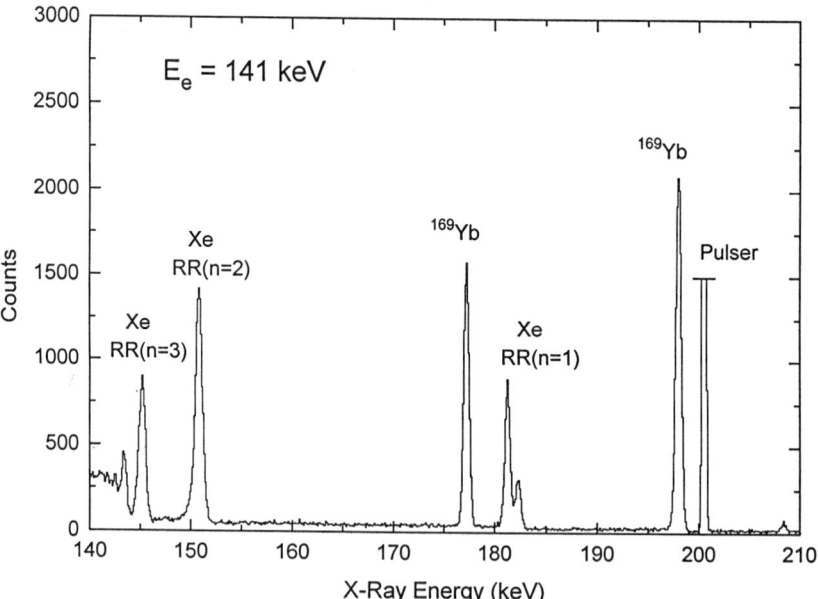

**FIGURE 5.** Example of spectra used to determine the ionization potential difference for hydrogenlike and heliumlike xenon. The $^{169}$Yb lines are from the calibration source.

**TABLE II.** The measured two-electron contributions to the ground-state energy of heliumlike ions compared to the relativistic MBPT calculations of Lindgren et al. [11] (in eV).

| Z | 2nd-Order Many Body | Two-electron Lamb Shift | Total Theory | Experiment |
|---|---|---|---|---|
| Ge(32) | -5.2 | -0.5 | 561.9 | 562.5 ± 1.6 |
| Xe(54) | -6.9 | -1.6 | 1028.1 | 1027.2 ± 3.5 |
| Dy(66) | -8.6 | -2.5 | 1336.6 | 1341.6 ± 4.3 |
| W(74) | -9.4 | -3.4 | 1574.6 | 1568 ± 15 |
| Bi(83) | -11.0 | -4.5 | 1882.7 | 1876 ± 14 |

Our results for the difference in the ionization potential between hydrogenlike and heliumlike ions are given in Table II. There is general agreement between our results and calculations using several different theoretical approaches. The results of one theory, a calculation by Lindgren et al. [12] using relativistic many body perturbation theory, are included in the table. The "2nd order many body" and "two-electron Lamb shift" contributions are tabulated separately so that they can be compared to our experimental uncertainties. The experimental uncertainties are smaller than the 2nd order many body contributions (except at higher Z), so the experimental results are sensitive to this part of the theory, but not yet to the two-electron Lamb shift. However, anticipated future improvements in the experimental precision could provide a significant test of the two-electron Lamb shift in heliumlike ions.

## Transition Energies In Few-Electron Ions

Accurate calculations of the transition energies in high-Z ions with more than two electrons require theoretical techniques different from those applied to heliumlike ions. Recent Super EBIT measurements of the $2s_{1/2}$ - $2p_{3/2}$ transitions in highly charged uranium [13] and thorium [14] ions provide an ideal data set for testing our understanding of multielectron ions because the measurements span eight ionization stages (lithiumlike through neonlike) in both elements, so the effect of a varying number of electrons can be studied.

The Super EBIT measurements used a high resolution Bragg crystal of the von Hamos type, arranged as shown in Fig. 6 [15]. A typical x-ray spectrum from thorium ions is shown in Fig. 7. Several charge states are present in the trap simultaneously, resulting in multiple but well separated lines in the spectrum. The charge state distribution was shifted to cover the full range of interest by adjusting the Super EBIT operating parameters. The spectrometer was energy calibrated with the Lyman series of x rays in hydrogenlike and heliumlike argon, as their energies are well known.

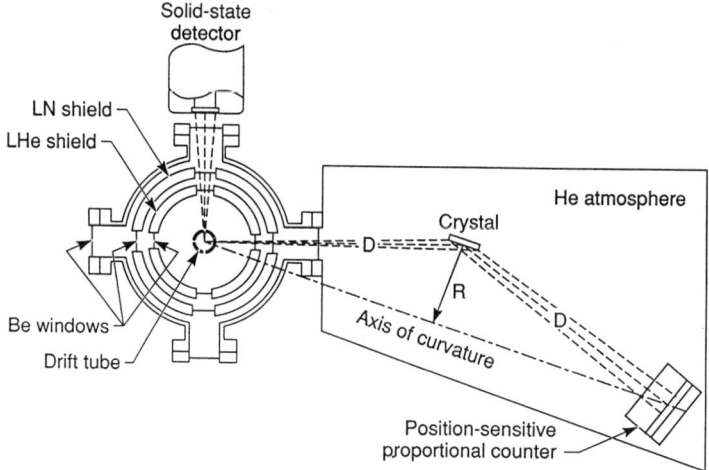

**FIGURE 6.** Diffraction plane layout of the high resolution von Hamos spectrometer used for the 2s-2p transition energy measurements. The electron beam direction is out of the page. The axis of curvature of the cylindrically bent crystal is the line joining the x-ray source and x-ray detector; the source-to-crystal and crystal-to-detector distances are equal. A solid state detector used for monitoring the EBIT x-ray emission is also shown.

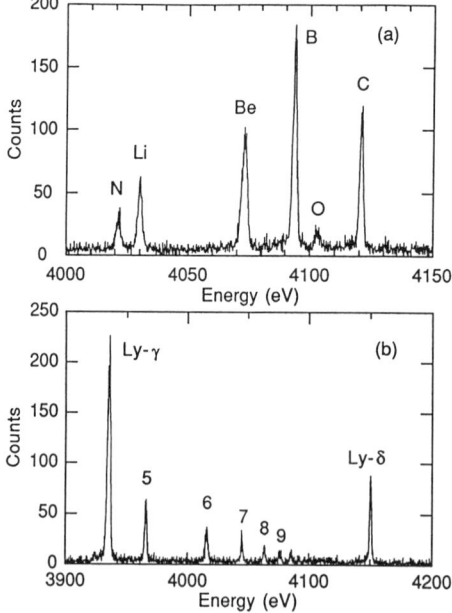

**FIGURE 7.** (a) The $2s_{1/2} - 2p_{3/2}$ transition in lithiumlike $Th^{87+}$ (labeled "Li") and other charge states of thorium excited at 100 keV electron energy. (b) The Lyman series in hydrogenlike and heliumlike argon used for wavelength calibration.

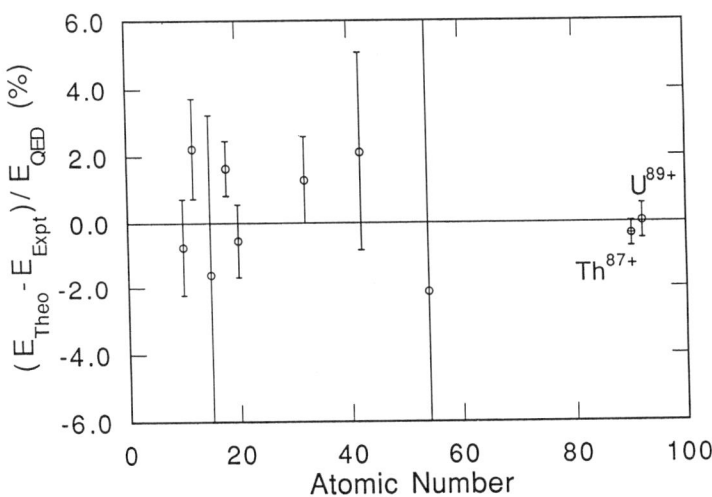

**FIGURE 8.** Difference between theory [16] and experiment for the $2s_{1/2}$ - $2p_{3/2}$ transition in the lithiumlike isoelectronic sequence expressed as a percentage of the theoretical QED energy. The uranium and thorium points are from the Super EBIT at LLNL.

In Fig. 8 the measured energies for the $2s_{1/2}$ - $2p_{3/2}$ transition in lithiumlike ions are compared to the theoretical QED energy [16] for this transition. Our Super EBIT results at high Z provide the best test of theory. For high-Z ions with several bound electrons, multiconfiguration Dirac-Fock (MCDF) theory has been a common tool for calculating energies. A comparison between MCDF theory and the Super EBIT results shows a systematic difference between theory and experiment that increases with the number of bound electrons [13,14]. This is attributed to the treatment of correlation energies in the MCDF calculations. More recently, notable progress has been made using many body perturbation theory [17] and configuration interaction techniques, resulting in better agreement with our Super EBIT measurements for the multielectron ions.

## NOVEL TECHNIQUES FOR SPECIAL MEASUREMENTS

### Rare Isotopes And Nuclear Sizes

Most EBIT and Super EBIT experiments have been done without regard to the isotopic composition of the element being studied, and large (gram) quantities of feed material with the natural isotopic abundance have been used for injection into the EBIT trap. However, several interesting experiments require isotopes or elements available only in trace quantities. Examples include $^{233}$U, $^{235}$U, and $^{249}$Cf. These isotopes cannot be injected into Super EBIT with the usual

MEVVA source [18] because the quantity of feed material required is too large. To solve this problem, we have developed a novel injection technique that enables Super EBIT operation with only nanogram quantities of feed material, 100-million-times smaller than previously possible [19].

The new technique works by mounting or plating a small amount of the desired rare isotope on the pointed end of a wire probe. As shown in Fig. 9, the wire probe is positioned near the compressed electron beam of our Super EBIT in the region where the beam passes through the ion-trap structure. The electron beam is surrounded by a cloud of trapped ions, some of which impact the sample and remove material from it by sputtering. On the order of 0.1% of the sputtered material comes off as positive ions and is captured in the EBIT trap. The rate of sputtering is controlled by the position of the probe and the bias voltage applied to it.

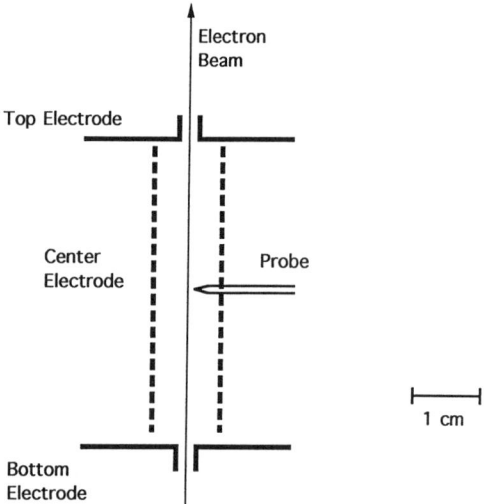

**FIGURE 9.** A schematic of the wire probe injector showing the placement of the probe with respect to the electron beam and trap electrodes. The probe location is controlled by a standard vacuum positioning device.

In one application of this new capability, we have measured the isotope shift of x-ray transitions in high ionization stages of three uranium isotopes, $^{233}$U, $^{235}$U, and $^{238}$U [20]. The results provide information on the nuclear charge radius for the different isotopes. A high-resolution Bragg-crystal spectrometer was used to record the $2s_{1/2}$ - $2p_{3/2}$ transitions in lithiumlike, berylliumlike, boronlike, and carbonlike uranium for the different isotopes. $^{233}$U and $^{235}$U were obtained from the rare isotope injector, and the more abundant $^{238}$U was obtained from a MEVVA injector.

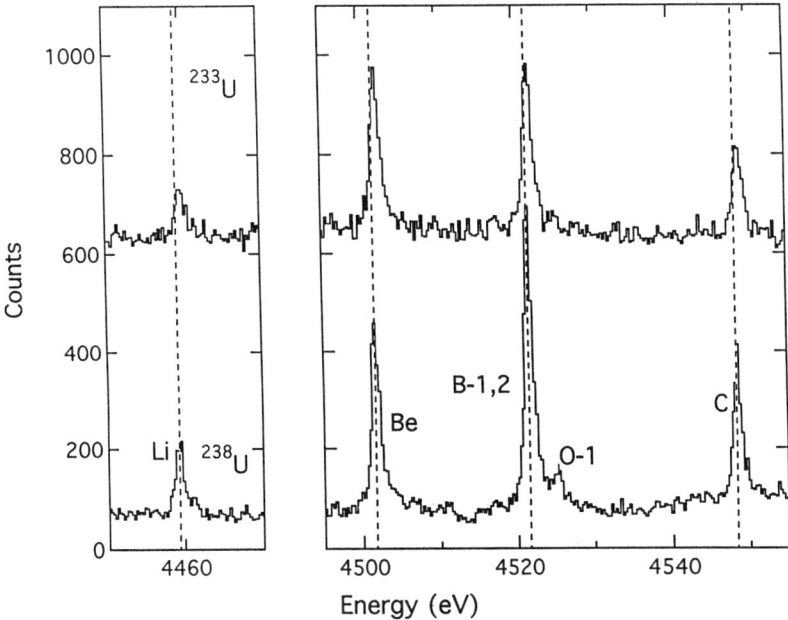

**FIGURE 10.** Comparison of $2s_{1/2}$ - $2p_{3/2}$ transitions for $^{233}$U and $^{238}$U. The x-ray lines are labeled with the chemical symbols for the different ionization stages.

A comparison of the spectra for $^{233}$U and $^{238}$U is shown in Fig. 10. The isotope shift of the transition energy can be clearly seen. The four different ionization stages give a consistent value of $\delta\langle r^2\rangle^{233,238} = -0.457 \pm 0.043$ fm$^2$ for the difference in root-mean-square nuclear charge radius between $^{233}$U and $^{238}$U. This may be compared with a previous value of $\delta\langle r^2\rangle^{233,238} = -0.383 \pm 0.044$ fm$^2$ derived from measurements of K$\alpha$ and optical transitions in neutral uranium [21].

## Cold Ions And X-Ray Line Widths

Observed x-ray lines from radiative transitions in highly charged ions are broadened by both the natural width of the transition and by the motion of the emitting ions. In measurements with accelerator beams, Doppler broadening (in detectors subtending a finite angle) can be very large because of the high velocity of the emitting ions. X-ray lines from (stationary) plasma sources are broadened by the thermal motion of the emitting ions. Ions confined in Super EBIT also have a thermal motion due to ion heating by the electron beam. However, the evaporative cooling process, always used to keep highly charged ions from boiling out of the Super EBIT electron beam, can be exploited to cool the trapped ions to surprisingly low temperatures. In fact, thermal broadening can be reduced to the point where the natural width of some transitions can be measured in Super EBIT if very high resolution spectrometers are used [22].

Bragg Crystal spectrometers arranged as shown in Fig. 6 can achieve very high resolving power with EBIT if the distances between the electron beam, Bragg crystal, and detector are made large. In one experiment, resolving powers as high as $\lambda/\Delta\lambda \approx 22,000$ were obtained for trapped heliumlike titanium ions, allowing the line broadening due to the thermal motion of the ions to be clearly seen [22]. The $1s^2(^1S_0) - 1s2p(^3P_2)$ transition in heliumlike titanium was selected because the natural width of this forbidden line is too small to contribute to the observed line width.

The temperature of the trapped heliumlike titanium ions was reduced primarily by lowering the axial trapping voltage, an effect loosely analogous to lowering the temperature of boiling water by reducing the pressure. This cooling technique has been used in Super EBIT to cool heliumlike uranium ions to an estimated temperature of roughly 2q eV [1], suggesting that a measurement of the natural width may be possible for certain short-lived E1 decays in highly charged ions.

## CONCLUDING REMARKS

The Super EBIT at LLNL supports an active and diverse research program with many notable accomplishments since the initial operation of Super EBIT in 1992. Recent work has focused on the few-electron ions of heavy elements, otherwise available only as high velocity beams from large accelerators. Looking to the future, it should be noted that the precision of many of the results reviewed above is limited by counting statistics; hence, future increases in x-ray intensity could substantially improve the experimental precision, probably to the point where even more subtle corrections to the structure of highly charged ions could be investigated. The variety of experiments will expand even further with the advent of Super EBIT devices under construction at other laboratories.

## ACKNOWLEDGMENTS

Experiments at the LLNL Super EBIT were done in collaboration with P. Beiersdorfer, J. Crespo, S. Elliott, D. Knapp, D. Schneider, Th. Stöhlker, and K. Widmann. This work was performed under the auspices of the U. S. Department of Energy by Lawrence Livermore National Laboratory under Contract No. W-7405-Eng-48.

## REFERENCES

[1] R. E. Marrs, S. R. Elliott, and D. A. Knapp, Phys. Rev. Lett. **72**, 4082 (1994).

[2] D. A. Knapp, R. E. Marrs, S. R. Elliott, E. W. Magee, and R. Zasadzinski, Nucl. Instrum. Methods A **334**, 305 (1993).
[3] J. H. Scofield, Phys. Rev. A **40**, 3054 (1989)
[4] N. Claytor, B. Feinberg, H. Gould, C. E. Bemis, J. G. Campo, C. A. Ludemann, and C. R. Vane, Phys. Rev. Lett. **61**, 2081 (1988).
[5] H. L. Zhang and D. H. Sampson, Phys. Rev. A **42**, 5378 (1990); K. J. Reed (private communication).
[6] D. L. Moores and K. J. Reed, Phys. Rev. A **51**, R9 (1995).
[7] C. J. Fontes, D. H. Sampson, and H. L. Zang, Phys. Rev. A **51**, R12 (1995).
[8] D. A. Knapp, R. E. Marrs, M. A. Levine, C. L. Bennett, M. H. Chen, J. R. Henderson, M. B. Schneider, and J. H. Scofield, Phys. Rev. Lett. **62**, 2104 (1989).
[9] D. A. Knapp, R. E. Marrs, M. B. Schneider, M. H. Chen, M. A. Levine, and P. Lee, Phys. Rev. A **47**, 2039 (1993).
[10] D. A. Knapp, P. Beiersdorfer, M. H. Chen, J. H. Scofield, and D. Schneider, Phys. Rev. Lett. **74**, 54 (1995).
[11] R. E. Marrs, S. R. Elliott, and Th. Stöhlker, Phys. Rev. A, in press.
[12] I. Lindgren, H. Persson, S. Salomonson, and P. Sunnergren, Proceedings of the Nobel Symposium on Trapped Charged Particles and Related Fundamental Physics, August, 1994 (to appear in Physica Scripta).
[13] P. Beiersdorfer, D. Knapp, R. E. Marrs, S. R. Elliott, and M. H. Chen, Phys. Rev. Lett. **71**, 3939 (1993).
[14] P. Beiersdorfer, A. Osterheld, S. R. Elliott, M. H. Chen, D. Knapp, and K. Reed, unpublished.
[15] P. Beiersdorfer, R. E. Marrs, J. R. Henderson, D. A. Knapp, M. A. Levine, D. B. Platt, M. B. Schneider, D. A. Vogel, and K. L. Wong, Rev. Sci. Instrum. **61**, 2338 (1990).
[16] S. A. Blundell, Phys. Rev. A **47**, 1790 (1993).
[17] W. R. Johnson, J. Sapirstein, and K. T. Cheng, Rev. A **51**, 297 (1995).
[18] I. G. Brown, J. E. Galvin, R. A. MacGill, and R. T. Wright, Appl. Phys. Lett., **49**, 1019 (1986).
[19] S. R. Elliott and R. E. Marrs, Nucl. Instrum. Methods B **100**, 529 (1995).
[20] S. R. Elliott et al., unpublished.
[21] A. Anastassov, Yu. P. Gangrsky, K. P. Marinova, B. N. Markov, B. K. Kul'Djanov and S. G. Zemlyanoi, Hyperfine Interactions **74**, 31 (1992).
[22] P. Beiersdorfer, V. Decaux, S. R. Elliott, K. Widmann, and K. Wong, Rev. Sci. Instrum. **66**, 303 (1995).

# Attaining low electron temperatures in electron coolers

## Håkan Danared

*Manne Siegbahn Laboratory at Stockholm University*
*S-104 05 Stockholm, Sweden*

**Abstract.** Some properties of electron coolers relevant to studies of ion-electron recombination are reviewed. We discuss the influence of the electron temperature on energy resolution and count rates in recombination experiments and how these have been increased by lowering the transverse electron temperature through adiabatic beam expansion. Possibilities to further reduce the electron temperature are mentioned.

## INTRODUCTION

Electron cooling was proposed as a method to cool and accumulate antiprotons for high-energy colliders [1]. For this purpose, stochastic cooling is now used. Instead, electron cooling has been implemented at about ten ion storage rings and is used not only for cooling and accumulation, but also for studies of recombination between atomic or molecular ions and electrons. In several of the rings (TSR in Heidelberg [2], ASTRID in Århus [3], and CRYRING in Stockholm [4]) atomic and molecular physics is the dominant activity, in others at least some of the time is devoted to this kind of physics. Electron coolers thus play a dual role in these rings. Cooling first reduces momentum spread and transverse dimensions of the stored ion beam, perhaps in combination with accumulation and other types of beam preparation. Then the cooler acts as an electron target for the actual recombination. In both cases, the temperature of the cooler's electron beam is an important parameter.

The use of storage rings with electron cooling for recombination experiments has many advantages compared to more traditional methods, such as drift experiments or single-pass merged- or crossed-beam techniques. The possibility to reuse ions that did not react during their first passage through the cooler is of course very important, since an electron beam is a very dilute target and the ions may be difficult to produce. Also the high energy that the particles have in a storage ring is advantageous, since background processes, for instance electron capture in rest-gas collisions, become much reduced. Further, there are many possibilities to prepare the ions before the measurements begin. The phase-space cooling improves precision and resolution in the experiments; the storage itself can allow time for internal cooling, i.e., relaxation of excited atomic or molecular states; specific states can be populated or depopulated through interaction with laser light; etc. In addition, the electron beam in a cooler has very well defined properties—the electron density is accurately known and the (anisotropic) electron temperature can be determined from independent measurements. These and other properties make storage rings with electron cooling an excellent tool for studies of the interaction between free ions and electrons, such as spontaneous and laser-induced radiative recombination, dielectronic recombination, three-body recombination, and dissociative recombination of molecular ions.

## THE COOLER/TARGET

An electron cooler is a device where an intense electron beam with a current of typically 0.01 − 1 A and a diameter of a few cm is guided from a gun to a collector by a magnetic field parallel to the electron beam. In the following, several examples will be taken from CRYRING, and the layout of its electron cooler is shown in figure 1. During cooling, the electrons are

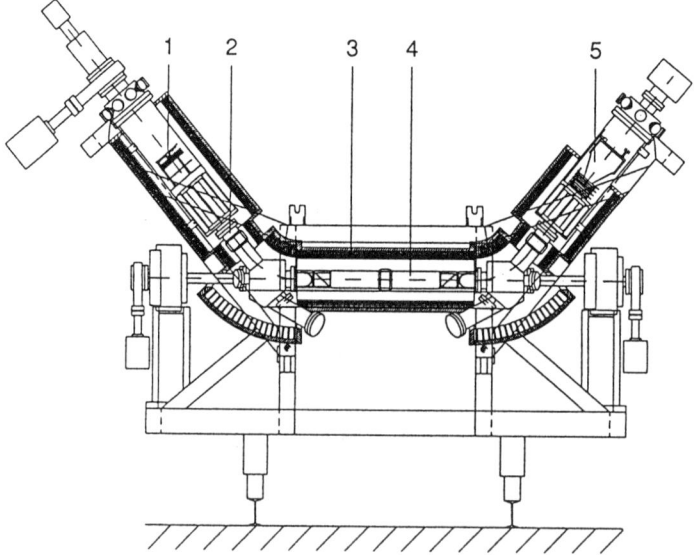

**FIGURE 1.** Schematic diagram of the CRYRING electron cooler with 1) electron gun, 2) beam expansion region, 3) magnet coils, 4) ion–electron interaction region, 5) electron collector.

given the same average velocity as the ions circulating in the ring, and in the interaction region where the electron beam is merged with the ion beam, heat is transferred from the ions to the continuously renewed cold electrons. Electron cooling has been used for ion energies up to 470 MeV per nucleon, but in the smaller rings for atomic physics, cooling is performed at ion energies in the range 0.3 to 30 MeV per nucleon, corresponding to electron energies between 0.16 and 16 keV.

After the ions have been phase-space cooled, the electron energy can be detuned from the cooling energy $E_{\text{cool}}$ in order to give the electrons an energy $E_d$ in the center of mass of the ions. The center-of-mass energy is much smaller than the energy change $\Delta E_{\text{lab}}$ in the laboratory and is (non-relativistically) given by

$$E_d = \frac{(\Delta E_{\text{lab}})^2}{4 E_{\text{cool}}} \quad (1)$$

when $\Delta E_{\text{lab}} \ll E_{\text{cool}}$.

As an example, cooling of an ion beam at 10 MeV per nucleon requires an electron energy of roughly 5 keV, obtained by applying this voltage to the cathode that the electrons are emitted from. Changing the cathode potential by 1 volt then gives a centre-of-mass energy of only $5 \times 10^{-5}$ eV. Measurements at such low center-of-mass energies are in practise very difficult, since, for example, the electrons tend to accelerate the ions rapidly, giving them the same average velocity as the electrons in a fraction of a second. A perhaps more serious problem, however, is that the velocity spread of the two particle beams no longer can be neglected.

After the ion beam has been phase-space cooled, its temperature will in general be of the same order of magnitude as that of the electrons. Since the ions are much heavier, this means that the electrons will have the largest velocity spread.

One usually assumes that, in the frame moving with the electrons, their velocity has an anisotropic Maxwellian distribution

$$f(\mathbf{v}_e) = \frac{m_e}{2\pi k T_{e\perp}} \left( \frac{m_e}{2\pi k T_{e\parallel}} \right)^{1/2} \exp\left( -\frac{m_e v_{e\perp}^2}{2 k T_{e\perp}} - \frac{m_e v_{e\parallel}^2}{2 k T_{e\parallel}} \right), \quad (2)$$

characterized by the transverse temperature $T_{e\perp}$ and the longitudinal temperature $T_{e\parallel}$. The transverse temperature is determined by the temperature of the cathode from which the electrons are emitted. All existing electron coolers use dispenser cathodes that are heated to 800 – 900 °C or more in order to emit electrons. The transverse electron temperature may become higher than this due to imperfections in the electron optics, space charge, etc., but at moderate energies, such as in CRYRING, one can get close to the cathode temperature, i.e., to $kT_{e\perp} = 100$ meV when the magnetic-field strength is constant throughout the cooler. The longitudinal temperature is much lower due to the acceleration of the electrons. According to expression (1) one would get that a longitudinal energy spread $kT_{\text{cath}}$ in the laboratory becomes

$$kT_{e\parallel} = \frac{(kT_{\text{cath}})^2}{4E_{\text{cool}}} \qquad (3)$$

in the moving system, or around a $\mu$eV at keV electron energies. However, the true longitudinal energy spread is set by relaxation processes within the electron beam. Each electron has a potential energy from the electrostatic field of the other electrons. If, after the acceleration of the electrons, this potential energy is larger than the spread in kinetic energy, the latter tends to increase until the two are in equilibrium [5]. Since the average potential energy is of the order $e^2/(4\pi\epsilon_0)n_e^{1/3}$, the resulting longitudinal energy spread could become $10^{-4}$ eV. Nevertheless, this is much less than the transverse energy spread.

## BEAM EXPANSION

In order to improve the properties of the CRYRING cooler when it acts as an electron target (and to increase the cooling rates), we have lowered the transverse electron temperature by expanding the electron beam in an adiabatically decreasing magnetic field [6, 7]. If the field changes over a distance that is long compared to the cyclotron wavelength (i.e., the axial distance an electron travels during the time it takes to make one turn in its cyclotron motion around the magnetic field lines), the ratio $W_\perp/B_\parallel$ is an adiabatic invariant. Here $W_\perp$ is the kinetic energy of the transverse motion, and $B_\parallel$ is the longitudinal magnetic field strength. Reducing the magnetic field by a factor of ten from the electron gun to the interaction region, as was done in CRYRING, thus lowers the transverse energy spread from 100 meV to 10 meV, provided that the adiabaticity condition is fulfilled. Since the cyclotron wavelength is

$$\lambda_c = \left(\frac{8\pi^2 m_e}{q} \frac{E_{\text{cool}}}{B_\parallel^2}\right)^{1/2}, \qquad (4)$$

one can easily see that the transition between the two fields indeed is adiabatic for a cooler such as the one in CRYRING, where $E_{\text{cool}}$ is around 10 keV or lower, $B_\parallel$ decreases from 0.3 T to 0.03 T, and the length of the transition is approximately 0.3 m

A reduction in transverse temperature by a factor of ten also means that the cross-sectional area of the electron beam increases ten times. At the same time as the field gradient is introduced, we therefore installed a new electron gun with a cathode area that is ten times smaller than that of the original gun. The new electron gun has exactly the same geometry as the old one, except that its linear dimensions are scaled down by a factor of $10^{1/2}$. Consequently the perveance did not change (the perveance $P$ of an electron gun is defined as $P = I_e/U_{\text{acc}}^{3/2}$, where $I_e$ is the electron current extracted from the gun, and $U_{\text{acc}}$ is the acceleration voltage, and it depends only on the shape of the gun, not on its size). The current density after the expansion is thus the same as in the original cooler design without expansion.

## RECOMBINATION SPECTRA

The electron temperature has a direct influence on energy resolution and reaction rates in recombination experiments. The reaction rate is obtained through

$$\alpha_r(v_d) = \langle \sigma v_{\text{rel}} \rangle = \int \sigma(v_{\text{rel}}) v_{\text{rel}} f(\mathbf{v}_e) d^3 v_e, \tag{5}$$

where $\sigma(v_{\text{rel}})$ is the cross section, $f(\mathbf{v}_e)$ is the electron-velocity distribution given by (2), $v_{\text{rel}} = [v_{e\perp}^2 + (v_d + v_{e\parallel})^2]^{1/2}$, and $d^3 v_e = 2\pi v_{e\perp} dv_{e\perp} dv_{e\parallel}$. As an illustration we can think of dielectronic recombination and look at an isolated transition approximated by a delta function of the relative energy with the resonance at $E_0$, i.e.,

$$\sigma(E_{\text{rel}}) = \sigma_0 \delta(E_{\text{rel}} - E_0). \tag{6}$$

Then, the integral (5) can be evaluated analytically [8, 9], yielding

$$\begin{aligned}\alpha_r(v_d) =& \frac{\sigma_0 v_0}{2\lambda k T_{e\perp}} \exp\left[\frac{-m_e}{2kT_{e\perp}}\left(v_0^2 - \frac{v_d^2}{\lambda^2}\right)\right] \\ & \times \left[\text{erf}\left(\sqrt{\frac{m_e}{2kT_{e\parallel}}} \frac{v_d + \lambda^2 v_0}{\lambda}\right) - \text{erf}\left(\sqrt{\frac{m_e}{2kT_{e\parallel}}} \frac{v_d - \lambda^2 v_0}{\lambda}\right)\right].\end{aligned} \tag{7}$$

Here, $\text{erf}(x)$ is the error function, $v_0 = (2E_0/m_e)^{1/2}$, $\lambda = (1 - T_{e\parallel}/T_{e\perp})^{1/2}$, and we assume $T_{e\parallel} < T_{e\perp}$.

Three different cases are of practical interest: $kT_{e\parallel} \ll E_0 \ll kT_{e\perp}$, $kT_{e\perp} \ll E_0 \ll kT_{e\perp}^2/T_{e\parallel}$, and $E_0 \gg kT_{e\perp}^2/T_{e\parallel}$. In the highest energy range, the peaks are symmetric and their widths are proportional to $(E_0 kT_{e\parallel})^{1/2}$ (full width at half maximum is $4(E_0 kT_{e\parallel} \ln 2)^{1/2}$). For lower $E_0$, approaching $kT_{e\perp}^2/T_{e\parallel}$, the peaks start to become asymmetric. On the low-energy side, the width is proportional to $kT_{e\perp}$, while on the high-energy side still to $(E_0 kT_{e\parallel})^{1/2}$. For even lower $E_0$, below $kT_{e\perp}$, the peaks extend all the way down to zero relative energy and appear more as "shoulders" in the spectra. Furthermore, the area under the peaks starts to drop and becomes inversely proportional to $kT_{e\perp}$. Thus, decreasing the transverse electron temperature not only improves the energy resolution for $E_0$ below $kT_{e\perp}^2/T_{e\parallel}$ but also increases the integrated rate below $kT_{e\perp}$. Both these effects are seen in figure 2, where the rate $\alpha_r(E_d)$ is plotted for three resonances at 10, 30 and 50 meV. The upper curve is for $kT_{e\perp} = 10$ meV and the lower curve for $kT_{e\perp} = 100$ meV. In both cases $kT_{e\parallel} = 0.1$ meV was used. Clearly, dielectronic recombination can be used as a rather accurate "electron thermometer", provided only that the resonances are reasonably well separated. The figure also shows that the maximum of a peak is shifted somewhat relative to the resonance energy $E_0$.

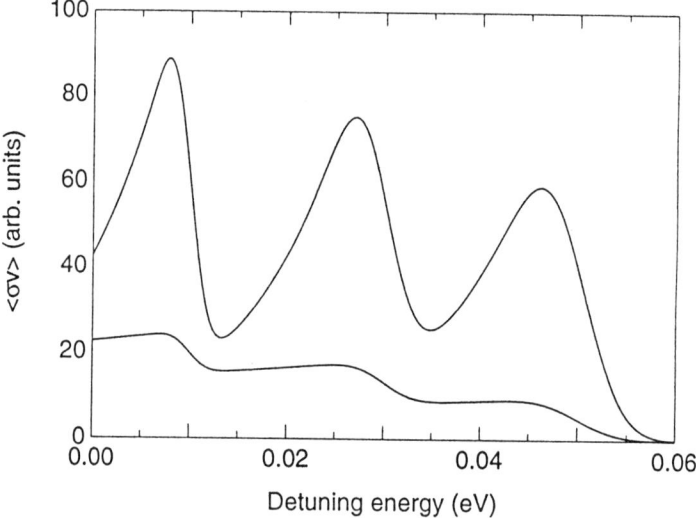

**FIGURE 2.** The recombination rate as a function of the detuning energy for a model cross section consisting of three $\delta$ resonances at 10, 30 and 50 meV. The upper curve was obtained with $kT_{e\perp} = 10$ meV and $kT_{e\parallel} = 0.1$ meV; the lower curve is for $kT_{e\perp} = 100$ meV and $kT_{e\parallel}$ still 0.1 meV.

Spectra of dissociative recombination can also have sharp features as was demonstrated at CRYRING, where HeH$^+$ ions were studied both with 10 and 100 meV transverse electron temperature [10, 11]. Figure 3 shows the recombination rate at low energies for both temperatures, and the difference between the two sets of data is in excellent agreement with what one can expect from the temperature difference.

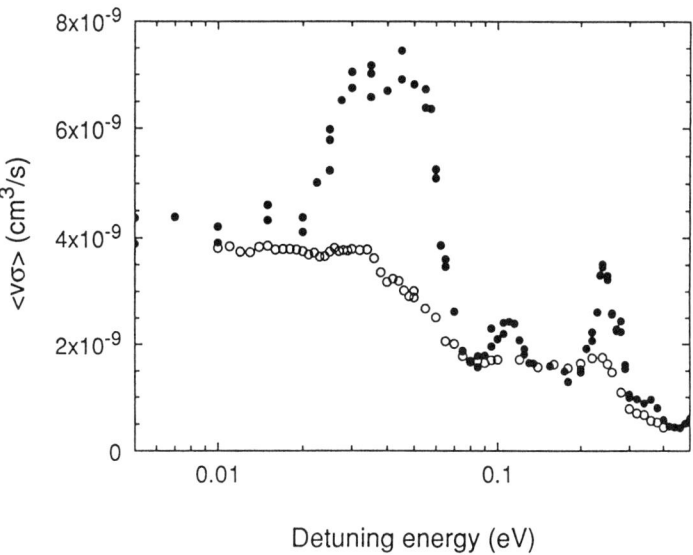

**FIGURE 3.** Dissociative-recombination rates for $^3\text{HeH}^+$ ions obtained at CRYRING [10, 11] with transverse electron temperature 10 meV (filled symbols) and 100 meV (open symbols).

## OUTLOOK

We have shown how a reduction in transverse electron temperature from 100 meV to 10 meV improves the energy resolution and reaction rates in recombination spectra at relative energies around 10 eV and below. The method of adiabatic beam expansion has been implemented at the electron coolers in CRYRING, TSR, ASTRID, and TARN II in Tokyo [12], with expansion ratios between 7 and 15, setting new standards for detailed recombination studies. Furthermore, it should be possible to extend this method to even higher expansion factors. Two practical limits are set by the maximum current density emitted from the cathode (up to 10 A/cm$^2$ for dispenser cathodes and substantially higher for other emitters, such as LaB$_6$, which, however, operate at higher temperatures), and by the maximum field ratio that can be achieved with present magnet technology (one should no go much below 0.05 T in the interaction region, and maximum field for standard superconducting solenoids is about 10 T or a little below). It is also difficult to accelerate and transport a very cold electron beam without heating it, particularly at high currents and/or high energies. Still, we believe that it is possible to reach transverse energy spreads in the order of 1 meV, using an expansion factor of 100, and we have recently ordered a 5 T superconducting gun solenoid for the CRYRING cooler.

Going from 10 meV to 1 meV will improve the resolution at relative energies of around 0.1 eV and below. This energy range is important for many different experiments. One example is the one of figure 3, showing peaks between 0.01 and 0.2 eV that Guberman [13] has identified with indirect recombination channels through Rydberg resonance levels in the neutral molecule, but where more structure is expected with higher experimental resolution. Another one is dielectronic recombination in many-electron ions [14], where resonances may exist so close to zero energy that they are difficult to separate from radiative recombination and hence may limit beam lifetimes during beam accumulation. We thus believe that electron coolers with transverse electron temperatures still lower than today's best values are within reach, and that ion–electron recombination studies of many types will benefit from this development.

## REFERENCES

1. G.I. Budker, *Proc. Int. Symp. on Electron and Positron Storage Rings (Saclay, 1966)*, eds. H. Zyngier and E. Crémieu-Alcan, Presses Universitaire de France, Paris 1967, p. II-1-1
2. E. Jaeschke, D. Krämer, W. Arnold, G. Bisoffi, M. Blum, A. Friedrich, C. Geyer, M. Grieser, D. Habs, H.W. Heyng, B. Holzer, R. Ihde, M. Jung, K. Matl, R. Neumann, A. Noda, W. Ott, B. Povh, R. Repnow, F. Schmitt, M. Steck, and E. Steffens, *European Part. Acc. Conf (Rome 1988)*, ed. S. Tazzari, World Scientific, Singapore 1989, p.365
3. R. Stensgaard, *Phys. Scr.*. **T22**, 315 (1988)
4. K. Abrahamsson, G. Andler, L. Bagge, E. Beebe, P. Carlé, H. Danared, S. Egnell, K. Ehrnstén, M. Engström, C.J. Herrlander, J. Hilke, J. Jeansson, A. Källberg, S. Leontein, L. Liljeby, A. Nilsson, A. Paál, K.-G. Rensfelt, U. Rosengård, A. Simonsson, A. Soltan, J. Starker, M. af Ugglas, and A. Filevich, *Nucl. Instr. Meth.*, **B79**, 269 (1993)
5. V.I. Kudelainen, V.A. Lebedev, I.N. Meshkov, V.V. Parkhomchuk, and B.N. Suchina, *Sov. Phys. JETP*, **56**, 1191 (1982)
6. H. Danared, *Nucl. Instr. Meth.*, **A335**, 397 (1993)
7. H. Danared, G. Andler, L. Bagge, C.J. Herrlander, J. Hilke, J. Jeansson, A. Källberg, A. Nilsson, A. Paál, K.-G. Rensfelt, U. Rosengård, J. Starker, and M. af Ugglas, *Phys. Rev. Lett.*, **72**, 3775 (1994)
8. L.H. Andersen and J. Bolko, *Phys. Rev.*, **A42**, 1184 (1990)
9. G. Kilgus, D. Habs, D. Schwalm, A. Wolf, N.R. Badnell, and A. Müller, *Phys. Rev.*, **A46**, 5730 (1992)
10. G. Sundström, S. Datz, J.R. Mowat, S. Mannervik, L. Broström, M. Carlson, H. Danared, and M. Larsson, *Phys. Rev.*, **A50**, R2806 (1994)
11. J.R. Mowat, H. Danared, G. Sundström, M. Carlson, L.H. Andersen, L. Vejby-Christensen, M. af Ugglas, and M. Larsson, *Phys. Rev. Lett.*, **74**, 50 (1995)
12. T. Tanabe, K. Noda, T. Honma, M. Kodaira, K. Chida, T. Watanabe, A. Noda,

S. Watanabe, A. Mizobuchi, M. Yoshizawa, T. Katayama, H. Muto, and A. Ando, *Nucl. Instr. Meth.*, **A307**, 7 (1991)
13. S.L. Guberman, *Phys. Rev.*, **A49**, R4277 (1994)
14. D.R. DeWitt, R. Schuch, S. Asp, C. Biedermann, H. Gao, W. Zong, *Nucl. Instr. Meth.*, in press; D.R. DeWitt et al., in preparation

# Production of bright, cold beams for atomic collision experiments

K.A.H. van Leeuwen*, E.J.D. Vredenbregt*, P.G.M. Sebel*,
J.P.J. Driessen*, M.D. Hoogerland[†], and H.C.W. Beijerinck*

*Physics Department, Eindhoven University of Technology
P.O. Box 513, 5600 MB Eindhoven, The Netherlands
[†] Atomic and Molecular Physics Laboratories
Research School for Physical Sciences and Engineering
Australian National University
Canberra ACT 0200, Australia

> Laser manipulation techniques for neutral atoms can be used to produce atomic beams which are more intense, brighter and colder than can be achieved by any other means. These beams will have a tremendous impact on the experimental study of low-energy atomic collisions. We first describe the design and operation of an intensifier for a thermal (axial velocity 600 m/s) beam of metastable neon atoms. The intensifier produces a gain in beam brightness of a factor 160 and a typical gain in usable flux of a factor 1400. Next, we discuss the design, construction and preliminary tests of a setup to produce an atomic beam which is slow and cold as well as bright and intense. The setup is expected to produce a beam of metastable atoms with an axial velocity of 100 m/s, a spread therein of 1.5 m/s, a diameter of 1 mm, a residual divergence of 1 mrad and a flux of $10^{12}$ atoms/s.

## INTRODUCTION

The study of atomic collisions has always constituted a formidable experimental challenge. One principal problem is the large number of parameters characterizing the collision process: in the ingoing channels, the electronic state and orientation of both collision partners, their relative velocity and impact parameter; in the outgoing channels, the state, orientation, direction and velocity (or energy) of the collision products. Obviously, a full characterization of any but the simplest collision process is next to impossible. With conventional techniques, in order to obtain the best characterization possible of the ingoing channels, the main sacrifice which has to be made is signal strength. To define the initial relative velocity, collisions between the atoms in colliding collimated atomic beams are studied. Precise knowledge of the relative velocity of the collision partners furthermore requires time-of-flight (TOF) velocity analysis of the atomic beams. Both the collimation (by inserting apertures) and the chopping of the beams, necessary for the TOF-analysis, cut down severely on the intensity of the atomic beams, and thus on the detection rate of the collision products. Depending on factors such as the initial beam intensity, the collision cross section, and the detection efficiency of the collision products, the expected count rates very quickly drop below acceptable levels.

A second problem is lack of control over the collision energy. Although the energy can be *analyzed* by time-of-flight techniques, it is conventionally only *controlled* by varying the source temperature. This control is very limited: the necessity of having a sufficient vapor pressure provides a limit far above room temperature for most elements. For rare gas atoms this limit is low. However, a discharge is normally used to excite the atoms to metastable states, as collisions with the inert ground state rare gas atoms are hardly interesting. The discharge heats up the source, thus effectively increasing the lower limit in source temperature to at least 100 K.

Unfortunately, some of the most fascinating collision physics is expected to show up in the very low energy regime ($T < 10$ K). Here, a semiclassical description of the collision process is not valid any more; the nuclear motion has to described in a fully quantum mechanical way (see, e.g., Ref. [1]).

Laser manipulation techniques provide a way both to achieve very low collision energies and to address the signal strength problem. Diverging atomic beams can be compressed without loss to small diameter, low divergence beams. This can easily lead to 6 orders of magnitude increase in detection rate in a typical crossed beam collision experiment. Atomic beams can also be slowed to a well defined, adjustable axial velocity. These "monochromatic and tunable" beams, to adopt the language of optics, obviate the need for TOF-analysis, which increases the detection rate by elimination of the beam chopper. The tunability of the beam velocity provides control over the collision energy.

Until now, ultracold atomic collisions have been studied by a number of groups, not in colliding beam experiments but in atom traps. For a recent theoretical paper with numerous references to experiments, see Ref. [2]. In these devices, slow atoms (either from a slowed atomic beam, or just from the extreme tail of a thermal Boltzmann distribution in a gas cell) are laser cooled to very low temperatures and trapped by applying a restoring force in three dimensions using magnetic or optical fields or a combination of these. In essence, the result of the cooling is a small, dense cloud of gas at a temperature which can range from a few mK down to the nK regime. Fascinating results have been obtained, perhaps culminating in the achievement of Bose-Einstein condensation of rubidium by the group of Cornell [3].

However fascinating, the trap experiments essentially constitute bulk phase collision experiments with all the well-known disadvantages: The collision energy is not well defined but Boltzmann distributed (albeit with a very low temperature) and the direction of the relative velocity is random. This severely restricts the information which can be obtained on the collision processes. Also, the atoms have to be quite cold before they can be trapped:

the usual depth of the trap is around 1 K. Thus, the transition of the "semi-classical" regime to the "quantal" regime cannot be studied by varying the temperature in the trap.

The work, which is described in this progress report, constitutes part of the GEMINI project. The final goal of this project is the construction of a bright and cold beam collider, in which two atomic beams are collimated to a small diameter, low divergence atomic beam without sacrificing intensity and slowed to a single axial velocity, which can be varied between 10 and 100 m/s. The two beams can be of a different atomic species. This beam collider will provide us with a unique tool to perform well-controlled collision experiments for a variety of systems. The average collision energy can be varied from 1 mK to 10 K, covering the transition from the semiclassical to the quantal regime. The project is a collaboration between our experimental group and the theory group of Verhaar in Eindhoven, and groups in Utrecht (Niehaus, Heideman, van der Straten, Rudolph).

In this report the work which has been done in our group is discussed. As a first step, we built and operated an intensifier for a thermal (axial velocity 600 m/s) beam of metastable neon atoms. This intensifier can be used by itself for a variety of collision experiments in the thermal energy range. The design and the measured performance of this setup will be discussed. Next, a prototype of a full GEMINI beam line has been constructed, which will produce a beam which is slow and cold as well as bright and intense. The design is similar to that of a bright and cold beam setup described by Scholz et al. [4], but incorporates several additional features to provide for a much higher beam flux and lower final divergence. We discuss the design, expected performance and preliminary tests of this beam line.

## THERMAL BEAM INTENSIFIER

The basic design of the beam intensifier consists of three stages of transverse laser cooling, using the closed-level Ne$\{3s\}^3$P$_2 \leftrightarrow \{3p\}^3$D$_3$ transition at $\lambda =$ 640 nm. A schematic view of the setup is given in Fig. 1.

In the first stage, we collimate the diverging atomic beam into a 20 mm diameter parallel beam. Second, this parallel beam is focussed to a 4 mm diameter spot using a two-dimensional magneto-optical trap [5] as the equivalent of a lens in optics. Third, near the focus of this lens, the atomic beam is recollimated into a narrow and nearly parallel beam.

The beam of metastable Ne$\{3s\}$ atoms originates from a discharge-excited supersonic source. The source is cooled with liquid nitrogen, which results in a supersonic axial velocity distribution centered at $v = 580$ m/s with a HWHM $\Delta v = 100$ m/s.

In the first stage we apply a two-dimensional optical molasses to collimate the beam. To achieve a large capture angle while keeping the cooling time as short as possible, it is desirable to use curved wavefronts for the molasses laser

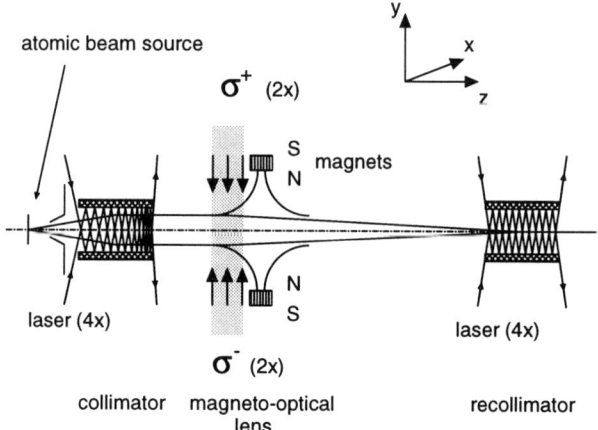

FIG. 1. Schematic view of the experimental setup. The total length scale from source to second collimator is 1 m.

beams [6,7]. For zero detuning of the laser, each atom will be locked to the curved wavefront once it comes into resonance, provided that the radiation pressure can match the centrifugal force. The trajectories with a large initial angle with respect to the beam axis are thus captured at the entrance of the collimator, those with a smaller angle further downstream. By using multiple laser beam reflections between two nearly parallel mirrors (see Fig. 1), both effectively curved wavefronts and a large interaction time are obtained with minimal laser power. The offset angle between the 150 mm long mirrors (60 mm apart) is $\alpha \approx 1.5$ mrad. We inject the laser light (from both sides, to obtain a complete coverage of the interaction region) at an angle $\beta_{coll}^0 \approx 100$ mrad with respect to the plane perpendicular to the atomic beam axis. With each reflection, the angle $\beta$ is reduced by an amount $\alpha$. This produces the effectively curved wavefronts. Since a zero exit angle would cause the laser beam to retrace its initial path and subsequently heat the atomic beam, a small but finite exit angle is used.

This principle is applied in two dimensions using two sets of mirrors. This way, we collimate an atomic beam with an initial HWHM divergence $\Delta\theta_0 \approx 64$ mrad, i.e., a solid angle of $1.3 \; 10^{-2}$ sr, to a well-collimated beam with a HWHM radius $\Delta r_1 = 10$ mm and a residual HWHM divergence $\Delta\theta_1 = 2$ mrad.

To focus the collimated atomic beam in the second stage we use the two-dimensional equivalent of a magneto-optical trap. In a quadrupole magnetic field perpendicular to the beam axis, of which the gradient varies with the axial position $z$ ($B_x = xG(z)$ and $B_y = -yG(z)$), we illuminate the atomic beam with counter-propagating laser beams with orthogonal circular polarization [8]. Because the characteristic cooling time of the transverse velocity is short

(1 μs) compared to the total interaction time in the lens (≈ 85 μs, determined by the interaction length of 50 mm), each atom is rapidly locked to a transverse velocity where the Doppler detuning equals the Zeeman shift of the magnetic field at its point of entrance in the lens. For the $x$-component this means $kv_x(z) = -xG(z)(\mu_B g_{eff}/\hbar)$, with $g_{eff}$ an effective $g$-factor and $\mu_B$ the Bohr magneton. Because the gradient $G(z)$ slowly increases with the axial position $z$ in the lens, from $G = 0.02$ T/m at the entrance to $G = 0.12$ T/m at the exit, the result is a converging trajectory with angle $\theta_2 = v_x(z_{exit})/v = -x/f$, where $f$ is given by $f = \hbar k v/\mu_B g_{eff} G(z_{exit})$. Consequently, for a parallel atomic beam this configuration mimics a lens in optics with focal length $f$. With $g_{eff} \approx 0.8$ for the transition used, the calculated focal length $f = 0.68$ m. The focal length of the lens is proportional to $v$: it thus has a "chromatic aberration". Consequently, for the velocity distribution of our liquid nitrogen cooled source, the size of the focus is limited to a HWHM radius $\Delta r = 1.7$ mm.

We use in-vacuum polarization-preserving mirrors to illuminate the atomic beam from four sides with circular polarization using a single input beam. The magnetic field is provided by ferrite permanent magnets outside the vacuum, at the downstream end of the interaction region: the gradient $G(z)$ in the interaction region increases approximately linearly with the axial position.

Near the focus, we recollimate the atomic beam with a mirror set which is identical to the first stage except for the angle $\alpha$ between the mirrors, which is chosen to be 0. Due to the small transverse velocity of the focussed atoms, curved wavefronts are not necessary.

To measure the performance of the setup, we use two scanning wire-detectors to record 1-D integrated beam profiles. A stainless steel wire is scanned through the atomic beam; at each transverse position the metastable atom-induced Auger emission is monitored by measuring the current from the wire.

**Results**

In Fig. 2(a) experimental results are shown for the atomic beam profile with only the first collimator stage in operation, at two positions $z = 210$ and 1040 mm downstream of the source ($z = 0$). The diverging atomic beam is collimated to a beam with a HWHM radius $\Delta r_1 = 10$ mm and a residual HWHM divergence $\Delta \theta_1 = 2$ mrad.

When the second, magneto-optic, stage is activated, the downstream atomic beam profile is focussed into a sharp peak. In Fig. 2(b) we show the HWHM of the atomic beam profile at position $z = 1040$ mm, measured as a function of the strength of the lens. This variation is achieved by partially blocking the downstream end of the laser beam entering the magneto-optical lens, thus varying the gradient $G(z_{exit})$ and thus the focal length. We observe a minimum in the atomic beam half-width as we go from an underfocussed to an overfocussed atomic beam. The minimum value of 1.8 mm is fully

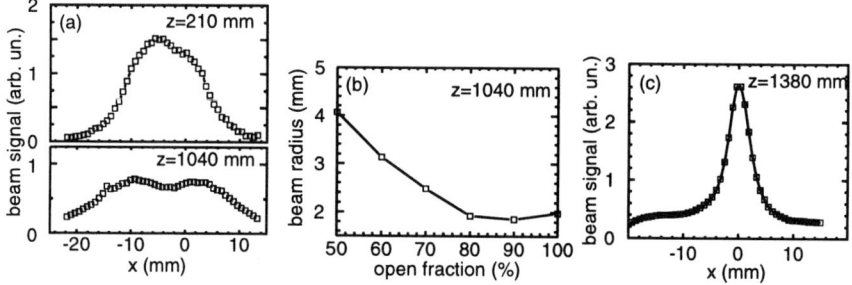

FIG. 2. (a) Atomic beam profile with only the first stage (collimator) activated, measured with the scanning wire detector at $z = 210$ mm (top) and at $z = 1040$ mm (bottom). (b) Atomic beam radius at $z = 1040$ mm with the first two stages active, as a function of the open fraction of the laser beam of the magneto-optical lens. (c) Atomic beam profile with all three stages in operation at $z = 1380$ mm, 340 mm downstream of the exit of the recollimator.

determined by the velocity spread of the atomic beam source. From these data we extrapolate the focal length of the lens without blocking, i.e., at maximum power of the lens, to be 0.7 m.

Finally, the third, recollimation stage at position $z = 960$ mm, i.e., centered approximately on the focus of stage two, is switched on. With a laser beam angle $\beta^0_{recoll} \approx 36$ mrad, the HWHM radius at $z = 1040$ mm is $\Delta r_3 = 1.8$ mm, equal to the minimum of Fig. 2(b). The beam profile at $z = 1380$ mm, 340 mm downstream of the exit of the recollimator, is shown in Fig. 2(c). The HWHM radius has increased to $\Delta r_3 = 2.5$ mm due to the residual beam divergence of $\Delta \theta_3 = 2.0$ mrad. By varying $\beta^0_{recoll}$, we can minimize the beam divergence to a value very close to the Doppler limit $\Delta \theta_D = v_D/v = 0.4$ mrad at the cost of an increase in beam width at $z = 1040$ mm to $\Delta r_3 = 2.5$ mm. This results in a minimum 4-D phase space volume of the atomic beam $(\Delta r_3 \Delta \theta_3)^2 = 1$ mm$^2$mrad$^2$. In comparison with the initial volume of $(\Delta r_0 \Delta \theta_0)^2 = (0.2$mm $64$ mrad$)^2 = 164$ mm$^2$mrad$^2$, with $\Delta r_0 = 0.2$ mm the HWHM virtual source radius [9] of the discharge-excited atomic beam source, the gain in phase space density (brightness) is 160. The gain in maximum flux of the final atomic beam is much larger, namely a factor of 1400. This gain in flux refers to a "Gedanken" experiment where we insert an aperture with a 1.8 mm radius in the beam at position $z = 1040$ mm and compare the laser-on and laser-off signals.

## GEMINI BRIGHT AND COLD BEAM LINE

A schematic view of the prototype GEMINI beam line is given in Fig. 3. The beam line, designed to produce a bright beam of metastable Ne$\{3s\}^3$P$_2$ atoms with a well defined axial velocity of 100 m/s, incorporates 5 different stages of

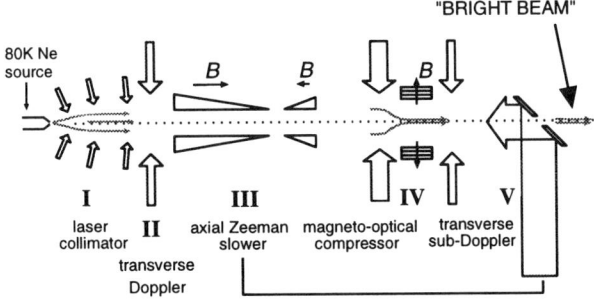

FIG. 3. Schematic view of the prototype GEMINI beam line.

cooling and slowing. The first stage is a 150 mm long nearly-parallel-mirror collimator, identical to the one discussed for the thermal beam intensifier.

To further decrease the residual divergence of this beam ($\approx 2$ mrad) to the Doppler limit of 0.4 mrad, a 50 mm section of pure transverse 2-D Doppler cooling is next. The arrangement of in-vacuum mirrors in this section to provide the transverse light field will be discussed further on in the context of the compression stage.

The third section is a conventional Zeeman-tuned slower [10], optimized to handle the large-diameter (15 mm) atomic beam exiting from the second section. In this stage, the atomic beam is decelerated by a counterpropagating, circularly polarized laser beam. The change of the Doppler blueshift, as the axial velocity of the atoms decreases, is compensated for by an axial magnetic field of which the strength changes with the axial position. The field is produced by a set of tapered solenoids producing fields of opposite sign (see Fig. 3). This configuration is chosen as it minimizes the maximum absolute value of the magnetic field, and thus the heat dissipation in the solenoids. The collimation of the atomic beam before the Zeeman slower drastically increases the throughput of the slower. Without collimation, only very few atoms would pass the slower; even a small transverse velocity will take an atom out of the acceptance area of the next stage. This is not only caused by the length of the Zeeman slower (1.2 m), but also by the fact that the slower acts like a divergence magnifier: As the atom is decelerated without changing the transverse velocity, the initial angle between the trajectory of the atom and the beam axis increases by a factor $v^{initial}/v^{final}$, i.e., a factor of 4.5 in our case.

A magneto-optical compressor forms the fourth stage. In essence, it is a two-dimensional magneto-optical trap like the one which is used as a lens for the thermal beam intensifier. However, due to the much longer interaction time (0.9 ms vs. 85 $\mu$s), the atoms are not only deflected in the interaction region; instead, their trajectories are actually funneled into the beam axis [4,8,11]. The compressor is designed for a large diameter incoming atomic beam (maximum of 30 mm). The optical interaction length is 90 mm. Using 4 separate laser beams to create the required transverse light field (counter-propagating,

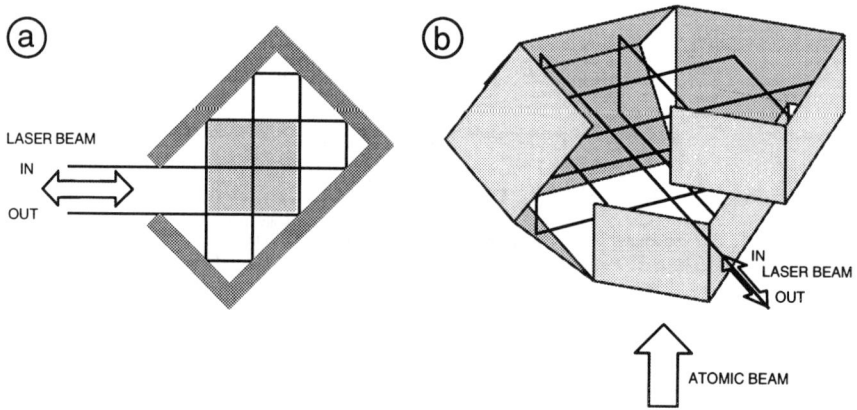

**FIG. 4.** Light recirculation schemes for magneto-optical stages. **a**: Two-dimensional recirculator used for the magneto-optical stage in the thermal beam intensifier. The shaded area is filled with light from four sides. The atomic beam runs perpendicular to the plane of the drawing. **b**: Three-dimensional recirculator used in the GEMINI beam line.

counter-rotating circularly polarized waves in two dimensions), would require more than a Watt of dye laser power.

As the light of one dye laser with a maximum output power of 0.3 W is used for all stages in the setup (shifted in frequency by acousto-optic modulators where needed), an efficient recirculation scheme is needed for the light. For the magneto-optical lens of the thermal beam intensifier, a fourfold recirculation scheme with 4 mirrors positioned in the vacuum (see Fig. 4a) was used. The dielectric mirror coating is designed for minimal phase difference for the reflection of s– and p–polarized light. Between two counter-propagating beams, there is an even number of reflections which ensures the change of the rotation direction of the light polarization.

For the compressor (as well as for the Doppler cooling second section) in the GEMINI beam line a different scheme was used (see Fig. 4b), recirculating the light 16 times. Thus, only 70 mW laser power is needed. The three-dimensional character of the mirror configuration is necessary to prevent the accumulation of polarization phase errors over the 32 reflections involved.

To decrease the residual divergence of the atomic beam, an extra stage of sub-Doppler transverse cooling is incorporated after the compressor. Here, polarization gradient cooling is used, which is calculated to reduce the transverse velocity spread to $\approx 10\,\text{cm/s}$. This limits the final divergence to 1 mrad.

We expect the final beam to have an axial velocity of 100 m/s with a FWHM spread therein of 1.5 m/s, a diameter of 1 mm, a residual HWHM divergence of 1 mrad and an atom flux of $10^{12}\text{s}^{-1}$. These numbers are based on the measured performance of the metastable atom source and a Monte-Carlo

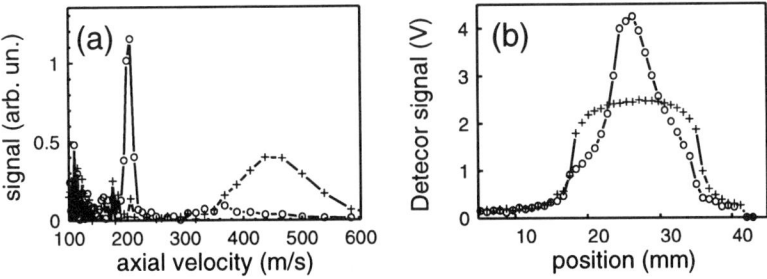

FIG. 5. a: Velocity distribution of the detected atoms with (circles) and without (crosses) the Zeeman slower in operation. b: Focussing of the unslowed beam by the magneto-optical compressor.

simulation of the atom trajectories through the whole setup.

### Preliminary tests of the setup

The setup has been assembled and presently the cooling stages are being tested separately. Testing of the Zeeman magnet requires time-of-flight (TOF) analysis of the velocity of the atoms. To achieve this, two apertures of 5 mm diameter are inserted in the atomic beam, one just after the Zeeman slower and one 700 mm downstream, in front of a surface ionization metastable atom detector. A separate laser beam, just after the first aperture, deflects the atomic beam through resonant radiation pressure so that it misses the second aperture if the laser beam is switched on. By periodically switching off the laser beam, the atomic beam is effectively chopped, enabling the TOF analysis.

In order not to distu rb the axial slowing laser beam, which for this test is sent in straight from the downstream end of the beam apparatus, the aperture substrates as well as the surface ionization detector are transparent. The detector has a transparent conductive coating. In Fig. 5a the results of the velocity analysis are shown, with the Zeeman slower tuned to 200 m/s final velocity. The unslowed part of the detected atoms, with a velocity distribution centered at 450 m/s, is due to the presence of $^{22}$Ne isotopes in the beam as well as $^{20}$Ne$\{3s\}^3$P$_0$ metastable atoms. The FWHM width of the slowed peak, which incorporates 75% of the atoms, is 10 m/s, well above the predicted width of 1.5 m/s. This has not been explained as yet.

The magneto-optical compression stage is tested separately by using it as a lens for the unslowed beam. The beam profile is recorded with a wire scanner, as described for the thermal beam intensifier. In Fig. 5b, the beam profile as passed through a 15 mm aperture just after the Zeeman slower is shown at 300 mmm from the exit of the compression stage, with and without the compressor in operation. The wire scanner is much closer than the focal length of the lens, so that the focussing is only partial. However, the measured profile conforms to the calculations.

At present we are proceeding with the testing of the operation of the compressor with the slowed beam, as well as the operation of the collimator and the transverse Doppler and sub-Doppler cooling stages.

## CONCLUSIONS

The thermal beam brightener which has been constructed and tested by us forms a very valuable tool for collision experiments at "thermal" energies. The device is reliable, and can be operated with a single laser with a modest output power (70 mW minimum). The achieved gain in typical usable beam flux is a factor 1400.

The prototype bright and cold beam line, which is constructed as part of the GEMINI twin cold beam collider project, is being tested. When fully operational, the bright and cold beam line will be a truly unique tool, specifically designed for precision studies of cold collisions in the range of 1 mK to 10 K. With a single beam line, the final beam brightness is large enough to allow the study of intrabeam collisions. The first application will be the study of collisions of metastable neon atoms in the presence of light exciting the $\{3s\}^3P_2 \leftrightarrow \{3p\}^3D_3$ transition ("optical collisions"). The full collider will allow a large range of collision experiments with heterogeneous as well as homogeneous collision partners.

## REFERENCES

1. B.J. Verhaar, AIP Conf. Proc. 323 (Atomic Physics 14), p. 351 (1995).
2. K.-A. Suominen, M.J. Holland, K. Burnett, and P. Julienne, Phys. Rev. A**51**, 1446 (1995).
3. E.A. Cornell, European Research Conference on Bose-Einstein condensation, Strasbourg, France (June 1995).
4. A. Scholz, M. Christ, D. Doll, J. Ludwig, and W. Ertmer, Opt. Commun. **111**, 155 (1994).
5. E. L. Raab, M. Prentiss, A. Cable, S. Chu, and D. E. Pritchard, Phys. Rev. Lett. **59**, 2631 (1987).
6. F. Shimizu, K. Shimizu, and K. Takuma, Chem. Phys. **145**, 327 (1990).
7. A. Aspect, N. Vansteenkiste, R. Kaiser, H. Haberland, and M. Karrais, Chem. Phys. **145**, 307 (1990).
8. J. Nellesen, J. Werner, and W. Ertmer, Opt. Commun. **78**, 300 (1990).
9. H. C. W. Beijerinck and N.F. Verster, Physica **111C**, 327 (1981).
10. W.D. Phillips and H. Metcalf, Phys. Rev. Lett. **48**, 596 (1982).
11. E. Riis, D. Weiss, K. Moler, and S. Chu, Phys. Rev. Lett. **64**, 1658 (1990).

# "Complete" Measurement of Molecular Coulomb-Explosions

U. Werner and H.O. Lutz

*Fakultät für Physik*
*Universität Bielefeld, Universitätsstraße 25, D-33615 Bielefeld, Germany*

---

The multiple ionization and fragmentation of small molecules, e.g. $H_2$, $D_2$, $H_2O$, and $CF_4$, by fast $H^+$, $He^+$, and highly charged $O^{q+}$-ions was studied utilizing a position- and time-sensitive multi-particle detector. The coincident measurement of the momenta of correlated fragment-ions yields a *kinematically complete* image of the molecular break-up process. Thereby, apart from relative cross-sections for specific reaction channels, the fragmentation energy as well as angular correlations can be derived for *each individual event*. Of special interest are "Coulomb-explosion" processes like $H_2 \rightarrow H^+ + H^+$ or $H_2O \rightarrow H^+ + H^+ + O^{n+}$. Whereas the $H_2$ and $D_2$ data are in good agreement with a pure Coulomb-explosion model, this model is insufficient to explain the detailed behaviour of more complex systems. In case of $H_2O$ better agreement is achieved with *ab initio* MCSCF-calculations of the intermediate $H_2O^{(n+2)+}$ parent-ion.

---

## INTRODUCTION

In contrast to studies of molecular multiple ionization and fragmentation by impact of electrons [1,2] and photons (see e.g. [3–5]) the ion impact-induced fragmentation has yet received comparatively little attention. Experiments in which *all* fragment ions emitted after a particular collision are detected in coincidence, can provide valuable information about the dissociation dynamics; under certain conditions even information about the geometric structure of the parent fragmenting system may be derived [6]. So far, most work on ion-induced fragmentation has concentrated on the observation of individual reaction products without further attention to the correlated behavior of the remaining fragments (cf. e.g. [7] and references therein). Exceptions can be found in dissociation studies of some diatomic molecules where the correlation of both fragments has been studied in detail (e.g. [8–10]) and in the "Coulomb explosion" studies of molecular ions (e.g. [6,11]). This latter work is of particular interest since all fragments from a certain break-up can be detected in coincidence: the molecular ion under study is passed at high velocities through a thin foil where the valence electrons are stripped off. The fragments from the ensuing Coulomb explosion (CE) are detected in a position- and time-sensitive detector. As powerful as it may be, this technique is in practice only

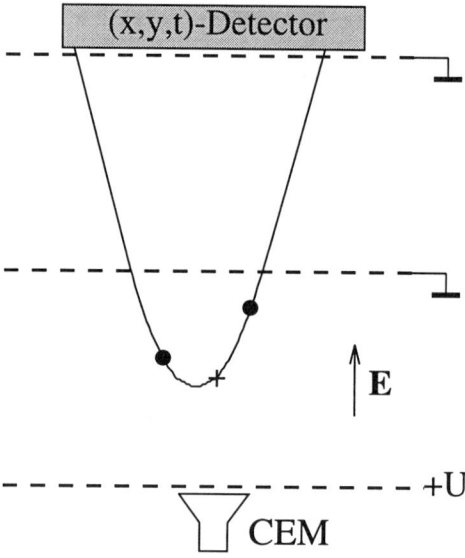

**FIG. 1.** Geometry of the fragmentation experiment. + is the locus of the Coulomb-fragmentation, — projection of correlated fragment-ion paths.

applicable to molecular ions with kinetic energies of at least several MeV, and the data analysis is complicated by the complex interactions in the foil. Many experimental situations (especially those with neutral molecules and clusters) require the handling of fairly low-energy particles; a typical experiment employs, after the dissociation, an acceleration of the charged reaction products and their detection in a time-of-flight (TOF) spectrometer. In our experiment we use a time- and position-sensitive detector to establish correlations between low-energy fragments from a particular molecular break-up, thereby avoiding complications caused by a dissociation foil.

## EXPERIMENTAL SET-UP

Collimated beams of $H^+$, $He^+$, $O^{6+}$, and $O^{7+}$ projectiles interact with a molecular gas target. $H^+$ and $He^+$ ions were produced at the ion accelerator in Bielefeld; highly charged $O^{q+}$ ions were provided by the electron cyclotron resonance ion source (ECRIS) of the KVI in Groningen. The slow ions and electrons generated in the collision process are separated by a weak homogeneous electric field perpendicular to the incident ion beam (Fig. 1). Electrons are detected in a channeltron (CEM) at one side of the interaction region; positive ions are accelerated towards the position- and time sensitive multi-particle detector [12] at the other side. The ions pass a field-free time-of-flight region before they hit the detector which is based on micro-channel-plates in

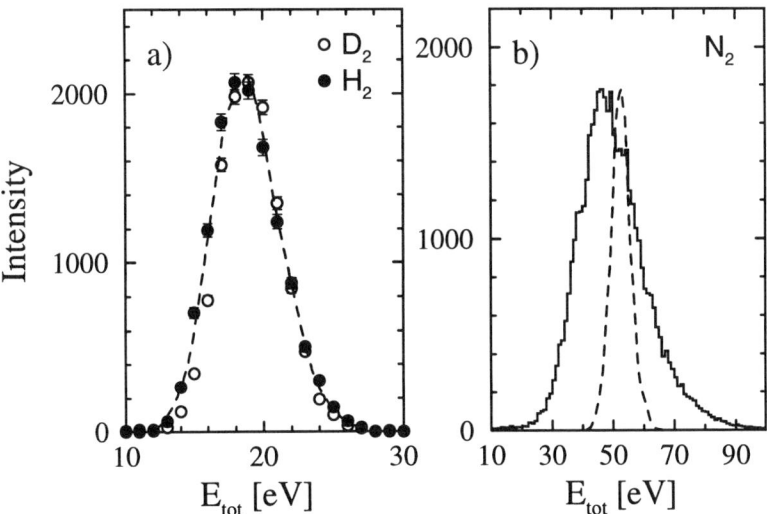

**FIG. 2.** a: Total kinetic energy of coincident fragments in collisions of 200 keV $He^+$ on $H_2$ and $D_2$. - - - result of a Franck-Condon calculation for the Coulomb-explosion of $H_2$ convoluted with the response function of the detector system. b: Total kinetic energy of coincident $N^{2+}$-$N^{2+}$ fragments produced in collisions of 200 keV $He^+$ on $N_2$. - - - is the prediction of a pure Coulomb explosion model.

combination with an etched crossed-wire structure consisting of independent "wires" in x- and y-direction. The first electron registered in the channeltron serves as a start pulse for the coincidence electronics which mainly consists of a specially developed time-to-digital conversion (TDC) system. Thereby, for each positive fragment the position $(x_i, y_i)$ on the detector and the time-of-flight $t_i$ relative to the start electron are recorded.

The present setup allows the simultaneous measurement of all reaction channels resulting in at least one electron and one or more positive fragments. Thus relative cross sections for the production of selected ions (as e.g. $H_2O^+$, $H^+$, or $O^{q+}$ in collisions with $H_2O$) and special processes (e.g. $H_2O \rightarrow H^+ + OH^+$ and $H_2O \rightarrow H^+ + H^+ + O^{q+}$) can be obtained [13]; furthermore, if all fragments of a particular break-up process are detected information about the fragmentation dynamics as well as the molecular structure can be derived. In the following we will concentrate on the latter application.

## DIATOMIC MOLECULES

The simplest and most frequently studied molecular target is $H_2$. Among the various reaction channels occuring in the collision processes we concentrate on those fragmentations leading to coincident proton pairs; these are widely called 'Coulomb explosions':

$$X^+ + H_2 \rightarrow X^+ + H^+ + H^+ + e + e \quad \text{double ionization}$$
$$\rightarrow X \ + H^+ + H^+ + e \quad \text{capture and ionization}$$
$$\rightarrow X^- + H^+ + H^+ \quad \text{double capture}$$

with $X^+$ the projectile ion. Since the detector system is currently triggered by an electron, only the first and second process are observed. For each individual event the positions $\vec{x}_i$ of the correlated protons at the detector and their flight-times $t_i$ are registered. These data allow an absolute determination of the dissociation energy by the use of classical mechanics [12].

Figure 2a shows the total kinetic energy distributions of coincident fragment ions from collisions of 200 keV $He^+$ with $H_2$ and $D_2$ molecules. The $H_2$ data are in good agreement with a Franck-Condon calculation [14] assuming a Coulomb-explosion process. The same holds for the isoelectronic $D_2$ which shows a somewhat narrower distribution as expected by a comparison of $H_2$ and $D_2$ ground state wavefunctions.

The situation changes in case of more complex diatomic molecules like $N_2$ and CO: whereas in case of $H_2$ and $D_2$ the final fragment ions are described by the simple Coulomb potential, there are in general many states of $(AB)^{(m+n)+}$ which finally result in $A^{m+} + B^{n+}$ pairs with different characteristic kinetic energies. Therefore an analysis in terms of a simple CE-model is often inappropriate (c.f. next section). As a typical example Fig. 2b shows the total kinetic energy distribution of coincident $N^{2+} + N^{2+}$ fragment ions in comparison to the prediction of a point-charge CE-model. Even more complicated is the Coulomb fragmentation of poly-atomic systems, a situation our detector was built for.

## COULOMB-FRAGMENTATION OF $H_2O$

As pointed out above even the collision-induced fragmentation of small polyatomic molecules may result in a large number of different reaction channels. As prototype of a triatomic system we choose $H_2O$ which is fairly simple in the following sense: it consists only of two kinds of atoms which can be easily distinguished due to their large mass difference, and only the O-atom may occur in different charge states. A coincidence map as shown in Fig. 3 gives an overview on the two-particle events detected in collisions of 742 keV $O^{7+}$ with $H_2O$. In this representation a coincidence between e.g. an $H^+$ on the right of the detector and an $O^{2+}$-ion on the left results in an event at $T_R \sim 560$ ns and $T_L \sim 200$ ns; the practically symmetric appearance of the diagram is caused by the detector symmetry. As a consequence of momentum conservation the complete fragmentation $H_2O \rightarrow H^+ + OH^+$ results in a narrow structure perpendicular to the diagonal. If one fragment is not detected the structure

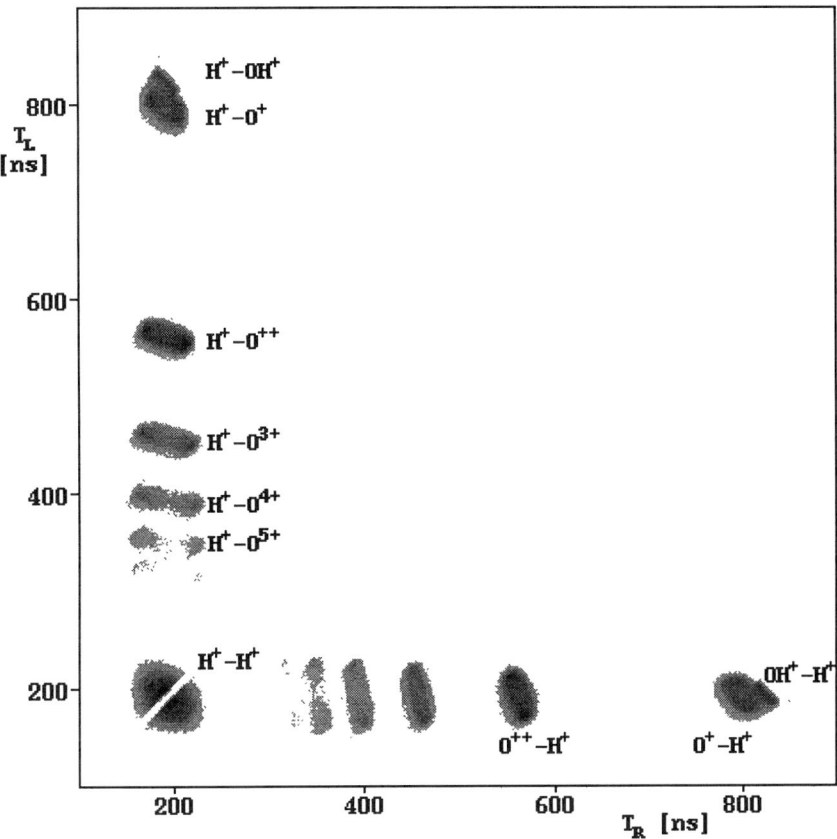

FIG. 3. Coincidence map of correlated positive fragment pairs from collisions of 742 keV $O^{7+}$ on $H_2O$. $T_L$ and $T_R$ are the flight times of the fragments which hit the detector at the left- and rightmost position.

is broadened due to the momentum carried away by the unobserved particle. In the case of $H_2O$ most channels can be separated and analyzed in great detail; in particular cross sections for the correlated production of selected ions can be derived. It should be pointed out that the data obtained by our detector contain also the correlations between more than two fragments, although there is no such intuitively understandable graphic representation in higher dimensions. As an application Fig. 4 shows cross sections for the complete fragmentation of $H_2O$ into three fragments in collisions with 100-350 keV protons. The measured relative data were normalized to absolute cross sections from Edwards et. al. [15–17] for $H^+ + H_2$ collisions using a mixed $H_2 + H_2O$ gas target. The rates of these reaction channels are surprisingly large, e.g. the cross section for $H_2O \rightarrow H^+ + H^+ + O^+$ is about one order of

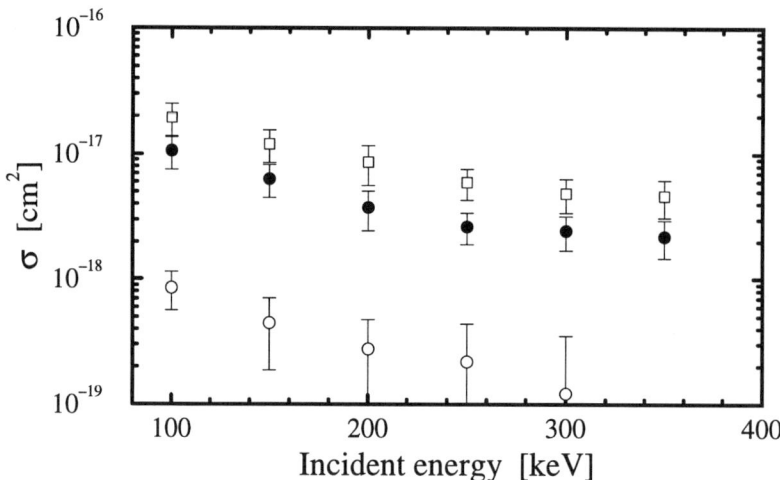

**FIG. 4.** Cross section for complete fragmentation in $H^+ - H_2O$ collisions: □ $H_2O \to H^+ + H^+ + O^{0,-}$, ● $H_2O \to H^+ + H^+ + O^+$, and ○ $H_2O \to H^+ + H^+ + O^{++}$.

magnitude larger than for $H_2 \to H^+ + H^+$ [18].

Such coincidence maps or similar techniques (as e.g. the covariance map which was successfully applied in photoionization measurements [3,4] and electron ionization studies [2]) are valuable tools for the analysis of the fragmentation of small molecules. However, one should be aware that they represent only part of the information contained in the fragmentation dynamics: a kinematically complete analysis involves the direct measurement of all momentum vectors. Among the various reaction channels occuring in ion-water collisions we will concentrate on complete fragmentations of the type

$$X^{p+} + H_2O \to H_2O^{(q+2)+} + X^{(p-m)+} + (q+2-m)e^-$$
$$\to H^+ + H^+ + O^{q+}$$

where $q + 2 - m \geq 1$. In the experiment these events appear as 4-fold coincidences between an electron and the three positive fragment ions. In collisions with $H^+$ and $He^+$ we observed processes with $q \leq 2$ [13]; in collisions with $O^{6+}$ and $O^{7+}$ complete fragmentations up to $q = 5$ were observed [19]. The fragmentation kinematics can be analyzed in terms of three independent parameters which are derived from the measured momentum vectors of coincident fragments. Besides the total kinetic energy of all fragments we choose the angle $\theta_v$ between the two $O^{q+}$-$H^+$ relative velocities $v_{OH}$, as well as the angle $\chi$ between the $H^+$-$H^+$ relative velocity $v_{HH}$ and the velocity of the $O^{q+}$-ion $v_O$ (see Fig. 5).

An important dynamical problem in multi-fragmentation processes is the question whether the participating bonds break simultaneously or in a step-

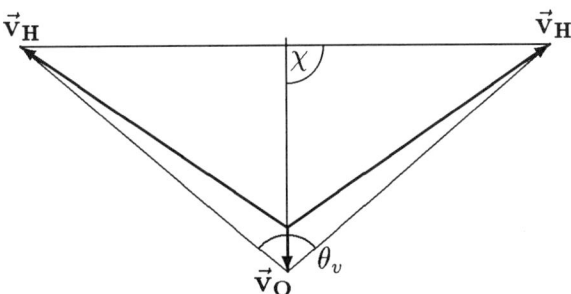

**FIG. 5.** Definition of characteristic angles for a complete fragmentation of $H_2O$ in velocity space.

wise fashion. In case of $H_2O$ the angle $\chi$ may be used as a criterion for the time-scale of the molecular break-up [20]: a break-up of both OH-bonds in a time short compared to the rotation and vibration periods of the system leads to a strong angular correlation between the corresponding velocities which shows up as a narrow peak in the $\cos\chi$-distribution. In case of a two-step process the 'intact' $OH^{(q+1)+}$ subsystem may rotate around its center of mass and the correlation would be lost resulting in a uniform $\cos\chi$ distribution. Figure 6 shows measured $\cos\chi$ spectra for 100 keV $He^+$ and 126 keV $O^{7+}$ impact together with a Monte-Carlo simulation of a simultaneous fragmentation into $H^+ + H^+ + O^+$ based on the MCSCF-calculations described below. The widths of experimental and calculated curves are in reasonable agreement indicating a practically simultaneous bond-breaking in the $(H_2O)^{3+} \rightarrow H^+ + H^+ + O^+$ fragmentation. Similar results hold in case of all observed $(H_2O)^{(q+2)+} \rightarrow H^+ + H^+ + O^{q+}$ processes in all collision systems studied.

A simultaneous break-up into positive fragment ions would suggest a simple Coulomb-explosion (CE) model: the electrons are removed from the molecule during a short collision time and the fragmentation dynamics is governed by the strong mutual repulsion of the generated positive ions. As a first approximation the kinetic energies and emission angles may be computed by assuming Coulomb forces acting between point charges. Figure 7 shows the prediction of this model in comparison to measured total kinetic energies of correlated $H^+ + H^+ + O^+$ fragments from collisions of 200 keV $He^+$ and 92.4 keV $O^{6+}$ with $H_2O$. Maybe surprisingly, the CE-model predicts a much narrower distribution shifted towards higher energies. Furthermore even a qualitative study of the energy spectra indicates that this model is insufficient in case of $H_2O$. For example, in a pure CE process the fragmentation energy and the angular distributions should be independent of the projectile type and its energy. The shapes of the measured spectra of $\theta_v$ and of the total kinetic energy (Fig. 7), however, clearly change with projectile; in case of

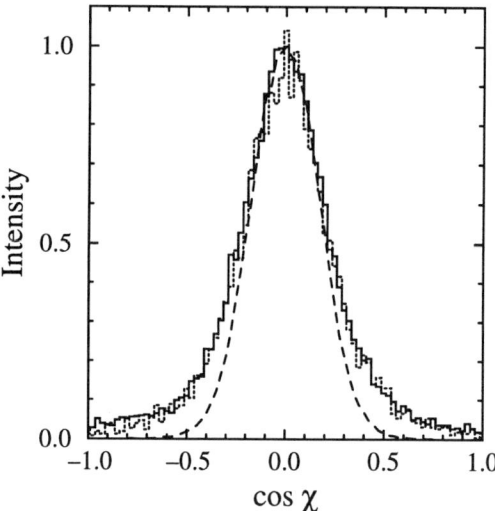

**FIG. 6.** cos $\chi$-distribution of $H^+ + H^+ + O^+$ from collisions of 100 keV $He^+$ (—) and 126 keV $O^{7+}$ (...). (- - -) Monte-Carlo simulation based on a MCSCF-calculation assuming a simultaneous break-up of both H-O bonds.

highly charged ions they even depend on the incident energy [19]. This is in contrast to the simple "point-charge" CE model: several competing processes which all result in three positive fragment ions must be involved to explain the observed behavior.

To account for the most important reaction channels we used the MOLPRO code [21,22] for an *ab initio* multi-configuration self-consistent field computation (MCSCF) of the nine lowest molecular states of the intermediate $H_2O^{3+}$ and $H_2O^{4+}$ ions [23]. For each of these potential surfaces the resulting total kinetic energy and angular distributions were calculated by Monte-Carlo techniques assuming a Franck-Condon transition from the $H_2O$ ground state to the particular dissociating $H_2O^{(q+2)+}$ state. The derived energy and angular distributions show a distinct dependence on the occupation of the $H_2O^{(q+2)+}$ orbitals. Since the transition strengths to the individual states are not known and a multi-parameter fit to the data gave ambiguous results we assumed a transition strength proportional to $1/E_i^2$ (with $E_i$ the excitation energy of the corresponding intermediate $H_2O^{3+}$ state), a scaling behavior which is well known e.g. in inner shell ionization. Of course a more detailed calculation would be highly desirable. Figure 7 shows the positions of the maxima and the weighted sum of the simulated energy spectra convoluted with the response function of the detector system; the calculated cos $\chi$-distribution (Fig. 6) was derived by a similar approach.

A comparison of the measured total kinetic energy of correlated $H^+ + H^+ +$

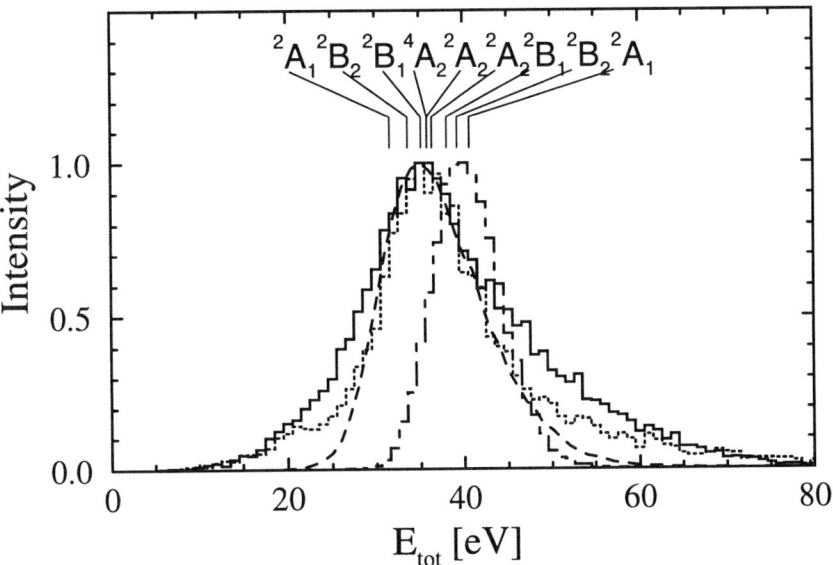

**FIG. 7.** Total kinetic energy distribution of coincident $H^+$-$H^+$-$O^+$ fragments from collisions of $H_2O$ with 250 keV $He^+$ (—) and 92.4 keV $O^{6+}$ (...). - - - is a MC-SCF-calculation taking into account the indicated molecular states of $H_2O^{3+}$ which are labeled by their symmetries. The prediction of a simple Coulomb-explosion model is shown as — -.

$O^+$ fragments to the MCSCF-prediction shows reasonable agreement (Fig. 7): for $He^+$ (and $H^+$) impact the deviations at the low- and high-energy end of the distribution may be attributed to neglected higher excited states in $H_2O^{3+}$ (which not necessarily lead to higher released kinetic energies) and to transition strengths different to the assumed $\propto 1/E_i^2$ scaling. The spectra obtained with highly charged primary O-ions show an even better agreement. According to the classical over-barrier model excited states are expected to be less important in such "gentle" collisions, in agreement with the experimental finding. A more detailed study reveals characteristic differences between individual collision systems indicating changes in the transition strength to particular molecular states of the intermediate $H_2O^{3+}$ ion.

## CONCLUSION

With a newly-developed multi-particle detection system we have studied the kinematically complete fragmentation of small molecules, as e.g. $H_2$ and $H_2O$. The derived data include relative cross sections for the channel-resolved multi-fragmentation as well as information on the dynamics of the initial multiple ionization process, e.g. dissociation energies and angular correlations. An analysis of the kinetic energy released in complete "Coulomb" fragmentations reveals a good agreement with the simple Coulomb explosion model for $H_2$ and $D_2$, whereas more complex systems require the consideration of the involved states of the intermediate multiply-charged ion. In case of $H_2O$ an interpretation in terms of the lowest molecular states of the intermediate $(H_2O)^{(q+2)+}$ indeed leads to a reasonable agreement with the experimental data. First measurements with more complex target molecules, as $CH_4$ or $CF_4$, indicate that molecules containing mainly H-atoms appear to be rather suitable for such detailed studies. So far, theoretical work on this subject is quite scarce and we hope that such detailed measurements will provide a stimulus to further studies.

## ACKNOWLEDGEMENTS

We wish to thank Prof. R. Morgenstern and the Groningen KVI group for the generous support during the experiments with highly charged ions, as well as for many stimulating discussions. This work was supported by the Deutsche Forschungsgemeinschaft (DFG) in Sonderforschungsbereich 216 and the EU-network CHRX-CT94-0643.

## REFERENCES

1. see e.g. R.N. Compton and J.N. Bardsley, in *Electron Molecule Collisions*, edited by I. Shimamura and K. Takayanagi, New York: Plenum Press, 1984, pp. 275-349.
2. M.R. Bruce, L. Mi, C.R. Sporleder, and R.A. Bonham, J. Phys. **B27**, 5773-5794(1994).
3. L.J. Frasinski, K. Codling, and P.A. Hatherly, Science **246**, 1029-1031 (1989).
4. L.J. Frasinski, P.A. Hatherly, and K. Codling, Physics Letters **A 156**, 227-232 (1991).
5. J.H.D. Eland and B.J. Treves-Brown, Int. J. Mass Spectrom. Ion Processes **113**, 167-176 (1992).
6. Z. Vager and E.P. Kanter, Nucl. Instrum. Methods **B33**, 98-101 (1988).
7. C.J. Latimer, Adv. At. Mol. Opt. Phys. **30**, 105-140 (1993).
8. D.P. de Bruijn and J. Los, Rev. Sci. Instrum. **53**, 1020-1026 (1982).
9. A.K. Edwards, R.M. Wood, and R.L. Ezell, Phys. Rev. **A31**, 99-102 (1985).
10. F.B. Yousif, B.G. Lindsay, and C.J. Latimer, J. Phys. **B21**, 4157-4164 (1988).
11. D.S. Gemmell, Chem. Rev. 80, 301-311 (1980).
12. J. Becker, K. Beckord, U. Werner, and H.O. Lutz, Nucl. Instrum. Methods **A 337**, 409-415 (1994).
13. U. Werner, K. Beckord, J. Becker, H.O. Lutz, Phys. Rev. Lett. **74**, 1962-1965 (1995).
14. K.E. McCulloh, J. Chem. Phys. **48**, 2090-2093 (1968).
15. A.K. Edwards, R.M. Wood, and R.L. Ezell, Phys. Rev. **A34**, 4411-4414 (1986).

16. A.K. Edwards, R.M. Wood, J.L. Davis, and R.L. Ezell, Phys. Rev. **A42**, 1367-1375 (1990).
17. A.K. Edwards, R.M. Wood, J.L. Davis, and R.L. Ezell, Phys. Rev. **A44**, 797-798 (1991).
18. U. Werner, K. Beckord, J. Becker, H.O. Lutz, to be published.
19. U. Werner, K. Beckord, J. Becker, H.O. Folkerts, and H.O. Lutz, Nucl. Instrum. Methods **B98**, 385-388 (1995).
20. C.E.M. Strauss and P.L. Houston, J. Phys. Chem. **94**, 8751-8762 (1990).
21. MOLPRO is an *ab initio* program written by H.J. Werner and P.J. Knowles with contributions from J. Almlöf, R. Amos, S. Elbert, K. Hampel, W. Meyer, K. Peterson, R. Pitzer and A. Stone.
22. H.J. Werner and P.J. Knowles, J. Chem. Phys. **82**, 5053-5063 (1985).
23. K. Beckord, J. Becker, U. Werner, H.J. Werner and H.O. Lutz, to be published.

# SELECTED TOPICS

## — PHOTONS —

Photoionization of Sr$^+$ Ions in the 3d Ionization Region ...... 755
    Y. ITOH, T. Koizumi, Y. Awaya, S.D. Kravis,
    M. Oura, M. Sano, T. Sekioka, and F. Koike

Angular Distributions and Retardation
in Photoionization of Two Electrons in Helium ................ 763
    M.A. KORNBERG and J.E. Miraglia

Experimental Separation of Photoabsorption
and Compton Scattering Contributions
to He Single and Double Ionization ........................... 773
    L. SPIELBERGER, O. Jagutzki, R. Dörner,
    J. Ullrich, U. Meyer, V. Mergel, M. Unverzagt,
    M. Damrau, T. Vogt, I. Ali, Kh. Khayyat, D. Bahr,
    H.G. Schmidt, R. Frahm, and H. Schmidt-Böcking

# Photoionization of Sr$^+$ Ions in the 3d Ionization Region

Y. Itoh [1], T. Koizumi [2], Y. Awaya [3], S. D. Kravis [2,3]*, M. Oura [3], M. Sano [2], T. Sekioka [4] and F. Koike [5]

[1] *Physics Laboratory, Faculty of Science, Josai University, Sakado, Saitama 350-02, Japan*
[2] *Department of Physics, Rikkyo University, Toshima-ku, Tokyo 171, Japan*
[3] *The Institute of Physical and Chemical Research (RIKEN), Wako, Saitama 351-01, Japan*
[4] *Himeji Institute of Technology, Himeji, Hyogo 671-22, Japan*
[5] *School of Medicine, Kitasato University, Sagamihara, Kanagawa 228, Japan*

**Abstract.** By measuring photoion-yield spectra for Sr$^{2+}$ and Sr$^{3+}$ from Sr$^+$ ions in the ground state, we have determined relative photoionization cross section of Sr$^+$ ions in the energy range from 135-210 eV. The 3d-ionization of charge-selected Sr$^+$ ions by photon-impact was studied for the first time. The yield-spectrum for Sr$^{2+}$ has no structure, while that for Sr$^{3+}$ shows prominent discrete peaks around 140 eV. For the assignment of discrete lines observed, we used a multiconfiguration Dirac-Fock calculation code. The calculation indicates very strong s-d mixing for the observed peaks. This is understood that after the creation of 3d-hole, the orbital energy of 4d-orbital almost coincides with that of 5s-orbital due to the collapse of the wavefunction.

## INTRODUCTION

Studying photoionization of free ions gives us a chance to examine ionic electronic structures systematically along iso-electronic, iso-ionic and iso-nuclear sequences. For this purpose, photoabsorption measurements using the laser-produced plasma technique[1-3], photoion measurements [4,5] and photoelectron measurements [6] have been done.

The photoionization of Sr$^+$ ions around the 4p-ionization region has been reported only by Lyon *et al* [7]. Very recently, McGuiness *et al* reported the 3d absorption spectra of the laser-produced plasma containing Sr and Sr$^{n+}$ ($n = 1$~3) [8]. Using a photon-ion merging-beam apparatus, Koizumi *et al* succeeded in measuring relative photoionization cross sections of Ba$^+$ ions in the 4d ionization region. [9].

In this article, we will report the results for the photoionization of charge-selected Sr$^+$ ions in the 3d ionization region [10] using the same apparatus reported by Koizumi *et al*. We have measured photoion-yields from photoionization of Sr$^+$ ( $3d^{10} 4s^2 4p^6 5s\ ^2S_{1/2}$ ) and compared the results with previous photoionization measurements of neutral Rb [11-13] ( in the same iso-electronic sequence as

© 1995 American Institute of Physics

$Sr^+$) in the same photon-energy region. Comparing the results with those of Rb, we could study the degree of the 4d-orbital collapse due to a one unit increase of nuclear charge.

## EXPERIMENTAL

A schematic diagram of the photon-ion merging-beam apparatus [ 9 ] is shown in Fig. 1. It consists of an ion-source, an electrostatic quadruple deflector (D), an interaction region (I), and an electrostatic parallel-plate charge-analyzer (A). A rotating chopper wheel (CP) driven by a stepping motor was placed in front of the deflector chamber to chop the photon-beam. A good vacuum is essentially required to reduce the background noise. Each chamber was evacuated differentially with a turbomolecular-pump (TMP). A glass-tube of 3mm$^\phi$ x 80 mm was set between the chopper-chamber and the deflector-chamber. This works as a differential wall between the beam-line of the synchrotron-radiation and the apparatus, and as a light-guide of the photon-beam also. To avoid the contamination of the interaction region by a bad vacuum of the ion-source region, which was about $2 \times 10^{-6}$ Pa during the operation, a stainless-tube of 5mm$^\phi$ x 50 mm was placed between the chambers. The background pressure of $2 \times 10^{-8}$ Pa at the interaction region was maintained during the measurements with this pumping system.

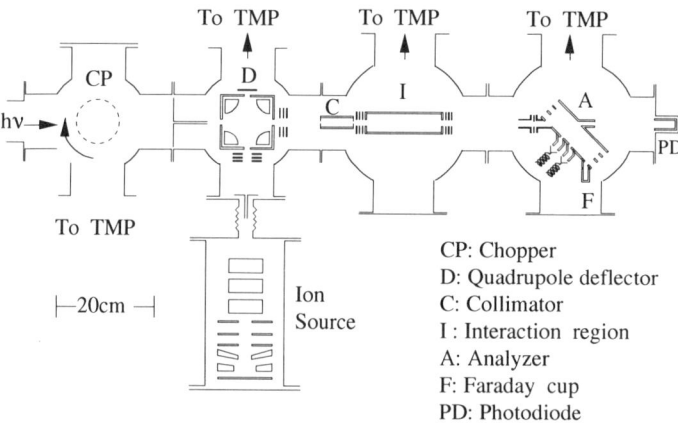

CP: Chopper
D: Quadrupole deflector
C: Collimator
I : Interaction region
A: Analyzer
F: Faraday cup
PD: Photodiode

**FIGURE 1.** Schematic diagram of the photon-ion merging-beam apparatus.

The energy level of the first excited state of the $Sr^+$ ion ($4p^6\ 4d\ ^2D$) is about 1.8 eV above the ground state. As we used a surface-ionization source, we can safely assume that the ions created by this source are all in the ground state. The ions were accelerated to 2 keV, collimated to 2 mm in diameter, then merged with a monochoromatized photon-beam from the BL-3B bending magnet [ 14 ] at the Photon Factory in the National Laboratory for High Energy Physics (KEK). The interaction region was biased to 800 V to separate the background $Sr^{2+}$ ions created outside the interaction region. Product ions were separated by the charge-analyzer and counted by two channel-electron-multipliers for doubly and triply charged ions. A small electron-gun set near the interaction region was used to check the analyzer by measuring $Sr^{2+}$ and $Sr^{3+}$ ions created by electron-impact.

The typical intensity of the ion-beam measured by a Faraday-cup (F) was 80 nA, and the photon-flux was estimated to be on the order of $10^{11}$ $s^{-1}$ with an energy resolution of $E / \Delta E = 450$. The accuracy of the energy scale of the present measurement is estimated to be $\pm 0.3$ eV around 150 eV.

The main component of the background noise, typically 200 $s^{-1}$, is considered to be due to the charge-stripping collision with the residual gas. The signal counting-rate was about 50 $s^{-1}$ for the intense peak at 140 eV when the resolving power was 200 ( see Fig. 2). The data accumulation time for each channel was 180 s. When the higher energy-resolution measurement was made, the signal intensity decreased to about 20 $s^{-1}$, therefore, the measuring time for each channel was increased to 350 s. The fluctuation of the data is estimated to be about 5% of the most intense line for the measurement with the $E / \Delta E = 200$, and about 10% of that with the $E / \Delta E = 450$ ( see Fig. 3).

## RESULTS AND DISCUSSION

Photoion-yield spectra of $Sr^{2+}$ and $Sr^{3+}$ from $Sr^+$ ions in the ground state as a function of photon energy are shown in Fig. 2. The band pass energy was about 0.7 eV at 150 eV. The yield-spectrum for $Sr^{2+}$ shows almost monotonic decrease towards the higher energy side, while that for $Sr^{3+}$ shows prominent discrete

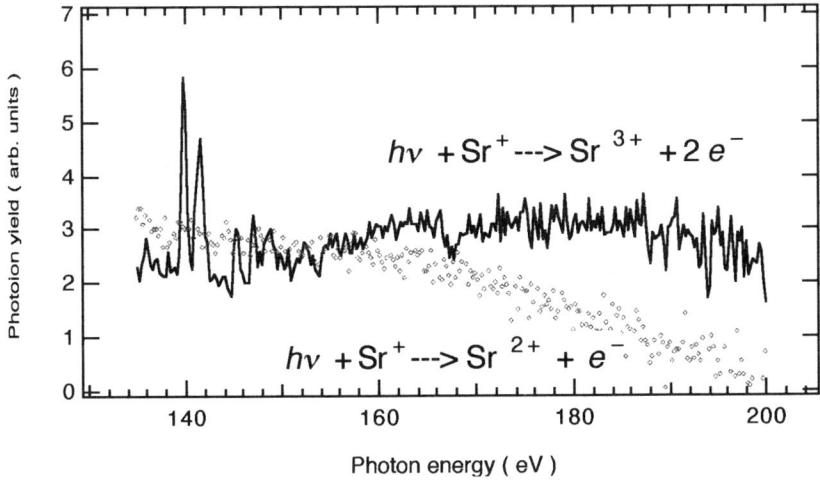

**FIGURE 2.** Relative photoion-yields from photoionization of $Sr^+$ ions.

peaks below 150 eV, and displays a broad structure from 155 eV to 200 eV. This broad structure is an indication of a giant resonance which corresponds to the 3d - $\varepsilon$ f transition. A window-type resonance is observed clearly at around 170 eV.

Considering the results of the photoabsorption measurements [15] and photoion-yield measurements [12, 13] for neutral Sr, we conclude that (i) the mechanism for $Sr^{2+}$ production ( a single ionization process) is the direct ionization of an outer 5s or 4p electron and (ii) the window type resonance is attributed to the simultaneous excitation of 3d and 4p electrons. (iii) $Sr^{3+}$ is produced by resonance-Auger or Auger processes after the 3d-electron is excited or ionized by photoabsorption. For example, below the 3d-ionization threshold, the following two-step process can be considered:

$$Sr^+(3d^{10}4s^24p^65s) + h\nu \longrightarrow Sr^{+*}(3d^94s^24p^65s\, np)$$
$$\text{step I} \longrightarrow Sr^{2+*}(3d^{10}4s\, 4p^55s\, np) + e_1 \quad (1)$$
$$\text{or}$$
$$\longrightarrow Sr^{2+*}(3d^{10}4s^2\, 4p^4\, 5s\, np) + e_{1'} \quad (1')$$
$$\text{step II} \longrightarrow Sr^{3+}(3d^{10}4s^24p^5) + e_2. \quad (2).$$

Koizumi et al [12] reported that discrete lines and giant-resonance structure were mainly observed in the $Sr^{3+}$-yield spectrum from the neutral Sr. They considered the following three-step process for triple-electron emission processes, for 3d-photoionization of Rb:

$$Rb(3d^{10}4s^24p^65s) + h\nu \longrightarrow Rb^*(3d^94s^24p^6\,5s\,np)$$
$$\text{step I} \longrightarrow Rb^{+*}(3d^{10}4p^65s\,np) + e_1 \quad (3)$$
$$\text{step II} \longrightarrow Rb^{2+*}(3d^{10}4s\, 4p^5\, 5s) + e_2 \quad (4)$$
$$\text{step III} \longrightarrow Rb^{3+}(3d^{10}4s^2\, 4p^4) + e_3. \quad (5).$$

Because $Sr^+$ ions have the same electron configuration as Rb atoms have, the similar process that is found in Rb may be considered also in $Sr^+$. If the Auger final state in step II ( for $Sr^{3+*}$ production in the present case ) is located higher than the ionization energy of $Sr^{3+}$, step III is possible to produce $Sr^{4+}$. Therefore, in our case, $Sr^{4+}$ ions are expected to be produced in considerable amounts. However, we can not currently measure $Sr^{4+}$ due to the limitations of the present apparatus.

In Fig. 3, $Sr^{3+}$ - yield spectrum measured with higher resolution, $E / \Delta E = 450$, is shown with a theoretical spectrum, which will be discussed later. The energy positions of the prominent peaks labelled as (a)-(e) are shown in Table. 1. McGuiness et al [8] analyzed the 3d-absorption spectra of the laser-produced plasma containing Sr and $Sr^{n+}$ ($n = 1\sim3$), and assigned 11 lines for the excited states of $Sr^+$ ions. The energy positions of the lines corresponding to the present results are also tabulated in Table 1. The agreement between their results and the present measurements is fairly good, especially for two "Clear" lines at 145.20 and 146.90 eV. However, we could not find remarkable lines at 137.91, 143.78, 143.92, and 144.52 eV.

For the assignment of the discrete peaks observed, we used the $GRASP^2$ code [16], a revised version of a multiconfiguration Dirac-Fock program by Dyall et al [17]. we calculated the energies of 3d-excited states and the oscillator strengths for the photoexcitation from the $Sr^+$ ground state. We included $3d^94s^24p^65s\,5p$,

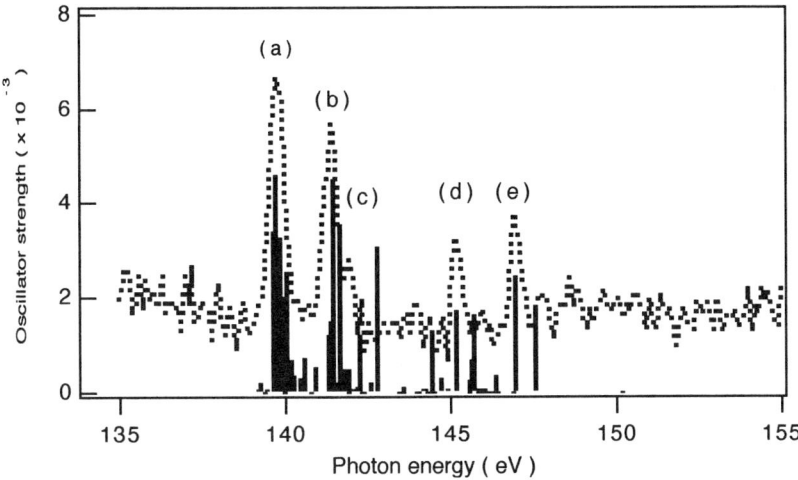

**FIGURE 3.** Sr$^{3+}$ -yield spectrum measured with a higher resolution, and theoretical spectrum.

3d$^9$4s$^2$4p$^6$ 4d5p, 3d$^9$4s$^2$4p$^6$ 5s6p and 3d$^9$4s$^2$4p$^6$ 4d6p for the excited state configurations. The oscillator strength in the velocity-form computed is already plotted in Fig. 3 as a function of transition energies from the ground state. Note that the theoretical spectrum is shifted by 0.5 eV to the higher energy-side to adjust to the experimentally observed intense peaks.

The computed spectrum is very complicated and shows that there exists almost no line that keeps single-configuration purity below 144.2 eV. Very strong s-d configuration mixing is seen for the 3d - 5p transitions. Some lines have a remarkable 3d$^9$ 4s$^2$ 4p$^6$4d 5p feature rather than that of 3d$^9$4s$^2$4p$^6$5s5p. In contrast to that, lines with relatively high single-configuration character are found for the 3d - 6p transitions. This tendency was also reported by Mansfield and Connerade [ 15 ] explaining that the degree of the collapse is controlled by the $n$ value of $n$p electron.

Though the peak labelled as (c) in Fig. 3 is not reproduced in the calculation, the agreement between measured and calculated results is very satisfactory for both

**TABLE 1.** Energy of the prominent peak in Fig. 3 ( in eV) and those reported by McGuiness *et al.*

| label | (a) | (b) | (c) | (d) | (e) |
|---|---|---|---|---|---|
| energy | 139.7 | 141.4 | 141.9 | 145.2 | 146.9 |
| McGuiness | 139.61 | 141.08 | 142.12 | 145.20 | 146.90 |

relative energy positions and intensities. The assignment and the percent composition of the electron-configurations for each line is following:

peak(a): $3d^9(^2D_{5/2})$ + 5s5p(33.2%) + 4d5p(66.3%) + 5s6p(0.4%) + 4d6p(0.2%),

peak(b): $3d^9(^2D_{3/2})$ + 5s5p(35.0%) + 4d5p(64.5%) + 5s6p(0.5%),

peak(d): $3d^9(^2D_{5/2})$ + 5s5p(1.1%) + 4d5p(0.3%) + 5s6p(97.6%) + 4d6p(1.0%),

peak(e): $3d^9(^2D_{3/2})$ + 5s5p(1.0%) + 4d5p(0.9%) + 5s6p(97.4%) + 4d6p(0.8%).

These percent compositions are determined by integrating over the theoretically obtained partial contribution from each line which is located under the experimental line. The averaged character of the two intense lines observed at 139.7 eV and 141.4 eV is seen to be dominated by $3d^9\ 4s^2\ 4p^6\ 4d\ 5p$ configuration.

A calculation of the same kind was performed for the isoelectronic element, Rb, applying the identical calculation procedure and configurations. This is to understand the origin of the very strong s-d configuration mixing seen in Sr $^+$. Theoretical results are shown in Fig. 4. In this figure, results for Sr$^+$ with no energy-shift is shown for comparison. We found that this theoretical spectrum for neutral Rb reproduces well the measured spectrum [ 11, 12 ] when the energy-scale is shifted about 1 eV to the higher energy-side. The theoretical spectrum for Rb is much simpler than that for Sr $^+$, because many lines keep single-configuration purity especially for the intense 3d - 5p transitions at 112.4 eV and 113.8 eV. This different degree of the s - d configuration mixing in the iso-electronic sequence may be understood from a difference of the orbital energies for each target.

The orbital energies for Rb * and Sr $^+$* after the 3d-hole creation are shown in Table 2. The order of 5s and 4d orbital-energy for Rb * is the same as the rare-gases, i. e., the energy of 4d orbital is located above that of 5s orbital. In Sr $^+$* case, the tendency is fully changed. Orbital energies almost coincide with each other and even the order is inverted. This is a clear example of the collapse

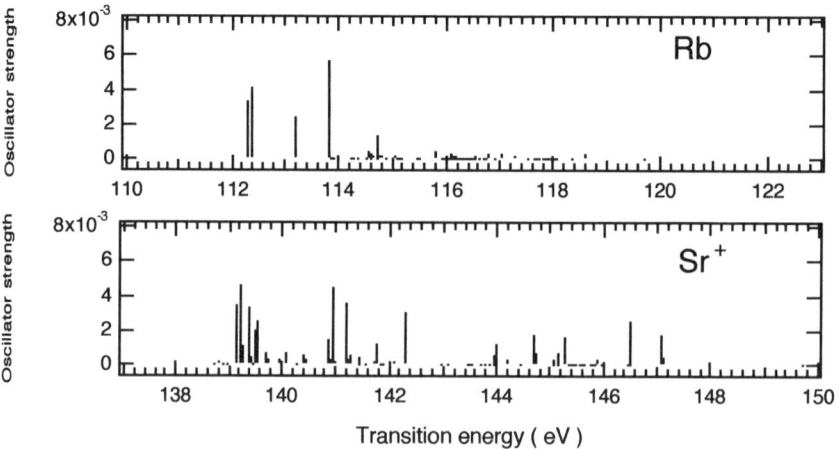

**FIGURE 4.** Theoretical spectra for neutral Rb and Sr$^+$.

**Table 2.** Orbital energies in eV for Rb * and Sr ** after 3d-hole creation.

| orbital | Rb* | Sr** |
|---|---|---|
| 4d($j$=3/2) | 5.374 | 15.018 |
| 4d($j$=5/2) | 5.289 | 14.850 |
| 5s | 7.197 | 14.319 |
| 5p($j$=1/2) | 2.871 | 8.398 |
| 5p($j$=3/2) | 2.758 | 8.203 |

of 4d orbital due to the increase of the effective nuclear-charge, which leads to very strong s-d mixing.

As a conclusion, our analysis for the 3d-excitation of Sr$^+$ shows that the intense lines observed in the photoion-yield spectrum are dominated by $3d^9 4s^2 4p^6 4d 5p$ configuration. Comparing the theoretical results for the 3d-excitation of Rb, we found that the collapse of 4d wavefunction of Sr$^+$ after the 3d-hole creation occurs very drastically, because of the increase of nuclear-charge.

## ACKNOWLEDGMENTS

This work was performed as an R&D program of the SPring-8 project, and partly supported by a Grant-in Aid from the JAERI-RIKEN SPring-8 Project Team. The theoretical calculation was partly supported by a Grant-in-Aid from the Ministry of Education, Science and Culture in Japan. We would like to thank Professor A. Yagishita and Dr. E. Shigemasa of the Photon Factory for helpful advice to perform the experiment. This experiment was done under approval of the Photon Factory of National Laboratory for High Energy Physics ( Proposal No. 92G302).


* Permanent address:  J. R. Macdonald Laboratory, Department of Physics,
Kansas State University, Manhattan, KS 66506, USA.

# REFERENCES

[1] Lucatorto, T. B., McIlrath, T. J., Suger, J., and Younger, S. M., *Phys. Rev. Lett.* **47**, 1124–8 (1981).
[2] Dunne, P., O'Sullivan, G., and Ivanov, V. K., *Phys. Rev.* A **48**, 4358–64 (1993).
[3] Jannitti, E., Gaye, M., Mazzoni, M., Nicolosi, P., and Villoresi, P., *Phys. Rev.* A **47**, 4033–41 (1993).
[4] Kravis, S. D., Church, D. A., Johnson, B. M., Meron, M., Jones, K. W., Levun, J., Sellin, I. A., Azuma, Y., Mansour, N. B., Berry, H. G., and Dretta, M., *Phys. Rev. Lett.* **66**, 2956–9 (1991).
[5] Dolder, K., *Electronic and Atomic Collisions*, Amsterdam: North-Holland, 1988, pp.549-56.
[6] Bizau, J. M., Cubaynes, D., Richter, M, Wuilleumier, F. J., Obert, J., Putaux, J. C., Morgan, T. J., Källne, E., Sorensen, S., and Damany, A., *Phys. Rev. Lett.* **67**, 576–9.
[7] Lyon, I. C., Peart, B., and Dolder, K., *J. Phys* B **20**, 1925–32 (1987).
[8] McGuiness, C., O'Sullivan, G., Carroll, P.K., Audley, D., Mansfield, M. W. D., *Phys. Rev.* A. **51**, 2053–62 (1995).
[9] Koizumi, T., Itoh, Y., Sano, M., Kimura, M., Kojima, T. M., Kravis, S., Matsumoto, A., Oura, M., Sekioka, T., and Awaya, Y., *J. Phys.* B **28**, 609–16 (1995).
[10] Itoh, Y., Koizumi, T., Awaya, Y., Kravis, S. D., Oura, M., Sano, M., Sekioka, T., and Koike, F., submitted to *J. Phys.* B (1995).
[11] Connerade, J. P., and Mansfield, M. W. D., *Proc. R. Soc. Lond.* A **348**, 539–52 (1976).
[12] Koizumi, T., Hayaishi, T., Itikawa, Y., Nagata, T., Sato, Y. and Yagishita, A., *J. Phys* B **20**, 5393-401 (1987).
[13] Koizumi, T., Hayaishi, T., Itikawa, Y., Itoh, Y., Matsuo, T., Nagata, T., Sato, Y., Shigemasa, E., Yagishita, A., and Yoshino, M. *J. Phys* B **23**, 403–15 (1990).
[14] Yagishita, A., Hayaishi, T., Kikuchi, T., and Shigemasa, E., *Nucl. Instr. and Meth.* A **306**, 578–83 (1991).
[15] Mansfield, M. W. D., and Connerade, J. P., *Proc. R. Soc. Lond.* A **342**, 421–30 (1975).
[16] Parpia, F. A., Grant, I. P., and Fischer., C. F. *private communication* ( 1992 ).
[17] Dyall, K. G., Grant, I. P., Johnson, C. T., Parpia, F. A., and Plummer, E. P., *Comput. Phys. Commun.* **55** 425–456 (1989).
[18] Mansfield, M. W. D., and Connerade, J. P., *J. Phys.* B **15**, 503–12, (1982).

# Angular distributions and retardation in photoionization of two electrons in helium

## M. A. Kornberg and J. E. Miraglia

*Instituto de Astronomía y Física del Espacio*
*Casilla de Correo 67, Sucursal 28, 1428 Buenos Aires, Argentina*

---

We study higher multipole corrections (retardation) to the dipole results for angular distributions in two-electron photoionization of the helium ground state.

The dipole asymmetry parameter is shown to have the well defined limits $\beta \to 0.0(2.0)$ as $\epsilon_1 \to 0(E_f)$, where $\epsilon_1$ is the energy of the observed electron and $E_f$ is the total final energy. This behaviour is reached at 3.0 keV.

Higher multipole effects require the inclusion of additional angular coefficients. A first-order retardation correction requires two additional coefficients; these coefficients are zero at $\epsilon_1 = 0$.

A simple calculation shows that the contribution of retardation on the total cross section $\sigma^{2+}$ is of the order $(v_m/c)^2$, where $v_m$ is maximum velocity of an electron, and so is the correction for $\sigma^+$. Therefore the ratio $R = \sigma^{2+}/\sigma^+$ gives the same value with or without retardation.

---

## INTRODUCTION

The photoionization of two electrons in helium has received a great deal of attention in the past few years. The center of the discussion has been devoted to the determination of the asymptotic value of the ratio $R = \sigma^{2+}/\sigma^+$, since experimental results for this quantity have become available with the advent of modern synchrotron light sources [1]. The accepted value for the asymptotic ratio is now 1.67 %. A typical energy range for these high-energy experiments goes from 1.0 to 12.0 keV. However, we note that above 6.0 keV, Compton scattering dominates over photoabsorption [2,3], so experiments should be able to distinguish between these two contributions [4].

The calculation of the ratio $R$ stems from work in the 1960s [5–7], using what is generally called the sudden approximation [8]. In these papers it was proved that in the dipole velocity or acceleration approximation, a simple formula for the ratio $R$ in the limit $E_\gamma \to \infty$ is given by

$$R = \frac{\int |\psi_i(\mathbf{r}_1, 0)|^2 d\mathbf{r}_1}{\sum_n |\int u_n^*(\mathbf{r}_1)\psi_i(\mathbf{r}_1, 0)d\mathbf{r}_1|^2} - 1, \quad (1)$$

© 1995 American Institute of Physics

where the sum runs over all bound $ns$ states $u_n$ of the residual ion He$^+$, and $\psi_i(\mathbf{r}_1, \mathbf{r}_2)$ is the initial ground state of He. The formula of Eq. (1) reflects that, as the photon energy goes to infinity, photoabsorption is determined exclusively by the region where one electron is at the nucleus. That is why the initial state should satisfy to a good degree of accuracy the so called cusp condition at the nucleus given by

$$R_{\text{cusp}} = \frac{(\partial \psi_i / \partial r_2)_{r_2=0}}{\psi_i(r_1, r_2 = 0)} = -2. \tag{2}$$

It is interesting to note that, in a configuration-interaction (CI) expansion, it is of primary interest to obtain a good description of the $ss$ part of the wave-function, the only part that enters in the calculation of the ratio $R$ in Eq. (1).

Other calculations for the energy dependence of $R$ have been performed in the dipole approximation, where in the velocity or acceleration gauges the total cross section $\sigma^{2+}$ falls off as $E_\gamma^{-7/2}$ [9]. It should be recalled that the gauge dependence for two electron transitions [10] makes that a strong discrepancy appears in using the length form of the dipole operator [6,11], and gives an erroneous $E_\gamma^{-5/2}$ dependence for the asymptotic cross section.

Beyond the theoretical calculation of the ratio $R$ in this high-energy regime, it is interesting to investigate the angular distributions of the electrons, which show us some of the physics underlying this process. In doing so, care must be taken that we are not losing a great deal of information when the dipole approximation is employed, since this approximation is not automatically justified at high photon energies.

We present in this work results for the asymmetry parameter $\beta$ using a correlated and an uncorrelated two-electron continuum wave function (C3 and C2 models, respectively). We investigate the effects of retardation on the cross sections and we calculate the first-order retardation correction which requires the inclusion of two additional angular coefficients. Finally, we calculate the contribution of retardation on the total cross section $\sigma^{2+}$ and we investigate the validity of Eq. (1) when retardation is considered.

## THEORY

The asymptotic theory for evaluating the ratio $R$ using Eq. (1) has been successful, even for obtaining the functional dependence of this ratio [12]. When we want to obtain angular distributions we must have a knowledge, not only of the initial ground-state $\psi_i(\mathbf{r}_1, \mathbf{r}_2)$, but also of the final double-continuum state wave function $\psi_f^-(\mathbf{r}_1, \mathbf{r}_2)$.

We consider the two electron photoionization of the He ground state by one linearly polarized photon with energy $E_\gamma$. The photon polarization $\hat{e}$ is assumed to be on the $z$-axis, while the photon momentum $\mathbf{k}_p$ is taken directed along the $x$-axis. We denote by $E_0$ the ground state energy of the helium

**TABLE 1.** Energy, cusp ratio and asymptotic ratio obtained using Eq. (1) of wave functions for the helium ground state.

| Wave function | $E_0$(a.u.) | $R_{cusp}$ | $R$ |
|---|---|---|---|
| GS1 | -2.875661 | -1.685 | 1.13% |
| GS2 | -2.901923 | -1.807 | 1.77% |
| SH | -2.902890 | -2.161 | 1.65% |
| Exact | -2.903724 | -2.000 | 1.67% |

atom and $\epsilon_1$ and $\epsilon_2$ the ejected electron energies. The total final energy of the process will be denoted $E_f = \epsilon_1 + \epsilon_2$.

The basic observable of the double photoionization process is the fivefold differential cross section (5DCS)

$$\frac{d^5\sigma^{2+}_{RET}}{d\epsilon_1 d\Omega_1 d\Omega_2} = \left(\frac{4\pi^2}{c}\right)\frac{k_1 k_2}{E_\gamma}|\hat{e} \cdot \mathbf{T}_{RET}(\mathbf{k}_1, \mathbf{k}_2|\mathbf{k}_p)|^2, \quad (3)$$

where $\mathbf{k}_1$ and $\mathbf{k}_2$ are the momenta of the electrons. Atomic units are used ($\hbar = m_e = a_0 = 1$). In Eq. (3) we have introduced the matrix element of the process including retardation (RET), given by

$$\mathbf{T}_{RET}(\mathbf{k}_1, \mathbf{k}_2|\mathbf{k}_p) = \langle \psi_f^- | e^{i\mathbf{k}_p \cdot \mathbf{r}_1} \nabla_1 + e^{i\mathbf{k}_p \cdot \mathbf{r}_2} \nabla_2 | \psi_i \rangle. \quad (4)$$

Formulas in the dipole approximation have been given in Ref. [11], however, the dipole velocity formulation is simply obtained from the preceding equations by setting $\mathbf{k}_p = 0$, and thus $\mathbf{T}_{DIP}(\mathbf{k}_1, \mathbf{k}_2) = \mathbf{T}_{RET}(\mathbf{k}_1, \mathbf{k}_2|\mathbf{k}_p = 0)$. In this article we will restrict our dipole calculations to the velocity form of the T-matrix, in which the error introduced in the final-state diminishes with increasing energy as $E_f^{-1}$, as argued by Dalgarno and Sadeghpour [13].

Integrating the 5DCS over the angles of electron 2 leads to the triple differential cross section (3DCS)

$$\frac{d^3\sigma^{2+}_{RET}}{d\epsilon_1 d\Omega_1} = \int d\Omega_2 \frac{d^5\sigma^{2+}_{RET}}{d\epsilon_1 d\Omega_1 d\Omega_2}, \quad (5)$$

which in the dipole approximation could be cast into the form [14]

$$\frac{d^3\sigma^{2+}_{DIP}}{d\epsilon_1 d\Omega_1} = \frac{1}{4\pi}\frac{d\sigma^{2+}_{DIP}}{d\epsilon_1}[1 + \beta(\epsilon_1)P_2(\cos\theta_1)]. \quad (6)$$

Clearly in this case the angular distribution of one electron could be specified by the unique parameter $\beta$, which has a simple physical meaning: (i) $\beta > 0$: electron ejected preferentially in the direction of $\hat{e}$, (ii) $\beta < 0$: electron ejected preferentially perpendicular to $\hat{e}$ and (iii) $\beta = 0$: isotropic electron.

In Table 1 we give some properties of the initial bound-state wave functions employed in this article. They are of the Hylleraas-type [15], denoted GS1 and GS2 (with 2 and 4 parameters, respectively), and of the CI-type [16], denoted SH (in a basis of *spdf* Slater-type orbitals).

Double-continuum wave functions are approximated by the product of three Coulomb waves (C3 model) and by the product of two Coulomb waves (C2 model). The C3 model satisfies the correct asymptotic boundary condition [17], it is given by:

$$\psi^-_{C3}(\mathbf{k}_1,\mathbf{k}_2|\mathbf{r}_1,\mathbf{r}_2) = \mathcal{P}_{12}\frac{1}{(2\pi)^3}e^{i\mathbf{k}_1\cdot\mathbf{r}_1+i\mathbf{k}_2\cdot\mathbf{r}_2}N(\xi_1)N(\xi_2)N(\xi_{12})F_1F_2F_{12}, \quad (7)$$

where $N(\xi) = \exp(-\pi\xi/2)\Gamma(1-i\xi)$ is the Coulomb factor and $F_j = {}_1F_1(i\xi_j, 1, -ik_jr_j-i\mathbf{k}_j\cdot\mathbf{r}_j)$ is the hypergeometric function. As usual, we denote: $\mathbf{r}_{12} = \mathbf{r}_1 - \mathbf{r}_2$, $\mathbf{k}_{12} = (\mathbf{k}_1 - \mathbf{k}_2)/2$, $\xi_1 = -Z_T/k_1$, $\xi_2 = -Z_T/k_2$, $\xi_{12} = 1/(2k_{12})$ and $Z_T = 2$ is the helium nuclear charge. In Eq. (7), $\mathcal{P}_{12} = (1+P_{12})/\sqrt{2}$, where $P_{12}$ is the exchange operator.

The C2 model is also referred as the independent particle approximation, because no interaction between the two electrons is incorporated, it is given by:

$$\psi^-_{C2}(\mathbf{k}_1,\mathbf{k}_2|\mathbf{r}_1,\mathbf{r}_2) = \mathcal{P}_{12}\frac{1}{(2\pi)^3}e^{i\mathbf{k}_1\cdot\mathbf{r}_1+i\mathbf{k}_2\cdot\mathbf{r}_2}N(\xi_1)N(\xi_2)F_1F_2. \quad (8)$$

## RESULTS AND DISCUSSION

We present in Fig. 1 our calculated asymmetry parameter $\beta$ at three different photon energies, $E_\gamma = 1.0$, 3.0 and 5.0 keV, using the C2 and C3 states to describe the double-continuum. We observed that, although not equal in the whole electron energy range, differences in the $\beta$ values are not very dramatic. This is due to the fact that, as the energy $E_\gamma$ increases, ground state correlation is able to provide a correct description of the process in the velocity gauge, as was pointed out in the previous section. We must point out that the differences observed at a given photon energy could not be adjudicated entirely to differences in C2 versus C3, and not in part due to differences in the ground state used, since we are not using the same ground state in both calculations. Note from Table 1 that the ground states employed, SH and GS2, have different characteristics and give different values for the asymptotic ratio.

It is interesting to see that we could characterize, in general, a slow electron as an electron emerging with $\beta < 0$, while a fast electron having $\beta > 0$.

At photon energies larger than 3.0 keV, the asymmetry parameter has the well defined limits: $\beta \to 0.0(2.0)$ as $\epsilon_1 \to 0(E_f)$. This implies that the slow electron with $\epsilon_1 \simeq 0$ is an isotropic electron, while the fast electron with $\epsilon_1 \simeq E_f$ is ejected mainly along the direction of the photon polarization. Since the fast electron has $\beta = 2.0$, it is clear that the angular distribution of this electron is the same as that of the single photoionization of He leaving

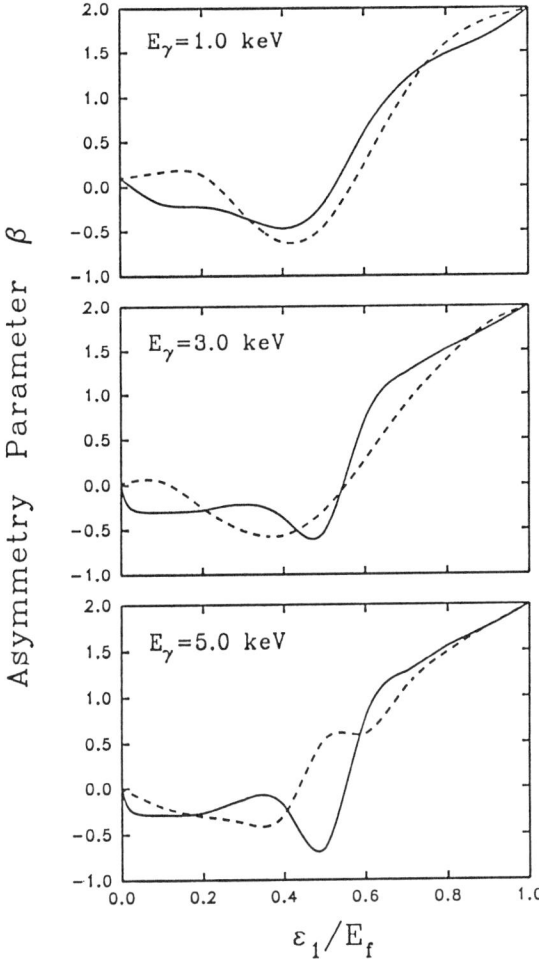

**FIG. 1.** The asymmetry parameter $\beta$ at three different photon energies. Solid line: calculation with C3-GS2 states. Dashed line: calculation with C2-SH states.

behind the He$^+$(1s) ground state. Note that when the single photoionization process leaves the He$^+$($ns$) excited state the angular distribution of the ejected electron has $\beta = 2.0$ as $E_\gamma \to \infty$ [18], which shows that the angular distribution of one electron escaping with almost all of the available energy is the same whether it is a single or a double photoionization process.

It is evident that the consideration of retardation will affect the angular distributions, and in this case the 3DCS will not be given by Eq. (6). Since our calculations include retardation to all orders in the multipole expansion, we can fit our results to a given formula. For the parametrization of our results, we have considered a first-order retardation correction, expected to

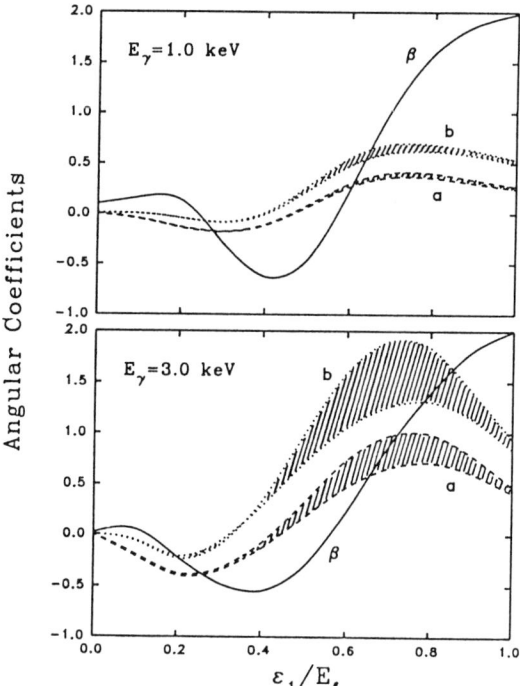

**FIG. 2.** Angular coefficients at two different photon energies, calculated with C2-SH states. The shaded area represents the uncertainty in the determination of the coefficients.

be valid for rather low energies, which includes two additional parameters $a(\epsilon_1)$ and $b(\epsilon_1)$, with the same form as the one that appears in the context of single photoionization [19]. It is given by

$$\frac{d^3\sigma_{\text{RET}}^{2+}}{d\epsilon_1 d\Omega_1} \simeq \frac{A(\epsilon_1)}{4\pi}\{1 + \beta(\epsilon_1)P_2(\cos\theta_1) + \sin\theta_1\cos\varphi_1[a(\epsilon_1) + b(\epsilon_1)P_2(\cos\theta_1)]\}. \tag{9}$$

Using the C2-GS1 states the parameters of this formula could be calculated analytically [20]. We note that $A(\epsilon_1) = d\sigma_{\text{DIP}}^{2+}/d\epsilon_1$ in the case Eq. (9) represents a good approximation for the 3DCS. One argument could be given to assert that the first-order retardation correction of the 3DCS is of the form given by Eq. (9). Since we have integrated over the directions of the electron 2, we have at our disposal to describe the angular distributions the vectors $\hat{e}$, $\mathbf{k}_1$ and $\mathbf{k}_p$, just as in the case of single photoionization, and as the correction $\sin\theta\cos\varphi[a + bP_2(\cos\theta)]$ is the more general one for a first-order in single photoionization, the 3DCS in double photoionization must have the same functional dependence.

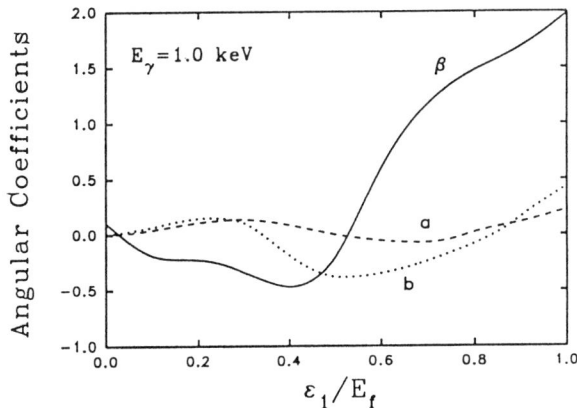

**FIG. 3.** Angular coefficients at the photon energy of 1.0 keV calculated with C3-GS2 states.

In Fig. 2 we present our calculated coefficients $a$ and $b$ using the models C2-SH for two photon energies 1.0 and 3.0 keV. Due to the fact that Eq. (9) could only be valid at rather low energies, a parametrization using different grid points gives different values for these coefficients; we indicate the error in the determination of the coefficients by a shaded area. We observe that, although for the case of 3.0 keV the formula is not very good, for electron energies less than 1.0 keV the error is very low and the parametrization given by Eq. (9) is reasonable.

In Fig. 3 we present the angular coefficients at 1.0 keV with C3-GS2 states. We note that, although they are not equal in the whole energy range with those calculated with C2-SH states, the limiting values at the electron energy endpoints are the same. We observe that $a = b = 0$ at $\epsilon_1 = 0$; the slow electron is an isotropic electron, even when we consider retardation. The slow electron is primarily produced by a shake-off process, so that retardation could not affect the angular distribution of this electron. Further, the production of the slow electron could be conceived in a physical picture as a transition from a high $n$-Rydberg level to a zero-energy continuum electron, since there is a continuity between $d\sigma^{2+}/d\epsilon_1)_{\epsilon_1=0}$ and the cross section $\sigma_n^+$ as $n \to \infty$ [21]. Clearly, in this transition, no trace of the photon momentum could be present.

For the fast electron ($\epsilon_1 \simeq E_f$) we obtain $\beta = 2$ and $b = 2a$, and in this case Eq. (9) adopts the form $\cos^2\theta_1(1 + a\sin\theta_1\cos\varphi_1)$, which shows that the retardation correction of the fast electron is that of an $s$ subshell electron in single photoionization [19].

We will finally discuss the contribution of retardation on the total cross section and on the ratio $R$. We have performed calculations of $\sigma^{2+}$ in the dipole approximation and with retardation into consideration [20] using the

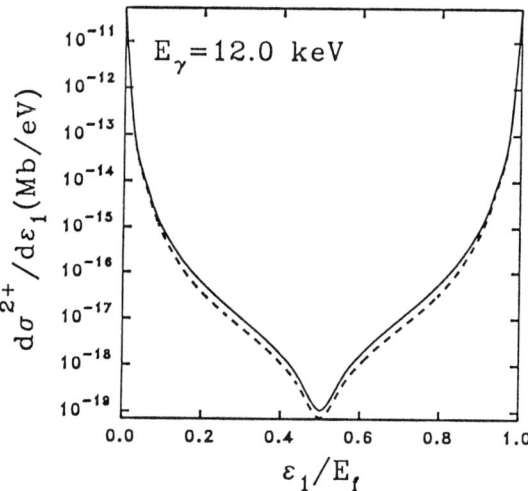

**FIG. 4.** The electron energy spectrum with retardation (solid line) and in dipole velocity approximation (dashed line) calculated using the states C2-SH at 12.0 keV.

C2-GS1 states. These calculations show that the contribution of retardation on the total cross section is of the order of $(v_m/c)^2$, where $v_m$ is the velocity of the fast electron. This shows that retardation gives a contribution, at the level of the total cross section, as if the process were that of single photoionization. Since the contribution of retardation on $\sigma^+$ is of the order $(v/c)^2$ [22], these contributions cancel in the ratio making $R_{\rm DIP} = R_{\rm RET}$. We are then lead to consider that Eq. (1) could be probably obtained considering the whole matter-radiation operator with retardation, although we do not have a proof on this subject.

A recent article [23] has claimed that there is a breakdown of the asymptotic ratio predicted by Eq. (1). This would imply that the sudden approximation, as we know it, should be reexamined. The author claims that the region of the spectrum where both electrons escape with nearly the same energy ($\epsilon_1/E_f \simeq 0.5$) tends to get importance as the energy increases, producing a breakdown of the asymptotic value of $R$. We have not observed this behaviour in our calculations. As an example, we have plotted in Fig. 4 the spectrum with retardation and in dipole velocity approximation for the states C2-SH at $E_\gamma = 12.0$ keV. We readily see that there is no such an effect in this region of the spectrum. So we disregard the analysis sustained in Ref. [23].

## ACKNOWLEDGMENTS

This work has been supported by CONICET. Discussions with R. H. Pratt and T. Surić are gratefully acknowledged.

# REFERENCES

1. J. C. Levin, D. W. Lindle, N. Keller, R. D. Miller, Y. Azuma, N. Berrah Mansour, H. G. Berry and I. A. Sellin, Phys. Rev. Lett. **67**, 968 (1991); J. C. Levin, I. A. Sellin, B. M. Johnson, D. W. Lindle, R. D. Miller, N. Berrah, Y. Azuma, H. G. Berry and D. -H. Lee, Phys. Rev. A **47**, R16 (1993).
2. J. A. R. Samson, C. H. Greene and R. J. Bartlett, Phys. Rev. Lett. **71**, 201 (1993).
3. K. Hino, P. M. Bergstrom and J. H. Macek, Phys. Rev. Lett. **72**, 1620 (1994).
4. L. Spielberger *et al.* Phys. Rev. Lett. **74**, 4615 (1995).
5. A. Dalgarno and A. L. Stewart, Proc. Phys. Soc. London **76**, 49 (1960).
6. F. W. Byron and C. J. Joachain, Phys. Rev A **164**, 1 (1967).
7. T. Åberg, Phys. Rev. A **2**, 1726 (1970).
8. T. Åberg, Ann. Acad. Sci. Fenn. AVI **308**, 1 (1969).
9. M. A. Kornberg and J. E. Miraglia, Phys. Rev. A **49**, 5120 (1994).
10. J. H. McGuire, Contributed Abstracts to ICPEAC XVIII, 1993, p. 50.
11. M. A. Kornberg and J. E. Miraglia, Phys. Rev. A **48**, 3714 (1993).
12. L. R. Andersson and J. Burgdörfer, Phys. Rev. Lett. **71**, 50 (1993).
13. A. Dalgarno and H. R. Sadeghpour, Phys. Rev. A **46**, R3591 (1992).
14. Z. -J. Teng and R. Shakeshaft, Phys. Rev. A **49**, 3597 (1994).
15. R. A. Bonham and D. A. Kohl, J. Chem. Phys. **45**, 2471 (1966).
16. N. Sabelli and J. Hinze, J. Chem. Phys. **50**, 648 (1969).
17. M. Brauner, J. S. Briggs and H. Klar, J. Phys. B **22**, 2265 (1989).
18. R. Wehlitz, B. Langer, N. Berrah, S. B. Whitfield, J. Viefhaus and U. Becker, J. Phys. B **26**, L783 (1993).
19. A. Bechler and R. H. Pratt, Phys. Rev. A **42**, 6400 (1990).
20. M. A. Kornberg and J. E. Miraglia, submitted to Phys. Rev. A.
21. Z. Fan, H. R. Sadeghpour and A. Dalgarno, Phys. Rev. A **50**, 3174 (1994).
22. R. H. Pratt, A. Ron and H. K. Tseng, Rev. Mod. Phys. **45**, 273 (1973).
23. E. G. Drukarev, Phys. Rev. A **51**, R2684 (1995).

# Experimental Separation of Photoabsorption and Compton Scattering Contributions to He Single and Double Ionization

L. Spielberger*, O. Jagutzki, R. Dörner, J. Ullrich[+], U. Meyer,
V. Mergel, M. Unverzagt, M. Damrau, T. Vogt, I. Ali,
Kh. Khayyat, D. Bahr[‡], H.G. Schmidt[‡], R. Frahm[‡], and
H. Schmidt-Böcking

*Institut für Kernphysik, Universität Frankfurt, D60486 Frankfurt/M., Germany,*
[+] *GSI, D64220 Darmstadt, Germany,* [‡] *HASYLAB am DESY, D22603 Hamburg, Germany*

---

We have experimentally separated the contributions of photoabsorption and Compton scattering to He single and double ionization for high-energy photon impact by measuring the full momentum vector of the recoiling $He^{1+,2+}$ ions. For recoil ions following photoabsorption large momenta and a distinct dipole emission pattern are observed. The ions produced by Compton scattering show small momenta. For the ratio of double to single ionization we find $(1.22 \pm .06)\%$ at $8.8^{+1.5}_{-1.65}$ keV for Compton scattering and $(1.72 \pm .12)\%$ at $7.0^{+2.1}_{-1.6}$ keV for photoabsorption. We compare our data with recent theories.

---

## INTRODUCTION

The interaction of a photon with an atom is described by a single particle operator. Thus double ionization is always mediated by electron–electron correlation. Therefore the description of He double ionization induced by photons is one of the most crucial tests of our understanding of these correlation effects. At small photon energies below 1 keV considerable success has been achieved in the realization of kinematical complete experiments and their theoretical description [1–3]. At these low energies absorption of the photon is the dominating ionization mechanism. In the high energy regime (above about 8 keV) Compton scattering is the leading process, the respective single ionization cross sections getting equal around 6 keV photon energy [4–8].

In the high-energy regime even the ratio of total cross sections for double to single ionization ($R = \sigma^{2+}/\sigma^{1+}$), which is predicted to converge with increasing energy to a constant value, is not yet well established [9]: Only a few experimental results on R have been reported [10,11]. Unfortunately these experiments were not able to distinguish between photoabsorption and

Compton scattering. From the theoretical side agreement has been achieved [12] in predicting the asymptotical value for $R_{Ph}$ to be 1.67% [13–15,4]. For Compton scattering different theoretical predictions for the high-energy limit as well as for the energy dependence of $R_C$ exist [8,12,16]. At 8 keV, where the present experiments were performed, the predictions for $R_C$ differ by more than a factor of two between 0.6% [8], 1.3% [12] and 1.65% [7].

Here we present an experimental approach which allows to distinguish both ionization mechanisms. Since ions from Compton events and photoionization are distinct in momentum space measuring the recoil momentum vector of the ions simultaneously with their charge state provides for the first time a separate determination of the ratios $R_C$ and $R_{Ph}$. We present these ratios determined at a photon energy of about 8 keV. We find a close connection between the Compton profile [17] and the single ionization recoil ion momentum distribution at low momenta.

## EXPERIMENT

In order to determine $R_{Ph}$ and $R_C$ separately not only the recoil ion charge state but also the recoil ion momentum was measured in the present experiment using COLTRIMS (COLd Target Recoil Ion Momentum Spectroscopy). The apparatus used is described in detail elsewhere [18–20]. It consists of a supersonic He gas jet, that provides a dense, localized and internally cold He target. The recoiling He ions created in the intersection volume of photon beam and target jet are extracted by a well defined homogeneous electrostatic field onto a position sensitive detector. Since *all* ions are projected on the detector the full $4\pi$ solid angle is obtained for both He charge states. The gas jet atoms have an offset momentum of 6 a.u. in y direction resulting from the supersonic expansion. The internal momentum spread of the target is below 0.2 a.u., which corresponds to a temperature below 0.1 K. The ion time of flight (TOF) is determined by measuring their timing signal with respect to the beam pulse provided from the storage ring. From the TOF we obtain the recoil ion charge state and the momentum component parallel to the field direction. The two momentum components perpendicular to the extraction field are calculated from the position on the channel-plate detector (see FIG. 1) and the TOF.

The experiment has been performed at the undulator beamline BW1 of HASYLAB at DESY in Hamburg. Two experiments have been carried out at the same time at this beamline. The main experiment used a small band pass of x-rays reflected out of the direct beam by a Be single crystal monochromator of 0.25 mm thickness operated in Laue transmission geometry. Our experiment was performed with all photons passing in forward direction through the Be crystal. Thus, the radiation used in this experiment was not monoenergetic. The photon spectrum is shown in FIG. 2. It was measured after the experiment with a Si (111) single crystal in $\Theta - 2\Theta$ geometry. The spectrum is dominated by the third harmonic of the undulator which was set to 9.2 keV. It is cut below 5 keV by the absorption of the beam line windows

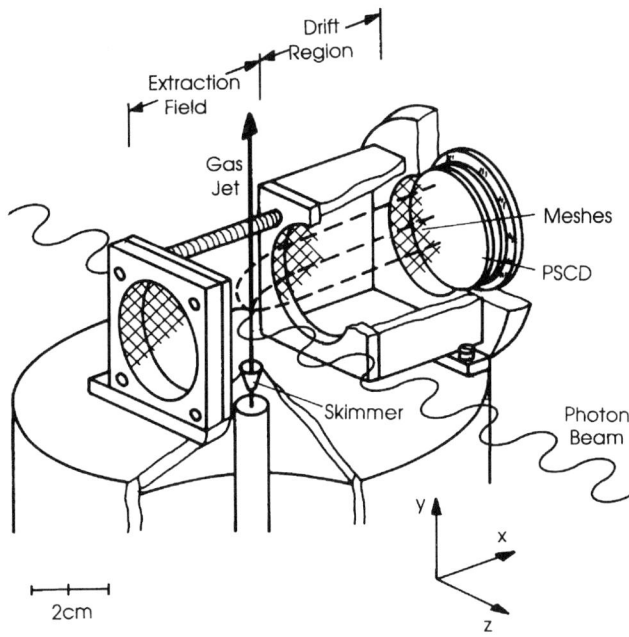

**FIG. 1.** Recoil-ion momentum spectrometer with supersonic He gas jet. The recoil ions have been measured in coincidence with the beam pulse, the channel-plate detector (PSCD) is two dimensional position sensitive with wedge and strip readout. The electric field vector of the linear polarized light is parallel to the extraction field.

(Carbon, Beryllium, and Aluminium) and the Be crystal, and on the high energy side at about 11 keV by two Au coated mirrors. These mirrors were used to focus the light in the vertical and horizontal direction. A beamspot of less than 1×2 mm was achieved by two sets of adjustable slits about 1 and 1.5 m upstream the collision region. The experiment was performed at a flux of about $1\times10^{14}$ photons/sec over the whole spectrum yielding about 80 $He^{1+}$ ions/sec.

## REMARKS ON THE RECOIL ION KINEMATICS

The momentum balances for the different reactions investigated are:

$$\vec{p}_\gamma + \vec{p}_{He} = \vec{p}_{He^{1+}} + \vec{p}_e \quad \text{(Photo, He}^{1+}\text{)}$$
$$\vec{p}_\gamma + \vec{p}_{He} = \vec{p}_{He^{2+}} + \vec{p}_{e1} + \vec{p}_{e2} \quad \text{(Photo, He}^{2+}\text{)}$$
$$\vec{p}_\gamma + \vec{p}_{He} = \vec{p}_{\gamma'} + \vec{p}_{He^{1+}} + \vec{p}_e \quad \text{(Compton, He}^{1+}\text{)}$$
$$\vec{p}_\gamma + \vec{p}_{He} = \vec{p}_{\gamma'} + \vec{p}_{He^{2+}} + \vec{p}_{e1} + \vec{p}_{e2} \quad \text{(Compton, He}^{2+}\text{)}$$

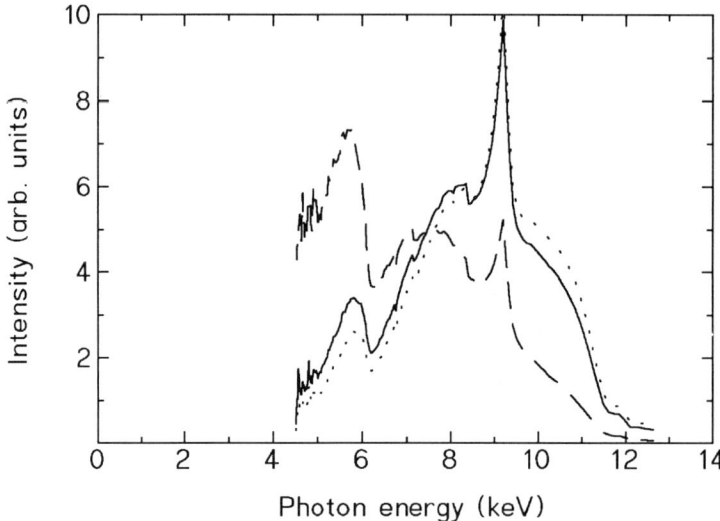

**FIG. 2.** Full line: photon energy distribution as used in the experiment. Dashed line: photon energy distribution folded with the $E^{-7/2}$ dependence of the $He^{1+}$ photoionization cross section [21,15], dotted line: photon energy distribution folded with calculated Compton scattering cross sections [4,5,7].

The incoming photon momentum, $\vec{p}_\gamma$, has a value of 2.3 a.u. at 8 keV energy, the momentum of the incoming target atom $\vec{p}_{He}$ is zero (within the above described properties of the supersonic gas jet). The final state momenta of the scattered photon, the emitted electrons and the recoil ion ($\vec{p}_{\gamma'}$, $\vec{p}_{e_i}$, $\vec{p}_{He^{1+,2+}}$) must compensate the incoming photon momentum.

For single ionization by photoabsorption at 8 keV the electron leaves with a momentum of 24 a.u.. Therefore the recoil ion momentum vector given by $\vec{p}_{He^{1+}} = \vec{p}_e - \vec{p}_\gamma$ ends on a sphere with radius of 24 a.u. shifted forward by $\vec{p}_\gamma$ in the laboratory frame. The recoil ion density distribution on the sphere must reflect the dipolar electron emission pattern due to the linear polarisation of the synchrotron radiation. According to theoretical investigations [22,23] higher orders than the dipole portion contribute only with less than 2% in this experiment.

In a Compton scattering process the nucleus acts a spectator, during the ionization process almost no momentum is transfered to the ion. Thus the recoil ion momentum represents essentially that of the nucleus in the initial state which is the compensated target electron momentum distribution. Therefore recoil ions following Compton scattering are expected with small final momenta of about 1 a.u.. Samson et al. have recently used this fact to measure the cross section for $He^{1+}$ production by Compton scattering [24].

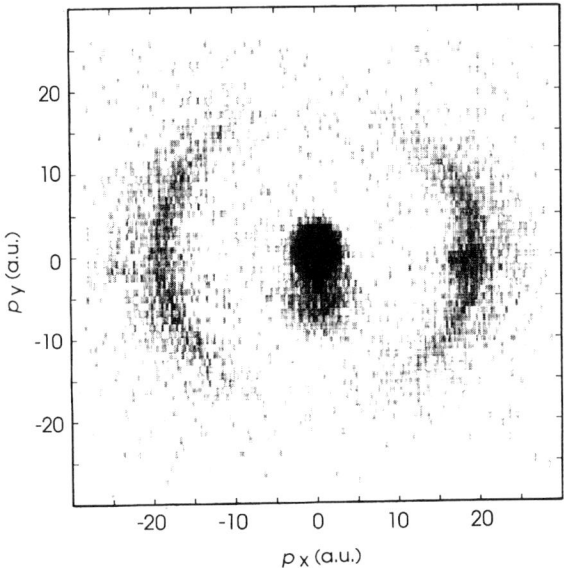

**FIG. 3.** Measured momentum distribution of He$^{1+}$ ions in the x-y plane, integrated over the momentum in z direction between -30 and 30 a.u.

## RESULTS

The measured momentum distribution of the He$^{1+}$ ions is shown in FIG. 3. It is plotted in the $p_x$-$p_y$ plane, where x is the direction of the electric field vector of the linear polarized light, y is the one of the He gas jet and z is the one of the photon beam. For the geometry see FIG. 1. The data of FIG. 3 are summed over momenta in z direction $-30$ a.u. $< p_z <$ 30 a.u. Due to the rotational symmetry of the system the distribution must be symmetrical with respect to the $p_x$-axis. Ions resulting from the residual He gas atoms can be seen in FIG. 3 at $p_y \approx -6$ a.u.

The distinct recoil ion momentum patterns described above can clearly be observed in this spectrum: the one close to zero momentum due to Compton scattering and the distribution along the sphere populated with a dipole distribution pattern from photoabsorption. The larger width of the sphere is mainly due to the energy spread of the incoming photons. The dashed line in FIG. 2 represents the photon energy distribution folded with the $E^{-7/2}$ dependence of the photoabsorption cross section [21,15]. After integrating over the full momentum sphere twice as many He$^{1+}$ ions from Compton scattering as from photoabsorption are observed in the experiment. This is in good agreement with the result calculated on the basis of the photon energy distribution and the calculated absolute cross sections for photoabsorption ($\sigma_{Ph}^{1+}$) and Compton scattering ($\sigma_{C}^{1+}$) [4,5,7].

Projections of the full momentum patterns on the x plane for single and

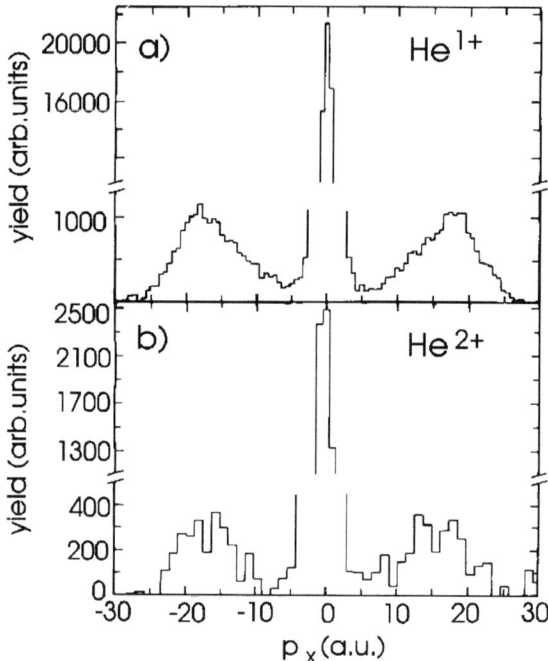

**FIG. 4.** Measured recoil ion momentum distributions in the x plane. (a) single ionization, (b) double ionization. Intergration over $p_z$ as in FIG. 3

double ionization are shown in FIG. 4. The $He^{2+}$ momentum distribution is found to be very similar to the one for single ionization. This is in agreement with previous findings that at this high energy the dominant $He^{2+}$ production process for photoabsorption is the emission of one very fast and one slow electron [23].

By integrating over the respective volumes in momentum space we obtain $R_C$ and $R_{Ph}$ separately. The results are given in Table I and displayed in FIG. 5. The respective energy is the mean value of the folded photon spectrum (see FIG. 2). The horizontal error bars indicate the spectral region from which 67% of all $He^{1+}$ ions originate. Due to the high photon flux the statistical error for $R_C$ is neglectibly small (0.01%). By our technique we are also able to definitely avoid the three major systematical errors of all former high-energy studies, which are small contributions from low-energy stray light,

|  | Photon energy | $R = \sigma^{2+}/\sigma^{1+}$ |
|---|---|---|
| Photoabsorption | $7.0^{+2.1}_{-1.6}$ keV | 1.72±.12% |
| Compton scattering | $8.8^{+1.5}_{-1.65}$ keV | 1.22±.06% |

**TABLE 1.** Ratios of double to single ionization.

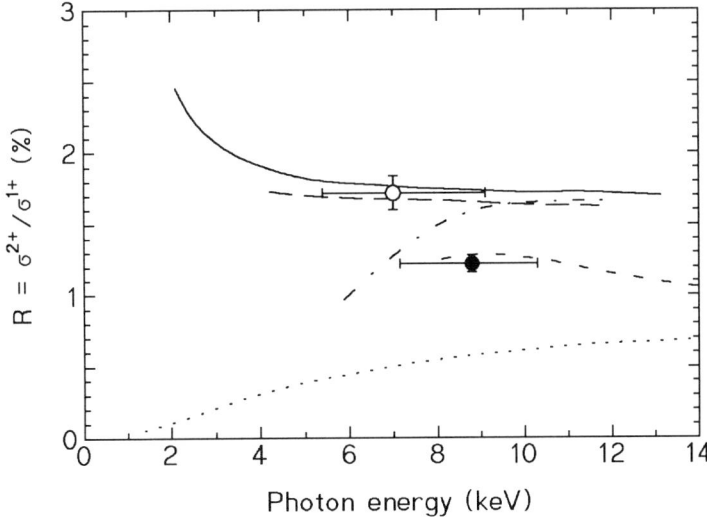

FIG. 5. Ratios of double to single ionization. Full circle: experimental ratio for Compton scattering ($R_C$), open circle: experimental ratio for photoabsorption ($R_{Ph}$). The data have been measured with light of an energy distribution as given in FIG. 2. The horizontal error bars indicate the energy region from which 67% of all He$^{1+}$ ions result (based on cross sections from [21,4,5,7]). Full line: $R_{Ph}$ from [4], broken line: $R_{Ph}$ from [7], dashed line: $R_C$ from [12], dotted line: $R_C$ from [8], dash-dotted line: $R_C$ from [7].

ionization by secondary electrons and charge exchange of the ions with the target gas [11,10]. Low energetic photons would cause photoabsorption and would yield smaller recoil ion momenta. Secondary electrons would not be restricted to the path of the photon beam and therefore result in He ions created along the gas jet and not only at the intersection point with the photon beam, which can be separated with the help of the detector position. The well localized gas jet and the good background pressure of $2 \cdot 10^{-7}$ hPa prevent secondary collisions of the ions resulting in charge exchange. Thus, the only remaining systematical error could be a charge state dependence of the detection efficiency of the channel-plate. This was checked two-fold: First the amplification of the channel-plate was reduced by a factor of two (the pulse height of the channel-plate signal is recorded for each event) by reducing the overall operation voltage by 100 V. Second, measurements were performed with two different postacceleration voltages of 2000V and 1000V. Both tests yielded the same $R_C$ within 5%. The error given in table I is the sum of our estimate of the systematical error based on these cross checks and the statistical error. The one of $R_{Ph}$ is somewhat larger and mainly due to the background subtraction for the He$^{2+}$ charge state. Ionization of

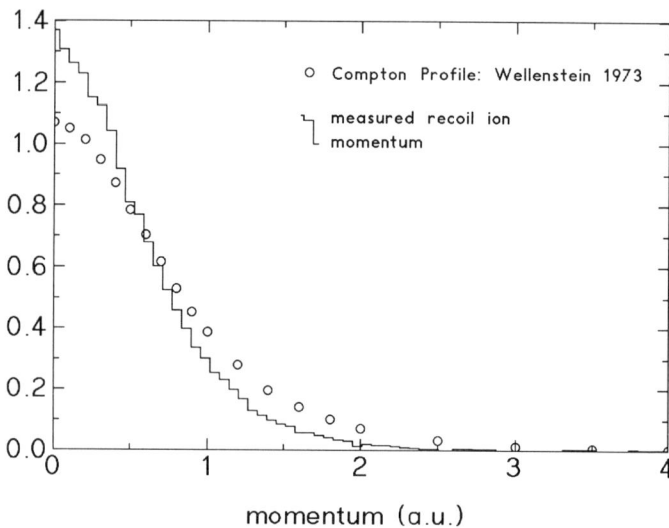

**FIG. 6.** $He^{1+}$ momentum distribution in the x plane around zero in comparison with the Compton profile [25]. The area under the recoil ion momentum distribution is normalized to that under the Compton profile.

the residual gas was the main contribution for the total recoil ion production, only about 4% of all events result from helium photoabsorption. These counts are distributed over a large volume in momentum space making background subtraction much more difficult than for the Compton effect.

For photoabsorption our experiment is in excellent agreement with the prediction of $R_{Ph}=1.67\%$ by different authors [13–15,4]. For Compton scattering our measured ratio is close to the prediction of Andersson and Burgdörfer and we can definitely rule out the values of $R_C=0.6\%$ as predicted by Surić et al. [8] and of $R_C=1.65\%$ as predicted by Hino et al. [7]. According to theoretical estimates [4] we have already reached the high-energy asymptotic value for $R_{Ph}$. This result is supported by the value $R=(1.6 \pm 0.3)\%$ found by Levin et al. [11] at 2.8 keV which agrees with our finding. This value represents only photoabsorption because at this energy the Compton cross section is in the order of 1% of the one for photoabsorption [4]. For Compton scattering the emitted electron energy is still significantly below 500eV and we expect not yet to have reached the asymptotic regime. Therefore, it cannot yet be determined if $R_C$ and $R_{Ph}$ have the same value in the high-energy limit as predicted by Amusia [16].

The $He^{1+}$ momentum distribution in x direction for Compton processes is presented in FIG. 6 in an enlarged scale in comparison with the Compton profile [25]. Since in this experiment only the momentum vector of the recoil ion is determined, the value of momentum transfer $K = \vec{p}_{\gamma'} - \vec{p}_{\gamma}$ is not known. Thus, the recoil ion momentum distribution represents the "Compton profile integrated over all momentum transfers $K$". In particular the included regime of small $K$ will affect the recoil ion momentum distribution. The efficient post collison interaction between the slow ejected electron and the ion will result in a shift of the momentum distribution towards smaller values, as it can be seen in our spectrum. A similar result was found for electron impact [26].

For a further investigation an experiment in the photon energy region from 50 to 100 keV is presently in preparation to reach the asymptotic limit for Compton scattering. A coincident detection of an emitted electron or the scattered photon will allow to determine $K$ and thus provide detailed information on the one-electron Compton profile and, in comparison to this, on the two-electron sum profile after double ionization.

In conclusion, we have experimentally demonstrated that photoabsorption and Compton scattering yield clearly distinct recoil ion momentum distributions. We have exploited this difference in momentum space to experimentally separate the processes for single and double ionization of He at a photon energy of about 8 keV. We find the ratio of double to single ionization by photoabsorption and Compton scattering to be different at this energy.

## ACKNOWLEDGMENTS

The work was financially supported by BMFT, DFG, and the EC. These experiments would not have been possible without the patient acceptance of us by the main synchrotron radiation users, namely K. Kjær, G. Brezesinski, L. Leiserowitz, and coworkers. We acknowledge helpful discussion with A. Lahmam-Bennani, C.L. Cocke, J. Burgdörfer, P. Bergstrom, T. Surić, M. Amusia, R. Dreizler, H.J. Lüdde and J.R. Samson. We received indispensable help in building the supersonic gas jet from U. Buck.

## REFERENCES

* e-mail: spielberger@ikf.uni-frankfurt.de
1. O. Schwarzkopf et al. *Phys. Rev. Lett.*, 70:3008, 1993.
2. A. Huetz et al. *J. Phys.*, B27:L13, 1994.
3. F. Maulbetsch and J.S. Briggs. *J. Phys.*, B26:L647, 1993.
4. L.R. Andersson and J. Burgdörfer. *Phys. Rev. Lett.*, 71:50, 1993.
5. P.M. Bergstrom, Jr., K. Hino, and J. Macek. *Phys. Rev.*, A51:3044, 1995.
6. J.A.R. Samson, C.H. Green, and R.J. Bartlett. *Phys. Rev. Lett.*, 71:201, 1993.
7. Ken-ichi Hino, P.M. Bergstrom, and J.H. Macek. *Phys. Rev. Lett.*, 72:1620, 1994.
8. T. Surić et al. *Phys. Rev. Lett.*, 73:790, 1994.
9. A. Dalgarno and H.R. Sadeghpour. *Comm. At. Mol. Phys.*, 30:143, 1994.
10. J.C. Levin et al. *Phys. Rev.*, A47:R16, 1993.
11. J.C. Levin et al. *Phys. Rev. Lett.*, 67:968, 1991.
12. L.R. Andersson and J. Burgdörfer. *Phys. Rev.*, A50:R2810, 1994.
13. F.W. Byron and C.J. Joachain. *Phys. Rev.*, 164:1, 1967.

14. T. Åberg. *Phys. Rev.*, A2:1726, 1970.
15. A. Dalgarno and H.R. Sadeghpour. *Phys. Rev.*, A46:R3591, 1992.
16. M. Ya. Amusia and A.I. Mikhailov. *J. Phys.*, B28:1723, 1995.
17. M. Inokuti. *Rev. Mod. Phys.*, 43:297, 1971.
18. J. Ullrich et al. *Comm. At. Mol. Phys.*, 30:285, 1994.
19. R. Dörner et al. *Phys. Rev. Lett*, 72:3166, 1994.
20. V. Mergel et al. *Phys. Rev. Lett*, 20:2200, 1995.
21. T. Ichihara, K. Hino, and J.H. McGuire. *Phys. Rev.*, A44:R6980, 1991.
22. R. H. Pratt and L. LaJohn. private communication.
23. M.Ya. Amusia et al. *J. Phys.*, B8:1247, 1975.
24. J.A.R. Samson et al. *Phys. Rev. Lett.*, 72:3329, 1994.
25. H.F. Wellenstein and R.A. Bonham. *Phys. Rev.*, A7:1568, 1973.
26. O. Jagutzki et al. accepted by *Z. Phys. D*

# SELECTED TOPICS

## – ELECTRONS –

Theoretical and Experimental Investigation of
Electron-Helium Scattering .................................... 785
    I. BRAY, D.V. Fursa, D.T. McLaughlin,
    B.P. Donnelly, and A. Crowe

Spin Effects in (e,2e) Collisions ................................ 795
    X. Guo, J. Hurn, J. LOWER, S. Mazevet,
    Y. Shen, I.E. McCarthy, and E. Weigold

Measurement of Exchange and Spin-Orbit Effects
and their Interference in Elastic e-Cs Scattering .............. 805
    M. TONDERA, G. Baum, P. Baum, L. Grau,
    B. Leuer, R. Niemeyer, and W. Raith

Studies of Electron-Molecule Scattering
at Microelectronvolt Energies
Using Very-High-$n$ Rydberg Atoms ........................... 815
    M.T. FREY, S.B. Hill, K.A. Smith,
    F.B. Dunning, and I.I. Fabrikant

On the Ionisation Mechanism of Reflection (e,2e) Events ..... 825
    S. IACOBUCCI, P. Luches, L. Marassi, R. Camilloni,
    B. Marzilli, S. Nannarone, and G. Stefani

Recombination of $H_3^+$ and $D_3^+$ Ions with Electrons ............ 835
    R. JOHNSEN, T. Gougousi, and M.F. Golde

# Theoretical and experimental investigation of electron-helium scattering

Igor Bray[1], Dmitry V. Fursa,

*Electronic Structure of Materials Centre, The Flinders University of South Australia, G.P.O. Box 2100, Adelaide 5001, Australia*

Damien T. McLaughlin, Brendan P. Donnelly and Albert Crowe

*Department of Physics, University of Newcastle upon Tyne, NE1 7RU, UK*

We present both calculations and measurements of the $3^{3,1}D$ state charge clouds after 30 eV electron-impact excitation of the ground state of helium. The measurements are obtained using the polarization-correlation technique. The theory is the convergent close-coupling (CCC) theory. Agreement between theory and experiment is very good indicating that the CCC theory is able to treat the exchange and continuum effects very accurately.

## I. INTRODUCTION

Electron-helium scattering is of interest in diverse applications and to fundamental scattering theory as one of the simplest examples of a Coulomb four-body problem. In order to be able to obtain useful scattering data we desire a general scattering theory that may obtain accurate results irrespective of the incident projectile energy or the scattering process of interest. Unlike hydrogen, the structure problem for helium is non trivial and requires careful consideration before attempting electron scattering calculations. From the experimental side, the helium atom is easier to work with than hydrogen, allowing for some of the most detailed measurements yet performed in electron scattering experiments [1–7]. These measurements are of paramount importance to the theorists since they serve as much needed stringent tests of the methods of calculation.

Until recently, the experimental information available for the electron-helium scattering problem invalidated all available scattering theories. Some of these such as the distorted-wave Born approximation (DWBA) [8] or the First Order Many-Body theory (FOMBT) [9] yield good cross sections for the dipole allowed excitation processes at intermediate and large energies, but

---

[1] electronic address: igor@esm.ph.flinders.edu.au

often have difficulties for exchange transitions and at low energies. R-matrix calculations e.g. [10] treat only the discrete spectrum with up to 29 states, and so are formally invalid above the ionization threshold, yielding identically zero for the total ionization cross section. Furthermore, inability to treat the target continuum implies an inability to reproduce the target dipole polarizability and hence the elastic cross section. The unitarity of the theory ensures that a wrong elastic amplitude is likely to yield an incorrect total cross section, which contributes to the errors in the magnitudes of individual integrated cross sections irrespective of the projectile energy. None of these deficiencies apply to the convergent close-coupling (CCC) method [11], which in our view is the only method that has not been invalidated by experiment. By treating both exchange and continuum effects accurately, a single CCC calculation for a given projectile energy yields elastic, excitation, ionization (singly-, doubly-, and triply-differential), and total cross sections. All of these may, and have been, tested against available experiment.

The purpose of this paper is to give an overview of the complementary theoretical and experimental work of our two research groups. The combination of the latest theoretical and experimental techniques has proved to be very fruitful in an effort to study electron-helium scattering.

## II. CCC THEORY

The CCC method was introduced by Bray and Stelbovics [12] in an effort to address the long-standing discrepancies between theory and experiment in the fundamental electron-hydrogen scattering problem. The discrepancy, in the language of Andersen, Gallagher, and Hertel [13], involves the description of the 2p charge cloud at large scattering angles after electron-impact excitation of the hydrogen ground state by 54.4 eV electrons. Unfortunately, the CCC theory yielded much the same results as most other theories. However, unlike other theories at the time the CCC method does not incorporate any substantial approximation for the e-H scattering problem. The method solves the non-relativistic Schrödinger equation governing the scattering to a precision determined by the convergence in the calculations with increasing number of states used in the close-coupling formalism. As these states are obtained from a truncated Laguerre basis we can be confident that completeness in the expansion of the total wave function is approached with increasing number of states. As such it is our view that on this occasion it is likely that theory is more accurate than the very difficult experiment.

Though the above-mentioned problem remains to this day, the CCC method has been able to claim successes in many other areas that support the validity of the CCC approach. These include the ability to obtain very accurate total ionization cross sections in the case of hydrogen [14], helium ion [15], and helium [11]. Even when the existing measurements didn't support the CCC theory, in the case of the sodium total ionization cross section [16], later measurements did so [17].

Prior to the application of the CCC method to helium the most spectacular demonstration of the ability of the theory to quantitatively describe the target charge cloud after electron-impact excitation was the application to sodium [18]. Development of the method for the helium target has enabled us to test the theory much more extensively than what was possible with the quasi one-electron targets. The latter applications have been reviewed in Ref. [19].

The details of the CCC theory for electron-helium scattering may be found in Ref. [11]. The extension to two-electron targets from one-electron targets [18] is quite substantial. It first involves the generation of the helium target states. This is done by taking an explicitly antisymmetric two-electron basis comprised of Laguerre functions. Diagonalizing the target Hamiltonian in this Sturmian-type basis results in both positive- and negative-energy states. These states span subsets of the helium true discrete and continuum spaces. The larger the Laguerre basis sizes the larger these subsets become. In the limit of infinite basis sizes the full target space is spanned by the two-electron Laguerre basis.

In the case of the helium target the discrete subspace contains only one-electron excitations, and is very well modeled by the frozen-core approximation [20], with the largest energy-level error being of order of 3% in the ionization energy of the ground state. Therefore, though we are able to treat the case of two electron excitation, we begin with the simple frozen-core model, where one electron is constrained to be the 1s orbital of $He^+$. This allows us to provide a simple convergence study by extending the target state description in only one-electron space. Fon et al. [10] critisize this model, yet their multi-configuration expansion leads to a worse description of the helium target structure (5% error in the ionization energy) than that of the frozen-core model [11]. Given this fact we suspect that the non-frozen-core configurations play an insignificant role in their structure calculation, while convergence in the frozen-core configurations has not been achieved. As a result we view their structure calculation as an approximation to the frozen-core model. In our view both structure approximations are sufficient for the description of electron-helium excitation at energies away from excitation thresholds. The issue then is how to correctly perform the scattering calculation.

The $N$ target states, obtained after diaginalizing the target Hamiltonian, are all square-integrable with the first few $\Phi_{nls}^N$ negative-energy states being good approximations to the true target eigenstates $\Phi_{nls}$. In our typical e-He calculations [11] this is the case for $n \leq 4$. The remaining negative-energy states would converge to the corresponding $\Phi_{nls}$ with increasing $N$, whereas the positive-energy $\Phi_{nls}^N$ provide a discretization of the true target continuum. In the limit of large $N$ the sum over the positive-energy $\Phi_{nls}^N$ is equivalent to integrating over the true target continnum [21].

The square-integrability of the $N$ states allows their incorporation within the close-coupling (CC) formalism. Formally, it doesn't matter how the close-coupling equations are solved. This may be done in configuration space, where the CC equations take the form of coupled integro-differential equations involving oscillatory wave functions, or in momentum space, where we have integral equations involving smoothly varying Born-like amplitudes [19]. We choose the latter approach as we are interested in coupling as many states as possible. The reason for this is simple. We need to establish convergence for each $ls$ symmetry. Within each symmetry we have to reproduce the effect of a sum over the infinite set of discrete states and an integral over the continuum. The existence of helium singlet ($s = 0$) and triplet ($s = 1$) $l$-states doubles the number of coupled states as compared to hydrogenic targets. For example, if we wish to include up to $F$ states in the CC formalism, taking just 10 states for each $ls$ symmetry, say 5 with negative energy and 5 with positive energy, results in as many as 80 states.

Thus it is clear that large-scale close-coupling calculations may be necessary when describing electron-helium scattering. In practice, the size of the calculation depends substantially on the scattering process of interest and the required accuracy. For example, elastic scattering does not require many $F$- or $D$-states in the close-coupling formalism. On the other hand, if we are interested in excitation of the ground state to a $D$-state, then we require a good representation of the $D$- and $F$-states. It is worthwhile noting that the size of the calculation varies with the number of channels, with a state of orbital angular momentum $l$ leading to $l+1$ channels. So inclusion of high-$l$ states rapidly increases the size of the calculations.

The formulation of the CCC method has reduced the electron-atom scattering problem to simply establishing convergence, with increasing number of states, in the scattering process of interest. The level of convergence indicates the accuracy of the calculations. Note that we establish convergence by simultaneously increasing the number of negative- and positive-energy states. This is not the same as establishing convergence in just the discrete subspace as is the case in standard close-coupling calculations.

## III. EXPERIMENT

The polarization correlation method is used in which the polarization (Stokes parameters) of the radiation from decay of the excited states is measured in coincidence with the scattered electrons which produced the same excited states. Hence for a particular scattering angle only radiation from identically prepared states is observed.

The apparatus is of the crossed electron atom beams type. A well defined beam of electrons ($10^{-7} - 10^{-6} A$) crosses a beam of helium atoms from a long narrow capillary. Scattered electrons which have excited the $n=3$ states are selected by a $180°$ hemispherical analyser and detected by a channel electron multiplier. The analyser is rotatable about the collision centre. Details of these aspects of the apparatus have been given by Crowe $et$ $al$ [22].

It is convenient to discuss the apparatus and the measurements performed in terms of the natural collision frame where the plane formed by the incident and scattered electron momenta is the $x - y$ plane, $x$ corresponding to the incident electron direction. Three Stokes parameters are measured for the visible radiation emitted perpendicular ($+z$-direction) to the scattering plane. Linear polarizations $P_1$ and $P_2$ defined by

$$P_1 I_z = I_z(0°) - I_z(90°) \tag{1}$$

and

$$P_2 I_z = I_z(45°) - I_z(135°), \tag{2}$$

where $I_z$ is the total photon intensity in the $z$-direction and $I_z(\alpha)$ the intensity transmitted by a linear polarizer, are measured. Addition of a quarter-wave plate enables the circular polarization $P_3$ to be measured, where

$$P_3 I_z = I_z(RHC) - I_z(LHC). \tag{3}$$

For the D states discussed here, a fourth Stokes parameter $P_4$ is measured corresponding to the linear polarization of the radiation emitted in the $y$-direction i.e. in the scattering plane and perpendicular to the electron beam direction

$$P_4 I_y = I_y(0°) - I_y(90°). \tag{4}$$

Narrow band interference filters are used to isolate the transitions of interest ($3\,^1D - 2\,^1P$ and $3\,^3D - 2\,^3P$). The radiation is detected by fast high-gain photomultipliers. A detailed discussion of the coincidence data accumulation method used, together with optical alignment and consistency checks are given in Refs. [23] and [4].

## IV. COMPARISON OF THEORY AND EXPERIMENT

A detailed comparison of the e-He CCC theory and various measurements and calculations at projectile energies of 1.5 to 500 eV may be found in Ref. [11]. In that work convergence was tested by presenting 69- and 75-state calculations. The former differered from the latter in that it had more $S$, $P$, and $D$ states but no $F$ states. Good agreement between the two indicated the accuracy of the CCC calculations. Here we take a slightly different approach to validating the CCC theory. Rather than applying the theory at many energies, we take only one, which we choose to be in the most difficult intermediate energy region. Of the many measurements available for comparison we take just those that are the most difficult for theory. These are the measurements of the charge-cloud distribution of the $3^{3,1}D$ state excitation. Good agreement with such data is only possible if issues such as treatment of exchange and the continuum are addressed correctly [24].

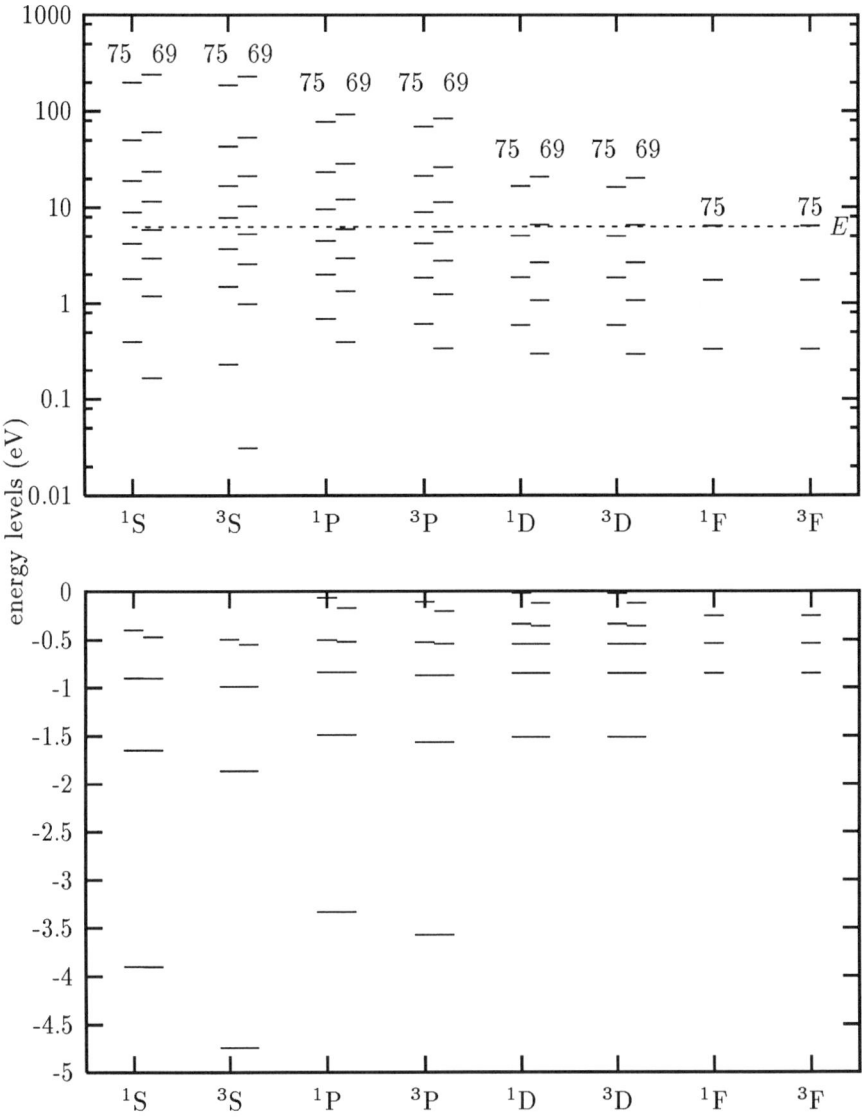

**FIG. 1.** Excited state energy levels in the CCC 75- and 69-state calculations. States with energies greater than the total energy $E$ lead to closed channels.

We present four different calculations in order to show the nature of convergence in more detail. The smallest calculation couples a total of 57 states, which comprise of 11, 10, 10, 10, 8, 8 states with symmetry $^1S$, $^3S$, $^1P$, $^3P$, $^1D$, $^3D$, respectively. The 63-state and 69-state calculations are obtained by progressively adding one extra state to each symmetry, with the 75-state calculation being obtained by adding 6 each of $^1F$ and $^3F$ states to the 63-state calculation. The distribution of the energy levels of the two largest calculations is given in Fig. 1.

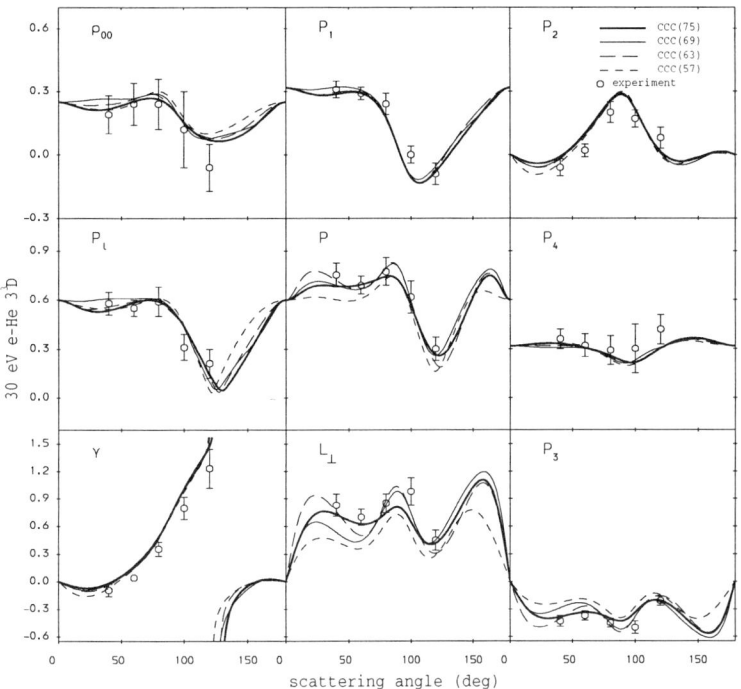

FIG. 2. Stokes and electron-impact coherence parameters for excitation of the helium $3^3D$ state by 30 eV electrons. Both the theory and experiment are previously unpublished.

In figure 2 we present the measured and calculated Stokes parameters $(P_1, P_2, P_3, P_4)$ and the derived electron-impact coherence parameters $(L_\perp, \gamma, \rho_{00}, P_\ell, P)$ for the $3^3D$ state excitation of the ground state of helium by 30 eV electrons. The general ideas relating to the description of an excited state charge-cloud have been given by Andersen et al. [13]. The specific relations pertinent here, which take into account fine structure depolarization, are given by Crowe et al. [25]. We see that all four calculations give the same qualitative picture, which is in accord with experiment. However, a careful examination indicates that the larger calculations are in slightly better agreement with experiment. We note that the effect of $F$ states is quite small, and indicates that there is no need to include $G$ states. This is an interesting result given that the integrated cross section for the $3^3D$ state drops upon introduction of $F$ states [11]. We are unaware of any other available measurements or calculations for the case presented.

In the next figure we present similar calculations and measurements for the $3^1D$ excitation by 30 eV electrons. Here we have the opportunity to compare with a distorted-wave calculation of Bartschat and Madison [8]. For the sake of clarity this is done at the expense of presenting the 63-state calculation, which happens to be barely distinguishable from the other CCC calculations. Once again we see that we have good convergence and agreement with experiment. This is not the case for the distorted-wave model, which is not surprising given the relatively low energy. It may seem odd to see the apparent disagreement with the $\gamma$ parameter at the large scattering angles. As $\gamma = \text{ATAN2}(P_1, P_2)/2$, and agreement with $P_1$ and $P_2$ is satisfactory, this indicates the sensitivity of this parameter whenever $P_1$ or $P_2$ are nearly zero, and underlines the importance of presenting the directly measured parameters as well as the derived ones.

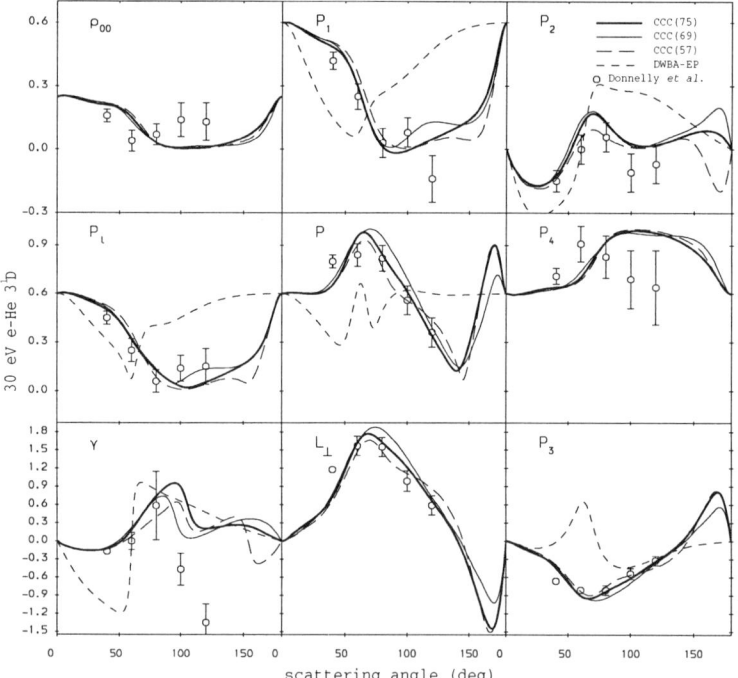

**FIG. 3.** Stokes and electron-impact coherence parameters for excitation of the helium $3^1D$ state by 30 eV electrons. The CCC(75) and CCC(69) theory is from Ref. [11]. The distorted-wave approximation using the excited-state distorting potential (DW-EP) is due to Bartschat and Madison [8]. The measurements are due to Donnelly, McLaughlin and Crowe [4].

## V. CONCLUSIONS

We have presented comparison of the most up-to-date theory and experiment for the description of the charge cloud of the $3^{3,1}D$ states of helium after 30 eV electron impact of the ground state of helium. The nature and detail of the measurements are such that only a theory that treats the exchange and continuum effects accurately is able to describe them adequately. As such, we believe that the presented measurements are able to invalidate all existing theories except for the presented CCC calculations. A future exciting direction for the development of the CCC theory is to application of differential ionization processes. Another direction is to apply it to two-electron excitation processes.

## ACKNOWLEDGMENTS

The authors are grateful for the support of the Australian Reseach Council and the United Kingdom EPSRC.

## REFERENCES

1. B. P. Donnelly and A. Crowe, J. Phys. B **21**, L637 (1988).
2. H. Batelaan, J. van Eck, and H. G. M. Heideman, J. Phys. B **24**, L397 (1991).
3. D. T. McLaughlin, B. P. Donnelly, and A. Crowe, Z. Phys. D **29**, 259 (1994).
4. B. P. Donnelly, D. T. McLaughlin, and A. Crowe, J. Phys. B **27**, 319 (1994).
5. D. T. McLaughlin, B. P. Donnelly, and A. Crowe, Phys. Rev. A **49**, 2545 (1994).
6. A. G. Mikosza, R. Hippler, J. B. Wang, and J. F. Williams, Phys. Rev. Lett. **71**, 235 (1993).
7. A. G. Mikosza, R. Hippler, J. B. Wang, and J. F. Williams, Z. Phys. D **30**, 129 (1994).
8. K. Bartschat and D. H. Madison, J. Phys. B **21**, 153 (1988).
9. D. C. Cartwright, G. Csanak, S. Trajmar, and D. F. Register, Phys. Rev. A **45**, 1602 (1992).
10. W. C. Fon, K. P. Lim, K. A. Berrington, and T. G. Lee, J. Phys. B **28**, 1569 (1995).
11. D. V. Fursa and I. Bray, to be published Phys. Rev. A **52**, (1995).
12. I. Bray and A. T. Stelbovics, Phys. Rev. A **46**, 6995 (1992).
13. N. Andersen, J. W. Gallagher, and I. V. Hertel, Phys. Rep. **165**, 1 (1988).
14. I. Bray and A. T. Stelbovics, Phys. Rev. Lett. **70**, 746 (1993).
15. I. Bray, I. E. McCarthy, J. Wigley, and A. T. Stelbovics, J. Phys. B **26**, L831 (1993).
16. I. Bray, Phys. Rev. Lett. **73**, 1088 (1994).
17. A. R. Johnston and P. D. Burrow, Phys. Rev. A **51**, R1735 (1995).
18. I. Bray, Phys. Rev. A **49**, 1066 (1994).
19. I. Bray and A. T. Stelbovics, Adv. Atom. Mol. Phys. **35**, 209 (1995).
20. M. Cohen and P. S. Kelly, Can. J. Phys. **44**, 3227 (1966).
21. I. Bray and A. T. Stelbovics, Comp. Phys. Comm. **85**, 1 (1995).

22. A. Crowe, J. C. Nogueira, and Y. Liew, J. Phys. B **16**, 481 (1983).
23. P. A. Neill, B. P. Donnelly, and A. Crowe, J. Phys. B **22**, 1417 (1989).
24. I. Bray, D. V. Fursa, and I. E. McCarthy, J. Phys. B **27**, L421 (1994).
25. A. Crowe et al., J. Phys. B **27**, L795 (1994).

# Spin effects in (e,2e) collisions

X.Guo, J.Hurn, J. Lower, S. Mazevet, Y. Shen,
I.E. McCarthy* and E.Weigold

*Research School of Physical Sciences and Engineering, I.A.S., A.N.U.,
Canberra, ACT 0200, Australia*
*\*Electronic Structure of Materials Centre,The Flinders University of South Australia,
Adelaide, S.A. 5001, Australia*

**Abstract:** The use of a polarised electron beam for a coplanar asymmetric (e,2e) experiment at intermediate energy on xenon is motivated by the possible influence of both exchange and spin-orbit interaction during the ionisation process. A preliminary experimental and theoretical investigation is presented.

## INTRODUCTION

The (e,2e) technique in electron atom collisions has considerably improved our understanding of the ionisation process during the past two decades. This technique, applied in different kinematic regions, gives access to different aspects of the problem. For incident electrons at high energy and out of plane symmetric detection of the outgoing electrons, one obtains important information on the structure of the target [1], while at lower energy the information extracted is more applicable to the study of the dynamics of the process.

From a theoretical point of view these ionisation processes are still a challenge as no theory can solve completely the boundary condition of three charged particles for the whole range of energies [2]. In addition, the success of methods such as Convergent Close Coupling for ionisation of hydrogen [3] suggests strongly that to provide a detailed description of this process one needs to include a large representation of the target state, which consists of an infinite number of bound states with a continuum spectrum.

Despite the fact that these problems are still not completely solved another challenge for both experimentalist and theorist is the use of polarised electrons and/or polarised atomic beams in ionisation processes. This experimental method has been successfully applied to both elastic and inelastic collisions and, combined with the density matrix formalism, has yielded important information on the dynamics of the processes [4]. The situation for ionisation is not so advanced and comparison between theoretical and experimental data requires in general an average over the spin direction. The use of polarised electron and/or polarised atomic beams suggests that, as for elastic and inelastic processes, our understanding of the role of spin interactions during ionisation will be improved.

# XENON (e,2e) COLLISIONS

## Spin interactions

In contrast with the two previous (e,2e) experiments performed with polarised electron beams, the use of xenon as a target at intermediate energy may involve the influence of both types of spin interactions found in e-atom collisions: namely the exchange and the spin-orbit interaction. The experiment performed by Baum et al [5] on lithium focuses on the exchange interaction only. The exchange interaction is formally a nonlocal interaction and is a consequence of the nature of the particles involved in the collision. For ionisation, the exchange between the two outgoing electrons is theoretically taken into account by a correct antisymmetrisation of the T matrix element.

The extension to the case of ionisation of the well-knowed "fine structure effect" [6,7] for inelastic collisions suggests that this may be an important contribution to the spin up-down asymmetry due to exchange. For a heavy target such as xenon the spin-orbit interaction within the atom is also important, producing a large fine structure splitting in the ion bound states. Further, electron-atom scattering at intermediate energy on a heavy target usually shows a significant contribution of the spin-orbit interaction also for the continuum electron. This interaction is understood as a change of spin direction for these electrons during the scattering without conservation of total spin for the e-atom system. Formally to account of this interaction one has to extend the scattering theory to the relativistic case (Mott scattering). This approach is required to describe collisions in the relativistic regime [8] as for the experiment performed by Prinz et al [9] who used a high-energy polarised electron beam in K-shell ionisation of silver to quantify a spin up-down asymmetry attributed to the $\vec{l}.\vec{s}$ interaction only. At intermediate regimes a complete relativistic treatment is not required and one usually considers the Breit-Pauli Hamiltonian for the electrons. The scattering wave function is in this case a two-component wave function describing an electron with its two possible spin directions in a spin-dependent potential, the Pauli potential. When one considers the elastic scattering by a heavy target such as xenon at intermediate energy one can quantify a rotation of the polarisation vector of an initially polarised electron beam due to this interaction. In this situation one observes a large spin up-down asymmetry due to this interaction near the minimum of the differential cross section. For ionisation of xenon one could therefore expect to see under the same conditions, in addition to the exchange effect, an important contribution from the $\vec{l}.\vec{s}$ interaction in the spin up-down asymmetry.

## Definition of the collision frame

An (e,2e) process is a reaction in which the kinematics is completely defined. This process is seen as a collision between an incident particle of a given energy and momentum with a bound electron ejected from the atom. The kinematics is shown in figure 1. Our (e,2e) experiment with polarised electrons on xenon has been performed with in-plane asymmetric kinematics $E_i=147$eV, $E_f=100$eV, $\theta_f=28$ degrees [10]. With this kinematics the triple differential cross section has a deep minimum in the binary lobe corresponding to a zero value for the recoil momentum of the ion.

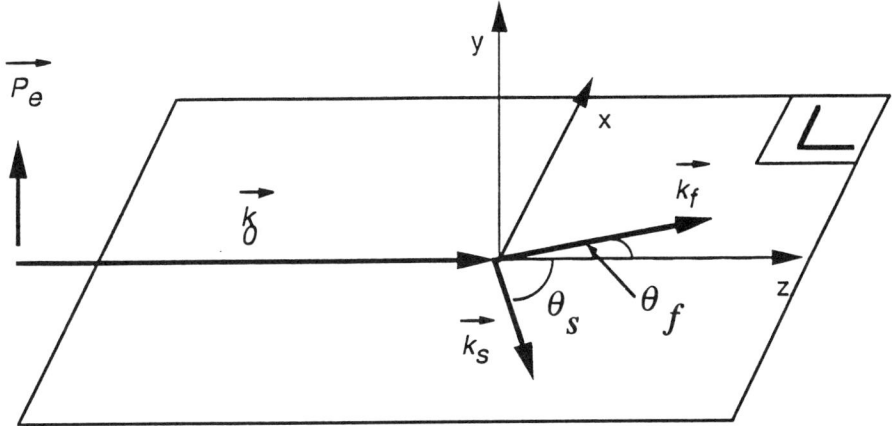

**FIGURE 1:** Present kinematic arrangement $\vec{k}_o$, $\vec{k}_f$ and $\vec{k}_s$ refer respectively to the momentum of the incident and the fast and slow outgoing electrons detected in coincidence.

The experimental spin up-down asymmetry is defined as

$$A = \frac{1}{P_e} \frac{\sigma\uparrow - \sigma\downarrow}{\sigma\uparrow + \sigma\downarrow},$$

where $\sigma\uparrow$ and $\sigma\downarrow$ refer to the triple differential cross section corresponding to an initial electron beam respectively polarised up and down. $P_e$ is the component of the polarisation of the electron beam perpendicular to the scattering plane and the spin-up direction is defined to be in the y direction as shown in the figure. The coordinate system is chosen to be in the collision frame, where the quantisation axis is along the incident direction and the fast outgoing electron is emitted to the left in the x-z plane. For a direct comparison between theory and experiment this frame will be used to express the (e,2e) collision parameters.

To include both spin-orbit and exchange interactions in the treatment of this asymmetry we use the density matrix formalism applied to the case of ionisation of a spinless target.

## DENSITY MATRIX FORMALISM

### Reduced density matrix of the initial states

The electrons have a spin s=1/2 and a particular momentum $\vec{k}$. The eigenvalues of these operators are denoted by $(k_x, k_y, k_z)$ for the components of the momentum $\vec{k}$ and $\nu=1/2$, $\nu=-1/2$ for the spin orientation of the electron. The state of an electron is a pure state and is described by the corresponding eigenvector $|n\rangle \equiv |k_x, k_y, k_z, \nu\rangle$.

An electron beam is a statistical mixture of pure states, each corresponding to the state of one electron present in this beam. We consider that all electrons of the beam have the same eigenvalues of momentum $\vec{k}$. The spin properties of the electron beam are characterised by the reduced density projector [11]:

$$\rho_e^{spin} = \sum_{\nu_0 \nu'_0} \langle \nu_0 | \rho_e^{spin} | \nu'_0 \rangle | \nu_0 \rangle \langle \nu'_0 |,$$

where $(\rho_e)_{\nu_0,\nu'_0} = \langle \nu_0 | \rho_e^{spin} | \nu'_0 \rangle$ are the reduced density matrix elements. In the $\{|\pm 1/2\rangle\}$ representation the explicit form of the reduced density matrix is

$$\rho_e = \frac{1}{2}\begin{pmatrix} 1+P_3 & P_1 - iP_2 \\ P_1 + iP_2 & 1-P_3 \end{pmatrix},$$

where $(P_1, P_2, P_3) \equiv (P_x, P_y, P_z)$ define the components of the polarisation vector along the x,y,z axes.

The atomic beam is described in the same way as the electron beam. A pure state of an atom is defined as $|\alpha'_0, J_0, M_0\rangle$ with $J_0, M_0$ respectively the total angular momentum of the atom and its projection, and $\alpha'_0$ is the set of all other quantum numbers required to define completely the atomic state. An atomic beam is a mixture of atoms in different pure states and is characterised by the density operator given by

$$\rho_a = \sum_{\substack{J_0,J_0' \\ M_0,M_0'}} |\alpha_0', J_0', M_0'\rangle\langle\alpha_0, J_0, M_0|(\rho_a)_{M_0 M_0'}^{J_0 J_0'}.$$

All the atoms before the collision are in their ground state defined by a given value $J_0$. In the present case, as the atomic beam is initially unpolarised, the density matrix $(\rho_a)_{M_0 M_0'}^{J_0 J_0'}$ is diagonal in the $2J_0+1$ subspace of a given value $J_0$ and its eigenvalues are equal to unity. The atomic density projector is then reduced to

$$\rho_a = \frac{1}{2J_0+1} I_{J_0} \sum_{M_0} |\alpha_0, J_0, M_0\rangle\langle\alpha_0, J_0, M_0|,$$

where $I_{J_0}$ is the $2J_0+1$ unity operator.

The electrons and the atoms are considered uncorrelated before the interaction begins. The density operator which describes the e-atom system before the collision is represented by the direct product of the two operators [4]

$$\rho^{in} = \rho_a \times \rho_e$$
$$= \frac{1}{2J_0+1} \sum_{\substack{M_0 \\ v_0',v_0}} |\alpha_0, J_0, M_0, v_0'\rangle\langle\alpha_0, J_0, M_0, v_0|(\rho_e)_{v_0,v_0'}$$

## Density matrix for the final state

The density matrix of the final sate is connected to the density matrix of the initial state by

$$\rho^{out} = T\rho^{in}T^\dagger.$$

where $T$ is the transition operator.

For an ionisation process the collision state is characterised by two continuum electrons and an ionic core in a given state $J_i, M_i$. The collision states are completely defined by the eigenvalues of the momentum and the spin operator for the scattered and ejected electron and the eigenvalues of the total angular momentum and its projection for the state of the ion. These states are

$$|\psi_{out}\rangle = |k_f, v_f; k_s, v_s; J_i, M_i\rangle.$$

In an (e,2e) experiment, since the kinematics are completely defined, the eigenvalues of the momentum of each electron are, in the limit of perfect experimental resolution, fixed. Using the relation between the initial and final density operators and the explicit expression for the initial reduced density projector, the reduced density matrix for the final state is

$$\langle v'_f, v'_s; J'_i, M'_i | \rho_e^{out} | v_f, v_s; J_i, M_i \rangle = \frac{1}{2J_0+1} \sum_{\substack{M_0 \\ v_0, v'_0}} f(v'_f, v'_s, M'_i, v'_0) f^*(v_f, v_s, M_i, v_0) (\rho_e)_{v_0, v'_0},$$

where the scattering amplitudes $f(v'_f, v'_s, M'_i, v'_0)$ and $f^*(v_f, v_s, M_i, v_0)$ are defined as

$$f(v_f, v_s, M_i, v_0) = \langle v_f, v_s, J_i, M_i | T | J_0, M_0, v_0 \rangle \quad ,$$

and * denotes the complex conjugate.

In an (e,2e) process only the subensemble of ions which gives two continuum electrons with the required energy are considered. For xenon, the $\vec{l}.\vec{s}$ interaction within the atom gives rise to a large fine-structure splitting between the ground state $^2P_{3/2}$ and the first excited state $^2P_{1/2}$ of the ion (1.3eV). In this case the experiment is able to distinguish between the two reactions corresponding to these two different value of $J_i$, the preceding reduced density matrix is therefore diagonal on the values of $J_i$.

Following the notation already used in the literature [12], the matrix elements are

$$\langle v'_f, v'_s, M'_i; v_f, v_s, M_i \rangle = \sum_{v_0, v'_0} f(v'_f, v'_s, M'_i, v'_0) f^*(v_f, v_s, M_i, v_0) (\rho_e)_{v_0, v'_0},$$

where the normalisation coefficient and the summation over $M_0$ cancel in the case of a closed shell atom.

As for the case of inelastic scattering with p shell atoms these matrix elements contain the maximum information on the scattering process if the initial and final polarisation of the projectiles and the final polarisation of the target are known while the target is initially unpolarised [4].

# Polarisation parameters for an (e,2e) experiment

The triple differential cross section, in the case where the polarisation of the ion and the outgoing electron beams are not detected, is given by the sum of the diagonal terms over the unobserved variables [13]

$$\frac{d^3\sigma}{d\Omega_f d\Omega_s dE_f} = \sum_{\substack{M_i \\ \nu_f, \nu_s}} \langle \nu_f, \nu_s, M_i; \nu_f \nu_s, M_i \rangle.$$

Using now the definition of the matrix element we obtain for the cross section

$$\frac{d^3\sigma}{d\Omega_f d\Omega_s dE_f} = (2\pi)^4 \frac{k_f k_s}{k_0} \sum_{\substack{M_i \\ \nu'_0, \nu_0 \\ \nu_f, \nu_s}} f(\nu_f, \nu_s, M_i, \nu_0) f^*(\nu_f, \nu_s, M_i, \nu'_0)(\rho_e)_{\nu'_0, \nu_0}.$$

By using the invariance of the scattering amplitude about the scattering plane the above expression can be written as

$$\frac{d^3\sigma}{d\Omega_f d\Omega_s dE_f} = \left(\frac{d^3\sigma}{d\Omega_f d\Omega_s dE_f}\right)_{unpol.} [1 + P_y \cdot A],$$

where we define the unpolarised differential cross section, the differential cross section corresponding to an initially unpolarised beam, as

$$\sigma_{unpol.} \equiv \left(\frac{d^3\sigma}{d\Omega_f d\Omega_s dE_f}\right)_{unpol.} = (2\pi)^4 \frac{k_f k_s}{k_0} \sum_{\substack{M_i \\ \nu_f, \nu_s}} f(\nu_f, \nu_s, M_i, \nu_0) f^*(\nu_f, \nu_s, M_i, \nu_0).$$

The asymmetry parameter is

$$A = \frac{-2(2\pi)^4}{\sigma_{unpol.}} \frac{k_f k_s}{k_0} \left\{ \sum_{M_i} (-1)^{M_i - J_i} \operatorname{Im}\left\{ f\left(\frac{1}{2}, \frac{1}{2}, M_i, \frac{1}{2}\right) f^*\left(-\frac{1}{2}, -\frac{1}{2}, -M_i, \frac{1}{2}\right) \right.\right.$$
$$\left.\left. - f\left(\frac{1}{2}, -\frac{1}{2}, M_i, \frac{1}{2}\right) f^*\left(-\frac{1}{2}, \frac{1}{2}, -M_i, \frac{1}{2}\right) \right\}\right\}$$

If the contribution of the $\vec{l}.\vec{s}$ interaction is now neglected, one obtains the expression for the asymmetry for a pure fine-structure effect in the collision frame, which is the contribution of exchange only to the spin-up down asymmetry.

The above expression does not contain at this stage any approximation. The approximations will appear only through approximation of the scattering amplitude.

## COMPARISON OF THEORY AND EXPERIMENT

A first attempt to evaluate the scattering amplitudes has been done with a DWBA in j-j representation. The distorted waves are computed with a real Pauli potential while the bound state of the atom is described by a Dirac-Fock wave function [14]. This method gives an evaluation of the potential matrix elements which do not conserve the spin even if the interaction is not spin dependent [15]. This is due to the inclusion of the Pauli potential in the calculation of the distorted waves and it is interpreted as a possible spin-flip for the continuum electrons in the entrance and the exit channel while the instantaneous transition due to the e-e interaction conserves the total spin of the system.

In figure 2a, 2b, 2c, this calculation for the asymmetry is compared with the experimental data for the final states of the ion $J_i=1/2$ and $J_i=3/2$ in a kinematic region where the incident fast and slow electron energies are fixed ($E_i$=147eV, $E_f$=100eV) and the scattering angle $\theta_f$ has respectively the values 15, 28 and 40 degrees. It appears that the description of this parameter is satisfactory for a scattering angle of 28 degrees for both final states of the ion, less so at 15 degrees and at 40 degrees the data are in strong disagreement. For the 40 degree scattering angle the shape of the asymmetry seems to be properly described if the signs are reversed. The contribution of the $\vec{l}.\vec{s}$ interaction at these three kinematical arrangement is quite insignificant, since the total asymmetry for the summed $^2P_{3/2}$ and $^2P_{1/2}$ transitions is neglible [15]. The disagreement of the calculated asymmetry could be attributed to the limited validity of the DWBA at such low energies, but could also be due to the use of a real potential in the calculation of the distorted wave. This approximation affects only the evaluation of spin-flip in the entrance channel, but as for elastic scattering [16] the inclusion of a complex potential could be required to describe properly the contribution of this interaction.

## CONCLUSION

This general approach using the density matrix formalism shows that the "fine structure effect" is the natural contribution of exchange to the asymmetry parameter when considering ionisation of a heavy atom at low to medium energies. The partial poor agreement of the DWBA calculations with the experimental data

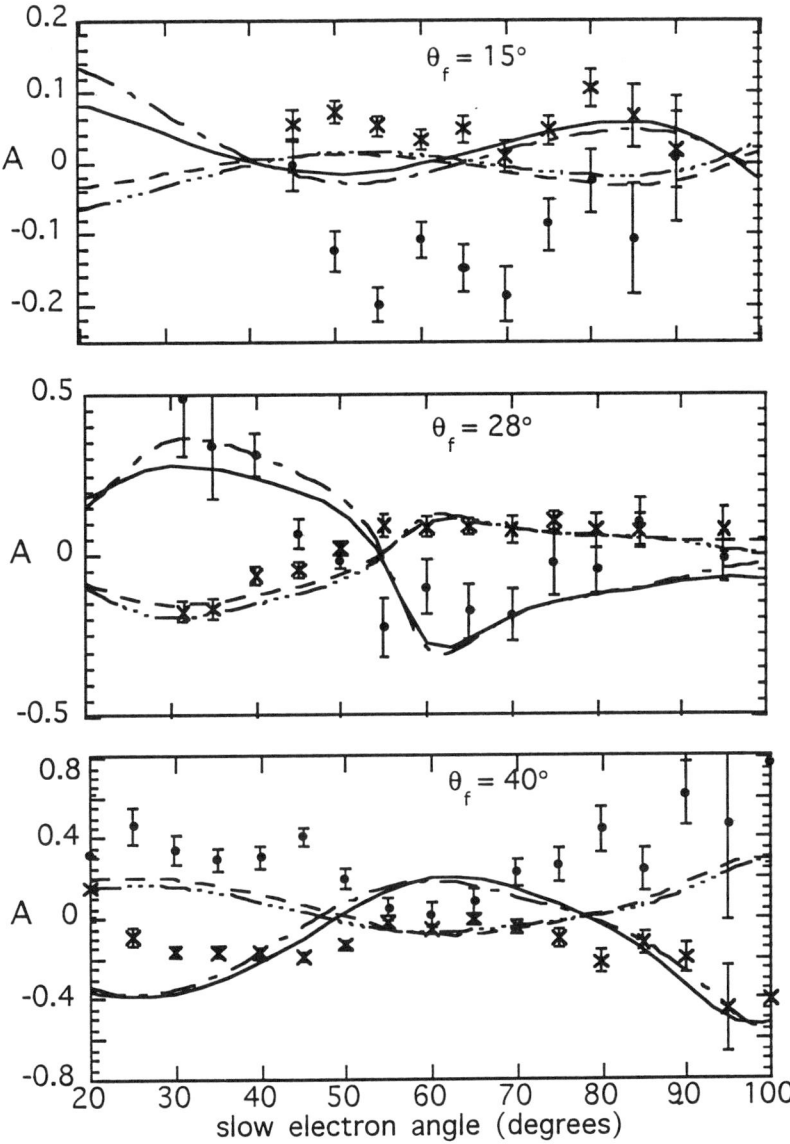

**FIGURE 2:** Comparison of calculated and experimental asymmetry parameters, A for an incident energy $E_i$=147eV and fast electron energy, $E_f$=100eV on a xenon target. Present calculations for J=1/2 —— and J=3/2 - - -, and Jones et al [7] for J=1/2 —— · —— and J=3/2 —— --- ——. Present measurements for J=1/2 (•) and for J=3/2 (×).

suggests that relativistic effects may have to be included more accurately and/or that more exact collision approximations may be needed. The inclusion of a complex potential in the DWBA calculations may be important and is under study.

## ACKNOWLEDGMENTS

We kindly thank Professor Madison for his results prior to publication.

## REFERENCES

1. McCarthy, I.E., and Weigold, E. :*Rep. Prog. Phys.* **54**, 789 (1991).
2. Klar, H., Konovalov, D., and McCarthy, I.E. : *J. Phys.* B **26**, L711 (1993).
3. Bray, I. : *Phys. Rev.* **A49**, 1066 (1994).
4. Bartschat, K., and Madison, D. : *J.Phys.* B **21**, 2621 (1988).
5. Baum, G., Blask, W., Freienstein, P., Frost, L., Hesse, S., Raith, W., Rappolt, P., and Streun, M. : *Phys. Rev. Lett.* **69**, 3037 (1992).
6. Hanne, G.F. : *Phys.Rep.* **95**, 96 (1983).
7. Jones, S., Madison, D., Hanne, F. : *Phys. Rev. Lett.* **72**, 2554 (1994).
8. Keller, S., Colm, T., Whelan, T., Ast, H., Walters, H., and Dreizler, R. : *Phys. Rev.* **A50**, 3865 (1995).
9. Prinz, H., Besch, K., and Nakel, W. : *Phys. Rev. Lett.* **74**, 243 (1995).
10. Guo, X., Hurn, J., Lower, J., Mazevet, S., McCarthy, I.E., Shen Y., and Weigold, E. : (*submitted to Phys. Rev. Lett.)* (1995).
11. McCarthy, I.E., and Weigold, E. : *Electron-atom collision,* Cambridge University Press, 1995.
12. Blum, K. and Kleinpoppen, H. : *Phys. Rep.* **95**, 251 (1983).
13. Kessler, J. : *Polarised electrons,* 2nd Ed., Springer Verlag, Berlin, 1985.
14. Grant, I., McKenzie, B., Norrington, P.H., Mayers, D.F., Pyper, N.C. : *Comp. Phys. Comm.* **21**, 207 (1980).
15. Guo, X., Hurn, J., Lower, J., Mazevet, S., McCarthy, I.E., Shen Y., and Weigold, E. : 1995 (in preparation).
16. Bartschat, K., and Burke, P. G. *Coherence in Atomic Collision Physics*, Plenum Press, 1988.

# Measurement of Exchange and Spin-Orbit Effects and their Interference in Elastic e-Cs Scattering

M. Tondera, G. Baum, P. Baum, L. Grau, B. Leuer, R. Niemeyer and W. Raith

*Fakultät für Physik, Universität Bielefeld, Postfach 10 01 31, D-33501 Bielefeld, Germany* [1]

Exchange and spin-orbit interactions are the spin polarization effects in the elastic scattering of electrons on high-Z one-electron atoms. Theoretical analysis showed that the presence of these two effects could lead to a detection of a special interference at low electron energies, measurable as a cross-section asymmetry ($A_1$) if unpolarized electrons are scattered on polarized cesium atoms. In a crossed-beam experiment with a highly polarized electron beam from a strained GaAs photocathode and a Cs atomic beam, polarized by optical pumping with two laser diodes, we measured simultaneously the relative differential cross section ($\sigma_0$), the special interference asymmetry ($A_1$), the spin-orbit asymmetry ($A_2$) and the double-spin asymmetry ($A_{nn}$) at 2 eV scattering energy. Our measurements can be compared with two different close-coupling R-matrix calculations, a Breit-Pauli approach, and a fully relativistic Dirac approximation.

## INTRODUCTION

We investigate spin-polarization effects in low energy elastic scattering of polarized electrons on polarized cesium atoms. For the electron scattering by high-Z one-electron atoms, two spin interactions occur, exchange and spin-orbit interactions. We observed an interference of both effects by measuring a distinctly non-zero interference asymmetry $A_1$ at 7 eV electron energy first [1]. At 20 eV and 13.5 eV this asymmetry was consistent with zero [2].

Exchange of identical particles occurs between the scattered electron and the cesium valence electron. If polarized electrons (polarization $\mathbf{P}_e$) are scattered by atoms with polarized valence electrons (polarization $\mathbf{P}_a$), exchange causes a parallel/antiparallel cross-section asymmetry described by

$$\sigma = \sigma_0(1 - A_{nn}\mathbf{P}_a \cdot \mathbf{P}_e), \tag{1}$$

---

[1]This work has been supported by the Deutsche Forschungsgemeinschaft in SFB 216.

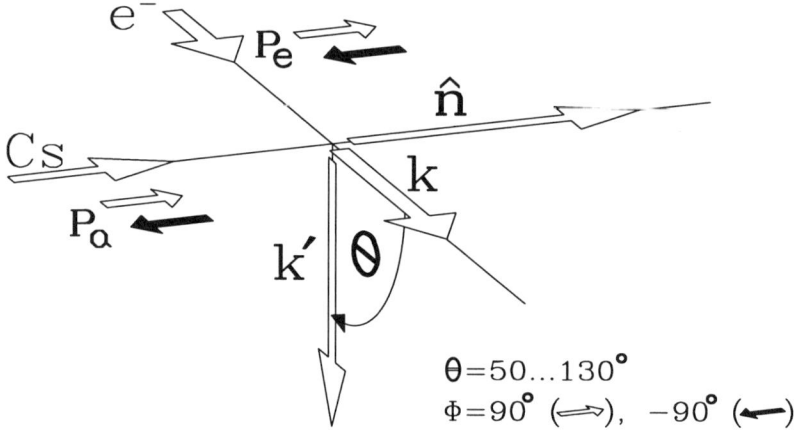

**FIG. 1.** Schematic of the scattering geometry ($\Theta$ is the scattering angle, $\Phi$ the azimuthal angle)

where $\sigma_0$ is the cross section for unpolarized particles and $A_{nn}$ is the exchange or double-spin asymmetry. In this experiment both spin directions are normal to the scattering plane (see Fig. 1). The minus sign in (1) results from the adoption of the definition for the asymmetry in electron–electron scattering [3].

The spin-orbit interaction is a magnetic interaction of the scattered electron's magnetic moment with the inhomogeneous magnetic field in its rest frame, produced by the charge of the atomic core moving around the electron. It is a relativistic effect, important for high-Z target atoms. The polarization effects of the so-called low-energy Mott scattering are most pronounced on both sides of deep diffraction minima of the cross section, where $|d\sigma_0/d\Theta|/\sigma_0$ is large [4]. If polarized electrons with spin perpendicular to the scattering plane are scattered by high-Z target atoms, the spin-orbit interaction produces an azimuthal dependence of the cross section according to

$$\sigma = \sigma_0 (1 + A_2 \mathbf{P}_e \cdot \mathbf{n}). \qquad (2)$$

$\mathbf{n}$ is the unit vector perpendicular to the scattering plane, pointing into the direction $\mathbf{k} \times \mathbf{k}'$ ($\mathbf{k}$, $\mathbf{k}' =$ electron momentum before and after scattering, respectively, see Fig. 1), $A_2$ is the spin-orbit asymmetry, known as Sherman function S in the literature of Mott scattering.

In 1974 Burke and Mitchell [5] gave a general theoretical analysis of relativistic electron scattering. They showed that this process involves more than the simple addition of exchange and spin-orbit interaction. The full

TABLE 1. Scheme of asymmetry evaluation

| $N_j$ | $N_k$ [a] | $\frac{N_j-N_k}{N_j+N_k}$ |
|---|---|---|
| ↑↓ + ↓↑ | ↓↓ + ↑↑ | $\mathbf{P}_a \cdot \mathbf{P}_e A_{nn}$ |
| ↑↑ + ↑↓ | ↓↓ + ↓↑ | $\mathbf{P}_a A_1$ |
| ↑↑ + ↓↑ | ↑↓ + ↓↓ | $\mathbf{P}_e A_2$ |

[a] $N_j, N_k$ : Signal Sums

theory depends on six complex scattering amplitudes that corresponds to 11 independent parameters, i.e. 6 moduli of amplitudes and 5 phase differences.

In discussing the consequences of Burke and Mitchell's analysis for experimental atomic physics Farago [6] drew attention to the fact that the full theory allows for a special interference of exchange and spin-orbit interaction. This interference should produce a left/right asymmetry $A_1$ (interference asymmetry) of the cross section if unpolarized electrons are scattered on polarized atoms with $\mathbf{P}_a$ perpendicular to the scattering plane according to

$$\sigma = \sigma_0(1 + A_1 \mathbf{P}_a \cdot \mathbf{n}). \tag{3}$$

This is an effect which neither exchange nor spin-orbit coupling alone could provide. For the light alkali atoms spin-orbit and interference effects are negligible. Investigations of e–Na scattering yielded interference asymmetries consistent with zero [7].

In our experiment both, the atom polarization $\mathbf{P}_a$ and the electron polarization $\mathbf{P}_e$ are oriented perpendicular to the scattering plane, parallel or antiparallel to the unit vector $\mathbf{n}$ which is also the direction of our atomic beam. The spin dependence of the cross section is described by superposition of (1) to (3):

$$\sigma = \sigma_0(1 - A_{nn}\mathbf{P}_a \cdot \mathbf{P}_e + A_1 \mathbf{P}_a \cdot \mathbf{n} + A_2 \mathbf{P}_e \cdot \mathbf{n}) \tag{4}$$

The cross section for unpolarized particles $\sigma_0$ and the three asymmetries depend on the electron energy E and the scattering angle $\Theta$.

Since our apparatus allows only measurements at scattering angles on one side of the primary electron beam, we cannot reverse the vector $\mathbf{n}$. But we can easily reverse both beam polarizations, see Fig.1, $\Phi = 90°$ (spin up) and $\Phi = -90°$ (spin down). The different combinations of beam-polarization directions lead to four different cross sections. We measure four different scattering intensities (counting rates) $N^{\uparrow\uparrow}$, $N^{\downarrow\uparrow}$, $N^{\uparrow\downarrow}$ and $N^{\downarrow\downarrow}$. The first arrow represents the direction of $\mathbf{P}_a$, the second of $\mathbf{P}_e$. By taking suitable combinations of these relative cross-section measurements we simultaneously obtain three independent cross-section asymmetries which are listed in Table 1.

## EXPERIMENT

The experimental arrangement is shown schematically in Fig.2. In a crossed-beam experiment we intersected a polarized electron beam from a GaAs source with a polarized Cs atomic beam. In previous publications we already described in detail the atomic system [8] and the polarized electron beam for measurements of ionization spin-asymmetries with high energy resolution [9].

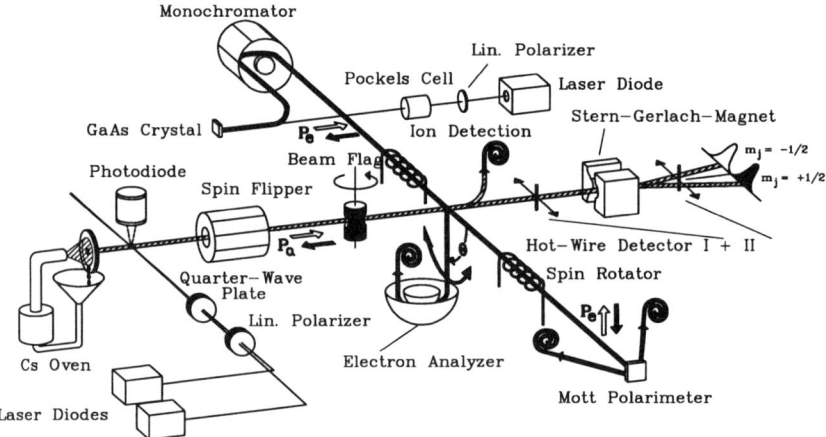

**FIG. 2.** Experimental arrangement

In this experiment the polarized electron source consists of a strained GaAs photocathode [10] prepared for negative electron affinity (NEA) by cesiating and oxidizing [11]. The photocathode provided electron currents of 5-10 $\mu$A. The lifetime of the NEA-layer amounted to several months. We continuously cesiated the GaAs crystal.

By using strained GaAs it is possible to excite all electrons into the conduction-band level with either $m_s = +1/2$ or $m_s = -1/2$ depending on the helicity of the polarized light, $\sigma^+$ or $\sigma^-$, respectively.

We used a laser diode that provided a wavelength of 850 nm and an output power of 40 mW. The electron polarization $\mathbf{P}_e$ can be reversed by changing the helicity of the laser-light with a Pockels cell. False asymmetries due to intensity changes associated with polarization reversals were found to be consistent with zero within the statistical uncertainties represented by the error

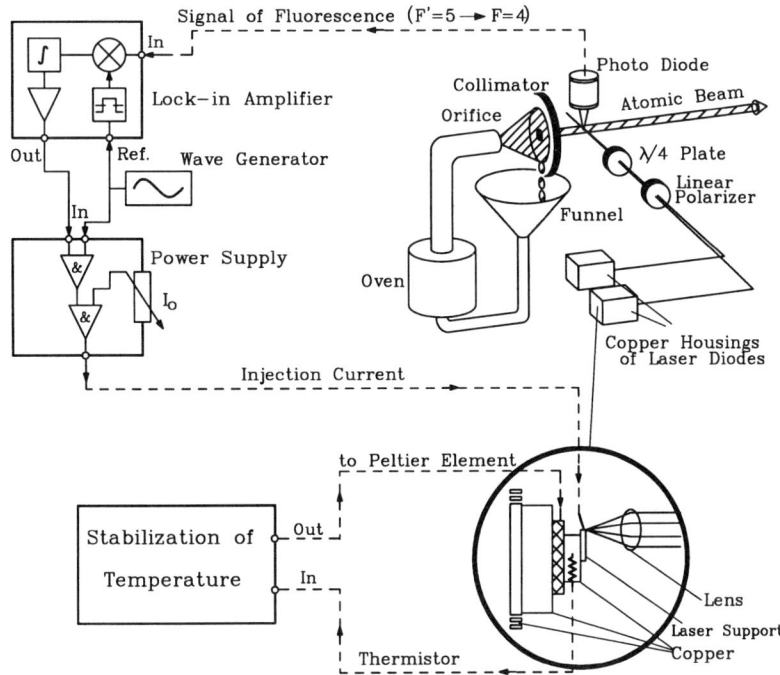

**FIG. 3.** Schematic view of laser frequency stabilization. In the upper right-hand corner the oven and the pumping arrangement with detection of the fluorescence radiation is shown. Fine tuning of the wavelength is accomplished with electronic regulation of the injection current (upper left) and the temperature of the laser diode (lower right)

bars in Fig. 5. The energy width of the electron beam amounts to $(0.3\pm0.05)$ eV. The electron beam intensity in the scattering region is about 150 nA.

We used a small-size retarding-field Mott polarimeter to determine the electron polarization. By using a thorium foil (thickness $12.5\mu m$) we obtained an analyzing power of $S_{eff} = -0.4$ at 45 keV electron scattering energy. The measured spin polarization varied from 0.6 to 0.8 with an accuracy of $\pm$ 0.03.

The rotatable detector for the scattered electrons covers the angular range of 50 - 130° with an angular acceptance of $\pm 3°$. In order to suppress detection of any inelastic scattered electrons, the energy resolution is reduced to 0.8 eV by means of a spherical deflector analyzer in front of the detector.

The cesium beam is produced by a recirculating oven system. By means of optical pumping with two laser diodes an atomic (valence electron) polarization of 0.9 is achieved for an atom density of $5 \cdot 10^9$ cm$^{-3}$ in the interaction region. We used two wavelength-selected GaAs laser diodes in order to pump both hyperfine-structure levels of the ground state $(6^2S_{1/2})$. The tuning was accomplished by temperature variation (Peltier element) and fine tuning by means of the injection current (Fig.3). The wavelength of each laser was stabilized by modulating the injection current which caused a wavelength modulation of the light beam. We measured the periodical change of the intensity of the resonance fluorescence radiation and employed feed-back circuits (lock-in amplifiers) for regulating the injection currents. The atom polarization $\mathbf{P}_a$ can be reversed quickly by means of a spin flipper [12] or slowly by turning the quarter-wave plate for the pumping light. We determine the atom polarization by evaluating the beam profiles produced by magnetic state separation in a Stern-Gerlach magnet [8].

## RESULTS

The measurements at 2 eV electron energy were performed in order to compare with two different close-coupling R–matrix calculations: first a Breit-Pauli (BP) approach (five- and eight-state), in which relativistic correction terms are added to the non-relativistic Hamiltonian, employed by Scott et al. [14] and subsequently improved by Bartschat [16] and secondly the fully relativistic Dirac approximation (five-state) of Thumm and Norcross [18] [15]. Both theoretical predictions are in excellent agreement if only the angle-differential elastic cross section is considered. Our data of the relative differential cross section $\sigma_0$ are displayed in Fig. 4. This measurement required an unpolarized atom and electron beam, easily established by turning off the laser diodes (optical pumping) and by using linear polarized laser light for the electron source. Alternatively, one can average over counting rates obtained with the four possible spin orientations. The values are normalized to the theoretical results (five-state BP) at the scattering angle $\Theta = 85°$. Our data agree very well with both theories and earlier measurements of Gehenn and Reichert [13].

The measured asymmetries $A_{nn}$, $A_2$ and $A_1$ are shown in Fig.5 together with the theoretical predictions. Eight-state Breit-Pauli and five-state Dirac calculations for the parameter $A_{nn}$ are not available. A five-state Breit-Pauli approach [17] for $A_{nn}$ exists and is displayed in the upper graph of Fig. 5.

Our data points were obtained with the following beam polarizations: $P_a = 0.86 \pm 0.035$ and $P_e = 0.68 \pm 0.03$. The systematic errors due to the beam-polarization uncertainties, common to all data points, amount to $\pm 6\%$ for $A_{nn}$, $\pm 4.5$ % for $A_2$ and $\pm 4$ % for $A_1$. All asymmetry data points are shown with statistical error bars corresponding to one standard deviation. Typical counting rates were 700 s$^{-1}$ in the maximum of the differential cross section. Background counting rates were about (50–200) s$^{-1}$ depending on the scattering angle. The 2 eV asymmetry data were obtained in a total running time of 140 hours. We calibrated the electron energy by measuring the ionization threshold. The uncertainty of the scattering energy is about $\pm 0.1$ eV.

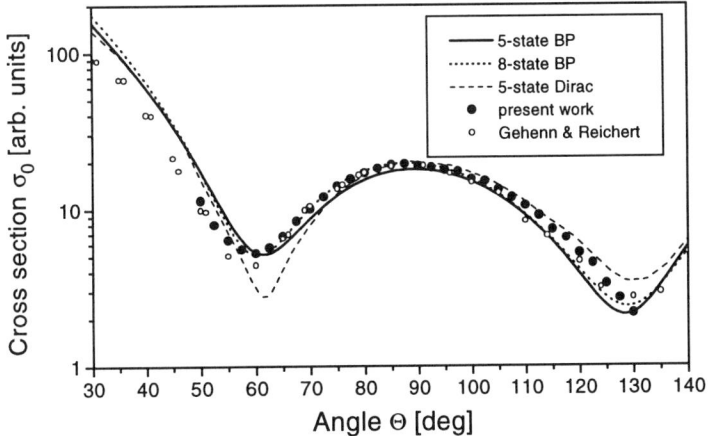

FIG. 4. Differential cross section $\sigma_0$ (relativ) at E = 2 eV

Our data points of $A_{nn}$ have a positive maximum of 0.21 ± 0.01 at the scattering angle $\Theta = 60°$. In the range from 60 to 130° the data show a nearly linear decrease and go through zero at the cross section maximum at $\Theta \approx 90°$. There is no good agreement with the five-state Breit-Pauli prediction.

The calculations for $A_1$ and $A_2$ shown in Fig. 5 predict dissimilar structures in particular for $A_1$. The pronounced structure of the Dirac approximation for the parameter $A_1$ disagrees extremely with our experimental results, whereas there is good agreement with the Breit-Pauli prediction within the angular range of 50–110°. At these angles the measured $A_1(\Theta)$ has a flat plateau of about +0.02. A zero transition occurs near 125°. Our $A_2$ data do not agree with either theory, but the five-state Breit-Pauli values seem to come somewhat closer.

The experimental results are averages over the energy width of the electron beam of about 0.3 eV. Unpublished five-state Breit-Pauli calculations [17] for energies near 2 eV indicate that the energy dependence of the asymmetries $A_{nn}$, $A_2$ and $A_1$ is sufficiently small to allow a meaningful comparison of our data with theoretical predictions at 2 eV. This can be seen by averaging the five-state Breit-Pauli calculations for energies from 1.8 to 2.2 eV with our energy resolution. These averaged curves are in excellent agreement with the theoretical asymmetries for 2 eV.

We are currently working on measurements with a reduced energy width and at lower energies in order to provide better and more data for critical comparison with theory.

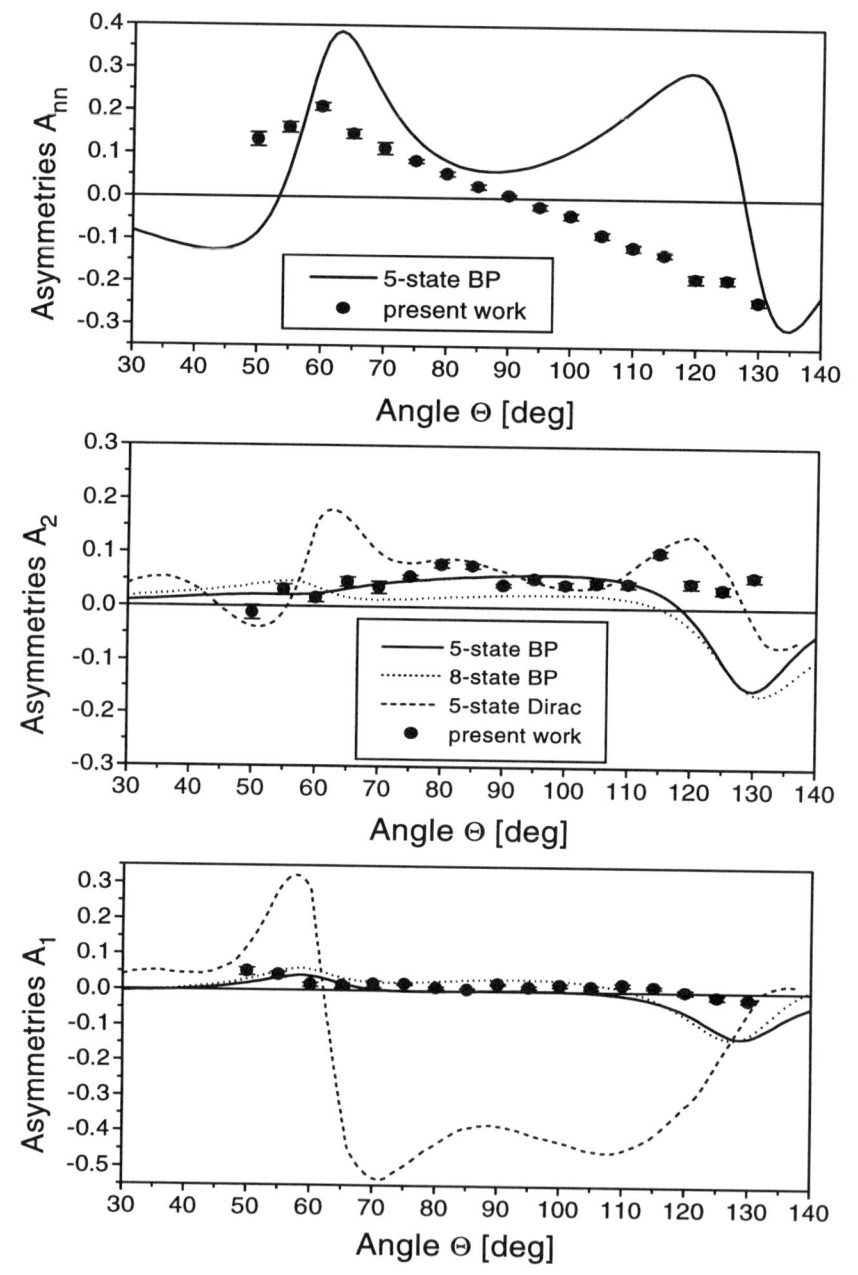

**FIG. 5.** Double spin asymmetry $A_{nn}$, spin-orbit asymmetries $A_2$ and interference asymmetries $A_1$ vs. scattering angle $\Theta$ at E = 2 eV

# REFERENCES

1. B. Leuer, G. Baum, L. Grau, R. Niemeyer, W. Raith, M. Tondera, Z. Phys. D **33**, 39 (1995).
2. W. Raith, Second SERC Workshop on Polarized Electron/Polarized Photon Physics at York, England, 15-16 April 1994; proceedings to be published by Plenum Press, London 1995.
3. P.S. Cooper, M.J. Alguard, R.D Ehrlich, V.W. Hughes, H. Kobayakawa, J.S. Ladish, M.S. Lubell, N. Sasao, K.P. Schüler, P.A. Souder, G. Baum, W. Raith, K. Kondo, D.H. Coward, R.H. Miller, C.Y. Prescott, D.J. Sherden, C.K. Sinclair, Phys. Rev. Lett. **34**, 1589 (1975).
4. J. Kessler : Polarized electrons, 2nd edn. (Springer Series on Atoms and Plasmas, vol. 1) Berlin, Heidelberg, New York: Springer 1985.
5. P.G. Burke and J.F.B. Mitchell, J. Phys. B **7**, 214 (1974).
6. P.S. Farago, J. Phys. B **7**, L28 (1974).
7. J.J. McClelland, S.J.Buckman, M.H. Kelly, R.J. Celotta, J. Phys. B **23**, L 21 (1990).
8. G. Baum, B. Granitza, S. Hesse, B. Leuer, W. Raith, K. Rott, M. Tondera, B. Witthuhn, Z. Phys. D **22**, 431 (1991).
9. G. Baum, B. Granitza, L. Grau, B. Leuer, W. Raith, K. Rott, M. Tondera, B. Witthuhn, J. Phys. B **26**, 331 (1993).
10. M. Chatwell, J. Clendenin, T. Maruyama, D. Schultz, eds., Proceedings of the Workshop on Photocathodes for Polarized Electron Sources for Accelerators, SLAC-432, Stanford, (1994).
11. D.T. Pierce and F. Meier, Phys. Rev. B **13**, 5484 (1976).
12. W. Schröder and G. Baum, J. Phys. E **16**, 52 (1983).
13. W. Gehenn and E. Reichert, J. Phys. B **10**, 3105 (1977).
14. N.S. Scott, K. Bartschat, P.G. Burke, W.B. Eisner, O. Nagy, J. Phys. B **17**, 3775 (1984b); J. Phys. B **13**, 4299 (1980);
15. U. Thumm, K. Bartschat, D.W. Norcross, J. Phys. B **26** , 1587 (1993).
16. K. Bartschat, J. Phys. B **26**, 3595 (1993).
17. K. Bartschat, private communication (1995).
18. U. Thumm and D.W. Norcross, Phys. Rev. A **45**, 6349 (1992), Phys. Rev. A **47**, 305 (1992).

# Studies of Electron-Molecule Scattering at Microelectronvolt Energies Using Very-High-$n$ Rydberg Atoms

M. T. Frey, S. B. Hill, K. A. Smith, F. B. Dunning, and
I. I. Fabrikant[†]

*Department of Physics and the Rice Quantum Institute,
Rice University, 6100 South Main Street, Houston, TX 77005-1892 USA*
[†]*Department of Physics and Astronomy, University of Nebraska, Lincoln, NE 68588 USA*

**Abstract.** Recent advances in experimental techniques now allow studies with atoms in very high-lying Rydberg states, $100 \lesssim n \lesssim 1100$. Such atoms provide a unique tool with which to investigate electron-molecule scattering at ultra-low electron energies and the new capabilities this affords are illustrated by considering collisions with molecules that attach free low-energy electrons, and with molecules that possess large electric dipole moments.

## INTRODUCTION

Atoms in which one electron is excited to a state of large principal quantum number, $n$, possess characteristics quite unlike those normally associated with atoms in ground or low-lying excited states. This is illustrated in Table 1 which lists a number of atomic properties, their dependences on $n$, and typical numerical values for these quantities at several different $n$. One striking feature of Rydberg atoms is their large physical size. Atomic dimensions increase rapidly with $n$ and at $n=1000$ the atomic diameter, ~0.1 mm, is larger than that of many microscopic organisms. The excited electron is so far from its associated core ion that it may be considered as an independent particle and the atom pictured using the Bohr model with the excited electron in a distant classical orbit about the core ion. For large values of $n$, the separation between the excited electron and core ion is much greater than the ranges typically associated with ion-neutral or electron-neutral interactions. Thus, in collisions with neutral targets, high-$n$ atoms behave not as an atom but rather as a pair of independent particles (1). Studies of Rydberg atom collision processes that are dominated by the (binary) electron-target interaction can therefore

© 1995 American Institute of Physics

**TABLE 1.** Properties of Rydberg Atoms.

| Property | n dependence | n=1 | n=100 | n=1000 |
|---|---|---|---|---|
| Bohr Radius | $n^2 a_0$ | $5.3 \times 10^{-9}$ cm | $5.3 \times 10^{-5}$ cm | $5.3 \times 10^{-3}$ cm |
| Binding Energy | $R/n^2$ | 13.6 eV | 1.36 meV | 13.6 µeV |
| Energy level Spacing | $2R/n^3$ | 10.2 eV | $2.7 \times 10^{-5}$ eV | $2.7 \times 10^{-8}$ eV |
| RMS velocity of Rydberg electron | $v_0/n$ | $2 \times 10^8$ cm s$^{-1}$ | $2 \times 10^6$ cm s$^{-1}$ | $2 \times 10^5$ cm s$^{-1}$ |
| Classical field ionization threshold | $1/16n^4$ | $3 \times 10^8$ V cm$^{-1}$ | 3 V cm$^{-1}$ | $3 \times 10^{-4}$ V cm$^{-1}$ |

provide information on electron-molecule scattering. (At high $n$ the atomic volume is such that even for low target gas densities, many target molecules are contained within the electron orbit, but these do not all interact simultaneously with the excited electron because of the limited range of individual electron-target interactions.) The time-averaged kinetic energy of the excited electron, which is equal to its binding energy, can be very small. For $n=1100$, the largest value of $n$ at which measurements have been undertaken so far, the mean kinetic energy is only ~11 µeV. Thus, very-high-$n$ Rydberg atoms provide a vehicle to study electron-molecule interactions at ultra-low electron energies. Such energies are of interest because the electron de Broglie wavelength becomes comparable to, or larger than, the ranges typical of most electron-molecule interactions. Also, the duration of the interaction is sufficiently long that rotation of the target molecule during the interaction is important.

This paper discusses recent advances in the study of collisions involving very-high-$n$ Rydberg atoms. The capability of Rydberg atom techniques is illustrated by considering collisions with molecules that attach free low energy electrons to form a negative ion, and molecules that possess permanent dipole moments.

## EXPERIMENTAL TECHNIQUES

The apparatus developed to study collisions of very-high-$n$ atoms is shown in the inset in Fig. 1 (2). Potassium atoms contained in a collimated beam are excited to selected high-lying state(s) using the crossed output of an intracavity-doubled Rh6G dye laser. To achieve stable excitation, the long term drift in laser output frequency is reduced to $\leq 2$ MHz per day using a scanning Fabry-Perot etalon and stabilized He-Ne laser (3). Excitation occurs near the center of an interaction region defined by three pairs of planar copper electrodes, each ~10 x 10 cm$^2$. The use of large electrodes well separated from the excitation region minimizes the effect of stray patch fields associated with non-uniformities in the electrode surfaces. However, even with all the electrodes grounded, stray fields of ~2 mV cm$^{-1}$ remain in the experimental volume which are sufficient to cause significant Stark mixing and, at very high-$n$, to even induce ionization. Thus the field in the experimental volume must be further reduced by application of small biases to the electrodes (4).

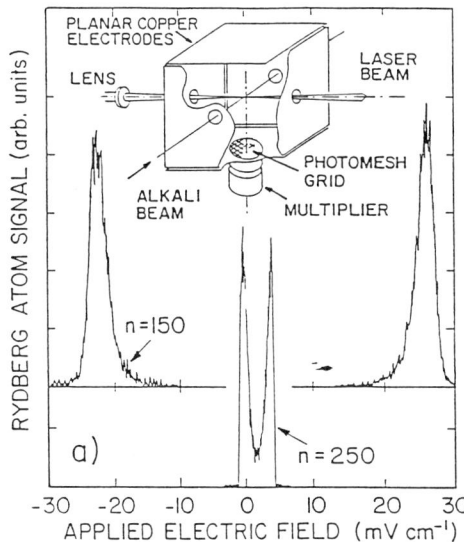

**FIGURE 1.** Rydberg atom signal observed as a function of applied electric field for excitation of Stark states with $n=150$ and 250. The inset shows a schematic diagram of the apparatus.

To determine these, the laser is detuned a few MHz to the red of a particular 4s-$n$p transition and the Rydberg atom production recorded as a function of the bias applied between each opposing pair of electrodes in turn. The bias superposes a variable component of electric field in the experimental region perpendicular to the plane of the biased electrodes. The dependence of Rydberg atom production on the applied field produced by biasing one particular pair of electrodes is shown in Fig 1 for excitation to states with $n=150$ and $n=250$. Two separate peaks are evident that correspond to the applied fields at which the absolute magnitude of the total field in the experimental region Stark shifts the $n$p state into resonance with the (fixed) laser frequency. (The separation of the peaks is governed by the laser detuning and the size of the Stark effect, which depends strongly on $n$.) Because the two peaks must be symmetrically positioned about zero, the midpoint between the peaks gives the external field that must be applied to neutralize the corresponding stray field component. Using this technique, the stray field in the excitation volume can be reduced to $\leq 50$ $\mu$V cm$^{-1}$ allowing excitation of individual $n$p states up to $n\sim 500$ as illustrated in Fig. 2. For larger values of $n$, Stark mixing induced by the residual field becomes important and results in the excitation of a narrow distribution of Stark states.

Rate constants for collisionally-induced state changing and/or Rydberg atom destruction are measured by operating the apparatus in a pulsed mode and observing the time evolution of the Rydberg atom population in the interaction region. The output of the laser is formed into a series of pulses of ~1-2 µs duration with a pulse repetition frequency of ~5 kHz using an acousto-optic modulator. Excitation occurs in the presence of target gas and collisions are allowed to occur for a predetermined time interval, typically 1-5 µs, whereupon the number and/or excited state distribution of Rydberg atoms remaining in the interaction region are determined using selective field ionization (SFI). A voltage ramp is applied to the

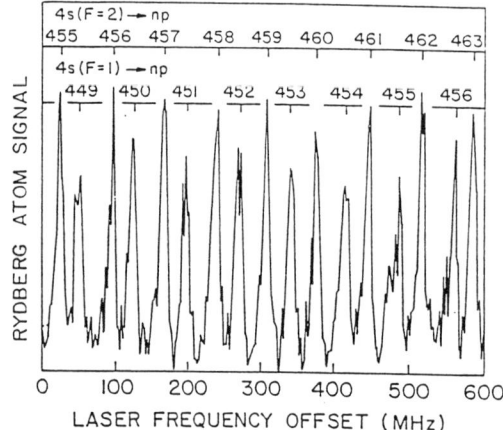

**FIGURE 2.** Rydberg atom production observed in the vicinity of $n=460$. Two well-resolved Rydberg series are evident that correspond to excitation from the ground-state F=1 and F=2 hyperfine levels.

lower electrode and the electrons resulting from field ionization are detected by an electron multiplier. Because atoms in different Rydberg states ionize at different applied fields, measurement of the field ionization signal as a function of applied field can provide a measure of the distribution of excited states present in the experimental volume immediately prior to application of the field. Rate constants for Rydberg atom destruction are determined by measuring the time evolution of the Rydberg atom population. The oscillator strengths for excitation to high-$n$ states are very small and the probability that a Rydberg atom is created during any laser pulse is low ($\leq 0.02$). The time development of the Rydberg atom population is therefore determined by accumulating data following many laser pulses.

## ELECTRON ATTACHMENT

Rydberg atom destruction in collisions with (non-polar) targets such as $CCl_4$ that attach low energy electrons is associated primarily with ionization through the electron transfer reaction

$$K(n\ell) + CCl_4 \rightarrow K^+ + CCl_4^{-*} \rightarrow K^+ + CCl_3 + Cl^- \qquad (1)$$

in which, in essence, the excited electron is captured by the target molecule. Measured rate constants $k_d$ for destruction of high-$n$ atoms in collisions with $CCl_4$ are shown in Fig. 3(a) as function of $n$ (5). The independent-particle model asserts that the rate constant for collisional destruction *via* such reactions should equal that for capture, by the target, of free electrons having the same velocity distribution as the Rydberg electron, i.e.,

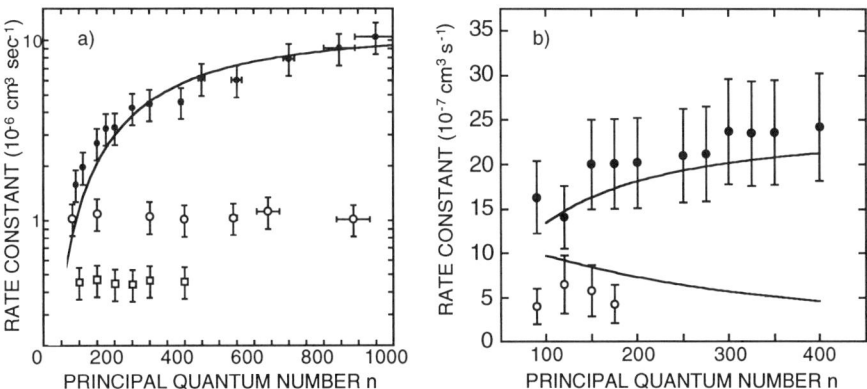

**FIGURE 3.** a) Rate constants for Rydberg atom destruction in collisions with CH$_3$Cl (●), CCl$_4$ (O), and H$_2$S (□). The solid line is a fit to the CH$_3$Cl data using Eq. (7) and assuming a virtual (or bound) state energy of 55 μeV. b) Rate constants for ℓ-changing (O) and collisional destruction (●) in K($n$p)-HF collisions. The solid lines show the results of model calculations.

$$k_d = \int_0^\infty v\sigma_e(v)f_R(v)dv \quad (2)$$

where $f_R(v)$ is the Rydberg-electron velocity distribution and $\sigma_e(v)$ the cross section for capture of *free* electrons with velocity $v$ (1). It is evident from Fig. 3(a) that at high $n$ the measured rate constants $k_d$ are essentially independent of $n$, i.e. independent of the Rydberg electron velocity distribution. Inspection of Eq. (2) shows that this requires that the attachment cross section $\sigma_e(v)$ vary as $1/v$, consistent with the Wigner threshold law for an s-wave inelastic process (6). The cross section $\sigma_e(v)$ inferred from the Rydberg atom data is presented in Fig. 4 together with velocity-averaged cross sections $\bar{\sigma}_e$ obtained using the expression $\bar{\sigma}_e = k_d / v_m$, where $v_m$ is the median velocity of the attached electrons, i.e., the Rydberg electron velocity such that integration of Eq. (2) from 0 to $v_m$ yields a value one-half that for integration form 0 to ∞. The median velocity can be calculated using the electron velocity distribution for $n$p states which is shown in Fig. 5(a) for $n$=1000 (7). The distribution is broad and peaks at low velocities, a consequence of the long time that the electron spends near the outer classical turning point of its (highly-elliptical) orbit. At very high $n$, however, the motion of the Rydberg atom relative to the target molecule becomes significant and influences the electron velocity distribution as observed in the rest frame of the target molecule. To illustrate the magnitude of this effect, Fig. 5 includes electron velocity distributions calculated including the effects of relative motion which sets a lower limit of ~1 μeV on the electron energies that can be investigated without cooling the collision partners. (At very high $n$, a mixture of Stark states is excited but when relative heavy particle motion is taken into account calculations indicate that the

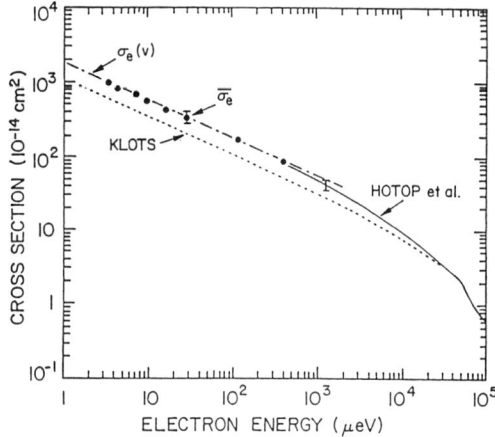

**FIGURE 4.** Cross section for electron attachment to $CCl_4$. ●, $\bar{\sigma}_e$ -K($np$); — - —, $\sigma_e(v)$ -K($np$); ———, photoelectron data (8); ------- theory (9).

median collision energies are similar to those for $np$ states.) At each $n$, the median collision energy $mv_m^2/2$ is significantly lower than the corresponding mean kinetic energy of the Rydberg electron. For comparison, Fig. 4 also includes the photoelectron data of Hotop and coworkers (8) and the cross section predicted by a theoretical expression given by Klots (9). The good agreement between the various data confirm the validity of the free electron model.

## COLLISIONS WITH POLAR TARGETS

In Rydberg atom collisions with polar targets, molecular rotational energy can be transferred to the excited electron leading to near resonant population of localized groups of higher-$n$ states through so-called $n$-changing reactions

$$K(n\ell) + ABC(J) \rightarrow K(n'\ell') + ABC(J-1) \qquad (4)$$

or, if the energy transfer is sufficient, to ionization

$$K(n\ell) + ABC(J) \rightarrow K^+ + ABC(J-1) + e^- \qquad (5)$$

At very high $n$, the electron binding energy is sufficiently small that nearly all rotational transitions will lead to ionization and measurements of the resulting Rydberg atom destruction can provide information on inelastic electron-dipole scattering. Quasi-elastic scattering of the electron by the target can also be important for strongly polar molecules and can lead to so-called $\ell$-changing reactions that populate states having a small range of $n$ close to the initial value but a near-statistical mix of $\ell$. The products of these state-changing reactions can be treated as a single "mixed" population which can also undergo collisional ionization (10). The electron velocity distribution $f_R(v)$ appropriate to such a population is

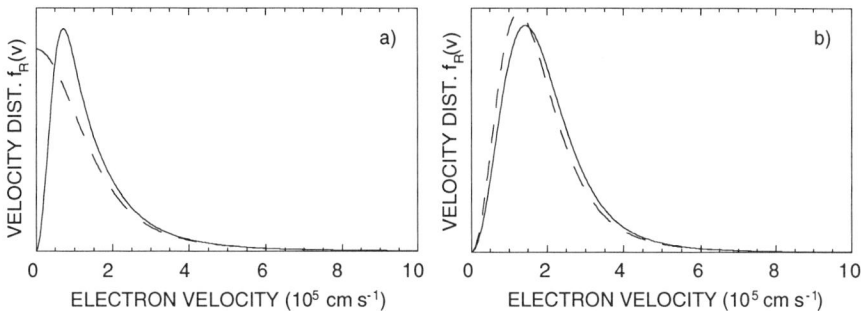

**FIGURE 5.** Rydberg electron velocity distributions $f_R(v)$ for $n$=1000 for (a) a p-state and (b) an $\ell$-mixed population both including (———) and excluding (— — —) the effects of relative motion.

plotted in Fig. 5(b) for $n$=1000 and is quite different from that for $n$p states (11). Furthermore, because it peaks at non-zero electron velocities, it is less strongly perturbed by relative heavy particle motion.

The Rydberg atom population in the interaction region may comprise both laser-excited "parent" atoms and the mixed population. The time dependence of the total Rydberg atom population is therefore governed by the rate constants for state changing, and for collisional ionization of parent atoms and of the mixed population. It is difficult to extract these separately from experimental measurements. However, two limiting cases can be identified that greatly simplify this analysis (12). If the rate constant for $\ell$-changing is small then for low target densities and short collision times the number of parent atoms will greatly exceed the number of state-changed atoms. In this case, Rydberg atom destruction results primarily from collisional ionization of parent atoms. Many strongly polar molecules however have relatively large rate constants for state-changing. For high target gas densities and long collision times the number of state-changed atoms will greatly outnumber the remaining parent population and Rydberg atom destruction is associated primarily with the destruction of mixed atoms.

Measured rate constants for Rydberg atom destruction in collisions with $H_2S$ and $CH_3Cl$ are shown in Fig. 3(a) (5,13). On the basis of the independent particle model, these rate constants should equal those for rotational deexcitation of target molecules by free electrons having the same velocity distribution as the Rydberg electron in accordance with Eq. (2), where $\sigma_e(v)$ is now the cross section for superelastic scattering of free electrons with velocity $v$. Since the measured rate constants for $H_2S$ are independent of $n$, i.e., independent of $f_R(v)$, this requires that $\sigma_e(v)$ again scale as $1/v$, consistent with the Wigner threshold law for an s-wave process. This is, at first sight, surprising because the Wigner threshold laws are not expected to apply for scattering by a $1/r^2$ dipole potential. The electron-target interaction lasts so long, however, that molecular rotation becomes important (14). $H_2S$ is an asymmetric top and, over sufficiently long times, its dipole moment will, as observed in a laboratory-fixed frame, average to zero.

Nonetheless, the electron can still induce a net dipole moment in the target, but since the resulting interaction potential will decrease faster than $1/r^2$, it is not unreasonable that the Wigner $1/v$ threshold law might apply. (For a static dipole, $\sigma_e(v)$ is expected to scale as $1/v^2$, which would predict a rate constant that increases linearly with $n$ (15).)

The behavior observed with $H_2S$ is quite different from that noted with $CH_3Cl$, which has a substantially larger dipole moment (1.89 D compared to 0.97 D for $H_2S$). For $CH_3Cl$, the rate constants for collisional destruction increase markedly with $n$. Since all rotational transitions in $CH_3Cl$ will, for the range of $n$ encompassed in Fig. 3(a), lead directly to ionization, the increase in $k_d$ at high $n$ cannot be attributed to an increase in the number of rotational transitions that can lead to ionization. $CH_3Cl$ is a prolate symmetric top and, unlike $H_2S$, the target dipole moment (as observed in a laboratory-fixed frame) will not average to zero over the duration of a collision. However, at room temperature rotational states with large $J$ but small $K$ tend to be populated so that the non-averaged component of the dipole moment is small. Based on studies of near-threshold photodetachment (15), detailed balance would suggest that under these conditions (and in the absence of other effects) $\sigma_e(v)$ should vary as $1/v^{2-x}$ where $x$ (~0.95 for $CH_3Cl$) is a parameter related to the non-averaged component of the dipole moment. Use of such a form for $\sigma_e(v)$ in Eq. (2), together with the electron velocity distribution appropriate to a statistical distribution of $\ell$ states ($CH_3Cl$ induces rapid $\ell$-changing), yields only a weak $n$-dependence in $k_d$ that is inconsistent with the observations (5).

The experimental observations, however, can be explained if dipole-supported bound or virtual states of near zero energy are important in electron-$CH_3Cl$ scattering. A static dipole with a dipole moment greater than 1.63D can support an infinite number of bound states (17). However, if the effects of molecular rotation are included this number becomes finite or zero (18). For example, HF, has a supercritical dipole moment, 1.83 D, but does not retain any bound states when rotation is included. Nonetheless, theory suggest that, at least for the ground rotational state, a virtual $HF^-$ state exists, although this result is tentative due to the very diffuse nature of dipole-supported states (14). Also, recent studies have shown that many molecules with somewhat larger dipole moments can form dipole-bound negative ions, even if rotationally-excited (19). The presence of dipole-supported bound or virtual states introduces resonances in the cross section for superelastic free electron scattering which, in the first approximation, assumes the form

$$\sigma(v) = \frac{a}{v^{2-x}(v^{2x} + \kappa^{2x})} \qquad (7)$$

where $-\kappa^2/2$ is the energy of the state (20). The best fit to the $CH_3Cl$ experimental data obtained using a cross section of this form (and $x=0.95$) in Eq. (2) is shown in Fig. 3(a) and corresponds to a virtual or bound state energy of ~55 μeV.

Recent measurements with HF offer further evidence of the importance of dipole supported states in electron-polar molecule scattering at low energies.

Measured rate constants for total Rydberg atom destruction ($k_d$), and for $\ell$-changing ($k_\ell$) in collisions with HF are shown in Fig. 3(b) which also includes the results of model calculations that take into account dipole-supported states. The electron-HF interaction is modeled by a combination of the dipole potential and a short-range interaction containing one free parameter that determines the value of the bound-or virtual-state energy of the negative ion at $J=0$, where $J$ is the total angular momentum. The system of rotational close-coupling equations corresponding to this interaction is solved by numerical integration for each $J$, neglecting all scattering channels where the electron angular momentum is higher than 2. The rate constants for ionization ($\ell$-changing) for each $J$ are determined from the calculated cross section for rotational deexcitation (elastic scattering) using Eq. (2). These rate constants are then averaged over the $J$-distribution appropriate to a thermal target. Reasonable agreement between experiment and theory for both $k_d$ and $k_\ell$ is obtained by assuming the existence of a virtual state of HF$^-$ ($J=0$) with an energy of ~1-1.5 meV. This agrees with earlier estimates which suggests that a virtual state exists with an energy higher than 0.37 meV (21,14). For $J>0$, due to rotational coupling, the virtual-state energy moves off the real axis and the corresponding singularities of the S-matrix manifest themselves as Feshbach or shape resonances rather than pure bound or virtual states.

The present work demonstrates that very-high-$n$ Rydberg atoms permit study of electron-molecule scattering in a new electron energy regime that extends down to a few µeV, equivalent to effective collision temperatures of below 100 mK. (Indeed, very-high-$n$ atoms may be viewed as a microscopic low-energy electron trap with the trapping potential provided by the core ion.) Measurements of a wide variety of elastic and inelastic processes appear possible and promise new insights into electron-molecule scattering at ultra-low electron energies.

This research is supported by the National Science Foundation and the Robert A. Welch Foundation.

## REFERENCES

1. For a further discussion of independent particle model see, for example, the articles by Matsuzawa M., Hickman A. P. and Olson R. E., Pascale J., in *Rydberg States of Atoms and Molecules*, edited by R. F. Stebbings and F.B. Dunning, New York: Cambridge University Press, 1983.
2. Ling X., Lindsay B. G., Smith K. A., and Dunning F. B., *Phys. Rev. A.* **45**, 242-6 (1992).
3. Lindsay B. G., Smith K. A., and Dunning F. B., *Rev. Sci. Instrum.* **62**, 1656-7 (1991).
4. Frey M. T., Ling X., Lindsay B. G., Smith K. A., and Dunning F. B., *Rev. Sci. Instrum.* **64**, 3649-50 (1993).
5. Frey M. T., Hill S. B., Smith K. A., Dunning F. B., and Fabrikant I. I., *Phys. Rev. Lett.*, in press.
6. Wigner E. P., *Phys. Rev.* **73**, 1002-9 (1948).
7. Gryzinski M., *J. Physique Lett.* **43** L425-30 (1982); Klar D., Mirbach B., Korsch H. J., Ruf M. -W., and Hotop H., *Z. Phys. D* **31**, 235-44 (1994).
8. Hotop H., private communication (1994).
9. Klots C. E., *Chem. Phys. Lett.* **38**, 61-4 (1976).

10. Kellert F. G., Smith K. A., Rundel R. D., Dunning F. B., and Stebbings R. F., *J. Chem. Phys.* **72**, 3179-90 (1980); Kalamarides A., Goeller L. N., Smith K. A., Dunning F. B., Kimura M., and Lane N. F., *Phys. Rev. A.* **36**, 3108-12 (1987).
11  McDowell M. R. C., and Coleman J. P., *Introduction to the Theory of Ion-Atom Collisions* Amsterdam: North-Holland, 1970, ch. 3, pp. 105-7.
12. Ling X., Frey M. T., Smith K. A., and Dunning F. B., *Phys. Rev. A.* **48**, 1252-6 (1993).
13. Frey M. T., Hill S. B., Ling X., Smith K. A., Dunning F. B., and Fabrikant I. I., *Phys. Rev. A.* **50**, 3124-8 (1994).
14. Fabrikant I. I., *J. Phys. B* **16**, 1269-81 (1983).
15. Fabrikant I. I., *J. Phys. B* **16**, 1253-67 (1983).
16. Engelking P. C., Phys. Rev. 26, 740-5 (1982).
17. Wightman A. S., *Phys. Rev.* **77**, 521-8 (1950).
18. Garrett W. R., *Chem. Phys. Lett.* **5**, 393-7 (1970); *Phys. Rev. A* **3**, 961-72 (1971); *J. Chem Phys.* **73**, 5721-5 (1980).
19. Defrançois C., Khelifa N., Lisfi A., and Schermann J. P., *Z. Phys. D*, **21** 177-84 (1994); Popple R. A., Finch C. D., and Dunning F. B., *Chem. Phys. Letts* **234**, 172-6 (1995).
20. Fabrikant I. I., *Sov. Phys. JETP* **46**, 693-97 (1977); *J. Phys. B* **11**, 3621-33 (1978).
21. Jordan K. D., and Wendoloski J. J., *Chem. Phys.* **21**, 145-54 (1977).

# On the Ionisation Mechanism of Reflection (e,2e) Events

S. Iacobucci[#+], P.Luches[*], L. Marassi[*], R. Camilloni[#], B. Marzilli[&],
S. Nannarone[*] and G. Stefani[&]

[#] Istituto di Metodologie Avanzate Inorganiche del CNR, Area della Ricerca di Roma, CP10 00016 Monterotondo Scalo, Italy

[&] Dipartimento di Fisica "E. Amaldi" and Unita' INFM, III Universita' di Roma, P.le A. Moro 2, 00185 Roma, Italy

[+]Laboratoire pour l' Utilisation du Rayonnement Electromagnetique, Centre Universitaire Paris-Sud, 91405 Orsay CEDEX, France

[*]Dipartimento di Fisica Universita' di Modena, via G. Campi 213A, 41100 Modena, Italy

**Abstract** The possibility of using the reflection grazing angle (e,2e) technique as a binding energy and/or momentum spectroscopy of surface states rests on the accurate knowledge of the ionisation process. Two possible mechanisms are envisaged that can generate pairs of correlated electrons in the reflection geometry: a single inelastic collision at large momentum transfer or a double collision (elastic plus inelastic). In this paper are presented the results of (e,2e) experiments that allow to elucidate the ionisation mechanism at intermediate energies (300 eV) and asymmetric kinematics. The measurements, performed on highly oriented pyrolitic graphite, also show that an overall energy resolution as good as 1.2 eV can be achieved.

## INTRODUCTION

Today most calculations of the electronic structure of solids are performed within density-functional theory. The approach relies on the existence of a model system of N non-interacting fermions, which has the same charge density as the actual interacting system. An effective one-particle potential exists, which generates this charge density, and it contains a term that is known only approximately. One of the most relevant feature for an interacting homogenous electron gas is that the occupation numbers in the momentum space are no longer described by a simple step function. In the presence of correlations there is a finite probability that the momentum states outside the Fermi surface are occupied. Hence, the expectation value of the momentum density is thought to be more sensitive than charge density or total energy to the presence of correlations in the ground state of the system. A sensitive experimental test for the momentum density should provide with a better knowledge of the correlation term for the aforementioned effective potential. This is just one example for the interest in directly testing momentum density in solids and surfaces in particular.

In spite of the large number of spectroscopies currently used to characterise the electronic structure of surfaces [1], none of them is capable of measuring the momentum distribution. To the best of our knowledge, only (e,2e) spectroscopy [2] or coincidence Compton scattering [3] can provide a direct determination of the electron momentum density of a quantum state. Both of these spectroscopies have already shown their capability to investigate bulk properties of solids. (e,2e) has already produced a fearly large body of investigation on thin films momentum distributions [4]. On the contrary, application to surfaces is in its infancy for (e,2e).

If momentum distributions can be measured only by coincidence experiments, binding energies for vanishing real momenta are measured with some difficulties by current surface spectroscopies, i.e. angle resolved photoelectron spectroscopy. This is specially true for disordered systems where, because of the low symmetry, it is not possible to reduce the real momenta to crystal momenta confined to the origin of the Brillouin zone. In photoionisation experiments this difficulty originates from the evanescence of the photon momentum that implies real momenta of the initial state which are directly proportional to the photoelectron energy. In electron scattering experiments this constraint on the momentum is not any more present in as much as for each value of energy transferred in the ionising collision does exist a continuum of kinematically allowed transferred momenta. To take full advantage of the kinematical flexibility of the electron collision processes, both free electrons produced by the ionisation are detected coincident in time thus allowing a complete balance of energy and momentum. For such experiments the acronym (e,2e) is usually adopted; they represent a unique tool for measuring the momentum distribution of an energy selected state as well as its dispersion in a zone of real momenta that extends all the way to zero value.

Even though the first application to surfaces was envisaged [5] shortly after the first (e,2e) experiment on a thin solid film [6], it has been only recently that the feasibility of such an experiment has been demonstrated.

The (e,2e) technique is not inherently surface sensitive. In order to achieve surface sensitivity, transmission mode experiments have been performed upon high energy asymmetric kinematics, relying on the shortness of the low ejected electrons mean free path [7]. Experiments performed in grazing angle reflection geometry are expected to be surface sensitive in as much as the penetration depth of the fast electrons, diffused and incoming, does not exceed the topmost layers of the sample. The existence of pairs of correlated electrons ejected from the surface under electron collision has been demonstrated by Kirschner et al. [8] at normal incidence, while Iacobucci et al. [9] have shown the feasibility of a binding energy spectroscopy with quasi-momentum resolution performed in grazing angle geometry.

Two main questions remain to be answered in order to fully exploit the potential capabilities of this latter spectroscopy: i) what is the limit for the overall energy resolution; ii) what is the dominant mechanism for production of the pair of ejected electrons. Both questions are addressed by this paper. It shows that energy resolution comparable to those characteristic of the best transmission (e,2e) experiments has been already achieved and that a mechanism of incoherent elastic reflection and inelastic ionising collision constitutes the framework for the interpretation of these experiments.

## REFLECTION (e,2e)

The (e,2e) spectroscopy consists of measuring simultaneously the energy $E_0$ of the incident electron, the energies $E_e$ and $E_s$ of the two final electrons, and the probability of their being emitted into solid angles $d\Omega_e$ and $d\Omega_s$ oriented along the directions $\theta_e$ and $\theta_s$ respectively (the five fold differential cross section). A schematic of the kinematics is given in figure 1.

Depending on the amount of momentum transfer $\vec{K} = \vec{K}_o - \vec{K}_s$, two approximations can be used to describe the (e,2e) process. In the dipolar limit, i.e. vanishing momentum transfer, the (e,2e) mechanism is equivalent to photoionisation [10] and binding energy ($\varepsilon(\vec{q})$) spectroscopy is possible. The dipolar (e,2e) differs from photoionisation in as much as the momentum associate with the transition, $\vec{q} = \vec{K}_e + \vec{K}_s - \vec{K}_o$, is not uniquely determined and can be changed by changing the geometry of the process while keeping fixed the energy balance $\varepsilon(\vec{q}) = E_0 - E_s - E_e$. In the impulsive (binary) limit (momentum transfer roughly equals the ejected electron momentum) the spectral momentum density of the electrons bound in the target can also be measured [11]. Several experiments performed on gaseous targets [12] and thin films have shown that the impulsive condition can be satisfied both in symmetric ($E_s = E_e$) and asymmetric conditions; hence both kinematics permit measurement of the momentum distribution. The potentiality of (e,2e) spectroscopy on surfaces was theoretically investigated by D'Andrea and Del Sole [5] under symmetric reflection kinematics. In a recent work, Iacobucci et al. [9] demonstrated the feasibility of such an experiment in asymmetric reflection kinematics and suggested that the ionisation mechanism amounts to an incoherent elastic reflection plus an

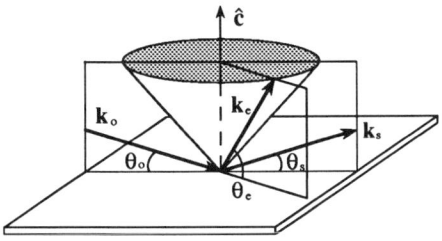

**FIGURE 1.** Kinematics of the (e,2e) experiment in grazing angle reflection geometry.

inelastic ionising collision rather than a single inelastic collision. Theoretical fundamentals to this double collision model can be found in Mills' calculation for the electron scattering cross-section at surfaces [13]. Experimental support comes from angle resolved electron energy loss (AREEL) measurements that have shown this mechanism to be the dominant one for valence [14] and core [15] ionisation, while a two-step process was also used to explain the electron impact induced spin polarization which was observed in secondary electrons emission experiments [16].

In the Plane Wave Impulse Approximation [17] and within the non interacting bound particle models, it is easily verified that the (e,2e) five fold differential cross section can be written (with implicit antisymmetrization ) [9] as:

$$\frac{d^5\sigma}{d\Omega_s d\Omega_e dE} \propto \frac{K_s K_e}{K_o} \times \left| \sum_t \left\langle \vec{K}_e \cdot \vec{K}_s \cdot \prod_{j=1, j\neq t}^{N-1} \Psi_q(\vec{r}_j) \middle| v_{o,t} \middle| \vec{K}_o \cdot \prod_{j=1}^{N} \Psi_q(\vec{r}_j) \right\rangle \right|^2$$

where E satisfies the energy balance, t and j are valence electron indexes and $\Psi_q(\vec{r}_j)$ is a one electron Bloch function with energy $\varepsilon(\vec{q})$ and crystal momentum $\vec{q}\,' = \vec{q} + \vec{G}$, $\vec{G}$ being a reciprocal lattice vector.

The main features of the five fold differential cross section are readily understood when a simplified description of the bound states is adopted. In the quasi-free electron model, $\Psi_q(\vec{r}_j)$ can be written as linear combination of plane waves. Furthermore, in the frozen core assumption the cross section becomes

$$\frac{d^5\sigma}{d\Omega_s d\Omega_e dE} \propto \frac{K_s K_e}{K_o} \cdot \frac{1}{K^4} \cdot \sum_{\vec{G}} \left| C_{\vec{q}\,'-\vec{G}} \right|^2 \qquad (1)$$

where the $C_i$'s are coefficients of the ground state wave function of the linear expansion. Therefore, the five fold differential cross section factorizes in a dynamic factor depending only upon the kinematics of the collision (first two terms of (1)), times a form factor which is solely determined by the target electronic structure. From equation (1) it can be also readily seen that the angular distribution of the scattered electrons is symmetric around the direction of the primary electron of the ionising inelastic collision.

Furthermore, whenever the ion recoil momentum $\vec{q}$ is identifiable with the single-electron crystal momentum (impulse approximation), the (e,2e) process provides a momentum spectroscopy of the target. The symmetry of the coincidence scattered electron distribution is therefore expected to be a sensitive test of the ionisation mechanism. Namely, if the double scattering model is dominant, the distribution will be symmetric along the direction of the specularly reflected beam ($\vec{K}'_0$) rather than along the direction of the primary beam ($\vec{K}_o$) and in equation (1) the momentum transfer $\vec{K}$ will be replaced by the specular momentum transfer $\vec{K}' = \vec{K}'_0 - \vec{K}_s$.

# EXPERIMENTAL

The experimental equipment is based on an Ultra High Vacuum (UHV) chamber containing an Electron Gun (EG), two electron analyzers - a Hemispherical Deflector Analyzer (HDA) and a Cylindrical Mirror Analyzer (CMA) -, and a sample holder mounted on a five-degrees of freedom manipulator. The permits to work in reflection geometry and at small grazing incidence ($\theta_0 = 0^o - 17^o$).

The EG is a commercial gun (Leybold EQ11/35) that has been modified in order to automatically scan the electron beam energy in the range 0-5000 eV under PC control.

The sample holder is mounted on a XYZ translator allowing a $\pm 180°$ rotation around the Z-axis and a $\pm 10°$ tilt of the sample to adjust the grazing angle $\theta_0$. The target temperature is controlled from -120 °C to 900 °C, thus permitting to condense gaseous and molecular specimen on the target surface.

The HDA consists of a dispersive hemispherical element equipped with a three-element electrostatic lens and it can rotate in the scattering plane ($\theta_s = 0^o - 12^o$). The analyzer has been optically aligned in air by using a laser beam and achieving an angular accuracy and repeatability of $\pm 0.2°$. Working in the constant pass energy mode the HDA features an energy resolution of 0.4 eV at 300 eV. It is used to analyze the fast scattered electrons. Their energy distribution is obtained by scanning the electron beam energy while keeping the HDA tuned at a fixed energy.

The CMA is a commercial (Riber) single-pass, modified to optimize the luminosity according to a numerical simulation. It is mounted with its main axis normal to target surface and it is used to detect Auger and low energy emitted electrons working in $\Delta E/E$ constant mode. Equipped with an exit slit of 1.5mm diameter it features an energy resolution of about 2% (FWHM), an angular acceptance of $3°$ and a field of view of about $\pm 1$mm.

The coincidence spectrometer has been characterized using gaseous targets by measuring (e,2e) spectra relative to the He 1S ionization. An (e,e'Auger) spectrum has also been measured by detecting the scattered electron after ionizing the Cl 2p core level in coincidence with the associated $L_{2,3}VV$ Auger electron in gaseous $CCl_4$ molecule.

Energy and angular resolutions of the two analyzers have been sacrified in order to obtain an optimization of the areas individually imaged by the two analyzers with that excited by the EG. This is of crucial relevance in order to maximize the overall luminosity of the coincidence spectrometer. In particular, for solids targets the HDA is operated at low-energy resolution (resolving power $\cong 3 \cdot 10^2$). The best overall energy resolution achieved is 1.2 eV and mostly limited by the HDA resolution.

The coincidence electronic chain is a conventional one [18]. The fast electronic pulses coming from the two single channel multipliers are fed through two fast amplifiers to constant fraction discriminators. The shaped pulses are then sent directly to the scalers to measure "single" count rates and, to establish time correlation between them, to a time multichannel analyzer (Lecroy 3001QVT) which yields the probability distribution of delay time between pulses (time spectrum). The time resolution of the spectrometer is about 12ns as measured on solid targets and it is largely due to the time spread of the trajectories in the two analysers. The *true* coincidence rate $I_t$ must be discriminate from the nearly flat background of the uncorrelated events (*accidental* coincidence) which occurs at a rate $I_a$. The optimal acquisition time is achieved when the ratio $I_t / I_a$, which depends on the incident current $I_o$, is about one [18]. In the best case we measured $I_t \cong 0.05$Hz with a ratio $I_t / I_a \cong 0.5$ for $I_o \cong 1$nA.

For the measurements of this paper a highly oriented pyrolitic graphite (HOPG) sample has been used. It was prepared according to UHV standard procedures by peeling in air and annealing at about 700°C at residual pressure of $1 \cdot 10^{-10}$ mbar. Working with 300 eV primary electrons at a typical grazing angle value of 5° a probe depth less than 1 Å is expected. Cleanliness and orientation of the surface were checked by Auger electron spectroscopy and AREEL measurements. The best monitor for surface cleanliness was the persistence of the sharp $\pi \rightarrow \pi^*$ transition at 6.2 eV in the AREEL spectrum and the presence of a narrow angular distribution (FWHM=2.3°) for the peak of specularly reflected elastic electrons. Reproducibility of these measurements over a period of six weeks has been assumed as a guarantee for the stability of the surface conditions and of the energy calibration during the long acquisition time needed for coincidence measurements.

## RESULTS

Two sets of measurements have been performed for two different grazing angles ($\theta_o$= 6.7° and 4.7°) always detecting 300 eV scattered electrons correlated in time with 8eV ejected electrons. The measurements performed under strict reflection geometry ($\theta_o = \theta_s$) were aimed at measuring the binding energy spectrum, while those performed at fixed binding energy and variable $\theta_s$ were aimed at measuring the symmetry of the five fold differential cross section.

In figure 2 is reported the HOPG binding energy spectrum as measured at $\theta_o = \theta_s$= 6.7°. The vacuum level is the origin for the binding energy scale which has been derived from the energy conservation law.

Under the assumptions used to derive the cross section in equation (1), the features in the binding energy spectrum are to be interpreted on the basis of the band structure of graphite. The chosen kinematics imply for the reconstructed momentum $\vec{q}$ non vanishing components ($q_p$ and $q_n$) both in the parallel and perpendicular directions to the HOPG $\hat{c}$ axis. Taking into account the finite analyser angular acceptances, the spectrometer work function and the surface potential barrier of the sample [9], the reconstructed values of $q_n$ range from nearly the middle to the boundaries of the first Brillouin zone in the $\Gamma$MK plane (from 0.65 to 1.53 Å$^{-1}$). The $q_p$ component ranges between 1.67 and 2.08 Å$^{-1}$, i.e. from

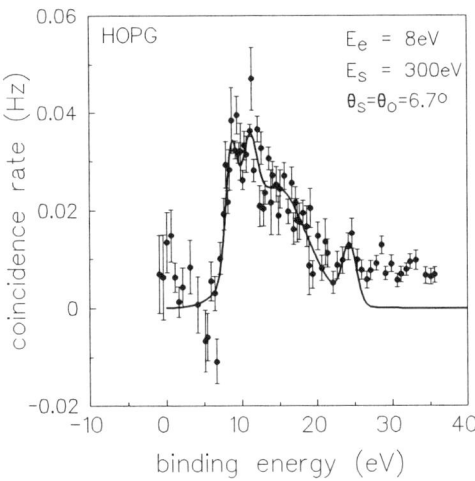

**FIGURE 2.** Coincidence spectrum from HOPG surface. Error bars represent one standard deviation for raw data. The continuous line is the best fit to the spectrum, see text for details.

bottom to middle of the third Brillouin zone along the $\Gamma$A direction.

According to valence band calculation [19] and on the basis of the volume of momentum space sampled by the experiment, four individual contributions from valence states are expected in the range of binding energies investigated. Consequently, the measured spectrum has been fitted by a least-square method, simulating the expected individual transitions by means of Gaussian functions of variable height, width and position. Experimental data for (e,2e) on solid samples are affected by multiple losses that result in enhanced transition intensities at larger binding energies. A previous work [9] has already shown that in grazing angle reflection geometry, binding energy and width of the individual transitions are essentially unchanged by multiple scattering effects that will only affect by a few percent the observed (e,2e) intensity, so it was not felt necessary to correct for multiple loss effects.

The improved energy resolution of this experiment (1.2 eV) over previous measurements (3 eV) [9], even though not yet satisfactory if compared to other well established techniques, is nonetheless sufficient to allow for a clearer interpretation of the (e,2e) spectrum of figure 2. In the binding energy region from 5 to 11 eV, the peak associated to the ionisation of the $\pi$ band from the $\sigma_3$ one is clearly resolved. According to the HOPG momentum density [20], in the volume of real momenta sampled by the present experiment the four main bands display similar values of the momentum probability; hence they should yield similar amplitude for the (e,2e) cross section. Consequently, based on the calculated dispersion band [19], four main components to the (e,2e) binding energy spectrum are to be expected roughly centered at 3.7, 6, 10 and 19.3 eV respectively for the $\pi$, $\sigma_3$, $\sigma_2$, and $\sigma_1$ bands. The best fit to the energy spectrum (continuous line of figure 2) locates the position of .the individually contributing peaks very close to the expected binding energy value. Furthermore, the $\sigma_2$ peak is broader than the others, this being in agreement with the large energy dispersion of the band in the momentum region investigated. The peak corresponding to the $\sigma_1$ band is deeper than expected by about 1 eV. This discrepancy is overcome if we assume for its dispersion curve the one measured in a transmission (e,2e) experiment [19] instead of the calculated one [18].

The angular distribution of the scattered electron has been measured tuning the spectrometer at the peak of the $\pi$ band feature of the (e,2e) spectrum. The measurement has been done at two grazing angles' $\theta_o$= 4.7º and 6.7º, and for the scattered angle variable around the specular reflection. The measure at the smaller grazing angle is shown in figure 3 and the symmetry of the five fold differential cross section around the specular reflection direction is evident. Similar measurements at 6.7º grazing angle, not shown in the figure, display similar

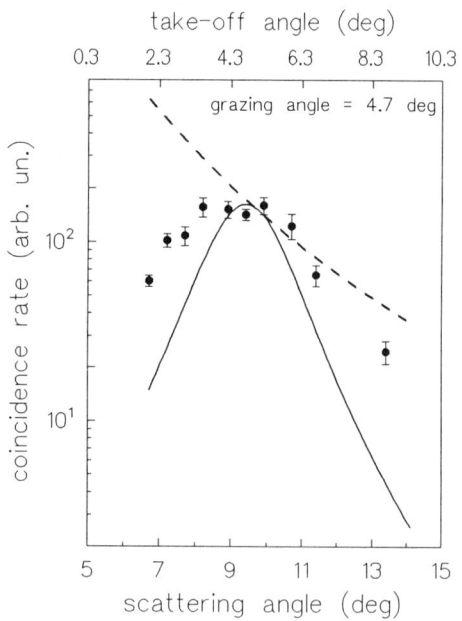

**FIGURE 3.** Angular distribution of the (e,2e) cross section at a grazing angle of 4.7º. The take off angle is $\theta_s$ while the scattering angle is $\theta_o + \theta_s$. Continuous and dashed lines are the kinematical factors for the double and single collision models respectively

symmetry around the elastically reflected beam. Regardless of the collisional model adopted, because of the CMA large accepted solid angle, as a first approximation, the form factor of (1) doesn't depend upon $\theta_s$.

On the other hand, the $\theta_s$ dependence of the kinematical factor drastically changes if a single or a double collision model is adopted. It can be readily seen if one explicitly looks at the momentum transfer dependence on the scattering angle: $K = [K_o^2 + K_s^2 - 2 K_o K_s \cos (\theta_o + \theta_s)]^{1/2}$ for the single-event model, which bears a symmetry for the cross section around the $\vec{K}$ direction (i.e. forward direction), and $K' = [K_o^2 + K_s^2 - 2 K_o K_s \cos (\theta_o - \theta_s)]^{1/2}$ for the double step model, which bears a symmetry around the $\theta_o = \theta_s$ direction, (i.e. specular reflection direction).

The relative cross section, as predicted by the two collisional models under the assumption of constant form factor is shown in figure 3. The shape of the measured cross section clearly supports the double collision hypothesis.

The experimental cross section is broader than the shape of the simple kinematical factor shape because the variation of the form factor is not taken into account in this calculation. It is finally to be mentioned that the measured coincidence angular distribution is very close to the correspondent non coincident energy loss distribution. This is a good evidence for a single dominant mechanism for all of the inelastic processes at glancing angle, i.e. the double collision one.

Further support to the double step over the single step model comes from the comparison between the reconstructed $q_p$ momenta in the two cases. If a single collision model is adopted $q_p$ should nearly range from 4.4 to 4.8 Å$^{-1}$ for the kinematics used to measure the spectrum of figure 2. This is a momentum space region where the graphite valence electrons momentum density is vanishing [19], so one should have no intensity at all in the (e,2e) spectrum.

## CONCLUSIONS

The present investigation confirms the possibility to use grazing angle reflection (e,2e) events to build up a binding energy spectroscopy with partial momentum resolution. The energy resolution achieved in this geometry is at least as good as the one obtained by the transmission (e,2e) experiments on thin films, i.e. an overall resolution on binding energy of 1.2 eV. The first necessary, (even though not sufficient) condition for realising a bound state momentum spectroscopy based on grazing (e,2e) experiments is the accurate knowledge of the mechanism that generates the pairs of coincident electrons detected by this spectroscopy. The present investigation clearly demonstrates that, at least under the kinematics investigated, the double collision mechanism prevails over the single inelastic collision one.

## ACKNOWLEDGMENTS

We are grateful to EEC Human Capital and Mobility, Contract No. ERBCHRXCT930359 and to Progetto Finalizzato Chimica Fine CNR, for partial support of the work. One of us (S.I.) is supported by the EEC Human Capital and Mobility, Contract No. ERBCHBICT940988.

## REFERENCES

[1] H. Lüth, *Surfaces and interfaces in solids*, Springer-Verlag, Heidelberg, 1993
[2] G. Stefani, L. Avaldi, R. Camilloni, *J. de Phys IV*, Colloq. C6/3, (1993) 1
[3] J. R. Schneider, F. Bell, Th. Tschhentschez, A. J. Rollason, *Rev. Sci. Instr.* 63 (1992) 1119
[4] P. Storer, this conference
[5] A. D'Andrea and R. Del Sole, *Surf. Sci.*, 71 (1978) 306
[6] R. Camilloni, A. Giardini Guidoni, R. Tiribelli and G. Stefani, *Phys. Rev. Lett.*, 29 (1972) 618
[7] Y.Q.Cai, M.Vos, P. Storer, A.S. Kheifets, I.E. McCarthy, E. Weigold, *Phys. Rev. B*, 51 (1995) 3449
[8] J. Kirschner, O. M. Artamov and A. N. Terekhov, *Phys. Rev. Lett.*, 69 (1992) 1711
[9] S. Iacobucci, L. Marassi, R. Camilloni, S. Nannarone, G. Stefani, *Phys. Rev. B*, 51 (1995) R10252
[10] A. Hamnett, W. Stoll, G. Branton, C. E. Brion and M. J. Van der Wiel, *J. Phys. B*, 9 (1976) 945
[11] V. G. Levin, V. G. Neudachin and Yu. F. Smirnov, *Phys. Stat.Sol.* (b), 49 (1972) 489
[12] L. Avaldi, R. Camilloni, E. Fainelli and G.Stefani, *J. Phys.B: At. Mol. Phys.*, 20 (1987) 4163
[13] D.L.Mills, *Surf. Sci.*, 48 (1975) 59
[14] S. Iacobucci, P. Letardi, M. Montagnoli, P.Nataletti and G.Stefani, *J. Electron Spectrosc. Relat. Phenom.*, 67 (1994) 479
[15] M. De Crescenzi, *J. Electron Spectrosc. Relat. Phenom.*, in press
[16] G. Ravano, M. Erbudak, *Solid State Commun.*, 44 (1982) 547
[17] I.E. McCarthy, E. Weigold, *Rep. Prog. Phys*, 54 (1991) 789
[18] G. Stefani, L. Avaldi and R. Camilloni, in *New Directions in Research with Third Generation Soft X-ray Synchrotron Radiation Sources*, NATO ASI E Vol. 254, A.F. Schlachter and F.J. Wuilleumier (eds.), KluweAcademic Publisher, 1994, p. 161, and references therein
[19] R. C. Tatar and S. Rabii, *Phys. Rev. B*, 25 (1982) 4126
[20] M.Vos, P. Storer, S.A. Canney, A.S. Kheifets, I.E. McCarthy, E. Weigold, *Phys. Rev. B*, 50 (1994) 5635

# Recombination of $H_3^+$ and $D_3^+$ Ions with Electrons

R. Johnsen, T. Gougousi,
Department of Physics and Astronomy, University of Pittsburgh,
Pittsburgh, PA 15260

and M.F. Golde
Department of Chemistry, University of Pittsburgh.
Pittsburgh, PA 15260

**Abstract.** Flowing-afterglow measurements in decaying $H_3^+$ or $D_3^+$ plasmas suggest that de-ionization does not occur by simple binary recombination of a single ion species. We find that vibrational excitation of the ions fails to provide an explanation for the effect, contrary to an earlier suggestion. Instead, we suggest that collisional stabilization of $H_3^{**}$ Rydberg molecules by ambient electrons introduces an additional dependence on electron density. The proposed mechanism would permit plasma de-ionization to occur without the need for dissociative recombination by the mechanism of potential-surface crossings.

## INTRODUCTION

It is usually taken for granted that typical plasma afterglow experiments yield the <u>binary</u> dissociative recombination coefficient for molecular ions and that, in principle, the measured values would be the same as those inferred from beam experiments. The assumption seems plausible since the capture of an electron into a repulsive state of the neutral is thought to induce rapid dissociation[1], leaving little time for third-body interactions. The argument, however, presupposes the existence of suitable dissociation channels. If such do not exist, then perhaps long-lived Rydberg molecules can be formed and third-body interactions may be more important. The experiments described here suggest that this situation may exist in plasmas containing $H_3^+$ and $D_3^+$ ions.

Theoretical arguments [2,3] indicate that the recombination

$$H_3^+ + e^- \rightarrow H+H+H$$
$$\rightarrow H_2 + H \quad (1)$$

for vibrationally cold $H_3^+$ (v=0) ions by direct dissociative recombination should be slow due to a lack of suitable crossings between the ionic and neutral potential surfaces. However, vibrationally excited $H_3^+$ ions in v≥3 are predicted to exhibit faster recombination. This expectation disagreed with the early microwave-afterglow experiments [4,5] which gave fairly "normal" recombination coefficients ($\sim 2 \times 10^{-7}$ cm$^3$/s) for $H_3^+$ ions believed to be in v=0. This "$H_3^+$ dilemma" has existed for about 20 years.

The structure and energy levels of $H_3^+$ and $H_3^*$ are quite well known. A brief introduction and extensive references can be found in the "thumbnail sketch of $H_3^{+}$" by Oka and Jagod [6]. The importance of $H_3^+$ in astrophysical environments is discussed in reviews by Dalgarno[7] and by Tennyson et al.[8]. A fairly complete account of the history of dissociative recombination work has been given by Bates et al. [9].

One possible resolution of the $H_3^+$ dilemma was offered by Adams et al.[10]. They interpreted their observation of an anomalous decay of a flowing afterglow plasma by assuming the presence of two species of $H_3^+$ ions, one of which recombined fairly fast while the other recombined slowly ($\alpha < 2 \times 10^{-8}$ cm$^3$/s). By identifying the fast and slowly recombining ions as, respectively, v>3 or v=0 ions, theory and experiment could be reconciled. This solution, while being in conflict with the microwave afterglow results, seemed quite plausible until Amano[11,12] showed that spectroscopically identified $H_3^+$ ions in v=0 recombined readily with electrons ($\alpha \sim 1.8 \times 10^{-7}$ cm$^3$/s). A repetition of the flowing-afterglow measurements by Canosa et al.[13,14] gave no evidence for the existence of two distinct types of $H_3^+$ ions.

Recombination <u>cross sections</u> for $H_3^+$ have been measured repeatedly in the merged-beam apparatus at the University of Western Ontario [15, 16, 17, 18]. The results showed that the cross section varied with conditions in the ion source and this was ascribed to vibrational excitation. In later work[16], the $H_3^+$ vibrational state was deduced from the threshold energy for electron-ion dissociative excitation. The result indicated that v=0 ions recombined only slowly with electrons.

When ion storage rings became available, they were very quickly applied to $H_3^+$ recombination. In these instruments, vibrationally excited ions have time to relax by radiative transitions (including the $v_1=1$ metastable level of $H_3^+$ with a lifetime[19] of 1.2 s). Larsson et al.[20] and Sundström et al.[21], using the CRYRING storage ring, obtained recombination cross sections for $H_3^+$ (v=0) over the wide energy range from 0.0025 to 30 eV. The analysis of the experimental cross sections yielded a recombination coefficient of $(1.15 \pm 0.13) \times 10^{-7}$ cm$^3$/s at an electron temperature of 300 K.

Since the storage-ring measurements also yielded fairly large recombination coefficients for v=0 ions, the problem of finding a recombination mechanism for v=0 ions became more acute. New mechanisms were proposed [22, 23] that do not require crossings between ionic and neutral potential curves. It is possible that such mechanisms may play a part in the recombination of $H_3^+$ ions, but no detailed calculations have been made.

Not all experimentalists were willing to accept a sizable recombination coefficient for $H_3^+$ in v=0. Smith and Spanel [24] repeated and refined the afterglow experiment of Adams et al.s' [10] and concluded that only $H_3^+$ ions in v>3 recombined efficiently with electrons. Again, those authors found that the decay of $H_3^+$ plasmas was anomalous in the sense that the recombination coefficient appeared to become smaller in the late afterglow. It was, in part, this work that motivated us to carry out an additional experiment. As will be seen, we largely reproduced Smith and Spanel's [24] results that there is something peculiar about the decay of $H_3^+$ plasmas but our interpretation of the experiment is quite different.

To avoid ambiguities, we will use the term "recombination coefficient" ($\alpha$) only for purely binary recombination. We will use the word "de-ionization coefficient" ($\beta$) to denote the quantity that is derived from plasma decay experiments. The two quantities may be equal, but this is not assumed from the onset.

A more complete description of the experiment, additional details of the analysis, and the interpretation will be given in a journal article [25].

## EXPERIMENTAL METHODS

Our flow tube is a stainless-steel tube of diameter 6 cm and length 36 cm. The carrier gas was helium (at pressures near 1 Torr and temperatures near 295 K). Flow velocities ranged from 5000 cm/s to 6000 cm/s. Reagent gases are added to the flowing plasma through several movable and fixed inlets. A movable Langmuir probe [26] serves to measure electron densities $n_e$ as a function of axial position z. A quadrupole ion mass spectrometer analyzes the ion composition at the downstream end of the flow tube.

### Plasma generation and ion conversion

The techniques of recombination measurements in flow tubes are quite well known and will not be discussed here. The method consists of creating a well-characterized plasma that contains only the desired ion species and measuring the rate at which free electrons disappear. The creation of the $H_3^+$ plasma involves several steps:

(1) The carrier gas (helium) is partially excited to the metastable state and (to a lesser degree) ionized by passing it through a microwave discharge.

(2) Argon (~15 % of helium) is added downstream to convert metastable helium atoms to $Ar^+$ ions by Penning ionization. After this, the plasma contains mainly $Ar^+$ ions with a small admixture of $He^+$ and $He_2^+$ ions (< 0.1 % of $Ar^+$). The electron density at this point is typically $4 \times 10^{10}$ cm$^{-3}$.

(3) To create $H_3^+$ ions, $H_2$ is added at densities from $10^{14}$ to $10^{15}$ cm$^{-3}$. The formation of $H_3^+$ via the reactions

$$Ar^+ + H_2 \rightarrow Ar H^+ + H \quad (+ 1.53 eV) \qquad (2)$$

$$ArH^+ + H_2 \rightarrow Ar + H_3^+ \quad (+ 0.57 eV) \qquad (3)$$

(see Radzig and Smirnov [27] for thermochemical data) takes place very rapidly compared to recombination since the rate coefficients for reactions (2) and (3) are large (~ $10^{-9}$ cm$^3$/s [28]).

### Vibrational excitation and quenching of the $H_3^+$ ions

If the entire energy of 2.1 eV that is released in reactions (2) and (3) were to appear as vibrational excitation, which seems unlikely, then $H_3^+$ could be produced in vibrational states up to v=5. However, $H_3^+$ ions in v>1 would be destroyed rapidly by proton transfer to Ar,

$$H_3^+ + Ar \rightarrow Ar H^+ + H_2 \qquad (4)$$

which is energetically possible for $H_3^+$ ions with internal energies above 0.57 eV. The non-reacting, surviving $H_3^+$ vibrational states are the ground state, the $v_2=1$ bending-mode vibration at 0.3126 eV, and the $v_1=1$ breathing mode vibration at 0.394 eV(see Lie and Frye [29] or Oka and Jagod [6]). At $[Ar] = 5 \times 10^{15}$ cm$^{-3}$, $v>1$ ions would be destroyed by proton transfer (k ~$10^{-9}$ cm$^3$/s) in less than 1 μsec, a time that is far shorter than the time scale of recombination (about 1 ms).

The radiative lifetimes of the $v_1=1$ and $v_2=1$ levels (1.2 s and 4 ms, respectively) [19] are too long for radiative relaxation to be important. Quenching of $v=1$ ions in collisions with $H_2$ is more effective. Amano's [30] absorption studies of the $v_2$ hot band indicate that the $v_2=1$ level is quenched by $H_2$ with a rate coefficient of approximately $3 \times 10^{-10}$ cm$^3$/s. This eliminates $v_2=1$ ions; they will be destroyed in 30 μs at $[H_2] = 1 \times 10^{14}$ cm$^{-3}$. On the other hand, there is evidence [31,32] for slower quenching (k~$10^{-12}$ cm$^3$/s) of other vibrational states. It seems very likely that the slowly-quenched ions are those in the $v_1=1$ level. A similar conclusion was reached by Bawendi et al. [33] who studied the absorption of $H_3^+$ hot bands in He/$H_2$ discharges.

The foregoing considerations indicate that only $H_3^+$ ($v_1=1$) ions live long enough to undergo recombination. Smith and Spanel [24] did not consider the destruction of the higher vibrational levels by proton transfer with argon and consequently believed that most $H_3^+$ ions would be in states $v \geq 3$. We do not agree with this conclusion; others have expressed the same criticism earlier [13,14]. This is an important point, since Smith and Spanel's proposed reconciliation of theory and experiment rests on the assumption that vibrationally excited ions dominate.

There is no *a priori* reason to expect that *para* and *ortho* $H_3^+$ ions should recombine equally fast. According to Oka [34], *ortho-para* equilibration by proton exchange with $H_2$ is rapid. We assume that the proton-transfer rate is sufficiently large (k>$10^{-10}$ cm$^3$/s would suffice at $[H_2] = 10^{14}$ cm$^{-3}$) to keep the para-to-ortho ratio at a constant value in the recombining plasma.

## DATA ANALYSIS

The most common method of deducing β from the slope of a graph of the reciprocal of the electron density $n_e(z)$ *vs.* axial position z (V being the plasma velocity),

$$\frac{1}{n_e(z)} = \frac{1}{n_e(z=0)} + \beta \frac{z}{V} \qquad (6)$$

provides quick, but only only approximate values of a z-averaged de-ionization coefficient. Since we are looking for deviations from a simple decay law, we were forced to employ a more rigorous approach (for details see ref. 25): It consisted of solving numerically the continuity equation for the electron density $n_e(r, z)$ in a flowing plasma with a parabolic velocity field. The computed solutions were then used to validate a computationally faster, approximate method that could be used for routine data analysis. The approximate method extracts a z-dependent de-ionization coefficient β(z) from the measured electron density $n_e(z)$. This form of analysis was shown to be sufficiently accurate by using the exact model to generate simulated $n_e(z)$ values and then analyzing the simulated data by the approximate method. In addition, tests were made on experimental data obtained in $O_2^+$ plasmas. As expected, no systematic variation of $\beta(z)$ with z was found and the measured $\beta$ agreed with the known recombination coefficient $\alpha(O_2^+)$.

## EXPERIMENTAL RESULTS

40 decay curves $n_e(z)$ were recorded for $[Ar]= 4 \times 10^{15}$ cm$^{-3}$ to $8 \times 10^{15}$ cm$^{-3}$, and $[H_2]=1 \times 10^{14}$ cm$^{-3}$ to $1.7 \times 10^{15}$ cm$^{-3}$. Even the simplest form of data analysis (based on Eq. (9)) showed that $\beta$ increased noticeably with increasing $H_2$ concentration (see Fig. 1). A similar, but less scattered graph was obtained when the average values of $\beta(z)$ in the first 4 cm were plotted as a function of $[H_2]$.

Figures 2 and 3 show examples of the electron density $n_e(z)$ and $\beta(z)$ for two different $H_2$ concentrations. It is apparent that $\beta$ is not constant but that it decreases with increasing z. This finding reproduces the observations of Smith and Spanel[24]; the decay of the plasma clearly does not follow a simple recombination law.

A smaller set of data was taken for $D_3^+$ ions. The observations largely mirrored those for $H_3^+$. Again, the recombination coefficient increased with increasing $D_2$ concentration, but the $D_3^+$ de-ionization coefficients were significantly smaller (from $0.7 \times 10^{-7}$ cm$^3$/s to $1.3 \times 10^{-7}$ cm$^3$/s in the range $[D_2] = 1$ to $14 \times 10^{14}$ cm$^{-3}$).

## INTERPRETATION OF RESULTS

The interpretation should explain the enhancement of $\beta$ upon hydrogen addition, the decline of $\beta(z)$ with increasing position z, and if possible, it should elucidate the recombination mechanism. We attempted to explain the peculiar decay of the plasma by two different models. The first model assumes the presence of $H_3^+$ ions in two different states. The second model invokes an entirely different explanation in terms of Rydberg molecules.

### Two - state model

The assumption that the z dependence of $\beta$ is caused by the presence of $H_3^+$ ions in $v=0$ and in $v_1=1$ leads to

$$\beta(z) = \alpha_0 f_0(z) + \alpha_1 f_1(z) \tag{6}$$

where $\alpha_0$ and $\alpha_1$ denote the recombination coefficients of $v=0$ and $v_1=1$ ions, and $f_0$ and $f_1$ are their fractional concentrations. Quenching of $v_1=1$ ions to $v=0$ by $H_2$ (rate coefficient $k_q$) and differing recombination losses make the concentrations z-dependent. It is not difficult to construct a kinetic model that yields a fit to the measured $\beta(z)$ curve at $[H_2] = 1.3 \times 10^{14}$ cm$^{-3}$. Figure (2) shows fits for $\alpha_0= 2.2 \times 10^{-7}$ cm$^3$/s, $\alpha_1= 4 \times 10^{-8}$ cm$^3$/s, $f_0 = 0.7$, and two quenching coefficients of $k_q=1 \times 10^{-12}$ cm$^3$/s and $k_q=1 \times 10^{-13}$ cm$^3$/s. However, when the same values are used to compute a fit to the data at $[H_2]= 1.7 \times 10^{15}$ cm$^{-3}$, the model (see Fig. 3) indicates that $\beta(z)$ should now increase with z. A decline of $\beta(z)$ can be forced by reducing the quenching rate to $k_q = 1 \times 10^{-13}$ cm$^3$/s, but the model fails to reproduce the observation that $\beta$ at z=0 (only 1 cm from the reagent inlet) is larger at the larger $H_2$ concentration. We eventually concluded that the two-state model does not provide a consistent explanation of the experimental findings.

## De-ionization by Rydberg molecule stabilization

We now return to the subject that was mentioned in the Introduction, namely de-ionization mechanisms which involve interactions with third bodies. The $\beta(z)$ data suggest that the de-ionization coefficient is larger at higher electron densities and higher $H_2$ densities. Is it possible that these particles affect the mechanism of de-ionization directly, rather than changing the state of the recombining ions? At first sight, this does not seem to be viable proposition. Electron-stabilized and neutral-stabilized recombination are known to be too slow (by several orders of magnitude)[35] to have an effect on recombination in low-density ($n_e < 10^{11}$ cm$^{-3}$) plasmas. Third-body effects are inefficient because electron capture requires energy transfer to a third particle during the brief electron-ion encounter.

In the following, we will argue that additional stabilization mechanisms exist in the case of molecular ions, especially those that do not undergo fast dissociative recombination. Theory indicates the $H_3^+$ ions may fall into this category. Non-dissociating, long-lived Rydberg states may be formed in collisions of electrons with molecular ions. These can be stabilized not only by energy-removing collisions but also by angular-momentum changing (*l*-mixing and possibly m-mixing) collisions.

The following sequence may occur: In the first step, an electron of low angular momentum (s or possibly p) is captured into a ro-vibrational level of a Rydberg state of $H_3$,

$$e^- + H_3^+ \rightarrow H_3^{**} (v=1, n, s). \tag{7}$$

The n=7 state should be favored since excitation to the $v_2=1$ state (0.3126 eV) in n=7, l=0 is resonant at an electron energy of +0.0309 eV (1.2 kT at 300 K).

We postulate that in the second step the angular momentum of the Rydberg molecule is randomized in collisions with ambient electrons (l-mixing and possibly m-mixing). It is known from studies of dielectronic recombination[36] and zero-electron-kinetic-energy (ZEKE) spectroscopy [37] that redistribution of angular momenta of Rydberg electrons (l-mixing) due to electric microfields and collisions can increase predissociation and autoionization lifetimes by several orders of magnitude. The basic reason is that the wave functions of high-l states do not interact with the ion core. Since the lifetimes of some $H_3^{**}$ autoionizing states are quite long (of the order of $10^{-7}$ s) [38, 39] and the l-mixing efficiency of thermal electrons is extremely large (rate coefficients $>10^{-3}$ cm$^3$/s) [40], l-mixing should occur in less than $10^{-7}$ s under the conditions of our experiment ($n_e = 4 \times 10^{10}$ cm$^{-3}$).

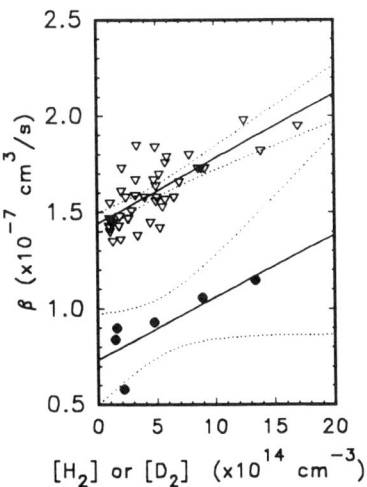

**FIGURE 1.** $\beta$-values in $H_3^+$ or $D_3^+$ plasmas (from the $1/n_e$ vs. z plots), as a function of $[H_2]$ (triangles) or $[D_2]$ (filled circles). Lines: Least-squares fit to the data. Dotted lines: 95% confidence limits the least-squares fit. From [25].

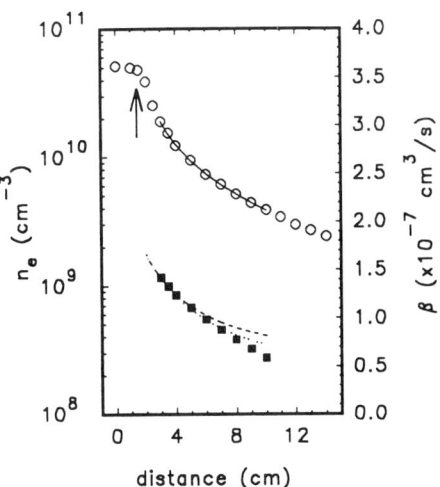

**FIGURE 2.** $\beta(z)$ values (filled squares) derived from $n_e(z)$ data (open circles) at $[H_2]=1.3\times10^{14}$ cm$^{-3}$. Lines through $\beta(z)$ points: Model calculations for $k_q=1\times10^{-12}$ cm$^3$/s (dashed) and $k_q=1\times10^{-13}$ cm$^3$/s (dotted). Arrow: Position of the $H_2$ gas inlet. From [25].

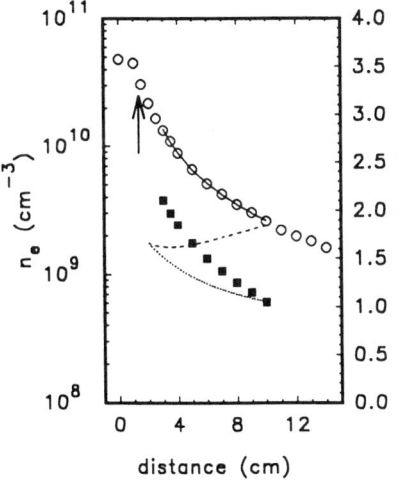

**FIGURE 3.** Same as Fig. 3, but for $[H_2]=1.7\times10^{15}$ cm$^{-3}$. From [25].

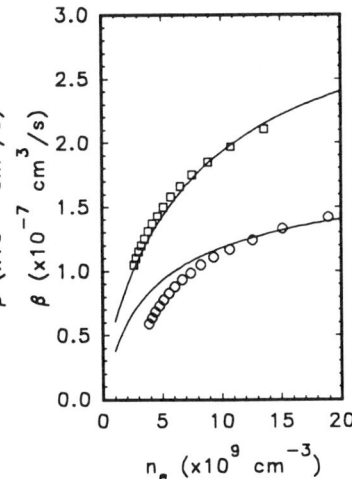

**FIGURE 4.** Comparison of observed values of $\beta(n_e)$ at $[H_2]=1.3\times10^{14}$ cm$^{-3}$ (circles) and $[H_2]=1.7\times10^{15}$ cm$^{-3}$ (squares) to the model calculations described in the text. From [25].

In the third step, the long-lived $H_3^{**}(n=7, v=1, l>1)$ can be returned to the autoionizing state (s, or p) by further collisions, they may be quenched to lower n-states by neutral atoms or electrons, they may undergo slow predissociation, radiate to a lower state, or react with an $H_2$ molecule. Some of the decay rates are known or can be estimated. Collisional energy loss (n-changing) with helium atoms is not important [41] compared to radiative decay. The rate coefficient for n-changing collisions of electrons with n=7 Rydberg atoms (electron quenching) has been measured by Devos et al. [43] to be $\sim 10^{-5}$ cm$^3$/s. Radiative decay rates for n=7 hydrogenic states are close to $1\times 10^6$ s$^{-1}$ for s-states, $2\times 10^7$ s$^{-1}$ for p-states, and about $2\times 10^6$ s$^{-1}$ for a statistical mixture of l-values up to l=6 [42]. The reaction with $H_2$

$$H_3^{**} + H_2 \rightarrow H_2 + H_3^* \tag{8}$$

provides the needed enhancement of the loss rate of electrons in the presence of $H_2$. It will compete with radiative decay if it occurs on a time scale of $<1\mu s$. At $[H_2] = 1\times 10^{15}$ cm$^{-3}$, this requires a rate coefficient of $\sim 1\times 10^{-9}$ cm$^3$/s for reaction (8), which is not unreasonable.

The set of reaction equations is easily solved to obtain an effective de-ionization coefficient. The capture and autoionization step

$$H_3^+ + e^- \leftrightarrow H_3^{**}(n=7, s,p) \tag{9}$$

is described by a capture coefficient $k_c$ and an autoionization rate $v_a$. Stabilization by l-mixing due to electrons

$$H_3^{**}(n=7, s, p) + e^- \leftrightarrow H_3^{**}(n=7, l>1) + e^- \tag{10}$$

and its reverse are described by rate coefficients $k_s$ and $k_i$, respectively. The following destruction reactions for $H_3^{**}$ were included:

| | | | |
|---|---|---|---|
| $H_3^{**}(n=7)$ | $\rightarrow H_3^*(n<7) + h\nu$ | (radiation, rate $v_{rad}$) | (11a) |
| $H_3^{**}$ | $\rightarrow H_2 + H$ (or 3H) | (predissociation, rate $v_p$) | (11b) |
| $H_3^{**}(n=7) + e^-$ | $\rightarrow H_3^*(n<7) + e^-$ | (rate coefficient $k_{qe}$) | (11c) |
| $H_3^{**} + H_2$ | $\rightarrow H_2 + H_3$ (low n) | (rate coefficient $k_H$) | (11d) |

To allow for the possibility that the rate coefficients for $H_3^{**}$ in the autoionizing states (s, p) differ from those in higher l-states, we treat their destruction separately and attach the subscripts "0" and "1" to the rate coefficients. After some algebra, the effective de-ionization coefficient can be put in the form

$$\beta = -\frac{1}{n_e^2}\frac{dn_e}{dt} = \frac{k_c}{1+\dfrac{1}{\dfrac{n_e k_s v_l}{(n_e k_i + v_l)v_a} + \dfrac{v_0}{v_a}}} \tag{12}$$

where $\quad v_0 = v_{rad,0} + v_{p,0} + k_{qe} n_e + k_H [H_2]$
and $\quad v_1 = v_{rad,1} + v_{p,1} + k_{qe} n_e + k_H [H_2]$.

It is possible to find a set of plausible rate coefficients so that Eq. (12) gives a reasonable fit to the data. Fig. 4 shows a comparison to experimental values of $\beta$ ($n_e$) for the following set of rate coefficients:

$k_c = 4.4 \times 10^{-7}$ cm$^3$/s,  $\nu_a = 1.3 \times 10^8$ s$^{-1}$,  $k_s = 1.25 \times 10^{-2}$ cm$^3$/s

$k_i = 3.1 \times 10^{-4}$ cm$^3$/s,  $\nu_{rad,0} = \nu_{rad,1} = 1.1 \times 10^6$ s$^{-1}$

$\nu_{p,0} = \nu_{p,1} = 0$,  $k_H = 4.5 \times 10^{-9}$ cm$^3$/s,  $k_{qe} = 1 \times 10^{-5}$ cm$^3$/s

We do not claim that this set of model parameters is unique or that the agreement with the data is sufficient proof for the validity of the model. More detailed studies would be required to prove that this model is valid. It is interesting, however, that the fitting parameters are quite close to those that were estimated earlier. If this model is correct, it would solve the theoretical dilemma of having to account for fast "binary recombination" of electrons with $H_3^+$ ions since this process would not be needed to explain the observed plasma de-ionization.

Our model is compatible with the most recent merged-beam observations of Mitchell and coworkers [43]. They found that the apparent $H_3^+$ recombination cross sections decreased by about a factor of five when a "deflection field" in the post-collision region was increased in magnitude. The effect is ascribed to field-induced re-ionization of autoionizing $H_3^{**}$ molecules that were formed in electron-$H_3^+$ collisions. The role of electric fields in such experiments is not yet perfectly understood, but the observations demonstrate that long-lived $H_3^{**}$ states can be produced in such collisions. The nature of these states and the reasons for their unexpectedly long life-time, however, are not at all clear.

## CONCLUSIONS

We reach the tentative conclusion that the recombination of $H_3^+$ plasmas does not exclusively occur by binary recombination. If this is true, then the results of afterglow measurements may be compatible with the theoretical expectation that binary recombination of $H_3^+$ ions in v=0 should be rather slow.

The "true" binary recombination coefficient of $H_3^+$ is needed for some applications, e.g. for models of the interstellar medium. It is conceivable that the recombination cross sections measured in merged-beam or ion-storage ring experiments are close to the true values, but the effects of residual electric fields in such experiments need to be examined more closely.

Acknowledgments: This work was, in part, supported by NASA

## References

1. J.N. Bardsley, J. Phys. B **1**, 365 (1968)
2. H.H. Michels and R.H. Hobbs, Ap. J. (Letters) **286**, L27 (1984)
3. K.C. Kulander and M.F. Guest, J. Phys. B **12**, L501 (1979)
4. M.T. Leu, M.A. Biondi, and R. Johnsen, Phys. Rev. A **8**, 413(1973)
5. J. A. Macdonald, M.A. Biondi, and R. Johnsen, Planet. Space Sci. **32**, 651 (1984)
6. T. Oka and M.-F. Jagod, J. Chem. Soc. Faraday Trans. **89**, 2147 (1993)
7. A. Dalgarno, Advances in Atomic, Molecular and Optical Physics, Vol. **32**, 57 (1993)

8. J. Tennyson, S. Miller, and H. Schild, J. Chem. Faraday Trans. **89**, 2155 (1993)
9. D. R. Bates, M.F. Guest and R.A. Kendall, Planet. Space Sci. **41**, 9 (1993)
10. N.G. Adams, D. Smith and E. Alge, J. Chem. Phys. **81**, 1778 (1984)
11. T. Amano, Ap. J. **329**, L121 (1988)
12. T. Amano, J. Chem. Phys. **92**, 6492 (1990)
13. A. Canosa, J.C. Gomet, B.R. Rowe, J.B.A. Mitchell, and J.L. Queffelec, J. Chem. Phys. **97**, 1028 (1992)
14. A. Canosa, B.R. Rowe, J.B.A. Mitchell, J.C. Gomet, and C. Brion, Astron. Astrophys. **248**, L19 (1991)
15. H. Hus, F.B. Yousif, A. Sen, and J.B. A. Mitchell, Phys. Rev. A **38**, 658 (1988)
16. J.B.A. Mitchell, J.L. Forland, C.T. Ng, D.P. Levac, R.E. Mitchell, P.M. Mul, W. Claeys, A. Sen, and J. Wm. McGowan, Phys. Rev. Lett. **51**, 885 (1983)
17. J.B.A. Mitchell, C.T. Ng, J.L. Forland, .R. Janssen, and J. Wm. McGowan, J. Phys. B **17**, L909 (1984)
18. F.B. Yousif, P.J.T. Van der Donk, M. Orakzai, and J.B. A. Mitchell, Phys. Rev. A **44**, 5653 (1991)
19. B.M. Dinelli, S. Miller, and J. Tennyson, J. Mol. Spectr. **153**, 718 (1992)
20. M. Larsson, H. Danared, J.R. Mowat, P. Sigray, G. Sundström, L. Broström, A. Filevich, A. Källberg, S. Mannervik, K.G. Rensfelt, and S. Datz, Phys. Rev. Lett. **70**, 430 (1993)
21. G. Sundström, J.R. Mowat, H. Danared, S. Datz, L. Broström, A. Filevich, A. Källberg, S. Mannervik, K.G. Rensfelt, P. Sigray, M. af Ugglas, and M. Larsson, Science **263**, 785, (1994)
22. S. L. Guberman, Phys. Rev. A **49**, R4277 (1994)
23. D. R. Bates, Mon. Not. R. Astron. Soc. **263**, 369 (1993) and D. R. Bates, Proc. Roy. Soc. Lond. A **443**, 257 (1993)
24. D. Smith and P. Spanel, Int. J. Mass Spec. Ion Proc. **81**, 67 (1987)
25. T. Gougousi, R. Johnsen, and M.F. Golde, Int. J. Mass Spectr. Ion Proc. (in print)
26. R. Johnsen, E.V. Shun'ko, T. Gougousi, and M.F. Golde, Phys. Rev. E **50**, 3994 (1994)
27. A.A. Radzig and B.M. Smirnov, Reference data on atoms, molecules, and ions. (Springer Verlag 1980)
28. D.L. Albritton, Atomic Data and Nuclear Tables **22**, 1 (1978)
29. G.C. Lie and D. Frye, J. Chem. Phys. **96**, 6784 (1992)
30. T. Amano, in "Dissociative Recombination:Theory, Experiment andApplications", edited by B.R. Rowe, J.B. Mitchell, and A. Canosa (Plenum Press, New York 1993)
31. H.S. Lee, M. Drucker, N.G. Adams, Int. J. Mass Spectrom. Ion Proc. **117**, 101 (1992)
32. C.R. Blakley, M.L. Vestal, and J.H. Futrell, J. Chem. Phys. **66**, 2392 (1977)
33. M.G. Bawendi, B.D. Rehfuss, and T. Oka, J. Chem. Phys. **93**, 6200 (1990)
34. T. Oka, Philos. Trans. R. Soc. London, A, **303**, 543 (1981)
35. M.R. Flannery, Advances in Atomic, Molecular and Optical Physics **32**, 117 (1994)
36. A. Müller, D.S. Belic', B.D. DePaola, N. Djuric', G.H. Dunn, D.W. Mueller, and C. Timmer, Phys. Rev. A **36**, 599 (1987)
37. F. Merkt and R.N. Zare, J. Chem. Phys. **101**, 3495 (1994)
38. V. Berardi, N. Spinelli, R. Velotta, M. Armenante, and A. Zecca, Phys. Rev. A, **47**, 986 (1993)
39. J.B.A. Mitchell, Bulletin of the American Physical Society Series II, 39, 1456 (1994)
40. F. Devos, J. Boulmer, J.-F. Delpech, J. de Physique 40, 215 (1979), and Phys. Rev. Lett. 39, 1400 (1977)
41. D.R. Bates and S.P. Khare, Proc. Phys. Soc. **85**, 231 (1965)
42. J. R. Hiskes and C.B. Tarter (Report UCRL-7088, Lawrence Radiation Laboratory, Livermore CA, (1964))
43. J.B.A. Mitchell, private communication

# SELECTED TOPICS

## – ATOMS –

Generalizations of Distorted-Wave Capture Theories
to Relativistic Atomic Collisions .................................. 847
      J. EICHLER and N. Toshima
X-Ray Emission in Relativistic Ion-Atom Collisions ........... 857
      J.F. McCANN, J.T. Glass, and D.S.F. Crothers
Measurements of Charge Transfer Cross Sections
for $Ar^{q+}$ (q=6,7,8,9 and 11) Collisions
with He and $H_2$ Targets at Low Energies ....................... 867
      K. OKUNO, H. Saitoh, K. Soejima,
      S. Kravis, and N. Kobayashi
Projectile Charge Dependence of Helium Single Excitation
at Medium and High Impact Energies ......................... 877
      F. Martín and A. SALIN
Hyperspherical Elliptic Coordinates:
New Approximate Symmetry of
Three-Body Coulomb Problem .................................. 887
      O.I. TOLSTIKHIN, S. Watanabe, and M. Matsuzawa

# Generalizations of distorted-wave capture theories to relativistic atomic collisions

J. Eichler[*][1] and N. Toshima[†]

[*] *Bereich Theoretische Physik, Hahn-Meitner-Institut, 14109 Berlin, Germany*
[†] *Institute of Applied Physics, University of Tsukuba, Tsukuba, Ibaraki 305, Japan*

---

The structure of various relativistic extensions of distorted-wave capture theories is analyzed in a unified manner and the relations between these approaches are discussed. Theoretical descriptions containing local potentials as residual interactions can be unambiguously generalized. Theories containing nonlocal residual interactions lead to difficulties, so that one is forced to confine oneself to semirelativistic generalizations.

---

## INTRODUCTION

Among the elementary processes occurring in atomic collisions, charge transfer is the most difficult one to describe theoretically. While already for nonrelativistic energetic collisions [1,2], no single theory has emerged as an undisputed standard, the problems increase for relativistic collisions, in which the velocities of the projectile and of the bound electrons approach the speed of light. We here analyze the various approaches that have been applied so far from a unified point of view.

Even in the simplest case of a transfer reaction, when an electron is captured by a bare projectile (with charge number $Z_P$) from a hydrogenic target atom (with charge number $Z_T$) one has to solve, in principle, a three-body Coulomb problem. Since this is not rigorously possible at present, one has to resort to approximate methods, in which the electron-target and the electron-projectile two-body problems are solved rigorously and the distortions caused by the third body are taken into account in some approximate fashion.

The first attempts to describe relativistic electron capture have neglected distortions altogether [3] by using the first-order Oppenheimer-Brinkman-Kramers (OBK1) approximation. Since this approach considerably overestimates experimental data, a simplified second-order version (OBK2) has been put forward which appeared to give agreement with the experimental data [4]. However, by way of examining these results, a rigorous numerical implementation of the OBK2 approach has been shown [5] to yield an even larger

---

[1] Also at Fachbereich Physik, Freie Universität Berlin, 14195 Berlin, Germany.

© 1995 American Institute of Physics

discrepancy with experiment than the OBK1 theory. As a result, it became clear that capture theories with undistorted waves are not suitable to describe experimental cross sections quantitatively. A realistic description of relativistic electron capture therefore requires distorted wave theories as is the case with nonrelativistic collision systems.

## NONRELATIVISTIC CHARGE TRANSFER

As a starting point for a relativistic formulation, we briefly discuss distorted-wave theories of electron capture in *nonrelativistic* collisions without attempting, of course, to be complete (for detailed reviews see [1] and [2]). Adopting the impact parameter picture, the exact nonrelativistic capture amplitude in the distorted-wave Born approximation can be written in the *post* form as

$$A_{fi}^{(+)} = -\frac{i}{\hbar} \int_{-\infty}^{\infty} dt \left\langle \left( H - i\hbar \frac{\partial}{\partial t} \right) \chi_f \bigg| \chi_i \right\rangle \tag{1}$$

and in the *prior* form as

$$A_{fi}^{(-)} = -\frac{i}{\hbar} \int_{-\infty}^{\infty} dt \left\langle \chi_f \bigg| H - i\hbar \frac{\partial}{\partial t} \bigg| \chi_i \right\rangle, \tag{2}$$

where $H$ is the total electronic Hamiltonian of the three-body system, and $\chi_i$ and $\chi_f$ are distorted waves. The physical meaning of the equations is the following. In the post form, the operator $(H - i\hbar\partial/\partial t)$ acting on $\chi_f$ yields the residual interaction *not* already included in the definition of the distorted wave $\chi_f$. In the prior form, the operator $(H - i\hbar\partial/\partial t)$ acting on $\chi_i$ yields the residual interaction *not* already included in the definition of $\chi_i$.

It is now the art of a theoretical formulation to invent distorted wave functions $\chi_i$ and $\chi_f$ that are physically reasonable, tractable and lead to realistic predictions. All approaches used so far, have been written in the general product form

$$\chi_i = U_i \, \psi_i(\mathbf{r}_T, t)$$
$$\chi_f = U_f \, \psi_f(\mathbf{r}_P, t), \tag{3}$$

where $U_i$ and $U_f$ are functions or operators yet to be specified and $\psi_i = \varphi_i(\mathbf{r}_T) e^{-iE_i t/\hbar}$ is the time-dependent target eigenfunction with $\mathbf{r}_T$ being the electron coordinate with respect to the target nucleus. The function $\psi_f$ is defined correspondingly with $\mathbf{r}_P$ being the electron coordinate with respect to the projectile nucleus. It also includes the translation factor [1,2] accounting for the Galileo transform from the moving projectile frame to the target frame.

By taking $U_i = U_f = 1$, we retrieve the undistorted Born approximation, OBK1, discussed in the Introduction. This formulation, besides predicting too large cross sections, does not have the proper asymptotic behavior imposed by the long-range Coulomb potential [1,2,6–8].

The simplest way of taking into account the Coulomb boundary condition, is to adopt the "boundary-corrected Born" (B1B) approximation [2,6], by choosing a phase distortion[2]

$$U_i^{\text{B1B}} = e^{-i\nu_P \ln(R-vt)}$$
$$U_f^{\text{B1B}} = e^{i\nu_T \ln(R+vt)}, \quad (4)$$

where $R$ is the internuclear separation, $v$ the projectile velocity, and $\nu_P = Z_P e^2/v\hbar$ and $\nu_T = Z_T e^2/v\hbar$ are the Sommerfeld parameters. This theory yields realistic predictions for experimental cross sections.

While in Eq. (4) the distorting phase factors only depend on time and on the impact parameter $b$ via $\mathbf{R} = \mathbf{b} + \mathbf{v}t$, a dependence on the electronic space coordinates is introduced if, instead, one adopts the eikonal phase factor. In the (asymmetric) eikonal approximation [9], one sets

$$U_i = 1, \quad U_f^{\text{eik}} = e^{i\nu_T \ln(r_T+z_T)} \quad \text{in the post form}$$
$$U_f = 1, \quad U_i^{\text{eik}} = e^{-i\nu_P \ln(r_P+z_P)} \quad \text{in the prior form.} \quad (5)$$

This approach, while not satisfying Coulomb boundary conditions and not being post-prior symmetric, is easily tractable and yields realistic results.

The symmetrized version (SE) of the eikonal approximation proposed by Maidagan and Rivarola [10] uses

$$U_i^{\text{eik}} = e^{-i\nu_P \ln(r_P+z_P)}$$
$$U_f^{\text{eik}} = e^{i\nu_T \ln(r_T+z_T)} \quad (6)$$

and leads to a residual interaction which is not simply given by a local potential as with the ansatz (4) or (5) but rather contains derivative terms and hence is nonlocal. The SE approximation is post-prior symmetric and satisfies Coulomb boundary conditions but yields unsatisfactory agreement with experimental data.

A widely applied and successful distorted-wave approach to electron capture is given by the "continuum distorted-wave" (CDW) approximation of Cheshire [11,12]. It uses the Coulomb factor modifying a plane wave in a Coulomb field for describing the distortion of the target (projectile) bound-state wave function by the projectile (target) nucleus. This implies the choice

$$U_i^{\text{CDW}} = e^{\frac{1}{2}\pi\nu_P} \Gamma(1-i\nu_P)\,_1F_1[i\nu_P, 1, iv(r_P+z_P)m_e/\hbar]$$
$$U_f^{\text{CDW}} = e^{\frac{1}{2}\pi\nu_T} \Gamma(1+i\nu_T)\,_1F_1[-i\nu_T, 1, -iv(r_T+z_T)m_e/\hbar] \quad (7)$$

which leads to post-prior symmetric transition amplitudes satisfying Coulomb boundary conditions. While the evaluation is not very simple and is marred

---

[2]Here and in the following phase factors, we discard irrelevant constant phases depending on the unit of length.

by a normalization problem [13], the calculated cross sections are in good agreement with experimental results.

There is a variety of multiple-scattering models using different guesses for the distorting functions $U_i$, $U_f$, and, in particular, there is a host of hybrid models rather arbitrarily combining $U_i$ functions taken from one approach with $U_f$ functions taken from another approach, see e.g. [2]. We here refrain from discussing this further.

For the purpose of later use, it is interesting to observe that the eikonal distortion functions $U_{i,f}^{\text{eik}}$ in Eq. (6) may be derived in two different ways.

(a) The distortion factor may be represented by the conventional eikonal integrals,

$$U_i^{\text{eik}} = \exp\left(-i\frac{e^2}{\hbar}\int_{-\infty}^{t}\frac{Z_P}{r_P}dt'\right),$$
$$U_f^{\text{eik}} = \exp\left(-i\frac{e^2}{\hbar}\int_{t}^{\infty}\frac{Z_T}{r_T}dt'\right), \qquad (8)$$

where an irrelevant constant phase factor has been dropped and where $r_T$ and $r_P$, respectively, are kept fixed when evaluating the integrals. This corresponds to the usual assumption that the collision time is short compared to the orbiting time of the electron.

(b) The second way to derive the eikonal distortion consists in taking the asymptotic limits of the full Coulomb distortion factors, that is

$$U_i^{\text{eik}} = \lim_{r_P \to \infty} U_i^{\text{CDW}} \qquad U_f^{\text{eik}} = \lim_{r_T \to \infty} U_f^{\text{CDW}}. \qquad (9)$$

Whereas in the nonrelativistic context it does not matter which point of view is taken, the relativistic (actually semirelativistic, see below) extension of Eq. (9) leads to a different result than Eq. (8) if one *first* performs the relativistic generalization and *then* takes the asymptotic limit (9). This point has not been properly recognized in the literature.

## RELATIVISTIC EXTENSIONS OF CHARGE TRANSFER

A relativistic extension of capture theories involves high projectile velocities up to values of the Lorentz factor $\gamma = (1 - v^2/c^2)^{-1/2} \gg 1$ as well as high electron velocities occurring for high-$Z$ nuclei. For a theory to be justifiable, one has to demand that it smoothly merges into the nonrelativistic limit $\gamma \to 1$, $\alpha Z \ll 1$, where $\alpha$ is the fine structure constant.

With these requirements, and adopting a description in the target frame, we can immediately generalize Eqs. (3) to (5) by the following replacements: (a) The target wave function $\psi_i$ is modified by the distortion operator $U_i$, which contains the coordinate $\mathbf{r}_P$. In a relativistic description, we measure coordinates with respect to the projectile in the projectile Lorentz frame (indicated by primed coordinates), i.e. $\mathbf{r}_P \to \mathbf{r}'_P$, $t \to t'$, and $R \to R'$, where the connection with the target frame is given by the usual Lorentz transformation (see e.g. [14,15]). (b) Correspondingly, the projectile wave function $\psi'_f$, constructed in the projectile frame, is modified by the distortion operator $U'_f$, which contains the target coordinate $\mathbf{r}_T$. Using the spinor transformation $S$ defined by [14,15]

$$\psi'_f(\mathbf{r}'_P, t') = S\,\psi_f(\mathbf{r}_T, t) \tag{10}$$

with

$$S = \sqrt{\frac{\gamma+1}{2}}\,(1 - \delta\alpha_z), \tag{11}$$

where

$$\delta = \sqrt{\frac{\gamma-1}{\gamma+1}}, \tag{12}$$

and $\alpha_z$ is the Dirac matrix in the direction $z$ of the projectile motion, we may transform the complete final-state wave function from the projectile frame to the laboratory frame.

We are now in a position to write down the capture amplitudes corresponding to the relativistic versions of the distortion operators (4) to (6). For the boundary-corrected Born approximation [14–16] we get, with Eqs. (2) and (4), the transition amplitude as a function of the impact parameter $b$ in the post form as

$$A^{(+)}_{\mathrm{B1B}}(b) = i\frac{e^2}{\hbar} \int_{-\infty}^{\infty} dt \left\langle S^{-1} U_f^{\mathrm{B1B}} \psi'_f \left| \frac{Z_T}{r_T} - \frac{Z_T}{R} \right| U_i^{\mathrm{B1B}} \psi_i \right\rangle. \tag{13}$$

and in the prior form as

$$A^{(-)}_{\mathrm{B1B}}(b) = i\frac{e^2}{\hbar} \int_{-\infty}^{\infty} dt \left\langle S^{-1} U_f^{\mathrm{B1B}} \psi'_f \left| S\left(\frac{Z_P}{r'_P} - \frac{Z_P}{R'}\right) S \right| U_i^{\mathrm{B1B}} \psi_i \right\rangle. \tag{14}$$

Here, the matrix $S^{-1}$ transforms the projectile wave function to the target system, while the combined operator $S^2$ in Eq. (14) does the same for the interaction, taking into account the commutation relation of $S^{-1}$ with the Dirac matrix $\gamma^0 = \beta$ occurring in the covariant formulation. The transition operator clearly exhibits the short-range residual interaction produced by the phase transformation (or gauge transformation) of Eq. (4).

For the (asymmetric) eikonal approximation [17], we have in the post form

$$A^{(+)}_{\mathrm{eik}}(b) = i\frac{e^2}{\hbar} \int_{-\infty}^{\infty} dt \left\langle S^{-1} \psi'_f \left| \frac{Z_T}{r_T} \right| U_i^{\mathrm{eik}} \psi_i \right\rangle, \tag{15}$$

and in the prior form

$$A_{\text{eik}}^{(-)}(b) = i\frac{e^2}{\hbar} \int_{-\infty}^{\infty} dt \left\langle S^{-1} U_f^{\text{eik}} \psi_f' \left| S \frac{Z_P}{r_P'} S \right| \psi_i \right\rangle, \tag{16}$$

with $\psi_f$ and $\psi_i$, respectively, remaining undistorted.

Equation (13) has been evaluated approximately [16] as well as by exact numerical computation [18], and yields good agreement with experimental cross sections and with the more rigorous relativistic two-center coupled-channel calculations of Toshima and Eichler [18]. Eq. (15) is in excellent overall agreement with a large body of experimental data [14]. It is important to observe that Eqs. (13) and (15) are applicable to extreme-relativistic collisions, $\gamma \gg 1$, as well as in the nonrelativistic limit $\gamma \approx 1$. Neither end shows any anomaly. What is common to both equations, is the fact that the nonrelativistic transition operators in Eqs. (1) and (2) are a local potentials.

The transition amplitudes for the symmetric eikonal (SE) approximation have a completely different structure. Inserting Eqs. (3) and (6) into Eqs. (1) and (2), Moiseiwitsch and Deco and Rivarola [19] obtain for the post form

$$A_{\text{SE}}^{(-)}(b) = i\frac{e^2}{\hbar} \int_{-\infty}^{\infty} dt \left\langle S^{-1} U_f^{\text{eik}} \psi_f' \left| \frac{Z_T}{r_T} \left(1 + \frac{1}{\beta} \frac{\boldsymbol{\alpha} \cdot \mathbf{r}_T + \alpha_z r_T}{r_T + z_T} \right) \right| U_i^{\text{eik}} \psi_i \right\rangle, \tag{17}$$

and for the prior form

$$A_{\text{SE}}^{(-)}(b) = i\frac{e^2}{\hbar} \int_{-\infty}^{\infty} dt \left\langle S^{-1} U_f^{\text{eik}} \psi_f' \left| \frac{Z_P}{r_P'} S \left(1 + \frac{1}{\beta} \frac{\boldsymbol{\alpha} \cdot \mathbf{r}_P' + \alpha_z r_P'}{r_P' + z_P'} \right) S \right| U_i^{\text{eik}} \psi_i \right\rangle, \tag{18}$$

where the transition operators originate from the kinetic energy term of the Hamiltonian. By construction, similarly as Eqs. (13) to (16), these expressions are applicable for $\gamma \gg 1$. However, remembering that the Dirac matrices $\alpha_x$ and $\alpha_y$ mediate spin-flip transitions, the occurrence of the factor $\beta = v/c$ in the denominators of the second terms suggests the existence of large spin-flip contributions at small velocities. Although the derivation of Eqs. (17) and (18) is a perfectly consistent procedure as long as exact relativistic bound-state wave functions are used, the resulting low-energy behavior is unrealistic, since spin-flip is caused by the motion-induced magnetic field of the projectile interacting with the magnetic moment of the electron. The anomalous behavior in the nonrelativistic limit has been analyzed in detail by Toshima and Eichler [20], both analytically and by exact numerical evaluation. Spin-flip contributions that are too large by many orders of magnitude show the non-existence of the nonrelativistic limit. It is hence unclear at which energies the SE approximation ceases to be applicable.

Recently, a distinctly different approach, also termed "symmetric eikonal approximation" has been advanced by Glass et al. [21–23] which, by construction, has the correct nonrelativistic behavior but ceases to be valid for high velocities and high-$Z$ nuclei. We come back to this approach in the following Section.

We finally turn to the CDW approximation with the distortion operators (7) representing nonrelativistic modifications of plane waves in a Coulomb field. Regarding a relativistic generalization, we observe two points. First, there is no relativistically exact closed-form Coulomb wave function for the continuum to replace Eq. (7): introducing factors of $\gamma$ in (7) is not sufficient. Since the Coulomb Dirac equation is not separable in parabolic coordinates, one has to resort to a partial-wave expansion. Second, irrespective of the non-existence of a closed form, a Coulomb wave function has a spinor structure and hence is not suitable, as it stands, to be identified with a distortion operator $U_{i,f}$ acting on another spinor wave function. In short, a direct, fully relativistic generalization of the CDW approximation for charge transfer is not possible. One may, however, resort to semirelativistic approximations.

## SEMIRELATIVISTIC EXTENSIONS FOR CHARGE TRANSFER

The basic equations (1) and (2) allow for a great deal of freedom in choosing the distorted waves. One may take advantage of this freedom in trying to find a symmetric eikonal or CDW approximation that has the proper nonrelativistic limit. Within a semirelativistic (or quasirelativistic) approximation, in which the Lorentz factor $\gamma$ does not deviate too much from unity and $\alpha Z \ll 1$, this has been done by Glass at al. [21–23]. In analyzing these approaches, it is advantageous to start with the semirelativistic CDW approximation (SCDW) and to proceed from there to the semirelativistic SE approximation (SSE).

The construction of approximate semirelativistic distorted waves $\chi_i$ and $\chi_f$ has three basic ingredients: (a) With the aid of the Darwin operator

$$U^D = 1 - i\frac{\hbar c}{2m_e c^2}\boldsymbol{\alpha}\cdot\nabla, \qquad (19)$$

semirelativistic bound-state wave functions to order $\alpha Z$ are constructed from the corresponding nonrelativistic wave functions multiplied with the basis spinors $\tilde{u}^{(+)} = (1,0,0,0)$ or $\tilde{u}^{(-)} = (0,1,0,0)$ for the positive or negative spin projections. (b) A nonrelativistic Coulomb distortion caused by the other collision partner (projectile or target, respectively) is produced by multiplying with the nonrelativistic operators $U_i^{\text{CDW}}$ or $U_f^{\text{CDW}}$ of Eq. (7), respectively, in which, however, the relativistic momenta $\gamma m_e \mathbf{v}$ enter as arguments. (c) Subsequently, the Sommerfeld-Maue operator

$$U^{\text{SM}} = 1 - i\frac{\hbar c}{2E}\boldsymbol{\alpha}\cdot\nabla, \qquad (20)$$

which is originally constructed to act on a relativistic plane wave $u^{(\pm)}$ [15], is applied to the resulting wave function in order to yield an approximate semirelativistic three-body wave function. All approximations are carried to the order $\alpha Z$.

The Sommerfeld-Maue modification is, of course, defined in the projectile frame for distorting the target wave function and in the target frame for distorting the projectile wave function. The operator (20), when acting on an exact relativistic plane wave, is known to yield approximate Coulomb continuum wave functions valid up to Lorentz factors $\gamma \gg 1$. In step (c), however, the operator acts on a distorted Darwin wave function whose small components are of the order of $\alpha Z$ (corresponding to $\gamma \approx 1 + \frac{1}{2}\alpha^2 Z^2$). Therefore, the sequence of operations (a) to (c) yields a distorted wave function justifiable only for $\alpha Z \ll 1$ and for $\gamma$ not too far from unity.

In assembling these operations, one has to neglect the second-order term in the derivatives, since it is of the same order as the terms already discarded. As a result, Glass et al. [22,23] obtain for the semirelativistic CDW approximation using Eqs. (3), (7), (19), and (20) the distorted SCDW wave function

$$\chi_i^{\text{SCDW}} = \left[ U_i^{\text{CDW}} \left( 1 - i\frac{\hbar c}{2m_e c^2} \alpha \cdot \nabla \right) - S \left( i\frac{\hbar c}{2E} \alpha \cdot \nabla \right) S^{-1} U_i^{\text{CDW}} \right] \psi_i^{\text{non}} u^{(\pm)}$$

(21)

and a corresponding term for $\chi_f^{\text{SCDW}}$. The first term in the expression (21) represents a nonrelativistic distortion by the projectile of a semirelativistic target wave function, while the second term is an addition to a semirelativistic distortion of a nonrelativistic target wave function. It is now a matter of straightforward algebra to calculate the residual interaction to be applied [22] in Eqs. (1) or (2). A precursor model for the SCDW approximation [24] contains inconsistencies in omitting important terms of the residual interaction, see [22], and uses various additional approximations. As has been shown by Bethe and Maximon [25], Sommerfeld-Maue wave functions are good approximations for partial waves with angular momenta $l \gg Z/137$. It is difficult to judge whether and when this condition is satisfied. The SCDW approach has the correct nonrelativistic limit, is post-prior symmetric, and yields promising agreement with experimental cross sections for not too high charges. The derivation of the SCDW approximation, Eq. (21), shows, however, that there is no unambiguous way to construct a semirelativistic CDW approximation. This should not be a serious drawback, as long as the residual interaction is derived consistently.

We finally return to the symmetric eikonal approximation. As is apparent from Eqs. (8) and (9), there are two ways to derive the distortion factor for the eikonal approximation. Equation (8) leads to an unambiguously defined consistently relativistic SE approximation, which, however, suffers from an unphysical behavior at low energies. If, instead, we start from Eq. (9), we can take advantage of the preceding development for the SCDW approximation and take the limit $r_T \to \infty$ at the end. The result is the semirelativistic symmetric eikonal approximation (SSE) proposed in [23] with a distorted-wave function $\chi_i^{SSE}$ obtained from Eq. (21) by simply replacing $U_i^{CDW} \to U_i^{eik}$ given in Eq. (6) and, similarly, $U_f^{CDW} \to U_f^{eik}$. By construction, the transition amplitude has the correct nonrelativistic limit and is post-prior symmetric.

If one wishes to relate the two types of symmetric eikonal approximations to the nonrelativistic CDW approximation, one may say that the SE is obtained by performing *first* the transition $r_{P,T} \to \infty$ and *then* the relativistic extension, while the SSE is derived by *first* taking the semirelativistic generalization and *then* the limit $r_{P,T} \to \infty$.

## SUMMARY

We analyze the structure of several relativistic extensions of distorted-wave capture theories in a unified fashion and discuss the relations between these approaches. The boundary-corrected Born approximation B1B and the (asymmetric) eikonal approximation, whose residual interactions are given by simple local potentials, allow for an unambiguous, fully relativistic extension. Their cross sections smoothly merge into the nonrelativistic limits. On the other hand, capture theories with nonlocal residual interactions lead to certain difficulties. While the direct generalization of the symmetric eikonal approximation (SE) leads to an unphysical behavior at low energies, a distinctly different semirelativistic symmetric eikonal approximation (SSE) derived via a semirelativistic continuum distorted wave (SCDW) approximation has the correct nonrelativistic limit. The SCDW approximation itself is based on a first-order treatment (in $\alpha Z$) of approximate Darwin wave functions with Sommerfeld-Maue type of distortions.

## ACKNOWLEDGMENTS

The present research is supported through the International Scientific Research Program by the Ministry of Education, Science and Culture of Japan. J.E. gratefully acknowledges the hospitality extended to him at the Institute of Applied Physics, University of Tsukuba, Japan.

## REFERENCES

1. B.H. Bransden and M.R.C. McDowell, *Charge Exchange and the Theory of Ion-Atom Collisions*, (Clarendon Press, Oxford 1992).
2. D.P. Dewangan and J. Eichler, Phys. Rep. **247**, 59 (1994)
3. M.H. Mittleman, Proc. Phys. Soc. (London) **84**, 453 (1964); R. Shakeshaft, Phys. Rev A **20**, 779 (1979); B.L. Moiseiwitsch and S.G. Stockman, J. Phys. B **13**, 2975 (1980)
4. B.L. Moiseiwitsch, J. Phys. B **21**, 603 (1988); Phys. Rev. A **39**, 5609 (1989)
5. F. Decker and J. Eichler, Phys. Rev. A **44**, 377 (1991)

6. D.P. Dewangan and J. Eichler, J. Phys. B **18**, L65 (1985); J. Phys. B **19**, 2939 (1986)
7. Dz. Belkic, R. Gayet and A. Salin, Phys. Rep. **56**, 279 (1979)
8. S.E. Corchs, L.J. Dubé, J.M. Maidagan, R.D. Rivarola and A. Salin, J. Phys. B **25**, 2027 (1992)
9. F.T. Chan and J. Eichler, Phys. Rev. Lett. **42**, 58 (1979); J. Eichler and F.T.Chan, Phys. Rev. A **20**, 104 (1979); J. Eichler, Phys. Rev. A **23**, 498 (1981)
10. J.M. Maidagan and R. Rivarola, J. Phys. B **17**, 2477 (1984)
11. I.M. Cheshire, Proc. Phys. Soc. **84**, 89 (1964)
12. D.S.F. Crothers and L.J. Dubé, in: Adv. Atom Mol. Phys., Vol 30, eds. D.R. Bates and B. Bederson (Academic Press, New York, 1993) p. 287.
13. D.S.F. Crothers, J. Phys. B **15**, 2061 (1982).
14. J. Eichler, Phys. Rep. **193**, 165 (1990)
15. J. Eichler and W.E. Meyerhof, *Relativistic Atomic Collisions* (Academic Press, San Diego, 1995)
16. J. Eichler, Phys. Rev. A **35**, 3248 (1987); Erratum: **37**, 287 (1988)
17. J. Eichler, Phys. Rev. A **32**, 112 (1985)
18. N. Toshima and J. Eichler, Phys. Rev. A **38**, 2305 (1988)
19. B.L. Moiseiwitsch, J. Phys. B **19**, 3733 (1986); J. Phys. B **20**, L171 (1987); G.R. Deco and R.D. Rivarola, J. Phys. B **20**, 5117 (1987)
20. N. Toshima and J. Eichler, Phys. Rev. A **41**, 5221 (1990)
21. J.T.Glass, J.F. McCann, and D.S.F. Crothers, J. Phys. B **25**, L541 (1992)
22. J.T.Glass, J.F. McCann, and D.S.F. Crothers, J. Phys. B **27**, 3445 (1994)
23. J.T.Glass, J.F. McCann, and D.S.F. Crothers, J. Phys. B **27**, 3975 (1994)
24. G.R. Deco and R.D. Rivarola, J. Phys. B **20**, 317 (1987)
25. H.A. Bethe and L. Maximon, Phys. Rev. **93**, 768 (1954)

# X-ray emission in relativistic ion-atom collisions

## J. F. McCann*, J.T.Glass †‡ and D.S.F. Crothers‡

\* *Physics Department, The University, Durham DH1 3LE, England.*
† *Institut für Theoretische Physik der Justus-Liebig-Universität,
39392 Giessen, Germany.*
‡ *Theoretical and Computational Physics Research Division,
Queen's University of Belfast, Belfast BT7 1NN, Northern Ireland.*

---

X-ray line-emission produced during collisions of relativistic multi-charged ions with atoms is discussed. A direct calculation for radiative electron capture (REC) rates is presented within the distorted-wave impulse approximation (DWIA). Simple formulae are obtained for the angular distribution of radiation, and closed analytic formula are derived for small $\alpha Z$. Reasonable agreement with recent experiments is found for photoemission rates and angular distributions. Our results indicate that the photon spectrum is rather insensitive to target structure, but dependent on multiple-scattering corrections, and this contradicts other theoretical predictions. However, in asymmetric collisions our results are in broad agreement with experiment and Relativistic Impulse Approximation (RIA) calculations. Magnetic transitions are also analysed in terms of the electron and photon polarization correlation parameters.

---

## INTRODUCTION

The high-frequency photoemission spectrum from relativistic ion-atom impacts is dominated by sharp lines due to collision-induced transitions [1]. This feature is accompanied by a background of electronic and nuclear *bremsstrahlung* and, in certain cases, an intense broad structure due to emission from quasimolecular states [2]. If the target atom is rather light and the projectile is highly-charged and very fast [3,4], then it is a suprising fact that electron capture via vacuum coupling is a stronger channel than potential scattering [5-9]. An important aspect of the collision is that strong spin-orbit coupling gives rise to magnetic (spin-flip) transitions during photoemission. However, this process only becomes important at very high energies, and in fact the *nonrelativistic* dipole approximation [10] is remarkably accurate even for high-$Z$ and kinetic energies up to 1 GeV/u [7]. Experiments have recently been performed to measure the rate of such processes and the frequency and angular distribution of photons emitted. Larmor's formula [11] dictates that the sudden acceleration along the beam axis of the electron leads to an intensity pattern that is peaked at right angles to the collision. However at relativistic energies retardation effects and magnetic transitions alter this pattern considerably. Recent measurements of this process have produced evidence of such effects [3,4] in confirming a skewed-lobe angular distribution.

In an asymmetric collision the process is dominated by the deep well of the highly-charged projectile/target. The electron tends to adjust diabatically during the transfer, and it is appropriate to use atomic orbitals. However, one should also take into account the influence of the fields in perturbing the atoms, through polarization and Stark shifting, to obtain accurate lineshapes and yields. One approach treats the process as inverse *photoionization*, with the target electron as almost free [5]. The target is treated in the *impulse approximation* [12,6,13,14] as the source of the electronic momentum distribution and some initial binding energy. Results obtained with this model have given good agreement with the angular distribution of the ejected photons [3,4], though the theory is not able to account for electron correlation effects or relativistic target structure. In addition, the theory is not designed to describe strong two-center effects.

In this paper we demonstrate the validity and accuracy of two-centered treatments of REC within the framework of the distorted-wave impulse approximation (DWIA), and compare with recent measurements for a variety of collision events. We also present photon polarisation correlation functions for this process.

## DISTORTED-WAVE IMPULSE APPROXIMATION

In the 'direct' approach one solves the relativistic field-collision problem. While vacuum coupling can be considered a small perturbation, the electron dynamics is a complicated electromagnetic tug-of-war. Nonetheless, one can try perturbation methods, but the size of the coupling strength ($\alpha Z$) may mean very slow convergence. The leading order is the Sauter formula, and for the total K-shell REC cross-section [15] can be written:

$$Q_S = \frac{4\pi Z_P^5 \beta \gamma}{c^8 (\gamma - 1)^3} \left\{ \frac{4}{3} + \frac{\gamma(\gamma - 2)}{(\gamma + 1)} \left( 1 - \frac{1}{2\beta\gamma^2} \ln \frac{1 + \beta}{1 - \beta} \right) \right\} \; , \tag{1}$$

where $\beta = v/c$, $\gamma = (1 - \beta^2)^{-\frac{1}{2}}$ and $Z_P$ is the projectile charge, $v$ is the collision velocity, and all quantities are in atomic units. The scaling ($Z_P^5$) arises from the atomic momentum probability; all other terms are kinematic factors. This elegant formula is rather limited, since it is based upon the leading term in $\alpha Z_P$, and the target atom parameters do not appear at all.

One method of 'summing' higher-order terms involves separation of the three-body scattering into coupled two-body scattering terms. This characterises approaches such as the Fadéev method, and the impulse and distorted-wave approximations [16]. We adopt a hybrid method (DWIA) [17] which

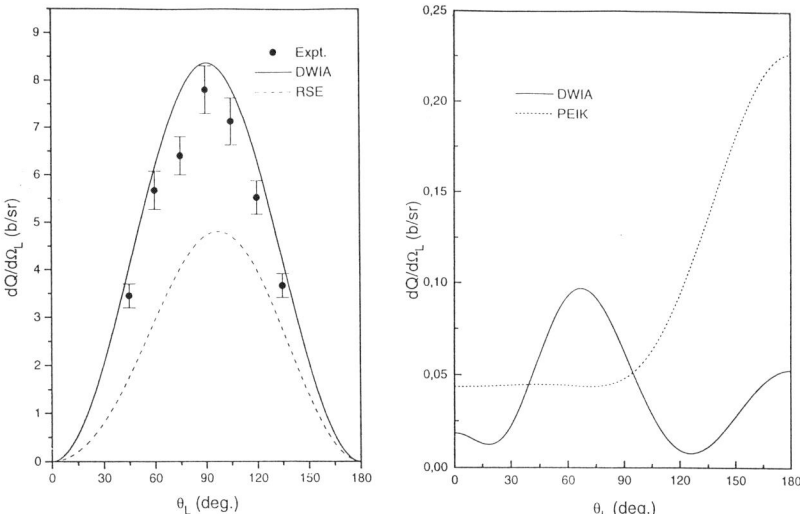

**FIG. 1.** Differential cross section, as a function of laboratory angle $\theta_L$, for *nonflip* REC within the DIWA and RSE approximations for Xe$^{54+}$ incident on Be at 197 MeV/u.

**FIG. 2.** Differential cross section, as a function of laboratory angle $\theta_L$, for *spinflip* REC within the DWIA and Eikonal approximations (PEIK) for Xe$^{54+}$ incident on Be at 197 MeV/u.

partially accounts for two-center dynamics. For a given impact parameter $b$, the first-order amplitude for a radiative transition is:

$$a_{fi}(\lambda, b) = -i \int d^4x' \ \Psi_f'^{\dagger} \alpha \cdot A_\lambda'^{*}(x'^{\mu}, k'^{\mu}) S \Psi_i, \qquad (2)$$

where $\Psi_f'$ and $\Psi_i$ are the scattering states. The matrix $S$ boosts the wavefunction from the target to the projectile frame [5]. All primed quantities refer to the inertial frame of the projectile, with origin at the projectile nucleus. In this frame the photon has a momentum $k'$, a polarization $\lambda = 1, 2$, and makes an angle $\theta'$ with the beam direction which also defines the electron spin-quantization axis. The vector potential $A_\lambda'$ is given by

$$A_\lambda' = \left(\frac{2\pi}{V\omega_p'}\right)^{\frac{1}{2}} e_\lambda \, e^{-ik' \cdot x'} \qquad (3)$$

and $k'^{\mu} = (\omega'/c, k')$. To be specific, the vector $e_2$ is parallel to $\hat{v} \times k'$, and $e_1 = \hat{k}' \times e_2$. The quantization of the field within a volume $V$ gives the density of states: $\rho_\gamma(\omega_p') = V\omega_p'^2 (2\pi)^{-3} c^{-3}$. The momentum-space amplitude can be defined as follows:

$$T_\lambda(\eta) = 2\pi i \gamma v \left(\frac{V\omega'}{2\pi c^2}\right)^{\frac{1}{2}} \int db \exp(-i\eta \cdot b) a_{fi}(\lambda, b). \qquad (4)$$

Thus the cross section for capture with emission of a photon of energy $\omega'_p$ into a solid angle $d\hat{\boldsymbol{k}}' = \sin\theta' d\theta' d\varphi'$ is given by

$$\frac{d^2 Q(\theta',\omega')}{d\hat{\boldsymbol{k}}' d\omega'} = \frac{\omega'}{(2\pi)^4 \gamma^2 v^2 c} \int d\boldsymbol{\eta} \sum_\lambda |T_\lambda(\boldsymbol{\eta})|^2, \qquad (5)$$

and the singly-differential cross section is then given by summing over all frequencies;

$$\frac{dQ(\theta')}{d\hat{\boldsymbol{k}}'} = \frac{1}{16\pi^4 \gamma^2 v^2 c} \int_0^{+\infty} d\omega'\, \omega' \int d\boldsymbol{\eta} \sum_\lambda |T_\lambda(\boldsymbol{\eta})|^2. \qquad (6)$$

The wavefunction of the target atom in the projectile frame is denoted as, $|\psi'_i\rangle = S\phi_T(\boldsymbol{x})e^{-iE_T t}$, and the impulse approximation may be succinctly written in the form:

$$|\Psi_i^{(+)'}\rangle = \sum_{\sigma=1}^{4} \int d\boldsymbol{p}' \Omega_P^{(+)}(E_{p'})|p'^\sigma\rangle \gamma^{-1} e^{i\boldsymbol{q}'\cdot\boldsymbol{b}+i\gamma \boldsymbol{q}'\cdot\boldsymbol{v}t'-i\gamma E_T t'} u_{p'}^{\sigma\dagger} S\tilde{\phi}_T(\boldsymbol{q}'). \qquad (7)$$

The argument of the Fourier transform ($\tilde{\phi}$) is defined as $\boldsymbol{q}' = \boldsymbol{p}'_\perp + (\gamma^{-1} p'_\parallel + v E_T/c^2)\hat{\boldsymbol{v}}$. The symbol $\Omega_P$ denotes the relativistic half-on-shell Coulomb scattering operator in the rest-frame of the projectile, and $\langle \boldsymbol{x}|p^\sigma\rangle = (2\pi)^{-\frac{3}{2}} e^{i\boldsymbol{p}\cdot\boldsymbol{x}} u_p^\sigma$ is the plane-wave spinor. In physical terms, equation (7) expresses the condition of the electron as simultaneously in a target bound-state and in the continuum of the projectile. The summation includes the negative energy states, though these are not found to be physically significant in this process.

The large discrepancy between plane-wave calculations and experiment can be remedied by multiple-scattering corrections, such as the Stöbbe formula [7]. More generally the transition amplitude for the DWIA is given by:

$$a_{fi}^{DW}(\lambda,\boldsymbol{b}) = -i \int dt' \int d\boldsymbol{x}' \phi_P'^\dagger(\boldsymbol{x}',t') L_{DW}'^\dagger c\boldsymbol{\alpha}\cdot \boldsymbol{A}_\lambda'^* \Psi_i^{(+)'}(\boldsymbol{x}',t') \qquad (8)$$

in which $L_{DW}$, the RCDW tensor [20], has been introduced in the final scattering channel, and $L'_{DW} = S L_{DW} S^{-1}$.

If the target is light, then one can simply set $L_{DW} = 1$ and regain the plane-wave impulse approximation (PWIA), for which:

$$T_{PW}(\lambda,\boldsymbol{\eta}) = (2\pi)^3 \sum_\sigma \langle \phi_P|\boldsymbol{\alpha}\cdot \boldsymbol{e}_\lambda^* e^{-i\boldsymbol{k}'\cdot\boldsymbol{x}} \Omega_P(E_{\bar{p}})|\bar{p}^\sigma\rangle u_{\bar{p}'}^{\sigma\dagger} S\tilde{\phi}_T(\bar{\boldsymbol{q}}) \qquad (9)$$

where $\bar{\boldsymbol{q}} = \boldsymbol{\eta} + (\gamma v)^{-1}(\gamma E_T - E'_P - \omega')\hat{\boldsymbol{v}}$ and $\bar{\boldsymbol{p}} = \boldsymbol{\eta} + (\gamma v)^{-1}(E_T - \gamma E'_P - \gamma \omega')\hat{\boldsymbol{v}}$. Thus the rate is governed by the the oscillator strength for radiative recombination, weighted by the target momentum distribution. And given

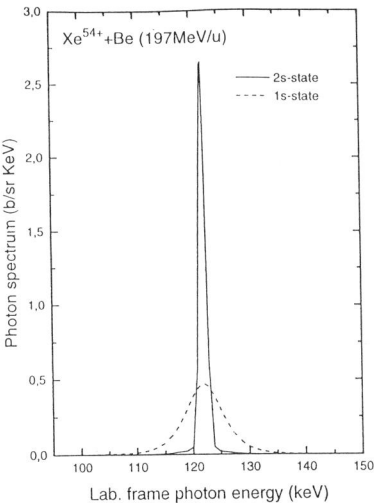

**FIG. 3.** Emission spectrum (lab. frame) for nonflip REC within the DWIA approximation for Xe$^{54+}$ incident on Be at 197 MeV/u.

that $\tilde{\phi}_T(\bar{q})$ is very small unless $\bar{q}_\parallel \approx 0$, the rate peaks at a frequency $\omega'_r = \gamma E_T - E'_P$. The fact that $\bar{q}_\parallel$ is large for nonradiative capture ($\omega' = 0$) at relativistic velocities, explains why capture mediated by photoemission is more likely. Consequently, when $\bar{q}_\parallel = 0$, then $\bar{p}_\parallel = -\gamma v(E_T/c^2)$ and so we can expect relativistic wavefunctions to be required in describing the electron-projectile scattering whereas *nonrelativistic* wavefunctions are adequate for the target. This breaking of the target-projectile symmetry was confirmed in detailed calculations. The PWIA result (Eq. 9), obtained by using the 'direct' approach, is similar to that found using the 'indirect' method given by Ichihara *et al* [6]. Small differences arise in the method adopted for relativistic velocity addition; in ref. [6] the target electron is assigned its rest energy only. Furthermore, the target electron was assumed to have a nonrelativistic momentum distribution, and spin-orbit effects were neglected. Clearly from our discussion above, these are reasonable assumptions for light targets.

## Light Targets

For light target atoms and nonrelativistic velocities, the REC cross section is insensitive to target structure [18,10]. We can show that this is also true for relativistic collisions. As mentioned above, in this limit the target function can be replaced by the Schrödinger approximation $\phi_T = \phi_{sT} u^i$ and we have, from (Eq. 9),

$$T_{PW}(\lambda, \eta) \approx (2\pi)^3 \sum_\sigma \langle \phi_P | \boldsymbol{\alpha} \cdot \boldsymbol{e}^*_\lambda e^{-i\boldsymbol{k}' \cdot \boldsymbol{x}'} \Omega_P(E_{\bar{p}}) | \bar{p}^\sigma \rangle u^\sigma_{\bar{p}'} S u^i \tilde{\phi}_{sT}(\bar{q}) \quad . \quad (10)$$

Independent of the photon polarisation, $\tilde{\phi}_{sT}$ is peaked very sharply around $\eta = 0$ *and* at the resonant photon frequency in the projectile frame, namely, $\omega'_r =$

**TABLE 1.** Differential cross sections for photons emitted at $\theta_L = 90°$ in b/sr for $Xe^{54+}$+Be 197 MeV/u.

| Theoretical model | Nonflip | | | | Flip | | Peaked |
|---|---|---|---|---|---|---|---|
| | 1s-states | | 2s-states | | 1s-states | 2s-states | 1s ≡ 2s |
| | Slater | RHF | Slater | RHF | Slater | Slater | ↑↑ / ↓↑ |
| DWIA | 4.15 | 4.16 | 4.18 | 4.18 | 0.0319 | 0.0316 | |
| PCDW | 4.17 | 4.18 | 4.18 | 4.18 | 0.0316 | 0.0316 | 4.1/0.0315 |
| TCDW | 11.2 | 11.2 | 11.3 | 11.3 | 0.682 | 0.680 | |
| Eikonal | 2.36 | 2.36 | 2.36 | 2.36 | 0.0243 | 0.0241 | 2.36/0.0240 |
| OBK1 | 11.2 | 11.3 | 11.3 | 11.3 | 0.683 | 0.680 | 11.3/0.678 |

**TABLE 2.** K-shell radiative electron capture cross sections. Collisions of fast, highly-charged projectiles with light targets.

| Collision Process | Kinetic Energy MeV/u | Theory : DWIA | | Experiment [4] barns |
|---|---|---|---|---|
| | | Dirac | Darwin | |
| $Pb^{81+}$(1s) + $N_2$ | 219 | 460 | 367 | 290 ± 87 |
| $Pb^{82+}$ + $N_2$ | 277 | 716 | 563 | 535 ± 161 |
| $Bi^{82+}$(1s) + C | 82 | 839 | 754 | 680 ± 136 |
| $Bi^{82+}$(1s) + C | 116 | 543 | 472 | 371 ± 74 |
| $Bi^{82+}$(1s) + C | 169 | 327 | 274 | 239 ± 48 |
| $U^{91+}$(1s) + $N_2$ | 295 | 460 | 374 | 290 ± 87 |
| $U^{92+}$ + $N_2$ | 295 | 942 | 745 | 569 ± 171 |
| $U^{92+}$ + Ar | 295 | 1290 | 1014 | 1069 ± 321 |

$\gamma E_T - E'_P$. If we approximate the oscillator strength and the external factors by their values at these two peaks and bring them outside the integrals we can invert the Fourier transform and find the peaked impulse approximation

$$\frac{dQ_\lambda(\theta')}{d\hat{k}'} \approx \frac{\omega'_r(2\pi)^2}{\gamma^2 v^2 c} \mid \langle\phi_P|\boldsymbol{\alpha}\cdot \boldsymbol{e}^*_\lambda e^{-i\boldsymbol{\kappa}'\cdot\boldsymbol{x}'} \Omega_P(E_s)|s^i\rangle u^{i\dagger}_s S u^i \mid^2 \quad (11)$$

where $\kappa' = \omega'_r/c$, and $s = -\gamma(E_T/c^2)v \approx -\gamma v$. For a target atom of atomic number $Z_T$ we multiply the single-electron cross section by $Z_T$, and hence the approximate scaling law follows: $Q \propto Z_P^5 Z_T$. It must be stressed that this two-body simplification for the cross section only arises through a peaking approximation to the full three-body expression. This approximation depends on the smallness of the parameter $Z_T/v$.

## SEMIRELATIVISTIC MODELS

When the charges involved in REC are not too large ($\alpha Z_P \lesssim 1$) then one can use approximate solutions of the Dirac equation. The Darwin and Sommerfeld-Maue solutions are the corrections to first-order in $\alpha Z$. Further simplifications such as the dipole approximation can also be made. The Darwin approximation is surprisingly accurate even in the high-momentum tail. We present results for REC using both models.

The target atom can be modelled with Slater functions or Roothan-Hartree-Fock orbitals [21]. We found very little difference for the frequency spectrum, though it was reported by other workers that the diffuse RHF orbitals gave rise to sharper lines [6].

For a semirelativistic REC process we considered the reaction

$$\text{Xe}^{54+} + \text{Be} \rightarrow \text{Xe}^{53+} + \text{Be}^+ + \gamma \qquad (12)$$

at 197 MeV/amu and compared with the data of Anholt et al [23]. This corresponds to $\gamma = 1.21$, ($v = 77.4$ a.u.) and the resonant photon energy in the target frame is $\omega_r = 121.8$ keV. A photon emitted at $\theta_L = 90^o$ has a wavelength of 0.100 Å (0.084 Å) in the target (projectile) frame.

Figure 1 shows the singly-differential cross section for DWIA for nonflip. We find excellent agreement with experiment [23] for the DWIA model. Most theoretical data in REC work is normalised to experiment and in the recent one-centered model of Ichihara et al this involves a reduction factor of 0.8. We would only require our data to be multiplied by 0.93 to agree with experiment at the central maximum. Figure 2, which is for the spin-flip process, highlights the differing behaviour for intermediate velocities. The eikonal approximation has become notoriously unreliable for spin-flip in nonradiative capture [17,20] , and while it is not quite as poor in REC, we still found that CDW-type theories were superior.

Table 1 highlights the target-parameter independence of the cross section. Note also the dramatic overestimation when projectile scattering is not included (OBK1). This is particularly evident in the flip cross sections. The last column shows that the peaked impulse approximation (Eq. 11) is extremely good for this asymmetric collision.

So how do the simple analytical models of the Darwin approximation compare with a full Dirac treatment? We have repeated the calculations with full Dirac bound states for the projectile and the results, again within the peaked DWIA, are given in Table 2. For these large projectile charges there is an increase of between 10 and 25% in the results. This is in line with the theory of Ichihara et al, though their predictions are in slightly better agreement with experiment. The addition of radiative corrections to the energy levels [2] increased the rate for $U^{92+} + N_2$ by 2.5 %. Fortuitously, it is the simple Darwin approach which gives the best agreement with experiment. The influence of two-center effects can be discerned in the last row in Table 2, where the target is *active* during the process. However, it is likely that electron correlation cannot be neglected in systems such as $N_2$, and thus the theoretical results obtained to date must be considered as rough estimates.

In a series of recent experiments Stöhlker et al [4] have measured and calculated L-shell REC for the first time. This enables more strenuous tests of theory and in particular the angular distribution is no longer so simple due to the loss of the retardation-aberration cancellation, a feature unique to K-shell.

## ASYMPTOTIC ENERGIES

We can separate the cross section for capture with and without spin-flip at asymptotic energies. It is straightforward to show for nonflip that, in the laboratory frame, for $v \approx c$,

$$\frac{dQ^{\uparrow\uparrow}(\theta_L)}{d\hat{k}} \simeq \frac{Z_P^5}{2\gamma c^8} \sin^2 \theta_L. \tag{13}$$

The cross section, in the asymptotic limit, is:

$$Q^{\uparrow\uparrow} \simeq \frac{4\pi Z_P^5}{3c^8 \gamma} = \frac{1}{3} Q_S. \tag{14}$$

where $Q_S$ is the Sauter cross-section.

For the flip process the asymptotic singly differential cross section can be shown to reduce to the simple form:

$$\frac{dQ^{\uparrow\downarrow}}{d\hat{k}_L} \simeq \frac{\sin^2 \theta_L}{(1-\beta\cos\theta_L)^2} \frac{dQ^{\uparrow\uparrow}}{d\hat{k}_L}. \tag{15}$$

This shows that the flip angular distribution has a strong anisotropy as $v \to c$. The total cross section is then given, in the asymptotic limit, by

$$Q^{\downarrow\uparrow} \simeq \frac{8\pi Z_P^5}{3c^8 \gamma} = \frac{2}{3} Q_S \tag{16}$$

It is interesting that the Sauter formula is dominated by spin-flip transitions, which explains the characteristic anisotropy [5]. Deco and Rivarola [9] showed how Coulomb wavefunctions modified the Sauter result.

## POLARISATION CORRELATION

To examine the role of electron polarization in more detail, we have calculated linear-polarisation correlation functions in which the electron's spin is also projected out:

$$P^{nf} = \frac{|T^{\uparrow\uparrow}_{\lambda=1}|^2 - |T^{\uparrow\uparrow}_{\lambda=2}|^2}{|T^{\uparrow\uparrow}_{\lambda=1}|^2 + |T^{\uparrow\uparrow}_{\lambda=2}|^2} \qquad P^{fl} = \frac{|T^{\uparrow\downarrow}_{\lambda=1}|^2 - |T^{\uparrow\downarrow}_{\lambda=2}|^2}{|T^{\uparrow\downarrow}_{\lambda=1}|^2 + |T^{\uparrow\downarrow}_{\lambda=2}|^2} \tag{17}$$

In their SPB model of REC, Hino and Watanabe [12] presented linear polarisation correlation functions as a function of the photon emission angle, having summed over electron spins. Broadly speaking, their results showed

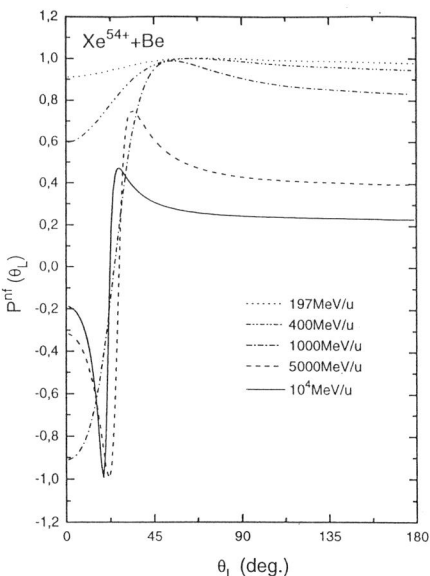

**FIG. 4.** Linear photon polarisation correlation function for nonflip $P^{nf}$ REC as a function of laboratory photon emission angle $\theta_L$, within the PWIA approximation for $Xe^{54+} + Be$.

that in the nonrelativistic limit, the photon is polarised in the scattering plane($P^{nf} = 1$), while at extreme relativistic energies both linear polarization states are equally probable ($P^{nf} = 0$). In Figure 4 we have shown the DWIA nonflip function $P^{nf}$ for $Xe^{54+}$+Be for a range of energies. As expected, for low energies the X-rays polarized in the scattering plane dominate, and for most emission angles $P^{nf} \to 0$ as $v \to c$. Interestingly there is a very dramatic effect for photons emitted close to the beam direction.

Using asymptotic analysis, we find that the amplitude for photon polarization in the scattering plane is zero when $\cos\theta' = -(\gamma - 3)(\gamma + 1)/\beta\gamma(\gamma - 1)$. Similarly the cross section for polarisation perpendicular to the scattering plane is zero at $\cos\theta' = -\beta\gamma(\gamma + 1)^{-1}$ which corresponds to $\cos\theta_L = \beta\gamma(\gamma + 1)^{-1}$ in the laboratory frame. This unusual behaviour might offer the first possibility of an experiment which could differentiate between capture with or without spin-flip, unlike nonradiative capture in which polarisation signatures are contained within projectile scattering. We add that these correlation results were also very sensitive to the model used, and it is hoped this work will stimulate further theoretical studies.

The polarisation correlation functions for flip $P^{fl}$ tend to zero asymptotically at all emission angles, with $\tilde{T}^{\uparrow\downarrow}_{\lambda=1}/\tilde{T}^{\uparrow\downarrow}_{\lambda=2} \simeq i(\gamma - 1)/\beta\gamma$. The very stable $P^{fl} \approx 0$ behaviour at high energies is in contrast to the nonflip behaviour, so much so that this may be of some interest experimentally. Finally, our summed polarisation correlation coefficents $P^S$ compare very well with the results of Hino and Watanabe [12].

## ACKNOWLEDGMENTS

This work was supported by the Department of Education for Northern Ireland, and the UK EPSRC. We are very grateful to Dr. Th. Stöhlker for very useful discussions and communications.

## REFERENCES

1. J. S. Greenberg, C. K. Davies and P. Vincent, Phys. Rev. Lett. **33**, 473 (1974).
2. W. Greiner, B. Müller and J. Rafelski, *Quantum Electrodynamics of Strong Fields*, Springer (1985)
3. Th. Stöhlker et al., Phys. Rev. A. **51** 2098 (1995).
4. Th. Stöhlker et al., Phys. Rev. Lett. **73** 3520 (1994).
5. J. Eichler, Phys. Rep. **193**, 165 (1990).
6. A. Ichihara, T. Shirai and J. Eichler, Phys. Rev. A **49**, 1875 (1994).
7. T. Stöhlker, C. Kozhuharov, A.E. Livingston, P.H. Mokler, Z. Stachura and A. Warczak, Z. Phys. D **23**, 121 (1992).
8. P.H. Mokler, T. Stöhlker, C. Kozhuharov, Z. Stachura and A. Warczak, Z. Phys. D **21**, 197 (1991).
9. G.R. Deco and R.D. Rivarola, Phys. Rev. A **39**, 5451 (1989)
10. R. Shakeshaft and L. Spruch, Rev. Mod. Phys. **51**, 369 (1979)
11. J.D. Jackson, *Classical Electrodynamics, 2nd Ed.* (Wiley) (1972).
12. K. Hino and T. Watanabe, Phys. Rev. A **39**, 3373 (1989).
13. D.H. Jakubassa-Amundsen and P.A. Amundsen, Z. Phys. A **298** 13 (1980).
14. M. Kleber and D. H. Jakubassa, Nucl. Phys. A **252** 151 (1974)
15. F. Sauter, Ann. Phys. **11**, 454 (1931).
16. B. H. Bransden and M. R. C. McDowell *Charge Exchange and the Theory of Ion-Atom Collisions* O.U.P. (1992)
17. J.T. Glass, J.F. McCann and D.S.F. Crothers, Z. Phys. D. *in press* (1995)
18. J.S. Briggs and K. Dettmann, J.Phys.B **10**, 1113 (1977).
19. J.S. Briggs and K. Dettmann, Phys. Rev. Lett. **33**, 1123 (1974).
20. J.T. Glass, J.F. McCann and D.S.F. Crothers, J.Phys.B **27**, 3445 (1994); J.Phys.B **27**, 3975 (1994).
21. E. Clementi and C. Roetti, At. Data Nucl. Data Tables **14**, 177 (1974).
22. P. Kienle, M. Kleber, B. Povh, R.M. Diamond, F.S. Stephens, E. Grosse, M. Maier and D. Proetel, Phys. Rev. Lett. **31**, 1099 (1973)
23. R. Anholt. S.A. Andriamonje, E. Morenzoni, C. Stoller, J.D. Molitoris, W.E. Meyerhof, H. Bowman, J.S. Xu, Z.Z. Xu, J.O. Rasmussen and D.H.H. Hoffmann, Phys. Rev. Lett. **53**, 234 (1984).

# Measurements of Charge Transfer Cross Sections for $Ar^{q+}$ (q=6, 7, 8, 9 and 11) Collisions with He and $H_2$ Targets at Low Energies

Kazuhiko Okuno, Hiroki Saitoh[1], Kouichi Soejima[2],
Scott Kravis[3]* and Nobuo Kobayashi

*Department of Physics, Tokyo Metropolitan University, Minami-Ohsawa, Tokyo 192-03*
*\*The Institute of Physical and Chemical Research (RIKEN), Hirosawa, Saitama 351-01*

**Abstracts**. The cross sections for single- and double-electron transfer, neglecting the transfer ionization processes, have been measured for $Ar^{q+}$(q=6, 7, 8, 9 and 11) with $H_2$ and He at low energies ranging from 0.075 to 225 eV/amu, systematically.

The single-electron transfer cross sections were found to distribute between the predictions of the scaling law by Müller and Salzborn and the absorbing model by Olson and Salop above about 1 eV/amu and to converge to the Langevin cross section at the low energy end measured. For the same projectile ion, the single-electron transfer cross sections for the $H_2$ targets are about threefold as large as those for the He target and both energy dependencies are very similar. The double-electron transfer cross sections are smaller than those for the single-electron transfer and they are characterized by a minimum structure near 1 eV/amu for He targets. All cross sections for the single- and double-electron transfer have a common signature increasing with decreasing collision energy at the low energy end measured. With analyses by simple models and the MCLZ calculation, the importance of an attractive induced-dipole polarization potential in the highly charged ion collisions is discussed.

## INTRODUCTION

The Coulomb electric field created by the ionic charge polarizes the target and an attractive induced-dipole polarization force bends the collision trajectory. In various thermal reactions of singly charged ions, it is very well known that almost all reaction rates are constant and their cross sections are enhanced in inverse proportion to the collision velocity. The signature is the so called "the orbiting effect" due to the attractive induced dipole polarization. By increasing the charge state of projectile, such signature is expected to be observed at higher energies than

---

Present address:1; H. Saitoh; Aloka corporation, Imai 3-2-7-19, Oume-shi, Tokyo 198.
2; K.Soejima; Photon Factory, National Laboratory for High-Energy Physics,Tukuba, Ibaraki 305.
3; S. Kravis; J. R. Macdonald Laboratory, Kansas State University, Manhattan, KS66506.

the thermal energy. However, although many investigations using highly charged ions have been made until now, only very few previous works[1,2,3] have found the enhancement of the charge-transfer cross section due to the induced-dipole attractive force.

Recently, single- and double-electron capture cross sections in $Ar^{q+}$(q=6, 7, 8, 9 and 11)-$H_2$ and He collisions have been measured at low energies ranging from 0.075 to 225 eV/amu, systematically. For $H_2$ targets, we have reported elsewhere.[4]

In this paper, the single- and double-electron transfer cross sections measured for $H_2$ and He targets are reported. Using the comparison with simple models, the Langevin cross section[5], a scaling law developed by Müller and Salzborn[6], the absorbing sphere model by Olson and Salop[7], and the analysis by the multi-channel Landau-Zener (MCLZ) calculation, the importance of "the orbiting effect" due to the attractive induced-dipole polarization potential in the highly charged ion collisions and factors which obscure such effect will be discussed.

## EXPERIMENTAL SETUP

The apparatus and technique used in the present cross section measurement were reported elsewhere in detail[8]. The schematic diagram of the apparatus used is shown in figure 1. For the low energy cross section measurements, there are some technical difficulties, especially in the preparation of a stable ion beam with a narrow energy spread and in the collection of large angle scattered ions. Here, these difficulties have been overcome by developments of a mini-EBIS (electron beam ion source), which can provide a multiply charged ion beam with a narrow energy spread typically 0.3q eV(WHM) in the DC ion-extraction mode[9], and of an OPIG (octo-pole ion beam guide), in which an r-f oscillatory electric field can accommodate product ions along the beam axis and make it possible to collect all product ions scattered into the forward direction in the laboratory systems.[8,10] Cross sections are determined from the initial growth of product ion intensity in increasing the target pressure.

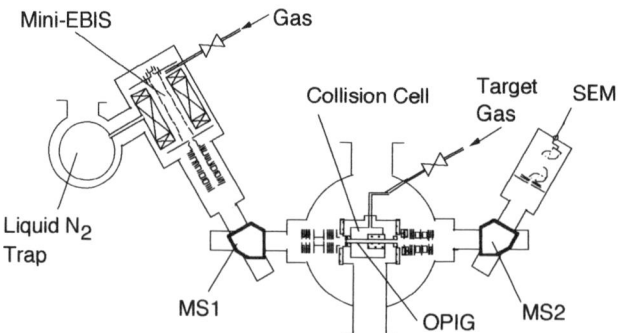

**FIGURE 1.** Schematic diagram of the apparatus used.

In the $Ar^{q+}$-$H_2$ and He collisions, single- and double-electron capture processes take place.

$$Ar^{q+} + H_2, He \rightarrow Ar^{(q-1)+} + (H_2)^+, He^+ \qquad \sigma_{q,q-1} \qquad (1)$$

$$\rightarrow Ar^{(q-2)+} + (H_2)^{2+}, He^{2+} \qquad \sigma_{q,q-2} \qquad (2)$$

$$\hookrightarrow Ar^{(q-1)+} + e \qquad \sigma_{TI} \qquad (3)$$

In our cross section measurements, neglecting the transfer ionization (3) following the double-electron capture (2), measured cross sections correspond to $\sigma_1 = \sigma_{q,q-1} + \sigma_{TI}$ and $\sigma_2 = \sigma_{q,q-2} - \sigma_{TI}$. The overall uncertainty in the measured cross sections is estimated to be ±20% and the ion energy uncertainty is found to be ±0.3q eV.

As is well known, the charge transfer process in highly charged ion collisions is an exothermic reaction with a large exothermic energy Q and the reduction of the collection efficiency for large angle scattered ions is troublesome, especially at low energies. However, in cases of the $Ar^{q+}$-$H_2$ and He collisions in which the mass of the projectile is much larger than that of targets, it is obvious from the consideration of kinematics that all product ions of $Ar^{(q-r)+}$ are scattered in the forward direction in the laboratory systems at the collision energies of $E_{amu} \geq Q/760$ for $H_2$ target and $E_{amu} \geq Q/360$ for He target. Thus all product ions of $Ar^{(q-r)+}$ in both collision systems can be detected in the whole collision energy range measured in this experiment using the OPIG.

## SIMPLE MODELS

*Langevin orbiting cross section.* As mentioned before, the attractive induced-dipole polarization potential of $V = -\alpha q^2/2R^4$ is significant in the low energy ion-neutral collisions, where $\alpha$ is the dipole polarizability of the target, q is the charge state of the projectile and R is the distance between the target and projectile. The projectile follows a circular orbit with a radius of $R_0$ in this potential at the critical impact parameter

$$b_0 = 2^{1/2} R_0 = [4 \alpha q^2 / \mu v^2]^{1/4}, \qquad (4)$$

where v is the relative velocity of the collision and $\mu$ is the reduced mass of the collision system. Whenever the impact parameter is smaller than $b_0$, the collision trajectory follows a spiral orbit falling into a region of small R. Therefore, if there is a crossing between diabatic potential-energy curves at an $R_x$ smaller than $R_0$, all collision trajectories with $b < b_0$ can reach $R_x$ and a probability of undergoing a charge-transfer at $R_x$ is enhanced. If the probability depends on a radial velocity of $v_r = v[1-b^2/R_x^2 - V(R_x)/(1/2)\mu v^2]^{1/2}$ at $R_x$, with decreasing the collision energy, $v_r$ is almost solely determined by the acceleration in the attractive potential and the probability becomes independent of the collision energy. Thus the low-energy charge-transfer cross section can be approximated as

$$\sigma = 2\pi \int P(v,b)b\,db \approx 2\pi P_{av} \int_0^{b_0} b\,db = P_{av}\pi b_0^2, \qquad (5)$$

where $P_{av}$ is a constant value of the averaged probability. $\sigma_L = \pi b_0^2$ is called the Langevin cross section[5] which gives an upper limit of the cross section in the low energy region where the orbiting radius $R_0$ is over the interaction region of $R_x$. Since the Langevin cross section is inversely proportional to the collision velocity and proportional to the charge state of the projectile ion, this model predicts that the orbiting effect enhances the cross section at low energies and the energy region where such signature is significant spreads to higher energies than near thermal energies with increasing the charge state of the projectile ion.

*Empirical scaling law.* Müller and Salzborn[6] developed an empirical scaling law for charge-transfer cross sections from 107 data for single-electron transfer and from 77 data for double-electron transfer, ranging in collision energy from 10 to 100 keV for various ions with $q \leq 10$. The scaling law is in terms of the projectile charge q and the ionization potential I (in eV) of the target,

$$\sigma_{ML} = A q^\alpha I^\beta. \qquad (6)$$

Fitting parameters of A, $\alpha$ and $\beta$ are $(1.43\pm0.76)\times10^{-12}$ cm$^2$, $1.17\pm0.09$ and $-2.76\pm0.19$ for single-electron transfer, respectively, and $(1.08\pm0.95)\times10^{-12}$ cm$^2$, $0.71\pm0.14$ and $-2.80\pm0.32$ for double-electron transfer. This empirical formula was confirmed to reproduce well single-electron capture cross sections in Kr$^{q+}$(q=7-25)-He collisions at 1q keV[11] and in I$^{q+}$(q=10-41)-He collisions at 1.25q keV.[12]

*Absorbing sphere model.* The absorbing sphere model was developed by Olson and Salop[7] in based on the Landau-Zener (LZ) theory described later. For single-electron transfer at a large number of curve crossings between a initial state and a band of final states inside some critical distance $R_c$, the charge-transfer probability inside this $R_c$ is assumed to be unity and the charge-transfer cross section is approximated as

$$\sigma_{OS} = \pi R_c^2. \qquad (7)$$

The Rc is determined from a following semiempirical expression in atomic units.

$$R_c^2 \exp(-2.648 R_c \alpha/q^{1/2}) = 2.864\times10^{-4} q(q-1) v_0/f \qquad (8)$$

Here, $\alpha = (I/13.6)^{1/2}$ with a ionization potential I in eV of the target, $v_0$ is the incident velocity of the projectile and $f$ is the Franck-Condon factor for specific vibrational transitions. This model is derived in the linear trajectory approximation neglecting the attractive induced-dipole polarization potential.

*Landau-Zener model.* When multiply charged ions collide with neutral targets, a variety of electron- transfer channels are possible. Multichannel Landau-Zener (MCLZ) calculation is useful to examine the contribution from each electron-transfer channels at diabatic potential curve crossings between an initial state and final states.

The single-crossing Landau-Zener (LZ) transition probability at a crossing distance $R_x$ is given by $p=\exp(-2\pi H_{12}^2/v_r\Delta F)$, where $H_{12}$ is the matrix element equal to half the energy splitting of the adiabatic potentials at $R_x$, the radial velocity is $v_r=v[1-b^2/R_x^2-V(R_x)/(1/2)\mu v^2]^{1/2}$ and $\Delta F$ is the difference in slopes of the diabatic potential curves. The cross section for the single-crossing is given by $\sigma=2\pi\int P(v,b)bdb$ with the single-crossing probability of $P=2p(p-1)$.

In order to demonstrate the effect of the attractive dipole-polarization potential, Landau-Zener calculations of some tentative single curve crossings with and without considering the target polarization are compared in figure 2.

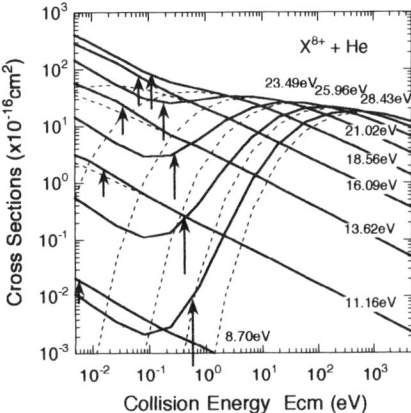

**FIGURE 2.** Landau-Zener calculation for tentative single-crossings labeled by $\Delta E$. Solid and dashed curves are with and without considering target polarization.

The dipole-polarization potential clearly enhances the cross section in the low energy region extending to a fairly high energy and the Langevin like cross section is characterized at energies lower than the arrow-marked position where the orbiting radius coincides with the crossing radius $R_x$.

In our MCLZ calculations for single-electron capture cross sections in $Ar^{8+}$-$H_2$ and He collisions, a semi-empirical formula of the coupling matrix element proposed by Olson and Salop[7] is used together with the correction factor $F_{nl}$ of Taulbjerg,[13]

$$H_{12} = F_{nl} 9.13 q^{-1/2} \exp(-1.324 \alpha R_x q^{-1/2}) \tag{9}$$

where 
$$F_{nl} = (-1)^{n+l-1}(2l+1)^{1/2}\Gamma(n) \left[\Gamma(n+l+1)\Gamma(n-l)\right]^{-1/2}. \tag{10}$$

Here $\alpha$ is $(I/13.6)^{1/2}$ with the ionization potential I (in eV) of the target, $n$ and $l$ are quantum numbers of the final state of the projectile to which the electron is captured.

## RESULTS AND DISCUSSION

The cross sections for single- and double-electron transfer in collisions of $Ar^{q+}$ (q =6, 7, 8, 9 and 11) with He and $H_2$, neglecting the transfer ionization processes following the double-electron capture, have been measured at low energies from 0.075 to 225 eV/amu, systematically. The numerical data for He targets are listed in Table 1 (data for $H_2$ targets were reported in ref.(4)). In figure 3, the measured charge-transfer cross sections of $\sigma_1$ and $\sigma_2$ for $H_2$ and He targets are shown together with previous data and the $\sigma_1$ cross sections are compared to three simple models of the Langevin cross sections $(\sigma_L)$,[5] the scaling law$(\sigma_{MS})$,[6] and the absorbing sphere model$(\sigma_{OS})$.[7] Calculated cross sections of the single- and double-electron transfer using a molecular orbital expansion method for $Ar^{6,8+}+H_2$ systems in the energy range from 6 to 2000 eV/ amu are in good agreements with the present data, especially in the energy dependence.[4]

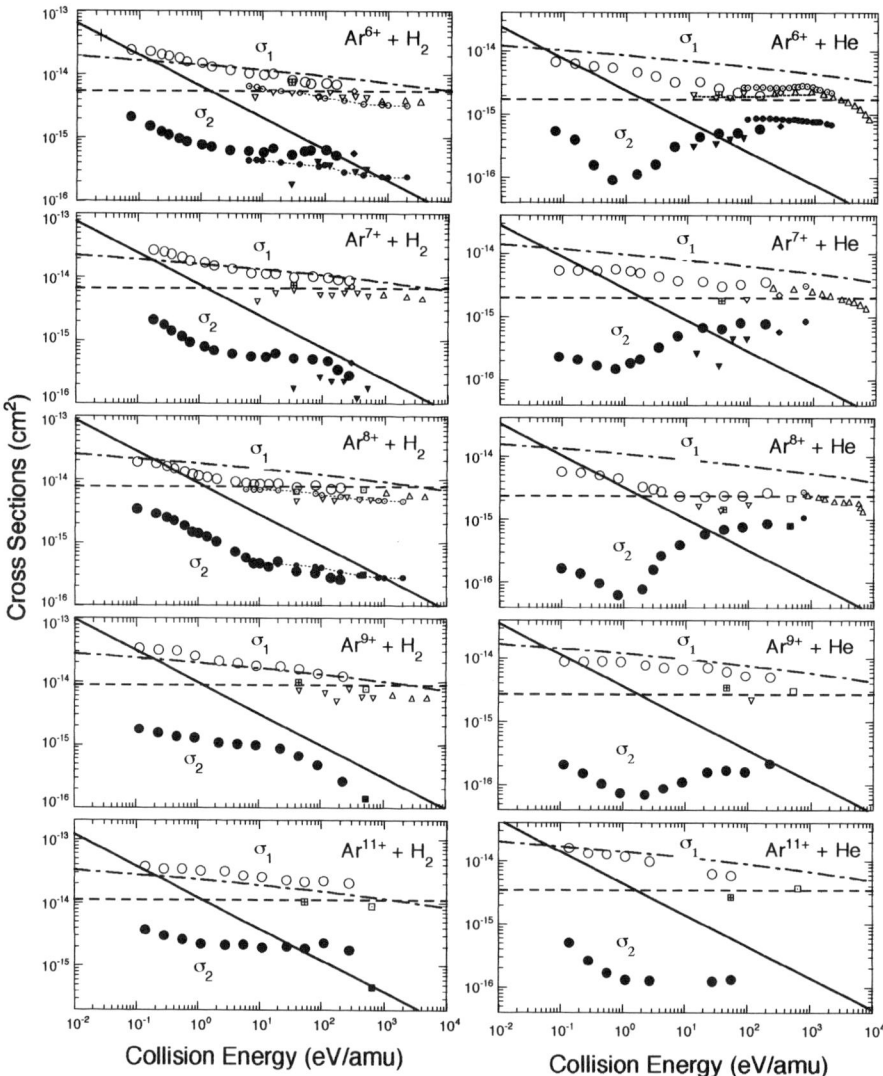

**FIGURE 3.** Single- and double-electron transfer cross sections for $Ar^{q+}$(q=6, 7, 8, 9 and 11)+$H_2$ and He collision systems. Legend; ○ for present $\sigma_1$, ● for present $\sigma_2$, —— for Langevin , - - - for Müller and Salzborn, ——— for absorbing sphere model, --○-- for molecular-orbital expansion $\sigma_1$, --●-- for molecular-orbital expansion $\sigma_2$.

**Table 1** Present experimental cross sections of $\sigma_1$ and $\sigma_2$ for single- and double-electron capture in collisions of $Ar^{q+}$ with He as a function of collision energy $E_{amu}$ over q. The cross sections are in $10^{-16}$ cm$^2$ and $E_{amu}/q$ is in V/amu.

| | Single electron capture ($\sigma_1$) | | | | | Double electron capture ($\sigma_2$) | | | | |
|---|---|---|---|---|---|---|---|---|---|---|
| $E_{amu}/q$ | q=6 | q=7 | q=8 | q=9 | q=11 | q=6 | q=7 | q=8 | q=9 | q=11 |
| 0.0125 | 68.5 | 54.2 | 56.4 | 90.4 | 161 | 5.36 | 2.34 | 1.66 | 2.10 | 5.07 |
| 0.025 | 64.5 | 53.9 | 54.5 | 88.2 | 132 | 3.99 | 2.14 | 1.38 | 1.53 | 2.65 |
| 0.050 | 58.4 | 54.6 | 49.8 | 90.3 | 127 | 1.57 | 1.71 | 0.98 | 1.05 | 1.70 |
| 0.10 | 55.6 | 56.6 | 45.2 | 88.7 | 118 | 0.92 | 1.52 | 0.63 | 0.75 | 1.32 |
| 0.175 | | 53.3 | | | | | 1.86 | | | |
| 0.25 | 47.2 | 49.9 | 33.2 | 78.3 | 98.0 | 1.14 | 2.16 | 0.79 | 0.70 | 1.28 |
| 0.375 | | | 30.5 | | | | | 1.60 | | |
| 0.50 | 40.5 | 43.9 | 28.3 | 71.9 | | 1.64 | 3.35 | 2.67 | 0.89 | |
| 1.00 | 33.0 | 37.7 | 23.4 | 67.6 | | 3.12 | 5.04 | 3.97 | 1.11 | |
| 2.50 | 33.1 | 36.3 | 23.2 | 71.9 | 63.0 | 4.43 | 6.85 | 5.80 | 1.59 | 1.24 |
| 5.00 | 26.0 | 30.6 | 24.0 | 61.8 | 58.9 | 4.99 | 6.52 | 6.98 | 1.71 | 1.34 |
| 10.0 | 22.7 | 32.9 | 24.1 | 53.2 | | 5.08 | 8.20 | 7.69 | 1.62 | |
| 22.5 | 20.4 | | | | | 5.88 | | | | |
| 25.0 | | | 36.0 | 26.8 | 50.7 | | | 7.92 | 8.65 | 2.19 |

As is seen in figure 3, the $\sigma_1$ cross sections fall between the scaling law($\sigma_{MS}$) and the absorbing sphere model ($\sigma_{OS}$) above about 1 eV/amu and converge to the Langevin cross section ($\sigma_L$). These simple models are found to be useful to know the general behavior of the single-electron transfer cross sections in the wide energy range. For the same projectile, the $\sigma_1$ cross sections for the H$_2$ target are almost threefold as large as those for the He target in the whole energy range measured and both energy dependencies are very similar. The cross section ratio $\sigma_1(He)/\sigma_1(H_2)$ for the same projectile is very close to $[I(He)/I(H_2)]^{-2.72}$ predicted by the scaling law and to $[\alpha(He)\mu_{H2}/\alpha(H_2)\mu_{He}]^{1/2}$ by the Langevin orbiting cross section. Although the scaling law has no energy dependence, it is very useful to discuss the charge dependence and the target dependence of the $\sigma_1$ cross section in the wide energy range. This agreement of the cross section ratio indicates that, with increasing with the charge state q of the projectile, the low

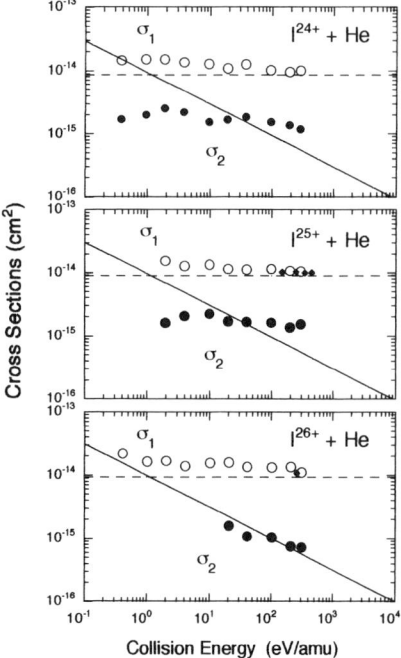

**FIGURE 4.** Single- and double-electron transfer cross sections for $I^{q+}$(q=24-26) + He collisions.

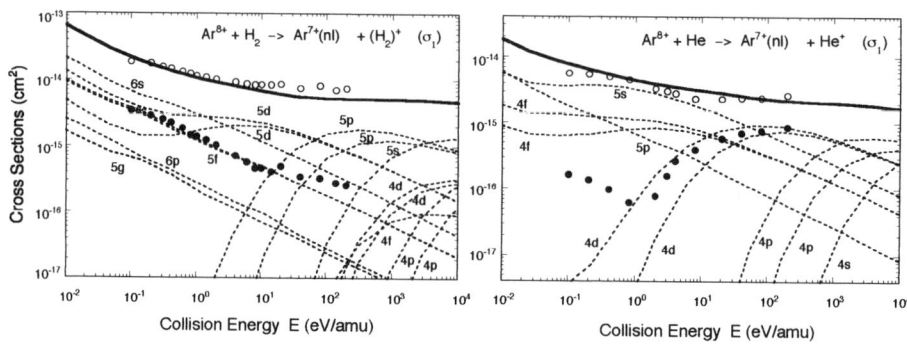

**FIGURE 5.** MCLZ calculations of partial and total single electron transfer cross sections for $Ar^{8+}+H_2$ and He collision systems. Legend ; MCLZ: — for total, --- for partial, present experimens: ○ for $\sigma_1$, ● for $\sigma_2$.

energy cross section of the Langevin type increases in proportion to q and that the high energy cross section also increases at a similar rate of $q^{1.17}$ according to the scaling law. The boundary energy position for both cross sections does not shift much to the high energy side even if q is much larger than q=11. Actually, as seen in figure 4 for the unpublished data for $I^{24,25,26+}$+He collisions, the situation of the energy dependence for large charge states from 24 to 26 is not so different from those for q=6-11. Thus, it is proved that the orbiting effect in the highly charged ion collision is masked by the spreading of the reaction window with increasing the charge state.

The MCLZ calculations for the single electron-transfer in $Ar^{8+}$ + $H_2$ and He collisions, neglecting double-electron transfer, are shown in figure 5. The calculated total cross sections reproduce well the present $\sigma_1$ data. It is clear that the partial cross sections are individually influenced by the induced-dipole polarization potential. However, the boundary on the energy dependence of the total cross section becomes indistinct with the overlapping of many channels having different energy dependence.

On the other hands, the double-electron capture cross sections of $\sigma_2$ are smaller than one tenth of the $\sigma_1$ cross sections and also tend to increase at the low energy end. However, $\sigma_2$ for the He target are very dependent on the collision energy and quite different from the corresponding ones for the $H_2$ target, and they are characterized by a minimum structure near 1 eV/amu. With increasing the charge state of the projectile, the minimum structure gradually shifts to the higher energy side and broadens. There are two possible ways of the direct path from an initial state to a final state and the two-step path through a single-electron capture path. However, since the two-step path through the single-electron transfer channel must pass over a Coulomb repulsive potential barrier, the path should produce a threshold structure on the cross section curve and can be ignored in the low energy range below the threshold energy. Thus the minimum structure in $Ar^{q+}$-He collisions can be expected to be caused from the combination of the two-step processes and the direct-double-electron transfer processes affected from the orbiting effect due to the attractive induced-dipole polarization potential.

Recently, in $C^{4+}$-Ne, Ar and Kr collisions, we have found analogous phenomena that target ions scattered into the backward direction appear and increase rapidly in the low energy region. For $C^{4+}$-Ar systems shown in figure 6, the backscattered target ions caused by the triple- and double-electron transfer are predominantly abundant at low energies and total cross sections for the multiple-electron transfer are revised upward by dot-dashed curves in the low energy side where the formation cross sections of target ions are roughly added to previous cross section data[14] measured from the detection of projectile $C^{4+}$ ions scattered in the forward direction. Also, the appearance energy for increasing target ions becomes large in multiplicative order of transferred electrons. This is consistent with the consideration of the orbiting radius covering the interaction region for the electron transfer. Thus, in the collisions of highly charged ions with heavy targets, the multi-electron transfer processes are supposed to be more strongly affected by the attractive induced-dipole polarization potential at higher energy and to be much more dominated by the large angle scattering at low energies.

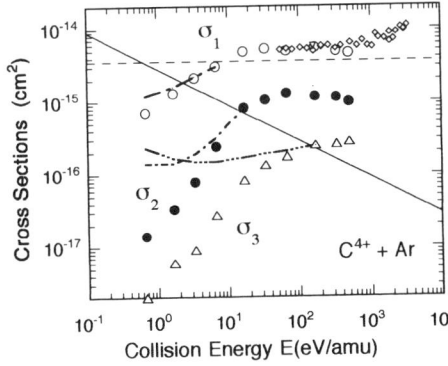

**FIGURE 6,** Single- and multiple electron transfer cross sections for $C^{4+}$-Ar, Forward; $\bigcirc(\sigma_1)$, $\bullet(\sigma_2)$, $\triangle(\sigma_3)$, total; dash-dotted lines.

## CONCLUSION AND SUMMARY

The single- and double-electron transfer cross sections in collisions of $Ar^{q+}$ (q=6, 7, 8, 9, and 11) with $H_2$ and He targets have been systematically measured at low energies ranging from 0.075 to 275 eV/amu. The presentation of these low energy data helps to understand the general behavior of the charge transfer processes taking place in the wide energy range.

The single-electron transfer cross section was found to converge to the Langevin cross section at the low energy end studied and to be reproduced by the scaling law and the absorbing model in the high energy side. The agreement of present data with these models proves the point that the low energy charge transfer reaction is characterized by the competition between the orbiting radius and the interaction distance and the cross section is enhanced at the energy where the orbiting radius exceeds the reaction distance. Also, by the MCLZ taking the attractive induced-dipole polarization potential into consideration, it was confirmed that each partial cross section is individually influenced from the polarization potential even at the energy much higher than the thermal energy, but the composition of each effect due to the attractive polarization force is obscured on the total cross section curve. The signature for single- and double-electron transfer cross sections enhanced at the low energy end studied was explained to be caused by the orbiting effect due to the induced-dipole polarization potential. Furthermore, in $C^{4+}$-Ne, Ar and Kr collisions, multi-electron transfer processes at low energies were found to be dominated by the large angle scattering due to the attractive polarization force.

Thus, this paper concludes that the orbiting effect due to the attractive induced-dipole polarization potential plays an important role in the slow electron transfer collisions of highly charged ions even at an energy much higher than the thermal energy and the low energy charge transfer is characterized by the competition between the orbiting radius and the interaction region.

Since the effect due to the attractive induced-dipole potential is broadened and obscured on the total cross section curve composed with many partial cross sections having an individual energy dependence, state-selective experiments on the energy dependence and the angular dependence are necessary to reveal more details of the orbiting effect in highly charged ion collisions.

## ACKNOWLEDGMENTS

The authors wish to express their thanks to Emeritus Professor Yozaburo Kaneko, who was the organizer of the laboratory when this study started, and for his continued interests in this study. This study was partially supported by a Grant-in-Aid for Scientific Research from the Ministry of Education, Science and Culture of Japan.

## REFERENCES

1. Okuno K., Koizumi T. and Kaneko Y., Phys. Rev. Lett. **40**, 1708-171 (1978); Koizumi T., Okuno K., Kobayashi N. and Kaneko Y., J. Phys. Soc. Jpn. **51**, 2650-2656 (1982); Koizumi T., Okuno K. and Kaneko Y., ibid. **53**, 567-573 (1984); Okuno K., ibid. **55**, 1504-1515 (1986); Okuno K., Kaneko Y., J. Mass Spectrom. Soc. Jpn. **34**, 351-365 (1986).
2. R. H. Neynaber R. H. and Tang S. Y., Chem. Phys. Lett., **92**, 556-559 (1982).
3. Havener C. C., Huq M. S., Schulz P. A. and Phaneuf R. A., Phys. Rev. A**39**, 1725-1740 (1989).
4. Kravis S., Saitoh H., Okuno K., Soejima K., Kimura M., Shimamura I., Awaya Y., Kaneko Y., Oura M. and Shimakura N., Phys. Rev. A **52**, 1206-1212 (1995).
5. Gioumousis G. and Stevenson D. P., J. Chem. Phys. **29**, 294-299 (1958).
6. Müller A. and Salzborn E., Phys. Lett. **62A**, 391-394 (1977).
7. Olson R. E. and Salop A., Phys. Rev. A **14**, 579-585 (1976).
8. Okuno K., Soejima K. and Kaneko Y., Nucl. Instr. Meth. Phys. Res. **B53**, 387-394 (1991).
9. Okuno K., AIP Conf. Proc. **188**, 33-44(1988); Jpn. J. Appl. Phys. **28**, 1124-1131(1989).
10. Okuno K., J. Phys. Soc. Jpn. **55**, 1504-1515 (1986).
11. Iwai T., Kaneko Y., Kimura M., Kobayashi N., Matsumoto A., Ohtani S., Okuno K., Takagi S., Tawara H. and Tsurubuchi S., J. Phys. B: At. Mol. Phys. **17**, L95-L99 (1984).
12. Tawara H., Iwai T., Kaneko Y., Kimura M., Kobayashi N., Matsumoto A., Ohtani S., Okuno K., Takagi S. and Tsurubuchi S., J. Phys. B: At. Mol. Phys. **18**, 337-350 (1985).
13. K Taulbjerg, J. Phys. B: At. Mol. Phys. **19**, L367-372 (1986).
14. Soejima K., Okuno K. and Kaneko Y., Organic Mass Spectrom. **28**, 344-348 (1993).

# Projectile Charge Dependence of Helium Single Excitation at Medium and High Impact Energies

Fernando Martín* and Antoine Salin[†]

*Departamento de Química, Facultad de Ciencias CIX,
Universidad Autónoma de Madrid, 28049-Madrid, Spain
[†]Laboratoire des Collisions Atomiques, Université Bordeaux I,
351 Cours de la Libération, 33405 Talence, France

---

We study the dependence of helium single excitation on projectile charge $Z_P$. For this purpose, it is essential to account for the competition with ionization, specially for large $Z_P$. Consequently, our close-coupling expansion includes a discrete representation of the continuum. We show that our method provides excitation cross-sections in good agreement with experiment and allows to represent not only the total ionization cross-sections but also the doubly differential ones. We study the Born series, the corresponding series expansion in $Z_P$ and conclude that their usefulness is rather limited. In particular the so-called $Z_P^3$ correction is not substantiated.

---

The theoretical knowledge on the excitation of atomic targets by multicharged ions has been rather scarce until recently. Efforts have been first directed toward the derivation of simple analytical behaviours in terms of the projectile charge $Z_P$. Many authors have used the Born series to introduce a power series expansion in $Z_P$ (see e.g. [1,2]) on which they base a qualitative discussion of various collision processes. Accordingly, it is traditionally claimed that the dominant correction to the first Born approximation is a term behaving as $Z_P^3$. This idea is used massively in stopping power works in particular to fit experimental results. However, it must be recognized that no information is available on the validity of this power series expansion and the main objective of the present contribution will be to bring quantitative evidence.

For a given impact parameter, the target is stripped of its electrons (by ionization or capture) when the projectile charge is large enough [3]. This must cause a related decrease in excitation probability, which is supported by recent theoretical evidence. Rodríguez and Salin [4] have solved numerically the time dependent Schrödinger equation by a finite difference method. Their results show the expected decrease in excitation probability at small impact parameters due to the competition with ionization. As a consequence, the dominant range of impact parameters for excitation shifts toward larger

© 1995 American Institute of Physics

values as the projectile charge increases. Accordingly, any solution of the excitation problem requires a treatment of the competition between excitation and ionization (or capture when the energy is small enough [3]). We have therefore developed a close-coupling solution of the problem which includes a realistic description of the continuum channels [5]. As we have used a one-center expansion, we expect this theory to be valid at least when capture can be neglected (though a more detailed study proves that its range of validity may be broader than expected). Our conclusions confirm those of [4].

A large body of close-coupling calculations represent the continuum through pseudo-states. The main drawback is that the only thing we know about pseudo-states is that they are not eigenstates of the target hamiltonian. Consequently, it is difficult to gauge the convergence and therefore the accuracy of calculations done with pseudo-states. Our approach is exactly the opposite: we want to include in our basis a representation of the real continuum states of the target, at least in a finite region of configuration space. As we show below, the realistic character of our discrete continuum states allows a determination of doubly differential ionization cross-sections. The discretization technique is a powerful tool to get accurate results in terms of surprisingly small basis sets. The case of the shape resonance in $H^-$ is a striking example [6]. On the other hand, the accuracy of the calculations can be gauged in a systematic way by decreasing the grain of the discretization.

Details of our theory have been given in [5,7]. Our basis sets include up to 123 one-center two-electron states. We have shown that our method allows to reproduce the resonance shapes in helium ionization by 0.1 to 3 MeV protons. As the latter property involves an interference between the continuum and doubly excited states, this is a strong indication on the validity of our description of the continuum channels.

## COMPARISON WITH EXPERIMENT
## FOR EXCITATION AND IONIZATION

Let us first compare our theory with experiment. In figure 1 we show our results for excitation of the He($4\,^1D$) state for a large number of $Z_P$ values and collision energies, plotted according to the Janev-Presnyakov scaling rule [8], together with the experimental data of [9]. Agreement is good down to a scaled energy of 50 keV/amu. In addition, we are able to reproduce the alignment of the final electronic density caused by the projectile charge. In figure 2 we give the percentage population of the He($4\,^1D$) substates. We observe an excellent agreement between theory and experiment [10] both in the magnitude and evolution of the alignment with the reduced energy of the projectile. For small reduced energies we have a dominant population of the $M = 0$ sublevel which reflects the alignment of the electronic cloud along the projectile velocity. At high energy, the alignment is governed by the usual rules in first order of perturbation theory.

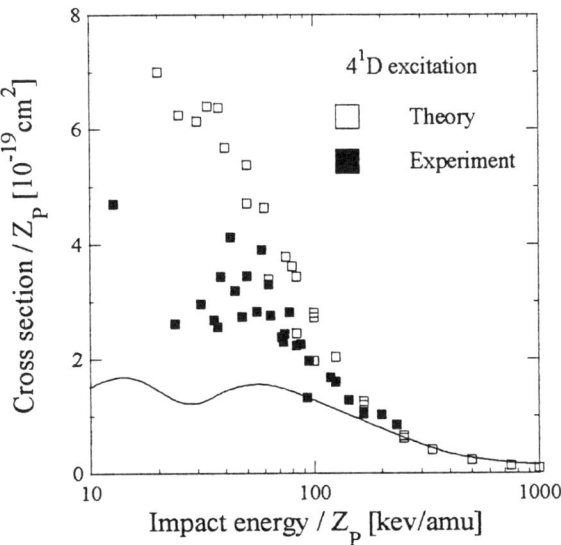

**FIG. 1.** Cross-section (divided by $Z_P$) for excitation of the He($4^1$D) state as a function of projectile energy divided by $Z_P$. Full symbols: experiment [9], open symbols: theory. The full curve gives the results for proton impact (quoted by [9]).

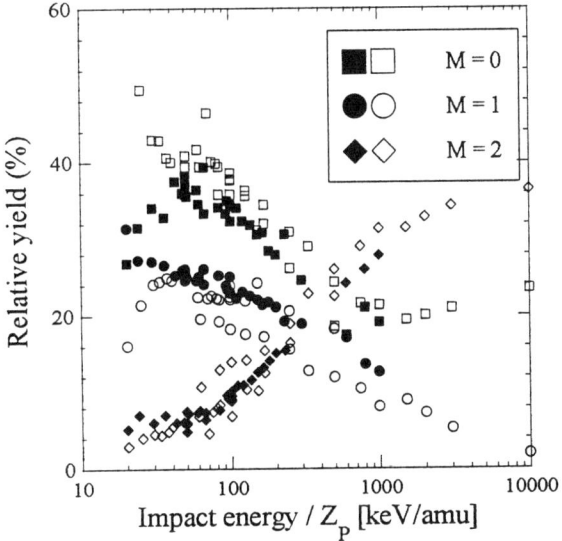

**FIG. 2.** Relative population of the He($4^1$D) substates as a function of projectile energy divided by projectile charge. Full symbols: experiment [10], open symbols: theory.

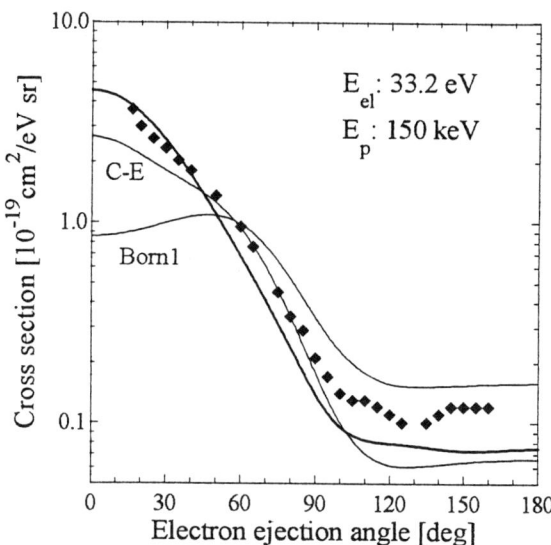

**FIG. 3.** Doubly differential cross-section for ionization of helium by 150 keV protons as a function of ejection angle. The ejected electron energy is 33.2 eV. Thick curve: present results, symbols: experiment [11], thin curves: CDW-EIS and first Born calculations.

The realistic character of our discrete representation of the continuum states is well illustrated by our results for the doubly differential ionization cross section as a function of ejected electron energy. This is shown in figure 3 for an electron energy of 33.2 eV. In the calculations, partial waves up to l=4 have been included for the ejected electron. The most striking feature is the ability of our one-center calculation to reproduce the two-center effect caused by the interaction of the electron with the projectile at the end of the collision. The agreement is very good with the CDW-EIS calculation that incorporates the same feature, in contrast with the Born approximation. We have obtained a similar agreement down to 1 eV. It should be noted that the alignment in excitation and the two-center effect in ionization have a common physical origin: the alignment of the electronic density caused by the projectile charge. This effect increases with projectile charge for a given projectile velocity. This is illustrated in figure 4 through the evolution of the doubly-differential cross-section with projectile charge. As expected, the ionisation cross-section becomes more and more forward peaked as the projectile charge increases.

## BORN EXPANSION AND POWER SERIES EXPANSION IN $Z_P$

A very important feature of the close-coupling method is that it allows a rigorous exploration of the Born series [5]. The close-coupling method is

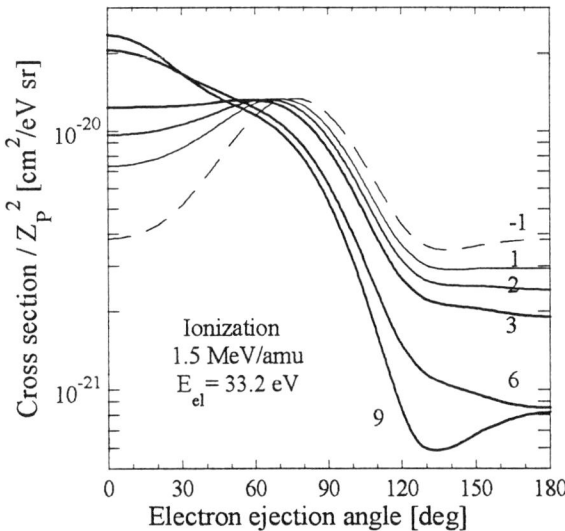

**FIG. 4.** Doubly differential cross-section (divided by the square of the projectile charge) for ionization of helium by 1.5 MeV/amu ions as a function of ejection angle. The ejected electron energy is 33.2 eV. The labels indicate the value of the projectile charge.

defined by a finite space $\mathcal{P}$ in which the problem is solved exactly. Of course, the choice of $\mathcal{P}$ must be realistic to get reliable results and this has been shown in the previous section (see also [5]) for our case. However, the important fact is that any approximation carried out within $\mathcal{P}$ can be gauged against exact results within this subspace since the latter are given by the solution of the close-coupling equations. This can be used for the Born series. All orders of the Born series can be calculated exactly within $\mathcal{P}$ because the basis set defining $\mathcal{P}$ is complete in this subspace. So we are in a position to compare exact calculations of the Born expansion within $\mathcal{P}$ with the exact solution of the problem within the same subspace.

The transition amplitude as a function of impact parameter for the Born approximation of order N can be cast into the form:

$$A_{IF}^{N} = \sum_{i=1}^{N} Z_{P}^{i}\, a_{IF}^{i} \tag{1}$$

where the quantities $a_{IF}^i$ are independent of $Z_P$ and the transition probability in the Nth order of the Born series is given by $P_{IF}^N = |A_{IF}^N|^2$. Now the terms in the Born series can be grouped in a different way to produce a power series in $Z_P$ for the excitation probability:

$$\Pi_{IF}^n = \sum_{i=2}^{n} Z_P^i \, \pi_{IF}^i \qquad (2)$$

where

$$\pi_{IF}^i = \sum_{k=1}^{i-1} \Re\left[(a_{IF}^k)^* a_{IF}^{i-k}\right] \qquad (3)$$

Here we have calculated all values of $a_{IF}^i$ up to $i = 4$ which allows to get terms in the series (3) up to $Z_P^5$, explicitely:

$$\begin{aligned}
\Pi_{IF}^5 = & Z_P^2 \mid a_{IF}^1 \mid^2 \\
& + Z_P^3 \left\{ 2\Re\left[(a_{IF}^2)^* a_{IF}^1\right] \right\} \\
& + Z_P^4 \left\{ \mid a_{IF}^2 \mid^2 + 2\Re\left[(a_{IF}^3)^* a_{IF}^1\right] \right\} \\
& + Z_P^5 \left\{ 2\Re\left[(a_{IF}^3)^* a_{IF}^2\right] + 2\Re\left[(a_{IF}^4)^* a_{IF}^1\right] \right\}
\end{aligned} \qquad (4)$$

The first term in the previous expression is the usual first Born term and the second one is the famous $Z_P^3$ term so often brought forward to explain deviations from the first order term with increasing projectile charge or when the sign of the latter is reversed. We consider explicitly the excitation of the $2\,^1P$ state by ions with energy 100 keV/amu. Our conclusions are changed only in minor way for other transitions.

We study first the Born series (1). Results for proton impact are given in figure 5. The first striking observation is that the second Born approximation does not bring any improvement over the first Born. A significant improvement is brought by the third and fourth Born approximation. Beyond first order, one gets increasingly unreasonable values at small impact parameters. Let us switch now to a projectile with charge $Z_P = 2$ (figure 6). A significant improvement is only obtained in fourth order. Furthermore the small impact parameter divergence appears already at 1 au. The results for $Z_P = 3$ are much worse: no improvement is obtained over the first Born by adding terms up to fourth order. Taking into account the divergence at increasingly large impact parameters, the Born series is no longer useful for $Z_P \geq 3$. As a conclusion, the Born series seems to be of very limited validity. The improvements eventually obtained at medium and large impact parameters by increasing the number of terms in the series are counterbalanced by the divergence at small impact parameters. Similar results are obtained at larger energies: there is a rather limited range of projectile charges over which one can improve over the first Born approximation.

Let us now consider the series in powers of $Z_P$ (2). The situation is already different from the previous series for $Z_P = 1$ (figure 7): there is little improvement over the first Born even by going to fifth order in $Z_P$. Results are given in figure 8 for $Z_P = 2$. The curves for the fourth and fifth order take negative values below those shown on the graph! The $Z_P^3$ term does

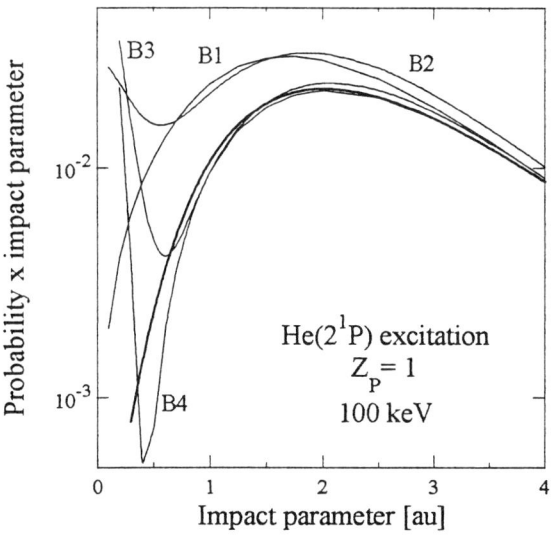

**FIG. 5.** Probability (times impact parameter) of exciting the $2^1P$ state of helium by 100 keV proton impact as a function of impact parameter. Thick curve: close coupling calculation. Thin curves: Born approximation of order N. The labels indicate the order of the Born calculation.

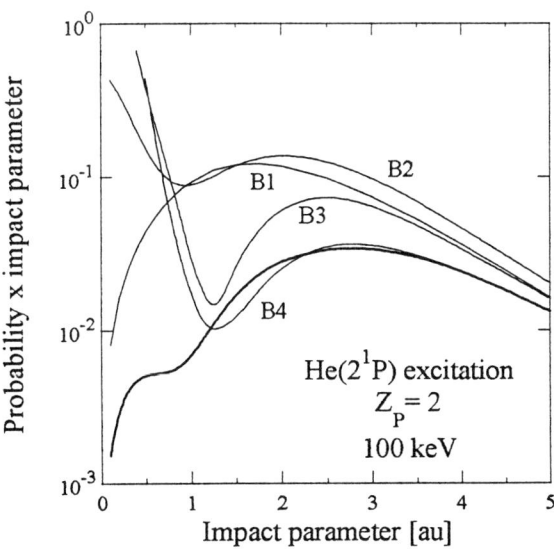

**FIG. 6.** Same as figure 5 for $Z_P = 2$.

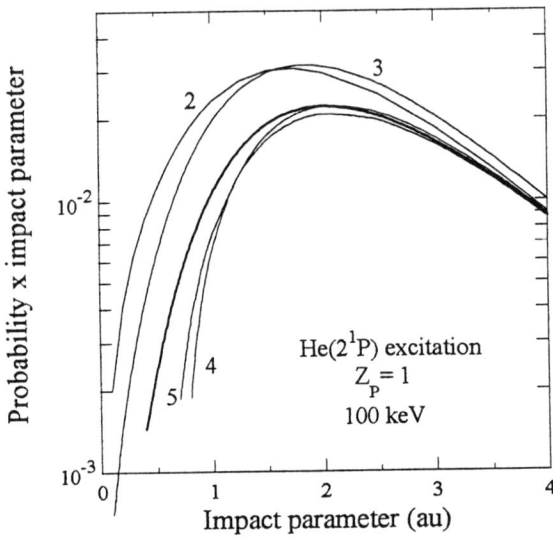

**FIG. 7.** Same as figure 5. The thin curves correspond to a calculation up to order $n$ (given by the label) in the power series in $Z_P$ (2).

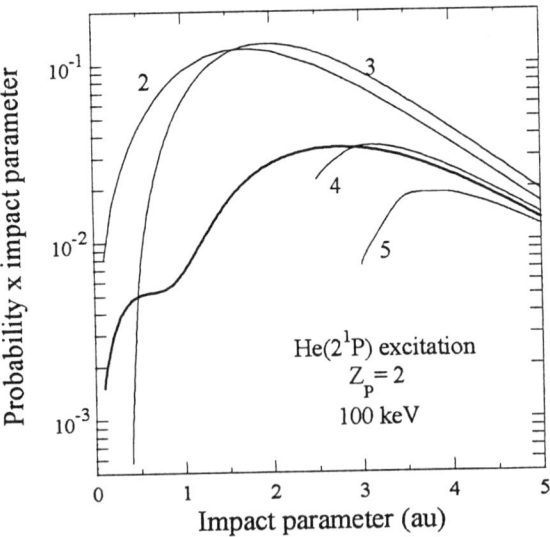

**FIG. 8.** Same as figure 7 for $Z_P = 2$.

not show any improvement over the $Z_P^2$ one and no significant improvement is reached by adding further terms in the series. The situation worsens for $Z_P \geq 3$. Our analysis is unchanged by increasing the energy. For example, for 1.5 MeV/amu, the $Z_P^2$ term gives an accurate value up to $Z_P \simeq 3$. The $Z_P^3$ contribution is negligible for any $Z_P$. For $Z_P \simeq 6$ the addition of further orders up to the fifth does not improve the results with respect to the $Z_P^2$ term. For $Z_P$ larger than 9, the agreement with the exact results deteriorates by introduction of the fourth and fifth order.

## CONCLUSIONS

The conclusion reached in the above cases is completely general. We think the evidence we have obtained raises serious doubts on the possibility to analyse the $Z_P$ behaviour in terms of a power series in $Z_P$. This may seem at first surprising since the usefulness of such an expansion has been taken as granted until now. However it must be recognized that most previous discussions remained at a purely formal level without support from any quantitative evaluation. It is in particular striking that the so called $Z_P^3$ correction is devoided of any justification. The consequence of this fact for the theory of energy loss in gases at medium energies is of particular importance.

Similar studies have been carried out by us for double excitation [7]. The situation is more complex than for single excitation. However we found a similar breakdown of the arguments based on a series expansion in $Z_P$.

We have stressed the increasing importance of competition between excitation and ionization (or capture) when the projectile charge increases. This is in contradiction with the concept of *saturation* [12] along which the excitation probability would go to a limit when the projectile charge increases [1]. Further experimental [9] and theoretical [13] evidences confirm the non validity of the saturation concept even at the level of total cross-sections.

## ACKNOWLEDGMENTS

We wish to thank Prof. K.-H. Schartner for useful correspondence. F. Martín has been partially supported by the DGICYT project PB93-0288-C02-01.

---

[1]The saturation obtained in [12] is a mere consequence of the particular choice of trial function in the application of the Schwinger variational principle [5].

# REFERENCES

1. H. Knudsen and J. Reading, Phys. Rep. A **212**, 107 (1992)
2. J.H. McGuire, Advances in Atomic, Molecular and Optical Physics **29** 217 (1992)
3. C.O. Reinhold, C.A. Falcón and J.E. Miraglia, J. Phys. B **20**, 3737 (1987)
4. V.D. Rodríguez and A. Salin, J. Phys. B **25**, L467 (1992)
5. F. Martín and A. Salin, J. Phys. B **28**, 671 (1995)
6. M. Cortés and F. Martín J. Phys. B **27**, 5741 (1994)
7. F. Martín and A. Salin, J. Phys. B **27** L715 (1994); **28**, 639 (1995); **28**, 1985 (1995); **28**, 2159 (1995); A. Bordenave-Montesquieu et al, J. Phys. B **28**, 653 (1995)
8. R.K. Janev and L.P. Presnyakov, J. Phys. B **13** 4233 (1980)
9. D. Detleffsen, M. Anton, A. Werner and K.-H. Schartner, J. Phys. B **27** 4195 (1994)
10. M. Anton, D. Detleffsen, K.-H. Schartner and A. Werner, J. Phys. B **26** 2005 (1993)
11. A. Bordenave-Montesquieu, P. Benoit-Cattin, A. Gleizes and H. Merchez, Atom. Data Nucl. Data Tables **17** 157 (1976)
12. R. Gayet and M. Bouamoud, Nucl. Instrum. Methods B **42** 515 (1989)
13. R.K. Janev, submitted for publication

# HYPERSPHERICAL ELLIPTIC COORDINATES: NEW APPROXIMATE SYMMETRY OF THREE-BODY COULOMB PROBLEM

O. I. Tolstikhin*, S. Watanabe, and M. Matsuzawa

*Department of Applied Physics and Chemistry
University of Electro-Communications
1-5-1 Chofu-ga-oka, Chofu-shi, Tokyo 182, Japan*

---

New approximate symmetry of the three-body Coulomb problem is discussed. The symmetry reveals itself on the level of the hyperspherical adiabatic Hamiltonian, giving two new quantum numbers $n_\xi$ and $n_\eta$. On the basis of these quantum numbers, we introduce the hyperspherical elliptic ansatz for representing a stationary state wave function of a three-body system with arbitrary masses, which gives direct generalization of the well-known Born-Oppenheimer formula. Such an approach provides a simple and accurate computational scheme for study of correlation and mass-polarization effects in a wide class of three-body systems.

---

## INTRODUCTION

Bound states and collisional reactions in a system of three particles interacting by Coulomb forces form the subject of the three-body Coulomb problem. The problem has four parameters, two mass ratios and two charge ratios, whose role should not be underestimated, for Nature provides us with systems where these parameters vary over *orders* of magnitude. Even restricting ourselves to electrons, muons, protons, and their charge-conjugated partners as constituents, we meet a broad family of systems embracing as different members as $eep$ ($H^-$) and $ppe$ ($H_2^+$), not to mention other more exotic ones, some of which are currently becoming available for laboratory observations (see Fig.1). Any approach, which reveals what is common rather than what is different in such a broad family, sheds some light on the three-body Coulomb problem as a whole, thus being of intrinsic interest from the theoretical point of view. If such an approach, in addition, yields an internally consistent and numerically reliable and accurate computational scheme, as does the hyperspherical method armed by the hyperspherical elliptic coordinates introduced in our recent paper [1], then its basic ideas deserve to be known by a wider audience of physicists such as the one we address here.

---

*Permanent address: P. N. Lebedev Physical Institute, Russian Academy of Sciences, Leninsky Prospect 53, 117924, Moscow, Russia

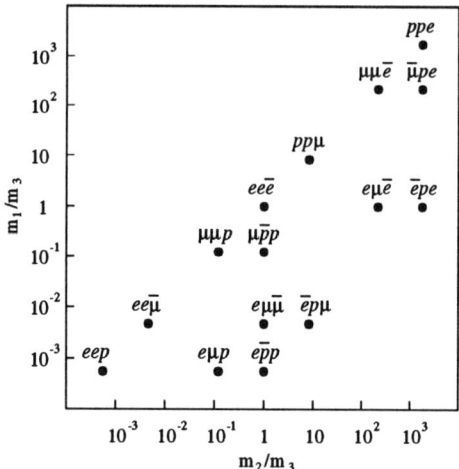

**FIG. 1.** Three-body Coulomb systems composed of electrons, muons, protons, and their charge-conjugated partners. Notations correspond to $Z_1 Z_2 > 0$, $m_1 \leq m_2$.

## HYPERSPHERICAL ELLIPTIC ANSATZ: INTRODUCTION AND COMPARATIVE DISCUSSION

Six variables are required to define a position of three particles in the center-of-mass frame. Different choices of variables lead to different schemes of decomposing the total wave function $\Psi$ into linear combination of some basic functions $\psi_a$, eventually emphasizing different aspects of the three-body dynamics. An important property of the functions $\psi_a$ is their separability — a condition, which is by no means necessary, though prevailing in practice. Thus, $\psi_a$ satisfy a set of separable equations, obtained in certain approximations from the original six-dimensional Schrödinger equation, and have a complete set of six quantum numbers $a$. These quantum numbers represent integrals of motion of the underlying model problem. If the given stationary state of the system can be represented by a single or few terms of the expansion, then the exploited model should be considered as reproducing correctly some dominant features of the system's dynamics. Inasmuch as the basic functions $\psi_a$ and the physical meaning of their quantum numbers result from the solution of *separable* equations, crucial importance of the choice of coordinate system is evident. Of course, the situation may depend on the total energy $E$: one may expect to have different $\psi_a$ for representing bound states, single-continuum states with negative energy $E < 0$, or double-continuum states with energy $E > 0$ above three-body break-up threshold.

We shall consider bound and single-continuum states only, to this end the following should be noted. In addition to total energy $E$ there are two more exact quantum numbers which can be used to label a stationary state of the three-body system — these are total orbital angular momentum $L$ and its projection $M$ along a space-fixed axis[1]. Bound states are not degenerate or have finite degeneracy in $L$, and it is natural to demand for $L$ and $M$ to be included into the set of quantum numbers $a$. For continuum states this is not generally the case: continuum states are infinitely degenerate in $L$ and appropriate linear combinations should be defined regarding asymptotic boundary conditions, so one may prefer to abandon $L$ and $M$ for the sake of individual momenta of colliding particles or other quantum numbers characterizing asymptotic behaviour of the wave function $\Psi$. Nevertheless, for low energy collisions, where the partial wave expansion converges rapidly, $L$ and $M$ again can be considered as "good" quantum numbers, and we shall adopt this approach further. Thus, a proper expansion is expected to have the form

$$\Psi_E^{LM} = \sum_{a'} C_{a'} \psi_a, \qquad a = \{a'LM\} \tag{1}$$

Note, that $L$ and $M$ represent a certain fundamental symmetry property (isotropy) of our *space* and have nothing to do with the nature of interparticle *interaction*. So, dynamic features of the system are presumably portrayed by the reduced set of four other quantum numbers $a'$.

Before discussing the general case with all three masses being arbitrary, it is worthwhile to remind the reader of what is encountered in two extreme limits of mass ratios, corresponding to systems with *one heavy - two light* and *two heavy - one light* particles, respectively. The following two formulas form the common language for interpreting a vast variety of phenomena in atomic and molecular physics:

the first corresponds to the one-center expansion

$$\psi_a = R_{\epsilon_1 l_1}(r_1) R_{\epsilon_2 l_2}(r_2) Y_{l_1 l_2}^{LM}(\hat{r}_1, \hat{r}_2), \quad a = \{\epsilon_1 l_1 \epsilon_2 l_2 LM\} \tag{2}$$

and the second is the Born-Oppenheimer ansatz

$$\psi_a = F_{\epsilon n_\xi n_\eta m \bar{L}}(R) f_{n_\xi m}(\xi) g_{n_\eta m}(\eta) Y_{\bar{L}\bar{M}}(\Theta, \Phi) e^{im\varphi}, \quad a = \{\epsilon n_\xi n_\eta m \bar{L}\bar{M}\} . \tag{3}$$

Consider firstly Eq.(2). It gives a solution of the problem, illustrated in Fig.2(a): two *noninteracting* particles 1 and 2 move in the *central* fields, produced by their interaction with infinitely heavy third particle, whose position is fixed in space. This problem splits into two 3D subproblems, each of them is completely separable in usual 3D spherical coordinates with $\epsilon_i$ and $l_i$ giving individual energies and angular momenta of the particles. An essential approximation of this model with respect to the original three-body Coulomb problem with the same masses and charges of the particles is that the 1-2 interaction is neglected or treated approximately by using the non-Coulombic 1-3 and 2-3 potentials. The only property of the model system, employed in (2) and represented by $l_1$ and $l_2$ quantum numbers, is that the 1-3 and 2-3 potentials are central - which is not something peculiar to the Coulomb field.

Next is the Born-Oppenheimer ansatz, Eq.(3). The model system here

---

[1] Discrete symmetries of the problem are not relevant to the present discussion.

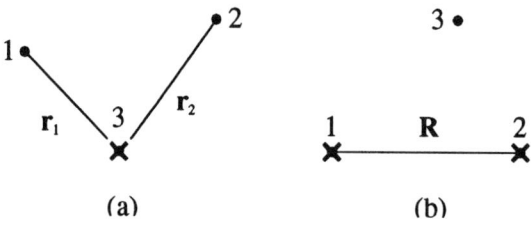

**FIG. 2.** Configurations of three particles, corresponding to (a) one-center expansion, Eq.(2); (b) Born-Oppenheimer expansion, Eq.(3).

contains two infinitely heavy particles 1 and 2, yet now they are not fixed in space but allowed to move *adiabatically*. The problem again splits into two 3D subproblems. The first is the *two-center Coulomb problem* with particles 1 and 2 being fixed at distance $R$ ("internuclear distance") at this stage, see Fig.2(b). This problem is completely separable in 3D spheroidal coordinates, defining the functions $f(\xi)$ and $g(\eta)$, and the adiabatic energy eigenvalue $U(R)$ ("electronic energy"). At the second stage the motion in $\mathbf{R} = (R, \Theta, \Phi)$ in the central potential field given by $U(R) + Z_1 Z_2/R$ is to be solved. The coordinate system, quantum numbers, and equations defining the functions $f(\xi)$ and $g(\eta)$ in (3) are explained in Table 1(a). It is well-known, that it is not a simple task to relate the model problem, which is solved by Eq.(3), with the original three-body Coulomb problem. However, we hope that Table 1(a) leaves no ambiguities in what is meant by Eq.(3). Two points should be emphasized. First, $\bar{L}$ and $\bar{M}$ are *not* the total angular momentum quantum numbers, and there is no finite linear combination of functions (3) restoring the exact $L$ and $M$. This is certainly a disadvantage of the form (3) comparing with (2). On the other hand, the two key quantum numbers $n_\xi$ and $n_\eta$ of Eq.(3) arise from the solution of the problem with *Coulomb* interaction. This problem has an additional integral of motion specific to the *Coulomb* field, corresponding to the separation constant $A(R)$ in Table 1(a). This observation gives an argument for (3) rather than (2) to be used as a starting point for understanding the general case.

One more comment before proceeding to our final goal of this section, Eq.(5). It is not only the masses but a proper combination of the masses *and* charges that determines applicability and success of each of the two forms of $\psi_a$ just discussed. Thus one-center expansion, undoubtedly applicable to two-electron atoms H$^-$, He ..., was recently demonstrated [2] to be surprisingly good in estimating properties of antiprotonic helium $e\bar{p}\mathrm{He}^{2+}$. However it is hardly expected to be suitable for the systems like $ep\bar{e}$ on account of the arrangement based on the $e\bar{e}$ bound states. A similar situation occurs for the Born-Oppenheimer ansatz: the formula (3), normally used for diatomic molecular ions H$_2^+$, HeH$^{2+}$ ..., may produce reasonable-looking results if applied, with a certain amount of imagination and corrections, to two-electron atoms [3], while for systems with two *attracting* heavy particles the Born-Oppenheimer approximation seems to require further justification [4].

**TABLE 1.** Coordinate systems, quantum numbers, and basic equations defining the Born-Oppenheimer and Hyperspherical Elliptic adiabatic wave functions.

| (a) | Born-Oppenheimer, see Eq.(3) and Fig.2(b) |
|---|---|
| $R$, $0 \leq R \leq \infty$ | Separation distance between particles 1 and 2 |
| $\xi$, $1 \leq \xi \leq \infty$ <br> $\eta$, $-1 \leq \eta \leq 1$ | Elliptic coordinates define the position of particle 3 with respect to particles 1 and 2 |
| $\Theta$, $0 \leq \Theta \leq \pi$ <br> $\Phi$, $0 \leq \Phi \leq 2\pi$ | Spherical angles define the orientation of the 1-2 axes |
| $\varphi$, $0 \leq \varphi \leq 2\pi$ | Angle of rotation of particle 3 around 1-2 axes |
| $\epsilon$ | Total energy may take discrete (vibrational states) or continuous values |
| $n_\xi$, $n_\xi = 0, 1, \ldots$ <br> $n_\eta$, $n_\eta = 0, 1, \ldots$ | Elliptic quantum numbers give the numbers of zeros of $f(\xi)$ and $g(\eta)$ |
| $\bar{L}$, $\bar{L} = 0, 1, \ldots$ <br> $\bar{M}$, $\bar{M} = -\bar{L}, \ldots \bar{L}$ | Angular momentum quantum numbers associated with rotation of the 1-2 axes |
| $m$, $m = 0, \pm 1, \ldots$ | Particle 3's angular momentum projection along 1-2 axes |

$$\left[\frac{d}{d\xi}(\xi^2 - 1)\frac{d}{d\xi} - \frac{m^2}{\xi^2 - 1} - Ra(\xi) + \frac{1}{2}U(R)R^2(\xi^2 - 1) - A(R)\right] f(\xi) = 0$$

$$\left[\frac{d}{d\eta}(1 - \eta^2)\frac{d}{d\eta} - \frac{m^2}{1 - \eta^2} - Rb(\eta) + \frac{1}{2}U(R)R^2(1 - \eta^2) + A(R)\right] g(\eta) = 0$$

where $a(\xi) = Z_3(Z_2 + Z_1)\xi$, $b(\eta) = Z_3(Z_2 - Z_1)\eta$

| (b) | Hyperspherical Elliptic, see Eq.(5) |
|---|---|
| $R$, $0 \leq R \leq \infty$ | Hyperradius |
| $\xi$, $2\gamma \leq \xi \leq 2\pi - 2\gamma$ <br> $\eta$, $-2\gamma \leq \eta \leq 2\gamma$ | Hyperspherical elliptic coordinates define the shape of the three-body triangle |
| $\Theta$, $0 \leq \Theta \leq \pi$ <br> $\Phi$, $0 \leq \Phi \leq 2\pi$ <br> $\varphi$, $0 \leq \varphi \leq 2\pi$ | Euler angles define the orientation of the body-fixed frame relative to the space-fixed laboratory frame |
| $\epsilon$ | Total energy may take discrete (vibrational states) or continuous values |
| $n_\xi$, $n_\xi = 0, 1, \ldots$ <br> $n_\eta$, $n_\eta = 0, 1, \ldots$ | Hyperspherical elliptic quantum numbers give the numbers of zeros of $f(\xi)$ and $g(\eta)$ |
| $L$, $L = 0, 1, \ldots$ | Total angular momentum |
| $M$, $M = -L, \ldots L$ | Its projection along a space-fixed axes |
| $m$, $m = -L, \ldots L$ | Its projection along a body-fixed axes |

$$\left[8\frac{d}{d\xi}(c - \cos\xi)\frac{d}{d\xi} - \frac{2s^2 m^2}{c - \cos\xi} - Ra(\xi) + U(R)(c - \cos\xi) - A(R)\right] f(\xi) = 0$$

$$\left[8\frac{d}{d\eta}(\cos\eta - c)\frac{d}{d\eta} - \frac{2s^2 m^2}{\cos\eta - c} - Rb(\eta) + U(R)(\cos\eta - c) + A(R)\right] g(\eta) = 0$$

where $c \equiv \cos 2\gamma$, $s \equiv \sin 2\gamma$, $\gamma$ is defined by Eq.(10)

Turning to the general case one should start from the re-examination of the choice of a coordinate system. Of major importance here is introduction of the hyperradius $R$ as one of the six independent variables - the step, firstly made by Fock [5]. Why hyperradius and not the interparticle distances? The "long range" argument comes first. Hyperradius $R$ defines the overall size of the system. In hyperspherical coordinates $R$ is the only unbound variable, while other five angles, denoted collectively by $\Omega$, are restricted to finite ranges. As far as one considers a system with negative total energy $E <$ 0, an exact solution has no more than one continuous quantum number, and one variable $R$ which can go to infinity from the mathematical point of view suffices to discribe the wave function's behaviuor in the whole configurational space, at the same time enabling a convenient separation in some moderate $R$ region, where simultaneous interaction of all three particles is important, and in the asymptotic region, where either the wave function decays exponentially, or the two-body interactions dominate the system's dynamics. The "short range" argument is probably even more important. As classical mechanics teaches us [6], a triple collision event causes an essential singularity in the classical equations of motion for the three-body system, and it is impossible to obtain a global solution unless one treats this triple collision point properly. This fact, whose importance may be argued by "practical" minds, has far reaching consequences: it was shown, for example, that though Hylleraas method permits one to calculate energy eigenvalues for the helium atom very accurately, the Schrödinger equation for helium *does not* have a solution of the assumed form. In the same paper[2] [5] Fock showed, that an exact solution *can* be constructed as an expansion in powers of $R$ and $\ln R$, thus demonstrating usefulness of $R$ as one of the coordinates.

Next step was done by Macek [7], who recognized the hyperradius $R$ as an adiabatic parameter for the Born-Oppenheimer-type expansion. For describing doubly-excited states of helium Macek introduced the hyperspherical adiabatic representation of the two-electron wave function

$$\psi_a = F_{\epsilon\nu L}(R)\Phi_{\nu LM}(\Omega), \qquad a = \{\epsilon\nu LM\} \qquad (4)$$

Here $\Phi(\Omega)$ is an eigenfunction of the adiabatic Hamiltonian $H_{ad}(R)$ (see Eq.(11) below), where $R$ acts as a parameter, and $F(R)$ describes the motion in $R$ in a way similar to the function $F(R)$ in (3). Index $\nu$ in (4) is not specified; it labels different eigenfunctions of the adiabatic Hamiltonian. Paper [7] gave birth to the new *hyperspherical* method of representing a statinary state wave function of a three-body system in the energy range $E < 0$. Due to the work of many researchers, during the last two decades this method was proved to be very effective and accurate in predicting different collisional and photo characteristics of two-electron atoms. However, in applications to systems with finite masses the difficulties in solving the adiabatic eigenvalue problem, caused by a sharp concentration of the adiabatic channel function $\Phi(\Omega)$ near

---

[2]O.I.T. thanks M.Inokuti for pointing out this particular side of Fock's paper.

the two attractive Coulomb singularities of the three-body potential, or in other words, by the essential nonseparability of the adiabatic Hamiltonian $H_{ad}(R)$ in the coordinate systems used so far, obscured the advantages of the hyperspherical method, making it implementable only by means of modern supercomputers.

Here we introduce the hyperspherical elliptic ansatz specifying Eq.(4) one step further, thus making all six quantum numbers explicit, namely

$$\psi_a = F_{\epsilon n_\xi n_\eta m L}(R) f_{n_\xi m}(\xi) g_{n_\eta m}(\eta) D^L_{Mm}(\Phi, \Theta, \varphi), \quad a = \{\epsilon n_\xi n_\eta m L M\} \quad (5)$$

This formula gives an approximate solution of the three-body Coulomb problem with arbitrary masses. All notations here are explained in Table 1(b). Notations are intentionally chosen in close correspondence with (3), for in the limit $\gamma \to 0$ Eq.(5) transforms to (3), otherwise giving a direct generalization of the Born-Oppenheimer formula. The hyperspherical elliptic coordinates used in (5) were introduced in [1]. In the next section we discuss the two key quantum numbers $n_\xi$ and $n_\eta$ of Eq.(5) in some more details. Because these quantum numbers account for the *relative* motion of the particles and do not concern the overall rotation of the three-body triangle, for the sake of simplicity we restrict ourselves here to the case of zero total angular momentum $L = 0$.

## BASIC EQUATIONS, APPROXIMATE SEPARABILITY

For a system of three particles of masses $m_i$ and charges $Z_i$, $i = 1, 2, 3$, we accept notations such that $Z_1 Z_2 > 0$ and $m_1 \leq m_2$. The Schrödinger equation for a state with $L = 0$ in the hyperspherical elliptic coordinates reads [1]:

$$\left[ -\frac{1}{2} \left( \frac{1}{R^5} \frac{\partial}{\partial R} R^5 \frac{\partial}{\partial R} - \frac{\Lambda^2}{R^2} \right) + \frac{C(\xi, \eta)}{R} - E \right] \Psi(R, \xi, \eta) = 0 \quad (6)$$

where $\Lambda^2$ is the square of Smith's grand angular momentum operator [8]:

$$\Lambda^2 = \frac{-16}{\cos\eta - \cos\xi} \left\{ \frac{\partial}{\partial \xi}(\cos 2\gamma - \cos\xi)\frac{\partial}{\partial \xi} + \frac{\partial}{\partial \eta}(\cos\eta - \cos 2\gamma)\frac{\partial}{\partial \eta} \right\} \quad (7)$$

and the effective charge $C(\xi, \eta)$ is given by:

$$C(\xi, \eta) = 4\frac{\cos(\xi/2) + \cos(\eta/2)}{\cos\eta - \cos\xi} \left[ q^+ \sin(\xi/4)\cos(\eta/4) + q^- \cos(\xi/4)\sin(\eta/4) \right]$$

$$+ \frac{\sqrt{2}q_3}{\sqrt{1 + p^+ \cos(\xi/2)\cos(\eta/2) + p^- \sin(\xi/2)\sin(\eta/2)}} \quad (8a)$$

$$q^\pm = Z_1 Z_3 \sqrt{\frac{m_1 m_3}{m_1 + m_3}} \pm Z_2 Z_3 \sqrt{\frac{m_2 m_3}{m_2 + m_3}}, \quad q_3 = Z_1 Z_2 \sqrt{\frac{m_1 m_2}{m_1 + m_2}} \quad (8b)$$

$$p^+ = 1 + \frac{2m_3}{m_1 + m_2}, \qquad p^- = \frac{m_1 - m_2}{m_1 + m_2} \qquad (8c)$$

The variables $R$, $\xi$, and $\eta$ are defined by the two sets of the mass-scaled Jacobi vectors $(\mathbf{x}, \mathbf{y})$ and $(\mathbf{x}', \mathbf{y}')$, where $\mathbf{y}$ and $\mathbf{y}'$ join the pairs of oppositely charged particles 1-3 and 2-3, respectively. If $y/x = \tan(\chi/2)$, $y'/x' = \tan(\chi'/2)$, then

$$R = \sqrt{x^2 + y^2} = \sqrt{x'^2 + y'^2}, \qquad \xi = \chi + \chi', \qquad \eta = \chi - \chi' \qquad (9)$$

$$0 \le R \le \infty, \qquad 2\gamma \le \xi \le 2\pi - 2\gamma, \qquad -2\gamma \le \eta \le 2\gamma$$

where $\gamma$ is a parameter of Smith's kinematic rotation between the two sets $(\mathbf{x}, \mathbf{y})$ and $(\mathbf{x}', \mathbf{y}')$

$$\gamma = \arctan \sqrt{\frac{m_3(m_1 + m_2 + m_3)}{m_1 m_2}}, \qquad 0 \le \gamma \le \pi/2 \qquad (10)$$

The first term in (8a) includes both attractive interactions; it has two singularities at $\xi = 2\gamma, \eta = \pm 2\gamma$, whose type and position coincide with those of two of the four singular points of the operator (7). This fact reveals the most essential property of the new coordinate system. The second term in (8a) corresponds to 1-2 repulsion; it is singular somewhere on the line $\xi = 2\pi - 2\gamma$. Note, that for symmetric systems $q^- = p^- = 0$.

In the hyperspherical method [7] the function $\Phi(\Omega)$ in (4) is defined as a solution of the adiabatic eigenvalue problem

$$H_{ad}(R)\Phi(\Omega) = U(R)\Phi(\Omega), \qquad H_{ad}(R) \equiv \frac{1}{2}\Lambda^2 + RC(\xi, \eta) \qquad (11)$$

In general, this equation is not separable. In [1] we have introduced the separable approximation to Eq.(11), replacing $C(\xi, \eta)$ by

$$C_s(\xi, \eta) = \frac{a(\xi) + b(\eta)}{\cos\eta - \cos\xi} \qquad (12)$$

Now substituting solution in the form $\Phi(\xi, \eta) = f(\xi)g(\eta)$ leads to the equations of Table 1(b), where $m = 0$ for the present case $L = 0$. This separable problem has an additional integral of motion, corresponding to the separation constant $A(R)$ in Table 1(b). Thus, in the separable approximation each solution of (11) can be labeled by the pair of eigenvalues $(U(R), A(R))$ or, equivalently, by the pair of quantum numbers $(n_\xi, n_\eta)$.

Our calculations show, that the asymptotically adapted definition of the potential functions $a(\xi)$ and $b(\eta)$ in (12) is most convenient in practice:

$$a(\xi) = \mathcal{C}(\xi, \pm 2\gamma) - b(\pm 2\gamma), \qquad b(\eta) = \mathcal{C}(2\gamma, \eta) \qquad (13)$$

where $\mathcal{C}(\xi, \eta) \equiv (\cos\eta - \cos\xi)C(\xi, \eta)$. Thus defined the separable basis enabled us to calculate bound state energies for different systems ranging from

**TABLE 2.** Several examples of the calculated bound state energies (a.u.). In all cases 40 functions (5) in the expansion (1) were used. $E_{1s}(ep) = 0.49972784$.

| System | State | Present | Most accurate [9] |
|---|---|---|---|
| $eep$ | $^1S^e$ | 0.5274 489 | 0.5274 4588 |
| $ee\bar{e}$ | $^1S^e$ | 0.2620 090 | 0.2620 0507 |
| $ppe$ | $^1S^e$, (0) | 0.5971 422 | 0.5971 3906 |
| $ppe$ | $^1S^e$, (10) | 0.5216 998 | 0.5216 9837 |
| $ppe$ | $^3S^e$, (0) | 0.4997 425 | 0.4997 4350 |

$eep$ to $ppe$ with the same relative accuracy $\sim 10^{-6}$, see Table 2. The same level of precision is expected for elastic and inelastic phase shifts; details of the calculations will be reported elsewhere. Here, in order to demonstrate quality of the separable approximation, we solved Eqs.(11), (12) with functions $a(\xi)$ and $b(\eta)$ defined self-consistently. Results for the $^1S^e$ symmetry are shown in Fig.3. The general conclusion, supported by Fig.3, is the following:
The adiabatic Hamiltonian (11) posesses an additional approximate symmetry. In the separable approximation, as well as in each of the limits $R \to 0$, $R \to \infty$, and $\gamma \to 0$, this symmetry is exact, giving $n_\xi$ and $n_\eta$ quantum numbers. Otherwise the symmetry is broken; breaking of the symmetry is most pronounced for systems like $eep$, in which limit $n_\xi$ and $n_\eta$ can be related to Lin's correlation quantum numbers [13], and its most severe effect can be seen in the appearence of avoided crossings of the adiabatic potential curves.

## PERSPECTIVES

The kind invitation to write this *pre*view has come while our work on the subject is in the state of fast progress. Accordingly, we believe it is approapriate here to mention some directions which could be pursued in the future.

i) Non-zero angular momentum $L \neq 0$. The Coriolis coupling which mixes different $m$-components of the symmetric-top decomposition (5) depends on the choice of a body-fixed frame. An exhaustive discussion on different choices of a body-fixed frame can be found in the literature (see [10] for example). However, in the new coordinate system the coupling terms have a different form, and the whole subject should be re-examined.

ii) Highly non-symmetric systems with $Z_1 = Z_2$ and $m_1 \ll m_2 \leq m_3$ are of special interest, with applications to the antiprotonic helium [2,4] in mind. Here a new kind of approximate decoupling scheme based on the $n_\xi$ and $n_\eta$ quantum numbers can be developed.

iii) The hyperspherical method is currently being extended to the ionisation problem, independently as a computational scheme [11] and in a form of a new refined theoretical approach [12]. In both cases, as far as finite masses of the particles are concerned, the hyperspherical elliptic coordinates may be found indispensable.

iv) The hyperspherical elliptic ansatz (5) could find its wide applications in theoretical study on atom-diatom chemical reactions [10].

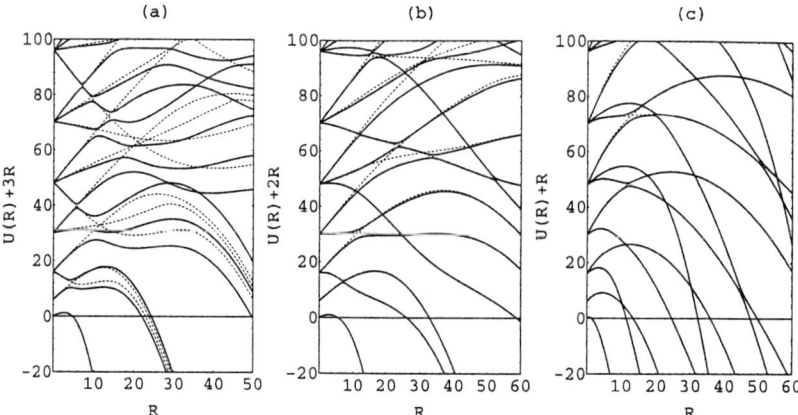

**FIG. 3.** Solid lines - solution of Eq.(11), dashed lines - self-consistent separable approximation. (a) $eep$, $\pi/2 - \gamma \approx 5.4 \times 10^{-4}$, (b) $ee\bar{e}$, $\gamma = \pi/3$, (c) $pp\mu$, $\gamma \approx 0.45$. For $ppe$, $\gamma \approx 3.3 \times 10^{-2}$, solid and dashed curves are indistinguishable to the eye.

## ACKNOWLEDGEMENTS

It is our pleasure to thank Drs. I. L. Beigman, L. P. Presnyakov, M. Inokuti, and J. McGuire for interesting discussions and encouragement. O.I.T. thanks all colleagues and friends in UEC for hospitality, warm attitude, and providing an excellent research environment. Work supported in part by Grant-in-Aid for Scientific Research on Priority Areas "Theory of Chemical Reactions" from the Ministry of Education, Science, and Culture of Japan.

## REFERENCES

1. O. I. Tolstikhin, S. Watanabe, and M. Matsuzawa, Phys. Rev. Lett. **74**, 3573 (1995).
2. T. Yamazaki and K. Ohtsuki, Phys. Rev. A **45**, 7782 (1992).
3. J. M. Feagin and J. S. Briggs, Phys. Rev. A **37**, 4599 (1988).
4. I. Shimamura, Phys. Rev. A **46**, 3776 (1992).
5. V. A. Fock, Izv. Akad. Nauk USSR, Ser. Fiz. **18**, 161 (1954) [Eng. Transl.: Kong. Norske Viden. Selsk. Forh. **31**, 183 (1958); **31**, 145 (1958)].
6. C. L. Siegel and J. K. Moser, *Lectures on Celestial Mechanics*, Springer, 1971.
7. J. Macek, J. Phys. B **1**, 831 (1968).
8. F. T. Smith, Phys. Rev. **120**, 1058 (1960).
9. A. M. Frolov, J. Phys. B **26**, L845 (1993); R. E. Moss, Mol. Phys. **80**, 1541 (1993).
10. R. T. Pack and G. A. Parker, J. Chem. Phys. **87**, 3888 (1987); Y.-S. Mark Wu, A. Kuppermann, and B. Lepetit, Chem. Phys. Lett. **186**, 319 (1991).
11. D. Kato and S. Watanabe, Phys. Rev. Lett. **74**, 2443 (1995).
12. J. H. Macek, S. Yu. Ovchinnikov, and S. V. Passovets, Phys. Rev. Lett. **74**, 4631 (1995).
13. C. D. Lin, Adv. At. Mol. Phys. **22**, 77 (1986).

# Author Index

## A

Abdoul-Carime, H., 599
Aguillon, F., 569
Ali, I. 773
Andric, L., 139
Arakaki, Y., 329
Arseneau, D. J., 413
Awaya, Y., 755
Aumayr, F., 631

## B

Bahr, D., 773
Baum, G., 805
Baum, P., 805
Beijerinck, H. C. W., 731
Bergmann, K., 247
Berlande, J., 657
Berrah, N., 117
Biquard, X., 657
Boesten, L., 279
Bordenave-Montesquieu, A., 537
Bray, I., 785
Breinig, M., 485
Burrow, P. D., 257

## C

Cai, Y. Q., 357
Camilloni, R., 825
Canney, S. A., 357
Caprari, R., 357
Chida, K., 329
Chung, Y.-S., 297
Clark, S. A. C., 357
Cocke, C. L., 495, 515
Crommie, M. F., 3
Crothers, D. S. F., 857
Crowe, A., 785

## D

Damrau, M., 773
Danared, H. 721
Datz, S, 485
Day, J. C., 609
DePaola, B., 609
Desai, D., 485
Desfrançois, C., 599
Deveney, E. F., 485
Djurić, N. 297
Donnelly, B. P., 785
Dörner, R., 495, 773
Drachman, R. J., 369
Driessen, J. P. J., 731
Dubois, R. D., 505
Dunn, G. H., 297
Dunning, F. B., 815

## E

Ehrenreich, T., 609
Eichler, J., 847
Eigler, D. M., 3
Errea, L. F., 445
Esaulov, V. A., 647

## F

Fabrikant, I. I., 815
Fleming, D. G., 413
Frahm, R., 773
Frey, M. T., 815
Fricke, B., 435
Fullerton, C. M., 201
Fursa, D. V., 785

## G

Gaveau, M. A., 657
Gianturco, F. A., 211
Gilbody, H. B., 19
Gordeev, Y. S., 579
Grau, L., 805
Greene, C. H., 127
Glass, J. T., 857
Golde, M. F., 835
Gougousi, T., 835
Guberman, S. L., 307
Guillemont, L., 647
Guo, X., 795

## H

Hansen, S. B., 609
Harel, C., 445
Haruyama, Y., 329
Hatanaka, K., 329
Hatano, Y., 67
Heller, E. J., 3
Hill, S. B., 815
Hitchcock, A. P., 89
Hoekstra, R., 547
Honma, T., 329
Hoogerland, M. D., 731
Horsdal-Pedersen, E., 609
Hosono, K., 329
Hotop, H., 267
Huetz, A., 139
Hurn, J., 795

## I

Iacobucci, S., 825
Itoh, Y., 755

## J

Jagutzki, O., 495, 773
Jean, A., 139
Johnsen, R., 835
Jones, N. L., 485
Jones, S., 341
Jouin, H., 445

## K

Kamegaya, H., 329
Katayama, I., 329
Keil, M., 247
Kernoghan, A. A., 397
Kessel, Q. C., 485
Khayyat, Kh., 773
Kheifets, A. S., 357
Khemliche, H., 495, 557
Klar, D., 267
Kobayashi, N., 867
Koike, F., 755
Koizumi, T., 755

Kornberg, M. A., 763
Kortyna, A., 247
Krause, H. F., 485
Kravis, S., 867
Kravis, S. D., 755
Kreil, J., 267
Krishnamurthy, M., 587
Krüdener, S., 677
Külz, M., 247
Kürpick, P., 435

## L

Lablanquie, P., 139
Lacombe, S., 647
Lallement, A., 657
Laricchia, G., 385
Leontiev, Y., 609
Lett, P. D., 667
Leuer, B., 805
Lin, C. C., 155
Lower, J., 795
Luches, P., 825
Lutz, C. P., 3
Lutz, H. O., 741

## M

MacAdam, K. B., 609
Macek, J. H., 347
Madison, D. H., 341
Marassi, L., 825
Marrs, R. E., 705
Martín, F., 877
Marzilli, B., 825
Mathur, D., 587
Matsuzawa, M. 887
Mayer, S., 163
Mazeau, J., 139
Mazevet, S., 795
McAlinden, M. T., 397
McCann, J. F., 857
McCarthy, I. E., 357, 795
McGuire, J. H., 515, 619
McLaughlin, D. T., 785
Mendez, L., 445
Mergel, V., 495, 773
Mestdagh, J. M., 657

Meyer, U., 773
Meyer, W., 247
Meyerhof, W. E., 515
Miraglia, J. E., 763
Mogensen, K. S., 609
Montenegro, E. C., 515
Moretto-Capelle, P., 537
Moshammer, R., 495
Müller, A., 317

## N

Nannarone, S., 825
Ngoc Tuan, V., 647
Niemeyer, R., 805
Noda, K., 329

## O

Ogurtsov, G. N., 579
Ohtani, S., 329
Okuno, K., 867
Olson, R. E., 495, 619
Oura, M., 755
Ovchinnikov, S. Yu., 347, 485

## P

Pan, J. J., 413
Passovets, S. V., 347
Pons, B., 445
Prior, M., 495
Prior, M. H., 557

## R

Raith, W., 805
Riera, A., 445
Ruf, M.-W., 267
Rühl, E., 89

## S

Saito, M., 329
Saito, N., 105
Saitoh, H., 867
Salin, A., 877
Sanders, J. M., 485
Sano, M., 755
Schermann, J. P., 599
Schmidt, H. G., 773
Schmidt-Böcking, H., 495, 773
Schmitt, W., 495
Schramm, A., 267
Schultz, D. R., 455
Sebel, P. G. M., 731
Sekioka, T., 755
Selles, P., 139
Senba, M., 413
Sepp, W.-D., 435
Shafroth, S. M., 485
Shen, Y., 795
Smith, K. A., 815
Soejima, K., 867
Solov'ev, E. A., 471
Sonntag, B., 39
Spielberger, L., 495, 773
Standage, M., 173
Stefani, G., 825
Stöhlker, Th., 525
Storer, P., 357
Sublemontier, O., 657
Suzuki, I. H., 105

## T

Takagi, H., 329
Tanabe, T., 329
Tanaka, H., 279
Tennyson, J., 233
Tolstikhin, O. I., 887
Tondera, M., 805
Toshima, N., 847

## U

Ullrich, J., 495, 773
Unverzagt, M., 495, 773
Uskov, D. B., 687
Utteridge, S., 357

## V

Van Leeuwen, K. A. H., 731
Vane, C. R., 485
Visticot, J. P., 657
Vogt, T., 773
Vos, M., 357
Vredenbregt, E. J. D., 731

## W

Wallbank, B. 189, 297
Walters, H. R. J., 397
Wang, J., 619

Watanabe, T., 329, 887
Weber, J. M., 267
Weigold, E., 357, 795
Werner, U., 741
Weyh, D., 247
Wu, W., 485, 495

## Y

Yoshizawa, M., 329

## Z

Zhaoyuan, L., 495

# AIP Conference Proceedings

| | Title | L.C. Number | ISBN |
|---|---|---|---|
| No. 214 | Beam Dynamics Issues of High-Luminosity Asymmetric Collider Rings (Berkeley, CA 1990) | 90-55857 | 0-88318-767-1 |
| No. 215 | X-Ray and Inner-Shell Processes (Knoxville, TN 1990) | 90-84700 | 0-88318-790-6 |
| No. 216 | Spectral Line Shapes, Vol. 6 (Austin, TX 1990) | 90-06278 | 0-88318-791-4 |
| No. 217 | Space Nuclear Power Systems (Albuquerque, NM 1991) | 90-56220 | 0-88318-838-4 |
| No. 218 | Positron Beams for Solids and Surfaces (London, Canada 1990) | 90-56407 | 0-88318-842-2 |
| No. 219 | Superconductivity and Its Applications (Buffalo, NY 1990) | 91-55020 | 0-88318-835-X |
| No. 220 | High Energy Gamma-Ray Astronomy (Ann Arbor, MI 1990) | 91-70876 | 0-88318-812-0 |
| No. 221 | Particle Production Near Threshold (Nashville, IN 1990) | 91-55134 | 0-88318-829-5 |
| No. 222 | After the First Three Minutes (College Park, MD 1990) | 91-55214 | 0-88318-828-7 |
| No. 223 | Polarized Collider Workshop (University Park, PA 1990) | 91-71303 | 0-88318-826-0 |
| No. 224 | LAMPF Workshop on $(\pi, K)$ Physics (Los Alamos, NM 1990) | 91-71304 | 0-88318-825-2 |
| No. 225 | Half Collision Resonance Phenomena in Molecules (Caracas, Venezuela 1990) | 91-55210 | 0-88318-840-6 |
| No. 226 | The Living Cell in Four Dimensions (Gif sur Yvette, France 1990) | 91-55209 | 0-88318-794-9 |
| No. 227 | Advanced Processing and Characterization Technologies (Clearwater, FL 1991) | 91-55194 | 0-88318-910-0 |
| No. 228 | Anomalous Nuclear Effects in Deuterium/ Solid Systems (Provo, UT 1990) | 91-55245 | 0-88318-833-3 |
| No. 229 | Accelerator Instrumentation (Batavia, IL 1990) | 91-55347 | 0-88318-832-1 |
| No. 230 | Nonlinear Dynamics and Particle Acceleration (Tsukuba, Japan 1990) | 91-55348 | 0-88318-824-4 |
| No. 231 | Boron-Rich Solids (Albuquerque, NM 1990) | 91-53024 | 0-88318-793-4 |
| No. 232 | Gamma-Ray Line Astrophysics (Paris-Saclay, France 1990) | 91-55492 | 0-88318-875-9 |
| No. 233 | Atomic Physics 12 (Ann Arbor, MI 1990) | 91-55595 | 088318-811-2 |

| Title | L.C. Number | ISBN |
|---|---|---|
| No. 234 Amorphous Silicon Materials and Solar Cells (Denver, CO 1991) | 91-55575 | 088318-831-7 |
| No. 235 Physics and Chemistry of MCT and Novel IR Detector Materials (San Francisco, CA 1990) | 91-55493 | 0-88318-931-3 |
| No. 236 Vacuum Design of Synchrotron Light Sources (Argonne, IL 1990) | 91-55527 | 0-88318-873-2 |
| No. 237 Kent M. Terwilliger Memorial Symposium (Ann Arbor, MI 1989) | 91-55576 | 0-88318-788-4 |
| No. 238 Capture Gamma-Ray Spectroscopy (Pacific Grove, CA 1990) | 91-57923 | 0-88318-830-9 |
| No. 239 Advances in Biomolecular Simulations (Obernai, France 1991) | 91-58106 | 0-88318-940-2 |
| No. 240 Joint Soviet-American Workshop on the Physics of Semiconductor Lasers (Leningrad, USSR 1991) | 91-58537 | 0-88318-936-4 |
| No. 241 Scanned Probe Microscopy (Santa Barbara, CA 1991) | 91-76758 | 0-88318-816-3 |
| No. 242 Strong, Weak, and Electromagnetic Interactions in Nuclei, Atoms, and Astrophysics: A Workshop in Honor of Stewart D. Bloom's Retirement (Livermore, CA 1991) | 91-76876 | 0-88318-943-7 |
| No. 243 Intersections Between Particle and Nuclear Physics (Tucson, AZ 1991) | 91-77580 | 0-88318-950-X |
| No. 244 Radio Frequency Power in Plasmas (Charleston, SC 1991) | 91-77853 | 0-88318-937-2 |
| No. 245 Basic Space Science (Bangalore, India 1991) | 91-78379 | 0-88318-951-8 |
| No. 246 Space Nuclear Power Systems (Albuquerque, NM 1992) | 91-58793 | 1-56396-027-3<br>1-56396-026-5 (pbk.) |
| No. 247 Global Warming: Physics and Facts (Washington, DC 1991) | 91-78423 | 0-88318-932-1 |
| No. 248 Computer-Aided Statistical Physics (Taipei, Taiwan 1991) | 91-78378 | 0-88318-942-9 |
| No. 249 The Physics of Particle Accelerators (Upton, NY 1989, 1990) | 92-52843 | 0-88318-789-2 |
| No. 250 Towards a Unified Picture of Nuclear Dynamics (Nikko, Japan 1991) | 92-70143 | 0-88318-951-8 |
| No. 251 Superconductivity and its Applications (Buffalo, NY 1991) | 92-52726 | 1-56396-016-8 |
| No. 252 Accelerator Instrumentation (Newport News, VA 1991) | 92-70356 | 0-88318-934-8 |
| No. 253 High-Brightness Beams for Advanced Accelerator Applications (College Park, MD 1991) | 92-52705 | 0-88318-947-X |

| | Title | L.C. Number | ISBN |
|---|---|---|---|
| No. 254 | Testing the AGN Paradigm (College Park, MD 1991) | 92-52780 | 1-56396-009-5 |
| No. 255 | Advanced Beam Dynamics Workshop on Effects of Errors in Accelerators, Their Diagnosis and Corrections (Corpus Christi, TX 1991) | 92-52842 | 1-56396-006-0 |
| No. 256 | Slow Dynamics in Condensed Matter (Fukuoka, Japan 1991) | 92-53120 | 0-88318-938-0 |
| No. 257 | Atomic Processes in Plasmas (Portland, ME 1991) | 91-08105 | 0-88318-939-9 |
| No. 258 | Synchrotron Radiation and Dynamic Phenomena (Grenoble, France 1991) | 92-53790 | 1-56396-008-7 |
| No. 259 | Future Directions in Nuclear Physics with $4\pi$ Gamma Detection Systems of the New Generation (Strasbourg, France 1991) | 92-53222 | 0-88318-952-6 |
| No. 260 | Computational Quantum Physics (Nashville, TN 1991) | 92-71777 | 0-88318-933-X |
| No. 261 | Rare and Exclusive B&K Decays and Novel Flavor Factories (Santa Monica, CA 1991) | 92-71873 | 1-56396-055-9 |
| No. 262 | Molecular Electronics—Science and Technology (St. Thomas, Virgin Islands 1991) | 92-72210 | 1-56396-041-9 |
| No. 263 | Stress-Induced Phenomena in Metallization: First International Workshop (Ithaca, NY 1991) | 92-72292 | 1-56396-082-6 |
| No. 264 | Particle Acceleration in Cosmic Plasmas (Newark, DE 1991) | 92-73316 | 0-88318-948-8 |
| No. 265 | Gamma-Ray Bursts (Huntsville, AL 1991) | 92-73456 | 1-56396-018-4 |
| No. 266 | Group Theory in Physics (Cocoyoc, Morelos, Mexico 1991) | 92-73457 | 1-56396-101-6 |
| No. 267 | Electromechanical Coupling of the Solar Atmosphere (Capri, Italy 1991) | 92-82717 | 1-56396-110-5 |
| No. 268 | Photovoltaic Advanced Research & Development Project (Denver, CO 1992) | 92-74159 | 1-56396-056-7 |
| No. 269 | CEBAF 1992 Summer Workshop (Newport News, VA 1992) | 92-75403 | 1-56396-067-2 |
| No. 270 | Time Reversal—The Arthur Rich Memorial Symposium (Ann Arbor, MI 1991) | 92-83852 | 1-56396-105-9 |
| No. 271 | Tenth Symposium Space Nuclear Power and Propulsion (Vols. I–III) (Albuquerque, NM 1993) | 92-75162 | 1-56396-137-7 (set) |
| No. 272 | Proceedings of the XXVI International Conference on High Energy Physics (Vols. I and II) (Dallas, TX 1992) | 93-70412 | 1-56396-127-X (set) |

| Title | L.C. Number | ISBN |
|---|---|---|
| No. 273 Superconductivity and Its Applications (Buffalo, NY 1992) | 93-70502 | 1-56396-189-X |
| No. 274 VIth International Conference on the Physics of Highly Charged Ions (Manhattan, KS 1992) | 93-70577 | 1-56396-102-4 |
| No. 275 Atomic Physics 13 (Munich, Germany 1992) | 93-70826 | 1-56396-057-5 |
| No. 276 Very High Energy Cosmic-Ray Interactions: VIIth International Symposium (Ann Arbor, MI 1992) | 93-71342 | 1-56396-038-9 |
| No. 277 The World at Risk: Natural Hazards and Climate Change (Cambridge, MA 1992) | 93-71333 | 1-56396-066-4 |
| No. 278 Back to the Galaxy (College Park, MD 1992) | 93-71543 | 1-56396-227-6 |
| No. 279 Advanced Accelerator Concepts (Port Jefferson, NY 1992) | 93-71773 | 1-56396-191-1 |
| No. 280 Compton Gamma-Ray Observatory (St. Louis, MO 1992) | 93-71830 | 1-56396-104-0 |
| No. 281 Accelerator Instrumentation Fourth Annual Workshop (Berkeley, CA 1992) | 93-072110 | 1-56396-190-3 |
| No. 282 Quantum 1/f Noise & Other Low Frequency Fluctuations in Electronic Devices (St. Louis, MO 1992) | 93-072366 | 1-56396-252-7 |
| No. 283 Earth and Space Science Information Systems (Pasadena, CA 1992) | 93-072360 | 1-56396-094-X |
| No. 284 US-Japan Workshop on Ion Temperature Gradient-Driven Turbulent Transport (Austin, TX 1993) | 93-72460 | 1-56396-221-7 |
| No. 285 Noise in Physical Systems and 1/f Fluctuations (St. Louis, MO 1993) | 93-72575 | 1-56396-270-5 |
| No. 286 Ordering Disorder: Prospect and Retrospect in Condensed Matter Physics: Proceedings of the Indo-U.S. Workshop (Hyderabad, India 1993) | 93-072549 | 1-56396-255-1 |
| No. 287 Production and Neutralization of Negative Ions and Beams: Sixth International Symposium (Upton, NY 1992) | 93-72821 | 1-56396-103-2 |
| No. 288 Laser Ablation: Mechanismas and Applications-II: Second International Conference (Knoxville, TN 1993) | 93-73040 | 1-56396-226-8 |
| No. 289 Radio Frequency Power in Plasmas: Tenth Topical Conference (Boston, MA 1993) | 93-72964 | 1-56396-264-0 |

| Title | L.C. Number | ISBN |
|---|---|---|
| No. 290 Laser Spectroscopy:<br>XIth International Conference<br>(Hot Springs, VA 1993) | 93-73050 | 1-56396-262-4 |
| No. 291 Prairie View Summer Science Academy<br>(Prairie View, TX 1992) | 93-73081 | 1-56396-133-4 |
| No. 292 Stability of Particle Motion in Storage Rings<br>(Upton, NY 1992) | 93-73534 | 1-56396-225-X |
| No. 293 Polarized Ion Sources and Polarized Gas Targets<br>(Madison, WI 1993) | 93-74102 | 1-56396-220-9 |
| No. 294 High-Energy Solar Phenomena A New Era<br>of Spacecraft Measurements<br>(Waterville Valley, NH 1993) | 93-74147 | 1-56396-291-8 |
| No. 295 The Physics of Electronic and Atomic Collisions:<br>XVIII International Conference<br>(Aarhus, Denmark, 1993) | 93-74103 | 1-56396-290-X |
| No. 296 The Chaos Paradigm: Developments an<br>Applications in Engineering and Science<br>(Mystic, CT 1993) | 93-74146 | 1-56396-254-3 |
| No. 297 Computational Accelerator Physics<br>(Los Alamos, NM 1993) | 93-74205 | 1-56396-222-5 |
| No. 298 Ultrafast Reaction Dynamics<br>and Solvent Effects<br>(Royaumont, France 1993) | 93-074354 | 1-56396-280-2 |
| No. 299 Dense Z-Pinches:<br>Third International Conference<br>(London, 1993) | 93-074569 | 1-56396-297-7 |
| No. 300 Discovery of Weak Neutral Currents:<br>The Weak Interaction Before and After<br>(Santa Monica, CA 1993) | 94-70515 | 1-56396-306-X |
| No. 301 Eleventh Symposium Space Nuclear<br>Power and Propulsion (3 Vols.)<br>(Albuquerque, NM 1994) | 92-75162 | 1-56396-305-1<br>(Set)<br>156396-301-9<br>(pbk. set) |
| No. 302 Lepton and Photon Interactions/<br>XVI International Symposium<br>(Ithaca, NY 1993) | 94-70079 | 1-56396-106-7 |
| No. 303 Slow Positron Beam Techniques for Solids<br>and Surfaces Fifth International Workshop<br>(Jackson Hole, WY 1992) | 94-71036 | 1-56396-267-5 |
| No. 304 The Second Compton Symposium<br>(College Park, MD 1993) | 94-70742 | 1-56396-261-6 |
| No. 305 Stress-Induced Phenomena in Metallization<br>Second International Workshop<br>(Austin, TX 1993) | 94-70650 | 1-56396-251-9 |
| No. 306 12th NREL Photovoltaic Program Review<br>(Denver, CO 1993) | 94-70748 | 1-56396-315-9 |

| Title | L.C. Number | ISBN |
|---|---|---|
| No. 307 Gamma-Ray Bursts Second Workshop (Huntsville, AL 1993) | 94-71317 | 1-56396-336-1 |
| No. 308 The Evolution of X-Ray Binaries (College Park, MD 1993) | 94-76853 | 1-56396-329-9 |
| No. 309 High-Pressure Science and Technology—1993 (Colorado Springs, CO 1993) | 93-72821 | 1-56396-219-5 (Set) |
| No. 310 Analysis of Interplanetary Dust (Houston, TX 1993) | 94-71292 | 1-56396-341-8 |
| No. 311 Physics of High Energy Particles in Toroidal Systems (Irvine, CA 1993) | 94-72098 | 1-56396-364-7 |
| No. 312 Molecules and Grains in Space (Mont Sainte-Odile, France 1993) | 94-72615 | 1-56396-355-8 |
| No. 313 The Soft X-Ray Cosmos ROSAT Science Symposium (College Park, MD 1993) | 94-72499 | 1-56396-327-2 |
| No. 314 Advances in Plasma Physics Thomas H. Stix Symposium (Princeton, NJ 1992) | 94-72721 | 1-56396-372-8 |
| No. 315 Orbit Correction and Analysis in Circular Accelerators (Upton, NY 1993) | 94-72257 | 1-56396-373-6 |
| No. 316 Thirteenth International Conference on Thermoelectrics (Kansas City, Missouri 1994) | 95-75634 | 1-56396-444-9 |
| No. 317 Fifth Mexican School of Particles and Fields (Guanajuato, Mexico 1992) | 94-72720 | 1-56396-378-7 |
| No. 318 Laser Interaction and Related Plasma Phenomena 11th International Workshop (Monterey, CA 1993) | 94-78097 | 1-56396-324-8 |
| No. 319 Beam Instrumentation Workshop (Santa Fe, NM 1993) | 94-78279 | 1-56396-389-2 |
| No. 320 Basic Space Science (Lagos, Nigeria 1993) | 94-79350 | 1-56396-328-0 |
| No. 321 The First NREL Conference on Thermophotovoltaic Generation of Electricity (Copper Mountain, CO 1994) | 94-72792 | 1-56396-353-1 |
| No. 322 Atomic Processes in Plasmas Ninth APS Topical Conference (San Antonio, TX) | 94-72923 | 1-56396-411-2 |
| No. 323 Atomic Physics 14 Fourteenth International Conference on Atomic Physics (Boulder, CO 1994) | 94-73219 | 1-56396-348-5 |

| | Title | L.C. Number | ISBN |
|---|---|---|---|
| No. 324 | Twelfth Symposium on Space Nuclear Power and Propulsion (Albuquerque, NM 1995) | 94-73603 | 1-56396-427-9 |
| No. 325 | Conference on NASA Centers for Commercial Development of Space (Albuquerque, NM 1995) | 94-73604 | 1-56396-431-7 |
| No. 326 | Accelerator Physics at the Superconducting Super Collider (Dallas, TX 1992-1993) | 94-73609 | 1-56396-354-X |
| No. 327 | Nuclei in the Cosmos III Third International Symposium on Nuclear Astrophysics (Assergi, Italy 1994) | 95-75492 | 1-56396-436-8 |
| No. 328 | Spectral Line Shapes, Volume 8 12th ICSLS (Toronto, Canada 1994) | 94-74309 | 1-56396-326-4 |
| No. 329 | Resonance Ionization Spectroscopy 1994 Seventh International Symposium (Bernkastel-Kues, Germany 1994) | 95-75077 | 1-56396-437-6 |
| No. 330 | E.C.C.C. 1 Computational Chemistry F.E.C.S. Conference (Nancy, France 1994) | 95-75843 | 1-56396-457-0 |
| No. 331 | Non-Neutral Plasma Physics II (Berkeley, CA 1994) | 95-79630 | 1-56396-441-4 |
| No. 332 | X-Ray Lasers 1994 Fourth International Colloquium (Williamsburg, VA 1994) | 95-76067 | 1-56396-375-2 |
| No. 333 | Beam Instrumentation Workshop (Vancouver, B. C., Canada 1994) | 95-79635 | 1-56396-352-3 |
| No. 334 | Few-Body Problems in Physics (Williamsburg, VA 1994) | 95-76481 | 1-56396-325-6 |
| No. 335 | Advanced Accelerator Concepts (Fontana, WI 1994) | 95-78225 | 1-56396-476-7 (Set) 1-56396-474-0 (Book) 1-56396-475-9 (CD-Rom) |
| No. 336 | Dark Matter (College Park, MD 1994) | 95-76538 | 1-56396-438-4 |
| No. 337 | Pulsed RF Sources for Linear Colliders (Montauk, NY 1994) | 95-76814 | 1-56396-408-2 |
| No. 338 | Intersections Between Particle and Nuclear Physics 5th Conference (St. Petersburg, FL 1994) | 95-77076 | 1-56396-335-3 |
| No. 339 | Polarization Phenomena in Nuclear Physics Eighth International Symposium (Bloomington, IN 1994) | 95-77216 | 1-56396-482-1 |

| | Title | L.C. Number | ISBN |
|---|---|---|---|
| No. 340 | Strangeness in Hadronic Matter<br>(Tucson, AZ 1995) | 95-77477 | 1-56396-489-9 |
| No. 341 | Volatiles in the Earth and Solar System<br>(Pasadena, CA 1994) | 95-77911 | 1-56396-409-0 |
| No. 342 | CAM -94 Physics Meeting<br>(Cacun, Mexico 1994) | 95-77851 | 1-56396-491-0 |
| No. 343 | High Energy Spin Physics<br>Eleventh International Symposium<br>(Bloomington, IN 1994) | 95-78431 | 1-56396-374-4 |
| No. 344 | Nonlinear Dynamics in Particle Accelerators:<br>Theory and Experiments<br>(Arcidosso, Italy 1994) | 95-78135 | 1-56396-446-5 |
| No. 345 | International Conference on Plasma Physics<br>ICPP 1994<br>(Foz do Iguaçu, Brazil 1994) | 95-78438 | 1-56396-496-1 |
| No. 346 | International Conference on Accelerator-Driven<br>Transmutation Technologies and Applications<br>(Las Vegas, NV 1994) | 95-78691 | 1-56396-505-4 |
| No. 347 | Atomic Collisions: A Symposium in Honor of<br>Christopher Bottcher (1945-1993)<br>(Oak Ridge, TN 1994) | 95-78689 | 1-56396-322-1 |
| No. 348 | Unveiling the Cosmic Infrared Background<br>(College Park, MD, 1995) | 95-83477 | 1-56396-508-9 |
| No. 349 | Workshop on the Tau/Charm Factory<br>(Argonne, IL 1995) | 95-81467 | 1-56396-523-2 |
| No. 350 | International Symposium on Vector Boson<br>Self-Interactions<br>(Los Angeles, CA 1995) | 95-79865 | 1-56396-520-8 |
| No. 351 | The Physics of Beams<br>Andrew Sessler Symposium<br>(Los Angeles, CA 1993) | 95-80479 | 1-56396-376-0 |
| No. 352 | Physics Potential and Development of<br>$\mu^+ \mu^-$ Colliders: Second Workshop<br>(Sausalito, CA 1994) | 95-81413 | 1-56396-506-2 |
| No. 353 | 13th NREL Photovoltaic Program Review<br>(Lakewood, CO 1995) | 95-80662 | 1-56396-510-0 |
| No. 355 | Eleventh Topical Conference on Radio<br>Frequency Power in Plasmas<br>(Palm Springs, CA 1995) | 95-80867 | 1-56396-536-4 |
| No. 357 | 10th Topical Workshop on Proton-<br>Antiproton Collider Physics<br>(Batavia, IL 1995) | 95-83078 | 1-56396-543-7 |
| No. 358 | The Second NREL Conference on<br>Thermophotovoltaic Generation of Electricity | 95-83335 | 1-56396-509-7 |
| No. 360 | The Physics of Electronic and Atomic Collisions<br>XIX International Conference<br>(Whistler, Canada, 1995) | 95-83671 | 1-56396-440-6 |